Handbook of Nuclear Medicine and Molecular Imaging for Physicists

Series in Medical Physics and Biomedical Engineering
Series Editors: Kwan-Hoong Ng, E. Russell Ritenour, Slavik Tabakov

Recent books in the series:

Advanced Radiation Protection Dosimetry
Shaheen Dewji, Nolan E. Hertel

On-Treatment Verification Imaging. A Study Guide for IGRT
Mike Kirby, Kerrie-Anne Calder

Modelling Radiotherapy Side Effects. Practical Applications for Planning Optimisation
Tiziana Rancati, Claudio Fiorino

Proton Therapy Physics, Second Edition
Harald Paganetti (Ed.)

e-Learning in Medical Physics and Engineering: Building Educational Modules with Moodle
Vassilka Tabakova

Diagnostic Radiology Physics with MATLAB®: A Problem-Solving Approach
Johan Helmenkamp, Robert Bujila, Gavin Poludniowski (Eds.)

Auto-Segmentation for Radiation Oncology: State of the Art
Jinzhong Yang, Gregory C. Sharp, Mark Gooding

Clinical Nuclear Medicine Physics with MATLAB: A Problem Solving Approach
Maria Lyra Georgosopoulou (Ed.)

Handbook of Nuclear Medicine and Molecular Imaging for Physicists – Three Volume Set Volume I: Instrumentation and Imaging Procedures
Michael Ljungberg (Ed.)

For more information about this series, please visit: www.routledge.com/Series-in-Medical-Physics-and-Biomedical-Engineering/book-series/CHMEPHBIOENG

Handbook of Nuclear Medicine and Molecular Imaging for Physicists

Instrumentation and Imaging Procedures, Volume I

Edited by
Michael Ljungberg

CRC Press
Taylor & Francis Group
Boca Raton London New York

CRC Press is an imprint of the
Taylor & Francis Group, an informa business

First edition published 2022
by CRC Press
6000 Broken Sound Parkway NW, Suite 300, Boca Raton, FL 33487-2742

and by CRC Press
2 Park Square, Milton Park, Abingdon, Oxon OX14 4RN

CRC Press is an imprint of Taylor & Francis Group, LLC

Library of Congress Cataloging-in-Publication Data
A catalog record has been requested for this book

ISBN: 978-1-138-59326-8 (hbk)
ISBN: 978-1-032-05868-9 (pbk)
ISBN: 978-0-429-48955-6 (ebk)

DOI: 10.1201/9780429489556

Typeset in Times
by Newgen Publishing UK

Access the Support Material: www.routledge.com/9781138593268

Contents

Preface

During the spring of 2017, I was writing a review of a proposal for a book to potentially be published by CRC Press. Upon closing the discussion with CRC Press regarding the result of this review, I was asked to be an editor for a handbook of nuclear medicine, with focus on physicists of this field. After spending the summer thinking about a relevant table of contents and related potential authors, I formally accepted the offer. I soon realized that the field of nuclear medicine was too extensive to be covered in a single book. After consolidating with the publisher, it was decided that instead of one book it would be best to develop three volumes with the titles, (I) *Instrumentation and Imaging Procedures*, (II) *Modelling, Dosimetry and Radiation Protection* and (III) *Radiopharmaceuticals and Clinical Applications.*

My vision was to create state-of-the-art handbooks, encompassing all major aspects relating to the field of Nuclear Medicine. The chapters should describe the theories in detail but also, when applicable, have a practical approach, focusing on procedures and equipment that are either in use today, or could be expected to be of importance in the future. I realized that the topic of each chapter would be broad enough, in principle, to lay the foundation for individual books of their own. As such, the chapters needed only cover the most relevant aspects of each topic. Therefore, this book series will, hopefully, serve as references for different aspects relating to both the academic and the clinical practice of a medical physicist.

I originally struggled with the definition of the word 'handbook'. I did not want the chapters to serve as point-by-point guidelines, but rather to function as independent chapters to be read more or less independently of one another. Consequently, there is some overlap in the content between chapters but, from a pedagogical point-of-view, I do not see this as a drawback, as repetition of key aspects may aid in the learning.

Volume I in the series of three books focuses primarily on the detection of radiation, beginning with an introduction to the history of nuclear medicine. This introduction is followed by chapters emphasising basic physics, interaction processes and radionuclide production, after which different types of detectors ranging from single ionisation chambers to complex PET/CT and PET/MRI system are described. To get a better understanding of image-based nuclear medicine systems characteristics, we then also describe properties of digital imaging, tools utilized for image access, processing, and reconstruction, and discuss how to calibrate systems and accurately process data and extract quantitative information. Due to the rapid increase in the use of machine learning, this topic is also covered in this volume. Lastly, since hybrid SPECT/PET and CT/MRI systems are oftentimes used in combination, and such systems are likely to become more widely used in the foreseeable future, two chapters are dedicated to describing the basic principles of CT and MRI.

These three volumes are the result of the efforts of outstanding authors who, despite the exceptional circumstances related to the COVID-19 pandemic, have managed to keep to the deadline of the project – although, I must admit, there were times when I questioned the feasibility of doing this. As COVID-19 hit, many of us were faced with unexpected tasks to solve: distance teaching, restrictions and changes in administration, and sometimes also rapid modifications to local procedures at departments and hospitals. Naturally, the combined effect of these interruptions impacted the time available to dedicate to writing. However, despite these many challenges, we all did our utmost to complete the chapters according to the deadline.

I would like to thank all authors for their contributions, which made these books possible. You have all done a phenomenal job, especially considering the extraordinary circumstances we are currently faced with, but also considering the fact that you all have other obligations of high priority. I would especially like to thank Professor Philip Elsinga, who initially helped me define the content of the radiopharmaceutical section being prepared for Volume III. This subtopic of nuclear medicine is the one I have the least knowledge of, and I am therefore very grateful for the kind support I received during the initial planning of Volume III.

I would like to thank CRC Press officials for entrusting me with the position as editor of this series of books. I would also like to thank Kirsten Barr, Rebecca Davies and Francesca McGowan, who have been the points of contact for me during these years.

It is also important to acknowledge two authors who are sadly no longer with us: Anna Celler, University of British Columbia, Vancouver, Canada, and Lennart Johansson, Umeå University, Sweden. Both were dear friends and great scientists. Throughout the years, their work has made a huge impact in their respective fields of research.

Finally, I would like to dedicate this work to my wife, Karin, as well as to my beloved daughter Johanna, who lives in Brisbane, where she is pursuing her PhD at the University of Queensland. Karin – I am so grateful for your patience, especially during the intense period around Christmastime right before the submission of the manuscript for this volume. I love you both very much.

Michael Ljungberg, PhD
Professor, Medical Radiation Physics, Lund
Lund University, Lund, Sweden

Access to colour images and support material: http://www.routledge.com/9781138593268

Editor

Michael Ljungberg is a Professor at Medical Radiation Physics, Lund, Lund University, Sweden. He started his research in the Monte Carlo field in 1983 through a project involving a simulation of whole-body counters but later changed the focus to more general applications in nuclear medicine imaging and Single Photon Emission Computed Tomography (SPECT). As a parallel to his development of the Monte Carlo code SIMIND, he started working in 1985 with quantitative SPECT and problems related to attenuation and scatter. After earning his PhD in 1990, he received a research assistant position that allowed him to continue developing SIMIND for quantitative SPECT applications and established successful collaborations with international research groups. At this time, the SIMIND program also became used worldwide. Dr. Ljungberg became an associate professor in 1994 and in 2005, after working clinically as a nuclear medicine medical physicist, received a full professorship in the Science Faculty at Lund University. He became head of the Department of Medical Radiation Physics in 2013 and a full professor in the Medical Faculty in 2015.

Besides the development of SIMIND to include a new camera system with CZT detectors, his research includes an extensive project in oncological nuclear medicine and, with colleagues, he developed dosimetry methods based on quantitative SPECT, Monte-Carlo absorbed dose calculations, and methods for accurate 3D dose planning for internal radionuclide therapy. In recent years, his work has focused on implementing Monte-Carlo based image reconstruction in SIMIND. He is also involved in the undergraduate education of medical physicists and biomedical engineers and is supervising MSC and PhD students. In 2012, Professor Ljungberg became a member of the European Association of Nuclear Medicines task group on dosimetry and served in that group for six years. He has published over a hundred original papers, 18 conference proceedings, 18 books and book chapters and 14 peer-reviewed papers.

Contributors

Gudrun Alm Carlsson
Department of Radiation Physics, Linköping University, Linköping, Sweden

Karl Åström
Centre for Mathematical Sciences, Lund University, Lund, Sweden

Tom Bäck
Medical Radiation Sciences, Institute of Clinical Sciences, Sahlgrenska Academy, University of Gothenburg, Sweden

Ronald Boellaard
Department of Radiology and Nuclear Medicine of the Amsterdam University Medical Centres, VU University Medical Center (VUMC), in Amsterdam, Netherlands

Anna Celler
Department of Radiology, University of British Columbia, Vancouver, Canada

Magnus Dahlbom
Ahmanson Translational Theranostics Division, Department of Molecular and Medical Pharmacology, David Geffen School of Medicine at UCLA, Los Angeles, USA

John Dickson
Institute of Nuclear Medicine, University College London Hospitals (UCLH) and University College London, United Kingdom

Kjell Erlandsson
Institute of Nuclear Medicine, University College London (UCL), United Kingdom

Jonathan Gear
Joint Department of Physics, The Royal Marsden National Health Service Foundation Trust (NHSFT), Sutton, United Kingdom

Mikael Gunnarsson
Radiation Physics, Skåne University Hospital, Malmö, Sweden

Johan Gustafsson
Medical Radiation Physics Lund, Lund University, Lund, Sweden

Charles Herbst
Department of Medical Physics, University of the Free State, Bloemfontein, South Africa

Brian F Hutton
Institute of Nuclear Medicine, University College London (UCL), United Kingdom

Mats Isaksson
Medical Radiation Sciences, Institute of Clinical Sciences, Sahlgrenska Academy, University of Gothenburg, Gothenburg, Sweden

Bo-Anders Jönsson
Medical Radiation Physics Lund, Lund University, Lund, Sweden

Andres Kaalep
Department of Medical Technology, North Estonia Medical Centre Foundation, Tallinn, Estonia

Chi Liu
Department of Radiology and Biomedical Imaging, Yale School of Medicine, New Haven, Connecticut, USA

Michael Ljungberg
Medical Radiation Physics Lund, Lund University, Lund, Sweden

Hans Lundqvist
Department of Immunology, Genetics and Pathology, Uppsala University, Uppsala, Sweden

Brian W. Miller
Department of Medical Imaging, University of Arizona, Tucson, Arizona, USA

Anders Örbom
Department of Clinical Sciences Lund, Oncology, Lund University, Lund, Sweden

Christopher Rääf
Medical Radiation Physics, Department of Translational Medicine, Lund University, Malmö, Sweden

Andrew J. Reader
School of Biomedical Engineering and Imaging Sciences, King's College London, United Kingdom

Per Roos
Environment, Safety, Health & Quality (ESH&Q) Division, European Spallation Source ERIC, Lund, Sweden

David Sarrut
Centre National de la Recherche Scientifique (CNRS), Université de Lyon, CREATIS lab; UMR5220, Inserm U1206, Institut national des sciences appliquées (INSA) -Lyon, Université Lyon and Léon Bérard Cancer Center, Lyon, France

Bernhard Sattler
Department of Nuclear Medicine, University Medical Centre Leipzig, Leipzig, Germany

Terez Sera
University of Szeged, Department of Nuclear Medicine, Szeged, Hungary and European Association of Nuclear Medicine Research Ltd.

Katarina Sjögreen Gleisner
Medical Radiation Physics Lund, Lund University, Lund, Sweden

Kris Thielemans
Institute of Nuclear Medicine, University College London (UCL), United Kingdom

Carlos Uribe
Functional Imaging, BC Cancer and Department of Radiology, University of British Columbia, Vancouver, Canada

Roel van Holen
Department of Electronics and Information Systems, Ghent University, Ghent, Belgium

Stefaan Vandenberghe
Medical Image and Signal Processing, Ghent University, Ghent, Belgium

Dimitris Visvikis
Director of Research Institut national de la santé et de la recherche médicale (INSERM), Faculte de Médecine Université de Bretagne Occidentale (UBO), Brest, France

Ronnie Wirestam
Medical Radiation Physics Lund, Lund University, Lund, Sweden

Jing Wu
Center for Advanced Quantum Studies and Department of Physics, Beijing Normal University, Beijing, China

Kristina Ydström
Radiation Physics, Skåne University Hospital, Lund, Sweden

Brian Zimmerman
National Institute of Standards and Technology, Gaithersburg, Maryland, USA

1 The History of Nuclear Medicine

Bo-Anders Jönsson

CONTENTS

This chapter provides a historic overview, primarily in chronological order, of those milestones and pioneer's research which have been relevant and important for the development of nuclear medicine and today's status. The content is not comprehensive, and a full review is beyond the purpose of this chapter. More detailed reviews are available elsewhere [1–9] as well as articles referred to therein.

1.1 1890–1930: THE RANDOM DISCOVERIES AND SYSTEMATIC RESEARCH

Although the discovery of X-rays on 8 November 1895 by Wilhelm Conrad Röntgen (1845–1923) is not directly associated with nuclear medicine, it is truly the starting point for using radiation in medicine. Both diagnostic and therapeutic use in medicine of the unknown radiation were applied shortly after the discovery. The first public radiographic exposure was demonstrated by Röntgen at a meeting of the Würzburg Physical Medical Society on January 23, 1896 [10, 11].

A few months later, another unknown type of radiation was discovered. On 26 March 1896, Antoine Henri Becquerel (1852–1908] accidentally discovered an unknown phenomenon when examining fluorescence from uranium salts. With the encouragement of his friend, Henri Poincaré, Becquerel attempted to determine if the rays were of the same nature as Röntgen's X-rays; however, he observed that the emitted radiation from the uranium penetrated black paper and blackened a photographic plate without having to be exposed to light in advance [1, 2]. This unknown radiation was first termed as "Becquerel rays", but its origin was established later by Paul Villard (1860–1934) in 1900 while he was studying radium salts. Villard's radiation was named gamma rays in 1903 by Ernest Rutherford (1871–1937).

Marie Sklodowska Curie (1867–1934) and her husband Pierre Curie (1859–1906) discovered the same type of penetrating radiation from uranium and named the phenomenon radioactivity in 1897. Furthermore, the Curie couple discovered the elements polonium (Z=84) and radium (Z=88), where ^{226}Ra for many years became a frequently used 'panacea' for various ailments, both in vivo and in vitro. Almost directly after these incredible discoveries, radiation from different constructed X-ray tubes and the gamma radiation from ^{226}Ra were used for various medical applications as well as for enjoyment for some decades. In medicine, radium sources were used for brachytherapy or teletherapy for almost the entire twentieth century. Röntgen was awarded the first Nobel Prize in Physics in 1901, while Becquerel and the Curie couple were the Nobel Laureates in Physics in 1903 (Figure 1.1). Other Nobel Laureates with special relevance to nuclear medicine are listed in Table 1.1.

DOI: 10.1201/9780429489556-1

FIGURE 1.1 Swedish stamps issued by the Swedish Post Office 1961 and 1963 in honour of the Nobel laureates 60 years earlier, Röntgen (left) and Becquerel and Curie couple (right). Swedish Post Office.

Other scientists who highly influenced the further development of physics and chemistry and were important for future applications were Joseph John Thomson (1856–1940), Ernest Rutherford, and Frederick Soddy (1877–1956). Advised by Thomson, Ernest Rutherford in 1896 investigated the effects of X-rays on electrical discharges in gases. The theories formulated based on his experiments enabled his colleague Thomson to discover the electron in 1897. Through his studies pertaining to cathode tubes, it was evident that the stream of free particles observed was electrons. As is well known, high-energy electrons that are slowing down constitute the basics for the generation of *bremsstrahlung*, that is, X-rays used for different applications depending on the accelerating voltage. The existence of isotopes was first suggested in 1913 by the radiochemist Frederick Soddy; however, Thomson was the first to use mass spectrometry and demonstrated the presence of isotopes, that is, variants of the same chemical elements but with different numbers of neutrons, in a non-radioactive substance.

An important chemist who pioneered the future of nuclear medicine was George de Hevesy (1885–1966), with his key role in the development of radioactive tracers being the study of chemical processes such as those pertaining to the metabolism of animals. George de Hevesy performed research at several scientific laboratories during his career [12, 13]. In 1911, de Hevesy worked with Rutherford at the Cavendish Laboratory, where Rutherford asked the young researcher to separate radium-D (^{214}Pb) that was present in large amounts of natural lead (Pb) in pitchblende. After lengthy trials he failed and realized that the task could not be accomplished. However, as a result de Hevesy suggested the ingenious idea that a radioactive element that is inseparable from a chemically identical element (isotope), can be used as an indicator for the same stable element.

In springtime 1920, de Hevesy went to the Niels Bohr Institute in Copenhagen and spent six years in an extensive research programme, discovering hafnium (Copenhagen in Latin). He first applied a radioactive isotope, thorium-B (^{212}Pb) in a study pertaining to the solubility of lead salts. George de Hevesy realized that the method could be used to study biological processes. At that time, experiments in vivo were difficult because only toxic substances were available, and he limited his first experiments for studying the uptake and distribution in a flowering plant, that is, the fava bean (Vicia faba). However, in 1924, together with J.A. Christiansen and Sven Lomholt, he presented the first results from animal studies; ^{210}Pb and ^{210}Bi were used. Additionally, this excellent scientist presented further discoveries when more human-friendly radio isotopes became available [1, 12, 13].

Some harmful effects of radiation on tissues became evident quite soon after X-rays and radium were introduced for medical use. However, biological effects were reliably explained only after several years. This also applies to methods developed for the measurement of absorbed energy by suitable equipment, as well as any versatile definition of quantities and units describing exposure to ionizing radiation. In the beginning of the 1920s, a Swedish physicist, Rolf Maximilian Sievert (1896–1966), also known as the father of radiation protection, became one of the world-leading pioneers in medical radiation physics. Sievert demonstrated characteristic scientific interest in radiation biology and radiation protection [14]. Beginning in 1921, in his small private physics laboratory (Figure 1.2) in the Radiumhemmet, founded 1910 as the first oncologic clinic in Sweden, and later at the Radiation Physics Department, Karolinska Institute, in Stockholm, Sievert developed numerous techniques and equipment used for measuring radiation fields. He designed the Sievert integral and a capacitor ion chamber that is also known as the Sievert chamber, among other inventions; furthermore, he established a mobile measurement division [15].

TABLE 1.1
Nobel Laureates with Relevance to Nuclear Medicine. "For the greatest benefit to humankind": Alfred Nobel (1833–1896)

Year	Laureate	Motivation[a]
1901	Wilhelm Conrad Röntgen	*"in recognition of the extraordinary services he has rendered by the discovery of the remarkable rays subsequently named after him"*
1903	Antoine Henri Becquerel	*"in recognition of the extraordinary services he has rendered by his discovery of spontaneous radioactivity"*
1903	Pierre Curie and Marie Curie, neé Sklodowska	*"in recognition of the extraordinary services they have rendered by their joint researches on the radiation phenomena discovered by Professor Henri Becquerel"*
1906	Joseph John Thomson	*"in recognition of the great merits of his theoretical and experimental investigations on the conduction of electricity by gases"*
1908	Ernest Rutherford	*"for his investigations into the disintegration of the elements, and the chemistry of radioactive substances"*
1911	Marie Curie, neé Sklodowska	*"in recognition of her services to the advancement of chemistry by the discovery of the elements radium and polonium, by the isolation of radium and the study of the nature and compounds of this remarkable element"*
1921	Albert Einstein	*"for his services to Theoretical Physics, and especially for his discovery of the law of the photoelectric effect"*
1921	Frederick Soddy	*"for his contributions to our knowledge of the chemistry of radioactive substances, and his investigations into the origin and nature of isotopes"*
1927	Arthur Holly Compton	*"for his discovery of the effect named after him"*
1933	Paul Adrien Maurice Dirac	*"for the discovery of new productive forms of atomic theory"*
1935	James Chadwick	*"for the discovery of the neutron"*
1935	Frederic Joliot and Irene Joliot-Curie	*"in recognition of their synthesis of new radioactive elements"*
1936	Carl David Anderson	*"for his discovery of the positron"*
1937	Clinton Joseph Davisson and George Paget Thomson	*"for their experimental discovery of the diffraction of electrons by crystals"*
1938	Enrico Fermi	*"for his demonstrations of the existence of new radioactive elements produced by neutron irradiation, and for his related discovery of nuclear reactions brought about by slow neutrons"*
1939	Ernest Orlando Lawrence	*"for the invention and development of the cyclotron and for results obtained with it, especially with regard to artificial radioactive elements"*
1943	George de Hevesy	*"for his work on the use of isotopes as tracers in the study of chemical processes"*
1944	Otto Hahn	*"for his discovery of the fission of heavy nuclei"*
1948	Patrick Blackett	*"for his development of the Wilson cloud chamber method, and his discoveries therewith in the fields of nuclear physics and cosmic radiation".*
1951	Sir John Douglas Cockcroft and Ernest Thomas Sinton Walton	*"for their pioneer work on the transmutation of atomic nuclei by artificially accelerated atomic particles"*
1951	Edwin Mattison, McMillan and Glenn Theodore Seaborg	*"for their discoveries in the chemistry of the transuranium elements"*
1977	Rosalyn Yalow	*"for the development of radioimmunoassays of peptide hormones".*
1979	Allan M. Cormack and Godfrey N. Hounsfield	*"for the development of computer assisted tomography"*

Note: [a] The Nobel Prize Foundation (www.nobelprize.org).

In July 1928, the Second International Congress of Radiology was held in Stockholm, where topics discussed included haematological diseases that frequently occurred among the X-ray personnel. At the meeting Sievert contributed significantly to the discussions regarding radiation protection recommendations and dose limits; consequently, the International Commission on Radiation Protection, ICRP, was established. Since its inception in 1928, the ICRP has published more

FIGURE 1.2 Rolf M. Sievert (1896–1966) in his private 5 m^2 laboratory at the Radium Hospital in Stockholm, 1923. Courtesy of Swedish Society of Radiation Physics.

than 145 (2020) publications with recommendations for radiation protection, including several for the nuclear medicine field. In 1928, a quantity that can characterize a photon radiation field in air, that is, the Exposure, with the unit Roentgen,[1] R, was introduced and became an important quantity in radiation protection for a long time to come.

Except for the radium emanation (i.e., radon therapy by radioactive drinking water), used for both medical purposes and in quackery, it was first of all external X-ray for diagnostics and therapy used in the half of the twentieth century. It was not until the 1950s when the first standardization of absorbed dose estimation from internal emitters was introduced. The use of radium water in vivo was halted, and regulations were quickly established after the tragic deaths of individuals who deliberately or unknowingly ingested radium. Examples include the radium dial painters [16, 17] and the quackery scandal that involved the death of prominent businessman Eben M. Byers by radiotoxicity after daily consumption of radioactive water, 'Radithor', which contained radium isotopes and was sold by the William Bailey Radium Laboratory in New Jersey [18].

1.2 1930–1950: DISCOVERY, PRODUCTION, AND DEVELOPMENT OF RADIONUCLIDES

The first artificially produced radionuclide was presented in 1934 by Irene Joliot-Curie (1897–1956) and Fredric Joliot (1900–1958) (Figure 1.3) after they have successfully produced Phoshors-30, ^{30}P, by irradiating an aluminium target with alpha particles from a radium-beryllium source.

$$\alpha + {}^{27}_{13}Al \rightarrow {}^{30}_{15}P + {}^{1}_{0}n \text{ followed by } {}^{30}_{15}P \xrightarrow[25\text{min}]{} {}^{30}_{14}Si + \beta^{+} + \nu$$

In addition to ^{30}P, the couple produced ^{13}N and ^{27}Si, which all demonstrated a continuous spectrum of positrons. However, the amount of activity produced was insufficient for any application. In the nuclear reaction, in addition to the ^{30}P, a

[1] Exposure (old unit: Roentgen) is only valid for photons in air and is the total charge (in Coulomb) produced per kilogram of air at NTP. 1 R in air is approximately equivalent to 0.01 Gy in water.

FIGURE 1.3 Irène and Fredric Joliot-Curie, the inventors of the first artificially produced radionuclide, ^{30}P, in their laboratory in the beginning of the 1930s. Courtesy of: Bibliothèque nationale de France.

neutron is emitted, which was at first misinterpreted as a gamma photon. After additional experiments, James Chadwick (1891–1974) presented a correct interpretation of these experiments in 1932, and the neutron was discovered. The discovery of the neutron changed the radionuclide research scene, since it enabled artificial radionuclides to be produced without an accelerator.

In the subsequent years, research intensified into fission and the neutrons emitted, in which both were misinterpreted but later rectified. In 1939, Otto Hahn (1879–1968) and Friedrich Wilhelm 'Fritz' Strassmann (1902–1980) reported that neutron irradiation of uranium resulted in the production of barium – the first evidence of fission. Finally, in 1939 Frederik Joliot, Hans von Halban (1908–1964), and Lew Kowarski (1907–1979) published evidence pertaining to a chain reaction by neutron multiplication due to fission. A few years later in 1942, Enrico Fermi (1901–1954) demonstrated the first controlled chain reaction, which demonstrated the emission of neutrons from ^{235}U undergoing fission, and the basics for nuclear reactors were established. When the Second World War began, nuclear physics became a military science, with its devastating consequences occurring at the end of the war. However, later fission resulted in extensive civil nuclear power applications, and the neutrons became an important component for scientific research at neutron facilities, as well as for the production of important radionuclides for various medical purposes, such as Cobolt-60 and Cesium-137 for external beam therapy; and Xe-133, Iodine-131 and later Molybden-99, which is used as the mother nuclide in a Technetium-99m generator, for nuclear medicine. The first reactor-produced radionuclides for clinical applications became available in the United States at the Oak Ridge reactor in Tennessee, soon after the end of Second World War in 1945, and at Harwell in the UK from 1947, which became important for Europe [6].

In 1930, a few years before the Curie–Joliot couple demonstrated the first artificially produced radionuclide, Ernest Orlando Lawrence (1901–1958) built the first successful cyclotron at the University of California, Berkeley. The cyclotron measured only 13 cm in diameter and accelerated protons to 80 keV. Together with Milton Stanley Livingston (1905–1986), Lawrence continued developing the circular-type accelerator; in 1938, they presented their fourth 37-inch cyclotron (Figure 1.4), which became another milestone for the production of 'medical' radionuclides, from the very beginning, such as ^{11}C and ^{32}P. The latter began to be used, and still is, for the treatment of polycythemia.

FIGURE 1.4 Stanley Livingston and Ernest Lawrence in front of their 27-inch cyclotron developed in 1934 at the University of California, Berkeley. Courtesy of Lawrence Berkeley National Laboratory. The Regents of the University of California, Lawrence Berkeley National Laboratory.

In 1935, after completing a professorship at the University of Freiburg (1927–1935), George de Hevesy went to the Niels Bohr Institute in Copenhagen for a second stint (1935–1943). His primary interest was to identify radionuclides that can be used as tracers in biological research. He managed to produce some kilobecquerels of ^{32}P using a strong Ra/Be source emitting slow-energy neutrons. George de Hevesy started a series of experiments in which different ^{32}P-labelled compounds were administered to animals to study the distribution and metabolism of the substances. This radiotracer principle is the foundation of all diagnostic and therapeutic nuclear medicine procedures, and de Hevesy is widely considered as the father of nuclear medicine. After the Nazi regime occupied Denmark, he felt unsafe and, in October 1943 he fled to Sweden, where he continued his research at the University of Stockholm (Figure 1.5). George de Hevesey was awarded the Nobel Prize in Chemistry in 1943. In 1948, he published a 556-page compilation about available isotopes and their applications for research, entitled *Radioactive indicators; their applications in biochemistry, animal physiology, and pathology* [19].

In 1937, the element technetium (Z=43) was discovered by Carlo Perrier (1886–1948) and Emilio Segré (1905–1989). In 1938, Glenn Theodore Seaborg (1912–1999) together with Segré identified the metastable isotope technetium-99m ($^{99}Tc^{m}$). However, it was not until 20 years later that $^{99}Tc^{m}$ became available for use in nuclear medicine diagnostics, and when the $^{99}Mo \rightarrow {}^{99}Tc^{m}$-generator was invented. Some other well-known radionuclides, that is, ^{60}Co (1937), ^{131}I (1938), and ^{137}Cs (1941) were discovered by Seaborg and colleagues.

A frequent user of the new Berkeley cyclotron was Joseph Gilbert Hamilton (1907–1957). For instance, he was the first to study the dynamics and excretion of radioactive sodium in the human body. In 1938, Saul Hertz (1905–1950), Arthur Roberts (1912–2004), and Robley D. Evans (1907–1995) administered ^{128}I to rabbits and found a significant uptake in their thyroid glands, more than nine times the concentration in the liver. From this study they predicted that when higher amounts of activity became available, radioiodine would be a suitable radionuclide for the diagnosis and therapy of the thyroid [20, 21].

Immediately after the discovery of ^{131}I, it became an important radionuclide in diagnostic nuclear medicine, both for examinations of the thyroid as well as a tracer in different iodine-labelled substances. The first paper pertaining to the

FIGURE 1.5 The father of radiation protection, Rolf Sievert (left), and the father of nuclear medicine, George de Hevesy (right) in Stockholm (ca. 1950). Courtesy of Swedish Society of Radiation Physics.

diagnostic use of ^{131}I in patients was published by Hamilton, Mayo Soley (1907–1949), and Evans, in 1939. The first thyroid treatment using ^{131}I was performed by Hertz in 1941 using a mixture of 90 per cent ^{130}I and 10 per cent ^{131}I, which was administered to a female patient with hyperthyroidism.

The therapeutic application of ^{131}I became prominent in the 1940s, resulting in a breakthrough in radionuclide therapy after its discovery and introduction in the late 1930s. The first major achievement occurred in 1946 when ^{131}I was discovered to be a suitable radionuclide for diagnosing thyroid diseases, and Samuel M. Seidlin (1896–1955), Leo D. Marinelli (1906–1974), and Eleanor Oshry (1920–2007) treated a patient with thyroid cancer using ^{131}I. In the next year, Benedict Cassen (1902–1972) used ^{131}I to examine whether benign thyroid nodules could be differentiated from malignant ones. The use of ^{131}I for the diagnosis and treatment of thyroid diseases gradually spread worldwide. In 1949 George Ansell and Joseph Rotblat (1908–2005) produced the first radioiodine scan of a patient with goitre in Liverpool University's physics department, using a collimated Geiger–Müller detector. Two years later, a scintillation detector designed for brain studies and using ^{131}I-fluorescein was developed by E.H. Belcher and H.D. Evans [6]. In Sweden, inspired by a visit to the United States, Bengt Skanse (1918–1963) and Jan Waldenström (1906–1996) introduced ^{131}I for the diagnosis and treatment of thyroid illness around 1950, when nuclear medicine began to be used at university hospitals [1, 22]. For examination of the localization of brain tumours prior to surgery, George Moore used ^{131}I- diiodofluorescein in 1947, whereas B. Silverstone used ^{32}P in 1949. Another radionuclide, ^{198}Au, was used by Müller and colleagues in 1945 for intracavitary purposes.

In 1949 Leonidas D. Marinelli (1906–1974) presented early formulas for internal dosimetry, whereby he described the relationship between radiation dosage (absorbed dose was not yet defined) and the concentration of a radionuclide. He introduced two quantities, 'equivalent roentgen' and 'differential absorption ratio', where the latter refers to the ratio between the concentration in a specific tissue to the average body concentration. Pertinent information was presented in tables, as were some clinical applications and formulas for some of the specific radionuclides available at the time [23].

1.3 1950–1970: FIRST IMAGING APPARATUS AND RADIOPHARMACEUTICALS

The 1950s was the decade when two ground-breaking workhorses in nuclear medicine were invented, that is, the rectilinear scintigraph and the scintillation camera. It was also the decade when many ^{131}I-labelled radiopharmaceuticals were developed.

The company Abbott Laboratory began selling ^{131}I-HSA 1950 and FDA-approved Na^{131}I (sodium-iodine) for patient use in 1951. In 1955, George V. Taplin used ^{131}I- labelled rose bengal for scintigraph liver imaging, and ^{131}I-hippuran to measure kidney function using scintillation detectors. Other applications included a test with vitamin B12 labelled with ^{60}Co. Subsequently in 1953, Robert F. Schilling used other cobalt isotopes for studies of blood diseases, and in 1957 H. Knipping used ^{133}Xe to measure lung ventilation.

For many years the only detector available for non-invasive gamma measurement was the Geiger–Müller detector. When collimated, it could be moved step-by-step over the thyroid for 'imaging' using ^{131}I. As this was an ineffective detection method, physicists were looking for more efficient detector materials and instruments. The photoelectron multiplier tube (PMT) was developed in 1940 by C.C. Larson and H. Salinger. In combination with a NaI(Tl) scintillation crystal it became the detector of choice for almost all measurements of gamma radiation in nuclear medicine and paved the way for further development of better detector systems. The first scintillation detector consisted of a CaW-crystal was constructed by Benedict Cassen and colleagues in 1950 and used for scintigraphy (imaging) of the thyroid after administration of ^{131}I. The detector was improved by the introduction of a larger NaI(Tl) crystal connected to a PMT. The first medical detection of positron emission, that is, annihilation photons detected by two of NaI(Tl) detectors connected in coincidence, was performed already in 1951 by Gordon Brownell (1922–2005) and William Sweet (1910–2001), using ^{64}Cu and ^{74}As to localize brain tumours.

The instruments and counting methods were successively developed, and in 1951 Cassen and colleagues presented the rectilinear scanner, that is, the scintigraph, which automatically positioned the detector and scan over an organ [24–26]. The rectilinear scanner became important equipment in nuclear medicine departments worldwide and was extensively used for imaging to produce a scintigram of organs such as the brain, lung, liver, spleen, and thyroid until the 1980s. In 1951, the first rectilinear scanning device in the UK was developed at the Royal Cancer Hospital London to image the thyroid gland [6]. In 1957, the first whole-body scintigraph was developed by a Swedish team comprising Lars Jonsson, Lars-Gunnar Larsson (1919–2009), and Inger Ragnhult (1925–2006) [1, 22, 27] (Figure 1.6). The primary radiopharmaceuticals that dominated were ^{131}I, ^{85}Sr and ^{198}Au. However, the acquisition time required for large organs was lengthy, which made dynamic studies impossible. A solution to this had already come in 1952, when Hal O. Anger (1920–2005) invented the pin-hole camera, which through a collimator projected an image on a NaI(Tl) crystal in optical contact with photographic film [28]. A similar method was developed by Sven Johansson (1923–1994) and Bengt Skanse (1918–1963) (Figure 1.7), but with a multiple parallel hole collimator, which increased the sensitivity, yet insufficient for medical purposes [29].

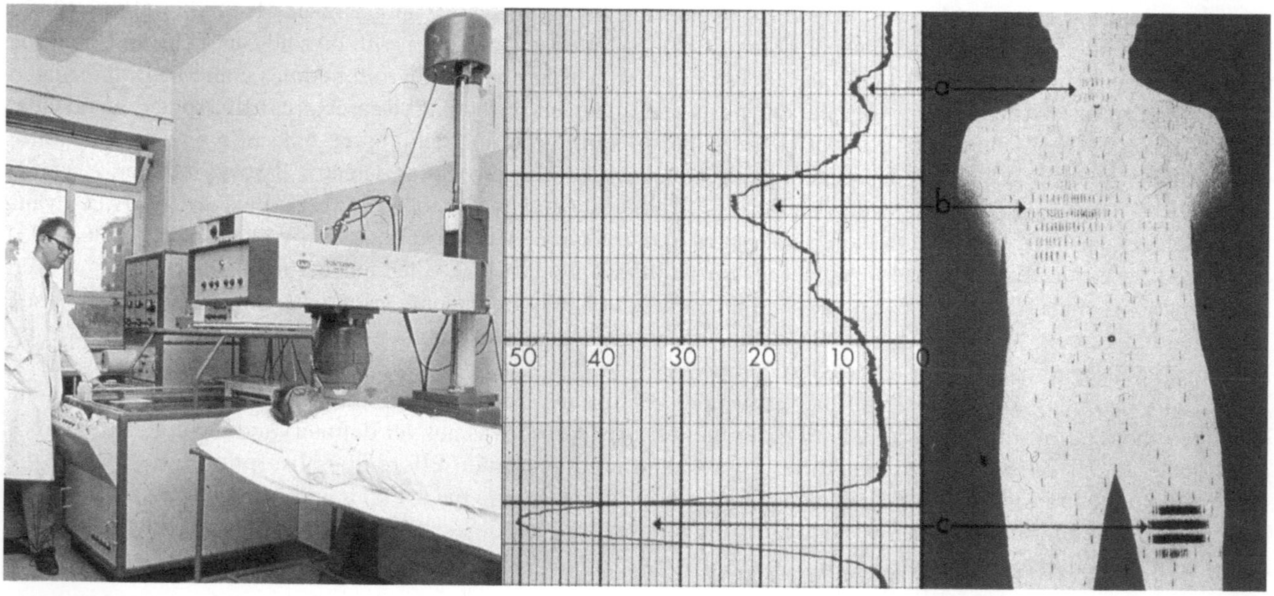

FIGURE 1.6 The whole-body scintigraph developed by Lars Jonsson and Inger Ragnhult at the Isotope Laboratory, Karolinska Hospital, Stockholm, which was later sold by LKB Ltd (left). An example of an ^{131}I scintigram with low activity in the thyroid (a) and metastasis in the right lung (b) after thyroid cancer treatment. The high activity in the left thigh (c) is an image artefact due to a contaminated handkerchief after the patient coughed up activity. Courtesy of Swedish Society of Radiation Physics.

FIGURE 1.7 The first scintillation camera: (a) parallel hole collimator, (b) scintillation NaI(Tl) crystal and (c) holder for a photographic plate developed by Lund University, Lund, Sweden.

A more significant step in the development of instrumentation in nuclear medicine was the development of the scintillation camera in 1958 by Hal O. Anger. He replaced the film with an array of PMTs mounted on a large single NaI(Tl) crystal. The origin of the registered gamma ray – that is, the scintillation in the crystal – was detected by weighting together the signals from the PMTs. This apparatus, also known as the Anger camera or gamma camera, revolutionized the field of nuclear medicine and was a real breakthrough for imaging, thereby enabling dynamic in vivo studies possible. The first Anger camera was installed at Ohio State University in 1962 by Nuclear Chicago and in subsequent years at many hospitals in the United States and Europe. The technique is still the basis used in today's single photon emission computed tomography (SPECT) cameras, although extensive improvement has been achieved since then [3, 30].

The ^{99}Mo → ^{99}Tc^{m} – generator (Figure 1.8) was developed in 1957 by Walter Tucker and Margaret Greene at the Hot Laboratory Division, Brookhaven National Laboratory, managed by Powell Richards (1917–2010). After a sluggish start, ^{99}Tc^{m} labelled to numerous substances became the world's most widely used radionuclide for in vivo imaging, both with planar scintillation cameras and SPECT. The mother-nuclide ^{99}Mo is produced in specialized fission reactors for radionuclide production. Other fission-produced medical nuclides include ^{32}P, ^{51}Cr, ^{90}Y, ^{133}Xe, ^{177}Lu and ^{186}Re.

In addition to this breakthrough and the subsequent development of instrumentation for imaging, the need for new suitable radionuclides for the gamma camera became obvious by production in a cyclotron. Since the pioneering days, many radionuclides have been developed mainly for diagnostics; many have disappeared completely, whereas others have been established in recent years. Examples include ^{57}Co, ^{67}Ga, ^{111}In, ^{123}I, and ^{201}Tl, which decay by electron capture (EC) and emits γ-radiation. Positron emitters utilized in positron emission tomography (PET) imaging are ^{11}C, ^{13}N, ^{15}O, ^{18}F, ^{64}Cu, ^{68}Ga, ^{110}In, and ^{124}I. Examples of radionuclides for therapy include β^--emitters such as ^{32}P, ^{90}Y ^{131}I, ^{153}Sm, ^{177}Lu, ^{198}Au, and ^{186}Re, as well as some α-emitting radionuclides, for example, ^{211}At and ^{223}Ra [31].

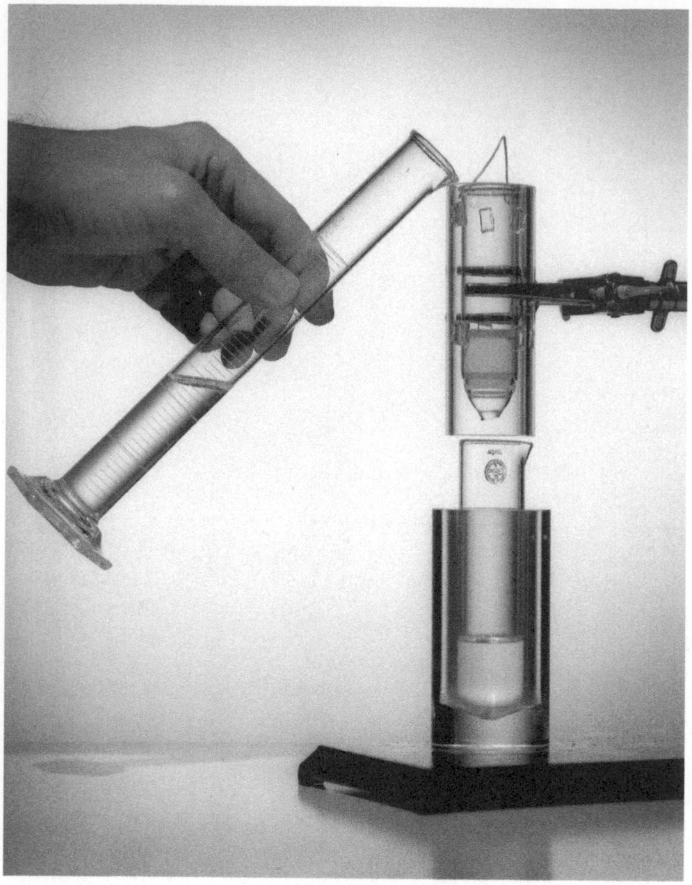

FIGURE 1.8 The first $^{99}Tc^{m}$-generator prototype $\left(^{99}Mo \rightarrow {}^{99}Tc^{m} \rightarrow {}^{99}Tc\right)$, invented in 1958 at Brookhaven National Laboratory, Upton, New York, by Walt Tucker, Powell Richards, and Margaret Greene. The $^{99}Tc^{m}$ pertechnetate solution is eluted as an ionic yield from ^{99}Mo bound to a substrate in the chromatographic column. Courtesy of Brookhaven National Laboratory.

The first medical cyclotron placed at a hospital for direct in-house production was built in 1955 at the Hammersmith Hospital, London, followed by the first installation in an American hospital at the Washington University Medical School in 1961. Subsequently, it was installed at many hospitals in the United States, and then worldwide. The number of cyclotrons existing currently is estimated to be more than a thousand.

The increasing use of radiopharmaceuticals during the 1950s necessitated the establishment of special radiopharmacy facilities at hospitals. Radioactive open sources were managed by hospital radiopharmacists, who used special equipment for dispensing owing to high activities and emitted radiation. At this time, medical physics progressed considerably, and a more formal academic society with a wider circle of members with interest in the field was necessitated, and many important societies were established. Many well-educated and trained clinical physicists at the hospitals contributed significantly to the development in the field of nuclear medicine in the subsequent decades.

A noteworthy aspect in nuclear medicine that is non-existent currently owing to newer fluorescent-antibody techniques, is radioimmunoassay (RIA), which was developed in 1959 by Rosalyn Yalow (1921–2011) and Saloman Berson (1918–1972) [32]. RIA is an in vitro nuclear medicine method for the quantitative measurement of small amounts of substances in plasma, such as hormones. The method mainly uses the characteristic X-ray-emitting and γ-emitting nuclide ^{125}I (discovered by Allen Reid and Albert Keston in 1946), but also β^- emitting nuclides such as ^{3}H, ^{14}C, ^{35}S, and $^{32,33}P$.

The 1950s was also the decade in which internal dosimetry evolved. In 1956, G.L. Brownell and Gerald J. Hine (1916–1987) together with Robert Loevinger (1916–2005) published a standard of absorbed dose estimations from internal emitters and described early studies of compartmental analysis of radiopharmaceuticals. In 1965, a committee named Medical Internal Radiation Dose (MIRD) was established within the Society of Nuclear

Medicine; its purpose was to develop standardized internal dosimetry procedures, improve published decay data for radionuclides and enhance the data on pharmacokinetics of radiopharmaceuticals. The first report, MIRD Pamphlet No. 1, was published in 1968 [33].

In 1961, David H. Ingvar (1924–2000) and Niels A. Lassen (1926–1997) introduced ^{133}Xe for quantitative measurements of regional cerebral blood flow (rCBF) using a system of multiple single NaI(Tl)-detectors [1]. In 1962, David Kuhl (1929–2017) introduced emission reconstruction tomography, which was later further developed and became the basis for image reconstruction in SPECT, PET and CT. Its reconstruction method was relatively simple, and the rapid development of computer tomography, CT, by Godfrey Hounsfield (1919–2004) benefitted PET significantly.

New radiopharmaceuticals and applications were continually developed. In 1963, Henry Wagner (1927–2012) introduced ^{131}I-labelled albumin aggregates for imaging of lung perfusion, which was performed on himself by his colleague Masahiro Lio. This was the first produced lung scan ever performed in a human being; subsequently, it was performed on a patient with acute pulmonary embolism [34]. Furthermore, ^{131}I-labelled albumin aggregates were also used to study of the macrophage system in the liver by George V. Taplin (1910–1979). Another scintigraphy study performed was ^{133}Xe ventilation for pulmonary embolism by Wagner and colleagues in 1968. Additionally, C.L. Edwards used ^{67}Ga for tumour scintigraphy.

1.4 1970–1990: TOMOGRAPHIC TECHNIQUES, RADIOIMMUNOLOGY, AND DOSIMETRY

In the early 1970s, a new radionuclide was introduced, that is, indium-111 (^{111}In), which significantly affected nuclear medicine imaging for many decades. Initially, ^{111}In-chloride was used for the diagnosis of tumours and for bone marrow scintigraphy. Subsequently, different ^{111}In-chelates, ^{111}In-oxine being the most prominent, were introduced for labelling different types of blood cells for imaging of suspected inflammations, which was first described by Thakur in 1977 [35]. During the 1980s, tumour imaging using labelled monoclonal antibodies progressed considerably, and ^{111}In became one of the most important radionuclides.

During the 1970s, Gopal Subramanian and John McAfee introduced ^{99}Tc^{m}-labelled phosphates for imaging of bone metastasis. David Kuhl performed the first human quantitative measurement of cerebral blood volume, and the first stress-test myocardial scintigraphy was implemented by H. William Strauss. The use of ^{99}Tc^{m} radiopharmaceuticals as instant kits was developed and increased during the 1970s. The first one was ^{99}Tc^{m}-DTPA for renography developed by W. Eckelman and P. Richards. Elliot Lebowitz introduced ^{201}Tl for perfusion studies of the heart wall. In 1973, David Goldenberg demonstrated that radiolabelled antibodies can target and image human tumours in animals. Those antibodies were further developed during the 1970s; their research was subsequently intensified by other researchers in the following decade. In 1976, N. Firusian began using ^{89}Sr for the palliative treatment of bone metastasis.

In 1975, Michel Ter-Pogossian (1925–1996) and colleagues presented a PET that used the filtered back projection method of reconstruction. In the following year, John Keyes developed the first general-purpose SPECT camera, and the first SPECT camera dedicated to brain studies was developed by Ronald Jaszczak. Subsequently, he developed a special phantom for the quality-control purposes of SPECT or PET cameras, known as the Jaszczak Phantom.

Robert Loevinger (1916–2005) and M. Berman presented the general S-formalism of the MIRD in 1976 [36], for the calculation of absorbed dose from internal radionuclides. Hitherto, 26 reports and 6 books have been published by MIRD, including the MIRD-primer 1991, and subsequent pamphlets are important manuals for internal dosimetry [33, 37].

In 1980, Stig Larsson (1943–2014), Karolinska University Hospital, Stockholm, developed the first complete SPECT rotating scintillation camera system, which was further developed and commercialized by General Electric (GE 400T) and became the technique routinely used in SPECT diagnostics [38].

In the 1980s, parallel to the continuing imaging instrumental development, a significant amount of research was performed to obtain new radiopharmaceuticals and develop new diagnostic methods. Some examples include the increased pursuit for the ‘magic bullet’, that is, to find suitable radionuclides and carriers for specific tumour localization, for both for diagnosis and therapy. Two pioneering works often referred to are that by Jean-Pierre Mach and colleagues (1981), who used radioiodine-labelled monoclonal antibodies for colon carcinoma, and that by Steve Larson and Jeff Carrasquillo (1982), who treated malignant melanoma with ^{131}I-labelled monoclonal antibodies.

One radiopharmaceutical that accelerated the acceptance and development of PET was ^{18}F-FDG (2-uorodeoxy-D-glucose), which, similar to many other radiopharmaceuticals, had been already developed at the Brookhaven National Laboratory in 1978. However, it became as important as ^{99}Tc^{m} in nuclear imaging only after several decades. Two successful radiopharmaceuticals for SPECT brain imaging that were introduced were ^{99}Tc^{m}-HMPAO (Ceretec®, Amersham) and ^{99}Tc^{m}-ECD (Neurolite). These complexes cross the intact blood brain barrier and are trapped within the brain parenchyma. In 1985, Peter Ell and colleagues published the world's first cerebral blood flow image using $^{99}Tc^{m}$-HMPAO. Because both are liposoluble, they became an alternative to ^{111}In-oxine for labelling of leukocytes for imaging of unknown inflammations, as shown by Peters and colleagues (1988) [39].

1.5 1990–2010: IMPROVED IMAGING BY MULTI-MODALITY SYSTEMS AND NOVEL MOLECULAR IMAGING

The SPECT systems were further developed and characterized as more robust rotation gantries and dual detectors. Dedicated scintillation cameras for smaller organs, particularly in high-resolution cardiac imaging, were constructed based on the new cadmium zinc telluride (CZT) solid-state detector technology. Methods for attenuation and scattering correction in SPECT have evolved, enhancing the image quality and quantification and hence the diagnostic quality. At the end of the 1990s, the SPECT systems were combined with low-dose and few-slice CT mounted on the same rotation gantry to perform patient-specific attenuation correction through transmission measurements. Subsequently, SPECT and fully complete CT were combined into hybrid SPECT/CT systems. Hence, the simultaneous acquisition of functional information from the SPECT and anatomical information from the CT could be accomplished and, after image processing resulting in superimposed images, so-called image fusion. The further development of SPECT in recent years in addition to the significant progress in computing power has resulted in the introduction of the CZT solid-state detectors, replacing the PMT and special multi-pinhole collimator, which is used in SPECT/MRI brain-imaging systems.

The progress of quantitative three-dimensional imaging using hybrid devices, that is, SPECT/CT, and PET/CT, continued in the beginning of the twentieth century, whereas PET/MRI has been further developed in recent years. Pre-clinical research has increased markedly, particularly that involving the use of small-animal imaging equipment as miniature hybrid system copies of those used for clinical imaging [40].

Whole-body PET scanning was introduced in the beginning of the 1990s and began to be accepted as an important imaging tool, which significantly affected nuclear-medicine imaging. In 1991, Ronald Nutt and David William Townsend proposed the use of simultaneous PET and CT imaging, which yielded the diagnostic information of both the organ physiology as well as its anatomy. However, their first prototype was first presented in 1998, and the first commercial PET/CT was launched in year 2000 by CTI. Subsequently, hybrid PET/CT-scanners quickly became highly valued imaging equipment for molecular imaging [41, 42]. In PET systems, the traditional scintillation material NaI(Tl) was replaced by new crystal block detectors, first with BGO (bismuth germanium oxide) and later by cerium-doped LSO (lutetium oxyorthosilicate) or LYSO (lutetium-yttrium oxyorthosilicate). The latter demonstrated much better performances in terms of energy, spatial, and time resolution.

Nuclear medicine is no longer limited to the research and development of new radiopharmaceuticals and clinical methods. By contrast, the field has progressed and now includes research in molecular biology and through labelled biomarker visualization of the cellular function and the subsequent molecular processes in living organisms, including human organs. The development of nuclear medicine and new modalities in molecular imaging has contributed to the improvement in radionuclide therapy through the optimization of pre-clinical and clinical developments [43].

During the last two decades, the number of PET examinations has increased significantly with ^{18}F-FDG being the most used radiopharmaceutical. In 1998, the first study using ^{18}F-FDG was conducted in PET imaging to analyse chemotherapy response to predict the response to subsequent high-dose chemotherapy [44]. Another positron-emitting radiopharmaceutical that is widely used is ^{18}F-NaF (sodium fluoride). This agent has become an alternative to $^{99}Tc^{m}$-phosphate substances for skeletal scintigraphy, but the latter is still the most typically used. The ^{18}F-NaF was developed and used occasionally already in the 1970s; however, because it was expensive and PET was rarely accessible at the time, $^{99}Tc^{m}$-MDP was the preferred choice for many years [45]. Other positron-emitting radionuclides that have been introduced after the millennium shift include ^{68}Ga. It is mainly produced in by the $^{68}Ge \rightarrow {}^{68}Ga$ generator system, and typically used for the diagnosis of neuroendocrine as ^{68}Ga-DOTATOC or ^{68}Ga-DOTATOC [46, 47]. This radiopharmaceutical is often used in combination with radionuclide therapy involving ^{90}Y- or ^{177}Lu- DOTATOC.

Patient dosimetry continued to develop parallel with the rapid growth of new instruments, radiopharmaceuticals, as well as diagnostic and therapeutic applications. Already in 1987, the ICRP published a report regarding the radiation doses of radiopharmaceuticals to patients with successive updates, as well as on reference computational phantoms in 2009 [9]. The ICRU published a report in 2002 regarding absorbed dose estimations in nuclear medicine [48], whereas the MIRD committee continued to publish a number of new pamphlets, including the recently introduced concept Biological Effective Dose, BED, in nuclear medicine [49, 50] and on the reliability of radiation dose estimations in internal radionuclide therapy [51].

With this chapter's modest overview of numerous milestones and breakthroughs in the field of nuclear medicine for more than a whole century, the author wishes upon the reader many interesting hours of reading of the continuing chapters of this book, and new self-improvements in the exciting field of nuclear medicine. Remember – our current actions will be history tomorrow.

REFERENCES

[1] S. Carlsson, "A Glance at The History of Nuclear Medicine," *Acta Oncologica*, vol. 34, no. 8, pp. 1095–1102, 1995, doi: 10.3109/02841869509127236.

[2] T. F. Buddinger and T. Jones, "History of Nuclear Medicine and Molecular Imaging," in *Comprehensive Biomedical Physics*, T. F. Buddinger, Ed., Vol. 1, ed. *Nuclear Medicine and Molecular Imaging*, A. Brahme, Ed. Amsterdam: Elsevier, 2014, pp. 1–37.

[3] B. F. Hutton, "The Contribution of Medical Physics to Nuclear Medicine: Looking Back – A Physicist's Perspective," *EJNMMI Physics*, vol. 1, no. 1, p. 2, 2014, doi: 10.1186/2197-7364-1-2.

[4] H. Scheida, Ed. *History of Nuclear Medicine in Europe*. Stuttgart: Schattauer GmbH, 2003.

[5] H. N. Wagner, "Nuclear Medicine: 100 Years in the Making 1896–1996." *Journal of Nuclear Medicine*, vol. 37, no. 10, pp. 18N–37N, October 1, 1996. [Online]. Available: http://jnm.snmjournals.org/content/37/10/18N.short. http://jnm.snmjournals.org/content/37/10/18N.full.pdf.

[6] *A History of Radionuclide Studies in the UK – 50th Anniversary of the British Nuclear Medicine Society*. Open Access: Springer International Publishing AG, 2016, p. 152.

[7] T. Jones and D. Townsend, "History and Future Technical Innovation in Positron Emission Tomography," *Journal of Medical Imaging*, vol. 4, no. 1, p. 011013, 2017, doi: 10.1117/1.jmi.4.1.011013.

[8] L. E. Williams, "Anniversary Paper: Nuclear Medicine: Fifty Years and Still Counting," *Medical Physics*, vol. 35, no. 7, Part 1, pp. 3020–3029, 2008, doi: 10.1118/1.2936217.

[9] S. Mattsson, "Patient Dosimetry in Nuclear Medicine," *Radiation Protection Dosimetry*, vol. 165, no. 1–4, pp. 416–423, 2015, doi: 10.1093/rpd/ncv061.

[10] A. Romer, "Accident and Professor Röntgen," *American Journal of Physics*, vol. 27, no. 4, pp. 275–277, 1959, doi: 10.1119/1.1934825.

[11] E. C. Watson, "The Discovery of X-Rays," *American Journal of Physics*, vol. 13, no. 5, pp. 281–291, 1945, doi: 10.1119/1.1990728.

[12] H. Levi, *George de Hevesy – Life and Work*. Rhodos, Copenhagen: Hilde Levi, 1985.

[13] W. G. Myers, "George de Hevesy: The Father of Nuclear Medicine," *Journal of Nuclear Medicine*, Proceedings of the 26th Annual Meeting vol. 20, no. 6, pp. 590–698, June 1, 1979. [Online]. Available: http://jnm.snmjournals.org/content/20/6/590.short. http://jnm.snmjournals.org/content/20/6/590.full.pdf.

[14] B. Lindell, *Pandora's Box. The History of Radiation, Radioactivity, and Radiological Protection. Part 1. The Time before World War II*. 1996 Bo Lindell and 2019 NSFS (Nordic Society for Radiation Protection), 1996 (translated into English in 2019), p. 202.

[15] M. Sekiya and M. Yamasaki, "Rolf Maximilian Sievert (1896–1966): Father of Radiation Protection," vol. 9, no. 1, pp. 1–5, 2016, doi: 10.1007/s12194-015-0330-5.

[16] H. S. Martland, "The Occurrence of Malignancy in Radioactive Persons – A General Review of Data Gathered in the Study of the Radium Dial Painters, with Special Reference to the Occurrence Of Osteogenic Sarcoma and the Inter-Relationship of Certain Blood Diseases," *The American Journal of Cancer*, vol. XV, no. 4, pp. 2435–2516, 1931, doi: 10.1158/ajc.1931.2435.

[17] R. B. Gunderman and A. S. Gonda, "Radium Girls," *Radiology*, vol. 274, no. 2, pp. 314–318, 2015, doi: 10.1148/radiol.14141352.

[18] R. M. Macklis, "The Great Radium Scandal," *Scientific American*, vol. 269, no. 2 (August), pp. 94–99, 1993.

[19] G. de Hevesy, *Radioactive Indicators. Their Applications in Biochemistry, Animal Physiology and Pathology*. New York: Interscience Publ., 1948.

[20] F. H. Fahey, F. D. Grant, and J. H. Thrall, "Saul Hertz, MD, and the Birth of Radionuclide Therapy," *EJNMMI Physics*, vol. 4, no. 1, 2017, doi: 10.1186/s40658-017-0182-7.

[21] B. Hertz, "Dr. Saul Hertz (1905–1950) Discovers the Medical Uses of Radioactive Iodine: The First Targeted Cancer Therapy," InTech, 2016.

[22] S. Mattsson, L. Johansson, H. Jonsson, and B. Nosslin, "Radioactive Iodine in Thyroid Medicine – How It Started in Sweden and Some of Today's Challenges," (in English), *Acta Oncologica*, Article; Proceedings Paper vol. 45, no. 8, pp. 1031–1036, Dec. 2006, doi: 10.1080/02841860600635888.

[23] L. D. Marinelli, "Dosage Determination in the Use of Radioactive Isotopes," *Journal of Clinical Investigation*, vol. 28, no. 6, pp. 1271–1280, 1949, doi: 10.1172/jci102194.

[24] B. Cassen, L. Curtis, C. Reed, and R. Libby, "Instrumentation for I-131 Use in Medical Studies," *Nucleonics*, vol. 9, no. 2, p. 46, 1951.

[25] W. H. Blahd, "Ben Cassen and the Development of the Rectilinear Scanner," *Seminars in Nuclear Medicine*, vol. 26, no. 3, pp. 165–170, 1996/07/01/ 1996, doi: https://doi.org/10.1016/S0001-2998(96)80021-3.

[26] W. H. Blahd, "Benedict Cassen: The Father of Body Organ Imaging," *Cancer Biotherapy and Radiopharmaceuticals*, vol. 15, no. 5, pp. 423–429, 2000, doi: 10.1089/cbr.2000.15.423.

[27] L. Jonsson, L. G. Larsson, and I. Ragnhult, "A Scanning Apparatus for the Localization of Gamma Emitting Isotopes in Vivo," *Acta Radiologica*, vol. 47, pp. 217–228, 1957.

[28] H. O. Anger, "Use of a Gamma Ray Pinhole Camera for In-Vivo Studies," *Nature*, no. 170, p. 200, 1952.

[29] S. A. E. Johansson and B. Skanse, "A Photographic Method of Determining the Distribution of Radioactive Material in Vivo," *Acta Radiologica*, vol. 39, no. 4, pp. 317–322, 1953, doi: 10.3109/00016925309136717.

[30] B. F. Hutton, "The Origins of SPECT and SPECT/CT," *European Journal of Nuclear Medicine and Molecular Imaging*, vol. 41, no. S1, pp. 3–16, 2014, doi: 10.1007/s00259-013-2606-5.

[31] F. F. Knapp and A. Dash, *Radiopharmaceuticals for Therapy*. India: Springer, 2016.

[32] S. J. Goldsmith, "Georg de Hevesy Nuclear Medicine Pioneer Award Citation-1986 Rosalyn S. Yalow and Solomon A. Berson," *Journal of Nuclear Medicine*, vol. 28, no. 10, pp. 1637–1639, October 1, 1987 1987. [Online]. Available: http://jnm.snmjournals.org/content/28/10/1637.short. http://jnm.snmjournals.org/content/28/10/1637.full.pdf.

[33] R. Loevinger, T. F. Budinger, and E. E. Watson, *MIRD Primer for Absorbed Dose Calculations*. New York: The Society of Nuclear Medicine, 1991.

[34] J. H. N. Wagner, *A Personal History of Nuclear Medicine*. London: Springer, 2006.

[35] M. L. Thakur, J. P. Lavender, R. N. Arnot, D. J. Silvester, and A. W. Segal, "Indium-111-Labeled Autologous Leukocytes in Man," *Journal of Nuclear Medicine*, vol. 18, pp. 1012–1019, 1977.

[36] R. Loevinger and M. Berman, *A Revised Schema for Calculation of the Absorbed Dose from Biologically Distributed Radionuclides. MIRD Pamphlet No. 1, Revised*. New York: Society of Nuclear Medcine, 1976.

[37] W. E. Bolch, K. F. Eckerman, G. Sgouros, and S. R. Thomas, "MIRD Pamphlet No. 21: A Generalized Schema for Radiopharmaceutical Dosimetry-Standardization of Nomenclature," (in English), *Journal of Nuclear Medicine*, vol. 50, no. 3, pp. 477–484, Mar 2009, doi: DOI 10.2967/jnumed.108.056036.

[38] S. Larsson and A. Israelsson, "Considerations on System Design, Implementation and Computer Processing in SPECT," vol. 29, no. 4, pp. 1331–1342, 1982, doi: 10.1109/tns.1982.4332190.

[39] A. M. Peters, "The Utility of [99mTc]HMPAO-leukocytes for Imaging Infection," *Seminars in Nuclear Medicine,* vol. 24, no. 2, pp. 110–127, 1994/04/01/ 1994, doi: https://doi.org/10.1016/S0001-2998(05)80226-0.

[40] J. A. Patton, D. W. Townsend, and B. F. Hutton, "Hybrid Imaging Technology: From Dreams and Vision to Clinical Devices," *Seminars in Nuclear Medicine*, vol. 39, no. 4, pp. 247–263, 2009, doi: 10.1053/j.semnuclmed.2009.03.005.

[41] M. Pizzichemi, "Positron Emission Tomography: State of the Art and Future Developments," *Journal of Instrumentation*, vol. 11, no. 08, pp. C08004–C08004, 2016, doi: 10.1088/1748-0221/11/08/c08004.

[42] B. F. Hutton, K. Erlandsson, and K. Thielemans, "Advances in Clinical Molecular Imaging Instrumentation," *Clinical and Translational Imaging*, vol. 6, no. 1, pp. 31–45, 2018, doi: 10.1007/s40336-018-0264-0.

[43] M. L. Thakur, "Genomic Biomarkers for Molecular Imaging: Predicting the Future," *Seminars in Nuclear Medicine*, vol. 39, no. 4, pp. 236–246, 2009, doi: 10.1053/j.semnuclmed.2009.03.006.

[44] M. D. Farwell, D. A. Pryma, and D. A. Mankoff, "PET/CT Imaging in Cancer: Current Applications and Future Directions," *Cancer*, vol. 120, no. 22, pp. 3433–3445, 2014, doi: 10.1002/cncr.28860.

[45] R. K. Kulshrestha, S. Vinjamuri, A. England, J. Nightingale, and P. Hogg, "The Role of 18F-Sodium Fluoride PET/CT Bone Scans in the Diagnosis of Metastatic Bone Disease from Breast and Prostate Cancer," *Journal of Nuclear Medicine Technology*, vol. 44, no. 4, pp. 217–222, Dec 1, 2016 2016, doi: 10.2967/jnmt.116.176859.

[46] T. D. Poeppel et al., "68Ga-DOTATOC Versus 68Ga-DOTATATE PET/CT in Functional Imaging of Neuroendocrine Tumors," *Journal of Nuclear Medicine*, vol. 52, no. 12, pp. 1864–1870, 2011, doi: 10.2967/jnumed.111.091165.

[47] M. M. Graham, X. Gu, T. Ginader, P. Breheny, and J. J. Sunderland, "68 Ga-DOTATOC Imaging of Neuroendocrine Tumors: A Systematic Review and Metaanalysis," *Journal of Nuclear Medicine*, vol. 58, no. 9, pp. 1452–1458, 2017, doi: 10.2967/jnumed.117.191197.

[48] ICRU, "International Commission on Radiation and Units. Absorbed-Dose Specification in Nuclear Medicine (Report 67)," *Journal of the ICRU*, Report vol. 2, no. 2, pp. 1–110, 2002.

[49] S. Baechler, R. F. Hobbs, A. R. Prideaux, R. L. Wahl, and G. Sgouros, "Extension of the Biological Effective Dose to the MIRD Schema and Possible Implications in Radionuclide Therapy Dosimetry," *Medical Physics*, vol. 35, no. 3, pp. 1123–1134, 2008, doi: 10.1118/1.2836421.

[50] B. W. Wessels et al., "MIRD Pamphlet No. 20: The Effect of Model Assumptions on Kidney Dosimetry and Response—Implications for Radionuclide Therapy," *Journal of Nuclear Medicine*, vol. 49, no. 11, pp. 1884–1899, 2008, doi: 10.2967/jnumed.108.053173.

[51] Y. K. Dewaraja et al., "MIRD Pamphlet No. 23: Quantitative SPECT for Patient-Specific 3-Dimensional Dosimetry in Internal Radionuclide Therapy," *Journal of Nuclear Medicine*, vol. 53, no. 8, pp. 1310–1325, 2012, doi: 10.2967/jnumed.111.100123.

2 Basic Atomic and Nuclear Physics

Gudrun Alm Carlsson and Michael Ljungberg

CONTENTS

2.1 THE ATOM AND ITS NUCLEUS

2.1.1 Understanding Radioactivity

It is necessary to gain the basic knowledge regarding the nature of atoms to understand the effects of ionizing radiation and its interactions with matter that are fundamental to the entire topic of nuclear medicine, and this chapter is a summary of the basic atomic and nuclear physics that are important for the further chapters in this book. More reading can be found in, for example, the books of Evans [1], Attix [2], and Podgorsak [3].

The atom (Greek *atomos* = indivisible) is the smallest part of an element that maintains the identity of a substance, that is, its characteristics have well-described chemical properties. Individual atoms are too small to be directly observed; however, by observing some of the characteristics of the atom, a simple yet useful and illustrative atomic model has been developed. With this model, it is possible to explain many experimental observations that are relevant for medical radiation physics.

DOI: 10.1201/9780429489556-2

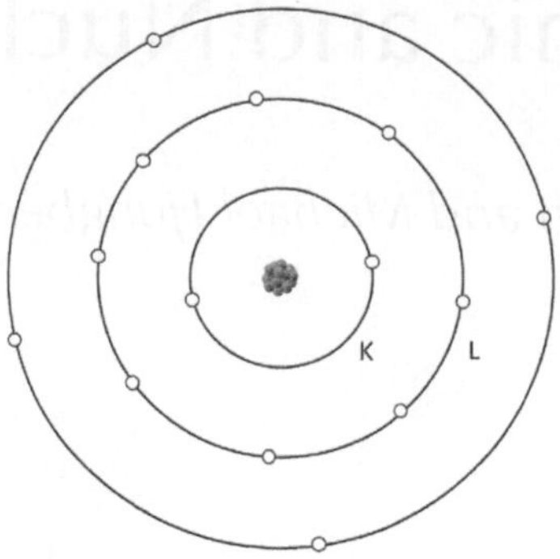

FIGURE 2.1 Schematic figure of the atomic structure. In this model, the electrons are circulating in different shells, labelled K, L, M, where each shell has well-defined binding energies.

TABLE 2.1
Some Data Characteristic for Electrons and Nucleons

Name	Symbol	Gram	Rest mass a.m.u.	MeV	Charge	Radii (cm)
Electron	e^-	$0.9108 \cdot 10^{-27}$	0.000549	0.511	-e	$2.82 \cdot 10^{-13}$
Proton	p	$1.6724 \cdot 10^{-24}$	1.00728	938.22	+e	$1.45 \cdot 10^{-13}$
Neutron	n	$1.6747 \cdot 10^{-24}$	1.00867	939.23	0	$1.45 \cdot 10^{-13}$

Note: One a.m.u. is equal to 1/12 of the atomic mass of ^{12}C.

An atomic model that today is widely accepted as a starting point is shown in Figure 2.1.

In this model, the atom consists of a central, electrically positively charged nucleus, around which a major part of the atomic mass is concentrated. Around the nucleus are circulating electrically negatively charged electrons, which will neutralize the positive charge of the nucleus when all charges are summed up; therefore, the atom when viewed externally seems electrically neutral.

The atomic nucleus comprises two elementary particles, protons and neutrons, which have approximately the same mass; however, the proton carries a positive charge, and its magnitude is equal to that of the electron's negative charge. The neutron is electrically neutral. A common name for both the protons and the neutrons is the 'nucleon'. The main characteristics of the electron, proton, and neutron are summarized in Table 2.1.

For comparison, the radius of the hydrogen atom is approximately 10^{-8} cm which, when considering the data in Table 2.1, indicates that the nucleus occupies a very small part of the overall atomic volume even though it virtually contains all the mass of the atom.

Currently, more complete models of the protons and neutrons and other particles have been developed. In the standard model of elementary particles, there exists a family of six particles, called quarks, that have the names 'up', 'down', 'charm', 'top', 'strange', and 'bottom'. The proton in this model consists of two 'up' quarks and one 'down' quark while the neutron consists of on one 'up' quark and two 'down' quarks. The quarks are kept together by the strong nuclear force (which is in the standard model described by an exchange of particles, called gluons). In addition, a family of six particles called leptons is also present in the standard model. These consist of the electron and five particles called tau, muon, electron-neutrino, tau-neutrino, and muon-neutrino. Among these leptons, the electron (and its anti-particle–the positron) are very important for the future text of this series of books. Quarks and leptons are termed fermions, and they are generally regarded as matter with a defined mass. Particles, that are active in interactions and exchange of forces are the photon and three gluons: the Z-, W- and the Higgs boson.

Three different types of forces are described by the standard model: (a) the strong force that holds the quarks together within the nucleus; (b) the electromagnetic force that results in the interactions between electrically charged particles; and (c) the weak force that interacts between subatomic particles and that can give rise to a radioactive decay of atoms. In addition to these three, there is also the force of gravity that brings objects with different masses together.

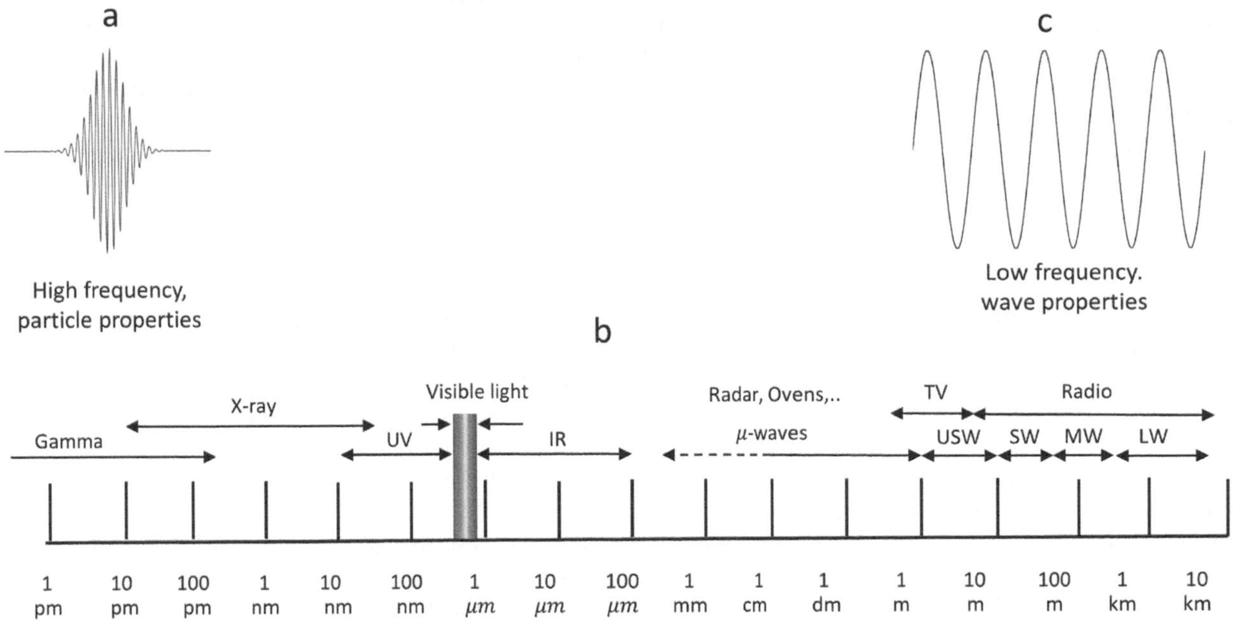

FIGURE 2.2 (a) Schematic figure of a wave package (photon) traveling in certain direction with an energy proportional to the frequency (E = hυ). (b) For lower frequencies, the electromagnetic radiation is often described as wave propagation instead of energy packages. (c) Spectrum of electromagnetic radiation, going from a very long wavelength commonly used for radio and TV communications (LW, MW, SW and USW) and passing µ-waves, infrared (IR) light, visible light, ultraviolet light (UV) up to the energetic x-ray and gamma radiation.

The equivalence between mass and energy of a particle is described by the Einstein relation $E=m_0c^2$, where m_0 is the rest-mass of the particle and c is the velocity of light as measured in vacuum. This equivalence has a direct implication in medical physics for Positron emission tomography (PET) when these systems measure dual photons emitted in opposite directions as a result of an annihilation of an electron and a positron. In this process, two masses are here converted to energy in the form of electromagnetic radiation.

The magnitude of the charge *e* of an electron, proton, or positron is approximately $1.602\ 10^{-19}$ C. The kinetic energy of a charged particle in motion is often measured in units of eV, which is defined as the kinetic energy that an electron will obtain when it is accelerated in an electrical potential of 1 V. One eV is, therefore, equal to $1.602\ 10^{-19}$ J.

The energy of a photon can be calculated from the equation $E = h\upsilon$, where υ denotes the frequency of the electromagnetic radiation and *h* denotes the Planck's constant. Electromagnetic radiation can be described both as a wave with an amplitude and a frequency (Figure 2.2), but it also has particle characteristics.

In a quantum mechanical description, a photon can be seen as a wave package travelling in a certain direction where the energy is defined by its frequency. In the energy range wherein most nuclear medicine applications reside, one can regard the photon as an energy quantum with particle properties more than having wave-properties; the opposite occurs as one goes down toward lower energies (i.e., longer wavelengths). A clear boundary between these two does not exist; however, generally visible light and ultra-violet light are described better as waves than particles (Figure 2.2).

2.1.2 Nuclear Physical Symbols and Notations

Each element has its special notations and labels (H, He, Li, Be, etc.). A compressed symbol language is used, which provides valuable information about the state of the element and the specific nucleus, and this is described by the following generic description.

$$ {}_{Z}^{A}X_{P}^{O} \tag{2.1} $$

Here, X, A, Z, O, and P denote a label of the element (H, He, Lu, Cs, etc.), mass number equal to the sum of protons and neutrons, atomic number equal to the number of protons, state of ionizations (2-, 3+, etc.) or excitation state of the nucleus, and molecular label equal to the number of atoms in a molecule (H_2, O_2, etc.), respectively. For metastable states, the position O should then be denoted by an 'm', for example, $^{99}Tc^{m}$. However, in many publications and textbooks ^{99m}Tc is also used.

In addition, the symbol N is often used to denote the number of neutrons in the nucleus. From the above, it can be seen that

$$N = A - Z \tag{2.2}$$

The term *nuclide* means a specific type of a nucleus that is uniquely determined by the mass number A and the atomic number Z. Nuclides with the same atomic number Z but with different mass numbers A are called *isotopes*. Because the chemical properties of an atom are determined by the structure of the electron cloud and how they are arranged, all atoms with the same atomic number Z but different A (isotopes) will express the same chemical properties; therefore, they are located on the same place in the periodical system, as is shown in Figure 2.3 (the name isotope originates from the two words (*iso* = equal and *topos* = place); *isotones* are nuclides with the same number of neutrons, and *isobars* are nuclides with the same mass number A.

The periodic table, first defined by Dmitrij Mendelejev in 1869, describes atomic elements according to their increasing atomic number Z and some of their chemical and physical properties along with their electron configuration in the outer electron shells. The table arranges elements with similar chemical properties in groups (column). The rows (period) define the elements with similar numbers of electrons in the outer shell. Within each period, the metals are in general found on the left side, while the non-metals are found on the right side. In the layout, shown in Figure 2.3, the

period	Group 1	Group 2	Group 3		Group 4	Group 5	Group 6	Group 7	Group 8	Group 9	Group 10	Group 11	Group 12	Group 13	Group 14	Group 15	Group 16	Group 17	Group 18
1	1.008 H 1																		4.003 He 2
2	6.94 Li 3	9.01 Be 4												10.82 B 5	12.01 C 6	14.01 N 7	16.00 O 8	19.00 F 9	20.18 Ne 10
3	22.99 Na 11	24.32 Mg 12												26.98 Al 13	28.09 Si 14	30.95 P 15	32.07 S 16	35.46 Cl 17	39.94 Ar 18
4	39.10 K 19	40.09 Ca 20	44.98 Sc 21		47.90 Ti 22	50.95 V 23	52.01 Cr 24	54.94 Mn 25	55.85 Fe 26	58.94 Co 27	58.71 Ni 28	65.54 Cu 29	69.72 Zn 30	69.72 Ga 31	72.60 Ge 32	74.91 As 33	78.96 Se 34	79.92 Br 35	83.8 Kr 36
5	85.48 Rb 37	87.63 Sr 38	88.92 Y 39		91.22 Zr 40	92.99 Nb 41	95.95 Mo 42	99 Tc 43	101.1 Ru 44	102.91 Rh 45	106.7 Pd 46	107.88 Ag 47	112.41 Cd 48	114.82 In 49	118.70 Sn 50	121.76 Sb 51	127.91 Te 52	126.91 I 53	131.30 Xe 54
6	132.91 Cs 55	137.38 Ba 56	138.9 La 57	58-71	178.58 Hf 72	180.95 Ta 73	183.86 W 74	186.22 Re 75	190.2 Os 76	192.2 Ir 77	195.05 Pt 78	197.0 Au 79	200.61 Hg 80	204.39 Tl 81	207.21 Pb 82	209.00 Bi 83	210 Po 84	(211) Ar 85	222 Rn 86
7	(223) Fr 87	228.05 Ra 88	227 Ac 89	90-103	(267) Rf 104	(268) Db 105	(269) Sg 106	(270) Bh 107	(277) Hs 108	(278 Mt 109	(281) Ds 110	(282) Rg 111	(285) Cn 112	(286) Nh 113	(289) Fl 114	(289) Mc 115	(293) Lv 116	(294) Ts 117	(294) Og 118

140.13 Ce 58	140.91 Pr 59	144.27 Nd 60	147 Pm 61	150.35 Sm 62	152.0 Eu 63	157.26 Gd 64	158.93 Tb 65	162.51 Dy 66	164.94 Ho 67	167.27 Er 68	168.94 Tm 69	173.04 Yb 70	174.99 Lu 71
232.05 Th 90	231 Pa 91	238.07 U 92	237) Np 93	242) Pu 94	243) Am 95	245) Cm 96	249) Bk 97	249) Cf 98	254) Es 99	253) Fm 100	256) Md 101	256) No 102	266) Lr 103

s block
p block
d block
f block

FIGURE 2.3 The periodic table of elements.

lower number is the atomic number, and the upper number is the mass number, which is presented as the average mass of the atoms of an element presented as the weighted average of all isotopes of the atomic element.

The atomic elements can be grouped in four blocks:

1. The s-block consists of alkali metals and alkaline earth metals (group 1,2) plus hydrogen and helium. In these atoms, the electrons with the highest energy in the ground state are of type s in the outer shell (s-orbitals).
2. The p-block consists of the atomic elements in groups 13–18 excluding helium. In these elements, the outermost electron shell contains electrons of type p (p-orbitals).
3. The d-block consists of elements in groups 3–12, that is, the transition metals. In the ground state of these atoms, the electrons with the highest energy are in the d-orbitals.
4. The f-block is a group of elements in which the highest-energy electrons are present in f-orbitals in the ground state. The f-elements in the 6th period are the *lanthanides* and those in period 7 are the *actinides*, and they are usually presented as two rows located under the main part of the periodic table.

2.1.3 Stable and Unstable Nuclides

In its ground state, not all combinations of protons and neutrons in a nucleus form a stable nuclide, that is, nuclides that do not change over time. Figure 2.4 shows a graph indicating the combinations of N and Z that provide stable nuclides.

For lighter elements with A< 20, it is required that Z ≈ N to obtain stability. However, for heavier elements, it is required that N>Z to ensure stability. The protons are positively charged and therefore repel each other by electrostatic forces. In a nucleus with more than ten protons, the electrostatic repulsion eventually becomes so strong that an excess of neutrons will be required to ensure that the nucleons hold together by direct nuclear forces as they possess no repulsive electrostatic forces; thus, this ensures stability. However, nuclear forces act within a very short range when

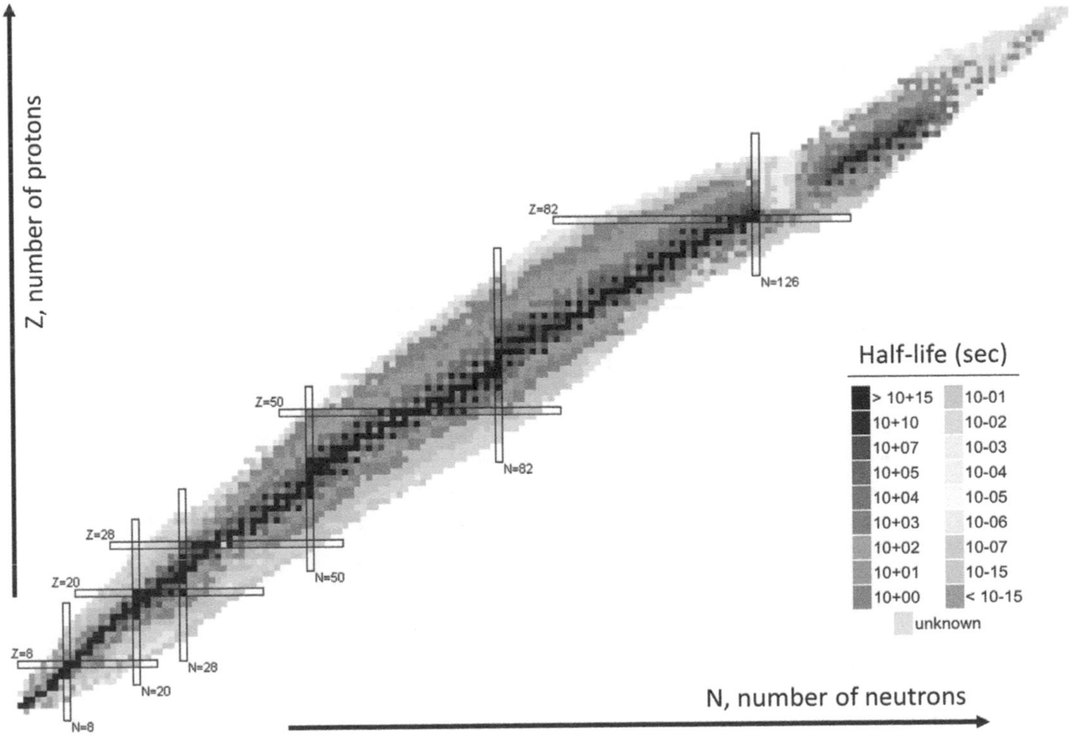

FIGURE 2.4 Neutron-proton diagram of stable nuclides. There exist no stable nuclides with Z = 43 or 61, with N = 19, 35, 39, 45, 61, 89, 115 or 126, or with A = Z + N = 5 or 8. All nuclides with Z > 83, N > 126 and A> 209 are unstable. The diagram is adopted with permission from NuDat2 database (www.nndc.bnl.gov/nudat2/), maintained by the National Nuclear Data Center, Brookhaven National Laboratory, Upton, New York.

compared to the electrostatic forces. The neutrons therefore interact only with the nearest nucleons in the nucleus while the protons interact electrostatically with each other over almost the entire nucleus. This limits the ability of neutrons to maintain a nucleus through nuclear forces for nuclides with a large number of protons. Consequently, the element ${}^{208}_{83}Bi$ therefore represents the boundary for stability.

Today, there are 252 stable nuclides and 34 radioactive nuclides with half-lives so long that these can be considered stable. All of them are distributed as isotopes of the different elements. For all other combinations of N and Z, stable nuclei are obtained from nuclei that are subsequently transformed and eventually end up into stable proton/neutron combinations. These unstable nuclei (radioactive nuclei) are known to be about 1400.

2.1.4 Electron Energy Levels

In the Bohr model, the electrons move in certain defined paths around the nucleus. An electron that moves in a certain path has a characteristic well-defined energy for that motion. Using quantum mechanical terminology (where the term 'path' does not really have a meaning), electrons can be in different shells around the core. Closest to the nucleus are the electrons located in the K-shell, and the electrons that appear with increasing distances are the L, M, N-, and so forth, shell. The closer the electrons are to the nucleus, the stronger they are energetically bounded to the nucleus by the electrical Coulomb force. Thus, the electrons located in the K-shell are hardest bounded. These therefore have the highest kinetic energy and binding energy and, therefore, that is all required from the main energy to remove the electron from the nucleus. In each of the main shells beyond the K-shell (L-, M, and so forth shells), the electrons can be distributed on different subshells, corresponding to different energy levels of electron pathways (circular or elliptical). While there is a relatively large difference in the binding energy between electrons located in different main shells, there is a smaller difference in the binding energies between electrons that are located in different subshells within the same main shell. Therefore, the binding energies of the electrons are often characterized by K-electrons, L-electrons, and so forth.

As stated above, in the model by Bohr, the electron was considered a particle that circulated around a nucleus. One of the features of a bounded system in quantum mechanics is that it can be seen as a series of quantum states with different discrete energies where the state of the lowest energy level is called the ground state. Other states with higher energies are excited states. Erwin Schrödinger (1887–1961) considered the electron as a wave motion, and the quantum mechanical description of an atom with many electrons is based on Schrödinger equations for the hydrogen atom, and it describes the state of an electron in an atom with four quantum numbers (**n**, **l**, **m**, and **s**) that can only have certain allowed values (Figure 2.5).

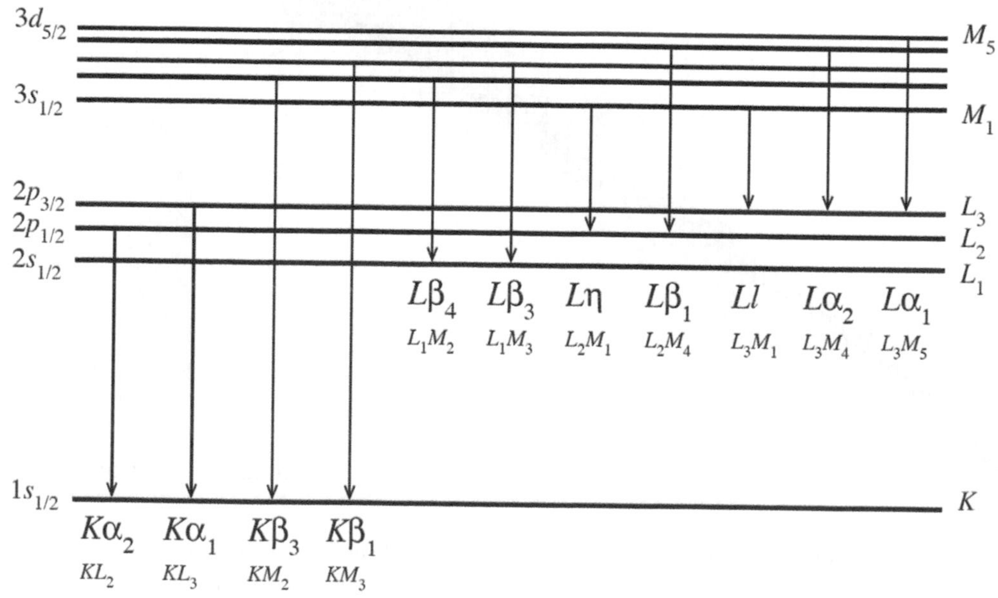

FIGURE 2.5 Energy levels of electrons in orbitals around the nucleus. Courtesy José M Fernández-Varea.

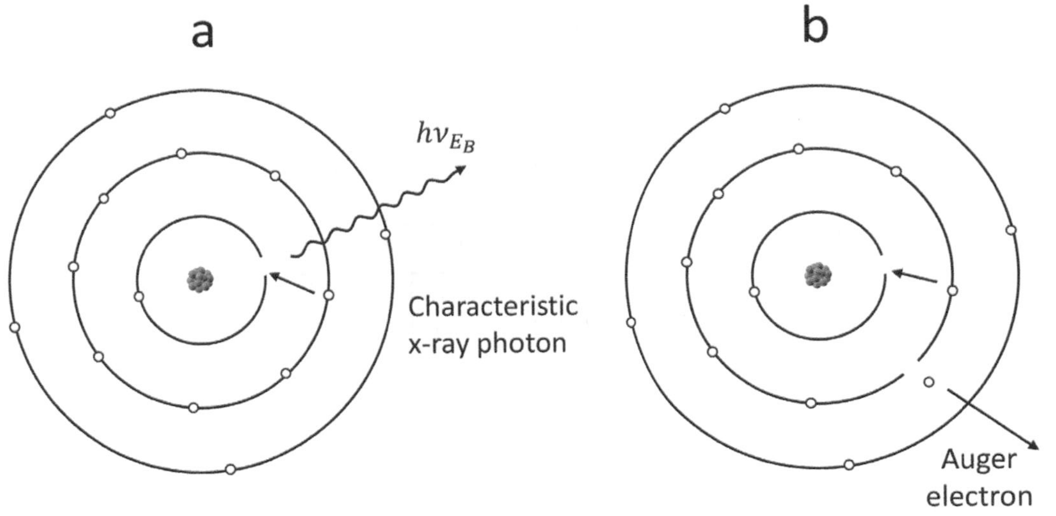

FIGURE 2.6 A vacancy in the K-shell is filled by an electron in the L-shell whereby an L-shell orbital electron is emitted with a kinetic energy = B_K-$2B_L$. Thus, in this process, two vacancies occur in the L-shell. If more Auger processes occur, the atom may be found to be in a highly ionized state at the end of the reorganization chain when it reaches its lowest energy.

The number **n** can have positive integer values (1, 2, 3, …) while number **l** only can have non-negative values up to **n**-1 (0, 1, 2, …, **n**-1). The quantum number **m** can have any value between (-**l**, -**l**+1 … 0 … **l**-1, **l**) and finally the number **s** can only have the values ± ½. The numbers **l** and **m** can be seen as describing the subshells of the shell **n** and the number of each subshell is determined by the value of **m**. It should here be mentioned that the terminology K, L, and M is equivalent to the quantum numbers; we will use this notation in the following text.

The number of electrons that can be in the same shell simultaneously is limited. For example, a K-shell can maximally contain two electrons while the L-shell can contain a maximum of eight electrons. In the ground state, the electrons are distributed over all available energy levels such that their total energy becomes as small as possible.

Figure 2.6 shows a neutral Si atom in its ground state, where the K- and L-shells are filled to their maximum number of orbital electrons, while four electrons are in the M-shell.

If energy is transferred to an orbital electron, this electron can then be elevated to a higher available energy level via an excitation process. The energy required to make this energetically possible needs to be at least the difference in binding energy for the electron and for the two energy levels. If an electron gains sufficient energy by some kind of interaction, the electron can also be completely released from the atom. The smallest energy required will be the binding energy of the particular shell electron. This event will thereby cause an ionization, that is, the atom that is electrically not neutral anymore and this state can then create disturbances, for example, in the bindings between atoms of a molecule, causing it to be split or modified. In either of the cases described above, a shell vacancy will often be created in one of the inner shells. This vacancy will strive to move the atom to a lower energy state by de-excitations resulting in the vacancy being filled with an electron from an outside shell. During this process, the excess energy will be released due to the change in the energy state of the vacant electron, and the released energy can be emitted in the form of either electromagnetic radiation, that is, photons or by so-called Auger electrons.

Photons emitted because of shell vacancy are called characteristic x-rays because the emission of the photons has a well-defined energy spectrum of the photons. If a K-shell vacancy is filled by an electron from the L-shell, the photon (K_α radiation) will have slightly less energy as compared to a vacancy filled with an electron from the M-shell (K_β radiation).

The larger the positive charge of the nucleus, the stronger the electrostatic force is on the orbiting electrons. Around a nucleus with a large number of protons (high Z atomic numbers) K-shell electrons circle closer to the nucleus, and they are bounded more strongly to the nucleus than to a nucleus with a fewer number of protons (lower Z atomic number). The energy of the electrons in different shells strongly depends therefore on the atomic number of the element. The energy of the emitted K, L, M and so forth, radiation varies with the atomic number of the element. For example, the K_α radiation from Lead (Pb) has a mean energy of 74 keV, while K_α radiation from Oxygen (O) has a mean energy of

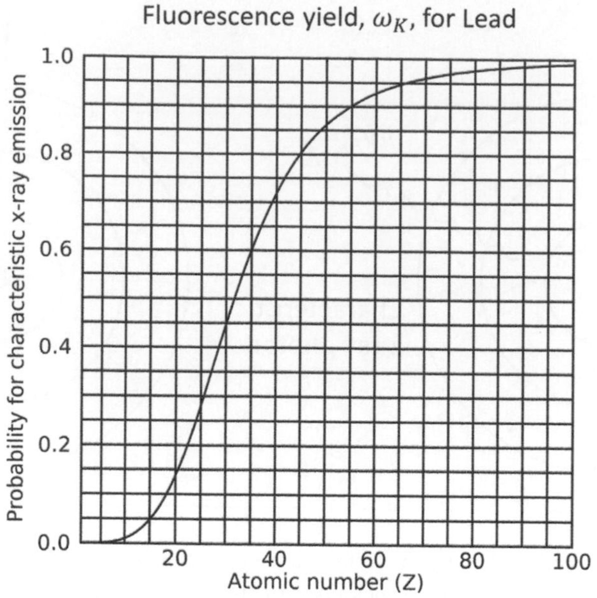

FIGURE 2.7 The probability for characteristic x-ray emission as a function of the atomic number Z. The curve has been calculated from the Eq. 2.1.

about 0.5 keV. It is therefore possible to identify different elements by measuring the energy of the emitted characteristic x-rays. This technique, called x-ray fluorescence, has been widely used to measure different elements in human tissues by exposing various parts of the body to photons and measuring the characteristic x-rays.

As mentioned above, when a vacancy in an inner shell is filled by an electron from an outer shell, the energy released can be transferred to one of the outer shell-electrons, which then then receive sufficient energy to escape the atom. These electrons are called Auger electrons, and usually they have a very low kinetic energy (in the order of keV or less). B_K and B_L are the binding energies of an electron in the K-shells and L-shells, respectively. Through this process, the vacancy is shifted to the L-shell and thus a cascade of processes is initiated until the atom finally reach its ground energy state.

The probability for the Auger-electron emission is higher for a low-Z material, and the characteristic x-ray emission dominates for high-Z materials. The relative probability for x-ray emission is called the fluorescence yield, $\omega_{K,L,...}$, and a function depending on the particular electron. The fluorescence yield for the K-shells, ω_K, as a function of the atomic number can be described by the equation

$$\omega_K = \frac{Z^4}{10^6 + Z^4} \tag{2.3}$$

is shown in Figure 2.7 for Lead.

2.1.5 Nuclear Energy Levels

It is accepted that the nucleus consists of A nucleons with Z protons and N neutrons. These are linked together into a bounded system by short-range nuclear forces, that can be described by quantum mechanics, as described briefly above. A nucleus that is in an excited state, as usually is the case for the daughter nucleus after a radioactive decay, strives to return to its ground state. If the excitation energy is sufficiently high, the nucleus is de-excited by the emission of a nucleon. However, if the excitation energy is not sufficiently high for the emission of a nucleon, the nucleus is usually de-excited by the emission of electromagnetic radiation. The electromagnetic radiation emitted because of the transition between two discrete energy states in the nucleus is in the form of a photon with an energy equal to the difference in energy between the two energy levels.

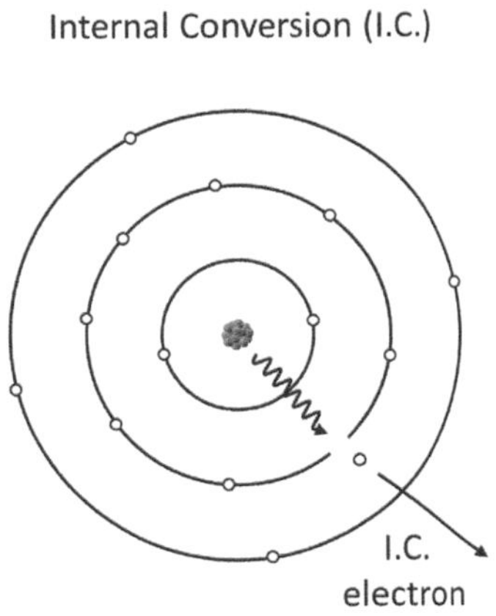

FIGURE 2.8 Schematic image of the internal conversion (I.C.) process.

Moving from an excited state to the ground state can take some several steps because of the lower available excitation energy levels. If de-excitation occurs in several steps, more than one photon can be emitted. A photon emitted upon de-excitation of an atomic nucleus is called a γ-quantum or γ-radiation, and these terms should therefore not be used to denote other photons besides those originating from the nucleus.

The de-excitation of an excited nucleus occurs immediately (prompt), which means so fast that the excited state has a very short lifetime. Sometimes, however, the likelihood of a transition from the excited state to a lower energy state may be such that the excited state has a measurable lifetime. One can then say that the excited state is metastable and that the excited nucleus by this is an isomer of the nucleus in the ground state. The transition from the metastable state to a lower energy state is called an isomeric transition (IT).

The de-excitation of the nucleus does not always occur because of the transmission of γ-radiation. The energy released at the transition between two energy states can instead be transferred to an electron in one of the surrounding shells (Figure 2.8). The electron is hereby released from the atomic shell with a kinetic energy equal to the energy released at the nuclear transformation minus the electron's binding energy in the atomic shell. This type of process is called *internal conversion* (I.C). For such a process to be energetically possible, the energy released at the nuclear decay must be greater than the binding energy of the electron. The I.C. is an important process in de-excitation of high-Z nuclides, when the energy released at the nuclear transition is immediately above the binding energy of an electron in the atomic scale and then the nucleus transition is isomeric.

For some transitions, the emission of γ-radiation is prohibited but the transition can then be made through I.C. In other cases, a nucleus transition may occur by issuing two γ-photons or by I.C. plus the emission of a γ-photon.

The I.C. will create a vacancy in the surrounding electron shell, resulting in a subsequent emission of a characteristic x-ray photon and/or Auger electrons. It is important not to mix γ-radiation with characteristic x-rays because if one does not consider the I.C. and its effect when evaluating a decay scheme, the misinterpretation of the detected radiation from the radioactive nuclides may occur.

The emission of γ-radiation and internal conversions are competing processes upon the de-excitation of a nucleus, and they are parallel to the competing processes of emission of characteristic x-rays and Auger electrons, respectively, upon the de-excitation of an excited electron shell.

The allowed discrete energy states of an atomic nucleus are characteristic of a nucleus with a certain number of protons and neutrons. The energy of a γ-photon, emitted from the nucleus, determined by the energy difference between two different discrete energy states, is also characteristic for the specific nucleus. Therefore, it is possible to identify the nuclei by studying the energy of the γ-photons emitted by the nucleus. While the line-spectrum of the characteristic x-rays is typical of a specific element, the line-spectrum of the γ-radiation is typical for a given nuclide. The isotopes of the same element emit different line-spectra of γ-radiation; however, it emits the same line-spectrum of the characteristic x-rays.

2.2 RADIOACTIVE DECAY

2.2.1 Mass–Energy Relationships

For understanding radioactive decay and related decay schemes, it is necessary to have knowledge of the mass–energy relationship. According to Einstein's special relativity theory, the mass of a particle is dependent on its velocity. As a consequence, the mass is equivalent to energy, and mass can be converted into kinetic energy and vice versa. The total energy, E, of a particle moving at the speed v is therefore given by:

$$E = mc^2 = m_o c^2 + T \quad (2.4)$$

$$m = \frac{m_o}{\sqrt{1-\left(\frac{v}{c}\right)^2}} \quad (2.5)$$

where m_0 denotes the particle mass in rest (rest- mass), T denotes its kinetic energy and c denotes the speed of light in vacuum. For a particle that moves in a potential field, where it also gain potential energy U, it applies

$$E = mc^2 = m_o c^2 + T + U \quad (2.6)$$

This means that the potential energy, U, can affect the mass. A nucleon in the nucleus has a potential energy because of the nuclear forces between the nucleons.

2.2.2 Nucleus Mass Defect and Bonding Energy

A precise measurement of the atomic weights shows that the nucleus mass is less than the sum of the individual masses of each nucleon if they were in a free unbounded state. The mass defect Δ can therefore be defined as

$$\Delta = Z \cdot m_p + (A - Z) \cdot m_n - M_Z^A \quad (2.7)$$

where m_n, m_p, and M_Z^A denote the rest-mass of the neutron, rest-mass of the proton, and the mass for a nucleus with Z protons and (A-Z) neutrons in its ground state, respectively. The mass defect can be explained by Einstein's mass–energy relation, and it is a striking proof of the equivalence between mass and energy. The nucleons in the nucleus are linked to each other by a bounded system of the attractive nuclear forces. To separate all nucleons in the nucleus from each other, an energy equivalent to the nuclei binding energy is required. The binding energy B of the nucleus is given by the mass-equal energy equivalent $B = \Delta \cdot c^2$. It then also follows that a nucleus in an excited state has a larger mass than the nucleus in its ground state.

2.2.3 Different Types of Instability

The ground state of the nucleus exhibits two different types of instabilities: dynamic instability and a β-stability. A dynamic instability exists if a nucleus can be split into two (or more) fragments such that the sum of the residual masses of each fragment is less than the mass of the rest of the original nucleus, that is, if the binding energy of the parent nucleus is less than the sum of the binding energies of each of the separated fragments. The dynamic instability is of importance to heavy nuclei. If the nucleus is split into two (or more) fragments of approximately the same mass, then the process is called spontaneous fission, and this process is possible for nuclides with a mass number A greater than about 80; however, the process has a very low likelihood to occur.

For a nucleus with mass number A greater than 150, another and more important form of dynamic instability will occur, called α-instability, leading to an α-decay where a helium nucleus $_4He^{2+}$ will be emitted from the parent nucleus. This is possible because of the relatively high binding energy of the $_4He$ nucleus. The emission of other lighter nuclei than the He is not energetically possible owing to the low binding energy of these potential candidates.

β-instability can occur if a nucleus can change its proton–neutron relation with retained mass number A and thus be converted to an isobar with a lower mass. β-instability is associated with a nucleon capacity to be transformed into each other. The following transformations can then occur.

$$n \rightarrow p + e^- + \bar{\nu}$$

$$p \rightarrow n + e^+ + \nu$$

$$p + e^- \rightarrow n + \nu \qquad (2.8)$$

The neutrino, υ, and its antiparticle, $\bar{\nu}$, are both elementary particles without an electric charge and with rest masses that are practically equal to zero. The positron, e^+, has the same rest-mass as the electron e^-, and a charge of the same magnitude but of an opposite sign. According to a stricter terminology, the term 'electron' holds true for both particles e^- and e^+, and wherein e^- instead is called a negatron.

Table 2.1 indicates that the mass of the neutral neutron is larger than the mass of the proton. This means that only the first transformation in Eq. 2.8 where a neutron transforms into a proton, a negatron and an antineutrino can occur for a free nucleon. The two other transformations are not energetically possible for a proton in a free state but can occur when the proton is bound in a nucleus. The transformations of nucleons in Eq. 2.8 correspond to three different types of β-decays, which will be discussed in more detail below.

2.2.4 Decay Scheme

The transformations (decays) that a nucleus in its ground state can undergo spontaneously without additional external energy is the α-decay and three types of β-decays. In all these cases, the decay involves a conversion of the nucleus by either a spontaneous emission of an electrically charged particle or from a captured electron by the nucleus. The emitted charged particles receive such high kinetic energy that they can further create ionizations. As a consequence of the primary decay of the parent nucleus, the daughter atom (the newly formed nucleus and its electron shells) emits additional ionizing radiation in the form of photons (γ-photons and characteristic x-rays) and/or electrons (internal conversion electrons and/or Auger electrons). Owing to the ionizing radiation emitted during the decay, the decay is called a radioactive decay. The decays are radioactive, and not the radiation itself; however, the radiation is ionizing. By radioactive nuclide (or a radionuclide), one refers to a nuclide that can undergo radioactive decay.

Important information of the transition for the radioactive decay is summarized in a so-called decay scheme, which is described schematically in Figure 2.9. The vertical direction provides information about the energy change and in the horizontal direction, the information is given about the change in the atomic number.

Some radioactive nuclides may collapse in a competing manner so that the daughter nuclei will involve different types. The decay scheme for such a branched decay scheme is as shown in Figure 2.10. However, the decay occurs such that the daughter nuclei only follow a single branch (one of the branches in Figure 2.9).

The vertical energy scale in the decay scheme is normalized so that the energy equivalence of the daughter-nucleus residual mass in its ground state is zero. The numerical value of the energy of a daughter atom with a nucleus in its excited state then directly indicates the excitation energy of the nucleus.

2.2.5 α-Decay

Heavy unstable nuclides pass spontaneously into lighter nuclei by the transmission of an α-particle, which is identical to a ^{4}He nucleus and contains two protons and two neutrons. By transmitting an α-particle, the nucleus reduces its atomic number Z and its neutron number with two units; the mass number A decreases with four units. The α-decay can be written in a decay diagram according to Figure 2.10.

An α-decay can be described according to the following:

$${}^{A}_{Z}X \rightarrow {}^{A-4}_{Z-2}X + \alpha + Q \qquad (2.9)$$

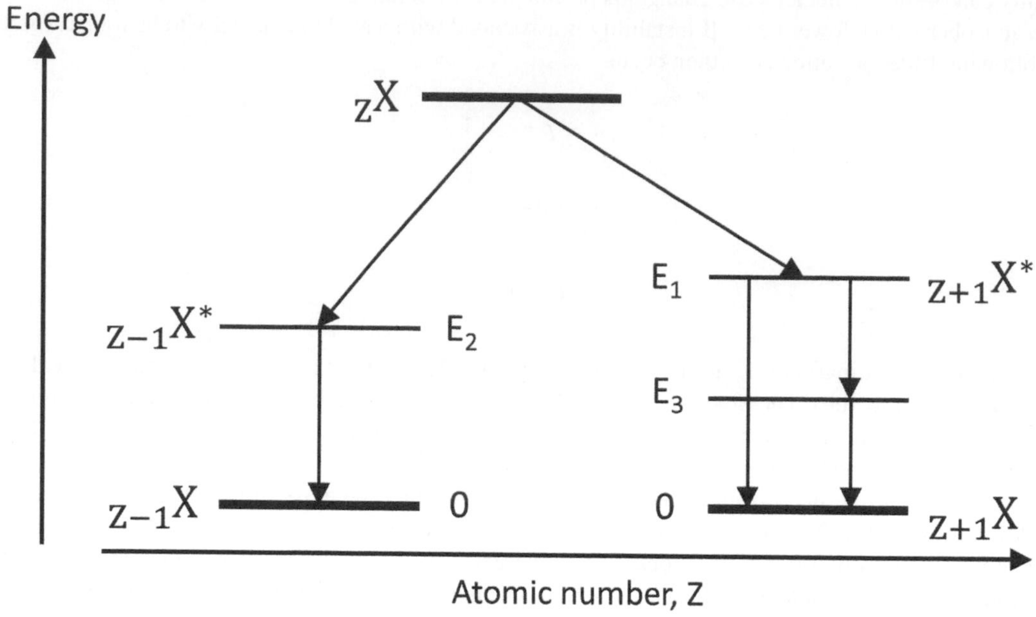

FIGURE 2.9 General description of a decay scheme. Horizontal lines indicate energy states relative to the ground state of the daughter nucleus. Arrows pointing down to the left/right indicate that the daughter's nucleus has a lower/higher atomic number than the parent nucleus. They also point to excited states (denoted by the star in the figure) for the nucleus after the decay where the vertical arrows then indicate the different de-excitation paths for the excited daughter nucleus.

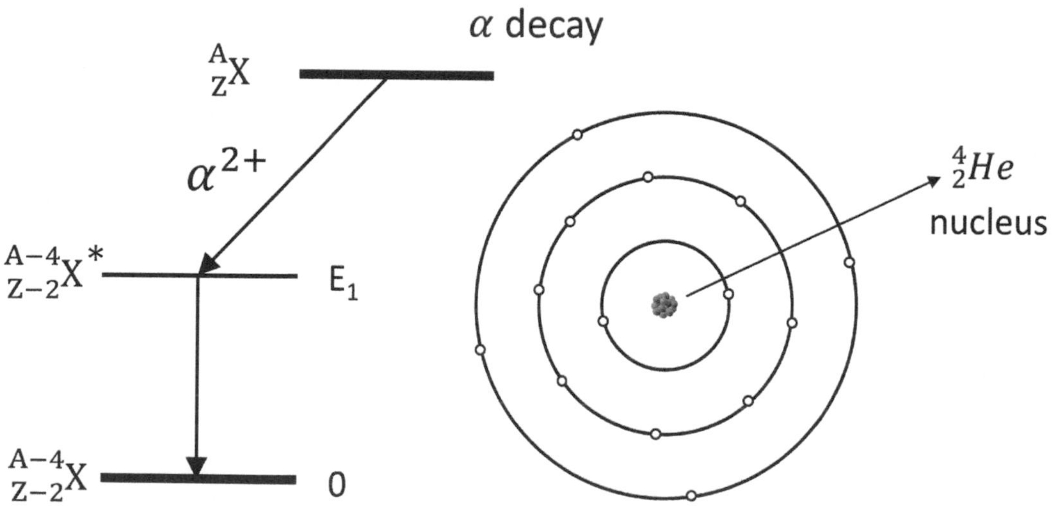

FIGURE 2.10 Scheme of a generic α-particle decay.

where Q denotes the sum of the released kinetic energy for the α-particle and the daughter nucleus, and the energy released by subsequent de-excitations. Because the radioactive decay is a spontaneous process without any external energy, it must hold true that $Q > 0$. This sets demands regarding the difference of mass between the parent and daughter nuclides. The decay's energy relation is given by

$$ {}_kM_Zc^2 = {}_kM_{Z-2}c^2 + m_\alpha c^2 + Q \tag{2.10}$$

where m_α denotes the α-particle rest mass, and ${}_kM_Z$ and ${}_kM_{Z-2}$ are the rest-masses of the parent and the daughters of their ground state, respectively. Thus,

$$ ({}_kM_Z - {}_kM_{Z-2} - m_\alpha)c^2 = Q \tag{2.11}$$

The requirement that Q must be greater than zero implies that the following condition must be true.

$$ {}_kM_Z - {}_kM_{Z-2} > m_\alpha \tag{2.12}$$

If instead of using nuclear masses ${}_kM_Z$, one can use the corresponding atomic masses ${}_aM_Z$. If so, the energy relations instead can be written as

$$ ({}_aM_Z - Z \cdot m_e)c^2 = ({}_aM_{Z-2}c^2 - (Z-2) \cdot m_ec^2 + m_\alpha c^2 + Q \tag{2.13}$$

where m_e denotes the rest mass for the electron. This results into

$$ ({}_aM_Z - {}_aM_{Z-2} - 2m_e - m_\alpha)c^2 = Q \tag{2.14}$$

If Q > 0 needs to be true, then this requires that

$$ {}_aM_Z - {}_aM_{Z-2} > 2m_e + m_\alpha \tag{2.15}$$

2.2.6 β⁻-DECAY

For a nuclide with an excess of neutrons in relation to the number of protons, a neutron can be transformed into a proton, negatron, and neutrino, according to

$$ n \rightarrow p + e^- + \overline{\nu} \tag{2.16}$$

At the transformation, the negatron and anti-neutrino are emitted from the nucleus. A negatron ejected from the nucleus is called a β^--particle and this notation implies that electrons that originate from the nucleus should be given this name. Through a β^--decay, the nucleus increases its proton number (atomic number Z) by one while the mass A remains the same. The decay is written in the decay scheme, as shown in Figure 2.11.

The β-particles emitted from multiple decays exhibit a continuous spectrum of kinetic energies, as is shown in Figure 2.12a. This energy spectrum occurs because the emitted β-particles share the released energy with the emitted neutrino. While the kinetic energy of the β-particle varies in different decays, the sum of the kinetic energies of the β-particle and neutrino remains the same in any decay.

The β^--decay can be written as

$$ {}^A_ZX \rightarrow {}^{A}_{Z+1}X + \beta^- + \overline{\upsilon} + Q \tag{2.17}$$

Here, Q denotes the released kinetic energy for the β^--particle and the anti-neutrino plus energy released during a de-excitation of a potential excited nuclei. The decay's energy excess is given by

$$ {}_kM_Zc^2 = {}_kM_{Z+1}c^2 + m_ec^2 + Q \tag{2.18}$$

$$ ({}_kM_Z - {}_kM_{Z+1} - m_e)c^2 = Q \tag{2.19}$$

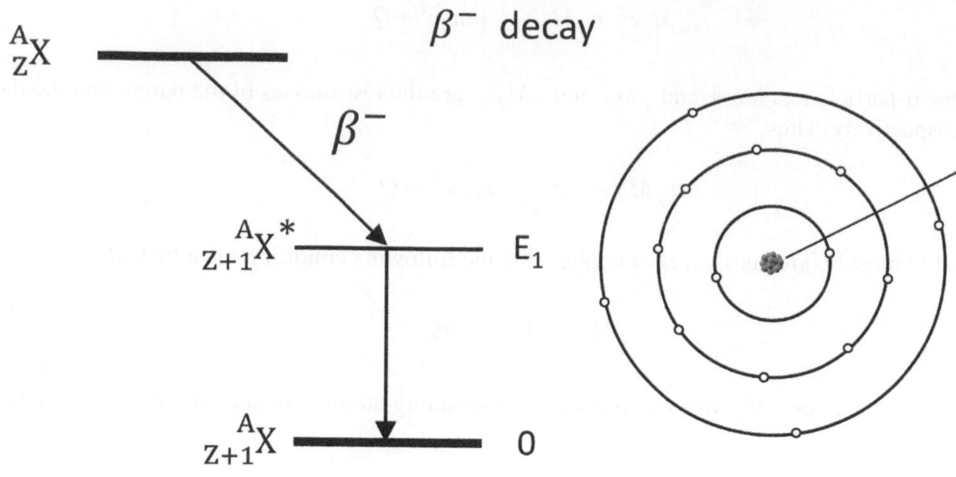

FIGURE 2.11 Scheme of a generic β-decay.

The requirement that Q is greater than zero implies that the following is true:

$$_{k}M_{Z} - {_{k}M_{Z+1}} > m_{e} \tag{2.20}$$

The energy relations can also be written as

$$({_aM_Z} - Z \cdot m_e)c^2 = ({_aM_{Z+1}}c^2 - (Z+1) \cdot m_e c^2 + m_e c^2 + Q \tag{2.21}$$

$$({_aM_Z} - {_aM_{Z+1}})c^2 = Q \tag{2.22}$$

If Q >0 this requires that

$$_{a}M_{Z} - {_{a}M_{Z+1}} > 0 \tag{2.23}$$

2.2.7 β⁺-DECAY

In a nucleus with an excess of protons in relation to the number of neutrons, the following transformation can occur:

$$p \rightarrow n + e^{+} + v \tag{2.24}$$

Here, a proton is transformed into a neutron, positron, and neutrino. When a positron originates from the atomic nucleus, it is called a β^+-particle.

The β^+-particles, that are emitted from different decaying nuclei, also exhibit a continuous range of kinetic energies as is shown in Figure 2.12b. However, the energy spectrum exhibits a smaller proportion of low-energy β^+-particles than can be seen from the energy spectrum of the β^- particles (Figure 2.12a). This is characteristic of β^+-particles spectra and is a consequence of the positive charge of the nucleus, which causes a transmitted β^+ particle to be repelled from the nucleus in the opposite direction to an emitted β^- particle that will be attracted to the nucleus. The nucleus reduces its atomic number to Z-1 while the mass number A remains unchanged. Figure 2.13 shows a generic decay scheme for β^+ particles. The arrow that breaks to the left in the decay scheme is explained by the mass condition of the β^+ decay.

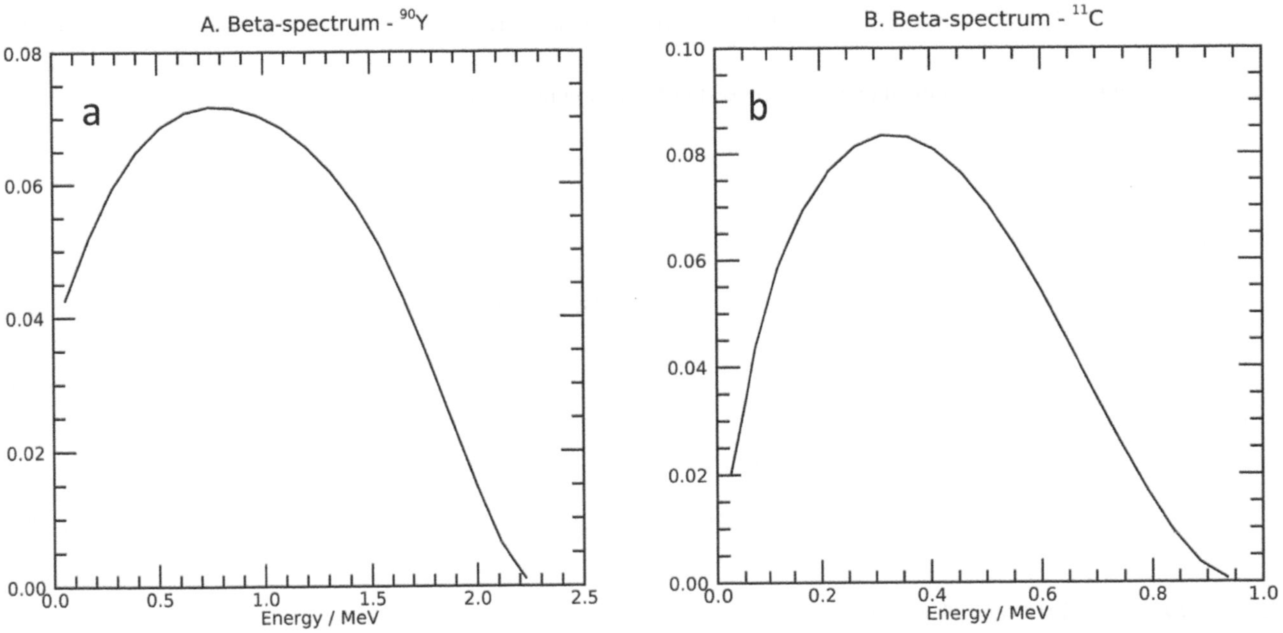

FIGURE 2.12 Spectrum of kinetic energies of β^--particles emitted from a ^{90}Y decay (a) and β^+-particles emitted a ^{11}C decay (b).

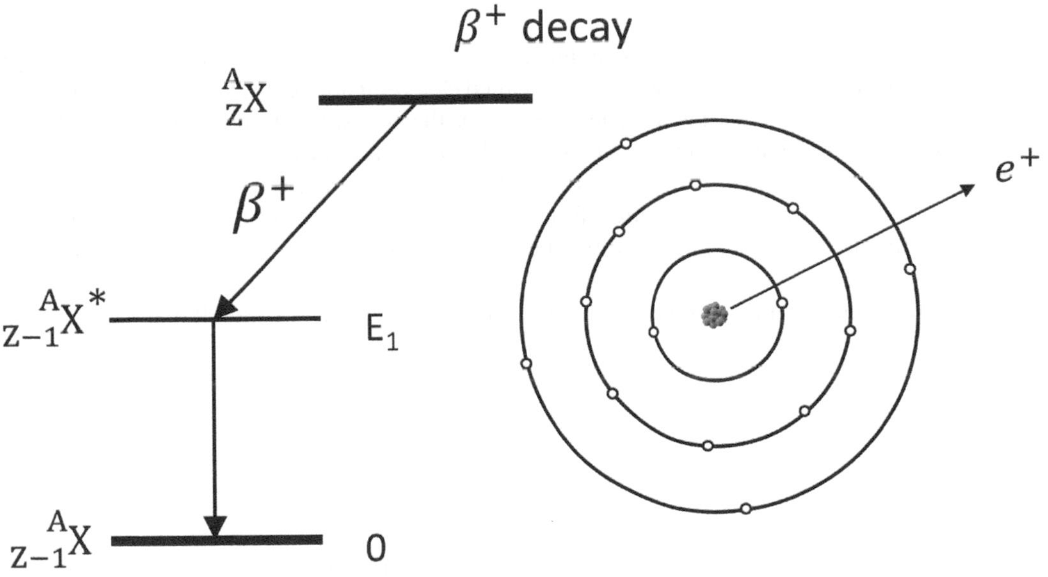

FIGURE 2.13 Decay scheme of a generic β^+ decay.

The β^+-decay can be written as

$$ {}^{A}_{Z}X \rightarrow {}^{A}_{Z-1}X + \beta^+ + \upsilon + Q \tag{2.25} $$

where Q is the released energy in the form of kinetic energy of the β^+ particle and neutrino, and the energy released by de-excitation of an excited daughter nucleus. As in the previous equations (Eqs. 2.18 and 2.19), the energy excess of the decay is given by

$$ {}_{k}M_{Z}c^2 = {}_{k}M_{Z-1}c^2 + m_e c^2 + Q \tag{2.26} $$

$$({}_kM_Z - {}_kM_{Z-1} - m_e)c^2 = Q \tag{2.27}$$

The requirement that Q is greater than zero implies that the following is true:

$${}_kM_Z - {}_kM_{Z-1} > m_e \tag{2.28}$$

The energy relations can also be written as

$$({}_aM_Z - Z \cdot m_e)c^2 = {}_aM_{Z-1}c^2 - (Z-1) \cdot m_ec^2 + m_ec^2 + Q \tag{2.29}$$

$$({}_aM_Z - {}_aM_{Z-1} - 2m_e)c^2 = Q \tag{2.30}$$

If the Q is larger than 0, then this requires that

$${}_aM_Z - {}_aM_{Z-1} > 2m_e \tag{2.31}$$

2.2.8 Decay by Electron Capture

Alternatively, for a β^+ decay in a nucleus with a proton excess, the following transformation can occur:

$$p + e^- \rightarrow n + \nu \tag{2.32}$$

An electron from the surrounding electron shell is captured by the nucleus, and it interacts with a proton, which results in the formation of a neutron and a neutrino. The process is called electron-capture (EC), and when capturing an electron from the K-shell, one defines this as a K-capture. The likelihood is the largest for electrons captured in the shells closest to the nucleus, especially those with the quantum number l=0. The nucleus reduces Z (number of protons) by one, while A remains unchanged. Electron capture is drawn as a decay diagram in Figure 2.14.

In an EC process, no charged particle is emitted from the nucleus. However, from the capture of a shell electron, there will be a vacancy in one of the shells that will be filled with a subsequent emission of characteristic x-rays and/or Auger electrons.

If the difference in mass between the parent atoms and their daughter atoms is so small that

$$\left({}_aM_Z - {}_aM_{Z-1}\right) \cdot c^2 < 2m_0c^2 \tag{2.33}$$

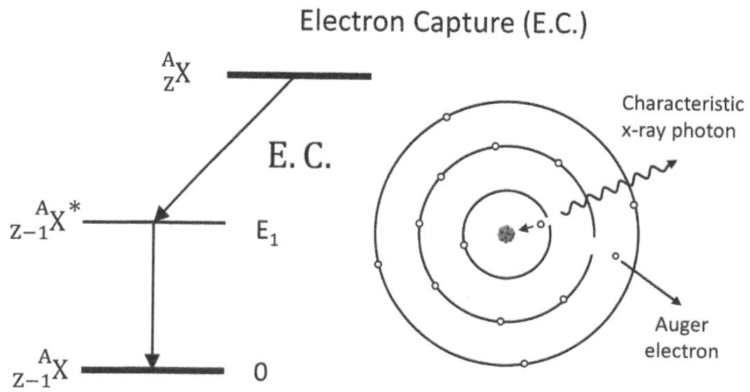

FIGURE 2.14 Decay diagram of a nuclear decay by an electron capture (E.C.).

then EC is the only possible approach for a nucleotide to reduce its number of protons in the nucleus. In all other cases, β^{+} decay and EC are two competing processes. For high-Z radionuclides, EC is more likely to occur than β^{+} decays because the innermost electrons circles in paths are closer to the nucleus, which favour the capture of the electron.

The EC-decay can be written as

$$ {}_{Z}^{A}X + e^{-} \rightarrow {}_{Z-1}^{A}X + \upsilon + Q \tag{2.34}$$

As with previous equations (Eq. 2.26 and Eq. 2.27), the decay's energy excess is given by

$$ {}_{k}M_{Z}c^{2} + m_{e}c^{2} = {}_{k}M_{Z-1}c^{2} + Q \tag{2.35}$$

$$ ({}_{k}M_{Z} - {}_{k}M_{Z-1} + m_{e})c^{2} = Q \tag{2.36}$$

The requirement that Q is greater than zero implies that the following is true:

$$ {}_{k}M_{Z} > {}_{k}M_{Z-1} - m_{e} \tag{2.37}$$

The energy relations can also be expressed with atomic masses as

$$ ({}_{a}M_{Z} - Z \cdot m_{e} + m_{e})c^{2} = ({}_{a}M_{Z-1}c^{2} - (Z-1) \cdot m_{e}c^{2} + Q \tag{2.38}$$

$$ ({}_{a}M_{Z} - {}_{a}M_{Z-1})c^{2} = Q \tag{2.39}$$

If Q > 0 then this requires that

$$ {}_{a}M_{Z} > {}_{a}M_{Z-1} \tag{2.40}$$

In the mass-calculations conditions, described above, the binding energy of the electron has been neglected. The mass of the nucleus is obtained from the atomic mass through the relation

$$ {}_{k}M_{Z}c^{2} = {}_{a}M_{Z}c^{2} - (Z \cdot m_{e}c^{2} - Z \cdot \overline{B}_{K,L,M...}) \tag{2.41}$$

where $\overline{B}_{K,L,M...}$ denotes the average of the binding energies for all shell electrons. The error from neglecting $\overline{B}_{K,L,M...}$ is, however, small. In the case of EC, the captured electron is usually a K-shell electron with a non-negligible binding energy, especially in high-atomic number elements. With a captured K-shell electron, the energy conversion is given by

$$ {}_{k}M_{Z}c^{2} + m_{e}c^{2} - B_{K} = {}_{k}M_{Z-1}\,c^{2} + Q) \tag{2.42}$$

B_K denotes the electron binding energy for the K-shell. If Q is greater than zero, then

$$ {}_{k}M_{Z} > {}_{k}M_{Z-1} - m_{e} + \frac{B_{K}}{c^{2}} \tag{2.43}$$

or if expressed as atomic masses

$$ {}_{a}M_{Z} > {}_{a}M_{Z-1} + \frac{B_{K}}{c^{2}} \tag{2.44}$$

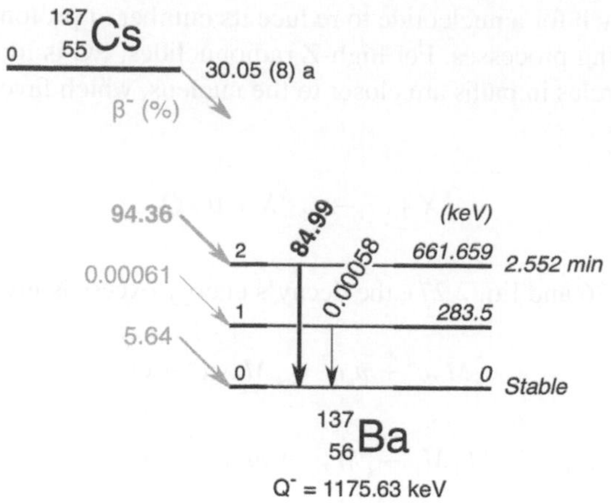

FIGURE 2.15 A chart of the ^{137}Cs decay scheme, reproduced with permission from a screen-capture from www.lnhb.fr/nuclear-data/module-lara/ web-page.

In a high-Z element, the binding energy B_K can go up to about 100 keV. If electron capture from the K-shell is not energetically possible, an electron from an outside shell (L, M, N, …) can be captured. The demand for the mass-difference between the parent and daughter's masses is not as strong for the capture of electrons from outside the K-shell.

2.3 INTERPRETATION OF DECAY SCHEMES

2.3.1 ^{137}Cesium

$^{137}_{55}$Cs decays by β^--decay to $^{137}_{56}$Ba (See Figure 2.15). After 94.7 per cent of the decays, the daughter nucleus remains in an excited condition with an excitation energy of 661.7 keV. This condition is meta-stable with a half-life of 2.552 min and the state is labelled $^{137}_{56}\text{Ba}^{\text{m}}$. The remaining 5.6 per cent of the decays goes directly to the ground level. The value Q_{β^-} is defined as the difference in energy between the masses of the parent nucleus and the daughter nucleus when they are at the ground level. For decays that go directly to the ground level, the maximum energy of the β^--particle will be 1175.63 keV. For the other β^--particles, the maximum energy will be the energy difference 1175.63 keV – 661.7 keV = 0.514 keV. In addition to the β^--particles, radiation is emitted as γ-photons and internal conversion (IC) electrons. When an IC electron is rejected from the shell, it creates a vacancy in that shell that will rapidly be filled by an electron. This results in the emission of either a characteristic x-ray photon or an Auger-electron. When the $^{137}_{56}\text{Ba}^{\text{m}}$ is in the excited state, the probability for IC is relatively high. It can be of interest to calculate the abundance between a γ-photon of 661.7 keV and an IC electron per unit decay. Using data obtained from the Laboratoire National Henri Becquerel webpage (www.lnhb.fr/nuclear-data/nuclear-data-table/), we can find that the number of K-electrons is 8.96 per cent per number of n_γ and the fraction $n_K / n_{L+M..}$ is 4.4755. For the decay scheme above, we see that

$$n_\gamma + n_K + n_{L+M+..} = 0.9436 \tag{2.45}$$

$$n_K = 0.0896 \cdot n_\gamma \tag{2.46}$$

$$n_{L+M+...} = \frac{n_K}{4.4755} = \frac{0.0896}{4.4755} \cdot n_\gamma \tag{2.47}$$

This gives that

$$n_\gamma + 0.0896 \cdot n_\gamma + \frac{0.0896}{4.4755} \cdot n_\gamma = 0.9436 \tag{2.48}$$

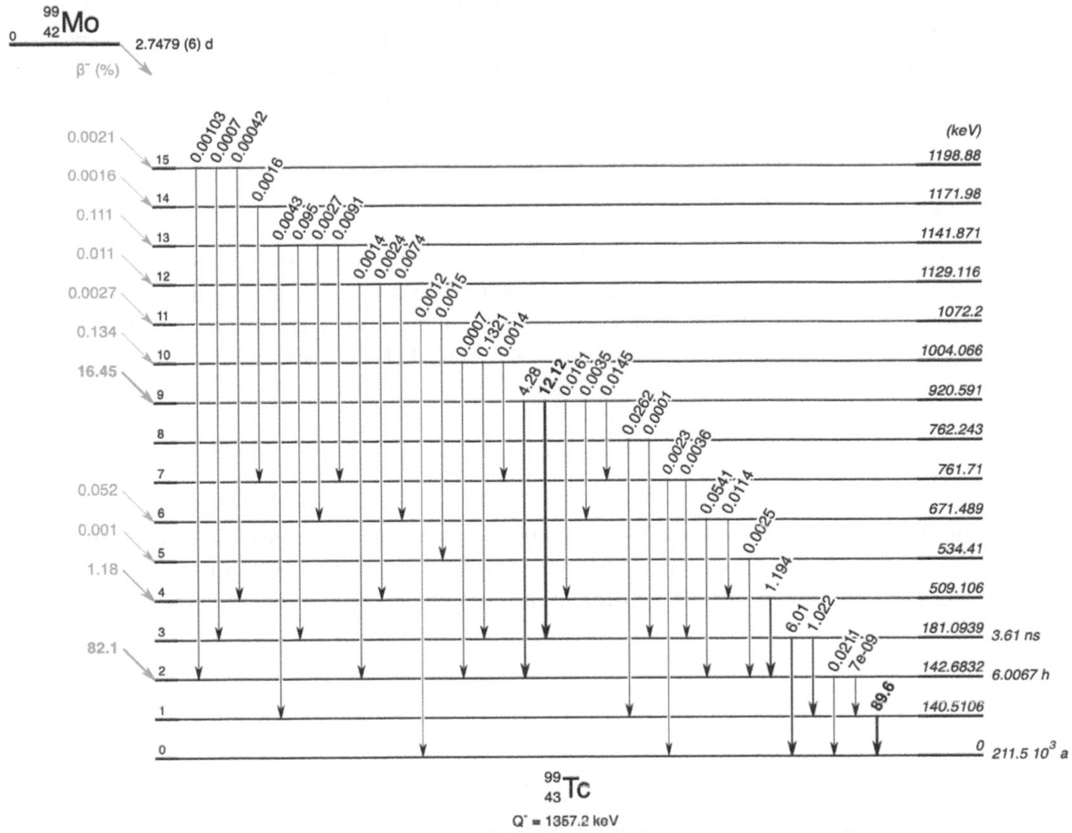

FIGURE 2.16 A chart of the ^{99m}Tc decay scheme, reproduced with permission from a screen-capture from www.lnhb.fr/nuclear-data/module-lara/ web-page.

This summarizes that we get (a) a β^- emission in 100 per cent of the decays, (b) a γ-photon of 661.7 keV in $n_\gamma = 85.0\%$ of the decays, (c) an IC electron from the K-shell for $n_K = 8.96\%$ of the decays (d) and $n_{L+M+..} = 2.00\%$ of the decays for the L+M+... shells and (e) subsequent emission of characteristic x-rays and Auger electrons. The fluorescence yield provides information about the relative probabilities between characteristic x-rays and Auger electrons. For Ba, the probability for x-rays for the K-shell is 86 per cent, which then means that characteristic x-rays are emitted for about 7.7 per cent of the decays with energy between 31.8 keV and 37.3 keV

2.3.2 ^{99M}Technetium

$^{99}_{43}Mo$ decays by β^- decay to $^{99}_{43}Tc$. In 82.1 per cent of all decays, the daughter nuclide is left in a metastable condition with a half-life of 6.007 h (See Figure 2.16). The remaining 17.9 per cent of the decays leave the daughter nucleus in several excited energy states. When these levels are de-excited, some will go through the metastable energy level. This mean that a total 89.07 per cent = 82.1 per cent + 1.18 per cent + (4.28/12.154) × 16.45 per cent of the decays will remain in the meta-stable condition.

Because the metastable condition has a relative long half-life, it is possible to extract the ^{99m}Tc radionuclide chemically for use as a source of γ-radiation because it is free from β-particle radiation, and it therefore produces a low radiation dose. Because the IC rate usually is high for metastable conditions, this advantage is generally not always useful. However, the metastable $^{99m}_{43}Tc$ of the de-excitation from the metastable energy level of 142.7 keV can be made by either one or two steps. When the de-excitation is made in one step, the probability for IC is high $(e/\gamma \approx 40.9)$. This also is true for the de-excitation from the level 142.7 keV to 140.5 keV. The level 140.5 keV is, however, not metastable and therefore, it has a relatively low conversion rate $(e/\gamma = 0.112)$.

When the de-excitation is made in two steps, a γ-photon is emitted with an energy of 140.5 MeV in 89.6 per cent of the de-excitations. In the other cases, IC electrons will be produced. For nearly 100 per cent of the de-excitations, a conversion electron is produced with a very low kinetic energy of ≤ 2 keV while for about 10 per cent, a conversion electron from the K-shell is produced with an energy of about 120 keV (140.5 keV − the binding energies 18–20 keV). Because of the vacancy in the shells, characteristic x-rays and Auger electrons will also be emitted.

2.4 RADIOACTIVE DECAY TIME

The radioactivity decay is a random Poisson-distributed process where the probability for a decay to appear, regarded at each time point, is a constant factor. This means that the probability for a decay does not depend on the age of the nucleus. Assume that there exist a large number of radionuclides N(t) at a given time point t, and that the variation in N only depends on the physical decay and that no new radionuclides will be introduced. Then, there will be a number of dN nuclei that decay between the time interval [t, t+dt]. Under the condition that dN is much less than N, the following relation is valid:

$$dN = -\lambda \cdot dt \cdot N(t) \tag{2.49}$$

The sign of dN is negative because the number of original nuclides is reduced. The product $\lambda \cdot dt$ here defines the probability that a nucleus decay within the time range [t,t+dt] and the decay constant λ is specific for each type of radionuclide. Equation 2.52 can then be rewritten as a differential equation and be solved according to

$$\frac{dN}{dt} = -\lambda \cdot dt \tag{2.50}$$

$$N(t) = N_0 \cdot e^{-\lambda t} \tag{2.51}$$

To calculate the number of radionuclides that have decayed, one should use

$$\Delta N = N_0 - N(t) = N_o\left(1 - e^{-\lambda t}\right) \tag{2.52}$$

Eq. 2.55 can introduce errors if the restriction dN<<N(t) is not fulfilled. The halftime of the radionuclide $T_{1/2}$ is the time that it takes to reduce the number of nuclides to $N_0/2$. When replacing N(t) with $N_0/2$ in Eq. 2.51, one obtains that that $\lambda = \ln(2)/T_{1/2}$

By measuring the emitted radiation from a radioactive substance with some type of detector, one can determine the number of decaying nuclides per unit of time at a specific time point. The activity A is then defined as

$$A = -\frac{dN}{dt} = \lambda \cdot N \tag{2.53}$$

and it is the ratio between the number of decays during a time interval and the time interval. The SI unit of activity is *becquerel* (Bq, s^{-1}), but often the unit *curie* is used. One Ci equals 3.7 10^{10} Bq and is the number of decays per second of 1g of ^{226}Ra. By multiplying both parts in Eq. 2.51 with λ, one obtains that

$$A(t) = A_0 \cdot e^{-\lambda t} \tag{2.54}$$

where A_0 denotes the activity at t = 0 and A(t) denotes the activity at a time point $t \geq 0$. It is also common to use the term activity for metastable radionuclides, such as $^{99}Tc^m$, where the metastable condition has a certain half-life.

2.5 DECAY CHAINS

In many cases, a radioactive decay will result in a new radionuclide that also will be radioactive. Consider a radionuclide X_1 that decays into X_2, that in turn, decays into the stable nuclide X_3. At a certain time point t, there exist $N_1(t)$, $N_2(t)$

and $N_3(t)$ nuclei of the X_1, X_2, and X_3 and where X_1 and X_2 have the decay constants λ_1 and λ_2, respectively. During the time interval [t, t+dt], the number of dN_1 nuclei decays according to

$$dN_1 = -\lambda_1 \cdot N_1(t) \cdot dt \tag{2.55}$$

when assuming that $dN_1 \ll N_1(t)$. The number of nuclei $N_2(t)$ for X_2 are changed because of the contribution from X_1 but also because of its own decay process. For this radionuclide,

$$dN_2 = \lambda_1 \cdot N_1(t) \cdot dt - \lambda_2 \cdot N_2(t) \cdot dt \tag{2.56}$$

For the radionuclide X_3, only a growth in $N_3(t)$ are expected, according to

$$dN_3 = \lambda_2 \cdot N_2(t) \cdot dt \tag{2.57}$$

We can then write these as three differential equations:

$$\frac{dN_1}{dt} = -\lambda_1 \cdot N_1(t) \tag{2.58}$$

$$\frac{dN_2}{dt} = \lambda_1 \cdot N_1(t) - \lambda_2 \cdot N_2(t) \tag{2.59}$$

$$\frac{dN_3}{dt} = \lambda_2 \cdot N_2(t) \tag{2.60}$$

The solution to the equations (2.58–2.60) for a time point t can be written as

$$N_1(t) = N_1(0) \cdot e^{-\lambda_1 t} \tag{2.61}$$

$$N_2(t) = \frac{\lambda_1 \cdot N_1(0)}{\lambda_2 - \lambda_1} \cdot \left(e^{-\lambda_1 t} - e^{-\lambda_2 t}\right) + N_2(0) \cdot e^{-\lambda_2 t} \tag{2.62}$$

$$N_3(t) = \frac{\lambda_1 \cdot \lambda_2 \cdot N_1(0)}{\lambda_2 - \lambda_1} \cdot \left(\frac{e^{-\lambda_2 t}}{\lambda_2} - \frac{e^{-\lambda_1 t}}{\lambda_1}\right) - N_2(0) \cdot e^{-\lambda_2 t} + N_1(0) + N_2(0) + N_3(0) \tag{2.63}$$

The activities for X_1 can be determined by

$$A_1(t) = \lambda_1 \cdot N_1(t) = A_1(0) \cdot e^{-\lambda_1 t} \tag{2.64}$$

$$A_2(t) = \lambda_2 \cdot N_2(t) = \frac{\lambda_2 \cdot A_1(0)}{\lambda_2 - \lambda_1} \cdot \left(e^{-\lambda_1 t} - e^{-\lambda_2 t}\right) + A_2(0) \cdot e^{-\lambda_2 t} \tag{2.65}$$

Note that $\lambda \cdot N$ always obtains the activity for a number of N nuclides. If the change of N is due only to physical decay, then dN/dt will also provide the activity. However, if the change of N over time is also a result of some contribution in addition to the physical decay, then the dN/dt will not describe the activity, since it is a unit to only describe the decay process.

Methods for using short-lived radionuclides are often based on the possibility to extract a short-lived daughter radionuclide from a parent nuclide with a longer half-life chemically. The parent radionuclide is here a generator of the short-lived radionuclide. If one extracts the daughter nuclide to be used in a patient study, it is of interest to evaluate the optimal time for performing this extraction. If one assumes a parent nuclide in its pure form, that is, the number of daughter nuclides $N_0 = 0$ at time $t = 0$, then the equations X and Y will provide the number of parent and daughter nuclides at a time t>0. If one is interested in

$$A_2(t) = \frac{\lambda_2 \cdot A_1(0)}{\lambda_2 - \lambda_1} \cdot \left(e^{-\lambda_1 t} - e^{-\lambda_2 t}\right) \tag{2.66}$$

$A_2\ (t)$ described a function that initially increased with time but reached the maximum and it declines with time.

2.5.1 Complex Decay Chains

In nature, there exist three complex decay chains originated from a radionuclide with a very high excess of energy, namely the Uranium-, Actinium- and Thorium-series. In addition to this, there is also Neptunium-series that originates from ^{237}Np, a nuclide created artificially by nuclear reactions. All decays start with an α-emission following by subsequent α- and β-emissions until stable Bi or Pb nuclides are reached. The half-life of the different radionuclides in the chains varies from milli-seconds to millions of years and, therefore, it is often very difficult to calculate the properties at a certain time point. However, in nuclear medicine there is an increased interest in using α-particles for therapy because of its high local energy absorption. Below is shown the Uranium series as an example.

$$\begin{matrix} & \alpha & & \beta^- & & \beta^- & & \alpha & & \alpha & \\ {}^{238}_{92}U & \rightarrow & {}^{234}_{90}Th & \rightarrow & {}^{234}_{91}Pa & \rightarrow & {}^{234}_{92}U & \rightarrow & {}^{230}_{90}Th & \rightarrow \\ & 4.510^9\ y & & 24.1d & & 1.17m & & 2.510^5\ y & & 7.510^4\ y \end{matrix}$$

$$\begin{matrix} & & \alpha & & \alpha & & \alpha & & \beta^- & & \alpha,\beta^- & \\ \rightarrow & {}^{226}_{88}Ra & \rightarrow & {}^{222}_{86}Rn & \rightarrow & {}^{218}_{84}Po & \rightarrow & {}^{214}_{82}Pb & \rightarrow & {}^{214}_{83}Bi & \rightarrow \\ & & 1600y & & 3.8d & & 3.1m & & 26.8m & & 19.9m \end{matrix}$$

$$\begin{matrix} & & \alpha & & \beta^- & & \beta^- & & \alpha & \\ \rightarrow. & {}^{214}_{84}Po & \rightarrow & {}^{210}_{82}Pb & \rightarrow & {}^{210}_{83}Bi & \rightarrow & {}^{210}_{84}Po & \rightarrow & {}^{206}_{82}Pb(stable) \\ & & 164\mu s & & 22.3y & & 5.0d & & 138.4d & \end{matrix}$$

2.6 RADIONUCLIDE DATA SOURCES

It is of most importance to have accurate and updated data for radionuclides and their decay properties and emissions [4]. There are several sources available that provide such nuclear-decay data. The National Nuclear Data Center (NNDC) at Brookhaven National Laboratory in Upton, New York, have compiled and maintained the evaluated nuclear structure data file (ENSDF) in a collaboration with the international nuclear structure and decay-data network [5]. The NNDC also maintain the NuDat2 database (www.nndc.bnl.gov/nudat2/) that is a Web interface to different types of databases and provides information about nuclear levels, gamma-radiation information, and other information about radiation following nuclear decays.

Nudat 2 also have a page for accessing decay data in the medical internal radiation dose (MIRD) format from the evaluated nuclear structure data file (www.nndc.bnl.gov/nudat2/mird/)

Another useful source of information is the decay-data evaluation project (DDEP) at the Laboratoire National Henri Becquerel [6]. The DDEP provides data specifically related to radioactive decay measurement, including x-ray and Auger emission energies and probabilities. Their web address is www.lnhb.fr and, from this page, the Nucléide – Lara application can be found (www.lnhb.fr/nuclear-data/module-lara/). This database interface provides searchable information about decay properties as well as decay schemes. There is also a page of useful collection of references about the process for determining recommended values, including an interactive periodical table with the option of downloading data in ENSDF, PenNuc, Lara ACSII-based format as well as beta-shape spectra – a feature useful for Monte Carlo simulations (www.lnhb.fr/nuclear-data/nuclear-data-table/.

The Nuclear Data Section of the International Atomic Energy Agency (IAEA) has a web page that provides nuclear data with over hundreds of databases for most nuclear applications. IAEA also provide an app for IOS and Android devices that includes more than four thousand nuclides and isomers, including nuclide's decay type, radiation and half-life.

REFERENCES

[1] R. D. Evans, *The Atomic Nucleus*. New York: McGraw Hill, 1955.

[2] F. H. Attix, *Introduction to Radiological Physics and Radiation Dosimetry*. New York: Wiley, 1986.

[3] E. B. Podgorsak, *Radiation Physics for Medical Physicists*, 3 edn. Switzerland: Springer, 2016.

[4] A. L. Nichols, "Recommended Decay Data and Evaluated Databases – International Perspectives," *Journal of Nuclear Science and Technology*, vol. 52, no. 1, pp. 17–40, 2015/01/02 2015, doi:10.1080/00223131.2014.929985.

[5] M. R. Bhat, "Evaluated Nuclear Structure Data File (ENSDF)," in *Nuclear Data for Science and Technology*, S. M. Qaim, Ed., (Research Reports in Physics. Berlin and Heidelberg: Springer, 1992.

[6] M. A. Kellett and O. Bersillon, "The Decay Data Evaluation Project (DDEP) and the JEFF-3.3 Radioactive Decay Data Library: Combining International Collaborative Efforts on Evaluated Decay Data," *EPJ Web Conference*, vol. 146, p. 02009, 2017. [Online]. Available: https://doi.org/10.1051/epjconf/201714602009.

3 Basics of Radiation Interactions in Matter

Michael Ljungberg

CONTENTS

3.1 INTRODUCTION

Most applications in nuclear medicine depend to some extent on the physics behind interactions between charged particles and the surrounding material or between electromagnetic radiation (photons) and charged particles, whether it is for radionuclide therapy, diagnostic imaging, or simply radiation detection. Solid knowledge of the underlying processes by which radiation interacts with matter is, therefore, of utmost importance in order to fully understand the potential and, more importantly, the limitations and risks for patients and personnel associated with the use of radiation.

DOI: 10.1201/9780429489556-3

The purpose of this text is to provide a basis for understanding the role of interactions in radiation dosimetry, radiation shielding, radiation protection, and radiation detection, and to provide further information on the ionizing radiation that occurs when the primary emitted radiation – for example, that emitted from a decaying radionuclide or its daughter products – interacts with its surrounding material.

The kinetic energy of ionizing particles is expressed by a special unit called the electron volt (eV), defined as the kinetic energy that a particle of unit charge (e.g., the electron) reaches when it is accelerated in a potential of 1 V. 1 eV = 1.602 × 10^{-19} J.

3.2 IONIZING RADIATION

As charged particles penetrate matter, they lose their kinetic energy via different types of interactions with nearby atoms and, as a result, create *ionizations* and *excitations* in those atoms and molecules along their path tracks. The process of excitation results in energy absorption by the atom, which leads to an excited state of electrons into higher energy levels. However, these excitations return more or less to the ground state by emission of electromagnetic radiation (photons). Ionization refers to situations in which an orbital electron will, during an interaction, receive a sufficient amount of energy so that it is energetically possible for the electron to completely leave its place in the atomic shell. The energy that is transmitted in this way to nearby atomic elements can give rise to a series of secondary interactions between the emitted electron and other atoms, and may result in potentially severe biological changes (e.g., damage to the DNA nucleus) when considering humans, or result in a measurable electric signal when considering a radiation detector.

The following definitions can be useful.

- Direct ionizing particles are charged particles $\left(e^-, e^+, p^+, \alpha^{2+}, \ldots\right)$, which have gained sufficient kinetic energy to subsequently produce ionizations through inelastic collisions along their path.
- Indirect ionizing particles are uncharged particles $(n, \gamma, \ldots)$ that, by their interactions, can either release directly ionizing particles or initiate nuclear conversion.

Ionizing radiation is therefore referred to as radiation that consists of both direct and indirect ionizing particles or a mixture of both.

The released energy from physical interactions can provide the ability to

1. estimate the original energies of the photons or charged particles that have deposited energy, thus enabling the identification of the radionuclide (gamma or charged-particle spectroscopy);
2. estimate the location of the radionuclide that has emitted radiation by forming an image of the interactions with an imaging system (scintillation cameras, SPECT, and PET); or
3. cause damage to tissues that, in some cases, can be very useful (as in radiotherapy with external beam, seal sources, or radionuclides) or harmful (side effects or late stochastic effects for patients and personnel).

It is therefore essential to understand the physical processes that occur in the volume of interest that is irradiated (e.g., in a detector, crystal, or a volume of tissue) because these processes will determine the accuracy and precision of the results of the applications for which we use radiation and radiation transport.

The total effect of radiation is determined by the properties of the particles (charge, type, mass, energy/velocity, etc.), but also on the number of particles per unit time and unit area. The *particle fluence* of a radiation beam is defined as the number of particles passing through a unit cross-sectional area (cm^{-2}). The two quantities related to fluence are the flux, defined as the fluence rate of particles, that is, the number of particles that pass through a unit cross section per unit time (cm^{-2} s^{-1}), and the energy fluence, which is the amount of energy that passes through a unit cross-sectional area (J cm^{-2}).

A beam of indirect ionizing radiation (e.g., X-rays, gamma rays, or neutrons) deposit their energy in a medium by a two-stage process whereby the energy carried by the photons or neutrons is transformed into the kinetic energy of charged particles. Direct ionization releases a charged particle (often an electron from its shell) and provides kinetic energy so that the particle can further excite to ionize other atoms. The quantity *kerma* is an acronym for 'kinetic energy released in matter', and is defined as the kinetic energy that is transferred to charged particles by indirect ionizing radiation.

It should also be mentioned here that the word 'photon' is a general name for the energy quantum that describes electromagnetic radiation. In quantum mechanics, every quantum of energy can be regarded as either a particle or a wave, that is, they can have the properties of both. For electromagnetic radiation, the properties of low-energy radiation are best described in terms of waves with a certain amplitude and frequency, while those of high-energy electromagnetic radiation are best described as independent particles that can be scattered in different directions. The particles in these energy ranges are called photons and have energy $E = h\upsilon$, where υ is the corresponding frequency and h is Planck's constant (6.62607×10^{-34} Js). The electromagnetic radiation used in medical physics applications is generally very high in energy because the particle properties best describe the effects and behaviour of the radiation.

There are also some differences in the naming of photons. Gamma (γ) rays are those photons emitted as a result of the decay of a nucleus, and which therefore originate from the nucleus. X-rays that are emitted as a result of energy loss when electrons, accelerated in the vicinity of the atomic nucleus, are named 'bremsstrahlung photons' (sometimes also 'braking radiation'). When a positron is close to an electron, and both particles are close to rest, a transformation of the two electron masses to energy in the form of electromagnetic radiation is possible with the result of two annihilation photons. A vacancy in an electron shell, caused by, for example, photoelectric absorption of an incoming photon, is called a 'characteristic X-ray' because the energies of these emitted X-ray photons are characteristic of the energy levels of the different electron shells and thus of the element. It is therefore recommended to follow these naming conventions because they provide information about the underlying interaction process.

3.3 SOLID ANGLE

A solid angle is a three-dimensional extension of a two-dimensional angle (Figure 3.1). It is essential to estimate the probabilities for different types of interaction; therefore, it is important to understand the basics of this type of angle.

The two-dimensional angle is defined by a part of the perimeter of the unit circle and is measured in units of radians, where a full length around the circle equals an angle of 2π radians. A solid angle (often denoted by Ω) is defined as a unit of area, A, on a unit sphere surrounding the midpoint of the sphere divided by r^2 (Figure 3.2). The unit of the solid angle is termed a *steradian*. The solid angle can then be calculated using the equation $\Omega = A / r^2$. For a unit sphere, the largest solid angle Ω is 4π.

The result of a particle interaction often is a change in direction relative to the incoming direction. In general terms, one can describe the number of scattered particles in a direction θ as

$$dn = f(\theta) \cdot d\Omega, \tag{3.1}$$

where $f(\theta)$ is an interaction-dependent function, and $d\Omega$ is the size of the solid angle. Note that the number of particles scattered exactly at angle θ is zero. For processes that are symmetrically distributed, one can sufficiently describe the number of particles only in terms of θ relying on the calculation of the surface element $dA = 2\pi \cdot \sin(\theta) d\theta = d\Omega$. The number of particles scattered in direction θ will then be

$$dn = f(\theta) \cdot 2\pi \cdot \sin(\theta) \cdot d\theta, \tag{3.2}$$

and from integration, one can obtain the total number of scattered particles according to

$$n = \int dn = 2\pi \cdot \int_0^{\pi} f(\theta) \cdot \sin(\theta) \cdot d\theta. \tag{3.3}$$

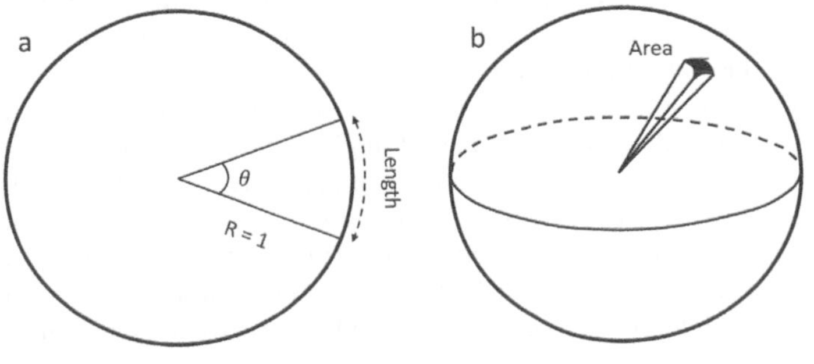

FIGURE 3.1 Definition of the radian (a) and steradian (b).

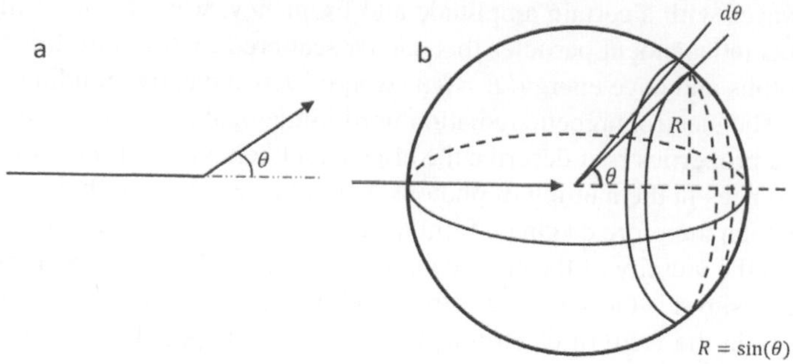

FIGURE 3.2 An element of area, dA, at an angle of θ (a) shaped as a small circular band on a sphere with radius 1 (b). The width of the band is dθ and the radius is sin (θ).

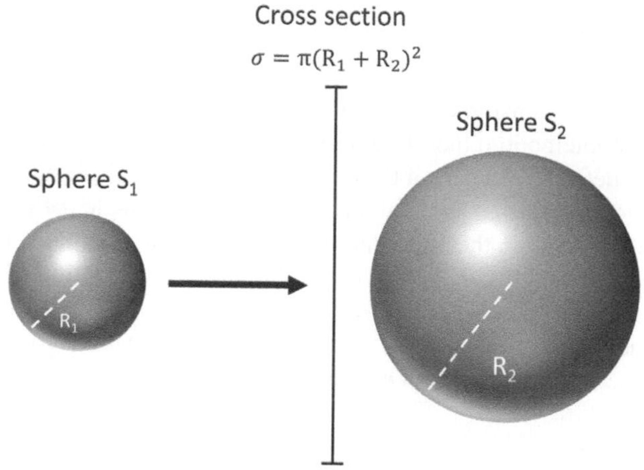

FIGURE 3.3 Interactions between two spheres, S_1 and S_2, each with radius R_1 and R_2. The cross section (i.e., a measure of the probability for collision) depends on the sum of R_1 and R_2.

3.4 CROSS SECTION

Cross sections are a fundamental property of the interaction process of photons, charged particles, and neutrons because they provide an estimate of the probability for a particular interaction. The unit of a cross section is that of area (for example, cm^2). To understand the definition of the cross section in this context, consider the example in Figure 3.3 as one homogeneous sphere S_1 of radius R_1 moves toward a second homogeneous sphere S_2 with radius R_2.

An interaction will occur only when the two spheres have a direct contact. The cross section then can be defined as the area that the centre of gravity for S_1 must hit to result in a collision (interaction). In the example, the cross section would then be equal to a circular area of $\sigma = \pi(R_1 + R_2)^2$. This means that the magnitude of the cross section depends on the radius for both spheres. If two different types of processes may appear at the interaction (for example, absorption versus scattering) with the probabilities $P_A + P_B = 1$, then the total cross section is $\sigma_A + \sigma_B = \sigma_T$. Thus, each process (e.g., photo absorption, Compton scattering) will have its own cross section and the probability that something will happen will be proportional to the sum of all separate cross sections.

In many cases, the direction of a particle exemplified by sphere S_1 will change direction according to some function $f(\theta)$, relative to its initial direction into a solid angle. This cross section is called the differential cross section and is described by $d\sigma_A(\theta) = \sigma_A \cdot f(\theta) \cdot d\Omega$. One can also calculate a differential cross section describing the probability for interaction A, where particle S_1 is scattered into solid angle $d\Omega$ around scattering angle θ, and the energy between E_1 and E_1+dE_1 can be calculated as $d\sigma_A(\theta, E_1) = \sigma_A \cdot f(\theta) \cdot h(\theta, E_1) \cdot d\Omega \cdot dE$. The general definition for a given interaction process can be described by $\sigma = P / \phi$, where P is the probability of the interaction process when one particle is irradiated by a specific type of radiation with a fluence of ϕ. The cross section is often given in units of barns (10^{-28} m^2).

3.5 PHOTON INTERACTIONS

Considering the energy interval relevant for most medical radiation physics applications (1 keV to 20 MeV), then photons passing through some material will interact with atoms through four major interaction processes, namely, photo absorption, Compton scattering, coherent scattering, and pair production. In the photo-absorption process, the energy of the photon is completely transferred to an orbital electron that will receive sufficient energy, which will be released from the atom shell. In Compton, or coherent, scattering, the photon is scattered in a new direction with or without a loss of energy, respectively. In the pair-production process, the photon interacts with the nucleus or an electron, resulting in the conversion of photon energy into a positron/electron pair that shares the residual kinetic energy. These four processes are now discussed in more detail.

3.5.1 Photon Absorption

The energy of the incoming photon is completely transferred to an orbital electron in one of the shells, following a very minor recoil energy. The process is schematically described in Figure 3.4. The conservation of energy can therefore be written as

$$h\nu = E_e + E_B, \tag{3.4}$$

where $h\nu$ is the incoming photon energy, E_e is the kinetic energy of the released orbital electron following the interaction process, and E_B is the binding energy of the orbital electron, that is, the energy that must be transferred to release the electron from its shell. The process is most likely to occur in the K-shell, but always results in a vacancy in the shell structure of the orbital electrons. The vacancy is rapidly filled by an electron from the outer shells. Because of the difference in the binding energies between the shells, the residual excess energy is emitted by either one or several characteristic X-ray photons or by the emission of Auger electrons. The process can only occur with an electron, bounded in a shell, as the atomic nucleus needs to receive some momentum.

The cross section per atom for photon absorption, $\sigma^a_{\gamma,e}$ has been formulated by Bethe based on quantum-mechanical methods, and can be described in a simple form by

$$\sigma^a_{\gamma,e} \sim k \cdot s(h\nu, Z) \cdot \frac{Z^5}{(h\upsilon)^{7/2}}, \tag{3.5}$$

where Z is the atomic number (number of protons), and $h\upsilon$ is the photon energy, expressed in units of keV. s is a complicated function that depends on Z and the incoming photon energy, $h\upsilon$, and can be written as

$$s = -0.18 + 0.28 \cdot {}^{10}log\left(\frac{h\nu}{Z^2}\right). \tag{3.6}$$

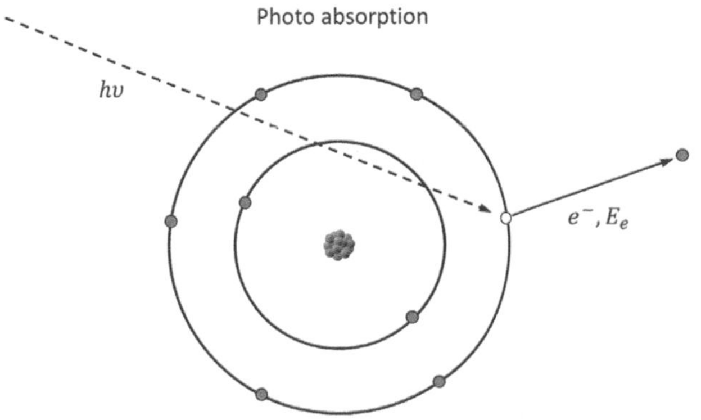

FIGURE 3.4 Schematic image of the photo absorption process. An incoming photon with an energy hυ is completely absorbed and its energy is transferred to an electron from one of the atomic shells. This electron is the released from the shell and with a kinetic electron E_e. A vacancy in the shell is created.

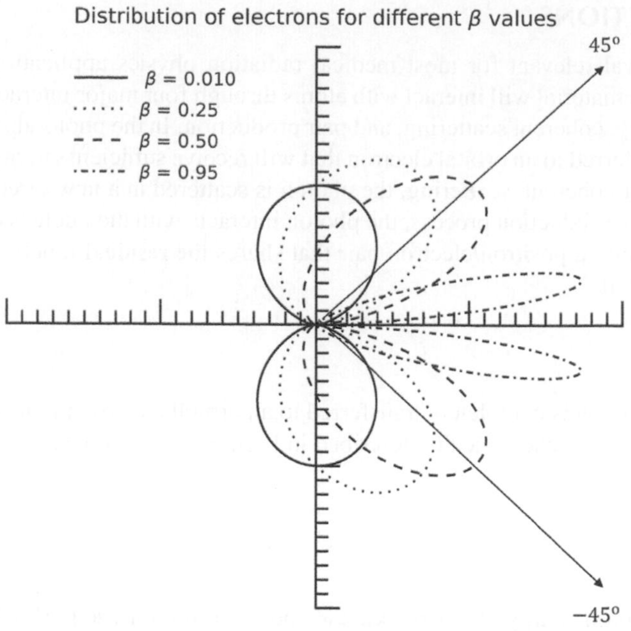

FIGURE 3.5 The angular directions of the ejected electrons from a photon absorption for different electron energy, expressed as β values, were $\beta = v / c$, where v is the velocity of the electron and c is the speed of light in a vacuum.

Equation 3.5 is valid only for photoelectrons that do not reach relativistic velocities, which holds true in the energy ranges for which this process is important. The energy and Z dependence can therefore be summarized approximatively by the following equation:

$$\sigma^{a}_{\gamma,e} \sim \frac{Z^5}{(h\nu)^{7/2}}. \tag{3.7}$$

The probability of photo absorption thus increases very rapidly with increasing atomic numbers and decreases with photon energy down to the K-absorption energy, where there is a discontinuity in the cross sections.

The direction θ of the emitted electron is given by

$$dN \sim \frac{sin^2\theta}{(1-\beta\cdot cos\theta)^4}\cdot d\Omega, \tag{3.8}$$

where $\beta = v / c$, v is the speed of the electron, c is the speed of light in vacuum. The distributions of the emitted electron for some incoming photon energies are shown in Figure 3.5.

The vacancy in the shell is filled by an electron in the outer shells, resulting in a subsequent emission of either Auger electrons or characteristic X-ray photons. The probability of an Auger electron emission is higher for low-Z materials, and the emission of characteristic X-ray emission dominates for high-Z materials. For more details, see Chapter 2.

3.5.2 Photon Scattering

3.5.2.1 Thomson Cross Section

Thomson scattering is a derivation based on classical electromagnetism of elastic scattering in the presence of an electromagnetic field and forms a basis for the derivations of the cross sections for coherent scattering. The term 'elastic' indicates that the photon will lose very little kinetic energy in the interaction. The theory is that the wave properties of the incoming photon cause an orbital electron to oscillate, and because of this, a photon will be emitted with the same wavelength as the incoming photon but heading in a different direction. The differential Thomson scattering cross section for an unpolarized incident photon is described by

$$\frac{d\sigma_{Th}}{d\Omega} = \left(\frac{e^2}{4\pi\epsilon_0 m_e c^2}\right)^2 \cdot \left(\frac{1+\cos^2\theta}{2}\right), \tag{3.9}$$

where e is the charge of an electron, ϵ_0 is the dielectrical constant and m_e is the rest-mass of an electron. The term in the first parenthesis is also called the classical electron radius, r_0. It can be seen that the differential cross section is a function of the scattering angle and that the intensity of the radiation is higher when θ are close to 0 or π (Figure 3.6).

3.5.2.2 Compton Scattering

In this process, the incoming photon interacts with a shell electron, resulting in a partial loss of energy, and the photon continues in a different direction with a lower energy. The electron will receive sufficient energy to leave the shell, creating a vacancy (Figure 3.7).

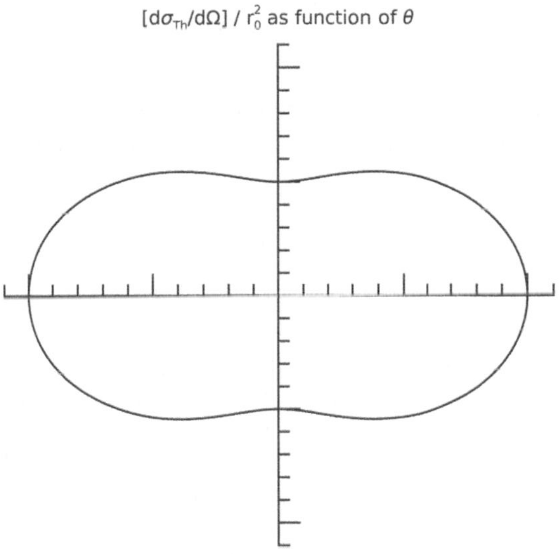

FIGURE 3.6 A polar plot of the Thompson cross section as function of different scattering angles θ. The values have been normalized to the square of the classical electron radius r_0^2.

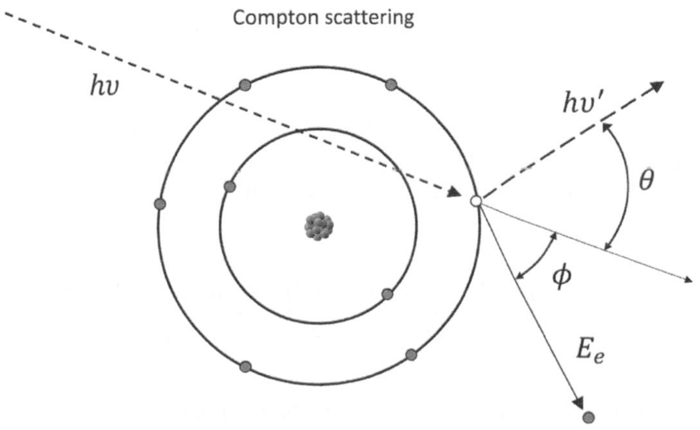

FIGURE 3.7 Schematic image showing a photon with an incoming energy $h\upsilon$ interacting with an orbital electron that will receive energy and be released in a direction ϕ with a kinetic energy of E_e. The photon will be scattered at an angle θ relative to the incoming direction and continue with a lower energy $h\upsilon'$.

If the energy of the incoming photon is much higher than the binding energy of the electron, the process can be regarded as a binary collision between two particles. Based on this assumption, the relation between the incoming photon energy hv and the energy of the scattered photon hv' can be derived as

$$hv' = \frac{hv}{1+\alpha(1-cos\theta)}, \tag{3.10}$$

where

$$\alpha = \frac{hv}{m_e c^2}. \tag{3.11}$$

An expression for the loss of energy for the photon is given by

$$hv - hv' = E_e = hv \cdot \frac{\alpha(1-cos\theta)}{1+\alpha(1-cos\theta)}. \tag{3.12}$$

From Eq. 3.10, it can be seen that the energy loss for an incoming photon with the energy hv, that is, the kinetic energy transferred to the released electron, is uniquely determined by the scattering angle θ of the photon. The angles θ and ϕ are related to each other by

$$\tan\phi = \frac{sin\theta}{(1+\alpha)(1-cos\theta)}. \tag{3.13}$$

If one of the quantities E_e, hv', θ, or ϕ is known, then all the other quantities can be determined. It can also be seen that for each scattering angle, the Compton photon energy approaches a limit with increased incoming photon energy. For large scattering angles, the energy hv' will approach the limit value for moderate photon energies. The above equations also show that, for incoming photons with high energies that are scattered with some degree or higher, most energy will be transferred to the electron. The largest energy transfer in a single Compton process is when the photon is scattered at 180°.

By using quantum mechanics theories, Klein and Nishina [1] derived an expression for the differential cross section per electron, $d\sigma^e_{\gamma,\gamma e}$, for a Compton process where an unpolarized photon is scattered an angle θ within a solid angle $d\Omega$.

$$d\sigma^e_{\gamma,\gamma e} = \frac{r_0^2}{2} \cdot \left(\frac{\alpha'}{\alpha}\right)^2 \cdot \left(\frac{\alpha}{\alpha'} + \frac{\alpha'}{\alpha} - sin^2\theta\right) \cdot d\Omega, \tag{3.14}$$

where α' is according to Eq. 3.11, but for the scattered photon energy. The subscript $\gamma, \gamma e$ can be understood as a photon that is absorbed, and photons and electrons that are emitted. Eq. 3.14 can be written in a more condensed way.

Integration of Eq. 3.14 over all scattering angles yields the total Klein–Nishina cross section in units of cm² per electron.

$$\sigma^e_{\gamma,\gamma e} = 2\pi r_0^2 \left\{ \frac{1+\alpha}{\alpha^2} \cdot \left[\frac{2(1+\alpha)}{1+2\alpha} - \frac{\ln(1+2\alpha)}{\alpha} \right] + \frac{\ln(1+2\alpha)}{2\alpha} - \frac{1+3\alpha}{(1+2\alpha)^2} \right\} \tag{3.15}$$

If the binding energies of all orbital electrons are negligible when compared to the incoming photon energy, the cross section will be the same for all electrons in an atom. Therefore, the differential cross section per atom can be written as

$$d\sigma^a_{\gamma,\gamma e} = Z \cdot d\sigma^e_{\gamma,\gamma e} \tag{3.16}$$

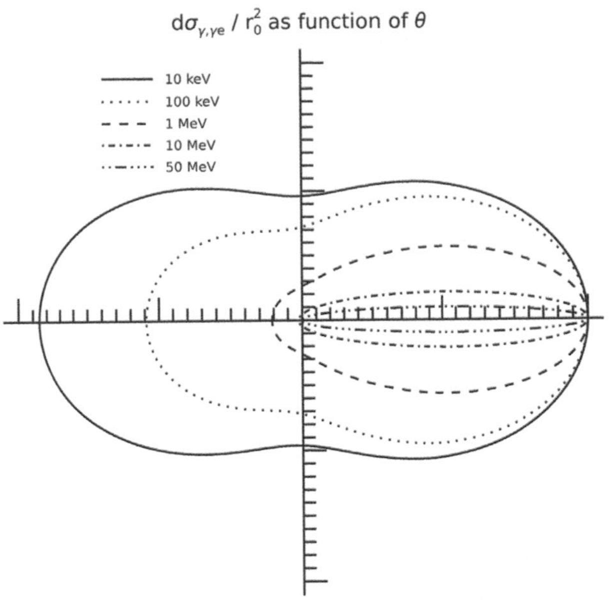

FIGURE 3.8 The angular distribution of scattered photons for some incoming photon energies is shown.

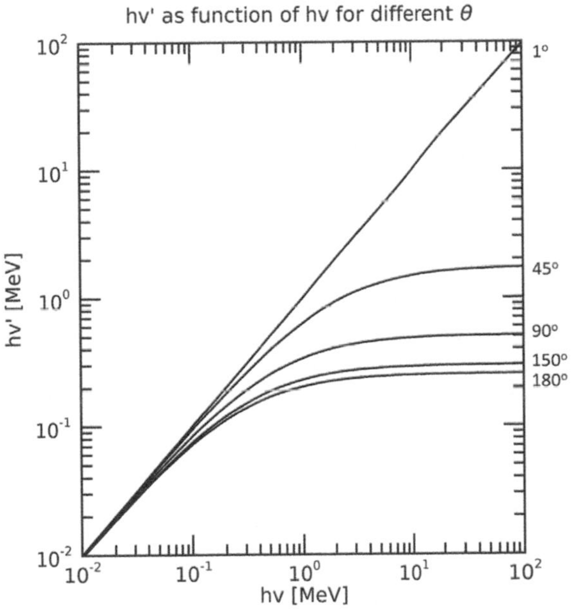

FIGURE 3.9 Graph showing the energy of the scattered photon hv' as a function of the energy hv of incoming photon for five different scattering angles θ.

The dependence on the scattering angle θ is best visualized when plotted in polar coordinates, as shown in Figure 3.8. It can be seen that the differential cross section approaches the same value when θ approach zero. For low energies, the distribution is symmetric and approaches the Thomson cross section, as shown in Figure 3.6. Figure 3.9 shows the relation between the incoming photon energy and scattered photon energy for some fixed scattering angles. It can be seen that the energy of the scattered photon converges to a maximum value when the incoming photon energy increases. Thus, at these energies, most of the energy is transferred to the released Compton electron.

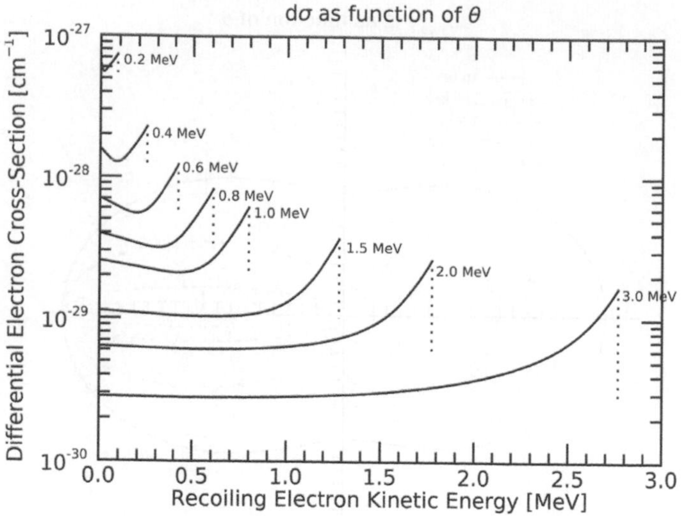

FIGURE 3.10 The energy distribution of recoiling electrons as a result of Compton interactions from incoming photon energies ranging from 0.02 to 3.0 MeV. The Compton edges (vertical dashed lines) define the maximum energy transfer to an electron that is possible in a single interaction, and the numerical values are also shown. In a scintillation detector (Chapter 6), it is the interactions and the deposit energy of these released electrons that form the measurable signal, so the Compton distributions demonstrated in this graph are often seen when detecting an incoming fluence of photons.

Also of interest is the expected energy distribution of photons and electrons after a Compton interaction, as shown in Figure 3.10.

$$\frac{d\sigma_{\gamma,\gamma e}}{dE_e} = \frac{\pi r_0^2}{\alpha h\nu}\left[2 - \frac{2E_e}{\alpha\left(h\nu - E_e\right)} + \frac{E_e^2}{\alpha^2\left(h\nu - E_e\right)^2} + \frac{E_e^2}{h\nu\left(h\nu - E_e\right)}\right] \tag{3.17}$$

When measuring photons using a detector, for example a scintillation detector, the signal that generates a pulse-height distribution for gamma spectroscopy (Chapter 8) comes from the interaction of released electrons, so we can expect peaks in the measurement, like those shown in Figure 3.10, that do not relate to a factual photon energy.

As stated above, the Klein–Nishina formula for Compton scattering from atomic electrons assumes that the electrons are free and at rest, which is a good approximation for photons with energies much larger than the binding energies of the K-shell. However, when the energy approaches the binding energies, the atomic cross section is reduced and is not directly a function of the atomic number Z. For a given scattering angle and incoming energy, the photon energy is not well defined but rather broadened, as can be seen in Figure 3.11.

This effect can be accounted for in the Waller-Hartree approximation [2] by multiplication with a function

$$\frac{d\sigma}{d\Omega} = \frac{d\sigma^{KN}}{d\Omega} \cdot S(x,Z) \tag{3.18}$$

The function $S(x,Z)$ is called the incoherent scattering function and is a function of the atomic number Z and the momentum transfer parameter x. This, in turn, is a function of the scattering angle θ and the photon wavelength λ:

$$x = \frac{\sin(\theta/2)}{\lambda} \tag{3.19}$$

Figure 3.12 shows examples of scattering functions for four different materials of various atomic numbers [3]. For higher photon energies (lower wavelengths), the influence of the binding energies decreases, and the numerical value of S(x,Z) converges to Z for the different elements.

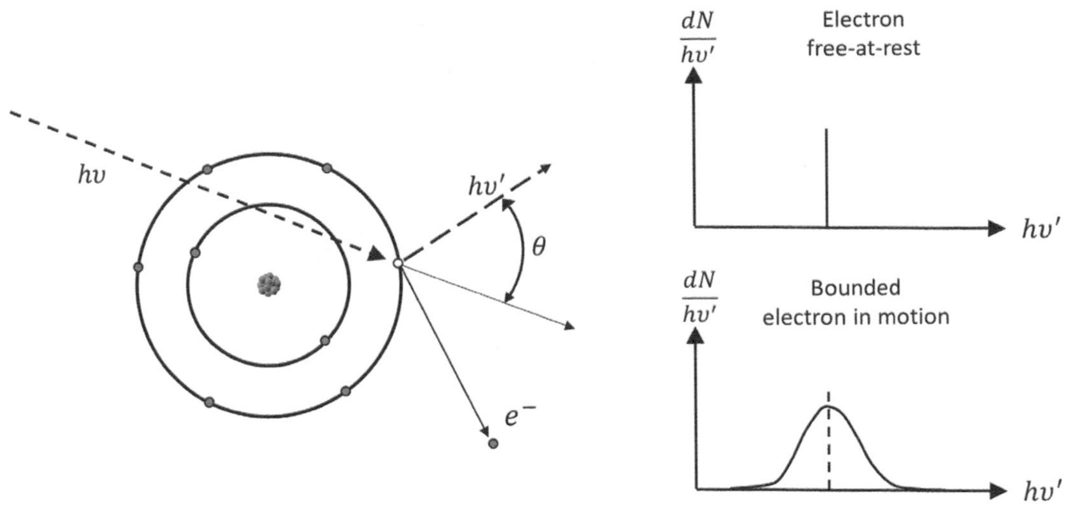

FIGURE 3.11 The effect of Compton scattering with a bound electron. The direct relation between the energy of the scattered photon and the angle is not well-defined but distributed.

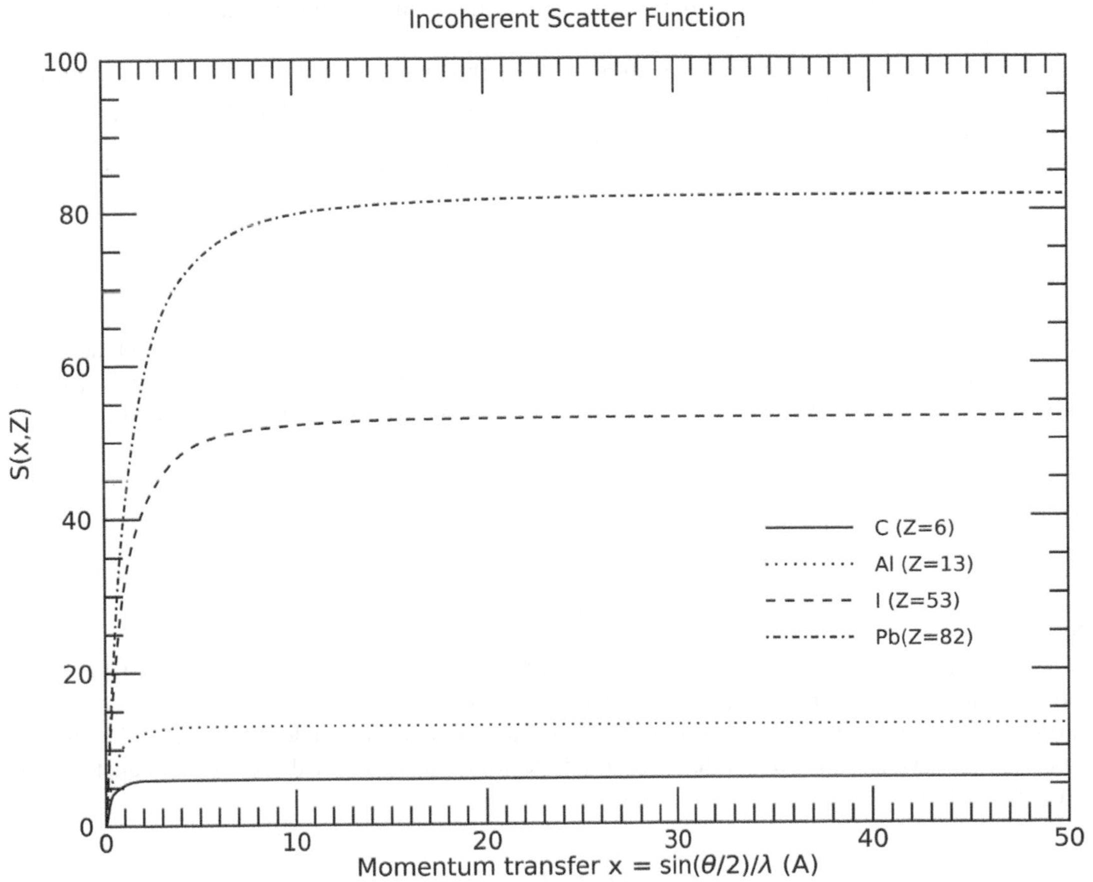

FIGURE 3.12 Plots of the incoherent scattering function S(x,Z) for different atomic elements. For higher photon energies, the effect of the binding of the electron on the atomic cross section becomes of less importance. The graphs have been based on data obtained from [3].

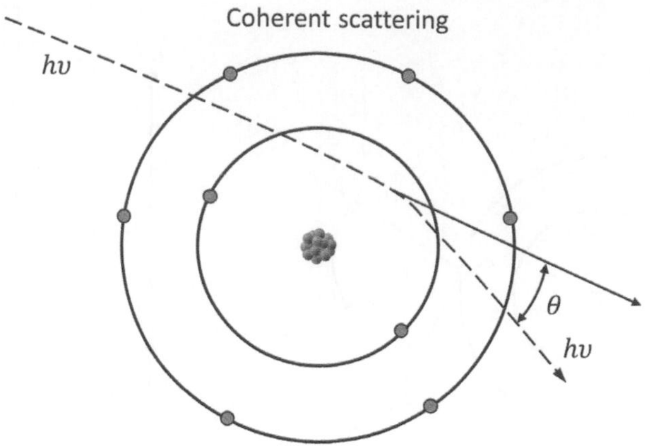

FIGURE 3.13 Schematic image of the coherent scattering process.

3.5.2.3 Coherent Scattering

For small photon energies, one cannot consider the shell electrons to act as if they are free and at rest. This increases the probability that the photon interacts with the whole atom. Because of the large mass of the atom, the recoil energy transferred from the photon will be very small, so this process essentially only results in a direction change (Figure 3.13). This process is called coherent scattering.

The cross section for this process increases with decreasing photon energy but when compared to the cross section for photo absorption, which increases by a power of approximately 3.5 for decreasing energies (Eq. 3.7), the contribution from coherent scattering can be neglected in some applications.

The differential cross section for coherent scattering and unpolarized photons can be described by

$$\frac{d\sigma_{coh}}{d\Omega} = \frac{r_0^2}{2}\left(1+\cos^2\theta\right)\cdot F^2\left(x,Z\right) \tag{3.20}$$

where the first part in the equation is the classical Thompson differential cross section (Eq. 3.9) and the function F(x,Z) is called the form factor [3]. Figure 3.14 shows the calculations using Eq. 3.22 for some materials and photon energies. It can be seen that coherent processes most likely result in small-angle scattering and is important for low photon energies.

3.5.3 Pair Production

If the incoming photon energy exceeds $2m_ec^2 = 1.022\,\text{MeV}$, it is energetically possible for the energy of the photon to be transformed into an electron–positron pair, where the residual energy from the photon is shared as kinetic energy between the two particles. Figure 3.15 shows a schematic of the procedure.

Pair production can only appear close to a charged particle that preserves momentum. Most likely, it will be a nucleus, and because of the large nucleic mass, the energy transfer is very small.

If pair production occurs with a shell electron, then this process is called triple production. The energy threshold is $4m_oc^2$ and an electron–positron pair plus a recoiling electron will be emitted. For pair production, we have the following energy relation:

$$hv = 2m_ec^2 + E_{e^+} + E_{e^-} + E_r. \tag{3.21}$$

In the following, the term 'pair production' refers to an interaction with an atomic nucleus (i.e., the recoiling energy $E_r \approx 0$). The kinetic energy is shared between the electron and positron in a random manner, but with a preference towards an equal share.

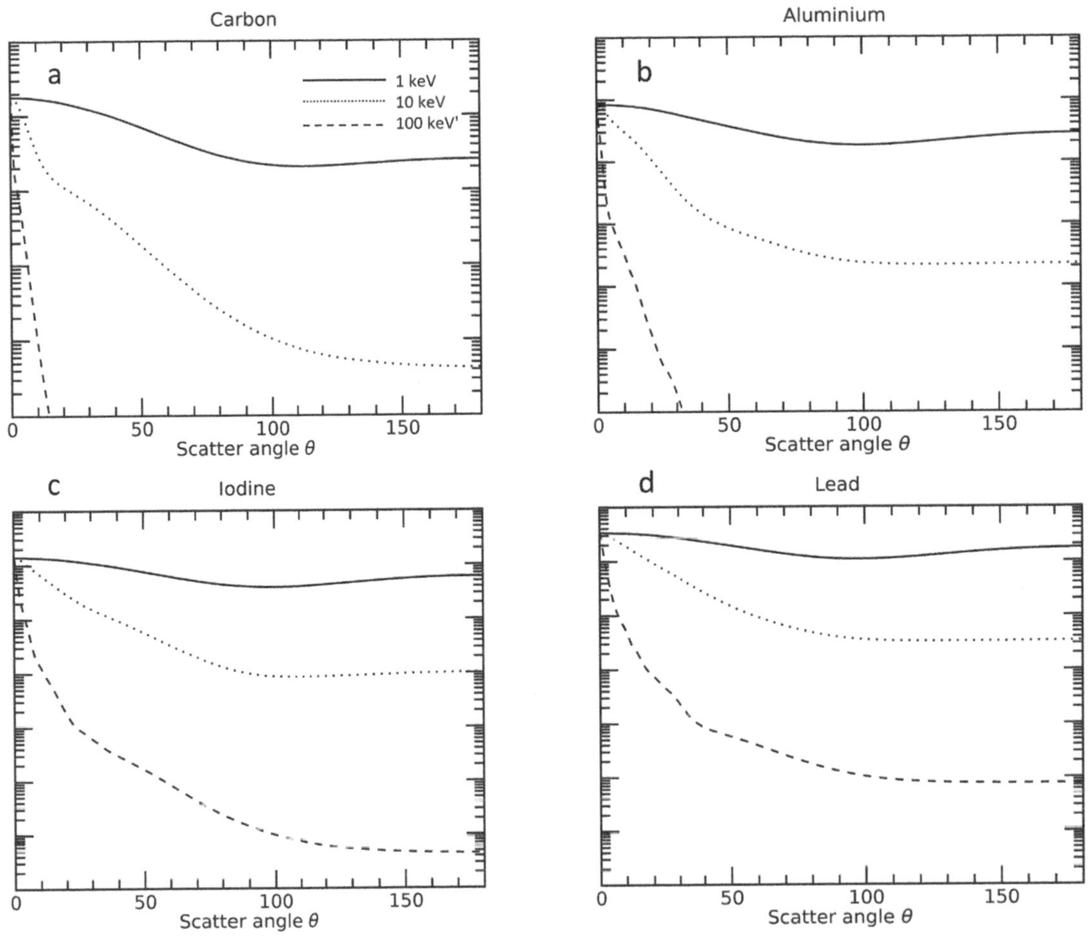

FIGURE 3.14 The cross section for coherent scattering as a function of scattering angle for three different photon energies (1,10 and 100 keV) for carbon (a), aluminium (b), iodine (c), and lead (d). Note that when the energy increases, the scattering angles are generally very small in the forward direction.

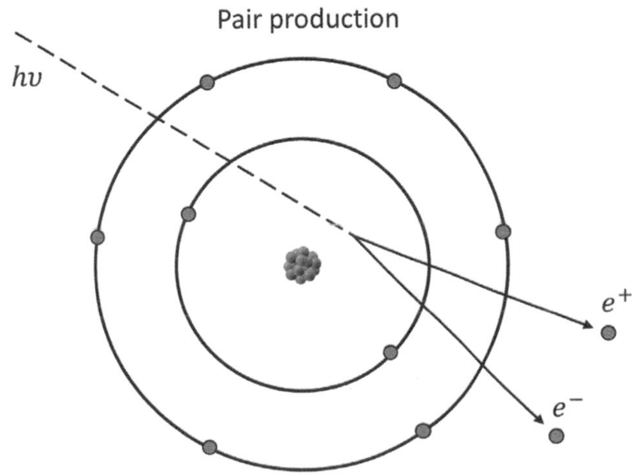

FIGURE 3.15 Schematic image of the pair production process.

The total cross section for pair production has been derived by Bethe and Heitler and is described by the following semi-empirical formula:

$$\sigma^{a}_{\gamma,e^-e^+} = \frac{1}{137}\left(\frac{q^2}{4\pi\varepsilon_0 m_e c^2}\right)^2 \cdot Z^2 \cdot \left(\frac{29}{9}\ln(2\alpha) - \frac{218}{27}\right). \tag{3.22}$$

Here, one can see that there is a Z^2 dependence for the atomic cross section and that the dependence on the incoming photon energy $(\sim \alpha)$ is low owing to the *ln* dependence. For very high energies, the photon-energy dependence is negligible; therefore, only the cross section depends on the material with which the photon interacts.

3.5.4 Photon Attenuation

3.5.4.1 Narrow-beam Geometry

Assume a beam of photons are impinging on some material (a radiation-protection shield or a volume of human tissue). An interesting question to be answered by a medical physicist is what will happen along the radiation beam; for example, how far will the photons reach an average and how much energy will be deposited along the path of the photons? To predict an answer, one can simplify the geometry to only consider those photons impinging in a narrow-beam geometry, that is, a situation in which a scattered photon is always deflected away from the beam and therefore cannot reach a volume of interest (for example a detector). Such a situation is exemplified in Figure 3.16a. If one considers an infinitely small layer of the material, with a thickness dx between the source and the detector, then the number of atoms per surface area of this layer will be

$$dn_a = \rho \cdot \frac{N_A}{M} \cdot dx, \tag{3.23}$$

where N_A is Avogadro's constant, ρ is the density of the material, M is the molar mass, and dx is the thickness. We also assume that every atom has a cross section for interaction, denoted σ^a_t, that is the sum of all possible photon interactions.

$$\sigma^a_t = \sigma^a_{\gamma,e} + \sigma^a_{\gamma,\gamma e} + \sigma^a_{\gamma,e^-e} + \cdots \tag{3.24}$$

Assume further that the material only consists of one chemical element, and that all photons have the same energy that together form a total cross section. Then, each unit of surface in our layer will then have a total cross section of $dn \cdot \sigma^a_t$

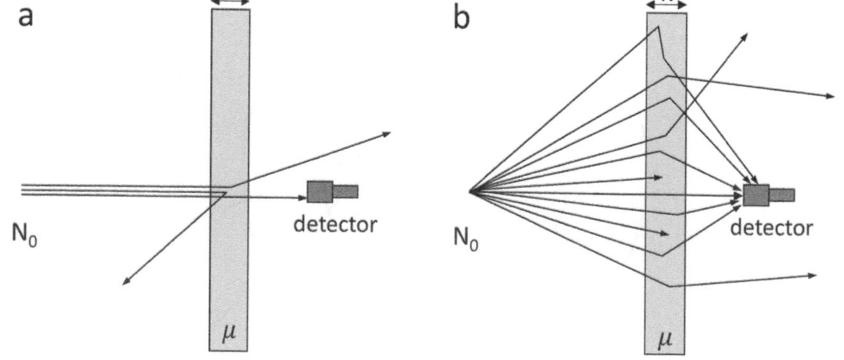

FIGURE 3.16 Left sketch (a) illustrates a narrow-beam geometry, where any interaction in the attenuator with thickness x will result in the photon being removed from the path that allows for detection, thus resulting in a lower detection rate. Right sketch (b) shows a broad-beam geometry, where photons emitted in broad angles may be scattered in the attenuator, resulting in a new direction that allows them to be detected. The broad-beam geometry increases the detected counts compared to those of a narrow beam.

for a single photon interaction. If we define the photon rate to be $\dot{N}$ before the layer and $\dot{N} - d\dot{N}$ photons after, then the following equation will be valid:

$$-\frac{d\dot{N}}{\dot{N}} = dn \cdot \sigma_t^a = \sigma_t^a \cdot \rho \cdot \frac{N_A}{M} \cdot dx. \tag{3.25}$$

We define the total linear attenuation coefficient, μ, as

$$\mu = -\frac{1}{\dot{N}} \cdot \frac{d\dot{N}}{dx}, \tag{3.26}$$

$$-d\dot{N} = \mu \cdot \dot{N} \cdot dx. \tag{3.27}$$

A comparison with Eq. 3.27 yields

$$\mu = \sigma_t^a \cdot \rho \cdot \frac{N_A}{M}. \tag{3.28}$$

If the photon rate of $\dot{N}_o$ is impinging on the material, the photon rate at a distance can be determined by the differential equation

$$\frac{d\dot{N}}{dx} = -\mu \cdot \dot{N}. \tag{3.29}$$

The solution of Eq. 3.31, with the boundary condition $\dot{N}(0) = \dot{N}_o$, will be

$$\dot{N}(x) = \dot{N}_o \cdot e^{-\mu x}. \tag{3.30}$$

The term $e^{-\mu x}$ here is the probability that a photon will pass through a material with thickness x without any interaction. Because of its exponential nature, this probability never reaches a value of zero. Therefore, it is not possible to define the maximum path length for photons, as is the case for charged particles. This also implies that it is not possible to construct a radiation shield that completely absorbs all impinging photons.

The mean free-path length, mfp, is defined as the depth at which, on average, one photon travels before interaction, or

$$mfp = \int_{x=0}^{\infty} x \cdot \left(\frac{-d\dot{N}}{\dot{N}_o} \right). \tag{3.31}$$

Solving Eq. 3.31 obtains

$$mfp = \mu \cdot \int_0^{\infty} x \cdot e^{-\mu x} dx = \mu^{-1}. \tag{3.32}$$

Thus, at a depth of one mfp, approximately 37 per cent of the photons in the beam still have not interacted with the material.

Narrow-beam geometry is in many situations an unrealistic geometry since most radiation fields have some kind of lateral extension that can cause a photon to be scattered back with a possible strike to a detector. Even scattered several times, a photon still can contribute to a detected signal. Therefore, we detect more events than the 'narrow-beam' geometry predicted by Eq. 3.35, as is seen in Figure 3.16b. As a consequence, the transmission equation in Eq. 3.30 must be complemented with a build-up factor, B(x), that is ≥ 1.

$$\dot{N}(x) = \dot{N}_o \cdot B(x) \cdot e^{-\frac{\mu}{\rho} \cdot \rho \cdot x} \tag{3.33}$$

The actual value of B(x) also depends on the photon energy, the composition and density of the material, and the detector geometry.

3.5.4.2 Mass-attenuation Coefficient

The mass-attenuation coefficient μ / ρ is defined as

$$\frac{\mu}{\rho} = \sigma_t^a \cdot \frac{N_A}{M} \tag{3.34}$$

and from this, Eq. 3.36 can be rewritten as

$$\dot{N}(x) = \dot{N}_o \cdot e^{-\frac{\mu}{\rho} \cdot \rho \cdot x} \tag{3.35}$$

The product $\rho \cdot x$ is often called the mass-thickness and has units of kg/m². The mass-attenuation coefficient now depends only on the atomic composition of the materials for a given photon energy. If one wants to show the contribution from the different processes, one can write

$$\frac{\mu}{\rho} = \sigma_{\gamma,e}^a \cdot \frac{N_A}{M} + \sigma_{\gamma,\gamma e}^a \cdot \frac{N_A}{M} + \sigma_{\gamma,\gamma}^a \cdot \frac{N_A}{M} + \sigma_{\gamma,e^-e^+}^a \cdot \frac{N_A}{M} + \ldots \tag{3.36}$$

Often, special notations are introduced for the different contributions.

$$\begin{aligned} \frac{\tau}{\rho} &= \sigma_{\gamma,e}^a \cdot \frac{N_A}{M} \\ \frac{\sigma_C}{\rho} &= \sigma_{\gamma,\gamma e}^a \cdot \frac{N_A}{M} \\ \frac{\sigma_R}{\rho} &= \sigma_{\gamma,\gamma}^a \cdot \frac{N_A}{M} \\ \frac{\kappa}{\rho} &= \sigma_{\gamma,e^-e^+}^a \cdot \frac{N_A}{M} \end{aligned} \tag{3.37}$$

Thus,

$$\frac{\mu}{\rho} = \frac{\tau}{\rho} + \frac{\sigma_C}{\rho} + \frac{\sigma_R}{\rho} + \frac{\kappa}{\rho}. \tag{3.38}$$

The dependence of μ / ρ on the photon energy is $\sim (h\nu)^{-7/2}$ in the energy range where photo absorption dominates; it decreases slowly in the energy range where the Compton process dominates and it increases slowly above the energy threshold where pair production starts to dominate. The dependence on the atomic number Z is $\mu / \rho \sim Z^4$ if the photon process dominates and μ / ρ is independent of Z when Compton interaction dominates, and μ / ρ is $\sim Z$ when pair productions dominate. For a composition of different chemical elements, the calculation of $\mu / \rho$is based on a weighted average of the cross sections for each element. Figure 3.17 shows partial attenuation coefficients and total attenuation coefficients for oxygen, copper, and lead.

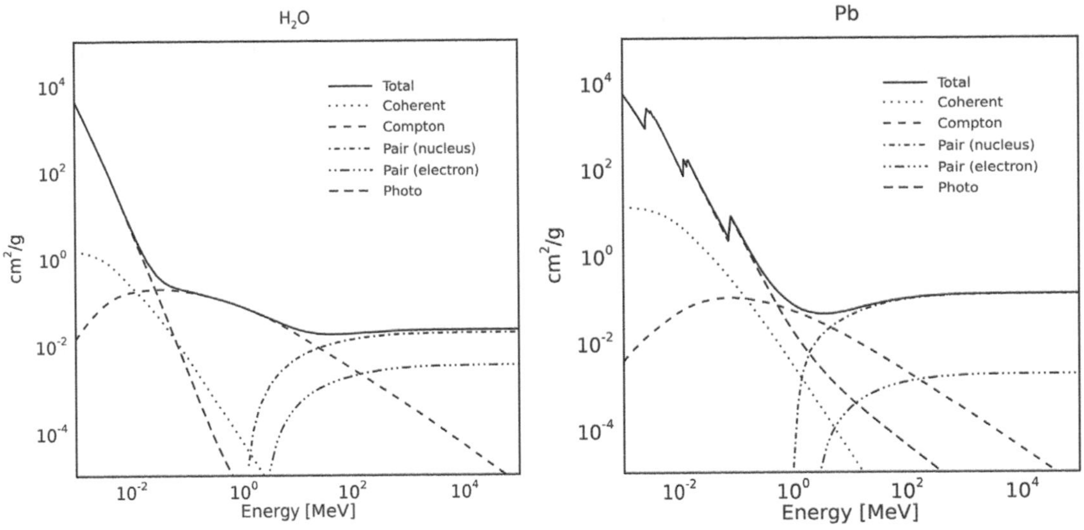

FIGURE 3.17 Partial interaction coefficients together with the total attenuation coefficient function of energy for three different materials. Note the discontinuities in the coefficients just below the binding energies of the shell electrons.

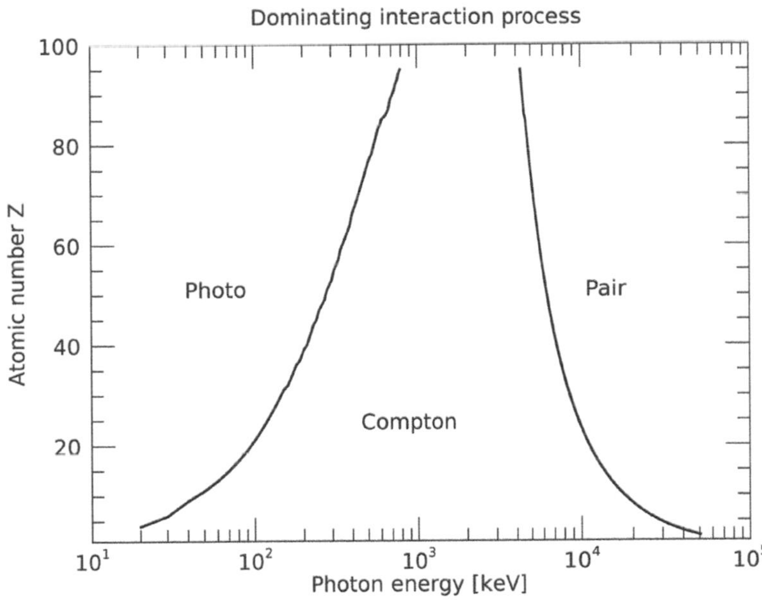

FIGURE 3.18 The graph shows the range of photon energy and atomic number where each of photo absorption, Compton scattering, and pair production dominates. Coherent scattering is not included, as it never dominates.

Figure 3.18 summarizes which of the three major interaction processes dominates for different photon energies and atomic numbers.

3.6 NEUTRON INTERACTIONS

Although neutrons are generally not directly used in nuclear medicine imaging or therapy applications, it is of interest to know about the properties of these particles because they are essential for the production of radionuclides that are important for daily work, such as ^{99}Mo, which is the mother radionuclide for ^{99m}Tc and is produced from fission of ^{235}U.

The neutron is an uncharged particle with a mass M_n=1.008962 u. When compared to the proton mass m_p, it can be seen that M_n = 1.00138 M_p. A free neutron is therefore expected to be unstable in terms of β-decay and decay to a proton, electron, and antineutrino according to

$$n \rightarrow p + e^- + \overline{\upsilon} + Q, \tag{3.39}$$

where Q is 782 MeV. The half-life for the decay of a free neutron is 12 min. When the neutron is in a bounded state as a member of the nucleus, it becomes stable with no half-life. In practical applications, the decay does not have any significance, as a free neutron only survives a very small time before being captured by an atomic nucleus. It has a spin of ½ and a magnetic momentum in the same order as the protons, indicating the presence of an internal structure. As a consequence of the lack of electrical charge, the neutron only has strong forces in the nucleus. This means that the cross section for interaction (per atom) is very small and varies strongly with the composition of the material and the kinetic energy of the neutron.

The cross section per atom is therefore equal to the cross section per nucleus. This can be useful to compare this cross section with the size of the geometrical cross section of the nucleus, $\pi \cdot R^2$, where R is the average radius of the nucleus, written as

$$R = R_0 \cdot A^{1/3}, \tag{3.40}$$

where R_o was measured as approximately 1.2 10^{-15} m and A is the mass number. The geometrical cross section, $\pi \cdot R^2$, is then

$$\pi \cdot R^2 \approx \pi \cdot R_o^{\ 2} \cdot A^{2/3}. \tag{3.41}$$

3.6.1 Interaction Processes

There are two main groups of interactions for neutrons: neutrons that change direction by scattering, followed by an interaction with the nucleus and neutrons that are absorbed by the nucleus (Figure 3.19).

When a neutron is absorbed, a compound nucleus is formed with a very small half-life into an excited state with rapid decay. The type of decay and cross section for absorption depends not only on the excitation energy, but also on the composition of the compound nucleus. The cross section becomes very large if the excitation energy comes close to certain energy levels of the nucleus, called resonances. A typical cross section curve therefore also includes several narrow peaks, where the number of peaks increases, and becomes broader as the excitation energy increases (Figure 3.20).

With kinetic energies below 500 keV, only a few different processes are energetically possible, namely, elastic potential scattering, elastic resonance scattering, and neutron capture. Elastic potential scattering can be described as an elastic bounce against the surface of the nucleus. Elastic resonance scattering occurs when the neutron is

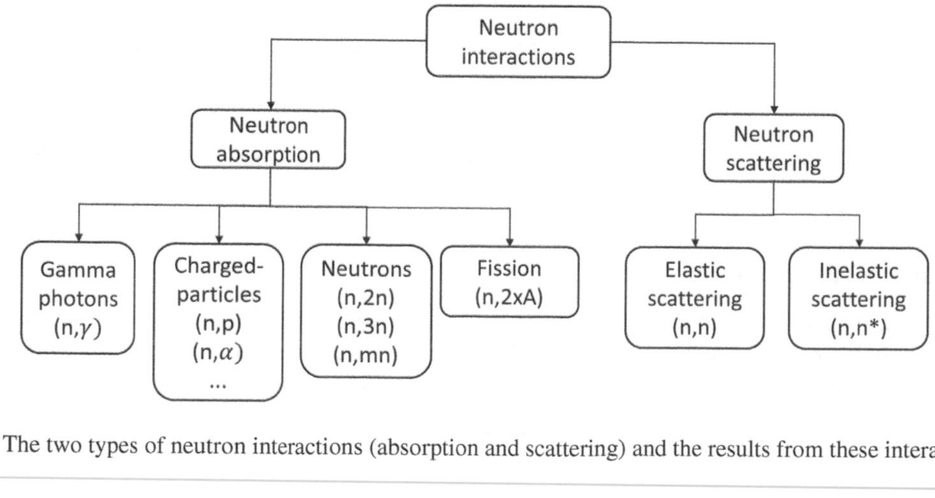

FIGURE 3.19 The two types of neutron interactions (absorption and scattering) and the results from these interactions.

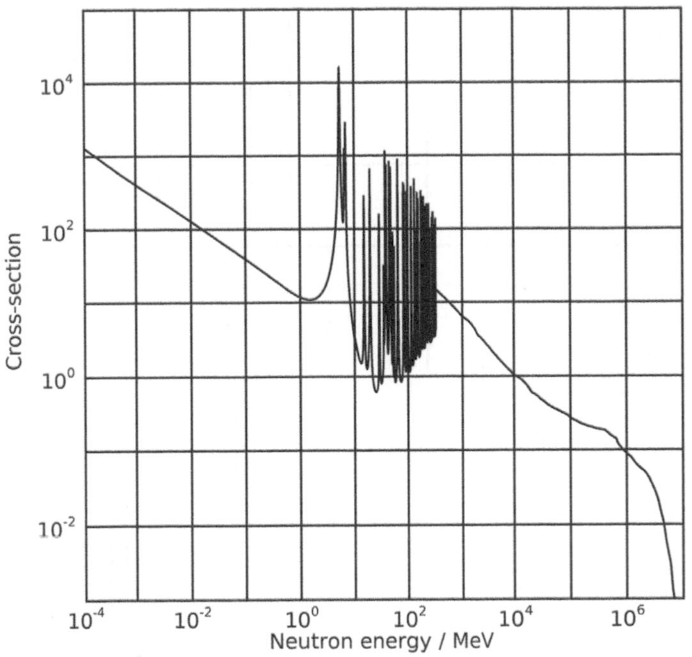

FIGURE 3.20 Example of a cross section for a neutron, where the general trend is a cross section proportional to $1/\sqrt{E_n}$ or $1/v_n$, where v is the velocity of the neutron. Large changes in cross section occur at certain energies called resonances.

absorbed by the nucleus, forming a compound nucleus following by a rapid decay with an emission of a neutron. The two processes *(n,n)* are different, but have the same resulting emission. The most common fate for a low-energy neutron is, however, a capture process (n,γ), by which the neutron is absorbed but the excitation energy is emitted as a γ-photon. The newly created nucleus often decays with β-emission due to excess neutrons. The cross section for each energetically possible process (excluding elastic scattering) can be regarded as proportional to $1/\sqrt{E_n}$ or $1/v_n$, where v is the velocity of the neutron. However, this is only valid for energies far from any resonance energies.

If no absorption process competes with elastic scattering, after a number of collisions the neutron will enter into a thermal equilibrium with the material (these neutrons are referred to as thermal neutrons having a motion in the material best described as diffusion). The most likely kinetic energy, E, of a thermal neutron is given by the Boltzmann distribution, $E = kT$, where k is the Boltzmann constant (1.38065×10^{-23} J/K) and T is the absolute temperature. At room temperature, E, is approximately 0.025 eV.

Elastic scattering is an important interaction process at all energies and for energies ranging from 30 eV to approximately 0.5 MeV, which is the dominating interaction process. Consequently, a typical scenario for a neutron that enters a material with an energy below 0.5 MeV is that it undergoes a number of elastic scatterings to finally be captured, before or after it has been slowed to thermal equilibrium with the material.

Most detector systems rely on the measurement of charged particles. This means that a neutron cannot be measured directly because it does not carry any charge. However, one can use some type of neutron interaction process with a large cross section that can create free charged particles, which can then be detected. Three commonly used processes are the (n,p) reaction in 3He, (n,α) in ^{6}Li, and (n,α) in ^{10}B. The last process has also been used clinically to treat cancer in the brain by a method called boron neutron-capture therapy (BNCT). For more information about this, see, for example, the literature review by Nedunchezhian and colleagues [4].

More types of interaction processes become possible as the energy of the neutron increases. One of the more important processes is the inelastic scattering $(n,n\gamma)$, in which the excess energy of a compound nucleus is shared between an emitted neutron and a γ photon, but also the emission of a charged particle (n,p) or an α-particle (n,α). To increase neutron energies, multiple particles can be emitted, such as (n,2n), (n,np), (n,n2p), $(n,n\alpha)$, (n,3n), and for very high energies, the incoming neutron can result in nuclear fission (Figure 3.21).

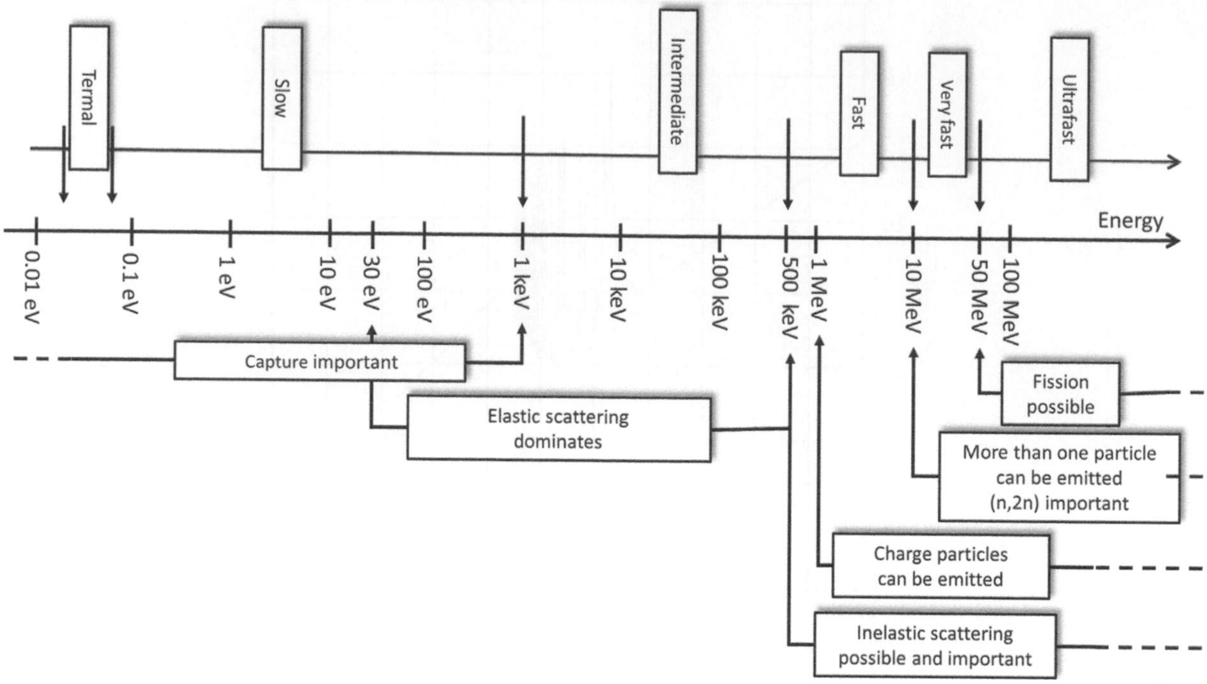

FIGURE 3.21 Sketch showing an energy scale where different interaction processes are possible.

3.6.2 Neutron Attenuation

An equation for the attenuation, similar to that for photons, can be derived for neutrons as

$$\dot{N}(x) = \dot{N}_o \cdot e^{-\mu x}, \tag{3.42}$$

where $\dot{N}(x)$ is the neutron rate in a beam of neutrons at depth x in the material, $\dot{N}_o$ is the incoming neutron rate, and μ is a linear attenuation coefficient for neutrons, defined as

$$\mu = \rho \cdot \frac{N_A}{M} \cdot \sigma_t. \tag{3.43}$$

Here, ρ is the density, N_A is Avogadro's constant, M is the molar mass of the material, and σ_t is the total cross section per atom for an interaction. One can also rewrite Eq. 3.44 as

$$\dot{N}(x) = \dot{N}_o \cdot e^{-\frac{\mu}{\rho} \cdot \rho \cdot x}, \tag{3.44}$$

where μ / ρ is the mass-attenuation coefficient for neutrons and ρ is the mass density. A difference when compared with the photon attenuation is that σ_t can vary for different isotopes of the same substance.

3.7 CHARGED-PARTICLE INTERACTIONS

The four charged particles that are often involved in medical physics applications are electrons, protons, α-particles, and other light ions.

1. **The electron,** e^-, is a negatively charged particle with rest mass m_e = 9.1094 10^{-31} kg. These particles can be accelerated up to very high energies and with velocities close to the speed of light. Electrons can also be produced

by photon interactions. The antiparticle of the electron is the positron, e^+, with the same properties as the electron but with a positive charge.

2. **The proton** p, is a positively charged particle with mass $M_p = 1836 \cdot m_e$. A large mass requires more complicated equipment to accelerate a positron. Free protons, such as alpha particles and heavy ions, can be produced when fast neutrons absorb in the material.
3. **The α-particle** is a ^4_2He nucleus consisting of two protons and two neutrons. It has a 2^+ charge and therefore is accelerated in a similar way as the proton. This particle is emitted due to the decay of some radioactive isotopes.
4. **Other light ions** are ionized atoms of heavy elements (in this context, atoms with Z > 3). These are positively charged ions with some or all orbital electrons removed.

For all charged particles passing through the material, electromagnetic interaction is the dominant process, especially the Coulomb force. The charged particles' energy and, especially for electrons, direction will change, and the charged particles give rise to ionizations and excitations along their paths, as is illustrated in Figure 3.22. The interaction processes between the charged particles and the material can, in basic terms, be divided into four groups.

1. **Inelastic collisions with atomic electrons.** This is the most dominant process in which a charged particle loses kinetic energy when passing through a material, and also the process that provides the largest energy loss for the particle. The result is that one or several atomic electrons are transferred to an excited state or liberated to a free electron. In this process, the created positively charged atom and the liberated orbital electron are called an ion pair. If the energy transferred to the electron is sufficiently high, then this will subsequently create excitations and ionization along its track. A number of ionizations, locally created within a small volume, is called a cluster. An orbital electron that receives so much energy that it can create multiple ionizations by itself is called a δ-particle.
2. **Acceleration in the vicinity of an atomic nucleus.** When the charged particle approaches the nucleus, interaction with the strong Coulomb field causes the charged particle to change direction. This can result in the emission of bremsstrahlung photons, which is very important for medical physics, especially in the field of radiology and radiotherapy (the major radiation therapy units are based on a linear accelerator producing electrons that crjeate high-energy bremsstrahlung radiation). In addition, nuclear reactions may occur, but these processes are very rare and not important for medical physics.
3. **Elastic collisions.** A deflection in direction when a charged particle is passing close to the nucleus can occur without energy losses or nuclear reactions. The incident particle only loses the energy needed to maintain momentum conservation. In particular, electrons have a high probability for this type of interaction. In the field from the atomic electrons, an incoming charged particle can be deflected with an energy loss smaller than the lower excitation potential of the atom. This means an interaction with the whole atom. The process is of importance only for incoming electrons with energies on the order of 100 eV or less, which makes this process not very important for medical physics applications.

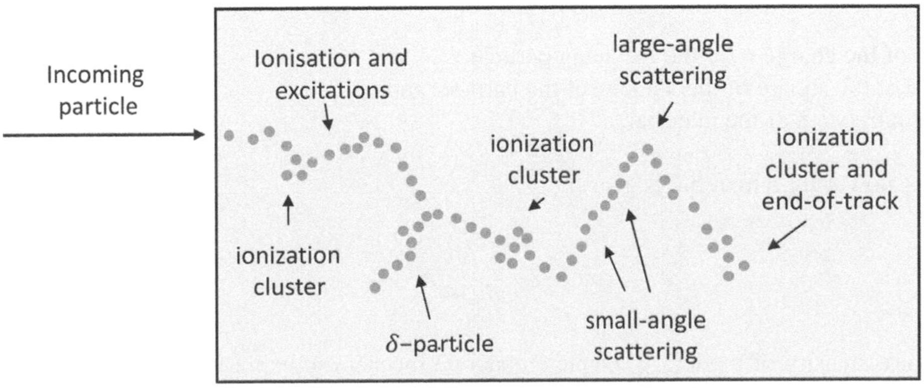

FIGURE 3.22 Schematic image of a particle track of ionization and excitations created from an incoming charged particle. Along the path of the particle are also randomly created clusters of ionizations and, in some cases, large energy transfers resulting in δ-particles that receive so much energy that they create ionizations of their own. Therefore, in some cases, these are treated as separate particles.

3.7.1 Inelastic Collisions with Atomic Electrons

With a straight collision between, for example, an α-particle with mass M_α and energy E_α, the maximum energy (Q_o) that can be transferred is $2mv^2$, where v is the speed of the α-particle. This can be written as

$$Q_o = 2mv^2 = E_\alpha \cdot \frac{4m}{M_\alpha}. \tag{3.45}$$

For a 5-MeV α-particle, $Q_o = 2.5$ keV, which implies that the α-particle loses its energy by small proportions via many collisions before coming to rest. This means that α-particles have a well-defined range with little statistical variation in the range between different α-particles with the same energy. However, the direction can be changed by an interaction with the nucleus (Rutherford scattering). If the incoming particle is an electron, the conditions will be different. In this case, the whole energy can be transferred to the interacting orbital electron in a single interaction. Generally, large energy transfers are common, and the range of the particle will be less well-defined. The electron can also be deflected close to the nucleus into large angles, but still maintains most of its energy.

In general, charged-particle energy transfer should be calculated based on quantum mechanical theories. However, a classical approach can be useful, especially for heavy particles. Here, one assumes that the electrons in the material are free electrons, that is, they receive so much energy in an interaction where the binding energy of the electron can be neglected. This reduces the event to be a simple Coulomb collision because the electromagnetic force will be the dominated force. The energy, dE, that a particle loses on average when passing through a distance dx (measured along the path) is used to characterize the material's ability to slow down the particle. The radio is called the total stopping power, or the total linear stopping power, if dE includes all types of energy losses. Here, one should use a subscript to make this clear, for example, $(dE/dx)_{tot}$. Often, the letter S is used to describe the ratio dE/dx. Because energy losses can occur partly through collisions (ionization and excitation) and partly through radiation, these two contributions to the total energy losses are often stated separately.

$$\left(\frac{dE}{dx}\right)_{tot} = \left(\frac{dE}{dx}\right)_{el} + \left(\frac{dE}{dx}\right)_{rad} \tag{3.46}$$

A classical derivation of the linear electronic stopping power (previously called collision stopping power) leads to the following dependence on the parameters:

$$\left(\frac{dE}{dx}\right)_{el} \sim \frac{e^2}{v^2} \cdot Z. \tag{3.47}$$

The energy loss is thus proportional to

(a) the square of the charge *e* for the incoming particle,
(b) the inverse of the square of the velocity of the particle, and
(c) the atomic number Z of the material.

The mass stopping power for a material is given by

$$\frac{1}{\rho} \cdot \frac{dE}{dx}, \tag{3.48}$$

where δ is the mass density of matter. A complete quantum mechanical expression is provided by the Bethe–Bloch formula:

$$\left(\frac{1}{\rho}\frac{dE}{dx}\right)_{el} \sim N_A \cdot \frac{Z}{A} \cdot \frac{z^2 e^4}{4\pi\varepsilon_o^2 m_e v^2} \cdot \left(ln \frac{Q_{max}}{I} - \ln\left(1-\beta^2\right) - \beta^2 - \frac{C}{Z} - \frac{1}{2} \cdot \delta \right), \tag{3.49}$$

where A is the molar mass of the material, z is the charge of the incoming particle, m_e is the electron rest-mass, v is the velocity of the incoming particle, I is the average ionization potential, C is a shell correction, δ is a correction term for the polarization effect, and β is the incoming particle relative velocity $(\beta = v/c)$. Q_{max} is the maximum energy that can be transferred with a straight collision and has a value that ranges from $mv^2/2$ for impinging electrons to $2mv^2$ for heavy particles (m assumes the relativistic mass).

The average ionization potential, I, is a weighted mean value over the atoms all excitation and ionization potentials and is therefore much higher than the smallest energy that is required to make an ionization. The value is determined from measurements of the stopping power or the range of heavy particles. The magnitude is approximately $13.5 \cdot Z$ eV. The stopping power depends on the logarithm of *I*, as can be seen from Eq. 3.51; the precise value of I is not that critical. The two terms that include the relative velocity, β, are only important for particles with speeds close to c.

The shell term C/Z is introduced to account for the fact that electrons will contribute less to the stopping power if the speed of the interacting electron is in the order of the speed of the orbital electrons. Furthermore, the theory of collision losses assumes the independence of the atoms. However, this holds true only for gases with low pressure. If the particles are passing dense materials, a correction term is needed to account for the density effect or the polarization effect. When the atoms are closely located, then the electric field between the incoming particle and the atom will be affected by fields from nearby atoms. The result is a field reduction that then reduces the stopping power. The material's dielectric and scattering properties determine the magnitude of the density effect. The effect is more pronounced at high kinetic energies and relativistic velocities. The density effect thus increases with the energy of the particle. Figures 3.23 to 3.25 show mass stopping-power for electrons, protons and α-particles as function of energy.

The total stopping power includes all interactions along the distance dx without any restrictions on where this energy is deposited. For an estimate of the local deposition along the path, one can define a restricted stopping power,

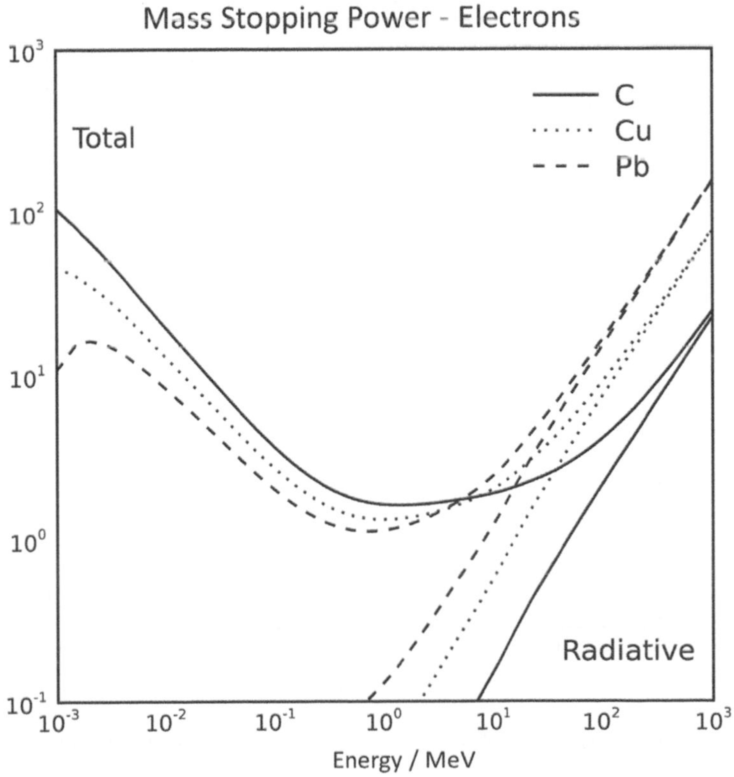

FIGURE 3.23 This figure summarizes the general trends in the stopping power curve as function of the energy of electrons. For low energies to the left of the peak, the stopping power decreases due to the decrease in the effective charge of the particle. The decrease to the right of the Bragg peak arises from the $1/v^2$ dependence. At relativistic velocities, the stopping power increases mainly due to the term $\ln(1-\beta^2)$. The shell correction is only relevant for high-Z materials, where the inner-orbital electrons have relativistic velocities.

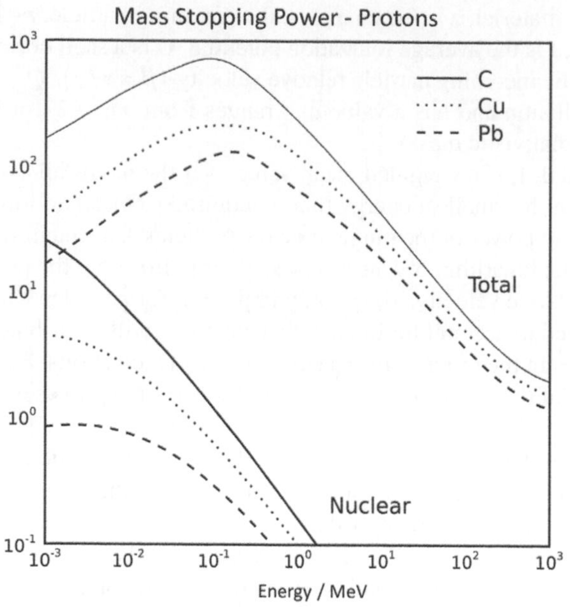

FIGURE 3.24 This figure summarizes the general trends in the stopping power curve as function of the energy of protons.

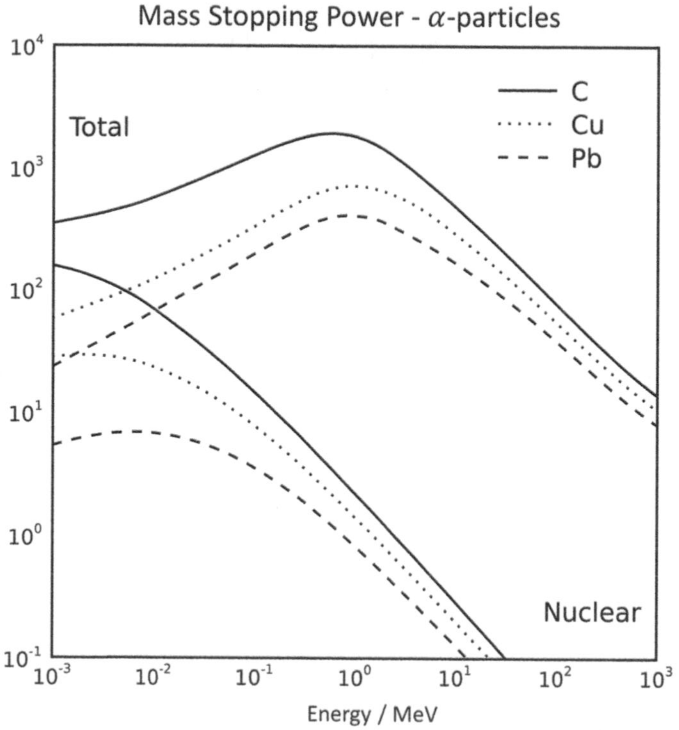

FIGURE 3.25 This figure summarizes the general trends in the stopping power curve as function of the energy of α-particles.

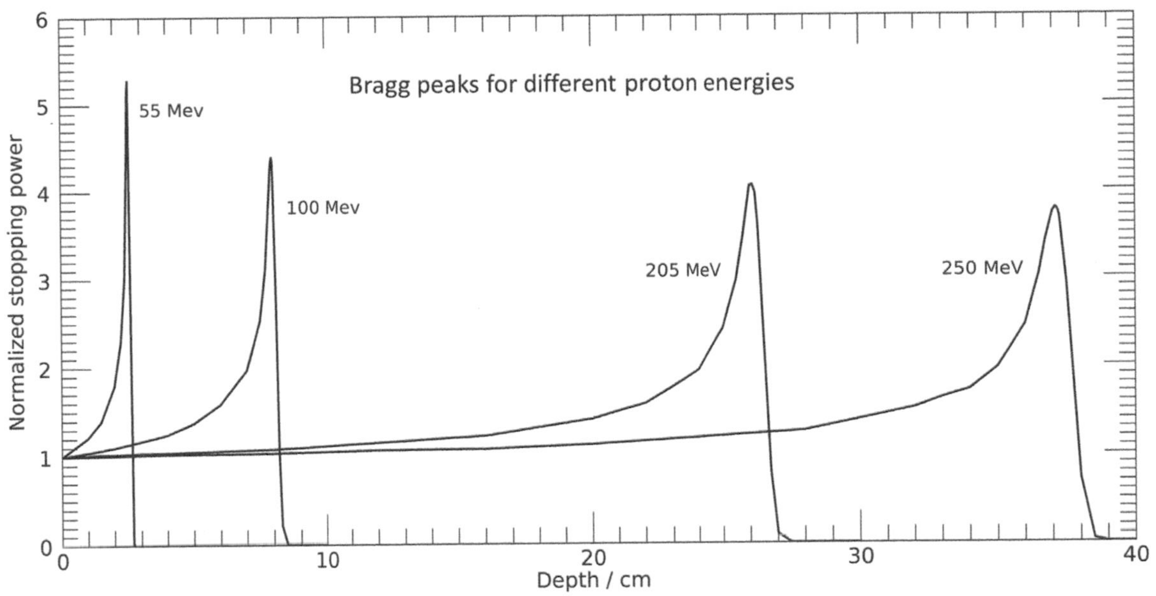

FIGURE 3.26 Normalized stopping power for protons with four different energies along their path. The protons deposit most of their kinetic energy per unit of length at the end of their well-defined range (the Bragg peak). This characteristic of the proton interaction is used in external proton-beam therapy where the energy of an incoming proton beam is adjusted so the protons deliver maximal energy in the target volume. Because the Bragg-peak may be too sharp in width relative to the volume for treatment, the proton beam is often scanned in a predefined pattern to create a uniform dose distribution. Data obtained with permission from NASA Space Radiation Laboratory, Brookhaven National Laboratory, Upton, New York (www.bnl.gov/nsrl/userguide/bragg-curves-and-peaks.php).

$(dE/dx)_\Delta$ which is often called the linear energy transfer (LET). The linear energy transfer is defined as the energy deposition per unit length created by collisions, each with an energy transfer less than Δ.

$$L_\Delta = \left(\frac{dE}{dx}\right)_{el,\Delta} . \tag{3.50}$$

This means that δ-particles with higher energies than Δ, but only these, are treated as new ionizing particles. L_Δ is thus essentially the electronic stopping power but without the contribution from of secondary electrons with a kinetic energy higher than Δ because these have such a high range that the energy cannot be considered locally absorbed. A typical value of Δ can be 100 eV. Because LET is a measure of the density of ionizations along the particle track, it is a very important concept in radiobiology as the biological effect of radiation is largely dependent on the LET value.

The specific ionization, that is, the number of created ion pairs per unit length, along a particle track will provide the so-called Bragg curve. This is closely related to the stopping power, dE/dx when the particle is slowing down and shows a peak at the end of the path where the specific ionization reaches a maximum (Figure 3.26).

Both excitations and ionizations occur along the particle track. To create an ion pair, one can calculate the average the energy $\overline{W} = E/N$, where N is the number of created ion pairs created by a particle with the kinetic energy E that is completely stopped. The value of $\overline{W}$ is often determined experimentally. To eject the loosely bound electrons, a typical value on the order of 10 eV is needed, and the average $\overline{W}$ is around 30 eV. $\overline{W}$ varies very little with the kinetic energy of the particle. It is higher for α-particles than for electrons. A useful value is $\overline{W}_{air}$ =33.97 eV.

3.7.2 Inelastic Collisions with Atomic Nucleus

According to quantum mechanics, there is a small but finite probability that a particle will emit bremsstrahlung radiation in connection with an acceleration or deflection. However, this probability is so small that in most cases no photon is emitted. However, when this actually occurs, a large fraction of the kinetic energy is emitted. The large number of small,

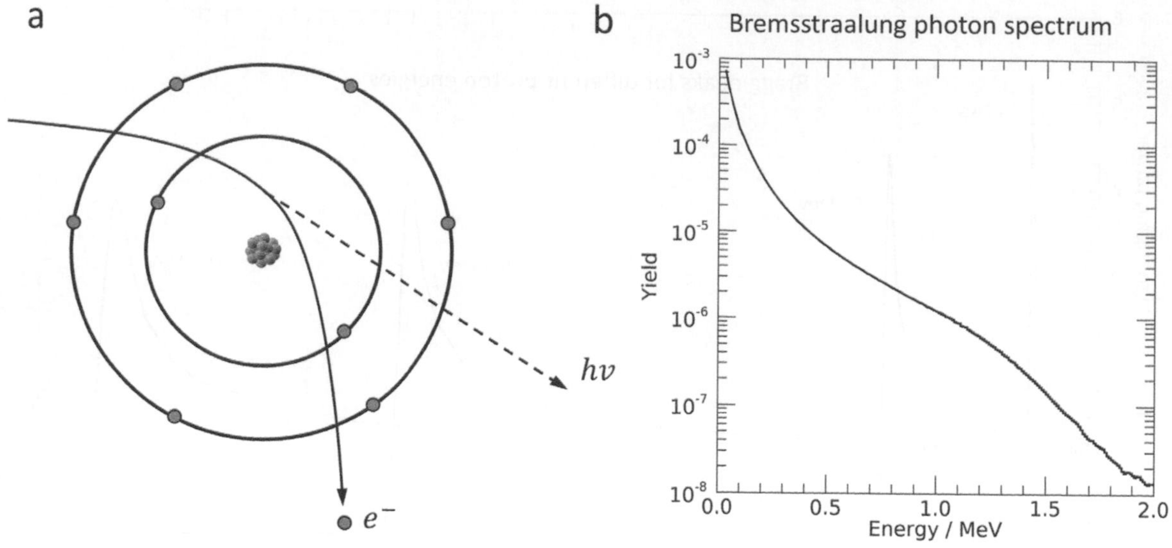

FIGURE 3.27 Schematic image of the bremsstrahlung process (a) where electrons slow down in the vicinity of the nucleus, and the kinetic energy is transferred to a photon. These energy losses for the electron are not discrete and well-defined, but follows a distribution, as can be seen from the energy distribution of bremsstrahlung created by interactions with a β-particle emitted from a ^{90}Y decay (b). The bremsstrahlung distribution closely depends on the velocity of the β-particle, so it is evident that most bremsstrahlung photons are of lower energies created at the end of the particle track.

but many, photon emissions, as predicted by the classical theory, is replaced with fewer emissions with large photon energies. The average emitted energy is the same for both, but the angular distribution differs. Figure 3.27a shows a schematic image of the interaction process.

The acceleration, a, that a charged particle experiences at a distance of r from a nucleus with charge Z can be expressed as

$$a = \frac{1}{4\pi\varepsilon_o} \cdot \frac{Z \cdot ze^2}{mr^2}, \tag{3.51}$$

where m is the mass of the particle. The radiation energy is then proportional to the square of the acceleration,

$$E \propto \frac{Z^2 \cdot z^2 e^4}{m^2}, \tag{3.52}$$

meaning that the total bremsstrahlung radiation per atom is proportional to the square of the atomic number for the material, Z^2, and inversely proportional to the square of the particle mass, m^2. Consequently, bremsstrahlung emission will be most important for electrons impinging on materials with a high Z. An electron can with a low probability transfer all of its kinetic energy, E, to a bremsstrahlung photon, or $hv_{max} = E$.

The radiative stopping power component of Eq. 3.48 depends, in addition to the parameters described above, upon

$$\left(\frac{dE}{dx}\right)_{rad} \propto \frac{EZ^2}{A} B(Z,E), \tag{3.53}$$

where B is a slowly varying function of Z and E. It can be observed that the radiative part of the stopping power depends on the square of the atomic number for the medium and depends linearly on the kinetic energy of the particle.

The radiation yield is the intensity of the radiation and can be calculated from the equation

$$Y(E_0) = \frac{1}{E_0} \int_0^{E_0} \frac{\left(\frac{dE}{dx}\right)_{rad}}{\left(\frac{dE}{dx}\right)_{el} + \left(\frac{dE}{dx}\right)_{rad}} dE. \quad (3.54)$$

Bremsstrahlung is the fundamental radiation in X-ray imaging and is generated by accelerating electrons in an electrical potential up to approximately 150 kV and the electrons impinging on a high-Z material (often tungsten that has a Z equal to 74 and mass-density of 19.25 g cm^{-3}).

In nuclear medicine, bremsstrahlung imaging has been used to determine the activity of pure β-electron emitting radionuclides, such as ^{32}P and ^{90}Y. The average kinetic energy of the β-emissions from ^{90}Y is 0.937 MeV is translated to a mean range of 2.5 mm in tissue-equivalent medial and a maximum track length of approximately 11 mm. However, the yield is very low for production in tissue-equivalent material, but if the activity is sufficiently high, images of bremsstrahlung can be produced. Figure 3.27b shows a bremsstrahlung photon energy spectrum [5]. It can be seen that most of the photons are of very low energies, but there is a small probability for photon emission up to the maximum energy of the β-particles. The high-energy bremsstrahlung photons are generated at the beginning of the β-particle close to the position of decay, and the opposite will occur for the low-energy bremsstrahlung photons. This means that in a SPECT imaging system, the resolution will depend on the possible range of the β-particle. For the particular ^{90}Y bremsstrahlung energy spectrum, shown in Figure 3.27b, the total yield for a photon to be emitted per β decay is 3.97 10^{-2}.

There is also a probability that a β-particle interacts with the nucleus from which it is emitted during the decay. This event is called the inner (or internal) bremsstrahlung. The loss of energy from the β-particles comes from interactions with the strong electric field close to the nucleus as the β-particles leave the nucleus. This process has often been neglected when modelling radiation transport, but recent studies have shown that it may have an impact on applications involving quantitative imaging of radionuclides based on bremsstrahlung imaging. In a study by Dewaraja and colleagues, [6] it was concluded that inner bremsstrahlung contributed significantly to the overall bremsstrahlung yield of ^{90}Y and that the process dominated over EB at energies above approximately 500 keV. When this effect was included, they found an improvement in the estimation of activity based on quantitative imaging methods.

3.7.3 Elastic Collisions

In an elastic collision with the nucleus, the particle path is deflected by Coulomb forces between the nucleus and the particle, but the energy loss is negligible, and no bremsstrahlung photons are emitted. However, the deflection angle and its probability are of interest. The process is most important for electrons traveling in a high-Z material, where this process is more likely than inelastic scatter. Very small deflections are the result of many small scatterings. A larger overall scattering angle is often caused by a single scatterer. The name of a small overall change in direction is called multiple Coulomb scattering. In addition, α-particles can be scattered into new directions by interaction with a nucleus, which is also associated with an energy loss. This phenomenon is called Rutherford scattering. Larger angle scattering is possible with a very small probability, and most scatterings result in a deflection angle of less than 1°.

In an elastic collision with an atomic electron, the direction of an incoming electron interacts with an atomic electron without energy loss but with a change in direction. The probability of elastic scattering increases with the atomic number Z. Most interactions only result in a very small change in direction, leading to a situation where the incoming particle travels roughly in the same direction even if a large number of elastic collisions occur.

Thus, for electrons that interact with the nucleus, this often results in a change in direction, while interactions with orbital electrons result in energy losses. The cross section for scattering with the nucleus increases proportionally to Z^2, while the inelastic cross section per atom is only proportional to Z. Therefore, the elastic process is more important for high Z. The probability for the two processes is equal, at Z = 30, and the inelastic process dominates for low-Z materials.

3.7.4 Path Ranges and Range Relations

Charged particles that pass through a material will undergo interaction, resulting in a loss of energy and deflections in the direction relative to the incoming path. The path length or range is used to describe how far the charged particle will travel before it completely stops – that is, before it has lost its entire kinetic energy. The range can be defined as the actual length a particle

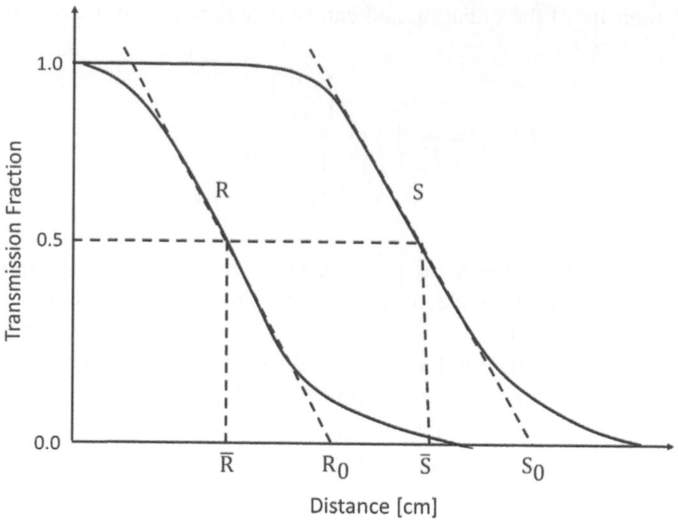

FIGURE 3.28 Range relations as defined by the ICRU report 'Linear Energy Transfer' [7].

has travelled (the track length) to the length projected along the incoming particle direction. For heavy particles, these two ranges are almost the same, but differ significantly for electrons due to statistical large-angle scatterings.

$$R_{CSDA} = \int_0^{E_0} \left(\frac{dx}{dE}\right) \cdot dE = \int_0^{E_0} \left(\frac{dE}{dx}\right)^{-1} \cdot dE. \tag{3.55}$$

The track length can be calculated using the stopping power values. The CSDA is an acronym for continuous slowing down approximation. R_{CSDA} are also sometimes called the average trajectory length or the average true range for the particle [7]. The range represents the path length that a particle will undergo with a total loss of E_o if the actual rate of energy loss along the track is equal to the average energy loss. Due to fluctuations in the energy loss, the particles achieve a distribution of ranges, even if the initial energy is the same. This distribution is called range straggling. For protons and α-particles, the range straggling is about a few percentages of the average energy, and for a measured distribution, the average of the ranges is slightly larger than R_{CSDA}. As stated earlier, heavy particles transfer only a small portion of their energy in each interaction, resulting in a well-defined range with little straggling. Electrons, on the other hand, can transfer large fractions in a single interaction and also deflect at large angles. Therefore, the range becomes less defined, and the projected range is less than the actual track length. Figure 3.28 illustrates the transmission curves for heavy particles and electrons. For comparison, the range of a beta particle from a radionuclide emission is also plotted. This curve differs in that the low-energy β-particles are absorbed much more, and therefore the slope of the transmission is steep. ICRU has also defined a set of definitions of particle ranges [7].

- $\bar{R}$ is the mean-projected range that is the thickness of the material that allow for a transmission of 50 per cent of the incoming particles.
- $\bar{S}$ is the mean path length, which is the range calculated by the CSDA approximation.
- R_0 is the extrapolated mean projected range, which is the range obtained by extrapolation of a line, along the tangent at a level of 50 per cent, down to the range axis.
- S_0 is the extrapolated path length.
- R_{max} is the absolute projected path length.
- R_q is the specified projected path length.

In dosimetry calculations for radionuclide-emitting β-particles, the actual range in mm of a charged particle may not be the best parameter to use in order to determine the absorbed energy from the emitted radiation. The X_{90} range measure has therefore been used to estimate the magnitude and distribution of the absorbed energy from interacting particles emitted by the radionuclide source. The X_{90} range, first introduced by Berger [8], defines the radius of a sphere

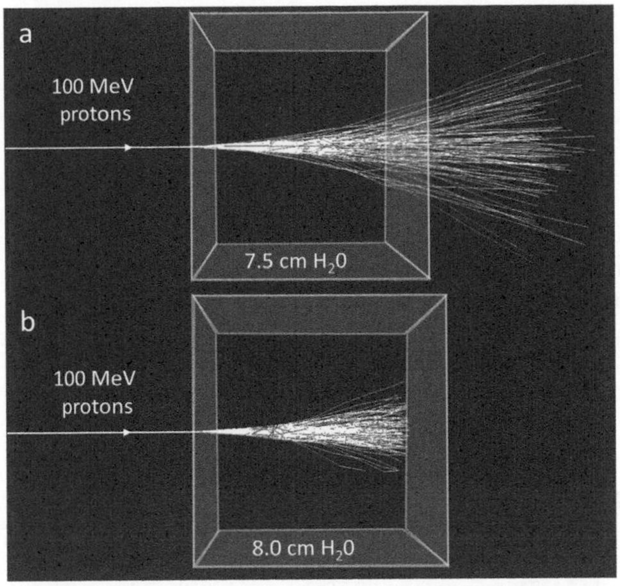

FIGURE 3.29 Visualization of a Geant4 Monte Carlo simulation showing protons with an energy of 100 MeV impinging on a slab of water with two different thicknesses. In example (a), the slab is 7.5 cm, and most protons (white tracks) pass through. However, if the slab thickness is increased to 8.0 cm (b), then all the protons are stopped because of their very well-defined range. Tracks from photons, generated from interactions along the proton track and emitted in all directions, can also be seen.

that surrounds the volume in which 90 per cent of the energy emitted from a point source located in the sphere centre is deposited. It is not possible to experimentally measure this range, so X_{90} needs to be obtained from Monte Carlo calculations of the deposited energy as a function of the radius from a point-source location for the relevant media.

As Eq. 3.47 shows, the energy losses for heavy-charged particles, such as protons and α-particles, are very small per interaction, and therefore the ranges are well-defined because the statistical fluctuations between the ranges of particles are very small. This is illustrated in Figure 3.29, where 100-MeV protons impinging on a water phantom with two different thicknesses are shown using a Monte Carlo simulation connected to a visualization tool. In Figure 3.29a, the ranges of the protons for the current energy are slightly larger than the phantom thickness and, therefore, all protons are passed through it. However, with an increase in thickness of 0.5 cm, we can see that all the protons are stopped (Figure 3.29b).

3.8 SOURCES FOR CROSS SECTIONS

3.8.1 Photons

A convenient computer program developed to generate cross sections for photons for single elements, as well as compounds and mixtures as needed, is XCOM [9]. This program is provided by the National Institute of Standards and Technology (NIST), and outputs photon cross sections for any element, compound, or mixture at energies between 1 keV and 100 GeV. The program includes a database of cross sections for the elements and provides interpolations and weights when creating cross sections for compounds or a mixture of compounds. XCOM data are also available from a webpage maintained by NIST [10].

3.8.2 Neutrons

The NGATLAS, (www-nds.iaea.org/ngatlas2/) maintained by the International Atomic Energy Agency (IAEA), contains numerical neutron capture cross sections in the range from 10^{-5} eV to 20 MeV for elements H to Cm.

3.8.3 Charged Particles

Stopping-power data for 74 materials in the range from 0.001 MeV to 10000 MeV can be downloaded from web pages for protons (PSTAR), electrons (ESTAR), and for α-particles (ASTAR) that are maintained by the National Institute of

Standard Technology (NIST). Here, for electrons, are provided the collision and radiative stopping power, density effect, CSDA range, and the fraction of kinetic energy of the primary electron converted into bremsstrahlung. For protons and α-particles, the projected range, stopping power, and CSDA range are provided. For all particles, calculations were made according to the methods described in ICRU 37 [11] and ICRU 49 [12].

REFERENCES

The content of this chapter is based on a Swedish report, written in the early 1980s by Lars Hallstadius, PhD, and Sven Hertzman, PhD, and is used here with permission from both. This excellent report has since that time been used in our Medical Physics education and we acknowledge their contribution. More readings can be found in, for example, the books of Evans [13], Attix [14], and Podgorsak [15], the chapter by Nilsson and Brahme [16], and the review by Turner [17].

[1] O. Klein and T. Nishina, "Über die Streuung von Strahlung durch freie Elektronen nach der neuen relativistischen Quantendynamik von Dirac," *Zeitschrift fur Physik*, vol. 52, pp. 853–868, 1929, doi: 10.1007/BF01366453.

[2] I. Waller and D. R. Hartree, "On the Intensity of Total Scattering of X-Rays," *Proc Royal Soc Lond. A*, vol. 124, pp. 119–142, 1929, doi: 10.1098/rspa.1929.0101.

[3] J. H. Hubbell, J. W. Veigle, E. A. Briggs, R. T. Brown, D. T. Cramer, and R. J. Howerton, "Atomic Form Factors, Incoherent Scattering Functions and Photon Scattering Cross Sections," *J Phys Chem Ref Data*, vol. 4, pp. 471–616, 1975.

[4] K. Nedunchezhian, N. Aswath, M. Thiruppathy, and S. Thirugnanamurthy, "Boron Neutron Capture Therapy – A Literature Review," (in English), *J Clin Diagn Res*, vol. 10, no. 12, pp. ZE01-ZE04, 2016, doi: 10.7860/JCDR/2016/19890.9024.

[5] X. Rong, Y. Du, M. Ljungberg, E. Rault, S. Vandenberghe, and E. C. Frey, "Development and Evaluation of an Improved Quantitative (90)Y Bremsstrahlung SPECT Method," (in English), *Med Phys*, vol. 39, no. 5, pp. 2346–58, May 2012, doi: 10.1118/1.3700174.

[6] Y. Dewaraja, R. Fleming, P. Simpson, S. Walrand, M. Ljungberg, and S. Wilderman, "Impact of Internal Bremsstrahlung on Y-90 SPECT Imaging," *Journal of Nuclear Medicine,* vol. 59, no. supplement 1, p. 577, May 1, 2018 2018. [Online]. Available: http://jnm.snmjournals.org/content/59/supplement_1/577.abstract.

[7] ICRU, "Linear Energy Transfer. Report 16." *International Commission on Radiation Units and Measurements*, 1970.

[8] M. J. Berger, "Distribution of Absorbed Dose Around Point Sources of Electrons and Beta Particles in Water and Other Media: MIRD Pamphlet No. 7," *Journal of Nuclear Medicine*, vol. 12, pp. 5–23, 1971.

[9] M. J. Berger and J. R. Hubbell, "XCOM: Photon Cross-sections on a Personal Computer," National Bureau of Standards, Washington, DC, 1987.

[10] M. J. Berger et al. "XCOM: Photon Cross Sections Database." www.nist.gov/pml/xcom-photon-cross-sections-database.

[11] M. J. Berger et al., "Report 37," *Journal of the International Commission on Radiation Units and Measurements*, no. 2, 1984, doi: 10.1093/jicru/os19.2.Report37.

[12] M. J. Berger et al., "Report 49," *Journal of the International Commission on Radiation Units and Measurements*, no. 2, 2016, doi: 10.1093/jicru/os25.2.Report49.

[13] R. D. Evans, *The Atomic Nucleus*. New York: McGraw Hill, 1955.

[14] F. H. Attix, *Introduction to Radiological Physics and Radiation Dosimetry*. New York: Wiley, 1986.

[15] E. B. Podgorsak, *Radiation Physics for Medical Physicists*, 3rd edn. Switzerland: Springer, 2016.

[16] B. Nilsson and A. Brahme, "Interaction of Ionizing Radiation with Matter," in *Comprehensive Biomedical Physics: Vol. 9: Radiation Therapy Physics and Treatment Optimization*, Ed. A. Brahme. Amsterdam and Heidelberg: Elsevier, 2014, pp. 1–36.

[17] J. E. Turner, "Interaction of Ionizing Radiation with Matter," *Health Phys,* vol. 88, no. 6, pp. 520–544, Jun 2005, doi: 10.1097/00004032-200506000-00002.

4 Radionuclide Production

Hans Lundqvist

CONTENTS

4.1 INTRODUCTION

With the 'Big Bang', the cosmic explosion about 14 billion years ago, the process to create all matter started. Particles like protons and neutrons, which form the building blocks of nuclei, appeared as free particles during the first seconds. Various types of nuclear reactions formed different combinations of protons and neutrons to create light elements as ^{4}He, ^{3}He, and ^{7}Li and later, in the stars, heavier elements. All matter around us was created in nuclear reactions as a mixture of stable and unstable (radioactive) combinations of protons and neutrons. Over time, the unstable combinations have undergone transformation (radioactive decay) to form stable combinations but some with exceptionally long half-lives (natural radioactivity) remain, potassium-40, lead-204, thorium-232 and the natural occurring isotopes of uranium. Some of these and their radioactive daughters were early applied in biology and medicine.

DOI: 10.1201/9780429489556-4

4.2 INDUCED RADIOACTIVITY

In the early 1930s humanity learned how to mimic nature to create radioactivity by nuclear reactions, first with energetic charged particles and later with neutrons both of low and high kinetic energy. The number of useful radionuclides for biology and medical applications increased considerably. Phosphorus metabolism was studied in rats injected with ^{32}P, and ^{128}I was applied in the diagnosis of thyroid disease. This was the start of the radiotracer technology in life science as we know it today.

In the beginning the cyclotron was the main producer of radionuclides. An early radionuclide of special importance was ^{11}C, since carbon is fundamental in life science, but the short half-life of 20 minutes was a drawback and limited the applications. After some years ^{14}C was discovered, with a half-life more suitable for laboratory work, as well as other isotopes of biological elements.

At first the nuclear reaction ^{13}C (d, p) ^{14}C was used, but ^{14}C produced this way is of limited use since the radionuclide cannot be separated from the target, meaning that stable carbon will dominate. However, a bottle of ammonium nitrate solution, standing close to the target, was at the same time irradiated by neutrons produced when deuterons were split in the target (d → p+n). By pure chance, it was discovered that this bottle also contained ^{14}C, which had been produced in the reaction ^{14}N (n, p) ^{14}C. Since no carbon was present in the bottle (except small amounts from solved airborne carbon dioxide) ^{14}C produced this way was of high specific radioactivity. It was also easy to separate from the target. In the nuclear reaction, a 'hot' carbon atom was created, which formed $^{14}CO_2$ in the solution. By simply bubbling air through the bottle, ^{14}C was released from the target.

Before the Second World War, the cyclotron was the main producer of radionuclides since the neutron sources at that time were very weak. But with the development of the nuclear reactor, a strong neutron source was readily available that easily could produce almost unlimited amounts of radioactive nuclides, including such biologically important elements as ^{3}H, ^{14}C, ^{32}P, ^{35}S and clinically interesting radionuclides such as ^{60}Co (for external radiotherapy) and ^{131}I for nuclear medicine. After the war, a new industry was born that could deliver a variety of radiolabelled compounds for research and clinical use at a reasonable cost.

However, accelerator-produced nuclides have a special character, which makes them differ from reactor-produced nuclides. Today, their popularity is increasing again. Generally, one may say that reactor-produced radionuclides are most suitable for laboratory work, whereas accelerator-produced radionuclides are more clinically useful. Some of the most-used radionuclides in nuclear medicine, such as ^{111}In, ^{123}I and ^{201}Tl, and the short-lived radionuclides (^{11}C, ^{13}N, ^{15}O and ^{18}F) used for positron emission tomography (PET) are all cyclotron-produced.

4.3 NUCLIDE CHART AND LINE OF NUCLEAR STABILITY

Chemists learnt during the late nineteenth century to organize chemical knowledge into the periodic system. Radioactivity, when it was discovered, conflicted with that system. Suddenly various samples, apparently with the same chemical behaviour, were found to have different physical qualities, such as radioactive decay with emission of particles carrying varying energies. The concept of isotopes or elements occupying the *same place* in the periodic system (which in Greek reads *isos topos meaning 'same space'*) was introduced by British scientist Frederick Soddy (1913), but a complete explanation had to wait for the discovery of the neutron by James Chadwick (1932).

The periodic system was organized according to the number of protons (atom number) in the nucleus, which is equal to the number of electrons to balance the atomic charge (Figure 2.3 in Chapter 2). The nuclide chart consists of a plot with the number of neutrons in the nucleus on the x-axis and the number of protons on the y-axis (Figure 4.1).

Figure 4.2 shows a limited part of the nuclide chart. The formal notation for an isotope and the relations of the numbers of protons and neutrons are described in Chapter 2 in this volume. It is important to understand that, whenever we use the expression isotope, we must always relate it to a specific element or a group of elements, for example, isotopes of carbon (e.g. ^{11}C, ^{12}C, ^{13}C and ^{14}C).

In the nuclide chart (Figure 4.1) the stable nuclides fall along a monotonically increasing line called the stability line. The stability of the nucleus is determined by competing forces; the 'strong force' that binds the nucleons (protons and neutrons) together and the Coulomb force that repulses particles of like charge, for example, protons. The interplay between the forces is illustrated in Figure 4.3.

For best stability, the nucleus prefers equal numbers of protons and neutrons. This is a quantum mechanics feature of bound particles and, in Figure 4.1, this is illustrated by a straight line. It is also seen that the stability line follows the straight line for the light elements but that there is a considerable deviation (neutron excess) for the heavier elements.

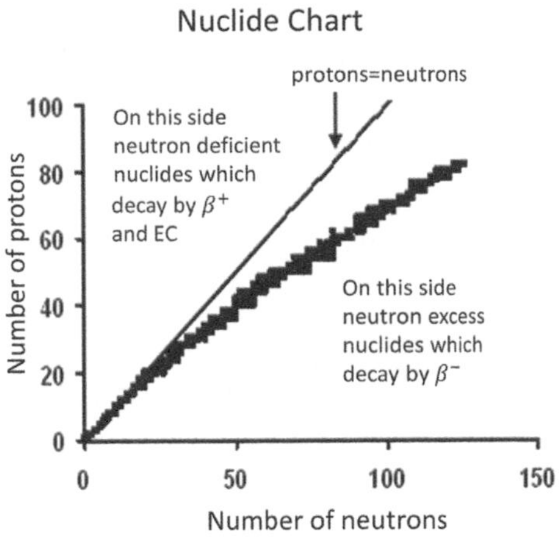

FIGURE 4.1 Chart of nuclides. The black dots represent 279 naturally existing combinations of protons and neutrons (stable or almost stable nuclides). Around this stable line are about 2,300 proton/neutron-combinations that are unstable.

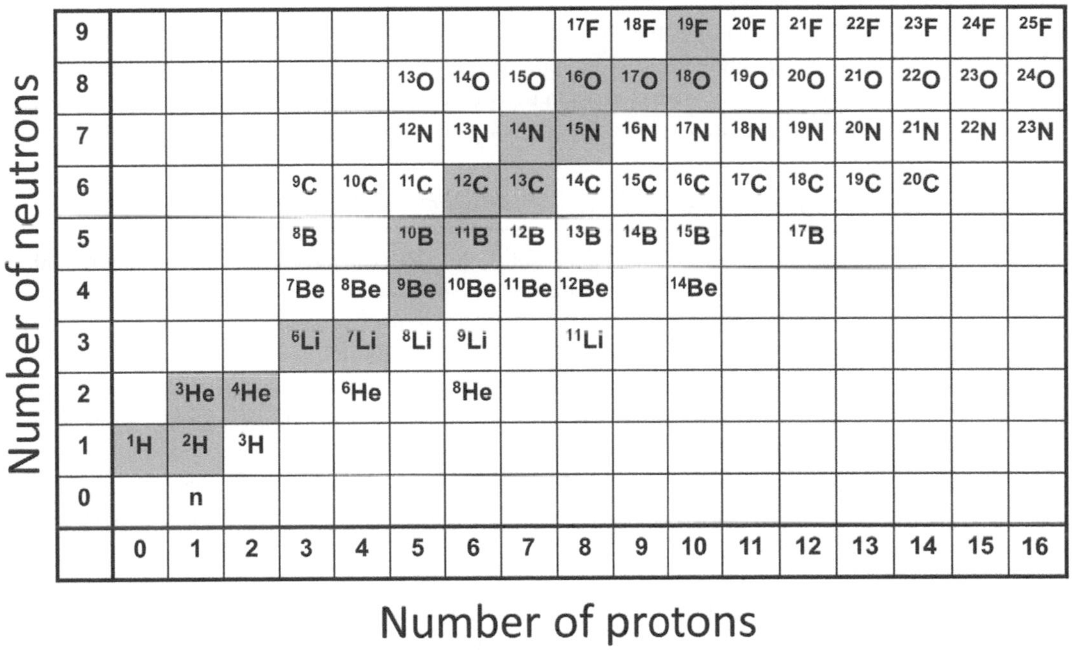

FIGURE 4.2 A part of the nuclide chart where the lightest elements are shown. The darkened fields represent stable nuclei. Nuclides to the left of the stable ones are radionuclides deficient in neutrons, and those to the right, rich in neutrons.

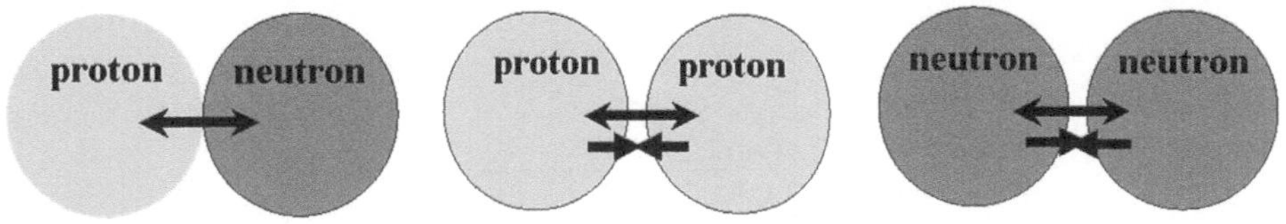

FIGURE 4.3 Between the proton and a neutron there is a nuclear force that amounts to 2.225 MeV. The nucleons are forming a stable combination called deuterium, an isotope of hydrogen. In a system of two protons the nuclear force is equally strong, but the repulsive Coulomb force is stronger. Therefore, this system cannot exist. Between two neutrons the nuclear force is equally strong and there is no Coulomb force. Still, this system cannot exist due to other repulsive forces, a consequence of the rules of pairing quarks.

The explanation is that, with many protons close together in the heavy elements, a large Coulomb force is created. When 'diluting the charge' by entering an excess of non-charged neutrons, the Coulomb force will decrease.

4.3.1 Binding Energy, Q-value, Reaction Threshold, and Nuclear Reaction Formalism

Between a free proton and neutron there are no barriers and no repulsive forces. They can fuse at low kinetic energies, forming a deuterium nucleus that has a weight somewhat smaller than the sum of the free neutron and proton weights. This mass difference can be converted into energy using the Einstein formula $E = mc^2$ and is found to be 2.2 MeV. This is also the energy released, as a gamma photon, in the reaction. If we would like to separate the two nucleons in deuterium nucleus at least 2.2 MeV has to be added to make it possible. The energy gained or lost in a nuclear reaction is called the Q-value. In a somewhat more complex reaction ^{14}N (p, α) ^{11}C the Q-value is calculated as the difference between the summation of the mass of the particles before the reaction (p, ^{14}N) from the mass of the particles after the reaction (α, ^{11}C). Using a Q-value calculator, that can be found on the net, for example, www.nndc.bnl.gov/qcalc/index.jsp, the Q-value for the reaction ^{14}N (p, α) ^{11}C is found to be -2921.92 keV. This means that the proton, when it reaches the ^{14}N nucleus, must have a kinetic energy of at least 2.92 MeV in order to make the reaction possible.

However, the proton needs to start with an even higher energy since, before it reaches the positively charged ^{14}N nucleus, it also has to overcome a repulsive Coulomb force. During this process the proton loses some energy and the starting value, called the threshold value, must then exceed the Q-value. The same calculator gives a threshold value of 3.13 MeV for the ^{11}C-production reaction.

The reaction energy (the 'Q-value') is positive for exothermal reactions (spontaneous reactions) and negative for endothermal reactions. Since all radioactive decays are spontaneous, they have positive Q-values. Some reactions used to produce radionuclides, mainly those that are based upon thermal neutrons, have positive Q-values but reactions based on positive particles usually have negative Q-values, that is, one need to add extra energy to make the reaction going.

4.3.2 Types of Nuclear Reactions, Reaction Channels and Cross Section

As seen in Figure 4.1 the radionuclides to the right of the stability line have an excess of neutrons compared to the stable elements, and they are preferably produced by irradiating a stable nuclide with neutrons. The radionuclides to the left are neutron deficient or have an excess of charge and, hence, they are mainly produced by irradiating stable elements by a charged particle. Although these are the main principles, there are exceptions.

Usually, the irradiating particles have a large kinetic energy that is transferred to the target nucleus to enable a nuclear reaction (the exception being thermal neutrons that can start a reaction by thermal diffusion). Figure 4.4 shows schematically the incoming beam incident upon the target, where it may be scattered and absorbed. It can transfer its energy totally or partly to the target nucleus and can interact with parts of or the whole target nucleus. Since the probability of interacting with a single atom is small, targets for radioactivity productions are usually thick, absorbing the whole or parts of energy of the irradiating particle.

In radionuclide production the nuclear reaction always involves a change in the number of protons or neutrons. Reactions that result in a change in the number of protons are preferable because the product becomes a different element, facilitating chemical separation from the target, compared to, for example, an (n, γ)-reaction, where product and target are the same element. Even low-energy neutrons down to thermal energies can enter the nucleus and cause nuclear reactions while charged particles need to overcome the Coulomb barrier (Figure 4.5).

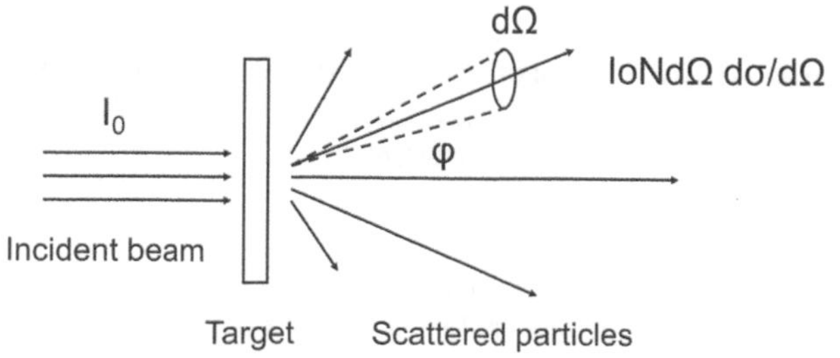

FIGURE 4.4 Target irradiation. A nuclear physicist is usually interested in the particles coming out, their energy and angular distribution, but the radiochemist is mainly interested in the transformed nuclides in the target.

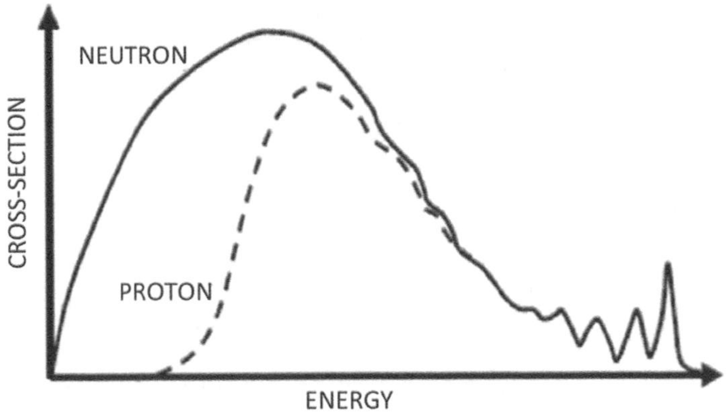

FIGURE 4.5 General behaviour of cross-sections as a function of the incident particle energy. For protons there is a threshold due to the Coulomb barrier that is not present for neutrons. Even low-energy neutrons may cause a nuclear reaction.

The parameter *cross section* [σ] is the probability for a certain interaction between incoming particle and the nucleus and is expressed as a surface. The geometrical cross section of a uranium nucleus is roughly 10^{-28} m^2, and this area has also been taken to define the unit for cross section, barn (b). This is not an SI unit but is commonly used to describe reaction probabilities in atomic and nuclear physics.

For fast particle reactions, the probability for a nuclear reaction is usually less than the geometrical cross-section area of the nucleus and the cross sections are in the range of mb. However, the probability of a hit is a combination of the area of both the nucleus and the incoming particle. The Heisenberg uncertainty principle states that the position and momentum of particles cannot be simultaneously known to arbitrarily high precision. This implies that particles of well-defined but low energy, like thermal neutrons, will have a large uncertainty as to their position. One may also say that they are increasing in size, and nuclear reactions involving thermal neutrons may have very large cross sections, sometimes in the order of several thousands of barns.

The general equation for a nuclear reaction is as follows:

$$a + A \rightarrow b + B + Q$$

where a is the incoming particle and A is the target nucleus in the ground state (the entrance channel). Depending on the energy and the particle involved, several nuclear reactions may happen, each with its own probability (cross section). Each nuclear reaction creates an outgoing channel, where b is the outgoing particle or particles, and B is the rest nucleus. Q is the reaction energy and can be both negative and positive.

A common notation of the nuclear reaction is A (a, b) B. If the incoming particle is absorbed we have a capture process type (n,γ) and in a reaction of type (p,n) we obtain charge exchange. If many particles are expelled, we can refer to the reaction, such as, for example, (p,3n). Each such reaction is called a reaction channel and is characterized by an energy threshold (energy that makes the nuclear reaction possible (*opens up the channel*) and a probability (cross section) varying with the incoming particle energy. A schematic illustration of different reaction channels opened in proton irradiation is given in Figure 4.6.

Different reaction mechanisms can operate in the same reaction channel. Here we differentiate between two ways:

- The formation of a compound nucleus
- Direct reactions

The compound nucleus has a large probability of being formed in a central hit of the nucleus and preferably at low energies close to the energy threshold of the reaction channel. Here, the incoming particle is absorbed, and an excited compound nucleus formed. This compound nucleus will rapidly (~10^{-19} seconds) undergo decay (fragmentation) with an isotropic emission of mainly neutrons and gamma. Direct reactions preferably occur at the edge of the nucleus and at high energies. The incoming energy is directly transferred to a nucleon (knock-on reaction) yielding two particles with high energy. The outgoing particles usually have high energy and are emitted in about the same direction as the incoming particle. The production of radionuclides is due to a mixture of these two reaction types. Their probability

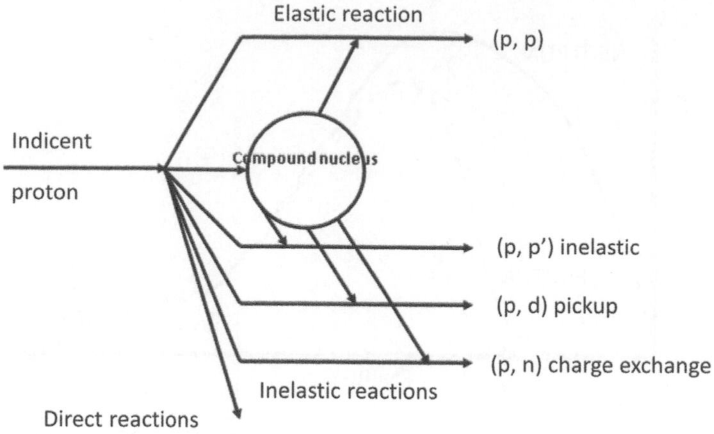

FIGURE 4.6 A schematic figure showing some reaction channels in proton irradiation.

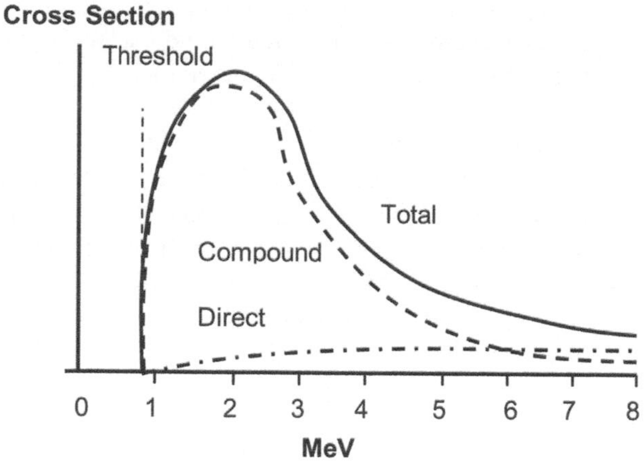

FIGURE 4.7 A schematic view of energy variation of cross sections for direct nuclear reactions and for forming a compound nucleus.

varies with energy in different ways. The direct reactions are heavily associated with the geometrical size of the nucleus and the cross section is usually small and fairly constant with energy. The compound nucleus has its highest probability to be formed just above the reaction threshold, as seen in Figure 4.7.

4.3.3 Relation Between Cross Section and Yield of Radionuclides

In a cyclotron the target is irradiated by a beam as shown in Figure 4.8. Since the target is assumed to be thin, only a small fraction of the beam particles interact, and the number of interactions is proportional to the number of incident particles. The probability to interact with the target is $n * x * \sigma$ and the activation rate $N_a = N * n * x * \sigma$.

Irradiation with neutrons in a reactor changes the geometry since the radiation is coming from all directions. The number of target nuclides (m) multiplied with the cross section gives the probability of interactions. The neutron can be described as a flux ϕ = neutrons/cm^2/s. The activation rate N_a can then be calculated as $N_a = \phi * m * \sigma$.

Na is the production rate of the radionuclides. During irradiation some of the produced radionuclides decay, and the formula $dM/dt = N_a - \lambda * M$ can be applied where M is the number of radionuclides and λ is the decay constant. The number of radionuclides as a function of time is then $M = N_a /\lambda * (1 - e^{-\lambda * t})$, and the activity, A, as a function of time is $A = N_a * (1 - e^{-\lambda * t})$.

In the calculations above it is assumed that the number of target nuclides is constant. This is true for nuclear reactions induced by high energy particles that generally have a cross section far less than one barn. In reactions involving thermal

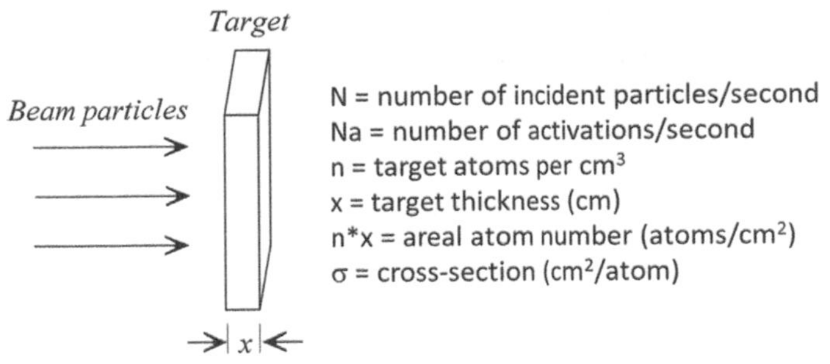

FIGURE 4.8 A schematic image of a thin target irradiated by a beam of ions (particles/second).

neutrons, cross sections can have values of several thousand barn that can lead to a substantial consumption of the target nuclides during irradiation. If needed, this must be accounted for to calculate the yield correctly.

In neutron irradiations the target usually does not affect the neutron flux, and all target nuclides are exposed to the same neutron number and energy. Only in large targets, or when the cross section is high, may neutron flux vary. In charged-particle reactions a thick target is used to maximize the activity yield (A_{int}), and as soon the particles enter, they start to degrade in energy. This means also that the cross-section value will vary. To calculate A_{int} one needs to integrate the formula $A = N * n * x * \sigma * (1 - e^{-\lambda * t})$ from the entrance of the target to the exit. Since only σ is varying, the integral can be written as

$$A_{int} = K\left(1 - e^{-\lambda t}\right) \int_{0}^{X_{max}} \sigma(x)dx \tag{4.1}$$

where K is a constant containing target material parameters and conversion factors between different units. Since the cross section usually is given as a function of energy it is an advantage to use energy as the integration variable and a variable transformation can be made using the stopping power S = dE/dx. The integration formula will then have following form:

$$A_{int} = K\left(1 - e^{-\lambda t}\right) \int_{0}^{E_{max}} \sigma(E)/S(E)dE \tag{4.2}$$

The integrations are usually made numerically.

4.4 REACTOR PRODUCTION

There are two major ways to produce radionuclides: using reactors (neutrons) or particle accelerators (protons, deuterons, alpha or heavy ions). Since the target is a stable nuclide, we generally obtain either a neutron-rich radionuclide (reactor produced) or a neutron deficient radionuclide (accelerator produced).

4.4.1 Principle of Operation and Neutron Spectrum

A nuclear reactor is a facility in which a fissile atomic nucleus such as uranium-235, plutonium-239 or plutonium-241 absorbs a low-energy neutron and undergoes nuclear fission. In the process, several fast neutrons are produced, with energies from about 10 MeV and downwards (the fission neutron spectrum). The neutrons are slowed down in a moderator (usually water) and the slowed-down neutrons start new fissions. By regulating this nuclear chain reaction there will be a steady-state neutron production and with a typical neutron flux in the order of 10^{14} neutrons/cm²/s.

Since neutrons have no charge, and are thus unaffected by the Coulomb barrier, even thermal neutrons (0.025 eV) can enter the target nucleus and cause a nuclear reaction. However, some nuclear reactions, depending upon the cross section, require fast neutrons (En < 10 MeV).

A reactor produces a cloud of neutron with varying energies. Usually, a pneumatic system is used to place the target at a predefined position where it is irradiated isotropically and with the wanted neutron energy. One must consider the heat that is evolved in the reactor core, since the temperature at some irradiation positions may easily reach 200°C. The reactor is characterized by the energy spectrum, the flux (neutrons/cm^2/s), and the temperature at the irradiation position. Most reactors in the world are for energy production and can, for safety reasons, not be used for radionuclide production. Usually only national research reactors are flexible enough for use in radioisotope production.

4.4.2 Thermal and Fast Neutron Reactions

The most typical neutron reaction is the (n, γ) reaction in which a thermal neutron is captured by the target nucleus forming a compound nucleus. The decay energy is emitted as a prompt gamma ray. A typical example is the reaction ^{59}Co (n, γ) ^{60}Co that produces an important radionuclide used in external therapy. However, since the produced radionuclide is of the same element as the target the specific activity, that is, the radioactivity per mass of the sample, is low. This type of nuclear reaction has little interest when labelling radiopharmaceuticals. In light elements, other nuclear reactions resulting from thermal neutron irradiation are possible, like (n, p). Table 4.1 lists possible production reactions for some biologically important radionuclides

For many reasons nuclear reactions with thermal neutrons are attractive. The yields are high due to large cross sections and the high thermal neutron fluxes available in the reactor. In some cases, the yields are sufficiently high to use these reactions as the source of charged secondary particles, for example, ^{6}Li(n, α)^{3}H for the production of high energy ^{3}H-ions, which can then be used for the production of ^{18}F by ^{16}O(^{3}H, n)^{18}F. The target used is ^{6}LiOH in which the produced ^{3}H-ions will be in close contact with the target ^{16}O. A drawback with this production is that when the target is dissolved the solution is heavily contaminated with ^{3}H-water that might be difficult to remove. Today, with an increasing number of hospital-based accelerators, there is little need of neutron produced ^{18}F.

Another reactor-produced neutron-deficient radionuclide is ^{125}I.

$$^{124}\mathrm{Xe}(\mathrm{n},\gamma)\,^{125}\mathrm{Xe}\left(\mathrm{T}\tfrac{1}{2}=17\,\mathrm{h}\right)\rightarrow\,^{125}\mathrm{I}\left(\mathrm{T}\tfrac{1}{2}=60\,\mathrm{d}\right)$$

This is the common way to produce high quality ^{125}I today. A drawback is that ^{124}Xe has a natural abundance of 0.1 per cent but the use of enriched targets increases the production yield. After irradiation, the target is stored for some days to let the relatively short-lived ^{125}Xe decay into the more long-lived ^{125}I. The expensive enriched ^{124}Xe-gas is carefully removed and used again. The target capsule is then rinsed with a suitable liquid that solves and removes the ^{125}I for further use.

Most reactor-produced radionuclides are not optimal for diagnostic since they, besides the useful γ-radiation, also emit high-energy β-particles that contribute to patient dose. However, a few will decay with no beta but mainly γ-rays since they are formed in exited states with long de-excitation times (metastable excited levels). Such radionuclides are

TABLE 4.1
Typical Nuclear Reactions in a Reactor for Radionuclide Production

Type of neutrons	Nuclear reaction	Half-life $T_{½}$	Cross-section (mb)
Thermal	^{59}Co (n, γ) ^{60}Co	5.3 a	2000
	^{14}N (n, p) ^{14}C	5730 a	1.75
	^{33}S (n, p) ^{33}P	25 d	0.015
	^{35}Cl (n, α) ^{32}P	24 d	0.05
Fast	^{24}Mg (n, p) ^{24}Na	15 h	1.2
	^{35}Cl (n, α) ^{32}P	14 d	6.1

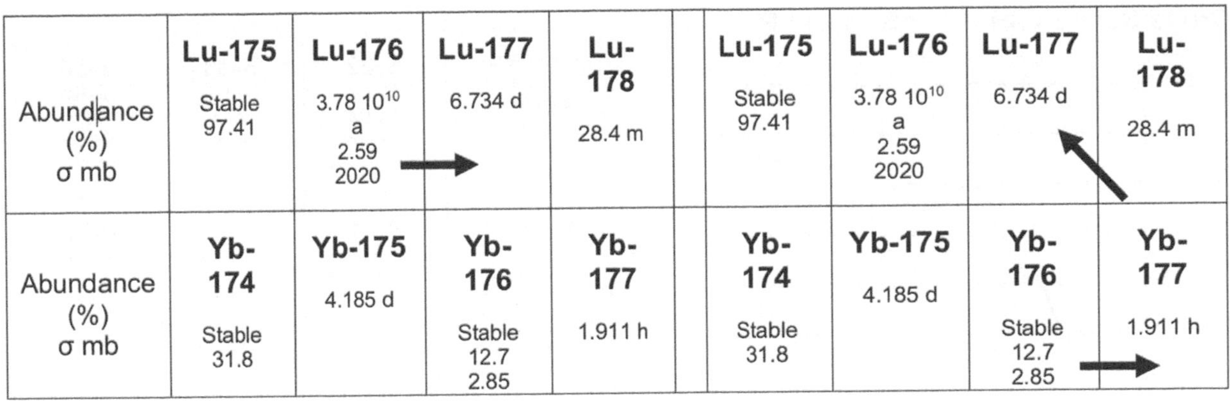
Abundance (%) σ mb
Lu-175
Stable 97.41
Lu-176
3.78 10^10 a 2.59 2020
Lu-177
6.734 d
Lu-178
28.4 m
Abundance (%) σ mb
Yb-174
Stable 31.8
Yb-175
4.185 d
Yb-176
Stable 12.7 2.85
Yb-177
1.911 h
Lu-175
Stable 97.41
Lu-176
3.78 10^10 a 2.59 2020
Lu-177
6.734 d
Lu-178
28.4 m
Yb-174
Stable 31.8
Yb-175
4.185 d
Yb-176
Stable 12.7 2.85
Yb-177
1.911 h

FIGURE 4.9 Production of ^{177}Lu from ^{176}Lu (left) and from ^{176}Yb (right).

ideal for nuclear medicine imaging, since they yield rather pure γ-radiation, with little charged radiation. The most used radionuclide in nuclear medicine, ^{99m}Tc, is of this type. The m after the atomic mass signifies that this is the metastable version of the radionuclide. For the production, see below.

In radionuclide therapy, in contrast to diagnostic applications, the emission of high energy beta radiation is desirable. Therefore, most radionuclides for radiotherapy are reactor produced. Examples include ^{90}Y, ^{131}I and ^{177}Lu. A case of interest to study is ^{177}Lu, which can be produced in two different ways using thermal neutrons. The most common production route is still the (n,γ) reaction on ^{176}Lu, which opposes two conventional wisdoms in practical radionuclide production for bio-molecular labelling.

1. Do not use a production route that yields the same product element as the target since it will negatively affect the labelling ability due to the low specific radioactivity.
2. Do not use a target that is radioactive.

However, ^{176}Lu is a natural radioactive isotope of lutetium with an abundance of 2.59 per cent and can be separated from the dominant ^{175}Lu to decrease the mass of the final product. Also, the high cross section (2,020 b) of ^{176}Lu results in a high fraction of the target atoms being converted to ^{177}Lu, yielding an acceptable specific radioactivity of the final product.

On the right of Figure 4.9, an indirect way to produce ^{177}Lu from ^{176}Yb is also shown. This method of production utilizing a generator nuclide ^{177}Yb, produced by a (n, γ) reaction, which then decays to ^{177}Lu. In principle, by chemically separating lutetium from ytterbium one would obtain the highest possible specific radioactivity. However, the chemical separation between two Lanthanides is not trivial and thus it is difficult to obtain ^{177}Lu without a substantial contamination of the target material Yb that may compete in the labelling procedure. Furthermore, the cross section for this reaction is almost 1000-fold lower resulting in a much lower yield of the product.

Reactions involving fast neutrons usually have cross sections that are in the order of mb, which coupled with the much lower neutron flux at higher energy relative to thermal neutron fluxes, leads to lower yields. However, there are some important radionuclides, for example, ^{32}P that must be produced this way. Figure 4.10 gives the details about this production.

4.4.3 Nuclear Fission and Fission Products

Uranium-235 is not just the fuel in a nuclear reactor but it can also be used as a target to produce radionuclides. ^{235}U irradiated with thermal neutron undergoes fission with a cross section of 586 b. The fission process results in the production of two fragments of ^{235}U nucleus plus several free neutrons. The sum of the fragments mass will be close to the mass of ^{235}U, but they will vary according to Figure 4.11.

The masses of the ^{99}Mo and ^{134}Sn produced in the reaction

$$^{235}\text{U}+\text{n} \rightarrow {}^{236}\text{U} \rightarrow {}^{99}\text{Mo} + {}^{134}\text{Sn} + 3\text{n}$$

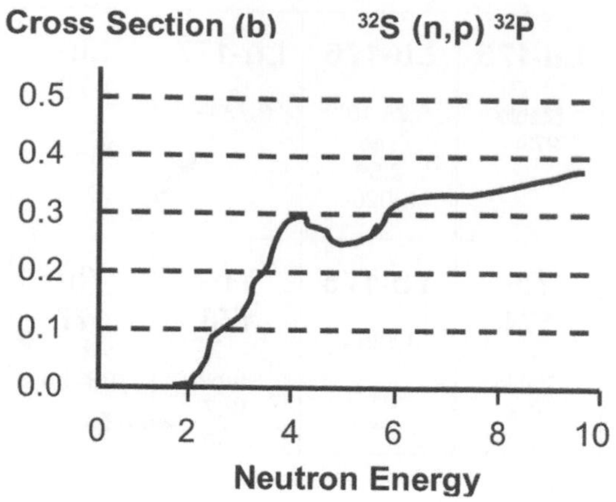

Abundance (%) fast n σ mb	**S-32** Stable 95.02 varying	**S-33** Stable 0.75 varying	**S-34** Stable 4.21 varying
Abundance (%) σ mb	**P-31** Stable 100	**P-32** 14.26 d	**P-33** 25.4 d

FIGURE 4.10 Data to produce ^{32}P in the nuclear reaction ^{32}S (n, p) ^{32}P. The reaction threshold is 0.51 MeV. From the cross-section data one can see that there is no substantial yield until a neutron energy of about 2 MeV. The yield is an integration of the cross-section data and the neutron energy spectrum. A practical cross section can be calculated to about 60 mb.

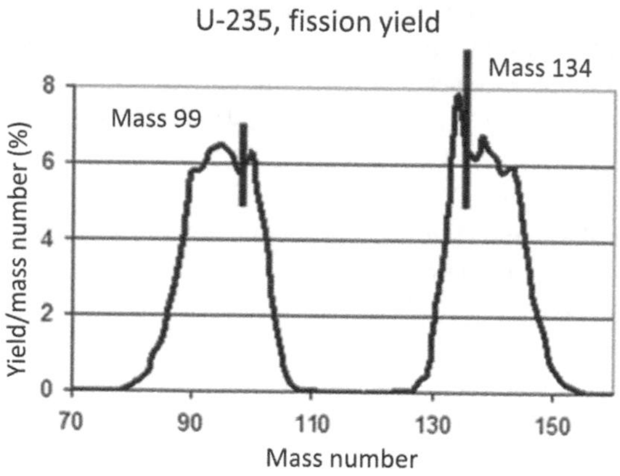

FIGURE 4.11 The yield of fission fragments as a function of mass.

are marked in Figure 4.11. Some medically important radionuclides are produced by fission like ^{90}Y (therapy) and ^{99m}Tc (diagnostic). They are not produced directly but by a generator system

$$^{90}\text{Sr (28.5 y)} \rightarrow {}^{90}\text{Y (2.3 d) and } {}^{99}\text{Mo (2.7 d)} \rightarrow {}^{99m}\text{Tc (6 h)}$$

The primary radionuclides produced are then ^{90}Sr and ^{99}Mo or more precisely the mass number 90 and 99.

Another important fission produced radionuclide in nuclear medicine, both for diagnostics and therapy, is ^{131}I. The probability of producing mass 131 is 2.9 per cent per fission. The practical fission cross section for this production is the fission cross section of ^{235}U * the fraction of fragments having mass 131 or 586 * 0.029 = 17 b. ^{131}I is the only radionuclide with mass 131 that has a half-life of more than 1 h meaning that all the others will soon have decayed to ^{131}I. To read more about reactor produced radionuclides, see reference [1].

4.5 ACCELERATOR PRODUCTION

Charged particles, unlike neutrons, are unable to diffuse into the nucleus, but need to have enough kinetic energy to overcome the Coulomb barrier. However, charged particles are readily accelerated to kinetic energies that open up more reaction channels than fast neutrons in a reactor. An example is seen in Figure 4.12 that also illustrates alternative opportunities with: p, d, ^3He and ^4He or α, to produce practical and economical nuclear reactions.

^{127}I	(p, 5n)	^{123}Xe	→	^{123}I
^{124}Xe	(p, np)	^{123}Xe	→	^{123}I
^{123}Te	(p, n)	^{123}I		
^{122}Te	(d, n)	^{123}I		
^{124}Te	(p, 2n)	^{123}I		
^{121}Sb	(^{4}He, 2n)	^{123}I		
^{121}Sb	(^{3}He, n)	^{123}I		
^{123}Sb	(^{3}He, 3n)	^{123}I		

FIGURE 4.12 Various nuclear reactions to produce ^{123}I. All reaction has been tried and can be performed at relatively low particle energies. The ^{123}Xe produced in the first two reactions decays to ^{123}I with a half-life of about two hours. In the first reaction the ^{123}Xe is separated from the target before it decays while in the second reaction the ^{123}I is washed out of the target container after decay.

TABLE 4.2
Characterization of Accelerators for Radionuclide Production

Proton energy (MeV)	Accelerated particles	Used for
<10	Mainly single particle, p or d	PET
10–20	Usually p and d	PET
30–40	p and d, ^{3}He and ^{4}He may be available	PET, commercial production
40–500	Usually p only	Often placed in national centres and have several users

An accelerator in particle physics can be huge as in CERN with a diameter of more than 4 km. Accelerators for radionuclide production are much smaller, needing to accelerate particles to much lower energies. The first reaction in Figure 4.12, where 5 neutrons are expelled, is the most energy demanding, since it requires a proton energy of about 5*10 = 50 MeV (rule of thumb states that about 10 MeV is required per expelled particle). All the other reactions require 20 MeV or less. A general characterization of accelerators used for radionuclide production is found in Table 4.2.

Another advantage with accelerator production is that it is usually easy to find a nuclear reaction where the product is a different element from the target. Since different elements can be chemically separated the product usually can be of high specific radioactivity, which is important when labelling biomolecules.

A technical difference between reactor and accelerator irradiation is that in the reactor the particles come from all directions but in the accelerator, there is a direction of the particles. The number of charge particles is often smaller and is usually measured as an electric current in μA (1 μA = $6*10^{12}$ protons/s but $3*10^{12}$ alpha/s due to the two charges of the α-particle).

A drawback in accelerator production is that charged particles are stopped more efficiently than neutrons, for example, 16 MeV protons are stopped in a 0.6 mm thick copper foil. A typical production beam current of 100 μA hitting a typical target area of 2 cm^2 will then put 1.6 kW in a volume of 0.1 cm^3, which will evaporate most materials if not efficiently cooled. Also, the acceleration of the beam is done in vacuum, but preferable the target irradiation is made outside in atmospheric pressure or in gas targets at 10–20 times over pressure. To separate the vacuum from the target the beam must penetrate foils that will steal some particle energy and they will also become strongly activated.

4.5.1 Cyclotron, Principle of Operation, Negative and Positive Ions

There are several types of accelerators and in principle, all can be used for radionuclide production. The dominant one for radionuclide production today is the cyclotron that was invented by Ernst O Lawrence in the early 1930s. Cyclotrons were first installed in hospitals in the 1960s, but during the last two decades, hospital based small cyclotrons yielding 10–20 MeV protons have become fairly common, especially with the increase of positron emission tomography (PET).

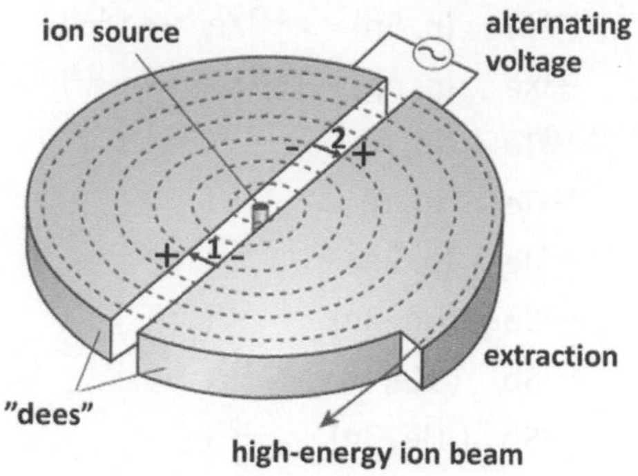

FIGURE 4.13 The cyclotron principle. In a homogeneous magnetic field two electrodes ('dees') are placed. An ion is injected into the gap between. An alternating electric field is applied across the gap (1) that causes the ion to accelerate. Inside the electrode and in the magnetic field the ion moves in a circular orbit reaching the gap again (2). During this time, the electric field is reversed in polarity, and the ion is accelerated again. The orbit radius increases with increasing ion energy until the ion finally is extracted.

A cyclotron is composed by four systems.

1. A resistive magnet that can create a magnetic field of 1–2 Tesla (T)
2. A vacuum system down to 10^{-5} Pa.
3. A high frequency system (about 40 MHz) providing a voltage with a peak value of about 40 kV, although these figures can vary considerably for different systems.
4. An ion source that can ionize hydrogen to create free protons as well as deuterium and eventually alpha-particles.

A view of the cyclotron principle is seen in Figure 4.13.

The ion source is usually placed in the centre and inside the vacuum (internal) but can in larger machines be external. The ions are then injected from the outside through a central hole in the magnet. The main idea of the ion source is to have a slow flow of gas that is made into plasma by an arc discharge. Through a collimator the wanted ion species are extracted and accelerated in a static electric field. There are several types of ion sources with different operating characteristics. In modern accelerators, usually negative ion (protons or deuterium with two orbit electrons) are used to facilitate extraction of the beam.

The ions leave the ion source with some velocity. Since the vacuum chamber is placed in a perpendicular magnetic field, the ions start to move in a circular orbit. Inside the vacuum chamber there are two electrodes, historically called the 'dees' since the first one's hade the shape of the letter D. These electrodes are hollow and so the ions can freely move inside (no electric field). If a voltage is applied with right polarity in the gap between the electrodes, the ions will experience an acceleration force when passing the gap. If the voltage polarity is switched at the correct rate, the ion will be continuously accelerated when crossing the gap resulting in an increase in energy and velocity. The radius of the ion orbit increases until the ion is extracted. The frequency of the electric field change across the 'dees' must be the same as the frequency of the circulating ions for a proper operation.

4.5.2 Commercial Production (Low and High Energy)

If the proton energy is > 30 MeV the particles tend to be relativistic, that is, their mass increases and their cycle time in the orbit increases. With a constant frequency of the accelerating electric field, the ions would soon come out of phase. This can be compensated either by increasing the magnetic field as a function of the cyclotron radius (isochronic cyclotrons) or by decreasing the radiofrequency during acceleration (synchrocyclotrons). Such accelerators tend to be more complex and expensive and for this reason 30 MeV is a typical energy for commercial accelerators that needs to have large beam currents and to be both reliable and cost effective.

Beam currents in commercial accelerators are usually in the order of several mA. Since it is technical difficult to extract such high beam currents due to heating problems in the separating foils, most commercial accelerators are using internal targets – that is, targets placed inside the cyclotron vacuum as shown schematically in Figure 4.14.

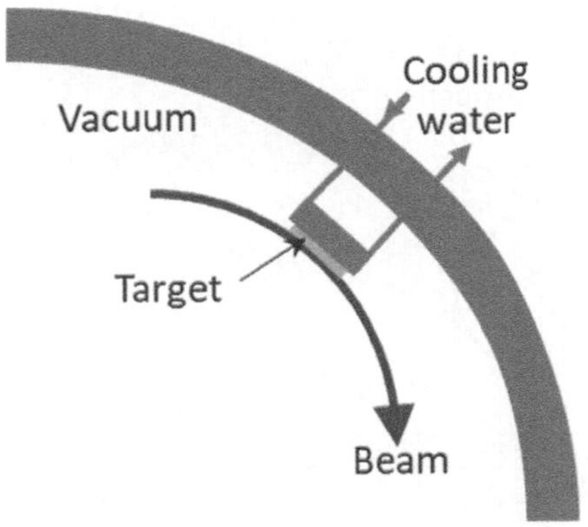

FIGURE 4.14 Schematic image of an internal target. The target material is usually thin (few tenth of µm) and evaporated on a thicker backing plate. The target ensemble is water cooled on the back. An advantage is that the beam is spread out over a large area which facilitates the cooling.

Most patients in nuclear medicine are still investigated by Single Photon Emission Computed Tomography (SPECT). Besides ^{99m}Tc (reactor produced), ^{67}Ga, ^{111}In, ^{123}I and ^{201}Tl (accelerator produced) are available for SPECT. Also, some accelerator produced radionuclides for PET like ^{124}I are becoming commercially available. The increasing demand for the generator ^{68}Ge/^{68}Ga has also started a commercial production of the cyclotron produced mother nuclide ^{68}Ge. Only few radionuclides of medical interest need production energies above 30 MeV. A limited number of high energy accelerators with high beam currents, usually at national physics laboratories, have the capacity to produce, for example, ^{52}Fe and ^{61}Cu and other isotopes used for research activities.

4.5.3 In-house Low-Energy Production (PET)

Accelerators dedicated to PET radioisotope production are limited both in energy (<20 MeV) and in beam current (< 200 µA). Many production routes utilize gases or water as target materials and therefore external targets are preferred. Due to the relatively low beam current, extraction is usually not a problem. Since internal targets need to be taken in and out of the cyclotron vacuum, they are not usually implemented in PET cyclotrons.

The modes of extraction depend upon whether positive or negative ions are accelerated. Extraction of positive ions is made by using a deflector that applies a static electric field that acts upon the particles when in the outer orbits. Some beam current is invariably lost in the process and the deflector often becomes quite radioactive. Modern proton/deuterium accelerators usually accelerate negative ions (p and d with two extra electrons) that are more easily extracted. When the ion reaches the outer orbits, it will hit a thin carbon foil that will strip away the electrons. Consequently, the particles suddenly change from negative to positive charge and are effectively bent out of the magnetic field with an almost 100 per cent extraction efficiency and with little activation.

The inner parts of a modern accelerator are seen in Figure 4.15. The accelerator is opened like a door with one of the magnetic poles sitting in the 'door'. The electrodes are differently shaped compared to Figure 4.13 but are still called 'dees'. In the figure is also seen the holder of the extraction foils. The beam pathway during extraction is indicated by the dashed curve.

The extracted beam can either be transported further in a beam optical transport system or it will hit a production target directly. Usually, the target is separated from the vacuum by two thin metallic foils that are strong enough to stand the pressure difference and the heat evolved from the absorbed beam. Two foils are used to facilitate the change of target without breaking the cyclotron vacuum and to enable cooling of the foils with a flow of He-gas. Helium is preferred as cooling medium since no induced activity will be produced in this gas.

The targetry of a modern 'In house accelerator' is seen in Figure 4.15. Five gas and water targets are mounted directly on the accelerator surface. The beam is directed toward the selected target by changing the position of the stripper foil. It is important that the target is sitting close to the extraction point since the beam is transported without any focusing and tends to spread somewhat.

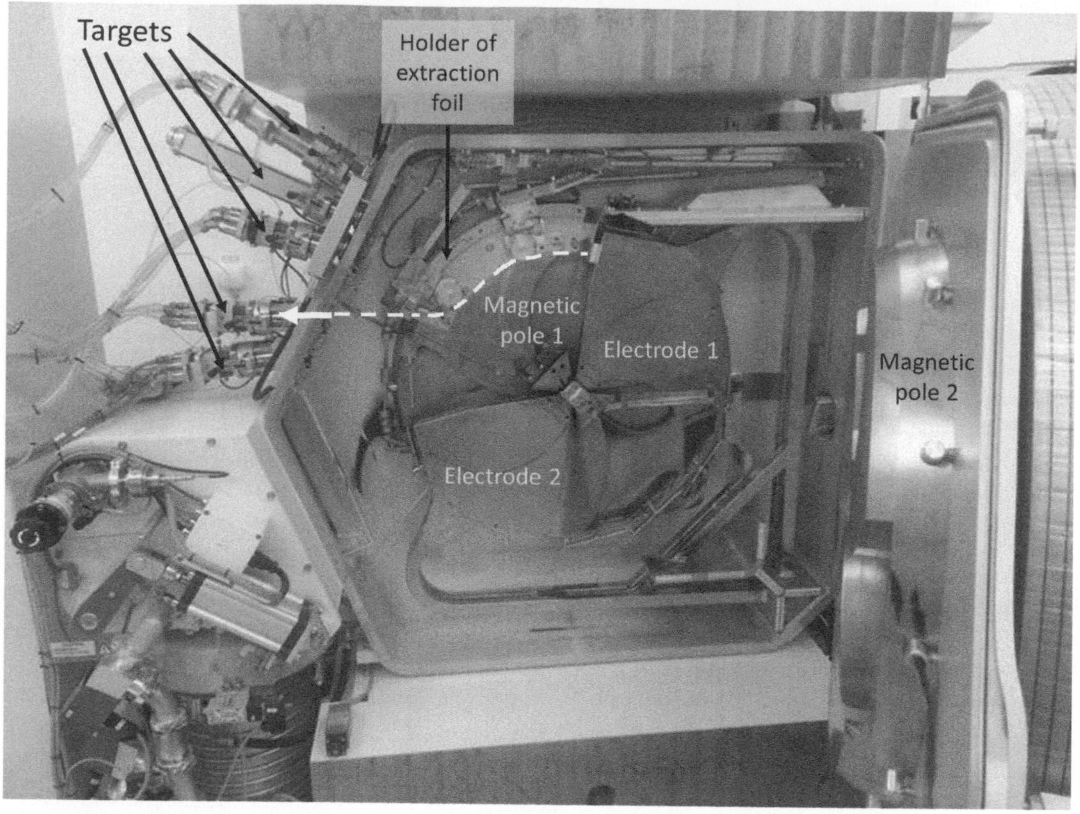

FIGURE 4.15 The inner parts of a modern cyclotron and the targetry. The accelerator is opened like a door showing the magnetic poles and the electrodes called the 'dees'. Also, the holder of the extraction foils is seen. The dashed curve shows the beam pathway during extraction. Five gas and water targets are mounted directly on the accelerator surface. The beam is directed toward the selected target by changing the position of the stripper foil. Courtesy Gustav Brolin, Skåne University Hospital, Lund, Sweden. Colour image available at www.routledge.com/9781138593268.

Choosing right reaction and target material is crucial and is illustrated by the production of ^{18}F (Table 4.3).

Not just the nuclear reaction is important, but also the chemical composition of the target. To irradiate ^{18}O as a gas would be the purest target (only target nuclide present) but the handling of a highly enriched gas is complicated, in addition to the hot-atom chemistry. Still, for some applications this might be the best choice. To irradiate ^{18}O as an oxide and a solid target is possible but the process after the irradiation to dissolve the target and to chemically separate ^{18}F is complex and the other element in the oxide can potentially contribute with unwanted radioactivity. The target of choice is ^{18}O-water, since ^{18}O is the dominating nucleus and hydrogen does not contribute to any unwanted radioactivity. There is usually no need of target separation, since the water containing ^{18}F often can be directly used in the labelling chemistry, for example, to produce ^{18}FDG. The target water can also, after being diluted with saline, be injected directly into patients, for example, ^{18}F-fluoride for PET bone scans.

Water targets are in some labelling chemistry not the best choice. An alternative production route is to use neon gas, $^{20}Ne(d,\alpha)^{18}F$. Adding fluorine-19 gas to the neon as carrier will yield $^{18}F_2$ that can be used for electrophilic substitution. However, adding carrier will lower the specific radioactivity of the labelled product.

A problem is the heat evolved when the beam is stopped in a few mL of target water. High pressure targets that force the water to stay in the liquid face can overcome some of these problems, but the productions are usually limited to beam currents < 40 μA. Gas and solid targets have the advantage since they can stand higher beam currents.

There also exist several options to produce ^{11}C. These include: $^{10}B(d,n)^{11}C$, $^{11}B(p,n)^{11}C$ or $^{14}N(p,\alpha)^{11}C$. The reactions on boron are made as solid target irradiations while the reaction on nitrogen is a gas target application.

Routine productions of the common positron emitters associated with PET are summarized in Table 4.4.

TABLE 4.3
Different Nuclear Reactions to Produce ^{18}F

Target	Reaction	Product	Comments
^{20}Ne	(d, α)	^{18}F	The nascent ^{18}F will be highly reactive. In the nobelgas Ne it will diffuse and stick to the target walls. Difficult to extract.
^{21}Ne	(p, α)	^{18}F	Same as above. Also, the abundance of ^{21}Ne is low (0.27%) and needs enrichment
^{19}F	(p, d)	^{18}F	Product and target are the same element. Poor specific radioactivity.
^{16}O	(α,d)	^{18}F	Cheap target, but accelerators that can accelerate α-particles to 35 MeV are expensive and not commonly available.
^{16}O	(d, γ)	^{18}F	Small cross-section and no practical yields can be obtained.
^{18}O	(p, n)	^{18}F	Expensive enriched target material, but the proton energy is low (low-cost accelerator), which makes this the nuclear reaction of choice.

TABLE 4.4
Routine Productions of the Common Positron Emitters Associated with PET

Radionuclides	Nuclear reaction	Target	Yield (GBq)
^{15}O	^{14}N (d,n) ^{15}O	gas	15
^{13}N	^{16}O(p, α) ^{13}N	liquid	5
^{11}C	^{14}N (p, α) ^{11}C	gas	40
^{18}F	^{18}O (p,n) ^{18}F	liquid	100

Oxygen-15 is produced by deuteron bombardment of natural nitrogen through the $^{14}N(d,n)^{15}O$ nuclear reaction. An alternative is the reaction $^{15}N(p,n)^{15}O$ if a deuterium beam is not available. In this case, the target needs to be enriched. In the nitrogen target, molecular oxygen ($^{15}O_2$) is produced directly. Direct production of carbon dioxide ($C^{15}O_2$) is possible by mixing the target gas with 5 per cent of natural carbon dioxide as a carrier. Water ($H_2{}^{15}O$) is preferable made by processing molecular oxygen-15.

Carbon-11 is produced by proton bombardment of natural nitrogen. By adding a small amount of oxygen to the target gas (<0.5%) carbon dioxide ($^{11}CO_2$) will be produced. Adding 5 per cent hydrogen in the target will produce methane ($^{11}CH_4$).

Liquid targets are today by far the most popular and widely used to produce ^{13}N. The reaction of protons on natural water produces nitrate and nitrite ions, which can be converted to ammonia by reduction. Water targets can also be used to form ammonia directly with the addition of a reducing agent, for example, ethanol or hydrogen.

4.5.4 Targetry, Optimizing the Production, Yield Calculations

When the nucleus is hit by an energetic particle, a complex interplay between physical and statistical laws determines the result. Important parameters are the entrance particle energy, the target thickness and the reaction channel cross sections for the particle energies in the target. Computer codes like ALICE and TALYS are available to calculate the size and the energy dependence of the cross section for a certain reaction channel but they are not easy to apply; hence caution should be exercised when interpreting the results from such codes. However, a rough estimation of the irradiating particle energy can be obtained using a well-known rule of thumb in radionuclide production illustrated in Figure 4.16 showing measured cross sections to produce various radioisotopes of Selene. The maximum cross sections are found at about 10, 30 and 40 MeV for the (p,n)-, (p,3n)- and (p,4n)-reactions, respectively. Thus, it takes about 10 MeV to kick out a nucleon, that is, a proton of 50 MeV can cover radionuclide productions that involve the emission of about 5 nucleons.

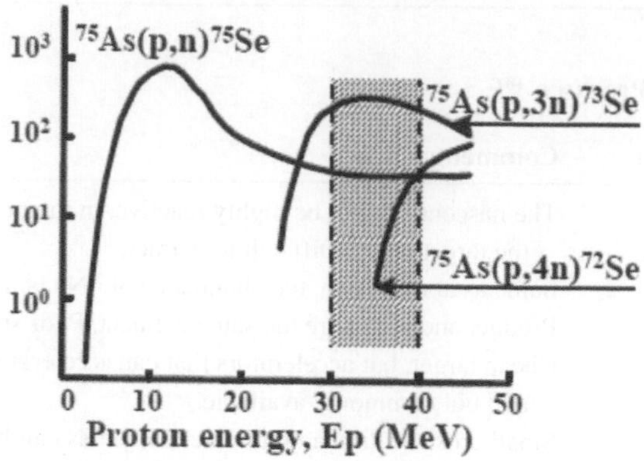

FIGURE 4.16 Excitation functions of ^{75}As (p,xn) ^{72,73,75}Se reactions. The optimal energy range to produce ^{73}Se is to use a proton energy of 40 MeV that is degraded to 30 MeV in the target. The figure is based on data taken from reference [2].

Figure 4.16 also highlights the fact that the production parameters are a compromise. The wanted radionuclide is ^{73}Se (T½ = 7.1 h) but it is not possibly to produce it without a contamination of the more long-lived ^{72}Se (T½ = 8.5 d) and ^{75}Se (T½ = 120 d). The practical setup when producing ^{73}Se is as follows. A suitable As-target is made and irradiated with 40 MeV protons since at this energy the contamination of ^{72}Se will be tolerable. The target thickness should decrease the proton energy down to 30 MeV, an energy where the ^{75}Se cross section starts to flatten out. The energy range 30–40 MeV will then use most of the maximum cross-section peak of ^{73}Se and will produce tolerable levels of ^{72}Se and ^{75}Se. This then gives a radioactivity yield of the desired radionuclide at end of bombardment mainly dependent upon the beam current and the irradiation time. The yield is usually expressed in GBq/μAh, that is, the produced radioactivity per time-integrated beam current. If possible, one tries to keep the radioactivity of the contaminants small (< 1 %). However, from the end of bombardment, the ratio of the product relative to any long-lived radio contaminants begins to decrease. To read more about cyclotron produced radionuclides see reference [3].

4.6 RADIONUCLIDE GENERATORS

Whenever a radionuclide (mother) decays to another radioactive nuclide (daughter) we call this a generator system. Most of the natural radioactivity is produced in generator systems starting with the uranium isotopes and ^{232}Th and involves about 50 radioactive daughters.

In nuclear medicine, we usually think of a special case in which the mother has a half-life long enough to enable delivery to far-away hospitals, and the daughter has a short half-life suitable for clinical investigations. A typical example is the ^{99}Mo/^{99m}Tc generator (Figure 4.17) producing the most-used radionuclide in nuclear medicine. The half-life of the mother (2.7 d) is adequate for transport and delivery, and the daughter has a suitable half-life (6 h) for patient investigations. The generator is used for a working week (2–3 half-lives of the mother) and is then renewed.

The ^{99}Mo/^{99m}Tc generator fulfils the requirement that the daughter is easily separated from the mother and that the product is pure and in a chemical form easy to use. Most commercial generators use column chromatography where ^{99}Mo is adsorbed onto alumina. Eluting this column with physiological saline yields the soluble ^{99m}Tc in a few mL liquids ready to use. In fact, most generators in nuclear medicine use ion exchange columns in much the same way due to its simplicity in handling.

Another generator of increasing importance is ^{68}Ge/^{68}Ga-system where ^{68}Ge, with a half-life of 271 days, produces a short-lived positron emitter, ^{68}Ga, with a half-life of 68 minutes. This is produced as an ion (+3) that can be rapidly labelled, using a chelating agent such as DOTA, to small receptor-binding peptides, for example, ^{68}Ga-DOTATOC. Due to the long half-life of the mother, the generator can be operated for up to two years and can be milked every fifth hour. One problem of such a long-lived generator is keeping it sterile, and that the ion-exchange material is exposed to high radiation doses that may reduce the elusion efficiency and the quality of the product.

The generator ^{90}Sr/^{90}Y is used to produce the therapeutic radionuclide ^{90}Y. This generator is not distributed to hospitals but is, due to the long half-life of the mother (28.5 a) and radiation protection considerations, eluted in special laboratories. The daughter ^{90}Y has a half-life of 2.3 days that is long enough to allow convenient hospital transportation.

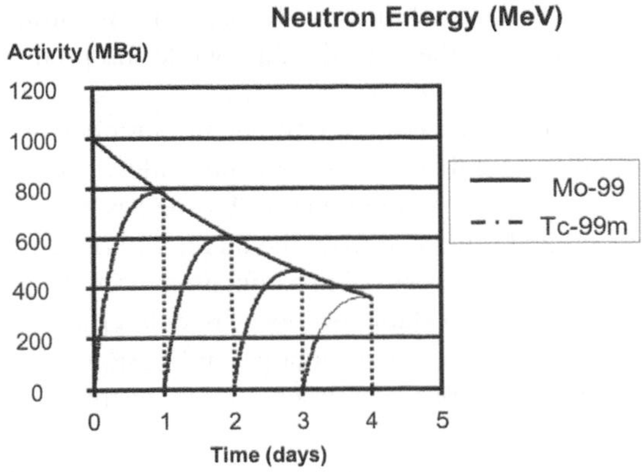

FIGURE 4.17 The elution of a ^{99}Mo/^{99m}Tc generator. The generator has a nominal activity of 1000 MBq at day 0 (Monday). It is 'milked' every day during the week yielding theoretically 1000, 780, 600, 470 and 360 MBq.

^{81}Rb (4.5 h)/^{81m}Kr (13.5 s) for ventilation studies and ^{82}Sr (25.5 d)/^{82}Rb (75 s) for cardiac PET studies are examples of other generators with a special requirement due to the extremely short half-life of the eluted product. Recently, generator systems producing alpha-emitters for therapy has become available, for example, ^{225}Ac (10 d)/^{213}Bi (45.6 m). Also, radionuclides used in therapy may by themselves be generators like ^{211}At (T½ = 7 h) decaying to ^{211}Po (T½ = 0.5 s) or ^{223}Ra, which generates a series of relatively short-lived radioactive daughters *in situ*. To read more about generator systems see reference [4].

4.6.1 Radiochemistry of Irradiated Targets

During a target irradiation, a few atoms of the wanted radionuclide are produced within the bulk target material. The energy released in a nuclear reaction is large relative to the electron binding energies and, therefore, the radionuclide is usually 'born' almost naked with no or few orbit electrons. This 'hot atom' will undergo prompt chemical reactions depending on the target composition. In a gas or liquid target, these 'hot atom' reactions may even cause the activity to be lost in covalent bonds to the target holder material. Also, during irradiation, the target is heated and may change structure and composition. A pressed powder target may be sintered and becoming more ceramic, making it more difficult to dissolve. The target may melt, and the radioactivity diffuses in the target and even possibly evaporates. In designing a separation method, all these factors must be considered. Fast, efficient, and safe methods are required to separate the few picograms of radioactive product from the bulk target material present in gram quantities.

The separation of the radionuclide starts already in the target as demonstrated in the production of $^{11}CO_2$. Carbon-11 is produced in a (p,α)-reaction on nitrogen gas. To enable the production of CO_2 some trace amounts of oxygen gas (0.1–0.5%) is added. However, at low-beam currents the absorbed energy may not heat the target gas enough and mainly CO will be formed. At high-beam currents, the CO will be oxidized to the chemical form CO_2. The separation, made by letting the target gas pass a liquid nitrogen trap, is simple and efficient. By adding hydrogen gas, 'hot atom' chemistry will produce CH_4 instead.

'Hot atom' chemistry is important to obtain a suitable chemical form of the radioactive product, especially when working with gas and liquid targets. Solid targets are usually dissolved and chemically processed to obtain the wanted chemical form for further labelling.

4.6.2 Carrier-free and Carrier-added Systems

The concept of specific activity (SA), that is the activity per mass of a preparation, is essential in radiopharmacy. A product that contains radioactive atoms only will have the theoretically highest specific activity. The relation between activity (A) in Bq and number of radioactive atoms (N) is then given by $N = A/\lambda$ and $SA = A/N = \lambda$ (the decay constant). For a short-lived isotope, the theoretical SA will be relatively large compared to a long-lived isotope. For example, the theoretical SA for ^{11}C (T½=20 m) is 1.5 10^8 times larger than that of ^{14}C (T½=5730 a).

The SA expressed in this way is a theoretical value that rarely is obtained in practical work. When producing ^{11}C, the target gas and target holder will contain stable carbon that will dilute the radioactive carbon. A more empirical way to define the SA is to divide the activity by the total mass of the element in consideration. This value will for ^{11}C will usually be a few thousand times lower than the theoretical value, while the production of ^{14}C can come closer to the theoretical SA.

In a labelled product, the SA is expressed as the activity per the number of molecules (a sum of labelled and unlabelled molecules). Instead of using the number of atoms or molecules, it is common to use the mole concept by dividing N by Avogadro's number ($6.022 * 10^{23}$). A common unit for the SA is then GBq/μmole.

If the radioactive atoms are produced and separated from the target without any stable isotopes, the product is said to be 'carrier-free'. If stable isotopes are introduced as being a contaminant in the target or in the separation procedure, the process is called 'no carrier added', that is, no stable isotopes are deliberately added. Both these processes usually give a high final SA. However, it may be necessary to use a target of the same element, or it may be necessary to add mass of the same element in order to have the separation process work. In this case 'carrier is added' deliberately and the SA will usually be low.

Note that a carrier does not necessarily need to be of the same element. When labelling a radiopharmaceutical with a chelator and metal ions, any ion fitting into the chelator will compete. An example is labelling a peptide with ^{111}In, where the activity usually is delivered as $InCl_3$ in a weak acid. By sampling the activity with a stainless-steel needle, Fe ions will be released and will most likely completely ruin the labelling process by outnumbering the ^{111}In atoms.

4.6.3 Separation Methods

After the radiation a small amount of desired radioactivity (≈ nanomole) needs to be separated from the bulk of the target in a suitable form for the subsequent labelling process and at high specific activity. The separation time should be related to the half-life of the radionuclide and should take at most one T½. Solid targets are usually dissolved, which is simple for salts like NaI but more complicated for Ni-foils, for example, where boiling aqua regia may have to be used. To speed up this process the Ni-foil can be replaced by a pressed target of Ni-powder that will increase the metal surface and will speed up the dissolving process.

In general, for separation two principles are used: liquid extraction and ion exchange. In liquid extraction two liquids that do not mix are used, for example, water and an organic solvent. The target element and the produced activity of another element should have different relative solubility in the liquids. The two liquids and the dissolved target are mixed by shaking, after which two faces are formed. The face with a high concentration of the wanted radioactive product is sampled and is usually separated again one or more times to reduce the target mass in that fraction. The relative solubility can be optimized by varying the pH or by adding a complexing agent.

In the ion exchange mechanism, an ion in the liquid face (usually an aqueous face) is transferred to a solid face (organic or ceramic material). To maintain the charge balance, a counter ion, for example hydrogen, is released from the solid face. The distribution ratio between liquid and solid face is often a function of pH. Furthermore, complexing agents can be used to modify the distribution ratio.

The dissolved target is adjusted to obtain the right separation conditions and is then put on to a column containing the ion-exchange material. The optimal separation conditions would be that the small mass of desired radioactivity sticks to the column, but the target material does not. Then, only a small amount of ion-exchange material is needed, and the wanted activity can be eluted from the column in a small volume, for example, by changing pH.

Often the two techniques are used together, for example, by using liquid extraction to reduce the target mass, after which ion exchange is used to make the final separation.

Occasionally, thermal separation techniques may be applied, which have the advantage that they do not destroy the target (important when expensive enriched targets are used) and that they lend themselves to automation. As an example of such dry methods the thermal separation of ^{76}Br (T½ = 16 h) is described. The target is $Cu_2{}^{76}Se$, a selenium compound that can withstand some heat. The nuclear reaction used is $^{76}Se(p,n)^{76}Br$.

The target is placed in a tube and heated, under a stream of argon gas, to evaporate the ^{76}Br activity by dry distillation (Figure 4.18). A temperature gradient is applied to separate the deposition areas of ^{76}Br and traces of co-evaporated selenide in the tube by thermal chromatography. The ^{76}Br activity deposited on the tube wall is dissolved in small amounts of buffer or water.

Separation yields of 60–70 per cent are achieved by this method, with a separation time of about one hour. Since dry distillation permits the extraction of radiobromine without destroying the target, the Cu_2Se targets are reusable. Considering the rather expensive ^{76}Se-enriched target material, this is a practical prerequisite for this type of production. The chemical form of the ^{76}Br activity after separation, analysed by ion exchange high-performance liquid chromatography (HPLC) and thin-layer chromatography (TLC), was almost exclusively found to be bromide.

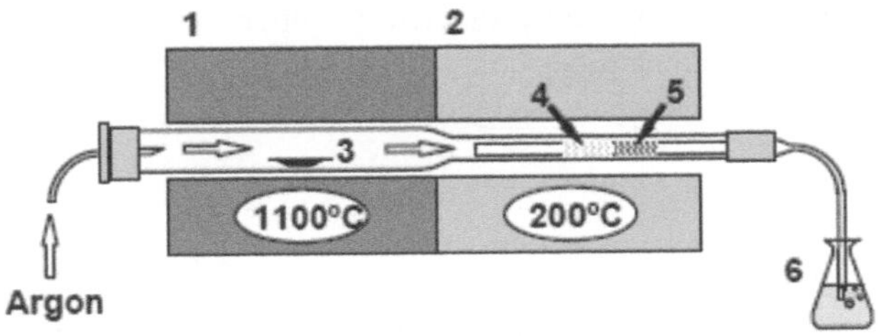

FIGURE 4.18 A schematic description of the equipment used to separate ^{76}Br by dry distillation: Furnace 1 of high temperature (1), furnace 2 at lower temperature (2), the irradiated target of $Cu_2{}^{76}Se$ placed in position (3), some selenium will also evaporate and is trapped in position (4) in the temperature gradient, ^{76}Br is deposit in position (5) mainly as bromide, eventual radioactivity passing through is trapped in the gas trap (6).

TABLE 4.5
Depending on the γ-radiation Abundance and Energy, the Dose Rate Varies for Different Radionuclides Determining the Amount of Lead Shielding.

Activity (TBq)	^{99m}Tc	^{111}In	^{18}F	^{124}I
	Dose rate (mSv/h) at 1 m			
1	18	81	135	117
	Thickness of lead shield (cm) giving 1 uSv/h			
0.1	0.28	1	5.8	20
1	0.36	1.3	7.1	22
10	0.43	1.6	8.5	27

Note: Calculations Made with RadProCalculator (www.radprocalculator.com/)

4.7 RADIATION PROTECTION CONSIDERATIONS

Besides the desired activity, the irradiated target usually contains several other radionuclides of varying elements, half-lives, and gamma energies. The presence of such contaminants needs to be considered when planning radiopharmaceutical labelling. An example is the production of ^{35}S using the reaction $^{35}Cl(n,p)^{35}S$. At first thought NaCl would be a suitable target due to the low atomic weight of sodium, one single isotope (^{23}Na) and a salt that is easy to dissolve. However, ^{23}Na has a huge thermal neutron cross section for producing ^{24}Na, which has a half-life of 15 h and abundant gamma energies up to 2.75 MeV. This target will be extremely hot, demanding lead protection more than 30 cm thick. If KCl is used instead, the emitted gamma radiation energy will be substantially lower and decay times shorter.

After irradiation the target usually is stored before processing to allow any short-lived radionuclides to decay. Depending on the half-life, this 'cooling period' can be from minutes to months but should not exceed one half-life of the desired radionuclide. The place for this depends upon the source activity and the energy and abundance of the gamma emissions. Separation of almost pure beta and alpha emitters may need just distance and some plastic shielding and can be performed in a standard fume hood while targets with a high gamma emission need significant lead shielding.

Handling reactor or accelerator produced radioactivity in the order of several hundreds of GBq requires adequate radiation protection usually in the shape of lead shields, hot boxes, lead shielded fume hoods and laminar air flow benches. Typical lead thicknesses required by common radionuclides are indicated in Table 4.5.

The radioactive target and the radionuclide separation are often the first steps in labelling a radiopharmaceutical. The shield then must fulfil the requirements both to protect the operator from the radiation and to protect the pharmaceutical from environment contamination. The first step usually implies a negative pressure hood to prevent eventual airborne radioactivity leaking out into the laboratory, while the second step implies the high positive pressure applied across the pharmaceutical to avoid contact with less-pure air from the laboratory. These contradictory conditions are usually handled by having a box in the box, that is, the pharmaceutical is processed in a closed facility at a high pressure placed in a hot-box having a lower pressure. The classical hot-box design with manipulators to manually process the radioactivity remotely is gradually replaced by lead-protected chambers housing an automatic chemistry system or a chemical robot making the pharmaceutical computer-controlled.

To read more about radiochemistry and irradiated targets, see references [1] and [3].

REFERENCES

[1] *Manual for Reactor Produced Radioisotopes*. Vienna: International Atomic Energy Agency, 2003.

[2] A. Mushtaq, S. M. Qaim, and G. Stöcklin, 'Production of 73Se via (p, 3n) and (d, 4n) reactions on arsenic,' *International Journal of Radiation Applications and Instrumentation. Part A. Applied Radiation and Isotopes*, vol. 39, no. 10, pp. 1085–1091, 1988/01/01/ 1988, https://doi.org/10.1016/0883-2889(88)90146-3.

[3] *Cyclotron Produced Radionuclides: Physical Characteristics and Production Methods*. Vienna: International Atomic Energy Agency, 2009.

[4] *Production of Long Lived Parent Radionuclides for Generators: 68Ge, 82Sr, 90Sr and 188W*. Vienna: International Atomic Energy Agency, 2010.

5 Radiometry

Mats Isaksson

CONTENTS

5.1 RADIATION DETECTORS IN GENERAL

The term 'detector' may refer to the part that is used to register the ionizing radiation (e.g. scintillation crystal, gas-filled cavity), but also to the whole measuring system, including electronic equipment and display. These radiometric systems can be divided into simple systems that essentially measure the intensity of the incident radiation (count-rate meters), and spectrometric systems, which are able to quantify the energy deposited by photons or charged particles. Some systems can also be used to visualize the spatial distribution of the radiation intensity (e.g. gamma cameras).

The purpose of a radiation detector is to provide a response that is proportional to the amount of energy deposited in the detector material. However, the nature of the response will depend on the detector material, and three main detection principles can be identified:

- Collection of electric charges.
  - Ionization of atoms in the detector material yielding a pulse or current of free charged particles that can be detected by electronic devices.

DOI: 10.1201/9780429489556-5

- Collection of light.
    - Prompt or delayed radioluminescence, that is, reemission of imparted energy by some atomic or molecular excitation-de-excitation process. This can also provide an electric signal by means of light-to-electric charge converters such as photomultiplier tubes (PMTs) or photodiodes.
- Measuring changes of chemical properties.
    - Radiation-induced chemical reactions causing permanent changes in the detector material, for example, darkening of photographic film.

5.2 GAS-FILLED DETECTORS FOR IONIZING RADIATION

Gas-filled cavities have for a long time been used to detect ionizing radiation by collecting the electric charges liberated by interactions between the detector gas and impinging photons or charged particles. This kind of detector consists of a volume of gas, which may be either enclosed or open. The active volume may contain air or mixtures of several gases, and to increase the probability of interaction the gas may also be pressurized.

In order to collect the charges produced when the ionizing radiation interacts with the detector gas, an electric field has to be applied over the detector volume. In cylindrical detectors, a thin metal wire, the anode, is connected to the positive terminal of a voltage supply, and the wall is connected to the negative terminal (or ground), the cathode. Ionizing radiation passing through the chamber will then create a track of positive ions and free electrons. The energy needed to ionize gas molecules and create electron-ion pairs is on the order of 30 eV at normal pressure and ambient temperature.

The ions and electrons will be attracted by the electric field and migrate to the cathode and anode, respectively, if the electric field is sufficiently high. If the electric field strength is weak, < 100 V cm^{-1}, the charges will not reach the electrodes before being absorbed in the gas (recombination). The charge collected at the electrodes can either be collected continuously as a current, or as a charge pulse during a specified time. The pulse can then be further analysed as a count rate or energy distribution of the radiation particles. However, the positive charge carriers (the ions) in gaseous detectors have low mobility, and it requires a comparably long time to collect those charges, which means that the time resolution is much poorer in most gaseous detectors than in semiconductor detectors. This can be mitigated by removing this slow component, thus shortening the rise time of the pulse, by using so-called gridded ionization chambers.

The detector response is highly dependent on the electric field strength in the gas, which in turn is affected by the applied voltage (bias) and the geometry of the gas volume. If the electrodes are two parallel plates, the electric field strength will be homogeneous in the gas volume. However, if the detector consists of a cylindrical cathode with a thin central wire as the anode, the electric field will decrease radially, being highest close to the anode wire. A sufficiently high electric field strength will cause the electrons to be accelerated to kinetic energies high enough to cause secondary ionization in the gas. These secondary electrons will in turn accelerate towards the anode and cause more ionization, eventually leading to an electron avalanche, and the charge reaching the anode can be a factor of 10^6 greater than the charge created by the primary ionization in the gas.

The features above are used to divide gas-filled detectors into three categories:

- ionization chambers,
- proportional counters, and
- Geiger-Müller (GM) counters.

The ionization chamber is characterized by an electric field strength lower than what is required to cause secondary ionization, usually operating at a voltage of around 100 V. Ideally, the total charge collected at the anode equals the charge of electrons liberated by ionizing radiation in the gas volume (in reality, however, electrons may recombine with ions before reaching the anode). If the total mass of the gas is known, the absorbed dose to the gas can be calculated from the detector signal. Since the number of ion pairs produced per unit energy of the ionizing radiation is low, the resulting current is very low (on the order of pA). The detection efficiency can be increased by pressurizing the gas (or replacing it by a liquid), but these chambers will be insensitive to irradiation with charged particles due to the thicker walls that are demanded.

Increasing the electric field strength in the gas volume, electron avalanches can be formed. For some field strengths – those that are highly dependent on the detector design (typically voltages on the order of 100–1000 V) – the detector signal is proportional to the voltage. A gas-filled detector operating in this mode is called a proportional counter. The charge collected after each primary event in a proportional detector will be on the order of 100 times higher than in an ionization chamber. Proportional counters can be used as contamination-control devices and for spectrometry.

By further increasing the applied voltage, the discharge avalanches may also result in emission of ultraviolet (UV) radiation that will traverse the gas and cause other avalanches distant from the primary avalanche, eventually involving the whole length of the anode wire. This will give rise to a continuous discharge that will cease when the charge concentration of positive ions is too high to sustain a sufficiently high electric field strength to enable further secondary ionizations. Detectors operating in this mode are called Geiger-Müller (GM) counters. Thus, the GM counter is insensitive to the energy of the incoming particle and each produced pulse will be of the same magnitude. The pulses are often large enough to be recorded without amplification, and GM counters are therefore often used in portable detector systems, such as count-rate meters.

Especially in GM counters, the formation of avalanches due to interaction between the positive ions and the cathode has to be reduced. When the ion is neutralized by an electron leaving the cathode material, energy is released, which can cause the emission of other electrons from the cathode. These electrons are accelerated by the electric field and may start new avalanches. To prevent this, a quenching gas is added to the gas volume. The molecules in this quenching gas have a lower binding energy than the molecules of the main detector gas and thereby less energy is released when they interact with the cathode. For example, if argon (Ar) is used as detector gas, methane (CH_4) can be added as quenching gas. The positive Ar-ions will then ionize the methane molecules instead of picking up an electron from the cathode. When the positively charged methane molecules are neutralized at the cathode, less energy is released, and the probability of emission of extra electrons from the cathode is reduced.

5.3 SCINTILLATION DETECTORS

Luminescent materials that are irradiated with ionizing radiation emit visible light in the form of photons in the wavelength region (1–5 eV; 250–1200 nm). When ionizing radiation interacts with a luminescent material, some of its energy can be imparted to the electrons in atomic or molecular states of the material, leading to excitation to higher energy levels. When these states decay back to their original level, energy is released in the form of light (luminescence). Depending on the details of the atomic or molecular energy levels, this de-excitation may be immediate (fluorescence) or delayed (phosphorescence). Luminescence can therefore be used for direct measurements of dose rate, or measurements of cumulated exposure (radiation dose). Both organic and inorganic materials may have the property of luminescence and are referred to as scintillators (see also Chapter 6).

5.3.1 Organic Scintillation Detectors

Organic materials consist mainly of carbon and hydrogen atoms and thus have an atomic composition close to that of tissue, which is useful when the detectors are used for dosimetry. Due to the approximate tissue equivalence, the cross section for photon interactions is similar to that in water or soft tissue. Many of these materials can also be mixed with liquids and plastics, which make it possible to produce detectors that can be fairly large. For example, plastic scintillators are often utilized in portal monitors that are used to monitor passing vehicles for radioactive elements. The large detector volume will then compensate for the comparatively low counting efficiency, due to the low-Z materials in the organic molecules.

Organic scintillators usually contain aromatic carbon compounds, consisting of carbon and hydrogen. In these compounds, the carbon atoms form benzene rings, each containing six carbon atoms and a carbon atom in the ring is thus bound to two carbon atoms and one hydrogen atom. This molecule is planar, that is, the orbitals of the bound electrons all lie in the same plane with a bond angle of 120°. However, one of the electrons in the carbon L-shell does not take part in the bonds, and these six electrons form orbitals that are orthogonal to the molecular plane. These π-electrons make it possible for the benzene molecule to bind to another benzene molecule. Absorption of energy may cause excitation of a π-electron, followed by emission of light upon de-excitation.

Luminescence is thus due to the de-excitation of the π-electrons and the process can be divided into three kinds, which differ regarding the timescale between excitation and de-excitation. These three kinds of luminescence are called *fluorescence*, *phosphorescence*, and *delayed fluorescence*. The total spin angular momentum of the excited molecule can either be 0 or 1 and, hence, two separate systems of energy levels are available for the excited molecule: singlet states and triplet states (Figure 5.1). If the molecule absorbs energy from a charged particle traversing the detector material (e.g. an electron liberated by photoelectric absorption or Compton scattering), it may be excited into one of the states depicted in Figure. 5.1. However, within the order of picoseconds, electrons excited to the higher energy states will eventually de-excite to the state $S_{1,0}$ by non-radiative transitions. Fluorescence then occurs in the transition from $S_{1,0}$ or $T_{1,0}$ to $S_{0,0}$ and is characterized by prompt (on the order of nanoseconds) emission of light, the intensity of which decays exponentially with time.

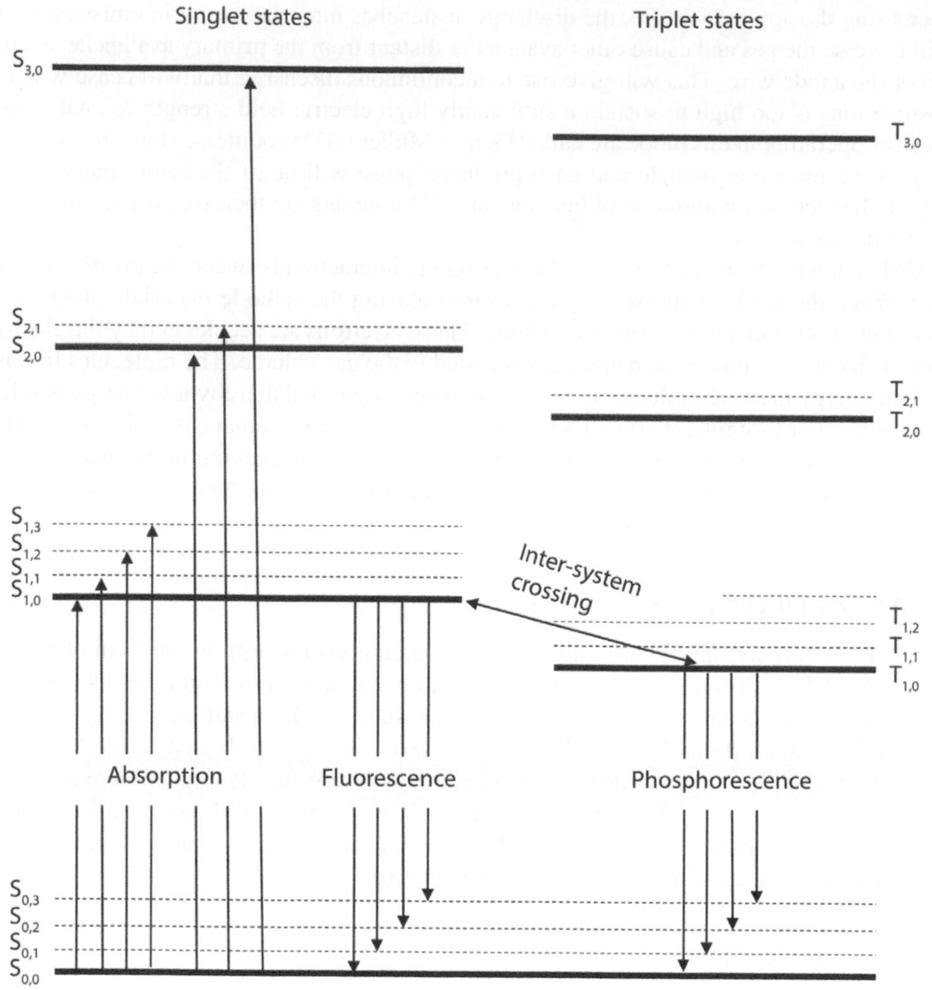

FIGURE 5.1 Molecular singlet and triplet states of a luminescent material showing possible paths of de-excitation. Electronic energy states are depicted by bold lines and thin lines depict vibrational energy states. Reproduced from [4].

Transitions from the lowest triplet state, $T_{1,0}$, to the lowest singlet energy state, $S_{0,0}$, are 'forbidden' by the quantum mechanical selection rule regarding the change of total spin angular momentum. In reality, this leads to the emission of light, phosphorescence, with a time constant of several microseconds. This is also called afterglow and can be a problem in some applications. Delayed fluorescence occurs when an electron in $T_{1,0}$ is thermally excited to $S_{1,0}$ and the luminescence occurs via the transition $S_{1,0} \rightarrow S_{0,0}$. The light that is emitted during this so-called *inter-system crossing* will have the same wavelength as the prompt fluorescent light. The processes involved in luminescence have been extensively discussed by Birks [1].

These kinds of detectors are not inherently transparent to their own light since the light emitted during luminescence may be reabsorbed. Therefore, a wavelength-shifting molecule can be added to the detector material. This molecule absorbs a luminescence photon and de-excites by emitting a photon of lower energy, that is, longer wavelength, than the luminescence light. Hence the emitted photon does not have sufficient energy to excite the scintillating material. However, self-absorption of the light might still be an issue, especially in larger detectors. The topic of transport of optical photons in organic scintillators have been discussed by Nilsson and colleagues [2].

5.3.2 Inorganic Scintillation Detectors

Inorganic materials, such as crystals of alkali metal halides (e.g. NaI, CsI), also show luminescence when irradiated with ionizing radiation. The atoms on these materials are arranged in a crystal lattice and this arrangement determines the energy states of the atomic electrons. The electrons in the so-called *valence band* are taking part in the bonds in the crystal lattice. If a sufficient amount of energy is transferred to an electron, it may become free to migrate in the crystal. These electrons are represented by the *conduction band*. Between the valence band and the conduction band is the *band gap*, representing energy states unavailable for the electrons in a pure crystal. However, by introducing impurities, so-called *activators*, in the crystal, energy states in the band gap can be made available for the electrons after being excited. Since the activator sites reside in the band gap, the energy released when these sites are de-excited back to the valence band is lower than the energy between the conduction band and the valence band. Hence, the photons emitted will lie in the visible range of the spectrum. One common activator is thallium, which is used both with sodium iodide and caesium iodide in the detector materials NaI(Tl) and CsI(Tl), respectively. CsI-detectors are also available with sodium as activator, CsI(Na).

When the crystal is irradiated by ionizing radiation, charged particles (electrons) will be liberated by ionization. A charged, and ionizing, particle will create a large number of electron-hole pairs when passing through the crystal, that is, electrons in the lattice are transferred from the valence band to the conduction band leaving a vacancy (a hole) in the lattice. The holes are also free to move within the crystal, which at first thought may seem rather strange. But consider a hole that is filled by an electron from a neighbour atom in the lattice. In effect, the hole has now moved to that neighbour atom. In this way, holes will migrate to the nearest activator site and capture the electron occupying the activator site (ionize the site) since these electrons are usually less tightly bound than electrons in the crystal lattice. The electron in the conduction band can then find an ionized activator site and neutralize it. This electron now occupies an energy state in the band gap from where it can de-excite to the valence band and produce luminescence. The half-lives of such excited states are typically 30–500 ns [3].

However, not all excitation events lead to fluorescence. The capture of the electron at the activator site may create an energy level from which transition to the ground state is forbidden. In order to de-excite, additional energy, for example, thermal energy, is required to 'lift' the electron to an energy level from which transition to the ground state is allowed. This will delay the emission of light, that is, phosphorescence. The de-excitation may also occur through radiationless transitions, in which case no luminescence is created. This effect is denoted, *quenching*.

In order for a scintillating material to be useful for spectrometry, some properties are desirable, based on density, effective atomic number, decay time, pulse rise time and photon yield:

- Efficient conversion of the kinetic energy of the charged particles into luminescence photons (high photon yield).
- A linear relation (proportionality) between the number of light photons emitted and the amount of kinetic energy imparted by the charged particles in the detector.
- The detector must be transparent to the scintillation light for efficient collection by a photomultiplier tube or photodiode.
- Prompt emission of fluorescence in spectrometers and fast count-rate meters (short decay constant).
- Low background from phosphorescence or delayed fluorescence (especially when measuring at high dose rates).
- Possible to grow the crystal material to sizes that are suitable for a γ-photon detector.

Properties of some scintillator materials are listed in Table 6.1. As an example of an organic scintillator, NE102A (a fast plastic scintillator) has a density of 1.03 g cm^{-3} and an effective Z of 4.5. The decay time is 0.0024 µs, pulse rise time (10–90%) 0.0006 µs, and the photon yield is 10000 MeV^{-1} [4]

5.3.3 Integrating Luminescent Detectors

The long-lived phosphorescence states, discussed above, is an important asset when measuring the accumulated dose to the detector. In contrast to prompt luminescent crystal materials, integrating luminescent detectors are designed to have long-lived energy states that can be deliberately de-excited by an external energy source. If the de-excitation is induced by thermal energy (heating the material), the process is called *thermoluminescence* (TL); if the excitation is induced by illumination with optical radiation, it is called *optically stimulated luminescence* (OSL). These detectors are often used as personal dosimeters to monitor occupational exposure, but also for environmental surveys or retrospective dose assessments. Figure 5.2 shows the energy levels and transitions in an integrating luminescent material.

A common material used in thermoluminescence dosimeters is lithium fluoride (LiF), which exhibits several desirable properties. The dose-response is linear or supra-linear over a wide range of absorbed doses (Table 5.1), and the sensitivity to γ-rays is high, giving detection limits on the order of a few tenths of a µGy. Also, the dose response is

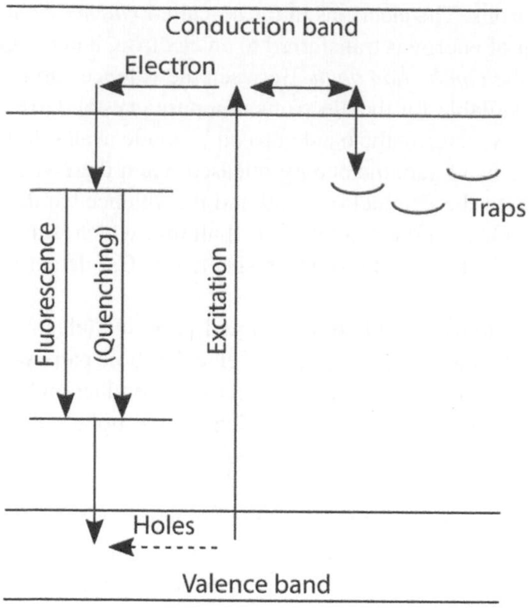

FIGURE 5.2 Illustration of the mechanism of delayed luminescence in an integrating luminescent detector material, induced by an external energy. Reproduced from [4].

TABLE 5.1
Applications of Integrating Luminescent Detectors in Personal Dosimetry, and their Typical Dose Ranges. Reproduced from [4]

Application	Dose range (Gy)	Uncertainty, 1σ(%)
Personal dosimetry	10^{-5}-0.5	-30 to +50
Environmental radiology	10^{-6}-10^{-2}	±30
Radiotherapy	10^{-1}-10^{2}	±3.5
Diagnostic X-ray examinations	10^{-6}-10	±3.5
Occupational exposure, industrial	10-10^{6}	±30

independent of the energy of the radiation and the long-lived energy states (i.e., latent luminescence) are stable. The stability of the long-lived energy states thus permits read-out of the detector signal even years after exposure.

If the detector is heated to 200–250 °C at a constant rate, the electrons occupying energy states in the band gap (electron traps) will attain sufficient energy to reach the conduction band from where they can return to the valence band at so-called recombination centres. The luminescence emitted will then be in the visible range and the light photons can be collected at the photocathode of a photomultiplier tube (PMT), see Figure 5.3a. Recording the intensity of the luminescence as a function of the temperature will result in a so-called glow curve, which is characteristic for the detector material and the absorbed dose to the detector (Figure 5.3b). The peaks in the glow curve reflect the energy required to excite an electron from a trap, that is, the temperature at which the thermal energy is sufficient to excite the electron from the trap. The half-life of the electron traps needs to be long to prevent the population of trapped electrons to decay, so-called fading. The detector should also exhibit a glow curve with at least one well-defined electron trap.

The use of TL dosimeters requires a given protocol in order to assure the quality of the dose estimations. A typical protocol for a TL-material that can withstand temperatures up to 600 °C is given below.

1. Individual calibration of each dosimeter. Although the detector material can be made highly homogeneous, the individual dosimeters vary in terms of their sensitivity, energy dependence, reproducibility and intrinsic background. Calibration is performed by irradiating each dosimeter to a well-determined absorbed dose prior to read-out.

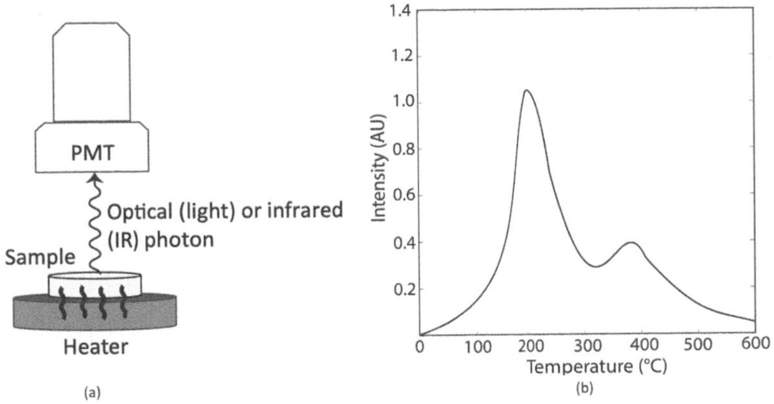

FIGURE 5.3 (a) Illustration of the read-out of a thermoluminescent dosimeter. (b) Typical glow curve for NaCl (ramping rate $\approx 2\ °C\ s^{-1}$). Reproduced from [4].

2. Annealing. Prior to use, all information from previous exposures must be erased from the dosimeter, that is, all electron traps must be emptied. This is achieved by heating the dosimeter in an oven to about 500 °C.
3. Cooling, possibly followed by calibration and annealing. If the sensitivity is not stable, steps 1 and 2 may have to be repeated.
4. Exposure. After completing steps 1–3, the dosimeter can be used as a personal dosimeter.
5. Pre-annealing. The contribution from low-energy traps to the read-out signal is less predictable than the contribution from the high-energy traps. The low-energy traps are therefore emptied before signal read-out by preheating the dosimeter.
6. Read-out. The luminescence is collected by a photosensitive device (e.g. PMT) during further heating to the optimal read-out temperature.
7. Determination of absorbed dose to the detector. The absorbed dose to the detector is calculated from the signal recorded during read-out using the calibration factor (sensitivity) obtained at calibration (step 1).

In recent years, materials that exhibit optically stimulated luminescence (OSL) have partly replaced TL dosimeters in personal dosimetry. An example of such a material is carbon-doped aluminium oxide, Al_2O_3:C [5], which can also be read out using heating. Also naturally occurring materials, for example, quartz and feldspar, can act as dosimeters and may be used in retrospective dosimetry. The registered dose can be as low as 10 mGy [6]. Integrating luminescent properties are also exhibited by materials that are commonly found in the environment, such as household salt (NaCl) and electrical components in mobile phones. Other examples are given in Table 5.2.

5.3.4 Chemical Detectors

As described in the beginning of this chapter, ionization and excitation can also cause reactions that change the chemical properties of the detector material. These changes can be utilized to assess the radiation dose, provided they can be related to the absorbed dose in a predictable and reproducible way.

One of the earliest types of detectors is photographic film, which is an example of a chemical detector. The active material of the film consists of grains of silver bromide (AgBr) embedded in a thin emulsion. This emulsion is placed on a transparent base and covered with a protective coating. Upon irradiation with ionizing radiation (or light, as in conventional photography), a latent image is created. This latent image is then made visible by exposing the film to a chemical (developing) that reduces the exposed silver halides to metallic Ag. In order to make the film insensitive to further exposure, the image is fixed by a chemical that produces soluble salts with Ag ions in the remaining silver halide crystals. These salts are then removed when the film is rinsed.

Photographic film has been extensively used in X-ray diagnostics, often in combination with a fluorescent material (e.g. $CaWO_4$) that converts the impinging photon energy to light (1–3 eV). In this way, the sensitivity of the film can be increased for X-ray photons in the energy range 10–140 keV. The combination of film and intensifying screen determines the relation between the optical density of the developed film and the absorbed dose to the film-screen

TABLE 5.2
Examples of Crystalline Materials with Known OSL Properties that can be Used for RetrospectiveD. Reproduced from [4]

Material	References (examples)
Quartz in building materials	Bøtter-Jensen et al. [6]; Thomsen et al. [7]
Household salt	Bailey et al. [8]; Bernhardsson et al. [9]; Christiansson et al. [10]; Spooner et al. [11]
Porcelain	Ramzaev et al. [12]
Mobile phone components	Beerten et al. [13]; Inrig et al. [14]; Woda et al. [15]; Bassinet et al. [16]
Dental repair material	Geber-Bergstrand et al. [17]
Dust (silicate)	Jain et al. [18]
Fingernails and toenails	Sholom et al. [19]
Banknotes and coins	Sholom and McKeever [20]
Salted snacks and nuts	Christiansson et al. [21]
Desiccants	Geber-Bergstrand et al. [22]

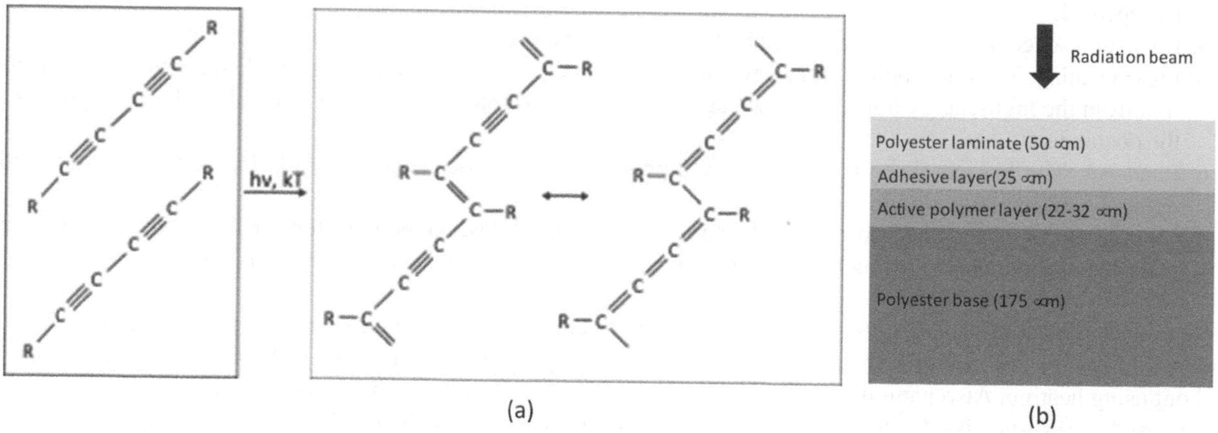

FIGURE 5.4 (a) Polymerization of monomers caused by irradiation. (b) Film structure in second-generation radiochromic film used, for example, for dose verification of diagnostic and therapeutic radiation beams. These films have also found use in environmental radiology as they change colour depending on the absorbed dose. Reproduced from [4].

system. However, most film detectors in X-ray diagnostics are now replaced by solid-state devices (digital imaging detectors). Also, in personal dosimetry films begin to be obsolete and are replaced by TL or OSL devices.

Another kind of film detector is radiochromic film. These contain a photosensitive emulsion that does not require processing or dedicated read-out equipment. These are used, for example, for measurements of dose distribution in radiotherapy. The chemical process caused by exposure to ionizing radiation is, in this case, polymerization of diacetylene, see Figure 5.4. Upon polymerization, the colour of the film will change in proportion to the accumulated energy imparted to the material.

In the 1920s, Fricke and Morse [23] developed a method to determine radiation dose, based on ionization (i.e., oxidation) of ferrous ions (Fe^{2+}) that produces ferric ions (Fe^{3+}). The presence of ferric ions in the solution causes the colour of the solution to change, and the absorbed dose can be determined by measuring the optical density. If ferrous sulphate is mixed in a gel [24], the distribution of ferric ions in the exposed gel will give three-dimensional information of the dose distribution. The exposed gel can be analysed using magnetic resonance tomography, utilizing that the magnetic properties of the gel change upon oxidation. However, the ferric ions formed tend to diffuse in the gel, thereby altering

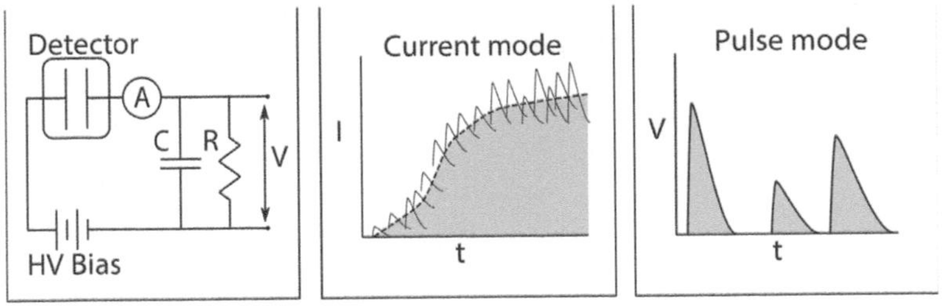

FIGURE 5.5 Schematic illustration of an ionization detector. Middle: In current mode the signal is the current of free charges created in the detector. Right: In pulse mode each pulse is amplified and presented separately. Reproduced from [4].

the initial three-dimensional dose distribution. To increase the stability, gel dosimeters based on polymerization of monomers can be used.

5.4 SPECIFIC FEATURES OF RADIATION DETECTORS

5.4.1 Time and Energy Resolution

The signal from the individual events taking place in the detector material can be analysed either by collecting the voltage pulses resulting from the charges liberated in the detector (pulse mode) or by integrating the charges (current mode, see Figure 5.5). The individual pulses, shown in the right panel of Figure 5.5, are characterized by a rise time and a decay time. The time between pulses is then determined by the dose rate and by the time constant of the detector electronics. If the detector is intended for spectrometry or as a survey meter (e.g. search instrument), both the detection time and pulse-processing time (including amplification of the pulse) need to be short in order to prevent pile-up, that is, the pulses need to be registered separately. The detection time, in this sense, is the combination of the time required to generate charges or luminescence photons in the detector material and the time required to collect the charges produced.

During the time required for the electronics to process an incoming pulse, the detector system is unable to process yet another pulse, and detector systems designed for spectroscopy handle this by increasing the acquisition time. If the operator has decided that the acquisition should last, for example, 3600 seconds, the actual time for acquisition of the spectrum may instead be 3800 seconds to allow the detector to be able to register pulses during a total time of 3600 seconds. The former, pre-set, time is called *live time* and the latter, which depends on the dose rate, is called *real time*. The time during which the detector system was unable to process pulses is referred to as *dead time*.

The value read-out from instruments designed for dose-rate measurements (dose-rate meters) is often a mean value from many consecutive pulses. This time-averaging of the signal is performed with the aim to reduce the statistical uncertainty of the measurement value and is determined by the *integration time*. The integration time should not be confused with the pulse-processing time, since the former is usually programmed in the detector software. For some instruments, it can be changed by the operator. The integration time could be several seconds long, which makes these kinds of instruments unsuitable for operation in environments where the dose rate could change significantly with time or position.

Energy resolution is a term used in spectrometry to describe the ability of the detector system to present the signals from incoming particles as a function of their energies (see also Chapter 6). A good energy resolution thus makes it possible to distinguish between particles whose energies are close to each other. The number of charge carriers produced when ionizing radiation interacts with the detector material in a solid-state or luminescent detector depends on the activator levels in the band gap. In a solid-state detector these energies are on the order of a few eV, while they are on the order of some tens on an eV in luminescence detectors. For example, the photon yield of NaI(Tl) is 38000 photons MeV^{-1}; it thus requires about 30 eV to generate a luminescence photon. In Ge, the energy required to generate an electron-hole pair is about 3 eV. Hence, ideally, the Ge-detector should be able to distinguish between photons whose energies differ by 3 eV.

The energy resolution of a spectrometer, where the result of a measurement is presented as a spectrum (pulse height distribution), is usually given as the full width at half maximum (FWHM) of a given peak in the spectrum (Figure 5.6), expressed as a percentage of the peak energy. Modern spectrometry systems based on solid-state detectors (e.g. Ge) have an energy resolution of about 0.1–0.2 per cent.

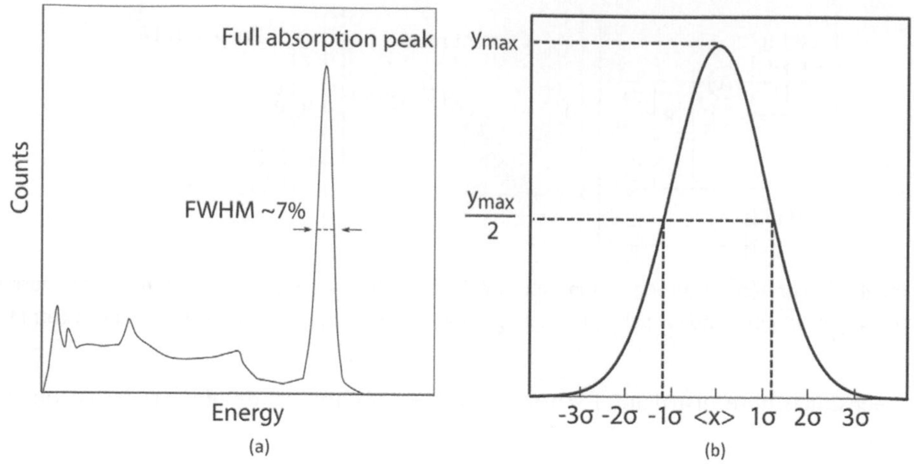

FIGURE 5.6 Typical pulse height distribution (a) recorded using a gamma detector (NaI(Tl) crystal), showing the Gaussian-shaped energy peak (b) with its centroid at the energy <x> and a full width at half maximum (FWHM) of 2.35 σ. Reproduced from [4].

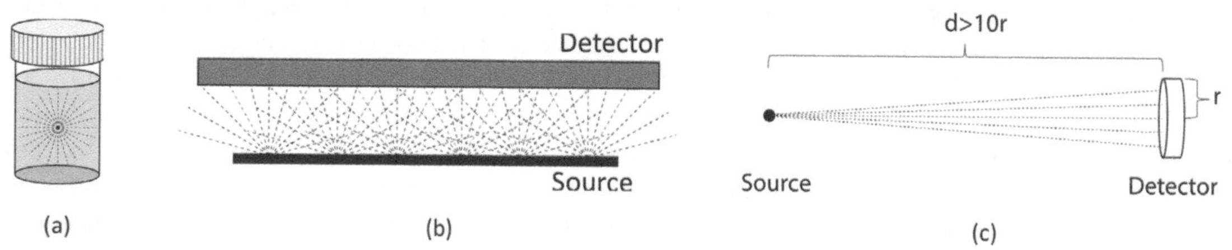

FIGURE 5.7 (a) 4π geometry with the sample dissolved in a liquid scintillator cocktail. (b) 2π geometry with a sample deposited on a surface parallel to a planar detector. (c) Point source geometry where the sample is at a distance at least ten times greater than the radius of the detector. Reproduced from [4].

5.4.2 Efficiency and Energy Dependence; Tissue Equivalence

The efficiency can generally be defined as the fraction of the emitted particles from a radiation source that results in a registration in the detector. For charged-particle detection, where almost all of the particle energy is deposited in the detector, the efficiency is often defined as the fraction of the emitted particles that deposit all of their energy in the detector. Liquid scintillation counting of alpha- or beta-emitting sources often has nearly 100 per cent efficiency, since the sample is dissolved in the scintillation material (Figure 5.7a). Figure 5.7b shows the situation in alpha spectrometry where the sample is distributed as a thin layer on a metal disc. In this case, nearly 50 per cent of the emitted alpha particles will be detected.

In γ spectrometry, total absorption of the photons in the detector is rarely the case. The efficiency is then a measure of the number of pulses recorded by the detector system per unit fluence rate from the source, at the detector surface and for a given photon energy. This quantity is called the *intrinsic efficiency*, and it is assumed that the source is sufficiently far from the detector so that the photons are normally incident on the detector, as shown in Figure 5.7c. However, for γ spectrometry, mostly the *absolute efficiency* is used. This quantity relates the *emission rate* of photons from the source to the number of pulses in a given full energy peak in the spectrum. The absolute efficiency thus depends on the actual source-detector configuration and also on photon energy. Measuring the absolute efficiency at various photon energies will result in an efficiency calibration curve, from which the activity of the measured sample could be calculated from the number of pulses in the full energy peak and the number of photons emitted per decay of the radionuclide in question.

The energy dependence of a detector is of concern not only in spectrometric applications. Also, for non-spectrometric instruments, such as dose and dose rate meters, it is important to consider. The factors mainly affecting the energy dependence of photon detectors are the thickness of the detector casing, atomic composition of the detector material and the detector size. The detector casing will cause attenuation of low energy photons, resulting in a low efficiency at

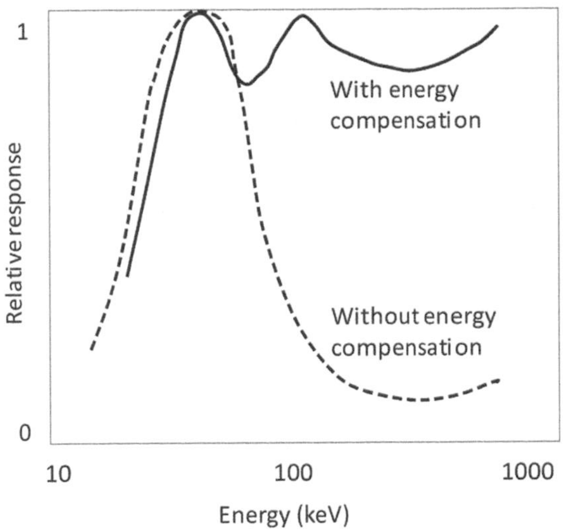

FIGURE 5.8 Schematic illustration of the energy response of an uncompensated GM tube and an energy-compensated GM tube. Reproduced from [4].

these energies. A detector designed for measuring dose rate will therefore underestimate the dose rate from low-energy photons, unless this is compensated for. For example, so-called energy compensated GM tubes are equipped with a metal filter to increase the attenuation of medium and high energy photons as well (Figure 5.8).

The atomic composition will determine the probability of interaction between the photons and the detector material, as well as which interactions will take place. Due to the decrease in interaction probability with photon energy, the efficiency decreases with photon energy. Detector size, and geometrical configuration, will therefore affect the efficiency since increasing the distance travelled by a photon will increase the probability of interaction in the detector material.

The absorbed dose from photons to the detector material depends on the energy fluence, that is, the radiation energy impinging on the detector per unit area, and the amount of that energy imparted in the detector material, expressed as the mass energy absorption coefficient, μ_{en}/ρ. The latter is energy dependent and therefore the absorbed dose is given as the integral over all energies present in the radiation field (E_γ), multiplied by $\mu_{en}/\rho(E_\gamma)$. The mass energy absorption coefficient also depends on the element composition, in terms of atomic number, Z, and detectors designed for dose measurements must be composed of materials with an effective atomic number, Z_{eff}, close to that of human tissue ($Z_{eff} \approx 7.8$). Air-filled detectors ($Z_{eff} = 7.6$) can be used for photons in the energy-range 60 keV to 3 MeV. However, care should be taken when using solid detectors, usually calibrated at 661.6 keV photons (^{137}Cs), since the higher atomic number ($Z > 14$) could lead to overestimation of the absorbed dose from photons of lower energy than those used at calibration. The exposure to photon energies higher than the calibration energy will, for the same reason, be underestimated.

5.4.3 Background and Radiation Quality

Measurements with a radiation detector is rarely performed in an environment where background radiation is absent. Cosmic radiation and ionizing radiation from naturally occurring radionuclides (e.g. ^{40}K and radionuclides in the U and Th decay series) will inevitably contribute to the detector signal. In addition, thermal excitation of charge carriers in the detector material will also contribute to pulses that are not related to the actual source being measured.

Excitation of charge carriers could also occur due to mechanical disturbances that cause vibrations in the detector material. These charge carriers will then contribute to the detector signal – for example, when recording spectra during mobile measurements. The detector signal could also be affected by defects in the detector and variations in the high voltage applied to permit collection of the charges liberated in the detector. In gamma-ray spectrometry, the full energy peaks are superpositioned on a background consisting of Compton scattered photons in addition to the environmental background.

In many applications, the detector system is designed to monitor one kind of radiation, for example, gamma-ray spectrometry. However, some applications require that more than one radiation quality can be monitored simultaneously.

This can be achieved by using two or more detectors in the same instrument housing – for example, combining a GM tube for gamma-ray detection with a Ag-doped zinc sulphide, ZnS(Ag), detector for detection of alpha particles.

The detector material itself could also be made sensitive to different radiation qualities. Thermoluminescence (TL) dosimeters of LiF could be enriched in ^{6}Li to increase the cross section for neutron interactions in the material by the ^{6}Li(n,α) reaction. Natural lithium contains 7.4 per cent ^{6}Li but in these materials it can be enriched to almost 96 per cent. Thus, ordinary LiF are sensitive to neutrons but this is further enhanced by the enrichment. On the other hand, separating out ^{6}Li, the other stable lithium isotope ^{7}Li can be enriched to almost 100 per cent, which makes the material less sensitive to neutrons. This is desirable when the dose from γ-photons is of importance, or when measuring the dose from both neutrons and photons in a mixed radiation field. The latter case requires two detectors, one that is insensitive to neutrons and one that has approximately the same sensitivity for neutrons and photons [25].

Enhancement of the sensitivity to a specific radiation quality could also be achieved geometrically. In proportional counters filled with ^{3}He gas, the cross section for neutrons is rather high due to the ^{3}He(n,p)^{3}H reaction, while the low density of the gas gives a low probability for photon interaction.

In liquid scintillation counting, the luminescence produced by an alpha particle will last longer than for a beta particle, for a given amount of deposited. This can be utilized in the pulse-shaping electronics to distinguish between events caused by alpha and beta particles. Another way to distinguish between these radiation types is the *phoswich* (phosphor sandwich) detector. These detectors are often found in hand-held instruments designed for simultaneous alpha and beta particle detection and are a combination of two scintillators, where the alpha particles are registered by a thin inorganic scintillator (e.g. ZnS(Ag)) and the beta particles are registered by a plastic scintillator. The luminescence emitted from the scintillators is then filtered to enable a discrimination between the light generated by alpha and beta particles, respectively.

5.4.4 General Features

Radiation detectors are used in a wide range of environments, ranging from prearranged laboratories to the harsh environment in outer space. Reliability demands that a detector must be able to work in the environment in which it is assumed to be used. This also applies to the electronics used for read-out of the detector signal.

Especially in outdoor use, humidity and moisture must be controlled since the crystals are often hygroscopic, for example, NaI(Tl). The crystal must therefore be protected by a cover that can withstand moisture. Temperature variations may, as previously noted, affect the excitation of electrons in the detector material, but can also be an issue for photomultiplier tubes and other kinds of light-to-charge converters.

As discussed above, mechanical disturbances that cause vibrations in the detector material could lead to spurious pulses. More severe mechanical stress could cause the detector crystal to crack. During the last decades, triggered by the 9/11 terrorist attacks in the United States in 2001, several kinds of portable detector systems intended for mobile measurements have been developed. These spectrometric detector systems were previously found almost only in prearranged laboratory environments but have now been equipped with protective covers in order to withstand outdoor environmental conditions and mechanical stress. The use of liquid nitrogen for cooling of high-purity germanium detectors have also been replaced by electrical cooling (e.g. Upp [26]).

Ageing of a detector system should also be considered since over time this will change the response. This effect could be due to radiation-induced damage or to the ageing of some components. Especially exposure to neutrons may cause changes in the crystal lattice, leading to reduced energy resolution. Solid-state detectors that are operated under vacuum conditions may lose vacuum due to ageing of guard rings losing their elasticity, which may also lead to moisture building up in the cavity between the casing and the detector crystal. The efficiency, especially at low photon energies (<100 keV), of these detectors may also be reduced if the detector is not kept cooled to 77 K. Temperature cycling between room temperature and cooling, may lead to a slow increase in the insensitive outer layer of the detector crystal (dead-layer), thereby increasing the attenuation of photons reaching the sensitive parts of the detector crystal.

Other examples of ageing phenomena are contamination of the detector from samples or the environment, degradation of light-transmitting materials connecting a scintillation detector and photomultiplier tube and increased quenching in liquid scintillators due to light-induced chemical reactions. All the ageing effects mentioned above will affect the performance of the detector, but can, in some cases, be remedied. A quality-control program is nevertheless necessary to assure the accuracy of the measurements.

5.5 EXAMPLES OF DETECTOR CONFIGURATIONS

5.5.1 Neutron Detectors

The cross section for interactions between neutrons and matter are highly energy-dependent, and a single detector material cannot be used in the whole energy range. Another complication is that the neutron energy fluence spectrum will change when the neutrons are traversing a medium. For detector applications, three interaction processes are of major interest: elastic scattering, nonelastic scattering, and neutron capture.

Elastic scattering between the neutron and light atomic nuclei, for example H, leads to a recoiling proton that causes ionization or luminescence in the detector material. Nonelastic scattering, in which the kinetic energy of the system is not conserved, leads to nuclear reactions, for example, the emission of protons or alpha particles. These charged particles will then act as ionizing particles in the material. The neutron may also be captured by the atomic nucleus, which will then emit a γ-photon. The energy of these photons depends on the nucleus and is, for example 2.2 MeV in the $^{1}H(n,\gamma)^{2}H$ reaction and 10.8 MeV in the $^{14}N(n,\gamma)^{15}N$ reaction.

The detection efficiency for neutrons of a wide range of energies is often improved by surrounding the sensitive detector volume by a moderating material in which the neutrons undergo elastic collisions that decrease their velocity to correspond to thermal energy (0.025 eV). However, the moderation will cause the information of the initial neutron energy to be lost. Examples of detector materials that can be used are solid plastic scintillators, liquid scintillators, and gas-filled ionization or proportional counters.

The sensitivity to photons must be taken into account when designing a neutron detector, since several of the neutron interaction processes will lead to the emission of photons. Given below are some factors that affect the relative counting efficiencies for photon and neutron interactions (Table 5.3).

- Passive shielding of gamma rays, for example, a lead shield will effectively absorb gamma photons without significantly absorbing fast (>1 MeV) neutrons.
- Some materials, for example, He- and B-based materials, have considerably different cross sections for thermal neutron absorption and gamma-ray interaction.
- The secondary particles generated by the interactions of neutrons are heavy charged particles, whereas photon interactions produce mainly electrons. The volume and pressure in gas-filled detectors can therefore be chosen to match the effective path length of the heavy charged particles.

TABLE 5.3
Examples of Neutron Detectors and their Detection Efficiency. Reproduced from [4]

Detector type	Size	Neutron active material	Incident neutron energy	Neutron detection efficiency[1)] (%)	γ-ray sensitivity[2)] ($mGy\ s^{-1}$)
Plastic scintillator	5 cm	^{1}H	1 MeV	78	$2.8 \cdot 10^{-5}$
Liquid scintillator	5 cm	^{1}H	1 MeV	78	$2.8 \cdot 10^{-4}$
Loaded scintillator	1 mm	^{6}Li	thermal	50	$2.8 \cdot 10^{-3}$
Hornyak button	1 mm	^{1}H	1 MeV	1	$2.8 \cdot 10^{-3}$
Methane (7 atm)	5 cm diam.	^{1}H	1 MeV	1	$2.8 \cdot 10^{-3}$
^{4}He (18 atm)	5 cm diam.	^{4}He	1 MeV	1	$2.8 \cdot 10^{-3}$
^{3}He (4 atm), Ar (2 atm)	2.5 cm diam.	^{3}He	thermal	77	$2.8 \cdot 10^{-3}$
^{3}He (4 atm), CO_2 (5%)	2.5 cm diam.	^{3}He	thermal	77	$2.8 \cdot 10^{-2}$
BF_3 (0.66 atm)	5 cm diam.	^{10}B	thermal	29	$2.8 \cdot 10^{-2}$
BF_3 (1.18 atm)	5 cm diam.	^{10}B	thermal	46	$2.8 \cdot 10^{-2}$
^{10}B-lined chamber	0.2 mg cm^{-2}	^{10}B	thermal	10	2.8
Fission chamber	2.0 mg cm^{-2}	^{235}U	thermal	0.5	$2.8 \cdot 10^{3}$

1) Interaction probability of neutrons impinging at right angles.
2) Approximate upper limit for usable output signals.

Source: Data from Crane and Baker [29].

- Differentiation between the signal from gamma rays and neutrons can be achieved by using pulse-shape analysis in the subsequent amplification.

Instruments used for the detection of neutrons within radiation protection are primarily designed for qualitative measurements, that is, to detect the presence of neutrons in the environment. Large gas-filled ^{3}He chambers could be used, for example, in vehicle-borne monitoring of the environment, or for searches of sealed radiation sources. These tubes can be configured as proportional counters and enclosed in a moderating material for the detection of for example, ^{239}Pu.

A combination of ^{6}LiF and ^{7}LiF tablets is commonly used in personal dosimetry at nuclear power plants. This combined detector will then be sensitive to neutrons, as well as to photons, which enables estimation of the dose contribution from neutrons in a mixed radiation field.

Hand-held instruments used for searches of radiation sources could be equipped with semiconductors containing an element with a high thermal neutron capture cross section. An example is adding Cd to ZnTe to form the detector material CdZnTe.

5.5.2 Electronic Personal Dosimeters

Personal dosimeters that continuously display the dose rate, as well as the radiation dose, have in recent years became common as a supplement to personal dosimeters based on TL or OSL. The latter materials are, however, still common since legislation often requires that a personal dosimetry system should have been approved for use as an official record of personal radiation doses.

The detector part of the dosimeter may consist of a semiconductor, often a PIN diode (see also Chapter 7). This silicon diode is a combination of a p-type and a n-type semiconductor with an intrinsic region in between. The intrinsic region is the radiation sensitive region and is made wider in the PIN diode, compared to an ordinary diode, in order to increase the sensitivity.

MOSFET (metal–oxide–semiconductor field-effect transistor) dosimeters is an alternative to TLD in radiation therapy. Irradiation of the MOSFET increases the intrinsic voltage shift due to trapped charges liberated by the radiation. These changes can persist for a very long time before read-out. MOSFET may also be used for direct reading of the dose rate.

5.5.3 Activity Calibrators (Dose Calibrators)

Dose calibrators are commonly used for verifying the activity of radiopharmaceuticals prior to administration. They are generally well-type cylindrical gas-filled detectors, operating as ionization chambers. The vial or syringe containing the radioactive source is placed on a so-called dipper and inserted into the well during measurement in order to simulate a 4π geometry. To further increase the efficiency, the detector is filled with a highly pressurized gas (for example Ar). Thus, the exposure rate, which is measured by the detector current, can be converted to activity.

The relation between exposure rate and activity depends on the energy and yield of the emitted radiation, that is, on the radionuclide in the sample. It also depends on the squared distance between sample and detector, and accurate positioning of the sample is therefore crucial. Dose calibrators are typically pre-calibrated for a number of radionuclides commonly encountered in nuclear medicine applications. However, corrections may be necessary if the geometry of the vials or fill volume deviate from the geometry and volume used when calibrating the dose calibrator.

In order to obtain a reliable measurement result, a systematic quality assurance protocol, as exemplified below, should be employed.

- Reproducibility – tested by repeated measurements with, for example, ^{99m}Tc of clinically relevant activity, where the dipper is lifted from the detector well between each measurement; the deviation between individual measurement results is compared to specifications.
- Constancy – tested by day-to-day measurements using, for example, a ^{57}Co or ^{137}Cs source to ensure that variations are within allowed limits (e.g. ±10%).
- Accuracy – test of the dose calibrator's ability to reproduce the activity of a radiation source with a certified and traceable activity.
- Linearity – test of the dose calibrator's ability to give accurate activity readings for different activities, within the specified limits; tested by varying the activity in, for example, a vial with ^{99m}Tc.

5.5.4 Whole-body Counters

Assessment of internal radiation dose requires knowledge of the incorporated activity, as well as of the radionuclide(s) in question. In occupational radiation protection, the radionuclide may be known from previous experience but in case of other emergencies, this may not be the case. Thus, the activity in the body and the type of radionuclide have to be determined by *in vivo* or *in vitro* measurements. *In vivo* measurements refer to estimation of the activity in the body, while *in vitro* measurements refer to analyses of, for example, excretion samples.

In vivo measurements can be made with various kinds of detector systems. For example, iodine in the thyroid can be measured by collimated NaI- or HPGe-detectors, but also with hand-held dose rate meters, provided they have been properly calibrated. However, measurements of the activity in the whole body are usually performed by some kind of whole-body counters, which consist of dedicated detector systems with heavy shielding for background radiation. The term 'counters' is still in use although modern detector systems are built on detectors and electronics for spectroscopy.

Stationary whole-body counting is usually performed in low-activity laboratories, which are highly shielded to optimize the signal-to-background. The shielding material has to be free from inherent radionuclides, such as ^{214}Bi in concrete, ^{60}Co in steel and ^{210}Pb in lead, and therefore have to be chosen carefully. The walls of the laboratory could typically be made from concrete containing low-activity sand. In this room, as an example, a steel chamber made of 10 cm-thick steel (smelted before 1945) with inner walls clad in old lead, houses the detector(s). The main purpose of the steel is to attenuate cosmic radiation, and the lead will attenuate secondary radiation caused by cosmic rays interacting with the steel. Build-up of radon daughters in the laboratory can be supressed by forced ventilation, causing an overpressure, and filtering the incoming air. Measurement times are typically 5–20 minutes, giving detection levels of 40 Bq for ^{137}Cs.

Whole-body counters used for occupational radiation protection are often calibrated for the most-commonly encountered radionuclides at the workplace. However, if other radionuclides are present, a specific calibration for each of these is required. The calibration factor will also depend on the geometry of the subject, that is, body weight and height, and calibration of whole-body counters are therefore performed using so-called phantoms made from a tissue-equivalent material and which mimic a human body. The availability of radiation sources is then a limiting factor, and whole-body counters are more and more calibrated by mathematical methods.

A mathematical calibration of a whole-body counter starts with a mathematical model of the detector system. This system can consist of stationary mounted scintillation or solid-state detectors, or else be set up to permit the detectors to move during measurement (scanning-bed system), see Figure 5.9. Then, a mathematical model of the subject is made, which 'contains' the radionuclides to be used in the calibration, as well as the distribution of the radionuclides in various body organs [27]. Such models for the reference man and reference woman have been developed by ICRP [28]. The detector response is then simulated by Monte Carlo methods.

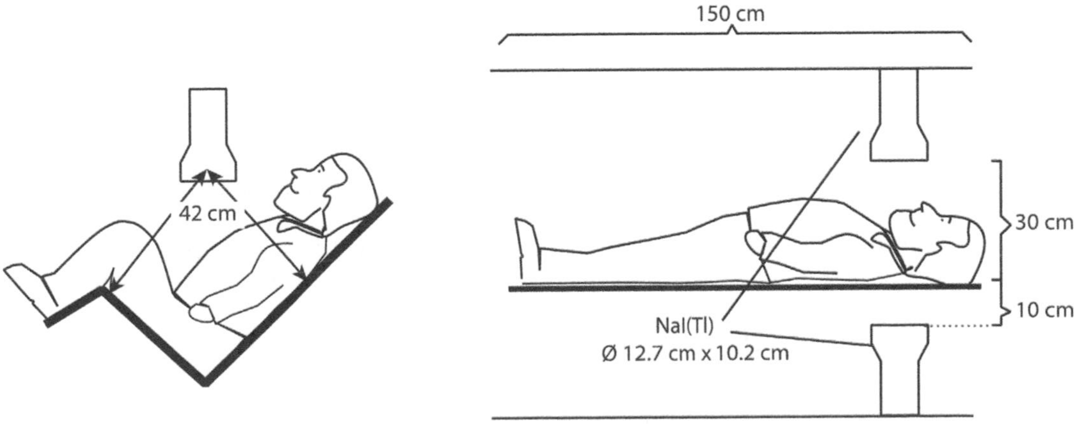

FIGURE 5.9 Left: Patient reclining in a chair with the gamma detector positioned so that the torso and legs are equidistant from the centre of the detector volume. Right: A dual detector scanning bed geometry with the patient lying between the upper and lower detectors. This system is less affected by inhomogeneous radionuclide distribution in the body. Reproduced from [4].

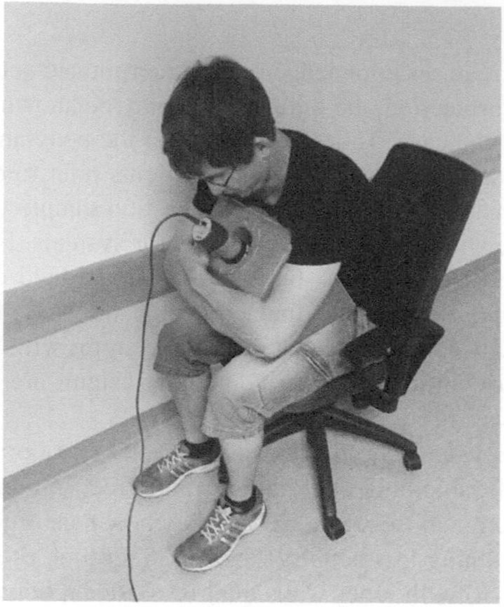

FIGURE 5.10 Mobile *in vivo* γ-ray spectrometry using the so-called Palmer geometry. Reproduced from [4].

Mobile *in vivo* measurements can be made using either shielded systems carried by, for example, a truck, or in an unshielded geometry. Commonly used for the latter case is the so-called Palmer-geometry, in which the subject leans over the detector (NaI or HPGe), see Figure 5.10. This kind of measurements has been used in screening of exposed individuals, for example, following the atmospheric nuclear weapons tests and the Chernobyl accident.

REFERENCES

[1] J. B. Birks, *The Theory and Practice of Scintillation Counting*. Oxford: Pergamon Press, 1964, p. 453.

[2] J. Nilsson, V. Cuplov, and M. Isaksson, "Identifying Key Surface Parameters for Optical Photon Transport in GEANT4/GATE Simulations," *Applied Radiation and Isotopes: Including Data, Instrumentation and Methods for Use in Agriculture, Industry and Medicine*, vol. 103, pp. 15–24, 05/22 2015, doi:10.1016/j.apradiso.2015.04.017.

[3] G. F. Knoll, *Radiation Detection and Measurement*. New York: Wiley, 2000.

[4] M. Isaksson and C. L. Rääf, *Environmental Radioactivity and Emergency Preparedness* (Series in medical physics and biomedical engineering). Boca Raton: CRC Press, Taylor & Francis Group, 2016.

[5] L. Bøtter-Jensen, N. Agersnap Larsen, B. G. Markey, and S. W. S. McKeever, "Al2O3:C as a Sensitive OSL Dosemeter for Rapid Assessment of Environmental Photon Dose Rates," *Radiation Measurements*, vol. 27, no. 2, pp. 295–298, 1997/04/01/ 1997, https://doi.org/10.1016/S1350-4487(96)00124-2.

[6] L. Bøtter-Jensen, S. Solongo, A. Murray, D. Banerjee, and H. Jungner, "Using the OSL single-aliquot regenerative-dose protocol with quartz extracted from building materials in retrospective dosimetry," *Radiation Measurements*, vol. 32, pp. 841–845, 12/01 2000, doi:10.1016/S1350-4487(99)00278-4.

[7] K. J. Thomsen, L. Bøtter-Jensen, A. S. Murray, and S. Solongo, "Retrospective Dosimetry Using Unheated Quartz: A Feasibility Study," *Radiation Protection Dosimetry*, vol. 101, no. 1–4, pp. 345–348, 2002, doi:10.1093/oxfordjournals.rpd.a005998.

[8] R. Bailey, G. Adamiec, and E. J. Rhodes, "OSL Properties of NaCl Relative to Dating and Dosimetry," *Radiation Measurements*, vol. 32, pp. 717–723, 12/01 2000, doi:10.1016/S1350-4487(00)00087-1.

[9] C. Bernhardsson, M. Christiansson, S. Mattsson, and C. Rääf, "Household Salt as a Retrospective Dosemeter Using Optically Stimulated Luminescence," *Radiation and Environmental Biophysics*, vol. 48, pp. 21–8, 10/01 2008, doi:10.1007/s00411-008-0191-y.

[10] M. Christiansson, C. Bernhardsson, T. Geber-Bergstrand, S. Mattsson, and C. Rääf, "Household Salt for Retrospective Dose Assessments Using OSL: Signal Integrity and Its Dependence on Containment, Sample Collection, and Signal Readout," *Radiation and Environmental Biophysics*, vol. 53, 05/09 2014, doi:10.1007/s00411-014-0544-7.

[11] N. Spooner, B. Smith, D. Creighton, D. Questiaux, and P. Hunter, "Luminescence from NaCl for Application to Retrospective Dosimetry," *Radiation Measurements*, vol. 47, pp. 883–889, 09/01 2012, doi:10.1016/j.radmeas.2012.05.005.

[12] V. Ramzaev and H. Göksu, "Cumulative Dose Assessment Using Thermoluminescence Properties of Porcelain Isolators as Evidence of a Severe Radiation Accident in the Republic of Sakha (Yakutia), Russia, 1978," *Health Physics*, vol. 91, pp. 263–269, 10/01 2006, doi:10.1097/01.HP.0000215837.68118.d3.

[13] K. Beerten, F. Reekmans, W. Schroeyers, L. Lievens, and F. Vanhavere, "Dose Reconstruction Using Mobile Phones," *Radiation Protection Dosimetry*, vol. 144, pp. 580–583, 11/01 2010, doi:10.1093/rpd/ncq343.

[14] E. L. Inrig, D. Godfrey-Smith, and C. L. Larsson, "Fading Corrections to Electronic Component Substrates in Retrospective Accident Dosimetry," *Radiation Measurements*, vol. 45, pp. 608–610, 03/01 2010, doi:10.1016/j.radmeas.2009.12.032.

[15] C. Woda, I. Fiedler, and T. Spöttl, "On the Use of OSL of Chip Card Modules with Molding for Retrospective and Accident Dosimetry," *Radiation Measurements*, vol. 47, pp. 1068–1073, 11/01 2012, doi:10.1016/j.radmeas.2012.08.012.

[16] C. Bassinet et al., "Retrospective Radiation Dosimetry Using OSL of Electronic Components: Results of an Inter-laboratory Comparison," *Radiation Measurements*, vol. 71, 03/01 2014, doi:10.1016/j.radmeas.2014.03.016.

[17] T. Geber-Bergstrand, C. Bernhardsson, S. Mattsson, and C. Rääf, "Retrospective Dosimetry Using OSL of Tooth Enamel and Dental Repair Materials Irradiated Under Wet and Dry Conditions," *Radiation and Environmental Biophysics,* vol. 51, 09/13 2012, doi:10.1007/s00411-012-0434-9.

[18] M. Jain, C. Andersen, L. Bøtter-Jensen, A. Murray, H. Haack, and J. C. Bridges, "Luminescence Dating on Mars: OSL Characteristics of Martian Analogue Materials and GCR Dosimetry," *Radiation Measurements*, vol. 41, pp. 755–761, 08/01 2006, doi:10.1016/j.radmeas.2006.05.018.

[19] S. Sholom, R. Dewitt, S. Simon, A. Bouville, and S. McKeever, "Emergency Optically Stimulated Luminescence Dosimetry Using Different Materials,", *Radiation Measurements*, vol. 46, no. 12, pp. 1866–1869, 2011, doi:10.1016/j.radmeas.2011.03.004.

[20] S. Sholom and S. W. S. McKeever, "Emergency OSL Dosimetry with Commonplace Materials," *Radiation Measurements*, vol. 61, pp. 33–51, 2014, https://doi.org/10.1016/j.radmeas.2013.12.008.

[21] M. Christiansson, T. Geber-Bergstrand, C. Bernhardsson, S. Mattsson, and C. L. Raaf, "Retrospective Dosimetry Using Salted Snacks and Nuts: A Feasibility Study," (in English), *Radiation Protection Dosimetry*, vol. 174, no. 1, pp. 1–5, Apr. 20, 2017, doi:10.1093/rpd/ncw044.

[22] T. Geber-Bergstrand, C. Bernhardsson, M. Christiansson, S. Mattsson, and C. L. Rääf, "Desiccants for Retrospective Dosimetry Using Optically Stimulated Luminescence (OSL)," *Radiation Measurements*, vol. 78, pp. 17–22, 2015/07/01/ 2015, https://doi.org/10.1016/j.radmeas.2014.11.002.

[23] M. Fricke and S. Morse, "The Chemical Action of Roentgen Rays on Dilute Ferrous Sulphate Solutions as a Measure of Radiation Dose," *The American Journal of Roentgenology Radium Therapy and Nuclear Medicine*, vol. 18, 1927, pp. 430–442.

[24] J. C. Gore, Y. S. Kang, and R. J. Schulz, "Measurement of Radiation Dose Distributions by Nuclear Magnetic Resonance (NMR) Imaging," *Physics in Medicine & Biology,* vol. 29, no. 10, pp. 1189-97, 1984, doi:10.1088/0031-9155/29/10/002.

[25] F. H. Attix, *Introduction to Radiological Physics and Radiation Dosimetry*. New York: Wiley, 1986.

[26] D. Upp, R. Keyser, and T. Twomey, "New Cooling Methods for HPGE Detectors and Associated Electronics," *Journal of Radioanalytical and Nuclear Chemistry*, vol. 264, pp. 121–126, 03/01 2005, doi:10.1007/s10967-005-0684-y.

[27] P. Cartemo, J. Nilsson, M. Isaksson, and A. Nordlund, "Comparison of Computational Phantoms and Investigation of the Effect of Biodistribution on Activity Estimations," *Radiation Protection Dosimetry*, vol. 171, no. 3, pp. 358–364, 2016, doi:10.1093/rpd/ncv415.

[28] ICRP, "Adult Reference Computational Phantoms," in 'ICRP Publication' *Annals of the ICRP*, vol. 110, 2009.

[29] T. Crane and M. Baker, "Neutron Detectors," in *Passive Nondestructive Assay of Nuclear Material*, D. Reilly, N. Ensslin, and H. Smith (Eds), pp. 379–406 1991.

6 Scintillation Detectors

Per Roos

CONTENTS

6.1 INTRODUCTION

Scintillation as a phenomenon to detect ionizing particles is one of the oldest methods. A high fluence of ionizing radiation can be seen directly as a glow when it interacts in the eye-bulb. This has been reported by several scientists in the early days of radioactivity and X-rays but also by members of space missions and accelerator scientists. The phenomenon has been summarized by Steidley [1].

Scintillation as a dedicated detection method played a central role both for Wilhelm Röntgen using scintillation screens of barium platinocyanide in the discovery of X-rays and for Ernest Rutherford in his formulation of a new atom model. Rutherford's data were largely based on observations made by Hans Geiger and Ernest Marsdon who, with great patience, observed the faint scintillations from a zinc sulphide screen bombarded by alpha particles deflected in a gold foil. At the time no technique was available to amplify the weak light flashes, so the human eye and a completely dark room were the prerequisites for these human scintillation counters. As Birks [2] states in his introduction chapter to *The Theory and Practice of Scintillation Counting* – these were truly men of vision.

The development of gas ionization chambers and associated electronic circuitry in the 1930s eventually made these men obsolete since it provided the users with a possibility not only to more precisely quantify the individual ionization events, but the measurements could also be performed at much higher count rates than what the human eye could

DOI: 10.1201/9780429489556-6

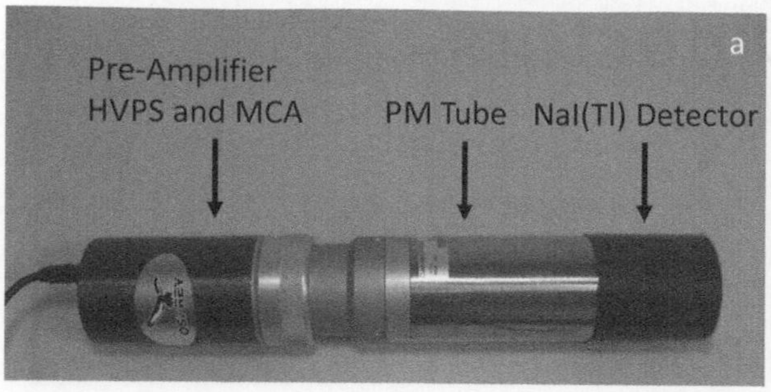

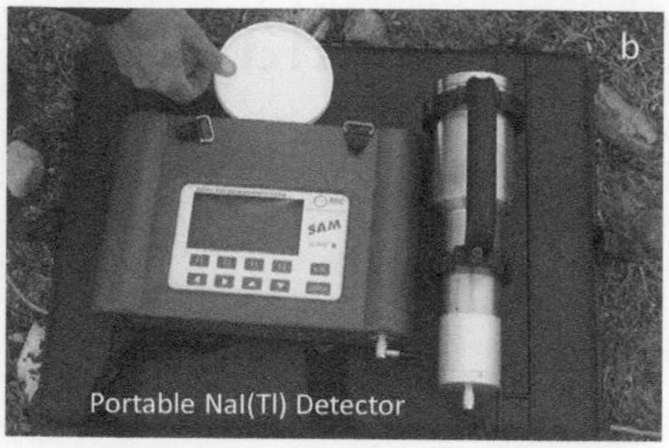

FIGURE 6.1 Two examples of NaI(Tl) scintillation detectors. To the left a 2"x2" NaI(Tl) detector connected to a PM tube and a module with integrated high voltage power supply (HVPS), amplifier and MCA ready to use following connection to the usb-port of a computer. To the right a 3"x3" NaI(Tl) for outdoor use with a battery-powered unit providing HVPS, amplifier, MCA, and a spectrum display.

accomplish and, more importantly, this development could detect not only alpha particles but also the weaker ionizing β-particles. Prior to 1930, scintillation screens of barium platinocyanide or artificially made willemite (a natural highly fluorescent zinc silicate mineral), were used as alternatives to activated zinc sulphide for visual scintillation counting. The disadvantage of these scintillators was that they were only available as small crystals or crystalline powders and the ZnS(Ag) scintillator had a severe shortcoming in that it could only be used in thin layers since it is almost opaque to its own scintillation light. The scintillations originating from weakly ionizing radiations on ZnS(Ag) could thus not be seen even by a trained human eye.

In the mid-1940s several groups [3–7] proved the possibility of converting the weak scintillation light, usually from ZnS-screens, to readable electric pulses by coupling the scintillator to a photomultiplier tube, thus enabling a revival for scintillators as radiation detectors. Kallmann in 1947 [8] and Deutsch in 1948 [9] showed that scintillations from gamma-irradiated naphthalene-blocks could be detected using photomultipliers, naphthalene thus being the first organic scintillator and the first large-volume scintillator. Eventually Hofstadter [10] discovered the (nowadays standard) thallium-doped sodium iodide crystal following numerous studies of the luminescent properties of the thallium ion in alkali halides. Although much focus on scintillation radiation detectors traditionally is on the properties of the scintillator material, it is important to remember that the development of the light-amplification techniques (PM tubes and photodiodes) also have played a crucial part in enabling today's radiation detection systems. Figure 6.1 shows two examples of modern scintillation detector designs (NaI(Tl)).

6.2 SCINTILLATION

The mechanisms by which various materials receive energy that is later emitted as visible light has given rise to a large number of names of phenomena, such as thermoluminescence (energy received due to thermal action and emitted differently than by black-body radiation); chemoluminescence (energy received through chemical reactions); or the common photoluminescence (energy received from visible or UV light). Radioluminescence or cathodoluminescence are phenomena whereby visible light is emitted following excitation by charged particles or only by electrons (cathodoluminescence). The overall term 'scintillation' roughly refers to the flashes of light that appear from a scintillator during de-excitation from a higher energy level and may thus to most readers be nearly identical to radioluminescence, a fluorescence or phosphorescence due to the absorbed energy from ionizing radiation. Scintillations are generally thought of as being the near prompt, fluorescent light, while the fraction of energy emitted as phosphorescence and delayed fluorescent light is at a minimum. Scintillation light may thus be treated as a subgroup of radioluminescence where timing of the emitted light is important. There may not be a complete consensus as to whether scintillation and radioluminescence cover the same physical phenomena, but this is out of the scope of this chapter. For a detailed description of the kinetics and mechanisms of scintillators please refer to [2] and [11].

The nature of scintillation differs between organic and inorganic materials. While the emissions in organic scintillators are due to the energy levels of individual molecules, the mechanism in inorganic solids are due to the energy levels of the crystalline structure. The physical state (gas, liquid, solid) of organic scintillators thus has little influence on the scintillator properties. For inorganic scintillation crystals the energy levels involved in the light emission must be contained in the forbidden energy band in order to avoid being reabsorbed, these are materials said to have a sufficiently large Stokes shift (difference in energy between absorbed and emitted energy) and the crystal is then said to contain luminescence centres.

Luminescence centres can be either intrinsic or extrinsic. Intrinsic luminescence centres can be due to defects in the lattice or due to the energy levels of individual atoms which possesses radiative transitions between an excited and lower energy state. In either case the transitions must be in the forbidden energy band. An example of a scintillator having intrinsic luminescence centres is bismuth germanate (BGO) where energy transitions of the Bi^{3+} atom is responsible for the scintillation light. Also, both undoped NaI and CsI act as intrinsic scintillators, although with weak emission. An example of the importance of the dopant ions in an extrinsic doped scintillator is the classical ZnS doped with Ag, Mn, or Cu, which then yields light emissions in blue, yellow, and green respectively [3]. The position of the activator energy levels relative to the valence and conduction band of the lattice is thus very important and determines both light yield, wavelength, and timing properties of the radioluminescence. Generally, the dopants are foreign ions that substitute for the cations or anions of the host material.

The interference between energy levels of the various luminescence centres, the crystal, and the Coulomb interaction of charged particles generally creates a more complex light emission in solid inorganic crystals compared with the light emission from organic molecules or single atoms. In the latter the emission is therefore in general much faster, follows a single exponential decay and with much less phosphorescence (popularly called *afterglow*) or delayed fluorescence. Increased phosphorescence can also be seen in organic molecules irradiated with high-LET particles due to the more complex damage of the individual molecules. The amplitude of the light emission from scintillators is frequently characterized by a time constant implying an exponential decline. The complex coupling between energy states of luminescence centres and the lattice in a solid inorganic crystal often creates a strong non-exponential decay, but it is nevertheless common to use time constant(s) to characterize the emitted light. This regulates to a large degree the choice of activator elements that can be used in a given matrix. The activator should also show charge stability to ensure stable energy levels during irradiation. This could for instance rule out the use of Fe as activator if the risk of oxidation ($Fe^{2+} \rightarrow Fe^{3+}$) or reduction ($Fe^{3+} \rightarrow Fe^{2+}$) exist during irradiation.

The scintillation process is usually characterized by three steps, (1) conversion, (2) energy transfer to luminescence centres and (3) luminescence. In the conversion step all the energy in the ionizing radiation event is transferred to the crystal until no electrons are left to ionize any atoms. In the transfer step the large number of electrons in the conduction band and holes in the valence band migrate through the material until their energy is transferred (phonon transport) to the luminescence centres in the forbidden energy band. Finally, in the last step, recombination of the electrons and holes in the luminescence centres generate the scintillation event. In order to be a fast emission, it should be an allowed transition of the electric dipole such as the 5d-4f transition for some rare-earth elements. The total time to transfer the energy to the luminescence centres ('rise time') is usually in the order of some picoseconds (ps) while the relaxation process can be from several ps to many μs or even seconds if phosphorescence is present. The transfer step (2) is normally considered the one most critical for the losses and trapping of electrons/holes due to the presence of lattice defects, ions in interstitial positions, boundaries, impurities, or traps generated due to the ionizing radiation (Frenkel defects). The fabrication process, the environment during fabrication, and the irradiation history may all cause such traps and thus influence the energy transport in a given scintillation crystal. Trapped electrons/holes usually relax their stored energy for a time long after the initial fluorescence or scintillation emission, and if this appears through radiative processes it is classified as phosphorescence or delayed fluorescence. The wavelength of the latter is identical to the scintillation light while it is longer for the phosphorescence light. Mechanisms to overcome afterglow phenomena and to speed up the fluorescence have been the subject of a large part of research around scintillator materials.

In order for a given scintillation material to work as a detector material for ionizing radiation – in this case primarily for gamma and X-ray photons – it is worth considering the important properties of such a material.

- The material should convert the kinetic energy of ionizing radiation into detectable light with a high scintillation efficiency (high light output per absorbed energy).
- High density and atomic number.
- This conversion should be linear – the light yield should be proportional to deposited energy over as wide a range as possible.
- The medium should be transparent to the wavelength of its own emission for good light collection.
- The decay time of the induced luminescence should be short so that fast signal pulses can be generated.
- The proportion of energy emitted as phosphorescence or delayed fluorescence should be minimal.
- The material should be of good optical quality and subject to manufacture in sizes large enough to be of interest as a practical detector (in case used for gamma or X-radiation).

- Its index of refraction should be that of glass (about 1.5) to permit efficient coupling of the scintillation light to a photomultiplier tube or other light sensor.
- The emitted light should be of wavelengths suitable for a light-amplification device such as a photomultiplier or a photodiode.
- Good mechanical stability to enable it to be used in rough environments (e.g. vibrations).
- Insensitive to radiation damage.
- Simple and cheap means of production.
- Minimum dependence of light yield and timing on temperature.

Several scintillator materials fulfil nearly all these requirements but are still not considered commercial materials due to manufacturing problems making it complicated to produce larger crystals, which is why the most common detector materials are not necessarily the same as the best scintillator materials. Requirements on a scintillator also depend on the objective. For some applications it may be an advantage to have a LET dependent light yield but in others not.

6.3 CLASSIFICATIONS OF INORGANIC SCINTILLATORS

During the past twenty to thirty years the field of scintillation detectors has expanded almost exponentially. From the early use of ZnS(Ag) and $CaWO_4$ during the investigations of Röntgen, Rutherford, and Crooke, there are now more than a hundred known scintillators that could be used as ionizing radiation detectors, although less than 10 per cent of these have become available commercially. The rapid development of scintillator materials during the last thirty years is mainly due to the requirements from high energy physics – for example, the CERN Large Hadron Collider (LHC) and the Crystal Clear Collaboration – but also to the need for improved detectors in the medical field (PET, SPECT, CT). The majority of scintillators developed during this period has been Ce^{3+} and Pr^{3+} doped materials due to their strong luminescence and fast decay. There are a number of reviews and books covering the development [11–20]. In order to provide an overview of the field it is useful to divide the development in phases, as suggested by Weber [13] and Dorenbos [12]. Figure 6.2 shows how these phases have appeared with time.

The first phase was the early use of ZnS, $CaWO_4$ and barium platinocyanide screens as scintillators by Röntgen, Crookes, Rutherford, and many others. The second phase was triggered by the introduction of the photomultiplier tube in the mid 1940s and the discovery of the NaI(Tl) [10, 21] and CsI(Tl) [22] scintillators in the beginning of the 1950s. In the third phase, crystal growth from high-melting point materials could be handled, and the use of rare earth materials (REE) as luminescence centres was initiated. The use of REE elements (mainly Ce and Eu) as dopants in various types of materials expanded heavily during the fourth phase, and a number of lanthanide-activated halides were discovered. The technique of co-doping (see below) and dedicated bandgap engineering improved various properties in already-existing scintillation materials and improved manufacturing technologies brought forward garnet compounds and ceramic scintillators. The fundamental understanding of the scintillation phenomena through both experimental and theoretical work enabled adjusting fundamental properties of already-existing scintillators through so called bandgap engineering. Current research (the fifth phase) focuses, from a medical point of view, on further improving energy resolution and developing very fast scintillators for TOF-PET.

A selection of some of the most common scintillator materials in medical service, together with their main characteristic, is given in Table 6.2. A great number of scintillator materials have been produced in the last decades, but very few have achieved becoming commercial products. Some scintillators in the table are only commercially available as relatively small crystals (few cm size), this is often due to crystal growth problems or that they are not sufficiently transparent for their emitted light (e.g. SrI_2:Eu). It is not straightforward to compare the various scintillators by just using the numbers given in the table. This goes, for instance, when comparing the light yield, which is the total integrated number of photons emitted during decay (apart from afterglow). A very fast scintillator emitting 10,000 photons in 1 µs may thus show the same temporal flux as a scintillator emitting 100,000 photons in 10 µs. When fast electronics with short shaping time are used the signal strength can thus be stronger with a fast scintillator with low light yield than with a slow scintillator having strong yield. The light yield and decay time should therefore be viewed in parallel; this is especially important when considering coincidence techniques where the time gate should be short, yet with high luminescence. Another important parameter is the wavelength of the emitted light, of which only that of maximum emission is tabulated. The wavelength distribution of the emitted scintillation light can be quite wide, for LuAG for instance the max emission is at 535 nm but emission is from 475–800 nm [www.advatech-uk.co.uk/luag_ce.html]. The light yield depends on how it is measured and, more importantly, on the size of the crystal. Several scintillators are not entirely transparent to their own emission. A classic example is ZnS(Ag), which has among the highest light outputs of all

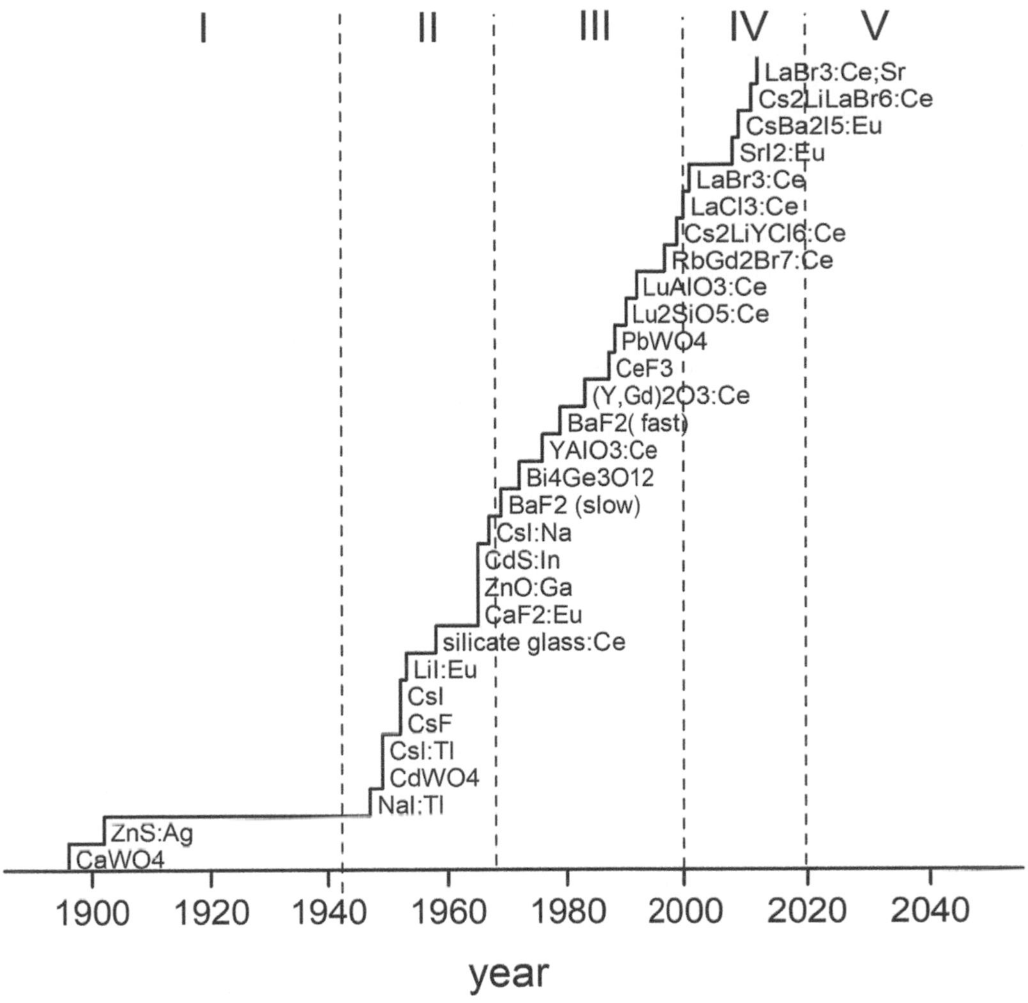

FIGURE 6.2 The various phases in the development of scintillation detectors. Reproduced from [12] with permission from the author.

scintillators, but which is almost opaque to the emitted light and therefore would have a tabulated light yield extremely dependent on size. As a consequence, energy resolution depends also on the detector size. Other parameters not listed in the table can be very important, the degree of afterglow perhaps being the most important. Other parameters are usually temperature stability and mechanical stability and ability to cleave the crystal material into thin planes and cost and manufacturing techniques as well as the ability to grow large crystals.

Scintillators can be classified in various ways. By functionality, speed, material morphology or their use, and so forth. Since the scintillation properties are very dependent on the composition of the inorganic material they are made of, the overview below of the various types of scintillators is divided accordingly.

6.3.1 Tl-doped Alkali-halides

The Tl-doped NaI and CsI have been studied in detail and the NaI(Tl) have for many years acted as standard for inorganic scintillation detectors in a number of fields. NaI(Tl) has an exceptional brightness (38,000 photons per MeV), a maximum emission at 415 nm and a refractory index of 1.85, both of which reasonably well fits to a standard bialkali photomultiplier tube with a glass (quartz) window. NaI(Tl) can be made relatively straightforward by mixing about 1 per cent Tl-powder with a NaI salt, heating it under vacuum to about 650 °C or above to allow for melting and mixing and then using a seed crystal to form a single-crystal ingot during cooling (Bridgman method [23]). The manufacturing of very large sizes (e.g. gamma cameras) at reasonable prices and a relatively good energy resolution at 662 keV (^{137}Cs) of about 6 per cent have been the major reasons for its popularity. As a radiation detector it is not without drawbacks: the material is highly hygroscopic and mechanically unstable and the light response to ionizing radiation is not perfectly linear. The fluorescent light has a decay time of around 230 ns but some 10 per cent of the total emission is emitted much later (phosphorescence or 'afterglow'). The light emission is highly temperature-dependent, with respect to both intensity and decay time. The abovementioned drawbacks combined with only a moderately high density (3.67 g/cm3) and average atomic number (51) have been the reasons for a search for other scintillator materials better suited for gamma spectrometry. The cousin CsI(Tl) is much less fragile and has both a higher density (4.51 g/cm3) and a slightly higher average atomic number (54), thus providing a better peak to Compton ratio than NaI(Tl). The light yield is higher (65,000 photons/MeV) but at a longer wavelength than the NaI(Tl) emission, which reduces the number of photoelectrons produced in a PM tube. The added non-linearity, similar to that of NaI(Tl), makes the energy resolution somewhat poorer than the corresponding energy resolution for NaI(Tl). The decay time of the emitted light is very long (~μs), and several per cent of the emitted light appears as phosphorescence or delayed fluorescence ('afterglow'). CsI is, however, a scintillator in widespread use due to its excellent material properties, high light output and low cost. The development of silicon photodiodes as photon detectors with a significantly better efficiency for longer wavelength offered the CsI(Tl) a position in early CT-scanners. However, with the increased scanning speed there followed a demand to suppress the residual afterglow, and the CsI(Tl) detectors were gradually exchanged for other scintillator materials. Due to the long decay time (several tens of milliseconds) for such afterglow the background intensity in a given detector continues to build up during a CT-scan. This results in a gradual poorer contrast and reconstruction artifacts in the image [24, 25].

A specific feature of CsI is the cubic crystalline structure, which allows fabrication into sub-millimeter microcolumnar films [26] which, together with large position sensitive SiPM arrays or CCD/microchannel image sensors, can be used to construct high-resolution digital X-ray screens or small field of view (SFOV) gamma cameras for SPECT [27]. The scintillator layer, made of a forest of thin (few μm) CsI(Tl) needles, can be grown directly from the vapour phase and is then optically coupled to a SiPM or a CCD/microchannel image sensor to be used in static or dynamic imaging [28, 29]. Each CsI(Tl)-needle is not only a scintillation detector but also a light guide that restricts the crosstalk between the CsI(Tl) columns, thus enabling a high spatial resolution. It is important to realize the need for scintillators with a high density and atomic number as well as good energy resolution to achieve high spatial resolution. The high density and atomic number maximize full energy absorption for a given detector volume, thus minimizing scattered Compton photons which, together with a poor energy resolution, otherwise would result in a poor spatial resolution. No other scintillation detector than CsI(Tl) can be constructed with the same number of crystals per unit area, thus providing a high spatial information [29].

6.3.2 Self-Activated Oxides

BGO, $CdWO_4$ and $CaWO_4$ (Scheelite – a natural mineral) constitute a group of scintillators that are attractive due to their high density and average atomic number. Their performances have been well studied, [30, 31]. The $CaWO_4$ scintillator was introduced as early as 1896 as a powder phosphor for X-ray detection in the first fluoroscope by Thomas Edison; the $CdWO_4$ in 1950 [30]; and the BGO in 1973 [31]. They are intrinsically doped in the sense that the WO_4^{2-} and the Bi^{3+} acts as luminescence centres in the crystals and thus have energy levels in the forbidden band gap. The two main advantages with these three scintillators are their very high intrinsic detector efficiency and very low phosphorescence (afterglow), which makes them suitable for use in CT where the irradiation intensity may change quickly during a scan. These self-activated oxides have further advantages in that they are mechanically rugged, not hygroscopic, have acceptable luminescence changes with temperature, and are only moderately sensitive to radiation damage. The BGO-scintillator has an acceptable long decay time and has been a popular detector in PET scanners because a high peak to Compton ratio can be obtained even for small detector crystals at medium gamma energies. If the spatial resolution in imaging detectors is to improve when making smaller detector elements (smaller pixel size), the peak to Compton, or at least the intrinsic detector efficiency, must increase accordingly. Otherwise, reduction of the detector element size may result in the opposite direction and deteriorate the spatial resolution since the detector elements mainly will experience Compton scatter to neighbouring detector cells. The $CdWO_4$ and $CaWO_4$ materials have a relatively long decay time (μs), which limits their use in PET scanners, where the coincidence time window should be small to minimize background. All three materials have a relatively low light output (10–20% relative to NaI(Tl)) and high refractive index, which results in a relatively poor energy resolution but which is of less importance in CT. The relatively poor energy resolution lowers the peak to Compton ratio, which affects the BGO negatively when used in PET. New detector materials with faster decay have gradually begun to replace these detectors both in CT and PET.

6.3.3 Ce-doped REE-halides and Eu-doped SrI_2

Most common detectors in this group are the $LaCl_3$:Ce, $LaBr_3$:Ce, $CeBr_3$ and SrI_2:Eu, all known for their extremely high luminescence and excellent energy resolution. For a long time it has been known that the rare earth elements (REE) show a very strong fluorescence (e.g. Tomaschek & Deutschbein [32]). It was thus not unexpected that they should be part of the search for new scintillator materials. In the mid-1990s attempts to use REE's as activator in various crystals were initiated, and it was soon realized that some Ce-activated halides like $LaCl_3$ and $LaBr_3$ showed an exceptional brightness [33] and an outstanding energy resolution of about 3 per cent together with a fast decay. This triggered an intensive phase in trying to find new scintillators and a large number of primarily Ce^{3+} doped materials were tested over the following decade [12]. Ce^{3+} was selected due to its relatively large Stokes shift when added to REE-halide crystals, and because the allowed 5d-4f transition is very fast. This allowed for very fast decay time and little self-absorption of the emitted light even in large crystals. In order to find detector materials with even better energy resolution, studies have been conducted to try Eu^{2+} doping of SrI_2 and BaI_2 [34, 35] or more general on Eu^{2+} doped alkaline earth halides [36]. Most of these resulted in strong luminescence and, in particular for SrI_2:Eu, an excellent energy resolution (3–4%), at least for small crystals. Unfortunately, the energy resolution becomes degraded with larger crystals due to a small Stokes shift in the Eu^{2+}-emission. There is thus a re-absorption/re-emission process which results in a fraction of the photons being permanently lost [37]. The same process also results in a decay time that is dependent on crystal size. Absorption of the emitted scintillation light also takes place in Ce-doped materials, but at a much lower probability, yet can be seen to increase with heavily Ce-doped materials. The $CeBr_3$ scintillator can be considered a 100 per cent doped $LaBr_3$:Ce material and the total light yield is reduced to about 70 per cent compared to a $LaBr_3$ doped with some mole per cent Ce. The choice between $LaBr_3$:Ce and SrI_2:Eu, both with excellent energy resolution, depends largely on the application. The $LaBr_3$:Ce has a better energy resolution and can be grown at larger sizes, but its most severe drawback is the inherent radioactivity from ^{138}La (0.089% abundance, half-life ~10^{11} y) and if not using very pure lanthanum salts in the manufacturing step also other long-lived rare earth elements will be present (e.g. ^{176}Lu). The purity of the ingredients was a problem in the first $LaBr_3$:Ce detectors where abundant ^{227}Ac with daughter isotopes could be present. This made low-level counting practically impossible. The advantage of SrI_2:Eu is that it is practically free of inherent radioactivity and thus would be a better choice for low-level counting if detectors could be grown at larger size. $LaBr_3$:Ce has its main advantage in measurements of relatively high activities or as anticoincidence guard where both detector size, energy resolution and fast decay is of importance. Both materials are very hygroscopic, which makes production expensive. The $LaBr_3$:Ce in addition suffers from ease of cracking during production unless cooling is properly allowed. This makes this material even more expensive to produce.

6.3.4 Ce-doped REE-oxyorthosilicates

This is one of several groups of oxide scintillators developed in attempts to replace other traditional scintillators, primarily CsI(Tl) and BGO having either significant afterglow or relatively slow decay, respectively. The REE-oxyorthosilicates (REE_2SiO_5) used in these materials almost exclusively consist of the rare earth elements Y, Gd or Lu, and the dopant is almost always Ce, although both Y_2SiO_5:Ce and Y_2SiO_5:Tb (YSO:Ce and YSO:Tb) have been developed. The choice of matrix materials and dopant ions is determined by the energy levels such that the luminescence centres (dopants) have energy levels in the forbidden band gap. The group has several attractive properties such as very fast decay, relatively good scintillation yield, little or no afterglow and, in addition, an emission wavelength and refractory index that relatively well match demands by most PM tubes. The Gd- and Lu-base materials also have high z_{eff} and densities. The decay time has been found to depend on the concentration of the Ce-dopant.

The material furthermore has a very good radiation resistance [38] which is of importance in CT where high doses can be accumulated in the detector elements. Both GSO and LSO have over the past two decades gradually replaced BGO as the preferred scintillators in PET. A comparison of several scintillator materials aimed for PET can be found in [39] Although GSO has a lower z_{eff}, and density than BGO, it is attractive in PET due to its much faster decay time (90% of the fluorescence in 56 ns compared to 300ns for BGO). In spite of having about twice the luminosity of BGO, the energy resolution is about the same due to severe non-proportionality of GSO [40, 41]. A comparison of the performance of YSO, GSO and LSO is presented in [42]. Perhaps the material of greatest interest in this group has been the LSO, discovered in 1992 [43], due to its potential in replacing BGO. It has approximately the same high linear attenuation coefficient as BGO but has a much faster decay making it attractive in PET scanners where it was first introduced in a small-animal Siemens PET scanner in 2001 [44]. LSO:Ce has a light yield, approximately 75 per cent that of NaI(Tl) and roughly 4 times that of BGO, but the energy resolution of LSO:Ce and BGO are similar, around 8–10 per cent, due to the severe non-proportionality of the LSO:Ce scintillator. By co-doping with Ca^{2+} it has been shown that both decay time and light yield can be improved [45, 46]. The major drawback with LSO, apart from its severe energy non-linearity, is the inherent radioactivity due to ^{176}Lu (2.6% abundance, 3.8 10^{10} y). Characteristic X-ray escape is not negligible from such high-Z materials, but the poor energy resolution does not enable a distinct escape-peak to be visible at the low-energy side of the full absorption peak although the energy difference is about 63 keV.

For the purpose of improving properties like decay time, scintillators made of mixed REE-oxyorthosilicates like LYSO ($Lu_{1.9}Y_{0.1}SiO_5$) and LGSO ($Lu_{0.4}Gd_{1.6}SiO_5$) have been developed. The two materials, both with high density and z_{eff} have sufficient different decay times such that they can be used as phoswich-detectors in PET, thus improving spatial resolution.

6.3.5 Ce-doped REE Aluminium Perovskites

This is the aluminium oxide version of the oxyorthosilicates; they have in general higher densities, higher indices of refraction, shorter decay times, shorter peak emission wavelengths, and lower light yields than their orthosilicate cousins. YAP:Ce ($YAlO_3$:Ce) is famous for being one of very few scintillation materials showing almost perfect energy linearity (see below). With an energy resolution around 4 per cent it is an interesting alternative to NaI(Tl) in applications not requiring high density or z_{eff} materials, for example, small dimension imaging detectors using 140 keV photons. The YAP:Ce is neither fragile nor hygroscopic and can relatively easily be machined in suitable sizes for instance to be used in array-detectors for positions sensing or as charged particle detectors [47]. It is also one of the fastest oxide scintillators. It has been observed that the nonproportionality of YAP is correlated to the Ce-concentration in the material [48], higher Ce-concentrations correlated both with improved nonproportionality and faster decay time.

The heavier LuAP:Ce ($LuAlO_3$:Ce) is a material which, just like LSO and LYSO above, is of interest in PET. LuAP:Ce has the highest linear attenuation coefficient of all scintillator materials, even larger than BGO, and with a decay time of only 17 ns it could be a very attractive scintillator material in PET. The LuAP has however had difficulties competing with the other Lu-based scintillators. Possibly due to manufacturing problems, the performance of LuAP has varied between suppliers. While some have reported poor energy resolution, low light output in combination with self-absorption of its scintillation light, and non-uniform crystals, some have reported energy resolutions of around 7 per cent in combination with excellent nonproportionality [49]. The difficulty in producing high quality crystals has resulted in a high price for this material. The mixed LuYAP:Ce (Lu,Y)AlO_3:Ce scintillator has been developed to improve phase stability during crystallization and thereby enable a commercial product [50].

6.3.6 Ce-doped REE Aluminium Garnets

The last class of oxide scintillators presented here is the cerium doped REE aluminium garnets. These garnets have earned their fame through the well-known YAG-laser ($Y_3Al_5O_{12}$:Nd). The YAG:Ce and LuAG:Ce materials have a distinct difference to the other Ce-doped oxide scintillators in that their emission spectrum is significantly shifted to longer wavelengths and, therefore, with Ce as dopant are less suitable to be used with ordinary PM tubes. The YAG:Ce has excellent mechanical properties that allow it to be cut in thin sheets. With the longwave emission it couples well to silicon photodiode detector readouts and is therefore common in thin position sensitive screens and cathode ray tubes. Research in particular in the LuAG material in the last decade has expanded the understanding of this material. Early reports on LuAG revealed a large fraction of slow, unexploitable, scintillation light [51] although the initial, bulk, light emission takes place with a fast component of about 20 ns. Thus, the steady-state light output from LuAG using infinite collection times (current mode) was found to be 7–8 times that of BGO, but the detector signal in pulse mode when using shaping times of a few μs was only about 1.5 times that of BGO. Shallow electron traps associated with misplaced Lu-atoms in the lattice were found to cause a delay for the migrating electrons on their way to the Ce^{3+} centres, resulting in a slow decay component – a problem also seen in YAG:Ce. Attempts to use Pr^{3+} as activator rather than Ce^{3+} shifted the emission from around 520 nm to around 310 nm, thus enabling connection to standard PM tubes, but with the same slow component in decay. Also light yield was reported to be somewhat lower than Ce-doped LuAG but due to improved energy-proportionality the energy resolution in Pr-doped LuAG could reach down to excellent 4.6 per cent as compared to 6–7 per cent in LuAG:Ce. Continued experiments in which Ga^{3+} was added to create $Lu_3(Ga_xAl_{1-x})_5O_{12}$:Pr resulted in creating radiation-free transitions for the trapped electrons and therefore almost completely removing the slow component in the light decay [52] but still with only an emission about 1.5 times that of BGO. The original LuAG was further developed as Ce^{3+} or Pr^{3+} doped mixed garnets of Lu^{3+}, Gd^{3+} and Ga^{3+} was manufactured, for example, $Lu_2Gd_1Ga_2Al_3O_{12}$:Ce or low-activity materials free of Lu like $Gd_3Al_2Ga_3O_{12}$:Ce or the GAGG:Ce scintillator. These scintillators all show very attractive attributes with light yields 5–8 times that of BGO, energy resolutions of 4–6 per cent, fast decay times, are mechanically robust and are non-hygroscopic. Their major disadvantage is that the single-crystal versions of these materials require expensive manufacturing. A review on the garnet scintillators can be found in [52].

6.3.7 Manufacturing

One major reason for that only a fraction of the many scintillator materials developed during the last decades has become commercially available is the cost of manufacturing. Manufacturing methods are perhaps not the first thing that comes to mind when comparing detector materials but the technology chosen to produce the various scintillators determines both their quality and cost and therefore the chances to compete with other materials on the market.

A rough comparison of manufacturing costs of various single-crystal scintillators is shown in Table 6.1.

Both the cheap halide scintillators NaI(Tl) and CsI(Tl) are usually grown by a directional solidification method, meaning the melted material in a sealed quartz ampoule under vacuum gradually moves from a hot zone into a colder one where solidification starts once a seed crystal is in place. The growth of the crystals in sealed ampoule under vacuum is an advantage with the hygroscopic NaI(Tl). Crystal growth of the halides and all the oxide materials can also be done by the well-known Czochralski method [53], where a seed crystal placed on a rod is dipped into the melt and then slowly pulled upwards to bring the growing crystal with it while rotating. The major disadvantage with this method is that only a fraction of the melt turns into crystal. The low fraction of melt turning into crystal means high cost for materials where ingredients are also costly (e.g. SrI_2:Eu). All the oxide type of scintillator materials, orthosilicates, tungstates, pyrovskites, and garnets can be manufactured using the Czochralski method, but temperatures of 2000 °C or more are required which necessitates inert expensive ampoule materials to be grown in, like platinum or iridium, which frequently requires being exchanged due to cyclic thermal and mechanical stress. Masses of such crucible can be several kilograms. Many materials require special inert atmospheres to avoid oxidation (e.g. Ce-doped materials to avoid oxidation to Ce^{4+}) or other unwanted reactions (e.g. avoiding water vapour due to the hygroscopic NaI(Tl) and $LaBr_3$:Ce). The high cost of some of the materials is associated with problems in obtaining pure enough materials to enable growing large crystals with high transparency, some materials require extreme care in cooling so as not to get brittle, and so forth. On the other hand, the Czochralski method is well optimized for the halides.

For small specimens of scintillator material, the so-called micro pulling down method can be employed. Again, a seed crystal is used to initiate the growth of the crystal when it is pulled out at a constant rate through a small hole at the bottom of the crucible. The advantage of this method is the fast growth rate (approximately ten times faster than the Czochralski growth). Long fibres (meters) from all the oxide scintillators can be manufactured in this way. The method has been used to produce crystals for testing, which is cheaper than using the full-scale techniques above. Micro-pulldown fibres can be from 100 um to 3mm, thus they can be sliced and used for pixelated detectors rather than cutting in expensive Czochralski-grown crystals. The material usage yield in micro-pulldown is nearly 100 per cent, which is good for expensive materials. μ-pull down technique has been used to manufacture scintillators from BGO, YAP, YAG, LSO, LYSO and LuAG.

Although single-crystalline scintillators are frequently used (e.g. the standard NaI(Tl)), there are alternative methods to produce the more expensive materials, for instance the garnets, which require very high temperatures in the Czochralski process. The technology of producing optical ceramic garnets rather than single crystals was triggered by the ceramic Nd-doped YAG-laser in the 1990s as a replacement for single-crystal lasers [54]. Larger crystals can also be made from many small nano- or micro-sized crystals, which are pressed together under high temperature and pressure (sintering). Such polycrystalline ceramic scintillators can be made of the majority of the oxide scintillators, and at a much lower price than the corresponding single crystals. The ceramic scintillators have found use as scintillator films and powders. A given scintillator material grown as single crystals for one purpose can instead be grown as micro- or nanocrystals and become sintered into a polycrystalline solid or a ceramic. The advantages of making transparent polycrystalline materials or ceramics out of many small crystals rather than using a single grown crystal are several. The manufacturing of single crystals requires melted material, and for this the Czochralski-method [53] is frequently used. This method requires high temperatures (for some materials, e.g. the oxide materials, above 2000 °C) and long periods (days) of growing, thus requiring substantial energy and adding to the cost of the material. The growth and sintering of micro-crystals are faster and thus cheaper, and approximately the same density can be reached in the sintered ceramic version as for a single crystal. It is also easier to control the amount of incorporated dopant ions and therefore adjusting the wanted properties like light emission or decay time and, in general, more complex compounds can be manufactured

TABLE 6.1
A Comparison of Manufacturing Costs for Various Scintillator Materials [96]

Scintillator material	Cost [US dollar per cubic centimetre]
$LaBr_3$:Ce	75–150
SrI_2:Eu	75–150
GAGG:Ce	250–600
NaI:Tl	2
CsI:Tl	4
LYSO:Ce	50–75
BGO	10–15

than with single-crystal growth. Ceramics are a benefit when a specific geometrical form of detectors are required. Ceramics manufacturing also has the benefit of selecting a wider combination of materials and activators, which may be complicated in single-crystal growth if it disturbs the growth process. More chemistry is simply allowed (several dopants) in ceramic manufacturing, which also results in cheaper materials. Ceramics usually have slower light decay and poorer resolution than their single-crystal versions, but some of the ceramic versions of the oxide materials (e.g. YAP:Ce [55]) have shown light yield and resolution even better than their single-crystal versions. Further advantage is that higher amount of Ce, resulting in higher light yield, can be added in ceramics compared to single-crystal growth [56]. The main disadvantage with ceramics is the usual inferior light transmission relative to the corresponding single-crystal version. Usually, the properties of the ceramic versions differ from that of the single-crystal versions. Often, ceramic scintillators are less transparent due to micropores or disturbances at grain boundaries. Fabrication of transparent ceramics of hygroscopic halides is challenging. Probably most gain is in oxides with a high melting point.

During the last decades (fourth phase above), two major discoveries were made in scintillation detector science; the possibility (understanding) of changing the main parameters in existing scintillators by so-called co-doping and the insight into what regulates energy resolution.

6.3.8 Co-doping

Crystalline materials to some extent always contain some defects. This may be lattice imperfections due to stress in the manufacturing process or impurities entering the crystal from all stages in the process from raw material to the final crystal. The crystal defects may be inhomogeneously distributed inside a given crystal and they may change from time to time as production methodologies and ingredients vary. These defects may act as hurdles or traps for the migrating electrons and holes, and they may therefore influence timing, intensity, and wavelength of the emitted light.

One way to overcome the problems has been to gain better knowledge of the role of manufacturing. Purification of the raw material, additional melt purification, and optimization of the crystallization process have been tried [11]. Optimization of the manufacturing technology is, however, not a general solution for all problems, and other ways have shown to be more effective. One of these methods is to dope the scintillator matrix material with an additional element other than the one constituting the recombination or luminescence centre: co-doping. Elements used for co-doping are not active as luminescence centres but have a positive effect on the scintillator properties such as improving light yield, reducing the afterglow, or improving the energy resolution. Although this so-called band-gap engineering is a relatively late technique, mainly appearing after around year 2000, observations of the positive effects of co-doping were noted earlier. In attempts to find better scintillator materials for CT, Yamada and colleagues at Hitachi laboratories [57] worked on the ceramic scintillator Gd_2O_2S:Pr (see Table 6.2), where the rare earth element praseodym (Pr) acts as a luminescence centre when added at some tens of ppm concentrations. Electron and hole traps in the host material created severe afterglow (defined as the fraction of light emitted some milliseconds after ionization) and prevented the material being used in CT-scanners. The addition of small amount of cerium as a co-dopant however had a dramatic effect on the afterglow, enabling it to be reduced to insignificant levels when adding Ce at permille concentrations. A similar effect was observed by adding fluoride which, in addition to lowering afterglow, also increased the light yield. Similarly to the development of this co-doped Gd_2O_2S:Pr, Ce, F scintillator, discoveries of the effect of co-doping on afterglow suppression was also used in the development of the $(Y,Gd)_2O_3$:Eu HiLight scintillator by General Electric

TABLE 6.2
Scintillator Materials in Medical Use. Data Collected from [11, 12, 16–20, 41, 97]. Afterglow Emission Is Not Included in Light Yield. Parameters Listed Are for Materials Without Co-doping

	Density [g/cm3]	Average atomic number (Z)	Max emission [nm]	Refractive index at max emission wavelength	Decay time [μs]	Light yield [photons/MeV]	Typical energy resolution at 662 keV [%]
Tl-doped alkali-halides							
NaI(Tl)	3.67	51	415	1.85	0.23	38000	6.7
CsI(Tl)	4.51	54	540	1.80	0.68 (64%), 3.34 (36%)	65000	6.6
Self-activated Oxides							
BGO ($Bi_4Ge_3O_{12}$)	7.13	75	480	2.15	0.3	8200	9
$CdWO_4$	7.90	64	470	2.3	1.1 (40%), 14.5 (60%)	15000	6.5
$CaWO_4$	6.1	64	420	1.94	8	15000	6.3
Ce-doped REE- halides							
$LaCl_3$:Ce	3.79	50	350	1.9	0.028	46000	3.3
$LaBr_3$:Ce	5.29	47	380	2.05	0.026	63000	2.9
$CeBr_3$	5.2	46	380	2.09	0.020	44500	4.1
Eu-doped halides							
SrI_2:Eu	4.55	49	435	2.05	1.2	115000	3.7
Ce-doped REE-oxyorthosilicates							
YSO (Y_2SiO_5)	4.54	39	420	1.80	0.07	24000	7.6
GSO (Gd_2SiO_5)	6.71	58	440	1.85	0.056 (90%), 0.4 (10%)	9000	8–9
LSO (Lu_2SiO_5)	7.4	75	420	1.82	0.047	25000	8
LYSO ($LuYSiO_5$)	7.1	60	420	1.81	0.041	20–30000	8–9
Ce-doped REE aluminium perovskites							
YAP ($YAlO_3$)	5.37	32	370	1.95	0.027	18000	4
LuAP ($LuAlO_3$)	8.4	65	365	1.94	0.017	17000	23
Ce-doped REE aluminium garnets							
YAG ($Y_3Al_5O_{12}$)	4.56	33	550	1.82	0.088 (72%), 0.30 (28%)	17000	6–7
LuAG ($Lu_3Al_5O_{12}$)	6.7	59	535	1.84	0.07	25000	6–7
GAGG ($Gd_3Al_2Ga_3O_{12}$)	6.63	54	520	1.9	0.05–0.15	40–60000	4–5
Ceramic scintillators							
GOS (Gd_2O_2S:Ce,Pr,F)	7.3	61	580		3	50000	

[58]. In the case of afterglow, the co-doping does not 'repair' the electron/hole traps but, due to their energy levels in the band gap, allows the trapped charge carriers to recombine in a non-radiative manner once released from the traps, thus preventing the afterglow. This shortcut in energy release may also affect the ordinary component of the scintillation light emission such that it is reduced. The amount of co-dopant added to the host matrix is therefore often a compromise between positive and negative effects (e.g. removal of afterglow at a cost of light yield). Starting around 2010 the methods of co-doping were empirically tested on a large range of scintillators. The energy resolution of the already excellently performing LaB_3:Ce could be further improved by co-doping with divalent alkaline earth elements such as Ca^{2+}, Sr^{2+} or Ba^{2+} [59–61]. Currently the best-ever energy resolution of 2 per cent (at 662 keV) was obtained by a small

specimen (mm-sized) of LaB_3:Ce, Sr [60], but energy resolutions of 2–2.6 per cent were later achieved in 3"x3" crystals [62]. The reason for the improved energy resolution when co-doping with alkaline earth elements is the improved linearity between detector signal and energy deposited in the crystal, which clearly can be seen for $LaBr_3$ with and without Sr^{2+} co-doping. Due to the specific energy levels of a given crystal material the effect of co-dopants is host-specific so that the changes that are enabled in one host material cannot necessarily be achieved in other materials.

Several attempts have also been made to reduce the severe afterglow in CsI(Tl) using co-doping. By co-doping the CsI(Tl) with Eu^{2+}, the afterglow in the time range of 10µs-100ms could be lowered by almost two orders of magnitude [63, 64]. Tests have also been done using Sm^{2+} [65–67], Tl^{+} [67], or Bi^{3+} [68] to establish non-radiative recombinations of the trapped charges. Although afterglow could be significantly reduced, in all cases it was at a cost of the scintillation yield or energy resolution. In 2014 it was shown that addition of small amounts of Yb^{2+} yielded a CsI(Tl):Yb crystal having both a suppressed afterglow, 0.035 per cent after 80ms, as well as a very high light yield, nearly 3 times the light yield of NaI(Tl) [56]. The afterglow problem was previously the largest problem for CsI(Tl) to be used in CT scanners, with very good mechanical properties and a fluorescence spectrum already very well suited to small-sized SiPMs, the CsI(Tl) detector is again a potential choice for use in CT.

It was not only for specialized scintillators that the effect of added co-dopants could create miracles. Even in the well-known standard NaI(Tl) scintillator improvements could be made in light output [69], energy resolution, and light output [70], non-proportionality, energy resolution, and decay time [71]. By applying co-doping the energy resolution could be improved from around 6.5 per cent to around 5 per cent.

Co-doping has not only been used to improve scintillation materials to become better radiation detectors but also to enable better manufacturing. As mentioned above, Ca co-doping of LSO:Ce scintillators have shown to have a desirable influence on both light yield and decay time [45, 46] but, unfortunately, it was observed that at optimum Ca-concentrations the crystal growth stability worsened due to the crystal melt sticking to the crucible being used in the Czoralski process [45]. In order to allow Ca co-doping during the crystal growth yet another co-dopant, zinc, was added to increase the surface tension of the melt so that it did not stick to the crucible during crystallization.

6.4 ENERGY RESOLUTION

Observed from a given scintillator system, the overall energy resolution, W_{tot}, is usually expressed as the squared sum of a number of contributing components.

$$W_{tot}^2 = W_{ph}^2 + W_{pc}^2 + W_{pm}^2 + W_x^2 \tag{6.1}$$

where W_{ph}, W_{pc} and W_{pm} is the standard deviation in the number of produced photons, produced photoelectrons at the photocatode and the gain in the PM tube respectively. W_x is the addition from all other sources.

The light collection and transport also introduce variance. If the scintillator is poorly transparent the fraction of collected scintillation photons will depend on the size of the detector element due to different pathlengths between the interaction point and the PM tube. The distance for the emitted photons to the PM tube also depends on the shape of the detector, which then also will influence the collected fraction. Energy resolution will thus depend on crystal size and shape. The shape also influences the amount of scatter at the scintillator surfaces where further losses may arise. The light collection process thus affects the energy resolution in at least two ways. First the absorption of light in the detector results in fewer photons reaching the PM tube, thus increasing the relative standard deviation in number of photons. Secondly, the uniformity with which light is collected depends on the position of the radiation interaction due to different path lengths of photons and the reflection at crystal surfaces and back. The spatial distribution of photons over the photocatode also contributes to the energy resolution if the optical coupling between scintillator and photocatode surface is poor or non-homogeneous, for example, remaining microscopic air bubbles in the optical coupling fluid. The scintillator refraction index is here also of importance to avoid back-scatter of light into the scintillator material. The thickness/efficiency of the thin photocatode likewise add to the peak broadening. Over prolonged measurements also other factors contribute to the energy resolution. The more common ones are the short-time variation in gains of the connected PM tube or SiPM from event to event, drift in associated electronics and, most importantly, temperature-induced variations in the scintillator and photon detector.

As mentioned above W_x is the added standard deviation from all other sources and includes, for instance, the light transport in the crystal and the amplification and shaping steps in the associated electronics. It also includes a term related to the non-linear energy response of the crystal. Usually the lowest number of 'information carriers' in the transport chain is at the photocatode in the total number of photoelectrons produced. If making a rough estimate of the standard deviation of the number of information carriers at this point from a NaI(Tl) irradiated by 1MeV photons and a typical PM tube with a photocathode quantum efficiency of 20 per cent, we would arrive at about square root of (38000 x 0.2) = 87 or a relative standard deviation of 87/(38000*0.2) = 1.1 per cent. This means a FWHM = 2.35 x 1.1 = 2.7 per cent. If this variation in the number of photoelectrons really would be the dominating contribution to the FWHM we would thus expect a FWHM of around 3 per cent for 1MeV photons. Commercial NaI(Tl) detectors are normally quoted as having energy resolutions for ^{137}Cs 662 keV of, at best, around 6 per cent. Commonly it is around 6–8 per cent. There are therefore other factors than the uncertainty in the number of photons and photoelectrons that dominate the energy resolution in NaI(Tl), as well as in many other scintillation detectors. It turns out that a factor that is of greater importance is the poor linearity between emitted light and absorbed energy in nearly all scintillation materials. Apart from affecting the energy calibration, this non-linearity severely affects the energy resolution.

A non-linear response between pulse-height and gamma photon energy in NaI(Tl) was observed already in 1950 [72], only two years after the introduction of this scintillator. Engelkemeir [73] summarized the early observations for NaI(Tl) in 1956 and Aitken [74] in 1967 showed that, apart from NaI(Tl), also CsI(Tl), CsI(Na), and CaF_2(Eu) showed a non-linear response. For many years this non-linearity was accepted, and little was done to investigate the underlying mechanism. It became however obvious that the limiting step in energy resolution was not the statistical broadening due to the number of photons since light yield and the photomultiplier tube (PMT) performance was already close to optimal. In order to improve the energy resolution, the focus was instead moved to the non-proportionality of scintillator materials. In connection with the strong development of new scintillator materials in late 1990, triggered largely by the CERN LHC, more thorough investigations on non-proportionality were initiated using Compton coincidence techniques [75–79] to study the detector response as a function of Compton electron energy. In this technique the actual energy deposition from individual gamma photons in the detector crystal can be determined by the difference between known incoming energy and measured energy of the scattered Compton photon. Further studies were conducted using monochromatic synchrotron X-ray irradiation [80, 81], and modelling work [82–85]. It was realized from these investigations that the non-proportionality does not depend, or depends very weakly, on the parameters that determine scintillators' quality, such as: impurities, lattice defects, activator concentration, manufacturing, afterglow, and so forth and, instead, that non-proportionality is an intrinsic property of the scintillator crystal material and structure. It was also realized that in some of the materials the non-proportionality could be altered by co-doping (see above).

The current understanding of the root cause for non-proportionality is to be found in the variations of light yield per energy deposition along the track of the electrons generated by the ionizing radiation. The luminosity is thus LET-dependent and, since a given gamma/x photon energy may give rise to a spectrum of secondary electrons in the scintillator (including generated Auger-electrons), the total amount of emitted light photons for a given gamma/x photon energy will depend on the spatial distribution of deposited energy from the electron energy distribution, which may change from gamma/x photon to photon although being from a monoenergetic source. When a typical 662 keV photon enters a NaI(Tl)-detector it does not necessarily interact with photoabsorption, delivering about 610–660 keV kinetic energy to the photoelectron. It can instead Compton scatter a few times before becoming absorbed through a photoabsorption event. The light emission from the sum of all Compton scattered electrons and final photoabsorption event differs relative to the light emission in the case of a single energetic photoabsorption event, thus generating two different sizes of the output signal, although in both cases the same energy (662 keV) was deposited in the crystal The variability in the secondary electron spectrum, and the spatial LET distribution, changes with photon energy (for instance, different cross sections for photoelectric absorption and Compton scatter), which is why deviation from linearity is observed to be photon energy dependent. The reason for the LET dependent emission is believed to be due to the competition between radiative and non-radiative processes, which depends on ionization density. Along a given particle track there is quenching of the light output due to radiation-less electron-hole recombination inside the volume of high ionization density, and in parallel diffusion of the charge carriers from the point of creation towards a volume of lower ionization density, eventually transferring the energy to luminescence centres. The faster the charge carriers escape the volume of high ionization density, the higher the probability of converting the energy of the carriers into optical photons. High-LET tracks will thus have a greater risk of causing recombination. A similar quenching behaviour is seen in organic scintillators [2]. Reviews on the non-linear behaviour of inorganic scintillators can be found in [12, 86–88]. Today it is accepted that nearly all scintillator materials show a more or less non-linear behaviour. Its contribution to the overall energy resolution can be exemplified by comparing the relatively common cerium doped yttrium aluminium perovskite scintillator ('YAP' or $YAlO_3$) with a standard NaI(Tl). YAP has a photon emission of only about 40–50 per cent to that of NaI(Tl) [89] but in spite of this has an energy resolution of about 4 per cent [89, 90] as

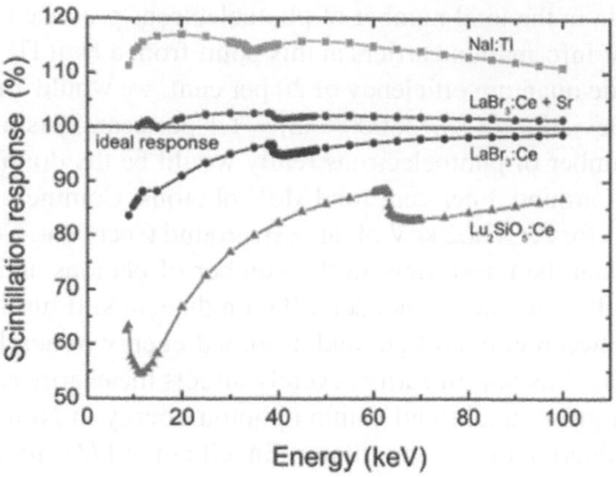

FIGURE 6.3 Scintillator response (photons per MeV) to monochromatic X-ray excitation relative to response for 662 keV photons. Although the figure seems to show a deviation in energy response only for low energies (<100 keV), the phenomena affect also the energy resolution at higher energies due to the energy distribution of generated Compton/photo electrons from monoenergetic gamma-rays. Reproduced from [60] with the permission of AIP Publishing.

compared to the 6–8 per cent of NaI. The YAP-scintillator is one of very few showing a good linearity between scintillator light output and electron energy (Figure 6.3). Table 1 in [88] lists a large number of scintillator materials in order of increasing non-proportionality.

6.5 SCINTILLATION DETECTORS IN MEDICAL USE

Naturally, the optimal scintillator for medical purposes depends on the purpose, and there is no ideal material that fits all applications. The bulk number of scintillators for medical purposes is for imaging using low- or medium-energy X-rays or gamma photons. A short mean free path in the detector material is then required, meaning detectors with high atomic numbers and density. These two parameters are not only of importance for the number of detected events (and thus patient doses) but also to minimize Compton scatter to nearby detector elements in a multidetector imaging device such as CT or PET. Detectors made of materials with high density and high atomic numbers can also be made physically smaller, thereby increasing spatial resolution. For imaging purposes, the energy resolution is also of great importance since this will determine how well the scattered photons can be discriminated when using energy window discrimination. The scattered photons can originate both in the detector and in the object. The peak-to-Compton value rather than detector efficiency and resolution alone is thus of importance to obtain high spatial resolution for a given detector volume. For PET, where coincidence technique is used, the length of the time gate also is of importance. A long gate allows for more background events, so detectors with a fast scintillation light decay enable a short time gate and thus a better signal to background ratio.

For low energy X-ray detectors scintillator density and atomic number are less important; the mean free path of low-energy photons in, for instance, yttrium-based YAP or YAG-detectors (see below) are similar to heavy bismuth-based BGO-detectors as long as the photon energy is below the K-edge of Bi (90 keV).

Among the most important parameters for perfect scintillator material for medical use is one with a good energy resolution, high atomic number and density, fast fluorescence decay with suitable wavelength, no afterglow, and with a material refractive index close to that of glass (1.5).

For planar scintigraphy or SPECT the large size of the detector (gamma camera) combined with the dominating radioisotope gamma energy (^{99m}Tc 140 keV) being used the most practical choice of detector material is NaI(Tl) with standard photomultipliers. The technique for growing large NaI(Tl) crystals is well developed and can be done at a reasonable price, and the density and average atomic number is sufficient for the 140 keV gamma photons. Since planar scintigraphy and SPECT are 'slow' techniques, they do not have any benefit from very fast scintillators or drawback with afterglow and, since the majority of investigations use ^{99m}Tc, the need for materials with higher density and atomic number is not needed. Better spatial resolution can, on the other hand, be accomplished by better energy resolution, and in 2000 the $LaCl_3$:Ce and $LaBr_3$:Ce materials with 3–4 per cent energy resolution at 662 keV were introduced. The price for these, however, was considerably higher than for NaI(Tl).

For computed tomography (CT) the spatial reconstruction of the scanned object requires many small detector elements and consequently one of the important parameters is the maximum absorption of the photons in each detector element. Scattered photons between detector elements would deteriorate the spatial resolution. Traditional scintillator materials used for this purpose has been $CdWO_4$ (CWO), $Bi_4Ge_3O_{12}$ (BGO) and CsI(Tl). The later coupled to silicon photodiodes in order to collect the long wavelength emission. The significant afterglow of CsI(Tl) hampered the use of fast CT-scans, and this material was therefore phased out from CT-scanners as the imaging speed was increased.

Compared to the standard NaI(Tl) scintillator, both CWO and BGO have higher densities (7.9 and 7.13 g/cm^3 respectively) and average atomic numbers (66 and 74 respectively) and have therefore frequently been used in CT. Both materials show the same degree of non-proportional energy resolution, but the light emission from CWO is slightly better than from BGO, resulting in better energy resolution. Both materials have a near absence of afterglow, which allows for fast scanning. The toxic Cd has been one of the reasons why BGO has been preferred ahead of CWO. In BGO it is the Bi^{3+} that acts as luminescence centre. In CWO it is the WO_4^{2-}.

The band-gap engineering has brought forward materials in which the fluorescent REE-elements have been incorporated depending on if their energy-levels fit relative to the band gap of the matrix. Several ceramic scintillator materials have been developed specifically for CT such as Gd_2O_2S doped with small quantities of Pr^{3+} as luminescence centre. Co-doping with Ce^{3+} and F^- enables some control of both light yield and afterglow.

In order to comply with requirements on optimum scintillator materials there has been a constant drive to find new and better scintillators

6.6 TIME-OF-FLIGHT PET

Ordinary PET depends on an image reconstruction from a number of events detected at each detector pair, which is proportional to the integral of the radioactivity along the line connecting the centres of the two detectors. The spatial resolution is typically around some millimetres and depends on several factors: for instance the detector size. An alternative approach is to use the difference in flight-time between the two 511 keV photons, which then makes it is possible to directly calculate the origin of the annihilation event, thus making image reconstruction obsolete. In order to compete with ordinary image reconstruction, PET – a time-of-flight PET (TOF-PET) system – requires the detectors to be able to distinguish events which appear within some tens or hundreds of picoseconds at the two detectors. A time resolution of 10 ps equals a spatial resolution of about 1.5 mm.

The major difficulties in achieving coincidence time resolutions (CTR) below about 100 ps is partly due to the nature of the scintillation light emission rate of inorganic scintillators and partly due to the time resolution of the connected photon detectors and associated electronics. The relatively slow light-emission rate results in even very bright scintillators emitting their light during a wide time window, resulting in only very few photons enabling the triggering of two detectors in a given 10 ps time window. For instance, the relatively bright cerium doped $LuYSiO_5$ (LYSO:Ce) scintillator, frequently used in PET, typically has a light emission of about 28,000 photons per MeV and a decay time of around 45 ns. Assuming a first-order decay pattern results in an initial light emission rate of less than one photon per ps for a 1MeV photon absorption event. In general, the CTR is proportional to the square root of the product of rise time and decay time of the emitted light. A fast rise and decay time is thus obviously important. While SiPM's and associated readout electronics continuously are developing, the bottleneck will thus be the rate of inorganic scintillation light emission. Although most *organic* scintillators have fewer photons emitted per MeV than *inorganic* ones, organic scintillators normally have a much faster light emission (in the order of a few ns) and would be ideal for coincidence-based techniques like TOF-PET if it were not for the very poor photo absorption capacity due to their low atomic number and density. The combination of an organic scintillator, allowing very fast timing, with a high-Z and high-density inorganic scintillator enabling high probability for photoelectric absorption is another way to approach the demands for fast timing in TOF-PET. An example of this is presented in Turtos and colleagues [91] in which pixel detectors (about 4x4x3mm^3) were constructed by 8 LYSO (or BGO) 200µm thick sheets intersectioned by 8 equally thick sheets of the ultra-fast (1.6 ns decay time) organic scintillator BC-422 (Saint-Gobain Crystals). The electrons produced through photoelectric effect or Compton scattering in the LYSO would continue into the organic scintillator thus providing the fast light output and acting as 'time-tagger'. When producing detector units like this the demands on the inorganic scintillator must not only be both high density and high atomic number but also an ability to be machined into suitable sizes and thus be mechanically stable and non-hygroscopic.

An alternative way to achieve a low CTR is by using the prompt-emitted Cerenkov light, emitted as charged particles at high-speed travel through media. Cerenkov light emission appears when charged particles have a velocity faster than c/n in a material where n is the refractive index of the material (depends on wavelength). The light emission is prompt and thus appears much faster than the scintillation light. Although the time-integrated Cerenkov photon flux per MeV energy deposited in a given scintillator is much lower than the scintillation light, it appears within a few picoseconds and the flux per picosecond is thus greater than for most scintillator materials. The classical BGO detector, previously a standard in PET, has returned as a potential detector in TOF-PET due to its high refractive index (2.15), which translates into Cerenkov photons being emitted from a larger energy range than in materials with lower refractive index. During the first picoseconds on an average 15 Cerenkov photons are emitted per 511 keV event in BGO. This should be compared to the <1 photons per picosecond from a LYSO:Ce scintillator in the example above. A further advantage of the BGO is the superior attenuation coefficient, much lower cost of production compared to LYSO, and the absence of internal background radioactivity.

Cerenkov emission intensity is inversely proportional to the square of the photon wavelength and is thus more abundant at lower wavelengths, which puts some demand on the detector material transparency at lower wavelengths as well as on the optical detector. While detectors providing fast Cerenkov emission with relatively good transmission properties also at lower wavelengths (e.g. BGO) may be utilized, the system bottleneck may rather be the time resolution of the optical detector. Current CTR's achieved by combining BGO crystals with SiPM detectors sensitive in the NUV-UV range are around 100–200 ps [92]. An interesting material not yet commercialized is TlBr (see Ch Semiconductor detectors), which is a semiconductor but with excellent properties for use with Cerenkov emission in TOF-PET.

The possibilities and current status of TOF-PET using Cerenkov emission is partly reviewed by Gundacker [92].

6.7 PHOTOMULTIPLIER TUBES

The sensitivity of the photocathode as a function of wavelength is very important. Some scintillators have a very strong light emission but, if not matched with a suitable light sensor, the advantage of the scintillator is lost. A typical example is CsI(Tl), which often is described as yielding a lower pulse-amplitude per MeV than the corresponding NaI(Tl) when connected to a standard bialkali PM tube. The total photon emission from CsI(Tl) is, however, more abundant than the NaI(Tl) emission, but it has a proportionally larger part occurring in the red/infrared not energetic enough to produce photoelectrons. At higher wavelengths the energy of the photon is not enough to create and eject a photoelectron from the photocathode. At lower wavelengths, around 350 nm, the glass obstructs the light. In some cases, glass can be substituted with quartz which enables transmission even in the UV-region, which is important, for instance for Xe.

The homogeneity of the thickness of the photocathode is very important for the energy resolution as is its ability to create the same number of photoelectrons per eV of photon energy, the quantum efficiency (Q). If these parameters differ over the surface of the photocathode the number of photoelectrons produced for a given input of light energy will depend on the size of the detector, since this partly determines the spatial distribution of the impinging photons.

There are two major types of photocathodes – multialkali (Na_2KSb) and bialkali (K_2CsSb). In NEA-dynodes (negative electron affinity) the lifetime of the created electrons is longer, which means a greater probability for escape from the surface before losing kinetic energy. The multiplication factor for NEA-dynodes is around 40–60 as compared with around 10 for normal dynode materials. The stronger emission of electrons is also important with respect to time resolution. The more abundant number of electrons means less relative variance and thus better-defined timing with less uncertainty. Normally the scintillator decay time determines the time resolution, but for fast scintillators it is the PM tube. Total time from photocathode to pulse out is about 50 ns.

The advantage with various types of photodiodes as light sensors is that they delivers the electrons faster. Instead of having a 50 ns delay the time between photon influx and electron output is about 1–2 ns.

Position-sensitive PM tubes can be constructed by using focused mesh or micro-channel plates, thus taking advantage of pixelated scintillators and obtaining a positioning similar to the Anger camera logic. The design would still be very sensitive to magnetic fields, thus preventing them from combined MR-PET instruments.

For detailed description of various types of PM tubes and their characteristics [93, 94] are excellent references.

6.8 PHOTODIODES

Although photomultiplier tubes still are very common as light sensors for scintillation detectors, the use of photodiodes in various forms offers some advantages apart from being smaller. Many scintillator materials (e.g. CsI(Tl), BGO and Gd_2O_5S (GOS)) have a significant photon emission in the long wavelength region not energetic enough to produce electrons in the photocathode of a PM tube, while photodiodes have a significantly better sensitivity for long wavelength photons. PM tubes are very sensitive to magnetic fields and must be shielded by a thin layer of mu-metal. This is however not enough when applied in strong magnetic fields that prevent them being used in coupled MR-PET systems or similar. Further advantages in using photodiodes are the higher quantum efficiency, lower power consumption, compact size, and the rugged design. All of which are factors of importance when designing compact, robust detector systems like hand-instruments or CT-systems. Due to their compact design, they are in general faster than PM tubes, which is an advantage in coincidence-based systems (e.g. PET).

The major disadvantage of photodiodes over PM tubes is the difficulty in producing large-area light detectors due to the much higher noise level per unit area compared to PM tubes.

There are three types of photodiodes used as readout detectors for scintillation light: Conventional photodiodes (usually called PIN photodiodes), avalanche photodiodes (APD), and array-constructed devices of avalanche photodiodes often called silicon photomultipliers (SiPM), multi-pixel photon counters (MPPC, the trademark of Hamamatsu photonics). The first type of detector (PIN) had no internal gain and operated at only a few volts. Since the total amount of photons produced in a scintillation event is not so large (for NaI(Tl) around 40,000 photons per MeV) the integrated charge collected without any internal gain (about one electron per incident photon in the PIN) is not so large. The amplitude above the noise level then becomes a major problem, especially at low-photon energies with few scintillation photons. The noise originates both from the PIN capacitance, which is proportional to the PIN area and inversely to its thickness (usually only in the order of 1mm thick), and the thermally generated electrons, which is proportional to the total volume of the PIN. For normal sized scintillation crystals of some centimetres the total noise level generated from a connected PIN therefore results in a much worsen energy resolution relative to when connected to standard low noise PM tubes. Cooling the PIN can dramatically reduce the thermal noise (dark noise) but the capacitance noise still remains. PIN photodiodes have their main advantage when applied as small size detectors in current mode for high photon flux applications like computed tomography (CT) with no direct need for energy resolution. The high photon flux at a small area PIN generates a sufficient high signal-to-noise ratio. The small size, fast response time and low cost make them attractive relative to standard PM tubes, especially when coupled to highly efficient Gd_2O_2S:Pr ceramic scintillators where the emission to a large extent is at longer wavelengths unavailable for PM tubes.

Gamma photons may pass through the scintillator and interact directly with the PIN photodiode. This may be minimized by having a thin PIN but then the capacitance generated noise will increase.

Silicon drift diodes or silicon drift detectors (SDD) are a further development of the standard PIN photodiode. Although developed to be used as a position-sensitive detector based on the time to collect charges (drift time), a consequence of this kind of photodiode with a small area anode is the much-lowered detector capacitance. This results in a significantly improved noise contribution and thus in a better energy resolution. These semiconductor detectors are used both as light readout devices when connected to scintillators and as stand-alone Peltier cooled X-ray detectors in, for instance, electron microscopes using energy dispersive X-ray fluorescence (EDX).

In the avalanche photodiode a higher voltage (a few hundred volts) is applied over the p and n contacts of the diode, sufficient to create an avalanche of electrons through the collisions of electrons accelerated through the diode. This results in a collected charge much larger than what is obtained through the PIN photodiode. The gain may reach a factor of hundred thus enabling better signal-to-noise ratios than for PIN photodiodes. The main disadvantage is the strong sensitivity of the gain to fluctuations in temperature and applied voltage. Analogy is often drawn to gas detectors operating as ion chambers, proportional counters, or in Geiger-Müller mode. The ordinary PIN photodiode is analogous to the ionization chamber while the avalanche photodiode mimics the behaviour in the proportional counter provided the voltage is not so high as to produce continuous avalanches, which would be the GM-mode of operation.

If in GM-mode of operation the gain is high but the proportionality to the amount of light collected is lost (just as in the GM-counter, each output pulse has equal size independent on energy of the incoming radiation) and no spectroscopy is possible. If instead creating an array of many very small (a few tens of μm) avalanche diodes connected so they can be read out in parallel, a SiPM or multi-pixel photon counters (MPPC) is designed. Since still operating in GM-mode, the size of individual avalanche diodes must be small enough so the probability to be struck by more than one scintillation photon must be nearly infinitely small. The total number of avalanche diodes in a SiPM can thus be in the order of 10^4–10^6, constructed on a single silicon chip. The total charge collected during a scintillation event will thus be the sum of all diodes output charge. The total gain can then reach around 10^6 or similar to the gain from an ordinary PM tube. The thermal production rate of electrons in a single avalanche PIN unit may be around 1–10 Hz at room temperature so the total background rate can approach MHz in a SiPM unit. Due to the fast discharges, from tens of picoseconds to less than a few nanoseconds, the individual charge pulses collected usually consist of only one or at most a few simultaneously discharged avalanche diodes. The total charge per pulse is thus not so great. The background due to thermally generated electrons can thus be removed by introducing a discriminator level that requires the total charge to be larger than what is combined from some tens of cells. In a typical scintillation event, some thousands of photons reach the SiPM in the first nanoseconds, thus creating a signal far above the discriminator level.

The physical sizes of commercial SiPM are not so large, in the order of 10x10mm typically holding some 50,000 pixels (individual APD's).

An important parameter of SiPMs is their photon detection efficiency (PDE). This is the product of the geometrical efficiency (how well the SiPM 'see' the scintillator), the quantum efficiency and the avalanche probability.

Today there is a gradual change going from the use of multi-anode PM tubes to SiPM based systems, partly due to need for detector systems capable of handling magnetic fields (such as in combined MR-PET), but also due to their smaller footprint and compactness [95].

Due to the individual avalanche PINs operate in Geiger mode in a SiPM, the avalanche will not stop by itself. For most devices, a quenching resistor is connected in series with each diode. The current output through the resistor will result in a voltage drop over the diode, thus reducing the electric field and terminating the avalanche.

REFERENCES

[1] K. D. Steidley, "The Radiation Phosphene," *Vision Research*, vol. 30, no. 8, pp. 1139–1143, 1990/01/01/ 1990, doi:https://doi.org/10.1016/0042-6989(90)90171-G.

[2] J. B. Birks, *The Theory and Practice of Scintillation Counting*. Oxford: Pergamon Press, 1964, p. 453.

[3] S. Ummartyotin, N. Bunnak, J. Juntaro, M. Sain, and H. Manuspiya, "Synthesis and Luminescence Properties of ZnS and Metal (Mn, Cu)-doped-ZnS Ceramic Powder," *Solid State Sciences*, vol. 14, no. 3, pp. 299–304, 2012/03/01/ 2012, doi:https://doi.org/10.1016/j.solidstatesciences.2011.12.005.

[4] F. H. Marshall, J. W. Coltman, and A. I. Bennett, "The Photo-Multiplier Radiation Detector," *Review of Scientific Instruments*, vol. 19, no. 11, pp. 744–770, 1948, doi:10.1063/1.1741156.

[5] F. H. Marshall, J. W. Coltman, and L. P. Hunter, "The Photomultiplier X-Ray Detector," *Review of Scientific Instruments*, vol. 18, no. 7, pp. 504–513, 1947/07/01 1947, doi:10.1063/1.1740988.

[6] M. Blau and B. Dreyfus, "The Multiplier Photo-Tube in Radioactive Measurements," *Review of Scientific Instruments*, vol. 16, no. 9, pp. 245–248, 1945/09/01 1945, doi:10.1063/1.1770379.

[7] S. C. Curran and W. R. Baker, "Photoelectric Alpha-Particle Detector," *Review of Scientific Instruments*, vol. 19, no. 2, pp. 116–116, 1948/02/01 1948, doi:10.1063/1.1741210.

[8] H. Kallmann, "Scintillation Counting with Solutions," *Physical Review*, vol. 78, no. 5, pp. 621–622, 06/01/ 1950, doi:10.1103/PhysRev.78.621.2.

[9] M. Deutsch, "Naphthalene as Scintillator," Institute of Nuclear Sciences, Massachusetts Institute of Technology, M.I.T. - I.N.S. Report No.3, Dec. 1947.

[10] R. Hofstadter, "Alkali Halide Scintillation Counters," *Physical Review – PHYS REV X*, vol. 74, pp. 100–101, 01/01 1948, doi:10.1103/PhysRev.74.100.

[11] P. Lecoq, "Development of New Scintillators for Medical Applications," *Nuclear Instruments and Methods in Physics Research Section A Accelerators Spectrometers Detectors and Associated Equipment*, vol. 809, 09/01 2015, doi:10.1016/j.nima.2015.08.041.

[12] P. Dorenbos, "(INVITED) The Quest for High Resolution γ-ray Scintillators," *Optical Materials: X*, vol. 1, p. 100021, 2019/01/01/ 2019, doi:https://doi.org/10.1016/j.omx.2019.100021.

[13] M. Weber, "Inorganic Scintillators: Today and Tomorrow," *Journal of Luminescence*, vol. 100, pp. 35–45, 12/01 2002, doi:10.1016/S0022-2313(02)00423-4.

[14] R. L. Heath, R. Hofstadter, and E. B. Hughes, "Inorganic Scintillators: A Review of Techniques and Applications," *Nuclear Instruments and Methods*, vol. 162, no. 1, pp. 431–476, 1979/06/01/ 1979, doi:https://doi.org/10.1016/0029-554X(79)90728-6.

[15] C. Melcher, "Perspectives on the future development of new scintillators," *Nuclear Instruments and Methods in Physics Research Section A: Accelerators, Spectrometers, Detectors and Associated Equipment*, vol. 537, pp. 6–14, 01/21 2005, doi:10.1016/j.nima.2004.07.222.

[16] P. Lecoq, A. Annenkov, A. Gektin, M. Korzhik, and C. Pedrini, "Inorganic Scintillators for Detector Systems: Physical Principles and Crystal Engineering," *Inorganic Scintillators for Detector Systems*, 01/01 2006, doi:10.1007/3-540-27768-4.

[17] G. F. Knoll, *Radiation Detection and Measurement*. New York: Wiley, 2000.

[18] M. Nikl et al., "Development of LuAG-Based Scintillator Crystals – A Review," *Progress in Crystal Growth and Characterization of Materials*, vol. 59, pp. 47–72, 06/01 2013, doi:10.1016/j.pcrysgrow.2013.02.001.

[19] C. Dujardin et al., "Needs, Trends and Advances in Inorganic Scintillators," *IEEE Trans Nucl Sci*, vol. 65, no. 8, pp. 1–1, 05/24 2018, doi:10.1109/TNS.2018.2840160.

[20] F. Maddalena et al., "Inorganic, Organic, and Perovskite Halides with Nanotechnology for High-Light Yield X-and gamma-ray Scintillators," *Crystals*, vol. 9, p. 88, 02/08 2019, doi:10.3390/cryst9020088.

[21] R. Hofstadter, "The Detection of Gamma-Rays with Thallium-Activated Sodium Iodide Crystals," *Physical Review,* vol. 75, no. 5, pp. 796–810, 03/01/ 1949, doi:10.1103/PhysRev.75.796.

[22] W. Van Sciver and R. Hofstadter, "Scintillations in Thallium-Activated CaI_2 and CsI," *Physical Review*, vol. 84, no. 5, pp. 1062–1063, 12/01/ 1951, doi:10.1103/PhysRev.84.1062.2.

[23] P. W. Bridgman, "Certain Physical Properties of Single Crystals of Tungsten, Antimony, Bismuth, Tellurium, Cadmium, Zinc, and Tin," *Proceedings of the American Academy of Arts and Sciences*, vol. 60, no. 6, pp. 305–383, 1925, doi:10.2307/25130058.

[24] J. H. Siewerdsen and D. A. Jaffray, "A Ghost Story: Spatio-temporal Response Characteristics of an Indirect-detection Flat-panel Imager," (in English), *Medical Physics*, vol. 26, no. 8, pp. 1624–41, Aug 1999, doi:10.1118/1.598657.

[25] CCC, "Crystal Clear Collaboration." [Online]. Available: https://crystalclear.web.cern.ch/crystalclear/

[26] V. V. Nagarkar et al., "High Speed X-ray Imaging Camera Using Structured CsI(Tl) Scintillator," in *1998 IEEE Nuclear Science Symposium Conference Record. 1998 IEEE Nuclear Science Symposium and Medical Imaging Conference (Cat. No.98CH36255)*, 8–14 Nov. 1998 1998, vol. 1, pp. 158–162 vol. 1, doi:10.1109/NSSMIC.1998.774828.

[27] V. V. Nagarkar, V. Gaysinskiy, I. Shestakova, and S. Taylor, "Microcolumnar CsI(Tl) Films for Small Animal SPECT," in *IEEE Symposium Conference Record Nuclear Science 2004*, 16–22 Oct. 2004 2004, vol. 5, pp. 3334–3337 Vol. 5, doi:10.1109/NSSMIC.2004.1466405.

[28] X. Deng et al., "Development of a Small Animal SPECT and CT Dual Function Imager with a Microcolumnar CsI(Tl) and CCD Based Detector," in *2012 IEEE Nuclear Science Symposium and Medical Imaging Conference Record (NSS/MIC)*, 27 Oct.-3 Nov. 2012, pp. 2742–2745, doi:10.1109/NSSMIC.2012.6551623.

[29] S. L. Bugby, L. K. Jambi, and J. E. Lees, "A Comparison of CsI:Tl and GOS in a Scintillator-CCD Detector for Nuclear Medicine Imaging," *Journal of Instrumentation*, vol. 11, no. 09, pp. P09009–P09009, 2016/09/22 2016, doi:10.1088/1748-0221/11/09/p09009.

[30] R. H. Gillette, "Calcium and Cadmium Tungstate as Scintillation Counter Crystals for Gamma-Ray Detection," *Review of Scientific Instruments*, vol. 21, no. 4, pp. 294–301, 1950, doi:10.1063/1.1745567.

[31] M. J. Weber and R. R. Monchamp, "Luminescence of Bi4 Ge3 O12: Spectral and Decay Properties," *Journal of Applied Physics*, vol. 44, no. 12, pp. 5495–5499, 1973/12/01 1973, doi:10.1063/1.1662183.

[32] R. Tomaschek and O. Deutschbein, "Fluorescence of Pure Salts of the Rare Earths," *Nature*, vol. 131, no. 3309, pp. 473–473, 1933/04/01 1933, doi:10.1038/131473a0.

[33] E. V. D. van Loef, P. Dorenbos, C. W. E. van Eijk, K. Krämer, and H. U. Güdel, "High-energy-Resolution Scintillator: Ce3+ Activated LaBr3," *Applied Physics Letters*, vol. 79, no. 10, pp. 1573–1575, 2001/09/03 2001, doi:10.1063/1.1385342.

[34] N. Cherepy et al., "Strontium and Barium Iodide High Light Yield Scintillators," *Applied Physics Letters*, vol. 92, pp. 083508–083508, 02/28 2008, doi:10.1063/1.2885728.

[35] J. Selling, S. Schweizer, M. D. Birowosuto, and P. Dorenbos, "Eu- or Ce-Doped Barium Halide Scintillators for X-Ray and y-Ray Detections," *IEEE Trans Nucl Sci*, vol. 55, no. 3, pp. 1183–1185, 2008, doi:10.1109/TNS.2008.922825.

[36] N. J. Cherepy et al., "Scintillators with Potential to Supersede Lanthanum Bromide," *IEEE Trans Nucl Sci*, vol. 56, no. 3, pp. 873–880, 2009, doi:10.1109/TNS.2009.2020165.

[37] J. Glodo, E. V. v. Loef, N. J. Cherepy, S. A. Payne, and K. S. Shah, "Concentration Effects in Eu Doped SrI $_{2}$," *IEEE Trans Nucl Sci*, vol. 57, no. 3, pp. 1228–1232, 2010, doi:10.1109/TNS.2009.2036352.

[38] M. Kobayashi and M. Ishii, "Excellent Radiation-resistivity of Cerium-doped Gadolinium Silicate Scintillators," *Nuclear Instruments & Methods in Physics Research Section B-beam Interactions With Materials and Atoms – NUCL INSTRUM METH PHYS RES B*, vol. 61, pp. 491–496, 10/01 1991, doi:10.1016/0168-583X(91)95327-A.

[39] S. Yamamoto, "Possibility Analysis of a GSO PET/SPECT Detector," in *2000 IEEE Nuclear Science Symposium. Conference Record (Cat. No.00CH37149)*, 15–20 Oct. 2000, vol. 2, pp. 14/17–14/20 vol. 2, doi:10.1109/NSSMIC.2000.950016.

[40] W. Mengesha, T. D. Taulbee, J. Valentine, and B. D. Rooney, "Gd2SiO5(Ce3+) and BaF2 Measured Electron and Photon Responses," *Nuclear Instruments and Methods in Physics Research Section A: Accelerators, Spectrometers, Detectors and Associated Equipment*, vol. 486, pp. 448–452, 06/21 2002, doi:10.1016/S0168-9002(02)00751-9.

[41] Y. Uchiyama et al., "Study of Energy Response of Gd2SiO5:Ce3+ Scintillator for the ASTRO-E Hard X-ray Detector," *Nuclear Science*, vol. 48, pp. 379–384, 07/01 2001, doi:10.1109/23.940084.

[42] M. Balcerzyk, M. Moszynski, M. Kapusta, D. Wolski, J. Pawelke, and C. L. Melcher, "YSO, LSO, GSO and LGSO. A Study of Energy Resolution and Nonproportionality," *IEEE Trans Nucl Sci*, vol. 47, no. 4, pp. 1319–1323, 2000, doi:10.1109/23.872971.

[43] C. L. Melcher and J. S. Schweitzer, "A Promising New Scintillator: Cerium-doped Lutetium Oxyorthosilicate," *Nuclear Instruments and Methods in Physics Research Section A: Accelerators, Spectrometers, Detectors and Associated Equipment*, vol. 314, no. 1, pp. 212–214, 1992/04/01/ 1992, doi:https://doi.org/10.1016/0168-9002(92)90517-8.

[44] M. Schmand et al., "Performance Evaluation of a New LSO High Resolution Research Tomograph-HRRT," in *1999 IEEE Nuclear Science Symposium. Conference Record. 1999 Nuclear Science Symposium and Medical Imaging Conference (Cat. No.99CH37019)*, 24–30 Oct. 1999, vol. 2, pp. 1067–1071 vol. 2, doi:10.1109/NSSMIC.1999.845845.

[45] M. Koschan, P. Szupryczynski, K. Yang, A. Carey, and C. Melcher, "Effects of Co-Doping on the Scintillation Properties of LSO:Ce," *Nuclear Science, IEEE Transactions on*, vol. 55, pp. 1178–1182, 07/01 2008, doi:10.1109/TNS.2007.913486.

[46] M. Giuseppina, G. Collazuol, S. Marcatili, C. Melcher, and A. Del Guerra, "Characterization of Ca Co-doped LSO:Ce Scintillators Coupled to SiPM for PET Applications," *Nuclear Instruments & Methods in Physics Research Section A-accelerators Spectrometers Detectors and Associated Equipment – NUCL INSTRUM METH PHYS RES A*, vol. 628, pp. 423–425, 02/01 2011, doi:10.1016/j.nima.2010.07.016.

[47] S. Yamamoto, K. Kamada, and A. Yoshikawa, "Use of YAP(Ce) in the Development of High Spatial Resolution Radiation Imaging Detectors," *Radiation Measurements*, vol. 119, 11/01 2018, doi:10.1016/j.radmeas.2018.11.001.

[48] S. Donnald et al., "Correlation of Non-proportionality and Scintillation Properties with Cerium Concentration in YAlO3:Ce," *IEEE Trans Nucl Sci*, vol. 65, no. 5, pp. 1218–1225, 04/19 2018, doi:10.1109/TNS.2018.2828428.

[49] M. Balcerzyk, M. Moszynski, Z. Galazka, M. Kapusta, A. Syntfeld, and J. L. Lefaucheur, "Perspectives for High Resolution and High Light Output LuAP:Ce Crystals," *Nuclear Science, IEEE Transactions on*, vol. 52, pp. 1823–1829, 11/01 2005, doi:10.1109/TNS.2005.856744.

[50] J. Trummer, E. Auffray, P. Lecoq, A. G. Petrosyan, and P. S. Roldan, "Comparison of LuAP and LuYAP Crystal Properties from Statistically Significant Batches Produced with Two Different Growth Methods," *Nuclear Instruments and Methods in Physics Research Section A Accelerators Spectrometers Detectors and Associated Equipment*, vol. 551, pp. 339–351, 10/01 2005, doi:10.1016/j.nima.2005.06.047.

[51] M. Nikl, "Energy Transfer Phenomena in the Luminescence of Wide Band-Gap Scintillators,"*Physica Status Solidi (A)*, vol. 202, pp. 201–206, 01/01 2005, doi:10.1002/pssa.200460107.

[52] M. Nikl et al., "Antisite Defect-free Lu3(GaxAl1−x)5O12:Pr Scintillator," *Applied Physics Letters*, vol. 88, no. 14, p. 141916, 2006/04/03 2006, doi:10.1063/1.2191741.

[53] A. Yoshikawa, V. Chani, and M. Nikl, "Czochralski Growth and Properties of Scintillating Crystals," *Acta Physica Polonica A*, vol. 124, pp. 250–264, 08/01 2013, doi:10.12693/APhysPolA.124.250.

[54] A. Ikesue, Y. Aung, T. Yoda, S. Nakayama, and T. Kamimura, "Fabrication and Laser Performance of Polycrystal and Single Crystal Nd:YAG by Advanced Ceramic Processing," *Optical Materials*, vol. 29, pp. 1289–1294, 01/01 2004, doi:10.1016/j.optmat.2005.12.013.

[55] T. Yanagida et al., "Evaluation of Properties of YAG (Ce) Ceramic Scintillators," *Nuclear Science, IEEE Transactions on*, vol. 52, pp. 1836–1841, 11/01 2005, doi:10.1109/TNS.2005.856757.

[56] N. Cherepy et al., "Cerium-Doped Single Crystal and Transparent Ceramic Lutetium Aluminum Garnet Scintillators," *Nuclear Instruments and Methods in Physics Research Section A: Accelerators, Spectrometers, Detectors and Associated Equipment*, vol. 579, pp. 38–41, 08/01 2007, doi:10.1016/j.nima.2007.04.009.

[57] H. Yamada, A. Suzuki, Y. Uchida, M. Yoshida, H. Yamamoto, and Y. Tsukuda, "A Scintillator Gd2 O 2 S: Pr, Ce, F for X-Ray Computed Tomography," *Journal of the Electrochemical Society*, vol. 136, no. 9, pp. 2713–2716, September 1, 1989 1989, doi:10.1149/1.2097566.

[58] S. Duclos et al., "Development of the HiLight (TM) Scintillator for Computed Tomography Medical Imaging," *Nuclear Instruments and Methods in Physics Research Section A: Accelerators, Spectrometers, Detectors and Associated Equipment*, vol. A505, pp. 68–71, 06/01 2003, doi:10.1016/S0168-9002(03)01022-2.

[59] K. Yang, P. Menge, J. Buzniak, and V. Ouspenski, *Performance Improvement of Large Sr2+ and Ba2+ Co-doped LaBr3:Ce3+ Scintillation Crystals*. 2012, pp. 308–311.

[60] M. S. Alekhin et al., "Improvement of γ-ray Energy Resolution of LaBr3:Ce3+ Scintillation Detectors by Sr2+ and Ca2+ Co-doping," *Applied Physics Letters*, vol. 102, no. 16, p. 161915, 2013/04/22 2013, doi:10.1063/1.4803440.

[61] M. Alekhin, S. Weber, K. Krämer, and P. Dorenbos, "Optical Properties and Defect Structure of Sr2+ Co-doped LaBr3:5%Ce Scintillation Crystals," *Journal of Luminescence*, vol. 145, pp. 518–524, 01/31 2014, doi:10.1016/j.jlumin.2013.08.019.

[62] G. L. Montagnani et al., "Spectroscopic Performance of a Sr Co-doped 3" LaBr3 Scintillator Read by a SiPM Array," *Nuclear Instruments and Methods in Physics Research Section A: Accelerators, Spectrometers, Detectors and Associated Equipment*, vol. 931, 04/01 2019, doi:10.1016/j.nima.2019.03.067.

[63] S. C. Thacker et al., "Low-Afterglow CsI:Tl Microcolumnar Films for Small Animal High-speed MicroCT," (in English), *Nucl Instrum Methods Phys Res A*, vol. 604, no. 1, pp. 89–92, 2009, doi:10.1016/j.nima.2009.01.036.

[64] C. Brecher et al., "Suppression of Afterglow in CsI:Tl by Codoping with Eu2+—I: Experimental," *Nuclear Instruments and Methods in Physics Research Section A: Accelerators, Spectrometers, Detectors and Associated Equipment*, vol. 558, pp. 450–457, 03/01 2006, doi:10.1016/j.nima.2005.11.119.

[65] S. R. Miller et al., "Reduced Afterglow Codoped CsI:Tl for High-Energy Imaging," *IEEE Trans Nucl Sci*, vol. 65, no. 8, pp. 2105–2108, 2018, doi:10.1109/TNS.2018.2807986.

[66] V. V. Nagarkar et al., "Scintillation Properties of CsI:Tl Crystals Codoped With ${\rm Sm}^{2+}$," *IEEE Trans Nucl Sci*, vol. 55, no. 3, pp. 1270–1274, 2008, doi:10.1109/TNS.2008.915689.

[67] E. Ovechkina, S. Miller, V. Gaysinskiy, C. Brecher, and V. Nagarkar, "Effect of Tl(+) and Sm(2+) Concentrations on Afterglow Suppression in CsI:Tl, Sm Crystals," *IEEE Trans Nucl Sci*, vol. 59, pp. 2095–2097, 10/01 2012, doi:10.1109/TNS.2012.2187072.

[68] Y. Wu et al., "Effects of Bi 3+ Codoping on the Optical and Scintillation Properties of CsI:Tl Single Crystals," *Physica Status Solidi (A) Applications and Materials*, 07/24 2014.

[69] A. Gektin et al., "Modification of NaI Crystal Scintillation properties by Eu-doping," *Optical Materials – OPT MATER*, vol. 32, pp. 1345–1348, 08/01 2010, doi:10.1016/j.optmat.2010.04.014.

[70] I. V. Khodyuk, S. A. Messina, T. J. Hayden, E. D. Bourret, and G. A. Bizarri, "Optimization of Scintillation Performance via a Combinatorial Multi-element Co-doping Strategy: Application to NaI:Tl," *Journal of Applied Physics*, vol. 118, no. 8, p. 084901, 2015/08/28 2015, doi:10.1063/1.4928771.

[71] K. Yang and P. R. Menge, "Improving γ-ray Energy Resolution, Non-proportionality, and Decay Time of NaI:Tl+ with Sr2+ and Ca2+ Co-doping," *Journal of Applied Physics*, vol. 118, no. 21, p. 213106, 2015/12/07, doi:10.1063/1.4937126.

[72] R. W. Pringle and S. Standil, "The Gamma-Rays from Neutron-Activated Gold," *Physical Review*, vol. 80, no. 4, pp. 762–763, 11/15/ 1950, doi:10.1103/PhysRev.80.762.

[73] D. Engelkemeir, "Nonlinear Response of NaI(Tl) to Photons," *Review of Scientific Instruments*, vol. 27, no. 8, pp. 589–591, 1956, doi:10.1063/1.1715643.

[74] D. W. Aitken, B. L. Beron, G. Yenicay, and H. R. Zulliger, "The Fluorescent Response of NaI(Tl), CsI(Tl), CsI(Na) and CaF2(Eu) to X-Rays and Low Energy Gamma Rays," *IEEE Transactions on Nuclear Science*, vol. 14, no. 1, pp. 468–477, 1967, doi:10.1109/TNS.1967.4324457.

[75] B. D. Rooney and J. D. Valentine, "Benchmarking the Compton Coincidence Technique for Measuring Electron Response Non-proportionality in Inorganic Scintillators," in *1995 IEEE Nuclear Science Symposium and Medical Imaging Conference Record*, 21–28 Oct. 1995, vol. 1, pp. 404–408 vol.1, doi:10.1109/NSSMIC.1995.504254.

[76] B. D. Rooney and J. D. Valentine, "Scintillator Light Yield Nonproportionality: Calculating Photon Response Using Measured Electron Response," *IEEE Transactions on Nuclear Science*, vol. 44, no. 3, pp. 509–516, 1997, doi:10.1109/23.603702.

[77] J. D. Valentine, B. D. Rooney, and J. Li, "The Light Yield Nonproportionality Component of Scintillator Energy Resolution," in *1997 IEEE Nuclear Science Symposium Conference Record*, 9–15 Nov. 1997, vol. 1, pp. 833–837 vol.1, doi:10.1109/NSSMIC.1997.672710.

[78] W. Choong et al., "Design of a Facility for Measuring Scintillator Non-Proportionality," *IEEE Transactions on Nuclear Science*, vol. 55, no. 3, pp. 1753–1758, 2008, doi:10.1109/TNS.2008.921491.

[79] P. B. Ugorowski, M. J. Harrison, and D. S. McGregor, "Design and Performance of a Compton-coincidence System for Measuring Non-proportionality of New Scintillators," *Nuclear Instruments and Methods in Physics Research Section A: Accelerators, Spectrometers, Detectors and Associated Equipment*, vol. 615, no. 2, pp. 182–187, 2010/04/01/ 2010, doi:https://doi.org/10.1016/j.nima.2010.01.028.

[80] I. V. Khodyuk and P. Dorenbos, "Nonproportional Response of LaBr3:Ce and LaCl3:Ce Scintillators to Synchrotron X-ray Irradiation," *Journal of Physics: Condensed Matter*, vol. 22, no. 48, p. 485402, 2010/11/17 2010, doi:10.1088/0953-8984/22/48/485402.

[81] I. V. Khodyuk, J. T. M. d. Haas, and P. Dorenbos, "Nonproportional Response Between 0.1–100 keV Energy by Means of Highly Monochromatic Synchrotron X-Rays," *IEEE Transactions on Nuclear Science*, vol. 57, no. 3, pp. 1175–1181, 2010, doi:10.1109/TNS.2010.2045511.

[82] S. A. Payne, S. Hunter, L. Ahle, N. J. Cherepy, and E. Swanberg, "Nonproportionality of Scintillator Detectors. III. Temperature Dependence Studies," *IEEE Transactions on Nuclear Science*, vol. 61, no. 5, pp. 2771–2777, 2014, doi:10.1109/TNS.2014.2343572.

[83] S. A. Payne et al., "Nonproportionality of Scintillator Detectors: Theory and Experiment. II," *IEEE Transactions on Nuclear Science*, vol. 58, no. 6, pp. 3392–3402, 2011, doi:10.1109/TNS.2011.2167687.

[84] S. A. Payne, N. J. Cherepy, G. Hull, J. D. Valentine, W. W. Moses, and W. Choong, "Nonproportionality of Scintillator Detectors: Theory and Experiment," *IEEE Transactions on Nuclear Science*, vol. 56, no. 4, pp. 2506–2512, 2009, doi:10.1109/TNS.2009.2023657.

[85] S. A. Payne, "Nonproportionality of Scintillator Detectors. IV. Resolution Contribution from Delta-Rays," *IEEE Transactions on Nuclear Science*, vol. 62, no. 1, pp. 372–380, 2015, doi:10.1109/TNS.2014.2387256.

[86] M. Moszynski et al., "Energy Resolution of Scintillation Detectors," *Nuclear Instruments and Methods in Physics Research Section A Accelerators Spectrometers Detectors and Associated Equipment*, vol. 805, pp. 25–35, 01/01 2016, doi:10.1016/j.nima.2015.07.059.

[87] W. W. Moses et al., "The Origins of Scintillator Non-Proportionality," *IEEE Transactions on Nuclear Science*, vol. 59, no. 5, pp. 2038–2044, 2012, doi:10.1109/TNS.2012.2186463.

[88] I. V. Khodyuk and P. Dorenbos, "Trends and Patterns of Scintillator Nonproportionality," *IEEE Transactions on Nuclear Science*, vol. 59, no. 6, pp. 3320–3331, 2012, doi:10.1109/TNS.2012.2221094.

[89] V. Baryshevsky et al., "YAlO3: Ce-Fast-acting Scintillators for Detection of Ionizing Radiation," *Nuclear Instruments and Methods in Physics Research B*, vol. 58, pp. 291–293, 05/31 1991, doi:10.1016/0168-583X(91)95605-D.

[90] M. Kapusta, M. Balcerzyk, M. Moszynski, and J. Pawelke, "A High-Energy Resolution Observed from a YAP:Ce Scintillator," *Nuclear Instruments and Methods in Physics Research Section A: Accelerators, Spectrometers, Detectors and Associated Equipment*, pp. 610–613, 02/01 1999, doi:10.1016/S0168-9002(98)01232-7.

[91] R. M. Turtos, S. Gundacker, E. Auffray, and P. Lecoq, "Towards a Metamaterial Approach for Fast timing in PET: Experimental Proof-of-concept," *Physics in Medicine and Biology*, vol. 64, no. 18, p. 185018, Sep 19, 2019, doi:10.1088/1361-6560/ab18b3.

[92] S. Gundacker et al., "Experimental Time Resolution Limits of Modern SiPMs and TOF-PET Detectors Exploring Different Scintillators and Cherenkov Emission," *Physics in Medicine and Biology,* vol. 65, no. 2, p. 025001, Jan 17, 2020, doi:10.1088/1361-6560/ab63b4.
[93] A. G. Wright, *The Photomultiplier Handbook.* Oxford University Press, 2017.
[94] Hamamatsi. "The Photomultiplier Tube." www.hamamatsu.com/resources/pdf/etd/PMT_TPMZ0002E.pdf.
[95] M. G. Bisogni, A. Del Guerra, and N. Belcari, "Medical Applications of Silicon Photomultipliers," *Nuclear Instruments and Methods in Physics Research A,* vol. 926, pp. 118–128, 2019, doi:10.1016/j.nima.2018.10.175.
[96] V. Taranyuk, "State of the Art of Scintillation Crystal Growth Methods." In: *Engineering of Scintillation Materials and Radiation Technologies*, Eds, Korzhik, M. and, Gektin, A., Springer Proceedings in Physics, 2019, pp. 147–161.
[97] C. W. E. v. Eijk, "Inorganic Scintillators in Medical Imaging," *Physics in Medicine and Biology,* vol. 47, no. 8, pp. R85–R106, 2002/04/05 2002, doi:10.1088/0031-9155/47/8/201.

7 Semiconductor Detectors

Per Roos

CONTENTS

7.1 INTRODUCTION AND HISTORICAL BACKGROUND

The first reported use of semiconductor materials as radiation detectors was by Stetter in 1941 [1], studying the response of ionizing radiation in diamonds, among other materials. Triggered by this work, Van Heerden [2] some years later proved that detection of alpha particles, electrons, and gamma rays could be done using AgCl crystals. A number of tests on various compound semiconductors (e.g. TlBr and GaAs) were performed in the 1940s and 1960s. Due to poor performance of new materials as compared to the recently discovered NaI(Tl) scintillator, future perspectives of these new type of 'crystal counters' was judged to be very limited. In the first edition of the radiation detection book by Price in 1958 [3], crystal counters were given a total space of about two pages among 'Photographic emulsions and other detection methods' (Scintillation detectors were given 48 pages). In the second edition in 1964 semiconductor radiation detectors had become a chapter of its own.

When characterizing detectors for ionizing radiation, at least four parameters are of importance: counting efficiency, energy resolution, time resolution, and linearity. Other factors, like price, the need for cooling and mechanical stability matter but depend very much on the application. Semiconductor detectors are roughly single-element devices (Ge, Si) or compound detectors (e.g. CdTe, HgI_2). Generally, detectors based on Ge and Si have their main advantages among the first four parameters above while compound semiconductors mainly have evolved due to a need to have detectors operating at room temperature and to a lesser extent being more compact and robust. The main advantage of semiconductor detectors is their superior energy resolution due to production of more information carriers than any other detector type for ionizing radiation. A visual Monte Carlo simulated example of the benefit of having good energy resolution is shown in Figure 7.1 where the simulated responses from a Ge- and NaI-detector are shown for a virtual ^{108m}Ag source [4]. A high energy resolution not only enables distinguishing interfering peaks in a spectrum but is also one of the major parameters of importance for the signal-to-noise ratio, which in gamma spectrometry, is referred to as the peak to Compton value and which depends on energy resolution and detector efficiency.

DOI: 10.1201/9780429489556-7

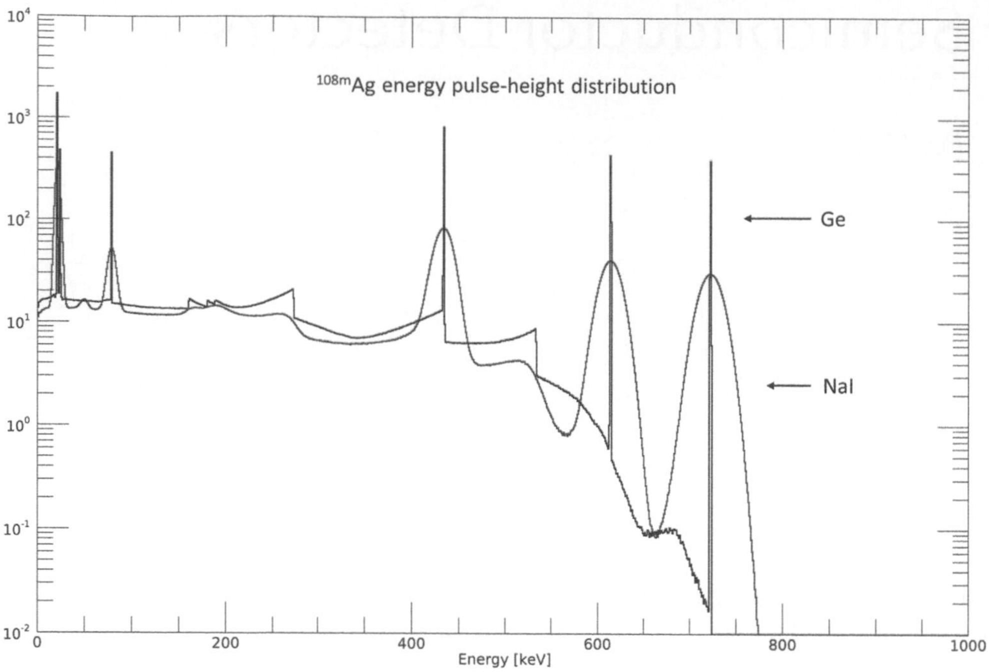

FIGURE 7.1 Monte Carlo simulated response of a Ge- and NaI(Tl)-detector to a virtual ^{108m}Ag-source.

In order to make a semiconductor material work as a spectrometric detector for ionizing radiation, it must have a low inherent production rate of charge carriers, that is, a high resistivity [Ωm] while at the same time allows charge carriers to move fast and uninterrupted once created. A suitable material must therefore be able to be produced without impurities and at sizes large enough to minimize secondary particle escapes. The material must accept being connected to an external voltage producing an electrical field high enough to sweep all charge carriers out of the volume before being trapped. The main obstacle in fulfilling these needs has been to produce high-purity materials with semiconductor properties. Materials of poorer quality have lower resistivity, and the crystal structure shows defects or inhomogeneous structures. Such materials produce higher background noise, less good charge carrier mobility, and cannot withstand applied high-voltage, resulting in poor spectroscopic performance.

In mid-twentieth-century, scintillation detectors, mainly NaI(Tl), were becoming the standard for γ-spectrometry. The first semiconductor detectors appeared approximately at the same time. Techniques for growing crystals of semiconductor materials were far from good enough and clean-room techniques were not yet implemented. The methodology to obtain high-resistivity material was therefore entirely based on the pn-junction technique, which was commonly used when constructing diodes and transistors. By combining p- and n-doped semiconductor materials (for instance boron-doped and phosphorus-doped silicon) at elevated temperatures, the diffusion of electron and holes at the junction forms residual immobile charged dopant ions thus creating an electrical field across the junction, which counteracted further diffusion of charges and a steady-state condition thus evolves. The electrical field (roughly 1 V over a distance of some tens of micrometres) is strong enough to swiftly sweep away any new mobile charges appearing due to thermal action or ionizing radiation. The region is therefore depleted in charge carriers relative to the rest of the material volume (which experience no electrical field) and is therefore called the 'depletion layer'. This construction is the basis for common semiconductor diodes and, if connecting such a diode to an external voltage in the forward direction (positive voltage to the p-end), a current will flow freely as soon as the voltage exceeds the junction potential. The current is made up by the excess electrons on the n-side being swept across the junction to the p-side, where the positive voltage is connected. By reverse biasing such a diode (negative voltage to the p-end) the inherent potential across the junction will instead increase and the depletion region expand. The magnitude of expansion depends mainly on the applied voltage and the resistivity of the material. The greater the resistivity and the voltage the larger the depletion region will extend. Due to the difficulty in making materials of sufficient resistivity (even when cooling) depletion depths in early detector materials were not large enough (some fraction of millimetres to some mm) to become attractive as γ-counters. Also, the practical diffusion length of doped materials into each other was limited. The initial use of such counters was therefore merely as detectors for charged particles, where smaller devices met the needs. The commercial names of such devices were, for instance, Diffusive Junction (DJ) detectors. Due to the inability to extend the depletion depth to the whole volume of the doped semiconductor in such DJ-detectors the entrance window to the detector became relatively thick, resulting in energy losses and peak broadening for charged particles. In addition, the leakage current of early DJ-detectors from minority carriers (impurities still present in the depleted layer) as well as poor contact connections resulted in poor energy resolution. Later techniques enabled the pn-junction to be formed by oxidative changes of the Si-surface (surface barrier detectors) and later by ion-implantation of doped material in silicon (PIPS – Passivated Implanted Planar Silicon) with orders of magnitude improvement in leakage current and thickness of entrance window resulting in better peak shape and energy resolution. These techniques all applied to the production of charged particle detectors since their limited dimensions (on the order of a few mm) was not sufficient to make them act as photon detectors. Detectors for charged particles are preferentially also based on Si rather than Ge due to the lower Rutherford back-scatter $\left(\sim Z^2\right)$ in Si relative Ge and due to the fact that Si-detectors can be operated at room temperature because of the larger band gap than Ge (1.12 eV vs 0.67 eV, respectively, at 300 K).

In order for semiconductor materials to meet the demands for γ-spectrometers, larger detector volumes were required; alternatively, the detector must be based on a material with higher mass-density and/or Z number. The two main disadvantages of using Si as material in γ-spectrometers are that high resistivity Si is not easily available and the combination of low mass-density and low atomic number does not favour the material to be used for γ-spectrometry at energies above some tens of keV. Germanium was a far better alternative, but impurity concentrations were still too high in the mid-1960s, even in several times re-crystalized (zone-refined) material. The first use of germanium as a radiation detector appeared in a report by McKay [5] where a surface barrier pn-junction was used to observe counts from a polonium alpha emitter. The detector was, however, too small to be used for γ-rays.

7.2 SINGLE-ELEMENT SEMICONDUCTORS

The discovery of Li-drifting Ge was the first technique that enabled large size γ-spectrometers for high-photon energies. The principles of Li-drifted detectors based on Si are described in [6–8] while Ge(Li) is described in [9]. The first Ge(Li) detectors were fabricated in 1962 while larger detectors of coaxial-type were first fabricated in 1965. Detailed descriptions on fabrication techniques and their use up to 1966 are provided in [10].

The Li^+ is a small ion (He-size) with large mobility in several materials. Li, in the form of oil or grease or as vapor can be applied on one side of a Ge-crystal and by applying a voltage of some hundred volts across the crystal and raising the temperature (~50 °C) for some days, the lithium ions can be made to diffuse deep into the Ge crystal. Since Ge materials at the time were slightly p-doped from the impurities in the processing, the addition of a n-type dopant like Li could act to compensate the Ge, thus producing high-resistivity material. Following the Li-drifting process the detector had to be cooled to liquid nitrogen temperatures (-196 °C) in order to prevent further Li migration, which otherwise would have ruined the detector. The Li-drifting enabled production of large-size Ge(Li) 'Jelly' detectors, which at the time became the state-of-the-art way in γ-spectrometry equipment. Due to shorter diffusion length of Li-ions in Si, the production of Si(Li) 'Silly' detectors resulted in smaller volume detectors more suited for low energy (<50 keV) photons. While Ge escape peaks complicates low energy spectra the absence of Si characteristic X-ray escape peaks in spectra obtained by Si(Li) provides this detector with a strong advantage in for instance X-ray fluorescence work. In 1980, further development of zone-refined Ge crystals resulted in manufacturing of the first high-purity Ge (HPGe) detectors which at the same time resulted in stopping further production of Ge(Li). Reviews on Ge-detectors are in [11, 12]. Figure 7.2 shows some of the principal components in a Ge(Li) and Ge solid-state detector system as well as an example of a portable Si(Li) detector.

Due to their superior energy resolution, the introduction of Si and Ge detectors radically took over from scintillator-based detectors in several research fields, but in medical imaging sciences energy resolution is not often the primary parameter. For that reason, Si- and Ge-based detectors have never played a major role in medical radiation physics. Imaging systems like Anger camera-based techniques and PET have relied on scintillator-based detector materials instead. In parallel to single-element semiconductors like Si and Ge, development of compound semiconductor detectors has been going on for more than fifty years. These types of detectors have had better success in medical radiation physics, and the rest of this chapter deals with them.

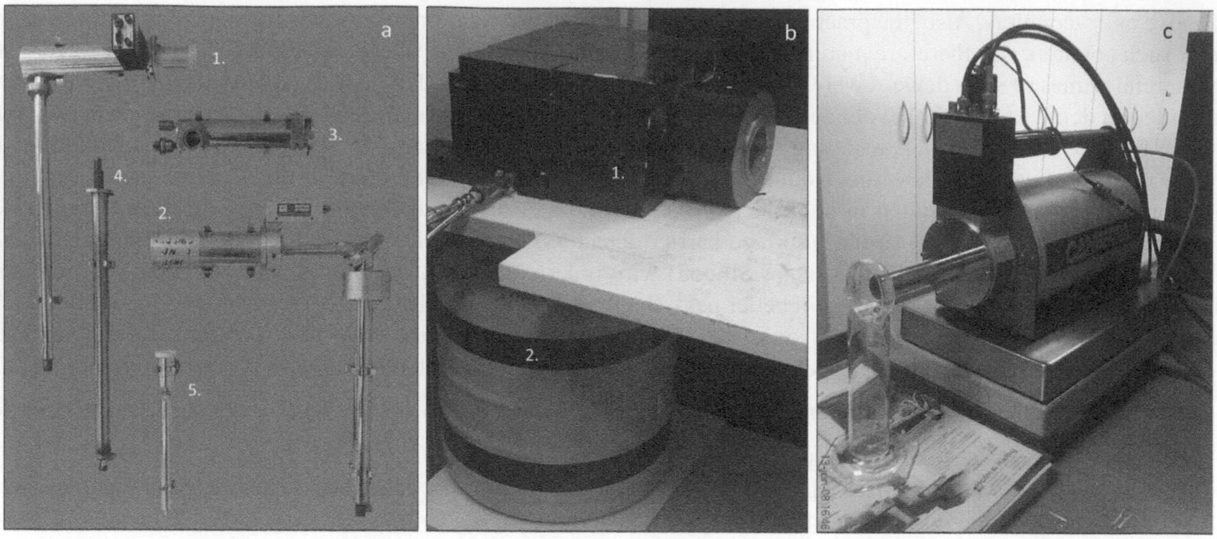

FIGURE 7.2 (a) Decommissioned Ge(Li) detectors with their associated aluminium cryostats. Items 1 and 2 show two different detectors in horizontal geometry with their preamplifier units still attached. Item 4 is the 'cold finger' providing thermal conductivity between detector element and liquid nitrogen thermos (the Dewar). Item 5 is a small vertical Ge(Li) detector. Both the detector cryostat and cold finger have internal vacuum when in operation. (b) High purity Ge detector surrounded by lead bricks (item 1) in order to lower influence from background radiation. The liquid nitrogen thermos (Dewar, item 2) contains typically some 20 litres of liquid nitrogen and is sufficient to cool a normal size Ge-detector for about a week. The metallic Dewar is both electrically and mechanically isolated from the floor below (plate under Dewar) in order to avoid electrical noise and microphonics. (c) A portable Si(Li) detector (about 15 mm diameter and 5mm thick) cooled with liquid nitrogen to enable measurements of low energy (~ 1 keV) characteristic X-rays or gammas. Contrary to Ge-detectors Si-detectors does not require cooling to operate but cooling provides lower spectrum noise, which enables spectrometry at lower energies.

7.3 COMPOUND SEMICONDUCTORS

Semiconductor detectors composed of Ge or Si show excellent energy resolution and linearity. They cover most applications for spectrometry of both charged particles and gamma/X-ray radiation. With respect to gamma/X-ray - spectrometry they have only two disadvantages, the need for cooling and the relatively low Z numbers, which lowers the photoelectric fraction thus requiring larger detector volumes to be effective for higher photon energies. It should be mentioned that the low Z number is also an advantage in low-energy gamma/X-ray spectrometry in that the fraction of escape peaks in the spectrum is significantly reduced due to lower fluorescence yields and lower energies of the characteristic X-rays, preventing them from escaping the detector volume except close to the detector surface. Nevertheless, applications where cooling is impractical or where more efficient detector material (higher Z number) are required give room for alternative detector materials apart from Ge and Si. Compactness, weight, spatial resolution, or operation at elevated temperatures may in several applications be more important than energy resolution, which has created a market for alternative semiconductor materials. The only additional valence 4 element apart from Ge and Si is carbon, which in the form of diamond has been used with success as charged-particle detector [13] in spite of being an insulator. One of the unique qualities with diamond is that it can be operated fully unprotected and exposed to sunlight due to its large band gap (5.5 eV), which is beyond UV.

Compound semiconductors are generally derived from mixtures of elements in groups II to VI of the periodic table. Most of these elements are soluble within each other as long as the Hume-Rothery rules are satisfied. In order to fill the outer electron shells of the elements in the crystal lattice, combination of elements take place between group II-VI, III-V and IV-IV so that the sum of the group numbers is always 8. Compound semiconductors can be made up of two or several elements and can thus be binary (e.g. CdTe), ternary (e.g. CdZnTe), or quaternary (e.g. InGaAsP) by alloying binary compounds together. In spite of that compound semiconductors were the first semiconductors to be experimented on, very few of them have been judged to fulfil the requirements for radiation detectors. Usually, difficulties appear when trying to grow chemically pure or perfect crystals, which often leads to charge trapping that prevents reasonable sizes of crystals to be grown without substantial loss of charge carriers. Many compounds suffer from being brittle, difficult

TABLE 7.1
Properties of Semiconductor Materials Used as Radiation Detectors.

	Si	Ge	CdTe	$Cd_{0.9}Zn_{0.1}Te$	$Cd_{0.7}Zn_{0.3}Se$	CdSe	TlBr	GaAs	PbI_2	HgI_2
Average Z number	14	32	50	49.1	38	41	58	31.5	63	62
Mass-density (300 K) [g cm^{-3}]	2.33	5.33	5.85	5.78	5.5	5.81	7.56	5.32	6.2	6.4
Band gap [eV]	1.12	0.67	1.44	1.57	2.0	1.73	2.68	1.43	2.32	2.15
Electron-hole creation energy [eV]	3.62	2.96	4.43	4.64	6.0	5.5	6.5	4.2	4.9	4.2
Electron mobility, μ [$cm^2 V^{-1} s^{-1}$]	1400	3900	1100	1000	NA	840	30	8000	8	100
Hole mobility, μ [$cm^2 V^{-1} s^{-1}$]	1900	1900	100	120	10	75	4	400	2	4
Electron lifetime, t [s]	$>10^{-3}$	$>10^{-3}$	$3x10^{-6}$	$3x10^{-6}$	NA	10^{-7}	$2.5x10^{-6}$	10^{-8}	10^{-6}	$3x10^{-6}$
Hole lifetime, t [s]	10^{-3}	10^{-3}	$2x10^{-6}$	$1x10^{-6}$	10^{-7}	10^{-6}	$3.7x10^{-5}$	10^{-7}	$3x10^{-7}$	$1x10^{-5}$
Electron μt product [$cm^2 V^{-1}$]	~1	>1	$3x10^{-3}$	$4x10^{-3}$	$\sim 10^{-4}$	$6x10^{-5}$	$5x10^{-3}$	$8x10^{-5}$	$\sim 10^{-5}$	$3x10^{-4}$
Hole μt product [$cm^2 V^{-1}$]	~1	>1	$2x10^{-4}$	$1x10^{-4}$	$\sim 10^{-6}$	$8x10^{-5}$	$2x10^{-4}$	$4x10^{-6}$	$3x10^{-7}$	$4x10^{-5}$

Note: Mobllities (μ) Are Given for 300 K, Except for Ge, Which Is for 77 K. NA=Not Available [12, 15]

to grow using standard techniques, or they show changing properties with time – for instance, due to charge build-up (polarization). Table 7.1 lists some of the compound semiconductor materials that have been investigated, together with Si and Ge as a comparison.

For some of the compounds (like TlBr), development in crystal production has developed very rapidly in recent years. On the other hand, some values in the table sometimes refer to only one, or very few, observations, making coming to conclusions on the degree of improvement questionable. Advances in various diagnostic techniques in solid-state physics constantly enable new identifications of the causes for poor material performance, access to and skills in handling such techniques can determine whether a given material can be improved or not. It should therefore be emphasized that the table does not show any static condition in the quality of the various materials. When comparing the materials listed in Table 7.1 it becomes clear that there are two major parameters distinguishing Si and Ge from the compound semiconductors, the superior mobility-lifetime product of charge carriers (μt, [$cm^2 V^{-1}$]) and their relatively similar mobilities of holes and electrons. The mobility-lifetime product determines to what degree the charge carriers will be trapped and lost before reaching the electrodes and will thus set an upper limit of practical detector volume. The importance of equal mobility of the charge carriers becomes evident when monoenergetic ionization events take place at different places in the detector volume, sometimes near the cathode, sometimes closer to the anode. Only if electrons and holes are collected with equal efficiency can artefacts like peak tailing be avoided in the spectrum. For all investigated compound semiconductor materials, these two parameters have been the Achilles heel that has prevented them becoming equally successful detector materials as Si and Ge. Figure 7.3 shows the attenuation (approximate detector efficiency) of 140 and 511 keV photons in various thicknesses of CdTe, Ge and Si. The significantly higher attenuation in CdTe, especially at 140 keV, is one of the advantages of this detector material. At this energy the size of an CdTe-crystal needs only to be about 20–30 per cent of the size of a Ge-crystal to obtain the same efficiency, which means that smaller detectors can be used and which therefore improves spatial resolution in imaging systems. The necessity of limiting the detector size for compound materials has made it even more important to consider the mass-density and average Z number of the material. It should however be remembered that average Z number does not correctly reflect the differences, since the photoelectric effect depends on the Z number raised to 3–4. Thus, the average Z number should be a weighted one; it is the element with the highest Z number that is the most important. As an extreme example, uranium hydride (UH_3) can be used. Examples of review-papers on compound semiconductors are [12, 14–18].

7.3.1 CdTe- and CdZnTe-detectors

Following a near total dominance of semiconductor detectors by Si and Ge materials in the early sixties it was gradually realized that there was a need for semiconductor detector materials capable of operating also at room temperature. CdTe was early one of the more attractive candidates due to its high average Z number with a cross section for photoelectric effect being roughly 4–5 times that of Ge and more than 100 times that of Si. CdTe crystals as potential γ-spectrometers were first experimented on in the 1960s while CdZnTe (CZT) crystals were grown for γ-spectrometry

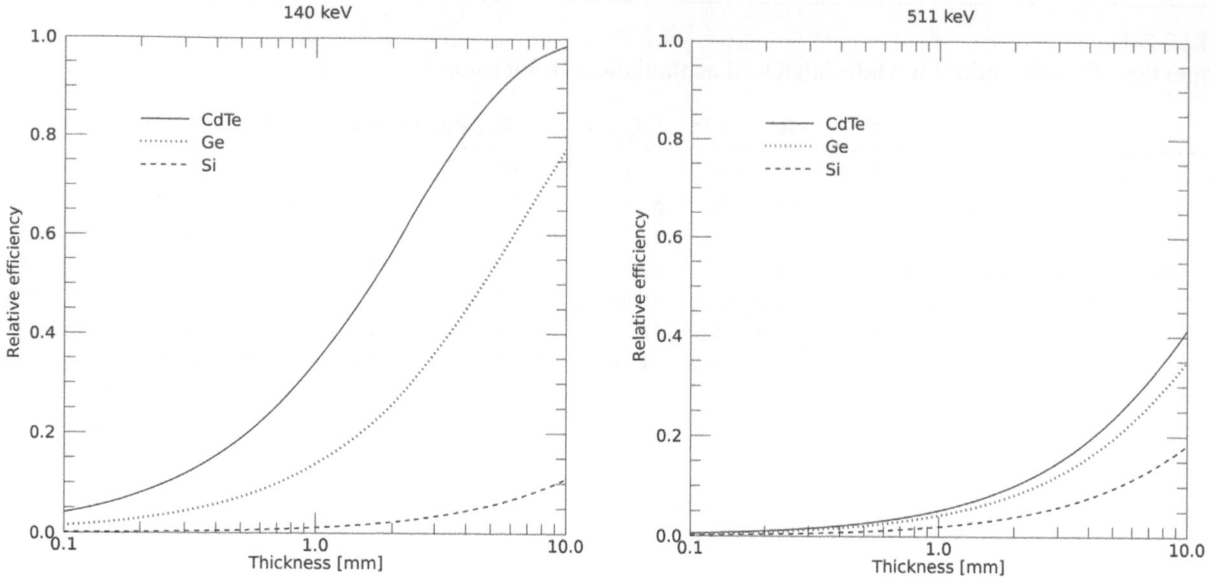

FIGURE 7.3 Detection efficiency for 140 and 511 keV γ-photons in various thickness of CdTe, Si and Ge. Attenuation coefficients obtained from [58].

first in early nineties. Both materials are here discussed together since their properties are very similar. In CZT detectors a fraction, x, of CdTe is exchanged with ZnTe in the manufacturing process, and the common designation is usually $Cd_{1-x}Zn_xTe$ with the fraction of ZnTe being between a few and up to 20 per cent and usually is around 10 per cent. When including Zn, the crystal band gap increases, resulting in higher resistivity and thus lower leakage currents that allow a higher voltage bias to be used, thus improving charge collection. While CdTe detectors often are cooled (-20 to -50 °C) to reduce polarization (see below), thermal noise, and thus leakage currents, CZT have less need for this and can be operated at room temperature due to the larger band gap. Due to the higher leakage current from CdTe, blocking (non-injecting) contacts are often used instead of metallic, ohmic, ones. The blocking contact is simply a highly p- or n-doped material with the p-type contact attached to the cathode. The high concentration of acceptors in the p-type contact prevents ('blocks') current, electrons, flowing in the external circuit through the detector. In earlier pn-junction detectors, this configuration was at place already during production due to the two p and n-type materials being a part of the process. Apart from influencing the leakage current the addition of Zn also has a general positive influence on the crystal manufacturing, lowering the dislocation density and strengthening the crystal lattice. One of the problems with CdTe detectors is the polarization phenomenon, which normally appears in insulators and which in semiconductor materials results in time-dependent (and integrated-exposure dependent) reduction of the charge collection efficiency and therefore also counting efficiency. These phenomena are due to trapping of electrons by deep acceptors within the material. The trapped electrons create a space charge that disturbs the electric field and eventually leads to a decreased depleted layer or active thickness of the detector. It is possible to reduce the polarization effect by increasing the applied bias and by operating the detector under cooled conditions [19]. CZT detectors suffer significantly less from polarization phenomena than CdTe. This is one of the major differences between CdTe and CZT. Both detector types are relatively expensive, primarily due to both CdTe and CZT showing severe problems in manufacturing of detector-grade crystal quality. Reviews of CdTe and/or CZT detectors can be found in [20–23].

It was early realized that there would be obstacles to solve in order to seriously compete with Si- and Ge-based gamma/X-ray detectors. In particular, it was difficult to obtain an energy resolution that was good enough to compete with existing Si- and Ge-based detectors. While the latter could excel with energy resolutions in the order of per mille (e.g. FWHM of 1–2 keV at 1000 keV) typical CdTe detectors commonly had energy resolutions that were a factor of ten worse. This can be explained by the poor charge collection in CdTe/CZT. To overcome the mechanisms responsible for this has proven to be difficult, and the detector types still today suffer from the same fundamental problem as when introduced. Several of the problems are associated with the manufacturing technique, the crystal-growth process, which has put constraints on production and therefore resulted in relatively costly detectors. Crystal growth and processing involves several steps, and each step may contribute with their individual obstacles leading to poor detector performance. Cadmium, for instance, is relatively volatile and is therefore removed from the melt to the overlying atmosphere during crystal growth, leading to deviations in stoichiometric ratios in the crystal. To compensate for this, excess Cd can be added to the melt before start. The different melting temperatures of Cd and Te also tend to create Te micro precipitations when the melt cools and Te solubility decreases, this can partly be solved by using long annealing times [24], using excess Cd in the vapor phase. The Te inclusions or precipitates are typically surrounded by a part of the crystal that shows reduced lifetime of the charge carriers. The first detector grade CdTe crystals were made of Cl-doped CdTe grown by the travelling heater method [25]. It was early realized that the crystal growth of CdTe is problematic in several ways. Due to the low thermal conductivity of CdTe, the classical Czochralski method, used to grow several types of scintillator crystal materials as well as Ge-crystals, cannot be satisfactorily used due to the decomposition of the crystal during pulling. A review on CdTe/CZT crystal growth is presented in [21].

The crystal growth demands, and the poor charge collection has also put upper limits on the size of CdTe/CZT crystals that can perform as radiation detectors with a reasonable energy resolution. Escape peaks are also a serious problem in these types of detectors, in particular since the detector volumes are made small so as not to have charge collection problems, which means that the surface-to-volume ratio is large, thus promoting escape of the characteristic X-rays around 25 keV emitted from Cd and Te. At the same time the high Z number enable even small (<1 cm^3) detectors to be relatively effective and, due to their small physical size and without strong demands of cooling, they can fill a niche where they have an advantage over Si and Ge detectors.

The poor charge collection in CdTe/CZT is exclusively due to trapping of the charge carriers. Traps appear due to crystal dislocations, impurities in the crystal or other inhomogeneities due to precipitates and inclusions of Te formed during the crystal-growth process. Microscopic tellurium inclusions or precipitates are well-known to exist in CdTe/CZT detectors [26]. The narrow band gap of Te (c:a 0.3 eV) relative to CdTe means that Te-enriched regions inside the crystal will have higher conductivity. This affects the leakage current and the electric field, thus disturbing the charge carrier transport. Shallow traps do not affect the mobility much, but de-trapping from deeper traps may have time-scales of μs to seconds [27] and the electronics time constant may thus not allow for the collection of the entire charge created in the detector, even if it can reach the electrodes. In this sense deep traps are the major reason for deteriorating charge transport, thus causing variability in the amount of charge collected from event to event which, in extension, worsens the energy resolution. Just as with ionization chambers, the output pulse is made up of the induced charge on the electrodes (anode and cathode) due to the movement of the charge carriers through the electric field in the detector. The total pulse is thus the sum of induced charge arising from both electrons and holes. In the ideal detector all electrons and holes created during ionization are collected swiftly and with no losses. If one of the charge carriers is partly lost on its way to the electrode, the total output pulse will have an amplitude less than if all charge carriers were collected. Although both electrons and holes are trapped to some extent, the major problem with current CdTe/CZT detectors and all other compound semiconductor materials, is the poor mobility of holes (Table 7.1). The combined poor mobility and relatively high trapping leading to short lifetimes result in pulse rise-time fluctuations and a total charge collection that is dependent on the spatial location of the initial ionization event. Due to the frequency distribution of charge carrier travel distances, depending on the location of the initial ionization event and the direction and interaction of the secondary electrons (e.g., Compton electron, photoelectron), a variable fraction of the charge carriers are being trapped on their way to the electrodes each time a primary photon hits the detector. This causes fluctuations of the total induced charge on the electrodes, which broadens the peak without significantly changing the peak position as long as the irradiation geometry remains constant. This also results in a spectrum with a low-energy tail where the fraction of the tail relative to the peak, as well as the shape of the tail, depends on several factors such as photon energy, divergence of the incoming photons (measurement geometry), applied voltage, detector shape, size and distance between contacts. Richter and Siffert [28] showed that there is a linear relation between the pulse rise time and the peak position. A longer pulse rise time means longer charge travel distances and thus a larger fraction of the charge being trapped, unable to contribute to the output pulse. Assuming a planar detector configuration, if the detector cathode faces incoming photons of low energy, interaction will take place near the surface, close to the cathode, and a large fraction of the holes will be efficiently collected due to the short distance to the cathode resulting in less tailing. If on the other hand the incoming photons face the anode side of the detector, the distance to travel for the holes will be longer, resulting in a larger fraction trapped and a pulse output that may be considerably lower in amplitude than in the previous case – both because of trapping and because of a time constant too small to allow the whole induced charge to be collected.

The problems associated with CdTe/CZT spectroscopy have resulted in parallel ways to solve them. On one hand, corrections techniques, principally for the poor hole collection, have been developed while simultaneously efforts to improve quality of materials, growth processes, purification, and detector manufacturing have taken place. It should be emphasized that improvement in charge collection would not only improve the energy resolution. The main reason preventing CdTe/CZT detectors to be built in larger volumes is the poor charge collection since spectrometric performance greatly deteriorates with larger detectors. Better charge collection is thus the key to both energy resolution and efficiency.

There have been several suggested remedies for the spectrometric effects of poor hole trapping. The first one follows from the example above by using the detector cathode as the detector surface [22]. This works best for low-energy photons. A second method involves pulse shape correction techniques which include both pulse shape discrimination methods [29–31] and pulse rise time compensation [29, 32, 33]. Both of these methods use hardware and software techniques to distinguish events with a large contribution arising from hole trapping. Such events can be identified by the slow rise of the output pulse, which is due to the slower hole movement in the electrical field. A drawback here is the elimination of pulses with a large contribution of the slow rise time. Events with a very large amplitude loss would be situated far from the full energy peak and would not affect the detector efficiency if being sorted away. However, events with only partial losses, appearing in the spectrum as part of the tail near the full energy peak, would also be sorted away provided the preset conditions allow for this. By elimination of slow and incomplete charge collection, the improved energy resolution thereby comes with a price in detector count rate or efficiency. Other ways have been tried to improve the energy resolution, but without significant drops in counting efficiency. Auricchio and colleagues [32] fed the detector output to two charge-sensitive preamplifiers with different shaping time constants to be able to collect both the fast component due to induced charge by moving electrons, thus dependent on the position of the primary interaction event, and the full collected charge composed by both electrons and holes. Using offline analysis and the magnitude of the fast electron pulse permitted them to estimate the loss of charge inside the crystal, thereby enabling reconstruction of the energy of the primary photon. Although these methods may be useful during certain conditions, the most effective remedy for compensating hole charge carrier losses are by configuring the electrical contacts in a way to minimize the material defects on energy resolution. Such contacts may be various types of the Frisch grid with single- or multi-anode contacts consisting of strip or pixelated design. Common for all these solutions is that the readout signal is only taken from the anode, opposite to what is used for Ge and Si detectors, where the signal usually is made up by the contribution from both hole and electron movement.

The Frisch grid was an early invention used in ionization chambers to overcome the spatial dependence of the initial ionization event on energy resolution. This appears due to the induced charge at the electrodes because of the moving charge carriers, in an ionization chamber made up by electrons and positive ions. For an acceptable energy resolution, the induced charge from both positive ions and electrons would have to be collected. Since the positive ions move very slowly relative to the electrons in an ion chamber, they are at greater risk to recombine and form neutral gas atoms, resulting in lost charge. The fraction that gets lost depends on their travelling distance to the cathode resulting in a collected sum charge that will vary depending on the interaction point in the gas volume. This is a situation analogous to most compound semiconductors, where the holes move much slower than the electrons. The slow output pulse is also difficult to handle due to its slow rise time. If only the electrons would to be considered in making up the pulse, the size of each pulse would vary due to the different distances travelled by the electrons through the gas volume (or the detector crystal). The insertion of a third electrode, the Frisch grid, aimed to solve this problem. The Frisch grid is placed in between the anode and cathode, relatively close to the anode. The potential at the grid is also set to a value between the anode and cathode. The chamber is thus divided in two compartments, cathode-grid, which is the main volume where ionization events mainly appear and the grid-anode, which is the part making up the output signal. During an ionization event in the cathode-grid volume electrons are swiftly swept to the anode, passing the thin wired grid with little loss, while the positive ion cloud essentially is immobile. If the output pulse is made up by the induced charge on the anode by electrons moving between grid and anode, the size of the output pulse becomes independent on the position of the initial ionization event. The size of the pulse will, however, be reduced relative to a non-gridded chamber due to the lower difference in potential between grid and anode, relative anode and cathode. In solid detectors it is not possible to place the Frisch electrode inside the detector, and different types of electrodes will instead have to be placed on the detector surface; the design is often referred to as a virtual Frisch grid (VFG). The VFG improves the energy resolution when measured as FWHM, and the low-energy tailing becomes less pronounced but is not completely eliminated. Although the design of the VFG electrodes results in a relatively small grid-anode volume, primary radiation interaction may still take place there, meaning that the drift of holes also will contribute to the collected charge. Several other reasons for the remaining tail, such as electron traps or poor electric field close to the surface [34], may also influence the magnitude of the low-energy tailing.

Multi anode contacts consisting of strips, grids or points (pixels) contacts made using photolithographic techniques are yet another way of improving the readout signal. Barrett and colleagues [35] showed that the deleterious effects of hole trapping were greatly reduced when the anode was divided in an array of pixels. This resulted in a significant reduction of the low-energy tailing. The results not only influenced energy resolution but also detector efficiency, since a larger fraction of the primary photon interactions could be included in the photopeak.

Furthermore, one of the obstacles in using larger single anode detectors was the poor hole collection over large drift distances, which previously required only using thin detectors. With pixelated anodes and reduced influence from poor hole transport larger detectors could be used. The positive effect on energy resolution through the division of the anode into small pixels have, in the literature, generally come to be called 'the small pixel effect'. However, the effect of charge sharing between the pixels, lowering signal to noise ratio in each individual pixel, influences the energy resolution in a negative way. The multi-anode contact also enables information on where in the crystal the interaction takes place by combining lateral information gained from the pixelated anode with depth information obtained from the relative proportions in signal from anode to cathode. Spatial 3D resolution at the mm scale is then achievable with an energy resolution of 1–2 per cent at 511 keV. This approach requires substantially more readout electronics coupled to the detector as compared to single anode contacts. This increases the complexity and price as well as the electronic noise. Designs using a pixelated anode usually have one single cathode while strip detectors (cross-strip detectors) use a strip-pattern for both anode and cathode to improve spatial resolution. CZT-detectors with strip-electrodes structure often requires less complex readout electronics than when using pixelated solutions for the same detector area. This is an important factor when considering complex detector systems like PET with a large number of detectors. Normally the multi anode design is combined with one or several VFG electrodes resulting in relatively complicated readout structures.

With developments in microelectronics and application specific integrated circuits (ASIC's) [36] applied to the output signal the digitized pulse shape, amplitude and timing, may be collected from all contacts, processed online or later. This enables removal of events contributing to a poor peak shape but at the cost of lower counting efficiency. Some advantages and disadvantages of the various detector designs [37] are listed in Table 7.2.

The possibility of improving the energy resolution in CdTe/CZT detectors have resulted in the raw detector signal rarely being displayed directly to users but rather allowed to go through some type of 'fix-filter' before being further processed. This filter can consist of both hardware and software components.

The poor charge collection of holes is not the only reason for the degraded energy resolution. Trapping of electrons has also been observed to influence the spectrometric performance of CdTe detectors. Zhang and colleagues [27]

TABLE 7.2
Summary of Different Contact Designs for CZT Detectors with Associated Advantages and Disadvantages (Redrawn from [37].

Geometry Type	Advantage	Disadvantage	Best performance in Energy Resolution (FWHM)
Planar electrode.	Simple structure.	Severe hole trapping problems.	-
Frisch strip and trapezoid prism electrode.	Simple structure.	Existing leakage currents between the grid and anode.	2.68 % at 662 keV
Insulating Frisch ring electrode.	Eliminating leakage currents between the grid and anode.	More complicated design and fabrication technique.	1.70 % at 662 keV
Pixelated electrodes.	Higher charge collection efficiency, suitable for imaging.	Charge sharing problems.	<3 % at 140 keV
Coplanar grid electrode.	Overcomes hole trapping more effectively.	Needs more readout electronics, more electronic noise.	1.3 % at 662 keV
Hemispherical electrode.	Uniform charge collection.	Complicated geometry design.	<1.9 % at 662 keV
Orthogonal coplanar strip electrode.	Less complexity for the device electronics.	Leakage current in the anode.	1.0 % at 662 keV
Charge-sharing strip electrode.	Simplified electronics and more effective non-collecting signal.	More electronic noise.	<6 % at 122 keV
Drift strip electrode.	The sensitivity to hole trapping is reduced due to the electrostatic shield to the readout anode.	More electronic noise.	0.8% at 356 keV

Note: Note that Only FWHM Have Been Included. Commonly Observed Low- energy Tailing May not Influence the FWHM, and Values on FWTM May Be More Indicative of the Improvements

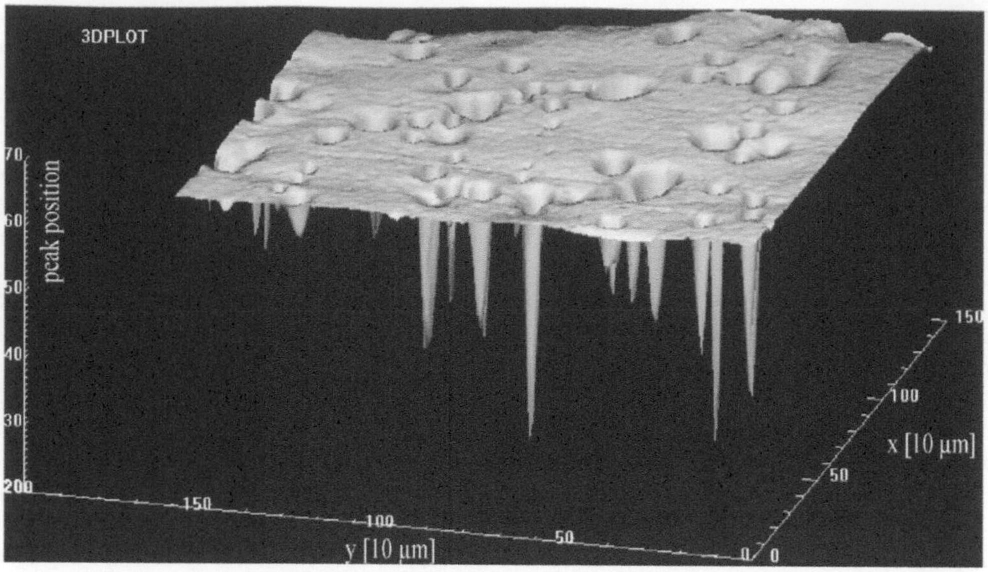

FIGURE 7.4 Three-dimensional sensitivity response (peak position) plot of a CdTe detector scanned by a 10 × 10 μm 30 keV monochromatic X-ray beam from a synchrotron. The sensitivity 'wells' correlate with Te-precipitate positions, which act as electron traps and therefore hinders the charge collection. The picture explains the low energy peak tailing in an energy spectrum. Reproduced from [38] with the permission of AIP Publishing.

identified at least five different types of traps in CZT crystals, with deep tellurium antisites being one of the more severe traps for electron charge carriers. Similarly, Carini and colleagues [38] showed a 100 per cent correlation between electron traps and tellurium precipitates (5–10 μm size) in the crystal (see Figure 7.4). In the case of randomly distributed electron traps throughout the detector the effect of charge trapping on energy resolution and peak position depend on the position of the interaction and the distance the charge carriers have to travel.

The effect on peak broadening and shift in peak position can then be corrected for using similar techniques as for holes, that is, depth-sensing using rise-time techniques (see below). In contrast to such randomly distributed traps, Te-precipitate regions are instead localized regions or non-homogeneously distributed 'hot spots' for electron trapping, where currently there are no existing correction techniques. The problems associated with electron trapping cannot be solved using contact design but must be sought in the material itself, thus requiring the crystal growth process of CdTe and CZT to improve or in one way or another modifying the material. The major remaining factor in fabricating larger CZT detectors today is, however, not the electron traps but rather the inhomogeneity of the crystal [39]. It was found that one approach of growing CdTe based materials with higher crystal uniformity was by addition of small amount of Mn or Se. The addition of selenium at the cost of Te appeared to inhibit the formation of sub-grains and reducing the density of Te inclusions in the crystal, thus ensuring a more homogeneous charge collection and an improved energy resolution. Roy and colleagues [40] reported an energy resolution of around 3.5 per cent at 662 keV for a VFG $Cd_{0.9}Zn_{0.1}Te_{0.985}Se_{0.015}$ detector and as low as 1 per cent for a VFG $Cd_{0.9}Zn_{0.1}Te_{0.98}Se_{0.02}$ detector [41].

7.3.1.1 Timing Properties of CZT-detectors for PET Applications

CZT detectors are currently the only semiconductor material that can compete with scintillator materials for use in PET systems. There is as yet no scintillator material available with a better energy resolution, and the attenuation coefficient for 511 keV photons is comparable with, for example, LYSO scintillators (about 20 per cent lower). The high spatial resolution due to 3D derivation of the impact point in the crystal in pixelated detectors adds to their attraction and CZT have been used in several PET applications. A drawback in using multi-anode equipped detectors is, however, the complex readout electronics, which becomes especially cumbersome when handling several detectors such as in PET. The timing properties of cross-strip or pixelated CZT detectors is also not as good as the fastest scintillator materials, and with a coincidence time resolution of several nanoseconds even for thin crystals (fast charge collection) they are not the ideal choice for TOF-PET. A review partly discussing the role of CZT-detectors for PET can be found in [42].

TABLE 7.3
Comparison of thallium bromide (TlBr) with LYSO and BGO with respect to important detector parameters

	LYSO	BGO	TlBr
Mass-density [g cm^{-3}]	7.1	7.13	7.56
Z_{eff}	66	73	74
Linear attenuation coefficient for 511 keV photons [cm^{-1}]	0.86	0.92	1.04
Mean free path 511 keV [cm]	1.16	1.09	0.97
Index of refraction at 570 nm	1.81	2.15	2.63

Note: LYSO is Lutetium-yttrium oxyorthosilicate, BGO = Bismuth germanate

7.3.2 Thallium Bromide (TlBr)

The TlBr is a compound semiconductor material that currently mainly exists at an experimental stage: commercial TlBr crystals are available but are not yet as ready-to-use detector systems. TlBr used as a spectrometric detector for γ-radiation was first reported in the mid-eighties, although research on mixed thallium bromide/iodide as detector for ionizing radiation was initiated much earlier [43]. TlBr is a competitive future low-cost alternative to CZT for room-temperature use. CZT currently being the only room-temperature operated high-resolution semiconductor detector for γ-spectrometry. Properties of TlBr and the competitive scintillator materials BGO and LYSO are listed in Table 7.3

With its higher mass-density and Z number, TlBr has a higher photoelectric and total attenuation coefficient than most other detector materials – scintillators as well as semiconductors – for energies above about 20 keV. The material can be grown at relatively low temperatures (about 480 C°) and, through repeated zone refinement, can be manufactured with high purity and thus high bulk resistivity. A comparison of crystal growth techniques for TlBr manufacturing is presented in Datta and colleagues [44]. The spectroscopic performance of semiconductors strongly depends on the raw crystal quality, which affects both charge transport and resistivity. Since its introduction in the mid-eighties the manufacturing of TlBr crystals has improved considerably: zone-refined material has enabled the mobility-lifetime product (μτ) to increase by almost three orders of magnitude and is now similar to that of CZT. The importance of the improved mobility-lifetime product is reflected in the manufacturing of larger crystals while maintaining good energy resolution. The band gap of 2.68 eV enables the material to operate at room temperature with a minimum dark current. An energy resolution similar to CZT detectors is achievable with values around 1 per cent for pixelated detectors [45]. Similar to CZT, TlBr shows approximately an order of magnitude better electron than hole mobility, which is why unipolar charge collection detectors like pixelated TlBr detectors show a benefit in energy resolution through the small pixel effect [35], since this minimizes the influence on signal from the migrating holes. This has also been shown for TlBr detectors equipped with Frisch grid electrodes to actively discriminate the hole migrating signal [46, 47]. A negative side of TlBr's high Z number is the severe escape peaks, especially in small detectors, from Tl (70–85 keV K_α-K_β, 10–15 keV L_x) and Br (12–13 keV K_x) characteristic X-ray emission.

Just as with CdTe detectors, deterioration in performance has been observed over time in TlBr detectors. The charge collection becomes incomplete or slower, making the semiconductor appear more as an insulator, the phenomenon usually termed polarization from disturbances of the electrical field. Polarization in CdTe is caused by a build-up of deep hole traps created during irradiation, which results in an increased number of sites with a high space charge thus disturbing the electrical field and preventing efficient charge collection. Since the polarization in CdTe depends on the number of deep hole traps formed, the polarization effect increases with the integrated irradiation and is thus a severe problem when used repeatedly – for instance in high-flux environments like X-ray imaging.

Polarization in TlBr is caused by another mechanism. It has been observed that in TlBr dissociation of the crystal gradually takes place, leading to ionic conductivity of Tl^+ and Br^- in the electrical field, which results in their build-up at the cathode and anode respectively, thereby degrading the charge-collection process through an increased space charge or polarization of the electrical field. Furthermore, the increased free ionic concentrations around the electrodes cause chemical reactions to take place that gradually corrode the electrode surfaces, causing the electrical field to be less stable. While for CdTe the remedy was the replacement of some of the Cd with Zn, creating CdZnTe, the suppression of polarization in TlBr has been successfully achieved: applying cooling [48], by replacing the gold anode contact with a thallium contact or by periodically switching the applied bias.

In order to operate a TlBr detector with a minimum of influence from the slow hole collection, the best energy resolution is obtained, just as with CZT, by operating the detector as an anode-only collector using a virtual Frisch grid (VFG). Of the ionic charges, just like the electrons, Br^- also accumulates at the anode contact [49]. The accumulated Br^- ions are therefore considered to be the primary cause for the TlBr polarization and anode degradation. Contrary to CdTe and CZT detectors where the polarization phenomena partially is reduced by using a VFG, the different mechanism behind polarization in TlBr requires additional measures to reduce it. Cooling reduces diffusivity, thus increasing the time the detector can be operative before sufficient Br^- accumulates at the anode. Cooling to about -20 °C appears to be sufficient for longer stable operations. The cooling also strongly influences the mobility-lifetime product of holes positively but does not seem to have any corresponding effect on electrons [50]. The requirements of cooling, however, introduces additional electronic complexity – for instance in PET-systems where a large number of detectors operate, and this removes one of the key arguments for using compound semiconductors, which is to be able to operate at room temperature. An alternative to reducing the influence of the migrating Br^- has been to apply anode contacts using Tl, which acts as a trap for the free Br^-, forming TlBr. The drawback here is the toxicity of metallic Tl and its quick oxidation compared to gold or platinum contacts but will allow stable operation of the detector at room temperature as long as the anode remains intact. This technique can be combined with a routine of reversing the applied bias when not using the detector. By reversing the bias, the direction of the migrating Br^- ions will change, thereby preventing polarization and metal degradation caused by Br^- accumulation and reactions at the anode.

7.3.2.1 Timing Properties of TlBr-detectors for PET Applications

The possibility to replace expensive CZT detectors with alternatives like TlBr is not only due to the price. The short mean free path of 140 keV photons in TlBr is below 2mm while for CZT it is approximately 5mm, which makes the TlBr an interesting alternative in SPECT. In general, due to the better photon capture per unit volume of TlBr, instrument and equipment for spatial resolution will gain from using TlBr in front of CZT due to their smaller size in multi-detector systems. Although energy resolution and efficiency are primary parameters in most systems, rather the timing properties are of utmost importance in TOF-PET. Normally semiconductors cannot compete with some of the very fast scintillators like LYSO, which are being used in state-of-the-art TOF-PET systems in spite of their poor energy resolution. The charge collection process in semiconductor devices is a relatively slow process, even if only electrons are considered; their mobility and the applied field set an upper limit for how fast the electrons can move. The collection process is also dependent on detector size and the applied field strength, which in turn depends on the resistivity of the material. For 511 keV photons the detector size and material need to be sufficient to enable full energy capture, thus a compromise between collection speed and photon capture exists for a given material. The required timing properties of TOF-PET scanners are often in the <0.5 ns range, while standard CZT detectors have achieved charge collection times of around 10 ns. For TlBr, Arino-Estrada and colleagues [51] measured a coincidence time resolution of 30–40 ns while Hitomi and colleagues [52] obtained a value of around 6 ns in a setup more optimized for fast timing.

Arino-Estrada [53] also examined the possibility of using the Cerenkov emission from TlBr rather than the traditional charge collection. Cerenkov light emission appears when charged particles have a velocity faster than c/n in a material where n is the refractive index of the material (depends on wavelength). The light emission is prompt and thus appears much faster than for instance scintillation light. Cerenkov emission is more abundant at lower wavelengths, which is why it often appears as blue when visible. Because of its wide band gap TlBr is transparent in blue light. Cerenkov photons emitted from fast-moving electrons in the TlBr crystal may thus be detected on the outside contrary to, for instance, CZT which has a smaller band gap (around 1.7 eV) and is only transparent in infrared, where not much Cerenkov intensity appears. The high index of refraction in TlBr (2.68) furthermore enables Cerenkov emission to take place from slower-moving charged particles, and thus over a larger energy interval than in materials with a smaller index of refraction. Arino-Estrada [53] measured a CTR FWHM of 430 ps from a 3x3x5 mm^3 crystal. In a follow-up paper the coincidence time resolution could be reduced to 330 ps [54], thus competing with the very fast LYSO scintillators currently in use in TOF-PET (see Table 7.3). A TlBr detector for TOF-PET would, however, require readout of both normal charge induction as well as Cerenkov light.

An advantage of TlBr compared to LYSO and BGO is the energy resolution, which in TlBr may reach down to around 1 per cent for 511 keV photons. The normal energy resolution at 511 keV for the LYSO and BGO detectors are around 10–15 per cent, which means that peak to background (or peak to Compton) values are substantially less than for a detector with better energy resolution, for example, the TlBr. The lower mean free path for 511 keV photons in TlBr than LYSO and BGO also means smaller detectors can be utilized without losses in count rate, thus improving the spatial information. Coincidence time resolution for LYSO using scintillation emission has been reported to around 100 ps [55], while for BGO the CTR using scintillation emission are around 1–1.5 ns [56] but can reach 0.3–0.6 ns when using Cerenkov light [57]. The higher index of refraction in TlBr compared to BGO, however, means that more Cerenkov photons will be produced for a given particle velocity in TlBr compared to BGO.

7.4 SUMMARY

In general, scintillator-based detectors have an advantage in medical applications due to their diversity, and they therefore provide the user a greater possibility of finding a material that fits a given application relative to what semiconductor-based detectors can provide. There are however applications where semiconductor-based detectors have an advantage – for instance when it comes to energy resolution. While improvement in energy resolution for scintillators requires modification of the material itself, corresponding improvement for compound semiconductor materials may, apart from growing purer crystals, also appear through improvement in the signal readout methodology. Faster, cheaper and more sophisticated ASIC's, real-time signal processing to address more complicated spectrum artefacts, will likely maintain these detectors as a competitive alternative in PET and possibly also in TOF-PET.

REFERENCES

[1] G. Stetter, "Durch Korpuskularstrahlen in Kristallen hervorgerufene Elektronenleitung," *Verhandl. Deutsche Physikalische Gesellschaft*, vol. 22, pp. 13–14, 1941.

[2] P. J. van Heerden, "The Crystal Counter, a New Instrument in Nuclear Physics," University of Utrecht, N.V. Noord Hollandishe, 1945.

[3] W. J. Price, *Nuclear Radiation Detection*. McGraw-Hill, 1958.

[4] M. Ljungberg and S. E. Strand, "A Monte Carlo Program for the Simulation of Scintillation Camera Characteristics," (in English), *Comput Methods Programs Biomed*, vol. 29, no. 4, pp. 257–272, Aug 1989. [Online]. Available: www.ncbi.nlm.nih.gov/pubmed/2791527.

[5] K. G. McKay, "A. Germanium Counter," *Physical Review*, vol. 76, no. 10, pp. 1537–1537, 11/15/ 1949, doi:10.1103/PhysRev.76.1537.

[6] E. M. Pell, "Ion Drift in an n-p Junction," *Journal of Applied Physics*, vol. 31, no. 2, pp. 291–302, 1960/02/01 1960, doi:10.1063/1.1735561.

[7] J. H. Elliot, "Thick Junction Radiation Detectors Made by Ion Drift," *Nuclear Instruments and Methods*, vol. 12, pp. 60–66, 1961/06/01/ 1961, doi:https://doi.org/10.1016/0029-554X(61)90114-8.

[8] N. A. Baily, R. J. Grainger, and J. W. Mayer, "Capabilities of Lithium Drifted p-i-n Junction Detectors When Used for Gamma-Ray Spectroscopy," *Review of Scientific Instruments*, vol. 32, no. 7, pp. 865–866, 1961/07/01 1961, doi:10.1063/1.1717543.

[9] D. V. Freck and J. Wakefield, "Gamma-Ray Spectrum Obtained with a Lithium-drifted p–i–n Junction in Germanium," *Nature*, vol. 193, no. 4816, pp. 669–669, 1962/02/01 1962, doi:10.1038/193669a0.

[10] *Lithium Drifted Germanium Detectors*. Vienna: International Atomic Energy Agency, 1966.

[11] D. Alexiev, M. I. Reinhard, L. Mo, A. R. Rosenfeld, and M. L. Smith, "Review of Ge Detectors for Gamma Spectroscopy," *Australasian Physical & Engineering Sciences in Medicine*, vol. 25, no. 3, pp. 102–109, 2002.

[12] G. F. Knoll, "Other Solid-state Detectors," in *Radiation Detection and Measurement*. New York: John Wiley, 2010.

[13] J. A. Duenas, J. de la Torre Perez, A. Martin Sanchez, and I. Martel, "Diamond Detector for Alpha-particle Spectrometry," *Applied Radiation and Isotopes*, vol. 90, pp. 177–80, Aug 2014, doi:10.1016/j.apradiso.2014.03.032.

[14] A. Mirzaei, J.-S. Huh, S. S. Kim, and H. W. Kim, "Room Temperature Hard Radiation Detectors Based on Solid State Compound Semiconductors: An Overview," *Electronic Materials Letters*, vol. 14, no. 3, pp. 261–287, 2018/05/01 2018, doi:10.1007/s13391-018-0033-2.

[15] A. Owens and A. Peacock, "Compound Semiconductor Radiation Detectors," *Nuclear Instruments and Methods in Physics Research Section A: Accelerators, Spectrometers, Detectors and Associated Equipment*, vol. 531, no. 1, pp. 18–37, 2004/09/21/ 2004, doi:https://doi.org/10.1016/j.nima.2004.05.071.

[16] S. Paul, "Cadmium Telluride and Related Materials as X-ray and Gamma-ray Detectors: A Review of Recent Progress," in *Proceedings of SPIE (Society of Photo-Optical Instrumentation Engineers)*, 1994, vol. 2305, doi:10.1117/12.187258. [Online]. Available: https://doi.org/10.1117/12.187258

[17] D. S. McGregor and H. Hermon, "Room-temperature Compound Semiconductor Radiation Detectors," *Nuclear Instruments and Methods in Physics Research Section A: Accelerators, Spectrometers, Detectors and Associated Equipment*, vol. 395, no. 1, pp. 101–124, 1997/08/01/ 1997, doi:https://doi.org/10.1016/S0168-9002(97)00620-7.

[18] P. Sellin, "Recent Advances in Compound Semiconductor Radiation Detectors," *Nuclear Instruments and Methods in Physics Research Section A: Accelerators, Spectrometers, Detectors and Associated Equipment*, vol. 513, pp. 332–339, 11/01 2003, doi:10.1016/j.nima.2003.08.058.

[19] M. Niraula, A. Nakamura, T. Aoki, Y. Tomita, and Y. Hatanaka, "Stability Issues of High-energy Resolution Diode Type CdTe Nuclear Radiation Detectors in a ong-term Operation," *Nuclear Instruments and Methods in Physics Research Section A: Accelerators, Spectrometers, Detectors and Associated Equipment*, vol. 491, pp. 168–175, 09/01 2002, doi:10.1016/S0168-9002(02)01175-0.

[20] S. D. Sordo, L. Abbene, E. Caroli, A. M. Mancini, A. Zappettini, and P. Ubertini, "Progress in the Development of CdTe and CdZnTe Semiconductor Radiation Detectors for Astrophysical and Medical Applications," *Sensors (Basel)*, vol. 9, no. 5, pp. 3491–526, 2009, doi:10.3390/s90503491.

[21] A. Zappettini, "8 – Cadmium Telluride and Cadmium Zinc Telluride," in *Single Crystals of Electronic Materials*, R. Fornari, Ed.: Woodhead Publishing, 2019, pp. 273–301.

[22] T. Takahashi and S. Watanabe, "Recent Progress in CdTe and CdZnTe Detectors," (in English), *IEEE Transactions on Nuclear Science*, vol. 48, no. 4, pp. 950–959, Aug 2001, doi:Doi 10.1109/23.958705.

[23] T. E. Schlesinger et al., "Cadmium Zinc Telluride and Its Use as a Nuclear Radiation Detector Material," *Materials Science and Engineering: R: Reports*, vol. 32, no. 4, pp. 103–189, 2001/04/02/ 2001, doi:https://doi.org/10.1016/S0927-796X(01)00027-4.

[24] H. R. Vydyanath et al., "Recipe to Minimize Te Precipitation in CdTe and (Cd,Zn)Te Crystals," *Journal of Vacuum Science & Technology B: Microelectronics and Nanometer Structures Processing, Measurement, and Phenomena*, vol. 10, no. 4, pp. 1476–1484, 1992/07/01 1992, doi:10.1116/1.586275.

[25] P. Siffert, "Current Possibilities and Limitations of Cadmium Telluride Detectors," *Nuclear Instruments and Methods*, vol. 150, no. 1, pp. 1–12, 1978/03/15/ 1978, doi:https://doi.org/10.1016/0029-554X(78)90450-0.

[26] R. B. James et al., "Material Properties of Large-volume Cadmium Zinc Telluride Crystals and Their Relationship to Nuclear Detector Performance," *Journal of Electronic Materials*, vol. 27, 06/01 1998, doi:10.1007/s11664-998-0055-x.

[27] J. Zhang et al., "Effect of Point Defects Trapping Characteristics on Mobility-Lifetime (μτ) Product in CdZnTe Crystals," *Journal of Crystal Growth*, vol. 519, pp. 41–45, 2019/08/01/ 2019, doi:https://doi.org/10.1016/j.jcrysgro.2019.04.026.

[28] M. Richter and P. Siffert, "High Resolution Gamma Ray Spectroscopy with CdTe Detector Systems," *Nuclear Instruments and Methods in Physics Research Section A: Accelerators, Spectrometers, Detectors and Associated Equipment*, vol. 322, no. 3, pp. 529–537, 1992/11/15/ 1992, doi: https://doi.org/10.1016/0168-9002(92)91227-Z.

[29] B. D. Keele, R. S. Addleman, and G. L. Troyer, "A Method to Improve Spectral Resolution in Planar Semiconductor Gamma-ray Detectors," *IEEE Transactions on Nuclear Science*, vol. 43, no. 3, pp. 1365–1368, 1996, doi:10.1109/23.507066.

[30] C. Bargholtz, E. Fumero, and L. Mårtensson, "Model-based Pulse Shape Correction for CdTe Detectors," *Nuclear Instruments and Methods in Physics Research Section A: Accelerators, Spectrometers, Detectors and Associated Equipment*, vol. 434, no. 2, pp. 399–411, 1999/09/21/ 1999, doi:https://doi.org/10.1016/S0168-9002(99)00546-X.

[31] J. Cardoso, J. Simoes, and T. Menezes, "CdZnTe Spectra Improvement through Digital Pulse Amplitude Correction Using the Linear Sliding Method," *Nuclear Instruments & Methods in Physics Research Section A-accelerators Spectrometers Detectors and Associated Equipment – NUCL INSTRUM METH PHYS RES A*, vol. 505, pp. 334–337, 06/01 2003, doi:10.1016/S0168-9002(03)01091-X.

[32] N. Auricchio et al., "Twin Shaping Filter Techniques to Compensate the Signals from CZT/CdTe Detectors," *IEEE Transactions on Nuclear Science*, vol. 52, no. 5, pp. 1982–1988, 2005, doi:10.1109/TNS.2005.856884.

[33] J. Tõke, M. Quinlan, W. Gawlikowicz, and W. Schröder, "A Simple Method for Rise-Time Discrimination of Slow Pulses from Charge-Sensitive Preamplifiers," *Nuclear Instruments and Methods in Physics Research Section A Accelerators Spectrometers Detectors and Associated Equipment*, vol. 595, 05/29 2008, doi:10.1016/j.nima.2008.07.024.

[34] A. E. Bolotnikov et al., "Performance characteristics of Frisch-ring CdZnTe detectors," *Nuclear Science, IEEE Transactions on,* vol. 53, pp. 607–614, 05/01 2006, doi:10.1109/TNS.2006.871509.

[35] H. H. Barrett, J. D. Eskin, and H. B. Barber, "Charge Transport in Arrays of Semiconductor Gamma-ray Detectors," (in English), *Physical Review Letters,* vol. 75, no. 1, pp. 156–159, Jul 3 1995, doi:10.1103/PhysRevLett.75.156.

[36] E. Vernon et al., *Front-end ASIC for Spectroscopic Readout of Virtual Frisch-Grid CZT Bar Sensors*. 2019.

[37] Q. Zhang, C. Zhang, Y. Lu, K. Yang, and Q. Ren, "Progress in the Development of CdZnTe Unipolar Detectors for Different Anode Geometries and Data Corrections," *Sensors (Basel)*, vol. 13, no. 2, pp. 2447–74, Feb 18 2013, doi:10.3390/s130202447.

[38] G. A. Carini, A. E. Bolotnikov, G. S. Camarda, G. W. Wright, R. B. James, and L. Li, "Effect of Te Precipitates on the Performance of CdZnTe Detectors," *Applied Physics Letters,* vol. 88, no. 14, p. 143515, 2006/04/03 2006, doi:10.1063/1.2189912.

[39] A. E. Bolotnikov et al., "High-Efficiency CdZnTe Gamma-Ray Detectors," *Nuclear Science, IEEE Transactions on,* vol. 62, pp. 3193–3198, 12/01 2015, doi:10.1109/TNS.2015.2493444.

[40] U. Roy, G. S. Camarda, Y. Cui, and R. B. James, "High-resolution Virtual Frisch Grid Gamma-ray Detectors Based on As-grown CdZnTeSe with Reduced Defects," *Applied Physics Letters*, vol. 114, p. 232107, 06/10 2019, doi:10.1063/1.5109119.

[41] U. Roy, G. S. Camarda, Y. Cui, and R. B. James, "Characterization of Large-volume Frisch Grid Detector Fabricated from As-grown CdZnTeSe," *Applied Physics Letters*, vol. 115, p. 242102, 12/09 2019, doi:10.1063/1.5133389.

[42] W. Jiang, Y. Chalich, and M. J. Deen, "Sensors for Positron Emission Tomography Applications," *Sensors (Basel)*, vol. 19, no. 22, Nov 17 2019, doi:10.3390/s19225019.

[43] R. Hofstadter, "Crystal Counters," (in English), *Nucleonics*, vol. 4, no. 4, p. 2; passim, Apr 1949.

[44] A. Datta, P. Becla, and S. Motakef, *Large Area Thallium Bromide Semiconductor Radiation Detectors with Thallium Contacts*. 2018, p. 37.

[45] H. Kim et al., "Thallium Bromide Gamma-Ray Spectrometers and Pixel Arrays," *Frontiers in Physics*, vol. 8, p. 55, 03/01 2020, doi:10.3389/fphy.2020.00055.

[46] K. Hitomi et al., "TlBr Capacitive Frisch Grid Detectors," *IEEE Transactions on Nuclear Science*, vol. 60, pp. 1156–1161, 04/01 2013, doi:10.1109/TNS.2012.2217155.

[47] A. Kargar et al., "Design of Capacitive Frisch Grid TlBr Detectors for Radionuclide Identification," in *2019 IEEE Nuclear Science Symposium and Medical Imaging Conference (NSS/MIC)*, 26 Oct.-2 Nov. 2019, pp. 1–5, doi:10.1109/NSS/MIC42101.2019.9059717.

[48] F. Costa, C. Mesquita, and M. Hamada, "Temperature Dependence in the Long-Term Stability of the TlBr Detector," *Nuclear Science, IEEE Transactions on*, vol. 56, pp. 1817–1822, 09/01 2009, doi:10.1109/TNS.2009.2024678.

[49] C. Rocha Leao and V. Lordi, "Ionic Current and Polarization Effect in TlBr," *Physical Review B*, vol. 87, p. 081202(R), 02/25 2013, doi:10.1103/PhysRevB.87.081202.

[50] T. Onodera, K. Hitomi, and T. Shoji, "Temperature Dependence of Spectroscopic Performance of Thallium Bromide X- and Gamma-Ray Detectors," *Nuclear Science, IEEE Transactions on*, vol. 54, pp. 860–863, 09/01 2007, doi:10.1109/TNS.2007.902703.

[51] G. Ariño-Estrada et al., "Development of TlBr detectors for PET Imaging," *Physics in Medicine and Biology*, vol. 63, 05/04 2018, doi:10.1088/1361-6560/aac27e.

[52] K. Hitomi, T. Onodera, and T. Shoji, "Influence of Zone Purification Process on T1Br Crystals for Radiation Detector Fabrication," *Nuclear Instruments & Methods in Physics Research Section A-accelerators Spectrometers Detectors and Associated Equipment – NUCL INSTRUM METH PHYS RES A*, vol. 579, pp. 153–156, 08/01 2007, doi:10.1016/j.nima.2007.04.028.

[53] G. Ariño-Estrada et al., "Towards Time-of-flight PET with a Semiconductor Detector," *Physics in Medicine and Biology*, vol. 63, 01/24 2018, doi:10.1088/1361-6560/aaaa4e.

[54] G. Ariño-Estrada et al., "First Cerenkov Charge-induction (CCI) TlBr Detector for TOF-PET and Proton Range Verification," *Physics in Medicine and Biology*, vol. 64, 07/25 2019, doi:10.1088/1361-6560/ab35c4.

[55] S. Gundacker et al., "State of the Art Timing in TOF-PET Detectors with LuAG, GAGG and L(Y)SO Scintillators of Various Sizes Coupled to FBK-SiPMs," *Journal of Instrumentation*, vol. 11, p. P08008, 08/08 2016, doi:10.1088/1748-0221/11/08/P08008.

[56] T. Szczesniak, M. Moszynski, M. Grodzicka, M. Szawłowski, D. Wolski, and J. Baszak, "MPPC Arrays in PET detectors with LSO and BGO Scintillators," *Ieee T Nucl Sci*, vol. 60, 06/01 2013, doi:10.1109/NSSMIC.2011.6152589.

[57] S. I. Kwon, A. Gola, A. Ferri, C. Piemonte, and S. Cherry, "Bismuth Germanate Coupled to Near Ultraviolet Silicon Photomultipliers for Time-of-flight PET," *Physics in Medicine and Biology*, vol. 61, pp. L38–L47, 09/02 2016, doi:10.1088/0031-9155/61/18/L38.

[58] M. J. Berger et al. "XCOM: Photon Cross Sections Database." www.nist.gov/pml/xcom-photon-cross-sections-database.

8 Gamma Spectrometry

Christopher Rääf

CONTENTS

DOI: 10.1201/9780429489556-8

8.1 BASIC PHYSICAL FEATURES OF A GAMMA DETECTOR

8.1.1 Configuration of a Gamma Spectrometer

Gamma (γ) spectrometry is used for detecting the presence of γ-emitting radionuclides in a sample. The aim is often to determine an activity concentration (Bq/kg) or (Bq/l) of the present γ emitters in the sample within a certain confidence limit. In most cases this requires that a sample-detector geometry is defined and calibrated so that the activity concentration in the examined samples can be assessed with a predetermined trueness, which is a term that describes how accurate, on average, a method used predicts the true value of the sample.

Gamma spectrometers used in nuclear medicine typically consist of a crystal composed either by a semiconducting diode (such as high-purity Germanium or Cadmium Zink Telluride, CdZnTe (abbreviated sometimes as CZT)) or scintillating materials, such as the NaI(Tl) and LSO (Lyso) crystals used for gamma camera imagining. Both detector types are described in more detail in Chapter 6 and 7. The sample is positioned in a well-determined fashion in relation to the detector crystal that in turn often is placed inside a shielding surrounding, both the crystal and the sample. As mentioned in previous chapters, the detector crystal is the unit where the essential interaction between the impinging γ-quanta and the detector material occurs that give rise to a characteristic pulse in terms of, for example, an electrical charge. The pulse is subsequently handled in a pulse process by being amplified and modulated to form an electronic pulse.

The first unit in direct connection with the crystal is the preamplifier that converts the original signal from one single interaction (or event, see The Composition of a γ-spectrum), into an electric pulse that can be amplified with minimum amount of distortion and noise. For traditional gamma camera imaging systems consisting of a scintillating layer as a gamma sensor, this preamplifier stage is represented by an array of photo-multiplier tubes (PMT) that convert the radiation-induced scintillation light into an electric charge. A modern variant of this light-to-charge converter are the Silicon photomultipliers (SiPM), which is becoming more used in gamma camera scintillator systems.

The next unit in the pulse-handling chain is the linear amplifier which essentially shapes the time pattern and amplitude of the pulse to fit the multichannel analyser (MCA). In this unit, the amplified pulses are ranged according to their amplitude or are rejected (discriminated) if the shape of an amplitude fails to meet certain criteria (e.g. too low or too high amplitude). The MCA is essentially a set of digital bins, spliced in equal size over a limited interval of pulse height, that registers the number of pulses the height of which falls within the interval of the particular bin. The final unit of a γ-spectrometer is the visualization interface, where the counts in the various bins of the MCA are displayed visually in the typical form of a spectrum, where a channel number represents an amplitude of the initial pulse in the crystal, which in turn translates into the energy deposited by the initial γ-interaction event. A schematic view of the event chain is given in Figure 8.1.

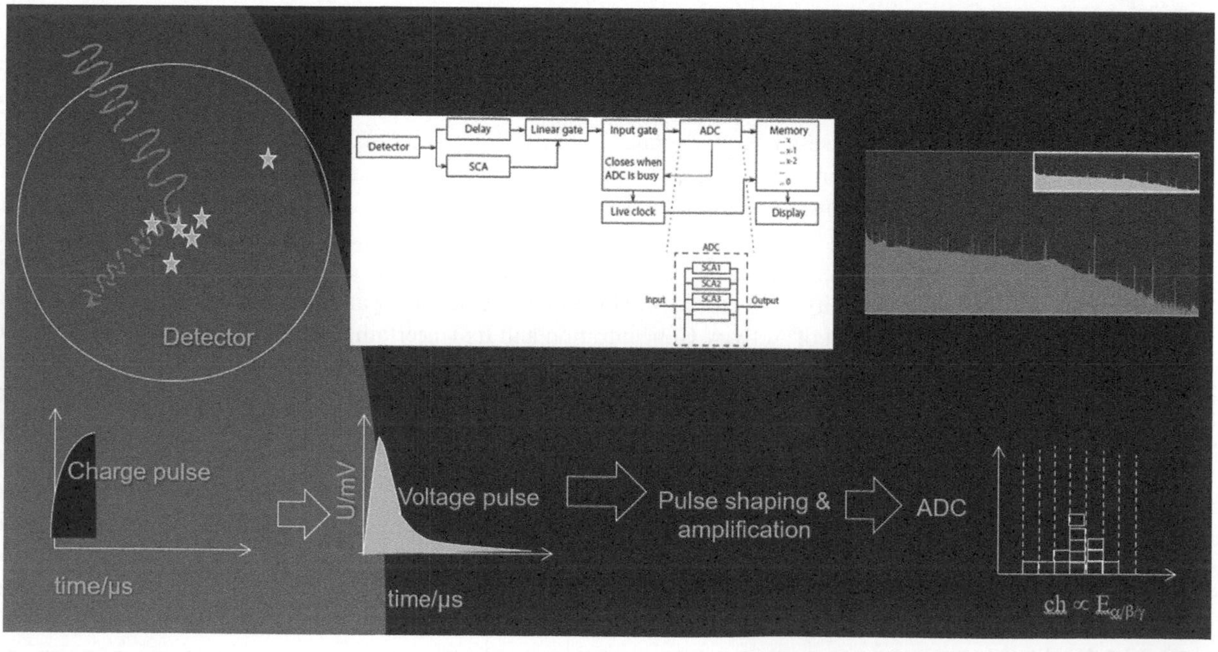

FIGURE 8.1 Event chain following a detected γ- in a detector, starting with the generation of the charge pulse, through amplification, multi-channel analysis and pulse storage.

Ideally, all energy of the impinging γ energy is deposited in the crystal in a single event and, hence, the initial pulse amplitude will have to be converted into a digital signal. In the classical constellation of γ-spectrometry electronics, this is done in a separate analogue-to-digital converter (ADC), and then the digitized signal is binned to a corresponding digital channel number in the MCA, representing the full energy of the γ-peak. Thus, when collecting the pulses of a succession of γ-events in the detector volume, the distribution in the number of counts in the bins will produce a pulse height distribution that is visually displayed by the software that emulates the MCA, and is often referred to as a spectrum (spectra in plural). If the pulse acquisition time is sufficient the presence of a γ emitter[1] in the source will then be manifested in terms of *full energy peaks* in the spectrum, which mainly reflect the quantum energies, E_γ (keV), of the γ-lines of the source. This spectrum will then provide a fingerprint of the radionuclide present in the sample and can be used for both the identification of the present γ emitters as well as quantification of these radionuclides in terms of activity concentration, A_x (Bq kg^{-1}).

Generally, some basic limiting conditions apply when selecting and using a γ-spectrometer for assessing the presence and the quantity of activity concentration of γ emitters in the samples. First, the type of detector material will determine the energy resolution, which is a measure of how well different full energy peaks can be separated and located in the spectrum. The energy resolution is defined as the Full Width of the photo peak at Half of the Maximum pulse height (FWHM). FWHM can be expressed either in terms of an energy width (keV), or as a percentage of the energy of the full energy peak (%). Scintillators, such as NaI(Tl), CsI, La_3Br and Cs_2LiYCl_6:Ce (CLYC), have relatively an energy resolution in the range between 4–8 per cent at $E_\gamma = 662$ keV, which is poor when compared with semiconductors such as high-purity germanium (HPGe) or silicon (Si) crystals. Semiconductors for γ-spectrometry have an FWHM of less than 0.25 per cent at 662 keV. Semiconductors are often much more expensive than scintillators of the same physical dimension. Furthermore, some semiconductors need to be cooled down to -196°C in order to function optimally, which requires either liquid nitrogen cooling (which in turn requires continuous maintenance and monitoring) or expensive electrical cooling that further increases the purchasing cost and maintenance. Therefore, the operator needs to determine the purpose of the γ-spectrometry assay: Is it to make regular constancy checks on the activity concentration of one or few radionuclides with well separated γ-lines? Then, a low resolution but robust NaI(Tl) crystal may be fit for purpose. Or is the aim instead a check for impurities and the presence of other unknown γ emitters in the sample? The importance of energy resolution becomes more imminent and high-resolution γ-spectrometers are required.

The physical dimension of the crystals available for γ-spectrometers in nuclear medicine varies from a few cm^3 (such as a 3.8·Ø3.8 cm^3 NaI scintillator for hand-borne γ contamination-control monitors), up to 0.95·2060 cm^2=1960 cm^3 crystals used for gamma camera imaging. The photon energy of the γ-emitting radionuclides of interest will in part determine which requirements must be put on the size of the crystal. The higher the energy of the γ-quantity to be detected, the larger the crystal needed to achieve efficient detection of the full energy peak. On the other hand, both for scintillators and semiconductors, larger crystals have the drawback of being more expensive and to a certain extent will also negatively affect the energy resolution with which different full energy peaks can be separated in the spectrum.

8.1.2 Common γ-Spectrometers Used in Nuclear Medicine

Gamma spectrometers dedicated to nuclear-medicine applications are most often optimized for the detection of photon energies within the energy range of ca. 30 to 600 keV. The γ emitters used are often short-lived ($T_{½,phys}$ < 30 d), as the combination of short-lived and low energetic γ emitters is less demanding in terms of patient-absorbed dose undergoing diagnostic imaging or radiopharmaceutical therapy.

For regular constancy controls of γ-sources, sometimes non-spectrometric systems, such as cylindrical ion-chamber systems, are still used. However, cheap and compact mobile scintillation crystals such as Sodium iodide (NaI(Tl)), Beryllium germanate (BGO), Cesium iodide (CsI), and CLYC (Cs_2LiYCl_6(Ce)) have made it possible to also utilize these for constancy checks. For delivery control of purchased radiopharmaceuticals and tests of radiochemical purity, high-resolution γ-spectrometry is more appropriate as many of these substances involve potential γ-emitting impurities with several photon energies that need to be resolved in the spectrum in order to be quantified.

[1] Strictly speaking, these are not γ emitters, but radionuclides used in positron-emitting tomography (PET), such as ^{18}F, since the two 511 keV photons emitted per decay are secondary particles caused by the annihilation process of the emitted positron, and not a primary photon from a nuclear or electro-magnetic transition in the nucleus. However, in the discussion of spectral properties of radionuclides in this chapter, no practical difference is made between these two types of radionuclides.

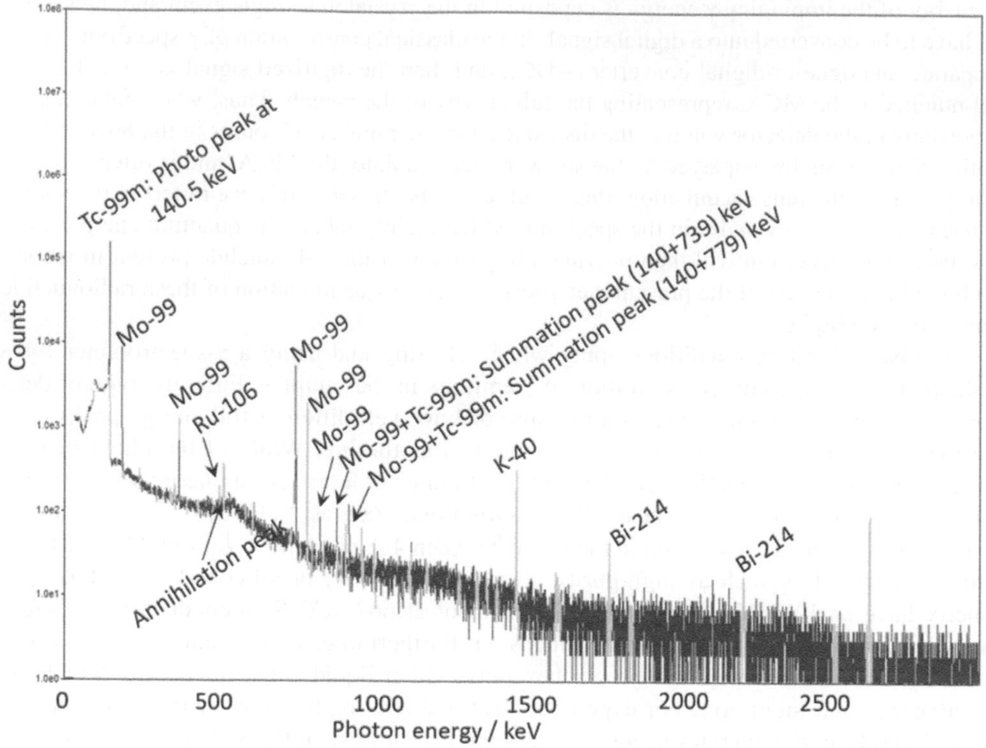

FIGURE 8.2 Pulse height distribution of a gamma emitting sample acquired by a 55 per cent HPGe detector. In addition to ^{99m}Tc, the gamma-emitting sample contains residues of ^{99}Mo and ^{106}Ru, together with so-called summation peaks. Peaks originating from the background of ^{40}K and ^{214}Bi are also visible.

Traditionally, large NaI(Tl)-crystal systems (typically 12.7 cm(Ø)·10.2 cm NaI(Tl)) have been used for monitoring of internal contamination of γ emitters among staff members. However, with the advent of compact electrically cooled high-resolution systems such as Low Energy HPGe-crystals (also referred to as LOAX crystals) – also for monitoring of γ-emitting contamination on working surfaces or external and internal contamination of staff members – medium-sized HPGe-detectors are preferably used. Even for imaging spectroscopy systems with spatial resolution, high-resolution systems such as HPGe and CZT crystals have been replacing scintillators. However, the emerging of semiconducting crystals in nuclear medicine for imaging is not driven by the spectroscopic property but rather by the improved spatial resolution (down to 0.5 mm) compared with traditional NaI(Tl) systems coupled with photomultiplier arrays. In this aspect, the semiconductors consisting of a combination of high atomic number, Z (or Z_{eff} for composite materials), elements Zn, Cd, Se and Te with high Z-values (~50) have an advantage before high-purity germanium (Z=32) and silicon (Z=14), since they can be operated at room temperatures and, due to their high atomic number, exhibit higher detection efficiency and contrast in imaging [1].

Figure 8.2 presents a typical high-resolution gamma spectrum from a sample containing residues of a ^{99m}Tc-solution, which is commonly used in nuclear medicine. In addition to a clearly distinguishable photo peak from the 140.5 keV gamma decay of the short-lived radionuclide, one also finds other peaks and formations in the spectrum by various origins. The origins of some of these peaks will be explained in more detail in Section 8.2.7.

8.1.3 Quality Parameters for a γ-Spectrometer

There are a number of characteristics of the detector material and the ambient conditions of the spectrometer set-up that must be considered in order to choose an optimal measuring system, which will be presented in the following sections.

8.1.3.1 Energy Resolution (FWHM)

The amount of signal quanta generated in the sensitive volume of the crystal per deposited energy determines the limit of how accurately the system can determine the deposited energy and the photon energies of the incident photons. The signal quantum in this first part of the pulse-generation process either consists of a charged particle released in an ionization detector such as an ion chamber or a solid-state semiconductor, or a photomultiplier or photocathode charge created by the luminescence that is generated by a scintillator when irradiated. The mean energy required to generate a signal quantum ranges between 1–3 eV in a semiconductor, 20–40 eV for gaseous ionization chambers, and more than 300 eV for solid state scintillators. In practice the energy resolution will be lowered by any contribution of noise generated in the detector or in the subsequent amplification of the signal. The electronic noise can be minimized by adjusting the bias, HV(V), over the semiconductor or the PMT and SiPM. The manufacturer of the γ-spectrometer often provides suggested values for HV. For scintillator systems the recommended HV-values range between 500 and 1000 V, whereas for high resolution semiconductor systems such as HPGe or Cadmium zink telluride (CZT), the HV values range between 2000 and 3000 V, depending on the size and geometrical configuration of the crystal.

8.1.3.2 Time Resolution

The signal generation caused by the interaction of the impinging photon, a process also referred to as an 'event', and the subsequent collection and amplification of the generated pulse need to be as fast as possible in order to avoid pulse pile-up that otherwise causes the system to erroneously represent the energy of the event. N.B. that this may also be a problem for non-spectroscopic systems, as the pulse pile-up will block the amplification and signal processing, resulting in a large dead-time during which the system cannot respond to additional incident photons. Time resolution for γ-spectrometers used in nuclear medicine mainly depends on the pulse rise time of the initial event, in combination with the time required for the electronics to process and amplify the signal. In the classical analogue set-up, the longest time-delay occurs in the ADC and MCA steps, and the original models of the MCA required up to 50 μs to process and correctly bin an incoming pulse. For both NaI(Tl)-scintillators and semiconductors, such as HPGe and CZT, the combined processing time is in the order of 1 to 10 μs.

Today, direct digital electronic circuits permit digitizing the signal output from the PMT or semiconductor and enables dead-time correction so that the pulses can be restored even at dead-times exceeding 10 per cent (e.g. [2]).

8.1.3.3 Energy Dependency in Detection Efficiency

The number of fully recorded events per unit impinging radiation quanta of a given energy, E_γ, will mainly depend on atomic composition (Z_{eff}) of the detector, the volume of the detector and the barrier thickness. A thicker NaI-layer than 1 cm is rarely considered relevant in nuclear medicine for gamma camera imaging, since (a) larger sizes will cause less precise localization of the interaction event by the PMT array, thus leading to poorer spatial resolution; and (b) photon energies up to about 600 keV are adequately recorded, and the use of higher γ-energies is rare. However, if using gamma camera systems for screening of internal contamination of high-energy γ emitters such as ^{60}Co (mean emitted γ-energy of 1250 keV), the thin crystal layers of the gamma camera system definitely limit the detection efficiency, as the probability of a detection per unit path length in the crystal decreases with increasing photon energy.

The energy dependence in the detection efficiency $\varepsilon(E_\gamma)$ in detectors designed for high-resolution γ-spectrometry is typically characterized by an increase at E_γ up to 130–150 keV due to the increasing penetration of the incident photons through the detector casing and inactive crystal layers in the entrance window of the detector. Since the crystal dimensions of such detectors for nuclear medicine are relatively limited (<100 cm^3), combined with the cross section for Compton scattering in the crystal becoming more predominant for higher photon energies, an increasing fraction of the deposited photon energy in the crystal will escape the detector and thus lead to decreasing counting efficiency. The energy interval around which counting efficiency peaks is referred to as a 'knee' (see also Figure 8.3). Conventional scintillation detectors for gamma camera imaging are often about 1 cm thick (3/8 inch or 0.95 cm), with a typical efficiency 'knee' located around the photon energy of ^{99m}Tc (140.5 keV). However, for imaging of ^{131}I therapies a thicker NaI(Tl)-crystal (5/8 inch or 1.6 cm) is preferable to use in order to increase the detection efficiency of the 364.5 keV photons, and to fully exploit the spatial information provided by the higher number of full energy γ counts.

8.1.3.4 Shielding and Background Suppression

In γ-spectrometry it is essential to depress unwanted detection of radiation particles coming from, for example, the presence of background radiation (such as ^{40}K in construction material and organic structures, ^{222}Rn daughters in air, ^{238}U and ^{232}Th daughters in ambient construction materials) or from incompletely registered events from cosmic radiation. Therefore, many detector systems need to be set up so that they are partly or completely shielded from these ambient radiation sources. Although many nuclear medicine applications involve the examination of substantially stronger γ-sources compared with, for example, environmental radioactivity – leading to the background radiation contribution

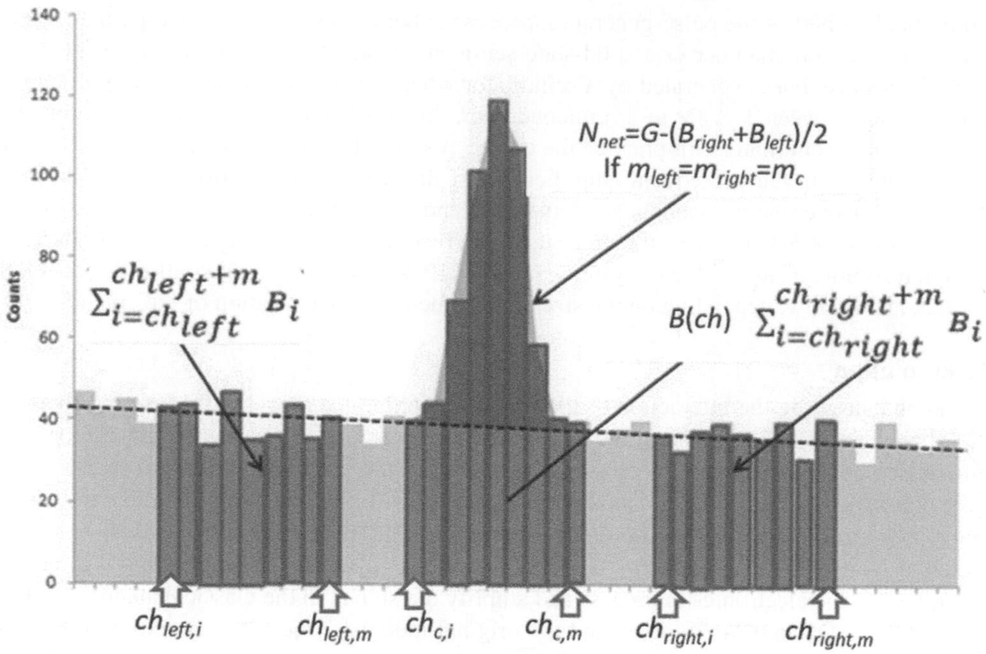

FIGURE 8.3 Example of a peak formation in a pulse height destitution where a background level of counts is subtracted, and the resulting net counts from the γ-source are used to calculate the centroid of the peak. All ROIs in this illustration have the same channel width of m channels.

being less important or even negligible – applications such as contamination controls and whole-body counting of staff members can still be considerably affected by the presence of ambient background radiation. Measurements of samples that are not intentionally radiolabelled or doped with radionuclides (such as swipe samples from working surfaces), must also be made in shielded environments. The background shield of the γ-spectrometer often consists of a lead-brick cave, where both source and detector are enclosed. Detection set-ups for such low-activity samples should also be done in a different locale than the one used for γ-spectrometry of radiochemical purity and other high-activity sources (> 1 MBq) in order to avoid so-called crosstalk and perturbation from the presence of radiolabelled substances.

Whole-body counting should also be done in a shielded environment, preferably in low-background rooms with construction materials dedicated for low presence of naturally occurring γ emitters (such as ^{238}U and ^{232}Th and their γ-emitting daughter nuclides) as this will suppress the ambient background. The detection level of the system, which is an important quality measure for various measuring applications, is mainly determined by how effectively the detector is shielded from the background radiation. Even if the γ-spectrometer is located in a shield environment, there will still be processes that contribute to perturbations and artifacts in the recorded γ-spectrum (addressed in the next section), which needs to be accounted for in the radionuclide assessment. The background and perturbations combined will essentially govern the detection limit (in terms of, for example, minimum detectable activity concentration, MDA (Bq kg^{-1}) of a γ-emitting radionuclide, and will be discussed more in Section Detection limits.

8.1.3.5 Ageing

The detector response may change over time due to irreversible fundamental detection processes, such as charge-particle interaction in thin semiconductors, or if subjected to neutron flux that results in inelastic scattering in the detector, altering the stoichiometry of the detector material). The latter will result in broadening of the FWHM and distorted shapes of the full energy peak [3].

8.2 QUANTITATIVE ASSESSMENT OF GAMMA SPECTRA

8.2.1 Basic Definitions

The pulse height distribution over the channel in the MCA in the recorded spectrum is used to define the locations of full energy peaks that can be attributed to a certain photon energy, E_γ. The centroid is the average channel number location of a peak, located between channel *i* and *m* and weighted by the individual number of counts, $N_{net,i}$, in each channel

(Eq. 8.1). When adjusting for the presence of background counts, the centroid, ch_c, is calculated only from the counts originating from the presence of the γ-emitting source.

$$ch_c = \frac{\sum_i^m N_{net,i} \cdot i}{\sum_i^m N_{net,i}} \tag{8.1}$$

Once the centroid of the peak, in terms of channel number or in photon energy (see next section) is computed, the sum of the net counts in the recorded peak needs to be computed as well. Due to the inherent broadening of the full energy peaks in all γ-spectrometry systems (Section 8.1.3.1), the channel interval cannot encompass the complete full energy peak. Instead, a limited channel range must represent the area over which the centroid and net counts are computed. The background counts in the pulse height distribution inside and outside the channel range of the centroid are used to estimate an average background count level around the location of the centroid. Often a linear slope is assumed described by the expression exemplified in (Eq. 8.2).

$$B(ch) = B(ch)_{left,i} \cdot \frac{\left(\frac{\sum_{i=ch_{right}}^{ch_{right}+m} B_i}{m}\right) - \left(\frac{\sum_{i=ch_{left}}^{ch_{left}+m} B_i}{m}\right)}{ch_{right,i} - ch_{left,i}} \cdot \left(ch - ch_{left,i}\right) \tag{8.2}$$

For simplicity it has been assumed that both the left and right background ROIs have the same channel widths, $m=(ch_{left,m}-ch_{left,i})=(ch_{right,m}-ch_{right,i})$ (see also Figure 8.3). The gross counts in each channel are then subtracted by this expression to obtain the estimated net counts in each channel, $N_{net}(ch)=G(ch)-B(s)$, and the centroid is then calculated as in (Eq. 8.3).

$$ch_c = \frac{\sum_i^m N_{net,i} \cdot i}{\sum_i^m N_{net,i}} = \frac{\sum_i^m \left(G(ch_i) - B(ch_i)\right) \cdot i}{\sum_i^m \left(G(ch_i) - B(ch_i)\right)} \tag{8.3}$$

Around this centroid, a region-of-interest (ROI) of channels can be defined by the software to represent the interval over which the counts emanating from a source is calculated. Usually, an ROI over a full energy peak is set to about 2.55·FWHM of the full energy peak, which then will cover almost 99 per cent of the theoretical number of counts under a Gaussian shaped pulse height distribution. The net counts, N_{net}, in a peak, over an ROI of m_c number of channels, can then be manually computed as (Eq. 8.4):

$$N_{net} = \sum_i^{i+m,c} G(ch) - \sum_i^{i+m,c} B(ch) = G_c - B_c \tag{8.4}$$

where G_c represents the number of gross counts summed over the whole region-of-interest, and B_c is the corresponding number for the background counts.

8.2.2 Calibration of a γ-Spectrometer

Several calibrations need to be done before any quantitative assessment can be done. These calibrations mainly involve three major types; (1) energy calibration, (2) Energy resolution calibration (FWHM), and (3) efficiency calibration. Most of these calibrations can be done in connection with the installation of the detector set-up by the aid of the vendor of the system, but some additional calibration (such as the relative counts in a full energy peak and other parts of the pulse-height distribution) and recalibration must be carried out by the operators themselves.

8.2.3 Energy Calibration

In the energy calibration a spectrum is recorded with a γ-emitting source with known composition of gamma emitters. The calibration source can be a cocktail of radionuclides such as ^{241}Am (432.2 y), ^{57}Co (271.8 d), ^{133}Ba (10.5 y), ^{113}Sn (115.1 d), ^{85}Sr (64.8 d) and ^{137}Cs (30.02 y), or it can be with a single radionuclide, such as ^{152}Eu (13.5 y) or ^{192}Ir (73.83

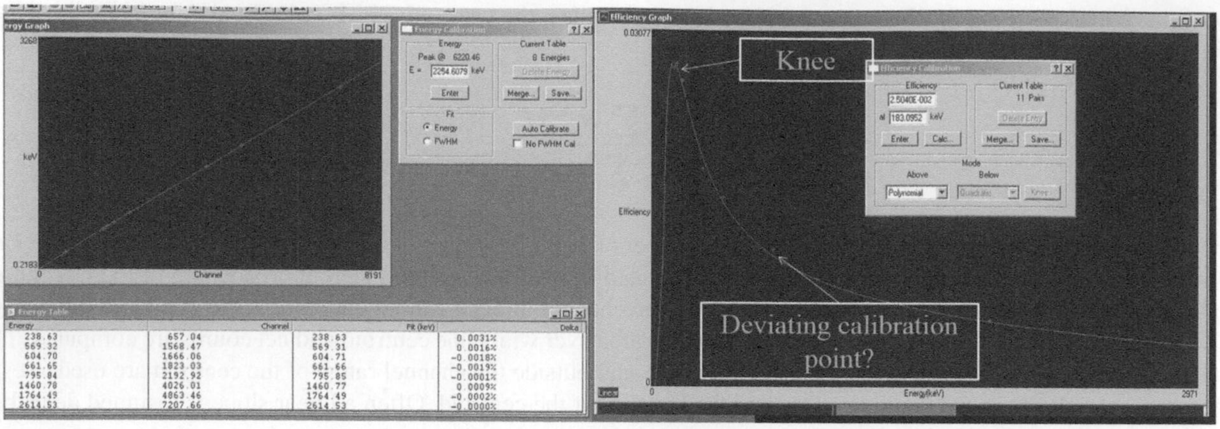

FIGURE 8.4 Left: Example of a software feature where the average channel number of a located peak (6220.46) can be associated with a γ-energy (2254.8 keV). Right: Example of a software feature where the efficiency of a high-resolution γ-spectrometer is calibrated.

d), which emits a number of prominent γ-lines in the energy range 30 keV up to more than 1000 keV depending on the application. For whole-body counting of staff members, a γ-energy range of up to 2000 keV should be used to cover the presence of naturally occurring radionuclides such as ^{40}K ($E\ \gamma$=1460.8 keV). The software controlling the multi-channel analyser (MCA) must then have a feature where the average channel location of the localized peaks (referred to as centroids, ch_c in the previous section) in the recorded spectrum can be associated with the correct γ-energy of the emitted photons (see Figure 8.4). The centroid can be computed manually, but often the software computes it automatically by the operator graphically marking the peak in the MCA-emulator. Usually, modern types of MCAs have a highly linear output in terms of E per channel number (Eq. 8.5). Small deviation from linearity can be compensated for by instead fitting a quadratic expression (Eq. 8.5: Alt 2).

$$Alt1: E_g = a_0 + a_1 \cdot ch$$

$$Alt2: E_g = a_0 + a_1 \cdot ch + a_2 \cdot ch^2 \tag{8.5}$$

A sufficient number of counts in the full energy peaks is required to obtain an accurate localization of the peak centroid (Eqs. 1–3). About 10,000 net counts are recommended.

8.2.4 FWHM Calibration

The energy resolution in terms of FWHM as a function of photon energy must be determined in many cases if automated analysis of the spectrum is to be used (discussed more later). This calibration is often done in connection with the energy calibration. The value of the FWHM is automatically calculated by most software (by e.g. marking the located peak in the MCA-emulator), but the computation of this parameter involves the number of counts in a number of single channels, which in turn leads to a higher tendency for statistical fluctuations. Hence, it is important that a sufficient number of counts are obtained in the full energy peak in connection with the calibration (at least 50,000 net counts) to suppress the uncertainty of the FWHM, otherwise the mathematical fit of FWHM as function of E_γ may be distorted and lead to biased assessments. FWHM is often related to E_γ in a predominantly square root relationship, although in some γ-spectrometric software a small linear contribution can be added in the mathematical fit to better accommodate the energy dependence in the charge collection of the semiconductor (Eq. 8.6).

$$\text{FWHM} = a_0 + a_1 \cdot E_\gamma + a_2 \cdot \sqrt{E_\gamma} \tag{8.6}$$

N.B., that it is not recommended to use the annihilation peak at 511 keV for FWHM calibration, as the FWHM is broader than the ones from an ordinary nuclear γ-transition line at the same energy range due to the so-called Doppler

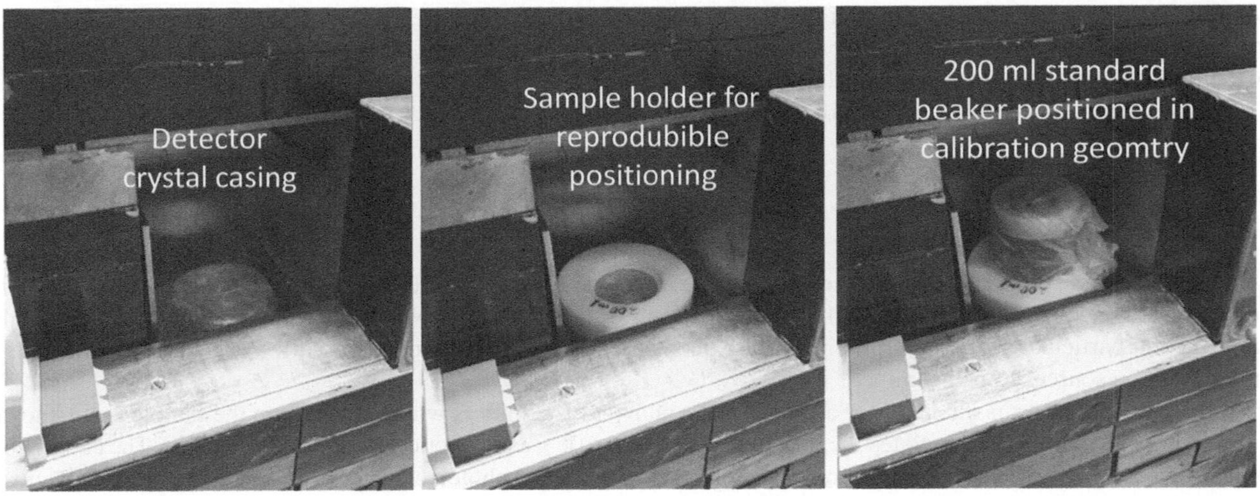

FIGURE 8.5 A calibration geometry for a 200 ml container placed in a sample holder on top of a 100 per cent (relative efficiency at 1.332 MeV) HPGe-detector-crystal that is surrounded by an 8 cm thick lead shield. The container is covered in a plastic bag to protect the detector from contamination.

effect from the non-zero kinetic energy of the annihilation particles (e^- and e^+). It is also important to ensure that the used full energy peaks in the FWHM calibration are clearly separated from nearby full energy peaks or interferences (such as escape peaks) that could otherwise perturb the computed FWHM values.

8.2.5 Efficiency Calibration

The efficiency calibration is a method of relating the measured number of net counts in a detected full energy peak with the activity of a γ-emitting source. The relation is specific for every source-detector geometry, and a calibration is needed for every geometry defined by the operator (see Figure 8.5). The obtained efficiency calibration is therefore essentially specified for a particular geometrical setting. Such a set-up can include a liquid sample in a beaker with a particular filling level and positioned at an appropriate distance from the detector surface. Any deviances in the source geometry, such as filling level, density or atomic composition of the sample beaker shape and material and distance to the detector surface, will lead to inaccurate results in the γ-spectrometry evaluation if these deviances are not appropriately corrected for. Such corrections can be made by mathematical corrections using software such as LABSOCS™ [4].

The calibration is carried out by using a standard source, with a well-determined (typically within 3 per cent relative uncertainty on a 95 per cent confidence level) activity of one or more γ emitters. To obtain metrological traceability of the activity measurement, the standard needs to be traceable to a reference. For γ-spectrometry, the most forward approach is to use a certified reference material. Care should be taken when purchasing radionuclide standards, since they involve relatively high costs and sometimes a lot of effort in administration. Due to standards in the market that are not certified nor traceable to international or national standards, it is important that thorough efforts have been put into specifying the standard before buying.

If the purpose of the γ-spectrometry is to only measure one radionuclide in a particular geometry, the standard may contain only that radionuclide. The activity concentration of the standard is defined at a certain reference date. The calibration factor is then obtained by measuring the standard and the related number of net counts in one of the peaks from the source, N_{net}, to the standard activity concentration, A_{st} (Bq kg^{-1}), to obtain a factor ε_r (cps/(Bq kg^{-1})) (Eq. 8.7).

$$\varepsilon_r\left(E_\gamma\right) = \frac{\left(\frac{N_{net}}{t_{acq}}\right)}{A_{st} \cdot n_\gamma} \cdot k_{decay} = \frac{\left(\frac{N_{net}}{t_{acq}}\right)}{S_{st}} \cdot k_{decay} \tag{8.7}$$

where t_{acq} is the pulse acquisition live time, n_g is the branching ratio of the γ-energy of interest. Some standard source manufacturers state the photon emission rate, S_{st}, instead of the activity, A_{st}, and is hence given by the factor $A_{st} \cdot n_g$. A factor, k_{dec}, must be done to correct the stated standard activity concentration, A_{st} (or γ-emission concentration rate, S_{st}) by the decay of the γ-emission rate of the standard source since its reference date:

$$k_{dec} = e^{\left(\frac{ln2}{T_{½}}\right)\cdot\left(t_m - t_{ref}\right)} \tag{8.8}$$

where t_m is the time of start measurement and t_{ref} is the reference date of the reference source. The difference between t_m and t_{ref} must be expressed in the same time units as the physical half-life, $T_{½}$. Note that the calibration factor, ε_γ, is only valid for the particular γ-energy, in the particular geometrical set-up and cannot be used to evaluate other potential γ emitters in the sources to be examined.

If the γ spectrometer is intended for evaluation of a number of γ emitters whose presence in the sample is not known beforehand, a calibration that relates the detection efficiency, $\varepsilon_r(E_\gamma)$, as a function of γ energy is needed. The calibration standard then needs to contain one or more γ emitters that have several photon energies over the energy range of interest, such as 30 to 600 keV for a radiopharmaceutical test, or between 30 to 1500 keV for contamination controls and whole-body counting. γ emitters such as ^{133}Ba, ^{134}Cs and ^{152}Eu are not suitable for sample geometries positioned close to the detector, as so-called true coincidence occurs (see Section 8.2.7) between several emitted gammas (such as the 604 keV and 795 keV photon being emitted from a ^{134}Cs decay and registered in the detector within less than 1 μs apart), which results in a given γ peak losing pulses that are positioned elsewhere in the pulse height distribution, and thus will make the detection efficiency at that energy appear lower than what it actually is. Therefore, standard 'cocktails' of γ emitters exist on the market that are composed of predominantly mono energetic radionuclides with no or less pronounced coincidence such as ^{241}Am, ^{57}Co, ^{113}Sn, ^{85}Sr and ^{137}Cs, that can be useful for efficiency calibration. However, if these cocktails also include ^{60}Co and ^{88}Y, coincidence loss correction may be needed for measurement geometries close to the detector. Calculation tools exist that can be used for true-coincidence summing (for example EFFTRAN [5]).

The standards are available in various constellations and in liquid or solid states. It is also possible to customize a standard source by submitting a beaker or vessel to the manufacturer, who can design a standard solution in that specific source constellation. An example of a certificate of a standard for a 200-ml beaker with unity density is given in Figure 8.6. The data sheet of the standard provides information regarding the photon emission rates of the γ lines of the radionuclides in the standard cocktail, $S(E_\gamma)$ (s^{-1}), and the mass of the standard, m_{st} (kg). N.B: that emission rate is not the same as activity, as most γ emitters have a branching ratio, $n\gamma$, of less than unity for their emitted γ lines (except for the 1173.2 and 1332.5 keV lines of ^{60}Co). The emission rate per unit mass of standard, can then be directly computed as $S(E_\gamma)/m_{st}$ (s^{-1} kg^{-1}).

Once procured, the standard is positioned in a given measurement geometry, and a spectrum is recorded with all the peaks from the γ emitters in the standard located. It is also important here to obtain at least 10,000 net counts in each of the peaks to be used in the calibration to ensure that the data point is obtained with sufficient statistical certainty. The *absolute efficiency*, $\varepsilon_r(E_\gamma)$ (cps s^{-1}) of a given energy is then given by Eq. 8.9.

$$\varepsilon_r\left(E_g\right) = (\frac{N_{net}\left(ROI\ of\ E_\gamma\right)}{t_m} \cdot \frac{1}{S_{st}\left(E_\gamma\right)} \tag{8.9}$$

A common software feature is to have a curve fit to all the recorded efficiency vs photon energy points, to obtain a continuous function of ε_r as a function of $E\gamma$. Typical curve fits are given in Eq. 8.10, where a_0, a_1, … a_n are fit parameters that are obtained by the software.

$$\ln\left(\varepsilon_r\right) = a_0 + a_1 \ln\left(E_\gamma\right) + a_2\left(\ln\left(E_\gamma\right)\right)^2 + \ldots a_n\left(\ln\left(E_\gamma\right)\right)^n ; n < 8$$

$$\ln\left(\varepsilon_r\right) = a_1 E_\gamma + a_2 E_\gamma^{-1} + a_3 E_\gamma^{-2} + \ldots$$

$$\varepsilon_r = 1/\left(aE_\gamma^{-x} + bE_\gamma^{-y}\right) \tag{8.10}$$

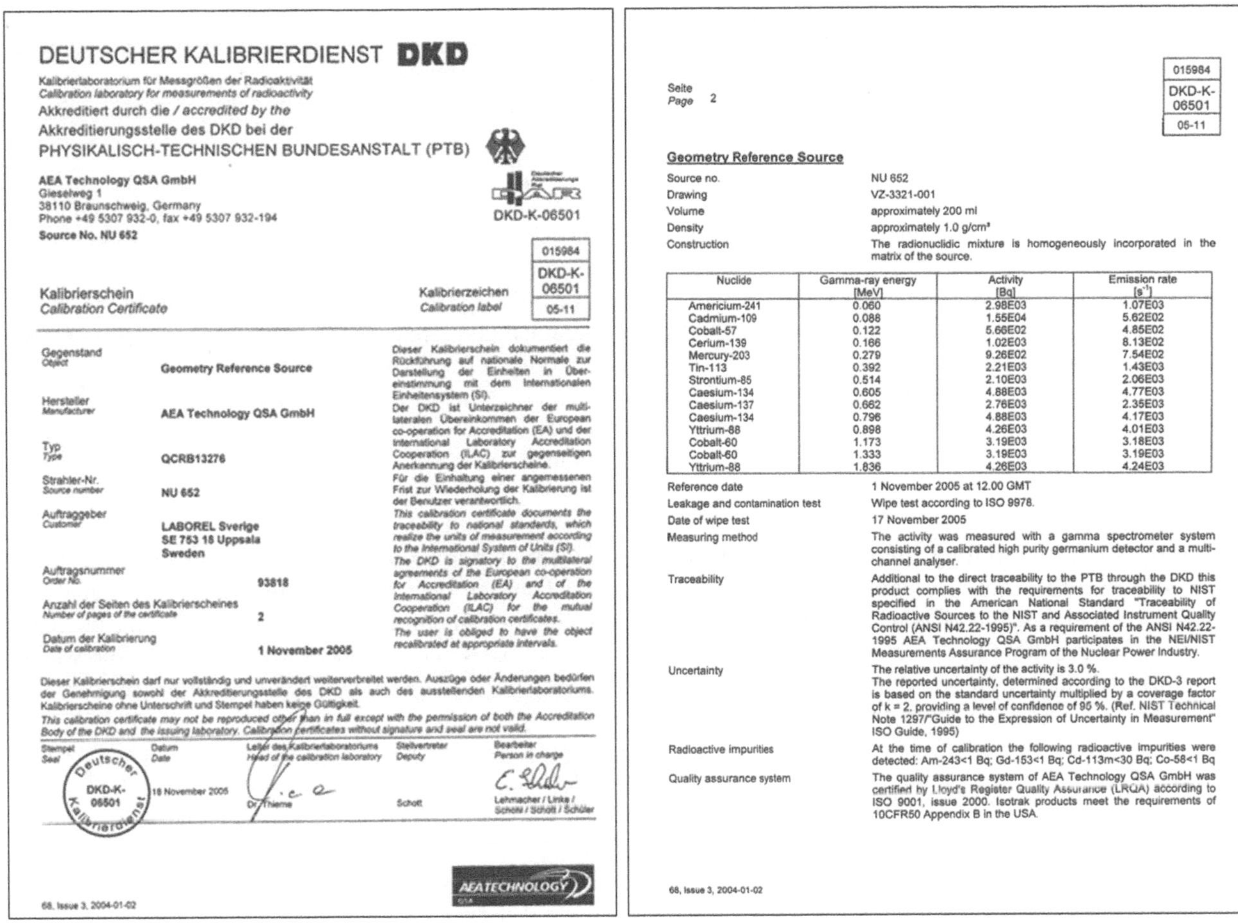

DEUTSCHER KALIBRIERDIENST DKD

Kalibrierlaboratorium für Messgrößen der Radioaktivität
Calibration laboratory for measurements of radioactivity
Akkreditiert durch die / *accredited by the*
Akkreditierungsstelle des DKD bei der
PHYSIKALISCH-TECHNISCHEN BUNDESANSTALT (PTB)

AEA Technology QSA GmbH
Gieselweg 1
38110 Braunschweig, Germany
Phone +49 5307 932-0, fax +49 5307 932-194
Source No. NU 652

DKD-K-06501

015984
DKD-K-06501
05-11

Kalibrierschein
Calibration Certificate

Kalibrierzeichen
Calibration label

Gegenstand *Object*	**Geometry Reference Source**
Hersteller *Manufacturer*	**AEA Technology QSA GmbH**
Typ *Type*	**QCRB13276**
Strahler-Nr. *Source number*	**NU 652**
Auftraggeber *Customer*	**LABOREL Sverige SE 753 18 Uppsala Sweden**
Auftragsnummer *Order No.*	**93818**
Anzahl der Seiten des Kalibrierscheines *Number of pages of the certificate*	**2**
Datum der Kalibrierung *Date of calibration*	**1 November 2005**

Dieser Kalibrierschein dokumentiert die Rückführung auf nationale Normale zur Darstellung der Einheiten in Übereinstimmung mit dem Internationalen Einheitensystem (SI).
Der DKD ist Unterzeichner der multilateralen Übereinkommen der European co-operation for Accreditation (EA) und der International Laboratory Accreditation Cooperation (ILAC) zur gegenseitigen Anerkennung der Kalibrierscheine.
Für die Einhaltung einer angemessenen Frist zur Wiederholung der Kalibrierung ist der Benutzer verantwortlich.
This calibration certificate documents the traceability to national standards, which realize the units of measurement according to the International System of Units (SI).
The DKD is signatory to the multilateral agreements of the European co-operation for Accreditation (EA) and of the International Laboratory Accreditation Cooperation (ILAC) for the mutual recognition of calibration certificates.
The user is obliged to have the object recalibrated at appropriate intervals.

Dieser Kalibrierschein darf nur vollständig und unverändert weiterverbreitet werden. Auszüge oder Änderungen bedürfen der Genehmigung sowohl der Akkreditierungsstelle des DKD als auch des ausstellenden Kalibrierlaboratoriums. Kalibrierscheine ohne Unterschrift und Stempel haben keine Gültigkeit.
This calibration certificate may not be reproduced other than in full except with the permission of both the Accreditation Body of the DKD and the issuing laboratory. Calibration certificates without signature and seal are not valid.

Stempel *Seal*	Datum *Date*	Leiter des Kalibrierlaboratoriums *Head of the calibration laboratory*	Stellvertreter *Deputy*	Bearbeiter *Person in charge*
Deutscher Kalibrierdienst DKD-K-06501	18 November 2005	Dr. Thieme	Schott	Lehmacher / Linke / Schönl / Schott / Schüler

AEATECHNOLOGY QSA

68, Issue 3, 2004-01-02

Seite
Page 2

015984
DKD-K-06501
05-11

Geometry Reference Source

Source no.	NU 652
Drawing	VZ-3321-001
Volume	approximately 200 ml
Density	approximately 1.0 g/cm³
Construction	The radionuclidic mixture is homogeneously incorporated in the matrix of the source.

Nuclide	Gamma-ray energy [MeV]	Activity [Bq]	Emission rate [s⁻¹]
Americium-241	0.060	2.98E03	1.07E03
Cadmium-109	0.088	1.55E04	5.62E02
Cobalt-57	0.122	5.66E02	4.85E02
Cerium-139	0.166	1.02E03	8.13E02
Mercury-203	0.279	9.26E02	7.54E02
Tin-113	0.392	2.21E03	1.43E03
Strontium-85	0.514	2.10E03	2.06E03
Caesium-134	0.605	4.88E03	4.77E03
Caesium-137	0.662	2.76E03	2.35E03
Caesium-134	0.796	4.88E03	4.17E03
Yttrium-88	0.898	4.26E03	4.01E03
Cobalt-60	1.173	3.19E03	3.18E03
Cobalt-60	1.333	3.19E03	3.19E03
Yttrium-88	1.836	4.26E03	4.24E03

Reference date	1 November 2005 at 12.00 GMT
Leakage and contamination test	Wipe test according to ISO 9978.
Date of wipe test	17 November 2005
Measuring method	The activity was measured with a gamma spectrometer system consisting of a calibrated high purity germanium detector and a multichannel analyser.
Traceability	Additional to the direct traceability to the PTB through the DKD this product complies with the requirements for traceability to NIST specified in the American National Standard "Traceability of Radioactive Sources to the NIST and Associated Instrument Quality Control (ANSI N42.22-1995)". As a requirement of the ANSI N42.22-1995 AEA Technology QSA GmbH participates in the NEI/NIST Measurements Assurance Program of the Nuclear Power Industry.
Uncertainty	The relative uncertainty of the activity is 3.0 %. The reported uncertainty, determined according to the DKD-3 report is based on the standard uncertainty multiplied by a coverage factor of k = 2, providing a level of confidence of 95 %. (Ref. NIST Technical Note 1297/"Guide to the Expression of Uncertainty in Measurement" ISO Guide, 1995)
Radioactive impurities	At the time of calibration the following radioactive impurities were detected: Am-243<1 Bq; Gd-153<1 Bq; Cd-113m<30 Bq; Co-58<1 Bq
Quality assurance system	The quality assurance system of AEA Technology QSA GmbH was certified by Lloyd's Register Quality Assurance (LRQA) according to ISO 9001, issue 2000. Isotrak products meet the requirements of 10CFR50 Appendix B in the USA.

68, Issue 3, 2004-01-02

FIGURE 8.6 Certificate of a standard sample containing a cocktail of gamma emitters such as ^{137}Cs (Cesium-137), ^{57}Co (Cobalt-57) and ^{113}Sn (Tin-113).

Typical features of such efficiency curves are the knee-shape around 100 to 150 keV, above which the efficiency decreases exponentially (Figure 8.4). The validity of the curve fit is best within the range of γ-line energies used in the calibration, and any extrapolation of this curve outside this range results in $\varepsilon_r(E_\gamma)$ values with higher relative uncertainties, $u(\varepsilon_r(E_\gamma))$, than within this range.

8.2.6 Required Features of γ-Spectrometry Software

A number of functions and parameters should be included in a γ-spectrometry software. In addition to being able to record and store the pulse height distribution, it also must be possible to add and store essential input parameters used in the evaluation of the activity concentration, A_x in an appropriate file format. These parameters include:

- the reference time, t_{ref}
- sample weight, m_s
- identity of energy calibration
- identity of efficiency calibration
- a *nuclide library* that contains data on half-lives and branching ratios of a set probable γ-emitting radionuclides present in the source

The software should also include algorithms that separate and localize the full energy peaks and their respective centroids and determine the background and net counts in these peaks. An algorithm for calculating the FWHM of the localized peaks is also necessary. These algorithms are often provided with the basic version of the spectrometry software.

However, for more complex assessments, where the composition of γ emitters in the source is not known beforehand, for example, in impurity checks, it is advisable to consult the manufacturers for whether more sophisticated software with more extensive analytical features is needed.

8.2.7 The Composition of a γ-Spectrum

For γ-emitting nuclides the energy deposited in the γ-detector should be translated into a spectrum with very sharp peaks/spikes, representing the energy of the incoming photons. For most spectrometry systems it is a challenge to accurately represent the fluence of emitted γ quanta impinging from the sources, due to the limited energy resolution. However, there are a number of factors that contribute to the shape of the pulse height destitution recorded by a small- or medium-sized γ-spectrometer that must be accounted for when making a quantitative assessment. Let us consider a source-detector constellation in lead shielding, where the γ-spectrometer is a medium-sized γ-crystal (such as a HPGe-detector (55 per cent relative efficiency @1.332 MeV with a crystal volume of c.a. 80 cm^3)), and with a mono energetic γ-emitting source positioned some distance away from the detector surface but still within the lead shielding. The following events can occur that will lead to a pulse being positioned at various energy intervals of the MCA (see also Figure 8.7).

1. **Complete absorption of incoming γ quantum:** The incident photon passes through the detector encapsulation (typically a few mm of aluminium) and other dead layers between the source and the sensitive volume of the detector, and then interacts with the atoms in the sensitive volume. The interaction event can be (a) the photoelectric effect, resulting in the release of a characteristic x-ray photon from one of the atoms in the detector, which is in turn completely absorbed by successive Compton scattering or photoelectric energy absorption; (b) successive

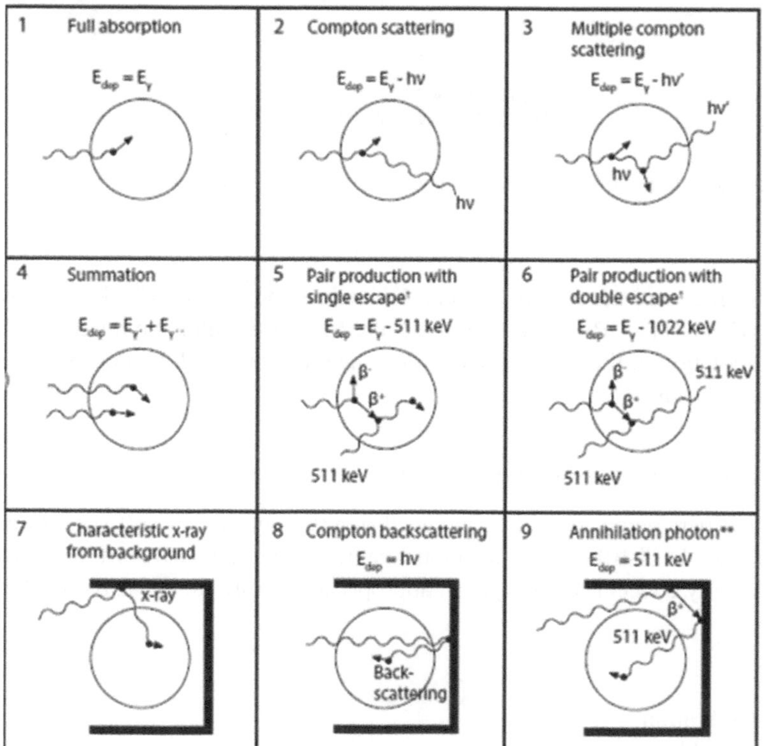

FIGURE 8.7 Schematic view of γ-interaction processes underpinning various components in a pulse height distribution recorded in a typical γ-spectrometer: 1. Complete absorption of incoming γ quantum. 2. Incomplete absorption: Single Compton escape. 3. Incomplete absorption: Multiple Compton scattering. 4. Summation (coincidence). 5. Incomplete scattering: Pair-production with single or double escape. 6. Incomplete scattering: Characteristic X-ray escape in the γ-detector. 7. Characteristic X-ray from shielding materials. 8. Compton back scattering from shielding materials. 9. Annihilation photon from background (Figure 8.7 taken from [25]).

Compton scattering, in which all the electrons released, and the scattered photons are completely absorbed within the sensitive volume of the detector. In both cases the deposited energy of the incident photon is completely absorbed and detected, albeit with some interference from detector noise. The MCA will then allocate the signal to a channel representing the full energy of the incident photon.

2. **Incomplete absorption – single Compton escape:** In this event a single Compton scattering of the incoming photon leads to the energy of the released electron being deposited in the sensitive volume of the detector. The scattered photon, however, escapes the sensitive volume, and the generated signal registered by the MCA represents the energy of the scattered electron which is related to $E_{\gamma'}=E_\gamma\cdot(1-1/(1+(E_\gamma/m_ec^2)(1-\cos\theta)))$ (Eq. 8.11). The term $m_0 \cdot c^2$ is the electron rest energy, which is equal to 511 keV, and θ is the scattering angle of the Compton photon. The maximum amount of energy transferred to the atomic electron, at θ=180°, leads to an energy of $E_\gamma\cdot(1-1/(1+2(E_\gamma/m_ec^2)))$. For a 511 keV primary photon this energy will be $E_g \cdot 2/3$ keV. For repeated such events, a continuum of pulses will be built up in the MCA from the low energy range up to this maximum energy, resulting sometimes in a distinct peak like shape.

$$E_{\gamma'} = \frac{E_\gamma}{\left(1+\left(\frac{E_\gamma}{m_ec^2}\right)\cdot(1-cos\theta)\right)} \tag{8.11}$$

3. **Incomplete absorption – multiple Compton scattering:** A similar event as above but here the scattered Compton photon interacts with an atom in the sensitive volume of the detector giving rise to a somewhat less escaped photon energy than in the previous case. In a detector crystal with lower atomic number (Z value) this procedure is much more common than for high Z value materials such as NaI(Tl). Instead of a Compton edge in the previous procedure the resulting pulses from repeated events of this kind will be located by the MCA as a continuous distribution of pulses.
4. **Summation (coincidence):** If two incident photons reach the detector within a sufficiently short time (in order of 10^{-9} s), the pulse shaping and amplification will not be able to resolve properly the two events, but instead the summed deposition results in a pulse being located at the summed energy of the two depositions. If the two depositions both are full energy absorption events, as in Case 1, the summed pulse is the sum of $E_{\gamma 1}$ and $E_{\gamma 2}$. However, in most cases the summing will be of fractions of $E_{\gamma 1}$ and $E_{\gamma 2}$ respectively, as the probability of Compton escape and other escape processes are just a likely as for a single photon impinging the sensitive volume, resulting in a pulse that is located somewhere between $E_{\gamma 1}$ and $E_{\gamma 1}+E_{\gamma 2}$, or $E_{\gamma 2}$ and $E_{\gamma 1}+E_{\gamma 2}$ depending on whether $E_{\gamma 1}$ is larger than $E_{\gamma 2}$. The higher the source strength, A_x, the higher the incident fluence of γ-photons, and subsequently there is a higher probability that a summing of two events will occur. Some gamma emitters decay in cascade passing through very short-lived meta-states and can thus emit two or more gammas in the short time in which both reach the detector, almost simultaneously, such as ^{134}Cs, where the 604 keV line and 795 keV line can be detected as a sum of the two energies, that is, at $E_{\gamma,tot}$=604+795~1400 keV. This process is referred to as a true coincidence, compared with a situation when the γ emitter coincides with an impinging photon from the background radiation or from two separate decays. Why make such a distinction? It is because true coincidence will be dependent on the source detector geometry and not on the count rate, or source strength, whereas the other type of coincidence process is directly related to the magnitude of fluence impinging on the detector. This means that true coincidence can, in relative terms, be just as important for a low-activity source as for a high-activity source of a γ emitter such as ^{134}Cs, and the only way to decrease this unwanted effect is to change the source-to-detector geometry. However, true coincidence can be corrected for mathematically, either using manually computed corrections for a single gamma line, E_{γ}, based on measured so-called total efficiency of the source-detector geometry (e_{tot}, described in Gordon and Gilmore [5]), or using pre-calculated codes based on Monte-Carlo simulations as in the aforementioned EFFTRAN [6].
5. **Incomplete scattering – pair-production with single or double escape:** For γ-photons with energies higher than $2\cdot m_0c^2$=1022 keV, the process of pair production and subsequent annihilation of the positron becomes possible. The higher the energy, E_γ, the more likely the process. In γ-spectrometry this process is often encountered when background radiation undergoes pair production in the sensitive volume, but either one or even two of the subsequent annihilation photons escape the volume, thus creating a pulse located at E_γ-511 or E_γ-1022 keV in the spectrum. For repeated events these two phenomena can give rise to shapes in the spectrum that can be mistaken for true full energy peaks of some γ emitter.

6. **Incomplete scattering – characteristic X-ray escape in the γ-detector**: If the number of counts of a distinct full energy peak is recorded long enough, a small peak can sometimes be discerned at the low-energy side of the peak, stemming from events of a photo-electric effect, but where the characteristic X-ray of the detector material, such as germanium or silicon, escapes the sensitive volume, and the remaining deposited energy is interpreted by the MCA as a pulse located at E_γ-$E_{X,K,det}$.
7. **Scattering in surrounding material – characteristic X-ray from shielding materials:** When a γ-photon originating either from the examined sample or from natural background radiation, such as the radon daughter ^{214}Bi, impinges on a high-Z material in the surround, such as the copper and lead shield, it may undergo a photoelectric process that results in a characteristic X-ray (of e.g. Pb or Cu) being emitted, that in turn can reach the sensitive volume of the detector. These peaks are therefore often clearly visible in most γ-spectrometry set-ups.
8. **Scattering in surrounding material – Compton back-scattering from shielding materials**: A similar event as the one previously described but with the difference that the γ-photon instead interacts in the surrounding material through Compton scattering resulting in a so-called back-scattered Compton photon reaching the sensitive volume and registered accordingly.
9. **Annihilation photon from background:** High-energy photons in the cosmic rays that impinge on the lead shield and other surrounding material, as well as the flux of some high-energy gammas from terrestrial natural background such as ^{208}Tl (^{232}Th-series) and ^{214}Bi (^{238}U-series) may give rise to pair-production processes that result in annihilation photons of 511 keV reaching the sensitive volume of the spectrometer. Often a distinct 511 keV peak is seen in most background γ-spectrometry measurements.
10. **Complete or fractional absorption of γ-quanta from background radiation:** Presence of background radiation, either as full energy peaks from γ emitted by the aforementioned radionuclides ^{40}K, ^{232}Th-daughters, and ^{238}U-daughters, or full energy absorption of cosmic γ-rays impinging the detectors will contribute to pulse counts by the MCA. For nuclear medicine applications, however, the source-to-background relationship is considerably higher than for low-activity measurements, except in whole-body counting and contamination-control measurements (Figure 8.8).

Another potentially important contribution to background and perturbations can be surface contamination of the detector or the surrounding material by γ-emitting radionuclides. The severity of this problem will be larger the longer the physical half-life, $T_{½}$, of the radionuclide. Even if most γ-emitting radionuclides within nuclear medicine are relatively short-lived, liquid materials should be handled with care as more long-lived impurities can contaminate the detector set-up and create artifacts and perturbing peaks in the subsequent measurements. In Figure 8.8 is given an illustration of how the aforementioned processes can be manifested in a γ-spectrum.

8.3 FROM SPECTRUM ACQUISITION TO A FINAL REPORT

8.3.1 Gamma Spectrometry Assessment: General

For routine measurement or industrial use of γ-spectrometry the evaluation of the examined sample is mostly based on a predetermined quality, in terms of overall uncertainty in the activity estimate, where the measurement geometry and pulse acquisition time may already be pre-set. The choice of measurement geometry is vital for the accuracy of the assessment, as any deviations from the one used in the calibration must in some way be corrected for, which in turn will lead to the introduction of uncertainties in the final assessment. Moreover, the acquisition time will greatly influence the stochastic uncertainty in the obtained pulse counts, and if the live time of the pulse acquisition must be set in advance to match time slots of detector systems, this pre-set must be adjusted so that it accommodates with other quality criteria set by the operator, such as a predetermined acceptable measurement uncertainty (e.g. 10% relative uncertainty) and detection limits L_D (see later).

Peak search in a recorded spectrum can be done either manually, by visual interpretation of the γ-spectrum, or automatically by various *analysis engines* of the software. For high-resolution γ-spectrometry, automated analysis engines are more often used since more nuclide-specific information can be extracted and assessed from the acquired spectrum than, for example, NaI(Tl)-spectra. Settings can be made in most software of how sensitive the chosen analysis engine is to suspected elevation of pulse counts in various parts of the pulse-height distribution. This analysis engine can be a combination of mathematical algorithms; one approach of an analysis engine is to first make a mathematically deconvolution of the pulse-height distribution of the spectrum, using a convolution function with a predetermined set of

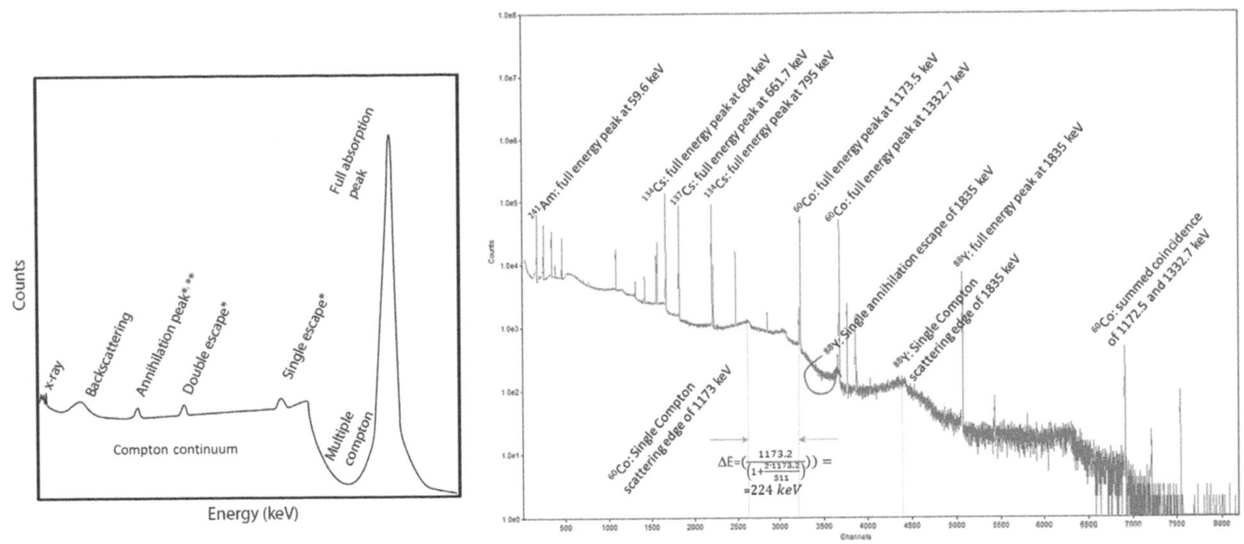

FIGURE 8.8 Gamma spectrum containing pulse contributions from processes 1 to 10 mentioned in Composition of a γ-spectrum. Left: Schematic illustration of peaks created by the processes 1–10. Right: An example of a real spectrum acquired by a 55 per cent (relative efficiency at 1.332 MeV) HPGe p-type crystal of a calibration sample containing a number of γ emitters, such as ^{88}Y and ^{60}Co.

γ-lines (taken from a nuclide library described below). This mathematical convolution function must account for the broadening of the peak width (FWHM) with increasing photon energy when computing the localization of the peak centroids After this step when the presence of all the predetermined peaks have been assessed, the remaining part of the spectrum is then analysed by applying the second derivate/difference and then locate additional potential centroids in the spectrum (mentioned in Calibration of a γ-spectrometer).

Once a peak is discerned and localized in terms of centroid, ch_c, it must also be identified in the sense that the peak is assigned to a full energy γ-peak from one or a set of likely γ-emitting radionuclides. The selection of nuclides is referred to as a *nuclide library* and may have been pre-selected and customized for the particular type of measurement series, such as impurity check of ^{99m}Tc. This selection is defined as a *working nuclide library*. An automatic assessment can include two nuclide libraries, one is selected to be customized by the operator him/herself, the *working library*, for the specific set of samples to be assayed, and one is the so-called *master library*, containing a more comprehensive set of nuclides. The master library is used to identify other peaks that have not been assigned any radionuclides in the working nuclide library and is intended for the user to have a suggested set of suspected radionuclides present in the sample that were not expected under normal conditions.

If the working nuclide library is too limited, only comprising the most likely encountered γ emitters, there is a risk of false negative findings. A *false negative* is the presence of a true γ emitter that is missed or misidentified in the assessment. However, if the working nuclide library includes too many γ emitters the probability of *false positive* identifications will increase, since there can be perturbations present (discussed in the previous section) that can be erroneously interpreted as full energy peaks. Moreover, the limited energy resolution may sometimes present a situation in which a number of candidate γ-lines can be assigned to a localized peak which also increases the probability of an erroneous identification and can be detrimental to the interpretation of the spectrum as it will require expensive operator time to resolve the misinterpretation. It is up to the user/operator to design and optimize the working library so that there will be a compromise between the probability of the two types of error.

To manage the assessment and minimize false positive results, customized nuclide sub-libraries need to be created from a master library so that both automatic and manual assessment can be made against a fit-for-purpose sets of possible γ-lines. A typical design of a nuclide library contains the following: If the aim of the assessment is to make impurity checks of ^{99m}Tc-generators, select from the master library provide by the software the following nuclides: ^{99m}Tc and ^{99}Mo, together with some radionuclides that are by previous experience known to be present in small amounts, such as ^{106}Ru. Then store the selected set of nuclides and assign this nuclide file as the working library of the search engine, so that this is library that is used in an automatic assessment.

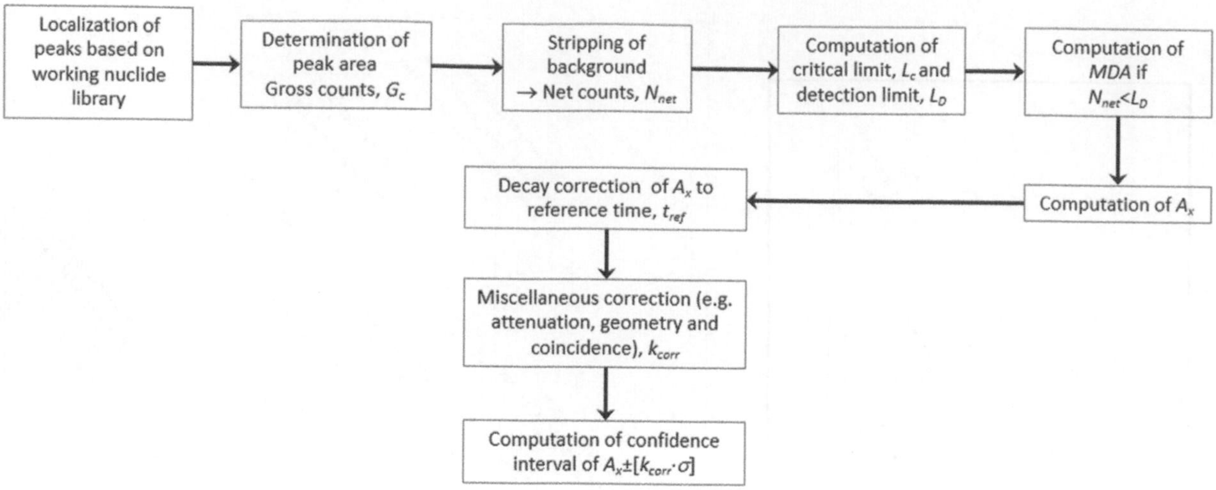

FIGURE 8.9 The various steps involved in γ spectrometry of a pulse height distribution with the aim of identifying and quantifying the γ-emitting radionuclides. Critical limit, L_c, and detection limit, L_D, are explained in Section 8.3.5.

The International Atomic Energy Agency (IAEA) has portals to links to various sources of updated nuclear data. Databases on nuclear data can be retrieved from a number of sources [7]. For nuclear medicine applications an ENSDF-database adapted for Committee on Medical Internal Radiation Dose (MIRD) [8, 9] format has been developed that may be the most convenient for medical applications.

The automatic analysis may be either nuclide-library driven or independent of any presumed presence of γ-peaks. For nuclear medicine applications it may be suggested to use library-driven automatic assessments, as the origin of the samples are relatively well known compared with environmental sampling with assays in more uncharted ecological systems. The procedure for a library-driven automatic assessment of a recorded spectrum generally follows a number of steps (Figure 8.9).

It is finally advised when using automated assessment to carefully read the software manual to select an analysis engine that best fits the purpose of the conducted measurements.

8.3.2 The γ-Spectrometry Report

The γ-spectrometry report should contain the following information to be used for the operator or the one in charge of reviewing the assessments:

- A list of all identified radionuclides
- A list of centroids, ch_c, of the located peaks
- A list of suspected radionuclides (retrieved from matching the spectrum with a more exhaustive nuclide library)
- Estimates of a quantification of the radionuclides in terms of activity concentration (or total activity, depending on which type of efficiency calibration is made) A_x±[confidence interval]
- The coverage factor used for the given confidence interval
- A list of minimum detectable activity concentration, MDA, for γ emitters in the working library for peaks where the computed net counts, N_{net}, was less than the detection limit, L_D.

In order to enable a swift/prompt quality assessment of the report, the following information should also be provided:

- Acquisition time
- Acquisition start time
- Sample identification (to check for mismatch between actual sample and the one listed in the additional spectrum information)
- Sample mass

- Reference date
- Efficiency calibration used in the assessment
- A list of corrections used in the analysis
- Which detector was used for the measurement
- Which operator took the measurement and did the evaluation

The software should provide a feature for providing Information on sample ID, sample mass, reference date, corrections to be used, and miscellaneous information regarding the origin of the sample, either in connection with the start of the pulse acquisition or in connection with the stopping and storing of the recorded spectrum. It is advised that the operator in charge of the γ-spectrometry determines whether the user can insert the sample information in connection with the start or storage of a pulse acquisition, in order to avoid mismatching and conflicting information.

8.3.3 Evaluation of Activity Concentration in the Examined Sample, A_x

Once the efficiency curve is established, the activity concentration in an examined sample, A_x, can be determined in the particular source detector geometry for all γ-energies within the range of which the calibration curve has been established (Eq. 8.12).

$$Ax = \frac{1}{n_\gamma} \cdot \frac{1}{m_{sample}} \cdot \left(\frac{\left(\frac{N_{net}}{t_m} \right)}{\epsilon_r \left(E_\gamma \right)} \right) \tag{8.12}$$

Additional corrections, k_{corr}, may be attached to Eq. 8.12. Often corrections for the decay, k_{dec}, between the reference time of the sample and the start of the pulse acquisition is needed, especially when measuring the relatively short-lived γ emitters used in nuclear medicine. For these types of γ emitters, there is also a need to correct for the gradual decay of the sources during pulse acquisition, $k_{dec,acq}$. Other corrections can be used as well, such as correcting for deviating atomic properties of the sample content from the one used in the calibration standard, k_{att}, mainly governed by the difference in density and atomic composition, which affects the self-attenuation of the γ emitters in the radioactive sample. For γ emitters prone to true coincidence, suitable correction factors for each γ-line, E_γ, are needed, k_{coin}. A number of miscellaneous correction factors may in some cases need to be attached. In many cases these correction factors may be relatively uncertain, especially concerning self-absorption and coincidence, as these are often extracted from a particular calibration event or from look-up tables requiring interpolation between different γ-energies. Consequently, each correction attached to the quantitative assay will also lead to an additional contribution to the overall uncertainty (discussed in Uncertainty of examined sample).

$$A_{x,corr} = \frac{1}{n_\gamma} \cdot \frac{1}{m_{sample}} \cdot \left(\frac{\frac{N_{net}}{t_m}}{\epsilon_r \left(E_\gamma \right)} \right) \cdot k_{dec} \cdot k_{dec,acq} \cdots k_{corr,i} \tag{8.13}$$

8.3.4 Uncertainty of Examined Sample

The confidence of the estimated activity concentration of the sample for the identified radionuclides is the most central part of the assessment, as it will provide a basis for a statement on how certain the user is of the obtained results. Especially in cases in which the aim of the assay is to ensure some kind of exemption criterion, that the observed concentrations of impurities do not exceed a certain intervention level, the confidence of the result is decisive for the statement of the results, and the continued procedures of the radioactive substances involved.

If recalling Eq. 8.13, one can use the formalism of the *Guide to the Expression of Uncertainty in Measurement* (*GUM*) [10] to express the uncertainty propagation of $A_{x,corr}$ for a given level of confidence, $u(A_{x,corr})$. If two simplifications are made, (a) there is no covariance between the independent variables of the function $A_{x,corr}$, and (b) all the independent variables have approximately Gaussian distributed probability density functions (pdf) – meaning that the standard deviation of each variable covers the same fraction of probability – is given by Eq. 8.14.

$$u^2\left(A_{x,corr}\left(N_{net},m_s,t_{acq},\epsilon_r\left(E_\gamma\right),\sum_{i=1}^{n}k_{corr,i}\right)\right)=\left(\frac{\partial A_{x,corr}}{\partial N_{net}}\right)^2\cdot u^2\left(N_{net}\right)+$$
$$\left(\frac{\partial A_{x,corr}}{\partial m_s}\right)^2\cdot u^2\left(m_s\right)+\left(\frac{\partial A_{x,corr}}{\partial t_m}\right)^2\cdot u^2\left(t_{acq}\right)+\left(\frac{\partial A_{x,corr}}{\partial \epsilon_r\left(E_\gamma\right)}\right)^2\cdot u^2\left(\epsilon_r\left(E_\gamma\right)\right)+ \tag{8.14}$$
$$\sum_{i=1}^{n}\left(\frac{\partial A_{x,corr}}{\partial k_{corr,i}}\right)^2 u^2\left(k_{corr,i}\right)$$

The uncertainties of the parameters, u(x), can be expressed in terms of fractions of one standard uncertainty of the variables, $k_a\cdot\sigma_x$, where k_a will represent the so-called coverage factor. For a 95 per cent confidence interval of a Gaussian distributed parameter, k_a=1.65, and for a 99 per cent confidence the corresponding value is k_a=1.96. Given the approximation that every pdf of the parameters is Gaussian distributed, the expression above can in turn be simplified to Eq. 8.15:

$$\left(\frac{u\left(A_{x,corr}\right)}{A_{x,corr}}\right)^2=\left(\frac{u\left(N_{net}\right)}{N_{net}}\right)^2+\left(\frac{u\left(m_s\right)}{m_s}\right)^2+\left(\frac{u\left(t_m\right)}{t_m}\right)^2+\left(\frac{u\left(\epsilon_r\left(E_\gamma\right)\right)}{\epsilon_r\left(E_\gamma\right)}\right)^2+\ldots+\left(\frac{u\left(k_{corr,i}\right)}{k_{corr,i}}\right)^2=$$

$$\left(\frac{\left(A_{x,corr}\right)}{A_{x,corr}}\right)^2=\left(\frac{\left(N_{net}\right)}{N_{net}}\right)^2+\left(\frac{\sigma\left(m_s\right)}{m_s}\right)^2+\left(\frac{\left(t_m\right)}{t_m}\right)^2+\left(\frac{\left(\epsilon_r\left(E_\gamma\right)\right)}{\epsilon_r\left(E_\gamma\right)}\right)^2+\ldots+\left(\frac{\left(k_{corr,i}\right)}{k_{corr,i}}\right)^2 \tag{8.15}$$

The net counts, N_{net}, may in many cases be retrieved directly from the software, or are considered directly in the automatic evaluation, but can also be a more complex function of the gross and background counts within and near the peak if for instance there is an existing peak in the background. However, as mentioned before, in nuclear medicine applications this latter case is not as common, provided that no external contamination of the detector set-up has occurred. Some sources of uncertainties can be handled as if they were a correction factor, k_{corr}, with a mean value of unity but with a standard uncertainty of $u(k_{corr})$ that will propagate as expressed in Eq. 8.15. An example of such a correction will be given in the next section.

8.3.5 Detection Limits

A definition of a set of signal levels is needed in γ-spectrometry (just as in radiometry and radiochemical analysis as well), to enable the user and operator to interpret the results and to make a statement of the results with a certain confidence. The following definitions were originally introduced by Currie in 1968 [11], and definitions given below are partly based on the work by Calmet and colleagues [12]. The first definition is the so-called *critical limit*, L_c, which is the level at which the background counts only exceed with a certain probability, α. In a situation when we know there is no presence of a certain radionuclide of interest, such as when measuring a blank sample, and computing the pulses in an ROI covering the full energy peak of this radionuclide (had it been present), we would still obtain counts emanating from other sources and from the background (discussed in the previous section). When repeating this measurement a number of times, we would in a·100 per cent of the measurement obtain a level exceeding L_c. Often a is set to 0.05 to represent the case when we can say with 1-α·100=95 per cent confidence that the background has exceeded L_c. This level is mathematically expressed in Eq. 8.16, with $K_{(1-\alpha)}$ being the quantile and σ_B is the standard uncertainty of the pulses acquired in the blank sample, or in a sample where the particular γ emitter is absent.

$$L_c = K_{(1-\alpha)}\cdot\sigma_b \tag{8.16}$$

The next situation of interest is when there actually is a presence of a γ emitter in a sample, but in such a small amount that the signal in terms of pulse counts in the ROI is so low that there is a risk with probability β of misinterpreting the signal as coming from the background. At which level of net counts, N_{net}, can the user then be confident on a certain confidence level, that N_{net} actually exceeds the counts with no presence of the gamma emitter (the situation represented by Eq. 8.16)? This situation is when the *detection limit*, L_D, of a γ emitter is defined and is given mathematically in Eq. 8.17.

$$L_D = K_{(1-\alpha)} \cdot \sigma_b + K_{(1-\beta)} \cdot \sigma_D \tag{8.17}$$

The factor $K_{(1-\beta)}$ is quantile for the probability distribution in the counts of N_{net}, and σ_D is the standard deviation of N_{net} at a certain level. The levels of α and β could be arbitrarily set by the user but is conventionally set so that α=β.This means that L_D is set at the level of which a measurement, on average, would give the same risk (=β) of interpreting the signal as below the critical limit, L_c, as the risk (=α) of erroneously interpreting a presence of a source if the sample were blank with a pulse count of L_c.

It is assumed that both gross, background, and net counts recorded by a γ-spectrometer are Poisson distributed, which means that the standard deviation of the parameter is equal to the square root of its mean (or expectancy) value; that is $\sigma_x^2=\mu_x$. Let us consider a measurement situation of a sample by a γ-spectrometer. The number of net counts, N_{net}, over a certain ROI, is given by Eq. 8.3:

$$N_{net} = G_c - B_c \tag{8.18}$$

where G_C is number of gross counts and B_C is number of counts emanating from the background. If we assume that there is no peaked background in γ-spectrometer, the background counts emanate from partially deposited γ-interactions from either cosmic radiation or from other radionuclides in the background or the sample. The background, B_c, is then calculated as described in Eqs. 2–3, using the pulse counts in adjoining ROI to estimate its value. Assume now that we have measured a sample that actually contains 0 activity, that is we have a 'blank' sample, but we do not know it in advance. We then compute N_{net} as usual. The definition of critical limit refers to this situation because it is an important reference to situations when one suspects the presence of a certain γ emitter in a sample. In the case of A_x=0 then N_{net} is by definition=0, which means that G_c=B_c. However, the variance (σ_x^2) and uncertainty in the computed value of N_{net} will not be zero as (Eq. 8.19):

$$\sigma^2(N_{net}) = \sigma^2(G_c) + \sigma^2(B_c) = G_c + B_c =$$

$$= (G_c)(= B_c) + B_c) = 2B_c \rightarrow \sigma(N_{net}) = \sqrt{2B_c} \tag{8.19}$$

The critical limit, L_c, on a 95 per cent confidence level (a=0.05) will then be given by the 95 per cent quantile for a Poisson distribution, $K_{0.95}$=1.645, and the standard deviation of a measurement with a blank sample:

$$L_c = 1.645 \cdot \sqrt{2B_c} = 2.33 \cdot \sqrt{B_c} \tag{8.20}$$

Using Eq. 8.20 and the approximation that for a non-blank sample, $\sigma_D \sim \sigma_B$ when the number of background counts is large compared with the net counts (which is the case when the number of N_{net} counts is small) the following expression for L_D is obtained (Eq. 8.21):

$$L_c = 1.645 \cdot \sqrt{(2B_c)} + 1.645 \cdot \sqrt{(B_G + B_c)} =$$

$$= 1.645 \cdot \sqrt{(2B_c)} + 1.645 \cdot \sqrt{2B_c} + 4.65 \cdot \sqrt{B_c} \tag{8.21}$$

N.B. that the above expression becomes somewhat different in cases where the background is much more well-determined. This can be the case if a long-time background measurement is conducted, t_{bkg}, which ascertains that the background count rate, $b=B_{bkg}/t_{bkg}$ (cps), is determined with high statistical certainty (if B_c is large$\rightarrow 1/\sqrt{B_c} \rightarrow 0$), and we have reasonable reason to assume that this background is unaffected in time, and relatively constant at a given full energy peak in time. In this special situation Eqs. 8.17–8.18 can instead for a blank sample be rewritten (Eq. 8.22):

$$
\begin{aligned}
N_{net} &= G - B_c; B_c = b \cdot t_{acq}; b = B_{bkg} / t_{bkg} \\
\sigma^2(N_{net}) &= \sigma^2(G_c) + \sigma^2\left(B_{bkg} \cdot \left(\frac{t_{acq}}{t_{bkg}}\right)\right) = \\
&= G_c\left(= B_c\right) + \left(\frac{t_{acq}}{t_b}\right)^2 \cdot \sigma^2(B_{bkg}) = \\
&= B_c + \left(\frac{t_{acq}}{t_{bkg}}\right)^2 \cdot \sigma^2(B_{bkg}) = B_c + \left(\frac{t_{acq}}{t_{bkg}}\right)^2 \cdot (B_{bkg})
\end{aligned}
\tag{8.22}
$$

Exploiting that $B_c=(t_{acq}/t_{bkg})\cdot B_{bkg}$, Eq. 8.22 can then be expressed as:

$$
\begin{aligned}
\sigma^2\left(N_{net\ blank}\right) &= B_c + \frac{t_{acq}}{t_b} \cdot B_c = \\
&= B_c\left(1 + \frac{t_{acq}}{t_b}\right) \sim B_c;\ if\ t_b \gg t_{acq}
\end{aligned}
\tag{8.23}
$$

When again considering a blank sample, the critical limit in a sample geometry with a well-determined background will then be L_c=1.645·σ(N_{net})~1.645· σ(B_c). Likewise, let us again consider a sample with an activity concentration A_x that gives rise to a net count, $N_{net=}G_D$-b·t_{acq}, just above the background level obtained if the sample were a blank, ·, the uncertainty of the net count rate in this sample is now determined by the standard uncertainty of the mean background level (Eq. 8.24):

$$
\begin{aligned}
\sigma(N_D)^2 &= \sigma(N_D)^2 + \left(\frac{t_{acq}}{t_{bkg}}\right)^2 \left(B_{bkg}\right) \\
&\rightarrow \sigma(G_D)^2; if\ t_{acp} \ll t_{bkg} \\
&\rightarrow \sigma\left(N_D\right) = \sigma\left(G_D\right)
\end{aligned}
\tag{8.24}
$$

Utilizing Eq. (8.23) the detection limit, L_D, will in this case be (Eq. 8.25):

$$
\begin{aligned}
L_D &= \sigma(N_{net}) + Lc = 1.645 \cdot \sqrt{B_c \cdot \left(1 + \left(\frac{t_m}{t_b}\right)\right)} + 1.645 \cdot \sigma(N_{net,D}) = \\
&\approx 1.645 \cdot \sqrt{B_c \cdot \left(1 + \left(\frac{t_{acq}}{t_{bkg}}\right)\right)} + 1.645 \cdot \sqrt{B_c} \approx 1.645 \cdot \sqrt{B_c} \approx 1.645 \cdot \sqrt{B_c} \\
&= 3.29 \cdot \sqrt{B_c}
\end{aligned}
\tag{8.25}
$$

The latter situation can be obtained for high-resolution γ-spectrometry in laboratories with relatively stable background conditions (that is, with a constant background count rate, b_C), but for nuclear medicine applications the relation in Eq. 8.21 is more common since the background in a ROI in a high-resolution γ-spectrometer will change from measurement to measurement due to the radionuclide composition in the sample. This in turn is because what was descried in Section 8.2.7 where the multiple Compton scattering leads to layers of background continuum (see Figure 8.8) that will vary with the relative composition of γ emitters in a sample, and the background cannot be derived from a long-term blank sample measurement.

Finally, the detection limit, L_D, can then be translated into a minimum detectable activity concentration, MDA (Bq kg^{-1}), by using Eq. 8.12 to obtain (Eq. 8.26):

$$MDA = \frac{L_D}{\varepsilon(E_g)\cdot t_{acq}\cdot m_s\cdot n_g} =$$

$$= 4.65\frac{\sigma_B}{\varepsilon(E_g)\cdot t_{acq}\cdot m_s\cdot n_g}; \textit{for arbitrary } b_c \textit{ if } t_{acq} = t_{bkg}$$

$$= 3.29\frac{\sigma_B}{\varepsilon(E_g)\cdot t_{acq}\cdot m_s\cdot n_g}; \textit{ if constanct } b_c \textit{ and } t_{acq} \ll t_{bkg} \tag{8.26}$$

where σ_B is the standard deviation of the background in the measured spectrum, which is obtained by, for example, the algorithm provided by the software.

An ISO standardized method of γ-spectrometry requires that also the uncertainty of the calibration $\varepsilon_R(E_\gamma)$ and other factors (such as in Eq. 8.13) be considered, and a more complex expression of MDA is then used (Eq. 8.27):

$$MDA_{ISO11929} = \frac{L_D}{\varepsilon(E_g)\cdot t_{acq}\cdot m_s\cdot n_g} =$$

$$= \frac{4.65\cdot\sigma_B + k(\sigma_\varepsilon^2, N_s)}{q(\sigma_\varepsilon^2)\cdot(\varepsilon(E_g)\cdot t_{acq}\cdot m_s\cdot n_g))} \tag{8.27}$$

where $k(\sigma\varepsilon_2, N_s)$ and $q(\sigma\varepsilon_2)$ are functions of the estimated variance of the efficiency, $\sigma^2(\varepsilon)$, and the net counts, N_{net}. For more detail refer to, for example, ISONORM [13].

8.4 EXAMPLES OF EVALUATION OF ACTIVITY CONCENTRATION, UNCERTAINTY ASSESSMENT AND DETECTION LIMITS

8.4.1 Evaluation of Assessment of Activity Concentration and Its Uncertainty in an Examined Sample

Assume a γ-spectrometer consisting of a HPGe-detector of 50 per cent relative efficiency at 1.332 MeV. The net counts of a localized full energy peak with a centroid located at 121.9 keV contains 17,749 pulses. The peak fit algorithm of the software estimates a stochastic pulse uncertainty of $\sigma(N_{net})$=145 counts. N.B. that this quantity is not directly given by the square root of the net counts as might be assumed from counting statistics of radiometry systems with no type of background whatsoever (see previous section for more detail). This will give a value of $(145/17{,}749)^2 = 6.67\cdot10^{-5}$ in the first term of the right-hand side of Eq. 8.15. The measurement time during the pulse acquisition, expressed in terms of live time as provided by the software, is 23,929.66 s. For live-times higher than ca. 100 s the uncertainty of t_{acq} can often be considered negligible, and thus the second term in the right-hand side of Eq. 8.15 is set to 0.

The efficiency is in this case taken from a measurement of a calibration standard containing a known activity of ^{57}Co, made prior to the measurement, being $\varepsilon(E_\gamma$ = 122 keV) = 0.02734 cps/(s^{-1}) for a 200 ml beaker positioned close to the HPGe-detector. The standard uncertainty of $\varepsilon(E_g$ = 122 keV) is estimated from the uncertainty of activity of the calibration standard, ($\sigma(A_{st})$). The calibration source contained a solution of ^{57}Co with specified estimated uncertainty of 3 per cent within a 99 per cent confidence level according to the manufacturer. This means that the relative standard deviation, σ_{st}, is 3.0%/1.96 (=coverage factor for 95 per cent confidence interval of a Gaussian distributed parameter) =

1.53 per cent of the stated activity concentrations in the standard. This gives a relative uncertainty of $\sigma(\varepsilon(E_\gamma = 122 \text{ keV}))$ = = 0.0153, which is the fourth term in Eq. 8.15.

The aliquot taken from a solution of ^{57}CoCl weighted 8.7 g, which was diluted with distilled water up to 200 ml in a similar type of beaker as the one used for the calibration standard. The sample mass, m_s is thus 0.0087 kg. With input of N_{net}, t_m, $\varepsilon(E_\gamma = 122 \text{ keV})$, m_s and with the knowledge of the nuclide being ^{57}Co with a branching ratio, $n_g = 0.856$ for $E_\gamma = 122$ keV, a value of $A_x = (1/0.856) \cdot (1/0.0087) \cdot (17{,}749/23929.66) \cdot (1/0.02734) = 3642.9$ Bq kg^{-1} is computed using Eq. 8.12.

The uncertainty in m_s can in many cases be non-normally distributed, especially when the sample weight is determined by an ordinary scale where the number of digits determine the precision of which the measured (in this case the mass in kg) is displayed. In this example we set that the sample of mass can vary from 8.65–8.75 g, and that the probability of for the accurate value within this range is uniform (also referred to as a rectangular pdf). A standard deviation in this case, σ_m, around a mean value at 8.7 g is $0.57 \times ((8.75\text{-}8.65)/2) = 0.028$ g. The factor $0.57 = (1/\sqrt{3})$ can be derived from the coverage interval obtained by computing the expectancy value of the standard deviation of a rectangular distributed parameter (20). Since the coverage interval of ±1 σ of Gaussian distributed parameters covers 68 per cent, whereas a corresponding interval for a rectangular distributed uncertainty only covers 57 per cent of the probability destitution, we cannot use σ_m in the uncertainty propagation given in Eq. 8.15 without adjusting for the non-matching coverage. To make $u(m_s)$ compatible with the other Gaussian distributed parameters we instead insert a value of $(68/33) \cdot 0.017 = 0.034$ and divide this with $m_s = 8.7$ g to obtain a value of $1.53 \cdot 10^{-5}$ for $(u(m_s)/m_s)^2$ as the second term in the right-hand side of Eq. 8.15.

The time between the start of acquisition and the reference date is 108 d. Given that the localized peak is identified as originating from ^{57}Co in the sample the decay correction factor becomes $k_{dec} = \exp(-(\ln(2)/T_{½,phys} = 270.9 \text{ d}) \cdot ((t_{acq,start} - t_{ref}) = 108 \text{ d}) = 0.759$. The uncertainty in this correction will stem from the uncertainty in the time, t_{ref}, and of the stated uncertainty of the tabulated physical half-time. However, assuming all laboratories use the same set of isotopic reference values, the constant $T_{½,phys}$ can be considered a well-known constant with negligible stochastic uncertainty and is thus disregarded. If t_{ref} is well-defined, the same apply for $u(t_{ref})$, and thus $u(k_{dec})$ is approximately 0.

However, the positioning of the sample in this specific set-up has been known to exhibit some irreproducibility. From a number of repeated constancy checks with an additional source, the uncertainty in the positioning will fluctuate. From 8 repeated measurements on different occasions, the peak count rate of the source of the constancy check, varied with a relative standard deviation around the mean of 1.7 per cent. A correction factor for the irreproducibility of the source positioning was thus introduced, with a value of $k_{rep}=1$, and with a standard deviation of 0.017. No additional corrections regarding attenuation and self-absorption have been considered as the standard source and the measured samples basically consist of water solutions with ^{57}CoCl in ionic state, assumed to be homogeneously distributed in the beaker. The resulting corrected activity concentration, $A_{x,corr} = (1/0.856) \cdot (1/0.0087) \cdot (17{,}749/23929.66) \cdot (1/0.02734) \cdot (1/0.759) \cdot 1 = 4799.6$ Bq kg^{-1}

The sum of the considered uncertainties in the right-hand side of Eq. 8.12 is then $u^2(A_{x,corr}) = 6.67 \cdot 10^{-5} + 0.000234 + 0 + 0.0153^2(=0.000234) + 0 + 0.017^2(=0.000289) = 0.000758$. The relative standard deviation $s(A_{x,corr})$ is thus the square root of this number, that is, 0.028. This means that the total relative uncertainty, expressed in one standard deviation, is 2.8 per cent. Expressed with on 99 per cent confidence level, the relative uncertain becomes $1.96 \cdot 0.028 = 5.4$ per cent. To communicate this result, the activity concentration value should thus be reported as being $A_{x,corr}(\pm 1.96 \cdot \sigma) = 4799.6 \pm 259$ Bq kg^{-1}.

8.4.2 Evaluation and Detection Limit and Minimum Detectable Activity Concentration

A spectrum is acquired for a lifetime Ll=1,000 s of a sample, measured in a standard geometry of 200 ml aqueous solution, positioned on top of a p-type HPGe-detector with 55 per cent relative efficiency at 1.332 MeV. In the beaker solution, a portion of 23.5 g sample solution to be determined, has been dissolved. A distinct full energy peak is found located around 122 keV. The software identifies this peak as the 122.1 keV γ-line from ^{57}Co, with $n_\gamma=0.856$. The curve fit of the absolute efficiency, ε_r (defined in Eq. 8.9) for this particular geometry has been determined to be 0.0273 cps s^{-1}. By marking the channels around the peak, covering about $2.5 \cdot \text{FWHM}(E\gamma=122.1 \text{ keV})$, the software computes a gross count, G=632 and N_{net}=404 counts. The estimated uncorrected activity, A_x (Bq), is computed according to Eq. 8.13 to be 404/1000/0.0273/0.856=17.3 Bq ^{57}Co. Dividing this value with the mass of 23.5 g radioactive solution solved into the beaker content, an activity concentration of 17.3/0.0235=736 Bq kg^{-1} is obtained.

Case 1: No ^{57}Co assumed in the background and presence of other radionuclides in the sample is found, for example, ^{51}Cr at 320 keV, which increases the background at 122 keV. What would have been the minimum detectable number of net counts, L_D, over the ^{57}Co peak at 122.1 keV? In this case Eq. 8.18 (Eqs. 18–20) should be used. The detection limit now refers to a hypothetical situation in which all other γ emitters are present except the one radionuclide of interest (^{57}Co in this case). In this situation the number of gross counts in the ROI would then be the number of background counts obtained in the ROI without the radionuclide present, that is B_m=G_m-N_m. The variance in the background found in the 1000 s measurement, that corresponds to a sample that is blank in respect to the particular radionuclide, will then have a pulse count of B_m of 632-404=228 counts, and a corresponding standard deviation of B_m is approximately given by √(632+404)=32.2. This gives a L_D of 4.65×32.2=149.7 counts, which in turn can be translated into an MDA of 149.7/(0.0273·1000·0.0235·0.853)=273 Bq kg^{-1}.

Case 2: No other evident γ emitters are found to be present. Given the single radionuclide composition in this sample, what is the detection limit, L_D, and what is the corresponding MDA of ^{57}Co? Comparing the spectrum with the corresponding spectrum from a blank sample (a beaker containing only distilled water of 0.200 kg) L_D is computed by Eq. 8.25. The background sample was measured for 511,036 s, and the number of gross counts G_B=17,523 over the same ROI (for ^{57}Co) as in the sample of interest. The software computes a small peak of net counts N_B=102 with a σ(N_B)=162. This weak peak is mainly originating presence of background radiation in the detector environment as discussed in 'The composition of a γ-spectrum'. This translates into a peak background count rate, b= 0.20·10^{-3} s^{-1}. If the sample of interest had been a complete blank sample, the measurement of 1000 s would have given a B_c of $b \cdot t_{acq}$(=1000 s)=0.2 counts. The standard deviation of these counts is then $\sqrt{0.2}$ =0.48 counts. According to Eq. 8.25 this translates into an L_D of 3.29·0.48= 1.45 counts. These number of counts is then used in Eq. 8.26 to obtain a corresponding MDA of 1.45/(0.0273·1000·0.0235·0.853)= 2.65 Bq kg^{-1}.

8.5 QUALITY ASSURANCE

It is suggested that already when the γ-spectrometry system is installed (including the fine-tuning of electronic settings by the vendor) all the initial settings should be saved, such as an amplification gain, number of channels in the MCA, optimal high voltage bias, HV, as well as generic detector characteristic properties provided by the vendor (relative efficiency at 1.332 MeV, dead-layer thickness, operating bias (HV), polarity, detector type: n or p-type, etc.). It is highly recommended to follow the manufacturer's instructions when performing the initial settings. If additional customized settings for non-accredited measurements are done, they should be documented, and it should be possible to easily reconfigure all settings to the one done by the vendors at the installation. The initial system performance should be carefully documented, including parameters such as energy resolution (FWHM) as a function of photon energy, peak shapes (which can be characterized by forming a ratio between the full peak width a 10 per cent maximum (FWHM) and FWHM, background count rate, etc.). In the following section, examples of such quality parameters are given.

Ambient settings, such as air venting and temperature conditioning of the laboratory, which may have influence on the measurements, should also be logged. It is also important that data files, containing either spectra or associated files used for the calibration and settings of the software, be arranged by the operator so that it is transparent and easily accessible for all the users of the system. The logbook, combined with a documentation structure are important when admitting other users to access the spectrometry system. For certain procedures it may be necessary to keep certain settings locked, so they only can be altered by the operator in charge, or by someone who has been authorized by the operator.

Once the γ-spectrometry system has been installed it is important that the γ-spectrometer system be regularly checked to ensure that the measured values are accurate within a predetermined confidence level. Certain parameters may be checked more often than others (e.g. daily, monthly or quarterly). Quality control of the system must also be done in connection with a bias voltage supply failure, after a repair of a component (e.g. replacing the preamplifier of the HPGe crystal and subsequent pumping of the vacuum) or a replacement of a component (e.g. swapping or replacing a MCA to the detector) has been done.

8.5.1 Constancy Tests for Quality Control of γ-spectrometry

A set of radiation sources is needed to perform regular constancy tests of the γ-spectrometry equipment and evaluation methods. It may be that different sources must be used depending on how the time slotting of the measurement times of the detector is optimized. Ideally, a γ-emitting source consisting of a cocktail of radionuclides that have physical half-lives long enough for the lifespan of the detector should be used. The activity of this ideal standard source should be high enough to yield at least 50,000 net counts in each of the γ-lines used in the evaluation, during a reasonable acquisition time, t_{acq}, such as one hour or less. The constancy tests mainly entail tests of efficiency and energy calibration, resolution (FWHM), and background radiation tests, described in the following sections.

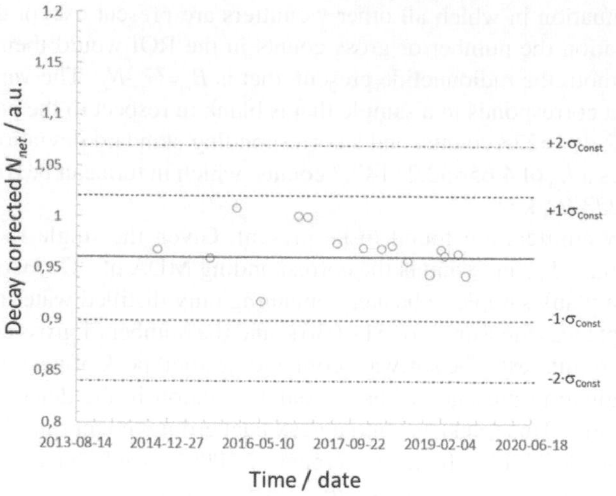

FIGURE 8.10 Time plot of constancy checks using ^{60}Co full energy counts in the net peak of 1332.5 keV. Values have been decay-corrected and normalized to the N_{net} at the initial measuring occasion.

8.5.1.1 Test of Efficiency

Regular measurements with a radiation source that can be assumed to be mechanically stable should be done to determine whether the efficiency, $\varepsilon_R(E_\gamma)$, and the method to extract the net counts, N_{net}, can reproduce the same number of counts when corrected for physical decay. The frequency of the efficiency test must be adapted to the estimated risk of obtaining biases or deviations in the assessments, giving rise to erroneous values (ranging from once a day to quarterly controls depending on, for example, the stability of the amplifiers). The constancy test source needs in turn be controlled for physical integrity by conducting smear tests to ensure that there is no leakage of the radioactive substance. Laboratories need to check the expiry date of the certificate of the manufacturer of the source. The expiry date represents the time over which the manufacturer can guarantee that the original specified activity is maintained, given that it is stored and treated with care and has not been subject to mechanical stress that may cause rupture of the source encasing. In Figure 8.10 is given an example of a plot of the net counts of some sources used for constancy checks of a HPGe-spectrometer, corrected for physical decay. Ideally the data point in this plot should be aligned within $\pm 2\sigma_{Const}$, with σ_{Const} being an uncertainty accounting for the stochastic uncertainty in the net counts, σ_N, and the relative positioning variability, σ_{Pos} (which should be less than 2–3%). If the result is outside another level, $\pm 3\sigma_{Const}$, it is usually recommended that a systematic trouble shooting is done. For definitions of different action levels for deviations in constancy test refer to, for example, Nordtest [14]).

As mentioned previously, it is preferable that the constancy test source consist of a nuclide with relatively long physical half-time, such as ^{241}Am (427 y), ^{137}Cs (30.2 y) or ^{60}Co(5.27 y). For nuclear medicine applications with the main energy of interest below 511 keV, sources such as ^{57}Co(270 d) or ^{133}Ba (10.51 y) can be used instead of ^{137}Cs and ^{60}Co.

If there is a high workload of the γ-spectrometer with many samples needing to be determined daily, the activity of the constancy test source should be high enough to enable at least 10,000 net counts within a pulse acquisition time of 1 h, in order not to occupy valuable detector time. A net count of 10,000 represents roughly a relative stochastic uncertainty of $1/\sqrt{(10000)}=0.01=1$ per cent. If the same acquired spectrum of the test is also used for energy resolution test (see next section), this number should be higher still, up to 50,000.

Ideally the test source should be in the same physical constellations as the regular sources used – however this may not be practical if a number of such standard source volumes and geometrical configurations are used. However, a shift or disruption from the $\pm 2 \cdot \sigma_{Const}$ interval, in the recorded N_{net} of the constancy test for a given measurement geometry, will strongly indicate a general deviation of the detector setting that needs to be investigated imminently.

8.5.1.2 Test of Energy Resolution, FWHM

The maintaining over time of the energy resolution (measured in terms of FWHM (see Section, Quality Parameters for a γ-spectrometer) is important and deviations of the FWHM value from the one specified by the detector manufacturer reflect changes in the detector set-up or deterioration of the physical conditions of the detector that need to be further investigated. The FWHM for a given full energy peak may be further reset during the installation of the detector by the vendors to optimize the performance (refer to e.g. Gordon and Gilmore [5]; Ortec [15]). An appropriate reference level for the FWHM of the detector is the values obtained during the installation. It is recommended that the FWHM of a full energy peak from the constancy test source be calculated from a peak containing more than 50,000 net counts (GammaVision 8). The value of the FWHM obtained by the software relies on pulse counts in individual channels and is therefore more sensitive to stochastic noise than when calculating N_{net}.

A gradual increase in FWHM over time indicates that the noise in the amplification has increased, which causes a broadening of the full energy peaks. For a semiconducting γ-spectrometer, such as HPGe, this noise consists of leakage current between the electrical contacts of the detector crystal, caused be gradual loss of vacuum in the high-resolution γ-spectrometer. A more sudden rise of FWHM indicates that increased noise is being consistently added to the pulses that enter the MCA, which may be a result from failure of one or more components in the detector and amplification.

For scintillator combined with photo-multiplier tubes (PMT) the resolution and energy calibration may be affected by deterioration of the optical conduction between scintillator and PMT-tube, or by gradually ageing dynodes that lowers their signal gain (so-called secondary emission gain [16]).

8.5.1.3 Background Levels

Repeated background counts of the detector are needed to verify that the shield and detector set-up have not been contaminated, or that no other ambient source of fluctuating background influences the measurements, such as increase in ^{222}Rn air concentrations in the laboratory locale. The frequency of such background measurements needs to be based on a risk assessment regarding factors that may influence the ambient conditions. The background can be measured by integrating the counts over the whole channel registry of the MCA, from the low energy end to the maximum channel number. The gross counts, $G_{bkg,tot}$, will then indicate any major change in the γ-radiation background of the detector. Irrespective of in which energy region a shift occurs. It is also beneficial to assess the background spectrum with a preset of ROIs that are usually used in the sample measurements (such as a 657 to 664 keV interval for the full energy peak of the 661.6 keV γ-line of ^{137}Cs). Alternatively, one can use the automatic peak search algorithm of the software to identify and quantify presence of full energy peaks in the background. Normally significant presence of ^{40}K, ^{228}Ac, ^{214}Bi, ^{214}Pb and ^{212}Pb is expected, as well as the annihilation peak at 511 keV (Figure 8.11).

8.5.1.4 Test of Energy Calibration

The accuracy of the energy calibration can be measured in terms of, for example, a relative shift in centroid compared with a reference value, ch_{ref}. The value ch_{ref} is preferably the centroid locations obtained when the calibration of the standards geometries was conducted. The relative shift of various full energy peaks can be summed to yield an overall measure of the energy calibration accuracy (Eq. 8.28).

$$\Delta = \sum_{i=1}^{n} \frac{ch_{meas,Ei} - ch_{ref,Ei}}{ch_{ref,Ei}} \tag{8.28}$$

Ideally, the energy calibration should be checked manually in connection with each measurement as a small drift in the amplification may cause misinterpretation of the results if the automated algorithm instead associates the centroid with a γ-line lying close to the one that was actually measured. The centroids should nevertheless be given in the report by the automated algorithm as a final check of the validity of the energy calibration. It is advised to adjust the amplifier gain so that the full energy peaks are always aligned as closely as possible to their original positions in the installation, $ch_{ref,Ei}$. It is therefore important to log the gain settings of the MCA if it needs to be replaced or repaired, in order to make it easier to reproduce the original energy calibration.

8.5.1.5 Test of Environmental Factors

It is important to monitor and document physical and environmental parameters that potentially can affect the validity of the γ-spectrometry results. Such factors can, for example, be the room temperature, humidity, liquid nitrogen vapor pressure, ambient ^{222}Rn concentration, or the presence of strong radiation sources in the adjacent localities of the laboratory. Related to this issue is also the control of access to the laboratory facility or other localities involving the preparation of the sample, to prevent contamination of surfaces that may influence the results.

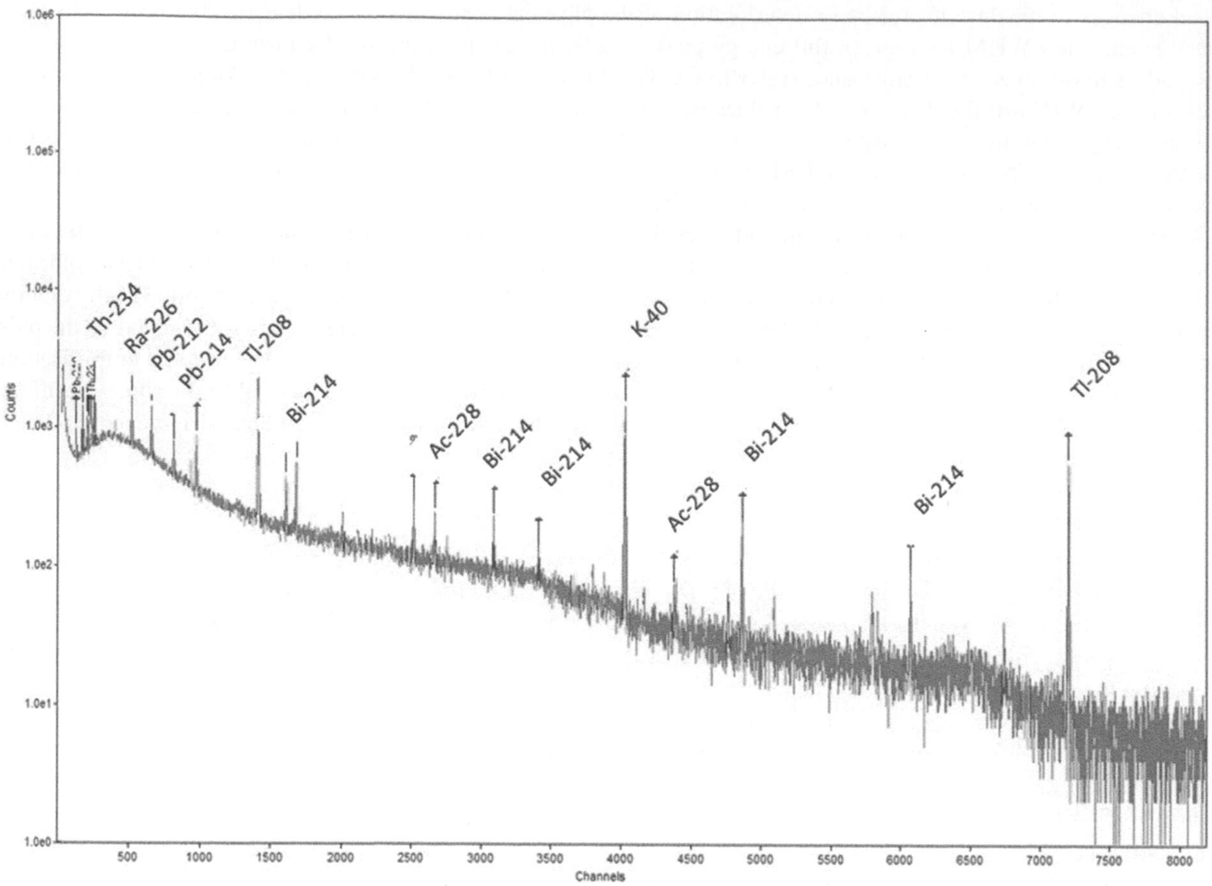

FIGURE 8.11 Typical spectrum acquired from a 55 per cent (relative efficiency at 1.332 MeV) HPGe-detector in the presence of a blank sample (200 ml distilled water in a beaker positioned close to the detector entrance surface).

8.5.2 Proficiency Tests

For a laboratory aspiring to maintain a quality assurance system, proficiency testing is a vital part of the quality assurance of an accredited assessment procedure. A proficiency test is a procedure in which a laboratory conducts an assessment of a reference sample to be reported for an inter-laboratory comparison. This reference sample is provided by an organization that compiles reported assessments from a number of other laboratories. The individual results from a given laboratory are then benchmarked against the mean of all participating laboratories or against a reference value determined by the organization that conducts the test. There are a number of proficiency tests carried out in γ-spectrometry by, for example, IAEA [17], NIST [18]) and PROCORAD [19]). However, few of these tests use samples that are appropriate for nuclear medicine, since the activity concentration in the environmental samples are often orders of magnitude lower than the relevant activity concentration values within nuclear medicine.

In the proficiency test, the reported value must be accompanied with an uncertainty estimate at a pre-set confidence level (e.g. ±2σ, that represents a 95 per cent confidence level for a normally distributed parameter).

There are several parameters that can be used to evaluate the participating laboratories' performance (e.g. [20]). One such parameter is the z-score, z_{score}, which measures the discrepancy between the reported activity concentration value and the target value by the proficiency test organizer is divided by a target standard deviation, σ_{target}, that is considered as acceptable by the organizer. A typical target standard deviation can be 10 per cent of the target value; that is $\sigma_{target}=0.1 \cdot A_{target}$ (Eq. 8.29).

$$z_{score} = \frac{A_{rep} - A_{ref/target}}{\sigma_{target}} \tag{8.29}$$

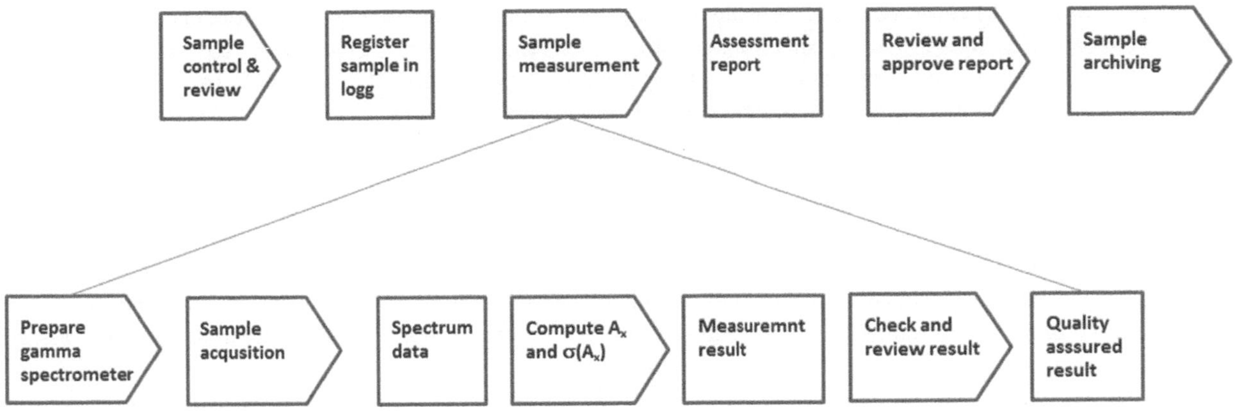

FIGURE 8.12 An example of a work-flow chart in a γ-spectrometry laboratory.

The organizer can define intervals of the z-score that represent an overall performance measure of the reported data. Typically, the score is deemed satisfactory if $|z_{score}|<2$, and questionable if a higher absolute value of z_{score} is obtained. The higher the z_{score} value the poorer the performance. Hence, if a z-score is judged unsatisfactory, the operator of the laboratory needs to revisit the reported data and identify which part of the assessment procedure given in Figure 8.12 contributed to the biased results.

To also take into consideration whether the uncertainties in the reported value are consistent, another score parameter, u_{test}, is more suitable (Eq. 8.30):

$$u_{test} = \frac{|A_{target} - A_{report}|}{\sqrt{u_{target}^2 + u_{report}^2}} \tag{8.30}$$

where u_{target} and u_{report} are equal to $k_{cov} \cdot \sigma_{target}$ and $k_{cov} \cdot \sigma_{report}$, respectively. The coverage factor, k_{cov}, specifies how large a confidence interval is to be reported, usually 95 per cent confidence, which means that $k_{cov} \sim 2$ for normally distributed parameter estimates. The coverage factor is in turn specified by the organizer of the proficiency test. This score is more appropriate in cases when a laboratory wants to test how well the uncertainty budget has taken all potential uncertainty and error sources into account.

8.5.3 Quality Management

One meaning of the term *good quality* of a γ-spectrometry can be that the demand of information regarding radionuclide concentration in a sample, requested by external or internal users of the organization, can be satisfactorily met by the laboratory. It is important that the party requesting information and the γ-spectrometry laboratory that is assigned to provide this information have a consensus on the expectations in terms of level of confidence in a reported activity concentration value. It may be that customers/task assigners have quite relaxed request regarding the confidence limits of the reported data, such as a 20 per cent relative overall uncertainty, whereas others need higher accuracy in the stated values of the laboratory. The assessment methods must then be fit for purpose so that the efforts are in proportion to the actual requested quality of the data. It is also important to know that, if the laboratory aspires for accreditation (according to ISO 2017 [21]) of their γ-spectrometry methods, a general system for quality management (quality system) is required.

A number of factors contribute to the risk of adversely affecting the quality of the data. These factors can be grouped in, for example; (1) *Human factors* (individual competence, stress, etc.); (2) *Management* (vision or lack thereof, planning or lack thereof, personal favouritism, communication skills, etc.); (3) *Equipment and instrumentation* (such as physical quality, lack of maintenance, non-availability of auxiliary parts, variability in performance); (4) *Methods* (conformity to specified procedures, definitions of procedures or lack thereof, inadequately defined procedures, conflicting requirements); and (5) the *environmental factors* (e.g. temperature, humidity, liquid nitrogen vapor pressure, ^{222}Rn concentration).

There are a number of references that provide useful information on how to manage the quality of a laboratory – references that can be easily applied to γ-spectrometry assessment (e.g. ISO17025 [21]or Isobudgets [22]). The following points will be useful when managing the quality of the γ-spectrometry assessment at the laboratory:

A *workflow description* is helpful, where processes are defined and described, and where responsibilities are clearly appointed. A workflow chart can illustrate the full chain of procedure of receiving a sample (assigning a task), measuring, assessing, and reporting of data. The laboratory should have *easy access to important reference documents* such as GUM [9] or standard methods suggested by IAEA (e.g. [23]).

A definition/statement of the *overall scope, or quality policy,* of the laboratory and the scope of the γ-spectrometry methods in particular should be stated and documented. The responsibility for the quality of the various procedures should preferably be proportioned between staff members based on their individual experience and competence regarding, for example, instrumentation or evaluation methods. In larger laboratories a technical leader can be assigned as well as a person in charge of the general quality assurance.

A *documentation plan* is recommended to maintain a structure that enables an overview of the management of vital documents, such as certificates, contracts and data reports, and a plan for the longevity of these documents. The plan can also outline selected users who have access to certain documents. All methods that are used should be described in well-documented *method descriptions* so that no method or procedure becomes dependent on one single user. Likewise, descriptions of how the methods have been *validated* should be done and logged. A comprehensive description of *the uncertainty budget* used for the confidence estimates in the reported values should be done, including a description of which factors have been considered in this budget.

Certificates of *traceable calibrations* of equipment should be listed and stored. As described previously, a laboratory should have *a control program* of instruments and equipment used in connection with the γ-spectrometry assessment (e.g. Uninterrupted Power Supply, Nitrogen cooling pumps, etc.), including the spectrometer itself. Regular *monitoring of ambient conditions*, such as air humidity or temperature, is often required for ISO-accreditation, if these parameters are considered to have a potential effect on laboratory performance. The outcome of previous participations in *proficiency tests* or other intercomparisons must be documented and be easily accessible to new and future operators of the system.

Regarding interaction with external task assigners, all *agreements and contracts* must be documented. Care must be taken to design the *data reports* so that all relevant information is provided for the end users. If any *complaints* or misunderstanding have arisen from the customers or end-user of the γ-spectrometry measurement results, these should be logged and documented. To ascertain the quality of the data output it is also recommended to request *feedback from the task assigner*.

To ensure maintenance of competence, *training programs and introductory programs* for newcomers must be established. Some methods may require mandatory training to be allowed access and handling of the γ-spectrometer. Which of these methods needing authorization must be defined by the responsible operator. It is also important that *a list of those staff members* who have been authorized must be logged, and it should also be clear for how long a certain staff member is granted approval to have a certain task. *Regular internal revision* of the methods and quality system should be done, in addition to *external audits* from, for example, an accreditation institute. The audits should be annual and are preferably divided into technical (equipment, instrumentation, etc.) and supportive (administration, reporting, communication with end-users) parts of the organization.

An important feature of the quality assurance is to have a *routine for dealing with non-conformance* in the laboratory (such as erroneous handling of data leading to inaccurate reported results), which may or could lead to an event that jeopardizes the quality of the reported data. Therefore, related to this, it is useful to have a plan for *preventive actions* to ensure minimizing the risk of perturbations or nonconformity. Regular *internal and external audits*, including both the technical and management aspects should be done.

Even if no accreditation of methods is aspired for, a laboratory can still assume a number of the aforementioned ambitions. The flexibility that comes with bypassing the formal demands of an accreditation must be used with precaution, and it is strongly advised to design and follow a quality-control program that is reviewed regularly.

Further reading can be found in [24].

ACKNOWLEDGEMENT

The author wishes to express his gratefulness for the valuable input from Sara Ehrs, Mattias Jönsson and Sigrid Leide-Svegborn.

REFERENCES

[1] D. G. Darambara and A. Todd-Pokropek, "Solid State Detectors in Nuclear Medicine,", *Quarterly Journal of Nuclear Medicine and Molecular Imaging,* vol. 46, no. 1, pp. 3–7, Mar 2002.

[2] S. M. Karabıdak, "Dead Time in the Gamma-Ray Spectrometry," in *New Insights on Gamma Rays*, 2017, Chapter 2.

[3] "ORTEC, Detecting High Energy Gamma Rays from Neutron Interactions: Neutron Damage and HPGe Detectors." www.ortec-online.com/service-and-support/library/tech-notes/neutron-damage.

[4] "LabSOCS (Laboratory Sourceless Calibration Software)." Mirion Technologies. www.mirion.com/detection-and-measurement.

[5] R. Gordon and G. Gilmore, *Practical Gamma-ray Spectrometry*, 2 ed. John Wiley, 2008.

[6] "EFFTRAN (Efficiency transfer and coincidence summing corrections for environmental gamma-ray spectrometry)." Home page for information regarding the code for EFFTRAN. www.efftran.com/.

[7] International Atomic Energy Agency, "IAEA Nuclear Data Services." www-nds.iaea.org.

[8] "Committee on Medical Internal Radiation Dose, MIRD. Nuclear Decay Data in the MIRD Format." www.nndc.bnl.gov/mird/.

[9] "Committee on Medical Internal Radiation Dose, MIRD." www.snmmi.org/AboutSNMMI/CommitteeContent.aspx?ItemNumber=12475.

[10] BIPM, IEC, IFCC, ILAC, IUPAC, IUPAP and OIMI, *Guide to the Expression of Uncertainty in Measurement* (*GUM*) JCGM 100:2008 GUM 1995, with Minor Corrections, 2008. www.bipm.org/utils/common/documents/jcgm/JCGM_100_2008_E.pdf.

[11] L. A. Currie, "Limits for Qualitative Detection and Quantitative Determination. Application to Radiochemistry," *Analytical Chemistry*, vol. 40, no. 3, pp. 586–593, 1968/03/01 1968, doi:10.1021/ac60259a007.

[12] D. Calmet, M. Herranz, and R. Idoeta, "Characteristic Limits in Radioactivity Measurement: From Currie's Definition to the International Standard ISO-11929 Publication," *Journal of Radioanalytical and Nuclear Chemistry,* vol. 276, no. 2, p. 299, 2008/05/04 2008, doi:10.1007/s10967-008-0502-4.

[13] "International Organization for Standardization, ISO 11929:2010: Determination of the Characteristic Limits (Decision Threshold, Detection Limit and Limits of the Confidence Interval) for Measurements of Ionizing Radiation – Fundamentals and Application," 2010.

[14] "Nordtest, Uncertainty from Sampling – A NORDTEST Handbook for Sampling Planners on Sampling Quality Assurance and Uncertainty Estimation (NT TR 604)," 2007.

[15] "ORTEC, GammaVision, Gamma-Ray Spectrum Analysis and MCA Emulator for Microsoft Windows 7, 8.1, and 10 Professional, A66-BW Software User's Manual," 2018.

[16] "Hamamatsu." (Online) www.hamamatsu.com/resources/pdf/etd/PMT_TPMZ0002E.pdf. Accessed in 2021.

[17] "IAEA, Interlaboratory Comparison." (Online) nucleus.iaea.org/rpst/referenceproducts/Proficiency Tests/index.htm. Accessed in 2021.

[18] "National Institute of Standards and Technology NIST." [Online]. Available: www.nist.gov/programs-projects/nist-radiochemistry-intercomparison-program-nrip. Accessed in 2021.

[19] "PROCORAD." www.procorad.org/en/register/Register-for-the-intercomparison.

[20] "International Organization for Standardization, Statistical Methods for Use in Proficiency Testing by Interlaboratory Comparison ISO13528:2015," 2015.

[21] "International Organization for Standardization, General Requirements for the Competence of Testing and Calibration Laboratories. ISO/IEC 17025:2017," 2017. [Online]. Available: www.iso.org/standard/66912.html

[22] "Isobudgets." www.isobudgets.com.

[23] *Measurement of Radionuclides in Food and the Environment*. Vienna: International Atomic Energy Agency, 1989.

[24] M. L'Annunziata, *Handbook of Radioactivity Analysis*, 3 ed. Oxford: Academic Press, Elsevier, 2012.

[25] M. Isaksson and C. L. Rääf, *Environmental Radioactivity and Emergency Preparedness* (Series in medical physics and biomedical engineering). Boca Raton: CRC Press, Taylor & Francis Group, 2016, pp. xxxv, 614 pages.

9 Properties of the Digital Image

Katarina Sjögreen Gleisner

CONTENTS

9.1 IMAGES FOR COMMUNICATION OF MEDICAL INFORMATION

Images have probably been used for communication throughout human history, with cave paintings an early from of art. Today the use of digital images for communication has literally exploded, entering our daily lives by different media such as cameras in our cellular phones. Images are extremely effective for human communication, providing a means to build mental images and to associate to our existing knowledge bank (Figure 9.1). Naturally, the words *imagine* and *image* are etymologically closely related.

The digital image is a carrier of information. It provides a link between the physical object, which in medical imaging is the interior of the examined patient, and the knowledge bank of the medical doctor (MD) who is assigned to interpret the image. Based on this information, the MD may make diagnostic decisions, such as from which disease the patient suffers, what kind of treatment should be prescribed or, for radiotherapy, how a radiation-treatment machine should be directed. It is evident that the image needs to convey information such that it corresponds to the physical object as exactly as possible, with minimal risk for misinterpretation.

Medical imaging covers a vast number of techniques that reflect different properties of the human body. The tissue properties that underlie the image formation differ between techniques, and may for example be

- for nuclear-medicine imaging: the uptake and accumulation of a radiopharmaceutical, in turn depending on physiological and cellular mechanisms;

DOI: 10.1201/9780429489556-9

FIGURE 9.1 Universally understandable communication?

- for X-ray imaging: the interaction and attenuation of photons in the kiloelectronvolt range, mainly governed by the mass density of tissues;
- for MR imaging: the concentration and rate of relaxation of hydrogen nuclei when these are placed in a strong magnetic field and exposed to a radio-frequency pulse – the relaxation rate is dependent on the tissue microenvironment and chemical composition;
- for ultrasound imaging: the reflection of an ultrasound wave when it propagates through tissues with different impedance, closely related to their mass density.

Together, these and other techniques form a tremendous arsenal for patient diagnostics. Depending on the question at hand and patient circumstances, the different techniques are useful for different purposes. Although the underlying physical and technical principles differ between modalities, there are some common principles in medical imaging, as described in this chapter.

9.2 FORMATION OF THE DIGITAL IMAGE

The formation of a digital image can be seen as a typical sequence of events, as summarized in Figure 9.2. The radiation source (a) may be placed outside the patient, as in X-ray imaging, or inside the patient, as in nuclear-medicine imaging. The correspondence in photographic imaging would be a light source such as the sun or a light bulb placed outside the object. In the detector system (b) the image is formed from the impinging radiation that interacts in the detector material, where the position, and sometimes also the frequency or energy of the interacting photons are recorded by formation of electronic pulses. In medical imaging theory, modelling of the image formation process includes an internal image plane (c) that represents the spatial distribution of the continuously varying signal generated in the detector system, which is then digitized into a matrix (d). Today, several detector systems rely on arrays of discrete detectors and the model does then not exactly represent the real, physical formation process. However, it is still useful for analysing the different components that affect the image quality, such as the limited spatial resolution of the imaging system, or effects of sampling. In most cases the image matrix is associated with a grey- or colour scale (e) enabling visual inspection and interpretation of the signal distribution on a computer screen, a projector, or a printer. The discrete matrix (d) is stored into the computer memory for later use and archiving.

9.3 ELEMENTS OF THE DIGITAL IMAGE

9.3.1 Sampling and Quantization

The process of digitization from the continuous signal distribution to an array or matrix with scalars (Figure 9.2, step from c to d) consists of two different processes, termed quantization and sampling (Figute 9.3).

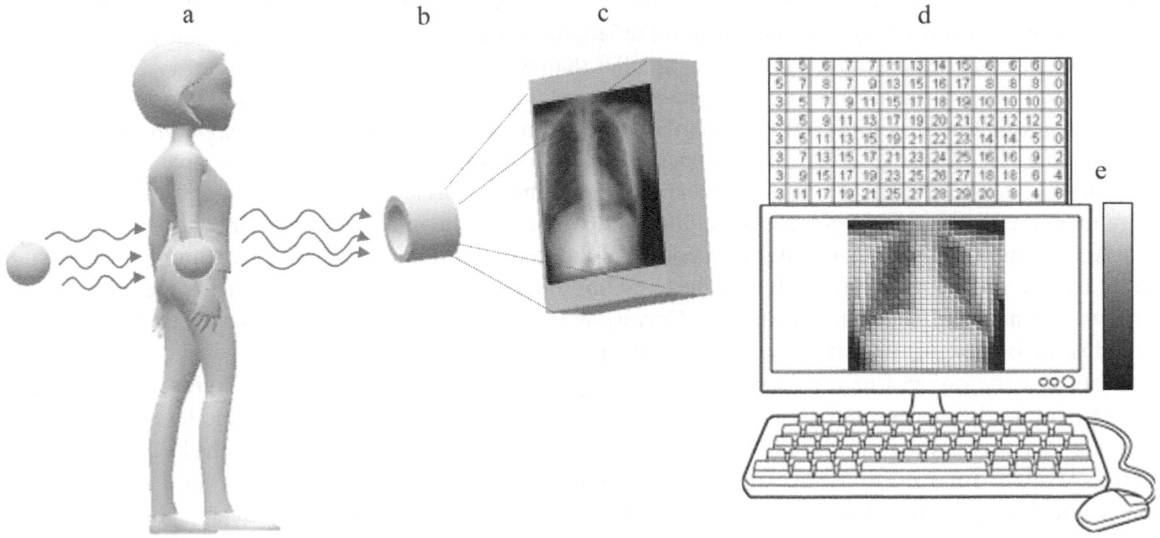

FIGURE 9.2 Schematic illustration of the formation of a digital image. Spheres (a) symbolize the radiation source which can be placed outside or inside the object (the patient). The detector system (b) in which electronic pulses are formed upon interaction with radiation emanating from the patient. The internal image plane (c) represents the continuously varying signal distribution generated in the detector system, which is then digitized to an image matrix (d). The matrix may be stored or presented on a computer screen using a grayscale (e).

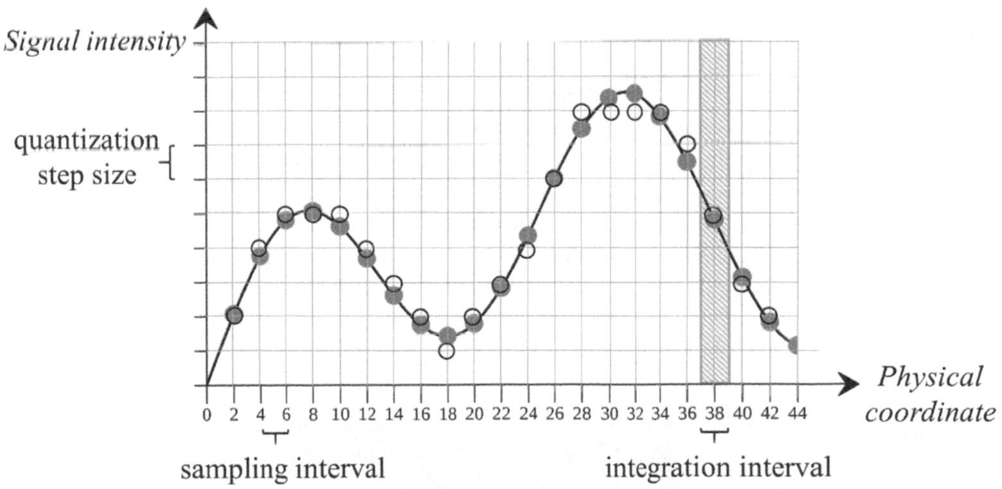

FIGURE 9.3 Schematic illustration of sampling and quantization. A one-dimensional continuously varying signal (solid curve) is sampled at regularly spaced positions along the horizontal axis (vertical lines) to give values indicated by filled circles. These values are then rounded off into the nearest quantization level (horizontal lines and open circles), where, for clarity, only a few levels are shown. Theoretically, sampling is described as the multiplication of a Shah-function and the underlying continuously varying function, while in practice, sampling is achieved by application of an integration interval (shaded area).

Sampling refers to the discretization along the horizontal direction, which in medical images represents a spatial coordinate, but may in other measurement setups represent time or some other coordinate along which the continuous signal propagates. The sampling interval is the space between each sample (in physical unit). In Figure 9.3 the vertical lines indicate sampling as a process where the signal amplitude is captured at regular spatial intervals. This is how sampling is described according to sampling theory, where the process is mathematically modelled as the multiplication of an infinite series of Dirac delta function (a Shah, or comb function) with the underlying continuously varying signal.

In practice, the amplitude of each sample is determined by integration of the signal within a pre-specified integration interval (Figure 9.3), followed by normalization to the length of that interval. The integration interval can be the same as the sampling interval but may also be larger or smaller than this length.

While Figure 9.3 illustrates sampling of a one-dimensional continuously varying signal, images represent signals that vary along two, three or more dimensions. For a two-dimensional signal, the corresponding sampling process would be modelled by the use of a two-dimensional distribution of delta functions (a stinger) separated by the sampling interval along two perpendicular directions (x and y). Each sample is associated with an image element, called pixel, for picture element. Correspondingly, for three-dimensional images, the image elements are called voxels, for volume elements. According to the Nyquist sampling theorem, if the sampling frequency is twice the highest spatial frequency that can be transferred by the system, that is, the sampling interval is the inverse of two times the highest frequency, then there will be no risk for frequency aliasing. Correspondingly, the highest spatial frequency that can be adequately represented is the inverse of two times the sampling interval.

Quantization is the process in which the continuous signal intensity is transferred to a finite set of equally spaced levels, most commonly represented as integer numbers. In Figure 9.3 this is illustrated as the levels along the vertical axis, with the step size indicating the intensity interval between consecutive quantization levels. The total number of quantization levels is referred to as the image depth or bit depth, given by 2^n where the power n is quoted. Common values in medical images are bit depths of 10 or 12, thus corresponding to 1024 or 4096 quantization levels. These are thus the number of levels available for representing the dynamic range of the signal that exists in the latent image in Figure 9.2c . In Figure 9.3, where only a few quantization levels are depicted, it is seen that signal amplitudes that fall between the levels suffer from the largest round-off errors at quantization. The bit depth together with the total number of samples, that is, the matrix size, essentially determines the memory space needed for storing the image, and compression methods can be used to reduce this space.

9.3.2 Image Coordinates and Spatial Scale

Once the digital image has been formed, the image matrix itself does not reveal any information about the spatial scale. For example, Figure 9.4 shows a microscopy image, a telescope image, and a medical image, each consisting of 512 columns and 512 rows of pixel values, but where the spatial scales are vastly different. To properly interpret the dimensions of the depicted objects an exact specification about the spatial scale is needed.

This information is generally supplied in header information that accompanies the image matrix, either embedded in the same computer file or supplied separately. For medical images, the DICOM standard (Digital Imaging and Communications in Medicine) encompasses a large amount of information required to transfer and read medical images, and the spatial scale is included among its header attributes. The sampling intervals between consecutive rows and columns are referred to as 'spacing between pixels', while the distance between consecutive image slices is termed 'spacing between slices'.

FIGURE 9.4 Images representing different spatial scales; a microscopy image of glioma cells (left), a telescope image of a galaxy (middle), and a human chest x-ray image (right). Each of these images consists of 512 columns and 512 rows of pixels, but the spatial dimensions of the imaged objects differ considerably. Information about the spatial scale is provided in the header information that accompanies the image matrix.

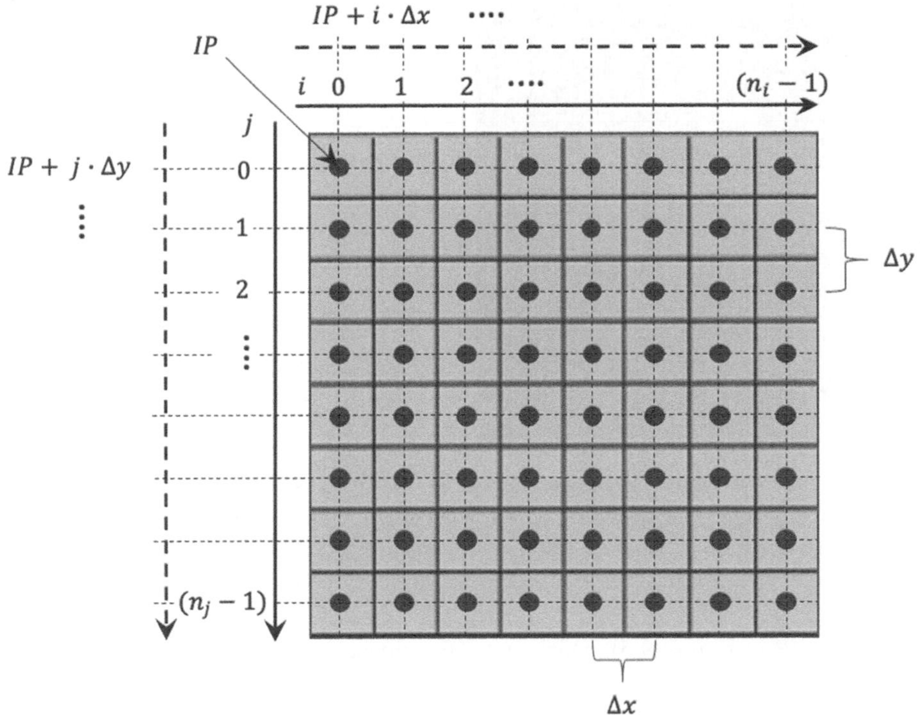

FIGURE 9.5 Image matrix consisting of 8 columns (n_i) and 8 rows (n_j), indexed with image coordinates i for column and j for row, as indicated by the black coordinate frame. According to the DICOM standard, pixels are associated with positions located centrally in the rows and columns (circles). The image coordinates are associated with physical coordinates in the Patient-based coordinate system, indicated by the dashed coordinate frame. This association is made by use of the attribute Image position (IP) that states the location of the origin of the image coordinates $(i = 0,\ j = 0)$ relative to the Patient system. The physical coordinates of remaining pixels are calculated from the image coordinates and the sampling intervals (spacing between pixels, Δx and Δy).

The interpretation of the coordinates of pixels or (voxels) in terms of physical units deserves some attention. In some applications the pixels are assumed to represent a distribution of patches, that is, small squares, or rectangles whose sides are determined by the sampling interval along the horizontal and vertical directions. In Figure 9.5 this corresponds to the square areas. Likewise, voxels are sometimes interpreted as a uniform signal distribution in the volume enclosed in the parallelepiped defined by the sampling intervals in three orthogonal directions. In principle, this is how pixel values are interpreted when shown on a computer screen or a projector by use of a grey- or colour-scale, or when printed.

However, when treating the image matrix with mathematical methods, the pixels are assumed to represent a mesh of points separated by the sampling intervals, represented by the dots in Figure 9.5. At application of image-processing methods or other image-based calculations, a convention then needs to be adopted for the exact location of these points, such that all methods in an image-processing system make the same interpretation. The convention adopted in the DICOM standard is illustrated in Figure 9.5, where it is assumed that the pixel values are located at the centres of rows and columns. Other systems may use the convention to associate pixel values with the lowest side of rows and columns, which in Figure 9.5 would correspond to the upper left corner of each grey square. If there are different conventions adopted for different steps in the same image-processing system, odd behaviours may arise such as small shifts when filtering is applied. Likewise, a convention needs to be adopted for the orientation of the coordinate system which, in Figure 9.5, increases from left to right and top to bottom, but there are also systems that have the vertical axis increase from bottom to top.

According to the DICOM standard, the translation from image coordinates to physical coordinates is made with reference to the so-called patient-based coordinate system. In its basic form this system is defined from a patient lying in supine position (on the back) with the head towards the gantry. The x axis then runs from the patients right to left, the y axis from the anterior to the posterior side (belly to back), and the z-axis along the inferior-superior direction (from feet to head). The position of transversal image slices relative to the patient-based coordinate system is specified by the

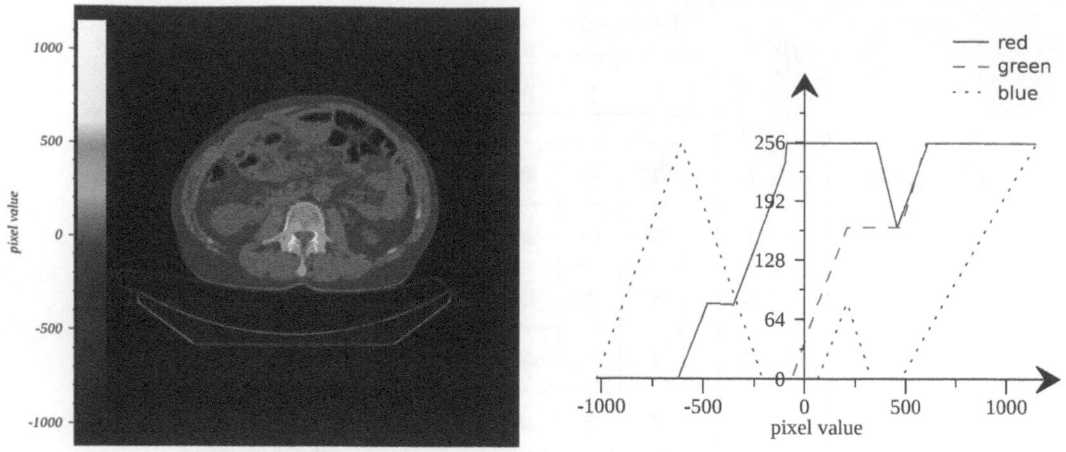

FIGURE 9.6 Left panel: A CT image whose pixel values range from -1024 to 1154. The CT image is displayed with a pseudo-colour scale, with mapping between different pixel values and colour is indicated by the colour bar. Right panel: graphical representation the LUT for the colour scale, consisting of the three channels red, green, and blue that each contributes to produce the colour that represents a given pixel value. Colour image available at www.routledge.com/9781138593268.

attribute 'Image Position' (*IP* in Figure 9.5), which states the physical x, y, and z coordinates of the upper left-hand corner of the image.

9.4 REPRESENTATION OF THE IMAGE INFORMATION

Provided that quantization has been made with a sufficient number of levels with respect to the dynamic range of the underlying object signal, the selection of a colour table that well reflects the image information is of some importance. The coupling between the pixel values and a specific colour- or grayscale is determined by the colour look-up-table (LUT), which can be stored on the graphics card of the computer, or as part of an image-processing program. For photographic images, which normally contain information about the intensity of a variety of colours, there are three separate planes, red, green, and blue (RGB), which are combined to reflect polychromatic colours. In medical applications usually only one plane is used for display, and pseudo colour-scales are constructed by combining three colour-channels, R, G, and B, into a single LUT. For an 8-bit colour graphics each channel consists of a 256-element array whose values define the intensity with which the channel contributes for the range of pixel values. Figure 9.6 illustrates a colour scale and the contribution from the respective channel for different pixel values.

Alternatively, surface plots or the extraction of profiles may be useful for examining the image information, where a profile is generally a one-dimensional graph of the pixel values along a chosen direction, for example along a row or a column of the image matrix.

9.5 IMAGE INFORMATION IN THE FREQUENCY DOMAIN

The discrete Fourier transform (DFT) is a central tool in image processing and is used for various operations such as filtering and modelling. The one-dimensional discrete Fourier transform of the data sequence $\{f_k\}$ consisting of N values, is defined as:

$$F_j = \frac{1}{N}\sum_{k=0}^{N-1} f_k \exp\left(-i2\,\pi\, j\, k / N\right)$$

$$f_k = \sum_{j=0}^{N-1} F_j \exp\left(i2\,\pi\, j\, k / N\right) \tag{9.1}$$

where the first equation is the forward DFT of the sequence $\{f_k\}$, and the second equation is the inverse DFT of the sequence of frequency components $\{F_j\}$. It should be noted that there are different variants of the DFT definition regarding the placement of the normalization factor $1/N$ and the sign of the arguments of the exponential functions.

The forward DFT decomposes functions of discrete variables into its constituent frequency components with trigonometric functions as basis functions. Applying the forward DFT to a data sequence yields a complex-valued sequence with the same dimensions, where each coordinate represents a frequency while the associated pixel value states the amplitude of a wave function with that frequency. Applying the inverse DFT synthesizes the original sequence from its frequency-domain representation.

While Eq. 9.1 shows the one-dimensional DFT with variables j and k to denote indices, for image processing we are mainly dealing with two or higher dimensions that then need to be included in the formulae. The physical spatial coordinates corresponding to the indices are obtained by multiplication to the sampling intervals Δx and Δy, as described above. In the frequency domain, the frequency interval between consecutive samples is given by $\Delta u = 1/(N\Delta x)$ and $\Delta v = 1/(M\Delta y)$, with N and M representing the number of columns and rows in the image, respectively.

The discrete complex exponential function $\exp(-i2\pi jk/N)$ has some interesting properties that have practical consequences for how the DFT should be interpreted and handled. It is periodic with a period of N, which means that it interprets any image to be Fourier transformed as one period of a periodic function. As a consequence, when performing filtering in the frequency domain based on the results of the convolution theorem, the image and filter first need to be zero-padded in order to avoid interference from adjacent periods. The exponential evaluates to a distribution that is conjugate-symmetric around zero, and the maximum frequency that can be represented is given by $j_{\max} = N/2$ (for even N). This means that the frequency components resulting from the forward transformation represents the physical frequencies u_j given by

$$u_j = \pm j/(N\,\Delta x); \quad j = 0..N/2 \tag{9.2}$$

The maximum frequency that can be represented is thus $|u_{max}| = 1/(2\Delta x)$. The corresponding relationships for other dimensions apply.

In practical applications the formulae in Eq. 9.1 are almost never used directly, since they are computationally inefficient for large N. Instead, an algorithm called the Fast Fourier Transform (FFT) is used, which is really only a way of rearranging the multiplications and sums in Eq. 9.1, using the properties of the exponential function, to

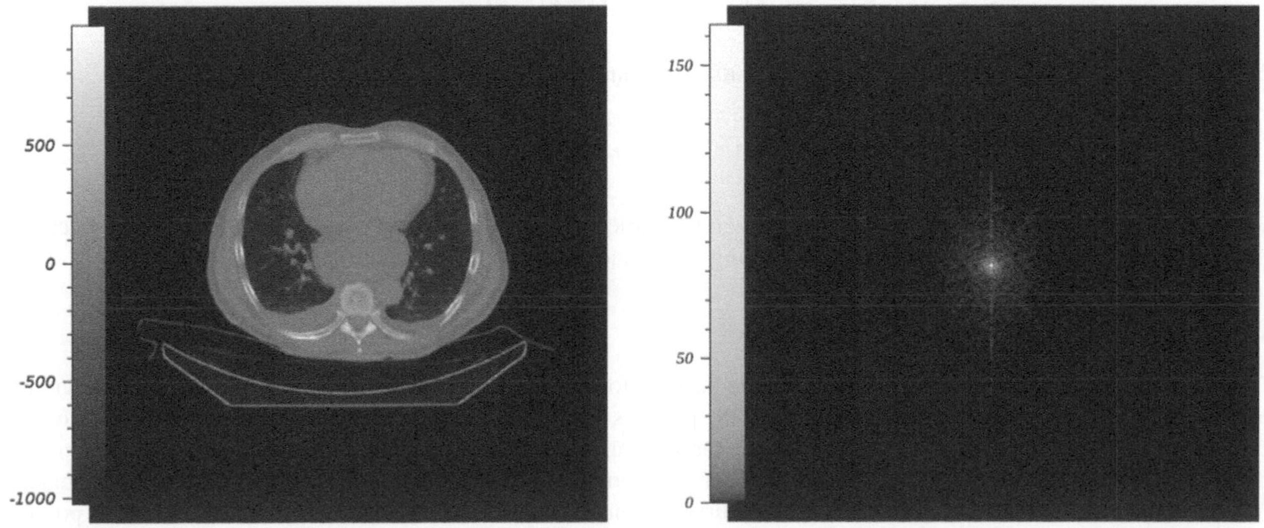

FIGURE 9.7 A CT image (left) and the absolute value of its discrete Fourier transform (right) with the origin placed in the centre. The left image is shown with a linear grey scale, while, for purpose of visibility, the right image is shown with a logarithmically increasing grayscale.

reduce the total number of arithmetic operations. The FFT has paved the way for a broad use of the DFT in medical imaging.

In many applications the image is acquired in the spatial domain and is real-valued, meaning that its Fourier transform are conjugate-symmetrically distributed around the origin. Figure 9.7 shows of a CT image over the chest along with its DFT in which the symmetric appearance is manifested. The frequency component at the origin represents the mean of the image values in the spatial domain and is sometimes called the DC component (stemming from electronics, 'direct current'). This component is usually considerably higher than the other frequency components, which is the reason for choosing a non-linear grayscale for displaying the image; due to the large dynamic range of the frequency components we would otherwise only see a spot in the middle of the image, that is, the frequency component at the origin. The appearance of the frequency-domain image in Figure 9.7 is also typical in the sense that amplitudes at higher frequencies, located towards the image periphery, are usually very low or even negligible. This property is used in image compression methods, such as those used for the JPEG format that includes a sibling of the DFT, the discrete cosine transform. This property is also taken advantage of in low-pass filtering, where it is assumed that the object signal is to a large extent centred close to the origin in the frequency domain, while components that arise due to noise are distributed along the entire frequency axes. Thus, by suppressing components at higher frequencies a less noisy image can be obtained.

9.6 FACTORS THAT AFFECT IMAGE QUALITY

For an ideal imaging system, the internal image plane would be an exact representation of the information that enters the detector system (the transitions a to b to c in Figure 9.2). However, any detector system suffers from some kind of imperfection, and it is important to be aware of the consequential limitations for the interpretation of the image information. The term *image quality* loosely refers to the accuracy with which an image reflects the physical object properties that underlie the image formation for the imaging modality. Especially three factors are used to describe image quality: spatial resolution, contrast resolution, and noise or signal-to-noise ratio. Knowledge of these factors is useful when comparing the information obtained from different imaging modalities, or for comparison of different system designs for a given modality.

In many cases it is useful to regard the imaging system as a *linear system*, which also gives an understanding of the image-degrading factors. Figures 9.2 (a to c) illustrate an imaging system as a 'black box' that transfers the signal that emanates from the object into the system, to the internal image plane that represents the output from the imaging system. For a two-dimensional imaging system that includes properties of: *i* yielding a linear and translation-invariant mapping, *ii* for which the system function $h(x,y)$ is defined as the impulse response, the relationship between the input signal distribution $f(x,y)$ and the signal distribution of the internal image plane $g(x,y)$ is given by the convolution

$$g(x,y)=f(x,y)\otimes h(x,y) \quad (9.3)$$

Eq. 9.3 can also be expressed in the Fourier domain, according to

$$G(u,v)=F(u,v)\cdot H(u,v) \quad (9.4)$$

where $F(u,v)$, $H(u,v)$ and $G(u,v)$ are the continuous Fourier transforms of $f(x,y)$, $h(x,y)$, and $g(x,y)$, respectively. Further details are given in the Appendix, including mathematical derivations of Eqs. 9.3 and 9.4.

9.6.1 Contrast

The perhaps most fundamental aspect of the image information is the image contrast, which is built up by the differences between pixel values. These differences make up the patterns that the observer can recognize and associate with objects that exist in the physical world. In terms of a linear system, the difference between the pixel values in the image is ideally linearly related to the signal difference from the corresponding parts of the imaged object. Practical imaging devices may not have an entirely linear response curve but may behave as a linear system over a certain range, possibly after some correction. When making measurements of the image contrast, the image values are generally evaluated after quantization. At evaluation of a system's capability to resolve a given signal difference in the object, quantization thus needs to be made with such a high image depth (low quantization step size) that the response curve is not substantially affected (Figure 9.2, c to d).

The image contrast may be defined in different ways, depending on the application and the image appearance. Measurement of the local image contrast involves the difference of pixel values, normalized to the background level. For example, for a high-intensity area located in a non-zero, low-intensity background, the contrast C may be defined according to:

$$C = \frac{\overline{g}_r - \overline{g}_b}{\overline{g}_b} = \frac{\Delta g}{\overline{g}_b} \tag{9.5}$$

where $\overline{g}_r$ represents the mean of the pixel values in a region delineated in the high-intensity area, and $\overline{g}_b$ the mean of the pixel values in the surrounding background.

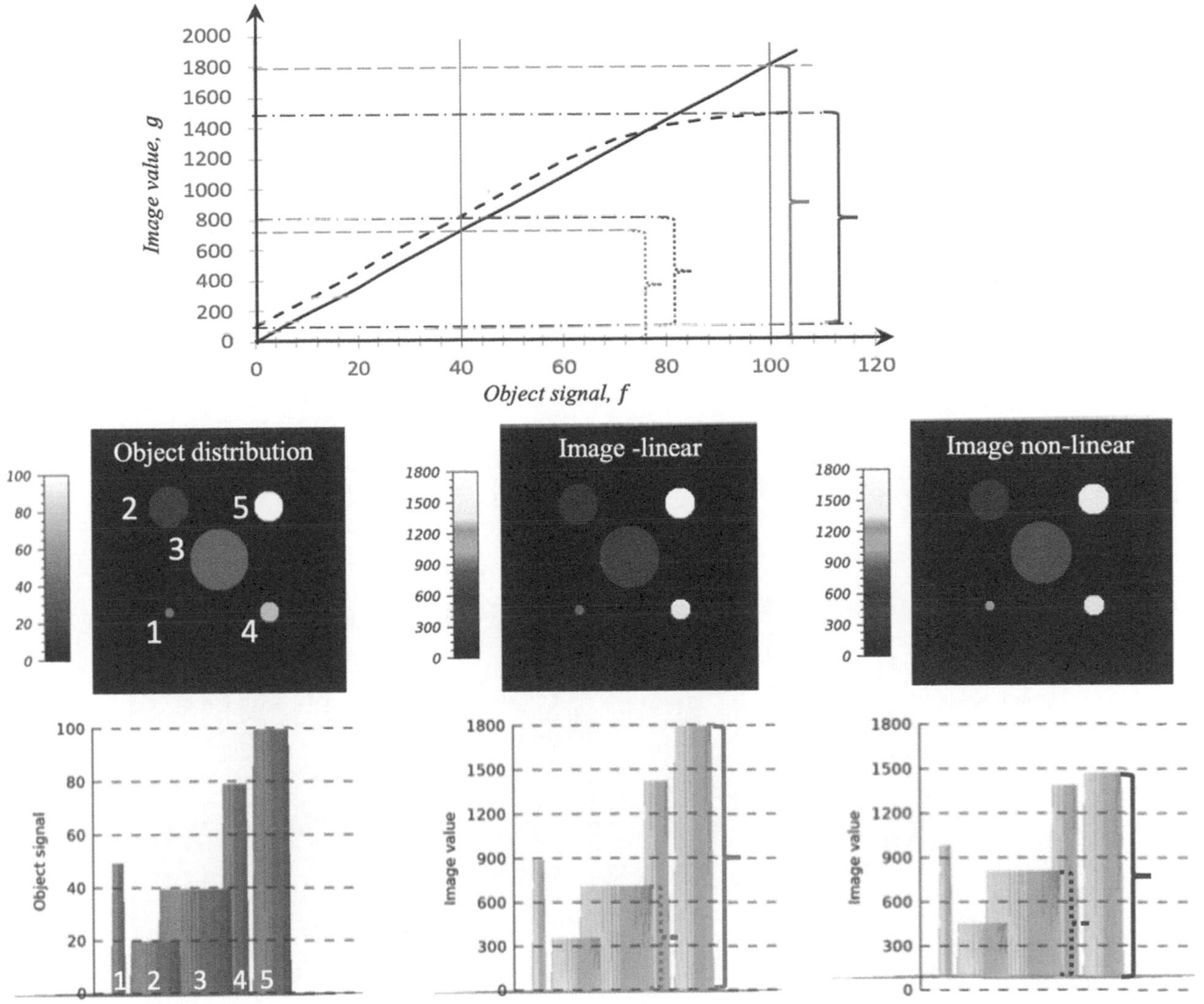

FIGURE 9.8 Top: Response curve for a linear system (solid line) with a linear relationship between the object signal and the image values, and for a system with deviation from linearity at the high end and including a background (dashed line). Vertical lines indicate two levels of the object signal (40 and 100) at which the image contrast is evaluated (brackets), also indicated in the bottom panel. Bottom panel shows images and surface plots for; the object distribution consisting of five circular discs (left); the image obtained for the linear system (middle), and the image obtained for the system that deviates from linearity (right). Colour image available at www.routledge.com/9781138593268.

Figure 9.8 illustrates the response curve for a hypothetical linear system. A second curve is shown as an example that deviates from linearity, due to saturation from pile-up and dead time at the high end, and addition of a background level that may be caused by background radiation or the instrument itself. A stylized object distribution consisting of five circular discs is also shown, and the image values that are obtained for the two systems.

For the linear system, the signal levels above background (indicated by brackets) are proportional to the object signal. For the non-linear system, evaluation at the lower object level (40) gives an image contrast that corresponds to the object contrast, while evaluation at the higher object level results in a biased image contrast. The image contrast thus becomes misleading in the sense that the high-intensity object is represented with a lower contrast than its real property.

The contrast resolution is essentially a metric of how well the image contrast reflects the underlying object contrast. In addition to any non-linearity in the response curve, the contrast resolution is strongly dependent on the modulating effect of limited spatial resolution and noise, as described later in this chapter.

9.6.2 Spatial Resolution

Spatial resolution, also termed detail resolution, is defined as the smallest distance in which two objects can be located with respect to each other and yet be identified as two separate objects in the image. Spatial resolution is governed both by detector properties and settings in the image acquisition and, depending on the imaging modalities, either of these factors may constitute the limitation in practice. The detector may have a limited ability to exactly determine the position of the signal interaction (in most cases photons) emanating from the object. As a result, the placement of the pulses in the internal image plane (Figure 9.2 c) suffers from a certain amount of uncertainty, which results in a blurred object representation (Figure 9.9).

In order to characterize the spatial resolution, an image of a point-shaped object can be acquired giving the point spread function (*PSF*). As an alternative, line-shaped objects may be used to estimate the line-spread function (*LSF*), which can be defined as the integral of the *PSF* along one direction, for example

FIGURE 9.9 Illustration of different levels of spatial resolution. As the spatial resolution becomes poorer, the overall features in the image are maintained while fine details are lost. For example, for the middle image the bricks on the ground, or details in the rear, are not well identifiable while the larger objects are still visible. For the right image it is barely possible to classify any features, and complementary information would be needed to interpret the image. Indeed, for medical imaging systems with poor spatial resolution, the middle or right image may be what we have at hand, and interpretation then inevitably needs to be accompanied by prior knowledge.

$$LSF(x) = \int_{-\infty}^{\infty} PSF(x,y)\,\mathrm{d}y \tag{9.6}$$

For some imaging systems, such as a gamma camera, LSF measurements need to be made with the line source rotated both along the horizontal and vertical axes of the imaging plane so as to assess the respective performance. For systems with a very good spatial resolution, an object with a sharp edge may be used to determine the edge-spread function (ESF), with the edge placed in a slanted direction with respect to the rows or columns of the imaging matrix to achieve a denser spatial sampling.

Quantitative analysis of the PSF is made by defining a profile through the centre of the point-shaped object and determining the full-width at half-maximum $(FWHM)$ as a metric of the spatial resolution (Figure 9.10). As a profile through a line is generally less sensitive to the exact location of the profile, the LSF is often the preferred method for determining the $FWHM$. For a Gaussian PSF insertion in Eq. 9.6 shows that the $FWHM$ of the LSF equates the $FWHM$ of the PSF. The ESF is generally analysed by extracting profiles perpendicular to the edge. Profiles are combined and then the first derivative (or local difference) with respect to the lateral position in the image is calculated to determine the LSF. Alternatively, an error function may be fitted to the combined profile.

The most basic model for representing the imaging system's impact on the spatial resolution is by means of convolution of an underlying object distribution by the PSF (Eq. 9.3), where the system function $h(x,y)$ thus represents the PSF. The absolute value of the system function in the Fourier domain (Eq. 9.4) normalized to the frequency component at zero frequency, that is, $|H(u,v)|/H(0,0)$, is called the modulation transfer function (MTF), revisited in Section 9.6.4.

Thus, for a linear and translation-invariant system, the impact of limited spatial resolution on the resulting image can be fully characterized by determining its PSF or MTF. For systems that do not fulfil the criteria of a linear system, the model is still often used as a starting point, with awareness of the approximations made. The broader the PSF, corresponding to a larger $FWHM$, the poorer the spatial resolution of the imaging system, with a more severe image blur as a result. Conversely, a broader MTF means a better spatial resolution as higher frequency components are passed through the system unmodulated, while a narrow MTF means that higher frequency components are dampened when they pass the detector system. It may be useful to consider the extreme cases of an *ideal* imaging system and a *useless* imaging system. The ideal imaging system is one where the PSF is a Dirac delta function, while the MTF is unity across all frequencies, implying no blur and all frequencies passed unmodulated. A useless imaging system is the opposite, that is, where the PSF has a constant value, while the MTF is a Dirac delta function, meaning that the resulting image will display one constant value (the mean pixel value) and all contrast is lost. Practical imaging systems lie somewhere between these two extremes.

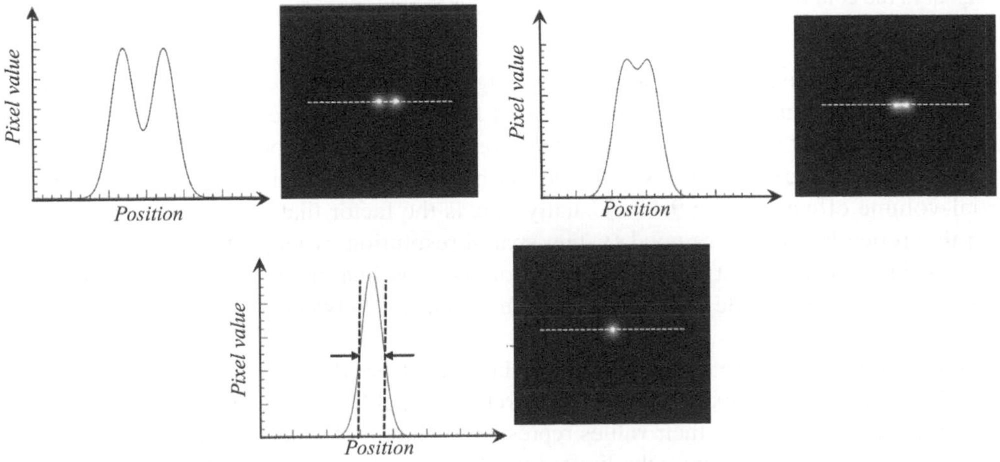

FIGURE 9.10 Top panel illustrates definition of spatial resolution as the smallest physical distance with which two objects can be placed and yet be identified as two separate objects in the image. Images and profiles of two point-sources with centre-to-centre separations of $1.5 \times FWHM$ (left) and $1.0 \times FWHM$ (right). Bottom panel illustrates how the spatial resolution is quantified from a profile through the centre of the PSF by determining the $FWHM$ (in physical unit).

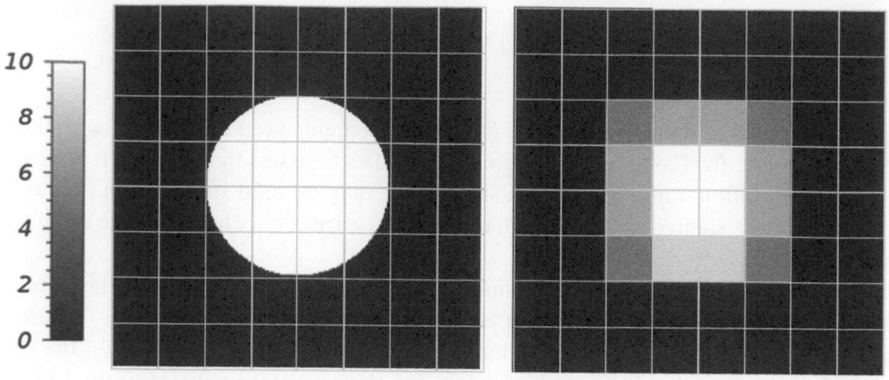

FIGURE 9.11 The image spatial resolution may be limited by the finite integration interval. Left image shows a disc-shaped object with a true signal level of 10 inside the disc and zero outside. White grid indicates the borders of the integration areas used for sampling for the right image, shown with the same colour scale as the left image. After this coarse sampling, the pixel values in the centre correctly reflect the true signal level, whereas at the object border the pixel values are decreased as they are formed partly by signal from outside the object.

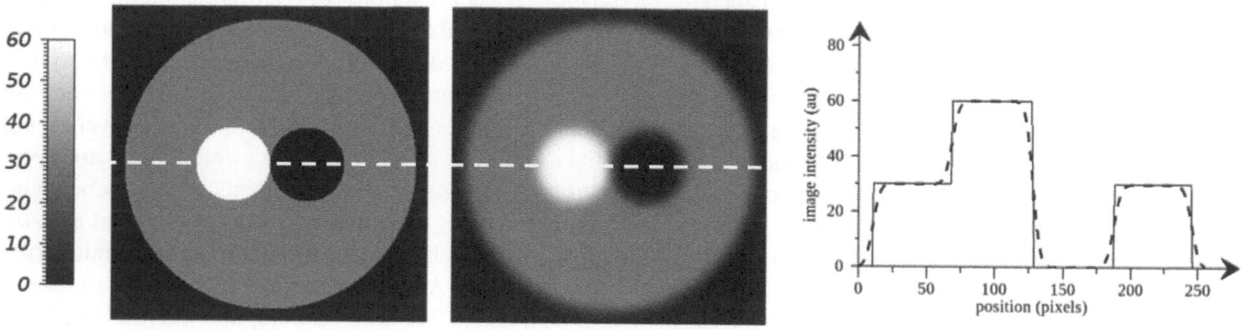

FIGURE 9.12 Left image shows an object with a uniform circular background (signal level 30), one hot object (signal level 60), and one cold object (signal level 0). The blur imposed by limited spatial resolution (middle figure) produces a spill-out of signal from the hot region and spill-in of counts to the cold region (profiles shown in right figure, at the position indicated by the horizontal line in the images). When a ROI is drawn along the border of the respective object, the mean signal in the hot object will be underestimated, while the mean signal in the cold object will be overestimated.

The sampling interval, that is, the pixel or voxel spacing, may also influence the spatial resolution of the image. As the integration interval (Figure 9.3) is by necessity of finite size, the pixel or voxel values may include signal that originate from a mixture of tissues. This means that even if the spatial resolution of the imaging modality would be ideal, the pixel values, especially regions at the border between different tissues, suffer from what is sometimes called the partial-volume effect (Figure 9.11). Usually, this is the factor that governs the spatial resolution in MR imaging, which theoretically has a very good system spatial resolution, whereas for nuclear medicine the imaging system's spatial resolution is the most important limitation. For any imaging technique, the choice of pixel or voxel size at acquisition needs to be made with regards to the trade-off between the signal-to-noise ratio and spatial resolution.

The partial volume effect refers to the mix of signal intensity from different objects due to limited spatial resolution (Figure 9.12). When pixel or voxel values not only reflect signal from a specific object coordinate, but also have contributions from nearby coordinates, their values represent a mixture of object intensity from different tissues. The cause of partial volume effects may be both the limited resolution of the scanner and limitations imposed by the sampling interval (pixel or voxel size).

9.6.3 Noise

Noise is a phenomenon that originates from randomness in the different processes involved in the generation and detection of the object signal. Such randomness may include, for example, the number of radioactive nuclei that decay during the acquisition time interval, the number of photons that interact with tissue on their passage towards the detector, the number of photons that interact in the detector, and the number of counts that are generated in response to these interactions. A noisy image has a grainy appearance, and this graininess is thus related to randomness in the processes involved in detection, rather than a true variability across the object distribution. As such, noise constitutes false information about the object, which needs to be considered at image interpretation.

Mathematically, noise is described in terms of stochastic variables. To illustrate how these relate to the image signal, we first consider a detector such as an ionization chamber or sodium-iodide crystal, and an experiment where repeated measurements are made under the exact same conditions. The number of detected counts will not be identical in each measurement but will be centred around a mean with a certain variation. By determining the frequency with which a given number of counts are obtained, the probability for obtaining this count number can be estimated. The probability distribution that governs the detector measurement can thus be estimated by constructing a frequency histogram. Similarly, for imaging devices we can, to a first approximation, view each pixel as a separate detector, meaning that for a 256×256 matrix we would have 65536 measured values. Assuming that the imaging device is exposed to a perfectly uniform object distribution (a flat field) and the imaging-system response is perfectly uniform over the detection area, each pixel will still not show an identical number of counts, due to noise. If the noise contribution is uncorrelated between pixels, the frequency histogram of the pixel counts may be used to estimate the probability distribution that governs the number of counts (Figure 9.13).

Random processes are mathematically described in terms of stochastic variables that follow some statistical distribution, and a valid model for counting statistics is in many cases the Poisson distribution. The probability p_g of obtaining a particular pixel value g when the expectation is λ, follows from:

$$p_g = \lambda^g \cdot \exp(-\lambda) / g! \tag{9.7}$$

This distribution has the property that the variance is equal to the expectation, such that if the expected number of counts is λ, then the variance in the number of counts also equates λ.

We can use this knowledge for estimating the number of counts that are required to arrive at a certain uncertainty in the pixel values. Considering the ideal imaging device outlined above, the standard deviation (*SD*) associated with a

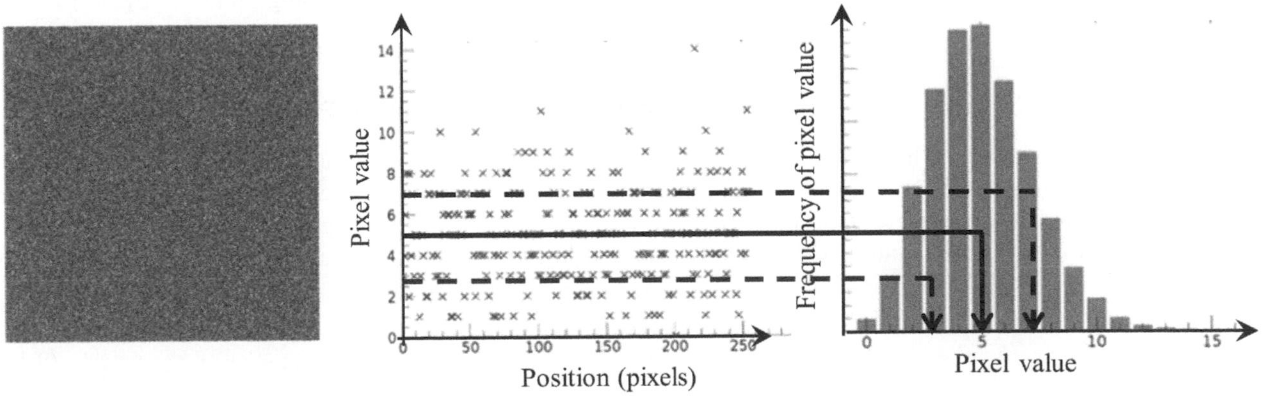

FIGURE 9.13 Assessment of the statistical distribution that governs the detected counts, by exposure to a flat field (left). Middle figure shows a profile through the image to the right (values along one matrix row). Right figure shows the histogram of the count distribution across the entire image, which in this case is represented by a Poisson distribution with a mean of 5 and standard deviation $\sqrt{5}$, represented by the solid and dashed lines, respectively.

pixel value g_i is $\sqrt{g_i}$. For a region-of-interest (ROI) encompassing n pixels with mean number of counts $\bar{g}_r$, the standard error of the mean $s(\bar{g}_r)$ is given by

$$s(\bar{g}_r) = \sqrt{\bar{g}_r / n} \tag{9.8}$$

since

$$s^2\left(\frac{1}{n}\sum_{i\in ROI} g_i\right) = \frac{1}{n^2}\sum_{i\in ROI} s^2(g_i) = \frac{1}{n^2}\sum_{i\in ROI} g_i = \frac{\bar{g}_r}{n} \tag{9.9}$$

where the first equality relies on the assumption that the values g_i are uncorrelated, and the second equality is valid for the Poisson distribution. It follows that if the relative *SD* in each pixel value g_i needs to be below say 5%, then the expected number of counts per pixel needs to be at least 400 $\left(s(g_i)/g_i = 1/\sqrt{g_i} = 1/\sqrt{400} = 0.05\right)$. If we delineate a ROI covering 100 pixels and aim for a relative *SD* in the ROI mean $\left(s(\bar{g}_r)/\bar{g}_r = 1/\sqrt{n\bar{g}_r}\right)$ of below 5%, then the mean pixel count needs to be higher than 4.

In the model for a linear system, (Eqns. 9.3 and 9.4), the contribution of noise can to a first approximation be incorporated as an additive term, following

$$g(x,y) = f(x,y) \otimes h(x,y) + n(x,y) \tag{9.10}$$

or expressed in the Fourier domain following

$$G(u,v) = F(u,v) \cdot H(u,v) + N(u,v) \tag{9.11}$$

where the term $n(x,y)$, with Fourier transform $N(u,v)$, is then modelling the position-dependent noise. It is sometimes of interest to simulate an imaging system that incorporates noise, which can be made following Eqs. 9.7 and 9.10. For each pixel position, a pseudo random number is calculated using the distribution in Eq. 9.7 with the expectation taken as the pixel value resulting from the convolution term in Eq. 9.10. This new value is placed at the corresponding position in a new image matrix, which, when all positions have been processed, represents the noisy image according to Eq. 9.10. It should be noted that the capability of such simulations to generate realistic noise distributions depends on the real origin of the noise, whether it arises before or after the signal enters imaging system. Moreover, distributions other than the Poisson may be more relevant, depending on the characteristics of the imaging system.

9.6.4 Interaction Between Image Contrast and Spatial Resolution, the MTF

As seen in Figures 9.9 and 9.12 limited spatial resolution introduces image blurring, which for small objects may result in a decreased image contrast. The modulation contrast C_M is a metric used when investigating the effects of limited spatial resolution for the imaging system. In practice, such investigation involves imaging of a bar pattern with alternating high- and low-intensity features with different spatial width, so called bar line pairs. The modulation contrast is defined as

$$C_M = \frac{g_{max} - g_{min}}{g_{max} + g_{min}} \tag{9.12}$$

where g_{min} and g_{max} are the maximum and minimum pixel values in the image for a given line pair. When the width of the line pairs decreases, the image contrast gradually decreases due to the blurring imposed by limited spatial resolution. The application of a bar pattern to determine the modulation contrast is illustrated in Figure 9.14, where, instead of line pairs with a rectangular profile, a sinusoidally varying intensity pattern with four different spatial frequencies is shown. Both the object distribution $f(x,y)$ and the resulting image $g(x,y)$ according to Eq. 9.3 are shown, along with profiles of the respective distributions. The bottom panel shows the obtained modulation contrast overlaid on the *MTF* determined from the Fourier transform of the *PSF* of the imaging system.

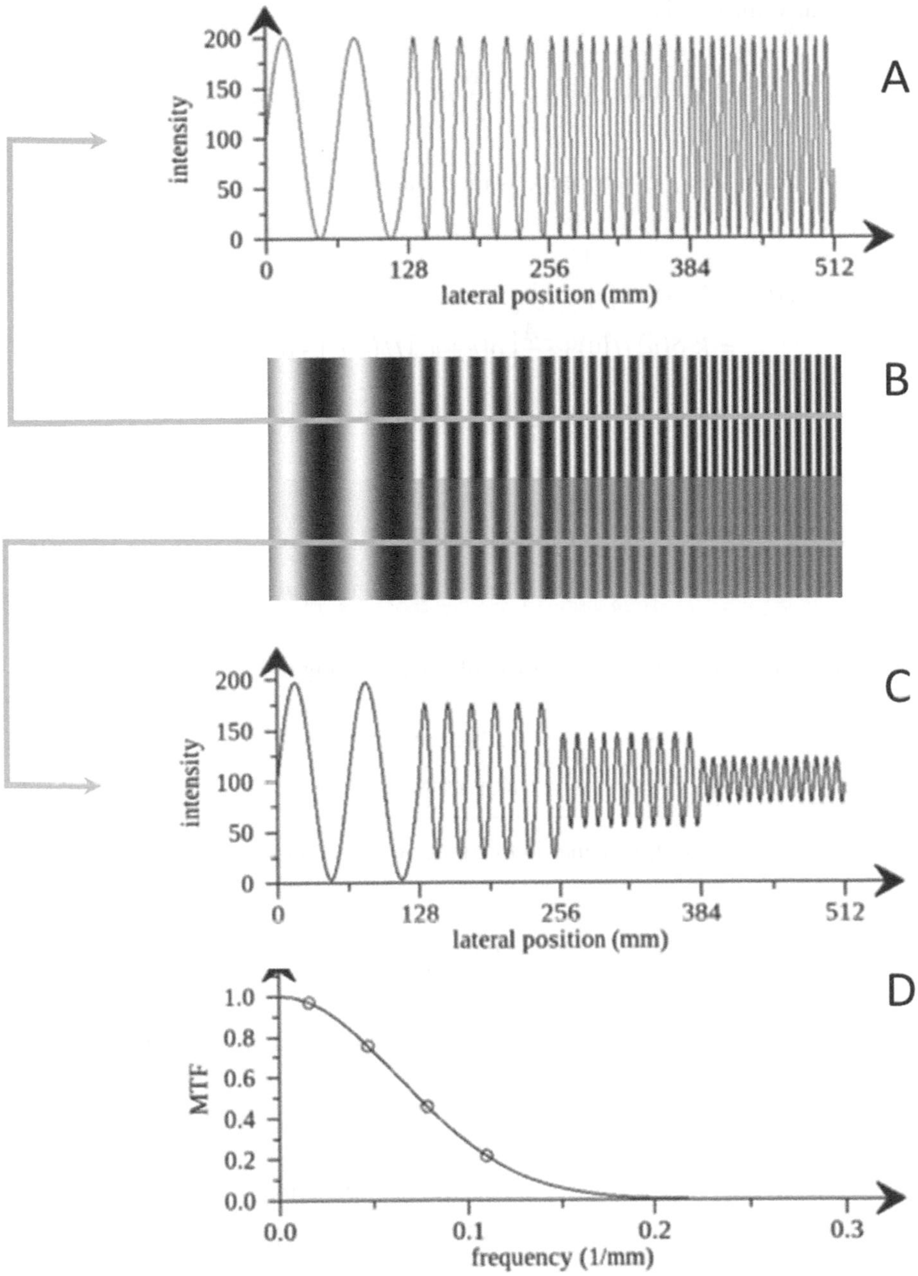

FIGURE 9.14 Panel (b) two rows of images of a sinusoidally varying intensity pattern with four different spatial frequencies. The upper row in (b) shows the object distribution, while the lower row shows the images when degraded by the limited spatial resolution. Profiles through the object distribution (a) and the image distribution (c). Panel (d) shows the *MTF* obtained from the modulation contrast C_M (circles) (Eq. 9.12) and using the Fourier transform of the *PSF* of the imaging system (solid line).

The relationship between the modulation contrast C_M and the *MTF* involves signal description in the Fourier domain. We consider a one-dimensional input object function $f(x)$ consisting of a sinusoidal variation with amplitude A, with frequency u_0, and oscillating around a constant level K

$$f(x) = K + A\sin(2\pi\, u_0 x) \tag{9.13}$$

The Fourier transform of this function is

$$F(u) = K\,\delta(u) + \mathrm{i}\frac{A}{2}\left[\delta(u+u_0) - \delta(u-u_0)\right] \tag{9.14}$$

When this function passes the system, its amplitude is modulated by the system function, $H(u)$, such that the output becomes

$$\begin{aligned} G(u) &= F(u)\cdot H(u) \\ &= K\,\delta(u)H(0) + \mathrm{i}\frac{A}{2}\left[\delta(u+u_0)H(-u_0) - \delta(u-u_0)H(u_0)\right] \\ &= K\,\delta(u)H(0) + \mathrm{i}\frac{A}{2}H(u_0)\left[\delta(u+u_0) - \delta(u-u_0)\right] \end{aligned} \tag{9.15}$$

where H is symmetric around zero. In the spatial domain this gives

$$g(x) = K\,H(0) + A\,H(u_0)\sin(2\pi\,u_0 x) \tag{9.16}$$

The modulation contrast for the object signal $C_{\mathrm{M},f}$ is calculated according to Eq. 9.12

$$C_{\mathrm{M},f} = \frac{K + A - (K - A)}{K + A + K - A} = \frac{A}{K} \tag{9.17}$$

while the modulation contrast for the output signal in the image $C_{\mathrm{M},g}$ becomes

$$C_{\mathrm{M},g}(u_0) = \frac{A\,H(u_0)}{K\,H(0)} \tag{9.18}$$

By the same reasoning this can be extended to any frequency u, and to two dimensions with spatial frequencies (u,v).

The modulation transfer function *MTF* is defined as the absolute value of the ratio $C_{\mathrm{M},g}\,/\,C_{\mathrm{M},f}$ as a function of frequency. For a two-dimensional description this becomes

$$MTF(u,v) = \frac{\left|C_{\mathrm{M},g}(u,v)\right|}{C_{\mathrm{M},f}} = \frac{\left|H(u,v)\right|}{H(0,0)} \tag{9.19}$$

When the *MTF* is the same along the horizontal and vertical directions, only one dimension is analysed. As noted, the *MTF* is defined to unity at a spatial frequency of zero. The *MTF* characterizes the decreased contrast obtained as the spatial frequency of the object signal increases, and as such, the effects of limited spatial resolution.

9.6.5 Interaction Between Image Contrast, Noise and Spatial Resolution

In addition to the object-signal dynamic range, the system response, the quantization, and limited spatial resolution, the image contrast may also be affected by noise, especially for small objects. The relationship between detectability, object size, and the noise level was early addressed by Albert Rose, giving name to the so-called Rose model. The signal-to-noise ratio, *SNR*, is used as a combined metric and describes the signal level relative to the background noise level. In medical imaging it is often calculated as the ratio of the mean of the pixel values in a region $\bar{g}_r$ and the standard deviation s, such that $SNR = \bar{g}_r\,/\,s(g_r)$. Closely related is the contrast-to-noise ratio, *CNR*, often calculated as $CNR = \Delta g\,/\,s(g_b)$, with notations as in Eq. 9.5. When detectability is in focus, such as in diagnostic imaging, it may be preferable to evaluate the *CNR*, whereas for quantitative imaging the *SNR* is more relevant. For both *CNR* and *SNR*, the region in which the pixel values are evaluated should be as uniform as possible. As the *SNR* generally depends on the spatial frequency it may also be characterized in the frequency domain by means of a noise-power spectrum. However, the above descriptions are often used for analyses in clinical applications, for example, for optimizing settings for the image acquisition.

Figure 9.15 illustrates the impact of noise and limited spatial resolution for stylized object distributions. Images have been generated by application of Eqs. 9.10 and 9.7 with Poisson distributed noise. The object contrast, that is, the object signal level above the background level, is identical for all objects within each image, and theoretically the *CNR* would thus be the same for all objects. However, even for an idealized spatial resolution (top row), the lower object contrast combined with a low *SNR* (right column image) make the smallest objects difficult to identify with certainty. This illustrates how noise can affect the identifiability of objects, depending on the object size and object-signal level above the background level. The bottom graph shows the measured *CNR* for objects with a signal level corresponding to a theoretical *CNR* of unity (middle column images). For repeated noise realizations, it is seen that when the spatial resolution is ideal, the measured *CNR* is on average equal to the theoretical one. However, the variability in the measured *CNR* increases as the object dimensions decrease, as reflected by the error bars. Recalling that $CNR = \Delta g / s(g_b) = (\overline{g}_r - \overline{g}_b) / s(g_b)$, this variability is caused by random uncertainties in the estimates of $s(g_b)$, $\overline{g}_r$ and $\overline{g}_b$. In combination with limited spatial resolution (middle row images), the smaller objects become unidentifiable for the lower object contrast combined with a low *SNR* (right column image), and the image contrast for the larger objects is also affected. As seen in the bottom graph, limited spatial resolution decreases the measured *CNR*, especially for the smaller objects. This is mainly a result of resolution-induced spill-out (Figure 9.12) that produces a decreased $\overline{g}_r$, in turn giving decreased Δg. Thus, the image quality, quantified in terms of the *CNR*, is affected by the signal level above the background, the presence of noise and limited spatial resolution, factors that have a more severe effect when the object dimensions are small.

9.7 SUMMARY

The digital image has become an important carrier of information that supports diagnostics and therapeutics in medicine. A basic understanding of the components of the digital image, how its information can be displayed and analysed, as well as its inherent limitations, is important for an adequate use. Today there are a number of different imaging modalities that each reflect different tissue properties and are based on different technical principles. The ambition to use a common terminology and knowledge base to describe basic image properties is regarded important to maintain a framework that enables translation of knowledge between the different specialties, of benefit for future development of the medical imaging field.

9.8 APPENDIX – LINEAR SYSTEMS THEORY

To describe the impact of the imaging system, we introduce the operator $\mathcal{H}$ to represent an initially unknown operation that maps the spatial signal-distribution put into the system, to an output signal-distribution. To make the notation shorter, the description is made for one-dimensional functions, but it is valid for higher dimensions as well. For a continuous input distribution $f(x)$ the output distribution $g(x)$ is given by

$$g(x) = \mathcal{H}\{f(x)\} \tag{9.20}$$

In principle this equation means that if we know the input distribution and the system operator, then we can predict the resulting output. Unfortunately, the application of Eq. 9.20 can become cumbersome without some further assumptions. There are especially two properties of $\mathcal{H}$ that simplifies its application: linearity and translation invariance.

A *linear* system preserves linear combinations of different input functions, that is, the operation $\mathcal{H}$ has properties of additivity and homogeneity. If we regard one input distribution $f_1(x)$ that when presented to the imaging system produces the output $g_1(x)$, and another input distribution $f_2(x)$ that produces the output $g_2(x)$, then for a linear system:

$$\mathcal{H}\{a_1 f_1(x) + a_2 f_2(x)\} = \mathcal{H}\{a_1 f_1(x)\} + \mathcal{H}\{a_2 f_2(x)\} =$$

$$a_1 \mathcal{H}\{f_1(x)\} + a_2 \mathcal{H}\{f_2(x)\} = a_1 g_1(x) + a_2 g_2(x) \tag{9.21}$$

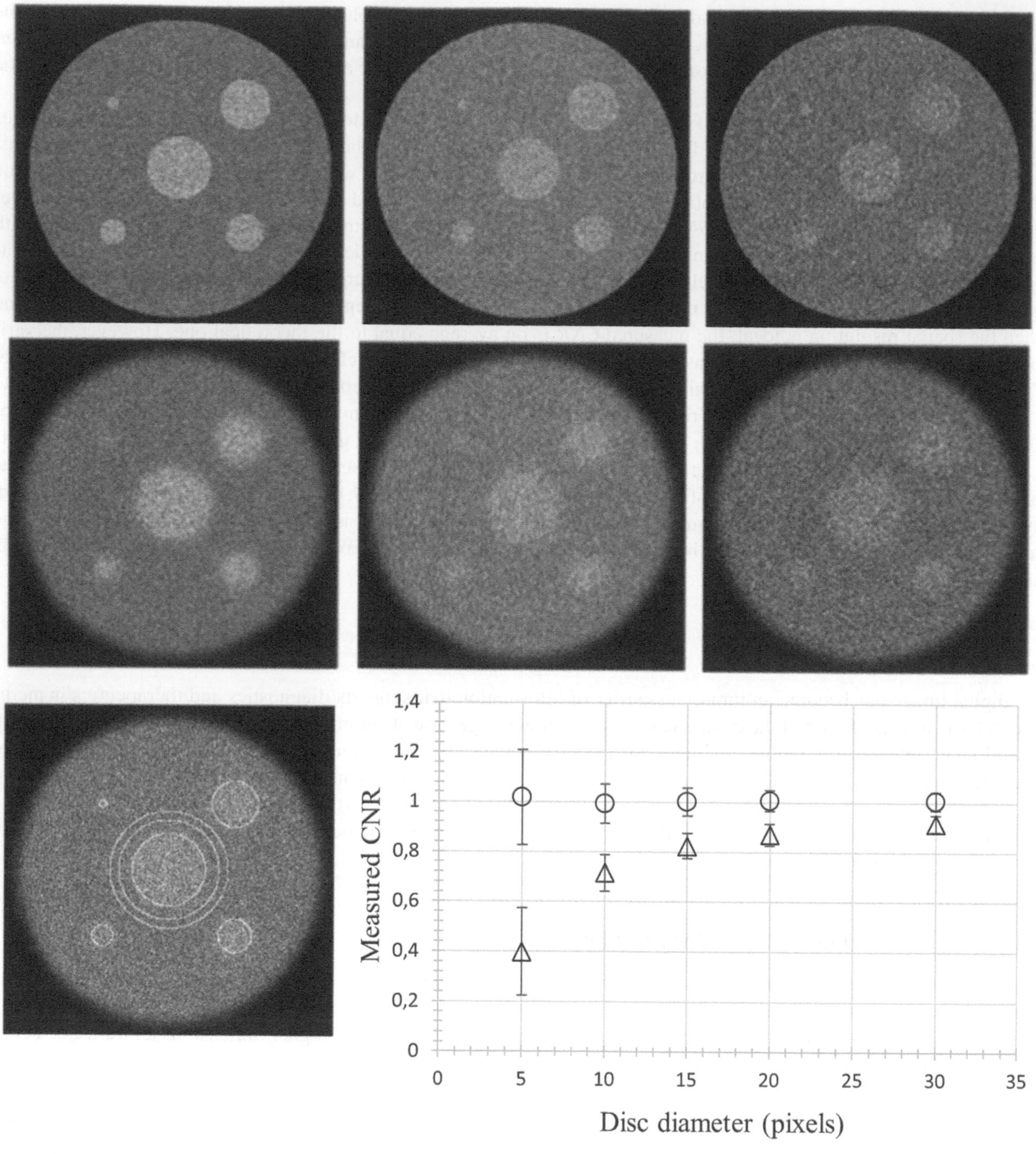

FIGURE 9.15 Disc-shaped objects with different diameters situated in a uniform background. Upper row represents images for an idealized spatial resolution, while middle row images have a limited spatial resolution. The object level above the background and the overall signal level have been varied between the images so as to obtain theoretical *CNR*s of 2, 1, and 0.7, from left to right, respectively. Bottom row (left) shows how the *CNR* has been measured using ROIs encompassing each separate object to quantify $\overline{g}_r$, and a band to quantify the background level $\overline{g}_b$ and standard deviation $s(g_b)$. The graph (bottom right) shows the measured *CNR* (mean and standard deviation) obtained from repeated noise realizations of images corresponding to the middle column (theoretical $CNR = 1$). The measured *CNR* is shown for images with idealized (circles) or limited (triangles) spatial resolution, respectively.

where a_1 and a_2 are scalars. In essence this means that the output obtained in response to the input from the combined objects is the same, whether they are imaged separately and the resulting outputs added together, or if the two objects are imaged simultaneously (in absence of noise). It also means that if the input signal from the object is amplified by some factor, a corresponding amplification of the resulting output image is obtained. Of note, the above definition is subtly different from a linear function in calculus since an intercept different from zero is not covered by a linear system, which is thus rather to be seen as following proportionality.

The above equation can be extended to an arbitrary number of terms, such that:

$$\mathcal{H}\left\{\sum_k a_k f_k(x)\right\} = \sum_k a_k \mathcal{H}\{f_k(x)\} = \sum_k a_k g_k(x) \tag{9.22}$$

which may be generalized to

$$\mathcal{H}\left\{\int_{-\infty}^{\infty} a_s f_s(x)\,\mathrm{d}s\right\} = \int_{-\infty}^{\infty} a_s \mathcal{H}\{f_s(x)\}\,\mathrm{d}s = \int_{-\infty}^{\infty} a_s g_s(x)\,\mathrm{d}s \tag{9.23}$$

where a_s are scalars and f_s are input functions that both vary over a continuum indexed by s.

For a *translation-invariant* system the response is not dependent on the coordinate of the input function. If we have the input distribution $f(x)$ that produces $g(x)$, then if $f(x)$ is displaced with a scalar α it is valid that

$$\mathcal{H}\{f(x-\alpha)\} = g(x-\alpha) \tag{9.24}$$

The *system function* $h(x)$ is defined as the impulse response, that is, the system response obtained for a Dirac delta function $\delta(x)$ placed in $x = 0$, such that when $f(x) = \delta(x)$ then the output is equal to the system function,

$$h(x) \equiv \mathcal{H}\{\delta(x)\} \tag{9.25}$$

We now regard our two-dimensional input distribution $f(x, y)$. Following the definition of the (two-dimensional) delta function, $f(x, y)$ can be written

$$f(x,y) = \int_{-\infty}^{\infty}\int_{-\infty}^{\infty} f(\alpha,\beta)\cdot\delta(x-\alpha, y-\beta)\,\mathrm{d}\alpha\mathrm{d}\beta \tag{9.26}$$

The function $f(x, y)$ is thus considered as a continuum of delta functions placed at positions given by the continuous α and β, weighted by the functional value of f.

Following Eqs. 9.24 and 9.25 the output from the displaced delta function becomes

$$\mathcal{H}\{\delta(x-\alpha, y-\beta)\} = h(x-\alpha, y-\beta) \tag{9.27}$$

The function $g(x, y)$ is then obtained following

$$\begin{aligned} g(x,y) &= \mathcal{H}\{f(x,y)\} = \mathcal{H}\left\{\int_{-\infty}^{\infty}\int_{-\infty}^{\infty} f(\alpha,\beta)\cdot\delta(x-\alpha, y-\beta)\,\mathrm{d}\alpha\mathrm{d}\beta\right\} \\ &= \int_{-\infty}^{\infty}\int_{-\infty}^{\infty} f(\alpha,\beta)\cdot\mathcal{H}\{\delta(x-\alpha, y-\beta)\}\,\mathrm{d}\alpha\mathrm{d}\beta \\ &= \int_{-\infty}^{\infty}\int_{-\infty}^{\infty} f(\alpha,\beta)\cdot h(x-\alpha, y-\beta)\,\mathrm{d}\alpha\mathrm{d}\beta = f(x,y)\otimes h(x,y) \end{aligned} \tag{9.28}$$

where linearity is assumed for the third equality and translation invariance for the fourth. The internal image plane obtained as output from the system (Figure 9.2) can thus be described as the convolution of the signal distribution from the object by the system function, provided that the imaging system can be assumed to be linear and translation invariant.

It is noted that when the input function $f(x,y)=\delta(x,y)$, then Eq. 9.28 gives

$$g(x,y)=f(x,y)\otimes h(x,y)=\int_{-\infty}^{\infty}\int_{-\infty}^{\infty}\delta(\alpha,\beta)h(x-\alpha,y-\beta)\mathrm{d}\alpha\mathrm{d}\beta=h(x,y) \tag{9.29}$$

which is thus consistent with Eq. 9.25. Following the convolution theorem, Eq. 9.28 can be expressed in the Fourier domain

$$G(u,v)=F(u,v)\cdot H(u,v) \tag{9.30}$$

with $F(u,v)=\mathcal{F}\{f(x,y)\}$, $H(u,v)=\mathcal{F}\{h(x,y)\}$, and $G(u,v)=\mathcal{F}\{g(x,y)\}$, where $\mathcal{F}$ symbolizes the operator for the continuous Fourier transform, and u and v represent spatial frequencies. In the Fourier domain Eq. 9.29 becomes

$$G(u,v)=\mathcal{F}\{\delta(x,y)\}\cdot H(u,v)=1(u,v)\cdot H(u,v)=H(u,v) \tag{9.31}$$

The assumption of a stationary system function may be sufficiently valid for planar images, such as obtained in nuclear medicine or X-ray imaging, for example. However, for tomographic images the system function is generally non-stationary.

It may also be of interest to consider functions of discrete variables, of particular relevance with regards to newer iterative techniques for tomographic reconstruction. Reverting to the one-dimensional notation, we may describe the relationship between the input signal $\hat{f}[n]$ and output signal $\hat{g}[m]$ following

$$\hat{g}[m]=\mathcal{H}\{\hat{f}[n]\} \tag{9.32}$$

with $n=0,..N-1$, and $m=0,..M-1$. By the discrete correspondence of Eq. 9.25, using the Kronecker delta function $\hat{\delta}$, the following expression for $\hat{g}[m]$ may be obtained:

$$\hat{g}[m]=\mathcal{H}\{\hat{f}[n]\}=\mathcal{H}\left\{\sum_k\hat{f}[k]\cdot\hat{\delta}[k-n]\right\}=\sum_k\hat{f}[k]\cdot\mathcal{H}\{\hat{\delta}[k-n]\} \tag{9.33}$$

Since translation invariance is not assumed, Eq. 9.33 does not result in a convolution. The expression may be written in a matrix form, such that

$$\hat{\boldsymbol{g}}=\boldsymbol{\mathcal{H}}\hat{\boldsymbol{f}} \tag{9.34}$$

where $\hat{\boldsymbol{g}}$ and $\hat{\boldsymbol{f}}$ are then column matrices of length M and N, respectively, and $\boldsymbol{\mathcal{H}}$ is the $M\times N$ matrix, commonly termed the system matrix. The system matrix essentially describes the fraction of the signal emanating from a given element in $\hat{\boldsymbol{f}}$ that is detected in a given element in $\hat{\boldsymbol{g}}$, where all elements in $\hat{\boldsymbol{f}}$ may contribute to any element in $\hat{\boldsymbol{g}}$. Although Eq. 9.34 may look straightforward, in practice the system matrix may become very large as is generally the case for three-dimensional tomographic imaging.

BIBLIOGRAPHY

Gonzales, R. C. and Woods R. E., *Digital Image Processing*, 4th ed. Hoboken, NJ: Pearson Education, 2018.
Bankman, I. N., *Handbook of Medical Imaging – Processing and Analysis*. San Diego: Academic Press, 2000.

Bracewell, R. N., *The Fourier Transform and Its Applications*. 3rd ed. Singapore: McGraw-Hill, 2000.

Brigham, E. O., *The Fast Fourier Transform and Its Application*. Englewood Cliffs, NJ: Prentice-Hall, 1988.

Rose, A., "Quantum and Noise Limitations of the Visual Process," *Journal of the Optical Society of America*, 43, 715–716 (1953).

IAEA, *Diagnostic Radiology Physics: A Handbook for Teachers and Students*. Vienna: International Atomic Energy Agency, 2014.

IAEA, *Quantitative Nuclear Medicine Imaging: Concepts, Requirements and Methods*. Vienna: International Atomic Energy Agency, 2014.

10 Image Processing

Johan Gustafsson

CONTENTS

DOI: 10.1201/9780429489556-10

10.1 INTRODUCTION

In image processing, various processing tools are applied in order to improve the interpretability or to extract information from images. Hence, it is a broad field that cannot be fully described within the scope of a single chapter, and the presentation given here will by necessity be limited to the most common operations. These selected subjects are colour-table transformations, filtering, spatial transformations and interpolation, and segmentation. The description is based on some mathematical prerequisites. In particular, basic knowledge of the Fourier transform, the discrete Fourier transform, and convolution are assumed. There are several good texts available for an introduction to these tools; see for example Bracewell [1] and Brigham [2].

Generally, the subject will be described in a one-dimensional (as a way to facilitate the description) or two-dimensional framework. In practice, nuclear-medicine images may also be three-dimensional or even four-dimensional. Typically, the generalization to a higher number of dimensions is straightforward.

10.2 COLOUR-TABLE TRANSFORMATIONS

Mathematically, a two-dimensional image can be represented as a matrix, that is, a rectangular scheme of numbers. Each element in the matrix is referred to as a picture element, or pixel. The most common way of extracting information from an image is through visual interpretation by a human observer. This requires conversion of pixel values into a colour. For this purpose, a colour look-up table is constructed, typically with 256 different colour levels, that maps from the range of pixel values in the image to a colour.

In the simplest case, the colour table is monochromatic, often mapping from black to white, as demonstrated in Figure 10.1A. In this case, the mapping is from black, for the lowest pixel value in the image, to white, for the highest pixel value, with a linear relationship in between. This mapping is referred to as a linear grayscale. However, the linear grayscale may not be the best choice for easy interpretation of a particular image. In principle, we have the freedom to choose a grayscale that suits our needs, but a very intricate mapping may of course become hard to interpret. Instead, there are a number of established techniques to modify the colour table in order to highlight some information in an image while at the same time suppressing other information.

From a mathematical perspective, the display of an image can be defined as a mapping from pixel values to 'whiteness'. Somewhat arbitrarily, the colour black will be represented by the value 0, while the colour white will be denoted by the number 1. In that case, the linear grayscale can be described as

$$g(p) = \frac{p - p_{\min}}{p_{\max} - p_{\min}}, \tag{10.1}$$

where p is the pixel value, $p_{\min}$ is the lowest pixel value in the image $I(x,y)$, and p_{max} is the largest pixel value in the image. The image I and the variable p refer to the same underlying pixel values, but for I we make the dependence on position in the image matrix explicit. Both kinds of notation will be necessary for the future discussion. In this case the grayscale has been adapted to fit the dynamic range of the current image. This assures that the full dynamic range of the display is used, which in turn maximizes the perceived contrast of the displayed image. However, it is not always desirable to adapt the colour table to the current image. For example, if the aim is to compare different images, the use of what will, in effect, be different mappings for the different images can be misleading and should be avoided.

10.2.1 Grayscale Transformations

A more formal perspective of displaying an image using a linear grayscale would be that two kinds of operators are applied to the image $I(x,y)$. The first operator $S(\cdot)$ maps the image values to the interval [0, 1], according to

$$S(I) = \frac{I - I_{\min}}{I_{\max} - I_{\min}} \tag{10.2}$$

and a linear grayscale operator $G(\cdot)$, which maps from black to white in the domain [0, 1] of pixel values. Hence, displaying an image with a lincar grayscalc as in Eq. 10.1 could alternatively be defined as applying the combined operator $G \circ S$ to the image, that is, first a scaling operation is applied followed by application of the grayscale operator.

FIGURE 10.1 Application of gray-scale transformations when displaying an image. (A) linear grayscale. (B) Windowed grayscale. (C) Gamma correction with $\lambda = 1.5$ (gamma expansion). (D) Histogram equalization. Plots of the mapping between pixel value and grayscale nuance is shown beneath each image. A higher gradient of the mapping results in a better contrast in the displayed image for that interval of pixel values.

One way of changing the way the image is displayed would be to apply the transform $T(\cdot)$ to the image prior to displaying with a linear grayscale, that is, to apply the operator $(G \circ S) \circ T$. Unfortunately, such an approach has the undesirable property of changing the underlying image data. If the only aim is to interpret the image visually, that is, the pixel values in themselves have no physical meaning, then this drawback is of little consequence. However, for cases where the actual pixel values have a quantitative meaning, changing the underlying data is cumbersome, at least from a theoretical perspective. A better method is then to consider the grayscale as the one being changed, that is, the linear grayscale is replaced for a non-linear one. In the remainder of this chapter we will apply transformation to the grayscale rather than the image. The exact same effects as those described could have been achieved by applying similar transformations to the image prior to display, but we wish to maintain the integrity of the image also at the time of display.

10.2.2 Windowing

On example of an alternative to a linear grayscale is to introduce two thresholds such that all pixel values below the lower threshold are mapped to the colour black while all pixel values above the threshold are mapped to the colour white, and a straight-line transition from black to white is used for pixel values in between the lower and upper thresholds. This operation is referred to as windowing, and the effect is to emphasize the visual contrast between pixel values between the thresholds, at the cost that all contrast below the lower threshold or above the upper threshold is lost. In mathematical terms, the grayscale becomes

$$g(p) = \begin{cases} 0, & p \le T_{\text{low}} \\ \dfrac{p - T_{\text{low}}}{T_{\text{high}} - T_{\text{low}}}, & T_{\text{low}} < p \le T_{\text{high}}, \\ 1, & p > T_{\text{high}} \end{cases} \tag{10.3}$$

where p is the pixel value, T_{low} is the lower threshold and T_{high} is the upper threshold. An example of the application of a windowed grayscale is shown in Figure 10.1 B.

10.2.3 Gamma Correction

Another technique to change the appearance of an image by colour table modification is gamma correction, where the grayscale is defined as

$$g(p) = \left(\frac{p - p_{\text{min}}}{p_{\text{max}} - p_{\text{min}}} \right)^{\gamma}, \tag{10.4}$$

where $\gamma > 0$ is an adjustable parameter. A value of $0 < \gamma < 1$ is referred to as gamma compression and a value of $\gamma > 1$ is referred to as gamma expansion. A value of $\gamma = 1$ makes the mapping equal to a linear grayscale. An example of gamma expansion is shown in Figure 10.1 C. The effect of the gamma expansion is to increase the gradient of the grayscale for high pixel values relative to a linear grayscale, while the gradient is decreased for low pixel values. This, in turn, amplifies the contrast between high pixel values in the displayed image while the contrast between low pixel values is suppressed. The effect would have been the opposite for gamma compression, that is, an amplification of the contrast between low pixel values and a suppression of contrast between high pixel values.

10.2.4 Histogram Equalization

A more refined way of adapting the grayscale to the properties of the image is histogram equalization [3]. The underlying theory is typically explained in the limit of an image of continuous variables. Hence, consider an image $I(x, y)$ of two continuous variables x and y (rather than the discrete x and y considered previously). Let the distribution of values in the image be $f(p)$. Consider the value of I in a random point and denote the corresponding stochastic variable ξ. The probability density function of ξ will then be $f(p)$. The cumulative distribution function of ξ is

$$F(p) = \int_{-\infty}^{p} f(p')\mathrm{d}p' = \Pr(\xi < p), \tag{10.5}$$

where $\Pr(\cdot)$ denotes probability.

Let a new image be formed according to $J(x,y) = F(I(x,y))$ and let the corresponding stochastic variable be $\eta = F(\xi)$. The cumulative distribution function of η has the form

$$G(p) = \Pr(\eta < p) = \Pr(F(\xi) < p) = \Pr(\xi < F^{-1}(p)) = F(F^{-1}(p)) = p, 0 \le p \le 1, \tag{10.6}$$

Hence, the probability density function for η, $g(p) = \frac{\mathrm{d}G}{\mathrm{d}p} = 1$, is uniform in the interval $[0,1]$, where the domain of definition for g and G stems from the fact that $0 \le F(p) \le 1$. So, by applying the transform F to the image an image with a uniform distribution of values is formed. In accordance with the previously stated preference of using a transformed grayscale rather than transforming the image itself, the same effect can be achieved by applying the transform to the colour table, that is, the cumulative distribution function is used as a colour table.

The effect of making the density function uniform is that the colour display is fully utilized since all grayscale levels are used equally often. The effect of histogram equalization is to give the colour table a high gradient in intervals where there are many pixel values, while the gradient is low in intervals where there are few pixel values in the image. This seems intuitively favourable, and the effect is an 'optimal' colour table (in some sense) for the current image.

In practice, digital images are discrete, and the probability density function is replaced by the histogram of the image, the cumulative distribution function is replaced by the cumulative histogram, and the (rescaled) cumulative histogram is used as grayscale. The property of a uniform histogram of grayscale values does not hold for the discrete case, but the improvement of contrast is still achieved. An example of histogram equalization is shown in Figure 10.1D, where the generally improved contrast compared with the linear grayscale in Figure 10.1A can be appreciated.

Histogram equalization is a special case of a broader range of techniques to make the displayed image follow a predefined distribution of grayscale values. For the continuous case, the desired histogram, having a cumulative distribution function T, can be achieved from an image I having cumulative distribution function F, by applying the transformation $T^{-1}\{F(I)\}$. The discrete case is more problematic due to the (general) non-invertibility of the cumulative distribution function for those images. Several methods to achieve approximate solutions have been proposed [4, 5].

10.3 A MODEL OF THE IMAGE-FORMATION PROCESS

The coming discussion about filtering will benefit from being viewed in the light of a model of the image-formation process. One simple, but useful, model is to consider the image-formation process as a linear translation-invariant process with additive noise, which can be formulated mathematically as

$$g(x,y) = h(x,y) * f(x,y) + n(x,y), \tag{10.7}$$

where * denotes convolution, $f(x,y)$ is the true distribution of the underlying object, $h(x,y)$ is the point-spread function (PSF) of the imaging system, that is, the response of the system to an impulse function, and $n(x,y)$ is the additive noise. Further details about the rationale for this model are given in Chapter 9. In the Fourier domain Eq. 10.7 corresponds to

$$G(u,v) = H(u,v) \cdot F(u,v) + N(u,v), \tag{10.8}$$

where $G(u,v) = \mathcal{F}\{g(x,y)\}$, $H(u,v) = \mathcal{F}\{h(x,y)\}$, $F(u,v) = \mathcal{F}\{f(x,y)\}$, and $N(u,v) = \mathcal{F}\{n(x,y)\}$. Low frequencies correspond to slow changes in the image, while high frequencies correspond to rapid changes in the image, for example, edges. Typically, $|F(u,v)\cdot H(u,v)|$ goes to 0 for large values of u and v while the same is not necessarily true for $|N|$. Hence, for low frequencies, that is, frequencies for which $\sqrt{u^2+v^2}$ is close to 0, $F\cdot H$ typically dominates over N, while the opposite is true for large values of $\sqrt{u^2+v^2}$.

10.4 FILTERING

One of the major degrading factors in an image is the presence of noise, that is, the term n in Eq. 10.7 and N in Eq. 10.8. Hence, an important image-processing task is to suppress noise in the image while preserving the signal from the object, thereby increasing the signal-to-noise ratio (SNR). Conversely, in some applications we are only interested in the part of the signal the describes the rapid changes in the image, that is, to highlight edges in the image. Both these aims can be achieved by the application of filters to the image, with so-called low-pass filtering used to suppress noise and so-called high-pass filtering used for edge enhancement.

Technically, the filtering can be performed either in the spatial domain or in the Fourier domain. Application of a filter in the spatial domain takes the form of a convolution, while application of a filter in the Fourier domain takes the form of a multiplication. The theoretical background of convolution and the Fourier transform is beyond the scope of this chapter. However, we will point out some important aspects of the convolution operation that is important to filtering.

Applications of filters in the spatial and Fourier domains are connected through the convolution theorem, so that convolution with a kernel in the spatial domain corresponds to multiplication with a filter in the Fourier domain and vice versa. Hence, the distinction between the two modes of application is, from a practical perspective, somewhat artificial. However, with respect to filter design, some filters are easier to understand in spatial domain and other filters are easier to understand in the Fourier domain.

10.4.1 Filtering in the Spatial Domain

10.4.1.1 Discrete Convolution

The mathematically strict definition of discrete convolution of two two-dimensional arrays f and h is

$$g(i,j) = f * h = \sum_{p=0}^{P-1}\sum_{k=0}^{K-1} f(k-i, p-j)\cdot h(k,p),\ i = 0,1\ldots N-1, \quad j = 0,1\ldots M-1, \tag{10.9}$$

where f is an image ($M \times N$ matrix) and h is a $K \times P$ kernel. This definition is a bit awkward for image-processing purposes. It is instead common to define the discrete convolution operation according to

$$g(i,j) = f * h = \sum_{p=0}^{P-1}\sum_{k=0}^{K-1} f\left(i+k-\frac{K}{2}, j+p-\frac{P}{2}\right)\cdot h(k,p), \tag{10.10}$$

The terms $K/2$ and $P/2$ make the convolution centred with respect to the kernel, that is, the origin is moved from the element $(0,0)$ to the element $\left(\frac{K}{2},\frac{P}{2}\right)$ in the kernel. The centring of the operation is natural since there is no reason to restrict the kernel to positive indices. Likewise, the arguments for f are $i+k$ and $j+p$, respectively, in contrast to the conventional definition of mirroring the function around the origin. Technically, this makes Eq. 10 a (centred) correlation operation rather than a convolution. However, since most kernels of practical use for filtering are symmetric, or in some cases antisymmetric, around the middle element, the distinction is of little consequence.

One may notice that, because of the centring, the filtering operation in this case makes most sense if the K and P are odd so that a well-defined centre element exists. For even values of K and P the result matrix will appear translated with respect to the original image. In order for the total sum of all pixel values not to change as result of the application of the filter, we also need to impose the constraint that $\sum_{p=0}^{P-1}\sum_{k=0}^{K-1} h(k,p) = 1$. Sometimes, an explicit normalization is included in the discrete-convolution formula, but we will consider such a normalization a property of the kernel rather than the filtering operation itself.

Another problem that needs to be handled when applying Eq. 10.10 is that the arguments for f, $i+k-\frac{K}{2}$ and $j+p-\frac{P}{2}$, for some values of i, k, j, and p will be out of bounds with respect to the matrix size of f. This is typically handled by either truncating the result matrix, in effect making the resulting image smaller than the original image, or by assuming out-of-bounds indices to correspond to elements with the value 0 (or some other value that is considered neutral for the particular application), which is referred to as zero-padding.

10.4.1.2 Low-pass Filtering

Assume that a two-dimensional image is convolved with the kernel

$$h = \frac{1}{9}\begin{bmatrix} 1 & 1 & 1 \\ 1 & 1 & 1 \\ 1 & 1 & 1 \end{bmatrix}. \tag{10.11}$$

The result is the moving average of the original image, as illustrated in Figure 10.2 B and the kernel is referred to as a 3×3 mean-value filter. Since the average tends to cancel out local variations in the signal, the SNR will improve at the cost of a deteriorated spatial resolution. A larger kernel, for example, 5×5, will result in a larger averaging effect, and thereby more noise reduction but also a further deteriorated resolution.

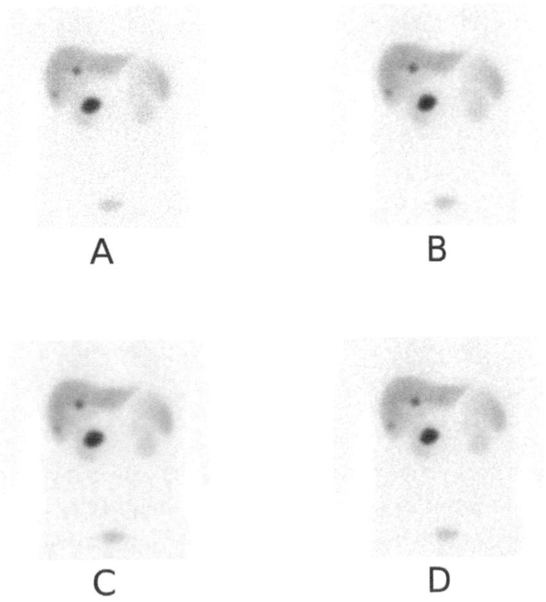

FIGURE 10.2 Examples of application of low-pass filtering in the spatial domain on a simulated planar image of a patient undergoing therapy with ^{177}Lu-DOTATATE. (A) The original image. (B) Image filtered with a 3 × 3 mean-value filter. (C) Image filtered with a 5 × 5 mean-value filter. (D) Image filtered with an NPS filter. Application of the kernels result in noise suppression, but also in loss of spatial resolution.

An alternative to a mean-value filter is to give a higher weight to the central element in the kernel, for example

$$h = \frac{1}{16}\begin{bmatrix} 1 & 2 & 1 \\ 2 & 4 & 2 \\ 1 & 2 & 1 \end{bmatrix}, \tag{10.12}$$

which is sometimes referred to as a nine-point-smooth filter. The smoothing effect is somewhat lower than for the corresponding 3×3 mean-value filter.

One property of the convolution-based filters, inherited from the convolution operation, is that the application of a filter is a linear operation, that is, if a filter is applied to the sum of two images, the result is equal to the sum of the two images filtered individually. This property greatly facilitates theoretical studies of the filter properties and adds to the predictability of the operation.

10.4.1.3 Edge Enhancement

10.4.1.3.1 Operators Based on the First Derivative

Rather than suppress local variation in the signal, filtering can also be used to enhance local variation. The typical aim is to enhance edges in the image. A natural strategy to detect changes in a signal is to consider the derivative of that signal. A high positive or negative value of the derivative of the signal indicates the transition between regions with different pixel values, that is, the absolute of the derivative indicates the positions of edges. In two, or higher, dimensions the absolute value of the derivative generalizes to the norm of the gradient. Furthermore, since digital images are discrete data the (partial) derivatives have to be approximated, or replaced by, finite differences. Hence, for an image $f(x,y)$ an approximation of the norm of the gradient becomes

$$G(x,y) = \sqrt{f_x^2 + f_y^2}, \tag{10.13}$$

where f_x and f_y are differences in the x- and y-directions, respectively. In a simple case,

$$f_x(x,y) = f(x+1,y) - f(x-1,y) \text{ and } f_y(x,y) = f(x,y+1) - f(x,y-1), \tag{10.14}$$

which corresponds to convolution with the kernels

$$g_x = \begin{bmatrix} -1 & 0 & 1 \end{bmatrix} \text{ and } g_y = \begin{bmatrix} -1 \\ 0 \\ 1 \end{bmatrix}, \tag{10.15}$$

so that

$$f_x = f * g_x \text{ and } f_y = f * g_y. \tag{10.16}$$

The main problem with derivation operators is their sensitivity to noise, since any form of local variation, random as well as systematic, will be amplified. One way of suppressing the noise-amplifying effect is to combine the filter for the finite differences with a low-pass filter in the orthogonal direction. Hence, more useful kernels for the differences in the x- and y-directions can be achieved as

$$g_x = \begin{bmatrix} -1 & 0 & 1 \\ -1 & 0 & 1 \\ -1 & 0 & 1 \end{bmatrix} \text{ and } g_y = \begin{bmatrix} -1 & -1 & -1 \\ 0 & 0 & 0 \\ 1 & 1 & 1 \end{bmatrix}. \tag{10.17}$$

Calculation of the magnitude of the gradient using these two kernels is referred to as application of the Prewitt operator [6]. The two filters in Eq. 10.17 have been applied to an image in Figures 10.3 B and 10.3 C and their combination using Eq. 10.13 is shown in Figure 10.3 D. An operator closely related to the Prewitt operator is the Sobel operator, which uses the kernels

$$g_x = \begin{bmatrix} -1 & 0 & 1 \\ -2 & 0 & 2 \\ -1 & 0 & 1 \end{bmatrix} \text{and } g_y = \begin{bmatrix} -1 & -2 & -1 \\ 0 & 0 & 0 \\ 1 & 2 & 1 \end{bmatrix} \tag{10.18}$$

to approximate the partial derivatives instead.

The Prewitt and Sobel operators are based on partial derivatives in the vertical and horizontal directions. These choices may seem natural, but for the aim of calculating the magnitude of the gradient, the direction in which differences are computed in principle should not matter as long as the directions are orthogonal to each other. This is explored in an operator referred to as Roberts cross operator [7], which calculates differences in the diagonal directions instead, that is,

$$g_x = \begin{bmatrix} 1 & 0 \\ 0 & -1 \end{bmatrix}, \ g_y = \begin{bmatrix} 0 & 1 \\ -1 & 0 \end{bmatrix}. \tag{10.19}$$

An example of application of Roberts cross operator is shown in Figure 10.3 E.

With respect to linearity, the calculation of partial derivatives is a linear operation, but the calculation of the norm of the gradient from these partial derivatives is a non-linear operation. This is in contrast to the low-pass filters discussed in the previous section.

10.4.1.3.2 The Laplacian Operator

A similar effect, that is, the highlighting of the transitions between regions, as with the first derivatives can be achieved using the second derivatives. For two and higher dimensions the partial second derivatives can be combined into a scalar-valued function using the Laplacian operator, that is,

$$\lambda(x,y) = \Delta f(x,y) = f_{xx}^2 + f_{yy}^2, \tag{10.20}$$

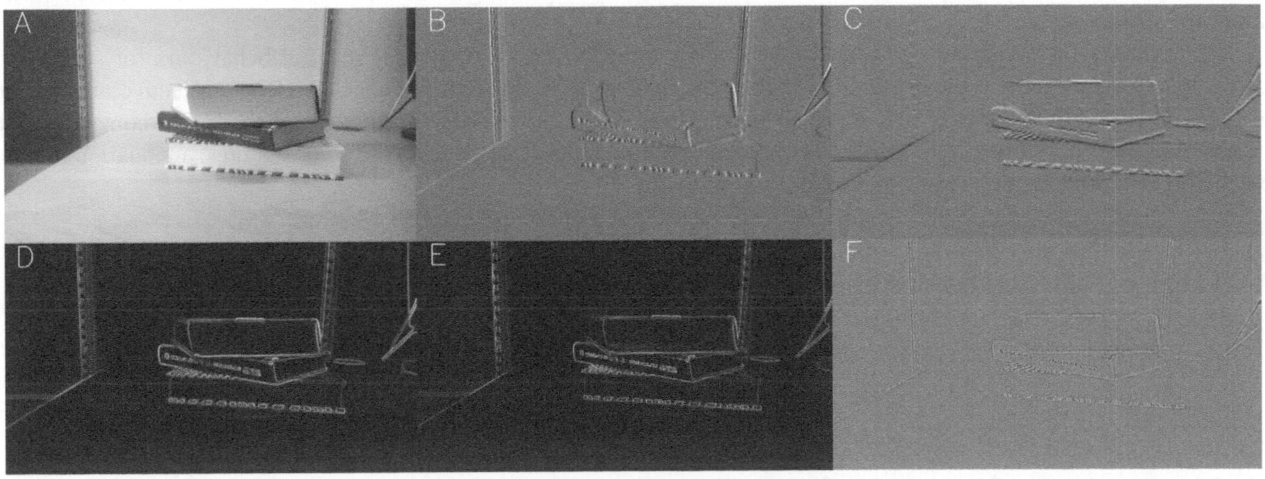

FIGURE 10.3 High-pass filtering in the spatial domain. (A) The original image. (B) Image filtered with a Prewitt filter in the horizontal direction. (C) Image filtered with a Prewitt filter in the vertical direction. (D) Image filtered with the Prewitt operator. (E) Image filtered with the Roberts cross operator. (F) Image filtered with the Laplace operator. The effect of all filters is to highlight edges in the images. For B and C, only edges in the horizontal and vertical directions, respectively, have been enhanced.

where f_{xx} and f_{yy} are the second order partial derivatives in the x- and y-directions, or, in practice, finite-difference approximations thereof. Such an approximation can be derived from simple Taylor expansions of f so that

$$\begin{aligned} f_{xx} &= f(x-1,y)-2f(x,y)+f(x+1,y) \\ f_{yy} &= f(x,y-1)-2f(x,y)+f(x,y+1) \end{aligned}. \tag{10.21}$$

Hence, application of the Laplacian operator can be achieved through convolution with the kernel

$$L = \begin{bmatrix} 0 & 0 & 0 \\ 1 & -2 & 1 \\ 0 & 0 & 0 \end{bmatrix} + \begin{bmatrix} 0 & 1 & 0 \\ 0 & -2 & 0 \\ 0 & 1 & 0 \end{bmatrix} = \begin{bmatrix} 0 & 1 & 0 \\ 1 & -4 & 1 \\ 0 & 1 & 0 \end{bmatrix}. \tag{10.22}$$

Sometimes, this kernel is modified to also be symmetric in the diagonal directions, leading to

$$L = \begin{bmatrix} 1 & 1 & 1 \\ 1 & -8 & 1 \\ 1 & 1 & 1 \end{bmatrix}. \tag{10.23}$$

An example of the Laplacian operator applied to an image is given in Figure 10.3 F.

As opposed to the edge-enhancement techniques based on first-order derivatives, application of the Laplacian operator consists only of convolution with a kernel. Hence, edge enhancement with the Laplacian is a linear operation.

10.4.2 Filtering in the Fourier Domain

Convolution in the spatial domain and multiplication in the Fourier domain are connected through the convolution theorem, stating that

$$\mathcal{F}\{f * g\} = F \cdot G \tag{10.24}$$

where F is the Fourier transform of f and G is the Fourier transform of g. The theorem holds both for the continuous and the discrete cases, possibly with some differences in scaling depending on the exact definition of the discrete Fourier transform used.

The rationale for constructing filters in the Fourier domain lies in the typical spectral behaviours for signal and noise, where the SNR is typically higher for low frequencies than for high frequencies. Hence, selective attenuation of high-frequency components will result in an improved SNR. In the same manner, since the high frequencies in a signal describes sharp transitions while low-frequency components describe the large-scale features, attenuation of low-frequency components will result in accentuation of edges.

10.4.2.1 Low-pass Filtering

An example of a filter that attenuates high frequencies while preserving low frequencies is

$$G(u,v) = \begin{cases} 1, & \sqrt{u^2+v^2} \le D_0 \\ 0, & \sqrt{u^2+v^2} > D_0 \end{cases}, \tag{10.25}$$

which is referred to as an ideal low-pass filter. The frequency D_0 is referred to as the cut-off frequency. Plots of the ideal low-pass filter in the Fourier domain and the spatial domain are shown in Figure 10.4A. Note the lobes in the spatial domain, which give rise to the ringing phenomenon after application of the filter. The oscillating behaviour of the two-dimensional ideal low-pass filter in the spatial domain is analogous to the well-known result that the Fourier transform of a one-dimensional box function is a sinc function and vice versa. The two-dimensional inverse Fourier transform of Eq. 10.25 is not a sinc function, but it has sinc-like properties in the sense that there is a large peak at the origin with oscillating tails that slowly go to zero. The ringing phenomenon resulting from filtering with an ideal low-pass filter is also referred to as the Gibbs phenomenon, Gibbs ringing, or Gibbs artefact [8, 9].

The result of applying the ideal low-pass filter in Figure 10.4A with respect to spatial resolution and SNR is as expected, that is, a deterioration in spatial resolution but an improved SNR. However, the ringing prevents the ideal low-pass filter from being practically useful and its utility comes more from theoretical considerations.

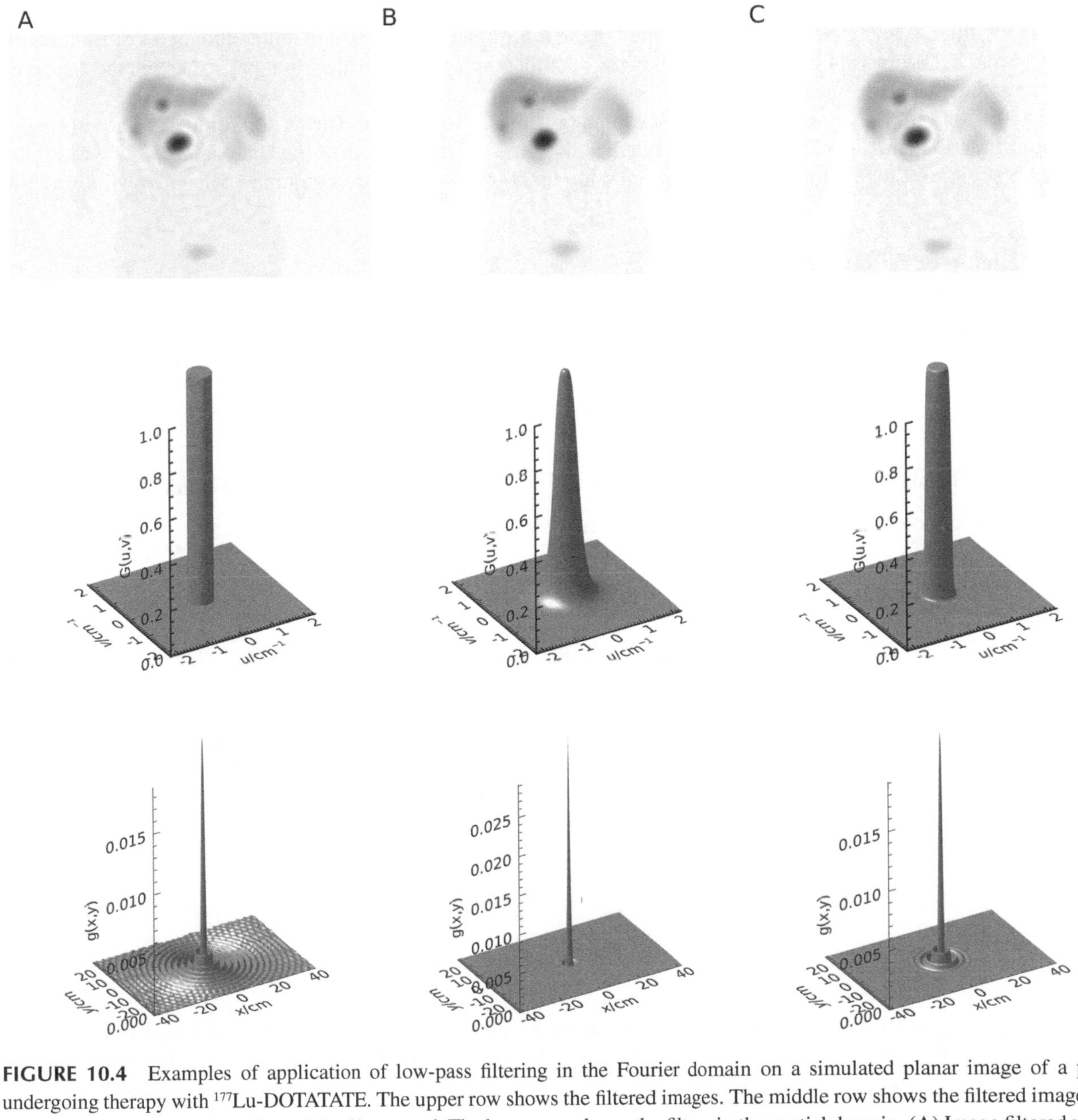

FIGURE 10.4 Examples of application of low-pass filtering in the Fourier domain on a simulated planar image of a patient undergoing therapy with ^{177}Lu-DOTATATE. The upper row shows the filtered images. The middle row shows the filtered images, and the lower row shows surface plots of the filters used. The lower row shows the filters in the spatial domain. (A) Image filtered with an ideal low-pass filter. (B) Image filtered with a Butterworth filter with order 2. (C) Image filtered with a Butterworth filter with order 10. The original image is shown in Figure 10.2A. Note the Gibbs artefacts visible in Figures 10.2A and 10.2C for example the wave pattern around the large tumour. The ringing artefacts are caused by the sharp transitions of the filters in the Fourier domain, resulting in oscillations of the filters in the spatial domain.

The problem of the Gibbs phenomenon is caused by the sharp transition from 1 to 0 in the ideal low-pass filter. Hence, a smoother transition from high to low amplitudes is desirable. This can be achieved in several ways, but one example is the so-called Butterworth filter [10]

$$G(u,v)=\frac{1}{1+\left(\frac{\sqrt{u^2+v^2}}{D_0}\right)^{2n}}, \tag{10.26}$$

where the cut-off frequency D_0 is the frequency where the filter is equal to $1/2$ and n is a parameter, referred to as the order of the filter, which determines the steepness of the transition from 1 to 0 of the filter. Examples of Butterworth filters are shown in Figure 10.4 C. As the order increases the filter approaches an ideal low-pass filter and the characteristic ringing phenomenon reappears.

One example of a filter that never exhibits Gibbs ringing is the Gaussian filter. The reason for the absence of ringing for this filter function is realized by considering that the inverse Fourier transform of the Gaussian filter will result in a Gaussian kernel in the spatial domain. Since the Gaussian function does not have oscillating behaviour, Gibbs ringing cannot occur.

10.4.2.2 High-pass Filtering

An ideal high-pass filter can be defined as

$$G_{\text{hp}}(u,v)=\begin{cases}0, & \sqrt{u^2+v^2}\le D_0\\ 1, & \sqrt{u^2+v^2}>D_0\end{cases}. \tag{10.27}$$

Alternatively, the ideal high-pass filter can be defined as $G_{\text{hp}}(u,v)=1-G_{\text{lp}}(u,v)$, where G_{lp} is an ideal low-pass filter. Accordingly, for an image $f(x,y)$ with Fourier transform $F(u,v)$

$$\mathcal{F}^{-1}\{F(u,v)\cdot G_{\text{hp}}(u,v)\}=\mathcal{F}^{-1}\{F(u,v)\cdot(1-G_{\text{lp}}(u,v))\}=$$

$$\mathcal{F}^{-1}\{F(u,v)\}-\mathcal{F}^{-1}\{F(u,v)\cdot G_{\text{lp}}(u,v)\}=f(x,y)-f(x,y)*g_{\text{lp}}(x,y). \tag{10.28}$$

Hence, application of an ideal high-pass filter is equal to subtracting a low-pass filtered version from the original image. Since the effect of low-pass filtering is to remove sharp edges in the image, those will be highlighted in the high-pass filtered image and the ringing problem of ideal low-pass filters is inherited by the high-pass version.

The high-pass version of the Butterworth filter can be defined as

$$G(u,v)=\frac{1}{1+\left(\frac{D_0}{\sqrt{u^2+v^2}}\right)^{2n}}, \tag{10.29}$$

which in turn is one minus the low-pass version of the filter. Examples of application of the ideal high-pass filter and the high-pass Butterworth filter are shown in Figure 10.5B. The suppression of the main signal and the accentuation of edges are clearly seen, as is the Gibbs ringing for the ideal filter.

10.4.3 Filtering in the Spatial Domain and in the Fourier Domain

Apart from the benefits of designing filters in the Fourier domain, there may also be advantages in applying filters originally designed in the spatial domain by first applying the Fourier transform to the filter and the image, multiplying the

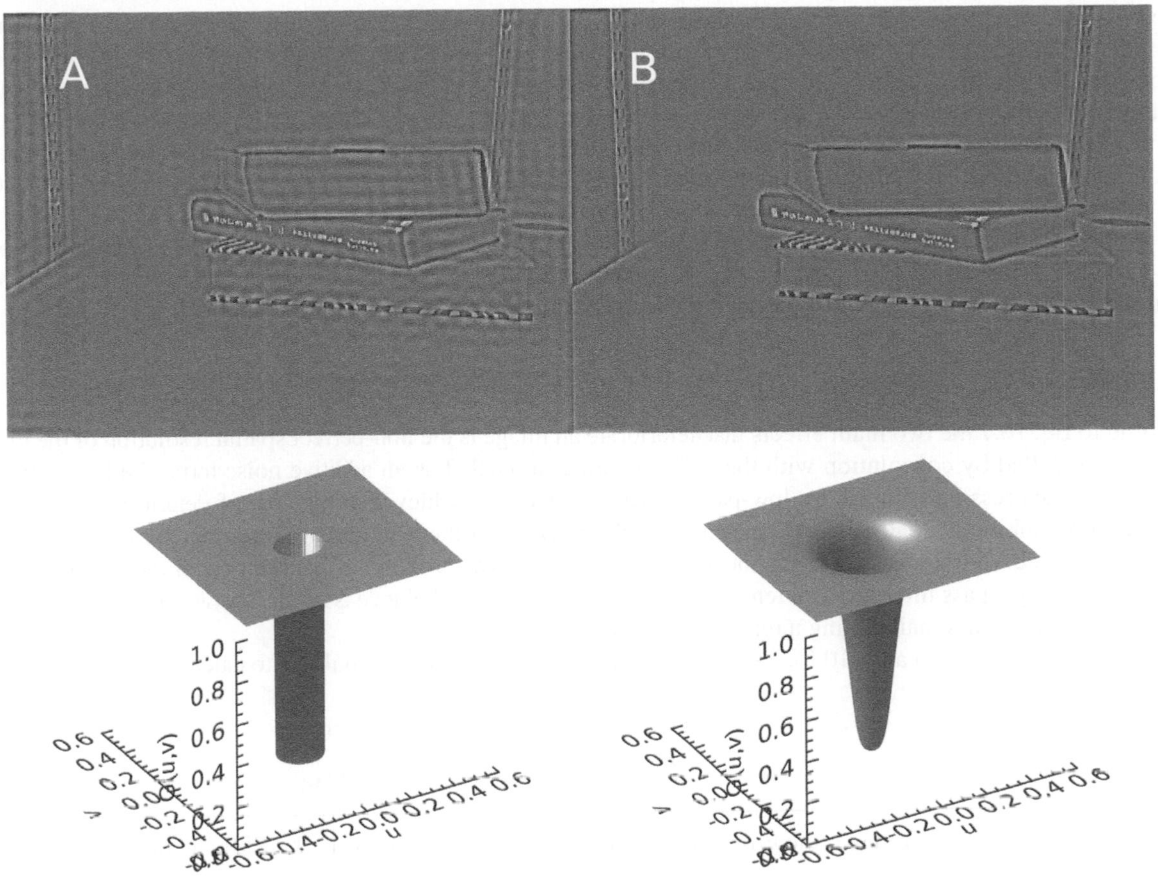

FIGURE 10.5 High-pass filtering in the Fourier domain. (A) Application of an ideal high-pass filter. (B) Application of a Butterworth high-pass filter with order 2. The original image is shown in Figure 10.3 A. Just as the ideal low-pass filter causes ringing in the filtered image, so does the ideal high-pass filter.

two and then performing an inverse Fourier transform. The reason for this procedure is that the discrete Fourier transform can be efficiently computed using the fast Fourier transform (FFT) algorithm [11, 12]. Hence, for some situations, especially for large images, using the route involving the Fourier transform might be a computationally more efficient approach than using discrete convolution directly.

The multiplication of matrices in the Fourier transform requires the matrices to be of the same size. Hence, if the kernel and image are not of the same size (which is often the case), they must be padded before Fourier transformation. Furthermore, a basic property of the discrete Fourier transform is that all arrays behave as if being periodically repeated in all directions. This makes convolution that is applied through the convolution theorem cyclical. This fact may lead to artefacts at the edges of the image matrix unless due caution is observed. The cyclic convolution can be exchanged for its zero-padded counterpart by zero-padding the matrices beyond the original size before Fourier transformation. If the original matrices are $N \times M$ and $K \times P$, zero padding to (a minimum of) $(N+K-1)\times(M+P-1)$ will make sure that the fold-over artifacts of circular convolutions do not affect the original $N \times M$ region, at the expense of computational time.

10.4.4 Non-linear and Adaptive Filtering

A problem with linear low-pass filters is the conflict between an effective noise suppression and the preservation of the spatial resolution of the image. Hence, a number of alternative filtration methods have been developed that aim to suppress noise while minimizing the deterioration of the spatial resolution.

A simple filtering method that is compatible with the existence of sharp edges is median filtering [13]. The value in each pixel is replaced with the median in its neighbourhood, for example, in a 3×3 region. The median filter is the best-known example of filters based on order statistics [14], other examples of such filters being various versions of trimmed means [15]. These kinds of filters often have good noise-removal properties for certain kinds of noise distributions. For example, the median filter is efficient in removing single pixel values with high or low values compared to their neighbourhood while at the same time preserving larger structures in the image. However, the median filter may also introduce artefacts in the image in the form of homogenous regions, sometimes referred to as streaking [16], and in general order based statistics filters are often harder to analyse theoretically than the more commonly used convolution-based filters, which is an unattractive feature.

A number of filtering methods try to adapt the smoothing effect to the image properties and the local noise properties of each pixel [17, 18]. By adjusting the amount of smoothing applied to each pixel, a better trade-off between noise and resolution can be achieved than when the same parameters are used throughout the whole image.

10.4.5 Wiener Deconvolution

According to Eq. 10.7 the two main effects that deteriorate an image is the non-perfect spatial resolution of the imaging system, as modelled by convolution with the PSF, and noise, modelled as an additive noise term. We have previously discussed the suppression of noise by low-pass filtering, which was achieved at the cost of deteriorated spatial resolution. A more ambitious goal would be to remove both the effect of the PSF and of noise, thereby retrieving a better estimate of the underlying object f from the measured image g. The problem is that removal of the effect of the PSF, in analogy with high-pass filtering, also tends to amplify noise. Hence, a balance is needed between the improvement of the spatial resolution and maintaining a reasonable SNR.

Starting from Eqs. (10.7) and (10.8), a naïve way of estimating f from g would be to calculate

$$\hat{f}(x,y) = \mathcal{F}^{-1}\left\{\frac{G(u,v)}{H(u,v)}\right\} = f(x,y) + \mathcal{F}^{-1}\left\{\frac{N(u,v)}{H(u,v)}\right\}. \tag{10.30}$$

Since Eq. 10.30 attempts to remove the effect of the convolution of the image with the PSF, the process is referred to as deconvolution. The problem with such an approach is the ratio $\frac{N(u,v)}{H(u,v)}$. Since $H(u,v)$ goes to 0 for large u and v while $\mathrm{E}(|N(u,v)|)$ is (often) approximately constant over u and v, the ratio goes to infinity rendering the result useless.

One way of avoiding the extreme noise amplification of Eq. 10.30 is offered by Wiener deconvolution [19]. The estimation of the underlying object is then computed as

$$\hat{f}(x,y) = \mathcal{F}^{-1}\left\{\frac{H^*(u,v)S_f(u,v)}{S_f(u,v)|H(u,v)|^2 + S_n(u,v)}G(u,v)\right\} = \mathcal{F}^{-1}\left\{\frac{H^*(u,v)}{|H(u,v)|^2 + \frac{S_n(u,v)}{S_f(u,v)}}G(u,v)\right\}, \tag{10.31}$$

where $S_n(u,v)$ is the noise-power spectrum, $S_f(u,v)$ is the power spectrum of f, which need to be estimated beforehand, and * denotes the complex conjugate.

Unfortunately, Weiner deconvolution requires substantial knowledge about the underlying object and the noise. As an alternative, a simplified version of Eq. 10.31 can be used according to

$$\hat{f}(x,y) = \mathcal{F}^{-1}\left\{\frac{H^*(u,v)}{|H(u,v)|^2 + K}G(u,v)\right\}, \tag{10.32}$$

where K is an empirically adjustable parameter rather than the theoretically derived function $S_n(u,v)/S_f(u,v)$ in Eq. 10.31.

10.5 SPATIAL TRANSFORMATIONS AND INTERPOLATION

A basic property of digital images is discretization, that is, that the values are only known at certain locations. However, for various operations the values at other locations than the original sampling points are needed. One example of such operations is spatial transformations of the image.

Examples of basic spatial transformations are translation, rotation, and scaling. These operations apply a mapping of the pixel coordinates so that $H\{(x,y)\}=(x',y')$. For example, the translation operator can be defined according to

$$T\{(x,y)\}=(x-x_0,y-y_0), \tag{10.33}$$

which translates the image a distance x_0 in the x-direction and a distance y_0 in the y-direction. Similarly, rotation can be expressed as

$$R\{(x,y)\}=(x\cdot\cos(\theta)+y\cdot\sin(\theta),-x\cdot\sin(\theta)+y\cdot\cos(\theta)), \tag{10.34}$$

where θ is the rotation angle, and scaling can be expressed as

$$S\{(x,y)\}=(a\cdot x,a\cdot y), \tag{10.35}$$

where a is a scalar.

The aforementioned transforms are examples of affine transformations, which can all be expressed in matrix form as

$$\begin{bmatrix} x' \\ y' \\ 1 \end{bmatrix} = \begin{bmatrix} h_{1,1} & h_{1,2} & h_{1,3} \\ h_{2,1} & h_{2,2} & h_{2,3} \\ 0 & 0 & 1 \end{bmatrix} \begin{bmatrix} x \\ y \\ 1 \end{bmatrix}. \tag{10.36}$$

The expansion of the point (x,y) with an extra dimension to $(x,y,1)$ is referred to as homogenous coordinates. Using homogenous coordinates, the affine transformations can be handled within a single framework.

In general, the coordinates (x',y') will not coincide with one of the original pixel coordinates in the image, and the values at point in between these original sampling points will be needed. This can be achieved as a two-step process. The first step is referred to as interpolation and aims, to the highest extent possible, to reconstruct the original function (of continuous variables) from its samples. The second step is referred to as resampling. Often, the whole process is, somewhat incorrectly, referred to as interpolation.

10.5.1 FORWARD AND BACKWARD INTERPOLATION

There are two fundamental strategies for interpolation and resampling when performing spatial transformations. The first option is to first transform the coordinates of the original sampling points according to the spatial transformation and then spread out the values of these points to the sampling points of the transformed image (according to some rule that has to be defined beforehand). This process is referred to as forward interpolation. The alternative is to first calculate the corresponding positions of the sample points in the transformed image in the original coordinate system, and the values at these points are subsequently found by interpolation. This process is referred to as backward interpolation. An illustration of the process is shown, for the example of a rotation, in Figure 10.6.

Forward interpolation is perhaps more intuitive than backward interpolation but has the disadvantage of not guaranteeing the same amount of data to contribute to all sampling points in the transformed image. Indeed, it is quite possible that some sample points in the transformed image get no contributions at all, thereby potentially leaving 'holes' in the result. Hence, backward interpolation is the preferred mode of operation for most applications. However, forward interpolation also has advantages that may be important under some special circumstances. For example, the total sum of all pixel values is preserved in forward interpolation while the same is generally not the case for backward interpolation. For some applications, the automatic preservation of the sum of pixel values is an important advantage [20]. In the rest of this description, only techniques for backward interpolation will be discussed since it is the standard mode of operation.

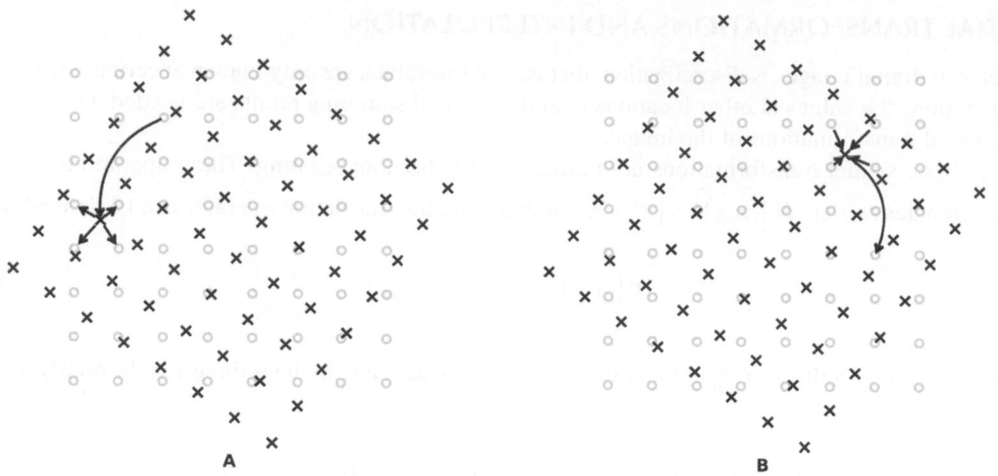

FIGURE 10.6 In forward interpolation (A), the point in the original grid is moved to another location by the transformation and the associated value spread out to locations in the grid of the transformed image. In backward interpolation (B), the locations of the sample points in the transformed image are first calculated and the value of the associated point calculated by interpolation from the original grid. In the shown example, the original and transformed grid coincide.

10.5.2 Ideal Interpolation

Interpolation in the strict sense, that is, the reconstruction of the function of continuous variables from samples, deserves a detailed discussion. In the following description, we will assume that the sampling distance in the signal is equal to 1. The results for sampling distances different from 1 may be obtained by scaling but will result in more elaborate formulas.

A well-known result by Shannon [21] is that a signal f may be fully recovered from its samples using the formula

$$f(x) = \sum_{i=-\infty}^{\infty} f(i)\operatorname{sinc}(x-i), \tag{10.37}$$

where the Fourier transform of f is assumed to have the property that $F(u) \equiv 0$ for $|u| > \frac{1}{2}$, that is, the function is band limited. Application of Eq. 10.37 is, in turn, equivalent with convolving with the kernel

$$g(x) = \operatorname{sinc}(x). \tag{10.38}$$

The underlying theorem is known as the sampling theorem, and the operation is referred to as ideal interpolation. Ideal interpolation will, in theory, restore the continuous signal, which in turn can be sampled with a sampling distance different from 1. Unfortunately, the condition that the function is band-limited is rather restrictive and, in practice, we cannot hope for Eq. 10.37 to hold exactly. The fact that the support for the kernel in Eq. 10.38 is infinite is also troublesome, partly because all signals that are handled in practice have a finite extension in time or space but also because of computational burden. If an interpolation kernel inspired by Eq. 10.38 were to be applied, a truncated version thereof would be the only practical solution. Unfortunately, it turns out that most of the good properties of ideal interpolation are lost in this truncation, and other finite-support kernels are better suited to act as approximations to Eq. 10.37.

A basic requirement for an interpolation kernel is that $g(0) = 1$ and $g(i) = 0$ for $i \neq 0$, thereby making the interpolated function equal to the sampled function at the sample points. A further desirable, but not necessary, property is that

$$\sum_{i=-\infty}^{\infty} g(d+i) \equiv 1, d \in [0,1), \tag{10.39}$$

which is referred to as the partition to unity condition. This makes the sum of the kernel at the sampling points equal and normalized for any position of the kernel relative to these points. For example, this assures that a uniform region in the original signal will remain uniform also after the interpolation.

There are several examples of interpolation kernels that satisfy the basic interpolation criterion as well as the partition to unity condition [22, 23]. Intuitively, we also want the approximative interpolation kernel to be similar to the sinc function within its domain of definition. This also applies to the kernel in the Fourier domain. The Fourier transform of the sinc function is the box function and, hence, we want the interpolation kernel to approximate an ideal low-pass filter in the Fourier domain.

10.5.3 Nearest-Neighbour Interpolation

The simplest approximation to ideal interpolation is nearest-neighbour interpolation, which, for the one-dimensional case, is equivalent to convolution with the kernel

$$g(x) = \begin{cases} 1, & -\frac{1}{2} \le x < \frac{1}{2} \\ 0, & x < -\frac{1}{2} \text{ or } x \ge \frac{1}{2} \end{cases}. \tag{10.40}$$

In practice, this means that the values at new sampling points are given the same value as the closest previous sampling-point. The kernel is illustrated in Figure 10.7A. The box function is a decent approximation to the sinc function close to the origin but is a poor approximation further away. There are also significant problems related to the behaviour of the kernel in Fourier space, with major deviations from the ideal response. The application of nearest-neighbour interpolation is illustrated in Figure 10.8B.

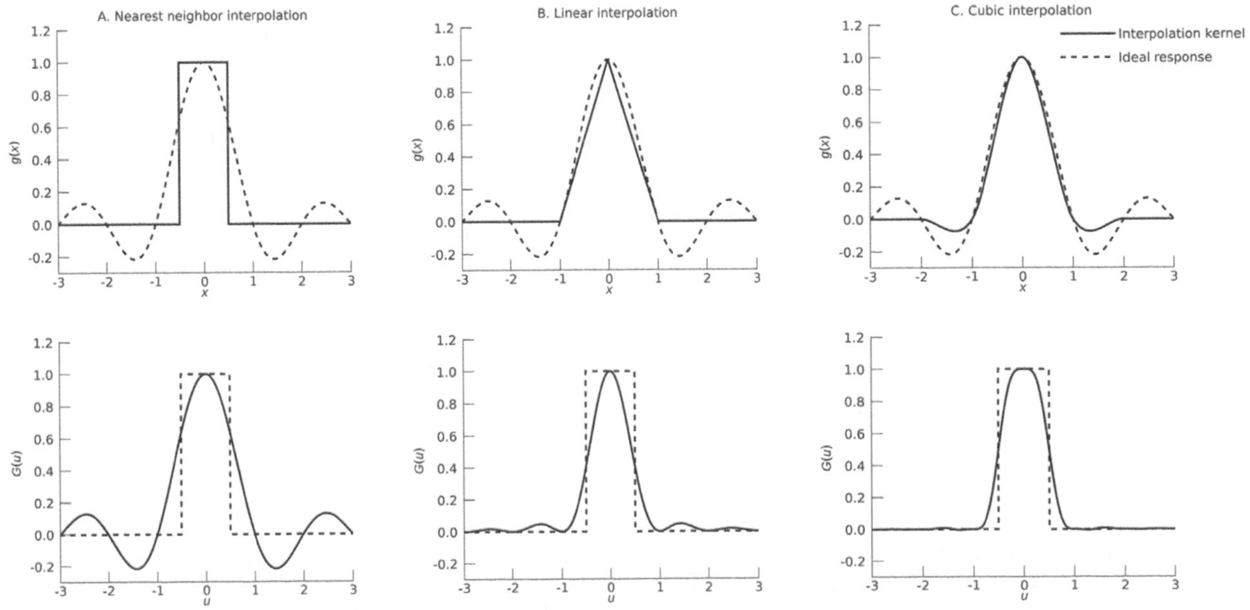

FIGURE 10.7 Kernels for different interpolation methods. (A) nearest-neighbour interpolation, (B) linear interpolation, and (C) cubic interpolation. The upper row shows the kernels in the spatial domain and the lower row shows the kernels in the Fourier domain. Nearest-neighbour interpolation is the worst approximation to an ideal filter, both in the spatial domain and the Fourier domain, while linear interpolation and cubic interpolation provide gradually better agreement.

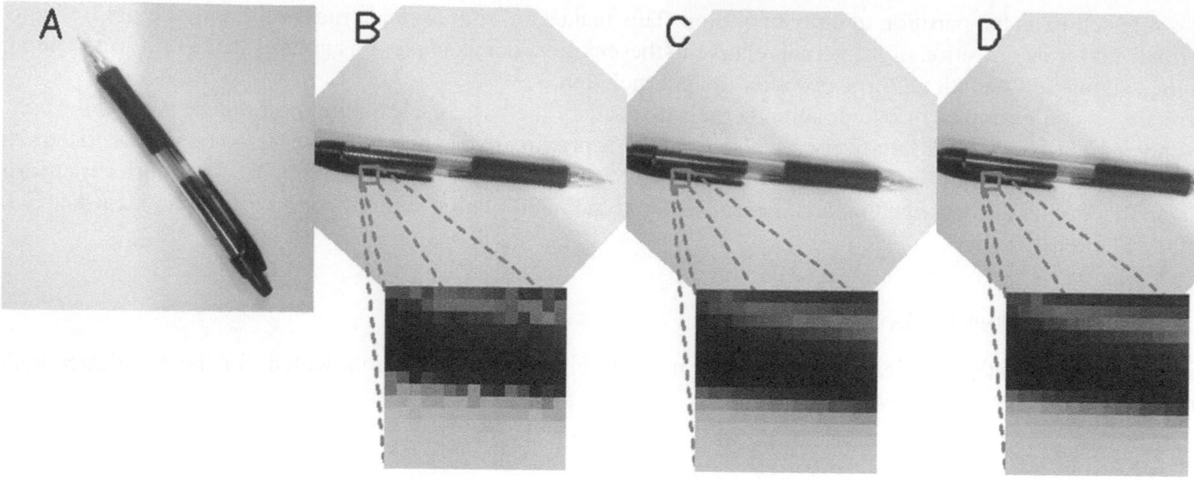

FIGURE 10.8 Effect of different interpolation methods. The original image is shown in A while the images B to D have been rotated. (B) Nearest-neighbour interpolation. (C) Linear interpolation. (D) Cubic interpolation. The differences in qualities between the interpolation methods are visible in the zoomed-in images.

10.5.4 Linear Interpolation

Linear interpolation can be achieved through convolution with the kernel

$$g(x) = \begin{cases} x+1, & -1 \le x < 0 \\ -x+1, & 0 \le x < 1 \\ 0, & x < -1 \text{ or } x \ge 1 \end{cases} . \tag{10.41}$$

The kernel is shown in Figure 10.7B and the improvement over nearest-neighbour interpolation is shown in Figure 10.8 C. From the perspective of frequency response, the Fourier transform approximates the box function reasonably well for large frequencies but has a relatively poor similarity close to the origin, leading to a deterioration of the spatial resolution.

An example of the application of bilinear interpolation is shown in Figure 10.8C. The zoomed-in edge gets a more natural appearance when linear interpolation is used compared with the use of nearest-neighbour interpolation.

10.5.5 Cubic Interpolation

Even further refinement compared to linear interpolation can be achieved by using cubic interpolation [24]. The interpolation kernel has the general form

$$g(x) = \begin{cases} (\alpha+2)|x|^3 - (\alpha+3)|x|^2 + 1, & |x| < 1 \\ \alpha|x|^3 - 5\alpha|x|^2 - 4\alpha, & 1 \le |x| < 2, \\ 0, & |x| > 2 \end{cases} \tag{10.42}$$

where α is a free parameter that can be tuned to the particular application. From a theoretical perspective, $\alpha = -0.5$ has been suggested as a good value [25].

A plot of the interpolation kernel for the case of $\alpha = -0.5$ is shown in Figure 10.7C. The similarity with the sinc-kernel for ideal interpolation can be noted. The application of bicubic interpolation to an image can be seen in Figure 10.8 D. Cubic interpolation is often considered a good compromise between accuracy and computational speed. However, it might be noted that opposed to nearest-neighbour and linear interpolation the cubic interpolation kernel shifts sign. This may be a disadvantage for applications where non-negativity of the interpolated image is a pre-requisite.

10.6 SEGMENTATION

Segmentation is the division of an image into two or more regions – for example an object region and a background region. This is in turn an important step for further analysis of for example the size and shape of the object.

In practice, the segmentation process can be divided into two subprocesses. The first step is to identify the object of interest in the image, and the second step is to define the border between that object and the background. These two processes are referred to as recognition and delineation, respectively [26]. Both these steps can be performed manually, automatically, or by a combination thereof. A segmentation process that combines manual and automatized steps, for example, manual recognition and automatic delineation – is sometimes referred to as semi-automatic segmentation.

The by far most-prevalent segmentation method, at least for medical images, is manual segmentation, that is, a manual operator performs both the recognition and the delineation tasks. The method is flexible, in the sense that an experienced operator is capable of several types of segmentation tasks and is able to combine data and prior knowledge.

The two main problems with manual segmentation are that the method is both time-consuming and suffers from poor reproducibility [27–29], and automatic or semi-automatic methods may be used to alleviate some of those problems [30]. A large number of (semi-)automatic segmentation methods are available, and application of such methods may be beneficial both in terms of the work time needed for the segmentation and with respect to reproducibility between operators. We will cover some of the classical techniques for image segmentation in the following section, but the list is by no means complete. Descriptions of some of the newer and more advanced segmentation methods that have been proposed for nuclear medicine, particularly for PET, can be found in Foster et al. [26] and Hatt et al. [31].

10.6.1 THRESHOLDING

The most common (semi)-automatic segmentation methods used in nuclear medicine are different forms of thresholding, that is, the classification of a voxel as part of the object or as part of the background is based on the voxel values directly. Thresholding methods have, in many cases, the appeal of simplicity, but often suffer from poor performance, in particular at low contrast-to-noise ratios.

10.6.1.1 Fixed Threshold

For a fixed-threshold technique all voxels with values above a predefined value will be classified as part of the object, while voxels with values below this threshold will be classified as background. Sometimes the threshold is fixed in an absolute sense, but even more common is to use a fixed value relative to the maximum value in an initial region defined by the operator. Often, a value of 42% is quoted as optimal [32], but the generality of such a number should be treated with caution. The 'best' threshold for a given object is affected by a number of properties of the object, for example, its shape and volume, and the environment of the object, as well as the contrast relative to the background.

10.6.1.2 Thresholding Based on the Image Histogram

A more refined approach than the use of a fixed threshold or a threshold fixed relative to the maximum value is to adapt the threshold based on the image histogram. One simple such method is referred to as the Otsu method [33], which partitions the image into two regions (classes) based on the principle of minimizing the within-class variance, that is,

$$\sigma_w^2 = \omega_0 \sigma_0^2 + \omega_1 \sigma_1^2, \tag{10.43}$$

where ω_0 is the fraction of voxels classified as background, σ_0^2 is the variance of the background, ω_1 is the fraction of voxels classified as object, and σ_1^2 is the variance of the object class. Minimizing the within-class variance is in turn equivalent to maximizing the between-class variance. The result is that the two classes, as determined from the adapted threshold, are maximally separated in histogram space, as illustrated in Figure 10.9.

10.6.1.3 Iterative Thresholding

Since it has been noted that the optimal threshold for segmentation depends on the size of the object of interest and its contrast with the background, one possible way of improving thresholding is to characterize this relationship in order to construct a calibration curve for the threshold to be used. Since the true size of the object and the true contrast between the object and the background are unknown for a real situation, the mode of operations becomes iterative, with the estimated volume and contrast used to pick a threshold which, in turn, is used to estimate the size and contrast. Many

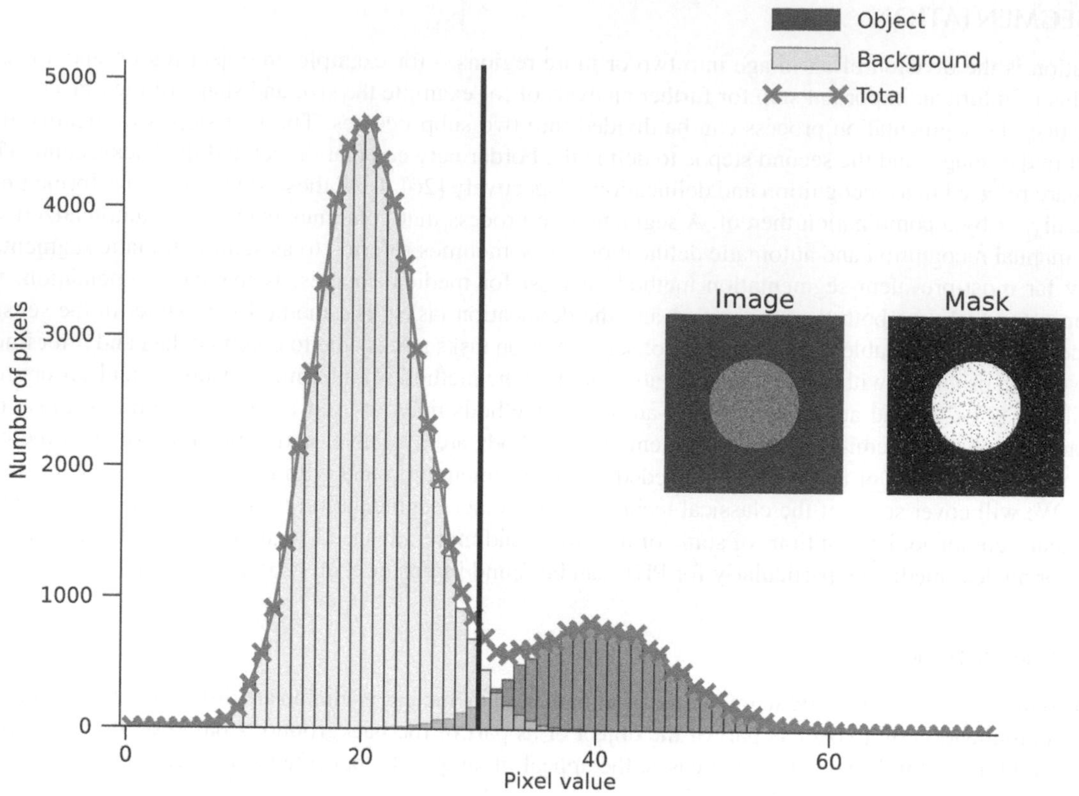

FIGURE 10.9 Segmentation of an image using the Otsu method. Underlying object is a circular disc with value 40 placed in a background with value 20. The noise in the image is Poisson distributed and hence the image histogram exhibits two peaks, one for the object and one for the background. The distribution of pixel values for the object and the background are shown separately together with their sum. The Otsu method tries to separate (the vertical line) the two classes of voxels as much as possible in the histogram. Note how the resulting mask suffers from inclusion of spurious pixels in the background and exclusion of spurious pixels in the object.

studies point to good results being achieved both for estimating the volume and the activity in regions using iterative thresholding [34–36], but the methods require calibration, which is a disadvantage.

10.6.1.4 Region Growing

One problem with thresholding techniques is that they only consider each voxel individually and do not consider the spatial relationship between voxels. Accordingly, there is no guarantee that the resulting region is connected, even if that is often a property that can be assumed about the underlying organ considered. A connected region can be enforced by a technique called region growing [37].

Region growing starts from one or more starting voxels, referred to as seeds. The neighbouring pixels are then considered and compared to some criterion for inclusion in the region. This criterion can for example be that the voxel value should be within a predefined window of values, or that the value should be within a certain interval of the average in the existing region. The process is repeated until the region has stabilized, that is, no more neighbouring pixels fulfil the criterion for inclusion. Since only neighbouring voxels are considered for inclusion in each step, the final region will automatically be connected.

10.6.2 Contour-based Methods

Rather than viewing the delineation problem as a classification problem of individual pixels as belonging to an object of interest or belonging to the background, the problem can be viewed as finding the boundary between the object and the background. Points inside the object boundary are then considered as belonging to the object and points outside the boundary are considered to belong to the background. Basic properties of such methods are what mathematical description is used for the contour or surface, what property of the image is used to attract the surface to the object boundary, and what method is used to optimize the surface parameters to conform to the object boundary as reflected by that image property.

10.6.2.1 Active Models

A classic example of a contour-based segmentation algorithm is so-called snakes [38], which is a contour that is allowed to evolve under the influence of external and internal forces. Hence, the problem of adapting the contour to the object of interest is transformed to an energy-minimization problem, with one external and one internal energy component. The external energy component makes the contour attracted to the boundary of the object of interest while the internal energy favours solutions with desirable properties, typically related to first and second derivatives of the contour, that is, the formation of discontinuities and edges are disfavoured in the minimization process. The external energy may, in principle, relate to any property of the image that is characteristic of the boundary of the object: for example the norm of the gradient of the image (i.e., a Sobel or Prewitt filtered version of the image).

A contour, or surface, that is gradually adapted to image features is referred to as active models, snakes being one of the best-known examples of active models. One of the most important properties of active models is the way the contour or surface is described. The most fundamental difference between contour descriptions is the difference between explicit and implicit surface descriptions. An explicit contour description has the form

$$C(s)=\big(x(s),y(s)\big), \tag{10.44}$$

that is, the contour is parametrized with a parameter s and the x- and y-coordinates explicitly described as a function of s. As opposed to explicit descriptions, an implicit contour description is of the form

$$C=\big\{(x,y);g(x,y)=0\big\}, \tag{10.45}$$

that is, the contour is defined as the set of points that fulfils a certain criterion. One example of an implicit curve description that has gained interest in the field of segmentation is level sets [30, 39], where the adaption of the contour to the object boundary is reformulated to the solution of a differential equation. One major advantage of level sets compared to explicitly defined contour descriptions is that the formulation intrinsically allows for topology changes as the contour evolves. Such automatic topology changes are difficult to accomplish with an explicit contour description.

10.6.3 Evaluation of Segmentation Methods

As for all method development and implementation, an important aspect of the development of segmentation methods is validation. The most realistic method evaluation is performed using real patient images, but the problem is that it is very difficult to establish a ground truth for those cases. Hence, evaluation based on patient images are often restricted to method comparisons, for example by comparing a new method to an established one, or studies of the variability between operators. As an alternative to patient images, phantom measurement or synthetic images are often used. For synthetic images, a ground truth of is often readily available, but the realism of the images can often be questioned.

10.6.3.1 Evaluation Metrices

A common metric for comparing regions, either between observers or with respect to a ground truth, is the volume agreement. The volume of a region is often a relevant parameter in itself but does not give a full description of the agreement between regions since it is quite possible for two volumes to agree while the underlying regions are quite different. Sometimes the centre-of-mass of regions are compared, thereby also indicating the global positions of the regions in combination with their size [40, 41]. While this is an improvement over using volume alone, it will still not reveal all the possible ways the two regions may differ.

A more refined measure for the agreement between two regions is the Dice Similarity Coefficient (DSC) [42], defined as

$$\mathrm{DSC} = 2\frac{|\mathrm{A} \cap \mathrm{B}|}{|\mathrm{A}| + |\mathrm{B}|}, \tag{10.46}$$

where A is the set of points corresponding to the first region, B is the set of points corresponding to the second region, and $|\cdot|$ denotes the cardinality of a set. The numerator is the volume of the overlap between the two regions and the denominator is the sum of the two regions individually. Hence, the DSC has a value between 0 and 1, where 0 indicates no overlap between the regions and 1 means that the two regions agree fully.

10.7 CONCLUDING REMARKS

While the current description of image processing is by no means complete, the hope is that it will serve as an introduction and overview of the aspects most important to nuclear-medicine imaging. The ultimate aim of the presented procedures is to improve the value of the acquired images, for example in terms of the ability for a human observer to confirm or dismiss the presence of a certain condition in a patient. The ability of a certain technique – for example, low-pass filtering – to provide such an advantage is not always straightforward, which requires dedicated study for the particular application.

REFERENCES

[1] R. N. Bracewell, *The Fourier Transform and Its Applications*, 3rd ed. Singapore: McGraw-Hill, 2000.

[2] E. O. Brigham, *The Fast Fourier Transform and Its Application*. Englewood Cliffs, NJ: Prentice-Hall, 1988.

[3] R. Hummel, "Image-Enhancement by Histogram Transformation," (in English), *Comput Vision Graph*, vol. 6, pp. 184–195, 1977, doi:10.1016/S0146-664x(77)80011-7.

[4] D. Coltuc, P. Bolon, and J. M. Chassery, "Exact Histogram Specification," (in English), *IEEE T Image Process*, vol. 15, pp. 1143–1152, May 2006, doi:10.1109/Tip.2005.864170.

[5] R. A. Hummel, "Histogram Modification Techniques," *Comput Vision Graph*, vol. 4, pp. 209–224, 1975.

[6] J. M. S. Prewitt, "Object Enhancement and Extraction," in *Picture Processing and Psychopictorics*, B. Lipkin and A. Rosenfeld Eds. New York: Academic Press, 1970, pp. 75–149.

[7] L. G. Roberts, "Machine Perception of Three-dimensional Solids," Doctoral thesis, Massachusetts Institute of Technology, 1963.

[8] H. Wilbraham, "On a Certain Periodic Function," *The Cambridge and Dublin Mathematical Journal*, vol. III, pp. 198–201, 1848.

[9] J. W. Gibbs, "Fourier's Series," *Nature*, vol. 59, p. 606, 1899.

[10] S. Butterworth, "On the Theory of Filter Amplifiers," *Exp Wirel Wirel Eng*, vol. 7, pp. 536–541, 1930.

[11] M. Frigo and S. G. Johnson, "The Design and Implementation of FFTW3," (in English), *P IEEE*, vol. 93, pp. 216–231, Feb 2005, doi:10.1109/Jproc.2004.840301.

[12] J. W. Cooley and J. W. Tukey, "An Algorithm for the Machine Calculation of Complex Fourier Series," *Math Comput*, vol. 19, pp. 297–301, 1965.

[13] B. R. Frieden, "A New Restoring Algorithm for the Preferential Enhancement of Edge Gradients," *J Opt Soc Am*, vol. 66, pp. 280–283, 1976.

[14] I. Pitas and A. N. Venetsanopoulos, "Order Statistics in Digital Image Processing," (in English), *P IEEE*, vol. 80, pp. 1893–1921, Dec 1992, doi:10.1109/5.192071.

[15] J. B. Bednar and T. L. Watt, "Alpha-Trimmed Means and Their Relationship to Median Filters," (in English), *IEEE T Acoust Speech*, vol. 32, pp. 145–153, 1984, doi:10.1109/Tassp.1984.1164279.

[16] A. C. Bovik, "Streaking in Median Filtered Images," (in English), *IEEE T Acoust Speech*, vol. 35, pp. 493–503, Apr 1987, doi:10.1109/Tassp.1987.1165153.

[17] J. S. Lee, "Digital Image Enhancement and Noise Filtering by Use of Local Statistics," (in English), *IEEE T Pattern Anal*, vol. 2, pp. 165–168, 1980, doi:10.1109/Tpami.1980.4766994.

[18] J. S. Lee, "Refined Filtering of Image Noise Using Local Statistics," (in English), *Comput Vision Graph*, vol. 15, pp. 380–389, 1981, doi:10.1016/S0146-664x(81)80018-4.

[19] R. C. Gonzales and R. E. Woods, *Digital Image Processing*, 4th ed. New York: Pearson, 2018.

[20] J. W. Wallis and T. R. Miller, "An Optimal Rotator for Iterative Reconstruction," *IEEE T Med Imaging*, vol. 16, pp. 118–123, Feb 1997, doi:10.1109/42.552061.

[21] C. E. Shannon, "Communication in the Presence of Noise," (in English), *P Ire,* vol. 37, pp. 10–21, 1949, doi:10.1109/Jrproc.1949.232969.

[22] T. M. Lehmann, C. Gönner, and K. Spitzer, "Survey: Interpolation Methods in Medical Image Processing," (in English), *IEEE T Med Imaging*, vol. 18, pp. 1049–1075, Nov 1999, doi:10.1109/42.816070.

[23] J. A. Parker, R. V. Kenyon, and D. E. Troxel, "Comparison of Interpolating Methods for Image Resampling," *IEEE T Med Imaging*, vol. 2, pp. 31–39, 1983.

[24] S. S. Rifman and D. M. McKinnon, *Evaluation of Digital Correction Techniques for ERTS Images; Final Report*, TRW Systems Group, Redondo Beach. TRW report: 20634-6003-TU-DO, 1974.

[25] S. K. Park and R. A. Schowengerdt, "Image Reconstruction by Parametric Cubic Convolution," (in English), *Computer Vision Graphics and Image Processing*, vol. 23, pp. 258–272, 1983, doi:10.1016/0734-189x(83)90026-9.

[26] B. Foster, U. Bagci, A. Mansoor, Z. Xu, and D. J. Mollura, "A Review on Segmentation of Positron Emission Tomography Images," *Comput Biol Med*, vol. 50, pp. 76–96, Jul 2014, doi:10.1016/j.compbiomed.2014.04.014.

[27] S. L. Breen et al., "Intraobserver and Interobserver Variability in GTV Delineation on FDG-PET-CT Images of Head and Neck Cancers," *Int J Radiat Oncol*, vol. 68, pp. 763–770, Jul 1, 2007, doi:10.1016/j.ijrobp.2006.12.039.

[28] H. Vorwerk et al., "The Delineation of Target Volumes for Radiotherapy of Lung Cancer Patients," *Radiother Oncol*, vol. 91, pp. 455–460, Jun 2009, doi:10.1016/j.radonc.2009.03.014.

[29] K. Gurleyik and E. M. Haacke, "Quantification of Errors in Volume Measurements of the Caudate Nucleus Using Magnetic Resonance Imaging," (in English), *J Magn Reson Imaging*, vol. 15, pp. 353–363, Apr 2002, doi:10.1002/jmri.10083.

[30] P. A. Yushkevich et al., "User-guided 3D Active Contour Segmentation of Anatomical Structures: Significantly Improved Efficiency and Reliability," (in English), *Neuroimage*, vol. 31, pp. 1116–1128, Jul 1, 2006, doi:10.1016/j.neuroimage.2006.01.015.

[31] M. Hatt et al., "Classification and Evaluation Strategies of Auto-segmentation Approaches for PET: Report of AAPM Task Group No. 211," (in English), *Med Phys*, vol. 44, pp. E1–E42, Jun 2017, doi:10.1002/mp.12124.

[32] Y. E. Erdi, B. W. Wessels, M. H. Loew, and A. K. Erdi, "Threshold Estimation in Single Photon Emission Computed Tomography and Planar Imaging for Clinical Radioimmunotherapy," *Cancer Res*, vol. 55, no. 23 Suppl, pp. 5823s–5826s, Dec 1 1995. [Online]. Available: www.ncbi.nlm.nih.gov/pubmed/7493353.

[33] N. Otsu, "A Threshold Selection Method from Gray-level Histograms," *IEEE T Syst Man Cyb*, vol. 9, pp. 62–66, 1979.

[34] J. Grimes, A. Celler, S. Shcherbinin, H. Piwowarska-Bilska, and B. Birkenfeld, "The Accuracy and Reproducibility of SPECT Target Volumes and Activities Estimated Using an Iterative Adaptive Thresholding Technique," (in English), *Nucl Med Commun*, vol. 33, pp. 1254–1266, Dec 2012, doi:10.1097/MNM.0b013e3283598395.

[35] W. Jentzen, L. Freudenberg, E. G. Eising, M. Heinze, W. Brandau, and A. Bockisch, "Segmentation of PET Volumes by Iterative Image Thresholding," (in English), *J Nucl Med*, vol. 48, pp. 108–114, Jan 2007. [Online]. Available: <Go to ISI>://WOS:000243306800018.

[36] M. Pacilio et al., "An Innovative Iterative Thresholding Algorithm for Tumour Segmentation and Volumetric Quantification on SPECT Images: Monte Carlo-based Methodology and Validation," (in English), *Med Phys*, vol. 38, pp. 3050–3061, Jun 2011, doi:10.1118/1.3590359.

[37] R. Adams and L. Bischof, "Seeded Region Growing," (in English), *IEEE T Pattern Anal*, vol. 16, pp. 641–647, Jun 1994, doi:10.1109/34.295913.

[38] M. Kass, A. Witkin, and D. Terzopoulos, "Snakes: Active Contour Models," *Int J Comput Vision*, vol. 1, pp. 321–331, 1988.

[39] V. Caselles, R. Kimmel, and G. Sapiro, "Geodesic Active Contours," (in English), *Int J Comput Vision*, vol. 22, pp. 61–79, Feb-Mar 1997, doi:10.1023/A:1007979827043.

[40] E. Berthelet, M. C. C. Liu, A. Agranovich, K. Patterson, and T. Currie, "Computed Tomography Determination of Prostate Volume and Maximum Dimensions: A Study of Interobserver Variability," (in English), *Radiother Oncol*, vol. 63, pp. 37–40, Apr 2002, doi:Pii S0167-8140(02)00026-9. doi:10.1016/S0167-8140(02)00026-9.

[41] A. M. van Mourik, P. H. M. Elkhuizen, D. Minkema, J. C. Duppen, and C. van Vliet-Vroegindeweij, and D. Y. Boost, "Multiinstitutional Study on Target Volume Delineation Variation in Breast Radiotherapy in the Presence of Guidelines," (in English), *Radiother Oncol*, vol. 94, pp. 286–291, Mar 2010, doi:10.1016/j.radonc.2010.01.009.

[42] L. R. Dice, "Measures of the Amount of Ecologic Association between Species," (in English), *Ecology,* vol. 26, pp. 297–302, 1945, doi:10.2307/1932409.

11 Machine Learning

Karl Åström

CONTENTS

Machine learning, a subfield of artificial intelligence, is the study of computer algorithms that improve automatically through experience. Although the term was coined in 1959, see [1], machine learning builds on questions/methods that were developed earlier in linear algebra, mathematical analysis, optimization and mathematical statistics.

In this chapter we will give a brief overview of machine learning. In particular, we will explain the main types of machine problems, provide a summary of some of the methods used, and how they can be applied to imaging problems.

One definition of machine learning, attributed to Tom M. Mitchell, is that 'A computer program is said to learn from experience E with respect to some class of tasks T and performance measure $\mathbb{P}$ if its performance at tasks in T, as measured by $\mathbb{P}$, improves with experience E', see [2]. This is a definition that works well, and it highlights that in order to use machine learning for a new task one has to work on defining what the task is, what experience E should be, and how the performance measure $\mathbb{P}$ should be defined.

The research in machine learning is vast. In order to keep the text in this chapter reasonably short, we concentrate on a brief overview of a few core topics – regression, classification, dimensionality reduction, and autoencoders. The chapter is structured as follows. First, we will briefly discuss a few basic regression and classification techniques, for example linear least squares regression (Gauss 1809). Then we introduce a few key concepts, such as Bayes Theorem (1753), prior and posterior distributions. Some machine-learning algorithms are based on estimating the likelihood function and using Bayes theorem to obtain the posterior distribution, but most can be viewed as trying to model the posterior distribution directly. This leads to linear logistic regressions, shallow neural networks, for example, the perceptron (1957), artificial neural networks (≈1980) and, finally, convolutional neural networks, (first used around 1990, but massive progress since 2011).

Machine-learning methods involve estimation of parameters, often using optimization. Determining what method to use and determining the so-called hyper-parameters require additional consideration in terms of over- and under-fitting

DOI: 10.1201/9780429489556-11

and the need to divide data into three parts denoted training, validation, and test. The progress within machine learning has been swift during the last decade. The techniques for designing computational architectures for solving different types of machine-learning problems have improved significantly, and researchers have shown considerable ingenuity on how to apply these new tools to a vast array of applications.

11.1 SUPERVISED LEARNING

In supervised learning the task T is to learn a function $f: X \mapsto Y$, from experience E = $(x_1,y_1),\ldots,(x_N,y_N)\}$ (or training data) consisting of many pairs of arguments x and values y. The aim is to develop methods that produce a function f so that $y_i \approx f(x_i)$, and so that the function generalizes well to unseen data, despite both uncertainties in the model and noise in the measurements. Therefore, supervised learning can be viewed as function approximation. There are two types of supervised learning, depending on the type of output space Y. Here **regression** is used for problems where Y is continuous and **classification** is used for problems where Y is discrete.

11.1.1 Least Squares Regression

For supervised learning, if the output space Y is continuous, the process is called regression. A simple example would be to find a function $f: \mathbb{R} \mapsto \mathbb{R}$.

Linear least squares regression uses a model function $f_w(x) = w_1 x + w_0$ and a performance (error metric) $\mathbb{P}(w) = \sum_{i=1}^{N} (f_w(x_i) - y_i)^2$. Minimizing $\mathbb{P}$ with respect to w is an ordinary least squares problem, which is a convex optimization problem [3, 4], for which there are efficient methods. The theory and methods for the least squares method, goes back to the developments of Gauss and Legendre, see [5, 6]. The model $f_w(x) = w_1 x + w_0$ is linear both in x and also in the unknown parameters $w = (w_0, w_1)$. The estimation of the parameter is easy if the model is linear in the parameter, even if it is not linear in x. Thus, it is possible to use a set of k possibly non-linear basis functions $\phi_j: X \mapsto Y$ and model $f_w(x) = \sum_{j=1}^{k} w_j \phi_j(x)$. Using the same performance metric above again leads to an ordinary least squares problem, which is convex and easy to solve. One choice here is to use a monomial basis, for example, choose $\phi_j(x) = x^{j-1}$, which tries to approximate the function f as a polynomial of order $k - 1$. Other popular choices are to use trigonometric functions. Linear least squares regression is optimal if the measurement model is $y_i = f_w(x_i) + \epsilon_i$, where the measurement errors are independent with Gaussian distribution of equal variance.

11.1.2 Classification and Bayes Theorem

Supervised learning is called classification if the output space Y is discrete. A simple example would be to classify into two classes based on one measurement. For a two-class case it is often convenient to denote the classes -1 and 1, that is, we are trying to find a classifier $f: \mathbb{R} \mapsto \{-1,1\}$.

We illustrate this with Example 1 – a two-class classification problem based on one real measurement. We use data from the classic paper by the statistician Ronald Fisher [7]. In the paper there is an example with a dataset of 50 samples of three kinds of flowers. For each flower one measures the length and width of both the sepal and petal and notes which species it is. In this illustration we study only one of the features (petal length) and we restrict ourselves to two of the species – Iris Setosa (class 1) and Iris Versicolor (class -1). In this case, the training data, or experience, consists of petal length measurements x_i and ground truth species y_i – coded as 1 or -1.

We assume that the **prior probability** is equal for two classes, that is, that $P(Y = 1) = P(Y = -1) = 0.5$. We are interested in estimating the **posterior probability** $P(Y = 1 | x)$ and $P(Y = -1 | x)$, that is, for each x what is the probability that it comes from class 1 or -1. Given this probability, we can choose to classify as class 1 if $P(Y = 1 | x) > 0.5$ and -1 otherwise. The posterior is often calculated using the **Bayes theorem**, see [8],

$$\underbrace{P(Y=1|x)}_{posterior} = \frac{\overbrace{P(Y=1)}^{prior}\,\overbrace{fx(x|Y=1)}^{likelihood}}{\underbrace{fx(x)}_{evidence}}. \tag{11.1}$$

The training data, as shown in Figure 11.1 can be used to estimate the probability density function (the likelihood) $f_X(x|Y = 1)$, that is, how probable it is that the petal length is x, given that the flower is of species $y = 1$ (Iris Setosa).

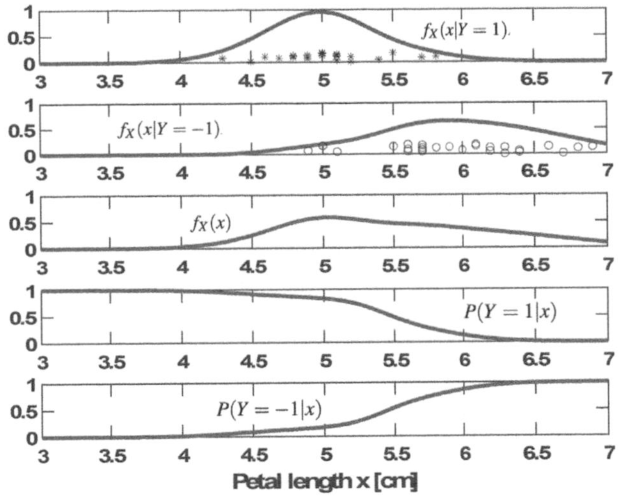

FIGURE 11.1 The figure illustrates the use of Bayes Theorem and probability density estimation to train a classifier. The two top rows illustrate data from the two classes and the estimated likelihoods. The middle row shows the total likelihood, and the two bottom rows show the estimated posterior distributions for the two classes.

This is illustrated in the top part of Figure 11.1, where such measurements are shown. There are many ways to estimate probability density functions, for example, (i) using histograms, (ii) parametric probability density estimation, and (iii) non-parametric probability density estimation. In this example we use a non-parametric method called kernel density estimation, see [9, 10]. The method is based on choosing a kernel, often a Gaussian kernel,

$$K(x) = \frac{1}{\sqrt{2\pi}} e^{-x^2/2}. \tag{11.2}$$

The kernel density estimation $\hat{f}_h$ of f is defined as

$$\hat{f}_h(x) = \frac{1}{n} \sum_{i=1}^{N} \frac{1}{h} K\left(\frac{x - x_i}{h}\right), \tag{11.3}$$

where the scale/width h is a so-called *hyper-parameter*. A larger h corresponds to more smoothing and regularization, which makes the estimate less sensitive to noise, but makes it difficult to capture details of the probability density. On the other hand, using a small h makes it possible to capture more details in the probability density estimation, but the method then becomes sensitive to noise. Typically, it would also need more training data.

Using kernel density estimation we obtain the estimate of $f_X(x|Y = 1)$ as shown in the figure. Using the same technique we have estimated $f_X(x|Y = -1)$ for Iris Versicolor. The analysis shows that petal length is often a bit shorter for Iris Setosa than for Iris Versicolor. By summing $f_X(x|Y = 1)P(Y = 1)$ and $f_X(x|Y = -1)P(Y = -1)$ we obtain the total probability or evidence $f(x)$. Using Bayes Theorem we then obtain $P(Y = 1|x)$ and $P(Y = -1|x)$ as illustrated in the figure. The analysis shows that we should classify the flower as Iris Setosa if the petal length is lower than 5.5.

11.1.3 Classification and Logistic Regression

Many classifiers do not explicitly model the likelihoods, that is, $f_X(x|Y = 1)$, but can be seen as modelling the posterior probabilities, that is, $P(Y = 1|x)$ directly. In the example of Figure 11.1 these posterior probabilities resemble smooth step functions. The method of linear logistic regression can be viewed as one such method, where one models the posterior as a combination of a linear function and the sigmoid (or standard logistic) function,

$$s(t) = \frac{1}{1 + e^{-t}}. \tag{11.4}$$

Using a vector $x \in \mathbb{R}^n$ as input, the model of the posterior probability is

$$P(Y=1,x) = f_{(w,w_0)}(x) = s\left(w^T x + w_0\right), \tag{11.5}$$

where the vector $w \in \mathbb{R}^n$ and the scalar bias w_0 are unknown model parameters. A maximum likelihood estimate of these parameters, using experience (or training data) $E = (x_1,y_1),\ldots,(x_N,y_N)\}$ leads to an optimization problem, where the loss

$$L\left(w,w_0\right) = \sum_{i=1}^{N} \log\left(1+e^{y_i\left(w^T x_i + w_0\right)}\right) \tag{11.6}$$

is minimized. For linear least squares problems, there exists a closed form solution for establishing the parameters, but here there is no closed form solution for (w,w_0). However, the optimization problem is convex, and there are fast and efficient techniques for calculating the optimum using optimization.

Sometimes a regularization term, for example, w^Tw, is added to the loss function. This gives smaller weights and in practice often better generalization to unseen data. Interestingly another machine-learning technique, the support vector machine [11], which uses the principle of maximal margin, gives a loss function w^Tw, if the two classes are separable. If they are not separable, one has to add a regularization term $max(0,y_i(w^Tx_i + w_0))$. Here the regularization term bears a slight resemblance to the original loss for the logistic regression and the original loss for support vector machine is the regularization term for the logistic regression. For support vector machines there is a rich theory showing how the methods generalize to unseen data [12].

For logistic regression and support vector machines, the final classifier is

$$f(x) = \begin{cases} 1 & \text{if } w^T x + w_0 >= 0, \\ -1 & \text{otherwise.} \end{cases} \tag{11.7}$$

Thus, the decision boundary, that is, the set $\{x|w^Tx_i + w_0 = 0\}$, which forms the boundary between the two classes, is a plane.

The support vector machine method can also be extended to non-linear classification, by using kernels, see [11, 12]. The main idea is to first map the original features x using a non-linear function $\phi(x)$, and then apply a linear map on the output. The classifier is then

$$f(x) = \begin{cases} 1 & \text{if } w^T \phi(x) + w_0 >= 0, \\ -1 & \text{otherwise.} \end{cases} \tag{11.8}$$

The history of linear logistic regression is described in [13]. Extensions of the linear logistic regression include (i) replacing the linear part with a non-linear function and (ii) increasing the number of classes. For more than two classes the sigmoid is replaced with the **softmax function**, $S: \mathbb{R}^n \mapsto \mathbb{R}^n$, where

$$S_i(x) = \frac{e^{x_i}}{\sum_{j=1}^{n} e^{x_j}}, \tag{11.9}$$

where S_i denotes element i of the vector S. Notice that for any input vector x, the output of the softmax function $S(x)$ has elements that are positive and sum to 1.

11.1.4 Classification Accuracy Versus Loss Function

There is one natural performance measure of a classifier, that is, the *classification accuracy* or the percentage of correctly classified examples. This involves the classifier

$$y = \arg\max_{y} P(Y = y, x), \tag{11.10}$$

where $P(Y = y,x)$ is modelled as

$$f(x, w), \tag{11.11}$$

where w are the unknown parameters. The performance $\mathbb{P}(w)$ in terms of how many correct classifications we make divided by the number of classifications, depends on $f(x,w)$, which depends on the parameters w. This **performance metric** $\mathbb{P}(w)$ is, however, difficult to use because the number of correctly classified examples is a discrete function. The performance measure is thus typically piecewise constant as a function of the parameters w. Most infinitesimal changes do not push any of the training examples across a decision boundary. This means that we cannot use derivatives of $\mathbb{P}$ with respect to w in traditional optimization to update the parameters. Therefore, it is common to use both a **performance metric** $\mathbb{P}(w)$ and a **loss function** $\mathrm{L}(w)$. The loss function $\mathrm{L}(w)$ is smooth and is usually directly connected to the posterior distribution $P(Y = y|x)$. For visualization purposes both loss and performance are sometimes plotted during training. We can think of the loss function L as a proxy for estimating the parameters and hope that by minimizing loss $\mathrm{L}(w)$ we will obtain a classifier that gives high performance $\mathbb{P}$.

11.2 ARTIFICIAL NEURAL NETWORKS

The Artificial Neural Network can be viewed as a generalization of the linear logistic regression. The main building blocks of an artificial neural network are

1. Affine functions $\mathrm{A}_{w,m,n}$ from $\mathbb{R}^m$ to $\mathbb{R}^n$. An affine function from $\mathbb{R}^m$ to $\mathbb{R}^n$ is $\mathrm{A}_{w,m,n}(x) = Ax + b$. The weights w consists of the $m \times n$ entries of the matrix A as well as the n entries of the vector b.
2. Activation functions T from $\mathbb{R}^n$ to R^n.
3. The softmax function S, given by Eq. 11.9.

The activation functions T have no parameters, but they are important since they introduce non-linearity to the classifier. There are many popular choices, for example, the sigmoid function from Eq. 11.4, Heaviside step function, the hyperbolic tangent activation function tanh, the rectified linear unit (ReLU) max(x,0). These are applied element-wise to the input.

A feed-forward artificial neural network consists of several layers of affine and activation functions followed by a softmax function. Assuming that there are m dimensions of the input data and n classes, an example of an artificial neural network is

$$f_w(x) = \mathrm{S} \circ \mathrm{A}_{w_3,n_2,n} \circ \mathrm{T} \circ \mathrm{A}_{w_3,n_1,n_2} \circ \mathrm{T} \circ \mathrm{A}_{w_3,m,n_1}, \tag{11.12}$$

where $\circ$ denotes function composition. For this network architecture there is first an affine mapping from m dimensions to n_1, followed by a non-linear activation function. The result is a vector with n_1 elements. Then follows another affine function from n_1 dimensions to n_2 dimensions, followed by a non-linear activation function. The final results are obtained by a third affine mapping from n_2 dimensions to n dimensions, followed by the softmax function. The output vector $f_w(x)$ consists of n positive numbers that sum to 1. The parameters w of the artificial neural network consists of all the parameters of the affine layers, that is, $w = (w_1, w_2, w_3)$.

Optimization of the parameters w is done by minimizing the loss function L. An example is to minimize the negative log likelihood of the correct class, that is, assuming training data $\mathrm{E} = \{(x_1,y_1),\ldots,(x_N,y_N)\}$, where $x_i \in \mathbb{R}^m$ and $y_i \in \{1,2,\ldots,n\}$.

$$\mathrm{L}(w, \mathrm{E}) = \sum_{i=1}^{N} -\log f_w(x_i)_{y_i}, \tag{11.13}$$

where $f_w(x)_j$ denotes element j in the output vector $f_w(x) \in \mathbb{R}^n$. The idea here is to view the output vector as a model of the posterior probabilities for the n classes, that is, $f_w(x)_j = P(Y = j|X = x)$. We would like to have a high probability for the correct class which translates to a low negative logarithm of the probability. Contrary to the previous examples, the optimization

$$\min_{w} \mathrm{L}(w, \mathrm{E}) \tag{11.14}$$

is typically non-convex. The function L often has many local minima and there are no guarantees that we will find the global optimum. Optimization is done by iterative techniques, for example, by starting with random weights and using variants of the steepest descent method.

11.3 CONVOLUTIONAL NEURAL NETWORKS

One problem with applying a general artificial neural network to the analysis of images is that such networks would have too many weights. Images often contain millions of pixel values. A general affine function from one million of features to one million of features requires one trillion weights, which is not feasible. Another issue is that often one would like to perform similar calculations for all parts of the image, that is, it is sometimes natural to have a linear operation that is also translation invariant. The so-called **Convolutional Neural Networks** can be seen as a generalization of the artificial neural network to images, where the affine functions above have been replaced with convolution and bias term. In a convolutional neural network, the data sent between the different computational parts are data blocks with two spatial dimensions and one channel dimension. Colour image is one example, which typically has three channels (red, green, and blue). We will use the notation $\mathbb{R}^{m \times n \times k}$ to denote a *m* by *n* image with *k* channels. The main building blocks of a convolutional neural network are

1. Convolutions that take a data block from $\mathbb{R}^{m \times n \times k}$ to $\mathbb{R}^{m' \times n' \times k'}$.
2. Activation functions. These are used element-wise on each element of the datablock. Thus, the spatial dimensions and the number of channels will not change. They map a block $\mathbb{R}^{m \times n \times k}$ to $\mathbb{R}^{m \times n \times k}$.
3. Max pooling functions. These return the same number of channels, but the spatial dimensions are typically halved. They map a block $\mathbb{R}^{m \times n \times k}$ to $\mathbb{R}^{m/2 \times n/2 \times k}$.
4. Softmax functions. These are usually applied along the channel direction. They map a block $\mathbb{R}^{m \times n \times k}$ to $\mathbb{R}^{m \times n \times k}$.

A typical classification network consists of a series of steps in which each step consists of a convolution, an activation function, and max pooling. After all of the steps, the output is of size $1 \times 1 \times n$. After a final softmax layer, these n numbers are again interpreted as posterior probabilities and used for classification. Each of these building blocks is usually called a *layer* of a network. We construct the architecture of a network by specifying the different layers. We illustrate this with two examples. Classification of grayscale digits are shown in Example 2 and of colour images for prostate cancer in Example 3.

Example 2.

MNIST classification.
The MNIST database (Modified National Institute of Standards and Technology database) is a set of images of handwritten digits that are often used for machine learning. The database is a continuation of an previous database from NIST. The database consists of 60,000 training examples and 10,000 test examples. Each example consists of an input image x_i of size $28 \times 28 \times 1$ and a label $y_i \in \{0,1,\ldots 9\}$. For this example, we use a deep-learning architecture similar to that presented in [14], that is, a series of convolution, activation function, max pooling followed by softmax and classification based on highest probability of the 10 output values. In Figure 11.2 (top-left) is shown 25 examples out of the 60,000 from the training set. Convolutional neural networks for classification often have a computational pipeline, where the spatial resolution becomes coarser and coarser. In LeCun's paper on MNIST digit recognition, the input was of size $28 \times 28 \times 1$ and in the different steps of the CNN the size of the data changes to $24 \times 24 \times 4$, $12 \times 12 \times 4$, $8 \times 8 \times 12$, $4 \times 4 \times 12$ and finally $1 \times 1 \times 10$. This architecture is visualized in Figure 11.2 (top-right). During optimization, at each iteration a batch of 2,000 examples are used to calculate the gradient of the loss function with respect to the parameters and then update those parameters. It thus takes 30 iterations

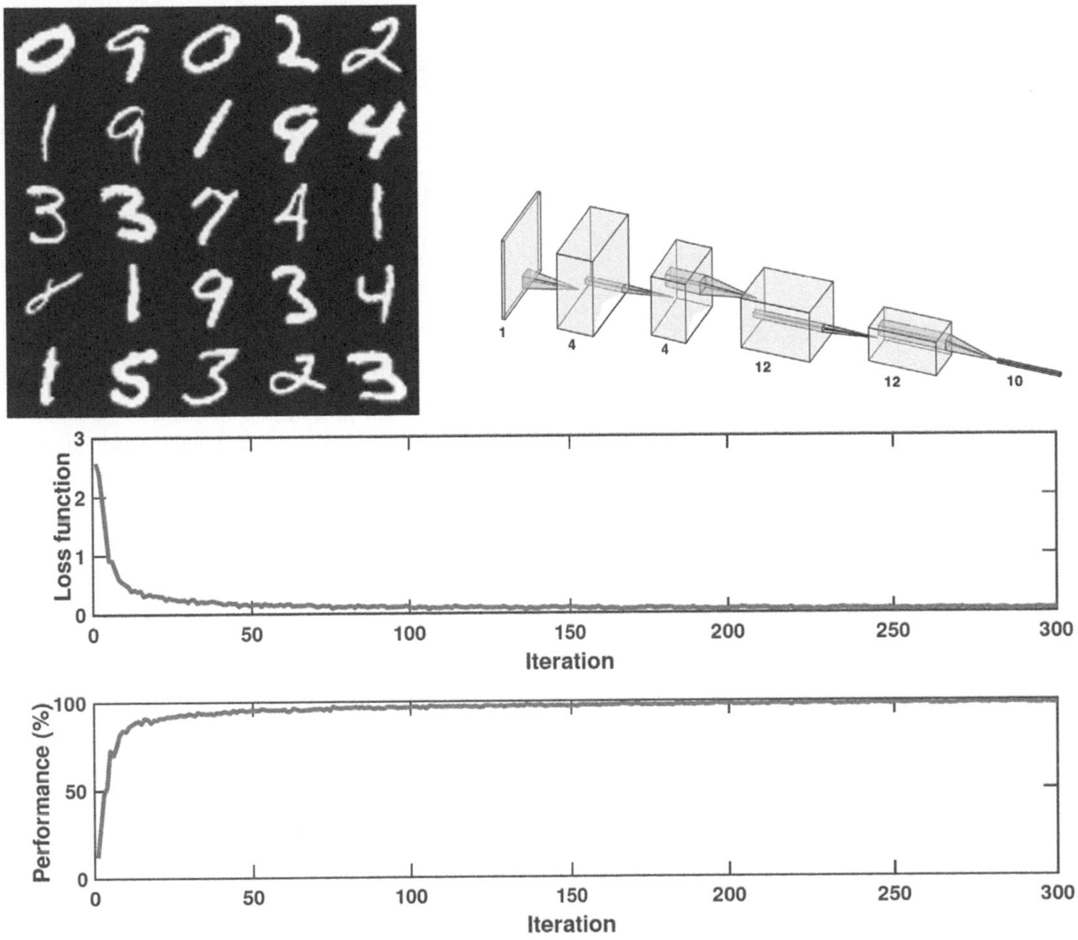

FIGURE 11.2 The figure shows 25 examples of 28 × 28 pixel grayscale images from the MNIST database (top left), the network architecture of the CNN (top right) and the progression of both loss L(w) and performance $\mathbb{P}(w)$ in per cent during optimization of parameters w during training (bottom).

to go through the whole training set, completing what is called an epoch. In this example we ran the optimization for 10 epochs. In the figure is shown the evolution of the loss function L, see Eq. 11.13, the evolution of the performance $\mathbb{P}$ of the machine-learning method (measured as the percentage of examples that are correctly classified) during these 300 iterations. The final classifier after 10 epochs of training made 98.4 per cent correct classifications on the training set and 98.4 per cent correct classifications on the test set. The fact that the performance on the test set is similar to that of the training set is an indication that there was not considerable over-training, see Section 11.4.

Example 3.

Prostate cancer grading
Cancer in the prostate gland is often detected through a process of screening, using for example prostate-specific antigen (PSA) testing, but the diagnosis is confirmed by pathologists based on ocular inspection of prostate biopsies in order to classify them according to the Gleason score, see [15]. In [16], a system for an automatic Gleason score is developed. The grading is based on digital images of prostate biopsies that have been stained to enhance certain structures in the biopsies. A convolutional neural network was trained on cut-outs with 318 × 318 pixels with three colour channels. An example is shown in Figure 11.3 (left). The architecture of the classifier is shown in the figure. The final classifier achieved about 93 per cent correctly classified samples.

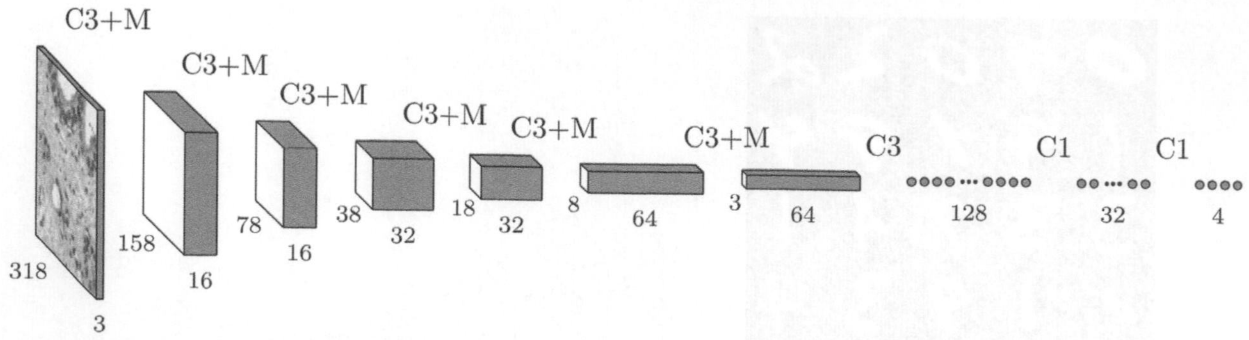

FIGURE 11.3 Network architecture of the CNN for the analysis of images of HEP-stained biopsies of the prostate. Similar to the MNIST example, we can see that the spatial dimensions become gradually coarser. The output are four numbers that we interpret as the posterior probabilities $P(Y = j|x)$ for the four classes (Benign, Gleason grade 3, Gleason grade 4 and Gleason grade 5).

11.4 THE BIAS–VARIANCE TRADE-OFF

A fundamental problem in machine learning is the bias-variance trade-off.

Bias/Underfitting: On one hand, a model with few parameters will not be able to capture the structure in the data even if there is a large training set. This error, the bias error, is caused by erroneous (simplified) assumptions in the learning algorithm. This can cause an algorithm to miss relevant relations between features and target outputs. This is often called underfitting.

Variance/Overfitting: On the other hand, if the model has too many parameters and if the training set is too small, the estimated model will be sensitive to small fluctuations in the training set. This can cause the machine-learning method to fit to noise in the data or to fit to the specific examples of the training data. This is the variance problem and is often called overfitting.

Ideally, one would like to have a method or model that generalizes well to unseen data by being capable of both capturing details in the machine-learning problem and also being insensitive to noise and limited training data. Unfortunately, this is impossible to achieve. There is a fundamental trade-off here between bias and variance.

Often there is a choice of many machine-learning methods or models. For each method there are parameters, w, that are optimized from training data, but there might also be hyperparameters. A common strategy is to divide data into three sets, the *training*, *validation*, and *test* sets. Here the training dataset is used to fit the parameters of a model, as described in the previous sections. This might be done for several choices of hyperparameters and/or models. These trained models are then evaluated on the validation dataset. It provides an unbiased evaluation of the model and can be used to select the model/hyperparameter that performs best on the validation dataset. This process results in a single model, which can be evaluated on the test dataset, sometimes also called the holdout dataset.

Several principles have been developed for selecting the model that makes a good trade-off for the bias-variance problem. See for example the Akaike information criterion, the Bayesian information criterion and also Minimum description length.

11.5 IMAGE-TO-IMAGE ARCHITECTURES

Convolutional neural networks for classification often have a computational pipeline, where the spatial resolution becomes coarser and coarser, as in Examples 2 and 3. There are many applications where it would be interesting if the output could have similar spatial dimensions as the input. To design such architectures, we need an upsampling layer. One example of upsampling is a process whereby, for example, each pixel could be replicated to four pixels. Thus, the output would have the double resolution in both spatial dimensions. A slightly more general upsampling can be achieved by the so-called **transposed convolution**, where a kernel with weights specifies how the value of each input pixel should be added to neighbouring pixels in the upsampled output. We illustrate this in the three following examples.

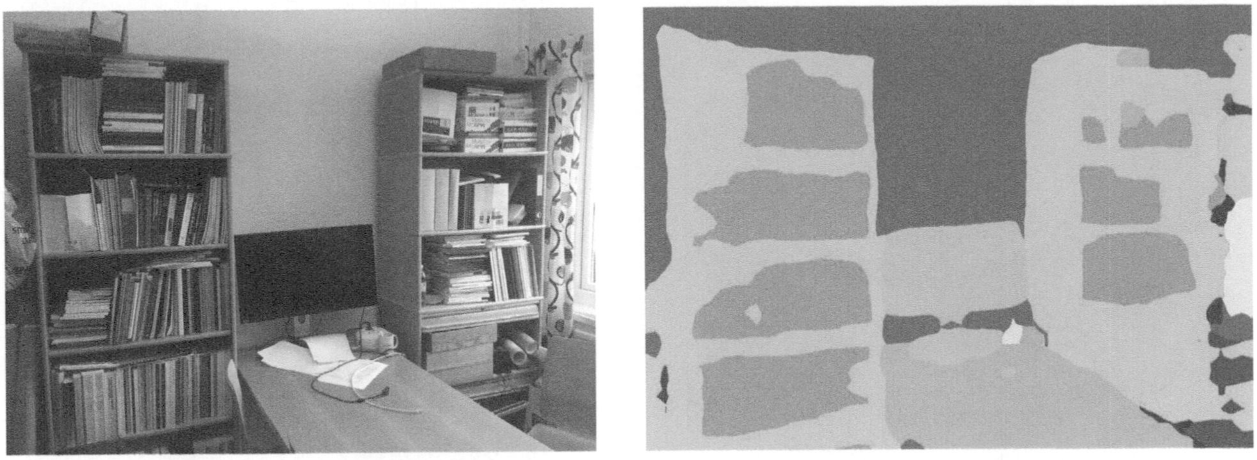

FIGURE 11.4 The figure shows input image (left) and output (right). Each pixel has been classified as one of 20 classes, for example: book, shelf, table, monitor, wall, and so forth. Colour image available at www.routledge.com/9781138593268.

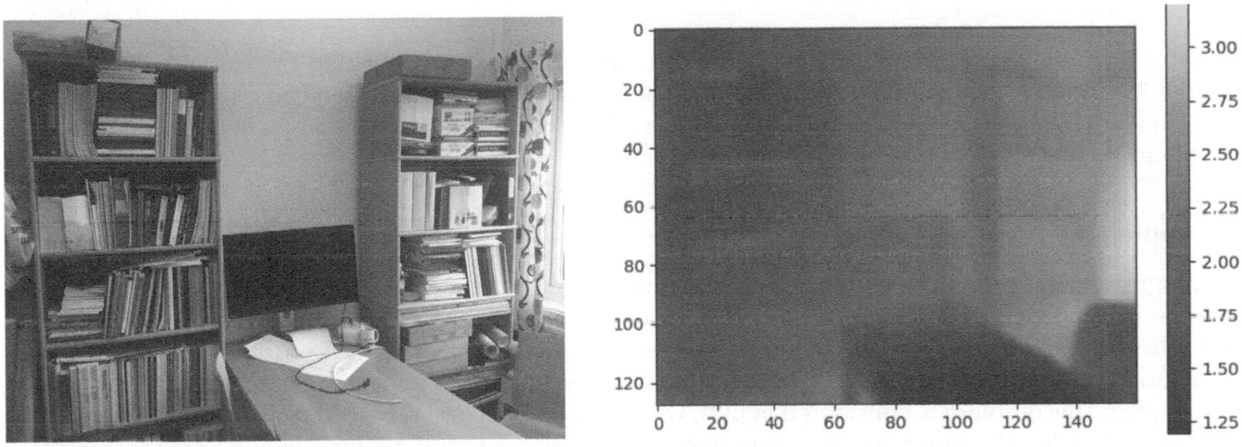

FIGURE 11.5 The figure shows input image (left) and output (right). The input is a colour image. The output is for each pixel a depth value (in meters). Colour image available at www.routledge.com/9781138593268.

Example 4.

Semantic Segmentation

Semantic segmentation is the process of classifying each pixel in one of a few semantic classes: see, for example [17, 18, 19, 20]. The input is thus a colour image as shown in Figure 11.4 (left). Many different architectures have been proposed for such a network. They typically involve several layers using convolution, activation function and downsampling followed by a series of layers that involve transposed convolution and activation function. The output is a classification of each pixel into a number of classes. In this example the network has been trained on indoor images, and pixel-wise classification has been provided for 20 classes. In Figure 11.4 an input image is shown to the left with corresponding semantic classification to the right.

Example 5.

Depth estimation from single views

Another machine-learning problem is the problem of guessing depth from a single view, see [21]. This should in principle be impossible, since a single colour image does not contain any information about

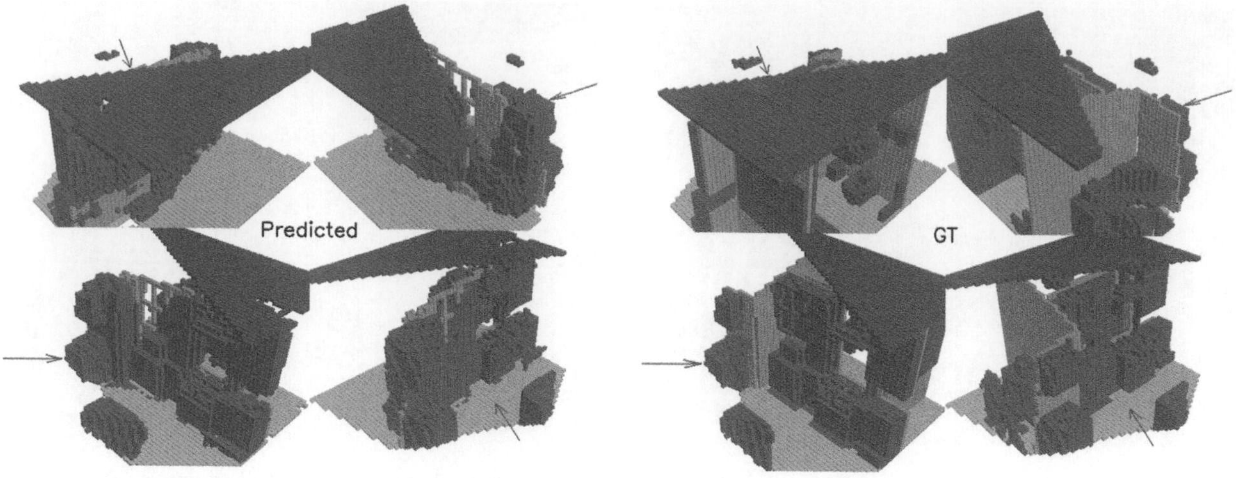

FIGURE 11.6 Here is one example output from test set for the scene completion experiments. To the left we have predicted labels and to the right ground truth labels. Colour image available at www.routledge.com/9781138593268.

depth. Nevertheless, humans can guess depth based on knowing what size objects usually have. In Figure 11.5 (left) the same office image is shown again. To the right in the same figure is shown the output of a machine-learning network. At each pixel there is an estimated depth value (in meters). The network has learned what typical depths are based on visual cues in the image.

Example 6.

Photographic image synthesis from semantic layouts
Semantic segmentation is the process of classifying each pixel in one of a few semantic classes. The input is a photographic image, and the output is a semantic layout of the scene. Is it possible to go in the other direction, that is, to have a semantic layout as input and synthesize a photographic image conditioned on this layout. This problem is considered in [22], where the architecture is based on a series of modules. The first module takes the input image, downsampled to 4 × 8 pixels and produces a number of features for each of these 4×8 pixels. Each subsequent module produces output with four times as many pixels, that is, the output of the second module is 8×16 and the ninth and last module, the output is 1024×2048. Each module takes the output of the previous module and a downsampled version of the semantic segmentation and outputs a number of features for each pixel with higher resolution.

11.6 ADVANCED NETWORK ARCHITECTURES

The machine-learning methods that we have described in the previous sections are constructed using layers (or computational blocks). These layers have parameters that are optimized using training data. One generalization is that the layers do not have to be assembled in a linear fashion. A network architecture can be constructed as any *directed acyclical graph.* A directed acyclical graph consists of vertices (the data) and directed edges (the computational processes). Another generalization is that any type of computational block can be used as long as it is possible to calculate the partial derivatives of the output, both with respect to the input and with respect to the parameters. There is currently rapid development of new types of computational blocks and new architectures as well as creative uses of such to solve problems in many domains. When communicating such architectures it is useful to find ways to visualize the architectures. One type of visualization concentrates on the data – that is, what is being sent between the computational blocks, as in Figure 11.2 top right. In this visualization we can clearly see the number of pixels and channels in the data, as computations flow from input to output. In another type of visualization, the focus is on the computational blocks, as in Figure 11.7.

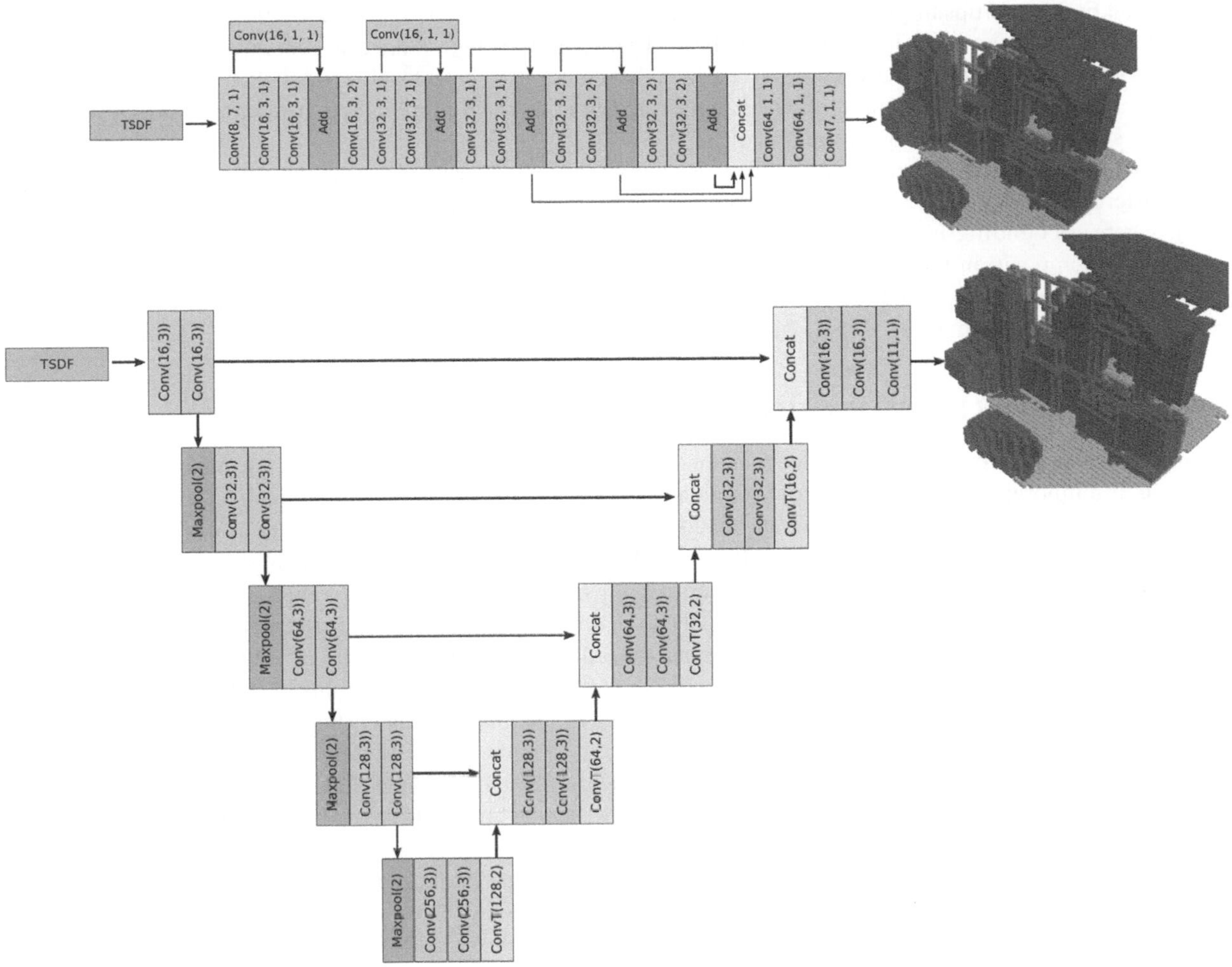

FIGURE 11.7 Architecture used in the scene completion experiments. Conv(d, k, l) stands for a 3D convolution filter stack of depth d and kernel size k and dilation l. ConvT(d, k) is the upsampling operation Transposed Convolution with depth d, kernel and stride k. Activation function is performed after every Conv layer. Softmax in the final layer. Colour image available at www.routledge.com/9781138593268.

Example 7.

Bayesian scene completion
Scene completion is the problem of both labelling 3D voxels that are visible, but also to label occluded voxels. The problem naturally arises when aiming to predict a 3D scene from a single view, but it can also be studied for multiple views. Predicting occluded areas can be of great help to autonomous vehicles during navigation and exploration. Especially for UAVs (Unmanned Aerial Vehicles), which navigate a 3D space where observations may be sparse, scene completion can be used for smoother trajectories during path planning. During exploration it can help the agent understand the likelihood of free space in occluded areas. In Figure 11.6 we see an illustration of the problem where the semantic labels are on the left and the ground truth is on the right. The figure is taken from [23], where a system for Bayesian Semantic Scene Completion (BSSC) is introduced. Along with the prediction scores the system also delivers an estimation of uncertainty. This is crucial for decision making during autonomous navigation and exploration, as it can help the agent understand when the data is new to the model and the prediction should not be trusted. It can also be used to understand what data should be added to the training to improve robustness.

One idea that utilizes the directed acyclic graph approach for semantic segmentation is the so-called *u-net* structure illustrated in Figure 11.7. It is similar to the architecture of Example 4. In this architecture we have several downsampling steps followed by several upsampling steps. After each downsampling step the spatial resolution is halved. In the U-net architecture, data from the early layers that have the highest spatial resolution are also sent directly to the latter layers of the same spatial resolution. These latter layers, thus, have access to both local, but detailed, information from the early layers as well as global information that has gone through both downsampling and upsampling. This approach has proved to be superior for many segmentation tasks.

The output of semantic segmentation, see Example 4, gives a class for each pixel, but the output does not distinguish between different objects of the same class. There are network architectures for object detections: see for example [24, 25], where a bounding box for each object is provided. The output of the architecture is a coarse sampling of the image. At each position the output consists of both classification (which object class it is) as well as information concerning the bounding box at that information. There are also network architectures that provide both a bounding box as well as pixelwise segmentation of that object.

11.7 DIMENSIONALITY REDUCTION

Machine learning and, in particular, deep-learning methods, have proved to be well suited for the analysis of signals that have spatial or temporal dimensions, for example, images (both 2D, 3D 4D), audio, radio. Such data often consists of a high number of variables (pixel, voxels, samples) and can thus be considered to be objects in a very high-dimensional space. A 10 mega-pixel colour image has 30 million variables and can be seen as an object in $\mathbb{R}^{30\,000\,000}$.

Dimensionality reduction techniques provide methods for mapping the data from a high-dimensional space into a low-dimensional space. The idea is to do this in a way so that the low-dimensional representation retains some meaningful properties (both local and global) of the original data. If data approximately lie on a lower dimensional manifold in the high-dimensional space, then dimensionality reduction techniques exploit this by estimating this manifold. Dimensionality reduction algorithms typically take a number of data in high dimensions $T = \{x_1,\ldots,x_N\}$, with $x_i \in \mathbb{R}^D$ as input and provide a low-dimensional representation $\{y_1,\ldots,y_N\}$, with $y_i \in \mathbb{R}^d$ as output. Some methods also provide mappings $f\colon \mathbb{R}^D \mapsto \mathbb{R}^d$ that can be used to map future data to the low-dimensional representation. Some methods also provide an inverse mapping $g\colon \mathbb{R}^d \mapsto \mathbb{R}^D$ that can be used to reconstruct the high-dimensional data from the low-dimensional representation.

Dimensionality reduction is useful in many ways. Often it is the low-dimensional representation in itself that is of interest. By reducing the dimensions to 1, 2, or 3, it is possible to visualize a full dataset in one single figure. This is useful for getting insight into the data, for example to understand if data cluster into a few distinct clusters. Working in a lower dimension is often easier because of the curse of dimensionality; see [26]. By first reducing the dimensions, machine-learning problems sometimes become easier to solve.

Principal Component Analysis [27] can be thought of as an affine dimensionality reduction technique. The mappings f and g are here affine, that is, $f(x) = Ax + b$ and $g(y) = Cy + d$. The problem of finding f and g can be posed as an optimization problem, where for every example x_i in the training set, one would like the reconstructed data $\hat{x}_i = g(f(x_i))$ to be as close to x_i as possible. The optimization problem

$$\min_{A,b,C,d} \sum_{i=1}^{N} \| x_i - g_{C,d}\left(f_{A,b}\left(x_i\right)\right)\|^2 \tag{11.15}$$

can be solved using singular value decomposition or eigen-decomposition. Once f and g have been calculated using this approach, the low-dimensional representation can be calculated using $y_i = f(x_i)$.

The *autoencoder* is somewhat similar to the principal component analysis. The idea is to represent both f and g using, for example, artificial neural networks or convolutional neural networks. The weights of both networks are optimized by minimizing a loss function, for example

$$\min_{w_f, w_g} \sum_{i=1}^{N} \| x_i - g_{w_g}\left(f_{w_f}\left(x_i\right)\right)\|^2, \tag{11.16}$$

although other loss functions could also be used. Once f and g have been calculated using this approach, the low-dimensional representation can be calculated using $y_i = f(x_i)$. Since both f and g are non-linear functions, it is possible for autoencoders to adapt to non-linearities in the data.

A so-called *denoising autoencoder* can be constructed by adding random noise to the input data

$$x_{i,noise} = x_i + \varepsilon_i, \tag{11.17}$$

where ε_j are simulated noise. By optimizing

$$\min_{w_f, w_g} \sum_{i=1}^{N} \left| x_i - g_{w_g}\left(f_{w_f}\left(x_{i,noise}\right)\right) \right|^2, \tag{11.18}$$

we hope that a noisy image $x_{i,noise}$ that is sent through both f and g will be close to the original (clean) image x_i. This has proven useful in many applications.

Multidimensional Scaling can be used for dimensionality reduction. The idea here is that one first calculates the pair-wise distances between points in the high-dimensional space,

$$d_{ij} = \left| x_i - x_j \right|, \tag{11.19}$$

for example, using the two-norm. Then one tries to place the points y_i in the lower-dimensional space so that their pair-wise distances are similar, that is,

$$d_{ij} \approx \left| y_i - y_j \right|. \tag{11.20}$$

By squaring both sides, we get

$$t_{ij} = d_{ij}^2 \approx \left(y_i - y_j\right)^T \left(y_i - y_j\right) = y_i^T y_i - 2y_i^T y_j + y_j^T y_j. \tag{11.21}$$

Without loss of generality, one may choose a coordinate system so that $y_1 = 0$. This means that $t_{1j} \approx y_j^T y_j$ and that

$$s_{ij} = \frac{t_{ij} - t_{1i} - t_{1j}}{-2} \approx y_i^T y_j. \tag{11.22}$$

By constructing a matrix S with the s_{ij} elements we notice that

$$S = \begin{pmatrix} s_{22} & s_{23} & \cdots & s_{2n} \\ s_{32} & s_{33} & \cdots & s_{2n} \\ \cdots & \cdots & \ddots & \cdots \\ s_{n2} & s_{n3} & \cdots & s_{nn} \end{pmatrix} \approx \begin{pmatrix} y_2^T \\ y_3^T \\ \cdots \\ y_N^T \end{pmatrix} \begin{pmatrix} y_2 & y_3 & \cdots & y_N \end{pmatrix}. \tag{11.23}$$

Matrix factorization, for example, the singular value decomposition, can be used to calculate y from S.

The distances d used in multidimensional scaling do not have to be Euclidian distances between points. In the *ISOMAP* method distances are computed in the following way. First, each point is connected to a few of the closest neighbouring points, for example the ten closest points. The distance between two points is then defined as the closest path, while moving along these edges in the graph. In this way the distance mimics the distance in the intrinsic manifold of the dataset. After calculating the distances in the high-dimensional space in this manner, we place the low-dimensional representation as was done for multi-dimensional scaling.

Perhaps the two most popular dimensionality-reduction techniques today are the *t-SNE* [28, 29] and the *UMAP* [30] methods. Similar to *ISOMAP*, the methods start with forming a graph by connecting neighbouring points. In the *t-SNE*, the following quantity

$$p_{j|i} = \frac{e^{-|x_i - x_j|^2 / (2\sigma_i)}}{\sum_{k \neq i} e^{-|x_i - x_k|^2 / (2\sigma_i)}} \tag{11.24}$$

is calculated. For the low-dimensional representation, we would like the similar quantity (based on the student t-distribution)

$$q_{j|i} = \frac{\left(1 + |y_i - y_j|^2\right)^{-1}}{\sum_{k \neq i} \left(1 + |y_i - y_k|^2\right)^{-1}} \tag{11.25}$$

to be similar to the ones in the high dimension. In the t-SNE method one optimize y_i so as to minimize the cost function

$$C(y) = \sum_i \sum_j p_{j|i} \log \frac{p_{j|i}}{q_{j|i}}. \tag{11.26}$$

In the *UMAP*, we calculate a weight between pairs of points

$$p_{j|i} = e^{\frac{-\max\left(0, d\left(x_i, x_j\right) - \rho_i\right)}{\sigma_i}}, \tag{11.27}$$

where, similar to $p_{j|i}$ in the t-SNE method, the weight is higher for points that are close to each other and lower for points further away. The parameter ρ_i is the distance to the closest point to x_i. This is set so that at least one other point has weight 1 for each i. The parameter σ_i is set so as to normalize according to

$$\sum_{i=1}^{k} p_{j|i} = \log_2(k). \tag{11.28}$$

From $p_{j|i}$ a symmetric weight is formed according to

$$p_{i,j} = p_{j|i} + p_{i|j} - p_{i|j} p_{j|i}. \tag{11.29}$$

For the low-dimensional representation we would like the similar quantity

$$q_{i,j} = \left(1 + a|y_i - y_j|^{2b}\right)^{-1} \tag{11.30}$$

to be similar to the ones in the high dimension. In the t-SNE method, one optimizes y_i so as to minimize the cost function

$$C(y) = \sum_i \sum_j \left(p_{i,j} \log \frac{p_{i,j}}{q_{i,j}} - p_{i,j} \log \left(\frac{1 - p_{i,j}}{1 - q_{i,j}} \right) \right). \tag{11.31}$$

Dimensionality reduction using principal component analysis, multi-dimensional scaling, t-SNE and UNET is shown in Figure 11.8.

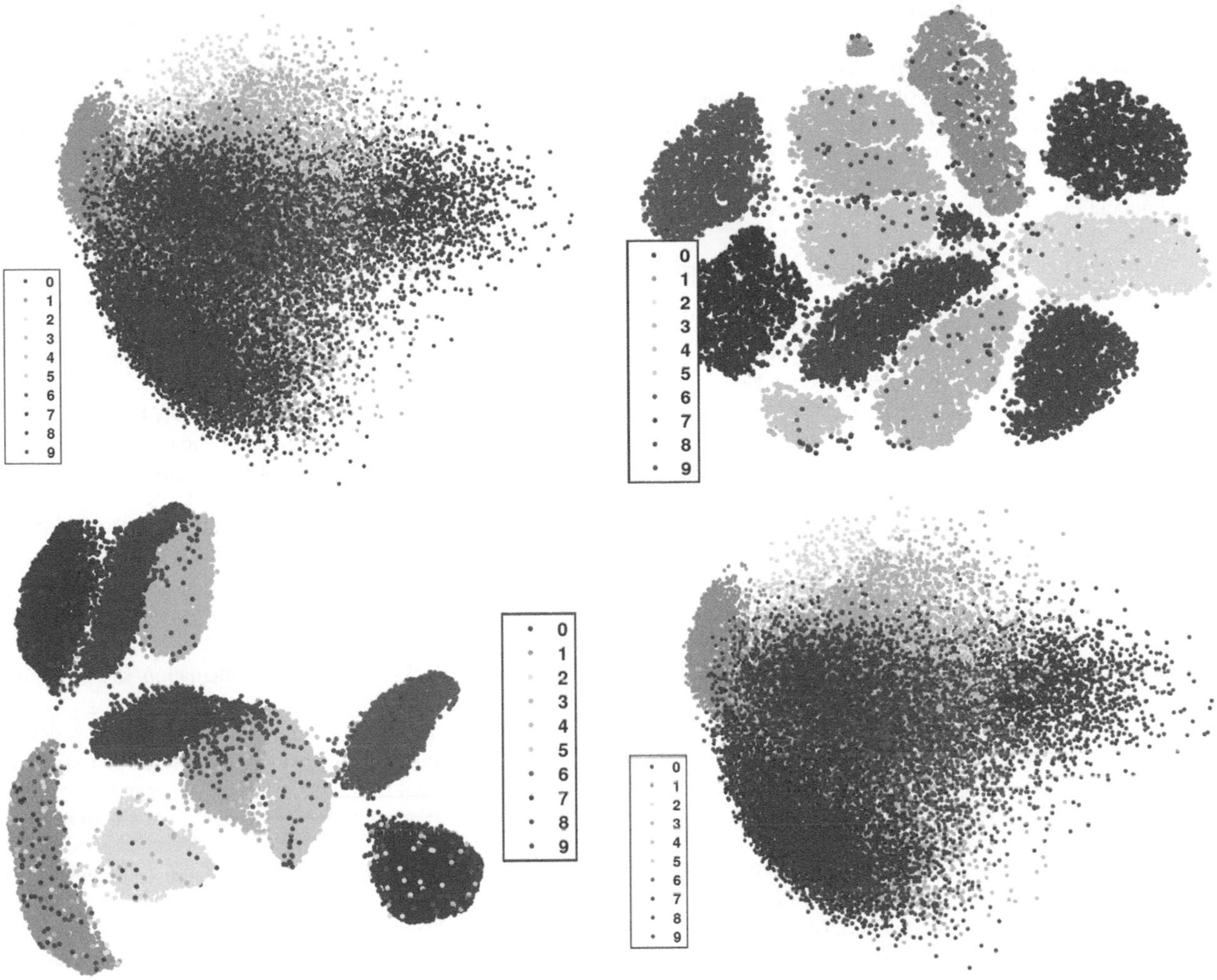

FIGURE 11.8 The figure shows the result of dimensionality reduction on the MNIST dataset. The 60,000 images in the dataset have 28×28 pixels, so each image have 784 dimensions. We use dimensionality reduction to find a representation in two dimensions. The results are 60,000 points in $\mathbb{R}^2$. For the MNIST dataset we also have the true label, but this is not used in the process. For visualization purposes we plot each point with a shade that corresponds to the label. In the figure we show the result of dimensionality reduction with PCA (upper left), t-SNE (upper right), UMAP (lower left) and multidimensional scaling (lower right). Colour image available at www.routledge.com/9781138593268.

11.8 SUMMARY

Machine learning builds on basic tools of mathematics, statistics, optimization and computation. The research area has evolved in waves, with several periods of intense research and interest. We are currently in a period where large databases, better computer hardware, and open-source programming packages make it possible to apply these methods with relative ease to new applications.

REFERENCES

[1] Arthur L Samuel. Some studies in machine learning using the game of checkers. *IBM Journal of Research and Development*, 3(3): 210–229, 1959.
[2] Tom M. Mitchell. *Machine Learning. McGraw Hill*, New York, USA, 1997.
[3] Stephen Boyd, Stephen P. Boyd, and Lieven Vandenberghe. *Convex Optimization*. Cambridge University Press, Cambridge, UK, 2004.
[4] Lars Hörmander. *Notions of Convexity*. Springer Science & Business Media, 2007.

[5] C. F. Gauss. *Theoria Motus Corporum Coelestium in Sectionibus Conicis Solem Ambientium*. Hamburg: Friedrich Perthes and I.H. Besser, 1809.

[6] Adrien Marie Legendre. *Nouvelles méthodes pour la détermination des orbites des comètes*. F. Didot, Paris, France 1805.

[7] Ronald A. Fisher. The use of multiple measurements in taxonomic problems. *Annals of Eugenics*, 7(2): 179–188, 1936.

[8] Thomas Bayes. An essay towards solving a problem in the doctrine of chances. By the late Rev. Mr. Bayes, F. R. S. communicated by Mr. Price, in a letter to John Canton, A. M. F. R. S. *Philosophical Transactions of the Royal Society of London*, 53: 370–418, 1763.

[9] Emanuel Parzen. On estimation of a probability density function and mode. *The Annals of Mathematical Statistics*, 33(3):1065–1076, 1962.

[10] Murray Rosenblatt. Remarks on some nonparametric estimates of a density function. *The Annals of Mathematical Statistics.*, 27(3): 832–837, (Aug) 1956.

[11] Vladimir Vapnik. Pattern recognition using generalized portrait method. *Automation and Remote Control*, 24: 774–780, 1963.

[12] N. Cristianini and J. Shawe-Taylor. *An Introduction to Support Vector Machines (and other kernel-based learning methods)*. Cambridge University Press, Cambridge, 2000.

[13] Jan Salomon Cramer. *The Origins of Logistic Regression*. Tinbergen Institute Working Paper no. 2002–119/4. 2002.

[14] Yann LeCun, Bernhard Boser, John Denker, Donnie Henderson, R. Howard, Wayne Hubbard, and Lawrence Jackel. Handwritten digit recognition with a back-propagation network. *Advances in Neural Information Processing Systems*, 2: 396–404, 1989.

[15] Jonathan I Epstein. An update of the Gleason grading system. *The Journal of Urology*, 183(2): 433–440, 2010.

[16] Anna Gummeson, Ida Arvidsson, Mattias Ohlsson, Niels Christian Overgaard, Agnieszka Krzyzanowska, Anders Heyden, Anders Bjartell, and Kalle Åström. Automatic Gleason grading of h and e stained microscopic prostate images using deep convolutional neural networks. In *Medical Imaging 2017: Digital Pathology*, 10140: 101400S. International Society for Optics and Photonics, 2017.

[17] Jonathan Long, Evan Shelhamer, and Trevor Darrell. Fully convolutional networks for semantic segmentation. In *Proceedings of the IEEE Conference on Computer Vision and Pattern Recognition*, 3431–3440, 2015.

[18] Guosheng Lin, Chunhua Shen, Anton Van Den Hengel, and Ian Reid. Efficient piecewise training of deep structured models for semantic segmentation. *Proceedings of the IEEE Conference on Computer Vision and Pattern Recognition*, 3194–3203, 2016.

[19] Panqu Wang, Pengfei Chen, Ye Yuan, Ding Liu, Zehua Huang, Xiaodi Hou, and Garrison Cottrell. Understanding convolution for semantic segmentation. In *2018 IEEE Winter Conference on Applications of Computer Vision (WACV)*, 1451–1460. IEEE, 2018.

[20] Hyeonwoo Noh, Seunghoon Hong, and Bohyung Han. Learning deconvolution network for semantic segmentation. In *Proceedings of the IEEE Conference on Computer Vision and Pattern Recognition*, 1520–1528, 2015.

[21] Bo Li, Chunhua Shen, Yuchao Dai, Anton Van Den Hengel, and Mingyi He. Depth and surface normal estimation from monocular images using regression on deep features and hierarchical CRFs. *Proceedings of the IEEE Conference on Computer Vision and Pattern Recognition*, 1119–1127, 2015.

[22] Qifeng Chen and Vladlen Koltun. Photographic image synthesis with cascaded refinement networks. *Proceedings of the IEEE International Conference on Computer Vision*, 1511–1520, 2017.

[23] David Gillsjö and Kalle Åström. In depth semantic scene completion. In *International Conference on Pattern Recognition, 2020*, 2020.

[24] Wei Liu, Dragomir Anguelov, Dumitru Erhan, Christian Szegedy, Scott Reed, Cheng-Yang Fu, and Alexander C. Berg. Ssd: Single shot multibox detector. *Lecture Notes in Computer Science*, 21–37, 2016.

[25] Joseph Redmon, Santosh Divvala, Ross Girshick, and Ali Farhadi. "You only look once: Unified, real-time object detection," IEEE Conference on Computer Vision and Pattern Recognition (CVPR), 2016, pp. 779–788.

[26] Richard Bellman. Dynamic programming. *Science*, 153(3731): 34–37, 1966.

[27] Karl Pearson. LIII. On lines and planes of closest fit to systems of points in space. *The London, Edinburgh, and Dublin Philosophical Magazine and Journal of Science*, 2(11): 559–572, 1901.

[28] Geoffrey E. Hinton and Sam Roweis. Stochastic neighbor embedding. *Advances in Neural Information Processing Systems*, 15: 857–864, 2002.

[29] Laurens van der Maaten and Geoffrey Hinton. Visualizing data using t-sne. *Journal of Machine Learning Research*, 9(Nov):2579–2605, 2008.

[30] Leland McInnes, John Healy, and James Melville. Umap: Uniform manifold approximation and projection for dimension reduction. *arXiv preprint arXiv:1802.03426*, 2018.

12 Image File Structures in Nuclear Medicine

Charles Herbst

CONTENTS

12.1 INTRODUCTION

This chapter is intended to provide a broad background to researchers who would like to use or develop custom software for extracting images or specific image information from Nuclear Medicine image files. It is not possible to cover all the complexities of the file structures, but a basic knowledge of the most common data and file structures will be a good starting point for programmers of any level of experience. In order to have a better understanding of the use of some of the file structures, a basic understanding of the binary information contained in an image is required as well as how this will impact on the display of an image. This will be addressed in the first section of the chapter.

12.2 IMAGE FILES AND IMAGE DISPLAY

All grayscale digital images may be regarded as a two-dimensional array or intensity function F(x, y) of pixel values where x and y represent the pixel coordinates in two dimensions, and F represents the image characteristic of the pixel at that point. The image characteristic represented by the pixel values may be the number of counts in a Nuclear Medicine image, or the relative amount of attenuation in an X-ray image, the CT number of a CT-image, the signal intensity generated by particular pulse sequence used in MR and so forth. The array size of the digital

DOI: 10.1201/9780429489556-12

image varies with the application but, in most cases, it will be a square array with sizes that are integer powers of two. The total file size will be determined by the array size as well as the maximum value required to represent the specific image characteristic.

Previously it was quite common to assume that the number of counts per pixel in a Nuclear Medicine image will not exceed 255, therefore it could be represented by 8 bits or a single byte. The CT numbers ranging from approximately -1000 to 3000 requires 12 bits, but most modern medical images will allow for 16 bits/pixel. The file size of a 16-bit, 1024 x 1024 image array will therefore be at least 2 Mbytes. Please refer to subhead 3.1 on general data structure for more detail.

The value of the image characteristic represented in the array is converted to an intensity value for display on a screen using a translation table. The maximum number of different screen intensity values is linked to the number of levels of grayscale that can be accommodated by the display system. Most medical display systems will be able to display 256 (8 bit) different grey levels, but systems with the ability to display 1024 (10 bits) grey levels are also available. Contrast and intensity scaling can be achieved by manipulating the translation table without changing the original data. Linear translation tables are most often used but it is also possible to use more elaborate scaling algorithms like histogram equalization to enhance particular aspects of the image.

Inherently all medical images are grayscale images, but these can be converted to pseudo colour images for display on colour screens by exploiting the fact that each pixel on a colour screen is able to display the three basic colours (red, green and blue) with different intensities. By changing the luminosity of the three basic colours individually, millions of different pseudo colours can be generated for each image pixel. Colour medical images may therefore be regarded as a grayscale image with additional display dimensions for the luminosity values of the three basic colours controlled by the RGB translation table. The three dimensions provide for the relative intensities of the three basic colours (RGB) but may be expanded to a fourth dimension if the relative opacity of the pixel is also added. The so-called RGBA colours (Red, Green, Blue, Alpha) therefore adds a new dimension, the so-called alpha channel, which can be set to determine the transparency of the image pixel as it is displayed on the screen. The latter is very helpful in superimposing displayed images on top of each other as may be required in overlaying PET data on top of a CT or MR image.

Image display software is available both as freeware as well as more comprehensive commercial software. Freeware includes ImageJ, Osiris, and Sante DICOM viewer Lite. It is not the intention to promote any providers of display software, as the list will certainly change in the future, but it may be worthwhile to also mention the independent commercial software, ANALYZE, as it has been around for quite a number of years. If one is challenged to make an informed choice on selecting between existing software applications or developing custom made software, the information in the following paragraphs will hopefully be helpful.

12.3 FUNDAMENTALS OF IMAGE ANALYSIS/DISPLAY SOFTWARE

12.3.1 General Data Structure

In the early computers all information, including the programming code as well as the data, was processed using electrical conductors. The conductor could have been switched on (represented later as '1'), or it could have been off (represented as a '0'). This was the basis for the binary (only two numbers or digits, called bits) coding system that is still applicable today. It is convenient to group the bits together in groups of 4 bits (called a nibble) as the bit information can then be conveniently contained in the shorter hexadecimal notation. The hexadecimal notation will be discussed in more detail later.

Two nibbles combined form a byte (8 bits) and in the older systems, a word consisted of two bytes. The concept of a word has been expanded for modern computers to indicate the fixed-sized piece of data that the processor handles. Today word lengths will therefore be 4 bytes for 32-bit processors and 8 bytes for 64-bit processors. Although outdated, all examples in the rest of this chapter will assume for the sake of convenience that a 16-bit processor is used, as it will be easy enough to apply the same principles to 32- and 63-bit processors.

In general, the information contained in the structure of a data file may be broadly described as a string of bytes that may represent either characters or numbers. Each character has a standard byte representation, but there is a difference in the way that bytes are combined to represent integer numbers and real or floating-point numbers. The different sets of bits associated with the different data types must therefore be linked because of the fact that they are all represented by the binary content of a byte. For example, the binary number 00110100 from a file may either be the character '4' or the decimal number 52. It is therefore important to have a working knowledge of both the data structure as well as the file structure in order to extract information correctly from a specialized file structure such as DICOM files.

Early computers with limited memory necessitated optimal use of every available bit. Today most of these older principles are still present in current file structures in order to ensure backwards compatibility. In this chapter we will firstly concentrate on the structure of data followed by a description of the structure of INTERFILE and DICOM files. Although the structure of the data may also influence the display and interpretation of the images, this dependence will only be mentioned in passing.

As already mentioned, a data file may include integer numbers, real numbers as well as characters. The correct interpretation of the bytes is the responsibility of the programmer. It is therefore also important to understand how the bytes are used to represent integers, characters and real numbers.

12.3.2 Number Representation

12.3.2.1 Integer Numbers

If a single byte is used to represent an integer, the8 bits will allow for 2^8=256 different combinations of 1s and 0s. The 256 different numbers will therefore cover all numbers from a minimum of 0 to a maximum of 2^8-1=255 (where all bits are set to '1'). Likewise, if only 4 bits (or a nibble) is used, the maximum number of different combinations will be 16, with a decimal range from 0–15. This combination of 4 bits or nibble is very useful in lower-level programming as a nibble can be represented by a single hexadecimal number. The 16 hexadecimal numbers are 0, 1, 2, 3, 4, 5, 6, 7, 8, 9, A, B, C, D, E and F. The hexadecimal or Hex representation allows for a much shorter and therefore less prone to error representation of the binary information. In order to avoid confusion with decimal or even octal numbers, the hexadecimal numbers are usually identified by either an 'h' after the number or a 0x in front of the number. The hexadecimal number 0x11 or 11h (equal to 17 decimal) will therefore be identified correctly as non-decimal. The binary and hexadecimal values of the first 16 decimal numbers are presented in Table 12.1.

If two bytes (equivalent to four nibbles) are combined, we can represent 2^{16}=65536 different numbers ranging from 0–65535. Please note that up to now we are restricting ourselves to positive integer numbers like the number of counts that will be present in a planar Nuclear Medicine image. If we want to include both positive and negative numbers as will be required to present the CT numbers in a slice, the first bit of the word is used as a sign bit. If the first bit is equal to one, then the following digits represent a negative number, while a first '0' will represent a positive number. As we now have only 15 bits available to represent a number, the maximum positive number will be 32767 and the minimum negative number -32767. In imaging processing, it is therefore important to know if you are working with signed integers (values ranging from -32767 to + 32767) or with unsigned integers (values ranging from 0–65535) in order to do the correct interpretation.

TABLE 12.1
Binary and Hexadecimal Representation of the First 16 Numbers

Decimal value	Binary value									Hex value	Observations
0	0	0	0	0	0	0	0	0		0h	2^0
1	0	0	0	0	0	0	0	1		1h	
2	0	0	0	0	0	0	1	0		2h	2^1
3	0	0	0	0	0	0	1	1		3h	
4	0	0	0	0	0	1	0	0		4h	2^2
5	0	0	0	0	0	1	0	1		5h	
6	0	0	0	0	0	1	1	0		6h	
7	0	0	0	0	0	1	1	1		7h	
8	0	0	0	0	1	0	0	0		8h	2^3
9	0	0	0	0	1	0	0	1		9h	
10	0	0	0	0	1	0	1	0		Ah	
11	0	0	0	0	1	0	1	1		Bh	
12	0	0	0	0	1	1	0	0		Ch	
13	0	0	0	0	1	1	0	1		Dh	
14	0	0	0	0	1	1	1	0		Eh	
15	0	0	0	0	1	1	1	1		Fh	
16	0	0	0	1	0	0	0	0	0	10h	2^4

12.3.2.2 Real Numbers

The structure of real numbers become more complex as fractions or decimal values must be represented. The following convention of the use of the bits in a 16-bit word will serve as an example of the structure of floating-point, or real numbers.

The first bit of the word is always used as a sign bit (as with integers), the next 4 bits are used as an exponent of ten and the rest as the mantissa (number following the decimal point). Thus, any real number can be presented as:

seeeemmmmmmmmmmm

with s the sign bit, e the exponential bits and m the values after the decimal comma. In summary, a number contained in a word of a 16-bit processor can therefore be in the form of:

seeeemmmmmmmmmmm	representing a real number, or
smmmmmmmmmmmmmmm	representing a signed integer number, or
mmmmmmmmmmmmmmmm	representing an unsigned integer.

From this example it is obvious that wrong answers will result if the programmer does not take care of the correct meaning of the different bits. It is also evident that the accuracy of a real number is restricted to 11 bits which is less than the accuracy of 16 bits of integer numbers. The convention is therefore to combine two or more words in order to increase the number of bits in the exponent as well as the mantissa. This larger number of bits assigned to the exponent will allow larger numbers to be accommodated, while the increase in the number of bits of the mantissa will allow for higher accuracy and therefore less rounding-off errors. It is also possible to combine two or more words in order to increase the accuracy of integer numbers. Combined words may be referred to a double precision, long integers, and so forth.

In summary, it has been pointed out that numbers can be presented as signed or unsigned integers, as well as signed floating (decimal, real) numbers and that all numbers can be either single precision or double precision. More information on the presentation of numbers can be found on the website of Comment Ça Marche [1]. The practical implications of the representation of numbers within a file will play an important role in the description of the DICOM file structure covered later in this chapter.

12.3.3 Character Representation

The original ASCII coding system was used to represent the 26 characters of the standard Roman alphabet used in the English language. Both capital characters as well as lowercase characters can be presented by the eight bits of a single byte. By convention, the five least significant bits of a byte was not used for characters. The decimal values ranging from 32–64 (bit 6 switched on) was used for special characters like !, @, #, $, %, and so forth. With bit 7 switched on, the resulting decimal numbers from 64–127 are used to represent the normal alphabetical numbers. Capital letters are represented by bit 7 switched on and bit 6 switched off (decimal 65 to 90), followed by the lower-case letters by switching on bit 7 as well as bit 6 (decimal 97–122). Switching bit 6 on in combination with bit 7 will therefore change from capital letters to lower-case letters. Readers are encouraged to test their skills at converting from ASCII to the binary and hexadecimal values using a converter available to the public [1].

Other characters are also included in the basic set, but the information above should be sufficient as a basis for understanding the basic ASCII character set. A more detailed explanation on the extension of the basic character set based on the full number of bits in a byte is beyond the scope of this chapter. However, more detail is available elsewhere for the interested reader [2, 3].

12.3.4 Little/big Endian

The two bytes of a 16-bit word may be referred to as a lower byte order if the byte containing the information from bits 1 to 8 (lower bits) is followed by the higher-order byte containing the information from bits 9 to 16. When a word is transferred using the TCP/IP protocol, the higher-order byte is always transferred first, followed by the lower-order byte. This is also the convention used in IBM mainframes for transferring bytes to the processor. However, the popular Intel processors requires that the lower-order byte be transferred first followed by the higher-order byte. Today, both conventions, 'lower order first' and 'higher order first', are in use. It is therefore very important to know which convention was used in generating and storing the data.

Endianness refers to the order of the bytes in a word. In the big-endian format the most significant byte will be first, followed by the lesser significant byte as in IBM mainframes, but the Intel microprocessors use the little-endian format where the less significant byte is first followed by the higher-order bytes. To illustrate the differences in character presentation and number presentation, as well as endianness, the different byte representation of the decimal number 4512 will be used as an example.

Firstly, the four characters of the number 4512 can be written to a file using the four bytes associated with the four different characters as illustrated in Table 12.2. Please note that four bytes are required to represent this number in ASCII format in memory.

If the same number is represented as an integer number, the binary value of 4512 must be determined. The binary value of the number 4512 is 1000110100000 and is represented as two separated bytes in Table 12.3. Both the binary as well as the hex values of these bits are illustrated in the table.

The big-endian order of the integer number 4512 will therefore be 11h, A0h and the little-endian order A0h, 11h. Note that only two bytes are required for binary coding.

There are significant advantages in keeping an integer number in the integer binary format in the computer, as the number can be used directly without conversions in calculations. Less memory is therefore required, but the binary value will have to be converted to the decimal equivalent by most people to correctly interpret the value (Please also note that the number 32767 still requires only two bytes in binary mode, but a minimum of six bytes are required in ASCII presentation!).

12.4 NUCLEAR MEDICINE IMAGE FILE STRUCTURES

With the collapse of specialist Nuclear Medicine image-processing units and the different gamma-camera vendors integrating their own acquisition and processing computers with their cameras in the beginning of the 1980s, a need arose for comparing the results from diverse algorithms used by the different vendors in processing functional organ data, for example, the ejection fraction of the left ventricle, kidney function analysis, and brain imaging [4]. In order to achieve these goals, a set of gamma camera images was sent to various institutions for evaluation using their standard protocols. This could only be achieved by developing an intermediate file format to convert to and from the propriety software of the main vendors. The Interfile file format was developed to achieve this and proved to be a great success over a number of years as the preferred means of transferring information between different systems [5].

A whole series of different permutations needed to be catered for in the intermediate file format as vendors often used different standards. It has been pointed out in the previous paragraphs that even a single number may be stored in a computer in different ways. A complicating factor was that the pixel data from a specific vendor could differ between diverse data-acquisition protocols, for example, it was possible to acquire static images in integer mode or byte mode.

TABLE 12.2
Coding Used for Representing the Characters of the Number 4512

Character	Decimal Character Code	Bit Value								Hex Value
4	52	0	0	1	1	0	1	0	0	34h
5	53	0	0	1	1	0	1	0	1	35h
1	49	0	0	1	1	0	0	0	1	31h
2	50	0	0	1	1	0	0	1	0	32h

TABLE 12.3
The Binary and Hex Values of the Number 4512

Byte Value								Hex Value	Description
1	0	1	0	0	0	0	0	A0h	Lower-order byte
0	0	0	1	0	0	0	1	11h	Higher-order byte

(If the number of counts per pixel was expected to be less than 255) A description of the way that the pixels values are represented in a file should therefore form an integral part of the file.

Additional required information such as the patient name and acquisition date should also be linked to the file. All the additional information pertaining to a file is regarded as metadata and can either be embedded as a header in the same file as the image information or it can be in a separate file that is associated with the image pixel information. In general, an image file therefore consists of at least two sections. In the metadata section, a description of the image data is provided. This will generally include information about the format of the pixel data as discussed above. Factors that are usually included will be whether the data is an integer or floating point number, the endianness, pixel depth, image dimensions, and so forth. The metadata therefore ensures that a software application will interpret and display the image data correctly.

Additional information about the patient and the modality through which it was produced may also be provided. For example, in Nuclear Medicine a set of reconstructed slices of a patient may contain information about the radionuclide used, number of slices, processing algorithm used, and so forth. The Nifti, Minc, and DICOM file structures are examples of embedded metadata, while Interfile and Analyze are two examples of separate files containing the image and metadata. Additional condensed information on the different file structures can be found in the article by Laborina and Loredana [6]. The files structures that are most frequently used today are the Interfile and DICOM standards and will be discussed in a little more detail in the next paragraphs.

12.4.1 Interfile Structure

It was pointed out that the Interfile file format was used for a number of years as the preferred means of transferring information between different imaging systems. Although use of the Interfile format has become less widespread over the last number of years, much can still be learned from the structure of this first across-the-board image file format – also to understand the more comprehensive DICOM file format that is in general use today.

There are two variants of the Interfile format. One variant includes the administrative or metadata and image information in the same file, but it is not recommended, although it is supported. The recommended variant places all the metadata in a text or ASCII file that contains a pointer to the file containing the binary image data. The binary image data file contains the 'raw' image data, which may consist of data in signed or unsigned integer or floating-point format. Bit format and ASCII format data are also supported, but seldom used. The byte order may either be big-endian (the default value) or little-endian and is applicable to both integer and floating-point data. The pixel data is ordered by column from left to right, followed by the row information from top to bottom.

The description of all the data is available in the associated administrative or metadata file. The administrative file consists of key-value pairs that can easily be read and edited by means of a standard text editor. This requirement necessitates that both the key as well the key value must be in ASCII. An example of some key-value pairs from an Interfile administrative file can be found in Table 12.4. A long list of keys is defined. Each key-value pair will be in a separate line in the metadata file, but the order of the key-value pairs is not important. Only a subset of the keys may be used, but this is not recommended as it is suggested that inclusion of optional keys may assist in the interpretation of the data. Keys are delimited by the characters':=' and the values of the keys by the characters '<cr><lf>', for example, the line containing the key 'patient name' with value 'John Smith' will be as follows:

patient name := John Smith<cr><lf>

Most of the keys are self-explanatory and therefore it is very easy to identify and change key values. Neither the keys nor the values are to be treated as case sensitive. Some of the keys are required to be specified, but others may be omitted. Required keys are preceded by an exclamation mark, for example,'!name of the data file'. It is important to note that when keys (either required or optional) are omitted, it will be equivalent to setting the value to null and therefore imply that the default value would be implemented if a default is specified.

Furthermore, some keys are associated with predefined values, for example, the key 'image data byte order' can only be 'BIGENDIAN', 'LITTLEENDIAN' or null. Predefined values must be as specified in the Interfile syntax declaration, but capitalization is not required. Null values are permitted, but when a null value is used, a default value will be invoked if a default is specified for the particular key. Default values can be defined in an external data dictionary used by the translation program containing a list of keys as well as default values. As an example, the default value for 'image data byte order' is 'BIGENDIAN'. Therefore, if this key is omitted or assigned a null value, all data will be regarded as big-endian.

TABLE 12.4
An Example of Some Key-value Pairs from an Interfile Administrative File (Information Extracted from [5])

```
!INTERFILE:=
!imaging modality:=nucmed
!originating system:=example
!version of keys:=3.3
!date of keys:=1992:01:01
;
!GENERAL DATA:=
data description:=static; this is optional
!data starting block:=0
!name of data file:=stat.img
patient name:=Joe Doe; this is optional
!patient ID:=12345; but this is required
patient dob:=1968:08:21
!study ID:=test
;
!GENERAL IMAGE DATA:=
!type of data:=Static
!total number of images:=1
!imagedata byte order:=BIGENDIAN
!number of energy windows:=1
```

Note that there are certain keys with no default value, for example, the matrix size of the image. In the event that the matrix size is not provided, nor is a null value, the image data will be uninterpretable. Also note that two different keys are available to indicate the offset of the image content in a combined file, namely 'data offset in bytes' and 'data starting block'. As the latter key assumes a block length of 2048 bytes, the use of 'data offset in bytes' may be more flexible.

It is possible to add comments to a key-value pair in order to improve readability by preceding the comment with a semicolon. It is strongly suggested for software developers to consult the full set of requirements for the Interfile syntax available within the original publication [5].

12.4.2 DICOM File Structure

DICOM is an acronym derived from 'Digital Imaging and Communications in Medicine' and was developed as a joint venture between the National Electronics Manufacturers Association (NEMA) and the American College of Radiology (ACR). The DICOM Standard defines both a format on which data associated to an object (often an image together with its administrative data) should be stored and transmitted between nodes in the imaging network as well as a protocol and description on what services can operate on the objects.

A challenge when designing a standard for a network system comprised of many different types of nodes (which is the case in medical imaging networks) are the large number of variations in data content, depending on what device a specific node represents. The data from a gamma camera compared to a sequence of ultrasound images differs in content when it comes to the physical nature of images, the technical and administrative data referring to the imaging device and the type of study. Consequently, besides the structure of the administrative data about the patient being examined, the files of data associated to the different types of devices will be very different in content. The solution has been to create a standard for images, and other types of generated data, a solution that is open and adapted to meet the requirements and developments in the medical imaging area.

The DICOM Standard defines the structure of Information Objects to each kind of data-generating device or modality. All defined Information Objects are described in part 3 of the DICOM Standard document (DICOM PS3.3). Following the development of new techniques and procedures, new parameters (attributes) to former objects, or entirely new objects are added to the DICOM standard continuously.

Besides the large number of defined Information Objects, different Services operating on the Information Objects have been defined. These services are generally described as Service Classes, for example, Storage Service Class, Print Management Service Class, and Media Storage Service Class. However, since the Information Object the services will operate on has a device specific content (e.g., a nuclear medicine (NM) image), the services need to be combined with the object as a Service Object Pair (SOP). These SOPs are described as SOP-classes, for example, Storage NM Image SOP-class. When a SOP-class is executed, a SOP-instance will be created which will be assigned a unique identification by its unique identifier (SOP-instance UID). Thus, each instance (image) can be traced to where it was created, at which institution and by what specific device. All service classes and SOP-classes are described in part 4 of the standard (PS3.4).

Since different nodes in an imaging network system have different roles designed for creating images, storing, printing, displaying and so forth, all nodes will naturally not have to meet the requirements of conformance in all possible aspects of the standard. For example, a gamma camera does not have to be able to store or display images created from an ultrasound imaging device. Hence, a single node will be designed to meet only a limited set of conformance in terms of services to be performed and objects to be handled. The set of conformance is described in a document, a DICOM Conformance Statement, which should be presented for each device and application by its manufacturer. The content and structure of this Conformance Statement is also described in the DICOM standard, in part 2 (PS3.2).

Creating DICOM files may be quite challenging because of the complexity of the Standard as pointed out above and is therefore beyond the scope of this chapter. The focus of this section will be on the format of the data associated with a Nuclear Medicine image, which is useful when investigating and extracting image data from a DICOM file. It is strongly advised to be more familiar with the Standard itself and its latest recommendations before starting a more complex project involving DICOM imaging. The Standard is available for free on the DICOM Standard website [7].

The lowest level of the DICOM information model consists of an image, report, waveform, or other data that hold the acquired information. Combining a group of images from the same protocol and piece of equipment will result in a *Series.* An example of a Series of images will be the dynamic series of images acquired during a gated blood pool study. A *Study* groups different *Series* to a requested procedure, for example, the two images (Series) acquired as an anterior and posterior whole-body scan will be combined into a single Study. A particular *Patient* may have many *Studies* that may even be acquired over different periods in time [8]. All this information that will allow linking the *Images*, Series and Studies to a particular Patient is contained in the DICOM file structure.

Each DICOM file consists of three major parts, as illustrated in Figure 12.1. The first part of the file (the 'header') consists of a 128-bytes preamble followed by a four-byte prefix containing the uppercase ASCII characters 'DICM'. There is no prescription for the 128 bytes and therefore it can be used for any information or it can be ignored when reading the file. The header is followed by the 'file meta-information header' that contains a dataset of information which specifies details such as the transfer syntax as well as other information regarding the storage of the data. The

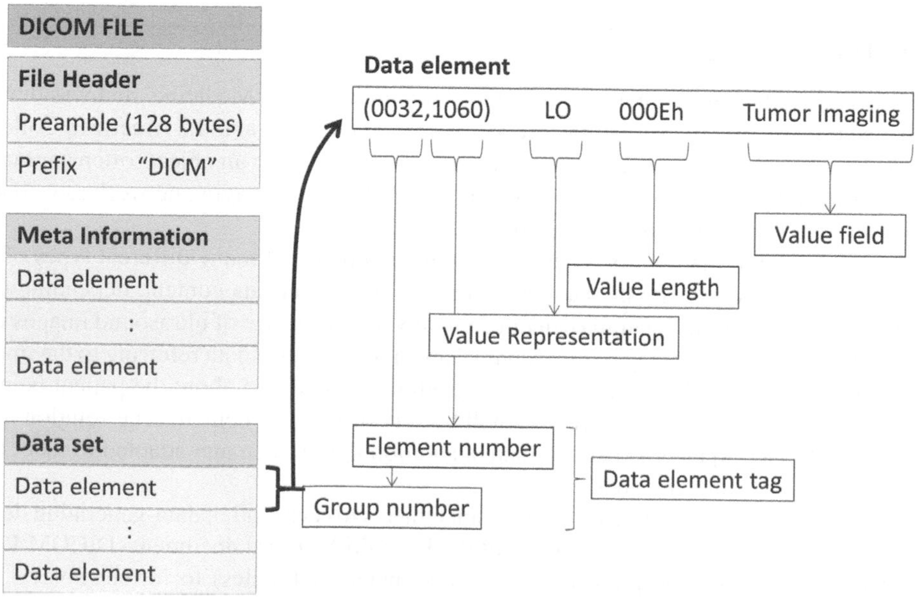

FIGURE 12.1 The relationship between the DICOM file, Data Set and Data Element structure.

third part of the file is the 'data object', or Data Set, containing information about the study, the series, and the patient as well as image information, including the scan position, image pixel data, and so forth.

The information in the 'file meta-information header' as well as the information in the Data Set section of the DICOM file are constructed as Data Elements. Each individual Data Element is uniquely identified by a Data Element Tag followed by three additional fields of information. The Data Element therefore usually consist of four different fields, namely a Data Element Tag composed of two 16-bit unsigned integer values representing the Group Number followed by an Element Number. The Data Element Tag is followed by the Value Representation (VR) consisting of two single byte characters. The VR of the Data Element is determined by, or dependent on, the Data Element Tag and can therefore be omitted (Implicit VR) in the Data Element or included (Explicit VR). VRs are not included as part of the Data Elements when the default transfer syntax is used. In this case the Data Element structure will contain only three different fields, namely the Data Element Tag, the Value Length and the value of the Data Element.

When the Group Number of the Data Element Tag is even, it is per definition a standard DICOM Tag. A full list of all the standard DICOM data elements including the descriptions and use of the tags can be found in the DICOM Data Dictionary, available in section 6 of the DICOM standard (PS3.6) [9]. Described in the DICOM Data Dictionary are thousands of different data elements that are updated a number of times per year. All previous standards are available, but most people will only be interested in the most recent publication that can be found on the DICOM Standard website [10]. The most recent dictionary at the time of writing was DICOM PS3.6 2019b – Data Dictionary.

Some implementations may require information that is not contained in the Standard Data Elements. Such implementations may make use of Private Data Elements. Private Data Elements have the same structure as that of the Standard Data Elements but with the provision that the Group Number of the tag shall be an odd number. More detail on the implementation of Private Data Elements is beyond the scope of this chapter. However, in section 7.8 of PS 3.5 [11], interested readers may find more detail on the provisions for avoiding collisions of private tags between different manufacturers as well as the mechanism by which a manufacturer can indicate that a block space has been reserved.

12.4.2.1 The DICOM Dictionary

In principle it is possible to link the DICOM Data Dictionary to the key and key value pairs used in Interfile. It was already mentioned that a long list of different Interfile keys has been defined. It was also mentioned that the key itself will always form part of the metadata information. The fact that all information in the Interfile metadata is available in ASCII allows for easy interpretation and change of the data.

The DICOM metadata includes thousands of different Data Element Tags (the equivalent of Interfile keys), but in the metadata of the DICOM file structure only the two 16-bit unsigned integer values of the Data Element Tag are included. The description of the tag uniquely defined in the Registry of the DICOM Data Elements is available in section 6 of the DICOM Data Dictionary and does not form part of the metadata. It is therefore only possible to make sense of the Data Tags if the information is read together with the Registry of the Data Elements. An example of some of the Tags and the associated attributes from the DICOM Data Dictionary [9] is given in Table 12.5. Please note that it is not a requirement that all tags be included in a particular DICOM file. However, it is required that the tags be ordered according to the tag values.

As mentioned previously, each and every Data Element will also have a Value Representation (VR) that describes the data type and format of the data element. This allows for the correct interpretations of the subsequent bytes. An explicit VR value can be used for the identification of the type of data in the Data Element, but implicit VR values are also allowed as pointed out above. With implicit definition of the VR value, the receiving application must use a DICOM

TABLE 12.5
An Example of Some Attributes from the DICOM Dictionary (Information Extracted from [9])

Tag	Name	Keyword	VR	VM	
(0020,0052)	Frame of Reference UID	Frame Of Reference UID	UI	1	
(0028,0030)	Pixel Spacing	Pixel Spacing	DS	2	
(0028,0104)	*Smallest Valid Pixel Value*	*Smallest Valid Pixel Value*	*US or SS*	*1*	*RET*
(0028,0105)	*Largest Valid Pixel Value*	*Largest Valid Pixel Value*	*US or SS*	*1*	*RET*
(0028,0106)	Smallest Image Pixel Value	Smallest Image Pixel Value	US or SS	1	
(0028,0107)	Largest Image Pixel Value	Largest Image Pixel Value	US or SS	1	

TABLE 12.6
An Example of Some Value Representations (VR) and their Descriptions (Information Extracted from [11])

VR Name	Definition	Character Repertoire	Length of Value
DS Decimal String	A string of characters representing either a fixed point number or a floating-point number. A fixed point number shall contain only the characters 0–9 with an optional leading '+' or '-' and an optional '.' to mark the decimal point. A floating-point number shall be conveyed as defined in ANSI X3.9, with an 'E' or 'e' to indicate the start of the exponent. Decimal Strings may be padded with leading or trailing spaces. Embedded spaces are not allowed. Note Data Elements with multiple values using this VR may not be properly encoded if Explicit-VR Transfer Syntax is used and the VL of this attribute exceeds 65534 bytes.	'0'-'9', '+', '-', 'E', 'e', '.' and the SPACE character of Default Character Repertoire	16 bytes maximum
UI Unique Identifier (UID)	A character string containing a UID that is used to uniquely identify a wide variety of items. The UID is a series of numeric components separated by the period '.' character. If a Value Field containing one or more UIDs is an odd number of bytes in length, the Value Field shall be padded with a single trailing NULL (00H) character to ensure that the Value Field is an even number of bytes in length. See section 9 and Annex B for a complete specification and examples.	"0"-"9", "." of Default Character Repertoire	64 bytes maximum
US Unsigned Short	Unsigned binary integer 16 bits long. Represents integer n in the range: $0 <= n < 2^{16}$.	not applicable	2 bytes fixed

Data Dictionary to look up the format of each data element. A short extract from the comprehensive list of VR values available in part 5 of the DICOM Standard is given in Table 12.6 [11].

Figure 12.2 illustrates a small section from a Data Set contained in a DICOM file. The first memory position in the file extracted is displayed in the block on the left, starting at memory position 500h in the file. The 16 subsequent memory positions are then from left to right, followed by the next 16 in the second row, and so forth.

Four Data Element Tags are highlighted in this figure. The first example to be explained in more detail is found at memory position 55Ch. According to the Registry of DICOM Data Elements, the attribute represented by the name of the tag (0020, 0052) is 'Frame of Reference UID' and the VR is 'UI'. Note that the information is extracted using an Intel processor and therefore the byte order of the two words comprising the Group Number and Element Number will differ from the default Big Endian byte order. The reverse order of the two bytes is, therefore (2000, 5200), as displayed in the figure. As the VR value (UI) is per definition an ASCII value, the order of these two characters (bytes) is not reversed. Also note that the name of the tag does not form part of the DICOM file and can only be identified using the Registry as pointed out above.

When you would like to extract particular information, you first have to identify the tag associated with that information. For example, if you would like to identify the pixel spacing of the image, you will have to consult the dictionary. Table 12.6 indicates that the name 'pixel spacing' is linked to tag (0028,0030) with a VR value of DS. The reverse byte order (2800,3000) can be found at memory position 602h. The ASCII information of the following two bytes ('DS') is the VR value associated with the particular tag and may be used to confirm that the correct tag has been identified. According to the definition of a decimal string ('DS') in Table 12.5, DS indicates that the decimal string will consist of a string of ASCII characters. The implicit value Multiplicity VM value associated with the tag (Table 12.5) is equal to 2 and therefore two different pixel spacing values will be included in the Data Element while the VL value of 2000h

MEM	0	1	2	3	4	5	6	7	8	9	A	B	C	D	E	F	0 1 2 3 4 5 6 7 8 9 A B C D E F
500h	55	49	38	00	31	2e	33	2e	31	32	2e	32	2e	31	31	30	U I 8 . 1 . 3 . 1 2 . 2 . 1 1 0
510h	37	2e	35	2e	36	2e	31	2e	36	39	33	39	31	2e	33	30	7 . 5 . 6 . 1 . 6 9 3 9 1 . 3 0
520h	30	35	30	31	31	35	30	39	30	32	31	31	35	36	30	32	0 5 0 1 1 5 0 9 0 2 1 1 5 6 0 2
530h	37	39	36	30	30	30	30	30	30	30	31	00	20	00	10	00	7 9 6 0 0 0 0 0 0 0 1 . . ┼ .
540h	53	48	02	00	32	20	20	00	11	00	49	53	04	00	31	30	S H ┐ . 2 . ◄ . I S ┘ . 1 0
550h	30	30	20	00	13	00	49	53	02	00	31	20	20	00	52	00	0 0 . ‼ . I S ┐ . 1 . R .
560h	55	49	38	00	31	2e	33	2e	31	32	2e	32	2e	31	31	30	U I 8 . 1 . 3 . 1 2 . 2 . 1 1 0
570h	37	2e	35	2e	36	2e	31	2e	36	39	33	39	31	2e	33	30	7 . 5 . 6 . 1 . 6 9 3 9 1 . 3 0
580h	30	35	30	31	31	35	30	39	30	32	31	31	35	36	30	32	0 5 0 1 1 5 0 9 0 2 1 1 5 6 0 2
590h	37	39	36	30	30	30	30	30	30	30	32	00	20	00	40	10	7 9 6 0 0 0 0 0 0 0 2 . . @ ┼
5A0h	4c	4f	08	00	75	6e	6b	6e	6f	77	6e	20	28	00	02	00	L O ◘ . u n k n o w n (. ┐ .
5B0h	55	53	02	00	01	00	28	00	04	00	43	53	0c	00	4d	4f	U S ┐ . . (. ┘ . C S ♀ . M O
5C0h	4e	4f	43	48	52	4f	4d	45	32	20	28	00	08	00	49	53	N O C H R O M E 2 (. ◘ . I S
5D0h	04	00	31	38	30	20	28	00	09	00	41	54	10	00	54	00	┘ . 1 8 0 (. . A T ┼ . T .
5E0h	10	00	54	00	20	00	54	00	50	00	54	00	90	00	28	00	┼ . T . . T . P . T . . (.
5F0h	10	00	55	53	02	00	80	00	28	00	11	00	55	53	02	00	┼ . U S ┐ . € . (. ◄ . U S ┐ .
600h	80	00	28	00	30	00	44	53	20	00	34	2e	37	39	35	31	€ . (. 0 . D S . 4 . 7 9 5 1
610h	39	39	38	37	31	30	36	33	32	5c	34	2e	37	39	35	31	9 9 8 7 1 0 6 3 2 \ 4 . 7 9 5 1
620h	39	39	38	37	31	30	36	33	32	20	28	00	51	00	43	53	9 9 8 7 1 0 6 3 2 (. Q . C S
630h	0e	00	55	4e	49	46	5c	55	4e	49	46	5c	55	4e	49	46	♬ . U N I F \ U N I F \ U N I F
640h	28	00	00	01	55	53	02	00	10	00	28	00	01	01	55	53	(. . U S ┐ . ┼ . (. U S
650h	02	00	10	00	28	00	02	01	55	53	02	00	0f	00	28	00	┐ . ┼ . (. ┐ U S ┐ . ☼ . (.
660h	03	01	55	53	02	00	00	00	28	00	06	01	55	53	02	00	└ U S ┐ . . . (. − U S ┐ .
670h	00	00	28	00	07	01	55	53	02	00	1e	00	28	00	50	10	. . (. • U S ┐ . . (. P ┼
680h	44	53	02	00	31	35	28	00	51	10	44	53	02	00	33	30	D S ┐ . 1 5 (. Q ┼ D S ┐ . 3 0
690h	29	00	10	00	4c	4f	12	00	53	49	45	4d	45	4e	53	20	) . ┼ . L O ↕ . S I E M E N S

FIGURE 12.2 An example of the hexadecimal data and character representation of the metadata from a DICOM File.

(correct order 0020h) indicates that the total number of characters in each of the two strings consist of 32 characters. The pixel spacing values are therefore the ASCII strings 4.7951998710632 and 4.7951998710632. Please note the slash separating the two values and also that a trailing undefined ASCII value is completing the full 32 characters.

Likewise, to determine the smallest image pixel value, the Tag associated with the name 'Smallest pixel value' must be determined from the Registry. The relevant Tag is (0028, 0106) with a VR value of US or SS and a VM value of 1. Searching for the reverse byte order (2800, 0610) results in memory position 668H. The VR value confirms to some extent the correct tag has been identified. The implicit VL value of one will result in only one value with a total length of 16 bits (0200) reversed to (0020). (Please note that US is by default fixed to two bytes). This is followed by the hex number 0000h indicating that the minimum pixel value in this Nuclear Medicine study is zero. (In this example it is also worth noting that a Tag Value of (0028, 0104) with the name 'Smallest Valid Pixel Value' is listed in the Registry, but the 'RET' indicates that this Tag Value has been retired and should therefore not be used in creating new files).

The last Data Element Tag in this example (0029,0010) is indicated at the bottom of Figure 12.2. As the Group Number of this Tag is an odd number, it is indicating that this is a Private Data Element with a VR value of 'LO' and that the total length of the data will be 18 (0012h) characters.

The example above should be sufficient to allow a fairly capable programmer to design software for reading a DICOM file. In short software for reading DICOM files will include the following general steps:

- Set up a list of the Tags that are essential for the information that you will require while considering the fact that the Tags in the file will always be in an ascending order.
- Read the total file and identify the position of the first Tag that is required.
- Use the VR value to confirm it is the correct Tag.
- Read the size of the data from the following bytes as specified in the implicit VL value as well as the Value Length following the VM value.
- Locate the data.
- Search for the next tag and repeat the procedure.

It is worthwhile to note that the image information will follow the tag (7FE0,0010) with a VR of OB or OW and a VL of 1. The principles set out above can then be used to extract the image information. Likewise, it is also possible to read the total file, search for the patient's name by using the correct tag and replace the name with a pseudonym preceded by the correct value representing the number of characters in order to anonymize the data.

The strategy described above will be quite effective if one would like to extract information linked to a particular Tag. However, it was pointed out above that there is no requirement that all Tags should be included in a DICOM file. The same strategy can be used to identify all the Tags in a file, but it may be worthwhile to have a look at some free software that is available to do this task. (Remember that it is not required to include all Tags in all files). Free software is available that will analyse a DICOM file, identify the tags and display the tag's position as well as the tag attributes. Free public domain software includes the DICOM Editor Tool [12], which is comprehensive and allows the user to read the different Tags, the description of the Tags, VR, VM and the value of the Tag of a specific file. Registration is required to use the software, but the there are no strings attached. If you are interested in the information of the DICOMDIR file, it may perhaps be more appropriate to use the free software from Sante [13]. Sante Dicom Hexviewer is not listed as a separate application for downloading on the Sante website, but it is still available elsewhere [14]. However, this very helpful application is incorporated in the free Sante DICOM Viewer Lite application [15]. Open any DICOM file for display and look for Hexadecimal viewer under 'view' when using Sante DICOM Viewer Lite to display the file.

Careful consideration should be given to additional factors if one would like to generate a true DICOM file – for example, a file containing an image generated using simulation software to be transferred to commercial systems for processing. In order to comply with the principals of interoperability and unique data transfer between systems without the possibility of clashes, DICOM files must always contain Unique Identifiers (UIDs) that will ensure global uniqueness across multiple countries, sites, vendors, and equipment.

All UIDs including Storage UIDs, Transfer Syntax UIDs, Study UIDs, Series UIDs, and so on, consist of two parts, namely an organization part <org root> and a suffix <suffix> as defined by the ISO/IEC 8824 standard and described in section 9 of PS 3.5 [11]. The <org root> and <suffix> are then combined to form the UID as <org root>.<suffix>.

The <org root> is usually issued by an ISO Member Body to a particular organization. For example, in the org root part of an UID equal to <1.2.840.xxxx> 1 identifies ISO, 2 the ISO Member Body, 840 the country code of a specific ICO Member Body, and xxxx the organization as identified by the ISO body. Ensuring the uniqueness of the suffix is the responsibility of the organization.

The suffix may include coding numbers for the system, study, series, image, and dates and times.

For example, in the UID 1.2.840.3434.3.152.235.2.12.187636473 the <org root> consists of the numbers given above and an organization number equal to 3434. The suffix <3.152.235.2.12.187636473> is constructed using 3 as representing the device type, 152 as the manufacturer serial number, 2 the series number, 12 the image number and 187636473 the time stamp. In this solution to ensure that the UID will be unique within the organization it is assumed that only one set of data can be acquired by a particular instrument at a particular time. Combining this unique suffix to the <org root> (which is already uniquely identified by the ISO Member Body) will ensure a unique UID that will have no duplicate in any other DICOM file. Note that different series of components may be used than those mentioned in this example as long as the organization ensure that the UID issued within the organization is unique.

A more thorough explanation of the description of the more comprehensive DICOM application is beyond the scope of this chapter. Support is available for programmers who would like to develop their own software in Python [16] or C++ [17]. Nevertheless, it might be worthwhile to cover, in passing, some aspects of the transfer of data between systems.

12.5 DATA TRANSFER BETWEEN SYSTEMS

The information provided above covers only the basic content and structure of DICOM files that may be interchanged between different imaging systems. However, network image management is by far the most widely implemented of the DICOM Standard.

The DICOM Message communication framework uses the TCP/IP protocol to communicate between DICOM compatible devices. The DICOM Transfer Protocol expands the generic file transfer protocols by including explicit semantics that will allow the server and client access to the information of the structured data on the other system. As the services and functions of the transfer communications are also shared, the receiving device will be aware of the information structure of the image data even before receiving it [18]. This allows for storage and retrieval of images and incorporating them into a patient-related information indexing system utilizing the patient related information from the

DICOM image file attributes. Fundamental to the sharing of information between systems is the definition and specification of DICOM Service Classes and services. For example, the Storage Service Class specifies the C-STORE service that enables a client to push a DICOM object to a server for storage. Likewise, the Query/Retrieve Service Class specifies the C-GET service that will allow an application to pull objects that match a set of key values.

Notwithstanding the detail available on the communications framework, it cannot be taken for granted that two systems claiming DICOM conformance will be able to communicate seamlessly with each other. The DICOM conformance document associated with a particular device will provide a list of the DICOM building blocks that the product supports as well as a description of the product-implementation details and behaviour. In order to determine the possibility for two devices to communicate with each other, it is necessary to have a complete description of the capabilities of each system in order to determine the level of possible communication. For example, a device may provide the C-STORE service and include that in the Conformance Statement, but the same device may exclude the C-GET service. The DICOM Conformance Statement will therefore allow a knowledgeable user to determine the possibility of two devices to communicate with each other as well as the level of possible communication, but it will not be enough to ensure interoperability at the level that may be required. However, the Conformance Statement will provide enough information to determine if it may be impossible for two systems to exchange data.

Ensuring seamless bidirectional communication between two systems is therefore no trivial task. However, some institutions with sufficient resources may like to develop their own networking infrastructure. A good starting point for the serious programmer involved in such a project may be the Python code based on pydicom/pynetdicom [16].

12.6 FUTURE TRENDS

The future of interoperable systems is determined to a great extent by the DICOM Standards Committee and the activities of the Working Groups of the Committee. The Committee meets three times per year overseeing the activities of the Working Groups. The Working Groups address revisions, including identification of deficiencies for specific modalities, clinical domains, and areas of technology. The manufacturing companies, service, and professional organizations and government agencies represented on the Committee and Working Groups are responsible for adopting these policies and also for revising the policies where necessary. They are therefore also responsible for identifying and accommodating future trends in the Standards [19].

Several challenges exist for the field of security as well as for distributed data and processing. These challenges are addressed by Working Groups 14 and 23 of the Committee. Among the areas that were identified was the fact that particular mechanisms may be appropriate for particular regulatory bodies but deemed inappropriate by another regulator. The seemingly conflicting regulations from different regulatory bodies in the various countries will become more difficult to accommodate as technology becomes obsolete and new technologies are introduced at a faster pace [20]. Cloud technology will definitely have a huge impact in this regard in the future.

The accommodation of artificial intelligence (AI) within DICOM is also quite challenging. The development of DICOM mechanisms to support AI workflows (including the interfaces between application software and the DICOM infrastructure) is the responsibility of Working Group 23 of the Committee. They are not only challenged with the same regulatory aspects pointed out above, but also with a technological landscape that is always rapidly changing. This working group has been creating examples of using DICOM Web services, including the Web-based Unified Procedure Step (UPS) to trigger a Web-based analytical service. With various tools falling in and out of favour, bulk analysis of large repository data for training deep-learning neural networks and potential intellectual property issues will also take a considerable amount of time to flesh out [21].

In order to facilitate communication needs resulting from clinical use of medical image distributed viewing and processing, Web-based extensions to DICOM need to be defined. These extensions should enable intra- as well as inter-enterprise communications and management of the medical image record as well as integration with other healthcare IT systems. Lastly, it can be mentioned that additional processing situations, including 3D compression of images, segmentation of images, and image overlay between different modalities will have to be addressed.

12.7 ACKNOWLEDGEMENT

I would like to thank Dr Tomas Kirkhorn, Lund University, Sweden, for his valuable insight, contribution and recommended changes to this chapter.

REFERENCES

[1] "Representation of Real Numbers and Integers." https://ccm.net/contents/62-representation-of-real-numbers-and-integers.
[2] "ASCII Code." https://ccm.net/contents/55-ascii-code.
[3] "Code Charts." www.unicode.org/charts/.
[4] P. S. Cosgriff, *COST B2 Final Report: Quality Assurance of Nuclear Medicine Software*, European Communities, Directorate-General Science, Research and Development, Luxembourg, Website EUR 17916 EN., 1998/01/27/ 1998. Accessed: 2019/06/12/13:16:48. [Online]. Available: https://publications.europa.eu/en/publication-detail/-/publication/430f7ce7-2860-44d6-9a8e-1da0b33d7784/language-en.
[5] A. Todd-Pokropek, T. D. Cradduck, and F. Deconinck. "A File Format for the Exchange of Nuclear Medicine Image Data: A Specification of Interfile Version 3.3." www.ncbi.nlm.nih.gov/pubmed/1448241.
[6] M. Larobina and L. Murino, "Medical Image File Formats," *J Digit Imaging*, vol. 27, no. 2, pp. 200–206, 2014/04// 2014, doi:10.1007/s10278-013-9657-9.
[7] "DICOM Standard."www.dicomstandard.org/.
[8] "Day1_S4-Medema-DICOM-Overview." www.dicomstandard.org/wp-content/uploads/2018/10/Day1_S4-Medema-DICOM-Overview.pdf.
[9] "PS3.6." http://dicom.nema.org/medical/dicom/current/output/html/part06.html#chapter_6.
[10] "Current Edition – DICOM Standard." www.dicomstandard.org/current/.
[11] "PS3.5." http://dicom.nema.org/medical/dicom/current/output/html/part05.html#PS3.5.
[12] "DVTk DICOM Editor." www.dvtk.org/dicom/editor/.
[13] "Sante DICOMDIR Viewer | Download." www.santesoft.com/win/sante-dicomdir-viewer/download.html.
[14] "Sante DICOM HEXViewer." https://download.cnet.com/Sante-DICOM-HEXViewer/3000-2192_4-75843799.html.
[15] "Sante DICOM Viewer Lite." www.santesoft.com/win/sante-dicom-viewer-lite/sante-dicom-viewer-lite.html.
[16] "A Python Implementation of the DICOM Networking Protocol: Pydicom/Pynetdicom." Dicom in Python. https://github.com/pydicom/pynetdicom.
[17] Anonymous, "The DICOM Landscape: Selecting the Right Software Development Kit," (in English), *MDDI Online*, 2015/06/17/T15:48:30-04:00 2015. [Online]. Available: www.mddionline.com/dicom-landscape-selecting-right-software-development-kit.
[18] W. D. Bidgood, S. C. Horii, F. W. Prior, and D. E. Van Syckle, "Understanding and Using DICOM, the Data Interchange Standard for Biomedical Imaging," *J Am Med Inform Assoc*, vol. 4, no. 3, pp. 199–212, 1997 1997. [Online]. Available: www.ncbi.nlm.nih.gov/pmc/articles/PMC61235/.
[19] "DICOM Standard Committee – DICOM Standard." www.dicomstandard.org/dsc/.
[20] "WG-14: Security – DICOM Standard." www.dicomstandard.org/wgs/wg-14/.
[21] "WG-23: Artificial Intelligence/Application Hosting – DICOM Standard." www.dicomstandard.org/wgs/wg-23/.

13 The Scintillation Camera

Jonathan Gear

CONTENTS

13.1 INTRODUCTION

The scintillation camera, commonly referred to as a gamma camera, is a medical device designed to image the distribution of radiopharmaceuticals *in vivo*. Its invention is attributed to Hal Anger at the Donner Laboratory at the University of California, Berkley, in the early 1950s. The appearance of the modern camera is quite different to that used to produce the first clinical images. However, the fundamentals of the technology used for image acquisition and formation remain largely unchanged. Figure 13.1a is an illustration of a scintillation camera, highlighting the key components required for clinical imaging. The primary components used for photon detection are located within a shielded housing (referred to as the detector or head), supported on a rotating gantry (Figure 13.1b). Gamma cameras usually have two detectors, although single- and triple-headed systems are available. A brief summary of the function of each component is summarized in Table 13.1 and discussed in more detail in this chapter.

13.2 GANTRY, HOUSING, AND IMAGING COUCH

As will be discussed in later chapters, it is necessary to orientate the detectors about the patient in a number of configurations to concentrate on the area of the body in which the pharmaceutical accumulates. The majority of investigations are carried out with the patient lying supine on the imaging couch. However, many systems allow imaging to be performed with the patient standing, sitting, or on a hospital bed.

Due to the quantity of lead used to collimate and shield the detector, the mass of each head can be in excess of 500 kg. The motors used to control the motion of such a weight are therefore very powerful and capable of delivering a large amount of force. Patient proximity sensors and touch-sensitive panels are often present to avoid a potential collision and reduce the risk of driving the detector into an object or patient. A commonly used proximity sensor is an infrared beam across the face of the detector which, if interrupted, alerts the system to the presence of the patient. This device may also be used to contour the detector more closely about the patient during a tomographic or whole-body acquisition.

DOI: 10.1201/9780429489556-13

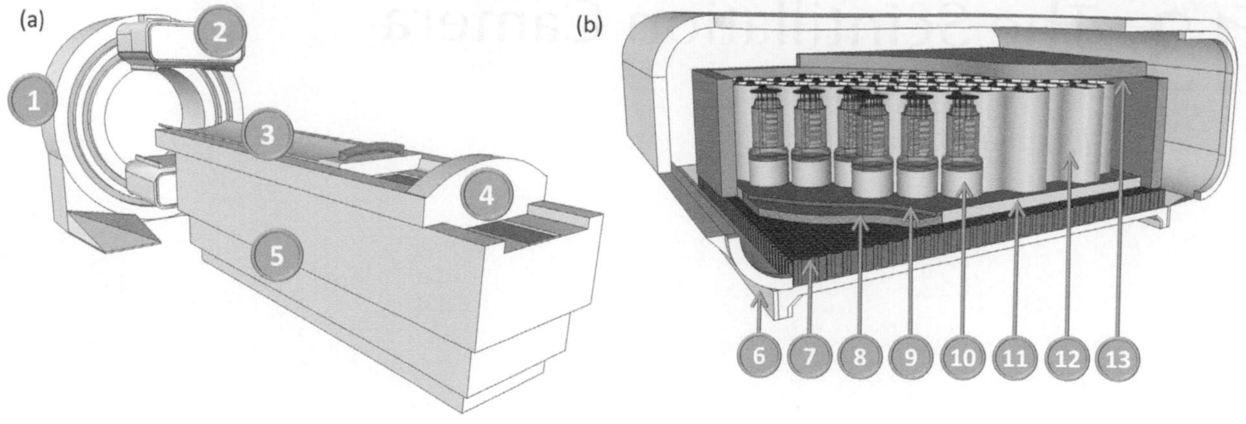

FIGURE 13.1 Diagram (a) depicts the basic components in a camera system, including the main gantry (1) the detectors (2), the patient pallet (3), the sub-pallet (4) and the main couch (5). Diagram (b) shows the components within the detector head, including a proximity sensor (6), collimator (7), NaI(Tl) crystal (8), light-guide (9), photomultiplier tube (10), aluminium casing (11), μ-metal shield (12), and lead shielding (13).

TABLE 13.1
A Summary of the Main Components of the Gamma Camera and their Basic Function

Nr	Component	Function
1	Gantry	Support and position detectors
2	Detector	photon collection
3	Patient Pallet	Low attenuation & positioning patient
4	Sub Pallet	Expending range of patient couch
5	Main Patient Couch	Positioning patient
6	Proximity sensor	Safety device to ensure detectors do not collide with patient
7	Collimator	Masks the detector to collimate photons traveling from a specific direction
8	Crystal	Absorb gamma emissions and emits scintillation photons
9	Light Guide	Ensures optical efficiency of photon transmission from crystal to PMT
10	Aluminium Casing	Haematically seals crystal and reflects scintillation photons back towards light guide
11	Mu-metal shield	Shield PMTs from earth's magnetic field
12	Photomultiplier Tube	Absorb scintillation photons and convert to electronic signal
13	Lead	Shield crystal from external ionizing radiation

A number of detector configurations and motions will often be pre-programmed for frequently used imaging acquisitions. The array of motions that the system is capable of are illustrated in Figure 13.2, and typical detector orientations are given in Figure 13.3.

The imaging couch is constructed from a material with low attenuating properties, such as carbon fibre, to ensure its presence does not adversely affect the scintigraphic image. The couch should be sufficiently strong to avoid sagging when fully extended. For hybrid systems containing both SPECT and CT imaging modalities, a sub-pallet is often required to ensure sufficient support is available to further extend the imaging couch beyond the gamma camera field of view and into the CT gantry. The sub-pallet does not have the same low-attenuation properties of the imaging couch, and care should be taken to ensure it is not present within the imaging field of view during an acquisition.

13.3 THE SCINTILLATION CRYSTAL

The scintillation crystal is the single most-expensive component of the camera. It is also arguably the most fragile, being sensitive to moisture, temperature, and physical shock. The function of the crystal is to absorb incident photon emissions and convert them to visible scintillation photons. A single scintillation crystal is used, which covers the entire

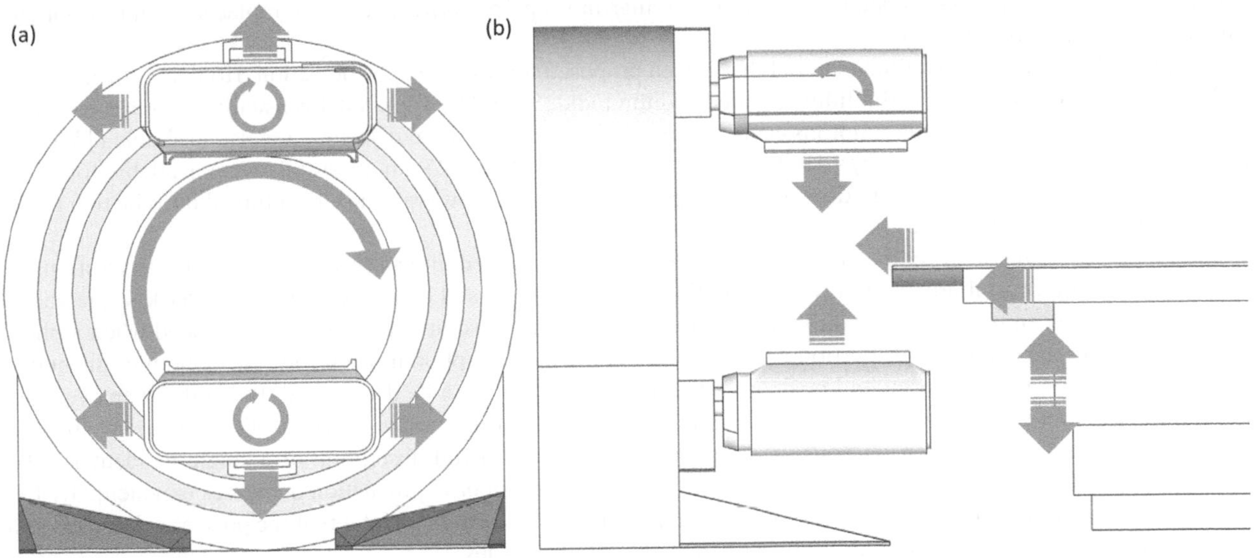

FIGURE 13.2 Diagram indicating the basic motions of gamma-camera system.

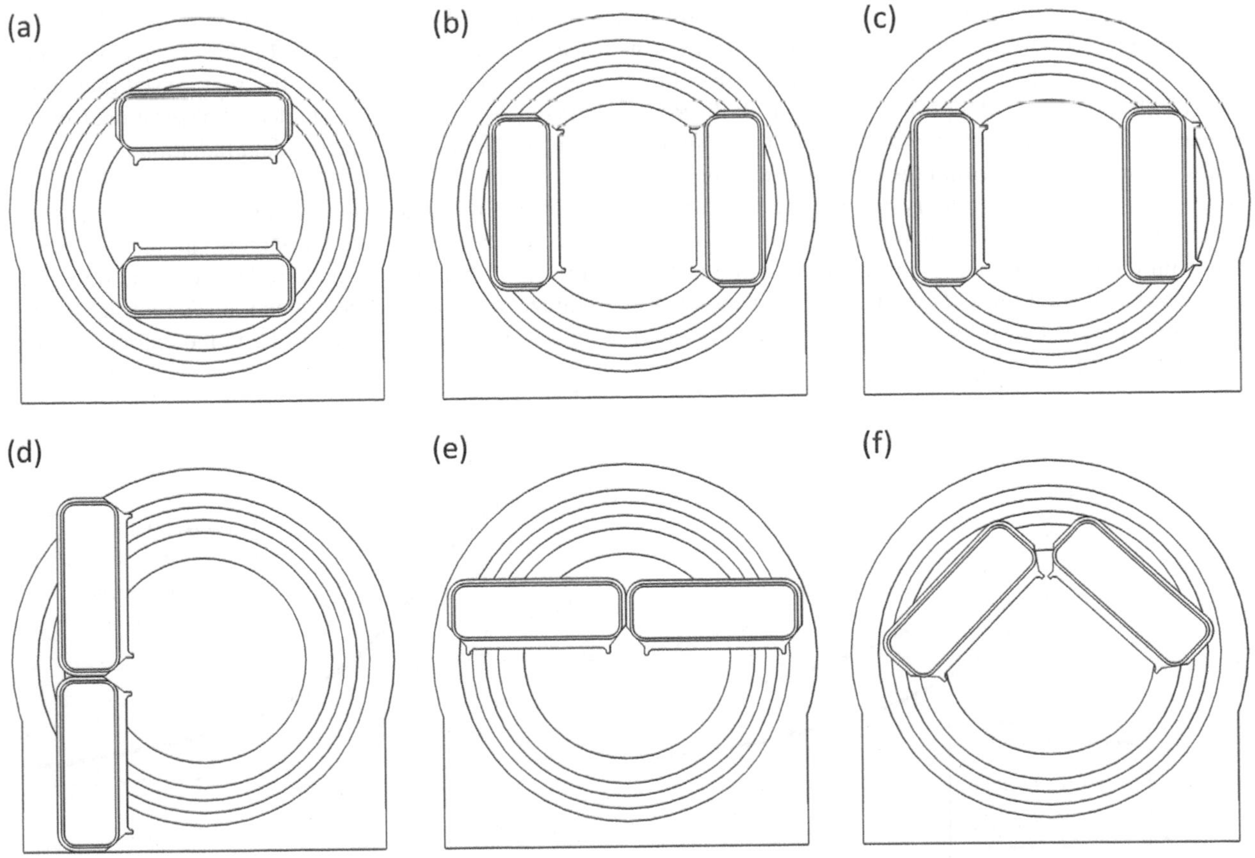

FIGURE 13.3 The basic detector orientations of a gamma camera. Detectors positioned at 180° for anterior and posterior imaging (a) and left and right lateral imaging (b). Detector facing out from gantry allowing posterior imaging with patient seated on a chair (c). Detectors orientated above each other allowing imaging with patient standing (e). Detectors orientated side-by-side allowing imaging of the patient on a hospital gurney (e). Detectors at 90° for tomographic imaging, common in cardiac imaging.

imaging field of view; in older systems these were circular in shape but have since been replaced with a rectangular geometry (generally 40 × 50 cm).

A number of inorganic crystalline materials have been proposed for use in photon detection. However, the most commonly used in gamma cameras is thallium activated sodium iodide, NaI(Tl). The crystal is usually 9.5 mm thick, which is ideal for absorbing the 140 keV photons emitted from ^{99m}Tc. Thicker crystal detectors are also available (15.4 mm), which are more suited to higher energy emitters such as ^{131}I. In the 1990s, 25.4 mm crystals were available for system capable of detecting 511 keV annihilation photons of positron emitters. However, since the clinical introduction of PET imaging these have become obsolete.

When a photon is incident on the scintillator, it causes ionization, predominantly via photoelectric absorption. The electrons within the crystal are excited into conduction bands of the crystal and, when de-excited back to the valence band, release energy in the form of scintillation photon. The number of scintillation photons produced when a photon is absorbed within the crystal is proportional to the energy of the incident photon. Therefore, by 'counting' the number of scintillation photons, the energy and position of the incident photons can be determined. For NaI(Tl) an average of one scintillation photon with a wavelength of 400 nm (blue optical light) is emitted for every 30 eV of energy absorbed. Approximately 4600 scintillation photons are produced as a result of a full absorption of a single incident 140 keV photon. This process is not instantaneous, and the rate of scintillation-photon production decreases exponentially from the time of initial ionization, with a decay constant of approximately $0.23\ \mu s$. Chapter 6 of the present volume provides more information about scintillation detectors and their specific properties.

Thicker crystals inevitably provide better absorption efficiency as the thicker material increases the probability of an interaction. However, for an incident photon absorbed at the front surface of the crystal, due to the inverse square law, the produced scintillation light will be more diffuse by the time the photons are detected at the back of the crystal, which can affect the precision with which the site of absorption is determined. Therefore, when selecting crystal thickness there is a trade-off between detection efficiency and intrinsic spatial resolution. Figure 13.4a shows calculated curves of the absorption probability of incident photons as a function of photon energy for three different crystal thicknesses. Figure 13.4b shows a similar probability curve for the 9.52 mm crystal separating out photoelectric absorption, Compton (incoherent) and Rayleigh (coherent) scattering. It can be seen that, as the energy of the incident radiation increases, the probability of scattering within the crystal also increases.

To protect the scintillation crystal, it is hermetically sealed within a thin aluminium cover with a glass window or 'light-guide' at the back to which the photomultiplier tubes are coupled. As scintillation photons are emitted isotropically from the point of energy absorption, a white reflective surface of the cover helps to reflect scintillation photons back towards the light guide (see Figure 13.5a) and thereby optimize scintillation light detection. The refractive index of the light-guide is carefully selected to reduce any internal reflection back into the crystal and the entrance window of the photomultiplier tube.

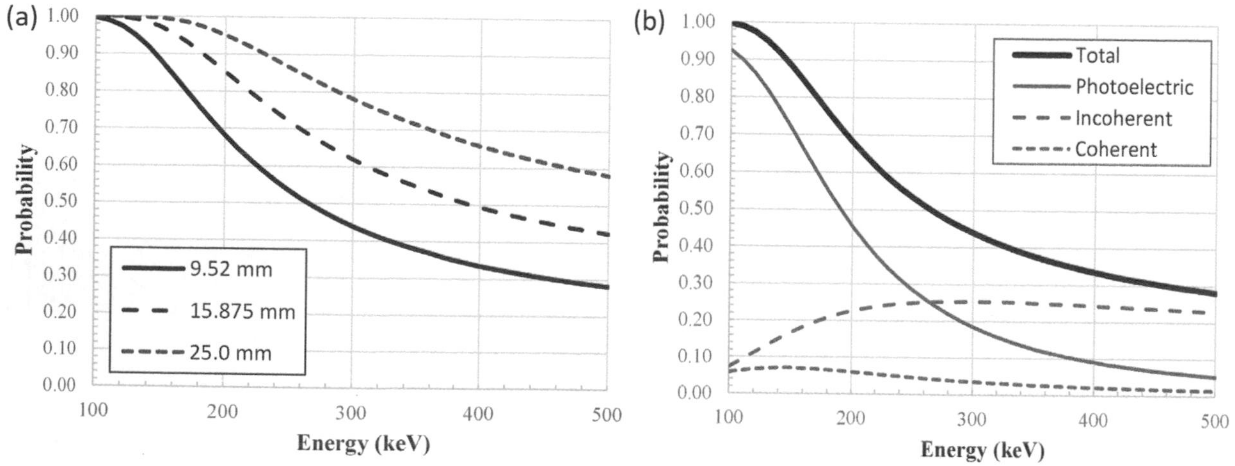

FIGURE 13.4 Absorption efficiency for three different crystal thicknesses (a). When working with radionuclides emitting photons with high energies, a too-thin crystal results in higher noise levels due to poor detection efficiency. Interaction probability of different interaction types for a 9.52 mm crystal (b). As photon energy increases Compton scattering begins to dominate.

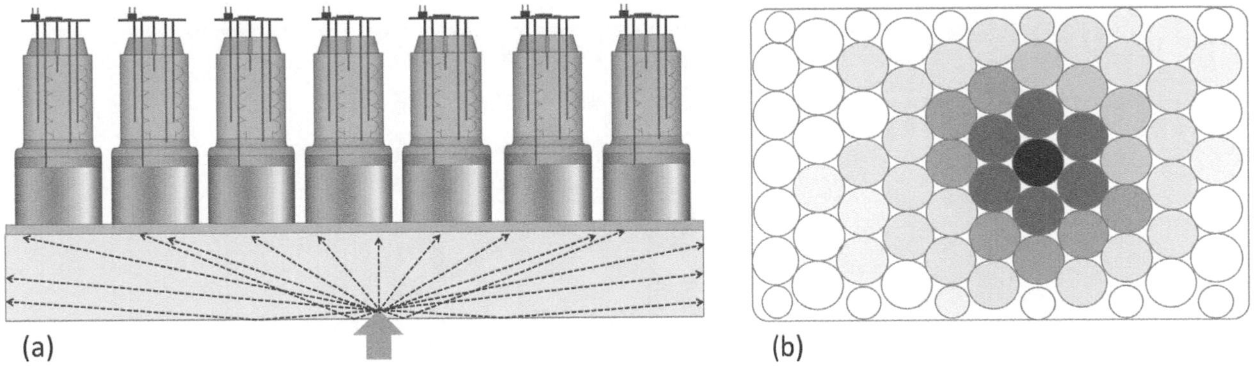

FIGURE 13.5 Diagrams showing the arrangement of PMTs at the back of the scintillation crystal. Scintillation light is isotropically emitted from the point of interaction (a). Conventionally, PMTs are arranged in a hexagonal pattern. The light incident on each PMT will decrease from the point of interaction, indicated by the shading in (b).

13.4 THE PHOTOMULTIPLIER TUBE

The purpose of the photomultiplier tubes (PMT) is to detect the scintillation photons emitted at the back of the crystal and convert them into a measurable electrical signal from which the deposited energy and location of the original incident photon is measured.

As discussed in Chapter 6, the PMT is an evacuated glass tube consisting of a front entrance window often coupled to the light guide with optical grease of suitable refractive index to reduce reflection of scintillation light away from the PMT. In simple radiation detectors, a single crystal/PMT arrangement is used to detect an event. However, for the determination of the interaction point with the crystal, a single PMT is insufficient. Therefore, between 50 and 70 PMT are contained within a gamma camera detector and are arranged across the back face of the crystal (see Figure 13.5). To optimize light detection, it is necessary to cover the highest surface area possible. Therefore, a high packing fraction of PMTs is required to reduce the dead-space between the PMTs. Hexagonal and squared shaped PMTs have been proposed. However, due to the more regular response across the entrance window, it is now more common to use cylindrical PMTs typically 40 to 60 mm in diameter, placed within a hexagonal arrangement. To further optimize the surface coverage smaller PMTs (20–30 mm diameter) are sometimes used to fill gaps at the periphery of the crystal as shown in see Figure 13.5b).

The inner surface at the entrance window of the PMT is coated with a semi-transparent bi-alkali photo-emissive substance (usually cesium antimony) called the photocathode. When scintillation photons enter the entrance window, interaction between the scintillation photons and the photocathode produces photoelectrons, which are produced with a conversion efficiency of approximately 2 to 3 electrons per 10 photons. A thin wire-focusing grid directs the photoelectrons towards a metal plate called a dynode, maintained at a high positive voltage (300 V) relative to the cathode. Photoelectrons accelerate from the photocathode towards the dynode (also coated with a photo-emissive material), which results in additional electrons being emitted from the dynode per incident photoelectron. Additional dynodes with increasing positive voltages are located down the length of the tube, and the number of electrons are multiplied at each dynode as the voltage increases. A series of approximately 10 dynodes are located in a single PMT, with a potential difference of 100 V between each. The total voltage across the entire PMT from photocathode to final diode is therefore approximately 1300 V and the number of electrons increases by a factor of 10^6. The electrons are collected at a final anode and recorded as a small electrical pulse of charge. For a single photon entering the PMT the anode pulse lasts for only a few nano seconds (ns), with a transit time (time from the scintillation photon entering PMT to output pulse) of between 10–50 ns depending on the supply voltage.

The size of the electrical pulse is highly sensitive to the voltage across each dynode and therefore a very stable high voltage to the PMT is required. For this reason, it is common practice to maintain power to the high voltage supply and leave sufficient time for the voltage to stabilize if power to the detector is inadvertently interrupted. Regular 'tuning' of the PMTs is often required to ensure identical output for a given input. This is achieved by making small adjustments to the supply voltage, dynode voltage or amplification of the output signal. The tuning process can be automatically controlled and may form part of the quality control program of the gamma camera. (see Chapter 23).

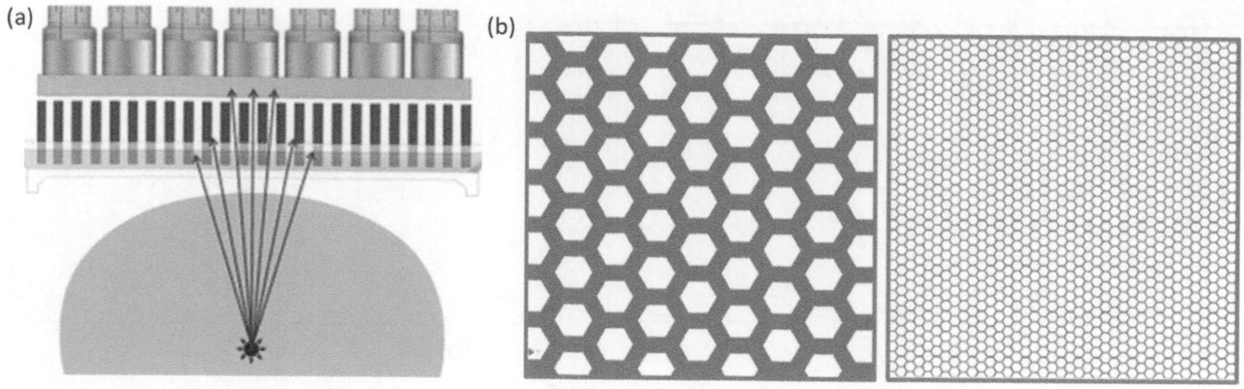

FIGURE 13.6 Cross-sectional image of the detector with a parallel hole collimator, showing how photons not perpendicular to the detector face are stopped by the septa, however some photons at small oblique angles may still pass through holes within the collimator (a). Different collimator designs are available depending on the imaging application and isotope being used. True-scale comparative designs for high and low energy collimators are shown in (b).

PMTs are also sensitive to magnetic fields – even those as weak as the earth's magnetic field – which are able to affect the trajectory of electrons between dynodes. Without sufficient shielding, the output of the PMT will vary when positioned in different orientations (a potential issue when placed in a rotating detector/gantry configuration). Shielding from magnetic fields is achieved by surrounding the PMT with a mu-metal casing, usually a nickel iron alloy.

13.5 THE COLLIMATOR

The overall aim of gamma camera imaging is ascertaining the origin of the emission within the patient. As it is not possible to ascertain any directional information of the photons that interact within the crystal, it is necessary to collimate the emissions before they reach the detector. This is achieved by placing a lead mask (referred to as a collimator) in front of the crystal. The collimator restricts the direction of incident photons so that they can only pass through holes within the lead, whilst lead septa separating the holes attenuate photons that are not travelling in the desired direction (see Figure 13.6a).

Most common collimator designs have septa and holes orientated perpendicular to the crystal face. This design is referred to as a parallel-hole collimator, as all photons that reach the detector are ideally parallel in direction, irrespective of their position across the detector face. The thickness and length of septa will determine the energy of the photons that it will be able to attenuate and, therefore, different collimator designs are required depending on the isotope being imaged. Figure 13.6 b illustrates the difference between a Low-energy collimator (used for ^{99m}Tc imaging) and a High-energy collimator (typically used for ^{131}I imaging). More detail on collimator designs and specification is given in Chapter 14.

13.6 READ OUT AND PULSE ARITHMETIC

The photomultiplier tube outputs an electrical charge in proportion to the number of scintillation photons received. To carry out energy and position analysis, the output current from the PMT is converted into a voltage by an integrating preamplifier. The rise and decay of this charge pulse is predominantly determined by the response time of the scintillator. A gated integration method is used to measure the cumulative charge from the PMT pulse, which is triggered by monitoring changes in output from the PMT dynodes.

Due to the small current produced from the PMT, it is necessary to perform preamplification as soon as the signal is received at the anode so that the signal is not lost to noise within the cabling. Early gamma cameras would use pulse arithmetic of these analogue voltage signals to determine the incident photons interaction position and energy. In modern systems analogue to digital conversion is applied at each PMT, which converts the output to a digital signal representing the amplitude of the pulse. The digital outputs from all PMTs are then passed to a microprocessor for signal arithmetic and image formation.

13.6.1 Energy Determination

If a photon, emitted via radioactive decay within a patient, undergoes Compton scattering within the body or in the detector, the assumption of a direct perpendicular trajectory from source to crystal is invalid, as demonstrated in Figure 13.7. If these events are included when forming the image, their apparent origin will be misrepresented.

By measuring the energy of the incident photons, most of those that have scattered can be distinguished from 'true' events as they will lose some energy during the scattering process. The energy of the photon can be determined by measuring the amount of scintillation light generated in the crystal when the photon is absorbed. In a simplified model, for a scintillation event at the front surface of the crystal, the distribution of scintillation photons from the light-guide will vary according to the inverse square law. The scintillation light incident on an individual PMT will therefore vary depending on the position of the scintillation event with respect to the PMT array (see Figure 13.5 and Figure 13.8a). The total output signal from a group of PMTs around the point of interaction (after some calibration) can be used as a means of determining the energy of the original gamma ray.

$$E = k \sum_{i=1...N} V_i \tag{13.1}$$

where k is a sensitivity factor and V is the signal output of PMT i.

The output signal from a single PMT can also vary with the position that the scintillation photons is incident on the photocathode. Although the focusing electrodes of a photomultiplier tube are designed so that electrons emitted from the photocathode are efficiently collected by the first or following dynodes, some electrons may deviate from their desired trajectories, resulting in a loss of collection efficiency. This loss varies with the position on the photocathode from which the photoelectrons are emitted and influences the spatial uniformity of a photomultiplier tube. In general, tubes especially designed for gamma-camera applications have excellent spatial uniformity. Figure 13.8b shows the typical sensitivity profile of a PMT array.

The total output signal from each PMT will therefore be a product of the scintillation light distribution and the sensitivity profile. This is demonstrated in Figure 13.9a, where the output signal of each PMT is represented by the shaded integral of the PMT signal and represented by V_i. It is evident that due to the inevitable gaps between the PMTs some

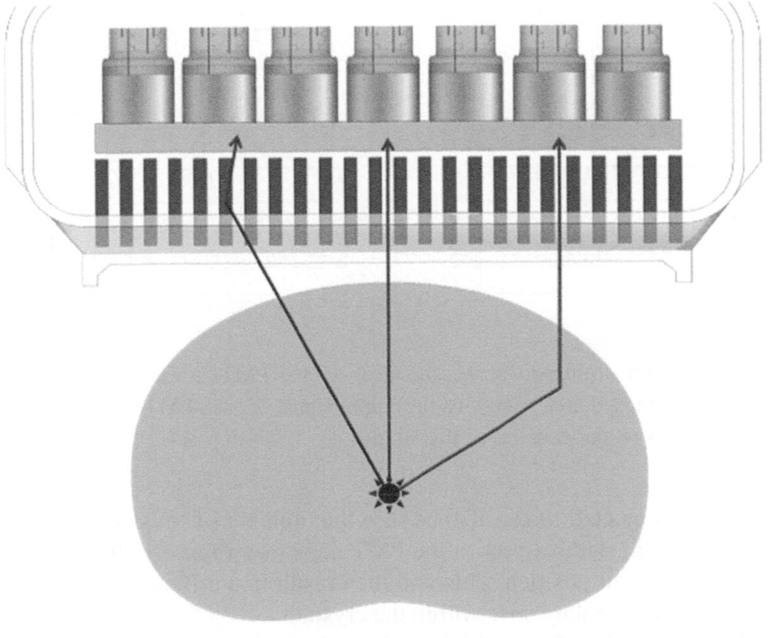

FIGURE 13.7 Photons can scatter within the patient, collimator or crystal. When doing so, the photon loses energy and the amount of scintillation light produced does not reflect the expected initial energy of the photon. When forming an image, energy discrimination can be applied to exclude photons that have undergone scatter.

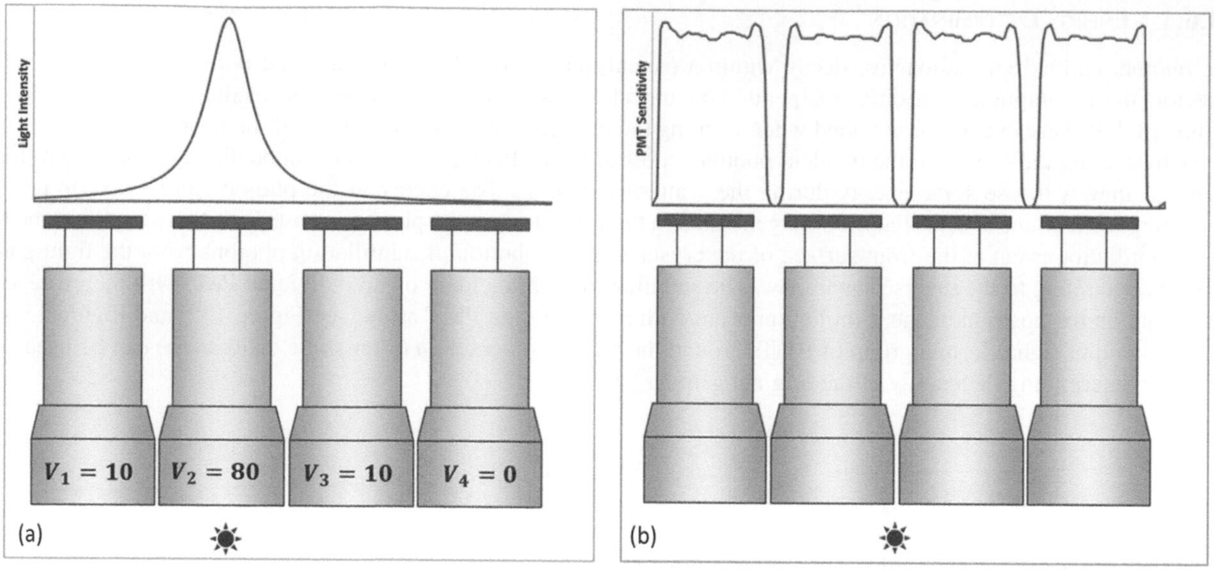

FIGURE 13.8 The distribution of light from a scintillation crystal modelled by the inverse square law and the relative signal from each PMT, V_i, given as the area under the intensity curve corresponding to the PMT (a). The sensitivity profile across an array of PMTs, showing a slight non-uniformity at the edges of the tubes (b).

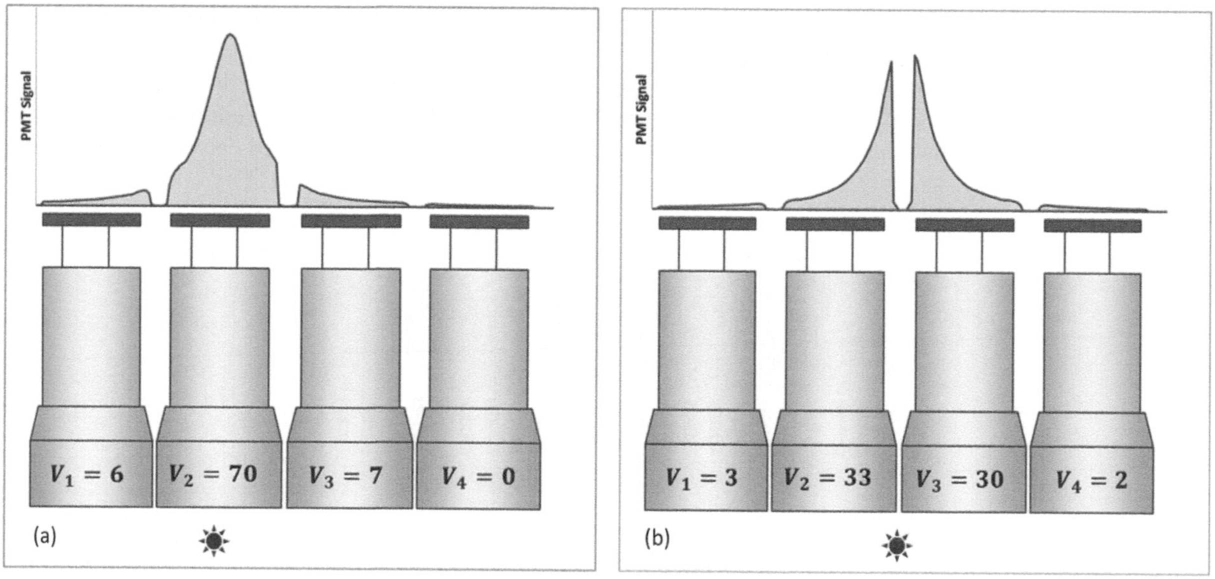

FIGURE 13.9 Position-dependent signal intensity across the faces of four PMTs for an interaction positioned centrally below a PMT (a) and between PMTs (b). The shaded area represents the output signal of each PMT, which are assigned an appropriate value of V_i. The total signal in scenario (b) is 68 per cent that of that in (a).

scintillation light photons are not detected. It can also be seen that number of 'missed' photons will vary depending on the position of the scintillation event with respect to the PMT gaps (see Figure 13.9b). The net effect is that the total output signal differs according to event position. This will then result in a different estimate of energy for scintillation events depending on the position where they arise within the crystal.

This effect is further demonstrated in Figure 13.10, where the total output signal across 7 PMTs is plotted as a function of event position. To overcome this issue a correction map is derived by measuring the observed offset and applying a position-dependent energy calibration, $k_{x,y}$

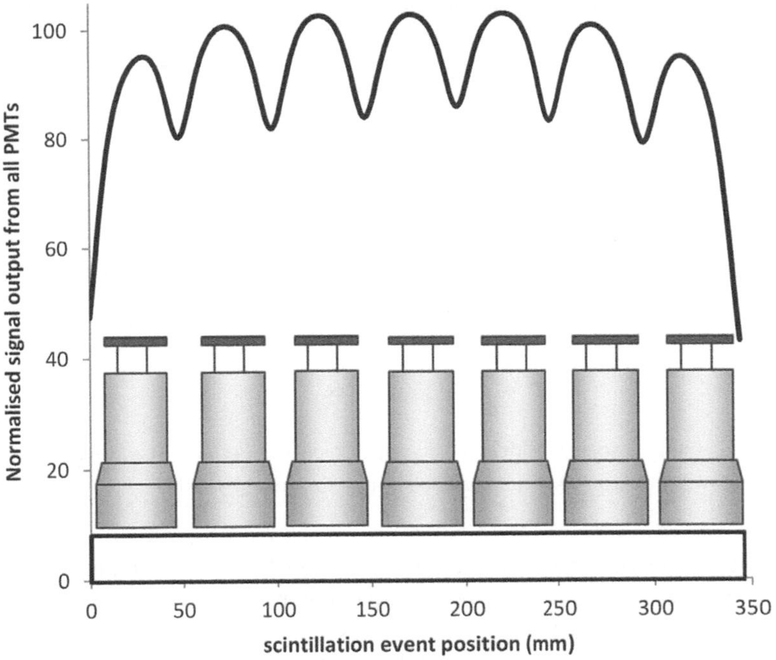

FIGURE 13.10 Normalized signal output from seven photomultiplier tubes for varying scintillation event positions relative to the PMT array.

$$E = \sum_{i=N} k_{x,y} V_i \tag{13.2}$$

The energy measurement process therefore relies on a number of factors, all of which are subject to random statistical errors. This in turn introduces uncertainties into the measured energy, the influential processes of which can be identified as:

- Variation in the amount of ionization within the crystal due to photoelectric absorption of the incident gamma ray.
- The probabilistic nature of Compton events within the crystal.
- The conversion efficiency of de-excited electron energy into scintillation photons.
- The transport of scintillation photons to each PMT (including scintillation light absorption, reflections, and crystal non-uniformities).
- The collection efficiency and non-uniformity of the photocathodes in each PMT.
- Photoelectron production at the photo-cathode.
- Photoelectron collection at the first and subsequent dynodes.
- Photoelectron multiplication at each dynode.
- Variations in high voltage and inter-PMT gain variation.
- Pre-amplification noise.
- Pulse arithmetic to determine energy.

The result is that for incident photons of a given energy, the measured energy of each photon will be normally distributed about an average value that should be equal to the true photon energy. The standard deviation of measured photon energy, σ, describes the statistical error of the measurement process. Often, one expresses the energy resolution in terms of Full-Width at Half-Maximum $\left(\mathrm{FWHM}=2\sqrt{2ln2}\sigma\right)$ at some reference energy, since this value often depends on the deposited energy. Common values of the energy resolution for scintillation cameras are between 8.5–10 per cent (the magnitude of the FWHM in units of keV divided by the photon energy).

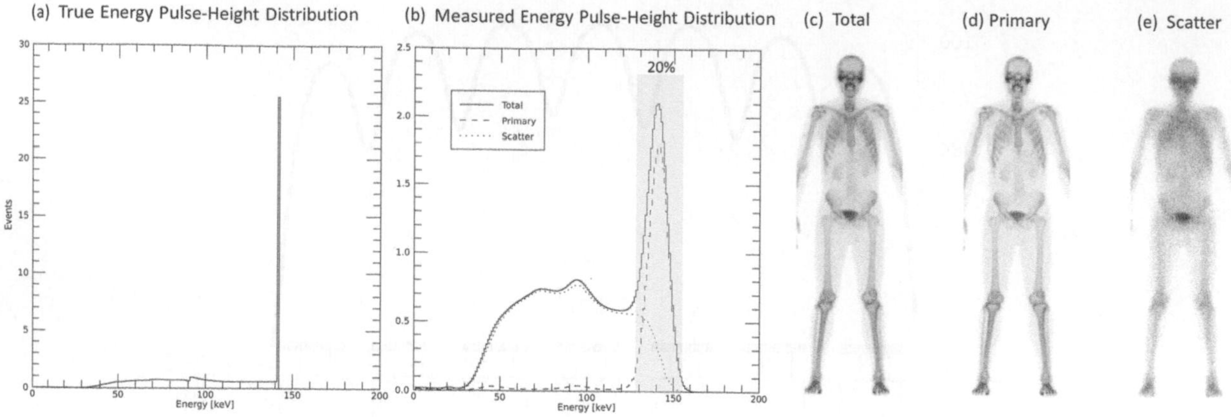

FIGURE 13.11 Modelled pulse-height spectra with typical scatter fractions. Perfect (a) and realistic (b) energy resolutions have been modelled demonstrating how scattered events are included within the energy window.

As a consequence of this imperfect energy resolution, an energy 'window' is required that discriminates incident gamma photons outside a specified energy range. Only gamma photons within the range are accepted as true events and used to form the image. Inevitably, there will be scattered events that, due to the finite energy resolution, still fall within the accepted energy window and will result in a positional error when included in forming the image. To illustrate this effect, a Monte Carlo simulation has been made for a typical 9.5 mm NaI(Tl) scintillation camera equipped with a low-energy, high-resolution collimator and a fictitious ^{99m}Tc distribution in a patient-like computer phantom. If the energy resolution of the detector had been perfect, the measured energy pulse-height spectrum would look like the graph in Figure 13.11a. However, once a representative energy resolution is simulated, an energy spectra of Figure 13.11b is observed. In order to include sufficient events, an energy window corresponding to twice the FWHM is often used. In addition, the components of the energy pulse-height spectrum that originated from scattered events have been included in these spectra.

13.6.2 Image Formation

An obvious necessity for image formation is the ability to determine the position within the crystal where the scintillation events occur. In addition, as demonstrated in the previous section, event position is also required to correct for the varying energy sensitivity across the detector field of view.

Event localization is achieved by using what is commonly referred to as 'Anger Logic' which is analogous to a centre-of-mass calculation. It follows that the PMT closest to the scintillation event will detect the largest number of scintillation photons and therefore output the largest signal (Figure 13.9). In addition, the neighbouring PMTs will detect a lower number of photons depending on their position relative to the scintillation event. Using a weighted average of the signal and the position of the centre of the PMT, the event location can be determined using

$$x = \frac{\sum_{iN} x_i V_i}{\sum_i V_i} \tag{13.3}$$

and

$$y = \frac{\sum_{iN} y_i V_i}{\sum_i V_i} \tag{13.4}$$

Where V_i is the output of PMT i and x_i and y_i are the Cartesian coordinates for the central position of PMT i.

For the PMT arrangement in Figure 13.9a, the coordinate of the scintillation event for a 50mm diameter PMT is;

$$x = \frac{6\times25+70\times75+7\times125+0\times175}{6+70+7+0} = 75.6\text{mm}, \tag{13.5}$$

and similarly, for Figure 13.9b,

$$x = \frac{3\times25+33\times75+30\times125+2\times175}{3+33+30+2} = 97.8\,\text{mm}. \tag{13.6}$$

13.6.3 Nonlinearity

Eqs. 13.3 and 13.4 work for events located at the exact centre or midpoint between PMTs. However, the number of detected scintillation light photons do not vary linearly with event position, and therefore the position estimate is exaggerated towards the centre of the PMT and away from the gaps. The 'nonlinearity' of the position estimate is demonstrated in Figure 13.12a for an array of 7 PMTs.

The result of this non-linearity on the image is a dramatic barrel distortion as shown in Figure 13.12b. A grid of evenly placed point sources across the image field of view are warped towards the centre of the PMTs. As with the energy correction, rather than correct for this effect using an analytical approach, it is conventional to measure the positional offset at different source positions and apply the offset to the image using a correction map. Indeed, the image in Figure 13.12b can be used for that purpose.

Once the corrected x and y coordinates of an event have been determined it is common to 'bin' the events into an image matrix. Each event is referred to as a 'count', and each element of the image matrix is referred to as a 'pixel' and assigned a different intensity or colour depending on the number of counts recorded in that pixel. In most cases a linear grayscale image is generated, with the darker parts of the image depicting regions of higher activity concentration. See Chapter 10 for more details about the digital image.

It is interesting to note that the energy correction of section 13.6.1 is applied before the linearity correction, so whilst the energy correction is position-dependent, it is applied using an uncorrected coordinate system. This may seem

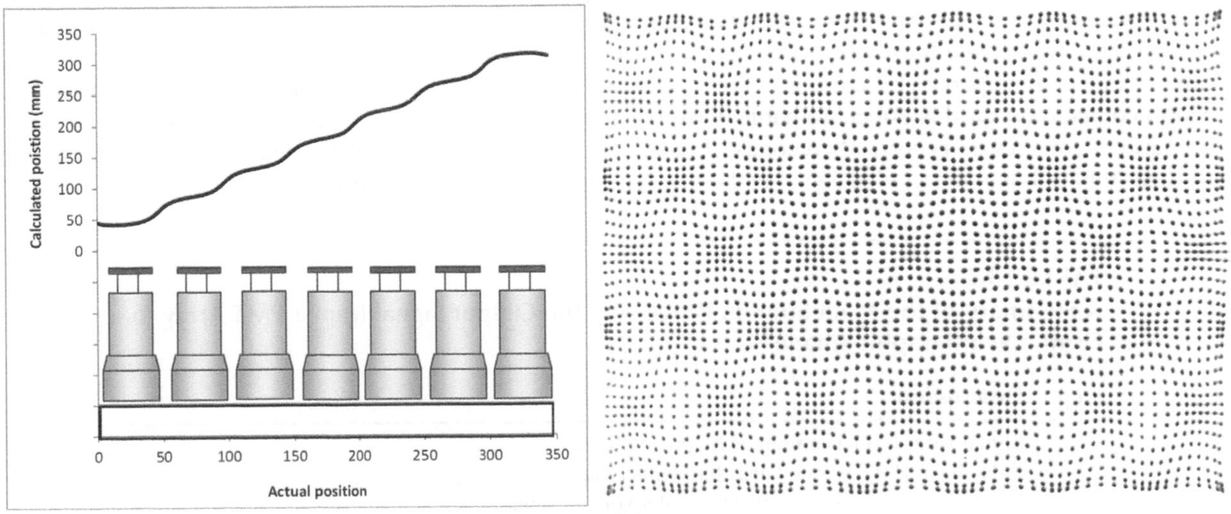

(a) (b)

FIGURE 13.12 A plot of position estimate against true source position, modelled using the inverse square law and typical sensitivity profiles of the PMTs (a). A distorted image of evenly spaced point sources across the imaging FOV. Positions are misplaced towards the centre of the PMT and away from the gaps.

counter intuitive. However, because the energy correction map is derived from a similarly 'warped' image the 'miscalculation' of position is negated at this stage.

13.6.4 System Sensitivity

Whilst energy and linearity corrections will correct for the majority of non-uniformities in the acquired image, it is inevitable that some additional variation in sensitivity will still be present. This can be due to variation in crystal thickness, density and other crystal and detector imperfections. Variation in detector response will also vary with photon energy. As higher energy photons inevitably penetrate deeper into the crystal, the depth of interaction will result in different distributions of scintillation light into the PMTs. Therefore, the energy and linearity correction are strictly only valid for the isotope used to generate the correction map. It is therefore common practice to apply a third correction map that is specific to the isotope being used. Non-uniformities may also be present due to construction tolerances in collimator manufacturing, and some vendors therefore recommend a collimator/isotope-specific uniformity correction.

Corrections for variation in detector sensitivity are made using a uniformity- correction map generated by irradiating the detector with a uniform flux of radiation. When including corrections for collimator effects, this will be acquired extrinsically (with the specific collimator in place) using a large flood source placed on the detector head and covering the entire field of view. If collimator effects are not being included, the map can be acquired intrinsically (without a collimator in place) and a point source positioned at a sufficient distance from the detector to produce a uniform flux across the detector. Specific details on this process are given in Chapter 23.

The image is acquired for a sufficient time period to negate effects of statistical noise. A smoothing filter is then applied, and the reciprocal of the image stored. When acquiring subsequent images, pixel data are multiplied with the reciprocal image and normalized to maintain event totality. A summary of the pulse-processing steps is given in Table 13.2.

13.6.5 Temporal Resolution

Proportionality is assumed between the amount of charge collected and the deposited energy of the incident photon. However, as previously stated, the scintillation process within NaI(Tl) continues for a few microseconds. For events that are separated by sufficient time, the integral of the current from the PMTs over an appropriate integration period can be used to determine the energy of each incident gamma photon (Figure 13.13a). If the flux of radiation is too high, the proceeding signal may commence before the current from the prior event has fully decayed (Figure 13.13b). In this scenario the signals from such events will overlap and result in a superposition of current. For the given integration period it is not possible to separate these two events, and a single event will be recorded with an energy equal to the sum of the two original gamma photons (Figure 13.13b). Most scintillation cameras have some kind of pulse pile-up rejection method to compensate for this effect. To partially overcome this problem, a shorter integration period can be used. In this scenario, the first event is correctly recorded (Figure 13.13c). However, the signal generated from the second event is still superimposed on the decaying signal of the first. The net effect of such occurrences is that events are potentially

TABLE 13.2
A Summary of the Main Steps in Processing the Output Signal from a PMT Array to Form a Scintigraphic Image

Step	Description
1	Signal from PMT is amplified by preamplifier
2	Pulse from PMT is gated and charge integrated
3	Integral of signal is converted to digital voltage signal
4	Total output from PMT group is used to estimate gamma energy
5	Weighting of signal from adjacent PMTs is used to estimate scintillation position
6	Correction of energy estimate is applied using energy correction map
7	Gating is applied to event if outside a user-defined energy window
8	Correction to event position is applied using a linearity map
9	Event position is saved within image matrix
10	Correction of additional inhomogeneities is applied using a uniformity/sensitivity map

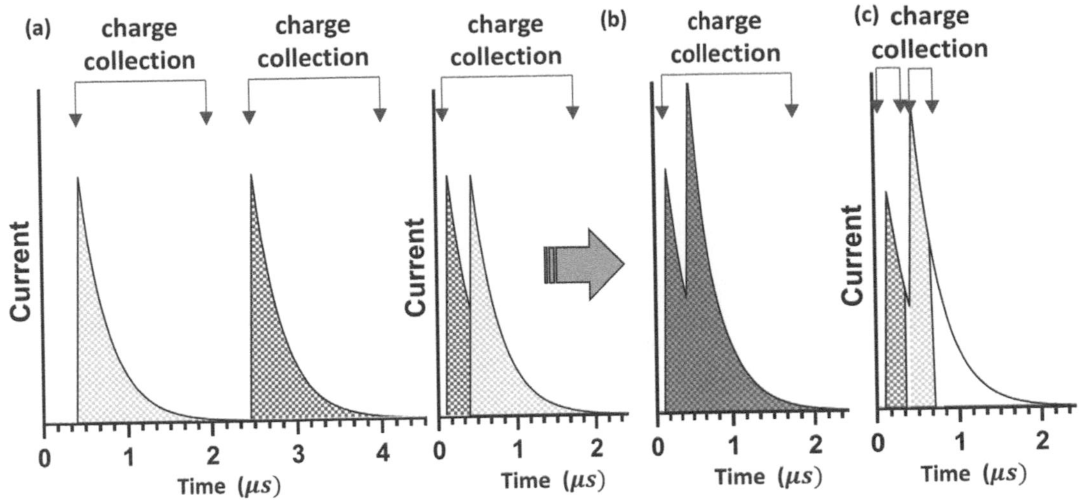

FIGURE 13.13 Graphs showing the shape of the charge pulse from the PMT. When count rates are adequately low, pulses can be separated (a). When count rates are too high signals from two events overlap and become superimposed (b). An incorrect energy is therefore assigned. A shorter integration period can be used to measure the pulse (c). However, the second pulse is still influenced by the first.

recorded with a higher energy than they actually possess. This can result in their falling outside of the energy-selection window and, hence, an overall loss of events. Alternatively, unwanted scattered events can be registered inside the energy window. The fraction of scatter in the energy window and in the image thus increases with count rate.

Since the position of an event is calculated from the energy estimate, superimposed PMT signals will result in a mispositioning of the event location. If there exist two sources of high activity, then events will often be registered somewhere along a line bridging the sources.

To overcome the misregistration of event position and energy, a period of dead time can be assigned after the collection of an event. This period can be fixed, or determined, based on the current from the PMT. During this time, the system is said to be paralyzed as no events are being recorded. Alternatively, the known decay time of the signal and the time difference between the two events can be used to model and correct for the contribution of the initial event on the later. Understanding how a system operates at high count rates is vital to ensuring consistent image quality. More details about the effects of system dead time are given in Chapter 19.

14 Collimators for Gamma Ray Imaging

Roel van Holen

CONTENTS

14.1 INTRODUCTION

In order to extract directional information from light photons – for example, for photography – one relies on the focussing properties of optical lenses. High energy γ-rays cannot, however, be focussed by lenses. Therefore, in order to extract directional information out of a beam of incoming γ-rays, collimators are used. Their task is to absorb all photons not traveling according to the collimator-imposed direction. As a result, only photons traveling in the required direction will pass the collimator and consequently be detected. In the rectilinear scanner, a single-hole collimator was first employed to accept photons only from the region in front of the scanner. Soon, the advantage of magnification induced by a pinhole collimator for imaging of small organs like the thyroid, was appreciated. In the Anger camera, single-hole collimation was replaced by parallel-hole collimation while the sensitivity gain by using multiple pinholes became clear in the early 1970s [1]. Focussed collimators, first used by Anger in the tomoscanner are now used in the form of fanbeam and conebeam collimators, mainly for brain imaging. All the aforementioned types of collimation suffer from the same trade-off between sensitivity, spatial resolution and FOV. A comparison of the most common collimators with respect to these three criteria will be discussed. For the derivation of the analytical formulas that will be used, the reader is referred to the overview papers by Wieczorek [2], Accorsi [3] and Moore [4].

14.2 PARALLEL-HOLE COLLIMATORS

A parallel-hole collimator can be regarded as a numerous amount of closely packed long, parallel holes through a slab of highly attenuating material such as lead or tungsten. Traditionally, it is fabricated by sheets of lead foil, folded in half-hexagonal holes precisely stacked together to form a honey-grate structure. Next to folded collimators exist cast collimators based on precise moulding techniques. The latter type of collimator is more precise and also can be fabricated with smaller thickness of the septa t. Besides the septal thickness, the most important design parameters of

DOI: 10.1201/9780429489556-14

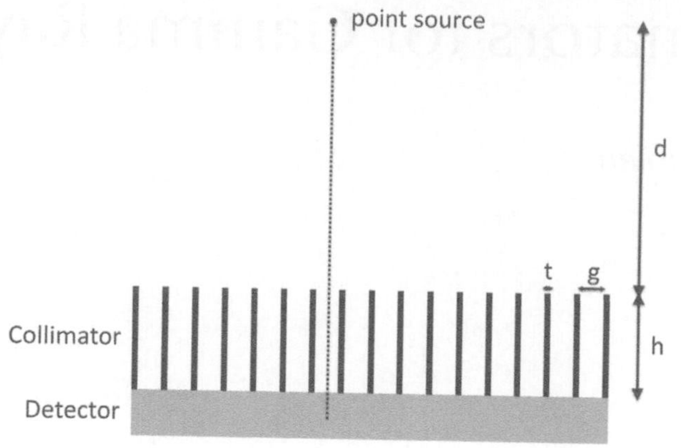

FIGURE 14.1 Section through a parallel-hole collimator with the indication of the notation. This schematic also represents an axial cut through a fanbeam collimator.

a parallel-hole collimator are its height h and hole size g (Figure 14.1). A parallel-hole collimator makes direct 1-to-1 projection images of the source on the detector. Therefore, the FOV of a parallel-hole collimator is equal to the size of the detectors, which makes this type of collimator generally applicable for all types of nuclear medicine examinations.

A critical parameter that determines the collimator design is the energy of the photons that need to be stopped. The septal thickness and the attenuation coefficient μ of the collimator material at the photon energy will determine the amount of septal penetration. This effect can be taken into account by defining an effective collimator height h_{eff}:

$$h_{eff} = h - \frac{2}{\mu} \tag{14.1}$$

Next, the desired sensitivity and resolution are the parameters determining the design of a collimator. A number of typical collimators, together with their most important geometric parameters and properties can be found in Table 14.1. In the next paragraphs, it will become clear that it is wise to use a different type of collimator when a large FOV is not required.

The most important property of a collimator is the trade-off between sensitivity and spatial resolution. The sensitivity of a parallel-hole collimator can be generally expressed as:

$$S_{paho} = \frac{g^2}{4\pi h^2} \tag{14.2}$$

Collimator resolution is given by

$$R_{c,paho} = g\frac{h+g}{h}. \tag{14.3}$$

Here, g is the hole size, h is the height of the collimator, and d is the distance from the collimator to the point where the sensitivity is evaluated. The complete system resolution of a gamma camera, R_s is however not only determined by the collimator. The detector resolution or intrinsic resolution. R_i, arising from the finite pulse discriminating capabilities of the detector also contribute to the system resolution.

This is important to mention here because the extent to which this intrinsic resolution contributes to the total resolution is determined by the type of collimator. For a parallel-hole collimators, assuming both collimator and detector PSFs are 2D Gaussians with respective FWHM R_c and R_i, the total system PSF is a 2D convolution of both PSFs resulting in a system FWHM of:

$$R_{s,paho} = \sqrt{R_{c,paho}^2 + R_i^2}. \tag{14.4}$$

TABLE 14.1
Properties of Some Typical Collimators.

	FOV@10cm (mm × mm)	g (mm)	h (mm)	t (mm)	R_{sys}@10cm[b] (mm)	S (cps/MBq)
Parallel-Hole (PH)						
General All Purpose (GAP)	240×400	1.57	25.4	0.24	8.3	229
High Resolution (HR)	240×400	1.40	27.0	0.18	7.2	168
Ultra-High Resolution (UHR)	240×400	1.40	34.9	0.15	6.2	104
Medium Energy (ME)	240×400	3.40	58.4	0.86	9.7	172
High Energy (HE)	240×400	3.81	58.4	1.73	10.8	160
Fanbeam[a] (FB)						
General All Purpose (GAP)	240×320	1.57	25.4	0.24	8.1	286
High Resolution (HR)	240×320	1.40	27.0	0.18	7.0	210
Ultra-High Resolution (UHR)	240×320	1.40	34.9	0.15	5.9	131
Conebeam[a] (CB)						
General All Purpose (GAP)	192×320	1.57	25.4	0.24	8.1	357
High Resolution (HR)	192×320	1.40	27.0	0.18	7.0	262
Ultra-High Resolution (UHR)	192×320	1.40	34.9	0.15	5.9	163
PinHole (PiH)						
General All Purpose (GAP)	218× 364	3.0	110		6.4	56
High Resolution (HR)	218× 364	2.0	110		5.0	25
Ultra-High Resolution (UHR)	218× 364	1.5	110		4.2	14

Notes: [a] Focus = 50 cm.
[b] Transaxial, calculated with R_{int} = 3 mm.
Source: Data from PRISM 3000XP SPECT Camera (Philips Medical Systems$_{TM}$)

Since the intrinsic resolution is usually small, and collimator resolution degrades with increasing collimator distance, it will be important only at small source-to-collimator distances; here it becomes the dominant factor in the system resolution. From the above formulas it can be read that, for a better spatial resolution (smaller R_c), a smaller hole size g or a larger collimator height h should be chosen. However, regarding the sensitivity, a linear change in g or h will result in a quadratically decreased sensitivity. The following far field ($d >> h$) relation between sensitivity and resolution shows the main limitation of a gamma camera, namely that resolution cannot be optimized without compromising sensitivity and vice versa:

$$S \sim R_c^2. \tag{14.5}$$

This relation can also be observed from Table 14.1: while the resolution improves with a factor 1.4, going from the GAP to the UHR collimator, sensitivity degrades by a factor 2.16. The trade-off is thus not exactly quadratic since parameters such as septal thickness also play a (limited) role in the above equations. More detailed expressions of sensitivity and resolution which take these effects into account can be found in [2].

14.3 FANBEAM AND CONEBEAM COLLIMATORS

In fanbeam collimators the holes are tilted toward a focal line (Figure 14.2a), while conebeam collimators have their holes focussed to a point (Figure 14.2b). Because the focal line of a fanbeam collimator is parallel to the axis of rotation of the camera, we only have a magnification effect in the transaxial direction and no magnification in axial direction. On the other hand, projections made with a conebeam collimator are magnified both in axial and transaxial directions.

The focal locus is usually chosen to lie at the opposite side of the patient or object under investigation in order to have magnification in the whole FOV. It has been shown, however, that in certain imaging conditions it can be better to choose the focal length inside the FOV – for example, for brain imaging [5]. Also, for cardiac imaging, asymmetric focussing collimators have been used. Compared to parallel-hole collimators, an extra parameter is introduced

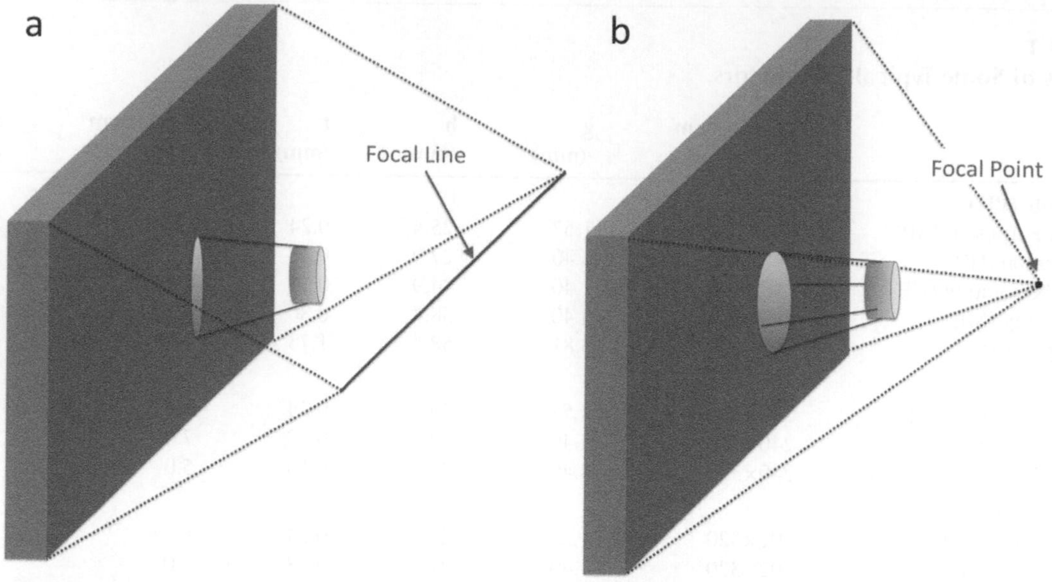

FIGURE 14.2 (a) In fanbeam collimators, all holes are directed to a line parallel to the axis of rotation of the scanner. (b) Conebeam collimators have all their holes directed to one single point.

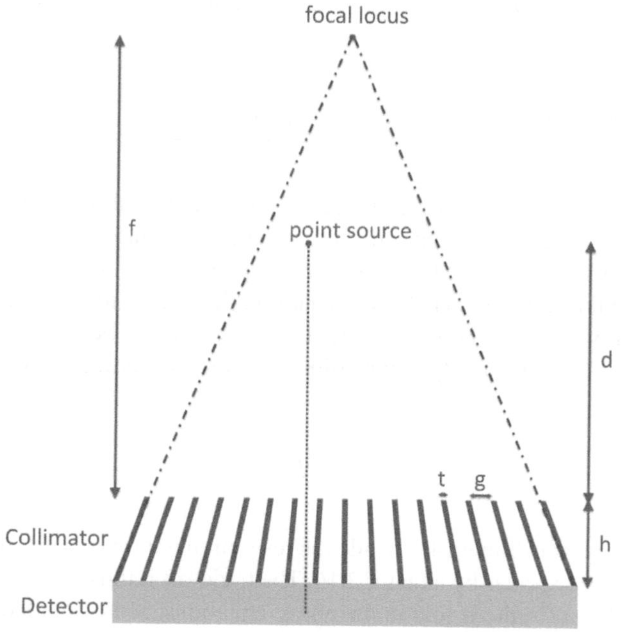

FIGURE 14.3 Transaxial cut through a fanbeam collimator or conebeam collimator. This schematic also represents the axial cut through a conebeam collimator.

in the sensitivity formula of converging-hole collimators: namely f . f is a design parameter representing the distance between the collimator surface and the focal locus of the collimators (Figure 14.3). The sensitivity of a fanbeam collimator is, compared to parallel-hole sensitivity, increased with a factor $\left(\frac{f}{(f-d)}\right)$. For a conebeam collimator sensitivity is additionally increased with $\left(\frac{f}{(f-d)}\right)$. This results in the general sensitivity expression:

$$S = \frac{g^2}{4\pi h^2}\left(\frac{f}{(f-d)}\right)^n, \tag{14.6}$$

where n respectively equals 1 and 2 for fanbeam and conebeam collimators, and d is the distance of the point in the FOV where the sensitivity is considered to the collimator surface. It is thus the ratio of f and the distance from the point to the focal locus (f-d) which determines the gain in sensitivity. There is no gain at the collimator surface (d = 0) while sensitivity gain increases the more the focal locus is approached. The sensitivity formula for parallel-hole collimators (Eq. 14.2) can be derived from Eq. (14.6) by assuming n = 0 but can also be obtained by the insight that a parallel-hole collimator is a converging collimator with focal distance f equal to infinity. Collimator magnification m is defined as:

$$m = \left|\frac{f+h}{f-d}\right|. \tag{14.7}$$

The collimator resolution of Eq. 14.3 has to be multiplied by a factor $\left(\frac{f}{(f+h)}\right)$ for the directions where there is focussing. The converging beam collimator resolution $R_{c,cb}$ becomes:

$$R_{c,cb} = g\frac{h+d}{h}\frac{f}{f+h}, \tag{14.8}$$

the intrinsic resolution is compromised by the magnification effect, and thus the system resolution becomes:

$$R_{s,cb} = \sqrt{R_{c,cb}^2 + \left(\frac{R_i}{m}\right)^2}. \tag{14.9}$$

Therefore, a slightly better resolution is obtained with converging collimators. The price paid for the better sensitivity and slightly better resolution is the smaller FOV_t which reduces to:

$$FOV = \frac{D}{m}, \tag{14.10}$$

with D the FOV of a parallel-hole collimator or, in other words, the size of the detector. The axial FOV of a fanbeam collimator remains unchanged compared to a parallel-hole collimator. A more detailed investigation of converging collimator characteristics can be found in [5–9] while the problem of singularity at the focal locus (S → ∞) of converging collimators is solved in [10] and [11]. A review on convergent beam collimation, including pinhole collimation, was provided by Gullberg [12].

14.4 PINHOLE COLLIMATORS

A pinhole is, as applied in the camera obscura, a very natural way of projecting (and mirroring) a large object (e.g. the outside world) to a small surface (e.g. a wall in a dark room). In nuclear medicine, pinhole collimators were used from the very beginning to make projections of small regions of interest to a larger detector surface. As opposed to the camera obscura, where objects are mapped to a minified projection image, a pinhole collimator exhibits the magnification effect by placing the object close to the pinhole. This magnification effect will again – as with converging hole collimators – minimize the effect of intrinsic detector resolution. This property, together with the high sensitivity in the near field, make collimation with pinholes an attractive alternative when imaging small organs like the thyroid [13], parathyroids [14], joints [15] or kidneys [16]. The magnification *m* of a pinhole collimator is defined as:

$$m = \frac{h}{d}, \tag{14.11}$$

with *h* the detector pinhole distance, and *d* the distance from the object to the pinhole plane (Figure 14.4).

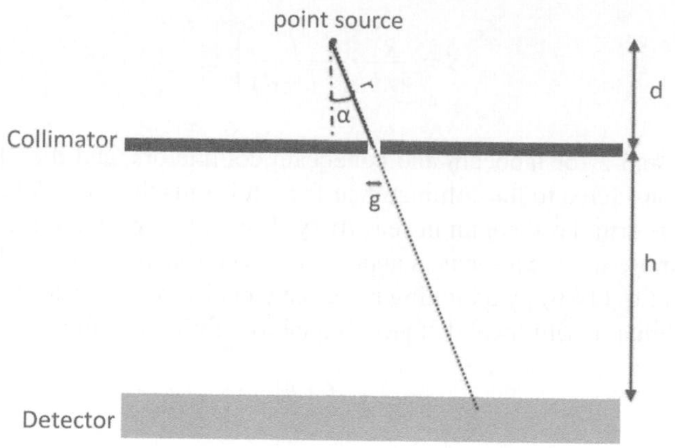

FIGURE 14.4 Cut through a pinhole collimator.

The sensitivity of a pinhole collimator is equal to the normalized solid angle subtended by the pinhole. For a round pinhole, sensitivity equals:

$$S = \frac{\pi \frac{g^2}{4} \cos(\alpha)}{4\pi r^2} = \frac{g^2}{16d^2} \cos^3(\alpha), \tag{14.12}$$

with $r = d / \cos(\alpha)$. A square pinhole has sensitivity equal to:

$$S = \frac{g^2 \cos(\alpha)}{4\pi r^2} = \frac{g^2}{4\pi d^2} \cos^3(\alpha), \tag{14.13}$$

Collimator resolution is equal to the resolution of a parallel-hole collimator:

$$R_{c,ph} = g\frac{h+g}{h}. \tag{14.14}$$

For pinholes, h however represents the detector-pinhole distance which is usually larger compared to the height of a parallel-hole collimator. Therefore, collimator resolution of a pinhole will be better. Total system resolution is given by:

$$R_{s,ph} = \sqrt{R_{c,ph}^2 + \left(\frac{R_i}{m}\right)^2}. \tag{14.15}$$

The FOV of a pinhole collimator gets smaller by a factor m. Thus, the closer to the pinhole, the smaller the FOV. Here the trade-off between sensitivity, spatial resolution and FOV again plays an important role. From sensitivity and resolution points of view, we would like to image as close as possible to the pinhole. However, the FOV reduces to zero when d = 0 (in reality, the FOV reduces to the size of the pinhole). In principle, when d would be larger than h, an FOV larger than the detector could be scanned. This is however never done in practice because the sensitivity decreases proportional to d^2 and already is too low at d = h.

14.4.1 Multi-pinhole Collimators

Pinhole collimators are mostly applied when a better spatial resolution is desired. When the object is small, and when the pinhole can physically be brought near the object, better resolution at equal sensitivity is possible. From the formulas it is clear that again, with better resolution (smaller pinhole diameter g), sensitivity is quadratically decreased.

Thus, when even higher resolution is desired, such as in preclinical imaging, sensitivity can be increased by directing multiple pinholes in the direction of the object. This approach was first introduced in nuclear medicine in the early 1970s by Wouters [1]. Multi-pinhole systems have had limited application in clinics but are still under active investigation and are promising for imaging of the heart [17–19], brain [20], kidneys [21], and bone [22]. On the other hand, multi-pinhole collimation has become the standard in small-animal SPECT imaging. Pre-clinical multi-pinhole systems can be classified according to their number of pinholes, but also as stationary or non-stationary. Also, the degree of overlap of projections from different pinholes, called multiplexing, is an important area of research. With these systems, resolutions down to 150 μm have been reported. Numerous different systems exist from which the most important have been developed by McElroy [23] (A-SPECT), Schramm [24] (HiSPECT), Beekman [25] (U-SPECT), Lackas [26] (T-SPECT) and Kim [27] (SemiSPECT). An overview of multi-pinhole collimation for pre-clinical imaging can be found in [28].

14.4.2 Sampling Completeness

One of the major problems in collimation with pinholes is the sampling completeness. Complete data are only recorded in the transaxial plane of the pinhole (which we will assume to be the central slice). Other transaxial slices suffer from incomplete data, which results in artefacts after image reconstruction. Typically, these artefacts become more pronounced the further we move away from the central slice. The extent of these artefacts are usually visualized by scanning a Defrise disk phantom (Data Spectrum, Inc.) which is a stack of axially repeated cold and hot slices. Typically, for a pinhole collimator, the more a slice is axially off-centre, the more artefacts it will show after reconstruction.

The first to address the issue of data completeness for parallel-hole collimation was Orlov [29]. For a point in the FOV of the camera, each possible photon detection path to the detector is intersected with a unit sphere, centred at the point considered. The union of points formed by the intersection of the projection lines and unit sphere defines a region Ω on the Orlov sphere. Orlov's condition for data completeness is fulfilled when any circle on the Orlov sphere intersects Ω. Sampling incompleteness becomes more severe if more great circles can be drawn without intersecting Ω.

For a parallel-hole collimator, Orlov's condition holds for all points in the FOV, when the detector is rotated for at least 180°. For points in the central slice in the FOV of a pinhole collimator, Ω is also a great circle or equator and thus data are complete (Figure 14.5a). The more one moves axially off-centre, the more the circle formed by the points Ω shifts toward the poles and more great circles can be found that do not intersect Ω (Figure 14.5b). Thus, the more we move off-centre, the less complete the data will be. Also, conebeam collimators suffer from data incompleteness toward the edge of the FOV, but there the effect is usually less pronounced since, in most cases, the focal distance is relatively large. Solutions for the problem of sampling incompleteness consist of scanning orbits different from circular ones (e.g. helical scans) [30, 31] or using multi-pinhole collimators [32].

14.5 SLIT-SLAT COLLIMATORS

Slit-slat collimators are based on a number of parallel slats that are placed on the detector (Figure 14.6). This slat collimation results in one-directional collimation as opposed to two-directional collimation in a parallel-hole collimation.

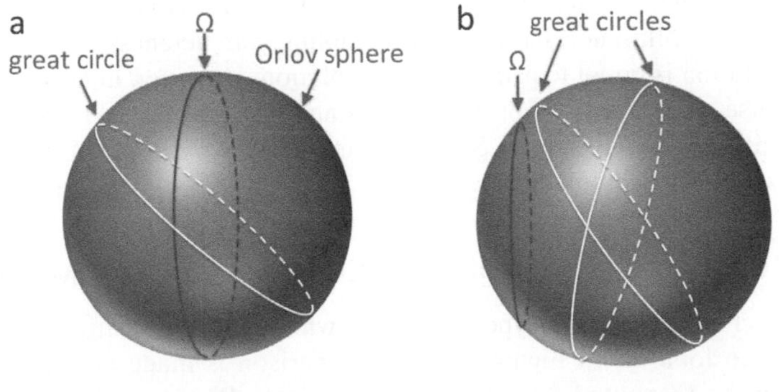

FIGURE 14.5 (a) Projections on the Orlov sphere for all points in the FOV of a parallel-hole collimator and the central transaxial slice in the FOV of a pinhole collimator. No great circle can be found that does not intersect Ω while in (b), showing the projections for an off-centre point in the pinhole FOV, a lot of great circles can be drawn that do not intersect Ω.

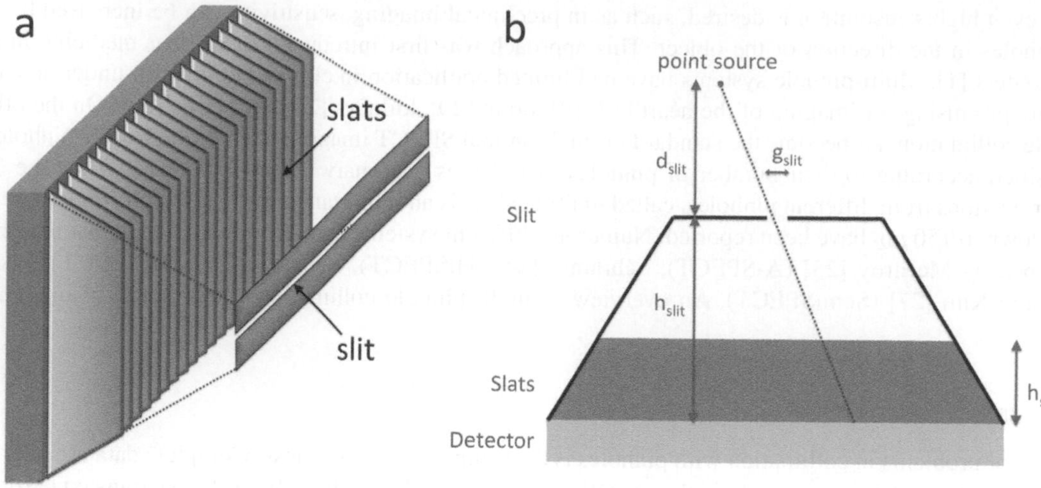

FIGURE 14.6 (a) A slit-slat collimator combines properties of a parallel-hole collimator in axial direction while exhibiting the high sensitivity properties of a pinhole collimator in transaxial direction. The transaxial geometry is clarified in (b).

However, in the direction perpendicular to the slats, photon selection is the same as with a parallel-hole collimator. Most commonly, the slats are oriented perpendicular to the axis of rotation (AoR), resulting in an axial collimation equal to parallel-hole collimation. To get information from the transaxial direction, a slit is placed above the slats parallel to the AoR. This results in pinhole collimation in the transaxial direction. Slit-slat collimation is thus nothing more than parallel-hole collimation in axial and pinhole collimation in a transaxial direction. The advantage of this collimator is that there will be no artefacts in the reconstructed image when axially moving away from the centre, and that a higher sensitivity can be obtained for equal (transaxial) resolution compared to a pinhole collimator. Resolution formulas can be borrowed from parallel-hole collimators for the axial direction and from pinhole collimation for the transaxial direction. Sensitivity calculation is, however, more complicated and for on-axis points has been shown to be equal to the geometrical mean of pinhole and parallel-hole sensitivity [33]. A more general expression of the sensitivity that is correct for all points in the FOV is given by Accorsi [34]:

$$S = \frac{g_{slit} \cdot g_{slat}}{4\pi d_{slit} \cdot h_{slat}} \cos^3(\alpha), \tag{14.16}$$

with g_{slit} and g_{slat} respectively the slit width and the slat spacing. d_{slit} is the perpendicular distance between the slit and the point where sensitivity is evaluated while h_{slat} is the slat height. When a smaller transaxial FOV is allowed, a slit-slat collimator can be used as an alternative, which is situated in between pinhole collimation and parallel-hole or fanbeam collimation. In transaxial direction, the slit-slat collimator inheres the resolution properties of a pinhole collimator – which is usually better due to larger h – while sensitivity is still high. Also, sensitivity does not drop as fast with collimator distance d compared to a pinhole collimator. Furthermore, sampling completeness is fulfilled for an orbital scan in all points of the FOV. Axial resolution is equal to parallel-hole resolution. However, in most designs, the object cannot be placed near to the slats because of the presence of the slit causing an offset in axial resolution (Figure 14.8b). Slit-slat collimators were first applied in the SPRINT brain SPECT camera design by Rogers [35, 36] and have been applied in cardiac studies [37] as well as in small-animal imagers [38].

14.6 OPTIMAL CHOICE OF COLLIMATOR AND COLLIMATOR OPTIMIZATION

To answer the question of which collimator type to use for which study, sensitivity, spatial resolution and FOV are the main characteristics to look at. In Figure 14.7, a comparison is made of the sensitivity of the previously described collimators. It can be respected that pinhole or slit-slat collimation should be used when the object is small and can be placed close to the detector. Further away, fanbeam and conebeam have higher sensitivity, at the cost of smaller FOV.

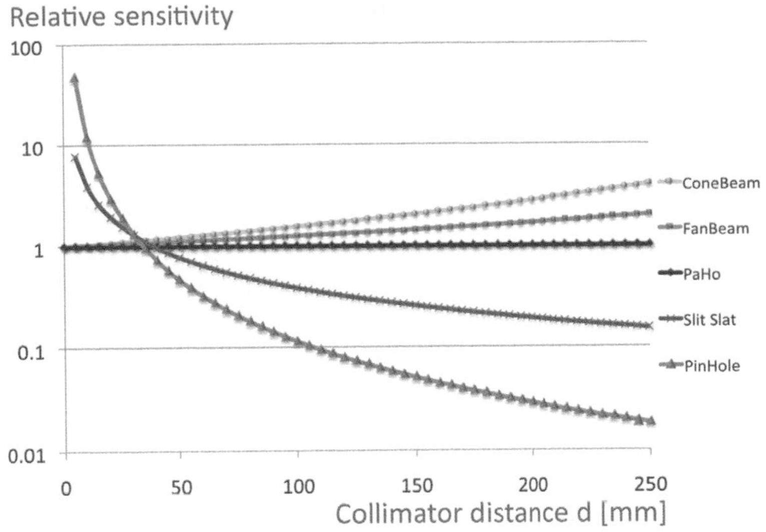

FIGURE 14.7 In this plot, the on-axis sensitivity is compared relative to the sensitivity of a parallel-hole collimator. Closer than 3 cm to the collimator, it is advantageous to use a pinhole or slit-slat collimator while for the converging hole collimators, sensitivity increases while moving further from the collimator. Design parameters for drawing these plots can be found in Table 14.1.

In Figure 14.8, collimators can be compared on the basis of respectively spatial resolution (a and b) and relative size of the FOV (c and d). Resolution is generally better with pinhole collimators. Slit-slat collimators offer better transaxial resolution while axial resolution is worse. This is due to the fact that the object cannot be placed close to the slats due to the slit, which usually is situated higher than the slats. The design parameters are based on clinically used collimators (Table 14.1). As a summary one could state that parallel-hole collimation is the best choice for whole-body imaging, which requires a large FOV for a reasonable scan time. Fanbeam and conebeam collimators are only advantageous when smaller organs are imaged at relatively large distance because their sensitivity increases with the collimator distance. When the organ of interest is small and can be placed close to the detector (small animal imaging), pinhole or multi pinhole collimation is preferable because for equal sensitivity, much better spatial resolution can be obtained. The slit-slat collimator should be situated in between parallel or fanbeam collimation and pinhole collimation. It combines the advantages of superior transaxial resolution and high sensitivity close to the slit. Furthermore, extended axial FOV and extended axial sampling completeness with respect to pinhole collimation is traded for axial resolution.

For a long time, there was only an optimization of collimators with respect to the allowable fraction of penetration of ^{99m}Tc photons and with respect to the desired resolution at a certain distance [39]. This ad-hoc approach does not take into account photon noise, nor background nor lesions variability, which influence both lesion detection and activity estimation. An optimization of parallel-hole collimators based on projection images has been made by Moore [40] and, more specifically, with respect to optimal detection and estimation for the special case of ^{67}Ga imaging in [41]. For tomographic imaging, an optimization of the collimator has been proposed by Zeng [42]. This study and [43] suggest that for SPECT, one should use larger collimator holes at the expense of spatial resolution, which can be recovered using an accurate system model during image reconstruction. Collimator optimization is especially important for isotopes that are not mono-energetic emitters. For instance, optimal collimator choice for ^{123}I imaging, hampered by the down-scatter and collimator penetration of high energy photons, has been the subject of a large number of studies [44–46].

Optimization of pinhole and especially multi-pinhole collimation is especially important for preclinical molecular imaging. A simulation study by Cao [47] finds the optimal number of pinholes to be used for a fixed geometry and the specific application of mouse brain imaging. An analytic technique proposed by [48] finds the optimal trade-off between sensitivity and spatial resolution through a feedback loop that varies the pinhole diameter for a fixed field of view. Yet another study proposes a theoretical method to optimize multi-pinhole imaging for the special case of post-smoothed ML-EM image reconstruction [49]. The parameters to be optimized in this study were hole size, collimator distance, collimator height, acceptance angle, position and number of pinholes.

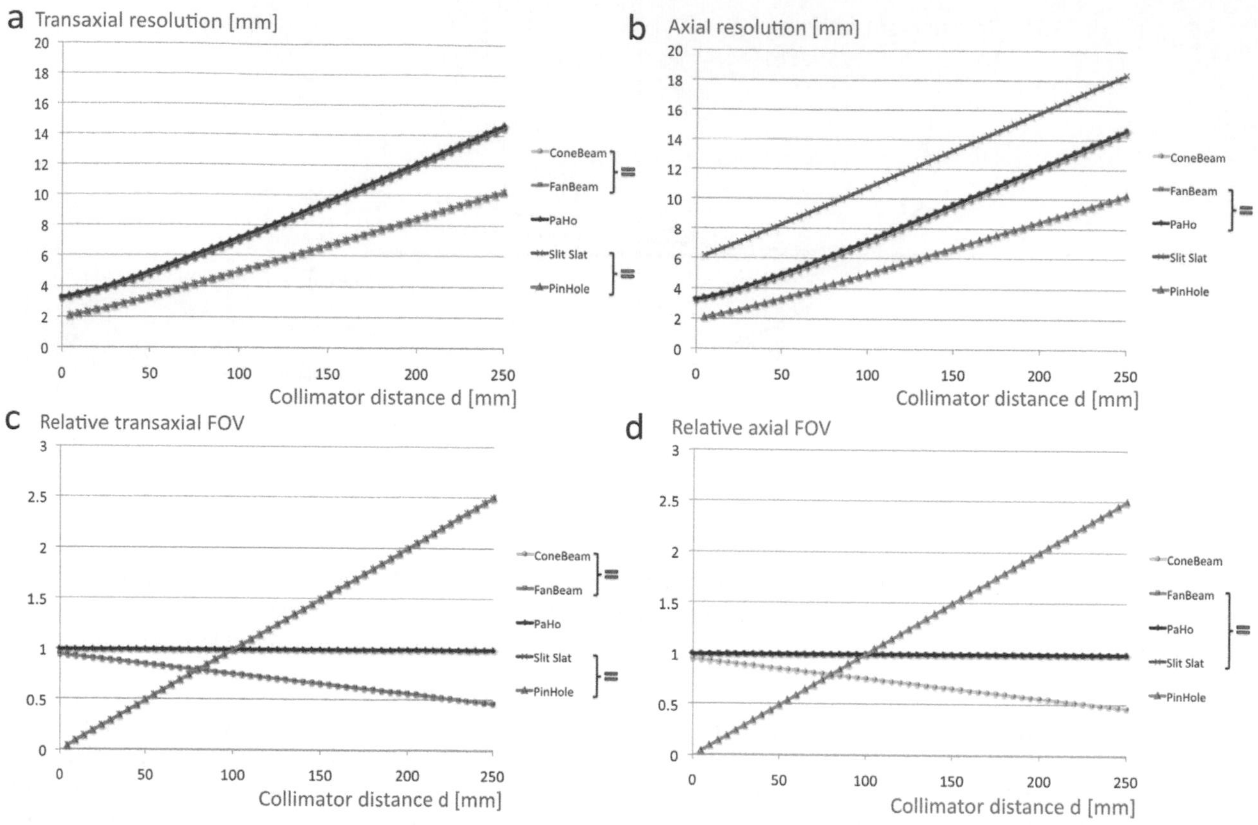

FIGURE 14.8 Plot (a) and (b) respectively compare the transaxial and axial resolution that can be obtained with the different collimators. Transaxial resolution is equal for fanbeam and conebeam and also for pinhole and slit-slat collimation. Axial resolution is equal for parallel-hole and fanbeam collimators. The axial resolution of a slit-slat is worse compared to parallel-hole because zero collimator distance is defined by the slit height h_{slit} and not by the slat height h_{slat}. Plot (c) and (d) respectively compare the transaxial and axial FOV, relative to the FOV of a parallel-hole collimator. Transaxial FOV is equal for fanbeam and conebeam and also for pinhole and slit-slat collimation. Axial FOV is equal for parallel-hole, fanbeam and slit-slat collimators. The design parameters used can be found in Table 14.1.

14.7 ROTATING SLAT COLLIMATORS

Rotating slat collimators, which are the subject of this dissertation, break with the traditional collimators described above. They fundamentally differ in the sense that instead of line integrals, they measure plane integrals. Furthermore, rotating slat collimators exhibit a different resolution/sensitivity relationship, which will be studied in the next chapter of this dissertation. This type of collimation has been used in combination with solid-state detectors in the SOLSTICE (SOLid STate Imager with Compact Electronics) design proposed by Gagnon [50, 51]. Due to the rotating slat design, only a limited number of detector elements is required to fill a strip area, which reduces the cost. A similar design has been published by Entine [52] about twenty years ago, combining a CdTe detector with a parallel-plate collimator. Before this the design of a linear detector has been proposed independently by Keyes [53] and Tosswill [54]. Traditional rectangular SPECT detectors have been studied in combination with rotating slat collimators by Webb, who found an increased sensitivity of about a factor of 40 for the rotating slat concept [55]. Due to the different nature of the data measured by this collimator, other reconstruction techniques have to be used to reconstruct images. Analytic reconstruction methods for planar and tomographic acquisitions have been derived by Lodge [56, 57]. The conclusions drawn from these studies were that, regarding signal-to-noise in the images, it is advantageous to use rotating slat collimators instead of parallel-hole collimators in the cases of hot spot imaging in small objects. A 3D iterative reconstruction algorithm for the data from the SOLSTICE camera has been proposed by Wang [58] and by Zeng [59]. In [58], next to better hot-spot contrast, also an increased contrast was found for cold lesions.

REFERENCES

[1] A. Wouters, K. M. Simon, and J. G. Hirschberg, "Direct method of decoding multiple images," *Appl Opt*, vol. 12, no. 8, pp. 1871–1873, 1973/08/01 1973, doi:10.1364/AO.12.001871.

[2] H. Wieczorek and A. Goedicke, "Analytical model for SPECT detector concepts," *IEEE Trans Nucl Sci*, vol. 53, no. 3, pp. 1102–1112, 2006, doi:10.1109/TNS.2006.874954.

[3] R. Accorsi, "High-efficiency, high-resolution SPECT techniques for cardiac imaging," in *Proceedings of Science*, pp. 1–12, 2006.

[4] S. C. Moore, K. Kouris, and I. Cullum, "Collimator design for single photon emission tomography," *Eur J Nucl Med*, vol. 19, pp. 138–150, 1992.

[5] M. A. Park, S. C. Moore, and M. F. Kijewski, "Brain SPECT with short focal-length cone-beam collimation," (in English), *Med Phys*, vol. 32, no. 7, pp. 2236–2244, Jul 2005, doi:10.1118/1.1929208.

[6] R. J. Jaszczak, C. E. Floyd Jr., S. H. Manglos, K. L. Greer, and R. E. Coleman, "Cone beam collimation for single photon emission computed tomography: analysis, simulation, and image reconstruction using filtered back-projection," *Med Phys*, vol. 13, pp. 484–489, 1986.

[7] S. R. Cherry, J. A. Sorenson, and M. E. Phelps, *Physics in Nuclear Medicine*, 4th ed. W. B. Saunders Co Ltd, 2012.

[8] M. Koole et al., "Modeling of the sensitivity of fan-beam collimation in SPECT imaging," in *2001 Conference Proceedings of the 23rd Annual International Conference of the IEEE Engineering in Medicine and Biology Society*, 25–28 Oct 2001, vol. 3, pp. 2375–2378 vol. 3, doi:10.1109/IEMBS.2001.1017255.

[9] D. Pareto, J. Pavia, C. Falcon, I. Juvells, A. Cot, and D. Ros, "Characterisation of fan-beam collimators," (in English), *Eur J Nucl Med*, vol. 28, no. 2, pp. 144–149, Feb 2001, doi:10.1007/s002590000436.

[10] R. Accorsi and S. D. Metzler, "Non-diverging analytic expression for the on-axis sensitivity of converging collimators: analytic derivation," (in English), *Phys Med Biol*, vol. 51, no. 21, pp. 5675–5696, Nov. 7, 2006, doi:10.1088/0031-9155/51/21/019.

[11] R. Accorsi and S. D. Metzler, "Non-diverging analytic expression for the on-axis sensitivity of converging collimators: experimental verification," (in English), *Phys Med Biol*, vol. 51, no. 21, pp. 5697–5705, Nov 7, 2006, doi:10.1088/0031-9155/51/21/020.

[12] G. T. Gullberg, G. L. Zeng, F. L. Datz, P. E. Christian, C. H. Tung, and H. T. Morgan, "Review of convergent beam tomography in single photon emission computed tomography," (in English), *Phys Med Biol*, vol. 37, no. 3, pp. 507–534, Mar 1992, doi:10.1088/0031-9155/37/3/002.

[13] J. W. Keyes, Jr., J. H. Thrall, and J. E. Carey, "Technical considerations in in vivo thyroid studies," (in English), *Semin Nucl Med*, vol. 8, no. 1, pp. 43–57, Jan 1978, doi:10.1016/s0001-2998(78)80006-3.

[14] A. Spanu et al., "The usefulness of neck pinhole SPECT as a complementary tool to planar scintigraphy in primary and secondary hyperparathyroidism," (in English), *J Nucl Med*, vol. 45, no. 1, pp. 40–48, Jan 2004.

[15] Y. W. Bahk, S. H. Kim, S. K. Chung, and J. H. Kim, "Dual-head pinhole bone scintigraphy," *J Nucl Med*, vol. 39, no. 8, pp. 1444–8, Aug 1998. [Online]. Available: www.ncbi.nlm.nih.gov/pubmed/9708525.

[16] K. E. Applegate, L. P. Connolly, R. T. Davis, D. Zurakowski, and S. T. Treves, "A prospective comparison of high-resolution planar, pinhole, and triple-detector SPECT for the detection of renal cortical defects," (in English), *Clin Nucl Med*, vol. 22, no. 10, pp. 673–8, Oct 1997, doi:10.1097/00003072-199710000-00002.

[17] R. A. Vogel, D. L. Kirch, M. T. LeFree, J. O. Rainwater, D. P. Jensen, and P. P. Steele, "Thallium-201 myocardial perfusion scintigraphy: results of standard and multi-pinhole tomographic techniques," (in English), *Am J Cardiol*, vol. 43, no. 4, pp. 787–793, Apr 1979, doi:10.1016/0002-9149(79)90079-1.

[18] T. Funk, D. L. Kirch, J. E. Koss, E. Botvinick, and B. H. Hasegawa, "A novel approach to multipinhole SPECT for myocardial perfusion imaging," (in English), *J Nucl Med*, vol. 47, no. 4, pp. 595–602, Apr 2006.

[19] P. P. Steele, D. L. Kirch, and J. E. Koss, "Comparison of simultaneous dual-isotope multipinhole SPECT with rotational SPECT in a group of patients with coronary artery disease," (in English), *J Nucl Med*, vol. 49, no. 7, pp. 1080–1089, Jul 2008, doi:10.2967/jnumed.107.040915.

[20] D. D. Patton, H. H. Barrett, and J. C. Chen, "FASTSPECT: A four-dimensional brain imager," (in English), 1994-05-01 1994. [Online]. Available: www.osti.gov/servlets/purl/197965.

[21] F. Forrer, E. Rolleman, N. Schramm, E. P. Krenning, and M. de Jong, "Is it possible to predict renal function in small animals using a multi-pinhole SPECT system," (in English), *Eur J Nucl Med Mol Imaging*, vol. 34, no. 7, pp. 1127–1128, Jul 2007, doi:10.1007/s00259-007-0445-y.

[22] B. Ostendorf et al., "High-resolution multipinhole single-photon-emission computed tomography in experimental and human arthritis," (in English), *Arthritis Rheum*, vol. 54, no. 4, pp. 1096–104, Apr 2006, doi:10.1002/art.21732.

[23] D. P. McElroy et al., "Performance evaluation of A-SPECT: a high resolution desktop pinhole SPECT system for imaging small animals," *IEEE Trans Nucl Sci*, vol. 49, no. 5, pp. 2139–2147, 2002, doi:10.1109/TNS.2002.803801.

[24] N. U. Schramm, G. Ebel, U. Engeland, T. Schurrat, M. Behe, and T. M. Behr, "High-resolution SPECT using multi-pinhole collimation," in *2002 IEEE Nuclear Science Symposium Conference Record*, 10–16 Nov. 2002, vol. 2, pp. 774–777 vol.2, doi:10.1109/NSSMIC.2002.1239437.

[25] F. J. Beekman et al., "U-SPECT-I: a novel system for submillimeter-resolution tomography with radiolabeled molecules in mice," (in English), *J Nucl Med*, vol. 46, no. 7, pp. 1194–200, Jul 2005.
[26] C. Lackas, N. U. Schramm, J. W. Hoppin, U. Engeland, A. Wirrwar, and H. Halling, "T-SPECT: a novel imaging technique for small animal research," *IEEE Trans Nucl Sci*, vol. 52, no. 1, pp. 181–187, 2005, doi:10.1109/TNS.2005.843615.
[27] H. Kim et al., "SemiSPECT: a small-animal single-photon emission computed tomography (SPECT) imager based on eight cadmium zinc telluride (CZT) detector arrays," (in English), *Med Phys,* vol. 33, no. 2, pp. 465–74, Feb 2006, doi:10.1118/1.2164070.
[28] F. Beekman and F. van der Have, "The pinhole: gateway to ultra-high-resolution three-dimensional radionuclide imaging," (in English), *Eur J Nucl Med Mol Imaging*, vol. 34, no. 2, pp. 151–161, Feb 2007, doi:10.1007/s00259-006-0248-6.
[29] S. Orlov, "Theory of three-dimensional image reconstruction: I Conditions for a complete set of projections," *Sov. Phys. - Crystallogr*, vol. 20, pp. 429–433, 1976.
[30] J. E. Bowsher, J. R. Roper, J. Peter, and R. J. Jaszczak, "Pinhole trajectories for SPECT imaging of the breast, axilla, and upper chest," in *2006 IEEE Nuclear Science Symposium Conference Record*, 29 Oct.–1 Nov., 2006, vol. 4, pp. 2387–2389, doi:10.1109/NSSMIC.2006.354393.
[31] S. D. Metzler, K. L. Greer, and R. J. Jaszczak, "Helical pinhole SPECT for small-animal imaging: a method for addressing sampling completeness," *IEEE Trans Nucl Sci*, vol. 50, no. 5, pp. 1575–1583, 2003, doi:10.1109/TNS.2003.817948.
[32] F. P. Difilippo, "Design and performance of a multi-pinhole collimation device for small animal imaging with clinical SPECT and SPECT-CT scanners," (in English), *Phys Med Diol*, vol. 53, no. 15, pp. 4185–4201, 2008, doi:10.1088/0031-9155/53/15/012.
[33] S. D. Metzler, R. Accorsi, J. R. Novak, A. S. Ayan, and R. J. Jaszczak, "On-axis sensitivity and resolution of a slit-slat collimator," *J Nucl Med*, vol. 47, no. 11, pp. 1884–1890, Nov 2006. [Online]. Available: www.ncbi.nlm.nih.gov/pubmed/17079823.
[34] R. Accorsi, J. R. Novak, A. S. Ayan, and S. D. Metzler, "Derivation and validation of a sensitivity formula for slit-slat collimation," (in English), *IEEE Trans Med Imaging*, vol. 27, no. 5, pp. 709–722, May 2008, doi:10.1109/tmi.2007.912395.
[35] W. L. Rogers et al., "SPRINT: A stationary detector single photon ring tomograph for brain imaging," (in English), *IEEE Trans Med Imaging*, vol. 1, no. 1, pp. 63–8, 1982, doi:10.1109/tmi.1982.4307549.
[36] W. L. Rogers et al., "Performance evaluation of SPRINT, a single photon ring tomography for brain imaging," *J Nucl Med*, vol. 25, no. 9, pp. 1013–1018, Sep 1984. [Online]. Available: www.ncbi.nlm.nih.gov/pubmed/6332182.
[37] W. Chang, H. Liang, and J. Liu, "Design concepts and potential performance of MarC-SPECT – A high-performance cardiac SPECT system," *J Nucl Med*, vol. 47, no. suppl 1, p. 190P, May 1, 2006. [Online]. Available: http://jnm.snmjournals.org/content/47/suppl_1/190P.3.abstract.
[38] S. Walrand, F. Jamar, M. de Jong, and S. Pauwels, "Evaluation of novel whole-body high-resolution rodent SPECT (Linoview) based on direct acquisition of linogram projections," *J Nucl Med*, vol. 46, no. 11, pp. 1872–1880, Nov 2005. [Online]. Available: www.ncbi.nlm.nih.gov/pubmed/16269602.
[39] E. L. Keller, "Optimum dimensions of parallel-hole, multi-aperture collimators for gamma-ray cameras," *J Nucl Med*, vol. 9, no. 6, pp. 233–235, Jun 1968. [Online]. Available: www.ncbi.nlm.nih.gov/pubmed/5647695.
[40] S. C. Moore, D. J. DeVries, B. Nandram, M. F. Kijewski, and S. P. Mueller, "Collimator optimization for lesion detection incorporating prior information about lesion size," *Med Phys*, vol. 22, pp. 703–713, 1995.
[41] S. C. Moore, M. Foley Kijewski, and G. El Fakhri, "Collimator optimization for detection and quantitation tasks: application to gallium-67 imaging," *IEEE Trans Med Imaging*, vol. 24, no. 10, pp. 1347–1356, Oct 2005, doi:10.1109/TMI.2005.857211.
[42] G. L. Zeng and G. T. Gullberg, "A channelized-hotelling-trace collimator design method based on reconstruction rather than projections," *IEEE Trans Nucl Sci*, vol. 49, no. 5, pp. 2155–2158, 2002, doi:10.1109/TNS.2002.803775.
[43] C. Kamphuis, F. Beekman, and B. F. Hutton, "Optimal collimator hole dimensions for half cone-beam brain SPECT," in *Proceedings 4th International Meeting on Fully Three-Dimensional Image Reconstruction in Radiology and Nuclear Medicine*, 1999, pp. 271–274.
[44] A. Cot et al., "Study of the point spread function (PSF) for123I SPECT imaging using Monte Carlo simulation," *Phys Med Diol*, vol. 49, no. 14, pp. 3125–3136, 2004/06/29 2004, doi:10.1088/0031-9155/49/14/007.
[45] F. D. Geeter, P. R. Franken, M. Defrise, H. Andries, E. Saelens, and A. Bossuyt, "Optimal collimator choice for sequential iodine-123 and technetium-99m imaging," (in English), *Eur J Nucl Med*, vol. 23, no. 7, pp. 768–774, Jul 1996, doi:10.1007/bf00843705.
[46] Y. Inoue et al., "Collimator choice in cardiac SPECT with I-123-labeled tracers," *J Nucl Cardiol*, vol. 11, no. 4, pp. 433–439, Jul-Aug 2004, doi:10.1016/j.nuclcard.2004.04.009.
[47] Z. Cao, G. Bal, R. Accorsi, and P. D. Acton, "Optimal number of pinholes in multi-pinhole SPECT for mouse brain imaging—a simulation study," *Phys Med Biol*, vol. 50, no. 19, pp. 4609–4624, Oct 7, 2005, doi:10.1088/0031-9155/50/19/013.
[48] M. C. Rentmeester, F. van der Have, and F. J. Beekman, "Optimizing multi-pinhole SPECT geometries using an analytical model," (in English), *Phys Med Biol*, vol. 52, no. 9, pp. 2567–2581, May 7, 2007, doi:10.1088/0031-9155/52/9/016.
[49] K. Vunckx, D. Beque, M. Defrise, and J. Nuyts, "Single and multipinhole collimator design evaluation method for small animal SPECT," (in English), *IEEE Trans Med Imaging*, vol. 27, no. 1, pp. 36–46, Jan 2008, doi:10.1109/tmi.2007.902802.

[50] D. Gagnon, G. L. Zeng, J. M. Links, J. J. Griesmer, and F. C. Valentino, "Design considerations for a new solid-state gamma-camera: Soltice," in *2001 IEEE Nuclear Science Symposium Conference Record (Cat. No.01CH37310)*, 4–10 Nov. 2001, vol. 2, pp. 1156–1160 vol.2, doi:10.1109/NSSMIC.2001.1009755.

[51] J. J. Griesmer, B. Kline, J. Grosholz, K. Parnham, and D. Gagnon, "Performance evaluation of a new CZT detector for nuclear medicine: Soltice," in *2001 IEEE Nuclear Science Symposium Conference Record (Cat. No.01CH37310)*, 4–10 Nov. 2001, vol. 2, pp. 1050–1054 vol. 2, doi:10.1109/NSSMIC.2001.1009733.

[52] G. Entine, R. Luthmann, W. Mauderli, L. T. Fitzgerald, C. M. Williams, and C. H. Tosswill, "Cadmium Telluride Gamma Camera," *IEEE Trans Nucl Sci*, vol. 26, no. 1, pp. 552–558, 1979, doi:10.1109/TNS.1979.4329689.

[53] W. I. Keyes, "Correspondence: The fan-beam gamma camera," (in English), *Phys Med Biol*, vol. 20, no. 3, pp. 489–493, May 1975, doi:10.1088/0031-9155/20/3/013.

[54] C. Tosswill, "Computerized rotating laminar collimation imaging system," US patent application 646 (granted December 1977), 917–967, 1977.

[55] S. Webb, D. M. Binnie, M. A. Flower, and R. J. Ott, "Monte Carlo modelling of the performance of a rotating slit-collimator for improved planar gamma-camera imaging," (in English), *Phys Med Biol*, vol. 37, no. 5, pp. 1095–1108, May 1992, doi:10.1088/0031-9155/37/5/006.

[56] M. A. Lodge, D. M. Binnie, M. A. Flower, and S. Webb, "The experimental evaluation of a prototype rotating slat collimator for planar gamma camera imaging," (in English), *Phys Med Biol*, vol. 40, no. 3, pp. 427–448, Mar 1995, doi:10.1088/0031-9155/40/3/007.

[57] M. A. Lodge, S. Webb, M. A. Flower, and D. M. Binnie, "A prototype rotating slat collimator for single photon emission computed tomography," *IEEE Trans Med Imag*, vol. 15, no. 4, pp. 500–511, 1996, doi:10.1109/42.511753.

[58] W. Wang, W. Hawkins, and D. Gagnon, "3D RBI-EM reconstruction with spherically-symmetric basis function for SPECT rotating slat collimator," *Phys Med Diol*, vol. 49, no. 11, pp. 2273–2292, 2004/05/20 2004, doi:10.1088/0031-9155/49/11/011.

[59] G. L. Zeng, D. Gagnon, F. Natterer, W. Wenli, M. Wrinkler, and W. Hawkins, "Local tomography property of residual minimization reconstruction with planar integral data," *IEEE Trans Nucl Sci*, vol. 50, no. 5, pp. 1590–1594, 2003, doi:10.1109/TNS.2003.817958.

15 Image Acquisition Protocols

Jonathan Gear

CONTENTS

15.1 BASIC ACQUISITION PARAMETERS

In this chapter we discuss the fundamentals of image acquisition and give examples of where different protocols and simple image-processing techniques can be beneficial in a clinical setting. The simplest form of scintigraphic imaging is the static acquisition, yet this forms the basis of the more complicated methods employed. Chapter 13 discusses how the gamma camera is used to detect the emitted gamma rays from the patient and convert them into detectable signals from which an image can be formed. However. there are still a number of user settings that must be understood to optimize image formation.

DOI: 10.1201/9780429489556-15

15.1.1 Flood Correction

As discussed in Chapter 13, a correction map is required to compensate for non-uniformities in the detector sensitivity. These sensitivities are dependent on the energy of the isotope being imaged, and in some cases, on the collimator fitted to the system. It is therefore important to select the correct flood for the required acquisition.

15.1.2 Matrix Size

The matrix size determines how many pixels will be used to form the image. With optical cameras or digital displays, it is often considered that more pixels give superior image quality. This is true as more detail is observed when the images are binned into finer pixels. The same is true for scintigraphy. However, there is an additional caveat regarding the introduction of noise into the data. Unlike optical imaging, which may use up to 10 trillion photons to form a photograph, in nuclear medicine, gamma photons are often limited to a few hundred thousand. Recording these events into smaller and smaller pixels will inevitably reduce the number of events recorded into each pixel and therefore the difference between the pixels with the highest and lowest number of pixels will reduce. This results in there being less contrast in the image and therefore the structure that was of interest becomes less visible with an increasing number of pixels. This is demonstrated in Figure 15.1 for a 3×3 pixel array. The central pixel has twice the signal than surrounding ones, and there is clearly a higher signal originating from this region of the array. If the matrix size is increased to 6×6 with the signal from each of the original pixels randomly assigned to the additional matrix elements, although the absolute signal values remain the same, the higher signal intensity from the inner section is more difficult to perceive.

This trade-off between contrast and sampling detail can be demonstrated using images of test objects – called phantoms, designed to illustrate these effects. Figure 15.2 is a set of acquisitions of the Williams phantom . The phantom is designed to replicate the size and shape of a liver containing regions of higher and lower activity concentration. In this example, a set of identical acquisitions have been performed and the matrix size altered from 1024 ×1024 to 32 × 32. The loss in detail can clearly be observed at the lower matrix sizes, but the additional noise in the higher matrices is also evident. The clearest image, where the smallest circular regions are most visible, is arguably the central image acquired on a 256×256 matrix size.

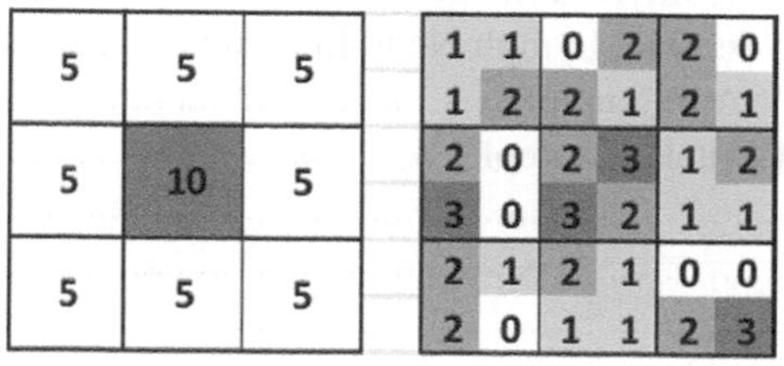

FIGURE 15.1 An illustration on the effect of voxel size. As events are shared between more pixels the difference between pixels decreases and contrast is reduced.

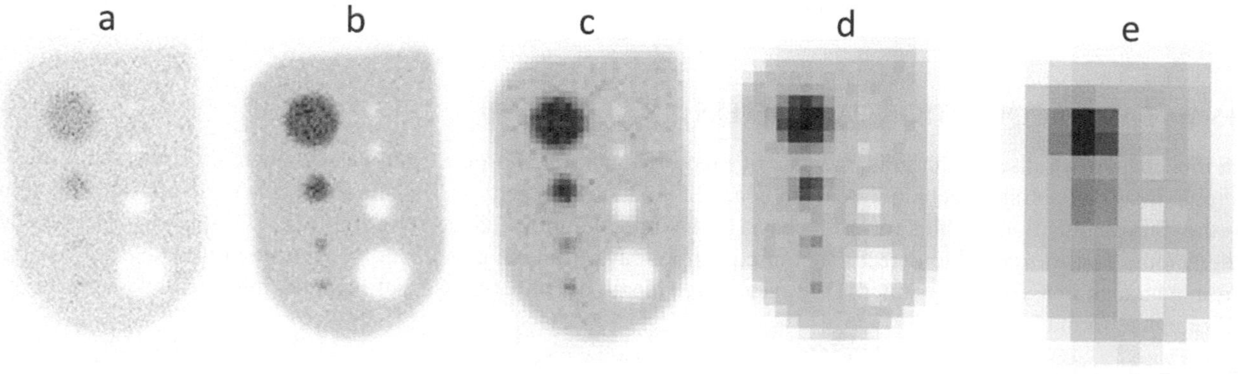

FIGURE 15.2 Scintigraphic images of a Williams phantom acquired at different matrix sizes (a – e) correspond to matrices of (1024, 512, 128, 64 and 32).

15.1.3 Field of View and Digital Zoom

A typical rectangular detector has a scintillation crystal of approximately 40 × 50 cm. Inevitably, image quality will be poorer at the edge of the crystal and the collimator masks events from the edges. The exposed area of the crystal, for which collimator holes are present, is called the geometric field of view (GFOV) and will vary depending on collimator type. The geometric field of view does not necessarily define the final image size, and a further electronic mask defining the useful field of view (UFOV) is often applied by manufacturers. A central field of view, with dimensions 75 per cent of the UFOV is also often defined, most commonly when reporting performance parameters, which are generally better in the central region of the detector. In many cases the anatomy being imaged will be smaller than the UFOV or CFOV of the detector. It is therefore unnecessary to acquire data covering the whole detector surface and a further digital zoom can be applied to mask out the unnecessary space. Whilst the physical resolution of the camera is not changed in this process, for the same matrix size, the pixel size will reduce.

15.1.4 Saturation Level and Behaviour

Numerical values for counts are stored digitally in the image matrix. Conventionally these are stored as 2-byte integers allowing the assignment of any value between 0 and 2^{16} (65,536). To allow for potential image processing where the handling of negative values may be necessary this range is often offset to -32,768 to 32,768. In the event that a value higher than this is obtained, the system may respond in a number of different ways. Either the acquisition will be stopped prematurely, the value of the saturated pixel will be reset to 0 or -32,768 or remain at the maximum integer value whilst the acquisition continues.

15.1.5 Collimator and Detector Distance

The choice of collimator is paramount for the intended application and isotope being used. The thickness and length of septa will determine the energy of gamma ray that it will be able to attenuate. In addition, there is trade-off between sensitivity and resolution. In acquisitions where a low activity is used, or the image needs to be formed quickly, it is often necessary to sacrifice image resolution for sensitivity. Figure 15.3 illustrates the difference between a number of different collimator choices.

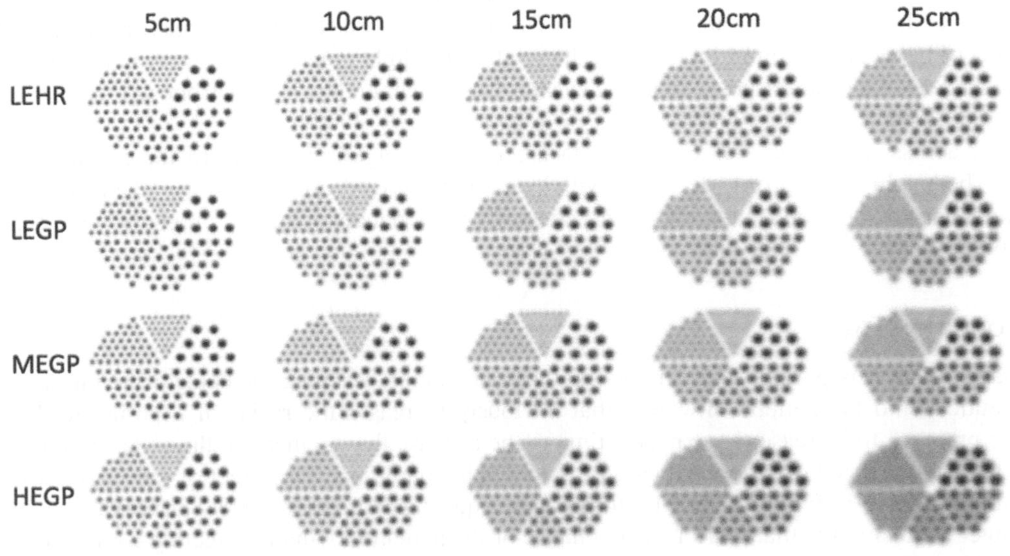

FIGURE 15.3 A thin plane of circles of different diameters position parallel to the surface of the collimator for different distances have been simulated for four types of collimator in order to illustrate the effect of source to collimator distance on the spatial resolution.

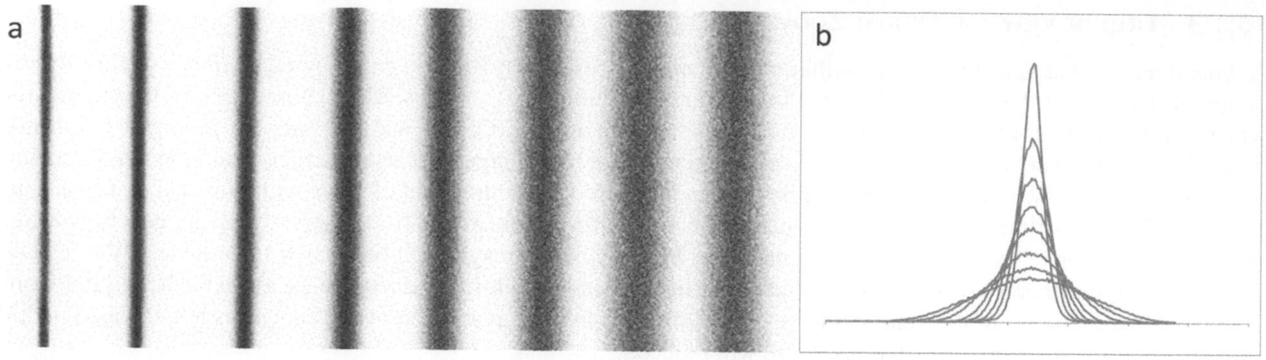

FIGURE 15.4 Static acquisitions of a line source of ^{99m}Tc, acquired at 0, 5, 10, 15, 20, 25, 30 and 35 cm from the detector surface (a). Count profiles across the individual line sources, the area under each profile remains the same (b).

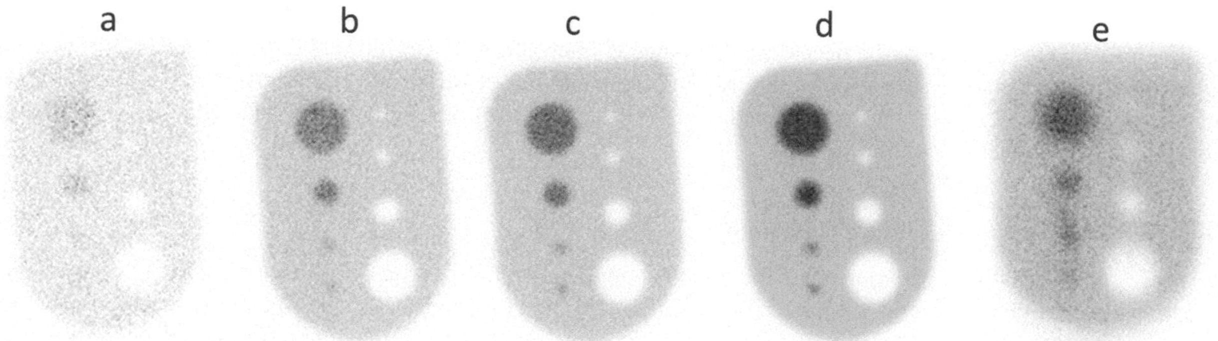

FIGURE 15.5 Images of the Williams phantom acquired at 50k (a), 500k (b), 1000k (c), and 5000k counts (d). In the far-right image (e), the phantom has been gently agitated to simulate effects of motion.

The effect of source distance on resolution is further demonstrated in Figure 15.4a. Images of line sources acquired at increasing distances clearly demonstrate a loss of resolution. It is also interesting to note that the resolution of the line source closest to the detector changes along its length. This is a result of a change in intrinsic resolution. When scintillation events occur between photomultiplier tubes, the scintillation light is more evenly spread across multiple PMTs. Therefore, the Anger logic used for positioning the event is less reliant on low signals from surrounding PMTs and less influenced by random noise.

Line profiles across the sources of Figure 15.4a are given in Figure 15.4b. Here the partial volume effect is demonstrated. The height of the profile decreases with increasing source distance. However, the total number of events remain the same and the area under the curves are constant. More details about the properties of collimators can be found in Chapter 14.

15.1.6 Acquisition Time

To reduce image noise, it is possible to increase the flux of radiation reaching the detector by injecting more of the radiopharmaceutical into the patient. However, that obviously increases the radiation exposure to the patient. It is therefore more practical to increase the exposure time to the camera, that is, increase the acquisition time. Analogous to increasing the shutter speed of an optical camera in low light conditions. The effect of longer acquisition durations on image quality is shown in Figure 15.5(a–d), for images acquiring 50k, 500k, 1000k, and 5000k counts. The disadvantage of longer acquisition is that there is more chance of the patient moving or fidgeting whilst the image is being acquired. The effect of motion during an acquisition is given in Figure 15.5e. The image duration and counts within the image are identical to that of Figure 15.5d, except the phantom was gently agitated during the acquisition. This results in an overall blurring of the image and the addition of reduced signal intensity and noise.

15.1.7 Spatial Resolution

It can be seen that the overall spatial resolution, R_{total}, of an image will be influenced by a number of factors, including the collimator design and source distance, $R_{collimator}$, the precision of determining an event position within the crystal, $R_{intrinsic}$, the choice of matrix size, R_{matrix}, and effects due to patient movement R_{motion}.

$$R_{total} = \sqrt{R_{collimator} + R_{intrinsic} + R_{matrix} + R_{motion}} \tag{15.1}$$

Where possible, to reduce these effects, the distance between the patient and detector should always be minimized, an appropriate matrix size should be selected such that the pixel size, Δr is $\leq$ FWHM/3 and acquisition time should be minimized to avoid patient discomfort.

15.1.8 Energy Window Selection

As the gamma camera is capable of imaging a variety of radioisotopes, each with different emission energies, the system must be correctly configured for the isotope being imaged. The energy window is used to discriminate events outside the required photopeak. It is therefore important to ensure that the window and photopeak are properly aligned prior to commencing an acquisition. Figure 15.6a is an example acquisition of ^{99m}Tc with a typical 20 per cent width energy window as shown in Figure 15.6b. As the energy resolution of the detector is finite, some scattered events are inevitably included within this window in addition to the true unscattered events. Figure 15.7a shows the estimated proportion of scatter that is included in each energy channel of the energy spectra. The proportion of scatter included in the image can therefore be reduced by narrowing the energy window, and the resulting image is shown in Figure 15.6b. However, the number of recorded events in this window is also dramatically reduced and, in fact, the image of Figure 15.6b took ten times longer to acquire as that of Figure 15.6a. Figure 15.6c demonstrates the effect of an overextended energy window. Obviously, in this situation scattered events outside of the energy window are also included. Figure 15.6d is the result of an offset energy window, which could be the result of the wrong isotope selection in the acquisition settings. In this instance a lower energy window has been selected, and only scattered photons are included. Had an isotope with a higher energy window (higher than the photopeak) been selected, then the result would have been a blank image with only very few events recorded due to background radiation.

15.1.9 Multiple Energy Windows

In some image acquisition protocols it may be beneficial to acquire images using more than one energy window. Events within these energy windows can either be combined or recorded into more than one dataset depending on need. Multiple photopeaks can arise either from imaging isotopes with more than one gamma emission, such as ^{111}In and ^{177}Lu, or when more than one isotope has been injected.

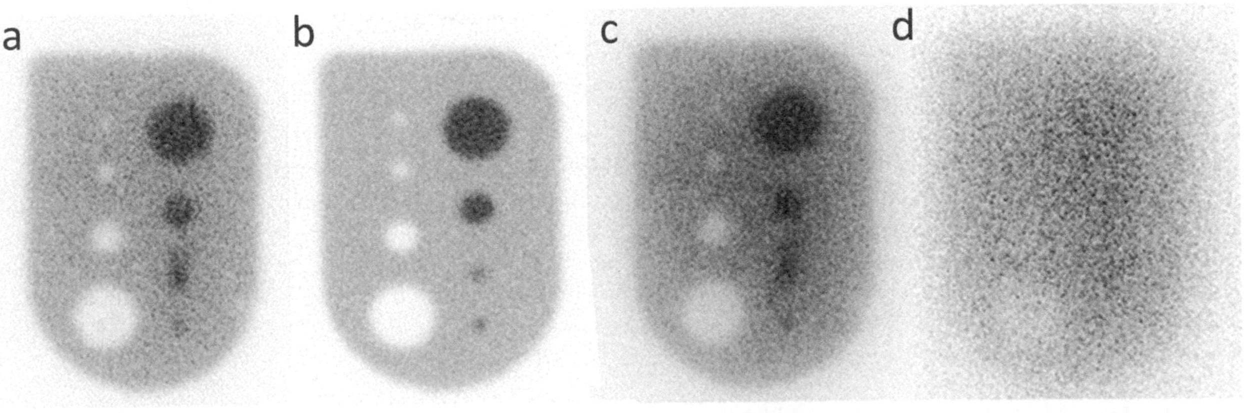

FIGURE 15.6 Effects of different energy window settings. Images acquired with a standard 20 per cent energy window (a), a 2 per cent energy window (b), a 50 per cent energy window (c) and an offset energy window (d).

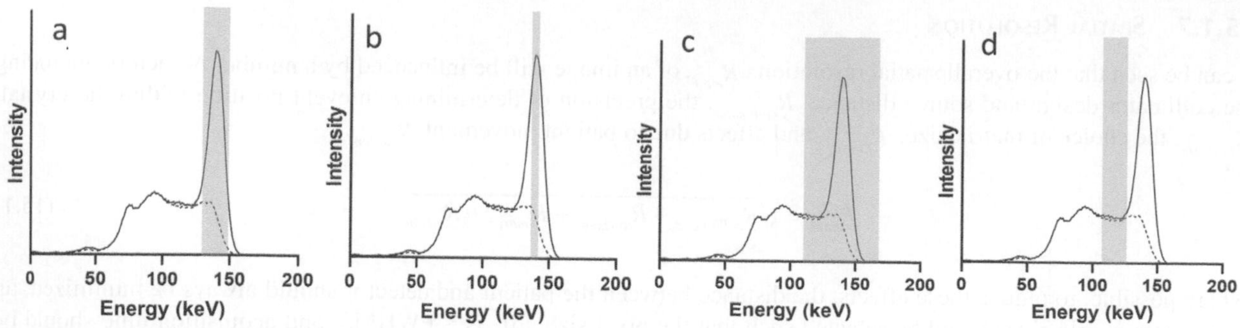

FIGURE 15.7 ^{99m}Tc spectra with illustrated scatter component (dotted lines). Energy widow range and positions are highlighted and correspond to that used to acquire images of Figure 15.6.

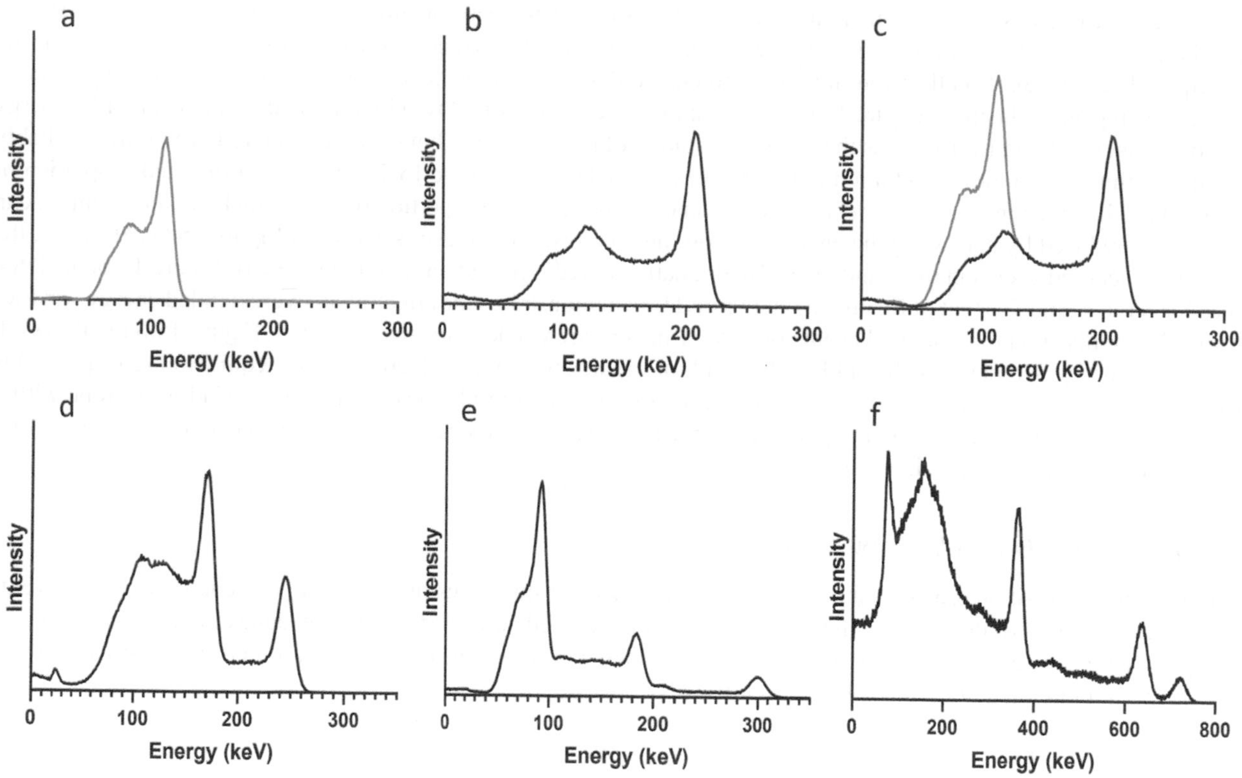

FIGURE 15.8 Energy spectra from the 113 keV emission of ^{177}Lu (a), the 208 keV emission from ^{177}Lu (b) and the combined spectate from both ^{177}Lu emissions (c). Typical energy spectra from ^{111}In (d), ^{67}Ga (e), and ^{131}I (f).

One of the difficulties with multiple energy acquisitions is the presence of scatter from the two peaks. Lutetium-177 has primary emissions at 113 and 208 keV with branching ratios of 0.10 and 0.06. Typical energy spectra from these individual emissions would be that of Figures 15.8a and 15.8b. However, as these two sets of emissions are concurrent, the spectra of the lower energy emission is superimposed on the scatter tail of the 208 keV events (Figure 15.8c). An energy window cantered on the 113 keV photopeak will therefore contain significantly more scatter events than if the 208 keV peak were not present. Methods to compensate for scatter events are discussed in Chapters 25 and 26. Spectra for other common multi-emission isotope shown in Figure 15.8 (d–f).

15.1.9.1 Example 1: Parathyroid Imaging

An example of the use of dual isotope imaging is for the diagnosis of parathyroid adenoma [1]. The parathyroids are small glands situated behind the thyroid. A parathyroid adenoma is a benign tumour on one of these glands that can

TABLE 15.1
Typical Acquisition Parameters for a Parathyroid Imaging Test

Parameter	Value
Radiopharmaceutical	^{201}Tl chloride & ^{99m}Tc Pertechnetate
Energy Window	70 keV for ^{201}Tl & 140 keV for ^{99m}Tc
Collimator	Low-Energy High-Resolution (LEHR) or pin-hole
Matrix Size	128 x 128
Injected Activity	40 MBq of each
Uptake time	20 min for ^{99m}Tc, ^{201}Tl injection on bed.
Acquisition time	300 s
Zoom	2x

TABLE 15.2
Typical Acquisition Parameters for a Lymphoscintigraphy Examination

Parameter	Value
Radiopharmaceutical	^{99m}Tc nano-colloid
Energy Window	140 keV
Collimator	Low-Energy High Resolution or pin-hole
Matrix Size	256 x 256
Injected Activity	20 MBq of each
Uptake time	injection on bed.
Acquisition time	300 s
Zoom	1x

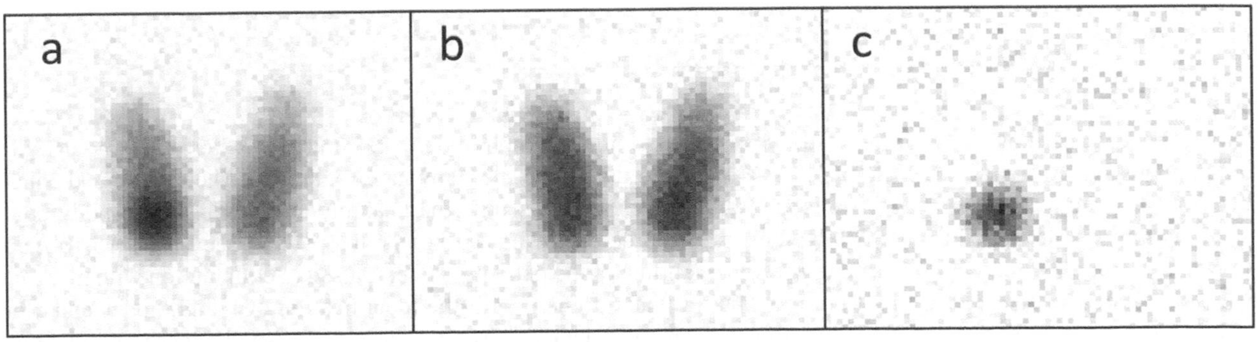

FIGURE 15.9 Typical images from a parathyroid test, showing ^{201}Tl uptake (a), ^{99m}Tc uptake (b) and a subtraction image (c).

result in the over expression of the parathyroid hormone. There is not a single radiopharmaceutical specific to parathyroid imaging. However, Thallium-201 is taken up by both parathyroid tissue and the thyroid. Unfortunately, as the parathyroid glands are situated behind the thyroid it is very difficult to distinguish the two on a single ^{201}Tl planar acquisition. Technetium-99m pertechnetate may also be administered in combination with the ^{201}Tl. The ^{99m}Tc is trapped by functioning thyroid tissue, but not the adenoma. A subtraction of ^{99m}Tc events from ^{201}Tl events therefore results in an image of just the adenoma. Typical imaging parameters for this test are described in Table 15.2 with example ^{201}Tl, ^{99m}Tc and subtraction images shown in Figure 15.9.

15.1.9.2 Example 2: Sentinel Lymph Node Imaging

Dual isotopes can also be used for combining emission and transmission imaging. One of the difficulties in interpreting functional imaging is that there is often no recognizable anatomical structure with which to compare the functional

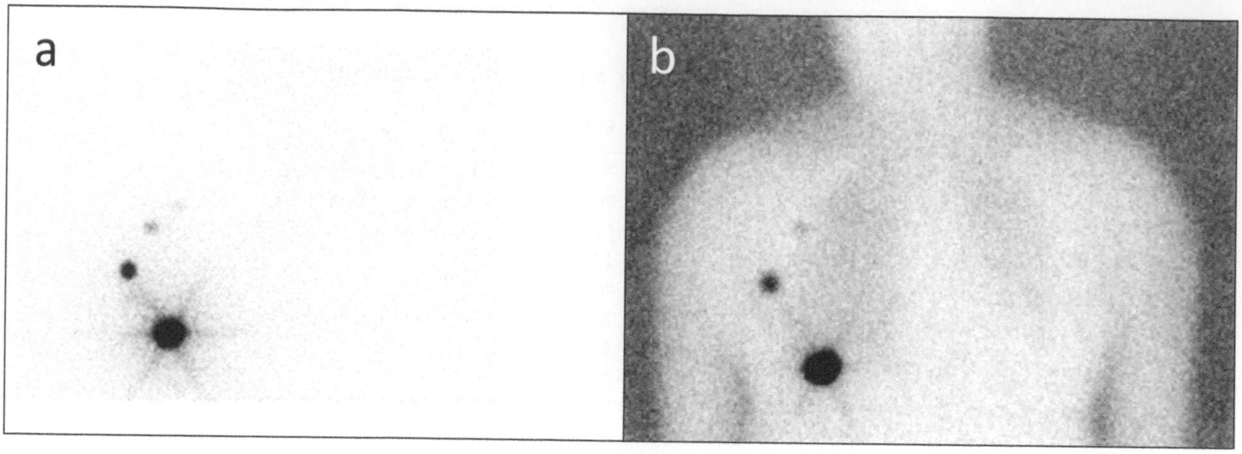

FIGURE 15.10 An example of a sentinel lymph node study without the presence of the transmission source the location of eth tracer is differ difficult to ascertain (a). When a ^{57}Co flood source is in place the outline of the patient is visible (b).

TABLE 15.3
Typical Acquisition Parameters for a Kidney Imaging Test

Parameter	Value
Radiopharmaceutical	^{99m}Tc DMSA (Dimercaptosuccinic acid).
Energy Window	140 keV with 20% width
Collimator	Low Energy High Resolution
Matrix Size	256 x 256
Injected Activity	80 MBq
Uptake time	2–3 hours post-injection
Acquisition time	600 s
Zoom	2x

radiopharmaceutical uptake. A prime example of this is sentinel lymph node (SLN) imaging where the drainage of a subcutaneous injection of ^{99m}Tc nanoparticles are used to determine the lymphatic route most likely to spread cancer cells from a tumour [2]. Figure 15.10a is an example SLN image where the anatomy and region of the body being imaged is impossible to determine. A large area ^{57}Co source can be placed behind the patient (see Figure 15.10b) and the events within the ^{57}Co window used to present an outline of the patient. ^{57}Co pens used in conjunction with markings on the skin can also be used to highlight areas on an image to help with the identification of lymph nodes and other anatomy. In reality the ^{57}Co and ^{99m}Tc photo peaks are sufficiently close in energy that a single energy window can be used to simultaneously detect events from both isotopes. Typical imaging parameters for this test are described in Table 15.2.

15.1.10 Geometric Mean (DMSA Example)

To assess the function of an organ or tissue it is sometime necessary to compare the relative pharmaceutical uptake in one region of the body with that of another. An example is the relative pharmaceutical uptake in the left and right kidney, which can be used as an indication of kidney health [3]. For two normal kidneys, the relative uptake should be between 45 and 55 per cent in each kidney. Typical acquisition details are given in Table 15.3.

A difficulty in assessing relative pharmaceutical uptake is that that the photons emitted from the organ are attenuated by the body and, therefore, organs at a greater depth of tissue will be attenuated more than those nearer the surface. This is also true for the kidneys, where it is relatively common for the left kidney to be more superiorly positioned than the right. This effect is demonstrated in Figure 15.11.

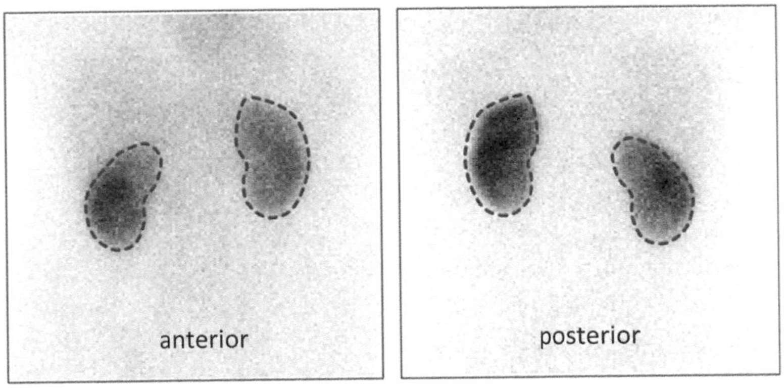

FIGURE 15.11 Anterior and posterior images of the kidneys from a DMSA study.

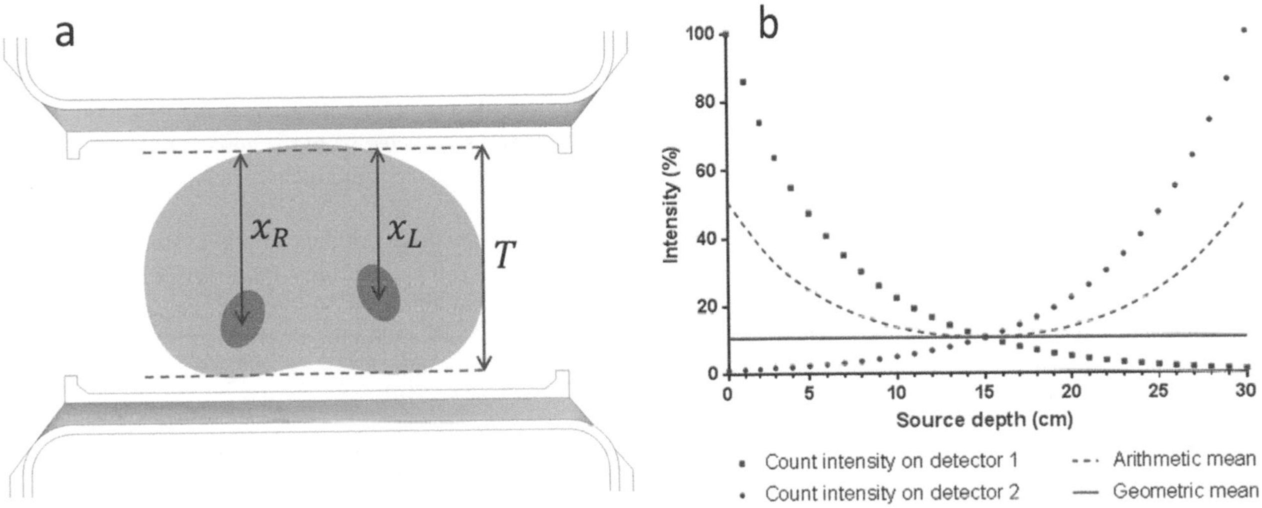

FIGURE 15.12 An illustration of source depth and patient size (a) and the change in count intensity as the distances vary (b).

It is therefore beneficial to acquire two opposing projection images, typically anterior and posterior. Counts, C_{Ant} and C_{Post} within regions of interest (ROIs) defined on the anterior and posterior images of the kidneys will vary as a function of kidney depth, x.

$$C_{Ant} = k \cdot A \cdot e^{-\mu \cdot x} \tag{15.2}$$

$$C_{Post} = k \cdot A \cdot e^{-\mu \cdot (T-x)} \tag{15.3}$$

Where, k, is a sensitivity constant, A, is the activity in the kidney, μ the linear attenuation coefficient for soft tissue and, T the thickness of the patient. A diagram showing these distances and a plot showing how these functions change with increasing source depth are given in Figure 15.12.

By taking the arithmetic mean of the anterior and posterior counts, the depth term, x, is still present,

$$\frac{C_{Ant} + C_{Post}}{2} = \frac{k \cdot A}{2} \cdot \left[e^{-\mu \cdot x} + e^{-\mu \cdot (T-x)} \right] \tag{15.4}$$

However, if the geometric mean is used, it is possible to obtain an expression for counts, C_{corr}, that is proportional to activity and independent of source depth:

$$C_{corr} = \sqrt{\frac{C_{\text{ant}} \cdot C_{post}}{e^{-\mu \cdot \text{T}}}} \tag{15.5}$$

Therefore, C_{corr} for kidneys at different depths can be directly compared, or the calculation performed pixel-wise and a geometric image formed. For more information about planar activity quantification, see Chapter 25.

15.1.11 Quantitation (Thyroid Uptake Example)

In some investigations, it is beneficial to be able to quantify the amount of the injected activity (%IA) that is taken up within an organ. Any variation outside of a normal range can then be used as an indicator of potential disorders. An example of this is thyroid pertechnetate imaging used for diagnosis of hyperthyroidism. The uptake and distribution of ^{99m}Tc can be used to distinguish Graves's disease from a toxic nodular goitre or to determine the therapeutic dosage for treatment, and to predict the outcome and potential side effects of therapy. Typical acquisition details are given in Table 15.4.

After acquiring the image, an ROI is drawn over the thyroid alongside a smaller background region. Counts within the background ROI are scaled to match the size of the thyroid ROI and subtracted from the counts within the thyroid ROI. Background subtracted counts within the thyroid ROI are then corrected for radioactive decay from the time of injection.

To convert the measured counts to activity, it is necessary to know the sensitivity of the camera. System sensitivity in this scenario is subject to acquisition protocol, set-up, and source geometry. It is therefore common to acquire an image of a phantom representing the thyroid. The phantom contains a known amount of ^{99m}Tc, and the scan is acquired and processed using the clinical protocol. A worked example calculating the %IA within the thyroid is given below for the patient of Figure 15.13a and phantom of Figure 15.13b. Normal reference ranges for %IA are generally between 1–5 per cent.

Activity in thyroid = unknown Thyroid ROI size =800 pixels Counts within Thyroid ROI = 186,076	Activity in phantom = 10 MBq Phantom ROI size =800 pixels Counts within Phantom ROI = 176,054
Background ROI size = 200 pixels Counts within background ROI = 2,985 BG counts scaled to thyroid = 11,940 Thyroid – BG= 174,136	Background ROI size = 200 pixels Counts within background ROI = 259 BG counts scaled to phantom = 1,036 Phantom – BG= 175,018
Decay corrected by 20 minutes =180,277 Decay corrected count rate = 751 cps Thyroid Activity = cps / sensitivity = 751/72.9 = 10.3 MBq	Decay corrected counts =175,018 counts Decay corrected count rate = 729 cps sensitivity = cps/activity = 729/ 10 = 72.9 cps/MBq
Percentage uptake = $A_{thy} / A_{inj} \times 100\%$ = 10.3/80 × 100% = **13 %**	

15.2 WHOLE-BODY IMAGING

As discussed, it is possible to reduce the FOV used by the system by applying a digital zoom during the acquisition. Conversely, it is possible to increase the field of view by moving the detector or patient during an acquisition. This type of acquisition protocol is generally referred to as whole-body imaging, as it is commonly used to image along the entire length of the patient. A clinical example where this is most common is for bone imaging, particularly for oncology where potential sites of disease can appear in any number of places within the skeletal anatomy. Typical imaging methods and patient preparation are summarized in Table 15.5.

TABLE 15.4
Typical Acquisition Parameters for a Thyroid Uptake Study

Parameter	Value
Radiopharmaceutical	^{99m}Tc Pertechnetate ($^{99m}TcO_4^-$).
Energy Window	140 keV with 20% width
Collimator	Low-Energy High-Resolution (LEHR)
Matrix Size	128x128
Injected Activity	80 MBq
Uptake time	20 min post-injection
Acquisition time	240 s
Zoom	2.5x

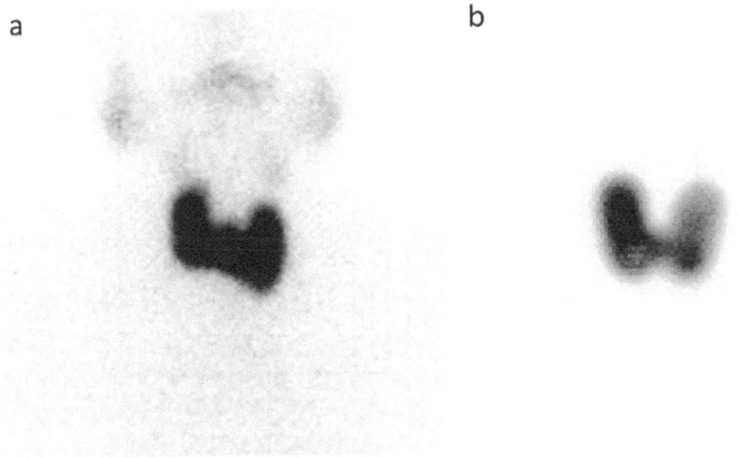

FIGURE 15.13 Image of thyroid pertechnetate scan (a) and thyroid phantom (b).

TABLE 15.5
Typical Acquisition Parameters for a Whole-body Bone Acquisition

Parameter	Value
Pharmaceutical	^{99m}Tc Phosphate
Energy Window	140 keV (optional scatter window)
Collimator	Low-Energy High-Resolution (LEHR)
Matrix Size	variable
Injected Activity	400 MBq of each
Uptake time	3–4 h
Acquisition time	300 s per bed position or 10 cm/min
Zoom	1x

An analogy with conventional photography would be panoramic imaging whereby the orientation of the camera is tracked during the exposure, and this motion used when the image is being generated. The same process is used with the gamma camera, except in most cases the patient is moved relative to the detectors. The imaging couch can be moved in one of two ways, 'Step and Shoot' or 'Continuous Motion'.

With 'step and shoot' mode the patient is positioned supine with the detectors in a posterior and anterior position. A standard static acquisition is performed for a duration set by the user. The couch will then move by a fixed distance (defined by the size of the detector) so that a lower part of the body is position under the detector and the process

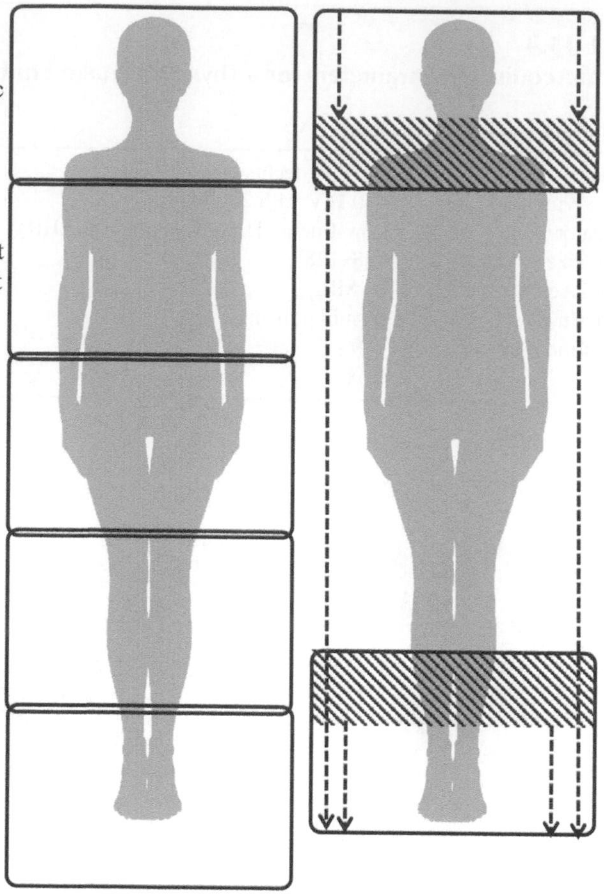

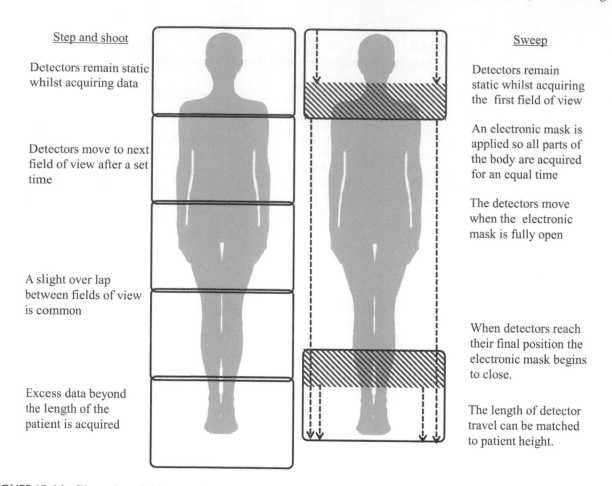

FIGURE 15.14 Illustration of different whole-body acquisition methods.

repeated until the required length of the patient is covered. Often, a slight overlap of a few pixels between couch positions will be applied to insure there is no risk of missing data.

For 'continuous' or 'sweep mode' the couch motion is continuous with a user-defined speed, typically 10–15 cm/min. This inevitably requires precise control of couch motion so that detector events can be binned into the correct pixel depending on the location of the couch at the moment the event is recorded. Continuous motion has the advantage that the detectors can potentially contour closer to the patient during the acquisition and improves image quality with a smoother transition between couch positions. In addition, the scan length does not have to be a multiple of detector places negating unnecessary imaging. However, to ensure each part of the patient is acquired for an identical duration, couch motion cannot be instantaneous. An electronic mask is applied across the detector, which gradually opens rows of pixels. The couch does not start moving until the mask is fully open. When the couch has completed its motion, the electronic mask will be applied again. The net result is that, for the same field of view and exposure time, a whole-body sweep with continuous motion will be one couch position longer in duration than a step-and-shoot protocol. A comparison of the two acquisition modes with an example image are given in Figure 15.14.

15.3 DYNAMIC IMAGING

The function of many biological processes cannot be adequately evaluated by taking a 'snap shot' of that function at a single time point. It is therefore often necessary to acquire multiple images over time to track the changing biological distribution of a radiopharmaceutical. This is achieved using dynamic imaging protocols analogous to time-lapse photography or a video recording [4]. The movement of the pharmaceutical through the body is measured by taking multiple static acquisitions at different points in time and combining those images into a single cine dataset. Frame duration (the time taken to acquire each static image) and the number of frames is usually based on the speed of the physiological process being monitored, and it is not uncommon to change the frame duration in the protocol as the physiological process slows (or speeds up). Each set of frames is referred to as a phase, and while each phase and frame usually starts immediately after the last, this does not have to be the case. In fact, even studies in which images are acquired with gaps of a few hours or days can still be considered as dynamic studies (such as those used for dosimetry).

It is evident that a dynamic acquisition requiring a fast-frame rate will inevitably be subject to low-count statistics and image noise. It is therefore common to use a high-sensitivity collimator and small matrix size during such acquisitions. The amount of activity administered in these studies is also generally higher.

15.3.1 Renogram Example

In addition to a visual representation of the pharmaceutical motion. Dynamic studies can be used to determine qualitative and quantitative measures of a physiological function. In the case of renography, the pharmaceutical (MAG-3 or similar) is excreted through the renal tubules. Typical acquisition setting for a dynamic renal study are given in Table 15.6.

Kidney function is assessed by measuring the rate that the MAG-3 is filtered from the blood and eliminated via the urinary tract. There are three distinct physiological phases that take place after administration of the radiopharmaceutical. An initial ventricular or perfusion phase, where the pharmaceutical is circulating throughout the body in the blood. An extraction phase when the kidney is actively filtering the MAG-3 from the blood, and an elimination phase where the pharmaceutical is released from the kidney into the bladder.

A renogram is a time-activity curve that can be used to describe each of the three physiological phases. The administration is given with the patient orientated within the gamma camera and posterior acquisitions commence a few seconds prior to a bolus injection. Due to the rapid nature of the initial phase, single second frames or less are usually acquired for the first few minutes. After this period, and during the elimination phase, frame duration is often increased to 10–20 seconds. Late acquisitions at 30 or 60 minutes may also be acquired for 5 minutes each. Typical dynamic frames are shown in Figure 15.15 and derived time activity curves in Figure 15.16.

The contribution of each kidney to total renal function can be determined by interrogating the renogram in a number of different ways. Methods include integration, convolution and Pat-Lak analysis. Measurements for the rate of transit of the radiopharmaceutical through the kidney include peak time, mean transit time, or renal outflow efficiency. More detail on renography and renogram analysis is given in Volume II of this book.

15.4 GATED IMAGING

For physiological processes that are very quick, it is often difficult to acquire sufficient events to form images of suitable quality, even with a high sensitivity collimator and highly administered activities. For processes that are repeated, such as the respiratory or cardiac cycles, events recorded over multiple cycles can be used to form a dynamic dataset that represents an average of that process [5]. To be able to add the frames from repeated parts of the physiological cycle, the system generally requires an input timing signal to which it can gate the events.

For imaging of the heart, this is achieved using an echocardiogram (ECG). Provided the output signal from the ECG is consistent, and the timing between R-waves predictable, events can be binned according to the time they were

TABLE 15.6
Typical Acquisition Parameters for a Dynamic Renal Study

Parameter	Value
Pharmaceutical	^{99m}Tc MAG-3
Energy Window	140 keV
Collimator	Low-Energy General-Purpose (LEGP)
Matrix Size	128 x 128
Injected Activity	100 MBq
Uptake time	Injected at start of acquisition
Frame duration	1 s then 10 s
Number of frames	60 then 174
Zoom	1x

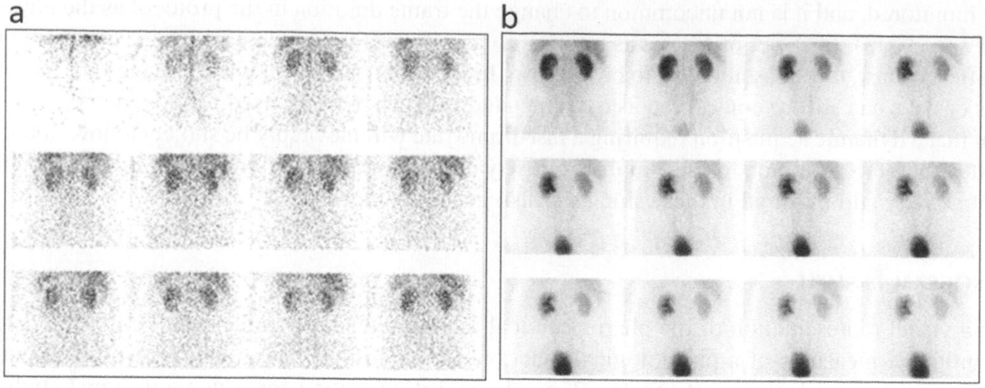

FIGURE 15.15 Dynamic images acquired during a DMSA study, images are shown for two phases acquired at 1 second per frame (a) and 10 seconds per frame (b).

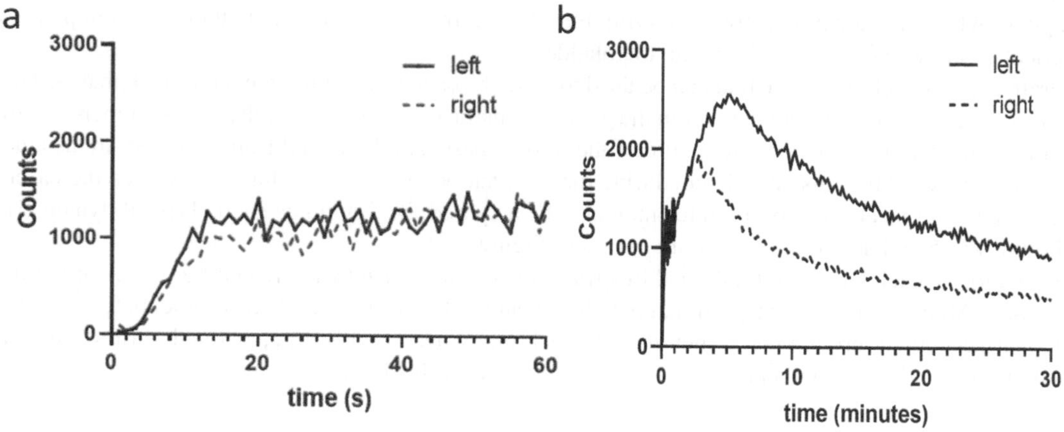

FIGURE 15.16 Time activity curves derived from a renogram study. Acquired with single second frames for the first minute (a) and then 10 second frames for thirty minutes (b).

detected after the last ECG R-wave. The system will monitor the RR interval for each heartbeat. If a significant variation is observed, events recoded during that interval will not be used in the image formation. Figure 15.17a is a typical ECG showing a single RR interval.

15.4.1 MUGA Example

A clinical example where this is implemented is radionuclide ventriculography, used to assess abnormalities in ventricular contractility and cardiac function. Red blood cells (euthrocytes) are labelled with ^{99m}Tc either in-vitro or in-vivo using a phyrophosphate agent. A typical imaging protocol is given in Table 15.7. Typically, up to 24 frames of 30 milliseconds duration are acquired for a sufficient number of heartbeats to record a hundred thousand to two hundred thousand counts per frame.

To get good separation of the left and right ventricles, right anterior oblique views are acquired, sometimes with a 20-degree craniocaudally tilt if the system is capable of it. Example images are given in Figure 15.17b

15.5 Tomographic Acquisitions

Thus far, only two-dimensional imaging has been discussed, whereby the gamma camera is used to take projected images of the radiation transmitted from the patient. However, in some cases it may be necessary to separate the activity within adjacent and superimposed tissues. To do so requires acquisition of data in three dimensions using

TABLE 15.7
Typical Acquisition Parameters for a Radionuclide Ventriculography Study

Parameter	Value
Pharmaceutical	^{99m}Tc Erythrocytes
Energy Window	140 keV
Collimator	Low-Energy General-Purpose (LEGP)
Matrix Size	64 x 64
Injected Activity	800 MBq
Uptake time	None
Acquisition time	16–24 frames per RR interval with 150 kcounts per frame (typically 15–20 minutes)
Zoom	2x

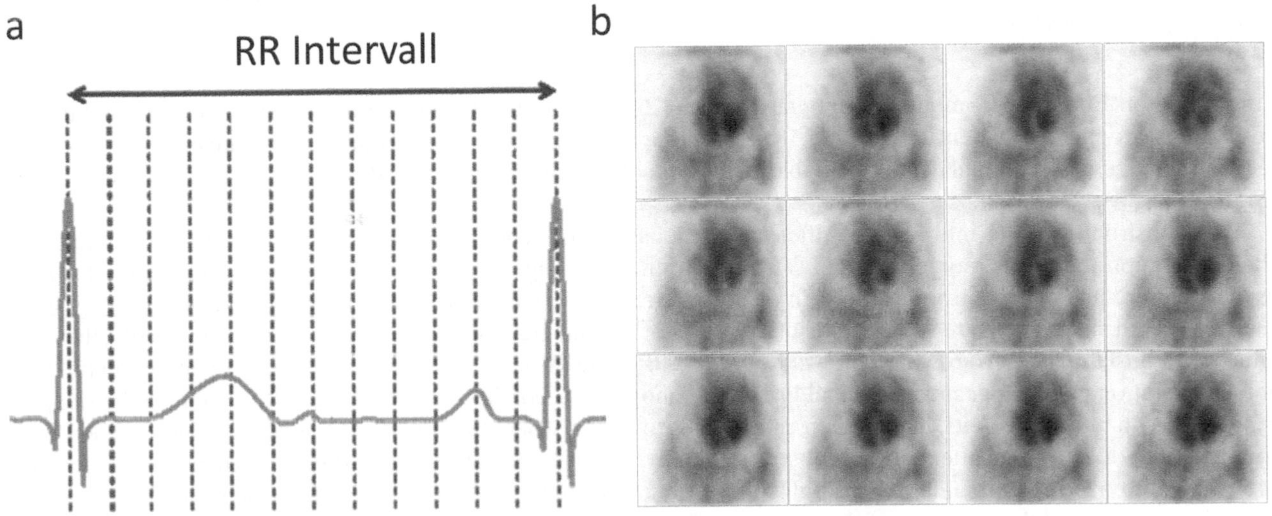

FIGURE 15.17 Typical ECG signal with appropriate frames used for gating (a). Typical gated frames generated during a ventriculography study (b). The change in volume of the left ventricle is clearly apparent.

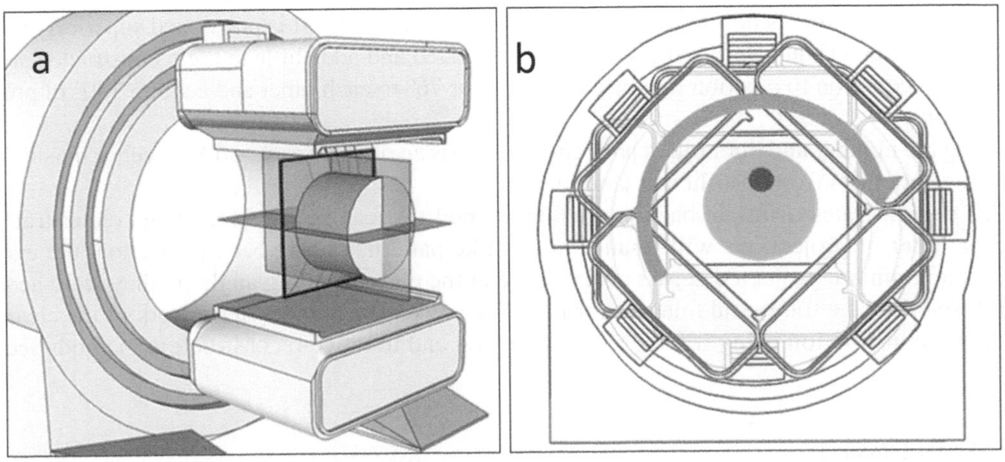

FIGURE 15.18 Illustrations of a gamma camera demonstrating rotation of detectors about an object (a) and illustration of reconstructed planes within the camera coordinate system (b).

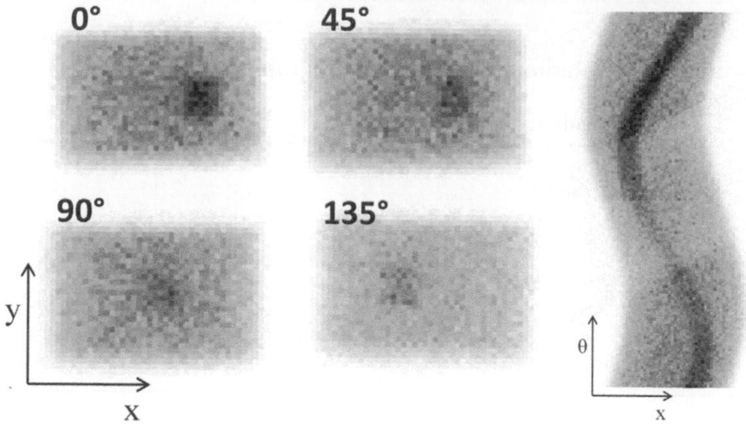

FIGURE 15.19 Example projection data at different angles about the object in Figure 15.18 and a sinogram representing a central slice through the projections.

Single Photon Emission Computed Tomography (SPECT). During this acquisition the detectors acquire multiple 2D projections around the patient (Figure 15.18a), which are reconstructed using Fourier or iterative techniques. Once reconstructed, cross-sectional data is available in the form of transaxial, sagittal, and coronal images through the patient (Figure 15.18b).

To reconstruct the data into 3D it is often conventional to first reorganize the 2D data according to the transaxial slices they represent. This reorganisation takes a row from each image projection and plots it as a function of the angle used to acquire it. The organisation results in what is referred to as a sinogram, and there will be a different sinogram for each transaxial slice. There are some interesting features of the sinogram in Figure 15.19 worth noting. Firstly, the sinusoidal pattern to the data, from which the sinogram gets its name. A point source positioned exactly at the centre of the detector rotation would appear as a vertical line on the sinogram. Secondly, a diagonal area at the centre of the sinogram is evident with reduced counts. This region corresponds to angles where the detector is under the imaging couch, and a small amount of additional attenuation is observed. Finally, this data is acquired with a dual-headed gamma camera with each detector covering 180°. There is a slight difference in sensitivity of the two detectors, which manifests as a slight reduction in counts at the bottom half the sonogram. To overcome this issue, some systems allow each detector to cover 360° and the projections acquired from each detector are summed before generating the sinogram.

With the introduction of a 3D rotation, there is the obvious need to consider the additional acquisitions parameters. These include the extent of rotation of each detector, the number of projections per rotation, and the period of each projection. In theory, it is not necessary to acquire opposing projections for image reconstruction, and therefore only 180° of detector coverage is required. In practise, due to effects of attenuation, it is advisable to acquire over the full 360°. The only application where this is not routine is cardiac imaging. As the heart is positioned superiorly on the patient's left, acquisition of data at a right inferior angle is highly attenuated and adds little to the reconstructed area of interest. In this situation, it is common to position the detectors at 90° or 76° to each other and acquire 180° of projections with fewer gantry motions.

When selecting the number and duration of projections, there is an obvious trade-off between acquisition time, image statistics and image artefacts from insufficient sampling.

Figure 15.20 demonstrates a simple back-projection method that can be used for image reconstruction and how an insufficient number of projections will result in a star-like pattern. During back projection, the events recorded in the row of a sinogram are projected across image space at the corresponding angle by they were acquired. As the projected events overlap, the transaxial image is formed. The greater the number of projections the clearer the image. For details about reconstruction using filtered back projection and iterative reconstruction methods see Chapters 16 and 20.

15.5.1 DaTScan Example

A clinical example that requires SPECT imaging is dopamine imaging, used to aid in the diagnosis of Parkinson's disease. [^{123}I]N-ω-fluoropropyl-2β-carbomethoxy-3β(4-iodophenyl)nortropane (^{123}I-FP-CIT), is a radiopharmaceutical

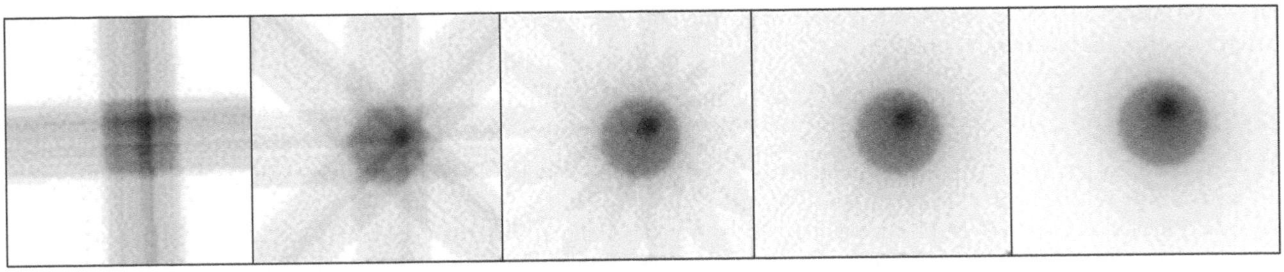

FIGURE 15.20 An illustration of a simple back projection technique to reconstruct emission data generated from the object of Figure 15.17. As more projections are back projected the image become clearer.

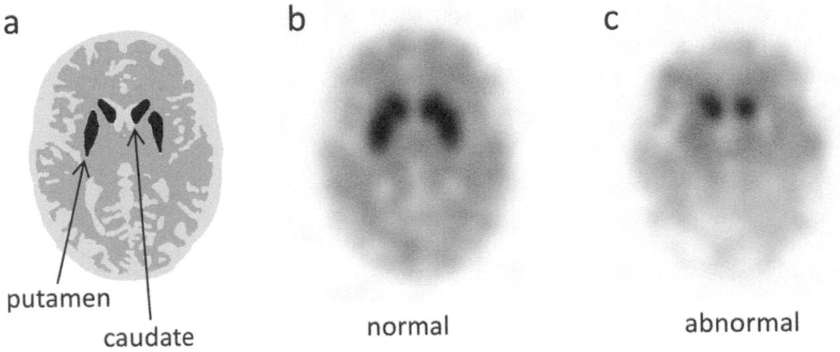

FIGURE 15.21 An illustration of striatum anatomy (a) and ^{123}I-FP-CIT uptake patterns typical of patients with essential tremor (b) and Parkinson's disease (c).

TABLE 15.8
Typical Acquisition Parameters for a Radionuclide Ventriculography Study

Parameter	Value
Pharmaceutical	^{123}I-FP-CIT
Energy Window	159 keV
Collimator	High-Energy (HE)
Matrix Size	128
Number of Projection	120
Injected Activity	185 MBq
Uptake time	3 to 6 hours
Acquisition time	30 s per projection
Zoom	1x

derived from cocaine and which binds to dopamine transporters (DaT) within the striatum [6]. A SPECT scan allows for visualization of the dopamine transporter function in the caudate and putamen. The main use of DaT scanning is in patients with a movement disorder to help differentiate essential tremors from tremor due to Parkinson's disease. A patient with an essential tremor will demonstrate high symmetric uptake in both caudate and putamen regions. A patient with Parkinson's disease will demonstrate reduced putamen activity and asymmetric caudate activity. An illustration of the relevant brain and striatum anatomy is given in Figure 15.21a, normal and abnormal uptake distributions are shown in Figures 15.21b and c.

Typical imaging parameters are given in Table 15.8. The patient is positioned supine on the imaging couch with the head extended off the couch, using a head holder, and with arms at the patient's sides. The patient's head is positioned straight and in the centre of the field of view. The detector radius is set as close as possible to the patient's head or auto-contouring is used, as excessive distance will affect image quality and, potentially, the ability to interpret the subtle differences in pharmaceutical uptake.

REFERENCES

[1] B. S. Greenspan et al., "Procedure guideline for parathyroid scintigraphy. Society of Nuclear Medicine," (in English), *J Nucl Med*, vol. 39, no. 6, pp. 1111–1114, Jun 1998.

[2] C. Bluemel et al., "EANM practice guidelines for lymphoscintigraphy and sentinel lymph node biopsy in melanoma," (in English), *Eur J Nucl Med Mol Imaging*, vol. 42, no. 11, pp. 1750–1766, Oct 2015, doi:10.1007/s00259-015-3135-1.

[3] A. Piepsz et al., "Guidelines for 99mTc-DMSA scintigraphy in children," (in English), *Eur J Nucl Med*, vol. 28, no. 3, pp. Bp37–41, Mar 2001.

[4] J. S. Fleming and P. M. Kemp, "A comparison of deconvolution and the Patlak-Rutland plot in renography analysis," *J Nucl Med*, vol. 40, no. 9, pp. 1503–1507, Sep 1999. [Online]. Available: www.ncbi.nlm.nih.gov/pubmed/10492372.

[5] B. Hesse et al., "EANM/ESC guidelines for radionuclide imaging of cardiac function," (in English), *Eur J Nucl Med Mol Imaging*, vol. 35, no. 4, pp. 851–885, Apr 2008, doi:10.1007/s00259-007-0694-9.

[6] J. Darcourt et al., "EANM procedure guidelines for brain neurotransmission SPECT using (123)I-labelled dopamine transporter ligands, version 2," (in English), *Eur J Nucl Med Mol Imaging*, vol. 37, no. 2, pp. 443–450, Feb 2010, doi:10.1007/s00259-009-1267-x.

16 Single Photon Emission Computed Tomography (SPECT) and SPECT/CT Hybrid Imaging

Michael Ljungberg and Kjell Erlandsson

CONTENTS

16.1 INTRODUCTION

Since the scintillation camera, in its essence, is a 2D imaging system that measures count projection generated by photons emitted from a 3D radionuclide distribution in a 3D object, there is no possibility of resolving the depth information of the distribution from a single projection image. In the past, different projections were measured, in order to account for this limitation, but this approach was really not a sufficient solution. By mounting the scintillation camera head on a gantry that could rotate around the patient, the development of SPECT (Single-Photon Emission Computed Tomography) followed the development of Computed Tomography (CT), since the problem with the depth information for planar X-ray imaging was essentially the same.

DOI: 10.1201/9780429489556-16

The development of SPECT and SPECT/CT is described in Chapter 1 in this book and more extensively in the excellent historical review article '*The origins of SPECT and SPECT/CT*', by Brian Hutton [1].

16.2 SPECT

As stated above, by mounting a scintillation camera head on a gantry that is able to rotate around the object it is possible to obtain more information of the radionuclide distribution, which then makes it possible to reconstruct a 3D image of the distribution. The general assumption is that the difference between the count distribution in the acquired images when rotating the camera is only dependent on the specific projection view, and that there exists only one possible activity distribution that can result in the acquired set of projection images. This is only true in the patient case when the radiopharmaceutical does not redistribute during the acquisition time or decreases by physical decay.

16.2.1 Image Reconstruction by Direct Back-projection

An image from a scintillation camera is, in most cases, obtained using a parallel-hole collimator that, due to its construction principle, only allows detection of photons traveling in a direction close to parallel to the hole axis. This procedure corresponds to a forward projection operation. One can mimic the opposite operation, that is, performing a back-projection of the measured data onto an image matrix along lines that are defined by a particular projection angle. Since it is not known where along the projection line the original emission point was located, the back-projection algorithm usually projects the measured data uniformly along the back-projection lines. Figure 16.1 shows schematically how forward projection relates to back-projection.

16.2.2 Image Reconstruction by Filtered Back-projection

The main problem with direct back-projection is that counts are back-projected uniformly along the lines of response, including areas outside the physical boundaries of the object. This results in a blurred image, as can be seen in Figure 16.2a. Since the back-projection operation is an additive process, the counts that are wrongly positioned will never be removed. The magnitude of the blur is proportional to the inverse of the radial distance from a particular source. It is therefore possible to account for this effect by applying a special type of filter, called a ramp filter. The filtering can be implemented as a multiplication in the Fourier frequency domain (Figure 16.2c) or as a convolution process in the spatial domain (Figure 16.2d).

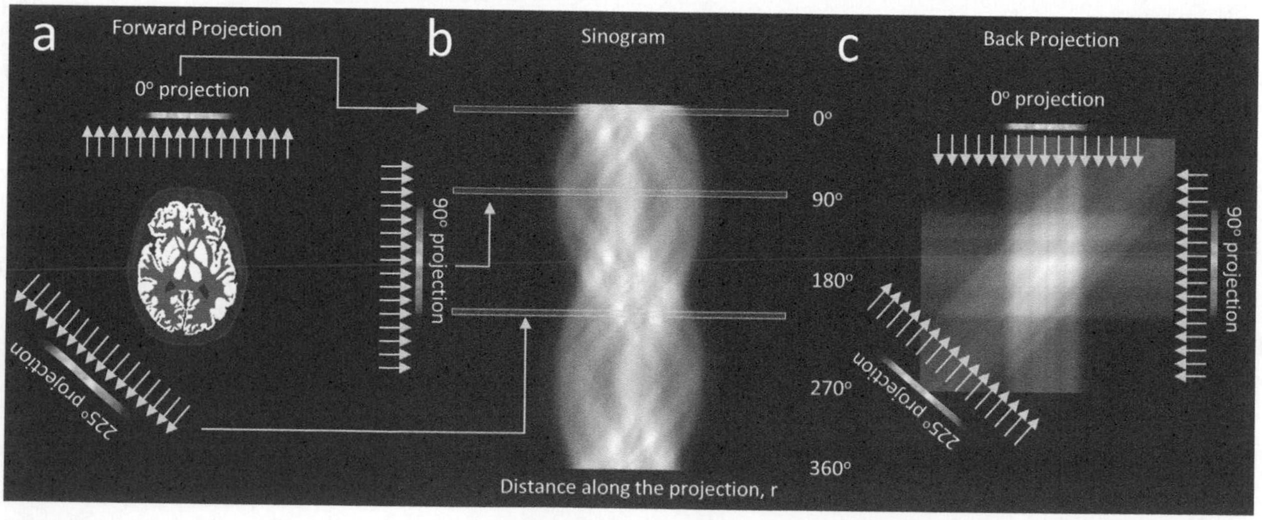

FIGURE 16.1 Three steps in the reconstruction of a tomographic image. (a) Photons emitted in parallel directions are passing though the collimator and generate a projection image. The camera is rotated to get a different view of the same distribution. (b) The data acquired are reorganized into a sinogram. (c) The data are back-projected into the object domain to generate an estimated activity distribution.

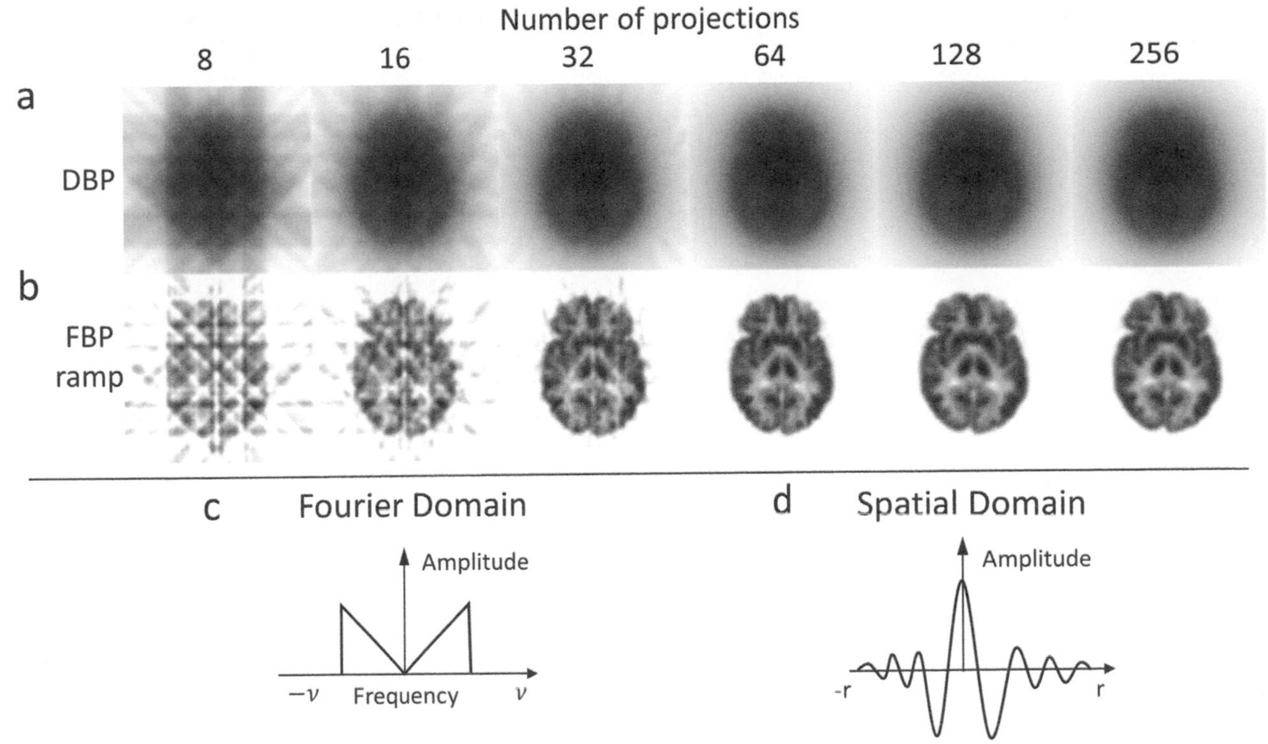

FIGURE 16.2 Example of images, reconstructed from a different number of projections ranging from 8 to 256 are shown in the upper row of (a) for direct back-projection (DBP) and in the lower row (b) for filtered back-projection (FBP). Figure (c) shows the ramp filter in the Fourier domain and its corresponding shape in the spatial domain (d).

The choice of filtering in the Fourier domain is preferable when filtering a large amount of data (large matrix sizes) because of better calculation efficiency.

Filtered back-projection (FBP) can be described mathematically by

$$f(x,y) = \int_0^{\pi} F^{-1}\left[|\upsilon| F\{p(r,\theta)\}\right] d\theta \quad (16.1)$$

where F and F^{-1} symbolizes the Fourier transform and the inverse Fourier transform operations, respectively, and $r = x\cdot\cos(\theta) + y\cdot\sin(\theta)$. One thus calculates the Fourier transform of the projection data $p(r,\theta)$, multiply the result with a ramp filter $|\upsilon|$, and then calculate the inverse Fourier transform. The result is then back-projected (Figure 16.2b).

16.2.3 Derivation of the FBP Algorithm

Here we will derive the formula on which the filtered back-projection (FBP) reconstruction algorithm is based. We start from the definition of the 2D Fourier transform (FT) of an image $f(x,y)$ and its inverse, which are used for moving from the spatial domain to the frequency domain and *vice versa*:

$$F(v_x, v_y) = \iint_{-\infty}^{\infty} f(x,y)\cdot e^{-i2\pi(xv_x + yv_y)} dxdy \quad (16.2)$$

$$f(x,y) = \iint_{-\infty}^{\infty} F(v_x, v_y)\cdot e^{i2\pi(xv_x + yv_y)} dv_x dv_y \quad (16.3)$$

where ν_x and ν_y are spatial frequency variables corresponding to spatial coordinates x and y, respectively. Setting ν_y =0 in Eq. 16.2, we get:

$$F(\nu_x,0) = \iint_{-\infty}^{\infty} f(x,y)\cdot e^{-i2\pi\cdot x\nu_x} dxdy = \int_{-\infty}^{\infty}\left(\int_{-\infty}^{\infty} f(x,y)dy\right)e^{-i2\pi\cdot x\nu_x}dx \tag{16.4}$$

Changing x to r and ν_x to ν_r, we can write:

$$= \int_{-\infty}^{\infty} p_\theta(r)\cdot e^{-i2\pi\cdot r\nu_r}dr\Big|_{\theta=0} = P_\theta(\nu_r)\Big|_{\theta=0} \tag{16.5}$$

where

$$p_\theta(r) = \int_{-\infty}^{\infty} f(x,y)ds\Big|_{x=r\cdot\cos\theta - s\cdot\sin\theta,\ y=r\cdot\sin\theta+s\cdot\cos\theta} \tag{16.6}$$

is the forward projection (FP) of $f(x,y)$ at angle θ and $P_\theta(\nu_r)$ its 1D-FT. The FP operator is also known as the Radon transform and can be used to describe the SPECT measurement process (ignoring photon attenuation and scattering as well as the distance-dependent resolution effect).

A result similar to Eq. 16.5 is obtained for $\theta = \pi/2$ with ν_x =0, and, due to rotational symmetry, the following general equality can be said to be valid for any angle θ.

$$P_\theta(\nu_r) = F(\nu_x,\nu_y)\Big|_{\nu_x=\nu_r\cdot\cos\theta,\ \nu_y=\nu_r\cdot\sin\theta} \tag{16.7}$$

In words, this means that the 1D-FT of the parallel projection at an angle θ of a 2D function is equal to a section through the origin at an angle θ of the 2D-FT of the same function. This is known as *the central section theorem* [2] and it is schematically described also in Figure 16.3.

The consequence of this theorem is that it is theoretically possible to reconstruct an image from its Radon transform. One way to do this could be to first construct the 2D-FT of the image and then apply Eq. 16.3. However, this would require interpolation of the data from polar to Cartesian coordinates in the frequency domain, which is doable but not straightforward.

An alternative approach is to rewrite Eq. 16.3 in polar coordinates (ν_r,θ), with $\nu_x = \nu_r\cdot\cos\theta$, $\nu_y = \nu_r\cdot\sin\theta$ and $\dfrac{d\nu_x d\nu_y}{d\nu_r d\theta} = \nu_r$:

$$f(x,y) = \int_0^{2\pi}\int_0^{\infty} F(\nu_r,\theta)e^{i2\pi(x\cdot\nu_r\cdot\cos\theta + y\cdot\nu_r\cdot\sin\theta)}\nu_r d\nu_r d\theta = \int_0^{\pi}\int_{-\infty}^{\infty} F(\nu_r,\theta)e^{i2\pi\cdot\nu_r(x\cdot\cos\theta+y\cdot\sin\theta)}|\nu_r|d\nu_r d\theta$$

where the change of the integration limits requires introducing the absolute value of ν_r, and now we can make use of the central section theorem (Eq. 16.7):

$$= \int_0^{\pi}\int_{-\infty}^{\infty}|\nu_r|\cdot P_\theta(\nu_r)e^{i2\pi\cdot r\nu_r}d\nu_r d\theta = \int_0^{\pi}\hat{p}_\theta(r)d\theta \tag{16.8}$$

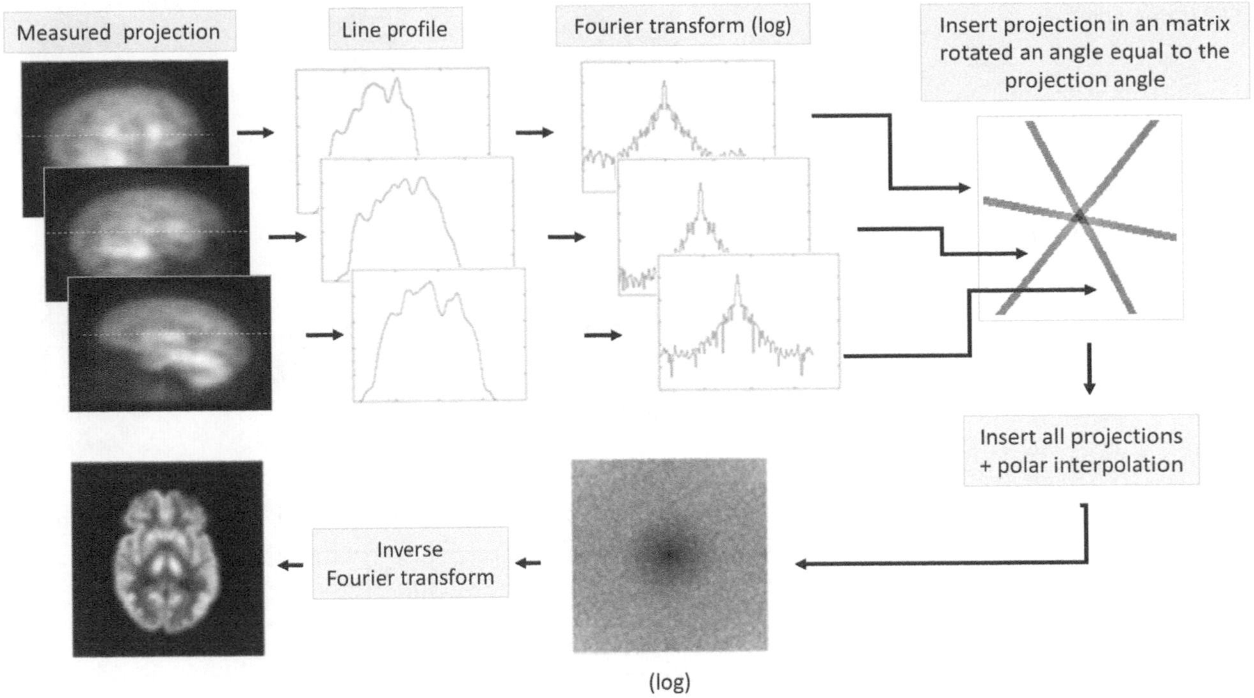

FIGURE 16.3 A schematic image showing the central section theorem that links acquired counts from a SPECT system, corresponding to a cross-section of an object, measured at different projection angles, to the spatial transversal distribution of activity in the particular cross-section.

where $\hat{p}_\theta(r)$ is the result after filtering the projection data $p_\theta(r)$ with the ramp filter, $R(v_r) = |v_r|$, in the frequency domain. The integral over the angle θ corresponds to the back-projection (BP) operator, which is the adjoint of the FP operator (Eq. 16.6).

Application of the ramp filter can intuitively be seen as a way of compensating for the non-uniform sampling in the frequency domain, where the sample density decreases linearly with increasing distance from the origin.

In real data, the signal-to-noise ratio typically decreases with increasing spatial frequency. Consequently, the effect of the ramp filter will be to amplify noise at high frequencies, where little or no actual signal is present. Therefore, the ramp filter is often combined with a low-pass filter in order to reduce the noise-amplification.

Equation 16.8 is valid for continuous variables only. In practice, however, the data obtained in a SPECT study are discrete, and the formula has to be appropriately adapted, involving approximations, which invariably lead to a more or less limited accuracy.

We should mention that it will be necessary to take into account the effects of photon attenuation and Compton scattering and, possibly, also distance-dependent resolution (Chapter 3 and 26).

16.2.4 FBP with Regularization

As can be seen in Figure 16.2c a ramp filter is a high-pass filter, so if noise is present in the projection data, these high-frequency components will be amplified, and this can be disturbing when viewing the tomographic image. A common method of reducing the noise is to apply a low-pass filter. This can be done (a) on the projection data prior to reconstruction by 2D pre-filtering, (b) as a post-reconstruction operation by applying a 3D filter on the noisy reconstructed image, or (c) within the FBP procedure as an addition multiplicative 1D filtering step, according to Eq. 16.9.

$$f(x,y) = \int_0^\pi F^{-1}\left[\, lp(\upsilon)\,|\upsilon|\; F\{p_\theta(r)\}\,\right] d\theta \tag{16.9}$$

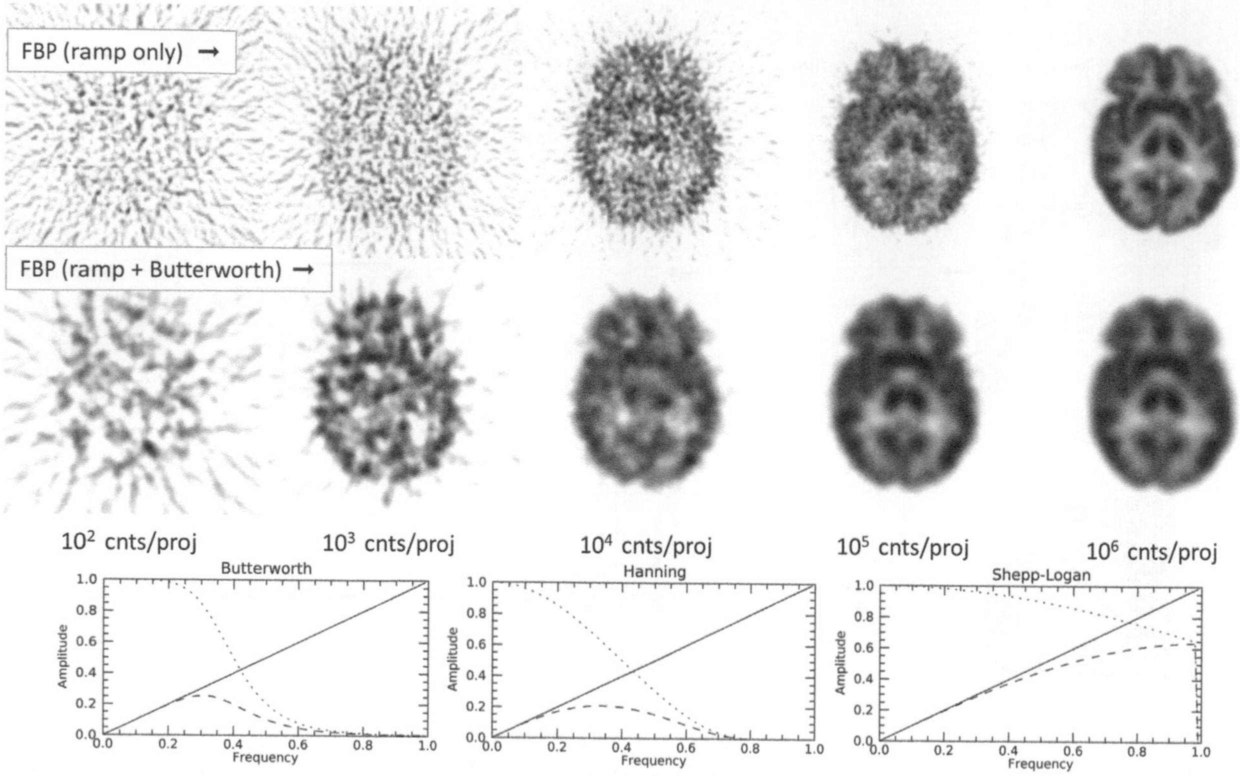

FIGURE 16.4 Examples of reconstructions of the brain example object with five different noise realizations applied to the projection data. Upper row show images reconstructed with a ramp filter only. Middle row shows the same images reconstructed with a ramp filter combined with a low-pass Butterworth filter. The three graphs, displayed in the lower row, show the Butterworth filter (left) together with two other low-pass filters (Hanning and Shepp-Logan) that have commonly been used in SPECT reconstructions. The solid line is the ramp filter, dotted lines in the low-pass filter and the dashed line is the product of the two curves. This is the curve that in practice is used when filtering in the Fourier domain.

where $lp(v)$ is the low-pass filter, defined in the frequency domain. The effect of including the low-pass filter can be seen in the middle row of Figure 16.4.

The example above is for a ramp filter modified with a 1D Butterworth low-pass filter. This filter is defined as $lp(\upsilon) = 1/\left(1 + (\upsilon/\upsilon_c)^{2n}\right)$ where υ_c is the critical frequency, corresponding to a filter amplitude of 50 per cent. The parameter n (the order) controls the steepness of the filter curve. These two parameters can be tuned by a user to optimize a certain acquisition protocol for a certain nuclear medicine study. It should also be mentioned that the ramp filter is essential for the FBP method, but if the noise level in the projection data is very low or absent, an additional low-pass filter step may not be needed, as it would only affect the spatial resolution. Therefore, the use of such a filter and the actual parameter values will depend on the particular study, and optimization of the filter is a common task for a medical physicist. In SPECT, several types of dedicated filters have been used [3]. Examples are the Hanning filter $lp(\upsilon) = 0.5 + 0.5 \cdot \cos(\pi\upsilon/\upsilon_c)$ and the Shepp-Logan filter $lp(\upsilon) = \sin(\pi\upsilon/2\upsilon_c)/(\pi\upsilon/2\upsilon_c)$

A SPECT study consists of a series of planar projections as a function of the projection angle around the patient. From Figure 16.2 one can see that a certain number of projections are required to reconstruct an image with acceptable quality. Given a total acquisition time, one can use this time to acquire a larger number of projections, each with a smaller acquisition time per projection, or vice versa. Since the Poisson noise in an acquired projection directly relate to the registered counts in the image, a too-low acquisition time will result in projections with a noise level that can create artefacts.

Figure 16.5 illustrates this by a simulation of a brain study with an activity of 150 MBq and a total acquisition time of 21 minutes (two detector heads). From left to right are shown images representing (a) 64×64 matrix, 64 projections, 40 sec, (b) 64×64 matrix, 128 projections, 20 sec (c) 128×128 matrix, 64 projections, 40 sec, and (d) 128×128, 128 projections, 20

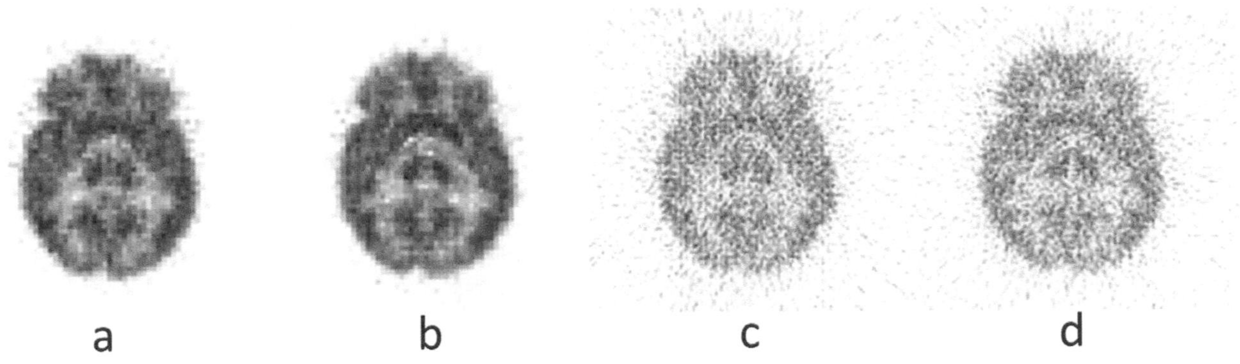

FIGURE 16.5 The simulated brain study with four different acquisition parameters. The matrix size for images (a) and (b) are 64×64 and 128×128 for images (c) and (d). Images (b) and (d) have been simulated with twice as many angles as in (a) and (c) and half the acquisition time.

sec, respectively. Most striking is the noisy appearance of the 128×128 images, which is due to the fact that the detection area represented by one pixel in the 128×128 is four times smaller than the 64×64 matrix pixel area. In addition, since the FBP algorithm reconstructs images slice by slice, the thickness of a 128×128 image is half that of a 64×64 image. Thus, on average only one eighth of the counts in a 64×64 voxel is used to determine the value in the 128×128 voxel. In theory, 128×128 projections should generate images with better spatial resolution compared to 64×64 projections, due to better sampling. However, depending on the observer the noisy image may require an additional low-pass filtering step, which has a negative effect on the spatial resolution. For the same voxel size (a, b) and (c, d), the image with more projections (b, d) but with lower acquisition time per projection appear to produce slightly better contrast in the images.

In theory, an angular sampling with smaller angular distance is always to be preferred but this will reduce the detected count level in a particular projection if a total acquisition time has been a constraint (which often is needed for patient comfort and patient throughput). Too-long acquisition times also increase the risk for motion artefacts. Furthermore, if image noise increases, and as a consequence a following need for post-filtering is required, the improvement due to small angular sampling may be lost due to the related degradation in spatial resolution by the low-pass filtering.

16.2.5 Image Reconstruction by Iterative Methods

Most SPECT systems today have the option of selecting statistical iterative reconstruction for creating tomographic images [4, 5] instead of FBP, and these methods are generally now more commonly used. An iterative procedure incorporates a computer model of the camera system and the patient being imaged with the purpose of calculating projections in a similar way as real projections are measured. The purpose of the reconstruction is to find the most probable estimate of the radionuclide distribution by comparing criteria – the calculated and measured projections. This is done by starting from an initial estimate of the unknown source distribution, calculating projection data using the camera/patient model, and modifying the image estimate until the difference is acceptably small. Thus, it is assumed that, if the calculated and measured projections agree, then the estimate of the source distribution represent the activity distribution in the patient. The most common iterative reconstruction algorithm implemented is Ordered-Subset Expectation-Maximization (OS-EM) [6]. Iterative reconstruction methods are discussed more in detail in Chapter 20.

It is important that the computer camera/patient model includes all relevant physical effects that can affect the image in a real measurement, such as non-homogeneous photon attenuation, the contribution of scatter, the blurring due to the collimator response, and potential septal penetration. Otherwise, the reconstructed image will not, even if reaching convergence (similarity between measured and calculated projection), represent the true distribution because of the difference between the computer-based modelling of the radiation transport and the actual radiation transport. Chapter 26 is discussing these necessary effects to consider in more detail.

16.2.6 Non-uniform Orbitals

The spatial resolution depends on the distance between the source and the camera. In previous, older cameras, a circular orbit was common, but such orbits were not optimal when imaging patients due to the difference in width and

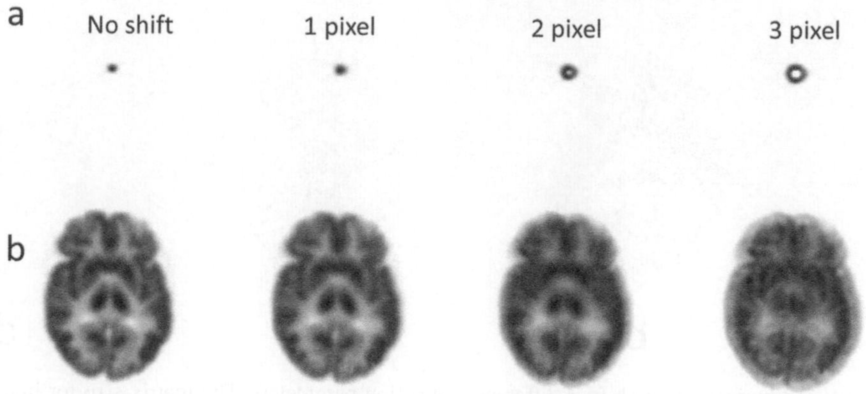

FIGURE 16.6 A simulated of a shift in the electronical centre-of-rotation (used in the image formation and reconstruction) relative to the mechanical centre-of-rotation for a point source (row a) and a distributed source (row b). For small shifts, no artefacts are created, but the spatial resolution will be reduced. For large pixel shifts, a clear change can be seen as a ring for the point source and in the superficial part of the object and the inner structures for the distributed source.

thickness and also because of the presence of the couch. It took a considerably long time to tune in the relation between the gantry angle and couch height, leading to an overall large overhead in time for the patient to stay on the couch before acquisition actually could start. Modern cameras nowadays have systems that automatically determine the orbit for both SPECT and for whole-body planar acquisition by the use of an array of detectors that adjusts the camera distance during the acquisition. This technology makes it both simpler and faster for the study to get started, leading to better comfort for the patient and allowing for longer acquisition times. The distance at different angles for the orbit can often be found in the Dicom header, allowing these values to be used in off-line reconstruction program that can compensate for collimator resolution.

16.2.7 The Centre-of-rotation

The physical camera has a central origin around which it rotates in a defined orbit. In order to achieve a proper reconstruction, the back-projection of the measured projection data must be made in such a way that there is an alignment between the mechanical centre point and the electronic centre point, which subsequently determines the midpoint in the image. If a point source is positioned exactly in the centre of rotation, a back-projection of a projection collected 0 degrees should follow the same path as a back-projection of data acquired at 180 degrees. However, if there is a mismatch, this will result in a degradation in spatial resolution and artefacts. Figure 16.6 shows a reconstruction simulation where an electronical mismatch has been introduced corresponding to 1, 2, and 3-pixels shift. For larger shift, the result will appear as an annulus, and the effect will create a deformation of the original shape of the source. COR correction is often made on a regular basis (Chapter 23, section 3)

16.2.8 Non-uniformities in the Camera

In a measured scintillation camera image, it is generally assumed that a variation in the counts over the whole field of view is caused by a variation in the radionuclide distribution or an effect related to photon attenuation and scatter in the object. Thus, for the case of a uniform fluence of photons impinging onto the camera, a measurement is expected to result in uniform image with a local variation only caused by counting statistics. However, non-uniformities can occur even for this scenario, and can be caused by non-linearities due to the PM tube positioning, difference in the gain of some PM tubes that results in an actual photopeak in the energy spectrum for the particular PM tube that is not centred within the energy window, and also local changes in the NaI(Tl) crystal itself due to aging, causing a regional attenuation of the scintillation light photons that subsequently result in a lower energy signal. This is an effect that is hard to see in a planar projection but can appear as a ring-shaped artefacts in reconstructed images. The shape of a ring appears since the problem that cause the non-uniformity is located on the same place in the FOV regardless of the projection angle of the camera. To illustrate this, a simulation was made whereby a cylindrical-shaped activity distribution without attenuation and scatter was simulated for 128 projection angles with a 360° rotation mode. A wrongly

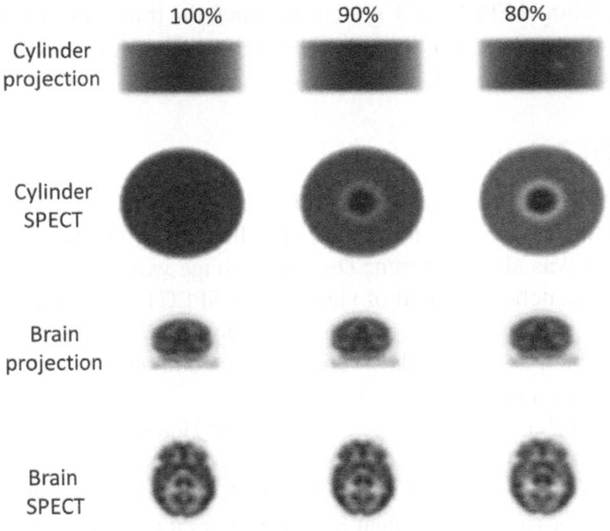

FIGURE 16.7 Simulation of an area on the crystal that arise to an inhomogeneity of 90 per cent (middle column) and 80 per cent (right column) relative to its surrounding part. Because this defect is located in fixed position on the crystal (regardless of the rotation) it will generate a circular artefact. In a symmetrically shaped source as is the case for the cylinder example, the artefact is clearly seen, but it is harder in a clinically realistic image, such as the reconstructed brain example.

tuned PM tube that resulted in a local decrease of signal to only 90 per cent and 80 per cent, relative to the surrounding tubes, was simulated by using a Gaussian function with a FWHM of 8 pixels (2 cm). The anterior projection and related reconstructed images are shown in Figure 16.7 for the two upper rows. The same defect was also applied on a brain phantom simulation, and these are shown in the two lower rows.

16.2.9 Multiple Bed-position

Most SPECT systems have a limited camera FOV of about 40 cm in the patient orientation. This means that when covering large areas, several bed positions need to be measured and combined to get a large FOV study or a WB SPECT study. Several vendors provide systems that have automatic procedures for doing this as conveniently and fast as possible, because this procedure in general extends the time the patient is on the couch and too much time increases the level of discomfort and potential risk for movement. In combining multiple bed-positions, it might be necessary to consider the fact that NaI(Tl)-based cameras have poorer image quality close to the edges of the camera due to fact that the Anger logic (positioning based on calculating the centroid of the scintillation light) does not work so well here. Therefore, some overlap may be needed. If equivalent overlap occurs in the CT measurement, then a larger absorbed dose, due to the CT exposure, might result.

16.3 SPECT/CT

Prior to SPECT/CT, most manufactures offered transmission imaging using either scanning-line sources, scanning-point sources, a single static line source or multiple static line sources. These systems were used to generate maps of linear attenuation coefficients for attenuation compensation [7], although the poor resolution and image quality prohibited more diverse applications. The need for faster and better attenuation maps, as well as anatomical localization capability, led to the development of SPECT/CT.

One application where anatomical information was found to be needed was in the diagnostic imaging of tumours. In many situations the tumour uptake of the radiopharmaceutical can be high compared to other organs, making it difficult to localize for potential future surgery. Anatomical landmarks in the background of the SPECT images were simply not present to accurately determine the locations of the malignancies. This was of course recognized, and research on rigid and non-rigid image registration between SPECT and CT was conducted, but the task of doing this accurately demanded sophisticated mathematical algorithms and, even when a diagnostic CT was available, the matching of the images was not always successful due to organ displacement between the studies. An obvious solution was therefore to investigate the possibility of designing a combined SPECT/CT system in which the transfer of the patient between the two modalities was accomplished in such a way as to minimize patient motion. These so-called, hybrid systems made fusion of SPECT and CT images possible in routine clinical procedures.

16.3.1 Commercial SPECT/CT Systems

The GE Millenium hybrid SPECT/CT camera equipped with the HawkEye single-slice CT was the first commercially available SPECT/CT system. Developed by General Electric Healthcare, Israel, the CT created images with a thickness of 1 cm and the spatial resolution was about 3.5 mm. The tube voltage was either 120 or 140 kV and the tube current was 2.5 mAs. An acquisition that matched the field of view for the SPECT cameras took approximately ten minutes to complete. A version of the system was also designed to be able to do coincidence PET by removing the collimators and was therefore equipped with one-inch crystals. This system was the commercial implementation of the very successful research by the late Bruce Hasegawa and colleagues [8, 9].

Current SPECT/CT systems – offered by major vendors, Philips Healthcare, General Electric Healthcare, Mediso Medical Systems, and Siemens Medical Solutions – all offer high-resolution diagnostic CT units with their SPECT systems. To describe the characteristics of each of these SPECT/CT systems is beyond the scope of this chapter, but it is worth mentioning that one vendor (Mediso Medical System) offers a triple SPECT/PET/CT system combined into one unit.

16.4 IMAGE RECONSTRUCTION AND QUANTIFICATION USING ANATOMICAL INFORMATION

16.4.1 Image Fusion

One of the most important applications for a combined SPECT and CT unit is the ability to fuse anatomy and morphology images. The is a great advantage, since in many cases the orientation and landmarks for the SPECT image are very difficult to use as guidance but, if the SPECT image can be overlaid on a CT image, the location of, for example a tumour, becomes much more evident. Figure 16.8 shows an example of such a combination. The example is a bone-scan, where there is an uptake within one of the vertebras, and the position is much more clearly defined when the SPECT image is fused with an anatomical CT image. Combined SPECT-CT is used for most nuclear medicine applications.

It should be mentioned that when performing a Dicom export of reconstructed SPECT images and corresponding CT images, they may not be an aligned to each other, making an external software necessary. However, Dicom have defined useful tags, such as 'Image Position', 'Pixel Width', and 'CT Slice Thickness', that together can make it possible to interpolate CT images to be able to fuse SPECT/CT images. These are also useful for concatenation of projections and related CT images obtained from several bed-positions.

16.4.2 Attenuation Correction

The information obtained from a CT system is turned into Hounsfield numbers defined as $HU = 1000 \cdot (\mu\text{-}\mu_{water}) / (\mu_{water}\text{-}\mu_{air})$, where μ is the linear attenuation coefficient, a function of photon energy, tissue composition, and tissue density. The baseline is water at standard pressure and temperature, while μ in turn can be expressed as a product of the mass-attenuation coefficient μ_m / ρ and the tissue density ρ and where μ_m is a function of the photon energy and the composition of the material. Mass-attenuation coefficients can be found in databases maintained by the National Institute of Standards and Technology [10]. Since μ_m reflects the electron density, a calibration is generally required against a phantom with different materials to scale to physical density (g/cm^3). It is important to remember that the scaling also depends on the X-ray tube voltage used in the given study. If the density distribution can be obtained with acceptable accuracy, this can then be used to create attenuation maps for use in attenuation correction. The scaling is straightforward, using a mass-attenuation coefficient for the current photon energy. However, several errors might occur in such as conversion. Firstly, it should be remembered that the HU numbers have been obtain from a system that uses a continuous energy spectrum of X-ray photons. In certain situations, with obese patients for instance, beam-hardening can occur, which corresponds to a change in the relative X-ray energy fluence along the transmitted projection bin through the patient duc to preferential absorption of lower energy X-ray photons. As a result, the HU units are different for the same tissue type compared to a smaller patient.

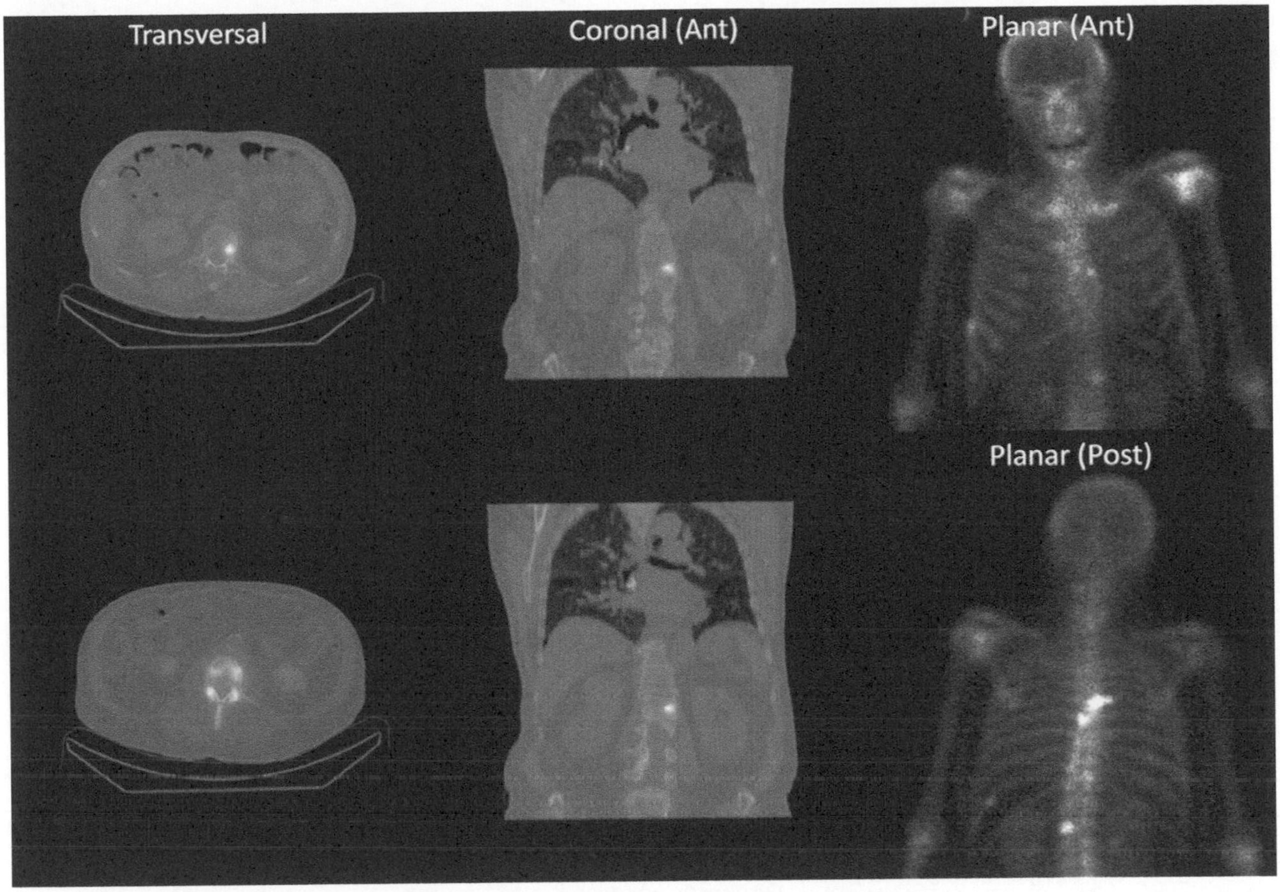

FIGURE 16.8 Example of a SPECT bone-scintigraphy images where SPECT and CT are fused to obtain location. Left column shows two transversal slices, Middle column shows coronal and sagittal images and right column shows anterior and posterior planar images. Image originally published in [21] and reprinted with permission. Colour image available at www.routledge.com/9781138593268.

16.4.3 Respiratory Motion

In most SPECT/CT system, the CT unit is a high-resolution diagnostic system that can perform very fast CT studies resulting more or less in snapshots of the patient's breathing-related geometry. However, the SPECT acquisition is made over a time frame that is relatively long and that necessarily includes breathing. This makes the attenuation, as 'seen' by the photons, into an average of the whole breathing cycle, and the resultant image is also affected by the breathing. Thus, if a 'snap-shot' CT study is used to be fusion with, or to correct, a SPECT study for attenuation, a mismatch may occur that can generate artefacts, especially close to boundaries between areas of different attenuation (lung/tissue or bone/tissue). Figure 16.9 illustrates this. The images have been generated from the XCAT phantom software [11], which includes very realistic models for organ motion. An CT image, averaged over the whole respiratory cycle, is shown together with images for the max exhalation and inhalations for the same position.

16.4.4 Scatter Correction

Most scatter-correction methods in clinical SPECT/CT systems are based on energy window methods, where the scatter contribution in the photo-peak window is estimated from a second energy window located below and, in some cases, also above the principal photon energy. It is assumed that the scatter in these windows after scaling represents the scatter in the main energy window. These methods are discussed further in Chapter 26. An alternative approach is that of calculating the scatter distribution from the Klein-Nishina probability for the first scatter order of photons. The principle is shown in Figure 16.10a. From a certain voxel, the probability (P_1) for a photon traveling distance L_1 is calculated.

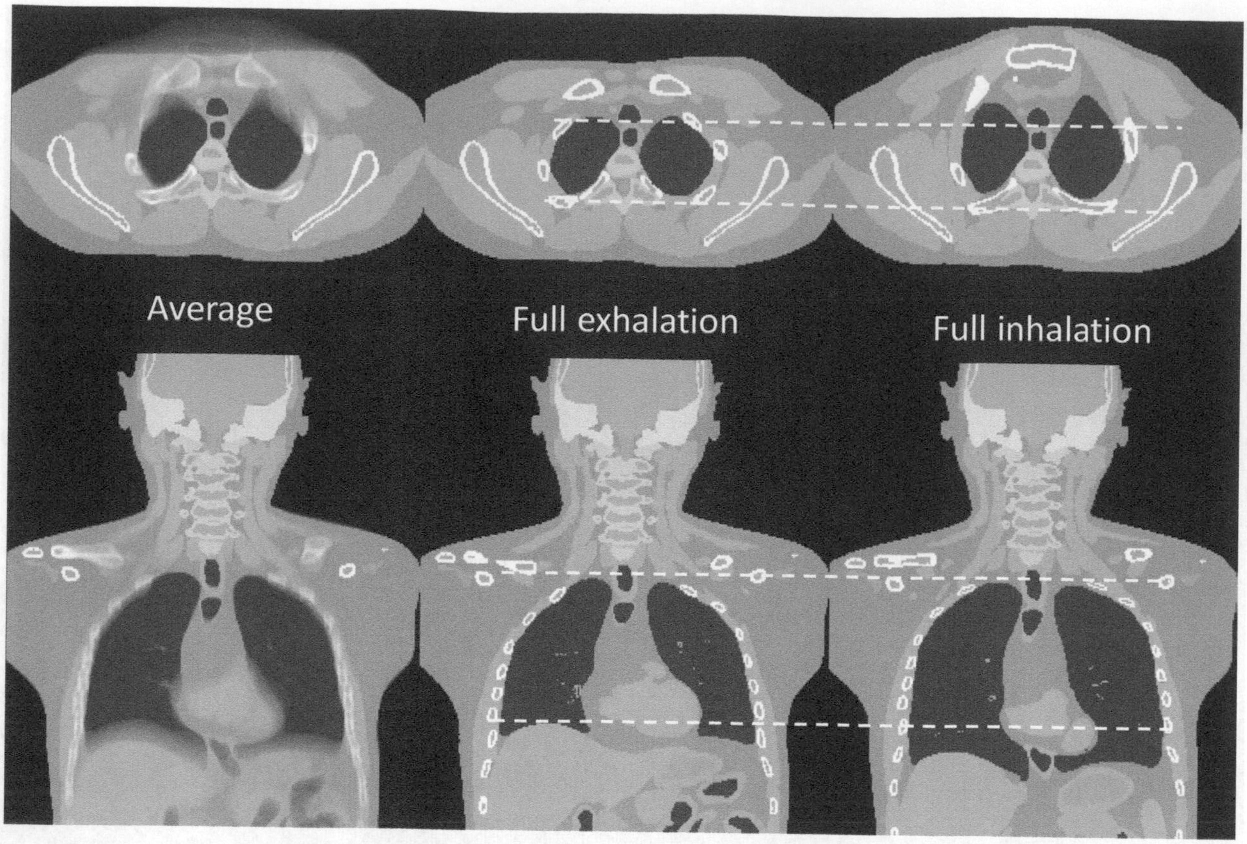

FIGURE 16.9 A simulation, using a computer phantom XCAT of the respiratory motion for full inhalation and full exhalation. The average represents the resulting blur that the motion is causing during an acquisition.

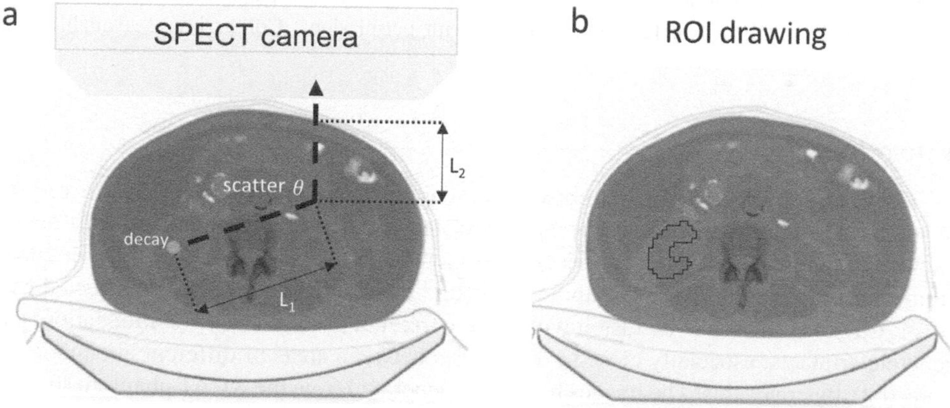

FIGURE 16.10 Figure (a) shows schematically how a CT image can be used to model scatter. The probabilities for photons traveling different paths in a CT volume is calculated and used to obtain an overall scatter projection. The right image exemplifies the usefulness of drawing volume of interests (VOI) from a CT and transferring this to calculate counts in reconstructed SPECT image. However, the difference in spatial resolution between the CT and the SPECT images needs to be accounted for.

Based on the initial direction, the scatter angle toward the camera is calculated, and the probability (P_2) for that scatter angle is calculated from the Klein-Nishina distribution (Chapter 3 in this volume). In addition, the probability (P_3) for an escape for the scattered photon with its new energy along the path length L_2 is calculated. The total probability for such a scattered photon impinging on the scintillation camera is then the product, $P_1 \cdot P_2 \cdot P_3$. A scatter profile is then created by considering all possible scatter locations and related angles. This method is based on the assumption of only one scatter order. A full Monte Carlo-based approach that includes also multiple scattered photons is possible to include in the reconstruction [12].

16.4.5 Outlining Volume-of-Interest

The CT images are also very useful in segmentation of volume-of-interest (VOI) since, because of its superior resolution compared to SPECT images, the outlines more reflect the actual volume of an organ or tumour (Figure 16.10b). However, when calculating activity and activity concentration within these CT-based VOI, some type of correction for the spill-in and spill-out of registered counts that should have been included if the resolution of the SPECT would be the same as for the CT. Correction for this is discussed in the next section.

Defining VOI from registered CT images may also be useful in situations where count levels in the SPECT images are so low that it is hard to define the VOI from the SPECT when using a linear grey-level scale. This is especially a problem if there are higher uptakes in other parts of the image that mask the boundaries of the VOI.

16.4.6 Partial-volume Correction

One of the problems with SPECT imaging using a scintillation camera with parallel-hole collimator is the limited spatial resolution. When drawing VOIs over an organ from reconstructed SPECT images, the boundaries of the organ can be hard to define due to the spatial resolution and to surrounding activity. If a registered high-resolution CT image is available, the physical boundary of the organ can be more accurately defined, but if this VOI is transformed and used on a set of SPECT images, some counts can be missed due to the spill-out effect, causing too-low counts. This effect is often called the partial-volume effect (PVE). Different methods to compensate for the PVE have been developed over the years, and a comprehensive review of these methods has been published by Erlandsson and colleagues [13]. One simple, but commonly used, method is to determine so-called recovery coefficient (RC). The principle here is to measure known activities in different source configurations by SPECT and apply correction for scatter, attenuation, and collimator response and then, by using a calibration factor, to estimate the activity in the sources from the quantitative SPECT images. This determination of the activity is defined from the physical outline of the sources. The ratio between the measured activity and the known true activity will then define the RC and describe the fraction of activity that results in counts appearing inside the defined VOI. For smaller source volumes, RC becomes smaller as the PVE becomes more apparent and, for larger source volumes, the RC approaches unity. A more comprehensive description of PVC methods is given in Chapter 26, Section 5.

16.4.7 Couch Bending

The SPECT/CT system are generally designed in such a way that the CT and SPECT can registered either directly with the clinical software or by external procedures, using the information of the Dicom tags for 'Image Position', 'Table Height' and 'Slice Thickness'. However, there is a potential problem in couch bending when imaging obese patients with large body masses. The CT are almost always behind the SPECT cameras relative to the couch and therefore the couch will extent out in the air with possible bending that will affect the A/P position of the CT image relative to the SPECT. This will affect both the image fusion and, if used, attenuation correction. The couch is constructed with materials that affect the measurements as little as possible in terms of photon attenuation when the camera head is on the posterior side of the patient. One way of measuring this effect is to fill line sources with both activity and contrast agents and, from a SPECT/CT measurement of this setup, to investigate the shift between the line source in a fused SPECT/CT as a function of the load and its position on the couch.

16.5 SPECT/CT in Radionuclide Dosimetry

It is well known that charge-particle interactions with surrounding molecules can cause damage and have biological effects on sensitive parts of the cell, such as the DNA, and this is used in radiotherapy with different types of radiation sources (accelerators, radioactive implants and administered radionuclides). The absorbed dose, D, is defined as the mean energy imparted, $d\bar{\varepsilon}$, to matter in an infinitesimal volume, with mass dm [14, 15], according to $D = d\bar{\varepsilon} / dm$. In practical situations, the mean absorbed dose within a volume (organs, tumours, etc.) is often used. The field of dosimetry is extensively covered for different clinical applications in Volumes II and III.

In nuclear medicine, dosimetry application, the emission of energy from a source volume, the following radiation transport, and the absorbed energy to a target volume is described by the MIRD formalism [16], according to the following equation

$$D = \frac{1}{M(r_T,t)} \sum_{r_s} \left[\int_0^{T_D} A(r_s,t)\,\mathrm{d}t \sum_i \Delta_i \cdot \phi(r_T \leftarrow r_S, E_i, t) \right], \tag{16.10}$$

where M is the mass of the target volume r_T, A is the activity in the source volume r_s, Δ_i I the energy release by particle i during a disintegration, ϕ is the fraction of energy from particle i that is absorbed in target volume r_T. In a clinical patient situation, the radiation transport, as described by the factor ϕ, is impossible to measure experimentally, so the only way to get an estimate of ϕ for a patient's geometry is to create a computer phantom of the patient from a CT study. Then this voxel-based estimate of the patient's geometry is used together with a quantitative set of registered SPECT images as input to a full Monte Carlo simulation of the radiation transport in order to score the energy deposited in the target volumes, which in this application is on a voxel level.

16.5.1 CT for Creating 2D Maps for Attenuation Correction

The method for performing dosimetry has, in many studies, been based on planar imaging using the conjugate-view method, a method that is described in detail in Chapter 25 of the present book. To quantify the activity in different organs, one needs to compensate for the attenuation. The properties of the conjugate-view method simplify in some way the attenuation correction to only require the knowledge of the total attenuation through the patient. This is similar to the attenuation in PET, where the attenuation is a function of the total distance for both annihilation photons through the patient. This attenuation can therefore be measured by an external source placed on the opposite side of the camera head.

According to the Beer-Lamberth law, the fractional number of photons transmitted in a narrow-beam geometry through a volume of material of thickness L is exponentially decreasing with L according to $e^{-\bar{\mu}L}$, where $\bar{\mu}$ is the average attenuation coefficient. If two measurements with a radionuclide source position of the other side of the patient relative to the camera are from one image – N(i,j), acquired with the patient *in situ*, and one reference image $N(i,j)_{ref}$ – a projection image of the fraction of transmitted photons can be derived from

$$\frac{N(i,j)}{N(i,j)_{ref}} = e^{-\bar{\mu}(i,j)L(i,j)} \tag{16.11}$$

to be used directly in the conjugate-view method (Section 3.3 in Chapter 25). Drawbacks with radionuclide-based flood sources used for transmission measurements are (a) poor spatial resolution in the attenuation maps determined by the selected collimator, and (b) image noise and related long acquisition time due to the low sensitivity of the gamma camera and limitations in possible level of activity in the flood source.

In a SPECT/CT system, a scout view (single X-ray projection) is used to outline the field of view for the CT measurement. This scout is in fact a transmission study performed according to Eq. 16.11 but created with a spectrum of X-rays instead. Therefore, this can be used in the conjugate-view method if a proper scaling is made, considering the fact that X-ray photons have a distribution of energies from the maximum energy, determined by the tube voltage down to very low energies. The group at Lund University have been using this method on two systems, namely the aforementioned HawkEye Discovery SPECT/CT system and a Discovery 670 SPECT/CT, as part of dosimetry calculation from planar whole-body images [17]. On the first system, it was possible to obtain the proper scaling factor from a system log file [18], but for the second system this was not possible, so a scaling method, based on the patient's weight, was developed [19]. An example of a scout measurement used for quantitative conjugate-view is shown in Figure 16.11 [20].

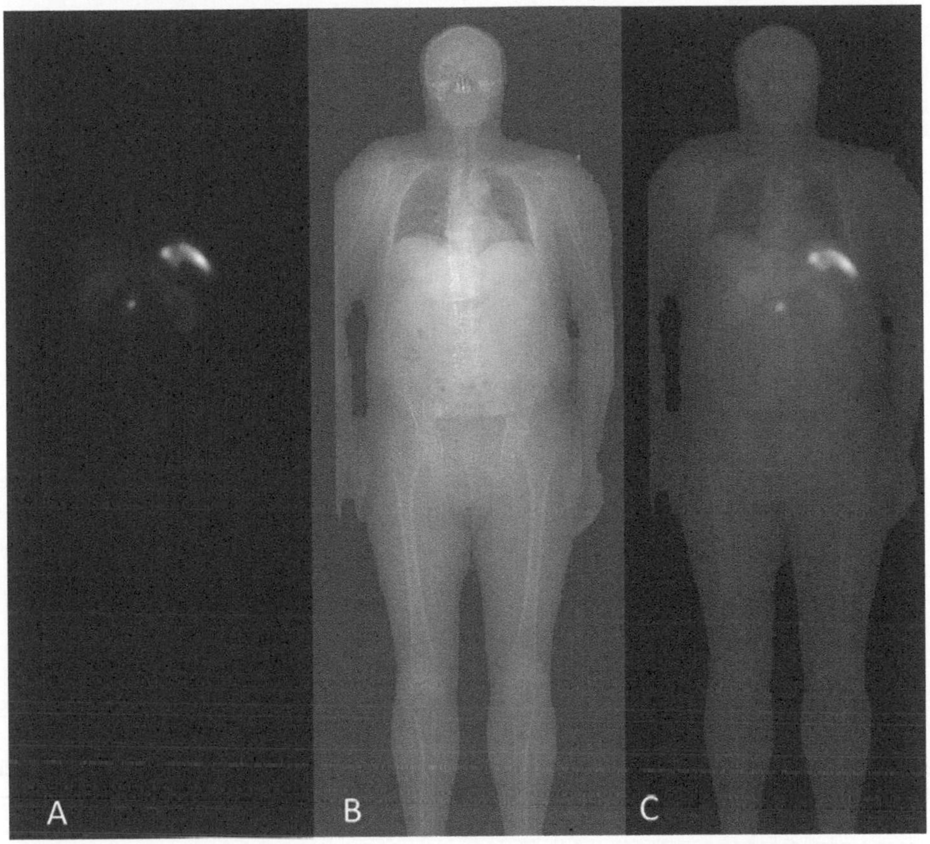

FIGURE 16.11 Example of a whole-body ^{177}Lu-DOTATATE study (left) with a matching scout-image acquired by the CT unit (middle). These two images are registered and used together (right). From these images, we can also understand that if the WB image and the scout images have been properly registered to match each other, then it is easier to define a region-of-interest from the scout image and use this on the WB image. Image originally published in [20] and reprinted according to Creative Common CC BY license (creativecommons.org). Colour image available at www.routledge.com/9781138593268.

16.5.2 CT for Creating 3D Maps for Dosimetry Calculations

The activity calculated from quantitative SPECT images is expressed in terms of activity per voxel volume. The images can therefore serve as input to dedicated absorbed-dose calculation programs, either based on assuming energy locally absorbed point-dose kernel convolution, or full 3D Monte Carlo simulation. In these voxel-based calculation procedures, the absorbed doses will be on a voxel level and the organ-absorbed dose can be obtained from a distribution of voxel-based absorbed doses using segmentation methods.

Full Monte Carlo simulations of the radiation transport provide the most accurate absorbed dose distribution but are also the most computer demanding method. It requires a registered SPECT activity and computer phantom in 3D, obtained by a CT study, that describes the geometry and composition of the patient. Decays are simulated according to the value in each source voxel (r_S), and particle emissions are sampled according to the decay scheme. The particles are followed within the 3D computer phantom, and the imparted energy from each particle interaction is scored in voxels, where interaction occurs making these target volumes r_T. When the particles are completely absorbed or have escaped the computer phantom, the absorbed dose for each voxel is calculated using the material map, knowing the density and the voxel volume to obtain $M(r_T)$ in Eq. 16.10. A schematic description of how image-based dosimetry methods works is shown in Figure 16.12. It should also be pointed out that Monte Carlo simulation is a statistical method based on random walks, which means that it requires a large number of histories to get a statistically accurate absorbed-dose image, even if the initial activity images are very well defined.

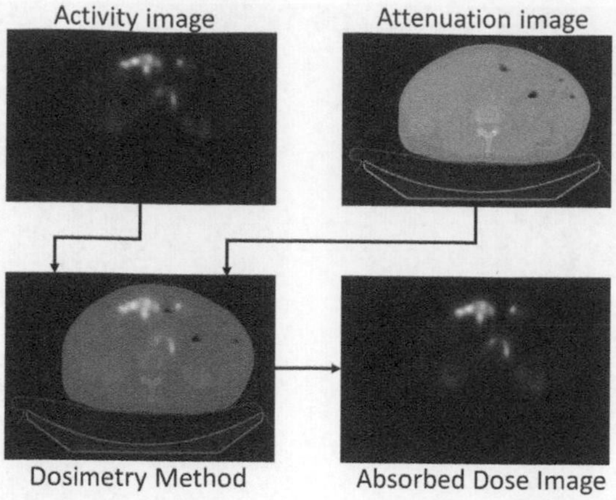

FIGURE 16.12 A schematic image showing the procedure where both SPECT images and CT images are used to calculate the absorbed dose on a voxel-by-voxel basis, meaning that each voxel is treated as a source independently from others and that the radiation transport from this source is simulated and the deposition of energy is scored for every other voxel, also including the source volume. Colour image available at www.routledge.com/9781138593268.

REFERENCES

[1] B. F. Hutton, "The origins of SPECT and SPECT/CT," *Eur J Nucl Med Mol Imaging*, vol. 41 Suppl 1, pp. S3–16, May 2014, doi:10.1007/s00259-013-2606-5.

[2] A. C. Kak and M. Slaney, *Principles of Computerized Tomographic Imaging* (Classics in Applied Mathematics). Society for Industrial and Applied Mathematics, 2001, p. 335.

[3] M. Lyra and A. Ploussi, "Filtering in SPECT Image Reconstruction," *Int J Biomed Imaging*, vol. 2011, p. 693795, 2011, doi:10.1155/2011/693795.

[4] P. P. Bruyant, "Analytic and iterative reconstruction algorithms in SPECT," (in English), *J Nucl Med*, vol. 43, pp. 1343–1358, 2002.

[5] P. Knoll et al., "Comparison of advanced iterative reconstruction methods for SPECT/CT," *Zeitschrift für Medizinische Physik*, vol. 22, no. 1, pp. 58–69, 2// 2012, doi:http://dx.doi.org/10.1016/j.zemedi.2011.04.007.

[6] H. M. Hudson and R. S. Larkin, "Accelerated image reconstruction using ordered subsets of projection data," *IEEE Trans Nucl Sci*, vol. 13, pp. 601–609, 1994.

[7] M. A. King, B. M. Tsui, and T. S. Pan, "Attenuation compensation for cardiac single-photon emission computed tomographic imaging: Part 1. Impact of attenuation and methods of estimating attenuation maps," *J NuclCardiol*, vol. 2, pp. 513–524, 1995.

[8] Y. Seo, C. M. Aparici, and B. H. Hasegawa, "Technological Development and Advances in SPECT/CT," *Semin Nucl Med*, vol. 38, no. 3, pp. 177–198, 2008, doi:10.1053/j.semnuclmed.2008.01.001.

[9] B. H. Hasegawa et al., "Dual-Modality Imaging of Cancer with SPECT/CT," *Technology in Cancer Research & Treatment*, vol. 1, no. 6, pp. 449–458, 2002.

[10] "Tables of X-Ray Mass Attenuation Coefficients and Mass Energy-Absorption Coefficients from 1 keV to 20 MeV for Elements Z = 1 to 92 and 48 Additional Substances of Dosimetric Interest." National Institute of Standards and Technology. www.nist.gov/pml/data/xraycoef/.

[11] W. P. Segars, G. Sturgeon, S. Mendonca, J. Grimes, and B. M. W. Tsui, "4D XCAT phantom for multimodality imaging research," (in English), *Med Phys*, vol. 37, pp. 4902–4915, 2010.

[12] J. Gustafsson, G. Brolin, and M. Ljungberg, "Monte Carlo-based SPECT reconstruction within the SIMIND framework," *Phys Med Biol*, vol. 63, no. 24, p. 245012, 2018/12/12 2018, doi:10.1088/1361-6560/aaf0f1.

[13] K. Erlandsson, I. Buvat, P. H. Pretorius, B. A. Thomas, and B. F. Hutton, "A review of partial volume correction techniques for emission tomography and their applications in neurology, cardiology and oncology," *Phys Med Biol*, vol. 57, no. 21, pp. R119–59, Nov 7, 2012, doi:10.1088/0031-9155/57/21/R119.

[14] I. C. o. R. U. a. Measurements, "Report 85: Fundamental quantities and units for ionizing radiation," *Journal of the ICRU*, vol. 11, no. 1, pp. 1–31, Apr 2011, doi:10.1093/jicru/ndr011.

[15] F. H. Attix, *Introduction to Radiological Physics and Radiation Dosimetry*. New York: Wiley, 1986.

[16] W. E. Bolch, K. F. Eckerman, G. Sgouros, and S. R. Thomas, "MIRD pamphlet 21: A generalized schema for radiopharmaceutical dosimetry – standardization of nomenclature," (in English), *J Nucl Med*, vol. 50, pp. 477–484, 2009, doi:10.2967/jnumed.108.056036.

[17] K. Sjogreen, M. Ljungberg, K. Wingardh, D. Minarik, and S. E. Strand, "The LundADose method for planar image activity quantification and absorbed-dose assessment in radionuclide therapy," (in English), *Cancer Biotherapy & Radiopharmaceuticals,* vol. 20, pp. 92–97, 2005, doi:10.1089/cbr.2005.20.92.

[18] D. Minarik, K. Sjogreen, and M. Ljungberg, "A new method to obtain transmission images for planar whole-body activity quantification," (in English), *Cancer Biotherapy & Radiopharmaceuticals,* vol. 20, pp. 72–76, 2005, doi:10.1089/cbr.2005.20.72.

[19] K. Sjögreen Gleisner and M. Ljungberg, "Patient-specific whole-body attenuation correction maps from a CT system for conjugate-view based activity quantification: method development and evaluation," *Cancer Biotherapy & Radiopharmaceuticals,* vol. 27, pp. 652–664, 2012.

[20] M. Ljungberg and K. S. Gleisner, "Hybrid imaging for patient-specific dosimetry in radionuclide therapy," *Diagnostics (Basel),* vol. 5, no. 3, pp. 296–317, 2015, doi:10.3390/diagnostics5030296.

[21] M. Ljungberg and P. H. Pretorius, "SPECT/CT: an update on technological developments and clinical applications," *Br J Radiol*, vol. 91, no. 1081, p. 20160402, Jan 2018, doi:10.1259/bjr.20160402.

17 Dedicated Tomographic Single Photon Systems

Jing Wu and Chi Liu

CONTENTS

The conventional SPECT scanners configured with one or more camera heads (typically two heads) using parallel-hole collimators and scintillation detectors are still the workhorse for clinical SPECT imaging. However, the image performance is limited by the low photon sensitivity and poor image resolution. For example, the tomographic count sensitivity is 130 count s^{-1} MBq^{-1} and the reconstructed spatial resolution is 15.3 mm as reported in [1] for the Symbia dual-head camera from Siemens Medical Solutions, using a low-energy, high-resolution (LEHR) collimator. A large number of dedicated SPECT systems have been reported in the recent years with improved sensitivity and image resolution for cardiac, brain, and breast imaging.

17.1 DEDICATED CARDIAC SPECT[1]

17.1.1 Commercial Systems with NaI(Tl) or CsI(Tl) Detectors

17.1.1.1 IQ-SPECT

The IQ-SPECT system [2–5] was released by Siemens in 2010, with multifocal SMARTZOOM collimators mounted on a conventional dual-head scanner with scintillation detectors, as shown in Figure 17.1. A cardio-centric rotational

---

[1] Part of the Dedicated Cardiac SPECT section is reproduced from J. Wu and C. Liu, "Recent advances in cardiac SPECT instrumentation and imaging methods", *Phys Med Biol,* vol. 64, no. 6, p. 06TR01, Mar. 13, 2019, doi: 10.1088/1361-6560/ab04de.

DOI: 10.1201/9780429489556-17

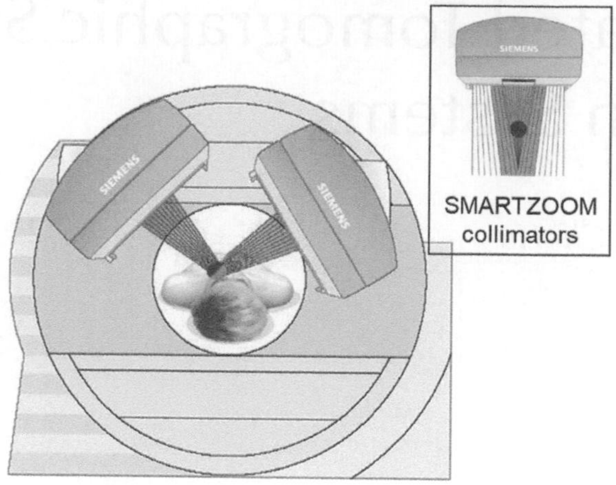

FIGURE 17.1 Schematic drawing of the Siemens IQ-SPECT scanner with SMARTZOOM collimators, reprinted from [2] with permission from Elsevier, COPYRIGHT © 2014.

orbit with scan radius of 280 mm is used for clinical cardiac imaging. The SMARTZOOM collimators have converging beams, which can focus on the heart without truncating the body to obtain four times higher sensitivity as compared to the parallel-hole collimators. A conjugate gradient method with accurate system matrix modelling that incorporates gantry deflections, collimator-hole angles and a system point spread function is used for image reconstruction to achieve high-quality cardiac images. Due to the improved sensitivity as compared to the conventional dual-head scanner, IQ-SPECT can provide comparable image quality but with only one-fourth the scan time or one-fourth the injection dose [6]. As reported in [1], the tomographic count sensitivity is 390 count s^{-1} MBq^{-1} and the reconstructed spatial resolution is 15 mm for IQ-SPECT scanner.

Phantom and human studies showed generally consistent performance between Siemens IQ-SPECT and conventional SPECT, in terms of segmental perfusion analysis and LV functions [7]. However, studies also identified substantial differences in the normal databases of ^{201}Tl myocardial perfusion imaging (MPI) between IQ-SPECT without attenuation correction and conventional SPECT [8]. Another study suggested that the summed scores were significantly higher, ejection fraction (EF) values significantly lower, and LV volumes significantly higher for the IQ-SPECT as compared to conventional SPECT, both without attenuation correction [9]. These studies might indicate that attenuation correction is preferred for IQ-SPECT.

In a phantom study with various correction methods [10], results suggested that one eighth of the acquisition time (6 s/view), compared with standard protocols with a full dose, or a lower dose at an acquisition time of 12 s/view, can be applied in IQ-SPECT MPI without the loss of diagnostic accuracy, leading to approximately 2.5–3 mSv effective dose to patients with ^{99m}Tc-labeled tracers. Subsequently, this group extended their study from phantom to human subjects. The study of 50 patients confirmed the findings in the phantom studies using ^{99m}Tc-labeled tracers on IQ-SPECT [11]. These studies used a scanning arc of 104° with 6° angular steps for each of the camera heads. For each detector, 17 views were acquired. Another investigation was performed for ^{201}Tl with both phantom and patient studies [12], which suggested 36 views/head with CT-based attenuation correction is more appropriate than 17 views/head for IQ-SPECT.

17.1.1.2 Digirad Cardius

Digirad Cardius series (Digirad Corporation, Poway, CA) uses one- to three-headed configurations with pixelated CsI(Tl) crystals and avalanche photodiodes (APDs) to enable a compact system design for dedicated cardiac imaging [2,13–16]. Figure 17.2 shows the Digirad Cardius 3 XPO system using a triple-head configuration. Each detector size is 21.2 cm by 15.8 cm with the crystal size of 6.1 mm by 6.1 mm. To improve the system sensitivity, the three detector heads are positioned towards the field of view (FOV) (15.8 cm by 21.2 cm) with an angular interval of 67.5° between heads. A scanning arc of 202.5° with 21–38 cm orbit radius is usually used for clinical data acquisition by only rotating an upright chair where the patient sits but keeping the detectors stationary. With improved detector count sensitivity, this triple-head scanner is reported to be able to achieve similar image quality as a dual-head scanner (CardioMD

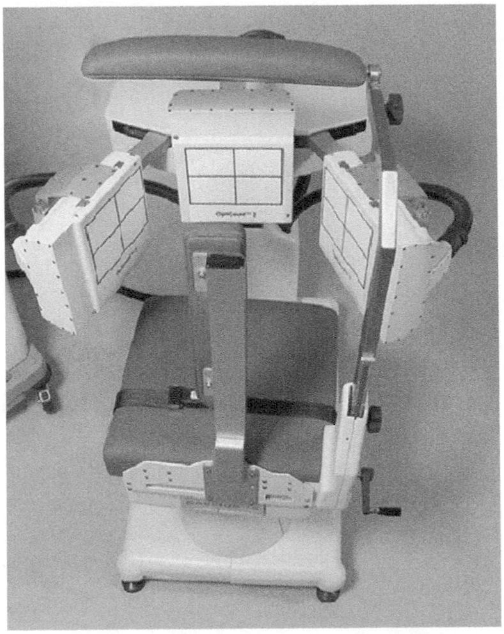

FIGURE 17.2 Digirad Cardius 3 XPO scanner, reprinted from [2] with permission from Elsevier, COPYRIGHT © 2014.

system – Philips Medical Systems, Milpitas, CA) with 38 per cent reduction in the scan time [15]. A novel 3D ordered subset expectation maximization (OSEM) algorithm (nSPEED) is used for image reconstruction, which uses a slab-by-slab blurring model as the distance-dependent geometric point response [17]. The nSPEED algorithm has been shown to achieve comparable image quality as the standard 2D OSEM algorithm for SPECT MPI with 50 per cent reduction in the scan time [18]. Similar to the system performance evaluation reported in [1], the tomographic count sensitivity is reported to be 324 count s^{-1} MBq^{-1} and the reconstructed spatial resolution is reported to be 9.2 mm at 20 cm orbit radius for the Cardius 3 XPO Digirad system [2].

17.1.2 Commercial Systems with Cadmium Zinc Telluride Detectors

17.1.2.1 D-SPECT

As shown in Figure 17.3, the D-SPECT system, from Spectrum Dynamics, Israel [19], consists of nine arrays of cadmium zinc telluride (CZT) detectors with parallel-hole collimators; each detector has a dimension of 4 cm by 16 cm. Each of the nine detectors rotates around the central axis independently during the data acquisition to cover the field-of-view (FOV). All the detectors are positioned towards the heart, leading to higher detection efficiency of photons emitted from the heart. In addition, the square-hole size (2.26 mm) of the tungsten parallel-hole collimators is larger, and the hole length (21.7 mm) is shorter than those of typical parallel-hole collimators, leading to the additional sensitivity gain [20]. The potential resolution loss due to larger collimator holes is compensated by the resolution recovery approach of modelling collimator-detector response during the reconstruction. According to the performance evaluation presented in [19], the D-SPECT system can achieve an average spatial resolution of 12.5 mm with standard clinical smoothing applied between OSEM iterations, which is comparable to the reconstructed resolution of 13.7 mm on the conventional GE Infinia scanner with LEHR collimators using the filtered back-projection (FBP) reconstruction and Butterworth post-filtering. The sensitivity in the ROI ranges from 647 to 1107 count s^{-1} MBq^{-1} for three different region of interest (ROI) sizes, which is 4.6–7.9 times the sensitivity of GE Infinia. The energy resolution is 5.5 per cent at 140 keV. The D-SPECT does not have the capability of performing transmission scans for attenuation correction.

In a multi-centre study with 276 patients, the agreement rates between D-SPECT and Anger SPECT, including GE DST-XL, GE Infinia, and Philips Cardio MD cameras with LEHR collimators were high for both ^{201}Tl and ^{99m}Tc tracers for various protocols [21]. For obese patients with much lower count levels, this study showed that the D-SPECT substantially enhanced the myocardial counts and reduced artefacts.

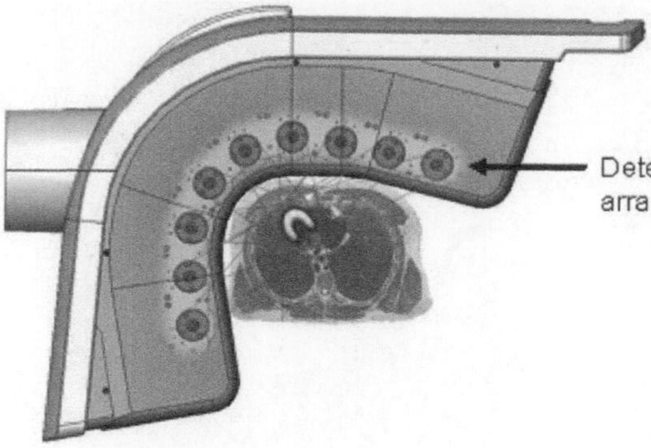

FIGURE 17.3 Schematic drawing of the D-SPECT scanner, reprinted from [19], © Institute of Physics and Engineering in Medicine. Reproduced by permission of IOP Publishing. All rights reserved.

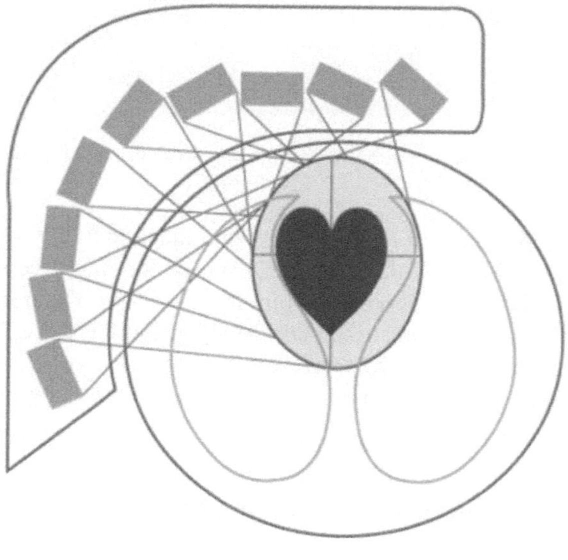

FIGURE 17.4 Schematic drawing of the GE 530c/570c scanner, reprinted from [22] with permission from Springer Nature: Springer, Journal of Nuclear Cardiology, COPYRIGHT © 2017.

17.1.2.2 Discovery NM 530c/570c

Another dedicated cardiac system with CZT detectors is the GE Discovery NM 530c/570c scanner as shown in Figure 17.4. It comes as either a SPECT-only system (GE 530c) or a SPECT-CT system (GE 570c). The 530c/570c system has 19 detector modules, each with a tungsten pinhole collimator. The system uses the same 4 cm by 4 cm CZT detector modules as the D-SPECT, with 4 modules per detector in a 2 by 2 array instead of a 1 by 4 array in D-SPECT. All the 19 detectors focus on collecting counts from the heart, leading to higher sensitivity. Different from other systems, none of the detectors move during the data acquisition.

According to a performance evaluation study [23], the system can achieve a central spatial resolution of 5.8 mm without using any post-filtering and an average sensitivity of 657 count s^{-1} MBq^{-1} with maximal rate of 848 count s^{-1} MBq^{-1} in the quality FOV, which is defined as the patient volume that is fully seen by all detectors. The energy resolution for 140 keV photons is 5.4 per cent. For pinhole collimator, both the resolution and the sensitivity are location-dependent, where the sensitivity changes as a function of the inverse-square of the distance from the source to the pinhole. A more comprehensive investigation of the location dependent resolution and sensitivity maps was performed [24]. As shown in Figure 17.5, there is a resolution and sensitivity gradient within FOV. The average resolution over thc entire FOV is about 7 mm but varies from 5 to 10 mm in the diagonal direction, as shown in

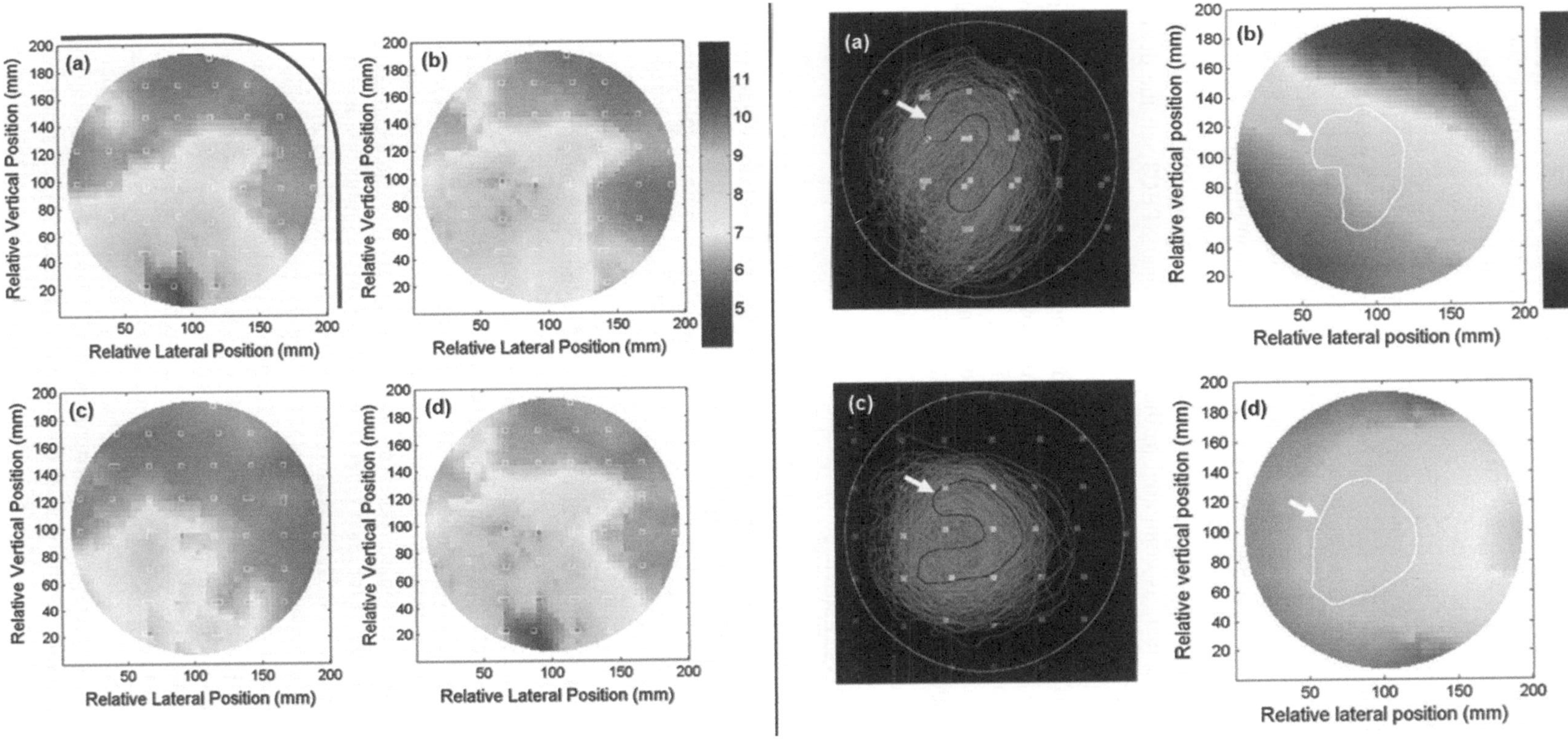

FIGURE 17.5 Left: transverse maps of iteratively reconstructed spatial resolution as measured by point sources in air with default reconstruction parameters in the three orthogonal directions: lateral (a), vertical (b), axial (c) and maximal value of all three maps (d). Maps were interpolated from 123 points in the 14 central transverse slices. The values are given in mm FWHM. Sample points are marked by white borders. Right: sensitivity measurements in the central transverse slice. (a) Plot of location of 64 measurements overlaid onto the log-intensity grayscale image of the myocardial contours of 266 clinical acquisitions. The white circle denotes the limit of the supported reconstruction region. Arrow denotes an example contour in black. (b) An interpolated map of the sensitivity in kc/s/MBq. Arrow denotes the contour of the average of all acquisitions. (c) and (d) corresponding coronal images based on 39 measurements. A diagram of the L-shape detector is overlaid on the sub-figure of left (a). Reprinted from [24] with permission from Springer Nature: Springer, Journal of Nuclear Cardiology, COPYRIGHT © 2014. Colour image available at www.routledge.com/9781138593268.

Figure 17.5 (left, d). The average sensitivity in the central transverse slice is 460 count s^{-1} MBq^{-1}, ranging from 200 to 910 count s^{-1} MBq^{-1} along the diagonal direction, as shown in Figure 17.5 (right, b). Therefore, patient positioning is critical to ensure the entire heart can be imaged with sufficient resolution and sensitivity. It is worth noting that the performance evaluation of spatial resolution for GE 530c/570c scanner described here cannot be directly compared with that reported in Section 17.1.2.1 for D-SPECT, as different experimental setup and reconstruction parameters were used.

In a comprehensive phantom study comparing the GE 530c camera with GE Infinia scanner with LEHR collimators [25], the full width at half maximum (FWHM) of line source of GE 530c was 1.7–3.5 mm in air and 3.9–6.6 mm in water, whereas the FWHM of the Infinia was 8.2–12.6 mm in air and 15.5–16.3 mm in water. While the resolution and contrast ratios were higher for 530c, the non-uniformity in the reconstructed transversal planes of cylindrical phantoms was larger for 530c, and truncation artefacts were also observed. Regarding EF as a function of acquisition time, it was found that 5-min acquisition was sufficient for accurate EF quantification with 530c using quantitative gated SPECT software, as no significant difference was observed among EF quantified using 5-min, 7-min and 10-min data.

A cross comparison study has been reported for conventional SPECT, IQ-SPECT, D-SPECT and GE 530c scanners [1]. The results showed that, as compared to the conventional SPECT, all the dedicated cardiac scanners led to improved sensitivity, while the dedicated scanners with CZT detectors further provided improved resolution.

The uniformity in polar maps was comprehensively investigated using multi-centre phantom studies for seven scanners, including both conventional and CZT scanners [26, 27]. As shown in Figure 17.6, the polar map patterns are quite different for various combinations of scanner and reconstruction methods. While superior performance can be observed using iterative reconstruction with scatter correction and attenuation correction, the residual differences among scanners suggest the need of adopting a normal reference database specific for each scanner system, regardless of the count density or noise level.

17.1.3 Systems under Research Development

In addition to the above-mentioned commercially available dedicated cardiac SPECT scanners, there are a number of new dedicated cardiac SPECT systems and collimator designs being developed by academic groups. Such new designs might provide additional improvement in resolution and sensitivity for clinical cardiac SPECT studies.

The new collimator designs that can be installed on any existing gamma camera all focus collimator holes to image the heart, in order to improve the sensitivity. A 20 pinhole design as shown in Figure 17.7 (a) was proposed to utilize as much of the detector areas as possible, leading to a 12.5 mm resolution and significant improvements in both signal-to-noise ratio and diagnostic sensitivity over typical LEHR parallel-hole collimators in a simulation study [28]. In a design for a stationary imager based on a dual-head gamma camera, a set of two different multi-pinhole collimators was proposed, taking into account the axis-dependent heart-size difference, as shown in Figure 17.7 (b). Both simulation and physical phantom studies [29, 30] demonstrated the feasibility of stationary imaging with improved sensitivity and resolution (~8 mm), compared to LEHR collimators. Another approach used multi-pinhole collimators mounted on a triple-head gamma camera for stationary data acquisition as shown in Figure 17.7 (c). A simulation study showed that more than 8 pinholes were needed for each detector to reconstruct an artefact-free image without gantry rotation [31]. Another novel detector design with pinhole collimator was proposed for stationary dedicated cardiac SPECT [32, 33]. A hemi-ellipsoid shaped detector was designed for each pinhole to increase the magnification and was demonstrated to improve the resolution over the flat-detector. By increasing the pinhole diameter in the Monte Carlo simulation study, the hemi-ellipsoid-detector system achieved ~3.13 times sensitivity improvement and similar resolution as the flat-detector system. In contrast to pinhole collimators, another design proposed to use multiple segment slant-hole collimators to achieve stationary data acquisition. As shown in Figure 17.7 (d), each detector contains seven segmented slant-hole sections that slant toward a common volume, to achieve a 5-fold sensitivity gain over LEHR collimators with similar spatial resolution [34]. Another design, referred to as C-SPECT, was proposed to improve the system sensitivity without sacrificing the resolution for cardiac imaging [35]. The main concept of this platform was to design a C-shaped stationary detector gantry wrapping around the heart with the ability of performing both high-quality emission and transmission scans. Based on this platform, a lab-prototype system was developed using slit-slat collimators, achieving a 2.5 times higher sensitivity and comparable resolution as a conventional dual-head SPECT system with parallel-hole collimator [36].

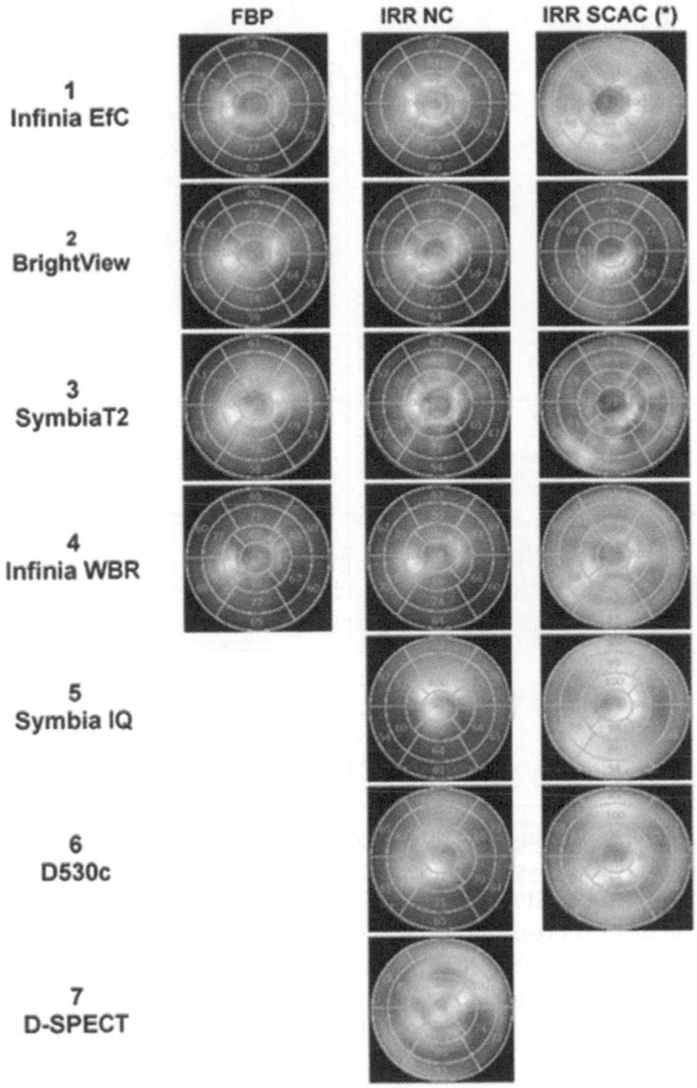

FIGURE 17.6 Phantom polar-map images obtained at reference count statistics: 6M counts in whole study for conventional systems (Camera 1-4) and 1M counts in the LV for advanced systems (camera 5-7). (*) An off-line CT was used to acquire the attenuation maps used to correct the emission data for Philips BrightView and D530c (i.e. GE 530c). No scatter correction was applied to the D530c data. No attenuation correction was applied with the D-SPECT system. IRR - iterative reconstruction with resolution recovery; NC - without any correction; SCAC - with scatter and attenuation corrections; Infinia WBR - Infinia wide beam reconstruction. Reprinted from [26] with permission from Springer Nature: Springer, Journal of Nuclear Cardiology, COPYRIGHT © 2016.

17.2 DEDICATED BRAIN SPECT

17.2.1 Commercial Systems

17.2.1.1 MILabs G-SPECT

A full-ring stationary SPECT scanner, G-SPECT-I from MILabs in the Netherlands, was launched in 2015 for dedicated brain imaging [37–40]. As shown in Figure 17.8, the scanner consists of 9 NaI(Tl) detectors, each with size of 595 × 472 × 9.5 mm^3. The nonagon-shape collimator is interchangeable, which contains a total of 54 pinholes with a 27° opening angle distributed over 3 rings. To achieve high system sensitivity, all the pinholes are focusing on the central FOV with the transaxial diameter of 100 mm and axial length of 60 mm. Three optical cameras are used for selecting the volume of interest on the user interface. The patient bed can be translated with an xyz-stage to obtain optimal sampling for the

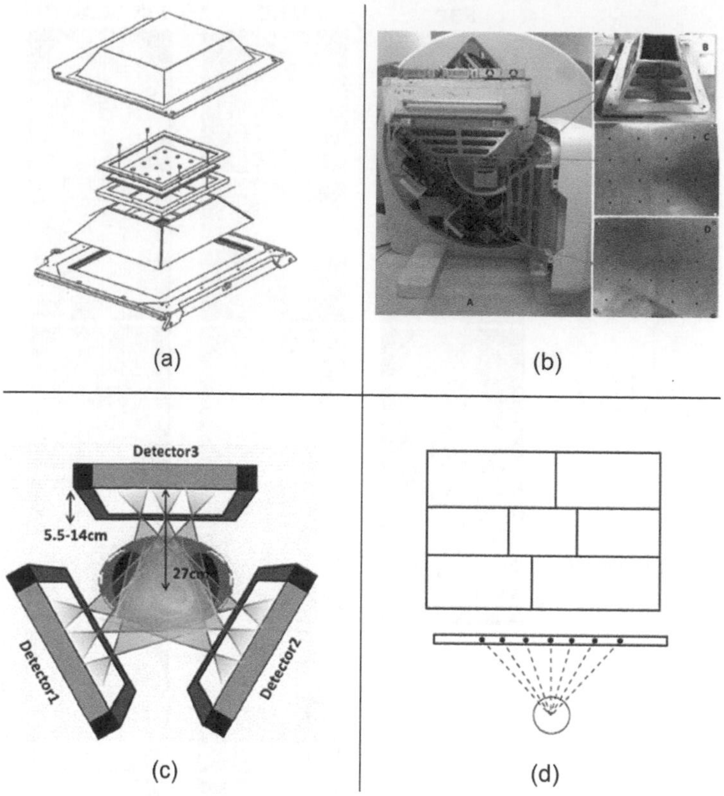

FIGURE 17.7 (a) A multi-pinhole collimator with 20 pinhole apertures designed for the GE Infinia Hawkeye 4. Reprinted with permission from [28], © Institute of Physics and Engineering in Medicine. Reproduced by permission of IOP Publishing. All rights reserved. (b) Prototype cardiac SPECT system based on a dual-head gamma camera with a set of two different multi-pinhole collimators. Reprinted with permission from [30], COPYRIGHT © 2015 IEEE. (c) Geometry of a triple-head myocardial SPECT with multi-pinhole collimators. Reprinted from [31] with permission from Springer Nature: Springer, Annals of Nuclear Medicine, COPYRIGHT © 2014. (d) Image configuration and the view-angles in the horizontal direction in the 2-3-2 configuration of multiple segment slant-hole collimators. Reprinted from [34] with permission from John Wiley and Sons, COPYRIGHT © 2015.

entire FOV. Then the 3D OSEM algorithm is used for image reconstruction using the projections from all bed positions with a scanning focus method [41]. By using pinholes with a 3-mm diameter, resolution phantom studies showed that the G-SPECT-I scanner can provide a resolution of 2.5 mm and a peak sensitivity of 630 count s^{-1} MBq^{-1} [42]. With its high sensitivity and full-ring stationary configuration, this SPECT scanner has the potential for dynamic imaging with a time frame of ~1 min for the entire brain [38]. Besides dedicated brain imaging, this scanner can be also used for pediatric imaging in clinic.

17.2.2 Systems under Research Development

A rotating full-ring SPECT scanner with three different multi-slit-slat system designs was proposed for clinical brain imaging with a FOV diameter of 200 mm [43]. To optimize the trade-off between sensitivity and resolution, the variable multiplexing strategy was used for each design. In the design of an asymmetric rotating collimator system shown in Figure 17.9 (a), the detector and slat collimator are stationary, while the slit collimator asymmetrically rotates around the FOV in a 'hula-hoop' manner with a fixed distance (x_c) between the FOV centre and the slit collimator centre. In the design of an asymmetric rotating detector system shown in Figure 17.9 (b), the slit collimator symmetrically rotates around the FOV, while the detector and the slat collimator with variable slat length asymmetrically rotate around the FOV, also in a 'hula-hoop' manner, with a fixed distance (x_d) between the FOV centre and the detector centre. In this design, the variation in the degree of multiplexing is achieved by the variable slat length. In the symmetric rotating

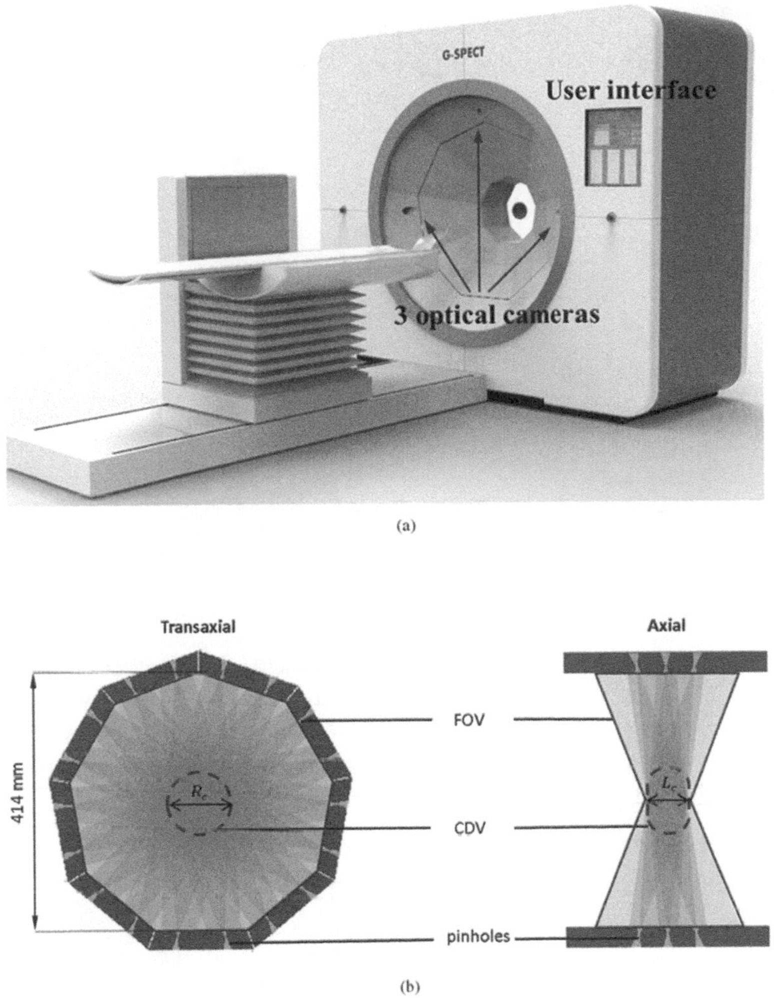

FIGURE 17.8 Illustration of G-SPECT-I scanner. (a) G-SPECT-I system with three optical cameras and a user interface for volume of interest selection; (b) multi-pinhole collimator of G-SPECT-I system. Reprinted with permission from [39], © Institute of Physics and Engineering in Medicine. Reproduced by permission of IOP Publishing. All rights reserved.

collimator system design shown in Figure 17.9 (c), the detector and the slat collimator are stationary, while the slit collimator symmetrically rotates around the FOV. In this design, the variation in the degree of multiplexing is achieved by unevenly distributed slits, where the densely distributed slits provide high degrees of multiplexing and the sparsely distributed slits provide no multiplexing. The analytical calculation showed that the symmetric rotating collimator system can achieve higher system sensitivity (510 count s^{-1} MBq^{-1}) than the other two systems (370 count s^{-1} MBq^{-1} and 420 count s^{-1} MBq^{-1} respectively) at the same average system resolution (6.0 mm).

A static full-ring SPECT system for brain imaging was designed with a multi-lofthole collimator insert on LaPET detector ring [44], as shown in Figure 17.10. The detector consists of 24 $LaBr_3$ (5% Ce) detectors, each including 27 × 60 crystals with the size of 4 × 4 × 30 mm^3. The lofthole has a circular aperture and a rectangular exit window that generates rectangular projection for optimal use of the detector area. The multi-lofthole collimator design is shown in Figure 17.11. The collimator contains a total of 128 loftholes with a 72.6° opening angle and 4 mm aperture distributed over 2 rings. The positions and tilt angles of the loftholes in each ring are carefully designed to sample one half of the transverse FOV. To obtain sufficient angular sampling but reduce the overlapping projections, a shutter mechanism is used for time-multiplexing. During data acquisition, 16 bed positions are used with 8-mm step size. Four different setups are acquired at each bed positions with 8 loftholes (from the same ring) opened simultaneously for 28.125 s in each setup. Based on this design, the sensitivity is reported to be 155 count s^{-1} MBq^{-1}, and 4 mm hot rods can be clearly identified on the reconstructed image.

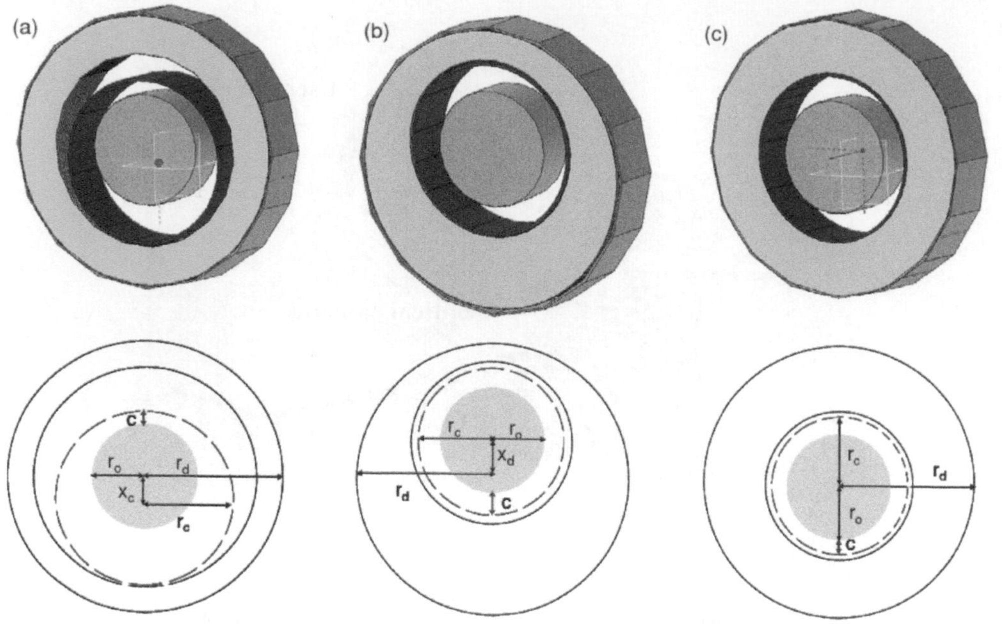

FIGURE 17.9 Asymmetric rotating collimator system design (a), asymmetric rotating detector system with variable slat length (b) and symmetric rotating collimator (c). The transverse drawing of each system is shown below the corresponding design; xc and xd represent the offset between the object centre for the ARC and the ARD systems and the centre of rotation of the collimator or detector, respectively. Reprinted with permission from [43], © Institute of Physics and Engineering in Medicine. Reproduced by permission of IOP Publishing. All rights reserved.

FIGURE 17.10 LaPET detector ring with collimator insert; and lofthole with a circular aperture and a rectangular exit window. Reprinted with permission from [44], © Institute of Physics and Engineering in Medicine. Reproduced by permission of IOP Publishing. All rights reserved.

To improve the SPECT system sensitivity for dedicated brain imaging, a custom-designed 20-aperture multi-pinhole collimator was mounted on a dual-head clinical SPECT/CT scanner, the GE Infinia Hawkeye 4, to replace the original LEHR collimator [45]. As shown in Figure 17.12, the tungsten collimator sheet size is 36 × 27 × 0.8 cm^3. A total of 20 pinholes are arranged to 4 × 5 array, with 6.2-cm distance between pinhole centres for the columns and 5.8-cm distance for the rows. Each pinhole has a 7.5-mm aperture diameter and a 65.4° opening angle. All the pinholes are positioned with tilt angles ranging from 6.2° to 29.4° between the pinhole axis and the normal to focus on the FOV with a 10.5-cm radius. Based on this design, the ideal sensitivity is 1,600 count s^{-1} MBq^{-1}, which is reported to be 22 times higher than the original LEHR collimator (72 count s^{-1} MBq^{-1}). However, the resolution degrades to 20.6 mm when compared with the LEHR collimator.

A triple-camera Triad-XLT SPECT system (Trionix Research Laboratories Twinsburg, OH) was proposed by using half-cone-beam collimators and circle-and-helix orbit [46]. The half-cone-beam collimators can provide a higher detection sensitivity due to their convergent geometry in both transaxial and axial directions, as compared to the fan-beam and the parallel-hole collimators that were also evaluated in this study. For the half-cone-beam, fan-beam and parallel-hole collimators investigated in the study, the hexagonal hole size was set to 1.45, 1.22 and 1.38 mm with the length of 4.00, 4.13 and 4.54 cm, where the septal thickness was set to 0.22, 0.15 and 0.16 mm. A novel circle-and-helix orbit

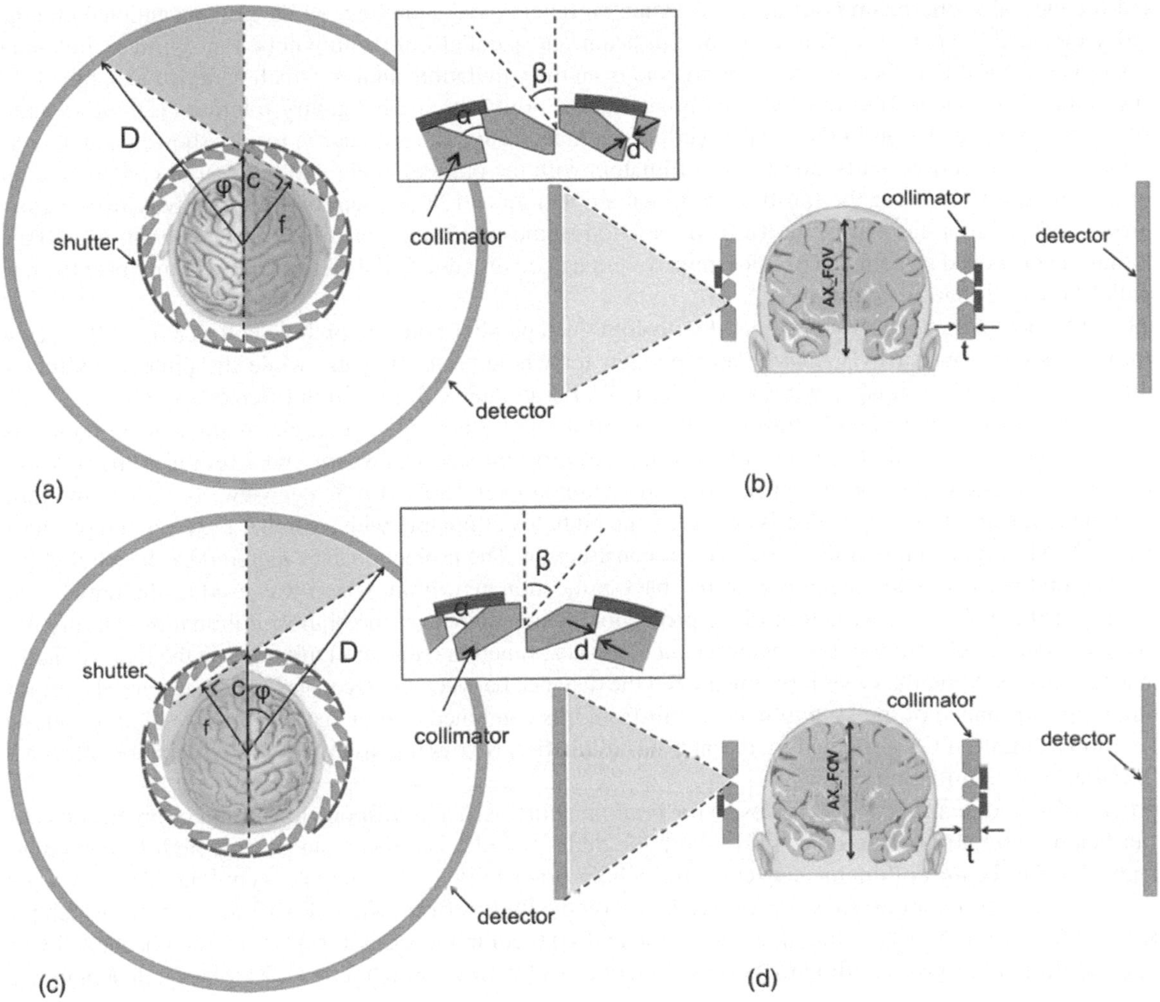

FIGURE 17.11 (a) Transverse cut through the first ring of loftholes with a close-up of the collimator. (b) Axial cut showing the two rows of loftholes on the collimator ring corresponding with the previous transverse view. Loftholes from the first ring are never opened at the same time as the loftholes from the second ring. (c) Transverse cut through the second ring of loftholes with a close-up of the collimator. (d) Axial cut showing the two rows of loftholes on the collimator ring corresponding with the previous transverse view. Reprinted with permission from [44], © Institute of Physics and Engineering in Medicine. Reproduced by permission of IOP Publishing. All rights reserved. Colour image available at www.routledge.com/9781138593268.

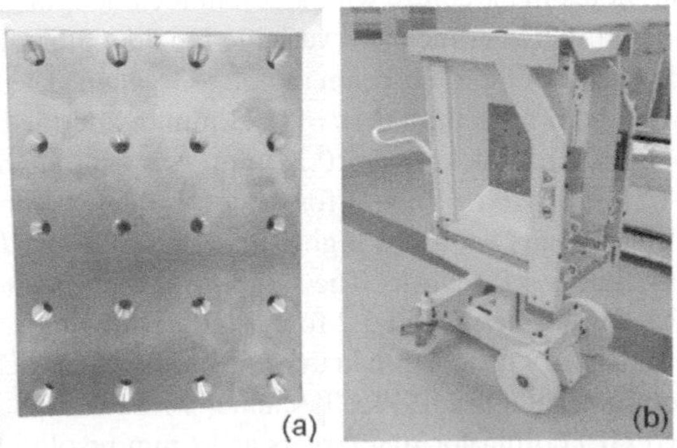

FIGURE 17.12 (a) Tungsten 20-pinhole aperture plate. (b) Assembled with collimator mount for use with a dual-head clinical SPECT scanner (GE Infinia Hawkeye 4). Reprinted from [45] with permission from John Wiley and Sons, COPYRIGHT © 2014.

was used for the half-cone-beam collimators to obtain sufficient axial sampling, while the conventional circular orbit with 360 views and 1° per view was used for fan-beam and parallel-hole collimators. An optimal circle-and-helix orbit was selected for the half-cone-beam collimators from the simulation studies, which included a circular orbit with 240 views and 1° per view followed by a combined helical orbit with a 240° gantry rotation (120 views and 2° per view) plus phantom translation in the axial direction simultaneously. The simulation results showed that at equal spatial resolutions, the proposed half-cone-beam collimators with the optimal circle-and-helix orbit can provide up to 26 per cent higher sensitivity over the fan-beam collimators, and up to 128 per cent higher sensitivity over parallel-hole collimators, both using a circular orbit. The experimental studies performed on a Hoffman 3D brain phantom further showed that the proposed system can provide improved image quality due to the increased sensitivity over the fan-beam and parallel-hole collimation systems.

A new multi-detector system was simulated with combined parallel-hole and pinhole collimators [47]. The parallel-hole collimators can cover a FOV that is large enough for whole brain imaging, while the pinhole collimators can obtain high resolution brain images within a smaller FOV. For pinhole collimation, the detector size is 22.5 × 22.5 cm^2 with 1.6 mm bin size. Each pinhole has a 0.5-mm aperture and a 60° opening angle. A total of 90 projections are acquired over 360°. For parallel-hole collimation, the detector bin size is 2.8 mm and 2.0 mm in the transaxial and axial directions, respectively, and the projections are acquired over 180° with 5° per view. To reduce the truncation artefact in the reconstructed images that is caused by the pinhole collimators with a smaller FOV, an iterative maximum a posteriori (MAP) algorithm was used for image reconstruction. The projection data acquired by the parallel-hole collimation was first reconstructed using the filtered back-projection algorithm, which was used as the initial image and prior image for the MAP reconstruction of the projection data acquired by the pinhole collimation. During the MAP reconstruction, two regularization terms were added to the cost function – one term minimizing the total variation in the reconstructed image, while the other term minimized the distance between the reconstructed image and the prior image. The imaging performance of the system may benefit from this combined design that can provide high resolution brain images within a smaller FOV provided by the pinhole collimation with compensation for the truncation effect provided by the parallel-hole collimation.

Another combined dual-head system design for brain imaging used a multi-pinhole collimator on one detector head and a fan-beam collimator on the other [48], where the multi-pinhole collimator can provide high resolution and sensitivity and the fan-beam collimator can cover the whole brain FOV with complete sampling. To demonstrate this design concept, a collimator design was proposed based on the Philips BrightView (BV) XCT scanner, which provided imaging detector size of 54 cm in the lateral direction and 40.6 cm in the axial direction for each head. A low-energy, ultra-high-resolution fan-beam collimator was used on one head with 1.4 mm hole size, 34.9 mm hole length, 0.15 mm septal thickness and 50 cm focal length. For multi-pinhole collimator design, a 3 × 3 pinhole array was selected with 2.2 mm aperture and 17.25 cm aperture to detector distance. This multi-pinhole system design resulted in a cylindrical FOV with 12-cm diameter and 8-cm height, which is suitable for the clinical DaTscan studies. Based on the analytical calculation, the multi-pinhole system can achieve a spatial resolution of 4.7 mm at the FOV centre that is much better than the fan-beam collimator (7.4 mm), and a sensitivity of 192 cpm/μCi that is equivalent to the fan-beam collimator.

A novel hybrid collimator design, as shown in Figure 17.13, was proposed for brain imaging with an ultrashort-focus cone-beam (USCB) collimator and a fan-beam collimator on a Siemens Symbia T6 SPECT/CT dual-head scanner [49]. The fan-beam collimator has a 39-cm focal length and can provide complete sampling of the FOV. The USCB collimator includes ultrashort cone-beam holes in the centre with a 20-cm focal length and slant-holes in the periphery with 37.7° tilt angle to the normal to the collimator surface and variable focal lengths. As shown in Figure 17.13, the collimator size is 53 × 39 cm^2, and the diameter of the cone-beam region is 36.4 cm. The fan-beam collimator has a hole size of 1.5 mm, a hole length of 34.9 mm and a septa thickness of 0.23 mm, while the USCB collimator has a hole size of 1.46 mm, a hole length of 37 mm and a septa thickness of 0.24 mm. After summarizing the total detected counts over 360° with 60 views, the sensitivity of the USCB collimator for the left caudate (near the centre of the brain) is 8.3 times higher than that of a fan-beam collimator, and 9.8 times higher than that of a LEHR collimator (hole size = 1.11 mm, hole length = 24.05 mm and septa thickness = 0.16 mm). The signal-to-noise for detecting a 5 per cent decrease in right putamen uptake is 7.4 for USCB, 3.4 for fan-beam, and 3.2 for LEHR collimators.

To achieve better resolution and sensitivity trade-off than the conventional parallel-hole collimator for brain imaging, a new design was proposed on a common dual-head SPECT scanner with two multi-pinhole collimators, where one is a multi-pinhole general purpose (MPGP) collimator that targets at 12 mm resolution, and the other is a multi-pinhole high-resolution (MPHR) collimator that targets at 8 mm resolution [50]. Each multi-pinhole collimator includes 9 pinholes, each with a 73° opening angle and being tilted towards the FOV centre, as shown in Figure 17.14. The pinhole aperture is 4.7 mm for MPGP and 2.8 mm for MPHR. The collimator to FOV centre distance is 167 mm, while the

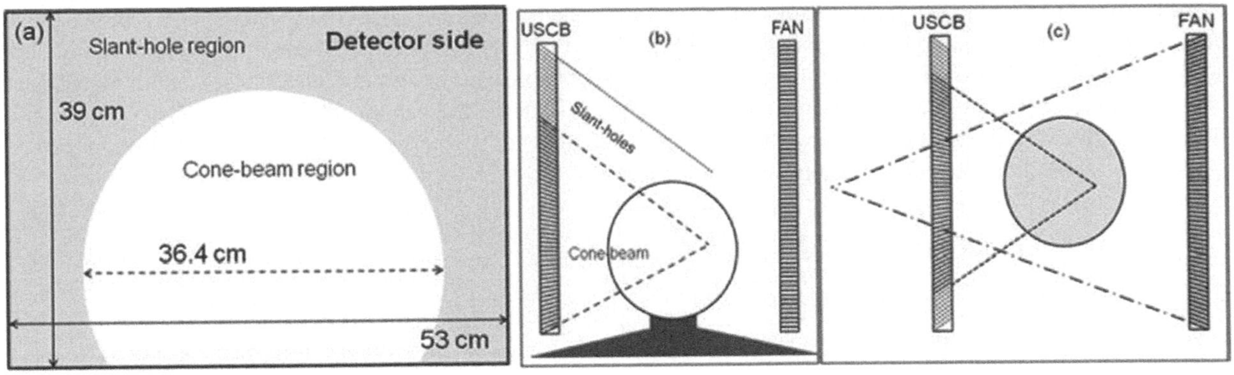

FIGURE 17.13 Design of the novel ultrashort cone-beam (USCB) collimator, (a) detector side of the collimator, (b) coronal view showing the collimator configuration for a brain SPECT acquisition using USCB and FAN collimators, and (c) transaxial view, indicating the differing focal lengths of the USCB and FAN collimators. Reprinted from [49] with permission from John Wiley and Sons, COPYRIGHT © 2016.

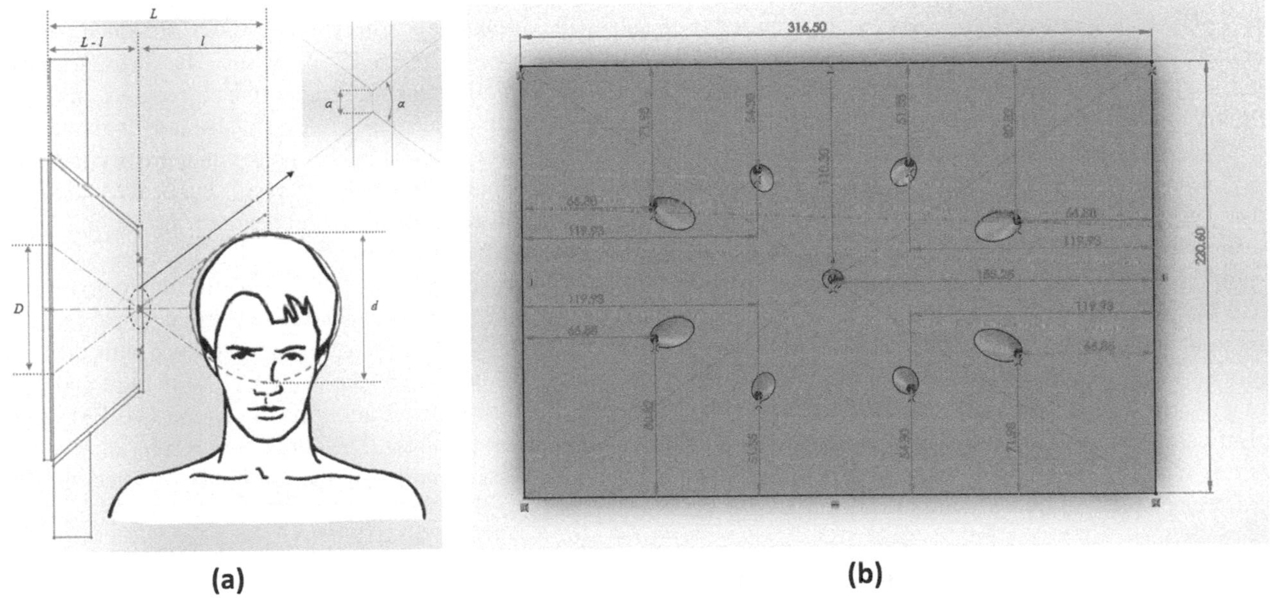

FIGURE 17.14 (a) Imaging geometry for the multi-pinhole brain scan. (b) Sample pinhole geometry pattern for the proposed 9-pinhole collimator in mm. Reprinted from [50] with permission from Springer Nature: Springer, Annals of Nuclear Medicine, COPYRIGHT © 2017.

collimator to detector distance is 127 mm. The analytical calculations showed that at the same resolution, the sensitivities of MPGP and MPHR are 270 per cent and 40 per cent higher than that of a LEHR collimator, while their resolution is the same and 33 per cent better. Superior bias-noise trade-off can be achieved for the multi-pinhole collimators than for the LEHR collimator. The Monte Carlo simulations indicated that 7.9-mm and 6.4-mm rods can be identified clearly by using the MPGP and MPHR collimators respectively.

To allow for static and dynamic brain imaging, a stationary brain-dedicated SPECT scanner, AdaptiSPECT-C, was proposed by using multiple detectors and multiple pinholes per detector with different pinhole configurations [51–53]. One of the current prototype designs [51] consists of 23 hexagonal NaI(Tl) detectors, which are arranged in 3 rings with 9 in the caudal ring, 9 in the middle ring, and 5 in the quasi-vertex ring. The additional quasi-vertex ring has been reported to improve the system sensitivity, spatial resolution, and image quality [52]. Each detector is configured with a knife-edge pinhole aperture with 4 mm diameter and 75° opening angle. All the pinholes are focusing on the centre of a spherical FOV with a 21 cm diameter, which can image the entire brain without truncation. The distances are 28.7 and

18.0 cm from the FOV centre to the detector surface and to the aperture centre, respectively. Studies have shown that this stationary scanner design can provide reasonable reconstructed images for brain perfusion and dopamine active transporter imaging. But additional axial and angular sampling may be needed as substantial enhancement in image quality has been observed with more axial and angular positions [51]. To avoid the bed translation and gantry rotation, adding additional apertures in the axial and lateral directions for each detector with applying the temporal shuttering technique can be used to provide additional axial and angular sampling [54].

17.3 DEDICATED BREAST SPECT

17.3.1 Systems under Research Development

For breast imaging, dedicated breast SPECT or molecular breast tomosynthesis systems have been proposed to improve lesion detection as compared to planar molecular breast imaging systems [55–59]. As the molecular breast tomosynthesis systems usually cover a limited angular range (less than 180°), this chapter will focus on the dedicated breast SPECT systems that can provide fully 3D images.

A fully 3D dedicated breast SPECT/CT scanner has been reported by using parallel-hole collimator and a CZT LumaGEM 3200S gamma camera (Gamma Medica in California) scanner [59]. The CZT camera with a 2.5 × 2.5 mm^2 crystal size can move in 3D around the breast translationally and rotationally to completely sample the breast FOV. The parallel-hole collimator has hexagonal holes with size of 1.22 mm, height of 25.4 mm and septa of 0.2 mm. To acquire fully 3D SPECT images, 128 views were acquired over 360° with vertical axis of rotation (VAOR), 30° tilted parallel beam (TPB), and 3-lobed projected sinusoidal (PROJSINE) ranging from 15° to 45° polar tilt. The SPECT images were reconstructed using OSEM algorithm with modelling of detector/collimator efficiencies and corrections for decay, attenuation, and scatter. A scaling factor was calculated based on experimental studies to obtain absolute quantification on SPECT images. Experimental results showed that the quantification error in large hot spots with activity concentration larger than 10 μCi/mL can be within 10 per cent. Regarding the acquisition trajectory, vertical axis of rotation and three-lobed projected sinusoidal can provide better quantification results than tilted parallel beam for breast imaging.

Besides the dedicated breast-imaging systems, dedicated cardiac SPECT systems with CZT detectors may have a potential to provide fully 3D breast images. The feasibility of breast imaging of sub-centimetre tumours with GE Discovery NM 530c/570c has been investigated using a simulation study [60]. NURBS-based Cardiac-Torso (NCAT) phantom [61] with breasts were used to simulate the SPECT images of ^{99m}Tc-tetrofosmin. Lesions with different diameters and contrast were inserted at several locations, including the centre and the chest wall of the left and right breasts. The SPECT images were reconstructed with maximum-likelihood expectation-maximization (MLEM) algorithm using the system matrix with modelling of the collimator-detector response. The simulation results showed that lesions with 8 mm diameter and contrast higher than 15:1 can be detected, indicating the feasibility of detecting sub-centimetre breast tumours using the GE 530c/570c system.

17.4 VERITON MULTI-PURPOSE SPECT

A whole-body CZT system, VERITON SPECT (or VERITON-CT hybrid SPECT-CT) by Spectrum Dynamics, is commercially available recently. As shown in Figure 17.15, VERITON has 12 CZT detector heads, each consisting of a 16 by 128 array of pixel units and equipped with parallel-hole collimators, which can be independently operated to provide full coverage of the body contour. Using an organ focus scanning mode, this system can be used for organ-specific acquisition by moving selected detectors and opens the opportunity for organ-specific dynamic SPECT imaging [62]. The VERITON SPECT system was reported to provide similar spatial resolution as the conventional system (Siemens Symbia with a LEHR collimator), where the tomographic resolution at the FOV centre was reported to be 7.72 mm and 5.92 mm for the VERITON system, and 7.75 mm and 7.72 mm for the conventional system, in the brain and body phantoms respectively [63]. For a point source in air, the tomographic sensitivity of the VERITON system with and without focus mode was reported to be 8 times and 1.6 times higher than the conventional system (1159 count s^{-1} MBq^{-1} and 236 count s^{-1} MBq^{-1} versus 144 count s^{-1} MBq^{-1}) [63].

17.5 SUMMARY

In this chapter, we summarized SPECT systems capable of providing fully 3D reconstructed images of dedicated organs of heart, brain, and breast. In contrast to conventional general purpose SPECT scanners with parallel-hole collimators,

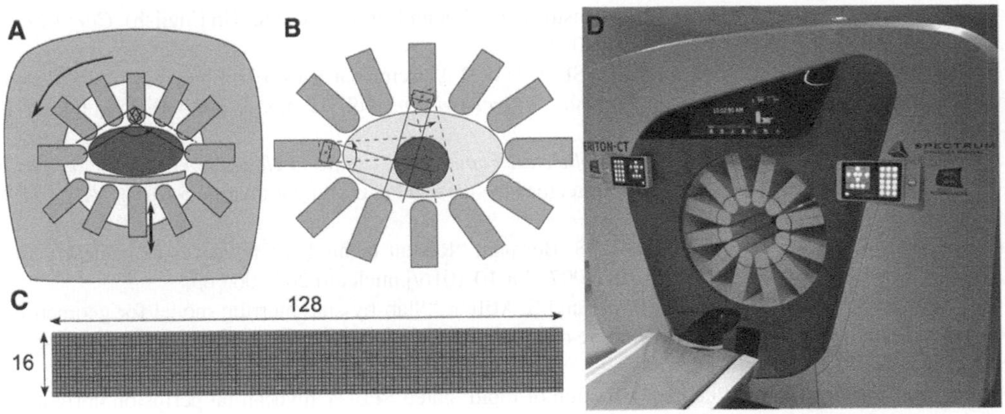

FIGURE 17.15 The VERITON SPECT scanner: general camera architecture showing the different movements of the 12 detectors (a), schematic principle of the focus mode showing the reduced swipe motion of detectors to a predefined region of interest shown in dark grey (b), design of the detection column consisting of an array of 16 by 128 pixel units (c) and picture of the system (d). Reprinted with permission from [63], COPYRIGHT © 2020.

the dedicated organ specific SPECT systems typically utilize innovative collimator design coupled with compact detectors to achieve improved trade-off between resolution and sensitivity. It is promising that such dedicated systems could eventually lead to improved diagnostic accuracy in various organ specific clinical applications.

REFERENCES

[1] L. Imbert et al., "Compared performance of high-sensitivity cameras dedicated to myocardial perfusion SPECT: a comprehensive analysis of phantom and human images," (in English), *J Nucl Med,* vol. 53, no. 12, pp. 1897–1903, Dec 2012, doi:10.2967/jnumed.112.107417.

[2] P. J. Slomka, D. S. Berman, and G. Germano, "New cardiac cameras: single-photon emission CT and PET," Semin Nucl Med, vol. 44, no. 4, pp. 232–251, 2014, doi: 10.1053/j.semnuclmed.2014.04.003.

[3] A. H. Vija et al., "A method for improving the efficiency of myocardial perfusion imaging using conventional SPECT and SPECT/CT imaging systems," in *IEEE Nuclear Science Symposium & Medical Imaging Conference*, 30 Oct. 6 Nov. 2010, pp. 3433–3437, doi:10.1109/NSSMIC.2010.5874444.

[4] R. Rajaram et al., "Tomographic performance characteristics of the IQ•SPECT system," in *2011 IEEE Nuclear Science Symposium Conference Record*, 23–29 Oct. 2011, pp. 2451–2456, doi:10.1109/NSSMIC.2011.6152666.

[5] J. Zeintl, T. D. Rempel, M. Bhattacharya, R. E. Malmin, and A. H. Vija, "Performance characteristics of the SMARTZOOM® collimator," in *2011 IEEE Nuclear Science Symposium Conference Record*, 23–29 Oct. 2011, pp. 2426–2429, doi:10.1109/NSSMIC.2011.6152660.

[6] S. M. S. USA. *IQ-SPECT Brochure*. (2014). [Online]. Available: www.siemens-healthineers.com/en-us/molecular-imaging/xspect/symbiatseries

[7] S. Matsuo, K. Nakajima, M. Onoguchi, H. Wakabayash, K. Okuda, and S. Kinuya, "Nuclear myocardial perfusion imaging using thallium-201 with a novel multifocal collimator SPECT/CT: IQ-SPECT versus conventional protocols in normal subjects," (in English), *Ann Nucl Med,* vol. 29, no. 5, pp. 452–459, Jun 2015, doi:10.1007/s12149-015-0965-7.

[8] K. Okuda et al., "Creation and characterization of normal myocardial perfusion imaging databases using the IQ.SPECT system," (in English), *J Nucl Cardiol*, Jan 3 2017, doi:10.1007/s12350-016-0770-2.

[9] M. Havel, M. Kolacek, M. Kaminek, V. Dedek, O. Kraft, and P. Sirucek, "Myocardial perfusion imaging parameters: IQ-SPECT and conventional SPET system comparison," (in English), *Hell J Nucl Med*, vol. 17, no. 3, pp. 200–203, 2014 Sep-Dec 2014. [Online]. Available: http://europepmc.org/abstract/MED/25526755.

[10] F. Caobelli et al., "IQ SPECT allows a significant reduction in administered dose and acquisition time for myocardial perfusion imaging: evidence from a phantom study," (in English), *J Nucl Med*, vol. 55, no. 12, pp. 2064–2070, Dec 2014, doi:10.2967/jnumed.114.143560.

[11] F. Caobelli, J. T. Thackeray, A. Soffientini, F. M. Bengel, C. Pizzocaro, and U. P. Guerra, "Feasibility of one-eighth time gated myocardial perfusion SPECT functional imaging using IQ-SPECT," (in English), *Eur J Nucl Med Mol Imaging*, vol. 42, no. 12, pp. 1920–1928, Nov 2015, doi:10.1007/s00259-015-3142-2.

[12] Y. Horiguchi et al., "Validation of a short-scan-time imaging protocol for thallium-201 myocardial SPECT with a multifocal collimator," (in English), *Ann Nucl Med*, vol. 28, no. 8, pp. 707–715, Oct 2014, doi:10.1007/s12149-014-0880-3.

[13] M. F. Smith, "Recent advances in cardiac SPECT instrumentation and system design," (in English), *Curr Cardiol Rep*, vol. 15, no. 8, p. 387, Aug 2013, doi:10.1007/s11886-013-0387-x.

[14] C. Bai et al., "Phantom evaluation of a cardiac SPECT/VCT system that uses a common set of solid-state detectors for both emission and transmission scans," (in English), *J Nucl Cardiol*, vol. 17, no. 3, pp. 459–469, Jun 2010, doi:10.1007/s12350-010-9204-8.

[15] H. Babla, C. Bai, and R. Conwell, *"A triple-head solid state camera for cardiac single photon emission tomography (SPECT)"*, Proc. SPIE 6319, Hard X-Ray and Gamma-Ray Detector Physics and Penetrating Radiation Systems VIII, 63190M (30 August 2006); https://doi.org/10.1117/12.683765

[16] J. A. Patton, P. J. Slomka, G. Germano, and D. S. Berman, "Recent technologic advances in nuclear cardiology," *J Nucl Cardiol*, vol. 14, no. 4, pp. 501–513, 2007/07/01 2007, doi:10.1016/j.nuclcard.2007.06.003.

[17] C. Bai, G. L. Zeng, G. T. Gullberg, F. DiFilippo, and S. Miller, "Slab-by-slab blurring model for geometric point response correction and attenuation correction using iterative reconstruction algorithms," *IEEE Trans Nucl Sci*, vol. 45, no. 4, pp. 2168–2173, 1998, doi:10.1109/23.708334.

[18] J. Maddahi et al., "Prospective multicenter evaluation of rapid, gated SPECT myocardial perfusion upright imaging," *J Nucl Cardiol*, vol. 16, no. 3, pp. 351–357, 2009/06/01 2009, doi:10.1007/s12350-009-9063-3.

[19] K. Erlandsson, K. Kacperski, D. van Gramberg, and B. F. Hutton, "Performance evaluation of D-SPECT: a novel SPECT system for nuclear cardiology," (in English), *Phys Med Biol*, vol. 54, no. 9, pp. 2635–2649, May 7 2009, doi:10.1088/0031-9155/54/9/003.

[20] S. S. Gambhir et al., "A novel high-sensitivity rapid-acquisition single-photon cardiac imaging camera," (in English), *J Nucl Med*, vol. 50, no. 4, pp. 635–643, Apr 2009, doi:10.2967/jnumed.108.060020.

[21] A. Verger et al., "Comparison between stress myocardial perfusion SPECT recorded with cadmium-zinc-telluride and Anger cameras in various study protocols," (in English), *Eur J Nucl Med Mol Imaging*, vol. 40, no. 3, pp. 331–340, Feb 2013, doi:10.1007/s00259-012-2292-8.

[22] J. S. Lee, G. Kovalski, T. Sharir, and D. S. Lee, "Advances in imaging instrumentation for nuclear cardiology," Journal of Nuclear Cardiology, vol. 26, no. 2, pp. 543–556, 2019, doi:10.1007/s12350-017-0979-8.

[23] M. Bocher, I. M. Blevis, L. Tsukerman, Y. Shrem, G. Kovalski, and L. Volokh, "A fast cardiac gamma camera with dynamic SPECT capabilities: design, system validation and future potential," (in English), *Eur J Nucl Med Mol Imaging*, vol. 37, no. 10, pp. 1887–1902, Oct 2010, doi:10.1007/s00259-010-1488-z.

[24] J. A. Kennedy, O. Israel, and A. Frenkel, "3D iteratively reconstructed spatial resolution map and sensitivity characterization of a dedicated cardiac SPECT camera," (in English), *J Nucl Cardiol*, vol. 21, no. 3, pp. 443–452, Jun 2014, doi:10.1007/s12350-013-9851-7.

[25] Y. Takahashi, M. Miyagawa, Y. Nishiyama, H. Ishimura, and T. Mochizuki, "Performance of a semiconductor SPECT system: comparison with a conventional Anger-type SPECT instrument," (in English), *Ann Nucl Med*, vol. 27, no. 1, pp. 11–6, Jan 2013, doi:10.1007/s12149-012-0653-9.

[26] O. Zoccarato et al., "Differences in polar-map patterns using the novel technologies for myocardial perfusion imaging," (in English), *J Nucl Cardiol*, vol. 24, no. 5, pp. 1626–1636, Oct 2017, doi:10.1007/s12350-016-0500-9.

[27] P. J. Slomka, M. Rubeaux, and G. Germano, "Quantification with normal limits: New cameras and low-dose imaging," (in English), *J Nucl Cardiol*, vol. 24, no. 5, pp. 1637–1640, Oct 2017, doi:10.1007/s12350-016-0563-7.

[28] J. D. Bowen et al., "Design and performance evaluation of a 20-aperture multipinhole collimator for myocardial perfusion imaging applications," (in English), *Phys Med Biol*, vol. 58, no. 20, pp. 7209–7226, Oct. 21, 2013, doi:10.1088/0031-9155/58/20/7209.

[29] H. Liu et al., "Feasibility studies of a high sensitivity, stationary dedicated cardiac SPECT with multi-pinhole collimators on a clinical dual-head scanner," in *2014 IEEE Nuclear Science Symposium and Medical Imaging Conference (NSS/MIC)*, 8–15 Nov., 2014 2014, pp. 1–4, doi:10.1109/NSSMIC.2014.7430982.

[30] H. Liu, J. Wu, S. Chen, S. Wang, Y. Liu, and T. Ma, "Development of stationary dedicated cardiac SPECT with multi-pinhole collimators on a clinical scanner," in *2015 IEEE Nuclear Science Symposium and Medical Imaging Conference (NSS/MIC)*, Oct. 31, 2015–Nov. 7, 2015, pp. 1–4, doi:10.1109/NSSMIC.2015.7582154.

[31] K. Ogawa and Y. Ichimura, "Simulation study on a stationary data acquisition SPECT system with multi-pinhole collimators attached to a triple-head gamma camera system," (in English), *Ann Nucl Med*, vol. 28, no. 8, pp. 716–724, Oct 2014, doi:10.1007/s12149-014-0865-2.

[32] J. Dey, "Improvement of performance of cardiac SPECT camera using curved detectors with pinholes," *IEEE Trans Nucl Sci*, vol. 59, no. 2, pp. 334–347, 2012, doi:10.1109/TNS.2011.2182660.

[33] K. Kalluri et al., "Multi-pinhole cardiac SPECT performance with hemi-Ellipsoid detectors for two geometries," in *2015 IEEE Nuclear Science Symposium and Medical Imaging Conference (NSS/MIC)*, Oct. 31, 2015-Nov. 7, 2015 2015, pp. 1–4, doi:10.1109/NSSMIC.2015.7582127.

[34] Y. Mao, Z. Yu, and G. L. Zeng, "Segmented slant hole collimator for stationary cardiac SPECT: Monte Carlo simulations," (in English), *Med Phys*, vol. 42, no. 9, pp. 5426–5434, Sep 2015, doi:10.1118/1.4928484.

[35] W. Chang, C. E. Ordonez, H. Liang, Y. Li, and J. Liu, "C-SPECT – a clinical cardiac SPECT/Tct platform: design concepts and performance potential," (in English), *IEEE Trans Nucl Sci*, vol. 56, no. 5, pp. 2659–2671, Oct. 6, 2009, doi:10.1109/tns.2009.2028138.

[36] W. Chang et al., "Final design of the C-SPECT-I lab-prototype," in *2015 IEEE Nuclear Science Symposium and Medical Imaging Conference (NSS/MIC)*, Oct. 31, 2015–Nov. 7, 2015, pp. 1–4, doi:10.1109/NSSMIC.2015.7582131.

[37] F. Beekman et al., "G-SPECT-I: a full ring high sensitivity and ultra-fast clinical molecular imaging system with< 3mm resolution," in *Eur J Nucl Med Mol Imaging*, vol. 42: Springer: New York, p. S209, 2015.

[38] M. Ljungberg and P. H. Pretorius, "SPECT/CT: an update on technological developments and clinical applications," *Br J Radiol*, vol. 91, no. 1081, pp. 20160402-20160402, 2018, doi:10.1259/bjr.20160402.

[39] Y. Chen, B. Vastenhouw, C. Wu, M. C. Goorden, and F. J. Beekman, "Optimized image acquisition for dopamine transporter imaging with ultra-high resolution clinical pinhole SPECT," *Phys Med Biol*, vol. 63, no. 22, p. 225002, 2018, doi: 10.1088/1361-6560/aae76c.

[40] Y. Chen, M. C. Goorden, B. Vastenhouw, and F. J. Beekman, "Optimized sampling for high resolution multi-pinhole brain SPECT with stationary detectors," (in English), *Phys Med Biol*, vol. 65, no. 1, p. 015002, Jan. 10, 2020, doi:10.1088/1361-6560/ab5bc6.

[41] B. Vastenhouw and F. Beekman, "Submillimeter total-body murine imaging with U-SPECT-I," *J Nucl Med*, vol. 48, no. 3, pp. 487–493, 2007.

[42] B. F. Hutton, K. Erlandsson, and K. Thielemans, "Advances in clinical molecular imaging instrumentation," *Clin Transl Imaging*, vol. 6, no. 1, pp. 31–45, 2018, doi:10.1007/s40336-018-0264-0.

[43] S. T. Mahmood, K. Erlandsson, I. Cullum, and B. F. Hutton, "Design of a novel slit-slat collimator system for SPECT imaging of the human brain," (in English), *Phys Med Biol*, vol. 54, no. 11, pp. 3433–3449, Jun. 7, 2009, doi:10.1088/0031-9155/54/11/011.

[44] K. V. Audenhaege, S. Vandenberghe, K. Deprez, B. Vandeghinste, and R. V. Holen, "Design and simulation of a full-ring multi-lofthole collimator for brain SPECT," (in English), *Phys Med Biol*, vol. 58, no. 18, pp. 6317–36, Sep. 21, 2013, doi:10.1088/0031-9155/58/18/6317.

[45] T. C. Lee, J. R. Ellin, Q. Huang, U. Shrestha, G. T. Gullberg, and Y. Seo, "Multipinhole collimator with 20 apertures for a brain SPECT application," (in English), *Med Phys*, vol. 41, no. 11, p. 112501, Nov 2014, doi:10.1118/1.4897567.

[46] R. Ter-Antonyan, R. J. Jaszczak, J. E. Bowsher, K. L. Greer, and S. D. Metzler, "Quantitative Evaluation of Half-Cone-Beam Scan Paths in Triple-Camera Brain SPECT," (in English), *IEEE Trans Nucl Sci*, vol. 55, no. 5, pp. 2518–2526, Oct. 1, 2008, doi:10.1109/tns.2008.2003255.

[47] Q. Huang, T. Zeniya, H. Kudo, H. Iida, and G. T. Gullberg, "High resolution brain imaging with combined parallel-hole and pinhole collimation," in *IEEE Nuclear Science Symposium & Medical Imaging Conference*, 30 Oct.–6 Nov. 2010, pp. 3145–3148, doi:10.1109/NSSMIC.2010.5874381.

[48] M. A. King, J. M. Mukherjee, A. Konik, I. G. Zubal, J. Dey, and R. Licho, "Design of a Multi-Pinhole Collimator for I-123 DaTscan Imaging on Dual-Headed SPECT Systems in Combination with a Fan-Beam Collimator," (in English), *IEEE Trans Nucl Sci*, vol. 63, no. 1, pp. 90–97, Feb 2016, doi:10.1109/tns.2016.2515519.

[49] M. A. Park et al., "Introduction of a novel ultrahigh sensitivity collimator for brain SPECT imaging," (in English), *Med Phys*, vol. 43, no. 8, p. 4734, Aug 2016, doi:10.1118/1.4958962.

[50] L. Chen, B. M. W. Tsui, and G. S. P. Mok, "Design and evaluation of two multi-pinhole collimators for brain SPECT," (in English), *Ann Nucl Med*, vol. 31, no. 8, pp. 636–648, Oct 2017, doi:10.1007/s12149-017-1195-y.

[51] N. Zeraatkar et al., "Investigation of Axial and Angular Sampling in Multi-Detector Pinhole-SPECT Brain Imaging," *IEEE Trans Med Imaging*, vol. 39, no. 12, pp. 4209–4224, 2020, doi:10.1109/TMI.2020.3015079.

[52] B. Auer et al., "Inclusion of quasi-vertex views in a brain-dedicated multi-pinhole SPECT system for improved imaging performance," *Phys Med Biol*, 2020, doi:10.1088/1361-6560/abc22e.

[53] B. Auer, J. D. Beenhouwer, K. Kalluri, J. C. Goding, L. R. Furenlid, and M. A. King, "Preliminary investigation of design parameters of an innovative multi-pinhole system dedicated to brain SPECT imaging," in *2018 IEEE Nuclear Science Symposium and Medical Imaging Conference Proceedings (NSS/MIC)*, 2018, pp. 1–3, doi:10.1109/NSSMIC.2018.8824691.

[54] K. S. Kalluri, N. Zeraatkar, B. Auer, P. H. Kuo, L. R. Furenlid, and M. A. King, "Preliminary investigation of AdaptiSPECT-C designs with square or square and hexagonal detectors employing direct and oblique apertures" (Proc. SPIE 11072, 15th International Meeting on Fully Three-Dimensional Image Reconstruction in Radiology and Nuclear Medicine, 1107215 (28 May 2019); https://doi.org/10.1117/12.2534885). SPIE, 2019.

[55] M. B. Williams, P. G. Judy, S. Gunn, and S. Majewski, "Dual-modality breast tomosynthesis," (in English), *Radiology*, vol. 255, no. 1, pp. 191–198, Apr 2010, doi:10.1148/radiol.09091160.

[56] O. Gopan, D. Gilland, A. Weisenberger, B. Kross, and B. Welch, "Molecular Imaging of the Breast Using a Variable-Angle Slant-Hole Collimator," *IEEE Trans Nucl Sci*, vol. 61, no. 3, pp. 1143–1152, 2014, doi:10.1109/TNS.2014.2322520.

[57] Z. Gong and M. B. Williams, "Comparison of breast specific gamma imaging and molecular breast tomosynthesis in breast cancer detection: Evaluation in phantoms," (in English), *Med Phys*, vol. 42, no. 7, pp. 4250–4259, Jul 2015, doi:10.1118/1.4922398.

[58] J. van Roosmalen, M. C. Goorden, and F. J. Beekman, "Molecular breast tomosynthesis with scanning focus multi-pinhole cameras," (in English), *Phys Med Biol*, vol. 61, no. 15, pp. 5508–5528, Aug. 7, 2016, doi:10.1088/0031-9155/61/15/5508.

[59] K. L. Perez, S. J. Cutler, P. Madhav, and M. P. Tornai, "Towards Quantification of Functional Breast Images Using Dedicated SPECT With Non-Traditional Acquisition Trajectories," *IEEE Trans Nucl Sci*, vol. 58, no. 5, pp. 2219–2225, 2011, doi:10.1109/TNS.2011.2165223.

[60] H. Liu et al., "3D molecular breast imaging using a high-resolution dedicated cardiac SPECT camera," in *2013 IEEE Nuclear Science Symposium and Medical Imaging Conference (2013 NSS/MIC)*, 2013, pp. 1–4, doi: 10.1109/NSSMIC.2013.6829097.

[61] W. P. Segars, G. Sturgeon, S. Mendonca, J. Grimes, and B. M. W. Tsui, "4D XCAT phantom for multimodality imaging research," *Med Phys*, vol. 37, no. 9, pp. 4902–4915, 2010, doi:https://doi.org/10.1118/1.3480985.

[62] B. F. Hutton, K. Erlandsson, and K. Thielemans, "Advances in clinical molecular imaging instrumentation," *Clin Transl Imaging*, vol. 6, no. 1, pp. 31–45, 2018, doi:10.1007/s40336-018-0264-0.

[63] C. Desmonts, M. A. Bouthiba, B. Enilorac, D. Agostini, and N. Aide, "Performance evaluation of a new multipurpose whole-body CzT-based camera: comparison with a dual head anger camera," *J Nucl Med*, vol. 60, no. supplement 1, p. 1384, May 1, 2019. [Online]. Available: http://jnm.snmjournals.org/content/60/supplement_1/1384.abstract.

18 PET Systems

Stefaan Vandenberghe

CONTENTS

18.1 INTRODUCTION

The major goal of a PET system is to make images of the distribution of positron emitters. There is a constant evolution in PET systems aiming at increased sensitivity, better count rate, and improved spatial and energy resolution. Although there are some important differences in the functioning of PET systems, they all share the same basic principles, these are first described. Hereafter, the more specific details about the different components of a system are described. Finally, an overview of the different generations of PET systems is given.

There are some major differences with SPECT systems. As the current systems consist of a full ring, there is no rotation required to obtain tomographic data. A SPECT system allows the making of multiple images at the same time by using isotopes with different energy. In PET the energy is always 511 keV, so no simultaneous imaging is possible. Modern PET systems are composed of pixelated detectors, while SPECT uses a continuous detector.

The most important difference is, however, the different collimation method, which results in PET having a much higher sensitivity than SPECT.

The use of positron emitters results in two opposed 511 keV photons at every decay. This allows for performing electronic collimation by the principle of coincidence detection. PET scanners are based on the assumption that the path of the photons and the annihilation point are at the same line (collinearity), and that both photons arrive around the same time on the detector ring (simultaneity). The PET scanner makes use of coincidence electronics: assume a positron annihilation – when one of the photons is detected by a crystal, the electronics receives a trigger, and a coincidence circuit is opened. If a second photon arrives within a very short time, the system has detected a coincidence (also called an *event* or *list-mode event*). By registering multiple of these coincident lines, projections of the emission distribution can be formed, and the distribution is calculated from these by image reconstruction techniques. This is illustrated in Figure 18.1.

The principle of coincidence detection, which lies at the basis of PET scanners, has a large potential, but its strength depends on the different components used to perform the coincidence detection. It requires a combination of accurate

DOI: 10.1201/9780429489556-18

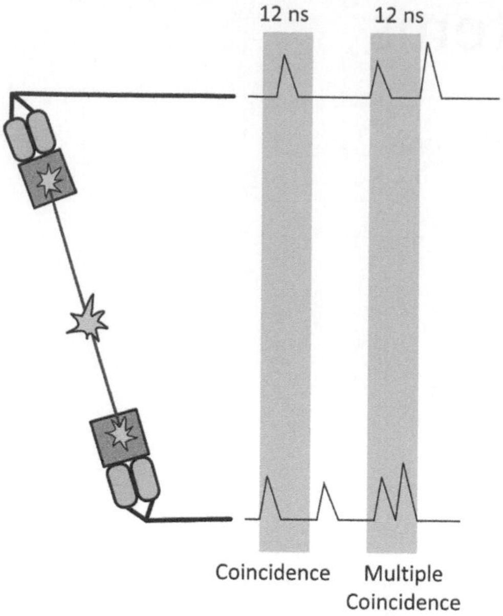

FIGURE 18.1 Use of a coincidence window to detect the lines of annihilation events in a PET scanner.

TABLE 18.1
Overview of Scintillator Materials Used for PET, and their Characteristics

Scintillators for PET	Attenuation coefficient (cm^{-1})	Decay time (ns)	Light output (photons/Mev)
BGO	0.95	300	7000
GSO	0.70	60	10000
LSO	0.86	43	26000
LaBr	0.47	15	60000

PMTs and electronics with a timing precision in the ns range, and the scintillators need to stop a large fraction of the incident 511 keV radiation. Preferably, these also have a good energy and spatial resolution. The different components in the detection chain are described below; emphasis is given to the important characteristics for PET.

18.2 SCINTILLATORS FOR PET

A scintillator in PET should, with a high probability, stop the incoming gamma ray and produce a light signal that allows for determining the energy, time, and position. During the early years of PET, several systems were based on NaI but, due to its low stopping power, this scintillator was suboptimal for PET. The crystal thickness needed to be 4–5 cm to achieve a good sensitivity. NaI was used as a continuous scintillator, but the current PET systems all make use of pixelated scintillators. Most systems during the 1980s and 1990s were using pixelated BGO (Bismuth Germanate) as a scintillator. BGO has excellent stopping power but releases a limited amount of light. This makes it difficult to identify crystals based on their light output when the crystals become quite small (below 6 mm). The limited amount of light also results in relatively poor energy resolution (20–25%). Therefore, most companies started using brighter (more light per 511 keV) scintillators like GSO or LSO in more recent higher resolution systems. LSO (or LYSO) has now become the new standard PET scintillator as it generates a lot of light, is quite fast, and has good stopping power. New scintillators are still being explored, and one of the most promising for PET is $LaBr_3$ due to its very high light output and fast response. Table 18.1 contains the performance values of the most common scintillators used in PET.

A human PET scanner has an axial dimension of 15–20 cm, and the radius of these systems is around 40–45 cm. To fill this cylinder with crystals of 4 x 4 mm (transverse x axial) more than 20000 crystals are needed. The crystal thickness is about 20–30 mm resulting in about 40–90 per cent crystal-detection efficiency for coincidence pairs.

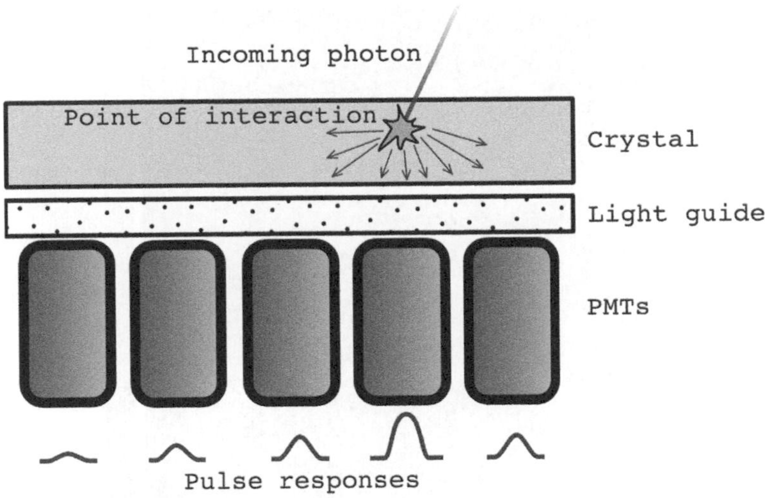

FIGURE 18.2 Light spread and PMT signals in a continuous PET detector.

18.3 DETECTORS FOR PET

A detector in PET is a device formed by the combination of scintillator, light conversion unit, and electronics. An ideal detector in PET should have the following properties: high stopping power for 511 keV, good energy resolution, good timing resolution, and good spatial resolution. Furthermore, the cost should be acceptable. The scintillators are discussed in the Section 18.2. The conversion of the output light into the electrical signal can be done by different devices (Photomultiplier tube, Position-sensitive Photomultiplier tubes, Avalanche Photodiode, Silicon Photomultiplier). As it is the most cost-effective way to achieve quite good performance, the previous generation of human PET systems are using Photomultiplier tubes (PMT) for this conversion step. These are characterized by a high-gain, low-noise, and fast response. The other devices are used in small animal PET systems or research prototypes to achieve better spatial resolution, count-rate capability, or because they can operate in an environment with high magnetic fields (MRI scanners). The focus in this part is on the use of PMTs for determining the position, energy, and time.

In the early PET systems, one PMT was used per scintillator, but this does not work when we want to use small crystals. While it is possible to use smaller PMTs, it is not quite effective to place one per scintillator element: the amount of electronics and PMTs would increase the cost and complexity of the PET system. Basically, there are two different detector designs in current PET systems: the continuous and the block detector.

The continuous detector is quite similar to the Anger logic approach used in SPECT. This allows the use of both a pixelated detector and a continuous detector (Figure 18.2). One of the important differences with SPECT is the much higher count rate in PET. Therefore, the light spread is limited by optimizing the thickness of the light guide, and only a limited number of PMTs (typically 7) are used to calculate the centroid and the total energy. This is possible because the PMT signals in modern systems are digitized, and these are used to calculate the position and energy. The local cluster is selected by selecting the PMT with the highest signal, and the neighbours are included in the position determination. This technique is called local centroid, and the advantage is that different spatially separated events can be processed simultaneously. The crystals are visible on so-called flood maps showing the distortion due to the PMTs. The result of Anger logic calculation will lie somewhere on this map and will be assigned to a certain pixel. The pulse coming out of the PMTs used as an input to the Anger logic calculation are shown in Figure 18.2.

Most systems, however, use block detectors. These blocks contain a limited number of crystals and PMTs: a typical example would be to couple 6 x 6 crystals to four PMTs. These four PMTs are placed in a rectangle above the crystals. Assume A and B are on top, and C and D at the bottom. The x and y positions (with origin the outer corner of the four PMTs) can be found from the PMT signals by using the Anger Logic technique. As we only have two positions of the PMT, this results in a simplified equation:

$$x = \frac{A+B}{A+B+C+D}$$

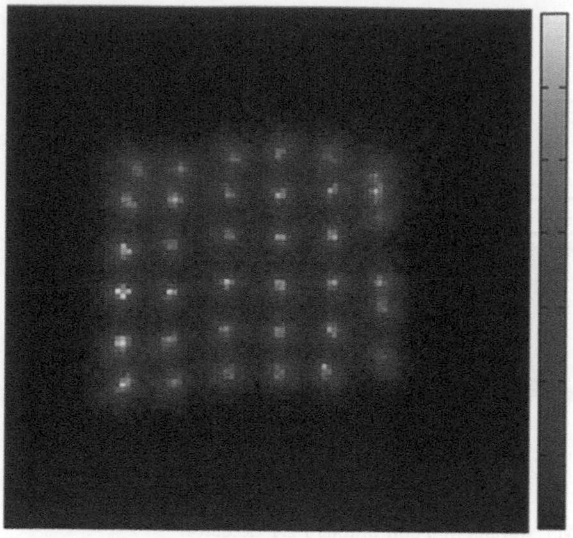

FIGURE 18.3 Example of a flood map. Colour image available at www.routledge.com/9781138593268.

$$y = \frac{A+C}{A+B+C+D} \tag{18.1}$$

Due to the non-uniformity of light spread and variation in PMT gains, this results in a so-called flood image (Figure 18.3). The isocontour lines are used to determine the region that belongs to a certain detector pixel. The total energy is found by summing the signal of the four PMTs.

The light, after an interaction in the scintillator, is released in a typical pattern: initially there is a rapid increase in the amount of light per time unit, followed by a gradual decrease. The time of each event is determined from the initial increase as measured by a group of PMTs (typically about twenty). In a human PET scanner this signal is first used to select the coincident events between different groups. The position and energy of each of the single events are then calculated from the digitized PMT signals (corrected for different gains).

18.4 GEOMETRY AND COLLIMATION OF PET SYSTEMS

There has been an evolution to more sensitive PET systems by optimizing the geometry (illustrated in Figure 18.4a–d), detection efficiency, axial FOV, and collimation (Figure 18.4e–i). While it is not necessary to use mechanical collimation in PET, in older PET systems the different slices are often collimated by lead septa. This makes it geometrically impossible for coincidences between different ring pairs to be measured. While it limits the sensitivity, it is used in order to reduce the count rate and limit the scatter, and it separates the reconstruction in different 2D reconstruction problems. When the septa are removed a system is said to be operating in fully 3D mode, meaning that all combinations between different rings are possible. An important difference between the 2D and fully 3D scanners is the different amount of wrong coincidences.

The first PET systems were composed of one or a limited number of axial rings. Some systems needed to rotate to obtain complete data. In the next-generation systems (second half of the 1980s) the axial extent of the scanners was increased to 10 cm in order to increase the sensitivity. Axial collimation was used. These scanners are called Multislice 2D PET scanners. The most common system at the second half of the 1990s was a multi-slice 2D PET scanner with pixelated BGO detectors. One fully 3D system based on continuous NaI was already produced and used clinically at the end of the 1990s. In the same period, different vendors also tried to produce low-cost PET systems: these were modified gamma cameras (capable of both SPECT and PET) or limited ring systems. Both types needed rotation to obtain complete tomographic data. These systems were characterized by suboptimal performance and lower reliability (due to mechanical rotation). Most of these systems had the option to operate either in 2D or fully 3D mode by using removable or retractable septa. In the period 2000–2005 most vendors started producing full-ring PET systems without any mechanical collimation, and operating exclusively in fully 3D mode. This evolution was made possible by faster scintillators (LSO and GSO) and more accurate electronics.

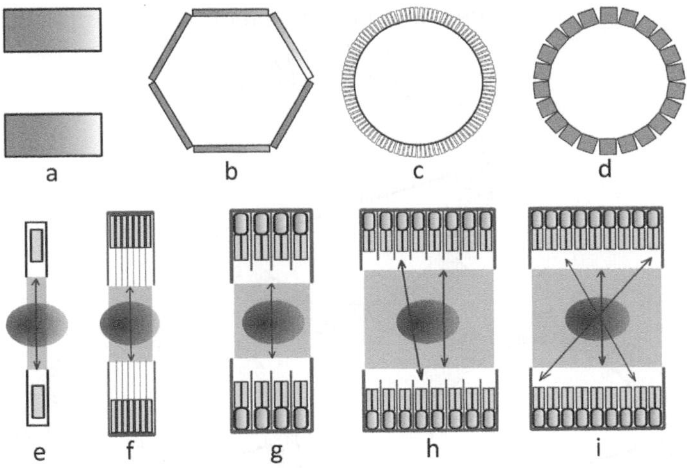

FIGURE 18.4 Evolution of the geometry (a–d) and the axial FOV and collimation (e–i) of PET scanners.

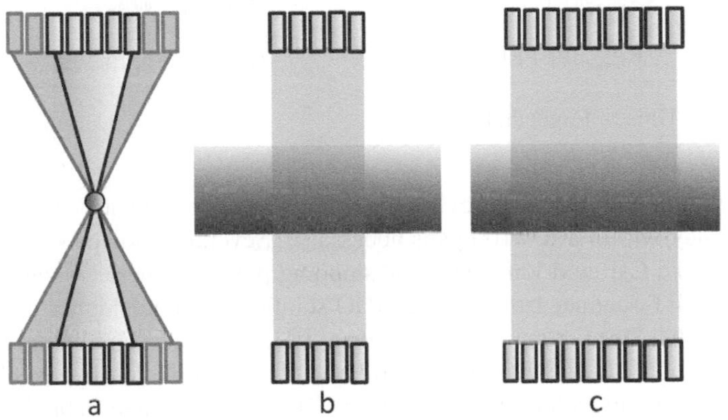

FIGURE 18.5 Effect of axial FOV on the point sensitivity (a) and the effect of axial extent on the imaged volume and sensitivity (b and c).

Practically all systems are now full-ring scanners (GSO or LYSO) operating in fully 3D mode. The axial dimension is about 15–20 cm and the radius of these systems is around 40–45 cm. The only remaining mechanical collimation is by lead shields at the axial ends to reduce effects from activity outside the field of view.

To further increase the sensitivity of PET systems, two approaches are followed. The first approach is to increase the axial extent of the PET scanner. This is a quite effective approach as it increases the sensitivity by two effects: a larger FOV is scanned at the same time (Figure 18.5a), and a higher sensitivity is obtained for a source in the FOV (Figure 18.5b–c). For example, doubling the axial extent of a PET scanner from Z to 2Z will increase the axial FOV by a factor of 2 and the sensitivity of point source in the centre will be doubled, as well as more coincidence pairs being 'captured'. The total sensitivity will therefore increase by a factor close to 4. Besides the evident increase in cost, there is also an increase in the amount of image-degrading effects. This increase can be minimized by faster scintillators with a better energy resolution.

The second approach is to measure more information about the coincidence itself. The current PET scanners are based on coincidence detection: when two events arrive within a short time frame at opposing detectors, the annihilation has occurred somewhere along the line connecting these detectors. The time difference information between the two detections can be used to determine the position of the annihilation (Figure 18.6). This extra information is used in reconstruction to improve image quality. Different papers have illustrated the high image quality gain that could be obtained with this principle. It was shown by different authors that the same image quality can be obtained in a much shorter acquisition time. This reduction in imaging time was predicted to be $D/\Delta x$, with D equal to the object diameter and Δx the spatial equivalent of the Time-of-Flight (TOF) Full Width at Half Maximum (FWHM). This idea was already proposed during the construction of the first PET scanners in the 1970s, but it is technically quite challenging to achieve

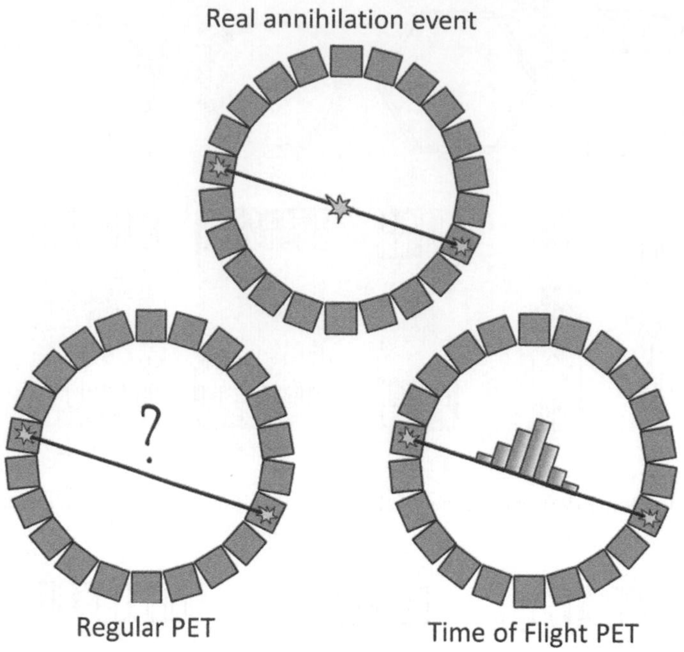

FIGURE 18.6 Regular PET and Time of Flight PET.

at a systems level. It requires much higher accuracy on the timing measurement, as we want to measure the difference in photon arrival time (accuracy within a few 100 ps is necessary). Several prototype scanners were constructed in the 1980s using fast scintillators like CsF and BaF_2. The low stopping power of these crystals for 511 keV photons could not compete with the classic PET scanner based on the BGO scintillator (much higher density) and, therefore, development of TOF-PET scanners has been stopped for more than a decade. New scintillators like LSO, LYSO and $LaBr_3$ were introduced in the second half of the 1990s. LSO and LYSO have a high density, excellent timing resolution, and good energy resolution, making them excellent candidates for TOF-PET scanners. This has relaunched developments in TOF-PET, and the first commercial system became available in 2006. The accuracy of the first systems is about 600 ps, which is equivalent to a Gaussian distribution of 9 cm FWHM. New systems have reported a timing resolution in the range of 200–400 ps.

18.5 PHYSICAL LIMITATIONS AND RELATED CORRECTIONS

Different effects due to physics or system performance degrade the image obtained by a PET scanner (See Figure 18.8a and 18.8b). While there are fundamental limitations imposed by the physics, these limits have not been reached, and there is a constant evolution in PET instrumentation to further reduce the amount of image-degrading effects.

The best obtainable spatial resolution in a PET scanner is limited by the physics of positron emission [1]. There are two major factors that limit the resolution: the positron range and the photon acollinearity. The positron range is the distance the positron travels through the object to lose enough kinetic energy before annihilation takes place. This range depends on the energy of the emitted positrons (different for each isotope) and on the material of the object. An isotope that emits low energy positrons will have a short positron range. Typically, in water 75 per cent of the annihilations occur within 1.2 mm for ^{18}F. Other isotopes have a longer positron range, for example. For ^{82}Rb this distance is even 12.4 mm. For ^{18}F this factor is not so important for human imaging as it is still dominated by other factors. Table 18.2 lists the major emitters in PET with their positron range, half-life, and maximum positron energy.

The effect of acollinearity is caused by the momentum of the positron and electron at annihilation. Due to the conservation of momentum, the resultant photons will not be emitted in exactly opposed directions. This small uncertainty is well modelled by a Gaussian distribution with a Full-Width of Half-Maximum (FWHM) of 0.23°. The absolute influence of this effect on the spatial resolution increases with increasing radius of the PET scanner. It is therefore the major physical limitation for human PET scanners: spatial resolution for a 90 cm diameter scanner is limited to 2.5–3 mm.

TABLE 18.2
Major Emitters in PET with their Positron Range, Half-life and Maximum Positron Energy

Isotope	Maximum positron energy (MeV)	Positron range in water (mm)	Half-life (min)
^{15}O	1.7	2.5	2.03
^{13}N	1.19	1.5	9.97
^{11}C	0.96	1.1	20.3
^{68}Ga	1.89	2.9	67.8
^{18}F	0.64	0.6	109.8
^{82}Rb	3.15	5.9	1.26

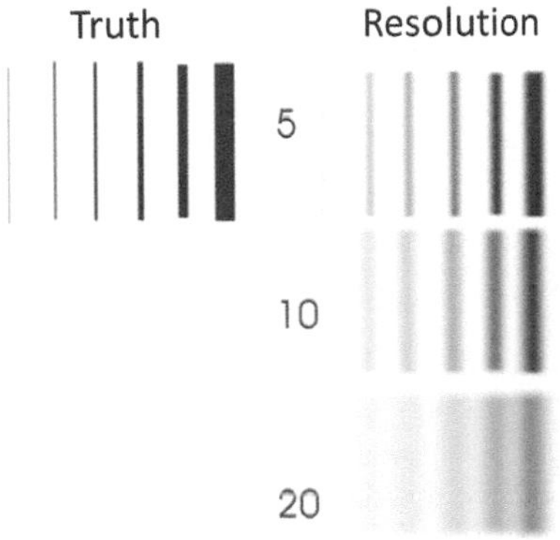

FIGURE 18.7 Loss in image quality in PET due to resolution loss.

Three different effects degrading the spatial resolution in PET loss are shown in Figure 18.7, the first two due to positron physics described above. The third (parallax effect) is described below. The dominant factor on the spatial resolution in current PET scanners is the detector itself. In pixelated detectors, the size of the detector is quite large (4–6 mm in human systems). The crystal Compton scatter and light spread can cause mispositioning (wrong crystal is identified) and further degrade the spatial resolution. Due to the high energy of the incoming photons (511 keV) a relatively large amount of interactions in the crystal will be Compton interactions followed by photoelectric effect. When these occur in different crystals, they can lead to mispositioned events. The light spread is the spread of the secondary photons that travel through the light guide to the PMTs.

Additionally, the parallax effect further degrades the spatial resolution (See Figure 18.8c). This is the result of the fact that the depth-of-interaction in the crystal is not measured. Due to the high energy, the photons are not directly stopped at the entrance surface of the detector. When the photons hit the crystal perpendicular on the surface this is not a problem. When they enter the crystals at oblique angles, it becomes a problem as it can lead to a different LOR (line of response) than the original one. The result of this effect is a resolution loss for oblique LORs. The effect can be minimized or compensated by crystals with different layers or by measuring the depth-of-interaction. Both approaches are, however, still challenging due to the complexity in crystal production and the increased demands in electronics.

All of the above effects lead to a degradation of image quality, which is called the Partial Volume effect. Due to the limited resolution, the activity of small hot structures will be underestimated, and the activity in small cold structures will be overestimated. This effect makes it difficult to follow up activity changes in small tumours as it is not clear whether a reduced activity comes from a reduction in size or in uptake. When anatomical data (CT or MRI) are available to derive the size of the region, the Partial Volume effect can be compensated for.

18.6 ATTENUATION

Because the 511 keV photons have to travel through the body before being detected, there is also a probability that they will undergo Compton scatter or Photoelectric effect in the body. These effects can lead to a decreased number of

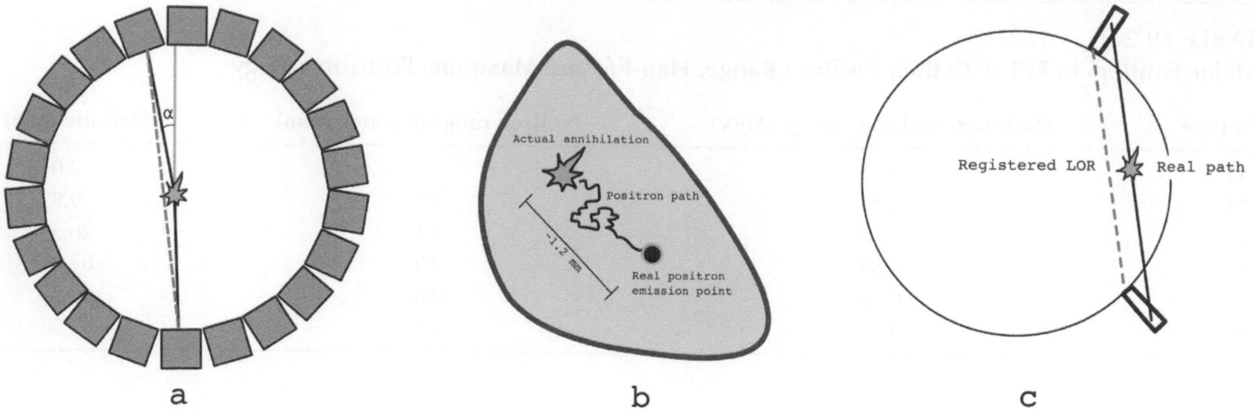

FIGURE 18.8 Spatial resolution loss due to acollinearity of both photons (a), positron range (b), and deviation from the exact LOR due to detector parallax error (c).

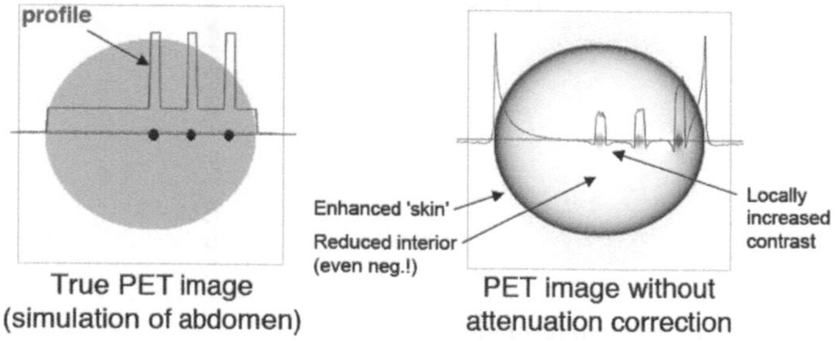

FIGURE 18.9 Attenuation effect in uniform phantom.

detected photon pairs along a given LOR. A photon pair is attenuated when one of the photons (or both photons) is not detected due to an interaction in the body. This can be due to both Compton scatter (deflection leading to non-detection) or photoelectric effect (absorption).

The attenuation of the material is characterized by the linear attenuation coefficient, which is the probability that a photon is attenuated while passing through a unit thickness of the material. It depends on the Z number of the material. For 511 keV the probability of absorption is relatively high, and the probability for scatter is somewhat lower. Due to the attenuation effect, annihilations from the centre of the body will have a much lower probability of escaping the body than will annihilations from the edge. This leads to the so-called cupping-effect figure.

Attenuation correction can be implemented in reconstruction when the attenuation map of the object is known. In stand-alone PET scanners this map was measured by a transmission source outside the object (with an energy close to or equal to 511 keV). As nowadays most scanners are combined with a CT, the attenuation map is derived from a low-dose CT (acquired before the PET scan).

18.7 TRUE, SCATTERED AND RANDOM COINCIDENCES

A measured coincidence that contains the annihilation point along its line (within the spatial resolution), is called a true coincidence. Due to limited energy and time resolution of PET scanners the measurement will, however, also contain a quite large fraction of incorrect coincidences.

The Compton interaction in the body not only leads to attenuated coincidences but can also lead to wrong coincidences. These are called scattered coincidences (illustrated in Figure 18.9). As one (or both) of the photons was deviated due to the Compton effect, the registered photon pair deviates from the correct one. As at least one of the photons has scattered, it is in theory possible to differentiate scattered events from the true events by measuring the energy of the photons. In practice the limited energy resolution (10–20%) of the current scintillators does not always allow us to differentiate 511 keV from lower energy photons. The compromise in current PET scanners is to accept only coincidences from a certain

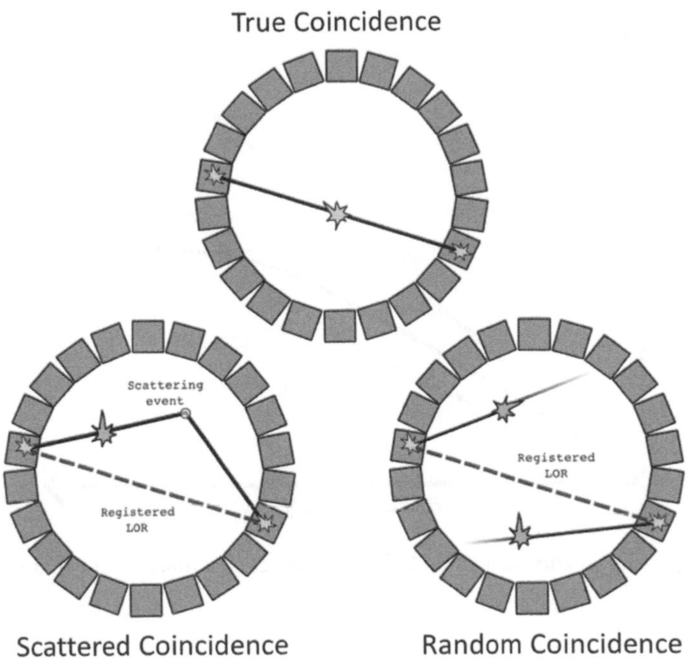

FIGURE 18.10 True, scattered, and random coincidences.

energy window around 511 keV (typically 420–650 keV). This accepts almost all true coincidences while minimizing the number of scattered coincidences. The scatter fraction of a system is the fraction of scattered coincidences from the total number. This is measured using a standard line source at low activities (in order to minimize randoms). 3D PET systems typically have a much larger scatter fraction (about 30–40%) than 2D PET systems (below 10%). As all systems are fully 3D scanners there is a continuing trend in further improving the energy resolution of PET systems. If scatter is not removed from the measured data, this will add a blurry background to the reconstructed images, which reduces the quantitative accuracy. Therefore, the scatter in the measurement is removed by combining estimation techniques and correction in iterative reconstruction.

In a PET scanner two photons are registered in coincidence if they are detected within a given coincidence time window τ. As the physical diameter of a human PET scanner is about 90 cm the maximum time difference between two photons is a about 3 ns. Because most PET detectors have limited timing resolution it is necessary to have a somewhat larger window (typically 4–8 ns) to avoid loss of true coincidences. This, however, also results in incorrect coincidences as two separate annihilations within the coincidence time window can each contribute a single detection. The combination of these two singles is called a random coincidence (illustrated in Figure 18.10). The randoms rate is proportional to the square of the singles rate on the detector. When the activity in the FOV of the scanner increases the amount of random, coincidences will also increase.

The amount of random coincidences in the measurement can be estimated indirectly from a singles measurement. A more common practice is to use a delayed coincidence window. When a single hit a detector only combinations with another single occurring in the time frame $[\tau,2\tau]$ are measured. This excludes all photon pairs coming from the same annihilation, so only random coincidences are measured. Not correcting for randoms adds a uniform background to the image. Therefore, the estimates obtained with the delayed window (or singles method) are also used in iterative reconstruction to correct for random coincidences.

18.8 COUNT RATE

During a PET exam the amount of radioactive tracer is in the same range as for a SPECT study (3–20 mCi). As no mechanical collimator is used in a PET scanner the detectors have to process the singles at much higher rates. This is a quite challenging task, and different components in the PET scanner can cause a loss in count rate. The process of scintillation and determination of the energy and position requires a certain integration time. This time is called deadtime (typically in the range of 30–200 ns). It means these detectors in the system should not process more singles during the

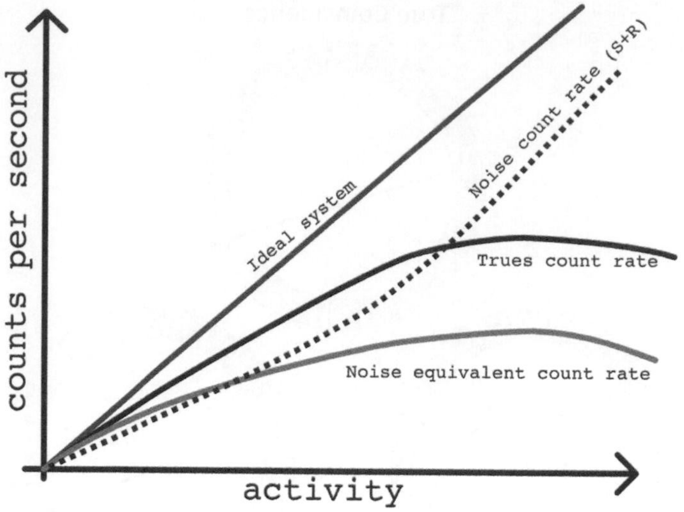

FIGURE 18.11 The number of trues, scattered, and random coincidences as a function of count rate; these are combined into the Noise Equivalent Count rate.

deadtime. If more singles arrive within the deadtime this will lead to pulse pile-up and the single will be registered as a higher energy photon. It will therefore fall outside the energy window and not lead to a coincidence detection. Due to this effect PET scanners tend to have typical count-rate curves (shown in Figure 18.11) that reach a maximum and then fall off. In clinical practice PET scanners are operating within the first half of the curve. If absolute quantification is important, it is necessary to correct for the count-rate loss. This can be done by applying count-rate loss factors based on the measured singles rate.

A measurement that gives insight into the quality of the PET scanner performance is the so-called Noise Equivalent Count rate [2]. It takes into account the deadtime, the randoms rate, and scatter and the ability to measure the true events count rate. NEC is calculated as

$$NEC = T2/(T + S + kR) \tag{18.2}$$

T being the trues count rate, S the scatter count rate, R the randoms rate. K is a factor that is determined by the used randoms correction and the variance. K=2 when real time subtraction is used, K=1 when the randoms estimate is first smoothed.

REFERENCES

[1] C. S. Levin and E. J. Hoffman, "Calculation of positron range and its effect on the fundamental limit of positron emission tomography system spatial resolution," *Phys Med Biol*, vol. 44, no. 3, pp. 781–799, Mar 1999, doi:10.1088/0031-9155/44/3/019.

[2] M. E. Daube-Witherspoon et al., "PET performance measurements using the NEMA NU 2-2001 standard," (in English), *J Nucl Med*, vol. 43, no. 10, pp. 1398–1409, Oct 2002.

19 Dead-time Effects in Nuclear Medicine Imaging Studies

Carlos Uribe and Anna Celler

CONTENTS

19.1 INTRODUCTION

The main application of nuclear medicine (NM) imaging is in the diagnosis of diseases. Nuclear medicine imaging modalities, positron emission tomography (PET) and single photon emission computed tomography (SPECT), use molecules labelled with radioisotopes to provide information about organ and tissue functions. Although interpretation of these images was traditionally mostly visual and qualitative, quantitative analysis gradually began to gain importance. Currently, in the majority of oncology PET, and in many modern SPECT studies, quantitative analyses, such as the determination of standardized uptake values (SUV) and measurements of radioactivity concentrations are performed.

In parallel, ever since targeted radionuclide therapies were introduced, efforts for image-based personalized treatment planning and treatment outcome assessments have been made. These require accurate estimation of the radiation dose delivered during treatments and its spatial distribution. All dosimetry assessments require performing quantitative imaging studies to determine biodistributions of radiotracers.

Generation of quantitative images of activity distributions inside the patient body involves performing image reconstruction with corrections for all image-degrading effects. For example, both PET and SPECT data need to be corrected for attenuation and scatter of photons. Additionally, corrections such as removing random coincidences must also be performed in PET. Furthermore, count losses due to dead time (DT) must be considered at high count rates.

DOI: 10.1201/9780429489556-19

19.2 DEAD-TIME EFFECT IN NUCLEAR MEDICINE IMAGING STUDIES

19.2.1 Sources of Dead Time

The term 'dead time' (DT) refers to the time during which the imaging system is busy processing previously detected events, and the new ones are not properly recorded. DT count losses are more significant at high count rates, and they result in changes of number of photons recorded in any energy window that is used for imaging. There are two mechanisms responsible for DTE:

1. Some of the counts are completely ignored (not recorded) due to the finite resolving or processing time of the camera electronics.
2. At high photon flux two or more photons arriving within a short time can be processed together. Consequently, they will be recorded at the energy equal to the sum of their individual energies. This energy summing (pile-up effect) alters the distribution of photons in the energy spectrum. It contributes to losses of primary (unscattered) counts in the photopeak window and increases the intensity of the higher energy part of the spectrum. Additionally, the positioning of the event is disturbed as it is now averaged from the interaction location of the two photons [1]. More importantly, however, camera exposure to very high flux irradiation may result in the loss of camera energy calibration, affecting its uniformity and linearity for an extended period of time (30 minutes or more) [1][2].

19.2.2 Models Used to Characterize Dead Time

Two models have been proposed [3] to describe the high count-rate performance of the radiation detecting systems, including PET and SPECT cameras:

I. **The Paralyzable system** where, after the first photon is recorded, the system is unable to process and record a second event unless it occurs at a time interval after the first event that is longer than the system resolving time τ. If a second event reaches the detector earlier (i.e., during the time interval τ), the resolving time extends again by τ resulting in a paralyzed system. The system remains paralyzed until a time longer than τ occurs between the subsequent photons. The following formula describes this behaviour, where N_o and N_t are the observed and true count rates, respectively, and τ is the system resolving time or system dead time.

$$N_o = N_t e^{-\tau N_t} \tag{19.1}$$

II. **The Non-paralyzable system** where, after the first photon is recorded, the system is unable to process the second event if it arrives within the time interval τ after the first event, but after this time has passed, the system is immediately ready for the next event. The formula for the non-paralyzable system is:

$$N_o = \frac{N_t}{1+\tau N_t} \tag{19.2}$$

Figure 19.1 shows the behaviour of the observed count rates versus true count rates for the paralyzable and non-paralyzable DT models.

It has been suggested that different components of the NM cameras may display paralyzable or non-paralyzable behaviour. In this case, the overall performance of the imaging system should be described by a combination of the two models. Indeed, early studies have shown that, at low count rates, the non-paralyzable component prevails, while at high photon fluxes the paralyzable model provides a good approximation for the camera behaviour. Therefore, most DT correction methods assume a paralyzable behaviour.

Electronic components of modern cameras may differ quite substantially from those used in the past, so their count-rate performances are also very different [2]. In particular, the pile-up rejection circuits are usually included. Nevertheless, the investigations of their count-rate performances shows that, although modern cameras may not follow the paralyzable model at all count rates, the agreement is good for clinically relevant count-rate ranges (300–400kcps), while significant deviations from this model at higher count rates have been observed [4].

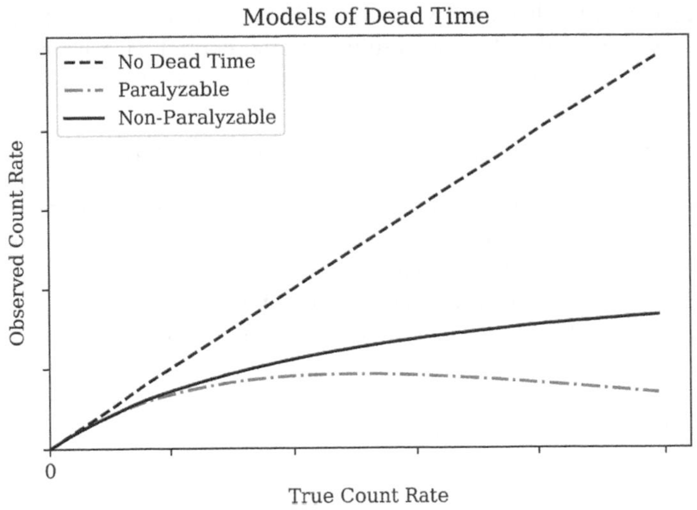

FIGURE 19.1 Behaviour of observed and true count rates for the different dead-time models. The dashed line represents the perfect system in which the observed count rate is exactly the same as the true count rate.

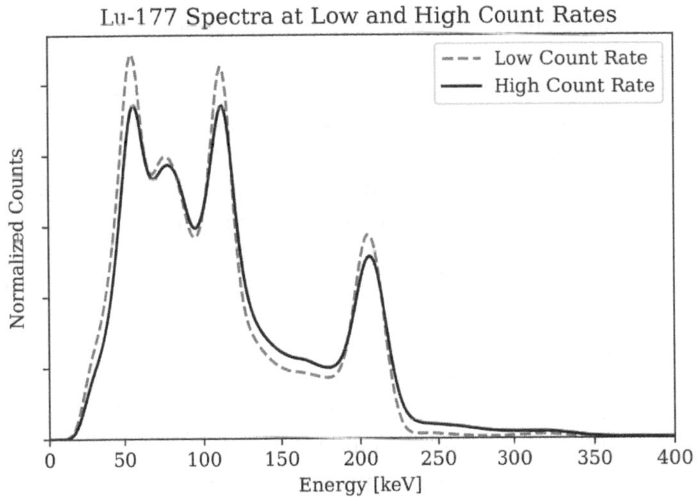

FIGURE 19.2 An example of the ^{177}Lu spectrum showing the pile-up effect as would be observed at high count rates (black line) compared to the spectrum corresponding to low count rates (grey dashed line). Both spectra have been normalized, that is, the area under the curves is equal to one.

19.2.3 Factors Influencing Dead Time

The factor responsible for creation of the dead-time effect in the imaging system is high photon flux. As already mentioned, when many photons reach the detector, some of them may be completely ignored and not be processed, while other photons may have the energy they deposit in the camera summed with that deposited by other photon(s), resulting in pile-up effect. Figure 19.2 illustrates this effect using an example of the ^{177}Lu spectrum (normalized such that the area under the spectrum is equal to one). Note that for the high count-rate, the emission peaks (72 keV, 113 keV, and 208 keV) have lower intensities than those for the low count rates. However, the intensity of the spectrum in the scatter regions (i.e., the regions between the photo peaks) has increased. Additionally, the region with energy above the 208 keV peak shows the pile-up effect particularly well as it has almost no counts at the low count rates and an increased number of counts at high count rates.

Correcting for the pile-up effect is especially important, as it affects the number of photons recorded in different energy windows. Fewer photons will be collected in the photopeak window. Additionally, if triple energy scatter correction is being used, the number of counts in the low-energy window will be decreased, while that in the high-energy window will increase. This effect may significantly affect the resulting scatter correction, especially since the dead-time effects are significantly influenced by the amount and distribution of scattering material in the object (patient) [5], [6].

19.3 CLINICAL SITUATIONS IN WHICH DEAD-TIME CORRECTIONS MUST BE APPLIED

In most clinical single photon imaging studies, both planar scintigraphic and tomographic (SPECT) scans, injected activities are low, thus the recorded count rates are also low. Therefore, corrections for camera dead-time effects need only rarely be applied. However, there are many exceptions such as, for example, cardiac first pass studies [7], cerebral blood-flow measurements using radionuclide angiography [8], dynamic renal studies [9], and imaging performed after administration of radionuclide therapy to verify radioisotope distribution and determine delivered doses [5], [10]–[13].

Quantitative image analysis is often required in PET studies. Since in most cases collimators are not used (resulting in high photon fluxes impinging on the detectors), dead-time corrections are routinely implemented by the software provided by the camera manufacturer.

19.4 DEAD-TIME CORRECTION METHODS

19.4.1 Corrections for Single Photon Imaging (Planar Scintigraphy and SPECT)

Analytic methods rely on a series of phantom experiments performed to determine the dead time correction factors (CF). Using several different activities, the relationship between the observed and the true count rates is established. These data are subsequently fit to the curve corresponding to the paralyzable model to determine the camera resolution time (i.e., τ). The experiments can be performed over the period of time spanning several half-lives of the investigated isotope, starting at very high source activity and repeating scanning until the source activity decreases to the level where dead-time losses become negligible. Alternatively, measurements can begin with a low activity source and continue, with gradually increasing source activity, until the required dead-time level is reached. The camera dead time determined from these experiments is subsequently used to calculate the DT correction factors corresponding to different count rates to be applied to the patient data, with the count rates corresponding to those observed in the phantom scans.

There are several challenges related to the analytic correction method. First, the DT depends on the distribution of the scattering medium in the object (patient). Therefore, when any given activity is placed in objects with different sizes, the camera resolving time will be different, as it depends on the objects' characteristics. Therefore, to provide accurate DT correction, the phantom used in the experiments performed to determine camera resolution time should closely model the patient body. Furthermore, there is a question as to which count rates should be used for the selection of the most appropriate CF. Since the quantification of the activity in the SPECT studies is performed using primary photons only (i.e., using images corrected for attenuation and scatter), the DT correction factor should be determined using primary photons only [6], and not counts recorded in the entire spectrum, as is often suggested. For detailed discussion of this issue please see Uribe and colleagues, 2018 [6].

Marker source methods employ a small source, assumed to create no dead time, positioned on the scanning bed beside the patient body, to estimate the dead-time losses due to high activities present in the patient. In this case, two scans are performed. During the first scan images of the source only (without the patient) are acquired, and then the second scan of the patient is performed with the source present. By comparing number of counts in the small region of interest (ROI) drawn around the image of source in these two scans, one without the dead time (source only) and the second with dead time (source + patient), the correction for DT count losses can be determined.

One limitation of this method is that it assumes the dead-time response of the camera is uniform and does not depend on the source position [8], which may not be true. An additional challenge is related to the fact that, to have negligible DT, the marker source must have low activity, which increases uncertainty of the count measurement in the small source ROI. Furthermore, the method will create problems if used in tomographic scans as, in some projections, the marker counts will overlap with patient counts. To solve this problem, planar acquisitions are recommended [12].

19.4.2 Correction Methods Used in PET

PET systems, as opposed to Anger cameras used in SPECT, deal with coincidence events. Two photons interact with two different detectors and, if this interaction occurs within the preset coincidence time window, the event is counted as a true coincidence. These two photons can come from the true annihilation event or they can originate from two different annihilation events. Therefore, in addition to the count losses of the single events (as in SPECT), PET must also account for the dead time added to the system by the coincidence electronics. The following summarizes different approaches that have been used to correct for DT in PET.

Hoffman and colleagues [14] presented a method for dead-time correction in a PET system used for brain imaging. Their method involved the analysis of triple coincidence rates (TC), which is the detection of a true coincidence event and a photon originating from another annihilation event. When this occurs, the true coincidence is rejected by the system. The relationship between the dead-time losses (DTL) and the triple coincidence rate was given by:

$$DTL = 1.58(TC) + 0.101(TC)^{3/2} \tag{19.3}$$

The total count rate was calculated as:

$$True\ Counts = Observed\ True\ Coincidences + DTL \times t \tag{19.4}$$

Where: t is the total acquisition time (in seconds).

A model that uses the total coincidence rate (i.e., true + random coincidences) was proposed by Mazoyer and colleagues [15]. A second coincidence window with the same duration as the one used to determine the true events (A) is set with a delay relative to the first window to measure only the random events (B). A is calculated by subtracting B from the total detected events. Their model assumed a paralyzable system with a single τ for the coincidence detectors. Mazoyer suggested that if N is the number of observed events in the coincidence window, and M is the number of events in the delayed coincidence window, then:

$$N = \frac{A+B}{f} \qquad M = \frac{B}{f} \tag{19.5}$$

Where: $f = \exp\left(\frac{\tau(A+2B)}{T}\right)$, for events that have been collected for an interval of time T, is the dead-time correction factor (see paralyzable model in Equation 19.1). To properly correct the data for dead time, it is important to have an estimate of the true count rate $\frac{(A+2B)}{T}$.

The method assumes that counts A and B follow an exponential decay with the decay constant λ of the radioisotope as:

$$A(t) = \alpha_0 e^{-\lambda t}$$

$$B(t) = \beta_0 e^{-2\lambda t} \tag{19.6}$$

If the count rate in the two windows is measured experimentally, using a series of scans as the source decays, α_0 and β_0 are the initial count rates for each of the windows, respectively. The true values of A and B at time t are calculated from these two equations. With the estimate of the true count rates, and the measured count rates, the resolving time τ is determined using the equation of the paralyzable model.

Yamamoto and colleagues [16] pointed out that the two previously discussed dead-time correction methods were dependent on the object size. They proposed a method in which single-event detection and coincidence detection should be considered independently because some components of the system work only when a coincidence occurs, while others work even on singles detection. The dead-time losses, however, are already included in the input to the coincidence-timing logic. The coincidence logic determines whether two events arrive within an accepted time window (on time) or at a higher time difference (off time), but the dead time of on-time and off-time events is the same (i.e., the same electronics is used in both situations). Similar to Mazoyer's method, the total coincidence rate is $A+2B$ where: A is the true coincidence rate and B is the random coincidence rate.

The observed singles rate is:

$$S_{obs} = \epsilon_s S_t \tag{19.7}$$

Where: ϵ_s is the efficiency of single-rate detection and S_t is the true single rate without any dead-time losses. The observed coincidence rate is given by:

$$C_{obs} = \epsilon_c \epsilon_s^2 \left(T_t + 2R_t\right) \tag{19.8}$$

Where: ϵ_s is the efficiency of coincidence detection, T_t is the true coincidence rate without dead-time losses and R_t is the true random coincidence rate without dead-time losses.

The observed random coincidence rate can be obtained in a similar manner as:

$$R_{obs} = \epsilon_c \epsilon_s^2 R_t \tag{19.9}$$

where R_t can be determined by extrapolation from a curve of randoms as a function of activity or time. The dead-time correction factor (DTCF) can then be calculated as:

$$DTCF = \frac{R_t}{R_{obs}} = \frac{1}{\epsilon_c \epsilon_s^2} \tag{19.10}$$

Subsequently, it can be applied to the true observed coincidence rate (T_{obs}) to determine the true coincidences

$$T_t = DTCF \times T_{obs} \tag{19.11}$$

Daube-Whithersppon and colleagues [17] proposed a model that took into account singles losses, coincidence losses and pile-up within a detector block. They modelled the singles rate S as a paralyzable system:

$$S_{obs} = S_{true} \exp\left(-\tau S_{true}\right) \tag{19.12}$$

but modified it to compensate for mispositioning of events caused by pile-up, using the following formula:

$$S_{obs} = S_{true_i} \exp\left(-\tau S_{true}\right)\left[1 + a_i \tau S_{true} - b_i \left(\tau S_{true}\right)^2\right] \tag{19.13}$$

Where: the subscript i represents the ring number in the scanner and a_i and b_i are the mispositioning parameters between rings. The different parameters of the equation are calculated from the decaying phantom study and are then applied to correct for other scans. The dead-time correction factor (DTCF) for the singles is:

$$\frac{S_{true}}{S_{obs}} = DTCF_s = \frac{\exp\left(\tau S_{true}\right)}{\left[1 + a_i \tau S_{true} - b_i \left(\tau S_{true}\right)^2\right]} \tag{19.14}$$

A $DTCF_s$ is calculated for each block and the coincidence $DTCF_c$ is the product of the singles $DTCF_s$ for each of the blocks where the coincidence is detected.

DeGrado and colleagues [18] presented yet another method and used it for calculating dead time in one of the first whole-body PET scanners. They measured the random coincidences using a delayed coincidence window. The sum of the true coincidences and the scatter rates can be calculated by subtracting the random coincidences rate from the total count rate. They measured the scatter fraction by scanning a line source at different positions within the axial FOV and using a 2 cm radius ROI around the source in the sinogram. The scatter events were considered to be uniform within the FOV and were estimated from the counts outside of the 2 cm region. The scatter fraction was calculated by summing up the scattered events and dividing them by the total detected counts. The percent dead time (%DT) was calculated as:

$$\%DT = 100 \times \left(1 - \frac{(T+S)_{meas}}{(T+S)_{extrap}}\right) \tag{19.15}$$

The phantom was measured over a wide range of activities (i.e., 14-hour scan). The values from low-activity measurements were extrapolated to find the expected sum of true and scattered coincidences at high count rates.

Eriksson and colleagues [19] proposed that two paralyzable dead times, one that is related to the detector system and another related to the data processing can model the count-rate performance of PET systems. They call the ratio of the observed count rate to the true count rate the 'live time fraction' (LTF), which is the inverse of the DTCF showed in Daube-Whiterspoon model. Two paralyzable models are chosen for the detector system and for the data processing part as

$$LTF_{detector} = e^{-NS_i \tau_{detector}}$$

$$LTF_{dataprocessing} = e^{-C_l \tau_{sys}} \tag{19.16}$$

Where: N is the number of detectors in one detector block, S_i is the average singles count rate for each detector, and C_l is the coincidence load that the data-processing components experience and is given by:

$$C_l = (True + Scatter + Multiples + Randoms) \times LTF^2_{detector} \tag{19.17}$$

The live fraction for the system is the product of the two LTF.

$$LTF_{system} = LTF^2_{detector} \times LTF_{dataprocessing} \tag{19.18}$$

From the true and observed singles in the detector, $\tau_{detector}$ can be determined, and from the corrected events in the detector that are seen by the data processing unit and the $LTF_{dataprocessing}$ the system dead time τ_{sys} is found. This model was validated with existing equipment but rather than being used to correct for dead time in existing scanners, it has been shown to be useful in Monte-Carlo simulations modelling the dead-time effect for the development of new PET scanners [20].

Wells and colleagues [21] used what is commonly known as 'the decaying source method' to measure the count-rate performance of a 3D PET camera. This method assumes that the background is negligible and uses a short-lived source (e.g. ^{18}F) with high activity. The initial N_i and subsequent count rates are measured over several half-lives of the isotope. The counts should follow the radioactive decay of the source such that

$$N(t) = N_i e^{-\lambda t} \tag{19.19}$$

where λ is the decay constant of the radioisotope. This equation can be combined with the paralyzable or non-paralyzable equations. Wells found that their camera could be well described using the paralyzable model:

$$N_{obs} = N_{true} e^{-\tau N_{true}} \tag{19.20}$$

If the natural logarithm is applied to both sides of this equation, then it becomes:

$$\ln(N_{obs}) = -\tau N_{true} + \ln(N_{obs}) \tag{19.21}$$

and following the radioactive decay:

$$\lambda t + \ln(N_{obs}) = -N_i \tau e^{-\lambda t} + \ln(N_i) \tag{19.22}$$

A plot of $\lambda t + \ln(N_{obs})$ as a function of $e^{-\lambda t}$ will be a line with the slope $N_i \tau$ and the intercept $\ln(N_i)$. They assume the model can be used for both singles and coincidences.

MacDonald and colleagues [22] used the paralyzable model to measure the lifetime of a system (similar to Eriksson's), assuming that it depends on the singles flux reaching the detector. Their procedure consisted of measuring the LTF from coincidence counts as a function of the activity concentration in a phantom

$$LTF(x) = \frac{N_o}{N_t} = e^{-\alpha x} \tag{19.23}$$

Where: x is the activity concentration in the phantom and α is a dead-time coefficient. They then simulated the experiment to determine the relationship between the flux of single photons Φ reaching the detectors and the activity concentration in the phantom. With this, it is possible to plot the coincidences LTF (measured as a function of activity distributions) as a function of Φ

$$LTF(\Phi) = e^{-\tau\Phi}$$

Where: τ is the dead-time coefficient measured in units of time. This function can be used to model new scanner configurations using Monte-Carlo simulations.

Vicente and colleagues [23] used the decaying source method and the non-paralyzable model to find dead-time losses and pile-up related effects. Following this model, the following equation is used:

$$N_{true}(t) = \frac{N_{obs}(t)}{1 - \tau N_{obs}(t)} \tag{19.24}$$

Where $N_{true}(t)$ and $N_{obs}(t)$ are the true and observed count rates at time t, respectively, and τ is the dead time. If the counts are corrected for the decay, then:

$$N_{true}(t=0) = \frac{N_{obs}(t)}{1 - \tau N_{obs}(t)} e^{\lambda t} \tag{19.25}$$

Rearranging this equation gives:

$$\frac{N_{obs}(t) e^{\lambda t}}{N_{true}(0)} = 1 - \tau N_{obs}(t) \tag{19.26}$$

Where the numerator $N_{obs}(t) e^{\lambda t}$ is the decay corrected observed count rate at t=0 and $N_{true}(0)$ is the expected count rate at t=0 that should be measured and corrected when the source has reached low activities.

The dead time factor accounts for the dead time pile-up effects:

$$\tau = \tau_{deadtime} + \tau_{ploss} - \tau_{pgain} \tag{19.27}$$

Vicente's group found that the relationship between τ and the singles to coincidences ratio (SCR) is linear. If all the singles are detected as coincidences, then SCR=2. The total τ is then a combination of the two effects which are related to the single-rate dead time τ_s; these that result in coincidences and those that are not measured as coincidences.

$$\tau = \tau_c + \tau_s (SCR - 2) \tag{19.28}$$

Here, τ_c is the coincidences' dead time. The SCR needs to be measured for a high SCR and a low SCR case.

Aykac and colleagues [24] discuss that dead-time correction methods, like those summarized previously, are typically applied at the detector-block (or group of detector blocks) level. They proposed a method to correct for DT at the crystal level. Similar to the decaying source method, they scanned a uniform phantom in list-mode and created the prompts and randoms sinograms at different activity concentrations. They used the randoms sinogram to estimate the singles count rate at the crystal level (a block detector has several crystals) for each of the acquisitions (or frames). Using each crystal's energy spectrum and the true coincidences they estimate the integrated crystal counts that represent the coincidence responses. The average integrated crystal counts (true coincidences) are plotted as a function of the average single count rates for each detector ring. They fit the data to a polynomial of order 8, and use a linear extrapolation from low count rates to determine the expected integrated counts at high count rates. They performed this procedure for the 3,200 crystals available in their scanner and calculated the dead-time value for each of them. The dead-time corrections for each crystal are included in the normalization of the system.

19.5 CONCLUSION

Dead time is caused by the inability of the detection systems to process new event(s) while the previous one is still being analysed. Two models, paralyzable and non-paralyzable, have been used to determine the resolving time of the systems. In SPECT, either an analytical approach or a marker placed beside a patient has been used to correct for count losses. In PET, although detectors can still be modelled as paralyzable or non-paralyzable, the fact that it uses detection of coincidences rather than single photons creates a new challenge when correcting for the dead-time count losses. This chapter summarized the methods used in both PET and SPECT for dead-time corrections. Dead time is particularly important at very high count rates for which not only counts are being missed but also the pile-up effect may significantly alter the shapes of the detected spectra. It is especially important to keep in mind the DT effects when performing radionuclide therapies.

19.6 GLOSSARY

NM	Nuclear Medicine
DT	Dead Time
DTCF	Dead-time Correction Factor
DTE	Dead-time Effect
DTL	Dead-time Losses
LTF	Live-time Fraction
PET	Positron Emission Tomography
SPECT	Single Photon Emission Computed Tomography
TC	Triple coincidence rate

In Memoriam: Anna Celler 1951–2020

Dr. Anna Celler was a Polish nuclear physicist who immigrated to Canada in 1984. In 1991 she joined Vancouver General Hospital (VGH) as a medical-imaging physicist and overviewed the quality-assurance program of 12 nuclear medicine departments around Vancouver, British Columbia. In 1996 she became a fellow of the Canadian College of Physicists in Medicine in recognition of her leadership and competence in physics as applied to medicine. Her passion for research led her to create the Medical Imaging Research Group (MIRG) at VGH and the University of British Columbia (UBC). She contributed 131 peer-reviewed manuscripts, 11 book chapters, 279 conference proceedings, and countless conference presentations and abstracts with topics ranging from development of Compton Cameras, quantitative imaging with SPECT and PET, dosimetry, and cyclotron production of ^{99m}Tc. Dr. Celler was considered a pioneer in quantitative and dynamic SPECT, as well as a leading expert in dosimetry for radiopharmaceutical therapies. During the last 3 decades, Dr. Celler supervised 13 post-doctoral fellows and 30 graduate students. Most of her trainees now have leading positions in academia, industry, and healthcare. Because of her contributions to the clinic, teaching, and research, in 2018 Dr. Celler was awarded the Gold Medal of the Canadian Organization of Medical Physicists (COMP), the highest distinction given by this organization. With all her awards and recognitions throughout the years, however, what she cared about most were the people around her. Sadly, Dr. Celler passed away during the final stages of the publication of this book after battling with cancer for two and a half years. I feel honoured by the opportunity of being mentored by Anna and for being able to work with her on this book chapter; a last one for her and a first one for me. She will be greatly missed, but her legacy will carry on for many years to come.

Carlos Uribe, PhD, MCCPM

REFERENCES

[1] R. Matheoud, F. Zito, C. Canzi, F. Voltini, and P. Gerundini, "Changes in the energy response of a dedicated gamma camera after exposure to a high-flux irradiation," *Phys. Med. Biol.*, vol. 44, no. 6, pp. N129–N135, 1999.

[2] T. K. Lewellen and R. Murano, "A comparison of count rate parameters in gamma cameras," *J. Nucl. Med.*, vol. 22, no. 2, pp. 161–168, 1980.

[3] J. Sorenson, "Deadtime characteristics of Anger cameras," *J. Nucl. Med. Off. Publ.* ..., vol. 16, no. 4, pp. 284–288, 1975.

[4] M. Silosky, V. Johnson, C. Beasley, and S. C. Kappadath, "Characterization of the count rate performance of modern gamma cameras," *Med. Phys.*, vol. 40, no. 3, p. 032502, Mar 2013.

[5] K. F. Koral, K. R. Zasadny, R. J. Ackermann, and E. P. Ficaro, "Deadtime correction for two multihead Anger cameras in 131I dual-energy-window-acquisition mode," *Med. Phys.*, vol. 25, no. 1, pp. 85–91, 1998.

[6] C. F. Uribe, P. L. Esquinas, M. Gonzalez, W. Zhao, J. Tanguay, and A. Celler, "Deadtime effects in quantification of 177Lu activity for radionuclide therapy," *EJNMMI Phys.*, vol. 5, no. 1, 2018.

[7] A. Dobbeleir et al., "Performance of a single crystal digital gamma camera for first pass cardiac studies," *Nucl. Med. Commun.*, vol. 12, no. 1, pp. 27–34, 1991.

[8] Y. Inoue, T. Ohtake, K. Yoshikawa, J. Nishikawa, and Y. Sasaki, "Estimation of deadtime in imaging human subjects," *Eur. J. Nucl. Med.*, vol. 25, no. 9, pp. 1232–1237, 1998.

[9] Y. Inoue, K. Yoshikawa, T. Ohtake, K. Ohtomo, and I. Yokoyama, "Deadtime correction in measurement of fractional renal accumulation of 99tcm-mag3," *Nucl. Med. Commun.*, vol. 20, no. 3, pp. 267–272, 1999.

[10] G. Delpon, L. Ferrer, A. Lisbona, and M. Bardiès, "Correction of count losses due to deadtime on a DST-XLi (SMVi-GE) camera during dosimetric studies in patients injected with iodine-131," *Phys. Med. Biol.*, vol. 47, no. 7, Apr 2002.

[11] J.-M. Beauregard, M. S. Hofman, J. M. Pereira, P. Eu, and R. J. Hicks, "Quantitative (177)Lu SPECT (QSPECT) imaging using a commercially available SPECT/CT system," *Cancer Imaging*, vol. 11, pp. 56–66, Jan 2011.

[12] A. Celler, H. Piwowarska-Bilska, S. Shcherbinin, C. Uribe, R. Mikolajczak, and B. Birkenfeld, "Evaluation of dead-time corrections for post-radionuclide-therapy (177)Lu quantitative imaging with low-energy high-resolution collimators," *Nucl. Med. Commun.*, vol. 35, no. 1, pp. 73–87, Jan 2014.

[13] K. R. Zasadny, K. F. Koral, and F. M. Swailem, "Dead time of an anger camera in dual-energy-window-acquisition mode," *Med. Phys.*, vol. 20, no. 4, pp. 1115–1120, 1993.

[14] E. J. Hoffman, M. E. Phelps, and S. C. Huang, "Performance evaluation of a positron tomograph designed for brain imaging," *J. Comput. Assist. Tomogr.*, vol. 7, no. 4, p. 743, 1983.

[15] B. Mazoyer, M. Roos, and R. Huesman, "Dead time correction and counting statistics for positron tomography," *Phys. Med. Biol.*, vol. 30, no. 5, pp. 385–399, 1984.

[16] S. Yamamoto, M. Amano, S. Miura, H. Iida, and I. Kanno, "Deadtime correction method using random coincidence for PET," *J. Nucl. Med.*, vol. 27, no. 12, p. 1925, 1986.

[17] M. E. Daube-Witherspoon and R. E. Carson, "Unified deadtime correction model for PET," *IEEE Trans. Med. Imaging*, vol. 10, no. 3, pp. 267–275, 1991.

[18] T. R. DeGrado, T. G. Turkington, J. J. Williams, C. W. Stearns, J. M. Hoffman, and R. E. Coleman, "Performance characteristics of a whole-body PET scanner," *J. Nucl. Med.*, vol. 35, no. 8, pp. 1398–1406, 1994.

[19] L. Eriksson, K. Wienhard, and M. Dahlbom, "A simple data loss model for positron camera systems," *IEEE Trans. Nucl. Sci.*, vol. 41, no. 4, pp. 1566–1570, 1994.

[20] C. R. Schmidtlein et al., "Validation of GATE Monte Carlo simulations of the GE Advance/Discovery LS PET scanners," *Med. Phys.*, vol. 33, no. 1, pp. 198–208, 2006.

[21] K. Wells et al., "Count rate performance and deadtime analysis of the new 3D PETRRA PET camera," *Penetrating Radiat. Syst. Appl. III*, vol. 4508, May, pp. 123–133, 2003.

[22] L. R. Macdonald et al., "Estimating live-time for new PET scanner configurations," *Ieee Nucl. Sci. Symp. Conf. Rec.*, pp. 2880–2884, 2007.
[23] E. Vicente et al., "Deadtime and pile-up correction method based on the singles to coincidences ratio for PET," *IEEE Nucl. Sci. Symp. Conf. Rec.*, pp. 2933–2935, 2011.
[24] M. Aykac, V. Y. Panin, and H. Bal, "Crystal-based deadtime correction for Siemens Next Generation SiPM based PET/CT Scanner," *2017 IEEE Nucl. Sci. Symp. Med. Imaging Conf. NSS/MIC 2017 – Conf. Proc.*, vol. 1, no. 865, 2018.

20 Principles of Iterative Reconstruction for Emission Tomography

Andrew J. Reader

CONTENTS

20.1 INTRODUCTION

This chapter will use the primary example of positron emission tomography (PET) to cover the principles of *iterative* image reconstruction for emission tomography in general, which includes the case of single photon emission tomography (SPECT). Image reconstruction concerns estimation of parameters, where the parameters are used to describe the radiotracer distribution (the object being imaged), which is inside the scanner's field of view (FOV). Strictly speaking, therefore, image reconstruction should be referred to as using the measured data to *estimate object representation parameters*. However, we colloquially refer to these estimated object representations – the radioactivity distributions – as *images* that have been 'reconstructed' (estimated) from the measured data.

The benefits of iterative reconstruction of a radiotracer distribution [1] [2] include more accurate modelling of three key components: (i) the discretization of the radiotracer distribution, (ii) the mean of the acquired data, and (iii) the noise in the acquired data. Non-iterative methods such as filtered backprojection (FBP) do not accurately model any of these key components, meaning that iterative methods often have lower reconstructed image noise and superior spatial resolution.

DOI: 10.1201/9780429489556-20

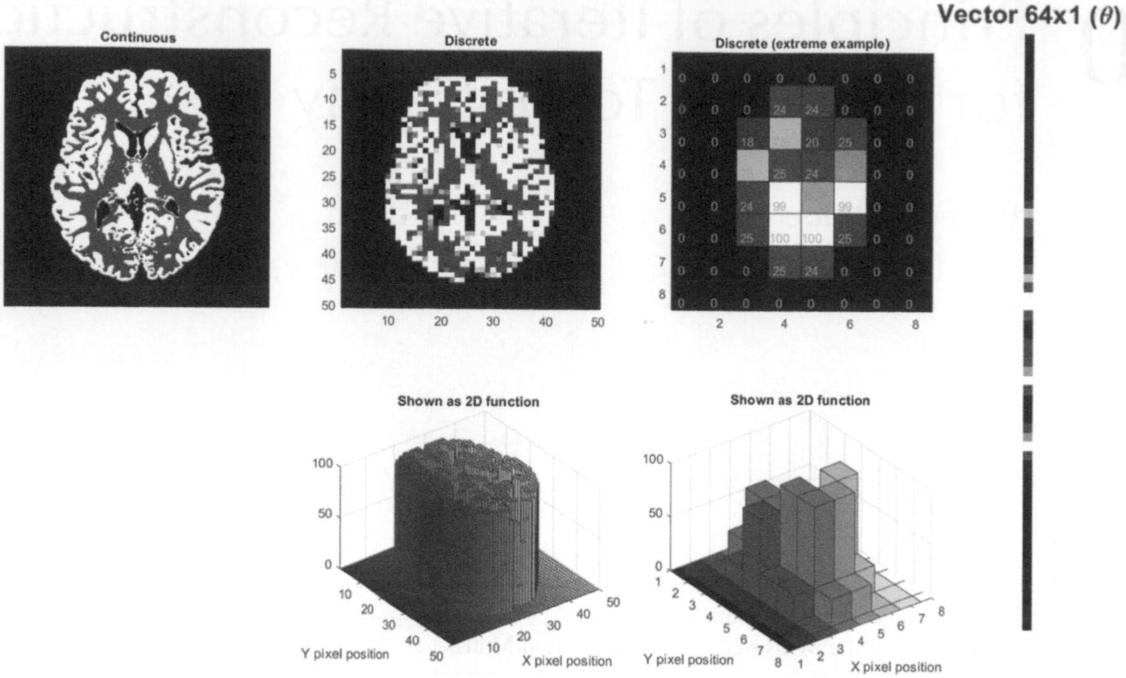

FIGURE 20.1 Using basis functions to represent the continuous distribution of a radioactive concentration $f(\boldsymbol{r})$. As a result, 'image reconstruction' just involves estimating a vector of parameters, $\boldsymbol{\theta}$, which are used in conjunction with basis functions (typically pixels or voxels) in order to represent the radiotracer distribution. We can place all these values, which are the amplitudes (often displayed as grey scale values for each pixel) into a single column vector $\boldsymbol{\theta}$, as shown on the right-hand side of this figure for the case of only 64 pixels (an 8×8 2D image).

20.2 PARAMETERS TO ESTIMATE AND BASIS FUNCTIONS

The parameters to estimate in image reconstruction are usually the values contained in pixels or voxels for 2D or 3D images, respectively. These values can be represented by grayscale shades, or colour maps, such as shown in Figure 20.1. We are therefore representing the distribution of radioactive concentration, f, as a function of spatial position, $\boldsymbol{r}$, simply as a collection of pixels or voxels (known as basis functions) as follows:

$$f(\boldsymbol{r}) = \sum_{j=1}^{J} \theta_j b_j(\boldsymbol{r}) \tag{20.1}$$

where $b_j(\boldsymbol{r})$ is a 2D rect or top-hat basis function corresponding to a given pixel index j, with a coefficient θ_j corresponding to the amplitude scaling of the basis function. This amplitude (or height) of the basis function is often displayed as a grayscale value, as also shown in Figure 20.1. It is important to note from Eq. 20.1 that it is only ever *approximating* the true radioactivity distribution $f^{TRUE}(\boldsymbol{r})$. The function $f^{TRUE}(\boldsymbol{r})$ has nearly infinitely high spatial resolution, which we can neither realistically measure nor reconstruct. Hence, the representation by basis functions is a simplifying approximation, leaving us with the simpler task of only estimating a coefficient θ_j for each one of the basis functions $b_j(\boldsymbol{r})$.

The fact that we cannot exactly recover $f^{TRUE}(\boldsymbol{r})$ combined with the fact that infinitely many different approximations are possible (e.g. different numbers and choices of basis functions), means that the reconstruction problem has many possible solutions according to these choices, corresponding to the approximation we choose to adopt. Image reconstruction, which concerns choosing basis functions and estimating their coefficients from measured data, is therefore known as an 'ill-posed' inverse problem, as there are many possible solutions.

As mentioned earlier, one of the benefits of image reconstruction by iterative methods is that the basis function approximation is explicitly accounted for, in contrast to analytic methods which are often discretized versions of analytic inversion formulae (e.g. FBP). Also, as we will see later, the function $f(\boldsymbol{r})$ can be used to indicate the *mean* number of counts emitted from any chosen subvolume of the object, as a function of position inside the scanner FOV, where this mean is the mean of a Poisson process at each position. Whenever we measure radioactivity by detection of counts, we have noisy data, and we always seek to estimate the mean values of the object that explain these detected counts, and not simply the number of counts emitted from the object during a single scan.

While pixels and voxels (2D or 3D rect functions) are chosen as the basis functions in the vast majority of cases in reconstruction, there have been many proposed alternatives including, for example, 'blobs' ([3]), through to the more recent proposal of basis function shapes derived from similarity of time-activity curves or from structural magnetic resonance (MR) images (e.g. the kernel method [4], which uses spatially-variant basis functions, with each one formed from a group of pixels or voxels). The choice of basis function can have a dramatic impact on spatial resolution and overall image quality [5] [6], as shown, for example, in Figure 20.2.

20.3 MODELLING THE MEAN OF THE MEASURED DATA: THE SYSTEM MATRIX

Eq. 20.1 captured what we wish to find: the coefficients, contained in a vector $\boldsymbol{\theta}$, for the basis functions, which when added together make up the approximate representation of the radioactivity distribution $f(\boldsymbol{r})$, as was shown in Figure 20.1.

However, the whole reason for image reconstruction is that when we do a PET or SPECT scan, we do not directly measure these values within $\boldsymbol{\theta}$. We could imagine situations where we do, though, such as autoradiography, whereby the radioactive distribution is considered to be in a slice only, and we directly measure radioactive counts from each area of the distribution corresponding to a pixel.

Instead, in emission tomography we usually measure projections at different angles external to the object that is being imaged. The projections are collated into *sinograms* – whereby we record counts measured externally to the slice, or volume, of interest, at a given angle, as depicted in Figures 20.3 and 20.4. It is this indirect, external, measurement of the parameters of interest that necessitates image reconstruction. Further note that these projections are usually regarded as parallel-ray projections, as encountered in the context of FBP reconstruction, but as will be seen later, this assumption is no longer required with the general capabilities of system modelling, which are enabled with iterative reconstruction.

To understand the nature of the sinograms that are actually measured in emission tomography, Figures 20.5 and 20.6 show simulation examples. We can see in Figure 20.5 that as more counts are detected, so a point source of radioactivity

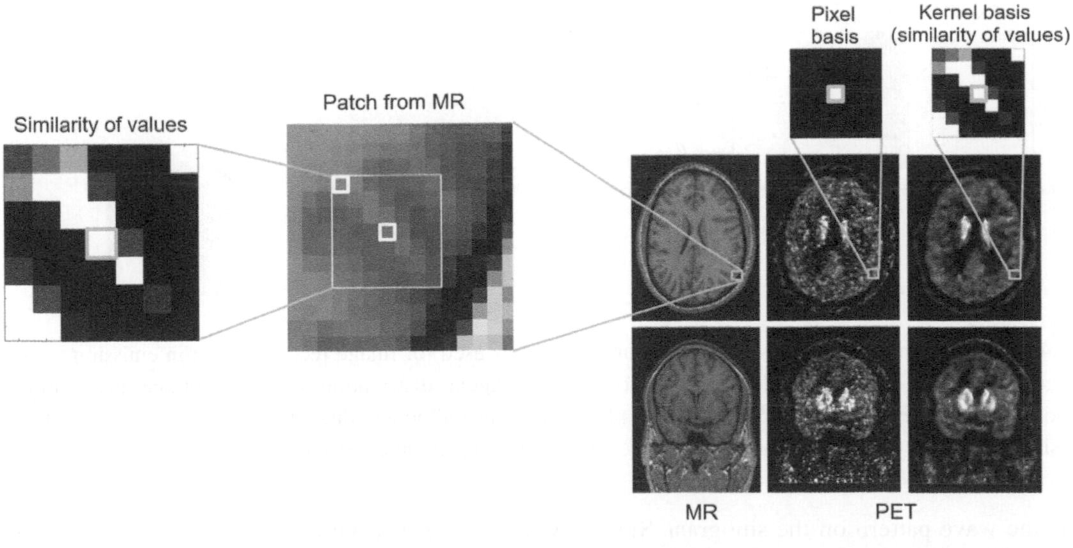

FIGURE 20.2 There are infinitely many different basis functions that can be used to represent the continuous radioactive concentration $f(\boldsymbol{r})$. Two examples are shown in this figure. Using pixels or voxels can result in noisy images, but using larger, overlapping basis functions, such as ones found from a registered MR image via the kernel method (kernel basis), can result in lower noise images. The 2D patch of pixels on the left side of this figure is an example of just one kernel basis function, composed of many pixels, where each pixel value is a measure of similarity between pixel values contained in an MR patch (shown in the centre of the figure). The images reconstructed with the kernel basis (right side of figure) are composed of numerous overlapping kernel basis functions, each with its own coefficient (amplitude) estimated from the measured data. Note further that these neighbourhood similarity measures can be used as weights to guide the smoothing used in regularization methods (see later Section 20.7). Figure is based on the work of Novosad and Reader [5].

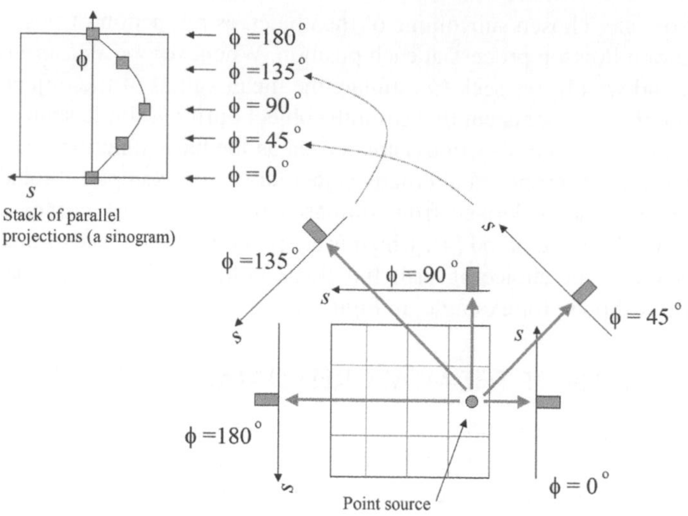

FIGURE 20.3 A sinogram is a 2D function consisting of a stack of 1D parallel projections. If the radioactive distribution was a point source, the 2D sinogram shows part of a sine curve. Note that in practice, with limited acquired counts in each projection, the projections will not be consistent with each other (e.g. one projection of the point source might have 10 counts, but a different projection angle will very likely have a different number of counts).

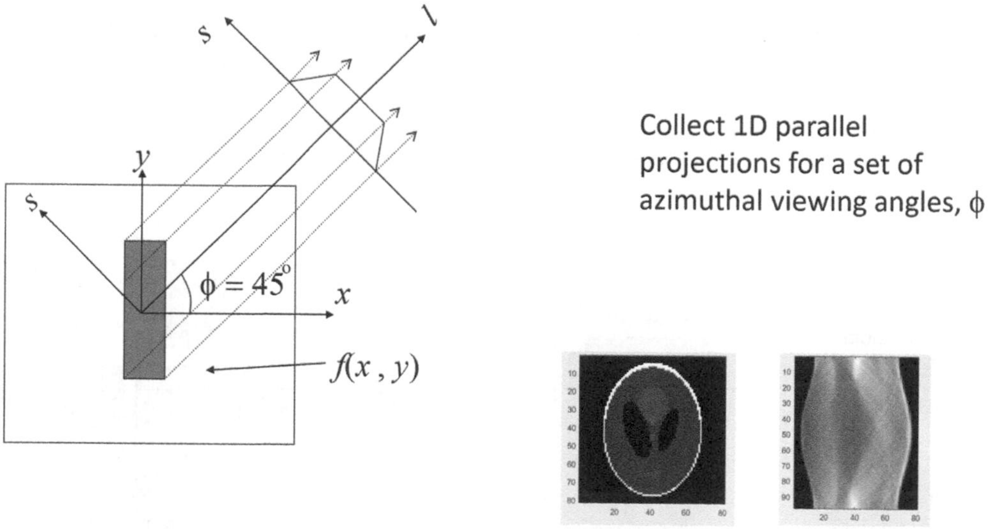

FIGURE 20.4 Conventional parallel-ray projections and sinograms used for image reconstruction in emission tomography. The left of the figure shows an example 45 degree projection of a rectangular distribution for the radiotracer distribution, $f(x,y)$. The lower right corner shows the well-known Shepp–Logan phantom for the radiotracer distribution, $f(x,y)$, and to its right the sinogram obtained by simple forward projection along parallel lines for a series of azimuthal viewing angles.

results in a sine wave pattern on the sinogram. Since any radiotracer distribution, such as that in Figure 20.6, can be regarded as a collection of point sources of differing amplitudes, so a general radioactive distribution will result in a summation of sine waves on a sinogram, each with an amplitude corresponding to each point source.

Now, in order to start estimating the object representation parameters $\boldsymbol{\theta}$ in Eq. 20.1, we need to know how to relate them to the measured sinogram data, $\boldsymbol{m}$. We do this by use of *a system matrix*, also known as the *forward model*, or the *system model*. We first stack our parameters of interest into a single column vector, as was shown previously in Figure 20.1. Similarly, we can consider our measured sinogram data as a column vector $\boldsymbol{m}$, as it is also composed of a large number of pixel values (these are usually called 'bins' for sinograms, as the detected photons are *histogrammed*, or 'binned' into these pixel elements).

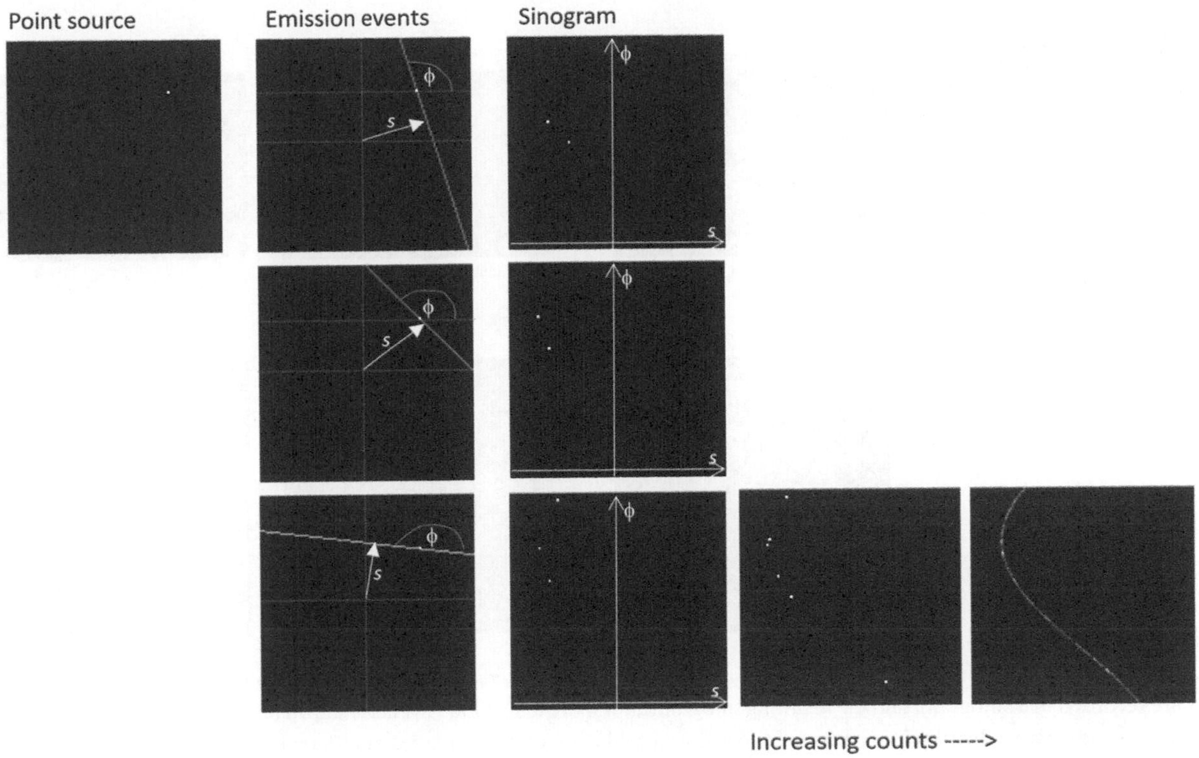

FIGURE 20.5 Emission tomography data are usually in the form of sinograms, but data are collected one count at a time. Top left: a point source positron emitter is shown as a white dot. Top centre: back-to-back photon pairs (for PET) are shown to be emitted as a line emanating from the point source. This line can be characterized by a distance s from the centre of the square (the field of view), and by an angle ϕ. The particular values of s and ϕ can be plotted as a dot (value = 1 count) on an s-ϕ coordinate system (top right of figure). A second event (pair of back-to-back photons) occurs, allowing a second line to be defined (by s and ϕ) and another point to be plotted (middle row of figure). The bottom row corresponds to three events (photon pairs), six events and after ~1000 events. Each pixel (each bin) of this sinogram contains the number of counts detected for a given line (s,ϕ) through the field of view.

We then require the mapping from vector $\boldsymbol{\theta}$ to vector $\boldsymbol{m}$. However, it is clear that a measured data set $\boldsymbol{m}$ is noisy, and so in most cases the data $\boldsymbol{m}$ are not consistent. If we consider a case of extremely low counts, it is easy to understand why the data are not consistent. For example, consider again Figure 20.3: a view angle of ϕ=0° could have a noisy projection that is not in agreement with the projection at ϕ = 90°, simply because a different number of counts have been measured at each different view angle of the sinogram. Even at normal count levels we cannot expect consistent data and, hence, we can never expect to exactly obtain a match for $\boldsymbol{m}$ from a single choice of $\boldsymbol{\theta}$.

As mentioned in Section 20.2, we seek instead something better: to estimate the *mean* number of emissions from the radioactivity distribution. We will therefore also model the *mean of the data*, which we will label as $\boldsymbol{q}$ (to distinguish it clearly from the noisy data $\boldsymbol{m}$), as follows. We will use a very large matrix, $\boldsymbol{A}$, which has as many columns as there are pixels or voxels (one for each element in $\boldsymbol{\theta}$), and which has as many rows as there are sinogram bins:

$$\boldsymbol{q} = \boldsymbol{A\theta} \tag{20.2}$$

A row of the matrix $\boldsymbol{A}$ reveals which pixels or voxels contribute to a given projection bin, and a column of the matrix $\boldsymbol{A}$ reveals how a given pixel or voxel contributes to the projection bins. In practice there is a very large number of rows / projection bins (> 100 million for 3D imaging) and a large number of columns / pixels or voxels (> 10 million for 3D imaging). This system matrix, $\boldsymbol{A}$, represents our PET or SPECT scanner (but, again, *without* the noise component that would normally arise due to limited counts/events being acquired in a single scan). An illustration of a system matrix for 2D reconstruction is shown in Figure 20.7. Each column corresponds to the mean data that would be measured from a point source contained in a pixel, and each row corresponds to an 'image' of a given sinogram bin's sensitivity to various

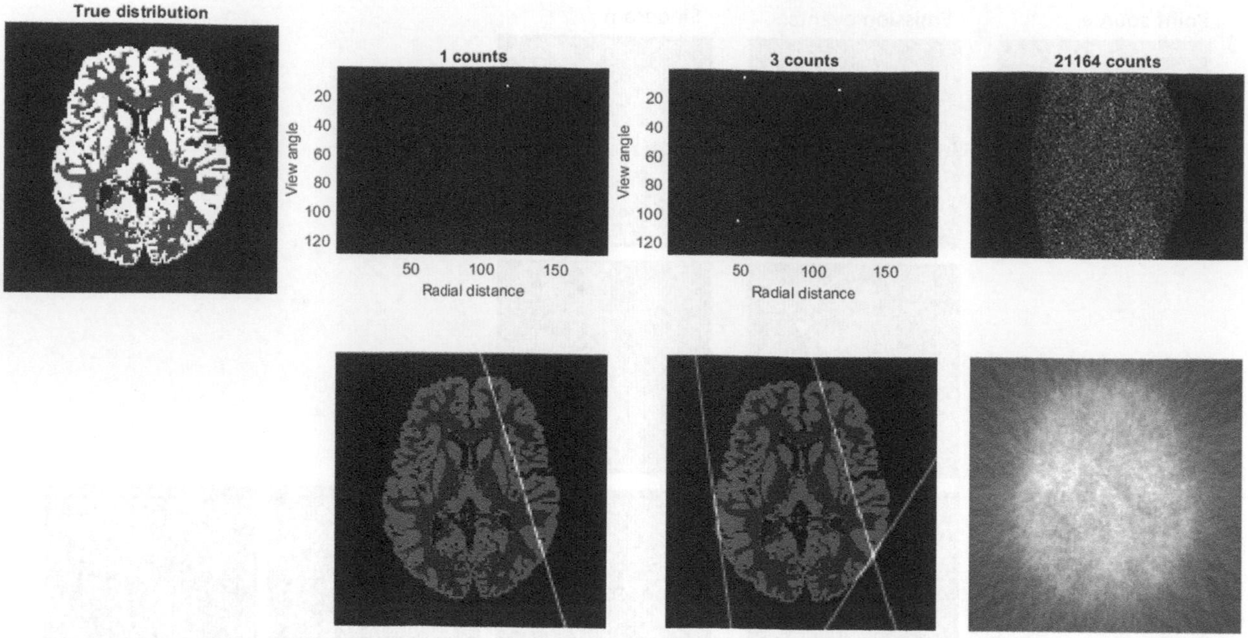

FIGURE 20.6 Example emission events (counts) are shown for a simulated radiotracer distribution in a slice of a brain [34]. The events are recorded in a sinogram, and the resulting sinogram after more than 20000 events is shown on the top right side of the figure. Note that any radioactive distribution can be considered as a collection of point sources of different intensities, and so the measured sinogram corresponds to a summation of noisy sinograms obtained for each point source in the scanner's field of view. The bottom row of the figure shows the lines along which the events were detected, and the bottom right shows the backprojection of the >20000 events.

pixels. Typically, a row of the system matrix $\boldsymbol{A}$ will correspond to a line 'image', showing the pixel values that need to be added together to model the mean value for that projection bin.

An important note on system modelling is that in practice, for emission tomography scanners, the system matrix can extensively model the physics of the imaging process, and this is achieved relatively simply by factorizing the matrix $\boldsymbol{A}$ as follows (see, e.g. [7]):

$$\boldsymbol{q} = \boldsymbol{A\theta} + \boldsymbol{s} + \boldsymbol{r} = \boldsymbol{NLX\theta} + \boldsymbol{s} + \boldsymbol{r} \tag{20.3}$$

where $\boldsymbol{N}$ and $\boldsymbol{L}$ are diagonal matrices modelling the normalization and attenuation losses, $\boldsymbol{X}$ is the line-integral (x-ray or Radon) transform, and $\boldsymbol{s}$ and $\boldsymbol{r}$ are the scatter and randoms coincidences (in the context of PET imaging). In the context of SPECT imaging, the attenuation of the single-emitted photons (rather than the pair of photons in PET) is depth dependent, and so the diagonal matrix factorization for $\boldsymbol{L}$ cannot be applied.

Note the crucial concept that is exploited in Eq. 20.3: decomposition of the system matrix $\boldsymbol{A}$ into a product of other matrices. We can go further with this to improve the system model, by including *resolution modelling*:

$$\boldsymbol{q} = \boldsymbol{NLXH\theta} + \boldsymbol{s} + \boldsymbol{r} \tag{20.4}$$

where, for example, $\boldsymbol{H}$ could be as simple as a convolution matrix, using a shift-invariant point spread function (PSF) [8]. It could however be used to model a shift-*variant* PSF as well. If such a model is used in an iterative reconstruction, the reconstructed image will effectively be compensated for that modelled spatial resolution (we achieve something comparable to a 'deconvolution' of the PSF during the image reconstruction steps, which are to follow below). In fact, if the overall model $\boldsymbol{A}$ is only a convolution matrix (i.e., a circulant matrix representing a linear shift-invariant system) containing shifted copies of the PSF in each column, iterative reconstruction algorithms that use such a model will correspond to doing deconvolution, although with the benefits of the improved modelling afforded by iterative reconstruction. One of these benefits will be mentioned in Section 20.4 – the benefit of accurate modelling of the noise in the data.

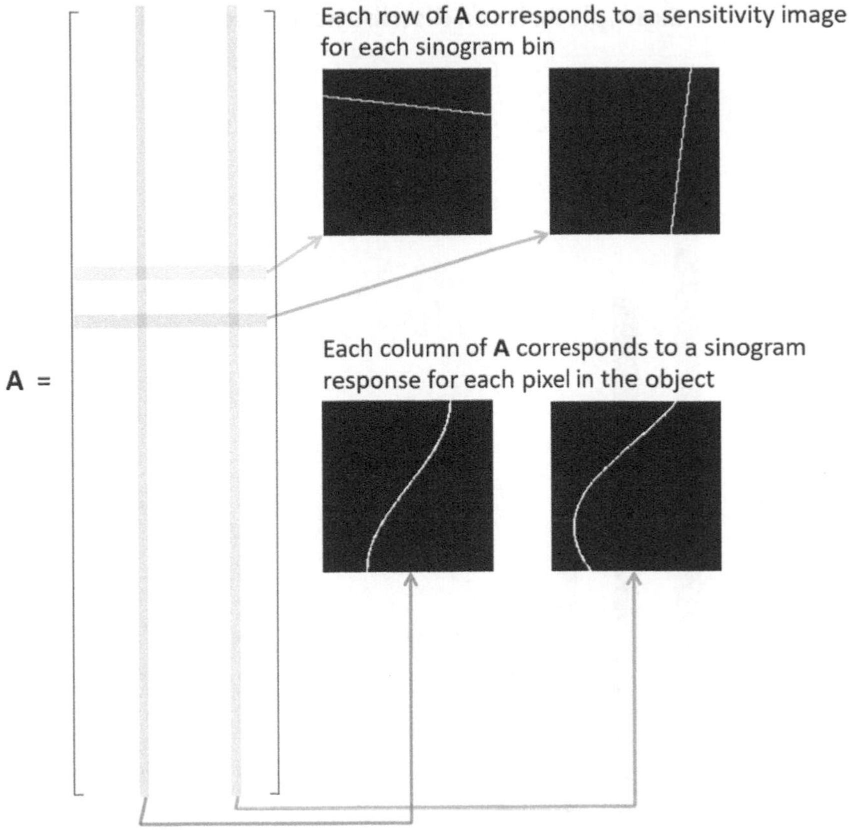

FIGURE 20.7 Visualization of the system matrix $\boldsymbol{A}$, which allows a vector $\boldsymbol{\theta}$ (modelling the mean radiotracer concentration in the field of view) to be mapped to a vector $\boldsymbol{q}$ (modelling the mean of each projection bin for one or many sinograms). For 3D emission tomography imaging, the vector $\boldsymbol{q}$ corresponds to many sinograms, and the system matrix is usually overdetermined, meaning there are many more rows (many more sinogram projection bins) than there are columns (where each column corresponds to the modelled response to a pixel of radioactivity in the 'image').

20.4 POISSON NOISE

Each projection bin i in the projection data, or sinogram data, collects m_i counts. Depending on how long the emission tomography scan lasts, the number of counts in each projection bin may be quite low, and the measured data suffer from noise arising from this limited number of counts. Collecting counts in projection bins is a *Poisson process*, other examples of which include counting the number of radioactive emissions from a source in a unit of time, as well as everyday examples such as the number of emails received during an hour of a normal working day. A Poisson process is one that follows the Poisson distribution, which was proposed in 1837 by French mathematician Siméon Denis Poisson (1781–1840).

Taking the example of emails: the number of emails received in any given hour will vary, but over a large number of hours, the average (or mean, or *expectation*) of the number of emails per hour can be estimated (= total emails / total hours). So, also, for a given projection bin in an emission tomography scan of a given radioactive distribution, if numerous scans were carried out, the mean number of counts (q_i) in bin i could be estimated. However, for any individual scan this would be a realization m_i of a random variable, which could be less than, equal to, or greater than the mean value q_i. So, we say that the measured number of counts m_i in projection bin i, if the mean is q_i, is distributed according to the Poisson probability distribution, which by definition is:

$$\Pr(m_i \mid q_i) = \frac{q_i^{m_i} \exp(-q_i)}{m_i!} \tag{20.5}$$

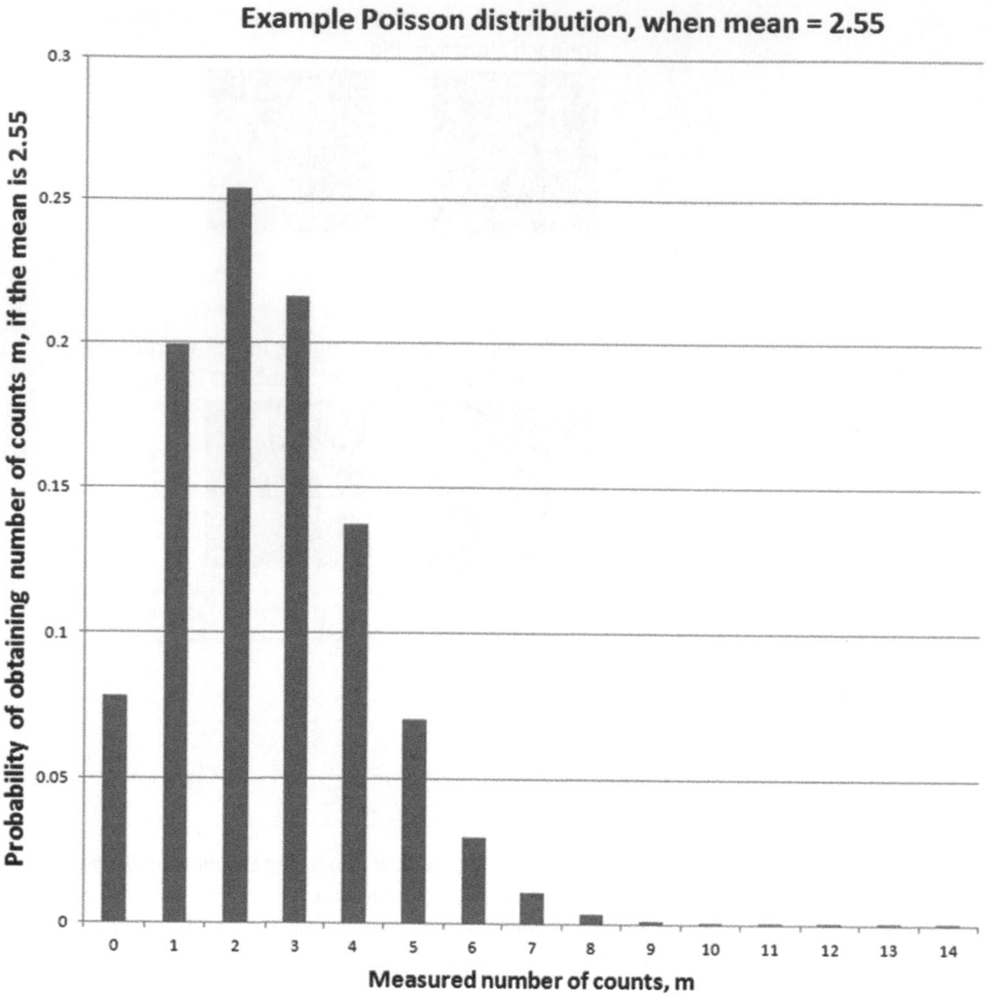

FIGURE 20.8 The Poisson distribution. This distribution simply gives the probability of obtaining the number of counts m, if the mean value is q. This graph was produced by substituting values of m=0, m=1, m=2 etc. into the expression for the Poisson distribution, holding q constant at 2.55. Note that the mean value is not an integer – it is unlikely that the mean value in a sinogram bin (i.e. when averaged across many, many identical scans) is an integer value. However, measured values are integers.

This equation gives the probability of obtaining m_i counts if (the '|' sign represents 'if') the mean number of counts in that bin (determined by repeating the experiment/scan under exactly the same conditions a large number of times) is q_i. Figure 20.8 visualizes the Poisson distribution.

An important further definition is that of the *likelihood*. We *define* the likelihood of the mean being q_i, if we measured m_i counts, according to the aforementioned probability:

$$l(q_i \mid m_i) = \Pr(m_i \mid q_i) = \frac{q_i^{m_i} \exp(-q_i)}{m_i!} \tag{20.6}$$

A notable subtlety is that Eq. 20.5 is a function of integer values m_i, for a fixed real q_i, whereas Eq. 20.6 is a function of continuous values q_i for a fixed integer value m_i.

Some characteristics of the Poisson distribution include:

1. For a low mean value q, there is a skew to the right (it has a long declining tail reflecting decreasing probabilities for larger and larger values of m).
2. The variance σ^2 is equal to the mean value q.

3. For large values of q (e.g. $q > 10$), the distribution closely approximates a Gaussian (or normal) distribution with mean=variance=q.

20.5 MAXIMUM LIKELIHOOD

We now consider how to go about estimating $\boldsymbol{\theta}$ in Eq. 20.1 from the measured data $\boldsymbol{m}$, assuming the measured data to be Poisson distributed (Eq. 20.5), with a mean modelled by $\boldsymbol{A\theta}$ (Eq. 20.2).

We therefore seek to find the vector $\boldsymbol{\theta}$, representing an object, which maximizes all of the individual likelihoods for all of the measured data bins, from i=1…I:

$$L(\boldsymbol{\theta} \mid \boldsymbol{m}) = \prod_{i=1}^{I} l(q_i(\boldsymbol{\theta}) \mid m_i) = \prod_{i=1}^{I} \frac{q_i^{m_i}(\boldsymbol{\theta}) \exp(-q_i(\boldsymbol{\theta}))}{m_i!} \tag{20.7}$$

where the capital pi, Π, denotes the product, and $L(\boldsymbol{\theta} \mid \boldsymbol{m})$ is the Poisson likelihood of the parameters $\boldsymbol{\theta}$, given the measured data $\boldsymbol{m}$. We are assuming the measured values $\{m_i\}$ are independent of one another. Intuitively it makes sense: we want the probabilities of each measured value in $\boldsymbol{m}$ to all be maximum, and if we take the product of all these probabilities, maximizing the product necessarily involves maximizing each of them individually. However, it is unlikely that all will actually be maximized, as a given $\boldsymbol{\theta}$ might not be able to fully satisfy all of the measurements simultaneously, due to the inconsistency of the noisy data as previously discussed.

The dependency of the Poisson likelihood on the parameter vector $\boldsymbol{\theta}$ is implicit due to the previous definition of $\boldsymbol{q}$ from Eq. 20.2, which in more explicit notation is given by:

$$q_i(\boldsymbol{\theta}) = \sum_{j=1}^{J} a_{ij} \theta_j \tag{20.8}$$

Our following analysis will consider optimizing just a single parameter value θ_j from the whole image $\{\theta_j\}$. Eq. 20.7 gives a single scalar value (L) for the likelihood if we feed into the equation a whole set of measured data values $\{m_i\}$ and a whole set of candidate parameter ('image') values $\{\theta_j\}$. First, to make the expression easier to manipulate, we will take the natural logarithm, being confident that the maximum value of (20.7) with respect to θ_j will occur at the same value of θ_j as when the logarithm of Eq. (20.7) is taken. (This is true, since the logarithmic function is monotonic – i.e., as x increases, so also log x increases.)

The Poisson log-likelihood is therefore

$$\ln L(\boldsymbol{\theta} \mid \boldsymbol{m}) = \sum_{i=1}^{I} \ln q_i^{m_i}(\boldsymbol{\theta}) - \sum_{i=1}^{I} q_i(\boldsymbol{\theta}) - \sum_{i=1}^{I} \ln m_i! \tag{20.9}$$

Simplifying gives

$$\Lambda(\boldsymbol{\theta} \mid \boldsymbol{m}) = \sum_{i=1}^{I} m_i \ln q_i(\boldsymbol{\theta}) - \sum_{i=1}^{I} q_i(\boldsymbol{\theta}) \tag{20.10}$$

where we have now labelled this expression as $\Lambda(\boldsymbol{\theta} \mid \boldsymbol{m})$, and we have ignored the last term in Eq. (20.9) since this is a constant value for any given measured data set $\{m_i\}$ and consequently has no impact on our seeking the maximum value of Eq. (20.7). We can rewrite (20.10) to explicitly show the components of $\boldsymbol{\theta}$:

$$\Lambda(\boldsymbol{\theta} \mid \boldsymbol{m}) = \sum_{i=1}^{I} m_i \ln\left(\sum_{b=1}^{J} a_{ib} \theta_b\right) - \sum_{i=1}^{I} \sum_{b=1}^{J} a_{ib} \theta_b \tag{20.11}$$

where the summation index has been changed to b (still summing over all the pixels b=1…J), as this will allow us to use j later on to refer to a specific pixel or voxel of interest.

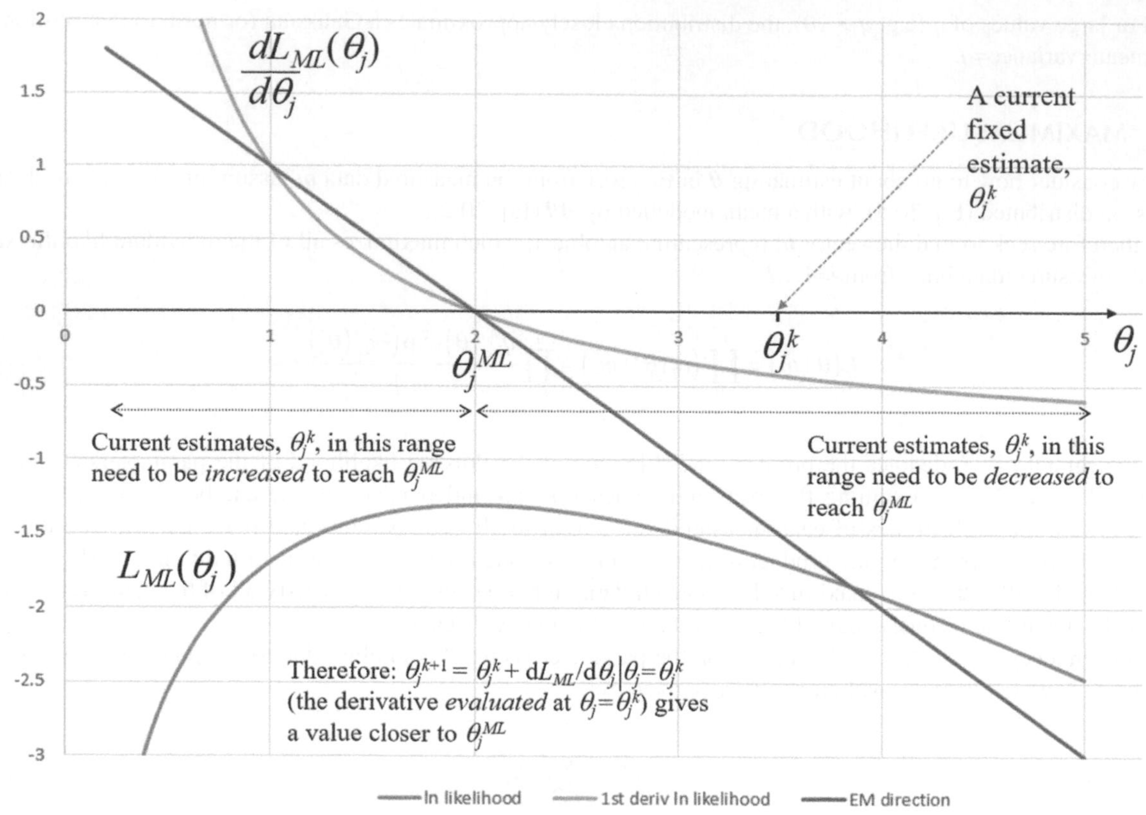

FIGURE 20.9 Visualization of the log-likelihood function for a 1D optimization of a single value θ_j, and how to improve upon a current fixed estimate of θ_j^k in order to get closer to the maximum of the log-likelihood. The EM algorithm modifies the gradient by multiplying by θ_j^k divided by the sensitivity image, to give the 'EM direction' shown on this graph.

As the values in the parameter vector $\boldsymbol{\theta}$ change, so the scalar value Λ will change. We require the optimum vector $\boldsymbol{\theta}^{ML}$, which, when substituted into Eq. (20.11) and evaluated, yields the maximum value of the log-likelihood ($\ln L$) – and, hence, is the maximum likelihood (ML) estimate of $\boldsymbol{\theta}$. So, we can formulate our problem as based on Eq. (20.10):

$$\boldsymbol{\theta}^{ML} = \arg\max_{\boldsymbol{\theta}} \sum_{i=1}^{I} \left[m_i \ln q_i(\boldsymbol{\theta}) - q_i(\boldsymbol{\theta}) \right] \tag{20.12}$$

Eq. 20.12 indicates that we seek a vector $\boldsymbol{\theta}$ that maximizes the entire expression on the right hand side of the equation, and that on finding that vector it will be the maximum likelihood estimate of the parameters, labelled $\boldsymbol{\theta}^{ML}$.

To conceptualize an approach to achieving this, we will consider just one pixel value θ_j, the log likelihood for different values of this pixel, along with the gradient of this log-likelihood function (Figure 20.9). We can see immediately that if we have a *fixed* current estimate θ_j^k, which is specifically labelled by superscript k to denote that it is a *specific fixed choice* of θ_j, then we can get closer to the optimal value by adding some scaled amount of *the derivative of the function*. This would yield a better estimate, which we will label with the superscript k+1: θ_j^{k+1}, which is closer to θ_j^{ML}.

Therefore, given a current fixed estimate θ_j^k for pixel j, we can obtain a better estimate θ_j^{k+1}, by adding a scaled amount of the gradient (the partial derivate of $\Lambda(\boldsymbol{\theta} \mid \boldsymbol{m})$ with respect to θ_j, which leaves a function of θ_j only) evaluated at our current fixed estimate θ_j^k:

$$\theta_j^{k+1} = \theta_j^k + \tau^k \left. \frac{\partial \Lambda(\boldsymbol{\theta} \mid \boldsymbol{m})}{\partial \theta_j} \right|_{\theta_j = \theta_j^k} \tag{20.13}$$

where the scaling τ^k can be used to modify the size of the step taken towards the maximum value, which is a critical choice if the algorithm is to converge. In a machine-learning context, this scaling or step size is called the learning rate, and often in that context only very small values are used.

We therefore need to find the partial derivative of $\Lambda(\boldsymbol{\theta} \mid \boldsymbol{m})$ with respect to a single image value θ_j. The log-likelihood, as a reminder, was, to within a constant:

$$\Lambda(\boldsymbol{\theta} \mid \boldsymbol{m}) = \sum_{i=1}^{I} m_i \ln\left(\sum_{b=1}^{J} a_{ib}\theta_b\right) - \sum_{i=1}^{I}\sum_{b=1}^{J} a_{ib}\theta_b \tag{20.14}$$

where, of course, our value of interest for a given fixed pixel j, θ_j, is just one of the terms being summed in the summations over the pixel index $b=1\ldots j\ldots J$.

Partial differentiation of Eq. 20.14 gives:

$$\frac{\partial}{\partial \theta_j}\Lambda(\boldsymbol{\theta} \mid \boldsymbol{m}) = f(\theta_j \mid \boldsymbol{m}) = \sum_{i=1}^{I} m_i \frac{a_{ij}}{\sum_{b=1}^{J} a_{ib}\theta_b} - \sum_{i=1}^{I} a_{ij} \tag{20.15}$$

where the single pixel j, and hence the value θ_j, is just one of the terms in the summation appearing in the denominator, and the function f of θ_j is shown to emphasize this is a function of θ_j only.

Evaluating this continuous function at our current fixed estimate θ_j^k, and then substituting into Eq. (20.13) gives the following pixel update step for each and every pixel in our current estimate $\{\theta_j^k\}$

$$\theta_j^{k+1} = \theta_j^k + \tau^k \left(\sum_{i=1}^{I} \frac{a_{ij}m_i}{\sum_{b=1}^{J} a_{ib}\theta_b^k} - \sum_{i=1}^{I} a_{ij}\right) \tag{20.16}$$

We can see that the gradient term in brackets could be negative and, depending on the choice of scale factor τ^k, this could result in a negative value for θ_j^k. We will see a theoretically motivated choice of τ^k in a later section, but for now we will motivate the choice by choosing its value in such a way as to avoid negatives, by setting it equal to the current image value θ_j^k, divided by the sensitivity of the scanner at that point $\left(= \sum_{i=1}^{I} a_{ij}\right)$:

$$\tau^k = \frac{\theta_j^k}{\sum_{i=1}^{I} a_{ij}} \tag{20.17}$$

Again, it will be seen in a later section why this value arises. For the present coverage of maximum likelihood we are seeking a simple gradient ascent-based understanding. With this choice of τ^k, Eq. 20.16 becomes

$$\theta_j^{k+1} = \theta_j^k + \frac{\theta_j^k}{\sum_{i=1}^{I} a_{ij}}\left[\sum_{i=1}^{I} \frac{a_{ij}m_i}{\left(\sum_{b=1}^{J} a_{ib}\theta_b^k\right)} - \sum_{i=1}^{I} a_{ij}\right] = \theta_j^k + \frac{\theta_j^k}{\sum_{i=1}^{I} a_{ij}}\sum_{i=1}^{I} \frac{a_{ij}m_i}{\left(\sum_{b=1}^{J} a_{ib}\theta_b^k\right)} - \theta_j^k \tag{20.18}$$

which gives a purely multiplicative and non-negative update, resulting in what is known as the expectation maximization (EM) algorithm for maximum likelihood (ML) estimation, EM-ML or ML-EM:

$$\theta_j^{k+1} = \frac{\theta_j^k}{\sum_{i=1}^{I} a_{ij}} \sum_{i=1}^{I} a_{ij} \frac{m_i}{\sum_{b=1}^{J} a_{ib}\theta_b^k} \tag{20.19}$$

This very well-known algorithm was first proposed by Shepp and Vardi in 1982 [9] and Lange and Carson in 1984 [10].

Provided that all the values in $\boldsymbol{m}$, $\boldsymbol{A}$, and $\boldsymbol{\theta}^k$ are all non-negative, we see that the values in the update image $\boldsymbol{\theta}^{k+1}$ will always be non-negative. This non-negativity constraint leads to noise reduction in low activity regions of the image, but for noisy data can lead to a positive bias in such regions as well.

It can be convenient to rewrite Eq. (20.19) in terms of matrices and vectors, as follows:

$$\boldsymbol{\theta}^{k+1} = \frac{\boldsymbol{\theta}^k}{\boldsymbol{A}^T\mathbf{1}} \boldsymbol{A}^T \frac{\boldsymbol{m}}{\boldsymbol{A}\boldsymbol{\theta}^k} \tag{20.20}$$

where ratios and products of vectors are understood to be, element-wise, consistent with (20.19), and matrix-vector multiplication remains conventional. This approach follows the notation introduced by Barrett [11].

Eq. 20.19 Eq. (or 20.20) is in widespread use in PET and SPECT imaging for iterative image reconstruction. It is in fact very simple to understand. We can express the ML-EM algorithm as a series of steps:

1. Start with an initial estimate of the radioactivity distribution.
2. Apply the system model to this estimate, in order to obtain a predicted set of projection data (this is the current model of the mean of the data).
3. Compare this predicted or calculated data with the data that were actually measured.
4. Correct the estimate of the radioactivity distribution, based on the discrepancy between what we actually measured with the PET scanner and what we predicted through our model (which used our estimate of the activity distribution).
5. Repeat steps 2 to 4 for each iteration, each time using the improved estimate of the radioactive distribution.

We will now spell out explicitly these steps.

Step 1: 'Guess' the radioactive distribution. Unless we really do have some prior information, we should make no strong assumptions here. So, we will say that the radioactivity is the same in every single pixel (a constant > 0). Therefore:

$$\theta_j^{(0)} = \text{constant} \quad \textit{for all} \quad j \tag{20.21}$$

The superscript 0, in brackets, is used to indicate that this is just a first estimate of the true θ. We will be iteratively improving the agreement of θ with the measured data $\boldsymbol{m}$, and so we will label successive estimates of the activity distribution with the number of iterations k that have been carried out. So, for this first step, $k = 0$, since no iterations have yet been carried out. Note that we will omit the brackets for the iteration number k.

Step 2: This is a crucial step: we must model our PET scanner as accurately as we can, using the system matrix method described previously. We have an estimate of the activity distribution, $\{\theta_j^k\}$, and we need to predict what the PET scanner would do if this were the actual distribution. So, we use the system of equations:

$$q_i^k = \sum_{j=1}^{J} a_{ij}\theta_j^k \text{ for } i = 1 \ldots I \tag{20.22}$$

which gives us the set of projection bin values that we need. This is essentially the same as Eq. 20.2, except that a superscript k is used (q_i^k is used instead of q_i). This reflects the fact that this is a calculation of $\boldsymbol{q}$ based on a current estimate of the underlying $\boldsymbol{\theta}^k$. The data set $\boldsymbol{q}^k$ is often referred to a model of the mean or *expectation* of the data. This model of the mean data, conceptually, would correspond to repeating precisely the same PET scan (with the identical level of radioactivity) an infinite number of times, and then taking the mean of all the acquired data sets.

Step 3: So now we have a prediction of the noise-free data $\boldsymbol{q}^k$ as well as the actual noisy measured data $\boldsymbol{m}$. In general, the two data sets $\boldsymbol{q}^k$ and $\boldsymbol{m}$ will be quite different, primarily due to the fact that our guess (estimate) of $\boldsymbol{\theta}^k$ is not yet accurate. A way of describing the difference is to look at the ratio of the data sets:

$$\frac{m_i}{q_i^k} \text{ for } i = 1 \ldots I \tag{20.23}$$

For a given projection bin i, this ratio will be greater than 1 if our calculated data value q_i^k is smaller than the noisy measured value m_i, and the ratio will be less than 1 if our predicted value is greater than the measured value. This ratio suggests a correction data set:

$$r_i^k = \frac{m_i}{q_i^k} \text{ for } i = 1 \ldots I \tag{20.24}$$

Step 4: We need to use this correction data set (contained in $\boldsymbol{r}^k$) to modify our current estimate $\boldsymbol{\theta}^k$. Since the correction values are for projection bin values, we need to map them to correction values for pixels. Thus, some form of *backward* mapping of the data is required. This is carried out using the system matrix again: we use the *transpose* of the matrix to place the correction values into pixel image space:

$$c_j^k = \sum_{i=1}^{I} a_{ij} r_i^k \text{ for } i = 1 \ldots I \tag{20.25}$$

Note the similarity between Eq. (20.25) and the forward model (20.22). Instead of mapping data from the image pixels to projection bins, we are mapping data from projection bins to image pixels. The forward model (forward mapping) is often called *forward projection*, and the application of the transpose of the system matrix (backward mapping) is known as *backprojection*. (To visualize backprojection, consider again Figure 20.6 and its caption, where the non-zero values in the top right sinogram can be backprojected along the lines through the FOV that they correspond to, giving the lower right image in that figure). It is very important to note that the transpose of the system matrix ($\boldsymbol{A}^{\mathrm{T}}$) is *not*, in general, equal to the inverse of the system matrix ($\boldsymbol{A}^{-1}$, which may not even exist). If the system matrix was orthogonal, then that would be case, but this is never the case for emission tomography imaging systems.

In order to further understand backprojection, we can look at the pixels that are related to a given fixed projection bin: that is, just scan across a row of the system matrix $\boldsymbol{A}$, and see which pixels are to be updated by the value we are backprojecting (see, again Figure 20.7). Such an approach is often what is used in practice, when storage of $\boldsymbol{A}$ is impractical, and on-the-fly calculation of the contents of $\boldsymbol{A}$ is used, such as via ray tracing methods.

So now that we have the backprojected correction ratio values in image pixel space, we can simply multiply our estimate of the radioactivity distribution with this correction image:

$$\theta_j^{k+1} = \theta_j^k c_j^k \text{ for } i = 1 \ldots J \tag{20.26}$$

We have updated the iteration number k to k+1 to reflect that we now have an improved estimate of the parameters representing the radioactivity (which is more in agreement with the measured data). This has now taken us through one iteration of the very popular iterative algorithm, called ML-EM, Eq. 20.20, as arrived at previously starting from the Poisson distribution. From Eq. 20.26, we can substitute for the correction image c_j^k using Eq. 20.25:

$$\theta_j^{k+1} = \theta_j^k \sum_{i=1}^{I} a_{ij} r_i^k \tag{20.27}$$

We can then substitute for the ratio, using Eq. 20.24:

$$\theta_j^{k+1} = \theta_j^k \sum_{i=1}^{I} a_{ij} \frac{m_i}{q_i^k} \tag{20.28}$$

A very necessary normalization term needs to be included as well:

$$\theta_j^{k+1} = \frac{\theta_j^k}{\sum_{i=1}^{I} a_{ij}} \sum_{i=1}^{I} a_{ij} \frac{m_i}{q_i^k} \tag{20.29}$$

where the normalization term is often referred to as the sensitivity image:

$$\sum_{i=1}^{I} a_{ij} = s_j \tag{20.30}$$

which remains a fixed definition for any given system matrix: it just backprojects a set of projection data filled with values of 1. This serves to account for the number of contributions of the correction ratio data set $\{r_i^k\}$ to each pixel, and so we view it as calculating the average correction value per pixel or voxel.

Eq. 20.29 is a standard way of writing the ML-EM image reconstruction algorithm but, for completeness, some papers write it in even more detail by substituting for q_i^k from Eq. 20.22:

$$\theta_j^{k+1} = \frac{\theta_j^k}{\sum_{i=1}^{I} a_{ij}} \sum_{i=1}^{I} a_{ij} \frac{m_i}{\sum_{b=1}^{J} a_{ib}\theta_b^k} \tag{20.31}$$

As previously briefly mentioned, when dealing with full-size PET scanners the system matrix $\boldsymbol{A}$ is impractical to store in memory. Instead, simplified factorized system models are used, using for example ray-tracing algorithms [12] to perform line integrals through the image estimate (matrix $\boldsymbol{X}$ in Eqs. 20.3 and 20.4). In this way, no matrix elements are explicitly stored in memory, but are in effect calculated as required by the reconstruction algorithm – since algorithms such as EM can be run just by accessing individual rows of the matrix $\boldsymbol{A}$ one (or many) at a time. Figure 20.10 illustrates the various components of the ML-EM algorithm in action for the simple case of reconstructing a 2D Shepp–Logan phantom from noise-free data. Figure 20.11 shows example 2D slices from 3D reconstructions when using the EM algorithm for real data, compared to conventional 3D filtered backprojection [13].

20.6 EXPECTATION MAXIMIZATION

The previous derivation of the ML-EM algorithm by a gradient method was informative, but it included a scaling of the gradient in a way that was only justified by retaining non-negativity of all values in the reconstructed image. This section will show how the ML-EM algorithm can be found via a different route, using the general method of expectation maximization [14]. Some of the previously discussed concepts will be repeated, but now in this different context.

20.6.1 One Pixel and One Projection Bin

First, consider just one single projection bin and one single pixel (2D), or voxel (for 3D). We will continue to use voxel and pixel interchangeably, knowing that the following can apply to either 2D or 3D image reconstruction. In anticipation of later using more than one projection bin and more than one pixel (as is normal for image reconstruction), we will refer to the projection bin with a label (or index) i and the pixel with label (or index) j.

Given a *measurement* of m_i counts in the projection bin i during a certain finite scanning time, we want to find the maximum likelihood estimate (MLE) of the parameter θ_j which specifies the mean number of counts emitted from the voxel j during that scanning time.

As always, we will model the number of counts in the projection bin as being a sample from a Poisson distribution. Furthermore, we will assume we know the probability that an emission of a photon-pair from voxel j will be detected in projection bin i (corresponding to a line of response (LOR) i of the PET scanner), we will label the probability a_{ij}.

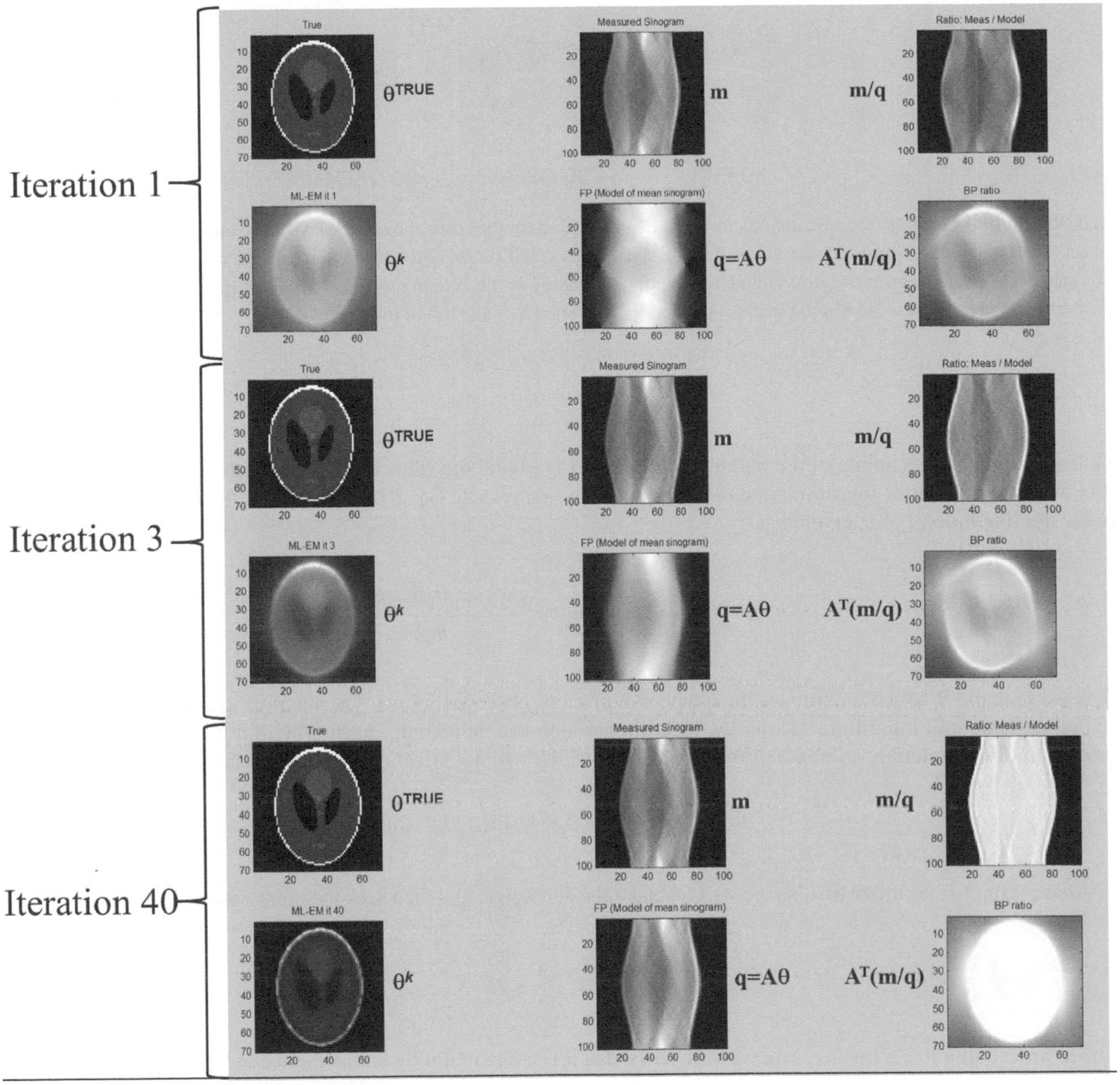

FIGURE 20.10 illustration of the ML-EM algorithm for iterations 1, 3 and 40. An example with noise-free measured data is shown. For each iteration, the true reference image and the fixed measured data are shown on the top row in each case. The bottom left image (row 2) for each iteration shows the current ML-EM iterate (reconstructed image), and its forward projection is shown to its right. On the far right hand side, for each iteration, the ratio of the measured data to the forward projected (modelled) data is shown, as well as its backprojection (transpose of matrix A applied to the element-by-element ratio $\boldsymbol{m} / \boldsymbol{q}$).

So, the probability of detecting m_i counts in projection bin i is:

$$\Pr(m_i \mid q_i) = \frac{\exp(-q_i)(q_i)^{m_i}}{m_i!} \tag{20.32}$$

where q_i is the model of the mean, or the expectation, of the number of counts in the projection bin i, based on a current estimate of θ_j:

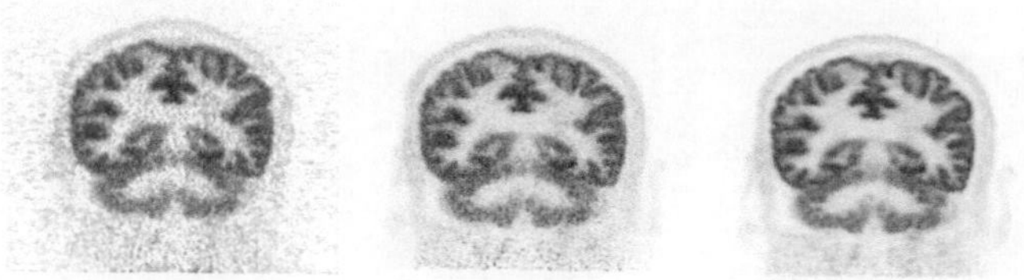

FIGURE 20.11 Example reconstructions from the EM algorithm (in the cases shown, 'OSEM' is given, corresponding to ordered-subsets EM, which is a simple acceleration of ML-EM as only a small subset of the projection data are considered in each update of the image [16]). The rightmost figure includes resolution modelling in the system model (such as that previously given in Eq. 20.4). For comparison purposes, 3D filtered backprojection (FBP) is shown on the left of the figure.

$$q_i = a_{ij}\theta_j \tag{20.33}$$

We can find the ML estimate of the parameter θ_j by seeking to find the value of θ_j, which maximizes the probability of obtaining the number of measured counts m_i. So, we can first rewrite Eq. 20.32 in terms of the desired parameter, and *define* the likelihood l of θ_j given m_i as:

$$l(\theta_j \mid m_i) = \Pr(m_i \mid \theta_j) = \frac{\exp(-a_{ij}\theta_j)\left(a_{ij}\theta_j\right)^{m_i}}{m_i!} \tag{20.34}$$

Now we seek the θ_j which maximizes the likelihood, given m_i observed counts. The maximization is much simpler if we first take the natural logarithm. Taking the logarithm (which is a monotonic function) will not alter the location of the maximum of the function, so it is a safe simplification which will not affect our maximization:

$$\ln l(\theta_j \mid m_i) = -a_{ij}\theta_j + m_i \ln\left(a_{ij}\theta_j\right) - \ln m_i! \tag{20.35}$$

We are seeking to maximize this, so we will first take the derivative, and then seek the point where the derivative is zero:

$$\frac{d}{d\theta_j}\ln l(\theta_j \mid m_i) = -a_{ij} + \frac{m_i}{a_{ij}\theta_j}a_{ij} = -a_{ij} + \frac{m_i}{\theta_j} \tag{20.36}$$

Before proceeding to set this to zero, we will verify that it is a maximum by taking the second derivative:

$$\frac{d^2}{d\theta_j^2}\ln l(\theta_j \mid m_i) = -\frac{m_i}{(\theta_j)^2} \tag{20.37}$$

Since the number of counts m_i is non-negative, we see that the second derivative must be negative for a finite (non-zero) number of counts, which indicates that setting Eq. 20.36 equal to zero will provide the value of θ_j, which gives the maximum of the log likelihood. Therefore, proceeding to set the derivative to zero gives

$$\theta_j^{ML} = \frac{m_i}{a_{ij}} \tag{20.38}$$

which is the ML estimate of the parameter θ_j. The result is intuitive: if, for example, the projection bin only has a 10 per cent chance of detecting an emitted count from the pixel (i.e., if a_{11}=0.1) then we need to scale up the number of measured counts by a factor of 10 (=1/0.1). We also see immediately that whilst m_i is an integer, it is highly unlikely that the estimated parameter θ_j (which is a mean) will be an integer. However, had our desired parameter to find simply been q_i in this case (i.e., if we had wanted to estimate the mean counts detected in the bin, rather than θ_j), then the ML estimate of q_i would be simply the single integer value m_i.

20.6.2 Many Pixels and Many Projection Bins

Next, let us consider the more complicated case, where we have multiple pixels or voxels (j=1…J) and multiple bins (i=1…I). Now we are seeking the most likely mean values in a vector $\boldsymbol{\theta}$, which gave rise to the measured data in vector $\boldsymbol{m}$. As we did in Section 20.5, we define the likelihood as simply the product of the individual likelihoods:

$$L(\boldsymbol{q} \mid \boldsymbol{m}) = \prod_{i=1}^{I} \frac{\exp(-q_i)(q_i)^{m_i}}{m_i!} \tag{20.39}$$

Where now the model of the mean in a given projection bin is the sum of all the contributions from all the pixels to that bin:

$$q_i = \sum_{b=1}^{J} a_{ib}\theta_b \tag{20.40}$$

Again, we wish to find the parameter vector $\boldsymbol{\theta}$ which maximizes the likelihood,

$$L(\boldsymbol{\theta} \mid \boldsymbol{m}) = \prod_{i=1}^{I} \frac{\exp(-\sum_{b=1}^{J} a_{ib}\theta_b)\left(\sum_{b=1}^{J} a_{ib}\theta_b\right)^{m_i}}{m_i!} \tag{20.41}$$

So, we take the natural logarithm, to make things easier again, to obtain the Poisson log likelihood:

$$\ln L(\boldsymbol{\theta} \mid \boldsymbol{m}) = \sum_{i=1}^{I} \ln\left(\frac{\exp(-\sum_{b=1}^{J} a_{ib}\theta_b)\left(\sum_{b=1}^{J} a_{ib}\theta_b\right)^{m_i}}{m_i!}\right) = \sum_{i=1}^{I}\left(-\sum_{b=1}^{J} a_{ib}\theta_b + m_i \ln\left(\sum_{b=1}^{J} a_{ib}\theta_b\right) - \ln m_i!\right) \tag{20.42}$$

which is:

$$\ln L(\boldsymbol{\theta} \mid \mathbf{m}) = \sum_{i=1}^{I}\left(m_i \ln\left(\sum_{b=1}^{j} a_{ib}\theta_b\right) - \sum_{b=1}^{j} a_{ib}\theta_b\right) + const. = \sum_{i=1}^{I}\left(m_i \ln q_i(\boldsymbol{\theta}) - q_i(\boldsymbol{\theta})\right) + const. \tag{20.43}$$

where the right-hand side of (20.43) is the well-known expression for the *Poisson log likelihood*. Then we can differentiate (20.43) with respect to a given parameter, for example, θ_j:

$$\frac{\partial}{\partial\theta_j} \ln L(\boldsymbol{\theta} \mid \boldsymbol{m}) = \sum_{i=1}^{I}\left(-a_{ij} + \frac{m_i}{\sum_{b=1}^{J} a_{ib}\theta_b} a_{ij}\right) \tag{20.44}$$

and set to zero for the ML estimate:

$$\sum_{i=1}^{I} a_{ij} = \sum_{i=1}^{I}\left(\frac{m_i}{\sum_{b=1}^{J} a_{ib}\theta_b^{ML}} a_{ij}\right) \tag{20.45}$$

If we had one voxel only, but still kept more than one projection bin, we would have:

$$\sum_{i=1}^{I} a_{ij} = \sum_{i=1}^{I}\left(\frac{m_i}{a_{ij}\theta_j^{ML}} a_{ij}\right) \tag{20.46}$$

which can be rearranged to find the maximum likelihood estimate as follows:

$$\sum_{i=1}^{I} a_{ij} = \frac{1}{\theta_J^{ML}} \sum_{i=1}^{I} m_i \tag{20.47}$$

then

$$\theta_j^{ML} = \frac{\sum_{i=1}^{I} m_i}{\sum_{i=1}^{I} a_{ij}} \tag{20.48}$$

The above equation is the MLE for one voxel, given two or more projection data bins. Returning back to Eq. 20.45 for multiple voxels, using matrix vector notation, we arrive at the following:

$$\mathbf{A}^T\mathbf{1} = \mathbf{A}^T\left(\frac{\mathbf{m}}{\mathbf{A}\boldsymbol{\theta}^{ML}}\right) \tag{20.49}$$

for which it is not possible to obtain a closed form solution for $\boldsymbol{\theta}^{ML}$.

20.6.3 Complete Data

With no closed form solution for Eq. 20.49, what then can be done?

First, we can imagine a situation where ML estimation becomes simple again, which was the case above, where we had just one pixel or voxel to consider. For that case, the key point was that we knew exactly which voxel was linked to the various projection bin measurements – as of course there was only one voxel! Imagine, therefore, if we had the following data set available to us, $\{z_{ij}\}$, which indicates exactly how many counts from a given voxel j were detected in a given projection bin i. For such a case of more *complete data*, ML estimation would become straightforward again. These Poisson-distributed complete data would relate to our observed Poisson-distributed data $\{m_i\}$ by a simple summation,

$$m_i = \sum_{j=1}^{J} z_{ij} \tag{20.50}$$

With this rich data set, we can also directly find the number of emissions from each voxel, n_j, just by summing all the counts that came from that voxel and were detected in the various projection bins:

$$n_j = \sum_{d=1}^{J} z_{dj} \tag{20.51}$$

Eq. 20.51 just reintegrates the detected counts back to the voxel j from which they came, and delivers a value n_j which is Poisson distributed. (A sum of Poisson distributed values is also Poisson distributed).

With this information, let's now revisit our ML estimation problem. First, revisiting again the case of just one voxel and just one projection bin we have:

$$l(\theta_j \mid z_{ij}) = \Pr(z_{ij} \mid \theta_j) = \frac{\exp(-a_{ij}\theta_j)(a_{ij}\theta_j)^{z_{ij}}}{z_{ij}!} \tag{20.52}$$

where we have modelled the mean, or expectation, of the number of counts in the projection bin i coming from the voxel j as $a_{ij}\boldsymbol{\theta}_j$:

$$\bar{z}_{ij} = a_{ij}\theta_j \tag{20.53}$$

We are seeking to maximize the likelihood of θ_j given z_{ij}, so we take the log and the derivative to seek the point where the derivative is zero:

$$\frac{d}{d\theta_j}\ln l(\theta_j \mid z_{ij}) = -a_{ij} + \frac{z_{ij}}{a_{ij}\theta_j}a_{ij} = -a_{ij} + \frac{z_{ij}}{\theta_j} \tag{20.54}$$

Setting this derivative to zero reveals the ML estimate to be

$$\theta_j^{ML} = \frac{z_{ij}}{a_{ij}} \tag{20.55}$$

which compares in an obvious way to the first case we considered, Eq. 20.38, as in fact the complete data are the same as the observed data for the case of having just one voxel and one projection bin.

Now then, for many projection bins and many voxels we can write the *complete-data Poisson likelihood* (we will reuse our label l, this time for the complete data likelihood):

$$l(\boldsymbol{\theta} \mid \mathbf{z}) = \prod_i \prod_b \frac{\exp(-a_{ib}\theta_b)(a_{ib}\theta_b)^{z_{ib}}}{z_{ib}!} \tag{20.56}$$

where we consider the voxel index b to range from 1 to J. (We use b to cover this range, as in the following we will be interested in considering one particular voxel, indexed by j, which is just one selection from this range of voxel indices.)

We then find the complete-data Poisson log likelihood:

$$\ln l(\boldsymbol{\theta} \mid \mathbf{z}) = \sum_i \sum_b \left(-a_{ib}\theta_b + z_{ib}\ln(a_{ib}\theta_b) - \ln z_{ib}!\right) \tag{20.57}$$

We can maximize by first partially differentiating with respect to one single parameter of interest θ_j (for one given voxel j):

$$\frac{\partial}{\partial\theta_j}\ln l(\boldsymbol{\theta} \mid \mathbf{z}) = \sum_i \left(-a_{ij} + \frac{z_{ij}}{a_{ij}\theta_j}a_{ij}\right) \tag{20.58}$$

and then setting the result to zero:

$$\sum_i a_{ij} = \sum_i \left(\frac{z_{ij}}{\theta_j^{ML}} \right) = \frac{1}{\theta_j^{ML}} \sum_i z_{ij} \tag{20.59}$$

Hence, the ML estimate can be found in a closed form expression (thanks to the use of z):

$$\theta_j^{ML} = \frac{\sum_i z_{ij}}{\sum_i a_{ij}} = \frac{n_j}{\sum_i a_{ij}} \tag{20.60}$$

where we recognize the numerator as being the number of counts emitted from voxel j (see 20.51), and the denominator is the *sensitivity* at that voxel (i.e., indicating the number of projection bins that can potentially receive contributions from that voxel, where each bin i in this sum is, of course, weighted by the probability that it will detect a photon from voxel j, hence, using a_{ij}, the probability of an emission from j being detected in bin i).

So Eq. 20.60 gives us the ML estimate of θ_j that we are seeking, and all values can be found in that manner to give $\boldsymbol{\theta}^{ML}$ (all the ML estimates of the mean emissions from each voxel – that is, the reconstructed image!).

There is a catch, of course: we are not in possession of the complete data z! That is the next problem to resolve.

20.6.4 Expectation of the Complete Data

First, we assume that we *do* know $\boldsymbol{\theta}$. Now we do actually know $\boldsymbol{m}$ (which we will now call the incomplete data, for obvious reasons, as it does not contain as much information as our desired data set z). With presumed knowledge of $\boldsymbol{\theta}$ (but in fact it is only an estimate, so we will label it $\boldsymbol{\theta}^k$) and with knowledge of $\boldsymbol{m}$, we can generate a complete data set z which would be consistent with $\boldsymbol{\theta}^k$ and $\boldsymbol{m}$ according to:

$$\frac{z_{ij}}{m_i} = \frac{a_{ij}\theta_j^k}{\sum_{d=1}^{J} a_{id}\theta_d^k} \tag{20.61}$$

Eq. 20.61 is simply an equality of fractions. It is presented here intuitively and requires that the fraction of the incomplete data for a projection, bin i, m_i, which arose from a given voxel j, (i.e., the fraction z_{ij} / m_i), be equal to the corresponding fraction of their means (their modelled means based on the current $\boldsymbol{\theta}^k$). Figure 20.12 illustrates this. This fractionation approach actually dates back to the late 1950s, when it was referred to as '*proportional allocation*' in the context of maximum likelihood estimation (see [15]), predating the EM formalism which was later introduced, in the late 1970s ([14]).

Therefore, each element of the complete data is given by

$$z_{ij}\left(\boldsymbol{\theta}^k, m_i\right) = \frac{a_{ij}\theta_j^k}{\sum_{d=1}^{J} a_{id}\theta_d^k} m_i \tag{20.62}$$

which can be shown to be the *conditional expectation of the complete data* given $\boldsymbol{m}$ and $\boldsymbol{\theta}^k$. For this reason, generation of the complete data by (20.62) is referred to as the expectation step (the expected complete data). Since the fixed current estimate $\boldsymbol{\theta}^k$ and the measured data are both constants, likewise then the generated complete data z can be regarded as a constant data set.

With this expression for the complete data, we can now use this in our complete-data Poisson log likelihood, which would therefore be called either 'the log likelihood of the conditional expectation of the complete data' or equivalently the 'conditional expectation of the log likelihood of the complete data':

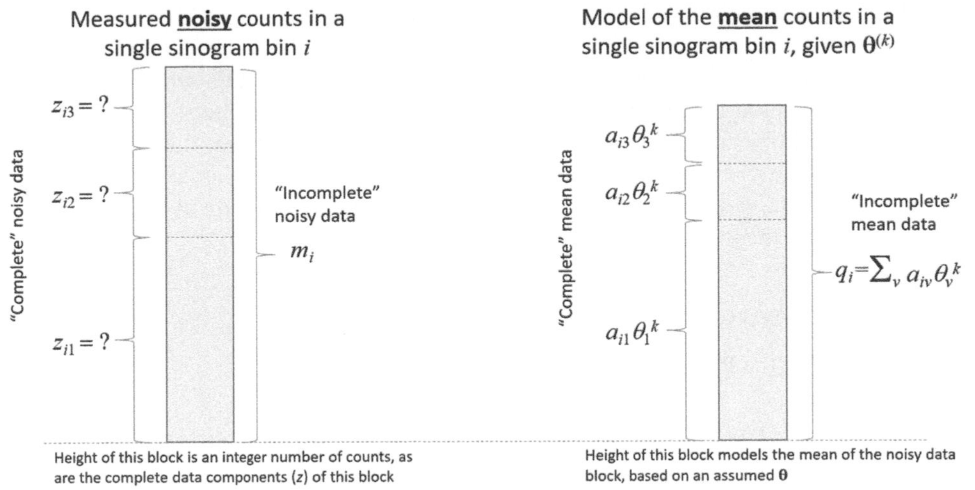

FIGURE 20.12 Illustrative explanation of how to obtain a complete data set z (for the simple case of an image containing just 3 voxels or pixels), given an example measured number of counts m for a bin i, and conditional on a current estimate of the parameters, θ^k. The complete data decomposes the measured incomplete data into its constituent parts, indicating which voxel v gave rise to each count in bin i. By examining the fractionation of the modelled mean counts, q_i, one can apply the very same fractionation to the actual measured counts m to obtain a data set z. Based on a from a figure from Reader and colleague [2].

$$\ln l\left(\boldsymbol{\theta} \mid z\left(\boldsymbol{\theta}^k, \boldsymbol{m}\right)\right) = \sum_i \sum_b \left(-a_{ib}\theta_b + z_{ib}\left(\boldsymbol{\theta}^k, m_i\right)\ln\left(a_{ib}\theta_b\right) - \ln z_{ij}\left(\boldsymbol{\theta}^k, m_i\right)!\right) \tag{20.63}$$

So, following what we had before (see Eqs. 20.57 to 20.60), after finding the conditional *expectation* of the complete data we proceed to the *maximization* of this expected log likelihood by partial differentiation of Eq. (20.63) with respect to a single choice θ_j and setting equal to zero to solve for θ_j:

$$\theta_j^{EM} = \frac{\sum_i z_{ij}\left(\boldsymbol{\theta}^k, m_i\right)}{\sum_i a_{ij}} = \frac{\sum_i \frac{a_{ij}\theta_j^k}{\sum_{d=1}^{J} a_{id}\theta_d^k} m_i}{\sum_i a_{ij}} \tag{20.64}$$

which delivers a parameter estimate (image estimate) that we label as EM (expectation maximization), as it is the *maximum* based on the current conditional *expectation* of the complete data, or more simply, Eq. 20.64 is:

$$\theta_j^{EM} = \frac{\theta_j^k}{\sum_i a_{ij}} \sum_i^I \frac{a_{ij}}{\sum_{d=1}^{J} a_{id}\theta_d^k} m_i \tag{20.65}$$

This now provides a new estimate of $\boldsymbol{\theta}$ (by considering every voxel), from which we can generate again a set of complete data (i.e., return to the expectation step – find the conditional expectation of the complete data given $\boldsymbol{m}$ and now the new estimate of $\boldsymbol{\theta}$). Then we can maximize the complete data log likelihood again, and so on.

This two-step procedure, which is entirely captured within the closed-form update step in Eq. 20.65, is the well-known EM-ML algorithm, or ML-EM, which we saw in the previous Section 20.5:

$$\theta_j^{k+1} = \frac{\theta_j^k}{\sum_i a_{ij}} \sum_i^I \frac{a_{ij}}{\sum_{d=1}^{J} a_{id}\theta_d^k} m_i \tag{20.66}$$

The use of the conditional expectation of the complete-data Poisson log likelihood, instead of the incomplete data Poisson log likelihood, as found by this EM approach, is an example of the more general concept of *optimization transfer*. With optimization transfer one seeks to find a surrogate objective function which is easier to maximize (or minimize) than the actually desired objective function. Maximization of the surrogate will result in closer proximity to the maximum of the desired objective, from which point a new surrogate objective function can be found and easily maximized, and so on. This permits arrival at maximization of the more difficult objective by decomposing the problem into a series of simpler optimizations. Hence, in the above, the conditional expectation of the complete-data Poisson log likelihood is a surrogate for the incomplete-data Poisson log likelihood.

20.6.5 Kullback–Leibler Divergence

Before going further, recall that the Poisson log likelihood was given by:

$$\ln L(\boldsymbol{\theta} \mid \boldsymbol{m}) = \sum_{i=1}^{I} \left(m_i \ln\left(\sum_{b=1}^{J} a_{ib}\theta_b \right) - \sum_{b=1}^{J} a_{ib}\theta_b \right) + Const. = \sum_{i=1}^{I} \left(m_i \ln q_i \boldsymbol{\theta} - q_i(\boldsymbol{\theta}) \right) + const. \quad (20.67)$$

It is very useful to note that the Poisson log likelihood is related to a discrepancy measure known as the *Kullback–Leibler (KL) divergence* between two vectors $\boldsymbol{m}$ and $\boldsymbol{q}$, *defined* by:

$$D^{KL}\left(\boldsymbol{q}(\boldsymbol{\theta}), \boldsymbol{m}\right) = \sum_{i=1}^{I} \left(m_i \ln \frac{m_i}{q_i(\boldsymbol{\theta})} + q_i(\boldsymbol{\theta}) - m_i \right) = -\ln L(\boldsymbol{\theta} \mid \boldsymbol{m}) + const. \quad (20.68)$$

Therefore, increasing the Poisson log likelihood (20.67) is equivalent to decreasing the KL divergence (20.68).

The ML-EM update algorithm of (20.66) gives a series of estimates of $\boldsymbol{\theta}$, which monotonically increase the Poisson log likelihood or, in other words, monotonically decrease the KL divergence.

20.6.6 Ordered Subsets EM

In general, the ML-EM algorithm can be slow. This is partly because it uses all of the measured data set $\{m_i\}$ in each and every iteration. The method can be accelerated by using just part of the data set in each iteration. This approach is called ordered subsets expectation maximization (OSEM) [16]:

$$\theta_j^{k,l+1} = \frac{\theta_j^{k,l}}{\sum_{i \in S_l} a_{ij}} \sum_{i \in S_l} a_{ij} \frac{m_i}{q_i^{k,l}} \quad (20.69)$$

where just one of the subsets S_l of the measured data $\boldsymbol{m}$ are used in a given update of $\boldsymbol{\theta}$. In machine-learning terminology, this is analogous to using 'mini batches', whereby a neural network's parameters are updated for each mini batch (subset) of the total training data. Typically, 10–30 subsets of data are used, and the subsets are chosen so as to select only a limited number of azimuthal angles in the sinogram data for each update. Hence, the algorithm performs approximately only a fraction of the amount of computation to obtain an image update (giving a speed-up of a factor approximately equal to the number of subsets). The method works well in early iterations, in effect carrying out ML-EM with just a subset of the data for each update. But as the algorithm progresses, the discrepancies between the different noisy subsets of data can lead to a limit cycle, where the algorithm seeks agreement of $\boldsymbol{\theta}$ with a series of inconsistent subsets of data.

20.7 MAXIMUM *A POSTERIORI*

To counteract noisy reconstructed images, we can use a maximum *a posteriori* (MAP) objective function. This allows us to model the probability of obtaining certain types of images so as to assign lower probabilities to images with greater noise levels. Often some energy function, U, is used to measure the noise level of a given image, and a Gibbs prior is used.

An example would be to consider a neighbourhood of pixels around a given pixel and to look at the square differences between the centre pixel and its neighbours. A uniform image, presumed to have low noise, would have a low sum of square differences (even zero for a perfectly flat and uniform region). However, a noisy image will have a large sum of square differences. So, a simple 'energy' or noise measure could be:

$$U(\boldsymbol{\theta}) = \sum_{c=1}^{J} \sum_{j \in N} w_{cj} \left(\theta_c - \theta_j\right)^2 \tag{20.70}$$

where U gives the energy (e.g. the noise level) for the image $\boldsymbol{\theta}$, by considering each and every pixel $c=1\ldots J$ in the image, and for each pixel looking at its neighbouring N pixels (e.g. a 3×3 neighbourhood) and adding up a sum of square differences with a weighting factor w_{cj} (e.g. inversely related to the distance of a pixel θ_j from the centre pixel θ_c under consideration). Importantly, the weighting factors can also be related to edge information from a guidance image (e.g. an MR image), by considering similarity of pixel values in a patch of the guidance image (similar to what was shown in Figure 20.2 for the kernel method).

We can then consider the probability of a given image $\boldsymbol{\theta}$ to be given by:

$$\Pr(\boldsymbol{\theta}) = \frac{1}{Z}\exp\left[-\beta U(\boldsymbol{\theta})\right] \tag{20.71}$$

where in this context the capital Z is just a normalizing constant to ensure the probability is between 0 and 1. Eq. 20.71 indicates that an image with a high value of U (high energy or high noise) will be assigned a small probability, and low values of U will be assigned higher probabilities. We can now make use of Bayes' law, which gives the so-called posterior probability:

$$\Pr(\boldsymbol{\theta} \mid \boldsymbol{m}) = \frac{\Pr(\boldsymbol{m} \mid \boldsymbol{\theta})\Pr(\boldsymbol{\theta})}{\Pr(\boldsymbol{m})} \tag{20.72}$$

where $\Pr(\boldsymbol{m}|\boldsymbol{\theta})$ is equal to the Poisson log likelihood. So we have the MAP objective function:

$$O_{MAP}(\boldsymbol{\theta} \mid \boldsymbol{m}) = \frac{1}{Z}\exp\left[-\beta U(\boldsymbol{\theta})\right]\left(\prod_{i=1}^{I} \frac{(q_i)^{m_i} \exp\left[-q_i\right]}{m_i!}\right) \tag{20.73}$$

Taking the natural logarithm:

$$\ln O_{MAP}(\boldsymbol{\theta} \mid \boldsymbol{m}) = \sum_{i=1}^{I} \left(\ln q_i^{m_i} - q_i - \ln m_i!\right) - \beta U(\boldsymbol{\theta}) - \ln Z \tag{20.74}$$

Further simplifying and removal of constants that have no impact on the subsequent maximization gives the *log posterior*, which needs to be maximized:

$$L_{MAP}(\boldsymbol{\theta} \mid \boldsymbol{m}) = \sum_{i=1}^{I} \left(m_i \ln q_i - q_i\right) - \beta U(\boldsymbol{\theta}) \tag{20.75}$$

Note the first term resulting from the Poisson log likelihood is maximized if $\boldsymbol{m}=\boldsymbol{q}$ and, as mentioned, this corresponds to the negative of an unnormalized version of the Kullback Leibler (KL) measure of difference between two vectors. We will follow the EM method to derive an algorithm to find a $\boldsymbol{\theta}$ that maximizes this log posterior.

We can now write out the complete-data MAP objective function – based on the complete data Poisson likelihood, and the Gibbs prior probability term. Compare with the previous Eq. (20.56) for reference:

$$O_{CD-MAP}(\boldsymbol{\theta} \mid \boldsymbol{m}) = \prod_{b=1}^{J}\prod_{i=1}^{I} \frac{(a_{ib}\theta_b)^{z_{ib}} \exp[-a_{ib}\theta_b]}{z_{ib}!} \exp[-\beta U(\boldsymbol{\theta})] \tag{20.76}$$

Ignoring constants (which will not impact on the maximisation task), the complete *log posterior* is:

$$O_{CD-MAP}(\boldsymbol{\theta} \mid \boldsymbol{m}) = \sum_b \sum_i (z_{ib} \ln(a_{ib}\theta_b) - a_{ib}\theta_b) - \beta U(\boldsymbol{\theta}) \tag{20.77}$$

At this point it should be noted that Eq. 20.77, when β=0, is very easy to solve in one single step (to give the ML estimate), which was the reason for using complete data. As before, we can find an estimate of the complete data based on $\boldsymbol{\theta}^k$ and $\boldsymbol{m}$ by:

$$z_{ij}(\boldsymbol{\theta}^k, \boldsymbol{m}) = a_{ij}\theta_j^k \frac{m_i}{\sum_{d=1}^{J} a_{id}\theta_d^k} \tag{20.78}$$

Returning to Eq. 20.77, which needs to be maximized, we can rewrite it as follows, using the plug-in value for the complete data:

$$O_{CD-MAP}(\boldsymbol{\theta} \mid \boldsymbol{m}) = \sum_b \left(\sum_i z_{ib}(\boldsymbol{\theta}^k, \boldsymbol{m}) \ln(a_{ib}\theta_b) - \theta_b \sum_i a_{ib} \right) - \beta U(\boldsymbol{\theta}) \tag{20.79}$$

By a very similar working to what was carried out before, this gives

$$O_{CD-MAP}(\boldsymbol{\theta} \mid \boldsymbol{m}) = \sum_b \left(\sum_i z_{ib}(\boldsymbol{\theta}^k, \boldsymbol{m}) \ln(\theta_b) + \sum_i z_{ib}(\boldsymbol{\theta}^k, \mathrm{m}) \ln(a_{ib}) - \theta_b \sum_i a_{ib} \right) - \beta U(\boldsymbol{\theta}) \tag{20.80}$$

Recognizing that the second term in the brackets is a constant that can be removed from consideration in our maximization, and factoring out the image, gives:

$$O_{CD-MAP}(\boldsymbol{\theta} \mid \boldsymbol{m}) = \sum_b \sum_i a_{ib} \left(\frac{\sum_i z_{ib}(\boldsymbol{\theta}^k, \boldsymbol{m})}{\sum_i a_{ib}} \ln(\theta_b) - \theta_b \right) - \beta U(\boldsymbol{\theta}) + \text{const.} \tag{20.81}$$

We will now simplify again by labelling the sensitivity image as s, and by identifying the EM-update image:

$$O_{CD-MAP}(\boldsymbol{\theta} \mid \boldsymbol{m}) = \sum_b s_b \left(\theta_b^{EM} \ln(\theta_b) - \theta_b \right) - \beta U(\boldsymbol{\theta}) \tag{20.82}$$

Eq. 20.82 is an important result: it shows that to perform MAP estimation an optimization in image space is all that is needed. This extension of the EM algorithm to accommodate MAP estimation can be referred to as generalized EM (GEM), for example [17]. Using the same definition of the complete data, the expectation step is the same as has been shown. However, the maximization step now has to maximize Eq.(20.82), that is, maximize the expectation of the complete-data log posterior. There is no tomographic aspect in the equation (the system matrix is not used). Eq. 20.82 can be compared to Eq. 20.75, where it can be seen that it has the very same form, this time just using the EM update image instead of the measured data $\boldsymbol{m}$, and the image values $\boldsymbol{\theta}$ instead of $\boldsymbol{q}$.

Eq. 20.82 needs to be maximized with respect to $\boldsymbol{\theta}$, and a good approach is to take the partial derivative with respect to a pixel value θ_j:

$$\frac{\partial}{\partial \theta_j} O_{CD-MAP}(\boldsymbol{\theta} \mid \boldsymbol{m}) = s_j \left(\frac{\theta_j^{EM}}{\theta_j} - 1 \right) - \beta \frac{\partial}{\partial \theta_j} U(\boldsymbol{\theta}) \tag{20.83}$$

where we note that the partial derivative of $U(\theta)$ is just a function of the pixel value θ_j, and so we can abbreviate it as follows:

$$\frac{\partial}{\partial \theta_j} U(\boldsymbol{\theta}) = U'(\boldsymbol{\theta}_j) \tag{20.84}$$

To maximize we set (20.83) to zero, to deliver the improved estimate $\boldsymbol{\theta}^{k+1}$:

$$0 = \frac{\theta_j^{EM}}{\theta_j^{k+1}} s_j - s_j - \beta U'(\theta_j^{k+1}) \tag{20.85}$$

We now consider four different cases of solving (20.85) for θ_j^{k+1}, to obtain the update of the MAP estimate (based only on the current conditional expectation of the complete data, which is already implicitly included in $\boldsymbol{\theta}^{EM}$). It is important to note that solving (20.85) for θ_j^{k+1} is not necessarily straightforward, as the energy derivative U' is a function of θ_j^{k+1}.

Case 1: No prior (β=0). In this case the value of θ_j that maximizes (20.85) is easy to find, remembering that it is just an update (so we label it k+1) as we did not actually have the complete data z, but only a current plug-in value based on $\boldsymbol{\theta}^k$ and $\boldsymbol{m}$. So, the update is:

$$\theta_j^{k+1} = \theta_j^{EM} \tag{20.86}$$

which, as expected, is just the EM update, ready for use in finding the next conditional expectation of the complete data.

Case 2: Use of a prior image. In this approach, the prior depends only on the local pixel value and not on any neighbours of a pixel. This is for the case where we might have some prior image $\boldsymbol{p}$, from which each pixel value should not differ too much. Often a quadratic (Gaussian) prior is used, as was first proposed by Levitan and Herman in 1987 [18]:

$$U(\boldsymbol{\theta}) = \frac{1}{2} \sum_{b=1}^{J} (\theta_b - p_b)^2 \tag{20.87}$$

Then, the partial derivative with respect to a specific pixel value θ_j is:

$$\frac{\partial}{\partial \theta_j} U(\boldsymbol{\theta}) = U'(\theta_j) = (\theta_j - p_j) \tag{20.88}$$

Substituting into the master Eq. 20.85:

$$0 = \frac{\theta_j^{EM}}{\theta_j^{k+1}} s_j - s_j - \left(\theta_j^{k+1} - p_j\right)\beta \qquad (20.89)$$

Rearranging into an explicit standard form quadratic in terms of the update θ_j^{k+1} , which we wish to solve for:

$$0 = \theta_j^{EM} s_j - \theta_j^{k+1} s_j - \left(\theta_j^{k+1} - p_j\right)\beta\theta_j^{k+1} \qquad (20.90)$$

$$0 = \theta_j^{EM} s_j + \left(\beta p_j - s_j\right)\theta_j^{k+1} - \beta\left(\theta_j^{k+1}\right)^2 \qquad (20.91)$$

This quadratic in θ_j^{k+1} has a solution (when using the less familiar Muller's method for the solution to a quadratic):

$$\theta_j^{k+1} = \frac{2 s_j \theta_j^{EM}}{\left(s_j - \beta p_j\right) + \sqrt{\left(s_j - \beta p_j\right)^2 - 4\beta\theta_j^{EM} s_j}} \qquad (20.92)$$

As will be covered further below, it turns out that embedding deep-learning methods into image reconstruction (e.g. for PET) makes use of this approach for including the prior information, whereby $\boldsymbol{p}$ has been obtained by a deep-learned denoised version of a previous iterate.

Case 3: To avoid the complication of the energy derivative U' depending on the very θ_j we are seeking – and so yielding a quadratic or higher-order polynomial for (20.85) – we can instead use the currently available image, and so approximate $U'(\theta_j^{k+1})$ by $U'(\theta_j^k)$ to obtain a linear equation:

$$0 = \frac{\theta_j^{EM}}{\theta_j^{k+1}} s_j - s_j - \beta U'\left(\theta_j^k\right) \qquad (20.93)$$

Thus, we can easily solve for θ_j^{k+1} to find the update:

$$\theta_j^{k+1} = \frac{s_j \theta_j^{EM}}{s_j + \beta U'\left(\theta_j^k\right)} \qquad (20.94)$$

which is the one-step late (OSL) update of Green 1990 [19]. Clearly it is 'one step late' as the derivative of the energy should be a function of the new estimate rather than the current one. A previously popular variation is to use the median root prior (MRP [20]) for the term U', and there has been a trend towards designing this derivative term to have desirable properties without necessarily being concerned about what the original function U would correspond to.

Case 4: Use a prior that depends on its neighbours. Returning to the case of a prior like (20.70), a scaled version:

$$U(\boldsymbol{\theta}) = \frac{1}{2}\sum_{c=1}^{J} \frac{1}{2}\sum_{b \in N_c} w_{cb}\left(\theta_c - \theta_b\right)^2 \qquad (20.95)$$

means we can use a substitution method for the optimization, a 'surrogate' (i.e. optimization transfer, as was done for the incomplete-data log likelihood). For the incomplete-data log likelihood we used the expectation of the complete data log likelihood as a surrogate. It is possible to devise a surrogate function for (20.95), as proposed by De Pierro in 1995 [21]:

$$U\left(\boldsymbol{\theta};\boldsymbol{\theta}^{k}\right)=\frac{1}{2}\sum_{c=1}^{J}\frac{1}{2}\sum_{b\in N_c}w_{cb}\left(2\theta_c-\theta_c^k-\theta_b^k\right)^2 \tag{20.96}$$

The genius of this surrogate approach is that we effectively now have a comparison of our image estimate $\boldsymbol{\theta}$ with a fixed prior image, $\boldsymbol{\theta}^k$, similar to case 2 above, but now defined over a neighbourhood. This permits a simple derivation of the update, comparable to what was carried out above in case 2. The outcome, unsurprisingly, is a similar expression to that obtained for case 2:

$$\theta_j^{k+1}=\frac{2\theta_j^{EM}}{\left(1-\beta v_j\theta_j^{SM}\right)+\sqrt{\left(1-\beta v_j\theta_j^{SM}\right)^2+4\beta v_j\theta_j^{EM}}} \tag{20.97}$$

The key difference, compared to case 2, is that we do not supply a prior image, but instead one that is derived from the previous/current estimate, as follows. First, we have a normalization image

$$v_j=\frac{\sum_{l=1}^{J}w_{jl}}{s_j} \tag{20.98}$$

and then we also have a smoothed version of the previous update:

$$\theta_j^{SM}=\frac{1}{2\sum_{l=1}^{J}w_{jl}}\sum_{l=1}^{J}w_{jl}\left(\theta_j^k+\theta_l^k\right) \tag{20.99}$$

This approach permits a solution in a very similar fashion to case 2 above, but importantly now, for the more general case of using Eq. 20.95 as the penalty function.

State-of-the-art MAP-EM methods (e.g. [22] [23]) use reconstruction algorithms such as the ones described in this section, in particular using MR images to guide the regularizing penalty, as demonstrated in Figure 20.13. Guided regularization can be achieved by using weights $\{w_{ij}\}$ in the quadratic penalty Eq.(20.95) that are related to the degree of similarity between pixels in a registered MR image (again, in a similar fashion to what was shown in Figure 20.2 for the kernel method).

20.8 CURRENT RESEARCH DIRECTIONS

Having laid down the fundamentals of iterative image reconstruction for emission tomography, this section now considers current research directions for reconstruction.

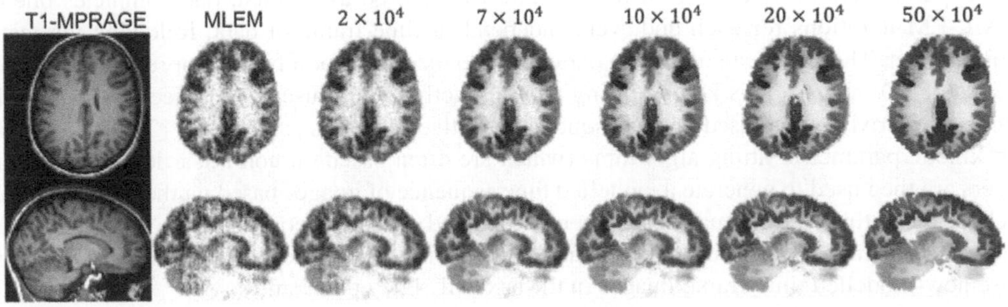

FIGURE 20.13 The impact of regularization parameter, β, on the quality of MAP-EM reconstructions of [^{18}F]FDG PET data using a penalty function which is guided by the MR, and in the case of this figure, also guided by the PET image during reconstruction. Impact of β is significant, and it is a topic of research (e.g. [35]). Figure is from Mehranian and colleagues [22].

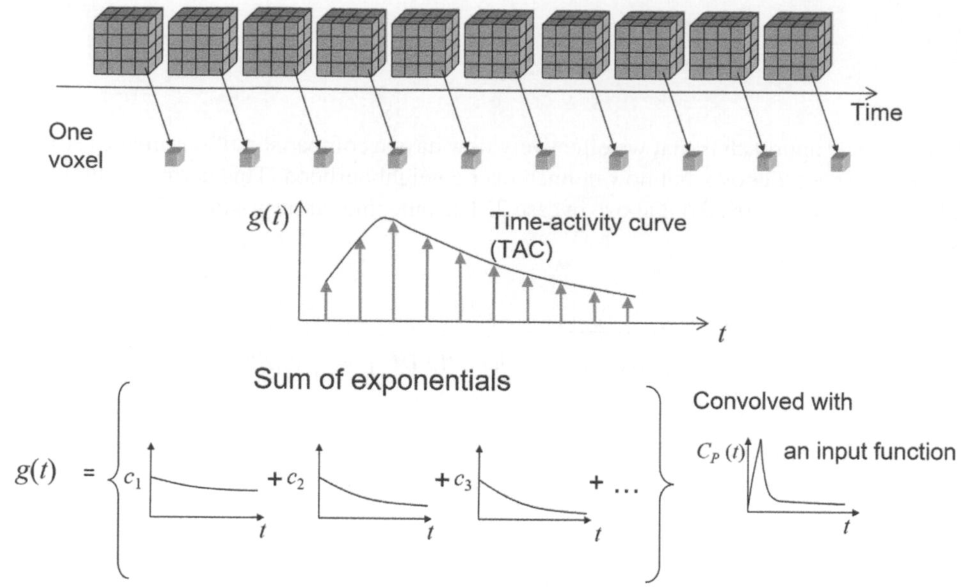

FIGURE 20.14 In 4D reconstruction, with a linear model, each time-activity curve in each voxel or region of the image is modelled as a summation of temporal basis functions. The reconstruction then estimates the coefficients for these basis functions for each voxel, which then provides a TAC model for each voxel. In the example shown here, there are a series of coefficients to estimate (c_1, c_2, c_3...) for spectral analysis basis functions, whereby each exponential basis function is convolved with the arterial input function.

20.8.1 4D Image Reconstruction

4D image reconstruction uses an explicit spatiotemporal model instead of Eq. 20.1. The advantage of a 4D reconstruction is that in principle every single 3D image time frame can benefit from all of the dynamically acquired data (which is needed when short time frames are used in conventional dynamic imaging, resulting in very low counts and noisy images). For linear models of the temporal variation in the radiotracer uptake – time-activity curves (TACs) with linear parameters – one can simply extend the system model used in ML-EM to include the temporal model as well (see [2]). An example linear temporal model is shown in Figure 20.14, where in that example we are estimating coefficients for spectral analysis basis functions, for each and every voxel in a 3D image.

For a case like Figure 20.14, with a summation of basis functions explaining a TAC, or related models like Patlak (which corresponds to modelling a TAC with just two basis functions), the model of the mean data used in EM algorithms would correspond to that shown in Figure 20.15, followed by standard modelling of the data previously covered, for each time frame.

Then the standard ML-EM can be used, with the transpose of the 4D system model. However, this can be very memory-demanding, and also very slow to converge [24], and so nested EM methods [25] have been proposed that accelerate the reconstruction by separating out the tomographic part of the reconstruction (which needs the sinogram/projection data) from the dynamic modelling (which can be done with the time-series of images).

For non-linear parameters of TACs, a nested procedure can also be used. First, one completes one or more conventional 3D ML-EM iterations for each and every independent time frame of data, following the methods already described in this chapter. Then, one can fit kinetic parameters to the TACs for all voxels or regions of interest as desired in this time series of 3D images. This kinetic fitting would strictly require use of an objective function involving the KL divergence, but approximation based on least squares objective functions can be used [26] [27], greatly facilitating use of existing kinetic parametric fitting algorithms (which are often based on non-linear least squares methods). The fitting parameters are then used to generate a modelled time sequence of images based on the current kinetic parametric estimates. Each of these time-frame images can then be updated using the conventional 3D ML-EM algorithm (i.e. using the tomographic data). The method then repeats by fitting the TACs in each voxel and/or region as before, and then using these new-modelled time-frame images in the next ML-EM update and so on.

Representative results from 4D reconstruction methods are shown in Figure 20.16. There have been a number of reviews on 4D image reconstruction [28] [29] [2] that expand considerably on the outline given here.

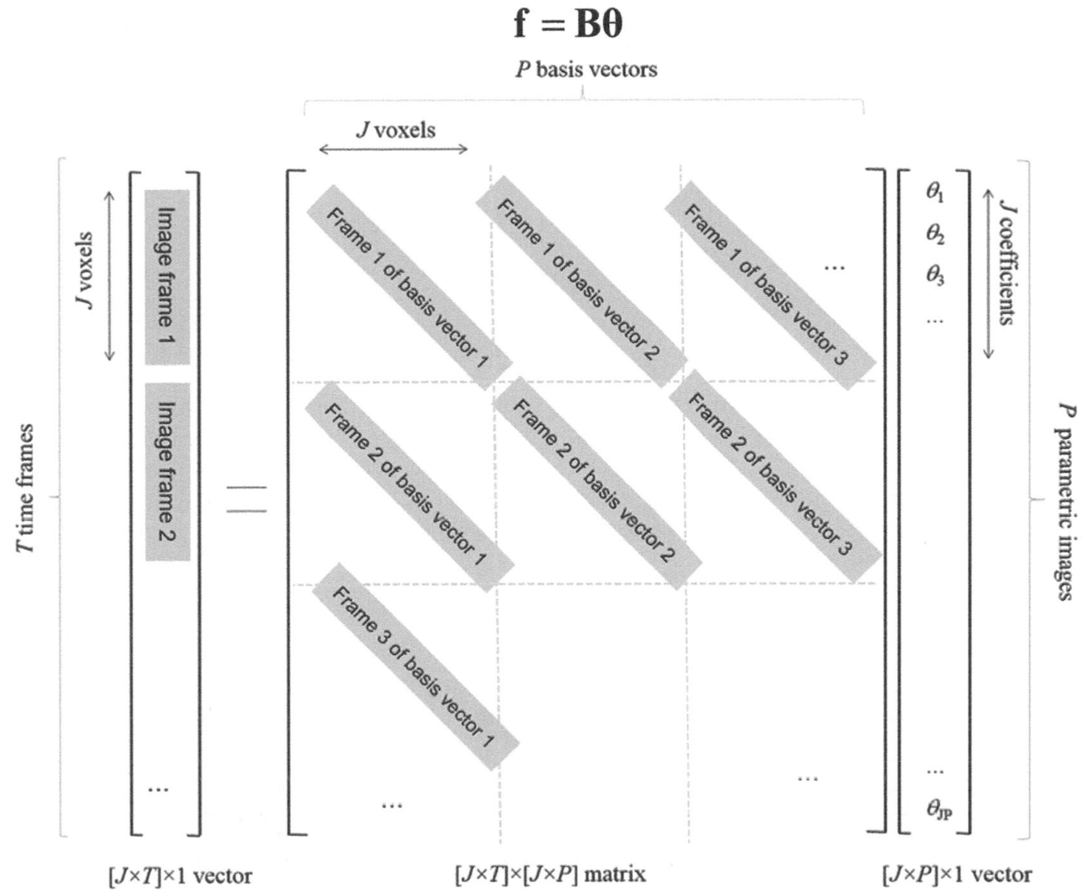

FIGURE 20.15 Example of how part of the system model would be modified for 4D reconstruction, where the parameter vector$\boldsymbol{\theta}$ now contains all the coefficients for all the basis functions for all the voxels (and is therefore a set of P parametric images). A time-series of T images, stored in $\boldsymbol{f}$, is generated from this matrix-vector multiplication, ready for the next stage of system modelling, such as that demonstrated earlier in this chapter for conventional 3D imaging. In the above, the large matrix $\boldsymbol{B}$ contains all the basis functions, or basis vectors, in a sparse matrix, which of course would not be stored in this way in practise but is shown here to demonstrate the ordering of data for 4D linear modelling of a time-series of 3D images.

20.8.2 AI/Machine Learning Within Image Reconstruction

Artificial intelligence (AI), in particular deep-learning, is having a significant impact on the field of iterative image reconstruction in emission tomography [30]. Whilst deep learning can in principle provide entire reconstruction mappings from measured data to reconstructed images (e.g. [31], [32]), one of the main aspects of reconstruction that deep learning can more practically help with is the choice of regularization prior for the MAP methods. In the earlier sections of this chapter, it was clear that simple analytic functions – for example, a quadratic like Eq.(20.70) – were used to penalize images that did not meet our prior beliefs. In contrast, deep learning can model the probability density function – $\Pr(\boldsymbol{\theta})$ in Eq.(20.71) – based on training data, and thereby offer superior regularization compared to the mathematically convenient methods previously presented.

The core idea in direct machine-learning reconstruction methods is to use training data composed of noisy data sets paired with a high quality reference data sets (found by simulation or by acquiring high-count scan data). The principle is then to find a mapping, a deep network, that can map any given noisy data set (or noisy current image estimate) to its respective high-quality reference. Figure 20.17 shows the principle. Having trained one or more such mappings, these mappings can then be used for newly acquired emission tomography data, with the expectation that the mappings learned should generalize well to new data, delivering training-data derived optimized denoised reconstructions and resolution recovery.

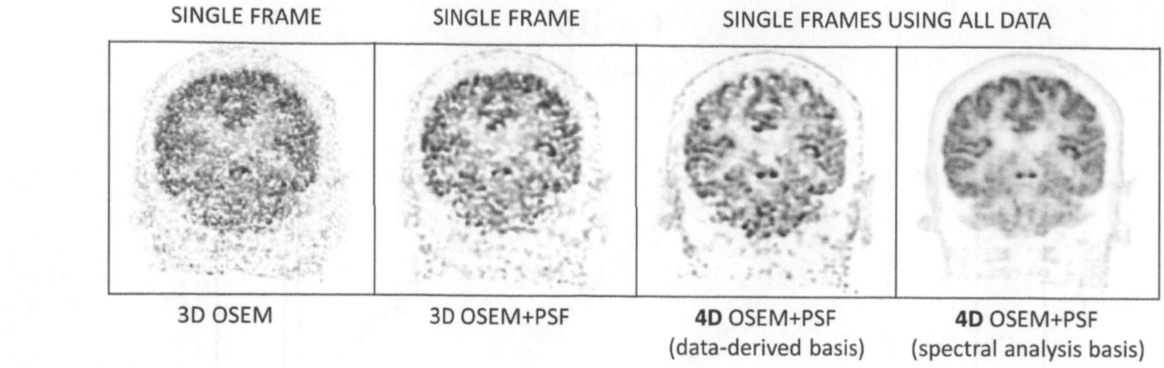

FIGURE 20.16 Representative results of 4D reconstruction for [^{18}F]FDG PET imaging [2]. On the left, single frame 3D reconstruction methods are shown (3D OSEM), and then on the right, 4D methods are shown, which benefit from all of the acquired data. The data-derived basis functions method [36] estimates both coefficients and temporal basis functions from the data, whereas the method on the far right side uses the prior information of a spectral analysis set of basis functions with a data-derived surrogate generating input function [37]. Figure is from Reader and colleagues [2].

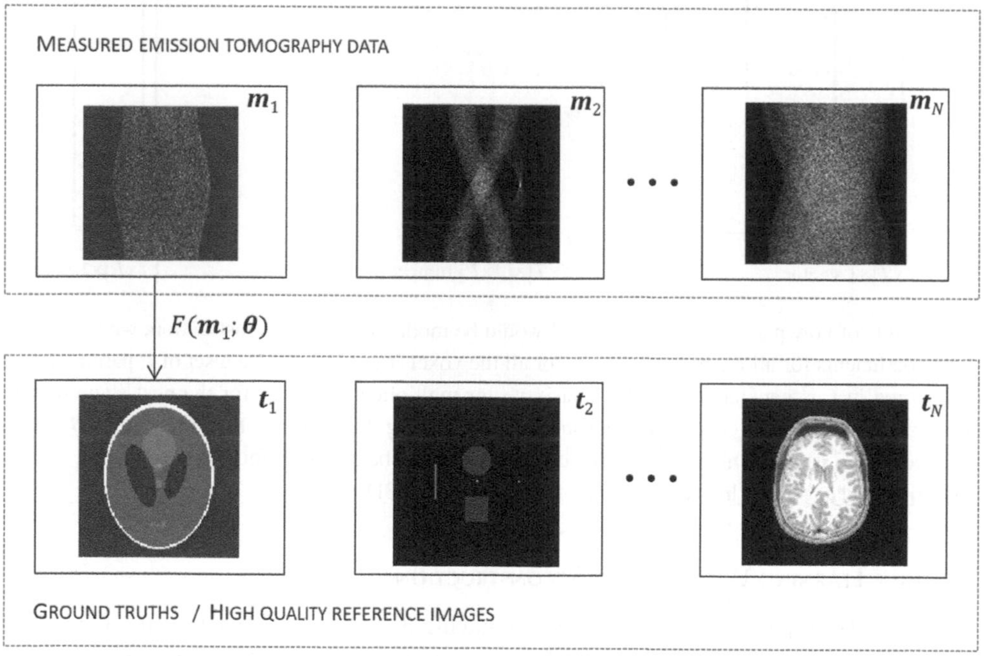

FIGURE 20.17 The AI paradigm for direct PET reconstruction: we estimate (or learn) the parameters $\boldsymbol{\theta}$ of a single mapping F which maps each data vector $\boldsymbol{m}$ to be as close as possible to its respective desired target vector $\boldsymbol{t}$. Supervised learning of the mapping needs example pairs of inputs and expected outputs (called targets or labels), which form the training data for the learning process. More advanced methods can learn how to map from one distribution to the other distribution, obviating the need for paired data vectors. Figure is from Reader and colleagues [30].

Direct deep-learning reconstruction methods have therefore been developed, mapping acquired sinogram data directly to images, but they typically require tens to hundreds of thousands of training-data pairs and are computationally demanding to train, and thus far have been relatively limited in the size of measured data and images they can handle (2D, rather than 3D). An early example of a network for PET image reconstruction is shown in Figure 20.18. In this convolutional encoder decoder (CED) architecture, the input 2D sinogram data is progressively downsampled spatially and converted into increasing numbers of feature maps, to the point that in the central part of the network the sinogram information is encoded in a latent space, with hardly any spatial samples, but with a large number of feature dimensions. This latent space representation of the PET data is then decoded by an image generator, which involves spatial upsampling

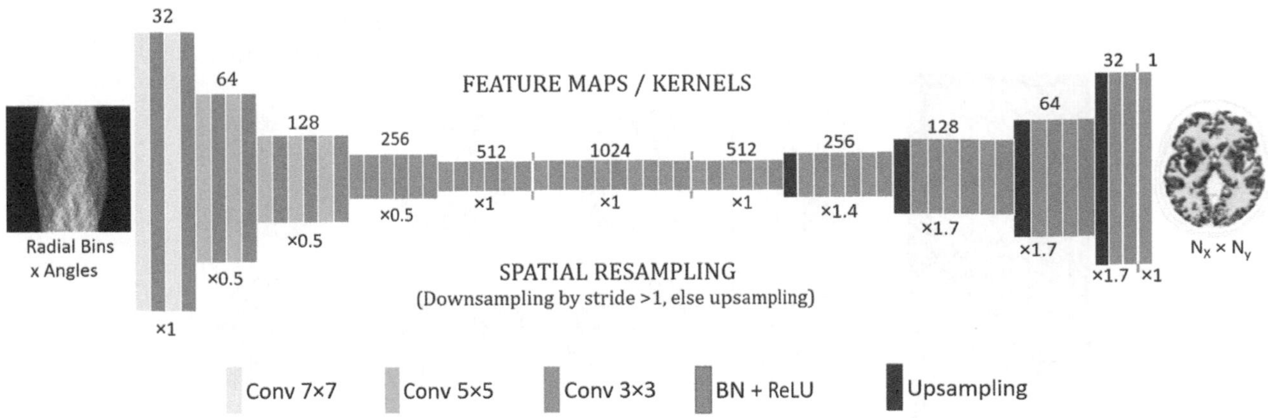

FIGURE 20.18 A convolutional encoder decoder (CED) architecture, representing a methodology proposed by Haggstrom and colleagues [32]. More than 200,000 training data pairs were required to learn the ~60 million parameters for this deep network. PET scan information expressed in the sinogram domain is progressively transformed by simultaneous reduction of spatial sampling and increasing of the number of feature maps, until a feature-rich latent space representation is obtained. This latent representation is decoded back out to an image-space representation of the same information, by increasing the spatial sampling and reducing the number of feature maps. Figure is from Reader and colleagues [30]. Colour image available at www.routledge.com/9781138593268.

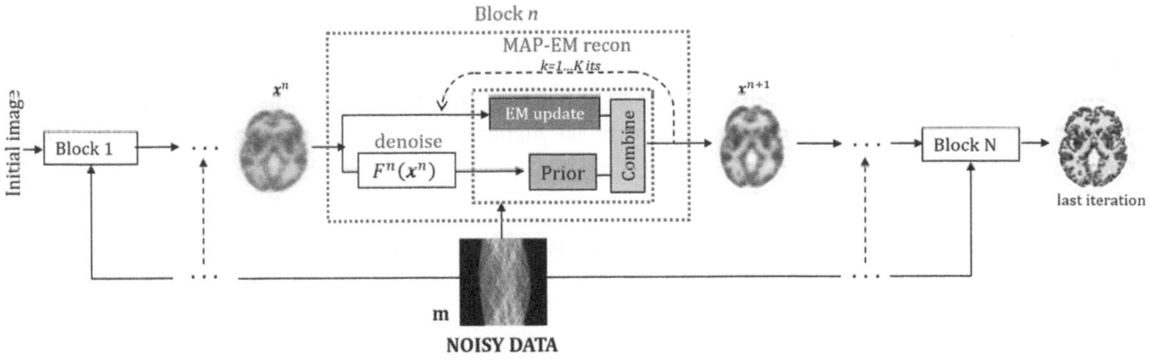

FIGURE 20.19 A framework for embedding deep learning as regularization into iterative image reconstruction in emission tomography. The unrolled series of updates is composed of n=1...N blocks or modules. Training can be done at the block level, where the goal is to denoise an update to make it best match a high quality (or true) reference. Or the training can be based on the very end image (last iteration), necessitating backpropagation through all N blocks during training in order to update the parameters for the denoiser network(s). The 'Prior' indicates a fixed image used in a quadratic penalty for the MAP-EM update, and the 'Combine' stage correspond to joining the EM update with the prior image, according to a method like that covered in Eq. 20.92. Figure is from Reader and colleagues [30].

and reduction of the number of features, until the end point of producing an image output, which is a different representation of the information contained in the PET sinogram.

A promising and more practical direction for using deep learning in PET reconstruction, readily suited to 3D reconstruction, is the use of unrolled networks. These methods unroll the iterative sequence of, for example, a MAP-EM iterative method. The denoising component – for example, essentially Eq. 20.99 in case 4 of the MAP-EM methodology presented earlier – is then replaced by a deep network that has been trained to denoise PET images based on training examples that pair noisy data with ideal (high-quality) data, available by simulations or by higher count acquisitions. Figure 20.19 shows a general framework for unrolled iterative PET reconstruction, embedding deep learning with standard reconstruction, and Figure 20.20 shows example results compared to conventional OSEM reconstruction as well as post-reconstruction denoising with a U-Net architecture [33]. The performance comparison between the unrolled method (FBSEM-Net in Figure 20.20) and the post-reconstruction method (U-Net in Figure 20.20) suggests that the potential advantage of embedding deep learning into reconstruction (compared to simply using it after reconstruction only) is perhaps still in need of a more convincing demonstration. Hence post-reconstruction deep-learning

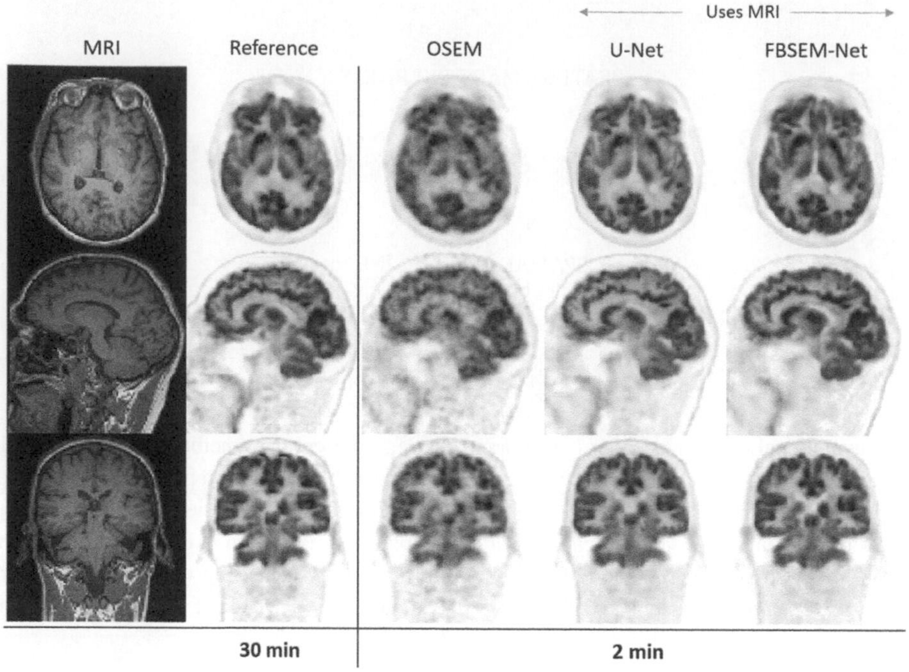

FIGURE 20.20 Example test results for real [^{18}F]FDG data for an unrolled iterative reconstruction method with deep learning included (FBSEM-Net [38], shown on the right), trained to match 30 minute reference data, when using 2 minute data along with a T1 weighted MR image for further information. FBSEM-Net is compared to conventional OSEM (no MRI benefit) and to a post-reconstruction denoised reconstruction using a U-Net supplied with MRI information. Figure is from Reader and colleagues [30].

methods may well be the most immediate way forward for using AI in the reconstruction pipeline for emission tomography.

20.9 SUMMARY

In summary, iterative image reconstruction for emission tomography allows considerable flexibility in accurate modelling of data acquired from a scanner. Combining accurate models of the data mean and the data noise with prior models of the reconstructed images delivers powerful MAP methods for regularized reconstruction. Furthermore, these methods can be upgraded still further with the use of spatiotemporal (4D) reconstruction and, in particular, through deep learning. Deep learning enables the prior probability of images, $\Pr(\boldsymbol{\theta})$, to be more accurately modelled based on training data, and these data-informed learned probability distributions can be embedded into the MAP reconstruction framework.

REFERENCES

[1] J. Qi and R. M. Leahy, "Iterative reconstruction techniques in emission computed tomography," (in English), *Phys Med Biol*, vol. 51, no. 15, pp. R541–R578, Aug. 7, 2006, doi:10.1088/0031-9155/51/15/R01.

[2] A. J. Reader and J. Verhaeghe, "4D image reconstruction for emission tomography," *Phys Med Biol*, vol. 59, pp. R371–R418, 2014.

[3] R. M. Lewitt, "Alternatives to voxels for image representation in iterative reconstruction algorithms," (in English), *Phys Med Biol*, vol. 37, no. 3, pp. 705–716, Mar 1992. [Online]. Available: <Go to ISI>://A1992HJ77500016.

[4] G. Wang and J. Qi, "PET image reconstruction using kernel method," *IEEE Trans Med Imaging*, vol. 34, no. 1, pp. 61–71, Jan 2015, doi:10.1109/TMI.2014.2343916.

[5] P. Novosad and A. J. Reader, "MR-guided dynamic PET reconstruction with the kernel method and spectral temporal basis functions," *Phys Med Biol*, vol. 61, no. 12, pp. 4624–4644, Jun. 21, 2016, doi:10.1088/0031-9155/61/12/4624.

[6] J. Bland et al., "MR-guided kernel EM reconstruction for reduced dose PET imaging," *IEEE Trans Radiat Plasma Med Sci*, vol. 2, no. 3, pp. 235–243, May 2018, doi:10.1109/TRPMS.2017.2771490.

[7] J. Qi, R. M. Leahy, S. R. Cherry, A. Chatziioannou, and T. H. Farquhar, "High-resolution 3D Bayesian image reconstruction using the microPET small-animal scanner," (in English), *Phys Med Biol,* vol. 43, no. 4, pp. 1001–1013, Apr 1998. [Online]. Available: <Go to ISI>://000073069000027.

[8] A. J. Reader, P. J. Julyan, H. Williams, D. L. Hastings, and J. Zweit, "EM algorithm system modeling by image-space techniques for PET \reconstruction," *IEEE Trans Nucl Sci*, vol. 50, no. 5 II, pp. 1392–1397, 2003. [Online]. Available: www.scopus.com/inward/record.url?eid=2-s2.0-0142126682&partnerID=40.

[9] L. A. Shepp and Y. Vardi, "Maximum likelihood reconstruction for emission tomography," (in English), *IEEE Trans Med Imaging*, vol. 1, no. 2, pp. 113–122, 1982, doi:10.1109/TMI.1982.4307558.

[10] K. Lange and R. Carson, "EM reconstruction algorithms for emission and transmission tomography," (in English), *J Comput Assist Tomogr*, vol. 8, no. 2, pp. 306–316, Apr 1984. [Online]. Available: www.ncbi.nlm.nih.gov/entrez/query.fcgi?cmd=Retrieve&db=PubMed&dopt=Citation&list_uids=6608535.

[11] H. H. Barrett, D. W. Wilson, and B. M. W. Tsui, "Noise properties of the EM algorithm. I. Theory," (in English), *Phys Med Biol*, vol. 39, no. 5, pp. 833–846, May 1994, doi:10.1088/0031-9155/39/5/004.

[12] R. L. Siddon, "Fast calculation of the exact radiological path for a 3-dimensional CT array," (in English), *Medical Physics*, vol. 12, no. 2, pp. 252–255, 1985, doi:10.1118/1.595715.

[13] P. E. Kinahan and J. G. Rogers, "Analytic 3D image-reconstruction using all detected events," (in English), *IEEE Tran Nucl Sci*, vol. 36, no. 1, pp. 964–968, Feb 1989. [Online]. Available: <Go to ISI>://A1989T377500191.

[14] A. P. Dempster, N. M. Laird, and D. B. Rubin, "Maximum likelihood from incomplete data via EM algorithm," (in English), *J Roy Stat Soc B Met*, vol. 39, no. 1, pp. 1–38, 1977. [Online]. Available: <Go to ISI>://A1977DM46400001.

[15] H. O. Hartley, "Maximum likelihood estimation from incomplete data," *Biometrics*, vol. 14, no. 2, pp. 174–194, 1958.

[16] H. M. Hudson and R. S. Larkin, "Accelerated image-reconstruction using ordered subsets of projection data," (in English), *IEEE Trans Med Imaging*, vol. 13, no. 4, pp. 601–609, Dec 1994. [Online]. Available: <Go to ISI>://A1994PZ97600004.

[17] T. Hebert and R. Leahy, "A generalized EM algorithm for 3D Bayesian reconstruction from Poisson Data using Gibbs Priors," (in English), *IEEE Trans Med Imaging*, vol. 8, no. 2, pp. 194–202, Jun 1989, doi:10.1109/42.24868.

[18] E. Levitan and G. T. Herman, "A maximum a posteriori probability expectation maximization algorithm for image-reconstruction in emission tomography," (in English), *IEEE Trans Med Imaging*, vol. 6, no. 3, pp. 185–192, Sep 1987, doi:10.1109/Tmi.1987.4307826.

[19] P. J. Green, "Bàyesian reconstructions from emission tomography data using a modified EM algorithm," (in English), *IEEE Trans Med Imaging*, vol. 9, no. 1, pp. 84–93, Mar 1990, doi:10.1109/42.52985.

[20] S. Alenius and U. Ruotsalainen, "Bayesian image reconstruction for emission tomography based on median root prior," (in English), *Eur J Nucl Med*, vol. 24, no. 3, pp. 258–265, Mar 1997, doi:10.1007/Bf01728761.

[21] A. R. De Pierro, "A modified expectation maximization algorithm for penalized likelihood estimation in emission tomography," (in English), *IEEE Trans Med Imaging*, vol. 14, no. 1, pp. 132–137, Mar 1995, doi:10.1109/42.370409.

[22] A. Mehranian et al., "PET image reconstruction using multi-parametric anato-functional priors," *Phys Med Biol*, vol. 62, no. 15, pp. 5975–6007, Jul. 6, 2017, doi:10.1088/1361-6560/aa7670.

[23] J. Bland et al., "Intercomparison of MR-informed PET image reconstruction methods," *Med Phys*, vol. 46, no. 11, pp. 5055–5074, Nov 2019, doi:10.1002/mp.13812.

[24] C. Tsoumpas, F. E. Turkheimer, and K. Thielemans, "Study of direct and indirect parametric estimation methods of linear models in dynamic positron emission tomography," (in English), *Med Phys*, vol. 35, no. 4, pp. 1299–1309, Apr 2008, doi:10.1118/1.2885369.

[25] G. Wang and J. Qi, "Acceleration of the direct reconstruction of linear parametric images using nested algorithms," (in English), *Phys Med Biol*, Research Support, N.I.H., Extramural vol. 55, no. 5, pp. 1505–1517, Mar. 7, 2010, doi:10.1088/0031-9155/55/5/016.

[26] J. C. Matthews, G. I. Angelis, F. A. Kotasidis, P. J. Markiewicz, and A. J. Reader, "Direct reconstruction of parametric images using any spatiotemporal 4D image based model and maximum likelihood expectation maximisation," (in English), *2010 IEEE Nuclear Science Symposium Conference Record (Nss/Mic)*, pp. 2435–2441, 2010. [Online]. Available: <Go to ISI>://000306402902131.

[27] G. I. Angelis, J. C. Matthews, F. A. Kotasidis, P. J. Markiewicz, W. R. Lionheart, and A. J. Reader, "Evaluation of a direct 4D reconstruction method using GLLS for estimating parametric maps of micro-parameters," (in English), *2011 IEEE Nuclear Science Symposium and Medical Imaging Conference (Nss/Mic)*, pp. 2355–2359, 2011. [Online]. Available: <Go to ISI>://000304755602118.

[28] C. Tsoumpas, F. E. Turkheimer, and K. Thielemans, "A survey of approaches for direct parametric image reconstruction in emission tomography," (in English), *Med Phys*, vol. 35, no. 9, pp. 3963–3971, Sep 2008, doi:10.1118/1.2966349.

[29] G. Wang and J. Qi, "Direct Estimation of Kinetic Parametric Images for Dynamic PET" *Theranostics*, doi:10.7150/thno.5130 2013.

[30] A. J. Reader, G. Corda, A. Mehranian, C. d. Costa-Luis, S. Ellis and J. A. Schnabel, "Deep Learning for PET Image Reconstruction," in *IEEE Transactions on Radiation and Plasma Medical Sciences*, vol. 5, no. 1, pp. 1–25, Jan 2021, doi: 10.1109/TRPMS.2020.3014786.

[31] B. Zhu, J. Z. Liu, S. F. Cauley, B. R. Rosen, and M. S. Rosen, "Image reconstruction by domain-transform manifold learning," *Nature*, vol. 555, no. 7697, pp. 487–492, Mar. 21, 2018, doi:10.1038/nature25988.

[32] I. Haggstrom, C. R. Schmidtlein, G. Campanella, and T. J. Fuchs, "DeepPET: A deep encoder-decoder network for directly solving the PET image reconstruction inverse problem," *Med Image Anal*, vol. 54, pp. 253–262, May 2019, doi:10.1016/j.media.2019.03.013.

[33] O. Ronneberger, P. Fischer, and T. Brox, "U-Net: Convolutional Networks for Biomedical Image Segmentation," (in English), *Med Image Comput Comput Assist Interv, Pt Iii*, vol. 9351, pp. 234–241, 2015, doi:10.1007/978-3-319-24574-4_28.

[34] C. A. Cocosco, V. Kollokian, R. Kwan, G. B. Pike, and A. C. Evans, "BrainWeb: Online interface to a 3D MRI simulated brain database," *NeuroImage*, vol. 5, p. 425, 1997.

[35] A. J. Reader and S. Ellis, "Bootstrap-optimised regularised image reconstruction for emission tomography," *IEEE Trans Med Imaging*, Jan. 14, 2020, doi:10.1109/TMI.2019.2956878.

[36] A. J. Reader, F. C. Sureau, C. Comtat, R. Trébossen, and I. Buvat, "Joint estimation of dynamic PET images and temporal basis functions using fully 4D ML-EM," *Phys Med Biol,* vol. 51, no. 21, pp. 5455–5474, 2006. [Online]. Available: www.scopus.com/scopus/inward/record.url?eid=2-s2.0-33751166570&partnerID=40.

[37] A. J. Reader, J. C. Matthews, F. C. Sureau, C. Comtat, R. Tre?bossen, and I. Buvat, "Fully 4D image reconstruction by estimation of an input function and spectral coefficients," in *IEEE Nuclear Science Symposium Conference Record*, Honolulu, HI, 2007, vol. 5, in 2007 IEEE Nuclear Science Symposium and Medical Imaging Conference, NSS-MIC, pp. 3260–3267. [Online]. Available: www.scopus.com/scopus/inward/record.url?eid=2-s2.0-48349134942&partnerID=40. [Online]. Available: www.scopus.com/scopus/inward/record.url?eid=2-s2.0-48349134942&partnerID=40.

[38] A. Mehranian and A. J. Reader, "Model-based deep learning PET image reconstruction using forward-backward splitting expectation maximisation," *IEEE Trans Radiat Plasma Med Sci*, vol. 5, no. 1, pp. 54-64, Jan. 2021, doi:10.1109/TRPMS.2020.3004408.

21 PET-CT Systems

Dimitris Visvikis

CONTENTS

21.1 INTRODUCTION

The development of multimodality imaging in the early 2000s has led to a real clinical revolution but has also presented numerous challenges and opportunities within the context of combining the anatomical and functional imaging modalities in the same scanner. Although the first-ever experimental multimodality device has been the SPECT/CT proposed by Bruce Hasegawa in the 1990s [1], the first multimodality device becoming a clinical reality has been the PET/CT [2]. One of the great advantages of combining PET and CT imaging within the same scanner has been the use of the acquired CT images for the correction of the attenuation effects in PET imaging [3]. Attenuation, caused by the photoelectric absorption of one or both of the 511 keV annihilation photons, is one of the most significant parameters influencing overall qualitative and, most importantly, quantitative accuracy in PET imaging (Figure 21.1). The development of PET/CT devices largely influenced the correction methodologies used for PET attenuation compensation because the acquired CT images offer clear advantages within this context [4].

21.2 ATTENUATION CORRECTION

Prior to the advent of PET/CT, PET attenuation correction has been based on the use of radioactive sources that were employed for the acquisition of patient-specific transmission maps. These were mostly acquired using $^{68}Ga/^{68}Ge$ line sources rotating around the patient. Transmission scans were acquired with and without the patient in the field of view, with the ratio of the two in the sinogram space allowing the calculation of the attenuation coefficient factors (ACFs) used to correct the PET data for the attenuation effects (Figure 21.2a). One of the issues associated with this approach concerned the low quality of the acquired transmission maps given the activity concentration of the line sources and the necessary compromise in terms of the overall acquisition times of a PET study. Within this context, the statistical noise present in the transmission maps propagated into the reconstructed emission data compromising the qualitative and quantitative accuracy of the PET emission images. Generally, the transmission scan acquisitions corresponded to half of the overall PET examination times (from 15 to 30 minutes).

Towards the latter part of the 1990s, two main technological and methodological solutions were put in place to facilitate the shorter transmission scan acquisition times. The first and most widely employed concerned the use of a segmentation of the transmission maps into three main tissue classes (bone, soft tissue, air) and assignment of associated attenuation coefficients, in order to reduce the impact of statistical noise ([3, 5] see Figure 21.3).

DOI: 10.1201/9780429489556-21

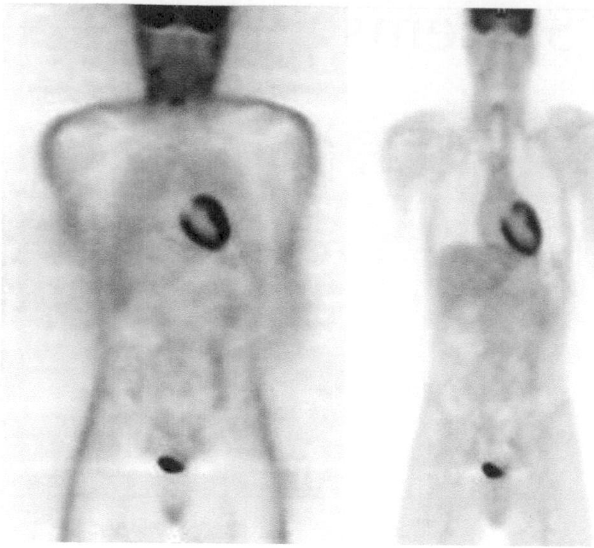

FIGURE 21.1 PET images (a) without and (b) with attenuation correction.

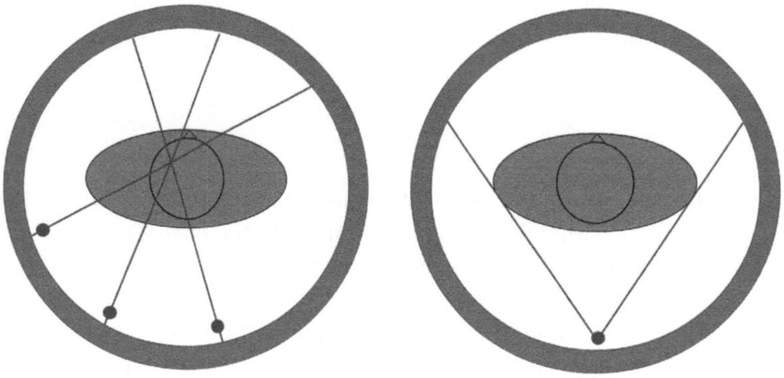

FIGURE 21.2 Transmission systems for PET attenuation correction using radionuclide sources: (a) positron emitters ($^{68}Ge/^{68}Ga$); (b). single photon emitters (^{137}Cs).

Although segmentation-based transmission scanning held the potential to reduce the impact of statistical noise, it also led, on the one hand, to an assumption concerning the accuracy of the segmentation algorithm in appropriately classifying the different tissue types considered while, on the other, ignoring the variable tissue types that could be present across the patient population. The alternative concerned the use of single photon rather than positron emitters for the acquisition of the transmission maps (Figure 21.2b). Most particularly, the use of ^{137}Cs was proposed, given that the energy of its emitted photons at 662 keV lead to a simple linear scaling correction for the energy differences relative to the 511 keV of the annihilation photons. One of the issues associated with the deployment of such a single photon transmission source was the mechanical system that had to be devised in order to facilitate the coverage of the whole scanner axial field of view [6], given that a point source format was used.

The advent of multimodality PET/CT clearly provided the potential to acquire very rapidly – and without any associated statistical noise – the transmission maps necessary for the characterization of the different body tissue attenuation properties [4]. Clearly the rapid acquisition of the transmission maps led to an increase by a factor of almost 2 on the number of patients that could undergo an attenuation corrected PET examination relative to the use of a radionuclide-based transmission scan. However, in order to ensure its accurate implementation, there have been two major issues. The first concerns the fact that the energy of an X-ray source is on average around 60–70 keV and, therefore, the use the attenuation maps acquired using a CT scanner requires a rescaling to the annihilation emission energy of 511 keV. It was rapidly shown that a bilinear relationship (see Figure 21.4) exists between the attenuation coefficients obtained at these two energies, one for Hounsfield units above 0 (above the

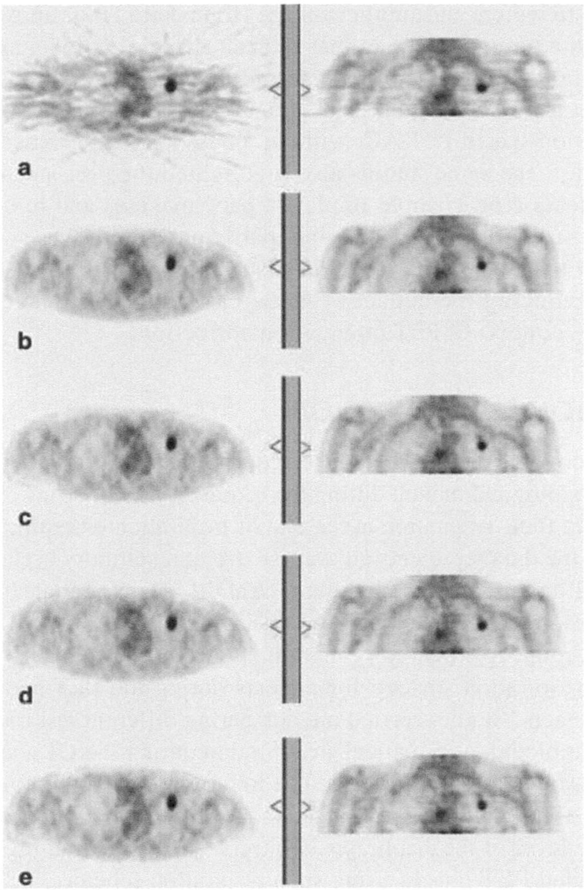

FIGURE 21.3 (from reference [5]): Coronal and transaxial slices from a patient with lung cancer. The slice location has been chosen to show the tumour site. Different reconstruction regimes are included, namely: (a) FBP with measured attenuation correction (15 min emission and transmission), (b) OSEM with segmented attenuation correction (SAC) (15 min emission and transmission), (c) OSEM with SAC (15 min emission and 3 min transmission), (d) OSEM with SAC (15 min emission and 2 min transmission) and, (e) OSEM with SAC (15 min emission and 1 min transmission).

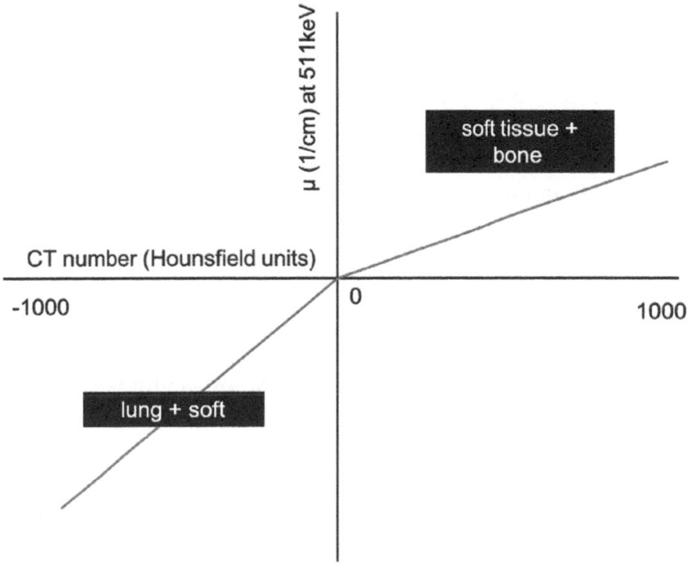

FIGURE 21.4 Bilinear relationship between the attenuation correction coefficients at 511 keV and the CT-based Hounsfield units.

attenuation level corresponding to water) and another for the Hounsfield Units above 0 [7]. Numerous studies were carried out in order to evaluate the potential impact that such an approximation may be entraining in terms of both qualitative and quantitative accuracy of attenuation corrected PET images. These studies globally concluded that CT-based attenuation correction for PET imaging leads to an increase in the overall signal to noise ratio of 20–30 per cent compared to segmentation-based PET AC, without an associated decrease at the level of the quantitative accuracy [3]. On the other hand, these same studies also highlighted the presence of artefacts associated with the occurrence of metallic components (for example implants, pace makers) and in certain cases in the presence of high-density oral contrast agents used during CT imaging, particularly in the case of Barium-based contrast agents [3, 8]. Certain developments in the field of iterative reconstruction algorithms in CT imaging over the previous decade have clearly helped in reducing the impact of metal-based image artefacts in CT images and, therefore, reducing their impact within the context of PET attenuation correction.

21.3 RESPIRATORY MOTION COMPENSATION

Another source of artefacts concerns the differences on the acquisition conditions between PET and CT scans, particularly considering the state of physiological motion during the acquisitions. CT scans over the thoracic zone are usually acquired with the patient blocking their respiration at the end of inspiration or expiration, while PET acquisitions last for a few minutes and, therefore, the data represents an average over a respiratory cycle. These differences at a patient's respiratory state during a PET/CT acquisition can, therefore, be at the origin of artefacts at the lower lung fields and the region of the diaphragm where respiratory motion has the larger influence.

There are different ways to account for respiratory motion effects in multimodality imaging. The first and simplest concerns the development of a respiration strategy for patients during the PET and CT acquisition, a strategy that reduces the presence of such artefacts. Studies carried out comparing different respiration schemes during the CT and PET acquisitions have shown that blocking the patient's respiration during the CT acquisition to the end of expiration leads to less artefacts compared to the end of inspiration [3]. In terms of clinical implementation in most cases today, both PET and CT acquisitions are mostly carried out under shallow breathing conditions, which also leads to a reduction of the visible artefacts at the level of the diaphragm. Conversely, despite limiting the qualitative inaccuracies and associated artefacts of reconstructed PET images using such respiration schemes, PET quantitative accuracy remains greatly reduced – and particularly for the middle and lower lung field regions [9].

Respiratory gating, involving the synchronization of the data acquisition with a respiratory signal recorded from the patient, can in principle reduce both qualitative and quantitative accuracies. In most cases the respiratory signal can be continuously monitored by using different external monitoring devices (see Figure 21.5) such as pressure belts ([9, 10], for example the Anzai respiratory belt (Anzai Medical, Japan), impedance plethysmography [11], the tracking of reflective markers stuck on the patient chest [12], for example, the Real-time Position Management system (RPM), by Varian Medical Systems (USA), or a spirometer [13]. A few studies comparing the performance of such external device-based respiratory monitoring devices have demonstrated variable correlation levels between the mono-dimensional respiratory signal obtained from these devices and the internal respiratory motion (at the level of organs and tumours) [14]. Although amongst all these devices a spirometer may provide the most accurate reflection of lung motion, it is more invasive and therefore difficult for patients to accept. More recent approaches concern the use of optical surface imaging devices that allow capturing the entire patient motion and, therefore, move beyond the use of a simple mono-dimensional respiratory signal, leading in turn to a superior correlation between external patient surrogate measures and corresponding internal structures' motion [15, 16].

An alternative approach is based on data-driven respiratory gating, where the respiratory signal is derived from either raw sinograms or reconstructed dynamic image series with acquisition frequency of <0.5s. Studies have shown that both approaches lead to the recovery of a data-driven respiratory motion signal equivalent to that of external respiratory monitoring devices [17–20]. The great advantage of using raw datasets to recover the required respiratory signal is that gating-based data binning can take place after the end of the acquisition, and it is totally transparent to the patient and radiographers, therefore making it much easier to accept and does not interfere with the overall acquisition procedure and associated protocol.

Once we are capable of synchronizing and bin data based on a patient's specific respiration pattern, the next step is to move forward with the reconstruction of respiratory-gated PET images. Data is binned based on the phase or the amplitude of the respiratory motion (see Figure 21.6a). The first option is a straightforward reconstruction of the data belonging to each position in the respiratory cycle and that for the whole of the PET data acquisition duration, resulting in a single PET image for each respiratory bin. The optimum number of respiratory bins, and therefore associated

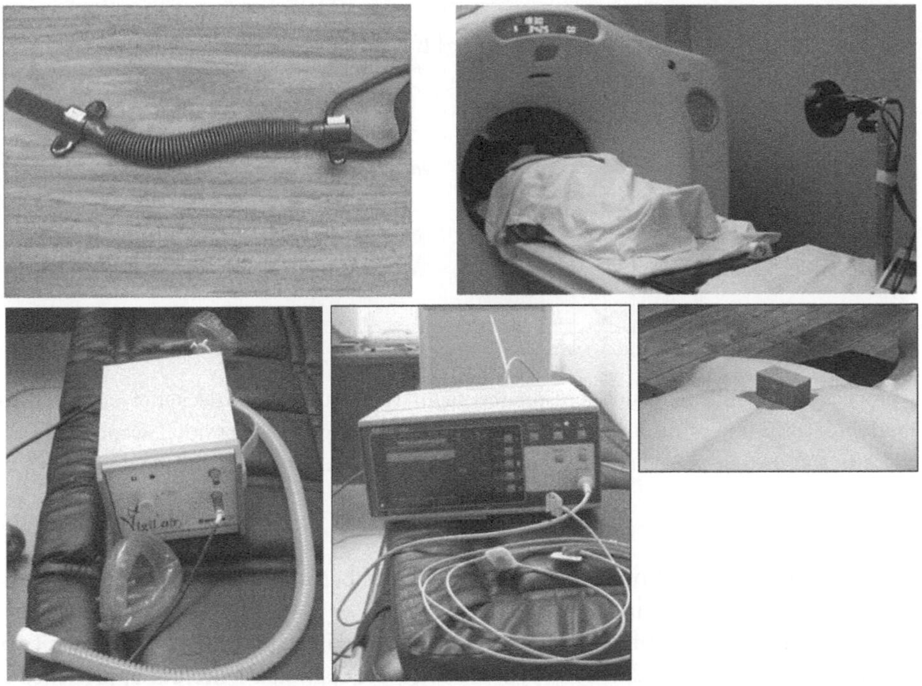

FIGURE 21.5 Different external respiratory monitoring devices; (a) pressure belt, (b) RPM system, (c) spirometer, (d) impedance plethysmograph

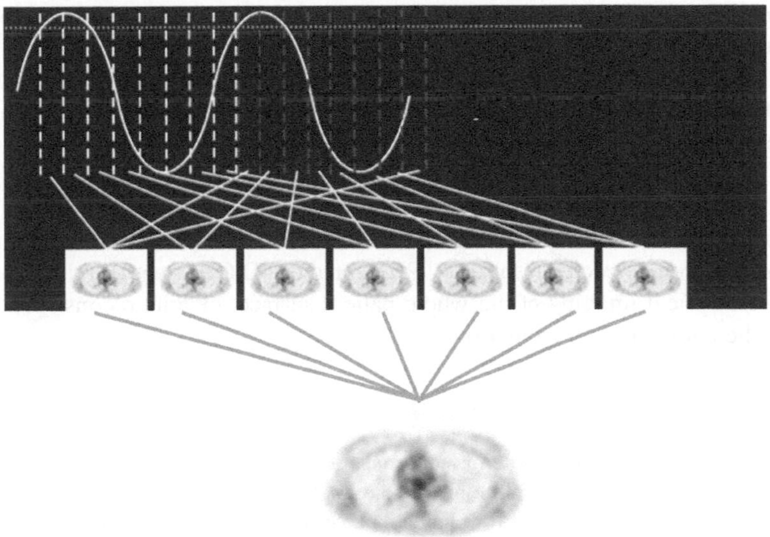

FIGURE 21.6 The principle of (a) respiratory motion gating based on the phase of the respiration data binning and reconstruction of individual respiration frames, and (b) the subsequent use of a motion model allowing the combination of the reconstructed gated images.

reconstructed images, has been determined to be between 6–8 given the frequency and variability associated with human respiration. One of the issues with respiratory gating is that the images of each respiratory frame suffer from reduced statistical quality given that only part of the data acquired is used in each reconstructed respiratory-synchronized gate [21]. Given that over 30 per cent of the data is acquired at the end of expiration phase of respiratory motion, one proposed approach is to use only the data acquired over this part of the respiratory cycle [22]. This reduces the influence

$$n^{k+1} = \frac{n^{k+1}}{S'} E_t^T A^T \frac{1}{A E^t n^k}$$

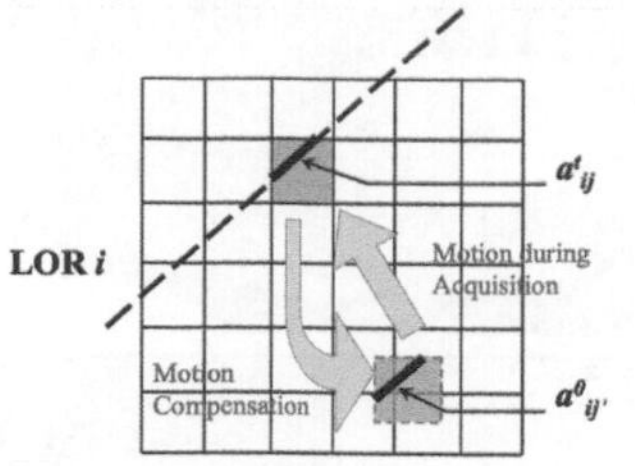

with

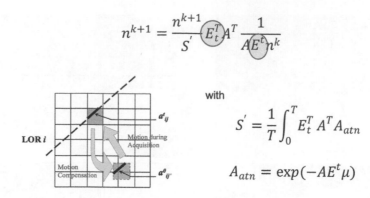

$$S' = \frac{1}{T}\int_0^T E_t^T A^T A_{atn}$$

$$A_{atn} = \exp(-AE^t\mu)$$

FIGURE 21.7 The process of the implementation of respiratory motion correction within the iterative image reconstruction process [24, 27]. The contribution of each line of response to a given voxel is translated to the position accounting for the motion effects by including the matrice E_t and its transpose during the projection steps. The sensitivity S' and attenuation A_{atn} corrections are also displaced accordingly.

of statistical noise relative to the acquisition of 6–8 respiratory gates but is still suboptimal in comparison to the use of the whole data acquired throughout a respiratory cycle [23].

An alternative is that being able to combine these individual frames in a single respiration-effects free image allows the realizing of all benefits of PET/CT respiratory-synchronized acquisitions. Different approaches can be envisaged for this combination step. The first solution is based on the registration of the individual respiratory-synchronized PET images. Different deformable models have been used during the years to realize this step [14]. Following this registration step, the individual images can be subsequently summed up (see Figure 21.6b). This image-based approach may be sub-optimal in terms of signal-to-noise ratio in comparison to a reconstruction-based approach. Such a methodology is based on incorporating the deformation matrices within the reconstruction process and, more specifically, during the forward- and back-projection steps. Given the deformable nature of the respiratory motion, the contribution of a line of response to each voxel is displaced to a new location based on these deformation matrices. At the same time, these deformations also have to be applied to all other data corrections such as normalization and attenuation correction steps (see Figure 21.7). Implementations to handle such reconstruction-based respiratory motion correction approaches have been proposed for both sinogram and list-mode datasets [24–26]. The majority of variants in terms of implementation concern the derivation of the necessary deformation matrices in order to transform the data to a single position in the respiratory cycle. Numerous approaches have been proposed throughout the years [14], including (i) the use of attenuation corrected or non-attenuation corrected respiratory PET-gated or 4D CT images, and (ii) the use of motion models based on external markers, optical imaging of the whole patient surface, or joint reconstruction approaches whereby motion fields are part of the parameters to estimation.

REFERENCES

[1] T. F. Lang et al., "Description of a prototype emission-transmission computed tomography imaging system," *J Nucl Med*, vol. 33, no. 10, pp. 1881–1887, Oct 1992. [Online]. Available: www.ncbi.nlm.nih.gov/pubmed/1403162.

[2] T. Beyer et al., "A combined PET/CT scanner for clinical oncology," *J Nucl Med*, vol. 41, no. 8, pp. 1369–1379, Aug 2000. [Online]. Available: www.ncbi.nlm.nih.gov/pubmed/10945530.

[3] D. Visvikis et al., "CT-based attenuation correction in the calculation of semi-quantitative indices of [18F]FDG uptake in PET," *Eur J Nucl Med Mol Imaging*, vol. 30, no. 3, pp. 344–353, Mar 2003, doi:10.1007/s00259-002-1070-4.

[4] J. S. Lee, "A review of deep learning-based approaches for attenuation correction in positron emission tomography," *IEEE Trans Radiat Plasma Medi Sci*, pp. 1–1, 2020, doi:10.1109/TRPMS.2020.3009269.

[5] D. Visvikis, C. Cheze-LeRest, D. C. Costa, J. Bomanji, S. Gacinovic, and P. J. Ell, "Influence of OSEM and segmented attenuation correction in the calculation of standardised uptake values for [18F]FDG PET," *Eur J Nucl Med*, vol. 28, no. 9, pp. 1326–1335, Sep 2001. [Online]. Available: www.ncbi.nlm.nih.gov/pubmed/11585291.

[6] F. Benard, R. J. Smith, R. Hustinx, J. S. Karp, and A. Alavi, "Clinical evaluation of processing techniques for attenuation correction with 137Cs in whole-body PET imaging," *J Nucl Med*, vol. 40, no. 8, pp. 1257–1263, Aug 1999. [Online]. Available: www.ncbi.nlm.nih.gov/pubmed/10450675.

[7] P. E. Kinahan, D. W. Townsend, T. Beyer, and D. Sashin, "Attenuation correction for a combined 3D PET/CT scanner," *Med Phys*, vol. 25, no. 10, pp. 2046–2053, Oct 1998, doi:10.1118/1.598392.
[8] S. A. Nehmeh et al., "Correction for oral contrast artifacts in CT attenuation-corrected PET images obtained by combined PET/CT," *J Nucl Med*, vol. 44, no. 12, pp. 1940–1944, Dec 2003. [Online]. Available: www.ncbi.nlm.nih.gov/pubmed/14660720.
[9] H. D. Kubo and B. C. Hill, "Respiration gated radiotherapy treatment: A technical study," *Phys Med Biol*, vol. 41, no. 1, pp. 83–91, Jan 1996, doi:10.1088/0031-9155/41/1/007.
[10] G. J. Klein, B. W. Reutter, M. H. Ho, J. H. Reed, and R. H. Huesman, "Real-time system for respiratory-cardiac gating in positron tomography," *IEEE T Nucl Sci*, vol. 45, no. 4, pp. 2139–2143, 1998, doi:10.1109/23.708323.
[11] C. J. Ritchie, J. Hsieh, M. F. Gard, J. D. Godwin, Y. Kim, and C. R. Crawford, "Predictive respiratory gating: A new method to reduce motion artifacts on CT scans," *Radiology*, vol. 190, no. 3, pp. 847–852, Mar 1994, doi:10.1148/radiology.190.3.8115638.
[12] S. A. Nehmeh et al., "Effect of respiratory gating on reducing lung motion artifacts in PET imaging of lung cancer," *Med Phys*, vol. 29, no. 3, pp. 366–371, 2002, https://doi.org/10.1118/1.1448824.
[13] J. Hanley et al., "Deep inspiration breath-hold technique for lung tumors: The potential value of target immobilization and reduced lung density in dose escalation," *Int J Radiat Oncol Biol Phys*, vol. 45, no. 3, pp. 603–611, Oct 1 1999, doi:10.1016/s0360-3016(99)00154-6.
[14] H. Fayad, F. Lamare, Merlin. T, and D. Visvikis, "Motion correction using anatomical information in PET/CT and PET/MR hybrid imaging," *Q J Nucl Med Mol Imaging*, vol. 60, no. 1, pp. 12–24, 2016.
[15] H. Fayad, T. Pan, O. Pradier, and D. Visvikis, "Patient specific respiratory motion modeling using a 3D patient's external surface," (in English), *Med Phys*, vol. 39, no. 6, pp. 3386–3395, Jun 2012, doi:10.1118/1.4718578.
[16] H. Fayad, T. Pan, J. F. Clement, and D. Visvikis, "Technical note: Correlation of respiratory motion between external patient surface and internal anatomical landmarks," (in English), *Med Phys*, vol. 38, no. 6, pp. 3157–3164, 2011, doi:10.1118/1.3589131.
[17] R. A. Bundschuh et al., "Postacquisition detection of tumor motion in the lung and upper abdomen using list-mode PET data: A feasibility study," *J Nucl Med*, vol. 48, no. 5, pp. 758–763, May 2007, doi:10.2967/jnumed.106.035279.
[18] C. Le Rest, O. Couturier, A. Turzo, and Y. Bizais, "Post-synchronization of dynamic images of periodically moving organs," (in English), *Nucl Med Commun*, vol. 21, no. 7, pp. 677–684, Jul 2000, doi:10.1097/00006231-200007000-00012.
[19] P. J. Schleyer, K. Thielemans, and P. K. Marsden, "Extracting a respiratory signal from raw dynamic PET data that contain tracer kinetics," *Phys Med Biol*, vol. 59, no. 15, pp. 4345–4356, Aug. 7, 2014, doi:10.1088/0031-9155/59/15/4345.
[20] H. Fayad, F. Lamare, T. Merlin, and D. Visvikis, "Motion correction using anatomical information in PET/CT and PET/MR hybrid imaging," (in English), *Q J Nucl Med Mol Imaging*, vol. 60, no. 1, pp. 12–24, Mar 2016.
[21] P. Jarritt et al., "Use of combined PET/CT images for radiotherapy planning: Initial experiences in lung cancer," *Brit J Rad*, vol. 78, pp. 33–40, 2005.
[22] A. Pépin, J. Daouk, P. Bailly, S. Hapdey, and M. E. Meyer, "Management of respiratory motion in PET/computed tomography: The state of the art," (in English), *Nucl Med Commun*, vol. 35, no. 2, pp. 113–122, Feb 2014, doi:10.1097/mnm.0000000000000048.
[23] M. Dawood, F. Büther, N. Lang, O. Schober, and K. P. Schäfers, "Respiratory gating in positron emission tomography: A quantitative comparison of different gating schemes," (in English), *Med Phys*, vol. 34, no. 7, pp. 3067–7306, Jul 2007, doi:10.1118/1.2748104.
[24] F. Lamare et al., "List-mode-based reconstruction for respiratory motion correction in PET using non-rigid body transformations," *Phys Med Biol*, vol. 52, no. 17, pp. 5187–5204, Sep. 7, 2007, doi:10.1088/0031-9155/52/17/006.
[25] F. Qiao, T. Pan, J. W. Clark, Jr., and O. R. Mawlawi, "A motion-incorporated reconstruction method for gated PET studies," *Phys Med Biol*, vol. 51, no. 15, pp. 3769–3783, Aug. 7, 2006, doi:10.1088/0031-9155/51/15/012.
[26] T. Li, B. Thorndyke, E. Schreibmann, Y. Yang, and L. Xing, "Model-based image reconstruction for four-dimensional PET," *Med Phys*, vol. 33, no. 5, pp. 1288–1298, May 2006, doi:10.1118/1.2192581.
[27] F. Lamare, T. Cresson, J. Savean, C. Cheze Le Rest, A. J. Reader, and D. Visvikis, "Respiratory motion correction for PET oncology applications using affine transformation of list mode data," *Phys Med Biol*, vol. 52, no. 1, pp. 121–140, Jan. 7, 2007, doi:10.1088/0031-9155/52/1/009.

22 Clinical Molecular PET/MRI Hybrid Imaging

Bernhard Sattler

CONTENTS

22.1 THE HISTORY OF HYBRID PET/MRI TECHNOLOGY

Undoubtedly, there is a multitude of advantages if combining a molecular imaging result with the superior high-resolution structural imaging that is possible with CT. There is lots of proofs of the utility of that combination both from a physical-technical perspective as well as in clinical routine. The combination of PET and CT is comparatively simple. The two modalities are mounted essentially in one and the same gantry with the fields of view axially centered. The same patient handling system is used for the patient to be positioned and axially moved first through the CT-system and then the PET-system. The patient is to not move during the whole image acquisition so the CT result can serve as an anatomical orientation and for attenuation correction of the PET as always required (see Chapters 18 and 26). More and more, also diagnostic CT imaging, that is, with higher doses than needed to just serve as a source for attenuation correction and anatomical orientation, and, moreover, with contrast enhancement

DOI: 10.1201/9780429489556-22

by CT contrast agents, has made its way into the diagnostic workups of molecular hybrid PET/CT imaging. Since its arrival in diagnostic workups two and a half decades ago it has been a huge success as a clinical tool, mainly in oncology. However, since CT has its limitations in imaging and to differentiate between soft tissues also, MRI always had to accompany most of those scenarios, particularly if soft tissue contrast or functional parameters are crucial. Already very soon after the advent of PET/CT in the clinical arena, there has been the wish to combine PET and MRI [1–6]. The combination of PET and MRI opens the field of cross-modality imaging to the marriage of the well-known, highly sensitive functional PET imaging with the superior soft tissue signal that MRI is able to add. Moreover, CT – even if carried out with intravenously applied contrast agents – is essentially a snapshot of the structure at the moment of imaging. Dynamic information, that is, data that is acquired in temporal resolution – except for some myocardial imaging protocols – are performed rather seldom. Here, MRI comes with a multitude of imaging sequences. Without any exposure to ionizing radiation, an ever-increasing plurality of information beyond just structural imaging can be acquired with a temporal resolution down into the orders of milli- or even microseconds. PET and MRI have been in the arena of clinical molecular tomographic imaging more than three decades each. While MRI originally aimed at imaging soft tissue structures, PET primarily aimed at imaging physiologic and pathophysiologic processes to image oncologic, neurologic, cardiologic, infection and inflammation diseases. It has been the aim of technical investigations and developments to investigate the combination of PET and MRI. After the proof of the operability of specific light detectors such as avalanche photo diodes (APD) [7, 8] and silicon photo multipliers (SiPM) [9–12] under the conditions of strong static and changing gradient magnetic fields [13] the way of combining PET and MRI into hybrid systems was paved. First, PET-detector inserts for preclinical [3, 7, 14–16] and clinical MRI-systems [4, 17, 18] have been developed to further prove the feasibility of combining the two imaging modalities without impairing the qualitative and quantitative imaging results of each other intolerably. In parallel, by using systems employing one and the same patient-handling system/table/bed, the clinical utility of having the image data of the involved modalities inherently co-registered to each other was thoroughly investigated. Using existing imaging systems, GE introduced a two-room-setup where a clinical PET/CT- and a MRI-System were installed in rooms right next to each other, and a unique patient-handling system designed connectable to the patient ports-bores of both of the systems [19, 20]. Philips came up with a comparable design, but having the two systems in the same room and using a bi-planar rotating patient table [21, 22]. The first system fully integrating the avalanche photodiode (APD) based PET detector system was manufactured by Siemens Healthcare and reached market maturity in 2010. Offering the option of a true simultaneous whole-body measurement of PET and MR signals, the limited timing resolution of the APDs, however, did not allow for time of flight (TOF) PET measurements [23, 24] as well known and utilized in photomultiplier-based PET-detector systems in hybrid PET/CT systems [25–27]. In 2013, GE healthcare introduced a TOF-PET-capable fully integrated whole-body PET/MRI-system employing SiPMs as light detectors in the PET-detector system [28]. Comprehensive overviews by Quick and Boellaard as well as Mannheim summarize the performance parameters of the PET/MRI-systems available for clinical use [29, 30]. For more information and explanations on the physical Basics of MRI the reader is referred to Chapter 32 where this topic is alluded to in full detail.

22.2 TECHNICAL CHALLENGES WITH INTEGRATING PET AND MRI

22.2.1 Design and Integration of Combined PET/MRI Systems

Whilst preclinical PET/MRI systems are employing MR-compatible PET-Detector setups that can be actually inserted into the bore of a high field preclinical MRI system to achieve the opportunity for simultaneous acquisition of the PET and the MRI signal, also the experiences in using PET/MRI clinically have shown that fully integrated clinical PET/MRI systems offering simultaneous PET and MRI acquisition have advantages over sequential measurements [31–35]. Whether true simultaneity is needed clearly depends on the clinical applications. There has been uncertainty about what clinical application will be key applications that are not available with other (sequential) hybrid imaging modalities, mainly PET/CT, and will be feasible and beneficial in the complex setting of diagnostic need, radiotracer availability, patient comfort, and relatively high cost of hybrid PET/MRI systems as compared to PET/CT. Another important consideration is constraints that come with the time that each of the acquisitions takes. While a CT takes just several seconds, even for larger body parts like the whole torso, MR sequences usually take several minutes and up to quarters of an hour or even longer, depending on the complexity of the MR-imaging sequences and protocols that have been chosen. Despite recent developments in advancing PET-Detector technology towards 3D-detection of events, time-of-flight detection (TOF) (see Chapters 6 and 18), the PET acquisition of one bed position today takes only 3–10 minutes

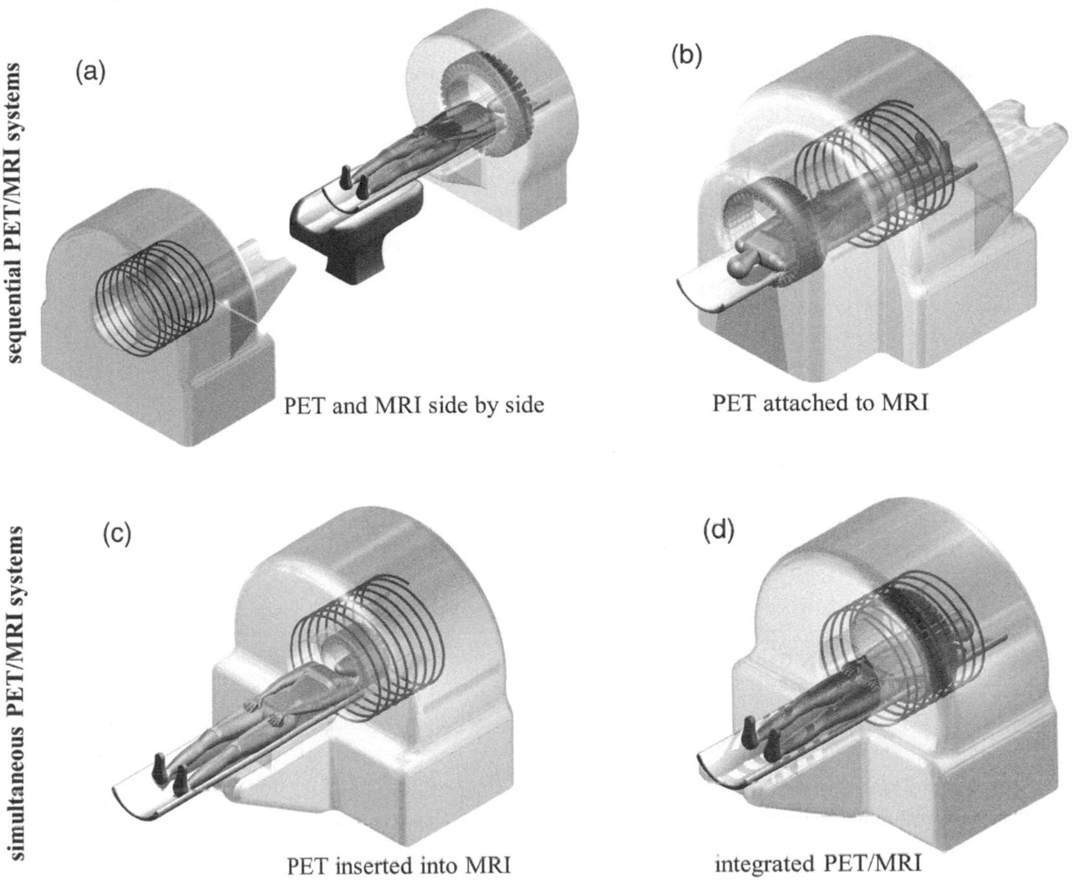

FIGURE 22.1 Design Concepts of combined PET/MRI systems. Sequential PET/MRI systems (a, b) allow for acquisition of the subject at the same axial position and the PET and MRI acquired consecutively. Systems with the PET detector system inserted/integrated into/with the MRI system (c, d) allow for simultaneous acquisition of PET and MRI with the subject at essentially the very same position/registration for the hybrid investigation (adapted from [40, 41]).

leading to a duration of a torso or whole body investigation of an adult of only about 10–20 minutes. The axial field of view of the PET-Detector system is usually not much longer than 25 cm in all hybrid PET/MRI systems that are currently available [24, 26, 30, 36, 37]. This limits true simultaneous acquisitions to areas that fit into that length (i.e. the brain or the heart). But, on the other hand, for investigations of larger body parts combining several of those bed positions, an integrated PET/MRI system allows for the acquisition of the PET data while MR-sequences are in progress at each of the bed position or 'stations', as they are usually called in the MRI world. Even if the resulting combined and fused volume data set is not truly simultaneously acquired anymore the acquisition time is no longer compared to MRI alone. If the PET and MRI gantries are placed apart from each other a patient handling system (bed) is needed that can place the patient in the very same axial position in the two gantries [38]. Moreover, such a design imposes the need for much more space, as this is actually a two- [39] or three-modality [20] setup with a footprint requiring more space than the usual scanner rooms offer, or have to be placed in adjacent imaging rooms. Ultimately, the PET and the MRI acquisitions have to take place one after the other and even more time is needed for a usual clinical PET/MRI investigation. Sequential PET/MRI systems had their place during the course of the clinical evaluation of combined PET and MRI. The mentioned drawbacks, however, lead to decreasing relevancy of such systems in the clinic.

22.2.2 Effects of MRI on the PET and Vice Versa

If already very complex molecular imaging technical modalities, one of which surrounded by strong magnetic fields and very sensitive against external electromagnetic influences and, the other, using ionizing radiation as a signal and detectors potentially sensitive against temperature shifts and magnetism are brought in close proximity or are even fully

TABLE 22.1
Impact of MR Features on the Performance of the PET Component

MR feature	Impact on PET	Solution	Consequences
B_0	Interaction with charged particles in the field (mainly distraction of electrons)	Use of solid state light sensors to detect scintillation light (APDs or SiPMs) instead of conventional PMTs	higher cost
B_1 and related RF-transmission	Heating, vibration interference with electronics	Re-engineering of system design subtle RF shielding of the PET detector and read-out electronics	more complexity more metal in the field →eddy currents and heating/vibrations

Source: (Adapted from [42])

TABLE 22.2
Impact of PET Features on the Performance of the MRI Component

PET feature	Impact on MRI	Solution	Consequences
Scintillator crystals	Static magnetic field inhomogeneity	Use of MRI compatible PET-Scintillators	higher cost
Gamma ray shielding	Static magnetic field inhomogeneity and gradient field distortion and non-linearity by eddy currents induced in this material	Modified (i.e.) multi-layer gamma-ray shielding material to interrupt eddy currents	more complexity, higher material cost
PET-power supply and electronics	Interference with the RF detection	RF-Shielding around PET electronics* and any cables	

Note: * This shielding can be the same as for the shielding of the PET detector and read-out electronics against RF-influences from the gradient- and transmission coils field.

Source: (Adapted from [42])

integrated, after proof of the general operability, it is essential to investigate the potential impact on the validity of the generated results. If we first focus on how the MRI system will impact the PET system, there are three main effects to be considered:

- Influence of the static magnetic field on electronic circuitry in its proximity
- Gradient magnetic fields (G_x, G_y, G_z) interference with any conductive material and electronic components inside the magnet bore
- RF transmission interfering with the electronic components of the PET detector system

On the other hand, we have to take into consideration how the PET component could impact the performance of the MRI system. There are basically two effects that have to be mentioned here:

- Influence on the homogeneity of the static magnetic field (B_0) and the linearity of the gradient field (B_1) by objects inside or close to the bore
- Interference on the detection/receiving of the very weak spin relaxation RF-signal after excitation

A summary of those impacts is shown in Tables 22.1. and 22.2

The ultimate (technical) goal of the integration of the PET and MRI hardware is to reduce the magnetic and electric influences of one to the other to a level so that the quality of the resulting outputs is not significantly impacted, and data comparable to the stand-alone systems can be generated at least. Undoubtedly, the ultimate key is to have PET detector technology available that can operate under the conditions of strong static and altering magnetic fields and, vice versa, is designed so that it does not influence the operation of the MRI system significantly. A more detailed description,

TABLE 22.3
Physical Features of Several Scintillator Materials Commonly Used in PET systems

Scintillator					
Feature	**NaI**	**BGO**	**LSO**	**LYSO**	**LuAP**
Wavelength of scintillation light (nm)	410	480	420	420	365
Light output R (10^3 photons/MeV)	41	9	30	30	12
Light intensity I_0 (photons/ns)	90	21	380	380	340
Light decay time τ (ns)	230	300	40	40	18
Density ρ (g/cm^3)	3.67	7.13	7.35	7.19	8.34
Effective Atomic Number Z_{eff}	50	65	64	65	33
attenuation length/stopping power $1/\mu$ (mm)	25.9	11.2	12.3	12.6	11.0

Note: Legacy materials: NaI thallium doped sodium iodide (NaI:Tl), BGO bismuth germinate ($Bi_4Ge_3O_{12}$); modern materials: LSO cerium doped lutetium oxyorthosilicate (Lu_2SiO_5:Ce), LYSO cerium doped lutetium-yttrium oxyorthosilicate ($Lu_{1.9}Y_{0.1}SiO_5$:Ce) and LuAP cerium doped lutetium-aluminum perovskite ($LuAlO_3$:Ce) (Reproduced and Adapted from [45]).

particularly on how the design of the PET detector system could be modified to further minimize its influence on the RF transmission is shown with an SiPM-based PET detector system that can be inserted into otherwise unmodified stand-alone MRI systems [43, 44]. A more general description of the influences of the sub-systems onto one another can be found in a topical review by Vandenberghe and Marsden [42].

22.2.3 PET Detectors in a Magnetic Field

First of all, the scintillator crystal material has to have some important features to be applied as detectors for 511keV gamma rays or photons. Besides all the features such as

- high light output (number of produced light photons per keV of the interacting gamma photon);
- high light intensity (number of produced photons per ns);
- a short scintillation light decay time;
- a short attenuation length or effective stopping power at 511keV (1/μ in mm, where μ is attenuation coefficient);
- no magnetic susceptibility.

Table 22.3 gives an overview of the PET scintillator materials that fulfil most of these requirements in comparison to the more legacy ones, the thallium doped sodium iodide (NaI:Tl) and the bismuth germanate (BGO).

Once the light is produced in the scintillators, this signal needs to be sufficiently detected and processed in a timely manner. Essentially, this means that, as much as possible, light photons should be detected with a timing resolution of at least some one digit nano-seconds to distinguish coincidences and, moreover, some several hundred piko-seconds to measure the time difference in detection of the photons belonging to the same line of response, that is, to enable time-of-flight PET (see Chapters 6 and 18). Now, as is physically obvious, light sensors that employ conventional photo-multiplier tubes (PMT) to detect the visible scintillation light created by the detector crystals upon interaction with the 511keV gamma ray photons cannot be used under the conditions of strong magnetic field influences. Therefore, in the beginning of the assessment of options for combining PET and MRI, the first direction of investigations was do develop PET detector systems with light guides of different lengths to guide the scintillation light generated by the scintillators inside the bore of the MRI system to conventional tube-based photomultipliers far enough away from the magnetic field influences of the MRI system [46, 47]. If the conventional PMTs were placed too close to the magnetic field, already comparatively small magnetic field strengths would totally disrupt their normal function by distracting moving charged particles, mainly electrons. Also, the application specific integrated circuit (ASIC) electronics would be heavily impacted in its operation, as was proven in comparative studies [8] and shown in Figure 22.2.

An intermediate state of further development was the application of shorter optic fibers guiding the scintillation light onto APDs, including application-specific circuitry contained in a shielding to resist the magnetic influences to a certain degree and, therefore, can be brought closer or even into the magnetic field of an MRI system [48]. The operability of APDs in PET-Detector modules has first been shown in PET-Systems [7]. Their application, including very short light

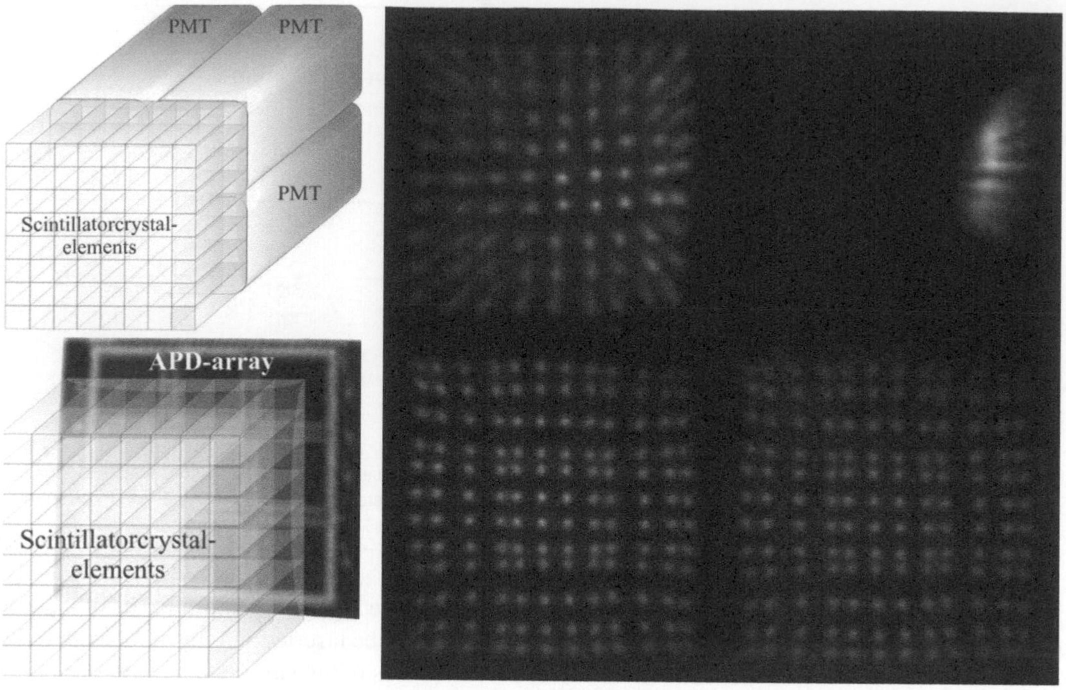

FIGURE 22.2 Principal PET detector designs. The upper row shows the conventional design with PMTs as light sensors the lower row the design with an avalanche photo diode (APD) array as the light sensor. The detector crystal read-out histograms show the impact of the magnetic field condition (middle B=0, right B≠0, adapted from [16]).

TABLE 22.4
Selected Key Features of Light Sensors for PET

Lightsensor Feature	Photomultiplier (PMT)	Avalanche-photodiode (APD)	Siliconphoto-multiplier (SiPM)
Size (diameter / active area)	∅10–100mm	1–100mm^2	~1mm^2
Gain	10^5–10^7	10^2	10^5–10^6
Signal rise time (ns)	<1	2–3	~1
Background (excess) noise factor	~0.1	>2	~0.1
Quantum efficiency	25%	60–80%	<40%
Dark current/count rate	<0,1nA/cm^2	10–100nA/cm^2	10^7–10^8Ips/cm^2
Bias Voltage	1000–2000	~100–1500	~50
power consumption	100mW/ch	1mW/cm^2	5mW/cm^2
Temperature dependence	<1%/°C	2–3%/°C	3–5%/°C
Magnetic susceptibility	Very high (mT)	No (tested up to 9.4T)	No (tested up to 15T)

Note: (Reproduced and Adapted from [45])

guides and the ASIC in a fully shielded module that can be inserted into a preclinical high fields MRI, followed [15]. In this setting their excellent read-out properties (see Table 22.4) have been evaluated in a ring setting.

Already here, the operability under strong magnetic field influences could be proven [49]. If comparing to the performance of conventional PMT-based PET-detector(ring)systems, APDs are not as fast as needed if time-of-flight is desired. Here, a timing resolution on the order of several hundred picoseconds is necessary, whereas APDs can perform 'only' down to several nanoseconds as needed to determine coincidences and, thus, valid lines of response (see Chapter 18). A decade ago, silicon photomultiplier (SiPM) cells were introduced as light sensors [50, 51]. Recently, they have been employed as digital scintillation light sensors within PET detector modules. While APDs work at operating

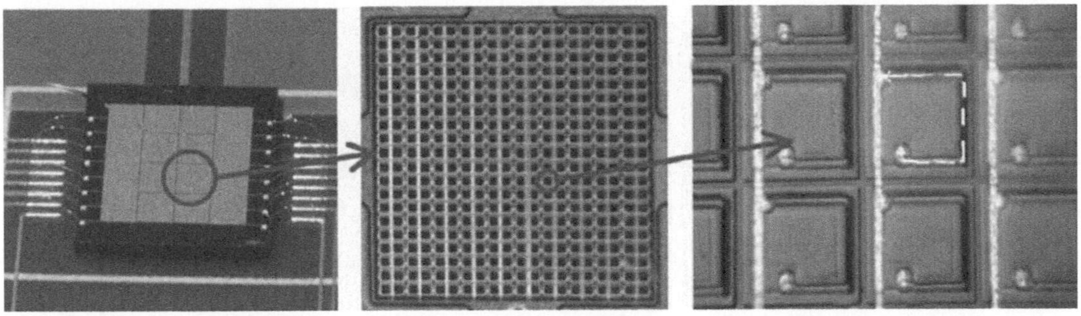

FIGURE 22.3 A SiPM array (left) is constructed of SiPM pixels (middle) that have dimensions of about 1x1 mm². The pixel consists of a multitude of individual SiPM-cells with dimensions from about 20x20 to about 100x100 μm². There are SiPM arrays available from 1x1 mm² (adapted from [11, 45]). They can be up-scaled by combining more pixels to lager arrays so they fit to the dimensions of the scintillator elements and can also be combined to modules reading out blocks of scintillator crystal elements of various dimensions as determined by the particular need and detector design.

voltages below their breakdown voltage where the obtained signal is proportional to the light photon energy, whole pairs generated as the photon is interacting, a SiPMs is essentially an APD working above the breakdown voltage, that is, in Geiger mode. Whenever a photon of visible light, generated in the scintillator material (crystal) is hitting the SiPM, it breaks down, and the response is a binary output signal (i.e. output=1 → visible light photon detected). Output=0 then means that no light photon was detected. So, the SIPM cell essentially generates a 'digital' output. Now enabled by the dimensions of the microcells of the SiPM, a considerable number of SiPM microcells can read out the scintillation light of one particular detector crystal. So, each crystal can be read out individually, eliminating the need for localization arrays, electronics and arithmetic within an 'Anger-mode' or the like detector block, as still widely implemented in state-of-the-art 'analog' clinical PET/CT systems. Basically being a solid state semiconductor working in Geiger mode, the SiPM is not sensitive to magnetic field influences up to 15T. Due to their favourable electrical and time resolution properties over PMTs, SiPMs have been introduced to be used also in digital PET detector systems that are employed in the current generation of hybrid PET/CT systems to further increase timing resolution and reduce the influence of noise by analog–digital conversion [52–54].

The energy discrimination – that is, determining the energy of the gamma ray causing the light pulse in the crystal material – is achieved by determining the intensity of the scintillation light. As the number of light photons in a pulse represents its intensity, counting the number of detected photons of visible light that is proportional to the number of 'fired' SiPM cells with output=1 (breakdown of the cell), enables the determination of the energy of the gamma ray. As there is no analog-to-digital signal conversion involved in this process, and photons of visible light lead to a rate of binary output of SiPM cells, this method can be called digital (light-) photon counting (DPC). More technical details and performance characteristics can be found in a nice review by Lecomte et al. [45] and an article by Seifert et al. [55].

It is the ultimate goal for hybrid PET/MRI systems to be able to acquire MR-signals at the very same time as the PET-signal is acquired, so a truly simultaneous measurement of them both is possible. This feature is needed for a very practical reason: first, MRI sequences take much longer to be acquired compared to a CT scan. A CT scan takes some seconds whilst a combination of desirable MRI-sequences can take up to an hour to be acquired. Moreover, the multitude of MRI-signals available (structural, molecular, in high temporal resolution) can be used for improvement of the PET-acquisition results. Due to their small dimensions compared to PMTs and their insensitivity to even very strong static and fluctuating magnetic fields, the availability of APDs and SiPMs was a key feature for simultaneous PET/MRI systems as they are available today. Beyond the rather practical technical reasons why one would aim for a simultaneous acquisition of PET along with MRI, there are a plurality of research and some clinical applications that only become possible, or at least profit, when PET and MRI are acquired simultaneously. The latter will be mentioned later in this chapter.

The slim design of the SiPM detectors (see Figure 22.3) allows the PET detector element and block design to be very compact compared to PMTs (see Figure 22.5) and the ring, composed of these detector blocks to be placed coaxial (see Figure 22.4) with the already very complex and large hardware of the MRI system, that is the static magnet (B_0), the gradient system (B_1) as well as the body transmission and receiver coil (see Figure 22.1d).

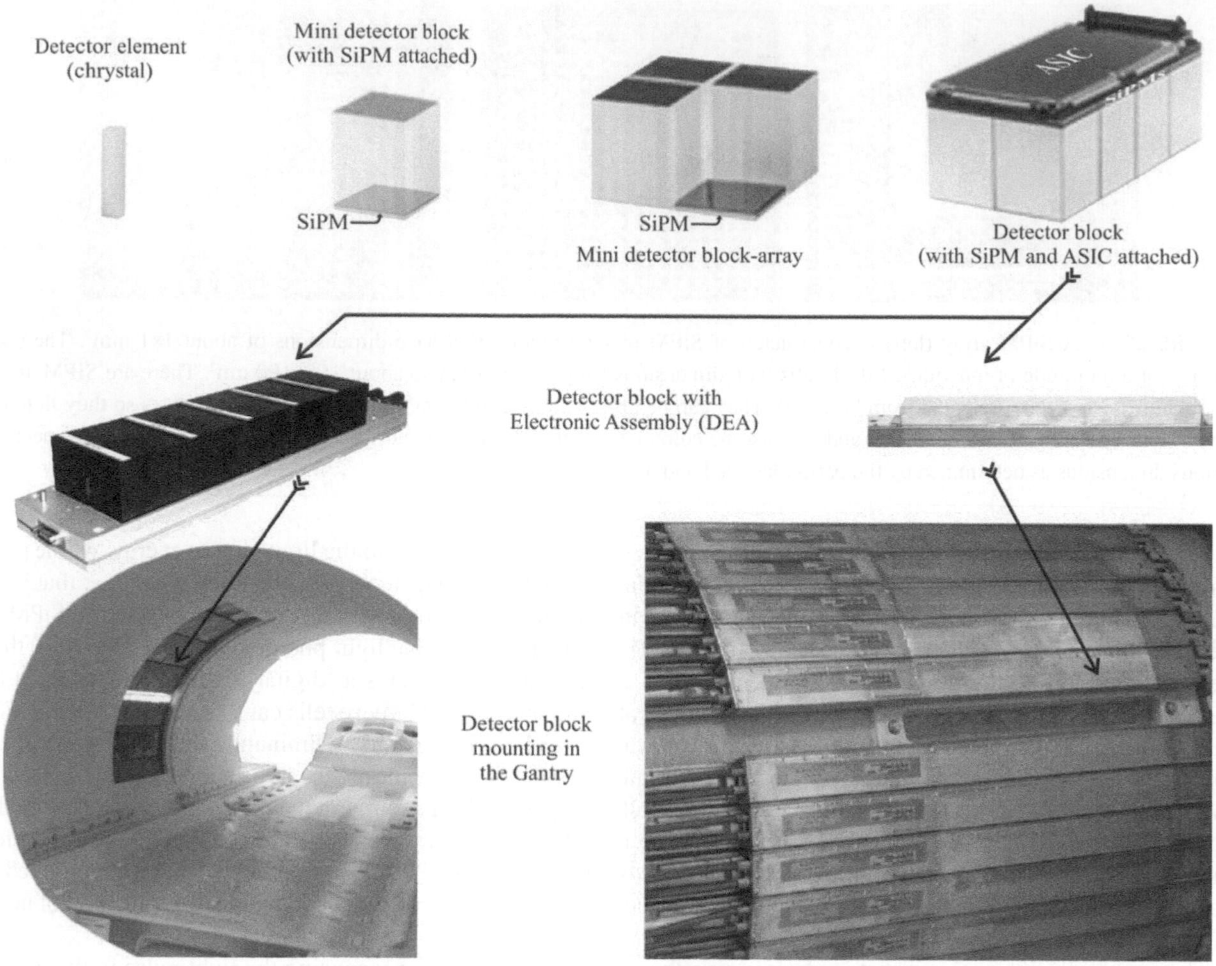

FIGURE 22.4 PET-Detector design beginning with the single LYSO detector crystal element, assembling a number of those to blocks, attaching the SiPM-arrays and the ASIC, combined with the Detector Electronic Assembly (DEA) to the circular mounting of the detector blocks into the Gantry forming the axial field of view of the detector (lower left: GE SIGNA PET/MRI; lower right: SIEMENS Biograph mMR, still with APDs as scintillation light sensors).

22.3 COMBINED PET/MRI SYSTEMS

22.3.1 Clinical PET/MRI Systems

22.3.1.1 Inserts into Stand-alone Standard MRI Systems

As shown in Section 22.2.2 (see Figure 22.1) there are four general ways of combining the PET and the MRI component. In this chapter emphasis will be on clinical, mainly integrated, simultaneous PET/MRI systems. Moreover, we will focus on the key features of the PET system and list the performance parameters of hybrid clinical PET/MRI systems as they are available from vendors today. To enable the proof of the clinical feasibility, more than a decade ago, a PET-insert to image the human brain was developed [56]. Subsequently, Siemens came up with the commercial clinical version, the brainPET detector system that could be inserted in an existing MRI system [17]. A very small number of these systems were manufactured for research purposes and to study the interplay with the MRI system. It had to be of a diameter to tightly fit into the bore of a commercial 3 Tesla Siemens MRI system. The experience in designing LSO/APD based PET detectors for preclinical PET inserts has been translated to develop the modules for this clinical system. It consisted of a 12 x 12 array of LSO crystals of $2.5 \times 2.5 \times 20$ mm^3 that are coupled via a short light guide to a 3×3 array of 5×5 mm^3 APDs. For the purpose of operation of the insert inside the MRI bore, the standard body coil of the MRI system was disabled as long as the insert was in place. Instead, a special transmit/receive head coil, designed

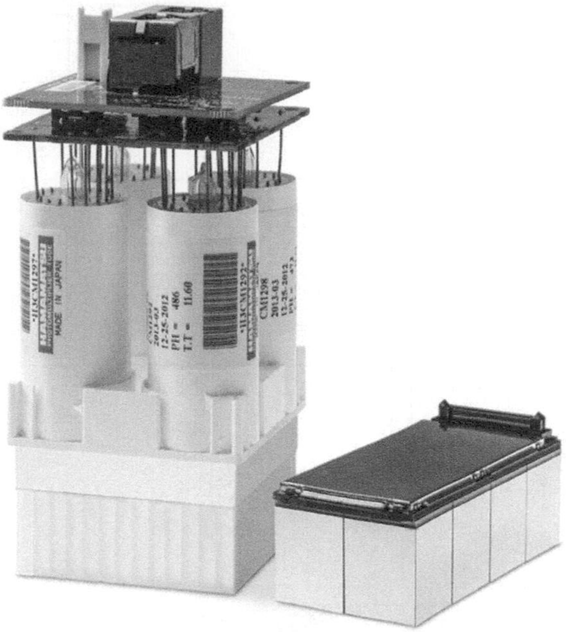

FIGURE 22.5 PET-Detector blocks. Right: Standard PMT based design; Left: SiPM based design. Due to the smaller dimensions of the PMTs the SiPM design allows for a much more compact design of the PET detector ring. (image reproduced with permission of SIEMENS Healthineers).

to minimize 511keV photon attenuation and to exactly fit into the PET field of view, was fixed to the patient handling system (bed) inside the gantry (see Figure 22.6, bottom right).

The system has an inner diameter of 376 mm and an axial length of 191 mm. This resulted in a comparatively high sensitivity of 7 per cent and a spatial resolution in the reconstructed PET data of about 3 mm full at half maximum, representing an excellent image quality. Obviously, this system was an optimal research tool to evaluate the many possible interferences of the systems with each other and artefacts such as, for instance, effects due to the presence of solid structures between the source of gamma emission and the PET detectors (photon attenuation by the transmit/receive coils), the design of the detector ring itself (gaps between the modules), or the effect of gradient switching on the PET detector ASIC electronics during simultaneous measurements. However, the latter turned out not to be of too much influence. The effect has been a reduction of the count rate of about -3 per cent as soon as fast switching gradients, such as during echo planar imaging (EPI) sequences, were transmitted. However, the influences on the accuracy of PET quantification – for instance by the aforementioned attenuation of the gamma rays by material and the fact that no tissue attenuation properties can be measured by means of an MR-measurement – were larger than the direct effects of electromagnetic influences of the systems onto each other's performance. Moreover, parallel to the development of whole-body integrated systems based on this experience, were investigations on:

- MRI-based methods for photon attenuation correction,
- the image quality and quantitative validity of the reconstructed activity concentrations in the PET data
- MRI-navigated motion correction of the PET data (gating)
- the optimization of clinically compliant workflows
- the further development, integration and optimization of data analyses and visualizations
- could be pushed forward to be consecutively incorporated in such systems. First clinical experiences such acquisitions of brain tumour patients imaged dynamically right after injection of 18F-fluoro-ethyl-tyrosine for 50 minutes [57]. Recently a further approach to develop a insertable PET detector ring has been undertaken, but this time with the goals to reduce cost for simultaneous PET/MRI again and to further investigate the influences of this SiPM-based, gaped PET detector system on the MRI performance of a standard stand-alone MRI system [37, 43, 44].

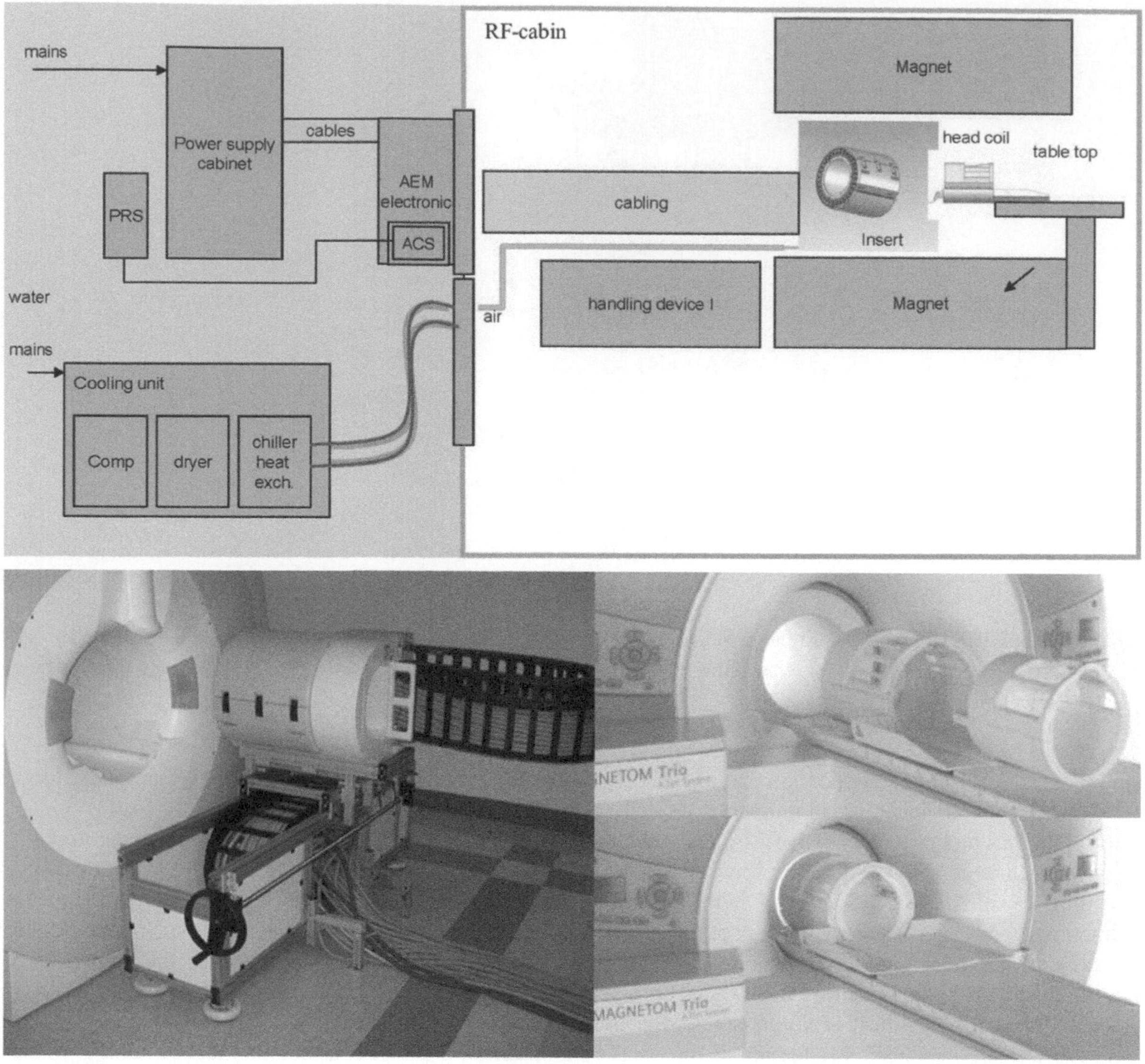

FIGURE 22.6 Top: Principal sketch of the brainPET insert with all required components (ACS: acquisition computer, PRS: reconstruction server, reproduced and adapted from [56] and with permission of SIEMENS Healthineers). Bottom: Physical layout of the brainPET insert (left [42]) with the RF transmit and receiver coils (right). Reproduced and adapted with permission from and SIEMENS Healthineers.

22.3.1.2 Sequential Whole-body PET/MRI Systems

Both Philips Healthcare and GE Healthcare came up with a sequential PET/MRI system (see Figures 22.7 and 22.8). Combined PET and MRI investigations were a close temporal correlation of the two modalities is needed was not possible with these systems. The important aims of these systems, however, were to explore potential clinical and research applications of hybrid PET/MRI and comparative studies to investigate and optimize MR-based attenuation correction methods without having to develop fully functional and regulatory approved integrated clinical PET/MRI systems. As PET/CT is essentially a sequential measurement, the clinical utility and value of sequentially acquired PET/MRI could be directly compared.

Obviously both systems had to ultimately rely on a sophisticated patient handling system capable of moving the patient from one modality to the other without, or with minimal, movement in between. The patient has to remain on the same bed-top throughout the transfer. If the MRI is performed first, all flexible receiver coils in place during this

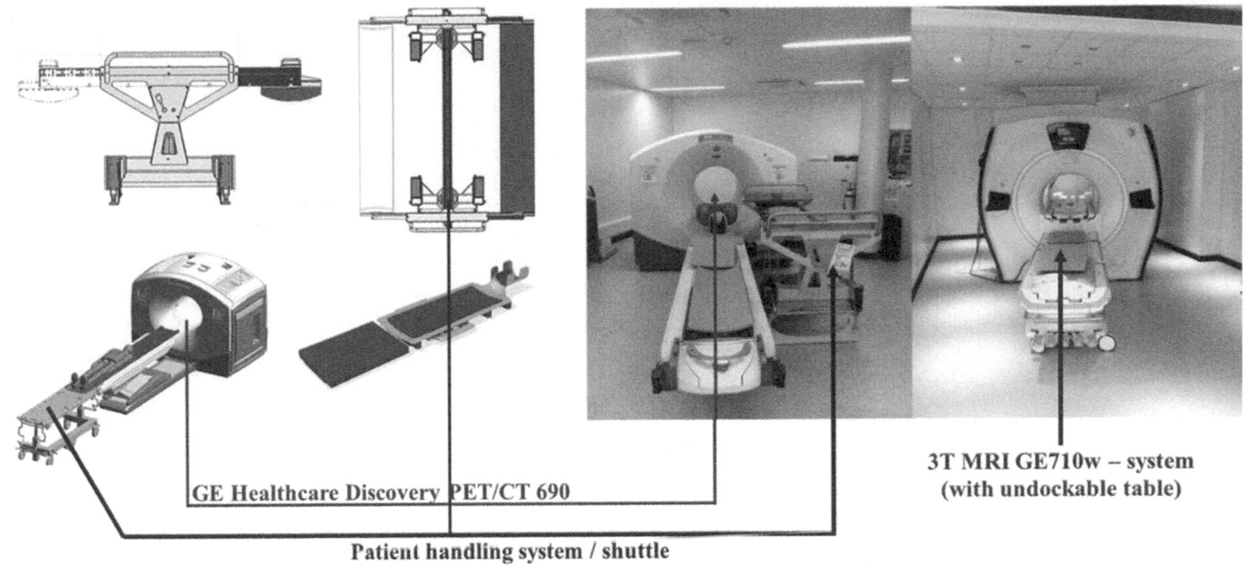

FIGURE 22.7 Sketch and setup of the GE-Healthcare tri-modality sequential PET/CT/MRI system. The PET/CT is situated in a room directly adjacent with the (RF-shielded) MRI-suite. The patient can be wheeled over from the one to the other modality without being moved from the relative position on the tabletop using a special patient handling system/shuttle. (image reproduced and adapted from [20]).

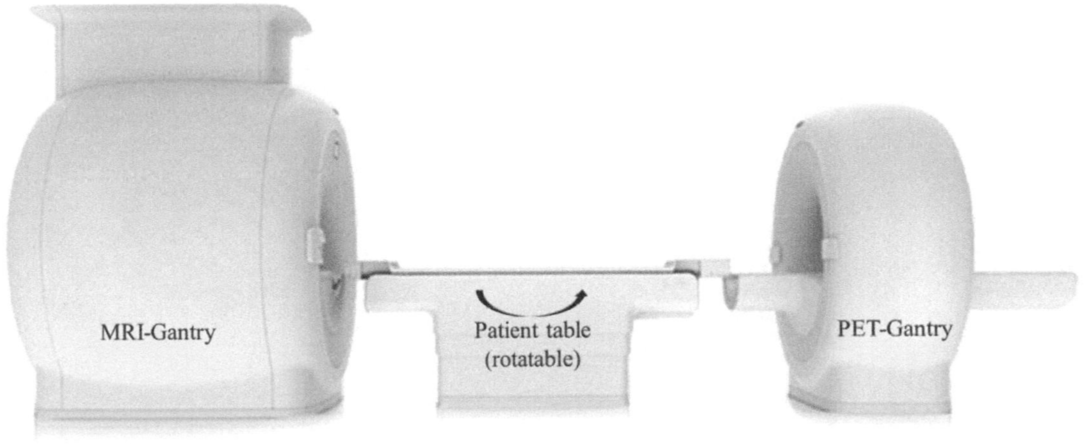

FIGURE 22.8 The Ingenuity TF PET/MRI of PHILIPS Healthcare. This was the first commercially available sequential clinical PET/MRI system. To move the patient between the modalities the table can be rotated around its vertical axis. (image reproduced and adapted with permission [21]).

investigation have to be carefully removed before repositioning of the patient in the PET(/CT) gantry. All rigid material that is reproducibly fixed to the patient couch/table and is present in the PET field of view again, must be represented with its linear attenuation coefficients and exact position relative to the patient-handling system to enable its involvement in the attenuation correction along with the reconstruction of the quantitative PET data.

22.3.1.3 Simultaneous Whole-body PET/MRI Systems

It has been stated that integrated systems with the capability of simultaneous acquisition of PET while several MRI sequences are running would be preferrable over sequential measurements. Figure 22.1 shows the considerations with respect to the combination of morphology and functional imaging. Moreover, and even if – due to the outcome requirements – a sequential measurement would suffice, it means a tremendous saving of time if the acquisitions of the

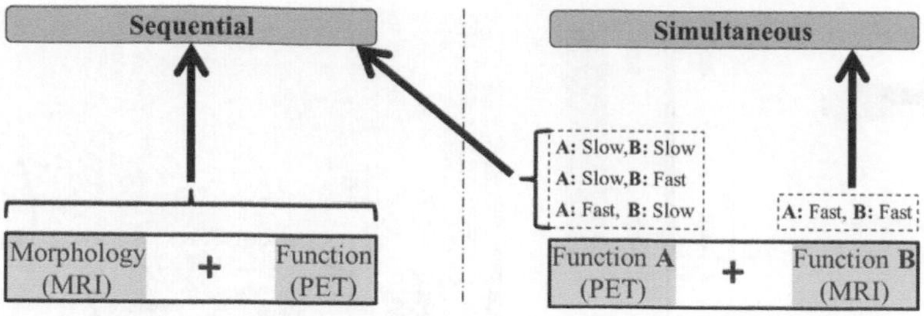

FIGURE 22.9 Morphology and Function. If we just consider these two, we need the simultaneity of the PET and the MRI measurement only in cases where we wanted to capture two fast functional processes or changes with PET and MRI at the very same time.

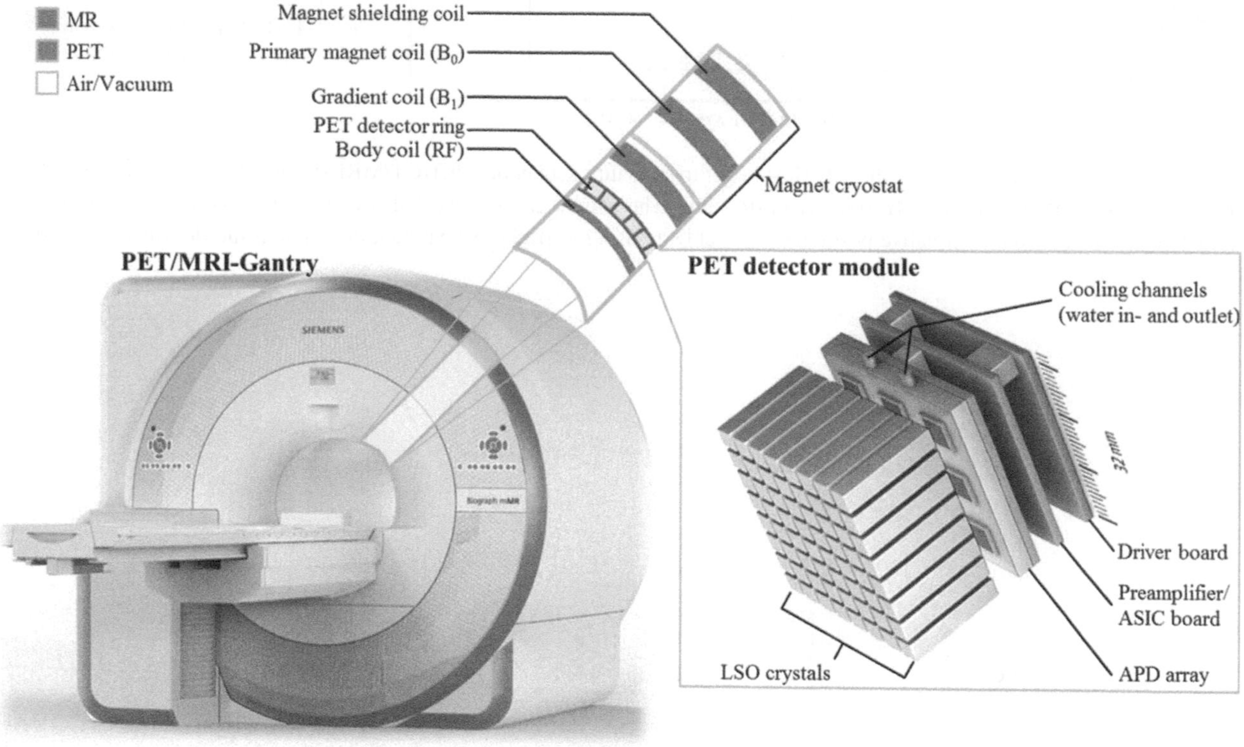

FIGURE 22.10 The SIEMENS Healthineers Biograph mMR simultaneous PET/MRI system. The left panel shows the stacking and Integration of the system components as necessary for operation of the MRI-System (green) integrated with the ring of APD-based PET detector modules (blue), kept in vacuum and cooled with liquid helium (cryostat). The right panel shows an explosion drawing of one of the PET detector modules (see also Figure 22.4, reproduced and adapted with permission of SIEMENS Healthineers).

two molecular imaging modalities can run simultaneously (see Figure 22.9). At first it was thought that PET would be the modality that limits the time needed for a whole simultaneous acquisition. But it turned out very soon that as soon as more than just a structural MRI (T1, T2 or the like) plus the MR-AC sequence is desired, the time for the MRI would exceed the time needed for the PET measurement to be completed at each station/bed position.

There is much more where simultaneity of the two is key, mainly in brain and cardiac molecular imaging and in situations where we want to utilize the high temporal resolution of MRI to observe involuntary cyclic or non-cyclic motion of the patient and use this information for gating and motion correction of the PET data. Last, but not least, the time saving is a major factor for patients when they have to undergo investigations in large-scale imaging systems with narrow bores and under varying conditions of morbidity [33].

Already, the brainPET as described above is an integrated system. Shortly after Philips came up with the Ingenuity TF PET/MRI (Figure 22.8), Siemens came up with the first integrated whole-body PET/MRI system, the Biograph mMR (see Figure 22.10). All experiences with the brainPET insert were thoroughly involved in this development. After thorough investigation of its technical performance [24] and sufficient accumulation of experience, it received CE and FDA approval in 2011. It is an APD-based system, whose detector assembly is installed between the gradient and body coils, leading to a tunnel diameter of about 60 cm. This is a bit less than the bore size of stand-alone Siemens 3 Tesla MRI systems but suffices for most of the investigational protocols that would be needed in clinical and research settings.

This system was the first commercially available fully integrated simultaneous whole-body clinical PET/MRI system and, thereby, is a milestone in the development of combining two major modalities of molecular imaging solving most of the technical challenges the modalities impose on each other. In 2017, major improvements of hardware and software were implemented in version E11, involving the results of the developments to improve the MR-based attenuation correction methods for PET as well as incorporating modern methods to apply motion correction of PET in the brain or other body regions using newly developed sequences to navigate and detect motion and apply the result for motion correction of the PET data. If listing performance parameters of the imaging system in the context of this book (see Table 22.5), we will of course mention basic features of the MRI system but otherwise we will primarily concentrate on the PET component.

TABLE 22.5
Specifications and NEMA Performance Parameters of the PET Component of the SIEMENS Biograph mMR (the Distance Column Gives the Distance to the Centre of the Field of View @ which the Value Is Obtained)

PET Specification and Parameter	Biograph mMR	
PET detector (crystal) material (see Table 22.3)	Lutetium oxyorthosilicate (LSO)	
Crystal element dimension	4 mm × 4 mm × 20 mm	
Number of detector elements	28672 (8×8/block, 56 blocks/ring, 8 rings	
Light sensor	APD	
Mode of acquisition	3D, delayed window random event estimation	
Field of View	25.8 cm (axial), 59,4 cm (transaxial)	
Detector ring diameter	65.6 cm	
Energy window / threshold	430–610 keV	
Coincidence window	5.9 ns	
Timing resolution (capability of TOF PET)	2930 ps (no)	
Spatial Resolution	*Distance*	
Transaxial	10 mm	4.2 mm (FWHM)
		8.8 mm (FWTM)
Axial	10 mm	5.7 mm (FWHM)
		11.5 mm (FWTM)
Transaxial (radial)	100mm	5.2 mm (FWHM)
		9.7 mm (FWTM)
Transverse (tangential)	100 mm	4.3 mm (FWHM)
		8.4 mm (FWTM)
Axial	100 mm	6.6 mm (FWHM)
		13.1 mm (FWTM)
Sensitivity	0 mm	15.0 kcps/MBq
	10 mm	13.8 kcps/MBq
Peak NECR		183 kcps @ 23.1 kBq/mL
Scatter fraction @ NECR peak		36.7%
Image quality parameters	*Contrast recovery*	*Background variance*
10 mm sphere	32.5 ± 5.1 %	5.3 ± 1.0 %
13 mm sphere	50.0 ± 9.2 %	4.8 ± 0.8 %
17 mm sphere	62.9 ± 7.2 %	4.1 ± 0.5 %
22 mm sphere	70.8 ± 6.0 %	3.7 ± 0.3 %
28 mm sphere (cold)	65.1 ± 1.2 %	3.3 ± 0.2 %
37 mm sphere (cold)	72.3 ± 1.1 %	3.0 ± 0.1 %

Notes: FWHM: Full Width at Half of Maximum; FWTM: Full Width at the Tenth of Maximum of the Point Spread Function. These Measurements Have Been Obtained Following the NEMA NU2- 2007 Standard [24, 58]

It is important to mention, that the image quality parameters are obtained using a µ-map of the Image Quality spheres phantom (see Chapter 21) that has been obtained from a CT-measurement for attenuation correction of the PET data. So, the image quality parameters obtained in this way do not represent the performance of the overall (i.e. combined) system, that is, involving an MR-based attenuation correction. Up to now this is the only option to involve phantoms or other equipment made of MR invisible material into the attenuation correction of PET in hybrid PET/MRI. The plastic housing of the phantom does not give any MRI obtainable signal that is related to its density. There has been research to optimize the kind of fluids used in the phantom [59], however the problem of MRI not being able to detect the phantom housing is still an obstacle if an image quality and quantification accuracy assessment is intended [60–62]. There is research underway trying to find materials that are MRI visible, ideally comparable to bone and 3D printable. A first study presenting a manually modelled head phantom with a central activity fillable compartment has shown the principle feasibility but challenges in constructing more subtle structures using 3D printing to reproducibly manufacture the fully MRI-visible phantom have to still be undertaken [63]. Usually, the NEMA NU2-2007 Standard for Performance Measurements of Positron Emission Tomographs [58] is sufficient guidance to assess the performance of PET detector system. The NU2-2012 version of this standard has only minor changes applied in contrast to the former version, mostly designed to make the tests easier to conduct, more reproducible, or more clearly defined. Further explanations of the changes can be found in the text of this standard [64]. The NU2-2018 version of this standard has two substantial additions incorporated now: A section describing how to assess the coincidence timing resolution of time-of-flight systems and a section describing how to assess the co-registration accuracy of hybrid PET(/CT) systems. Otherwise, there are some minor adaptions made, for instance to the type of radioactive sources that can be used in the tests and the way of assessing the image quality with the spheres phantom implemented. More detailed explanations of the changes can be found in the text of this version of the standard.

In 2014, GE came up with a fully integrated simultaneous system also, the SIGNA PET/MRI. It is based on a 3T MRI system also, the Discovery 750w. The major difference to the Siemens system is in the PET detector system. Utilizing SiPMs as light sensors coupled to L(Y)SO comparable lutetium-based scintillators, it is time-of-flight PET capable. The SIGNA PET/MRI is the first commercially available TOF-PET/MRI (Figure 22.11).

Since the appearance of the system on the market in 2014, it has been continuously investigated to test the interoperability of the components and to improve the hardware and software features. Recently, new reconstruction options using a Bayesian-penalized reconstruction iterative technique – well known from the PET/CT systems of GE – as well as Zero-time-Echo MR sequences to improve the attenuation correction by MRI-based detection and segmentation of bone, complement the formerly implemented atlas-based segmentation methods as well as the maximum likelihood reconstruction of attenuation and activity (MLAA) based quantification methods using the TOF information. No significant impact of the MRI- on the PET-performance and vice versa was found [65], not even when aggressive MR pulsing

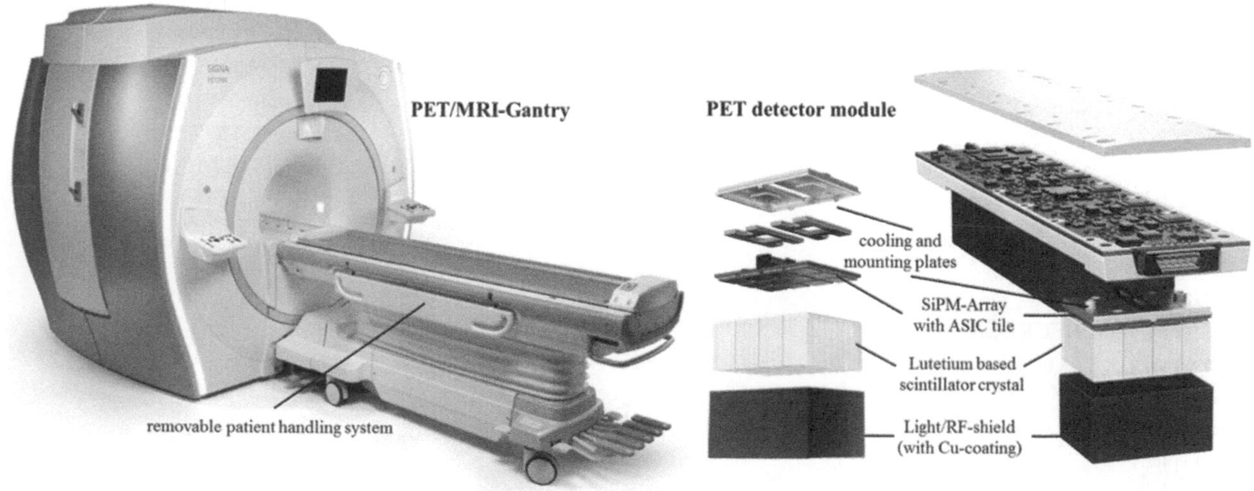

FIGURE 22.11 The GE Healthcare SIGNA simultaneous TOF PET/MRI system. The left panel shows the Gantry combining the PET and the MRI gantry as well as the removable patient couch. The right panel shows an explosion drawing of one of the PET detector modules and blocks (see also Figure 22.4, reproduced and adapted with permission of GE Healthcare).

TABLE 22.6
Specifications and NEMA Performance Parameters of the PET Component of the GE SIGNA TOF PET/MRI (the Distance Column Gives the Radial Distance to the Centre of the Field of View @ which the Value Is Obtained)

PET Specification and Parameter	SIGNA PET/MR	
Pet detector (crystal) material (see Table 22.3)	Lutetium based (L(Y)SO comparable)	
Crystal element dimension	4 mm × 5.3 mm × 25 mm	
Number of detector elements	20160 (8×8/block, 56 blocks/ring, 8 rings)	
Light sensor	SiPM	
Mode of acquisition	3D	
Field of View	25 cm (axial), 60 cm (transaxial)	
Energy window / threshold	425–650 keV	
Coincidence window	4.57 ns (±2.29ns)	
Timing resolution (capability of TOF PET)	<400 ps (yes)	
Spatial Resolution	*Distance (mm)*	
Transaxial	10	3.5 mm (FWHM, filtered TOF) 4.4 mm (FWHM, FBP)
Axial	10	4.7 mm (FWHM, filtered TOF) 5.3 mm (FWHM, FBP)
Transaxial (radial)	10	3,6 mm (FWHM, filtered TOF) 5,8 mm (FWHM, FBP)
Transverse (tangential)	100	3.5 mm (FWHM, filtered TOF) 4.4 mm (FWTM, FBP)
Axial	100	5.1 mm (FWHM, filtered TOF) 6.7 mm (FWTM, FBP)
Sensitivity	0 10	23.3 kcps/MBq 22.5 kcps/MBq
Peak NECR		218 kcps @ 17.8 kBq/mL
Scatter fraction @ NECR peak		43.6%
Image quality parameters (ToF)	*Contrast recovery*	*Background variance*
10 mm sphere	35.2 ± 2.0 %	4.9 ± 0.6 %
13 mm sphere	48.9 ± 2.2 %	4.5 ± 0.5 %
17 mm sphere	59.9 ± 2.6 %	3.2 ± 0.4 %
22 mm sphere	68.6 ± 2.2 %	2.7 ± 0.4 %
28 mm sphere (cold)	79.2 ± 1.9 %	2.2 ± 0.2 %
37 mm sphere (cold)	87.4 ± 0.7 %	1.9 ± 0.2 %

Notes: FWHM: full width at half of maximum; These measurements have been obtained following the NEMA NU2-2012 standard and under MR idle condition (no MR pulsing) using both filtered back projection (FBP) and iterative reconstruction involving ToF information [37, 64, 65]

such as fast-recovery, fast-spin echo MR sequences (high RF power) and echo planar imaging sequences (heat-inducing gradients) were selected. Based on the aforementioned NEMA standards, the performance of the system has been thoroughly investigated [37, 66]. A comprehensive summary of the specifications and of the performance measurements are comprised in Table 22.6.

Recently, to study the influence of the physical decay parameters on the performance parameters, investigations with ^{68}Ga and ^{90}Y have been carried out in comparison to 18F data [66]. The sensitivity investigation at 10 cm off the centre of the FoV results in 21.8, 20.1, and $0.653 \cdot 10^{-3}$ kcps/MBq for ^{18}F, ^{68}Ga, and ^{90}Y, respectively. The spatial resolution for ^{68}Ga and ^{90}Y is slightly degraded compared to ^{18}F. In the axial direction, the relative difference of FWHM values at 1 cm off centre compared to ^{18}F values is 17.8 per cent for ^{68}Ga and -1.3 per cent for ^{90}Y. However, as the NEMA tests are designed to characterize the detector – predominantly suggesting ^{18}F and ^{22}Na as radionuclides to be used – these results basically show the consequences of higher energy positron emission energies (^{68}Ga) and lower branching ratios (^{90}Y) of the physical decay.

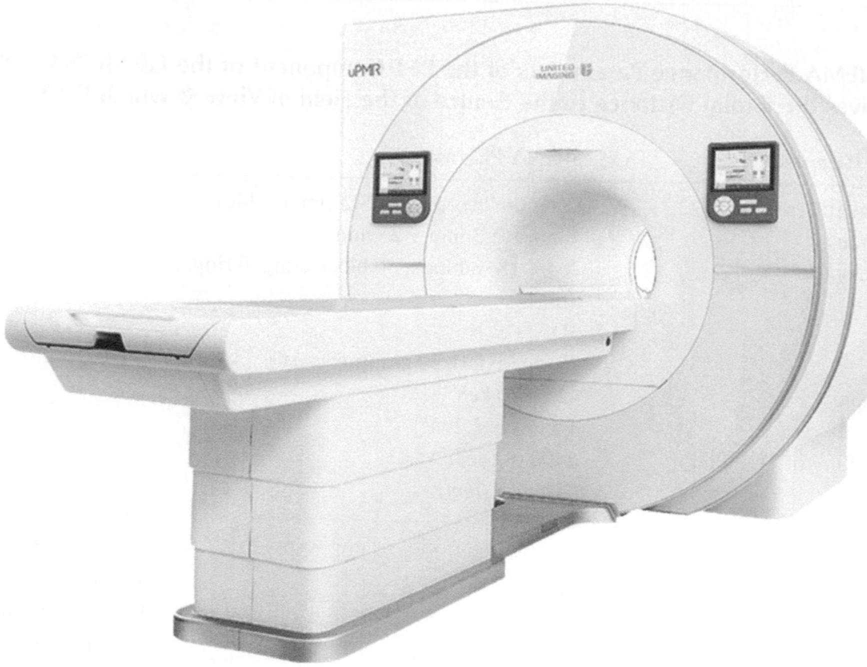

FIGURE 22.12 The United Imaging Healthcare uPMR 790 PET/MR simultaneous TOF PET/MRI system [67]. (with permission of UIH Healthcare).

Meanwhile, there is a third simultaneous TOF PET/MRI system commercially available, the uPMR 790 PET/MR system (see Figure 22.12) by United Imaging Healthcare, Shanghai, (UIH). As with the aforementioned systems, this system is based on a 3T MRI system of UIH also. The PET detector ring sits between the RF body coil and the gradient system. It utilizes LYSO/SiPM detectors assembled to a ring system covering an axial FoV of 32 cm and a transaxial FoV of 60 cm.

One of the first systems of the uPMR is installed in the Zhongshan Hospital, Fudan University, Shanghai. The physical–technical performance (see Table 22.7) as well as the clinical utility of the system have recently been investigated [67–71].

22.4 PET QUANTIFICATION IN HYBRID PET/MRI

22.4.1 Attenuation Correction of the PET – an Introduction

Essentially, a PET measurement aims at determining the three-dimensional distribution of activity in a given volume. For the quantitative validity of the PET measurement, this means that the concentration of activity (Bq/ccm) in the respective volumes and sub-volumes, and the respective physiological or pathological distribution of a given radiotracer, are to be determined as accurately as possible. Therefore, the energy degradation of the 511keV photons, as they travel through tissue until they reach the detector system, needs to be involved in the reconstruction of the emission data set. The half-value layer in tissue with properties similar to water is about 75 mm. The attenuation and scatter of photons is mainly determined by the electron density of the material they pierce through and interact with. If the trajectory of a photon is deflected, and it still happens to interact with the detector ring, it is considered a scattered event. If the deflection causes the photon not to reach the detector, it is irrecoverably lost. Attenuation of the annihilation photons to be detected by the detector ring, either scattered or even lost, does increase the background noise level in reconstructed PET data, causing a more or less blurred representation of the true activity distribution. Left uncorrected, this will lead to overestimations, if not to severe alterations of the reconstructed distribution of activity concentrations in the field of view under consideration, particularly on the body surface or regions with low attenuation, such as lungs or air cavities. Consequently, as much as possible, accurate scatter and attenuation correction is key in order to be able to reconstruct as much accurate activity concentrations as possible in critical structures of physiologically normal, as well as

TABLE 22.7
Specifications and NEMA Performance Parameters of the United Imaging Healthcare uPMR 790 PET/MR Simultaneous TOF PET/MRI System (the Distance Column Gives the Radial Distance to the Centre of the Field of View @ which the Value Is Obtained)

PET Specification and Parameter	uPMR 790 PET/MR	
PET detector (crystal) material (see Table 22.3)	Lutetiumoxyorthosilicate (LYSO)	
Crystal element dimension	2.76 mm × 2.76 mm × 15.5 mm	
Number of detector elements	78400 (7×8/block, 70 blocks/ring, 14 rings)	
Light sensor	SiPM	
Mode of acquisition	3D, delayed window random event estimation	
Field of View	32 cm (axial), 60 cm (transaxial)	
Energy window / threshold	430–610 keV	
Coincidence window	4 ns	
Timing resolution (capability of TOF PET)	480 ps (yes)	
Spatial Resolution	*Distance*	
Transaxial	10 mm	2.83 mm (FWHM)
Axial	10 mm	2.80 mm (FWHM)
Transaxial (radial)	100 mm	3,54 mm (FWHM)
Transverse (tangential)	100 mm	3.09 mm (FWHM)
Axial	100 mm	3.06 mm (FWHM)
Sensitivity	0 mm	15.4 kcps/MBq
	10 mm	15.3 kcps/MBq
Peak NECR		~125 kcps @ 11.5 kBq/mL
Scatter fraction @ NECR peak		~38%
Image quality parameters	*Contrast recovery*	*Background variance*
10 mm sphere	54.5 %	7.4 %
13 mm sphere	77.1 %	6.6 %
17 mm sphere	89.4 %	5.6 %
22 mm sphere	91.9 %	4.4 %
28 mm sphere (cold)	81.9 %	3.5 %
37 mm sphere (cold)	87.7 %	3.0 %

Notes: FWHM: Full Width at Half of Maximum. These Measurements Have Been Obtained Following the NEMA NU2-2012 Standard and under MR Idle Condition (No MR Pulsing during PET Acquisitions) Using Filtered Back Projection Reconstruction (FBP)

pathologically altered, uptake of the particular radiotracer. Now, the electron density can be directly found by obtaining transmission measurements either by using radioactive sources emitting 511keV photons or the X-rays as emitted by the tubes of CT systems. The reconstructed transmission volume data sets can be used directly if obtained from 511keV transmission measurements or after a (bi-linear) calibration/scaling of the linear attenuation coefficients (LAC) at X-ray energies to those at 511keV [72], the energy of the photons as emitted after annihilation of a positron with an electron. In contrast, a MR-signal basically represents the proton density in a tissue. This is by no means representative for the attenuation behaviour of material against gamma radiation. Thus, several approaches have been established to identify the different tissue classes using the MR signal. Basically, these tissue classes are fat, soft tissue, air, fat and lung. Depending on imaging circumstances (i.e. the imaged body part, position of the patient in the system, patient comfort and preparation, etc.) the detection of these tissue classes succeeds – or perhaps does not. Meanwhile there are methods capable of assigning continuous linear attenuation coefficients to the tissue classes rather than assigning a constant linear attenuation coefficient, as was the practice in the early days of using MRI-signals to segment the tissue classes. The so-constituted segmented μ-map is used for attenuation correction of the PET emission data. In the following chapters we will systematically allude to the various options of implementing attenuation correction of PET using data available in a hybrid combined PET/MRI imaging setting.

22.4.2 Approaches of Attenuation Correction for PET/MRI

We must confess that the linear attenuation coefficients of the tissue cannot be measured and obtained directly by a transmission measurement such as CT commonly done in the hybrid PET/CT setting. So, we have to use the signals available in the PET/MRI setting. To begin with, methods deriving the attenuation properties of tissue from MR signals will be mentioned and explained.

22.4.2.1 MRI-based Attenuation Correction

Still these algorithms suffer from being insufficient to detect bone and air. In their recent implementations most of the vendors use ultrashort and zero echo time MR sequences to detect bone and, thus, improve the performance of the tissue class segmentation. These methods are combined with methods of μ-map generation from MR data that use structural (i.e. T1- or T2 weighed) MR data sets in combination with CT atlas-based information of a particular part of the body to generate a more realistic map of linear attenuation coefficients, including bone [73–78]. As anatomical atlases are the basis for the assignment of LACs, those methods start to fail if the anatomy of the particular subject under investigation differs from the circumstances in the atlas. Ladefoged presented a method called Resolute [79]. The method was extensively validated for the head and neck region and proven to be very stable [80] in comparison with the CT as a ground truth. Its accurate performance could be shown under more difficult anatomic circumstances and alterations as might be present after surgical interventions also [81]. In recent research settings, neuronal network artificial intelligence approaches are employed to train algorithms using CT and 511keV data to learn generating continues-valued maps of LACs on the basis of structural MR data sets. Using the aforementioned methods, and depending on the body part under investigation, the accuracy of the PET measurement in hybrid PET/MRI settings reaches the order of that in PET/CT settings [80]. All the hardware in the trajectory of the gamma rays needs to be taken into account as it also degrades/attenuates the PET signal. The (flexible or rigid) MR signal receiver coils and the patient table are either implemented by CT-measured maps of LACs or designed so that the attenuation of the PET signal by this material is negligible. Most of the harmonization procedures of quantitative PET as known from PET/CT are based on the measurement of known phantom structures filled with watery solutions of radioactivity containing different fillable subvolumes and, thereby, representing known activity concentrations in volumes of different sizes in either a cold or hot background. Firstly, being constructed mainly of plastic, the structure of those phantoms cannot be detected sufficiently by MRI. Secondly, larger volumes of pure water in the MRI field of view causes major distortions of the MR signal. This topic has been addressed by searching for alternative liquids to fill the phantom [59]. Current approaches to use activity-fillable phantoms in hybrid PET/MRI, however, employs the implementation of CT-generated μ-maps of the particular phantom to account for the attenuation of the PET signal. Thus, inter-system quantitative comparisons give the comparability of just the quantitative performance of the PET detector system. If the clinical settings for attenuation correction – that is, the MR-based μ-map – is used for attenuation correction of phantom measurements, considerable deviations of accuracy of the PET measurement is found [61, 82]. After this rather quick compilation of the methods available, let us systematize the MRI based attenuation correction methods.

Segmentation-based methods make use of several structural MR images to identify anatomical structures and regions. Depending on the capability of the MRI sequences, up to five classes of different tissues, such as soft tissue, pure air, lung, bone and fat, can be distinguished. The labelling and identification of those tissues is based on parameters such as the signal intensity yielded by the various MR imaging sequences, the shape as well as the location of the particular structure, organ or system of structures or organs. Once identified, these structures are assigned predefined fixed linear attenuation coefficients. Usually, values of attenuation coefficients of 0.1 cm^{-1} (soft tissue, water), 0.085 cm^{-1} (fat), 0.023 cm^{-1} (lung) and 0.25 cm^{-1} (bone) are then assigned [78, 83]. To distinguish between watery tissue and fat, the Dixon-Sequence has proven to be sufficient [84]. This sequence delivers in-phase and out of-phase fat and water images that allow for serving as an anatomical reference to locate radiotracer uptake. Moreover, the segmentation of the lungs from these sequences is possible. Its accuracy, however, seems to depend on the presence and amount of adipose tissue in the proximity of the lungs [61]. However, due to the short relaxation time of bony structures in MR signals, it is not possible also to detect bone with the Dixon method. Meanwhile there is a variety of ultra-short echo time (UTE) [85–87] or zero echo time sequences [74, 88] that do account for that fact and are able to detect bone and, thus, facilitate the segmenting of bony structures and the linear attenuation coefficient be assigned, respectively. When assigning the aforementioned fixed linear attenuation coefficient for bone, however, this introduces errors as bones have varying linear attenuation coefficients, from spongy bone (around 0.1 cm^{-1}) to compact bone (0.25 cm^{-1}). This topic has been addressed in several studies, and sufficient methods to derive continuous valued linear attenuation coefficients from MR signals have been proposed [79, 89]. Ultra-short or even zero-echo-time sequences do prolong the MRI scanning time by about 3 to 5 minutes per imaging station / bed position. Moreover, if larger bone structures are considered, or if they are in close proximity to air cavities (e.g. in the case of facial bones), considerable errors in detection and, consequently, segmentation of bone do appear. Also, here, efforts have been made to address this issue, and methods that mostly are template-supported have been introduced [73, 90, 91]. Atlas- and template-based methods use anatomical priors to guide segmentation-based methods. Thereby, long acquisition times can be circumvented. Structural datasets such as T_1- and T_2 MR images as well as CT datasets do then serve as references in a standardized anatomical space. This atlas is then transformed to an MRI of the patient representing the individual anatomical structures. Templates are mostly formed from averages of anatomically normal structural CTs but also from more complex approaches using combinations of structural MRI and CT data, and sometimes involving individual information from the patients structural MRI are employed.

Without any longer acquisition times than is common in UTE and ZTE sequences, this method is potentially capable of involving more anatomical detail, including bone. It is immanent with atlases and templates generated from mostly anatomically regular and normal structures that methods relying on those are limited to patients without severe anatomical alterations in the imaging field of view or the structures represented in the atlas. The advantage in using supporting atlases and templates is that much more anatomical detail, including continuous values of LACs, for instance of bones, can be involved. The appropriate anatomical structures that the μ-values are assigned to are those selected on the basis of the individual MRI of the subject or patient under investigation.

Artificial intelligence (AI)-based methods to derive the map of linear attenuation coefficients have been introduced in the last decade. Combinations of segmentation and atlas-based methods form approaches where convolutional neuronal networks using deep-learning algorithms are trained to generate the μ-values from structural MRI datasets using training datasets of CT images as a reference [77, 89, 92–94]. In Figures 22.13 and 22.14 examples of this method are shown. Burgos and colleagues (see Figure 22.14) have compared their algorithms to generate pseudo CTs from structural MRI data. They propose a CT synthesis method based on pre-acquired MRI-CT image pairs. The validation consists of comparing the pseudo CT images synthesized with the proposed AI method based on a multi-atlas information propagation scheme [95] extended by involving multi-contrast MRI data [77] to the original CT images of patients.

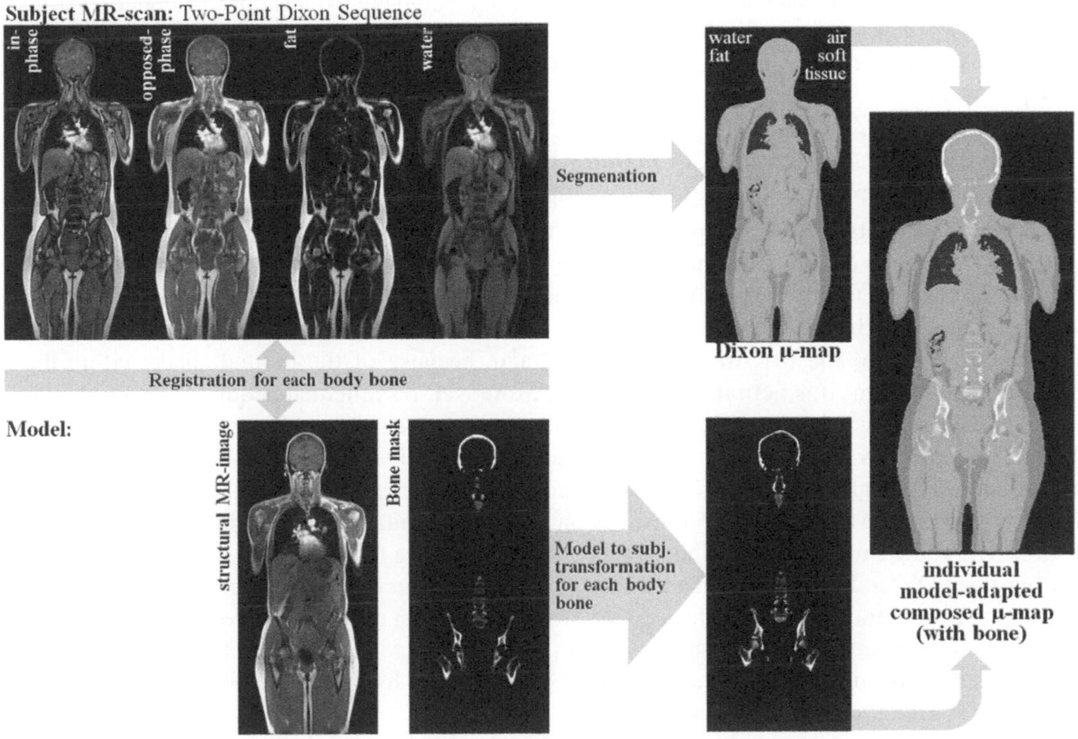

FIGURE 22.13 Template/Model/Atlas-based μ-map generation using anatomical priors to compose a map of linear attenuation correction values from a DIXON-based μmap (upper row) and a structural MRI of the individual patient (lower row). The bone mask/template is transformed involving the individual structural (i.e. T_1-MRI) and integrated into the DIXON-generated μ-map resulting in a map that represents water, fat, air, soft tissue and bone. (reproduced and adapted from [78]).

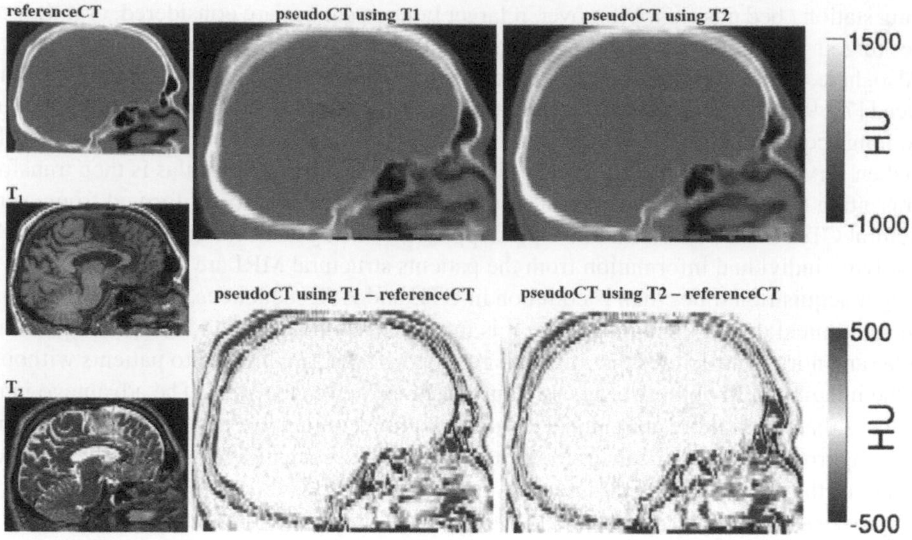

FIGURE 22.14 AI based method to generate pseudo CTs from structural MRI data based on a multi-atlas information propagation scheme [95]. Left column (small images): reference CT, T_1 and T_2 image of an example patient case. Upper row (large images): generated pseudo CT using T_1 and T_2 information. Lower row (large images): HU-difference images to the reference CT. Colour image available at www.routledge.com/9781138593268.

The success of these methods strongly depends on the diligence they are trained with. It is crucial that the training data set and – if applicable – the supporting templates and atlases are composed with the due diligence to be applicable for the intended result. Otherwise, the best neuronal network and deep-learning based approach can produce completely useless, fallacious, and misleading results. Up-to-date and complete explanations of both the basics of and the application of AI in medical imaging, particularly in the field of molecular imaging, is given in the recently published *European Nuclear Medicine Guide* [96] and several other very recent publications focusing on its application in PET/MRI [97–100]

22.4.2.2 PET Data-based Attenuation Correction

As the acquisition times of PET are typically shorter than those of MRI imaging sequences, it might be interesting to investigate deriving the attenuation properties of the tissue. Several methods that rely on emission as well as on transmission data have been developed. The acquisition of transmission data as has been standard with early stand-alone PET systems using rotating radioactive line- or point-shaped sources rules out as this would both introduce moving rotating parts into the FOV of the MRI system that, given a simultaneous acquisition of the both, would introduce severe artefacts to the MR image acquisition and, because of the limited space within the integrated bore of the combined system, it is actually not possible. Moreover, a sequential acquisition would add on the duration of the overall imaging procedure and, thus, is not suitable either. Recently, groups in Vienna (Austria) and Leipzig (Germany) have developed a phantom as well as pre-clinically tested head MRI receiver coil that involves a rotating, non-metallic, fillable point source that helically revolves around the coil while the MR signal is acquired [101, 102]. Using radioactive sources for transmission has the advantage of acquiring attenuation coefficients at the energy of 511keV, the same as the annihilation photons originating from the activity taken up by the tissues and volumes within the FOV and the patient. The main challenge with such transmission measurements simultaneous to the emission measurement, however, is to account for the separation of emission from transmission events by the detectors and for count-rate or dead-time problems in the PET detector system due to the necessarily rather high radioactivity concentrations in those sources. The latest generation of hybrid PET/MRI systems is capable of Time-of-Flight (TOF) PET signal detection [37, 67].

This information can be used in combination with algorithms of Maximum Likelihood reconstruction of Attenuation and activity [82, 103–107], as have been developed before PET/MRI became available. These methods have been adapted for PET data acquisition within combined PET/MRI systems. Moreover, these methods can be used to compensate for missing information about the body contour and fill information in this truncated region [108]. Figure 22.15 shows an example of a typical truncation artefact due to the limited transaxial field of view. For emissions-based

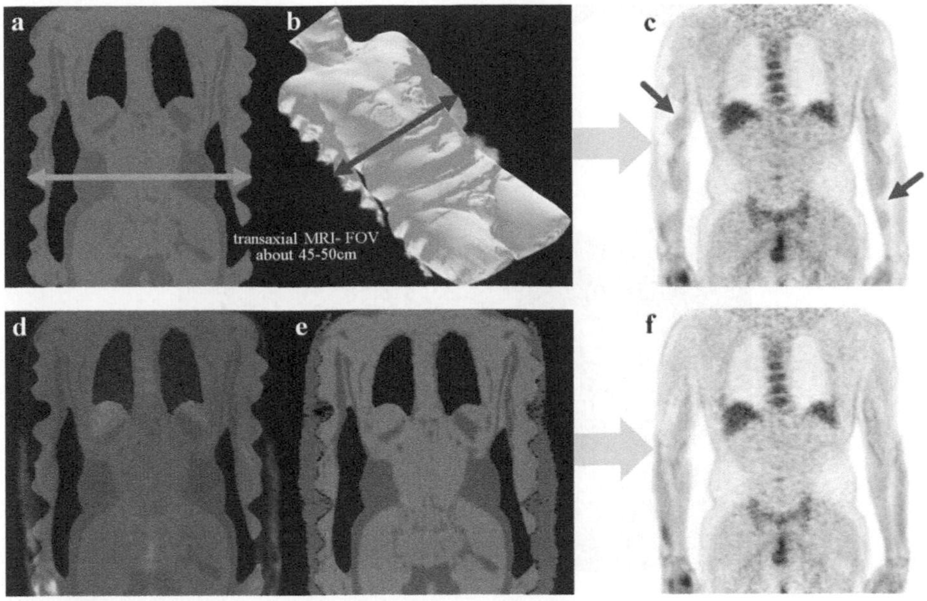

FIGURE 22.15 DIXON based μ-map derived from MRI with the arms down. Due to the limited transaxial field of view of the MRI (a,b), information of the periphery of the body is truncated. Left uncorrected, this translates into artefacts if this μ-map is used for attenuation correction of the PET data (c). Due to the larger diameter of the PET detector ring and if the tracer in use distributes peripheral (as for instance FDG does) the contours of the body can be derived from the non-attenuation corrected PET scan (d) and the missing μ-values can be filled into this area (e). Let alone that also with this approach no μ-values of the large arm bones are introduced, the resulting attenuation corrected PET scan (f) is still considerably improved to enable a visual assessment of these regions at least. (Reproduced and adapted from pictures with courtesy of Ambros Beer, Department of Nuclear Medicine, University medical centres Munich and Ulm). Colour image available at www.routledge.com/9781138593268.

attenuation correction methods it is very important that the tracers used do distribute also in the periphery of the body such as FDG and the like.

22.4.2.3 Attenuation Correction of Hardware Components

The correction of scatter and attenuation is more difficult if attenuating objects are without, or not surrounded by radiation emission, that is, outside the body. In the case of rigid material, pre-measured fixed attenuation maps of those (see Figure 22.16) either generated by 511keV transmission measurement or by scaled CT measurements have to be involved in the reconstruction or this material and shall cause only negligible attenuation of the radiation emitted by the patient.

Flexible MR-receiver coils to be applied in hybrid PRT/MRI are particularly designed to be very light so as to contain as minimal as possible metallic parts to be placed in the FOV of the PET detector system. The interested reader finds more details in several publications about the consideration and correction of hardware components in PET/MRI settings. This topic becomes of even more importance if PET/MRI serves as a source for external beam radiotherapy planning [91, 109].

None of the methods for attenuation correction in hybrid PET/MRI settings, may be MR-based, transmission-based or emission-based, do reach the overall accuracy, as was possible in the past with stand-alone PET systems using 511keV sources, or is possible with state-of-the-art CT-based attenuation correction of PET today. For the brain region, however, this issue has been widely solved by combinations of the described methods. In a large multi-centre study, Ladefoged could show that accuracy as we know it from standard PET/CT systems can be reached for this body part if the most advanced methods involving the AI technology, are applied [80, 110]. The error of quantifying various activity concentrations in various places in the brain is on the order of what we know from PET/CT. In regions close to the skull or air cavities near facial bones, this error slightly increases but, with methods using atlas and (bone-) template-supported AI-approaches it is still tolerably low (smaller than 3%). As shown in Figure 22.13, this development is underway for other regions of the body [78, 91] too, but due to the – with respect to attenuation – more complex and varying structures, there is more work to be done here. Meanwhile, some of those methods are implemented in the current clinical PET/MRI systems.

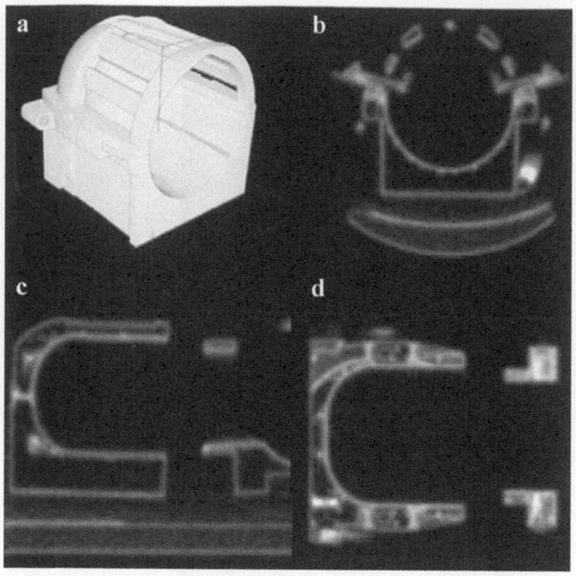

FIGURE 22.16 Head Coil of the SIEMENS Biograph mMR (a) and related hardware component μ-maps. Exemplary transaxial (b), sagittal (c) and coronal slice (d). (reproduced and adapted from pictures with courtesy of Ambros Beer, Department of Nuclear Medicine, University medical centres Munich and Ulm).

22.4.3 Image Reconstruction in PET/MRI

Generally, we can separate the algorithms into algebraic and iterative. Algebraic algorithm such as filtered back-projection needs sufficient angular equidistant projection data. This method is prone to noise artefacts when the count statistics in the projections is poor and decreasing with increasing the number of projections. Moreover, in case of missing data, that is, if metal implants influence the projections, severe artefacts are introduced into the reconstructed activity distribution. Contrary, iterative reconstruction techniques are much more robust in the presence of those fairly frequently occurring circumstances. The main advantage of iterative reconstruction is that the physical performance of the event detection and even the attenuation correction can be directly involved in the loop of image generation. Moreover, iterative methods can flexibly be optimized in speed and/or output quality, for instance splitting the iteration loops into subsets and/or involving a priori information about both properties of the object and the origin of the event, respectively. For more details of image reconstruction in general and particularly for emission tomography, please refer to Chapter 26 of this book or to the respective chapter by Belcari and colleagues in the *Textbook of Nuclear Medicine* [111]

The same PET reconstruction techniques applicable to emission data acquired in state of the art hybrid PET/CT systems can be employed in hybrid PET/MRI – particularly those methods that involve partial volume correction based on a perfect co-registration of the emission data to structural data profit from the capability of simultaneously acquired and, thus, inherently co-registered structural MR images with much better soft tissue contrast compared to CT. Moreover, MRI systems are capable of acquiring this structural information with a superior temporal resolution. This opens up opportunities to greatly improve the involvement of motion correction into the PET reconstruction. The extent and amplitude of patient motion can be of different origins. On the one hand, we have regular cyclic physiological motion, that is, of the heart and due to respiration. On the other hand, there is involuntary, irregular motion of patients such as tremor or which is due to the relatively long acquisition time of combined PET/MRI. These different kinds of motion are best considered entirely independent. While fully extended motion correction is technically complex and not always straightforward to integrate into clinical routine, attenuation correction involving motion-correction is relatively simple to implement and apply [112]. Indeed, the MR sequences used for attenuation correction can be readily modified to yield four-dimensional datasets representing, for instance, the respiratory cycle. With these at hand, regular PET gating techniques can be used to match emission and attenuation datasets [113].

22.4.4 Standardization of PET/MRI Imaging

As an accurate quantification of PET images is of utmost importance in research and also in clinical settings, where there need to be standards on how to acquire the data and quantify the results [114]. First of all, all centres/

departments where patients/subjects with a certain disease shall be PET/MR imaged shall use comparable, ideally the same protocols. The imaging procedure shall be standardized. Moreover, the image analysis shall be performed using comparable methodology and, thus, shall deliver comparable, ideally the same results across centres/departments. For PET/CT there are number of programs featuring and maintaining such standardization. Usually, there are preset procedures to follow, starting with simple phantom measurements, to check the overall calibration of the PET detector system including attenuation correction. Further, more complex cross-calibration measurements, including all systems measuring activity such as the activity meter (dose calibrator) and – at the lower end of activity concentrations – well counters to measure small volume samples taken from a phantom (aliquots) or the patient (mostly urine or blood). The idea behind cross-calibrations is that all activity-measuring systems that are involved in the protocol shall give the same results upon presentation of the same activity concentrations. For one of the available integrated PET/MRI systems the cross-calibration and long-term stability of the activity calibration has been found to be very satisfactory [86, 115]. One of the simplest, but very important, checks is to observe that all of those systems run at the same system time. Considering the half-lives of the usual radionuclides, for instance a difference in system times of only ten minutes between the PET imaging system and the activity meter will introduce a quantification error of some one-digit percent – additionally to unavoidable other errors introduced by other factors. Finally, there are more sophisticated protocol measurements, where more complex phantoms mimicking cold and hot lesions as well as background, and no activity structures are being filled with various activity concentrations. The latter measurements are intended to monitor the image quality in various situations. Some of these measurements are part of the acceptance tests of the PET system following the NEMA standards [64, 116]. Now, all the requirements of quantitative accuracy and image quality shall of course be fulfilled in one centre and/or PET imaging facility internally to ensure state-of-the-art clinical care. In multi-centre research projects, comparable image quality and quantitative accuracy are key if comprehensive data pooling is intended. Many of protocols of those trials have a requirement that centres that intend to participate have to prove the quality of their PET imaging by providing results of measurements as mentioned above, both before the start of the trial, as a proof of the required imaging practice, quality and quantification accuracy, and in preset intervals throughout the duration of the year.

For PET/MRI some of those requirements are challenging to fulfil. Particularly, this is the case because of the fact that the phantoms that are usually made of PMMA, which gives no signal in an MR measurement. Moreover, large water volumes are prone to introduce severe distortion artefacts into the MR images. Hence, the overall system performance including the MRI-based attenuation correction is hard to prove. Initially, these problems are circumvented by implementing pre-measured μ-maps with CT to be used for attenuation correction of the PET. Now, this delivers the performance of the PET part of the hybrid system *only*. Actually, one would want to know the *overall* performance of the *whole* hybrid system, that is, in the case of PET/MRI, including the MRI-based attenuation correction. To date, this feature is not available, mainly due to the lack of appropriate phantoms having MRI-compatible signalling features. The topic of phantom fluids has been addressed, and the suitability of various fluids was rated based on the ^{18}F-FDG solubility RF homogeneity in MR at 3.0 Tesla (MR signalling) and practicability [59]. The topic of PMMA not giving a signal in MRI has been addressed by various authors searching for such material, and it has been surveyed [117] and a first prototype for the brain has been produced [63]. The latter, being a kind of an anthropomorphic phantom, mitigates one more very important limitation: Most of the template-supported attenuation correction methods will work with this prototype, while for the standard phantoms coming from the stand-alone PET or PET/CT field, no templates exist and, thus, all the aforementioned methods of attenuation correction using templates cannot work. Manufacturers are aware of this limitation and provide template-based solutions for the most commonly used phantoms. As the development of anthropomorphic PET/MRI compatible phantoms is just beginning, at the present time the only way of testing a PET/MR system as a whole, or its MRAC, is by scanning an actual human subject, with all the limitations that entails [114, 118]. One of the examples to test the overall performance of MR-based attenuation correction of three of the four commercially available PET/MRI systems is a study in which three human subjects have been measured on the three systems [61] complimented by some phantom measurements [119] using commonly available phantoms.

22.5 CLINICAL APPLICATIONS FOR PET/MRI

In the scope of this book being primarily dedicated to readers from the field of (medical) physics and technology and engineering, the clinical applications of PET/MRI will just be named rather than being explained in much detail. First of all, it is very important to mention that PET/MRI is not intended to replace PET/CT. PET/CT will always be the primary hybrid modality in the fields where it is known to be applied. These are oncology, infection medicine, cardiology, and neurology. PET/MRI has set out to complement PET/CT where it provides superior features over PET/CT.

Generally, the quality of PET systems integrated with PET/MRI is comparable with those in PET/CT systems. It is the combination and full integration of the structural imaging modality, delivering a superior soft tissue contrast over PET/CT with the functional PET imaging so that a simultaneous measurement is possible (see Figure 22.9.). Moreover, MRI is also capable of functional imaging on a molecular level. The difference to PET, however, is the sensitivity being in the micromolar order of magnitude whereas – depending on the radiopharmaceutical used – PET reaches a nano-, sometimes picomolar sensitivity. Combining these features makes PET/MRI the milestone of molecular imaging of the last decade, enabling multi-parametric imaging for tissue characterization and deeper understanding of disease biology. The following list is shows the most frequent applications of PET/MRI but is not limited to these.

In the diagnostic imaging workup of neurologic, particularly *neuropsychiatric*, disorders, structural and functional, MRI has a key role. Combining these features with the strengths of PET gives a large variety of new diagnostic opportunities, beginning with simple things such as all the necessary imaging can be obtained in a single investigation, saving a lot of time and easing the burden on the patients. Particularly with patients suffering from dementia, this is a key factor. Further, the combination of PET with multi-parametric imaging along with the superior soft tissue contrast in structural MRI enables subtle studies and differentiation between various causes of dementia or – depending on the tracer in use – at least, exclusion of various types of dementia and/or neurodegeneration. This 'one-stop-shop' approach has proven to have great potential to improve not only diagnostic accuracy but also patients' and doctors' comfort [120].

Another important field of clinical application of PET/MRI is, of course, *oncology*. Again, it is of particular importance, where soft tissue contrast is needed. Here, prostate imaging combined with the different, now available mainly Ga^{68} but also F^{18}-labeled prostate specific membrane antigen (PSMA) radiopharmaceuticals must be mentioned. Where PET/MRI is available, it more and more steps into the place of PET/CT. Another very exciting application in oncology is the hybrid breast PET/MRI. Here, dynamic contrast enhanced MRI of the breast has been shown to be highly sensitive in assessing primary breast cancer. PET can provide important information regarding locoregional staging in locally advanced breast cancer. Moreover, PET/MRI advances monitoring treatment response to neoadjuvant chemotherapy and shows potential in the initial staging of breast cancer, particularly of distant metastases and further surveillance. For neuro oncology, PET/MRI opens up new diagnostic opportunities. Here it is the tracers F^{18}-Fuorethythyrosine (FET), ^{11}C-Methionine (MET), ^{18}F-Fluordopa (FDOPA) and, of course, ^{18}F-FDG that are most applied [121]. Recently, F^{18}-Fluciclovine entered the arena for imaging glioma, but having originally started out in prostate cancer, it needs more evaluation. Likewise, ^{18}F-MISO has been applied in neuro oncologic PET/MRI imaging of hypoxia. Imaging of cell oxygenation can be of great help in target definition and for the tailoring of external beam radiotherapy.

Another important feature of PET/MRI is that the structural imaging gets along without any additional exposure to ionizing radiation other than by PET. This is of utmost importance for imaging children and adolescents. In places where PET/MRI is readily available, almost all indications where formerly PET/CT was indicated, pediatric PET imaging is shifted to PET/MRI, particularly in situations where children and adolescents need to undergo such investigations for staging, surveillance, and treatment response assessment regularly.

Not to leave them out, but less frequent are applications of PET/MRI in inflammatory, cardiac, and respiratory diseases [122–124]. The latter greatly profiting – again – from the superior soft tissue contrast but much more from the temporal resolution of MRI. As aforementioned in the context of attenuation correction, the PET imaging for the myocard and heart function can be gated by timing information that is readily available from MRI [125].

Out there in the literature, there is much more detail about key, current, potential, and emerging clinical applications of PET/MRI, and the interested reader is referred to obtain much more information and scientific evidence in these sources [126–129]

22.6 PRECLINICAL PET/MRI

All technical operability assessments have first been carried out in preclinical settings. Starting with the testing of the interoperability of PET detectors with MRI and imaging protocol consideration towards the search for potential clinical key applications, preclinical PET/MRI has been, and is, key in further optimization of the main features of the hybrid imaging modality. Particularly in radiopharmaceutical development, animal models play an important role in the investigation of new radiopharmaceuticals. Different settings of preclinical PET/MRI systems are commercially available. First, Bruker Biospin (United States) provided a PET-insert to their high field MRI system. This system has the advantage of very high strength and a homogeneous static magnetic field, but it is a very heavy system requiring a large footprint and extensive peripheral facilities including, but not limited to, the cooling of the magnet with liquid helium. To circumvent these 'disadvantages' Mediso (Hungary) has presented another approach using a permanent magnet, enabling a very small and light system that, apart from power, actually does not require any special peripheral supply [131]. However, with permanent magnets the static field strength is limited to one Tesla and the homogeneity in the bore is not quite the axial extent of a small animal. Recently, MR Solutions (UK) came up with a series of systems combining a cryogen-free cooling of a superconducting magnet, higher field strengths of up to 9.4 Tesla, and a modular relatively small footprint system that can be installed on higher floors requiring only one small additional room for peripheral installations required mainly to keep the noise level down to not disturb sensitive investigations. Last, but not least, it remains to be mentioned, that the PET detector systems of the modular devices are always kept on the state of the art, today consisting of very small LSO crystals, all combined with silicon photomultipliers. Due to the small size of the animals, MR-based attenuation correction is not so much of an issue as known from clinical PET/MRI so that in some of the preclinical investigations it is not done at all.

22.7 SITING, STAFFING AND REGULATORY CONSIDERATIONS

Depending on the legislation and organizational requirements and settings, PET/MRI systems should usually be installed in and managed by nuclear medicine, radiology, or the collaboration of both departments. If it is integrated in a full hybrid imaging molecular imaging PET centre including a radiopharmaceutical production site and a cyclotron on site, the system should be located in close proximity to this facility. However, satellite PET/MRI units can also be installed at specialized clinical sites such as paediatric, psychiatric, or neurology departments if the appropriate staffing level is available [23].

REFERENCES

[1] Gaa J, Rummeny EJ, Seemann MD (2004) Whole-body imaging with PET/MRI. *Eur. J. Med. Res.*, 9(6), 309–312, PMID:15257872.

[2] Seemann MD (2005) Whole-body PET/MRI: The future in oncological imaging. *Technol. Cancer Res. Treat.*, 4(5), 577–582, doi:10.1177/153303460500400512, PMID:16173829.

[3] Judenhofer MS, Wehrl HF, Newport DF, Catana C, Siegel SB, Becker M, Thielscher A, Kneilling M, Lichy MP, Eichner M, Klingel K, Reischl G, Widmaier S, Rocken M, Nutt RE, Machulla HJ, Uludag K, Cherry SR, Claussen CD, Pichler BJ (2008) Simultaneous PET-MRI: a new approach for functional and morphological imaging. *Nat. Med.*, 14(4), 459–465, PMID:18376410.

[4] Schlemmer HP, Pichler BJ, Schmand M, Burbar Z, Michel C, Ladebeck R, Jattke K, Townsend D, Nahmias C, Jacob PK, Heiss WD, Claussen CD (2008) Simultaneous MR/PET imaging of the human brain: Feasibility study. *Radiology*, 248(3), 1028–1035, PMID:18710991.

[5] Antoch G, Bockisch A (2009) Combined PET/MRI: a new dimension in whole-body oncology imaging? Eur. *J. Nucl. Med. Mol. Imaging*, 36 Suppl (1), S113–20, doi:10.1007/s00259-008-0951-6, PMID:19104802.

[6] Bolus NE, George R, Washington J, Newcomer BR (2009) PET/MRI: The blended-modality choice of the future? *J. Nucl. Med. Technol.*, 37(2), 63–71; quiz 72-3, doi:10.2967/jnmt.108.060848, PMID:19447852.

[7] Ziegler SI, Pichler BJ, Boening G, Rafecas M, Pimpl W, Lorenz E, Schmitz N, Schwaiger M (2001) A prototype high-resolution animal positron tomograph with avalanche photodiode arrays and LSO crystals. *Eur. J. Nucl. Med.*, 28(2), 136–143, PMID:11303882.

[8] Pichler BJ, Swann BK, Rochelle J, Nutt RE, Cherry SR, Siegel SB (2004) Lutetium oxyorthosilicate block detector readout by avalanche photodiode arrays for high resolution animal PET. *Phys. Med. Biol.*, 49(18), 4305–4319, PMID:15509067.

[9] Li X, Lockhart C, Lewellen TK, Miyaoka RS (2011) Study of PET detector performance with varying SiPM parameters and readout schemes. *IEEE. Trans. Nucl. Sci.*, 58(3), 590–596, doi:10.1109/TNS.2011.2119378, PMID:22685348.

[10] Eraerds P, Legré M, Rochas A, Zbinden H, Gisin N (2007) SiPM for fast photon-counting and multiphoton detection. *Opt. Express.*, 15(22), 14539–14549, doi:10.1364/oe.15.014539, PMID:19550733.

[11] Llosá G, Barrio J, Lacasta C, Bisogni MG, Del Guerra A, Marcatili S, Barrillon P, Bondil-Blin S, La Taille C de, Piemonte C (2010) Characterization of a PET detector head based on continuous LYSO crystals and monolithic, 64-pixel silicon photomultiplier matrices. *Phys. Med. Biol.*, 55(23), 7299–7315, doi:10.1088/0031-9155/55/23/008, PMID:21081823.

[12] Dey S, Lewellen TK, Miyaoka RS, Rudell JC (2012) Highly-Integrated CMOS Interface Circuits for SiPM-Based PET Imaging Systems. *IEEE. Nucl. Sci. Symp. Conf Rec.*, (1997), 3556–3559, doi:10.1109/nssmic.2012.6551814, PMID:24301987.

[13] Zaidi H, Del Guerra A (2011) An outlook on future design of hybrid PET/MRI systems. *Med. Phys.*, 38(10), 5667–5689, doi:10.1118/1.3633909, PMID:21992383.

[14] Pichler BJ, Kolb A, Nagele T, Schlemmer HP (2010) PET/MRI: paving the way for the next generation of clinical multimodality imaging applications. *J. Nucl. Med.*, 51(3), 333–336, PMID:20150252.

[15] Judenhofer MS, Catana C, Swann BK, Siegel SB, Jung WI, Nutt RE, Cherry SR, Claussen CD, Pichler BJ (2007) PET/MR images acquired with a compact MR-compatible PET detector in a 7-T magnet. *Radiology*, 244(3), 807–814, PMID:17709830.

[16] Pichler BJ, Wehrl HF, Kolb A, Judenhofer MS (2008) Positron emission tomography/magnetic resonance imaging: the next generation of multimodality imaging? *Semin. Nucl. Med.*, 38(3), 199–208, doi:10.1053/j.semnuclmed.2008.02.001, PMID:18396179.

[17] Kolb A, Wehrl HF, Hofmann M, Judenhofer MS, Eriksson L, Ladebeck R, Lichy MP, Byars L, Michel C, Schlemmer HP, Schmand M, Claussen CD, Sossi V, Pichler BJ (2012) Technical performance evaluation of a human brain PET/MRI system. *Eur. Radiol.*, 22(8), 1776–1788, PMID:22752524.

[18] Quick HH (2014) Integrated PET/MR. J. Magn Reson. *Imaging*, 39(2), 243–258, PMID:24338921.

[19] Schulthess GK von, Kuhn FP, Kaufmann P, Veit-Haibach P (2013) Clinical positron emission tomography/magnetic resonance imaging applications. *Semin. Nucl. Med.*, 43(1), 3–10, PMID:23178084.

[20] Veit-Haibach P, Kuhn FP, Wiesinger F, Delso G, Schulthess G von (2013) PET-MR imaging using a tri-modality PET/CT-MR system with a dedicated shuttle in clinical routine. *MAGMA.*, 26(1), 25–35, PMID:23053712.

[21] Herzog H, van den Hoff J (2012) Combined PET/MR systems: An overview and comparison of currently available options. *Q. J. Nucl. Med. Mol. Imaging*, 56(3), 247–267, PMID:22695336.

[22] Schramm G, Langner J, Hofheinz F, Petr J, Beuthien-Baumann B, Platzek I, Steinbach J, Kotzerke J, van den Hoff J (2013) Quantitative accuracy of attenuation correction in the Philips Ingenuity TF whole-body PET/MR system: A direct comparison with transmission-based attenuation correction. *MAGMA.*, 26(1), 115–126, doi:10.1007/s10334-012-0328-5, PMID:22923020.

[23] Sattler B, Jochimsen T, Barthel H, Sommerfeld K, Stumpp P, Hoffmann KT, Gutberlet M, Villringer A, Kahn T, Sabri O (2013) Physical and organizational provision for installation, regulatory requirements and implementation of a simultaneous hybrid PET/MR-imaging system in an integrated research and clinical setting. *MAGMA.*, 26(1), 159–171, PMID:23053713.

[24] Delso G, Furst S, Jakoby B, Ladebeck R, Ganter C, Nekolla SG, Schwaiger M, Ziegler SI (2011) Performance measurements of the Siemens mMR integrated whole-body PET/MR scanner. *J. Nucl. Med.*, 52(12), 1914–1922, PMID:22080447.

[25] Surti S, Kuhn A, Werner ME, Perkins AE, Kolthammer J, Karp JS (2007) Performance of Philips Gemini TF PET/CT scanner with special consideration for its time-of-flight imaging capabilities. *J. Nucl. Med.*, 48(3), 471–480, PMID:17332626.

[26] Jakoby BW, Bercier Y, Conti M, Casey ME, Bendriem B, Townsend DW (2011) Physical and clinical performance of the mCT time-of-flight PET/CT scanner. *Phys. Med. Biol.*, 56(8), 2375–2389, doi:10.1088/0031-9155/56/8/004, PMID:21427485.

[27] Oprea-Lager DE, Yaqub M, Pieters IC, Reinhard R, van Moorselaar RJA, van den Eertwegh AJM, Hoekstra OS, Lammertsma AA, Boellaard R (2015) A Clinical and Experimental Comparison of Time of Flight PET/MRI and PET/CT Systems. *Mol. Imaging Biol.*, 17(5), 714–725, doi:10.1007/s11307-015-0826-8, PMID:25690949.

[28] Iagaru A, Mittra E, Minamimoto R, Jamali M, Levin C, Quon A, Gold G, Herfkens R, Vasanawala S, Gambhir SS, Zaharchuk G (2015) Simultaneous whole-body time-of-flight 18F-FDG PET/MRI: a pilot study comparing SUVmax with PET/CT and assessment of MR image quality. *Clin. Nucl. Med.*, 40(1), 1–8, doi:10.1097/RLU.0000000000000611, PMID:25489952.

[29] Mannheim JG, Schmid AM, Schwenck J, Katiyar P, Herfert K, Pichler BJ, Disselhorst JA (2018) PET/MRI Hybrid Systems. *Semin. Nucl. Med.*, 48(4), 332–347, doi:10.1053/j.semnuclmed.2018.02.011, PMID:29852943.

[30] Boellaard R, Quick HH (2015) Current image acquisition options in PET/MR. *Semin. Nucl. Med.*, 45(3), 192–200, PMID:25841274.

[31] Bailey DL, Antoch G, Bartenstein P, Barthel H, Beer AJ, Bisdas S, Bluemke DA, Boellaard R, Claussen CD, Franzius C, Hacker M, Hricak H, La Fougere C, Guckel B, Nekolla SG, Pichler BJ, Purz S, Quick HH, Sabri O, Sattler B, Schafer J, Schmidt H, van den HJ, Voss S, Weber W, Wehrl HF, Beyer T (2015) Combined PET/MR: The Real Work Has Just Started. Summary Report of the Third International Workshop on PET/MR Imaging; February 17–21, 2014, Tubingen, Germany. *Mol. Imaging Biol.*, 17(3), 297–312, PMID:25672749.

[32] Bailey DL, Pichler BJ, Guckel B, Antoch G, Barthel H, Bhujwalla ZM, Biskup S, Biswal S, Bitzer M, Boellaard R, Braren RF, Brendle C, Brindle K, Chiti A, La Fougere C, Gillies R, Goh V, Goyen M, Hacker M, Heukamp L, Knudsen GM, Krackhardt AM, Law I, Morris JC, Nikolaou K, Nuyts J, Ordonez AA, Pantel K, Quick HH, Riklund K, Sabri O, Sattler B, Troost EGC, Zaiss M, Zender L, Beyer T (2018) Combined PET/MRI: Global Warming-Summary Report of the 6th International Workshop on PET/MRI, March 27–29, 2017, Tubingen, Germany. *Mol. Imaging Biol.*, 20(1), 4–20, PMID:28971346.

[33] Bailey DL, Pichler BJ, Guckel B, Barthel H, Beer AJ, Bremerich J, Czernin J, Drzezga A, Franzius C, Goh V, Hartenbach M, Iida H, Kjaer A, La Fougere C, Ladefoged CN, Law I, Nikolaou K, Quick HH, Sabri O, Schafer J, Schafers M, Wehrl HF, Beyer T (2015) Combined PET/MRI: Multi-modality Multi-parametric Imaging Is Here: Summary Report of the 4th International Workshop on PET/MR Imaging; February 23–27, 2015, Tubingen, Germany. *Mol. Imaging Biol.*, 17(5), 595–608, PMID:26286794.

[34] Hofmann M, Pichler B, Schölkopf B, Beyer T (2009) Towards quantitative PET/MRI: a review of MR-based attenuation correction techniques. *Eur. J. Nucl. Med. Mol. Imaging*, 36 Suppl 1, S93–104, doi:10.1007/s00259-008-1007-7, PMID:19104810.

[35] Herzog H, Lerche C (2016) Advances in Clinical PET/MRI Instrumentation. *PET. Clin.*, 11(2), 95–103, doi:10.1016/j.cpet.2015.09.001, PMID:26952724

[36] Delso G, Khalighi M, Hofbauer M, Porto M, Veit-Haibach P, Schulthess G von (2014) Preliminary evaluation of image quality in a new clinical ToF-PET/MR scanner. *EJNMMI. Phys.*, 1(Suppl 1), A41, PMID:26501629.

[37] Grant AM, Deller TW, Khalighi MM, Maramraju SH, Delso G, Levin CS (2016) NEMA NU 2-2012 performance studies for the SiPM-based ToF-PET component of the GE SIGNA PET/MR system. *Med. Phys.*, 43(5), 2334, PMID:27147345.

[38] Kolthammer JA, Su K-H, Grover A, Narayanan M, Jordan DW, Muzic RF (2014) Performance evaluation of the Ingenuity TF PET/CT scanner with a focus on high count-rate conditions. *Phys. Med. Biol.*, 59(14), 3843–3859, doi:10.1088/0031-9155/59/14/3843, PMID:24955921.

[39] Zaidi H, Ojha N, Morich M, Griesmer J, Hu Z, Maniawski P, Ratib O, Izquierdo-Garcia D, Fayad ZA, Shao L (2011) Design and performance evaluation of a whole-body Ingenuity TF PET-MRI system. *Phys. Med. Biol.*, 56(10), 3091–3106, doi:10.1088/0031-9155/56/10/013, PMID:21508443.

[40] (2017) *Encyclopedia of Spectroscopy and Spectrometry*, ISBN: 9780128032244.

[41] Judenhofer MS (2017) *Combined Positron Emission Tomography–Magnetic Resonance Imaging*, 334–340, doi:10.1016/B978-0-12-409547-2.12142-0.

[42] Vandenberghe S, Marsden PK (2015) PET-MRI: a review of challenges and solutions in the development of integrated multimodality imaging. *Phys. Med. Biol.*, 60(4), R115–54, doi:10.1088/0031-9155/60/4/R115, PMID:25650582.

[43] Grant AM, Lee BJ, Chang C-M, Levin CS (2017) Simultaneous PET/MR imaging with a radio frequency-penetrable PET insert. *Med. Phys.*, 44(1), 112–120, doi:10.1002/mp.12031, PMID:28102949.

[44] Lee BJ, Grant AM, Chang C-M, Watkins RD, Glover GH, Levin CS (2018) MR Performance in the Presence of a Radio Frequency-Penetrable Positron Emission Tomography (PET) Insert for Simultaneous PET/MRI. *IEEE. Trans. Med. Imaging*, 37(9), 2060–2069, doi:10.1109/TMI.2018.2815620, PMID:29993864.

[45] Lecomte R (2009) Novel detector technology for clinical PET. Eur. *J. Nucl. Med. Mol. Imaging*, 36 Suppl 1, S69–85, doi:10.1007/s00259-008-1054-0, PMID:19107476.

[46] Shao Y, Cherry SR, Farahani K, Meadors K, Siegel S, Silverman RW, Marsden PK (1997) Simultaneous PET and MR imaging. *Phys. Med. Biol.*, 42(10), 1965–1970, doi:10.1088/0031-9155/42/10/010, PMID:9364592.

[47] Shao Y, Cherry SR, Farahani K, Slates R, Silverman RW, Meadors K, Bowery A, Siegel S, Marsden PK, Garlick PB (1997) Development of a PET detector system compatible with MRI/NMR systems. *IEEE. Trans. Nucl. Sci.*, 44(3), 1167–1171, doi:10.1109/23.596982.

[48] Catana C, Wu Y, Judenhofer MS, Qi J, Pichler BJ, Cherry SR (2006) Simultaneous acquisition of multislice PET and MR images: initial results with a MR-compatible PET scanner. *J. Nucl. Med.*, 47(12), 1968–1976, PMID:17138739.

[49] Pichler BJ, Judenhofer MS, Catana C, Walton JH, Kneilling M, Nutt RE, Siegel SB, Claussen CD, Cherry SR (2006) Performance test of an LSO-APD detector in a 7-T MRI scanner for simultaneous PET/MRI. *J. Nucl. Med.*, 47(4), 639–647, PMID:16595498.

[50] Young K (2010) *Advances in Optical and Photonic Devices*, ISBN: 978-953-7619-76-3, doi:10.5772/127.

[51] Saveliev V (2010) *Silicon Photomultiplier – New Era of Photon Detection*, doi:10.5772/7150.

[52] Rausch I, Ruiz A, Valverde-Pascual I, Cal-González J, Beyer T, Carrio I (2019) Performance Evaluation of the Vereos PET/CT System According to the NEMA NU2-2012 Standard. *J. Nucl. Med.*, 60(4), 561–567, doi:10.2967/jnumed.118.215541, PMID:30361382.

[53] van Sluis J, Jong J de, Schaar J, Noordzij W, van Snick P, Dierckx R, Borra R, Willemsen A, Boellaard R (2019) Performance characteristics of the digital biograph vision PET/CT system. *J. Nucl. Med.*, 60(7), 1031–1036, doi:10.2967/jnumed.118.215418, PMID:30630944.

[54] Pan T, Einstein SA, Kappadath SC, Grogg KS, Lois Gomez C, Alessio AM, Hunter WC, El Fakhri G, Kinahan PE, Mawlawi OR (2019) Performance evaluation of the 5-Ring GE Discovery MI PET/CT system using the national electrical manufacturers association NU 2-2012 Standard. *Med. Phys.*, 46(7), 3025–3033, doi:10.1002/mp.13576, PMID:31069816.

[55] Seifert S, van der LG, van Dam HT, Schaart DR (2013) First characterization of a digital SiPM based time-of-flight PET detector with 1 mm spatial resolution. *Phys. Med. Biol.*, 58(9), 3061–3074, PMID:23587636.

[56] Schmand M, Burbar Z, Corbeil J, Zhang N, Michael C, Byars L, Eriksson L, Grazioso R, Martin M, Moor A, Camp J, Matschl V, Ladebeck R, Renz W, Fischer H, Jattke K, Schnur G, Rietsch N, Bendriem B, Heiss W-D (2007) BrainPET: First human tomograph for simultaneous (functional) PET and MR imaging. *J. Nucl. Med.*, 48(supplement 2), 45P–45P.

[57] Herzog H (2012) PET/MRI: challenges, solutions and perspectives. *Z. Med. Phys.*, 22(4), 281–298, doi:10.1016/j.zemedi.2012.07.003, PMID:22925652.

[58] National Electrical Manufacturers Association (2007) *NEMA Standards Publication NU 2-2007: Performance Measurements of Positron Emission Tomographs.* 1300 N, 17th Street, Suite 900, Rosslyn, VA 22209.

[59] Ziegler S, Braun H, Ritt P, Hocke C, Kuwert T, Quick HH (2013) Systematic evaluation of phantom fluids for simultaneous PET/MR hybrid imaging. *J. Nucl. Med.*, 54(8), 1464–1471, doi:10.2967/jnumed.112.116376, PMID:23792278.

[60] Ziegler S, Jakoby BW, Braun H, Paulus DH, Quick HH (2015) NEMA image quality phantom measurements and attenuation correction in integrated PET/MR hybrid imaging. *EJNMMI. Phys.*, 2(1), 18, PMID:26501819.

[61] Beyer T, Lassen ML, Boellaard R, Delso G, Yaqub M, Sattler B, Quick HH (2016) Investigating the state-of-the-art in whole-body MR-based attenuation correction: an intra-individual, inter-system, inventory study on three clinical PET/MR systems. *MAGMA.*, 29(1), 75–87, PMID:26739263.

[62] Øen SK, Aasheim LB, Eikenes L, Karlberg AM (2019) Image quality and detectability in Siemens Biograph PET/MRI and PET/CT systems-a phantom study. *EJNMMI. Phys.*, 6(1), 16, doi:10.1186/s40658-019-0251-1, PMID:31385052

[63] Harries J, Jochimsen T, Scholz T, Schlender T, Barthel H, Sabri O, Sattler B (2020) A realistic phantom of the human head for validation of attenuation correction methods in PET-MRI. *J. Nucl. Med.*, 61(Suppl. 1), 389.

[64] National Electrical Manufacturers Association (2012) *NEMA Standards Publication NU 2-2012: Performance Measurements of Positron Emission Tomographs.* 1300 N, 17th Street, Suite 900, Rosslyn, VA 22209.

[65] Deller TW, Khalighi MM, Jansen FP, Glover GH (2018) PET imaging stability measurements during simultaneous pulsing of aggressive MR sequences on the SIGNA PET/MR system. *J. Nucl. Med.*, 59(1), 167–172, doi:10.2967/jnumed.117.194928, PMID:28747522.

[66] Caribé PRRV, Koole M, D'Asseler Y, Deller TW, van Laere K, Vandenberghe S (2019) NEMA NU 2-2007 performance characteristics of GE Signa integrated PET/MR for different PET isotopes. *EJNMMI. Phys.*, 6(1), 11, doi:10.1186/s40658-019-0247-x, PMID:31273558.

[67] Chen S, Gu Y, Yu H, Chen X, Cao T, Hu L, Shi H (2021) NEMA NU2-2012 performance measurements of the United Imaging uPMR790: an integrated PET/MR system. *Eur. J. Nucl. Med. Mol. Imaging*, doi:10.1007/s00259-020-05135-9, 03.01.2021, PMID:33388972.

[68] Liu G, Cao T, Hu L, Zheng J, Pang L, Hu P, Gu Y, Shi H (2019) Validation of MR-based attenuation correction of a newly released whole-body simultaneous PET/MR system. *Biomed. Res. Int.*, 2019, 8213215, doi:10.1155/2019/8213215, PMID:31886254.

[69] Yu H, Chen S, Liu H, Cao T, Hu L, Shi H (2019) A deep learning based technique for truncation completion in PET/MR. *J. Nucl. Med.*, 60(supplement 1), 2028.

[70] Zhao J, Shou Y, Jiang J, Song J (2019) Initial experience of domestic simultaneous PET/MR in the detection of metastases and/or recurrence. *J. Nucl. Med.*, 60(supplement 1), 1285.

[71] Chen S, Hu P, Gu Y, Pang L, Zhang Z, Zhang Y, Meng X, Cao T, Liu X, Fan Z, Shi H (2019) Impact of patient comfort on diagnostic image quality during PET/MR exam: A quantitative survey study for clinical workflow management. *J. Appl. Clin. Med. Phys.*, 20(7), 184–192, doi:10.1002/acm2.12664, PMID:31207077.

[72] Carney JP, Townsend DW, Rappoport V, Bendriem B (2006) Method for transforming CT images for attenuation correction in PET/CT imaging. *Med. Phys.*, 33(4), 976–983, PMID:16696474.

[73] Leynes AP, Yang J, Shanbhag DD, Kaushik SS, Seo Y, Hope TA, Wiesinger F, Larson PE (2017) Hybrid ZTE/Dixon MR-based attenuation correction for quantitative uptake estimation of pelvic lesions in PET/MRI. *Med. Phys.*, 44(3), 902–913, PMID:28112410.

[74] Sekine T, Ter Voert EE, Warnock G, Buck A, Huellner M, Veit-Haibach P, Delso G (2016) Clinical evaluation of Zero-Echo-Time Attenuation Correction for brain 18F-FDG PET/MRI: Comparison with Atlas Attenuation Correction. *J. Nucl. Med.*, 57(12), 1927–1932, PMID:27339875.

[75] Yang J, Wiesinger F, Kaushik S, Shanbhag D, Hope TA, Larson PEZ, Seo Y (2017) Evaluation of Sinus/Edge-Corrected Zero-Echo-Time-Based Attenuation Correction in Brain PET/MRI. *J. Nucl. Med.*, 58(11), 1873–1879, PMID:28473594.

[76] Burgos N, Cardoso MJ, Modat M, Punwani S, Atkinson D, Arridge SR, Hutton BF, Ourselin S (2015) CT synthesis in the head & neck region for PET/MR attenuation correction: an iterative multi-atlas approach. *EJNMMI. Phys.*, 2(Suppl 1), A31, PMID:26956288.

[77] Burgos N, Cardoso MJ, Thielemans K, Modat M, Dickson J, Schott JM, Atkinson D, Arridge SR, Hutton BF, Ourselin S (2015) Multi-contrast attenuation map synthesis for PET/MR scanners: assessment on FDG and Florbetapir PET tracers. *Eur. J. Nucl. Med. Mol. Imaging*, 42(9), 1447–1458, PMID:26105119.

[78] Paulus DH, Quick HH, Geppert C, Fenchel M, Zhan Y, Hermosillo G, Faul D, Boada F, Friedman KP, Koesters T (2015) Whole-Body PET/MR Imaging: Quantitative evaluation of a novel model-based MR attenuation correction method including bone. *J. Nucl. Med.*, 56(7), 1061–1066, doi:10.2967/jnumed.115.156000, PMID:26025957.

[79] Ladefoged CN, Benoit D, Law I, Holm S, Kjaer A, Hojgaard L, Hansen AE, Andersen FL (2015) Region specific optimization of continuous linear attenuation coefficients based on UTE (RESOLUTE): application to PET/MR brain imaging. *Phys. Med. Biol.*, 60(20), 8047–8065, PMID:26422177.

[80] Ladefoged CN, Law I, Anazodo U, St Lawrence K, Izquierdo-Garcia D, Catana C, Burgos N, Cardoso MJ, Ourselin S, Hutton B, Merida I, Costes N, Hammers A, Benoit D, Holm S, Juttukonda M, An H, Cabello J, Lukas M, Nekolla S, Ziegler S, Fenchel M, Jakoby B, Casey ME, Benzinger T, Hojgaard L, Hansen AE, Andersen FL (2017) A multi-centre evaluation of eleven clinically feasible brain PET/MRI attenuation correction techniques using a large cohort of patients. *Neuroimage.*, 147, 346–359, PMID:27988322.

[81] Ladefoged CN, Andersen FL, Kjaer A, Hojgaard L, Law I (2017) RESOLUTE PET/MRI Attenuation Correction for O-(2-(18) F-fluoroethyl)-L-tyrosine (FET) in brain tumor patients with metal implants. *Front Neurosci.*, 11, 453, PMID:28848379.

[82] Boellaard R, Hofman MB, Hoekstra OS, Lammertsma AA (2014) Accurate PET/MR quantification using time of flight MLAA image reconstruction. *Mol. Imaging Biol.*, 16(4), 469–477, PMID:24430291.

[83] Martinez-Möller A, Souvatzoglou M, Delso G, Bundschuh RA, Chefd'hotel C, Ziegler SI, Navab N, Schwaiger M, Nekolla SG (2009) Tissue classification as a potential approach for attenuation correction in whole-body PET/MRI: Evaluation with PET/CT data. *J. Nucl. Med.*, 50(4), 520–526, doi:10.2967/jnumed.108.054726, PMID:19289430.

[84] Dixon WT (1984) Simple proton spectroscopic imaging. *Radiology*, 153(1), 189–194, doi:10.1148/radiology.153.1.6089263, PMID:6089263.

[85] Sauter AW, Wehrl HF, Kolb A, Judenhofer MS, Pichler BJ (2010) Combined PET/MRI: One step further in multimodality imaging. *Trends. Mol. Med.*, 16(11), 508–515, doi:10.1016/j.molmed.2010.08.003, PMID:20851684.

[86] Benoit D, Ladefoged C, Rezaei A, Keller S, Andersen F, Hojgaard L, Hansen AE, Holm S, Nuyts J (2015) PET/MR: improvement of the UTE μ-maps using modified MLAA. *EJNMMI. Phys.*, 2(Suppl 1), A58, doi:10.1186/2197-7364-2-S1-A58, PMID:26956317.

[87] Keereman V, Fierens Y, Broux T, Deene Y de, Lonneux M, Vandenberghe S (2010) MRI-based attenuation correction for PET/MRI using ultrashort echo time sequences. *J. Nucl. Med.*, 51(5), 812–818, doi:10.2967/jnumed.109.065425, PMID:20439508.

[88] Leynes AP, Yang J, Wiesinger F, Kaushik SS, Shanbhag DD, Seo Y, Hope TA, Larson PEZ (2018) Zero-Echo-Time and Dixon Deep Pseudo-CT (ZeDD CT): Direct Generation of Pseudo-CT Images for Pelvic PET/MRI attenuation correction using deep convolutional neural networks with multiparametric MRI. *J. Nucl. Med.*, 59(5), 852–858, doi:10.2967/jnumed.117.198051, PMID:29084824.

[89] Juttukonda MR, Mersereau BG, Chen Y, Su Y, Rubin BG, Benzinger TLS, Lalush DS, An H (2015) MR-based attenuation correction for PET/MRI neurological studies with continuous-valued attenuation coefficients for bone through a conversion from R2* to CT-Hounsfield units. *Neuroimage.*, 112, 160–168, doi:10.1016/j.neuroimage.2015.03.009, PMID:25776213.

[90] Paulus DH, Braun H, Aklan B, Quick HH (2012) Simultaneous PET/MR imaging: MR-based attenuation correction of local radiofrequency surface coils. *Med. Phys.*, 39(7), 4306–4315, PMID:22830764.

[91] Paulus DH, Quick HH (2016) Hybrid Positron Emission Tomography/Magnetic Resonance Imaging: Challenges, methods, and state of the art of hardware component attenuation correction. *Invest. Radiol.*, 51(10), 624–634, PMID:27175550.

[92] Han X (2017) MR-based synthetic CT generation using a deep convolutional neural network method. *Med. Phys.*, 44(4), 1408–1419, doi:10.1002/mp.12155, PMID:28192624.

[93] Ladefoged CN, Marner L, Hindsholm A, Law I, Højgaard L, Andersen FL (2018) Deep learning based attenuation correction of PET/MRI in pediatric brain tumor patients: Evaluation in a clinical setting. *Front Neurosci.*, 12(), 1005, doi:10.3389/fnins.2018.01005, PMID:30666184.

[94] Arabi H, Zaidi H (2016) One registration multi-atlas-based pseudo-CT generation for attenuation correction in PET/MRI. *Eur. J. Nucl. Med. Mol. Imaging*, 43(11), 2021–2035, PMID:27260522.

[95] Burgos N, Cardoso MJ, Thielemans K, Modat M, Pedemonte S, Dickson J, Barnes A, Ahmed R, Mahoney CJ, Schott JM, Duncan JS, Atkinson D, Arridge SR, Hutton BF, Ourselin S (2014) Attenuation correction synthesis for hybrid PET-MR scanners: application to brain studies. *IEEE. Trans. Med. Imaging.*, 33(12), 2332–2341, doi:10.1109/TMI.2014.2340135, PMID:25055381.

[96] Deandreis D, Krizsán ÁK, Mirzaei S, Prior J, Sattler B, Castellucci P (2020) European Nuclear Medicine Guide. European Association of Nuclear Medicine, www.nucmed-guide.app/#!/startscreen, last accessed 12/18/2020.

[97] Mehranian A, Reader AJ (2020) Model-Based Deep Learning PET Image Reconstruction Using Forward-Backward Splitting Expectation Maximisation. *IEEE Trans. Radiat. Plasma Med. Sci.*, 1, doi:10.1109/TRPMS.2020.3004408.

[98] Papp L, Spielvogel CP, Grubmüller B, Grahovac M, Krajnc D, Ecsedi B, Sareshgi RAM, Mohamad D, Hamboeck M, Rausch I, Mitterhauser M, Wadsak W, Haug AR, Kenner L, Mazal P, Susani M, Hartenbach S, Baltzer P, Helbich TH, Kramer G, Shariat SF, Beyer T, Hartenbach M, Hacker M (2020) Supervised machine learning enables non-invasive lesion characterization in primary prostate cancer with 68GaGa-PSMA-11 PET/MRI. *Eur. J. Nucl. Med. Mol. Imaging*, doi:10.1007/s40336-014-0064-0, PMID:33341915.

[99] Liu F, Jang H, Kijowski R, Bradshaw T, McMillan AB (2018) Deep Learning MR Imaging-based Attenuation Correction for PET/MR Imaging. *Radiology*, 286(2), 676–684, doi:10.1148/radiol.2017170700, PMID:28925823.

[100] Gong K, Yang J, Kim K, El Fakhri G, Seo Y, Li Q (2018) Attenuation correction for brain PET imaging using deep neural network based on Dixon and ZTE MR images. *Phys. Med. Biol.*, 63(12), 125011, doi:10.1088/1361-6560/aac763, PMID:29790857.

[101] Renner A, Rausch I, Cal Gonzalez J, Frass-Kriegl R, Lara LN de, Sieg J, Laistler E, Glanzer M, Dungl D, Moser E, Beyer T, Figl M, Birkfellner W (2018) A head coil system with an integrated orbiting transmission point source mechanism for attenuation correction in PET/MRI. *Phys. Med. Biol.*, 63(22), 225014, doi:10.1088/1361-6560/aae9a9, PMID:30418935.

[102] Renner A, Rausch I, Gonzalez JC, Laistler E, Moser E, Jochimsen T, Sattler T, Sabri O, Beyer T, Figl M, Birkfellner W, Sattler B (2021) Back to 511 keV attenuation correction: A head coil system with orbiting transmission source for PET/MRI validated in a porcine study. *Med. Phys.*(submitted].

[103] Nuyts J, Dupont P, Stroobants S, Benninck R, Mortelmans L, Suetens P (1999) Simultaneous maximum a posteriori reconstruction of attenuation and activity distributions from emission sinograms. *IEEE. Trans. Med. Imaging*, 18(5), 393–403, doi:10.1109/42.774167, PMID:10416801.

[104] Benoit D, Ladefoged CN, Rezaei A, Keller SH, Andersen FL, Højgaard L, Hansen AE, Holm S, Nuyts J (2016) Optimized MLAA for quantitative non-TOF PET/MR of the brain. *Phys. Med. Biol.*, 61(24), 8854–8874, doi:10.1088/1361-6560/61/24/8854, PMID:27910823.

[105] Lougovski A, Schramm G, Maus J, Hofheinz F, van den Ho J (2014) Preliminary evaluation of the MLAA algorithm with the Philips Ingenuity PET/MR. *EJNMMI. Phys.*, 1(Suppl 1), A33, doi:10.1186/2197-7364-1-S1-A33, PMID:26501620.

[106] Cheng J-CK, Salomon A, Yaqub M, Boellaard R (2016) Investigation of practical initial attenuation image estimates in TOF-MLAA reconstruction for PET/MR. *Med. Phys.*, 43(7), 4163, doi:10.1118/1.4953634, PMID:27370136.

[107] Heußer T, Rank CM, Berker Y, Freitag MT, Kachelrieß M (2017) MLAA-based attenuation correction of flexible hardware components in hybrid PET/MR imaging. *EJNMMI. Phys.*, 4(1), 12, doi:10.1186/s40658-017-0177-4, PMID:28251575.

[108] Schramm G, Langner J, Hofheinz F, Petr J, Lougovski A, Beuthien-Baumann B, Platzek I, van den Hoff J (2013) Influence and compensation of truncation artifacts in MR-based attenuation correction in PET/MR. *IEEE. Trans. Med. Imaging*, 32(11), 2056–2063, doi:10.1109/TMI.2013.2272660, PMID:24186268.

[109] Paulus DH, Oehmigen M, Grüneisen J, Umutlu L, Quick HH (2016) Whole-body hybrid imaging concept for the integration of PET/MR into radiation therapy treatment planning. *Phys. Med. Biol.*, 61(9), 3504–3520, doi:10.1088/0031-9155/61/9/3504, PMID:27055014.

[110] Øen SK, Keil TM, Berntsen EM, Aanerud JF, Schwarzlmüller T, Ladefoged CN, Karlberg AM, Eikenes L (2019) Quantitative and clinical impact of MRI-based attenuation correction methods in 18FFDG evaluation of dementia. *EJNMMI. Res.*, 9(1), 83, doi:10.1186/s13550-019-0553-2, PMID:31446507

[111] Belcari N, Boellaard R, Morrocchi M (2019) *PET/CT and PET/MR Tomographs: Image Acquisition and Processing*, 47, 199–217, doi:10.1007/978-3-319-95564-3_9.

[112] Lassen ML, Rasul S, Beitzke D, Stelzmuller ME, Cal-Gonzalez J, Hacker M, Beyer T (2017) Assessment of attenuation correction for myocardial PET imaging using combined PET/MRI. *J. Nucl. Cardiol.*, PMID:29168158.

[113] Delso G, Nuyts J (2018) *PET/MRI: Attenuation Correction*, 53–75, doi:10.1007/978-3-319-68517-5_4.

[114] Aide N, Lasnon C, Veit-Haibach P, Sera T, Sattler B, Boellaard R (2017) EANM/EARL harmonization strategies in PET quantification: from daily practice to multicentre oncological studies. *Eur. J. Nucl. Med. Mol. Imaging*, 44(Suppl 1), 17–31, PMID:28623376.

[115] Keller SH, Jakoby B, Svalling S, Kjaer A, Højgaard L, Klausen TL (2016) Cross-calibration of the Siemens mMR: easily acquired accurate PET phantom measurements, long-term stability and reproducibility. *EJNMMI. Phys.*, 3(1), 11, doi:10.1186/s40658-016-0146-3, PMID:27387738.

[116] National Electrical Manufacturers Association (2018) *NEMA standards publication NU 2-2018: Performance Measurements of Positron Emission Tomographs*, 1300 N, 17th Street, Suite 900, Rosslyn, VA 22209.

[117] Valladares A, Beyer T, Rausch I (2020) Physical imaging phantoms for simulation of tumor heterogeneity in PET, CT, and MRI: An overview of existing designs. *Med. Phys.*, 47(4), 2023–2037, doi:10.1002/mp.14045, PMID:31981214.

[118] Delso G, Ziegler S (2009) PET/MRI system design. *Eur. J. Nucl. Med. Mol. Imaging*, 36 Suppl 1, S86–92, doi:10.1007/s00259-008-1008-6, PMID:19104809.

[119] Boellaard R, Rausch I, Beyer T, Delso G, Yaqub M, Quick HH, Sattler B (2015) Quality control for quantitative multicentre whole-body PET/MR studies: A NEMA image quality phantom study with three current PET/MR systems. *Med. Phys.*, 42(10), 5961–5969, PMID:26429271.

[120] Drzezga A, Barthel H, Minoshima S, Sabri O (2014) Potential clinical applications of PET/MR imaging in neurodegenerative diseases. *J. Nucl. Med.*, 55(Supplement 2), 47S–55S, doi:10.2967/jnumed.113.129254, PMID:24819417.

[121] Law I, Albert NL, Arbizu J, Boellaard R, Drzezga A, Galldiks N, La Fougère C, Langen K-J, Lopci E, Lowe V, McConathy J, Quick HH, Sattler B, Schuster DM, Tonn J-C, Weller M (2019) Joint EANM/EANO/RANO practice guidelines/SNMMI procedure standards for imaging of gliomas using PET with radiolabelled amino acids and 18FFDG: version 1.0. *Eur. J. Nucl. Med. Mol. Imaging,* 46(3), 540–557, doi:10.1007/s00259-018-4207-9, PMID:30519867.

[122] Lassen ML, Slomka PJ (2020) Cardiac PET/MR: Are sophisticated attenuation correction techniques necessary for clinical routine assessments? *J. Nucl. Cardiol.*, doi:10.1007/s12350-020-02057-9, PMID:32034663.

[123] Lassen ML, Manabe O, Otaki Y, Eisenberg E, Huynh PT, Wang F, Berman DS, Slomka PJ (2020) 3D PET/CT 82Rb PET myocardial blood flow quantification: comparison of half-dose and full-dose protocols. *Eur. J. Nucl. Med. Mol. Imaging*, doi:10.1007/s00259-020-04811-0, PMID:32372228.

[124] Kuttner S, Lassen ML, Øen SK, Sundset R, Beyer T, Eikenes L (2020) Quantitative PET/MR imaging of lung cancer in the presence of artifacts in the MR-based attenuation correction maps. *Acta. Radiol.*, 61(1), 11–20, doi:10.1177/0284185119848118, PMID:31091969.

[125] Hirsch FW, Sattler B, Sorge I, Kurch L, Viehweger A, Ritter L, Werner P, Jochimsen T, Barthel H, Bierbach U, Till H, Sabri O, Kluge R (2013) PET/MR in children. Initial clinical experience in paediatric oncology using an integrated PET/MR scanner. *Pediatr. Radiol.*, 43(7), 860–875, PMID:23306377.

[126] Volterrani D, Erba PA, Carrió I, Strauss HW, Mariani G (2019) *Nuclear Medicine Textbook*, ISBN: 978-3-319-95563-6, doi:10.1007/978-3-319-95564-3.

[127] Iagaru A, Hope T, Veit-Haibach P (2018) *PET/MRI in Oncology*. Springer International Publishing, ISBN: 978-3-319-68516-8.

[128] Umutlu L, Herrmann K (2018) *PET/MR Imaging: Current and Emerging Applications*, ISBN: 978-3-319-69641-6, doi:10.1007/978-3-319-69641-6.

[129] Carrio I, Ros P (2014) *PET/MRI, ISBN: 3642406912*, doi:10.1007/978-3-642-40692-8.

[130] Nagy K, Tóth M, Major P, Patay G, Egri G, Häggkvist J, Varrone A, Farde L, Halldin C, Gulyás B (2013) Performance evaluation of the small-animal nanoScan PET/MRI system. *J. Nucl. Med.*, 54(10), 1825–1832, doi:10.2967/jnumed.112.119065, PMID:23990683.

23 Quality Assurance of Nuclear Medicine Systems

John Dickson

CONTENTS

DOI: 10.1201/9780429489556-23

23.1 INTRODUCTION

A key element of any department's quality management system (QMS) is confirmation that equipment meets its expected level of performance. Specifically, at installation the equipment should perform at the level defined in the systems specification; it should be configured to meet its clinical needs; and it should perform throughout its lifecycle at a consistently high level. To achieve this, systems must follow a quality assurance (QA) programme that consists of a set of activities that ensures the proper outcome of the imaging process. There are three main activities in this quality assurance programme (Figure 23.1).

Acceptance testing. This is a series of tests that are performed to ensure the equipment meets its published performance specification, confirming that the equipment is fit for purpose. An example might be to check that a scanner equals or betters the advertised spatial resolution. These types of tests ensure that the equipment meets the expected level of performance.

Commissioning. This step prepares the system for clinical use. Following acceptance testing, it is the time where, for example, a PET system is calibrated for activity concentration and checked for its accuracy. Such tasks are not performance tests as such, but preparation of the system to scan patients. It is also at this time that baseline performance measurements will be made for periodic quality control tests such as uniformity, sensitivity, and so forth, and when clinical protocols will be optimized.

Quality control (QC) tests. These are periodic tests performed throughout the lifespan of the equipment to ensure that the system has a consistent and acceptable level of performance. QC tests are performed to find defects or reductions in the performance of a system, making reference to the baseline tests performed at commissioning when the scanner was in peak condition. Successful quality control results are a mark of a high performing, well-maintained scanner.

Many national and international standards are available that define acceptance testing, commissioning, and quality control. Equipment specifications for nuclear medicine imaging systems are almost exclusively set out by the National Electrical Manufacturers Association (NEMA). This US-based organization sets standards for defining equipment performance to allow comparison of imaging systems. Although many of the tests defined in these documents are intended for system manufacturers to be performed in a laboratory environment, tests can be exactly or closely replicated in a nuclear medicine department to assess individual equipment performance. Commissioning typically follows manufacturer guidance and will often be performed with support from manufacturer application specialists and medical physics experts (MPE). For quality control, there are many documents providing guidance from both national [1–4], and international organizations [5–8].

In addition to imaging equipment, dose calibrators (activity meters) also perform a key role in nuclear medicine. In the imaging process, dose calibrators determine the radiation dose we give to our patients and also help ensure that we give an appropriate activity of radiopharmaceutical to achieve a diagnostic image. It is essential therefore that these instruments are part of the quality assurance process.

The aim of this chapter is to give a useful handbook to be used 'in the field'. In the following sections the 'why, when, how and what' of quality assurance activities for gamma camera, SPECT and PET systems will be described together

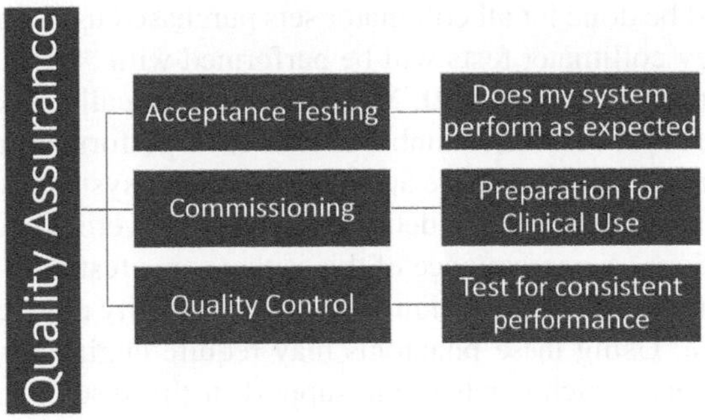

FIGURE 23.1 The key areas of quality assurance.

TABLE 23.1
Planar Acceptance Tests for Gamma Cameras

Intrinsic Tests	System Tests	Whole-body mode tests
Spatial Resolution	Spatial resolution without scatter	Spatial resolution without scatter
Spatial Linearity	Spatial resolution with scatter	
Energy Resolution	Sensitivity & Collimator Penetration / Scatter	
Uniformity	Detector Shielding	
Multiple Window	Count-rate performance with scatter	
Spatial Registration		
Count-rate Performance in Air		
Spatial resolution at 75 kcounts / s		
Uniformity at 75 kcounts / s		

Note: kcounts = 1000 counts. 75kcounts / s = 75,000 counts per second which is a higher-than-normal count-rate measurements.

with a suggested periodicity of these tests and acceptable levels of performance. The chapter will continue with QA activities for multimodality SPECT/CT, PET/CT and PET/MR systems before finishing with the quality assurance programme required for dose calibrators.

23.2 GAMMA CAMERA QUALITY ASSURANCE

This section focuses primarily on traditional gamma cameras comprised of single or multiple 'head' scintillation detector arrays using traditional Anger logic to determine positional information. Although much of what is described is also relevant to pixelated semiconductor detector-based systems, areas of divergence and extra requirements will be highlighted at the end of this section.

23.2.1 Acceptance Testing

Given that gamma camera manufacturers specify performance according to that defined by the National Electrical Manufacturers Association (NEMA), acceptance tests of this equipment follow the same imaging and analysis protocols wherever possible. The publication that specifies these tests is NU 1-2018: Performance Measurements of Gamma Cameras [9]. Tests are split into three sections for planar imaging: intrinsic (without collimator) measurements; system (with collimator) measurements; and whole-body measurements. A list of the tests specified in the document is given in Table 23.1.

The list of acceptance tests is quite exhaustive, and sites may not choose to do all these tests, depending on their expected scanner usage. As an example, high-count (75 kilocount per second) measurements of uniformity and spatial resolution are only really important for those performing first-pass or early post-molecular radiotherapy imaging.

All system measurements should be done for all collimator sets purchased using the radionuclide appropriate for that collimator: for example, low-energy collimator tests will be performed with ^{99m}Tc and/or ^{201}Tl and/or ^{123}I, while high-energy collimator tests should be performed using ^{131}I. Manufacturers typically make recommendations of the energy range for each collimator to assist this process. The number of tests to be performed can become quite extensive so, once more, sites may choose to concentrate on tests that are appropriate to their system use.

Finally, as mentioned in the introduction, tests defined by NEMA were designed to be performed by system manufacturers in laboratory conditions. A consequence of this is that some tests can be difficult to perform without the support of the system manufacturer. So, for example, intrinsic spatial linearity requires heavy, specialized slit phantoms that are placed on the detector face. Using these phantoms may require engineering safety overrides, and the use of expensive, rarely employed phantoms, which is where the support of the system manufacturer can be helpful. Given that acceptance tests are typically performed only once, and that they may require specialist software and hardware support, the details of each test will not be explained. However, some of the tests are repeated as periodic quality control measurements and so will be discussed more in the following sections. For those who want an in-depth understanding of acceptance testing, the NEMA documentation gives details of how these tests are performed.

23.2.2 Quality Control Tests

When considering the quality control tests to perform to ensure that the system has not become or is not becoming defective will depend on each site's usage. Nevertheless, there are certain key elements to system performance that will be consistent for all sites (Figure 23.2). These can be considered as the core components of a gamma camera quality control programme.

Non-uniform response of the detector. Gamma cameras are designed to image features that may be represented by high or low areas of radiopharmaceutical uptake in comparison to surrounding tissue. To accurately represent these features, it is important to have a spatially uniform detector response. A non-uniform response may hide, or indeed exaggerate, differences in radiopharmaceutical uptake. Given that a gamma camera detector is made up of multiple components, there is a real risk of a spatially varying response within the detector. It is essential therefore that quality control tests should assess uniformity of detector response.

Poor scatter rejection. A gamma camera imaging system uses energy windowing to preferentially select un-scattered photopeak photons for the imaging radionuclide being used. If this energy window is incorrectly set – either globally or on independent detector components, or if individual detectors have poor energy resolution – there is a risk that images will have an inferior image quality, or that the detector will have a spatially invariant response. The energy response of the system as a whole should therefore be tested as part of a routine quality control programme.

Poor detection or reduced contrast of features. The ability of the gamma camera to accurately represent objects, particularly small objects, is dependent on the spatial resolution of the system. Spatial resolution can also affect contrast and, therefore, also the detectability of features. For this reason, it is an important attribute to assess as part of a quality control programme.

Detector sensitivity. A key feature of any system including a gamma camera is its sensitivity to the incoming imaging information. Systems with low or reduced sensitivity can lead to less signal and consequently noisier data unless the time of acquisition or the level of radioactivity is increased. Measurements of sensitivity are consequently an important part of a quality control programme.

All gamma camera imaging is performed with a collimator and, because of this, any test of overall image quality should be performed with the appropriate collimators in place. However, because of the inherent spatial resolution of the

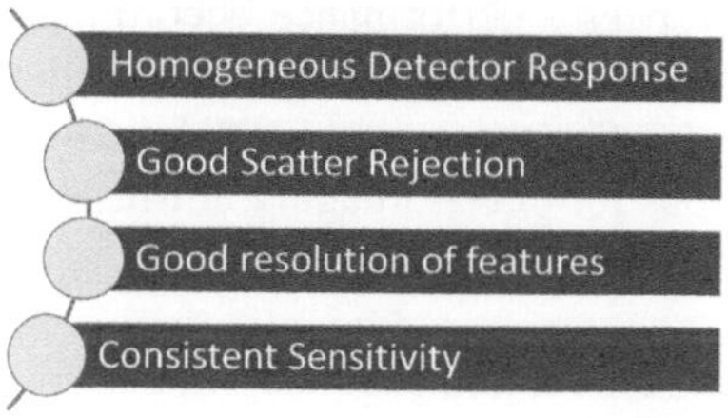

FIGURE 23.2 Essentials for good image quality.

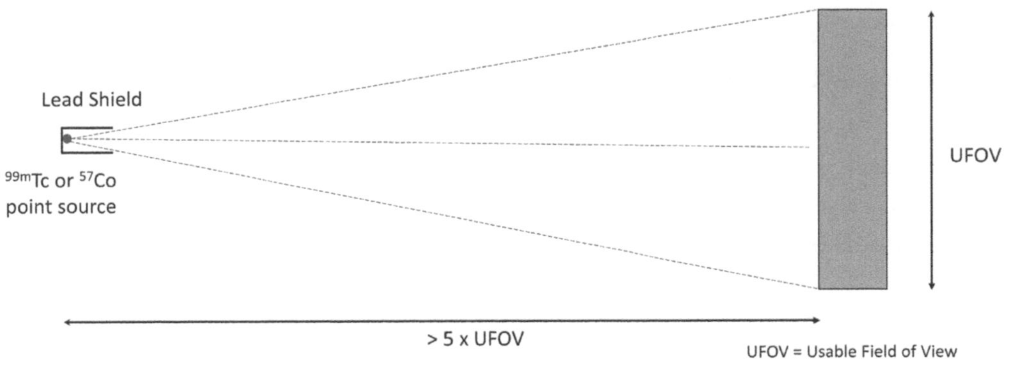

FIGURE 23.3 Experimental arrangements for intrinsic gamma camera measurements.

collimators themselves, reductions in homogeneous detector response or uniformity may be masked by a system measurement that includes the collimators. In addition, when problem solving, the combination of an intrinsic and system measurement can help determine where any loss of performance comes from. It is for these reasons that some quality control measurements are performed both with and without collimators.

As with acceptance testing, the choice of collimator used for any system test should be appropriate for the radionuclide being employed, for example, ^{99m}Tc for low-energy collimators and ^{131}I for high-energy collimators. Intrinsic tests are typically performed using the arrangement shown in Figure 23.3. The use of at least a five usable field-of-view distance between the source and the detector allows a uniformity of flux < 1 per cent. This allows a simple single source to be used for testing.

In addition to collimator and radionuclide choices, appropriate matrix/pixel size choices for each quality control test must also be considered. Low noise acquisitions such as that for uniformity measurements will require large pixel sizes and relatively long acquisition times while measurements that rely on good spatial resolution will use a small pixel size to achieve adequate spatial sampling but in doing so, may compromise on noise. Imaging times can be increased to help reduce noise, however, while such an approach is appropriate during acceptance testing, where more in-depth testing is justified, some compromise may be required for routine quality control testing.

The periodicity of quality control tests is an area with some uncertainty. While testing uniformity every day prior to imaging to check that all detectors are functioning and that the system has uniform response are considered to be essential, the periodicity of other tests is less clear. It is wise therefore to work with national or international guidelines on periodicity until it is proven that the results of the test are stable for your systems so that the local decision on the periodicity of a test can be extended. Alternatively, if test results are frequently changing from acceptable to borderline unacceptable, then more frequent testing of this parameter may be required.

In the following sections, the core quality control tests defined in Figure 23.2 and other quality control tests will be described. The reason for doing the test will be defined, together with possible areas of failure, and the suggested periodicity based on international IAEA guidelines [7] will be given. Typical expected values will also be described where appropriate.

23.2.2.1 Uniformity

Why? Uniformity is measured because of the key requirement of the gamma camera to be able to accurately represent an activity distribution. Major inhomogeneities in detector performance may mask or indeed exaggerate radiopharmaceutical uptake, so it is important that detector uniformity is acceptable.

How? *Imaging protocol.* Low noise, rather than detailed spatial sampling, is important for this acquisition to capture the true variation in signal rather than any variation from image noise. Given the Poisson nature of noise in counting statistics, this low-noise data requires acquisitions with high-count densities. Most nuclear medicine studies are performed with ^{99m}Tc. System uniformity measurements are therefore typically made using a ^{99m}Tc filled flood phantom or a long-lived ^{57}Co source. ^{57}Co has a similar energy emission (122 keV) to ^{99m}Tc (140 keV) and is often used as a more convenient alternative to regular filling of a flood phantom. Intrinsic measurements of uniformity should be performed using ^{99m}Tc, although less-regular measurements of other commonly used radionuclides, for example, ^{131}I, ^{111}In should

TABLE 23.2
Acquisition Parameters for Uniformity Measurements

Collimator	Intrinsic: None System: A single low-energy collimator should be used in most instances to allow trend analysis of results.
Pixel size	~6.4 mm to achieve low noise data.
Energy Window	Use manufacturer recommended settings. ^{99m}Tc or ^{57}Co for regular testing. Window relevant to radionuclide for other checks.
Counts	Daily QC (Intrinsic or System): 5–10 million counts Detailed Intrinsic: 50 million counts

also be performed to test the accuracy of the correction maps at these photon energies. Further details are given in Table 23.2.

Analysis

Analysis of uniformity is typically done for the usable field of view (UFOV) and the central field of view (CFOV). The usable field of view is the complete exposed detector area with the outer 5 per cent of the detector omitted to remove inhomogeneous responses caused by the edge of scintillation crystal and PMT packing at the edge of the detector. Assessment of uniformity is also performed for the central field of view, which is defined as the central 75 per cent of the usable field of view. This is the area that will be used for all imaging. CFOV uniformity should demonstrate better performance than the usable field of view because the signal has contributions from more photomultiplier tubes, and because it avoids any issues from any edge-of-detector effects.

Uniformity is assessed in three ways:

1. Visually;
2. Evaluating global non-uniformities using a measurement of integral uniformity;
3. Evaluating regional non-uniformities using a measurement of differential uniformity.

Integral and differential uniformity look at the difference in maximum and minimum pixel values using the equation:

$$Uniformity = \frac{(Max - Min)}{(Max + Min)} \times 100 \tag{23.1}$$

Integral uniformity identifies maximum and minimum pixel values within the whole field of view (UFOV or CFOV) and therefore assesses global non-uniformities, while differential uniformity identities the worst areas of local non-uniformity by identifying the biggest difference in maximum and minimum pixel values in a 5-pixel range for both x and y directions.

What could go wrong? A gamma camera detector is made up of a collimator, scintillation crystal, and an array of photomultiplier tubes (PMTs), with a series of corrections applied to ensure that the output from the detector is uniform across the field of view. A failure or defect in any one of these elements could cause non-uniform detector response. For example, a poor correction could give linearity issues, which would be reflected in the final image; the outputs from the multiple PMTs could be out of balance, which could give positional errors; there could be an imperfection in the fragile Sodium Iodide crystal caused by moisture ingress or mechanical damage; or there could be poor optical coupling between the scintillation crystal and the PMTs. In the collimator, there could be uneven hole size or angulation errors or, indeed, damaged collimator lead septa from general daily use. Because failures could occur within the detector or in combination with the collimator, both system tests (with collimator) and intrinsic tests (without collimator) should be performed. The intrinsic test is most likely to show issues because of the multiple points of failure that could occur within the detector, and because the acquired data will not be blurred by the response of the collimator. System

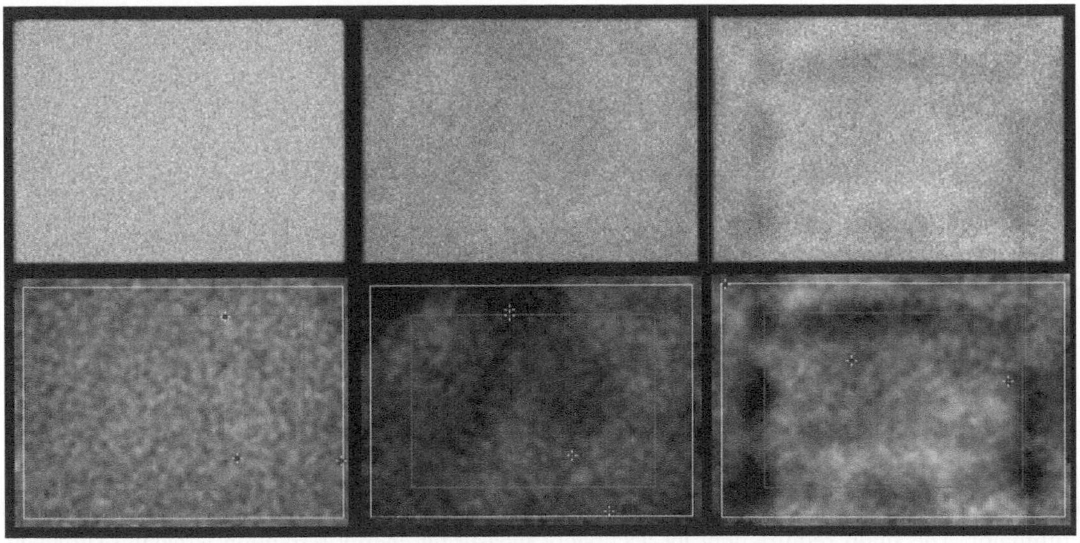

FIGURE 23.4 Top row shows three images from uniformity tests. The bottom row shows the same images *thresholded* to enhance the image contrast with usable field of view and central field of view shown, and the images pixel size are increased to calculate uniformity.

measurements are more relevant to clinical situations, but there is the possibility that a problem seen with an intrinsic measurement may not be noticeable in a system measurement which might allow continuing clinical use. Nevertheless, the underlying problem should still be dealt with in a timely manner, given that there will still be an effect on system performance.

When? Routine uniformity (system or intrinsic) – daily; intrinsic uniformity for ^{99m}Tc – weekly; intrinsic uniformity for other isotopes – half yearly.

Acceptable values. Collimated integral uniformity for ^{99m}Tc or ^{57}Co should be less than 5 per cent for the UFOV and less than 4 per cent for the CFOV. For intrinsic measurement with ^{99m}Tc, values < 3 per cent (UFOV) and < 2 per cent (CFOV) should be expected. Other radionuclides might display higher uniformity measurements than ^{99m}Tc, but uniformity values less than 5 per cent (UFOV) and 4 per cent (CFOV) should still be achievable. Differential uniformity results are always equal or less than that for integral uniformity, due to the smaller area tested.

Examples. Figure 23.4 shows three examples of degraded uniformity. While the top left image may appear to have good uniformity when windowed using the typical zero-maximum count windowing, lifting the lower threshold to around 50 per cent of the maximum pixel values (lower left image), we see lower outputs at the edge of the field of view. Note how the software has also increased the pixel size to calculate uniformity (bottom row), as per NEMA recommendations. A more drastic issue can be seen in the middle two images of Figure 23.4. In this instance the non-uniformities can be seen in the raw un-thresholded images and show severe differences in output across the field of view. To resolve these issues, a full calibration of the detector (energy, linearity, uniformity) is required. The right-most images show a distinctive pattern of non-uniformity, which could be a result of an underlying issue, for example, a power supply. All image data had uniformity measurements higher than expected for the detector.

23.2.2.2 Energy Resolution

Why? Energy resolution is a measure of the system's ability to determine photon energy. Given that scattered photons are rejected by using energy windowing, systems with poor or degrading energy resolution will be inferior at rejecting scattered photons compared to systems with good energy resolution. As a result, the inclusion of more scattered photons will lead to poorer image contrast.

How? *Imaging protocol.* There is no specific imaging protocol required to measuring energy resolution. Instead, it is typically measured at the same time as when performing uniformity measurements, both with and without collimators,

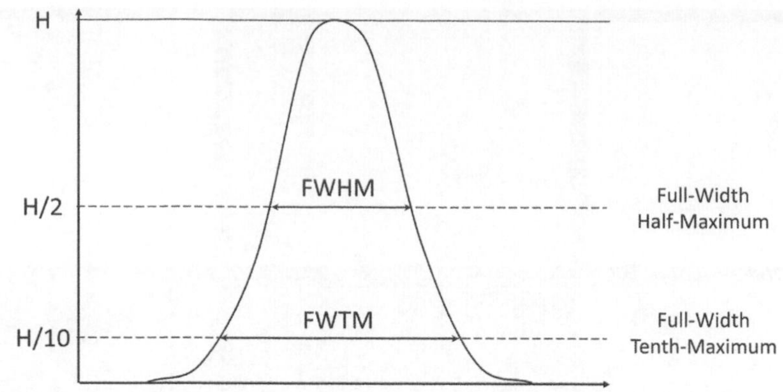

FIGURE 23.5 Definition of Full-Width Half Maximum (FWHM) and Full-Width Tenth Maximum (FWTM).

although intrinsic measurements are preferred because they have a sharper energy spectrum that is unaffected by scatter within a collimator. The main requirement for accurately measuring energy resolution is having sufficient acquired counts to produce a low-noise energy spectrum.

Analysis. On modern gamma camera systems, energy resolution measurements are performed for the user. A gaussian fit is made to the energy spectrum, with the energy resolution derived in terms of the Full-Width Half-Maximum (FWHM) and Full-Width Tenth-Maximum (FWTM) of the Gaussian fit (Figure 23.5).

What could go wrong? A gamma camera is made up of a series of approximately 60 photomultiplier tubes, which are tuned and calibrated to give a maximum and balanced response, and equal energy value when irradiated by a flux of photons. Degrading energy resolution could come from a number of sources. Drifting performance in photomultiplier tubes may lead to suboptimal calibration of the tubes which could lead to a larger spread of energy values across the detector. The number of light photons detected by the photomultiplier tube will also affect energy resolution, so a poorly performing PMT or indeed crystal inhomogeneities or coupling between the crystal and PMT may also lead to inferior energy resolution. If a PMT or a series of PMTs are performing badly, there could be underlying problems, such as an issue with power supplies. Although it may not be the most sensitive test of degrading system performance, an assessment of energy resolution is nevertheless an important and easy test to perform.

Acceptable values. Energy resolution for ^{99m}Tc is typically less than 10 per cent for FWHM. With increasing energy, FWHM energy resolution will decrease in percentage terms.

23.2.2.3 Spatial Resolution and Spatial Linearity

Why? The ability to resolve detail in a gamma camera image depends strongly on the spatial resolution of the system. With poor spatial resolution, uptake or absence of radiopharmaceutical uptake may be missed, so it is important that consistently good spatial resolution is achieved. The assessment of spatial linearity, which can be performed at the same time as spatial resolution, is important to ensure features are accurately placed in an image.

How? *Imaging protocol.* Good spatial sampling is required for this test, which leads to the use of small pixel sizes (large matrix sizes) with sufficient counts acquired to get good profiles from the images. The measurement of spatial resolution and spatial linearity can be made intrinsically, typically using slit masks in the x and y direction. However, more commonly, to mimic the situation seen clinically, spatial resolution is measured with collimators in place using line sources placed in the x and y direction 10 cm away from the collimator face. This distance is introduced to avoid geometric effects from the sources being placed over collimator septa. Line sources typically use ^{99m}Tc. An alternative for assessing spatial resolution and spatial linearity is to use a bar phantom (Figure 23.6) with a ^{57}Co source on top of the phantom to provide a radiation source. A bar phantom that typically has bar thicknesses of 2–5 mm can also be used for both intrinsic and system measurements. Further details on acquisition parameters are given in Table 23.3.

Analysis. Profiles are drawn through the line source, a Gaussian fit is made, and resolution is reported in terms of Full Width Half Maximum (FWHM) and Full Width Tenth Maximum (FWTM). System measurements can frequently use

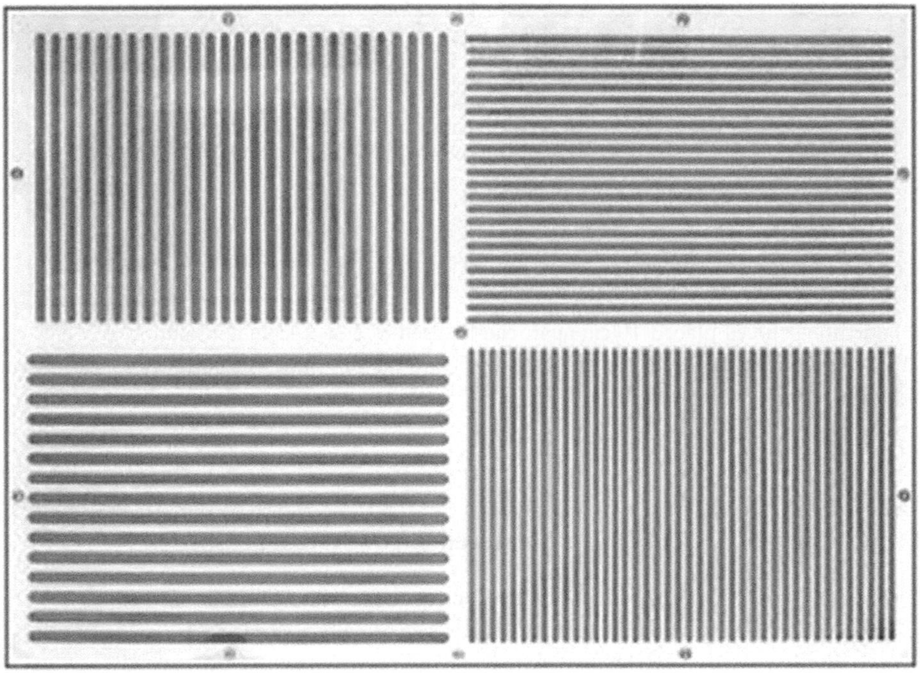

FIGURE 23.6 A quadrant bar phantom, which can be used to assess spatial resolution/ linearity.

TABLE 23.3
Acquisition Parameters for the Measurement of Spatial Resolution and Spatial Linearity

Collimator	Intrinsic: None System: Typically, periodic testing should be performed with the same Low-Energy High Resolution collimator. For radionuclides other than ^{99m}Tc/^{57}Co, the appropriate collimator should be used.
Pixel size	< 2mm. Spatial sampling requires a pixel size of at least ½ and ideally less than the object size and typical spatial resolution is ~ 4mm.
Energy Window	Use manufacturer recommended settings. Relevant to radionuclide although ^{99m}Tc window typically used
Counts	Intrinsic: Maximum pixel value in profile should be 10,000 counts. System with line sources: 1 million counts System with Bar Phantom: 4 million counts

four line sources, forming a 10-cm square. If this approach is taken, the distance in pixels between parallel line sources can be used as to convert the calculated FWHM and FWTM in pixels to millimetres. Spatial linearity is assessed using a visual assessment of the line source – possibly placing a guideline onto the workstation display or, quantitatively, by looking at the deviation from a fitted line through the line that is imaged. Measured in both the usable and central fields of view, quantitative spatial linearity is quoted as the standard deviation of the difference between the measured and fitted peaks (differential spatial linearity), and the maximum difference between the two peak locations (absolute spatial linearity).

What could go wrong? Assessments of system spatial resolution that show a localized degradation of performance may be caused by mechanical damage to the collimator holes. Other issues could come from poor balancing of PMT outputs which, in turn, could degrade positional information that would affect spatial resolution. This type of effect can also come with spatial linearity issues. A poor or corrupt linearity correction can also affect both spatial resolution and spatial linearity.

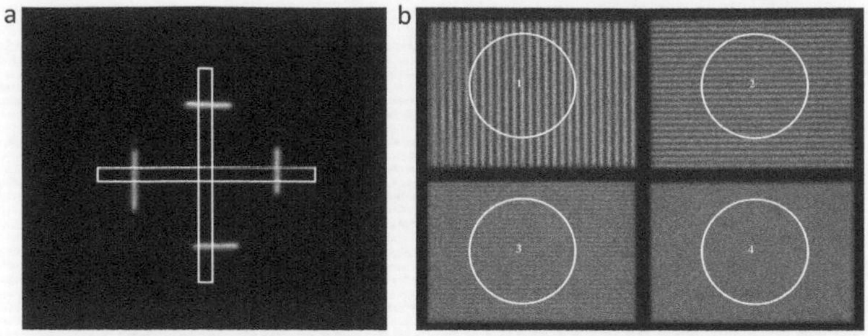

FIGURE 23.7 (a) Four line-sources placed 10 cm apart can be used to assess spatial resolution in x and y directions. The bar phantom, shown in (b), is typically used to qualitatively assess spatial resolution and spatial linearity but can also produce quantitative values of spatial resolution.

TABLE 23.4
Acquisition Parameters for the Measurement of System Sensitivity

Collimator	System: Should be performed for all collimators used clinically
Pixel size	Not important. A pixel size of ~4mm is acceptable.
Energy Window	Relevant to radionuclide. Use manufacturer recommended settings.
Counts	1 million counts

When? Six-monthly.

Acceptable values. Intrinsic FWHM resolution is typically around 4–5 mm. Collimated resolution will inevitably be dependent on the collimator but is typically around 10 mm at 10 cm distance, although this will be slightly better at around 7–8 mm for low-energy high-resolution collimators. Spatial linearity typically has values < 0.5 mm for absolute linearity and < 0.1 mm for differential linearity.

Examples: Figure 23.7 (left) shows how four line sources can be used to calculate spatial resolution in x and y directions. Placing the sources at a distance of 10 cm allows the pixel size to be calculated by determining the distance in pixels between the maximum of the two profiles. The fit of Gaussian curves to the line sources of several rows/columns allows the calculation of FWHM/FWTM in pixels and then millimetres. Figure 23.7 (right) shows the output from a bar phantom acquisition. Typically, the number of bar quadrants visible gives a qualitative assessment of spatial resolution, which can be followed over time, although regions in these quadrants can be used to determine the Modulation Transfer Function (MTF) and then the spatial resolution [10]. The lines in the quadrant can also be used to visually assess spatial linearity.

23.2.2.4 Sensitivity

Why? The sensitivity of the gamma camera determines the level of noise in the image, or the duration of scan we need to get an acceptable level of noise in our clinical data. Sensitivity measurements can also be used to calibrate the scanner to convert image counts into activity concentration.

How? *Imaging protocol.* A phantom, typically a large petri dish with a known amount of activity is imaged 10 cm away from the collimator using an acquisition of sufficient duration to achieve a precise value of measured counts in the phantom. Sometimes a background acquisition is also performed to subtract any ambient background signal in the sensitivity measurement. There is no need to for spatial detail in this acquisition, and given that the counts are averaged over a large area, noise is not normally an issue, either. Details on acquisition parameters are given in Table 23.4.

Analysis. Knowing the activity in the phantom, a region of interest is placed around the imaged phantom to determine the counts in the phantom and ultimately the sensitivity of the system in terms of counts per second (of acquisition) per MBq. The time between measuring the activity in the phantom and the acquisition of the image should be noted so that the decay-corrected activity in the phantom at the time of imaging can be calculated. If a background image is acquired, the count concentration in the region of interest of the background image should be subtracted from that in the phantom image.

What could go wrong? Damaged collimators could affect the sensitivity of the system as could outputs from photomultiplier tubes: for example, if they were not tuned correctly. Problems with the crystal could also affect system sensitivity, although this should be seen in uniformity assessments.

When? Six-monthly.

Acceptable values. Sensitivity values are completely dependent on the collimator being used. Sites should refer to their system specification to determine an acceptable value for this test. Equally as important is the fact that sensitivity remains stable over time. Degrading performance may be a consequence of slow detector failure.

23.2.2.5 Others

What has been described above can be considered the core quality control assessments of a gamma camera. Other assessments could include:

Multi-window spatial registration. This test applies to imaging radionuclides where multiple photopeak energies are used for imaging. The equation to determine the positional information of a scintillation in a gamma camera has a denominator that is the sum of all PMT outputs which, in turn, gives the energy of the incoming photon. If imaging with multiple photopeak energies, there is the potential that the denominator term at different energies positions the scintillation at different points. By imaging a point source in the centre, and at varying positions at the edge of the field of view, it is possible to calculate any differences in the position of the source at different energies. This is the measurement of Multi-window spatial registration, which should be performed on a six-monthly basis, according to IAEA recommendations [11].

Count-rate performance. As the flux of radiation hitting a gamma camera detector increases, the output increases in a linear manner until dead-time and pulse pile-up issues arise. This results in loss of counts and mispositioning of counts. Count-rate performance is a test that can be performed with collimators and a scatter environment, or intrinsically with a decaying source to assess the point at which non-linearity between input radiation flux and output count rate occurs.

Whole-body mode tests. Modern gamma cameras can perform sweeps of patients in so-called whole-body mode, where the detectors slowly move over the patient to produce an image larger than the detector field of view. This test requires software ramping and good mechanical control of the patient couch or detector, together with good positional information from the system to ensure that dwell times per row of the image are equal, and to ensure that the integrity of the spatial resolution in the image is maintained. Tests on whole body mode constancy to assess dwell time per row, and whole-body mode spatial resolution to assess feature positioning and detectability should be performed on an annual basis.

Software tests. Correct timing of computer systems performing nuclear medicine tests is essential for many quantitative nuclear medicine tests. For dynamic studies the timing of frames should reflect what is set on the computer so that, for example, two 10-second frames in a dynamic scan should have exactly the same duration. Similarly, with ECG gating in MUGA scans, for example, we would expect that binned frames in the cardiac cycle be split equally between the two triggering QRS complexes. Any errors with computer timings could lead to incorrect time-activity curves and possibly incorrect diagnosis. Computer software tests can be performed with stop clocks and analysis of time activity curves using a phantom with a fixed activity concentration. Such tests should be performed on a six-monthly basis.

23.2.3 CZT-based Gamma Cameras

In recent years, Cadmium Zinc Telluride (CZT) based gamma camera systems have become more common in nuclear medicine. Instead of using traditional scintillation detector arrays, these systems use pixelated arrays of CZT semiconductor detectors. Because of the different system architecture there can be some differences in the quality control assessments of such devices.

Uniformity is frequently measured with collimators in situ because the collimators are difficult to remove or are an integral part of the detector. Furthermore, given the geometry of some systems, the sources used to measure uniformity frequently require non-standard source positioning with inbuilt software corrections for geometry. The results from uniformity tests should also be interpreted differently from standard Anger gamma cameras. Non-uniformity is on a pixel-by-pixel basis and therefore frequently uses different metrics of uniformity, for example, coefficient of variation. Furthermore, on occasion dead pixels (with no signal) can occur and can be turned off so that they do not contribute to the final image, with a form of interpolation applied to cover the missing signal. If a large number of contiguous pixels are malfunctional, this may require a replacement detector module.

The FWHM energy resolution of CZT systems is very good – far superior (~5 per cent) to that achievable with traditional Anger camera systems (~9 per cent). However, because of incomplete charge collection the energy spectrum below the photopeak energy can have quite a long tail, which has a detrimental effect on FWTM energy resolution [12].

Spatial resolution is always measured with the collimator in place. Given that the intrinsic resolution is defined by the size of the pixelated semiconductor detector, and the system resolution should be stable so long as the collimator is not damaged, this measurement does not need to be performed as frequently as it does with traditional gamma cameras. There is also no need to measure spatial linearity, given that there is no linearity correction of mixed PMT outputs to affect this measure. The only contribution to poor spatial linearity is the physical array of pixelated detector elements, or possibly a damaged collimator.

Sensitivity is dependent on detector output and, given that such detectors may require cooling, sensitivity may change over time and should therefore be measured with the same frequency as for a traditional gamma camera. The source used for this test may depend on the geometry of the system.

The count-rate performance of CZT based systems is very good, being linear to much higher count rates than seen in an Anger gamma camera. Multi-window spatial registration is not required for such systems, given that it does not depend on the Anger logic used in traditional systems to define positional information.

23.3 SPECT QUALITY ASSURANCE

23.3.1 Acceptance Testing

The acceptance tests for SPECT and SPECT/CT as defined by NEMA in NU 2-2018 are given in Table 23.5.

Once more, sites may choose only to perform tests that are relevant to local circumstances. For example, SPECT/CT co-registration will not be relevant for standalone SPECT-only systems, and absolute quantification accuracy would be more appropriate for hybrid SPECT/CT systems, where quantification is performed, although the test does provide useful information about the overall performance of SPECT systems. These tests should always be performed using ^{99m}Tc (^{57}Co can be used as an alternative, where appropriate), although sites that use other radionuclides may choose to examine spatial resolution, sensitivity, and contrast and quantification accuracy for other radionuclides relevant to their practice.

23.3.2 Quality Control Tests

An essential prerequisite for good-quality SPECT imaging is good-quality planar imaging. Good planar uniformity is essential for good SPECT uniformity and good planar spatial resolution can translate to good SPECT spatial resolution. Indeed, one could argue that SPECT uniformity and SPECT spatial resolution are not essential tests for ensuring SPECT performance. The translation between good planar and SPECT parameters is influenced by three factors:

1. The reconstruction parameters used in tomographic reconstruction. Spatial filtering is commonplace in tomographic reconstruction, with filters such as Butterworth and Gaussian employed to achieve the optimal balance

TABLE 23.5
SPECT and SPECT/CT Acceptance Tests Defined in NEMA NU 1-2018

System Alignment	Detector-Detector Sensitivity Variation
SPECT Spatial Resolution with and without scatter	Tomographic Contrast and Absolute Quantification Accuracy
System Volume Sensitivity	SPECT/CT Co-registration Accuracy

TABLE 23.6
Acquisition Parameters for Assessing the Centre-of-rotation Correction

Collimator	Given the different weights and therefore mechanical stresses of different collimators. Tests should be performed with all collimators available on site.
Pixel size	Results depends on centroids. Adequate sampling but not fine sampling is required. A pixel size of ~4 mm is acceptable.
Energy Window	Standard settings for ^{99m}Tc are appropriate.
Time per projection	Sufficient time to get a well-defined point. Depends on activity of source(s), but 20 seconds per projection is normally sufficient to get the required 10,000 counts per projection
Angular sampling	6–12 degrees with each detector acquiring over a 360-degree arc is adequate.
Radius of rotation	Follow manufacturer guidelines, but typically two radii of, for example, 30 cm to simulate body studies and 15 cm to simulate brain studies is sufficient.

between the noise and spatial resolution in the final image. Inevitably the filter used will affect parameters such as uniformity, which is based on the variability of individual voxel values, and SPECT spatial resolution, which can be blurred or enhanced with filtering. For tests of spatial resolution or general SPECT performance where the highest-possible spatial resolution is required, a ramp filter or no filter is applied while, for other tests, a clinically appropriate filtering regime is suggested. The reconstruction algorithm, that is, filtered back-projection or iterative reconstruction (OSEM/MLEM), and the corrections used (attenuation, scatter, resolution modelling) can also affect SPECT performance. For spatial resolution measurements, filtered back-projection is often defined because it has no convergence issues [13], while for other tests, the algorithm and corrections used clinically should be selected. For some tests such as SPECT uniformity and general SPECT performance, reconstruction should always include some form of attenuation correction.

2. Detector-to-detector sensitivity. Balanced detector sensitivity is essential to avoid image artefacts. If, for example, one detector has a sensitivity one half of the second detector, the signal will be lower for all projections acquired with the detector with the lower sensitivity. In the final reconstructed image of a uniform volume source, this can result in a non-uniform image. Differences between detector sensitivity are rare, but should be checked at acceptance testing and tracked during routine planar uniformity measurements.
3. System alignment / centre of rotation. This test examines the relationship between the reconstructed centre of the field of view with the mechanical centre of the field of view. As shown in Figure 23.8, while with detectors positioned directly anterior and posterior to the patient/phantom, the two centres correspond, with the detectors in a lateral position gravity acts on the detectors, which pulls the mechanical centre below the reconstructed centre. For each projection angle the effect of gravity will be different, and also the effects of different collimators on the system will also be different. Typically, a centre-of-rotation correction is applied prior to reconstruction to correct for the mismatches of mechanical and software centres of the field of view. As part of a routine quality control programme the adequacy of these corrections is tested. A similar test performed at the same time is multiple head registration. This test ensures that multiple head systems image the same volume, that is, that the volumes imaged on both heads are not shifted in some way. Of all quality control tests performed on SPECT systems, this assessment can be considered the most important.

In the following sections, standard and optional quality control tests will be described. The reason for doing the test will be defined, together with possible areas of failure and the suggested periodicity (based on IAEA guidelines [7]) will be given.

23.3.2.1 Centre of Rotation/System Alignment/Multiple Head Spatial Registration

Why? Gravity acting differently on detectors at different angular positions can cause a misalignment between the software and mechanical centres of the field of view (Figure 23.8). Normally, this effect is mitigated by applying a correction to projection data prior to reconstruction to realign the two centres. A poor correction will provide insufficient or incorrect realignment, resulting in ring artefacts in the final image (Figure 23.8).

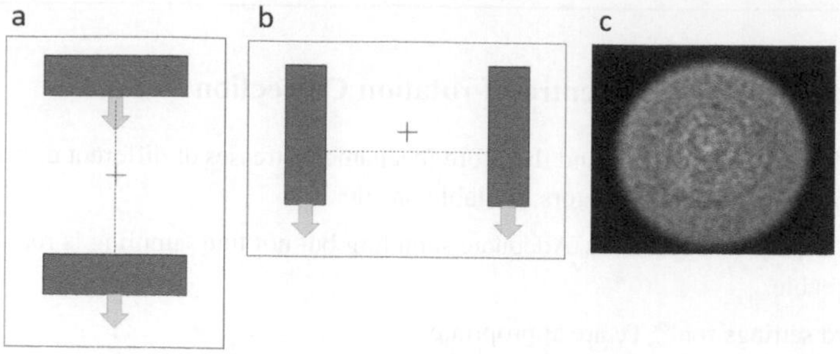

FIGURE 23.8 Effects of gravity on gamma camera detectors. In (a) the mechanical and software centres of rotation are the same but at lateral projections; (b) the effect of gravity creates a mismatch between the two centres; (c) Ring artefact seen when there is a poor centre-of-rotation correction.

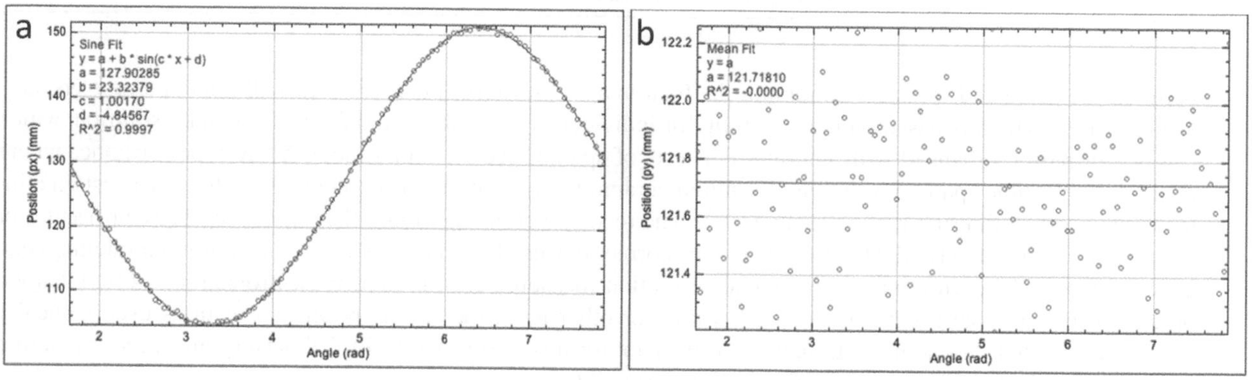

FIG. 23.9 Centre-of-rotation results showing the offset at each angle from a fitted line in x (a) in y (b) directions.

How? *Imaging protocol.* A point source, or several point sources along the axial field of view (going into the scanner bore), are placed centrally in the SPECT field of view. Some manufacturer software will ask for specific positioning of these sources, although a 5 cm off-centre position is adequate. A SPECT acquisition with each head travelling 360 degrees of arc is required to assess the centre-of-rotation correction Details on the acquisition parameters are given in Table 23.6.

Analysis. A line (or sinusoid if the source is off-axis) is fitted to the centroid position of the source for each projection. This is done in both x and y directions. Results are reported as the maximum offset from the fit.

What could go wrong? The QC test assesses the effectiveness of the centre-of-rotation correction. Poor performance in this test is mostly caused by a failure of the correction – possibly because of mechanical settling or shift of the detector holding point.

When? This test should be performed on a monthly basis but should be performed for all available collimators. For systems with multiple geometries, for example, 90-degree and 180-degree modes, centre-of-rotation QC tests should also be performed in both these modes, although 90-degree modes are normally only used for low-energy collimators.

Acceptable values. Fitting with sampling theory, acceptable results are for the deviation from the fit to be within half of the pixel size for a typical SPECT. So, for example, with a matrix size of 128 x 128 and a pixel size of 4.4 mm, a deviation of at most 2.2 mm would be acceptable. However, limits are typically set much lower, at around 0.5 mm, given that they are easy to achieve and facilitate earlier corrective intervention.

Figure 23.9 shows an example of centre-of-rotation analysis. In Figure 23.9a we see the points of the centroid of the source in the x-direction for each angle forming a sinusoid, with a line fitted through these points showing the deviation for each angle. In Figure 23.9b the fit this time is a line, and the distribution of values is tighter (see the y-axis scale). The maximum deviation in this instance (not shown) is 0.77 mm (x-direction) and 0.54 mm (y-direction) – well within the 1.1 mm acceptable for this 2.2 mm pixel size.

TABLE 23.7
Acquisition Parameters for Assessing General SPECT Performance

Collimator	Typically, low-energy high resolution or equivalent. The collimator used should be appropriate for the radionuclide being assessed.
Pixel size	Given that SPECT spatial resolution is approximately 8–10 mm, a pixel size of 2–4mm is appropriate to achieve images with acceptable levels of noise.
Energy Window	Standard settings for ^{99m}Tc are appropriate. If using other radionuclides, the appropriate window for that radionuclide should be chosen.
Time per projection	A high-count low noise acquisition is required. Typically, >800 kilocounts per projection is required.
Angular sampling	~3 degrees. If using a dual detector system, 180 degrees of arc per detector is acceptable.
Radius of rotation	A consistent value such be used that is relevant to clinical imaging. Ideally detector phantom distance should be minimized as much as possible to achieve optimal performance.
Reconstruction	Attenuation correction should be applied (Chang or CT). Use Filtered Back Projection with Chang Attenuation Correction OR Iterative reconstruction with a high number of updates (e.g. 10 iterations and 10 subsets) using CT attenuation correction. No filtering or a filter with the highest acceptance of all spatial frequencies should be used to show maximum performance.

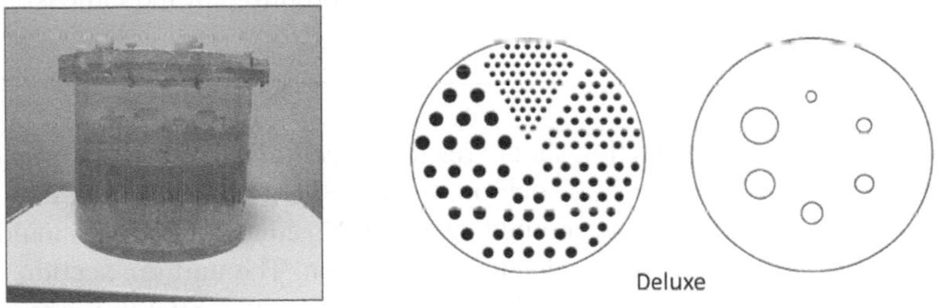

FIGURE 23.10 The Deluxe Jaszczak phantom. Solid sphere diameters are 31.8 mm, 25.4 mm, 19.1 mm, 15.9 mm, 12.7 mm, 9.5 mm. Rods have diameters of 12.7, 11.1, 9.5, 7.9, 6.4 and 4.8 mm.

23.3.2.2 General SPECT Performance

Why? Using one phantom such as a Jaszczak phantom (Figure 23.10), it is possible to assess general SPECT performance, such as uniformity, contrast, and spatial resolution. Typically, this assessment is performed qualitatively, although it is possible to make quantitative assessments. While good planar uniformity and spatial resolution should translate into good SPECT performance, reconstruction filters and SPECT-specific artefacts can affect SPECT image quality and, therefore, General SPECT Performance should be assessed.

How? *Imaging protocol.* A Jaszczak or equivalent phantom is typically filled with ^{99m}Tc solution. A high-count, low-noise acquisition (>800 kcounts per projection) is performed to allow the assessment of image performance potential and to spot SPECT artefacts. Other radionuclide solutions can also be used in the phantom to assess performance for applications that do not use ^{99m}Tc but are typically not performed as a routine quality control test. The acquisition parameters for this test are given in Table 23.7.

Analysis. Visual assessment of the uniform section of the phantom is acceptable and can be used to spot artefacts from, for example, centre-of-rotation or attenuation correction issues (see Figure 23.8c). The visible number of ball-and-rod features in the data can also be used to check the consistency of SPECT image contrast and spatial resolution. Slices should be summed to reduce noise and assist with visual assessment, assuming that the phantom alignment is straight. Although dependent on slice thickness, 5 slices typically can be summed in the uniform section, 3 slices in the ball section and 8 slices in the rod section. Quantitative assessments can be made in a manner like that suggested by the IAEA [7], [14].

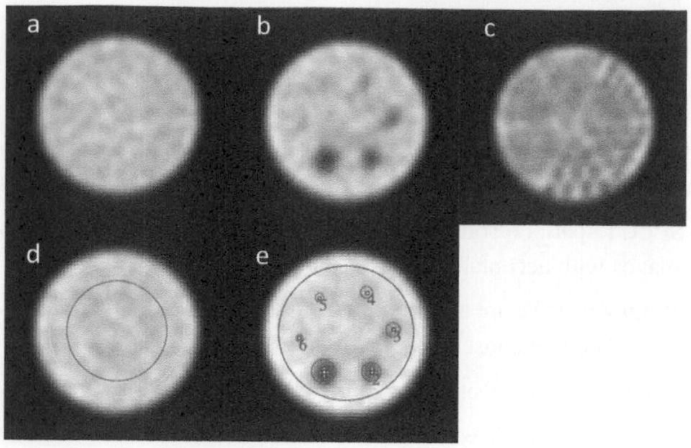

FIGURE 23.11 Jaszczak phantom images showing (a) uniform, (b) sphere and (c) rod sections, with (d) the border and central region shown for summed uniform slices and (e) contrast features for the solid spheres.

What could go wrong? An assessment of SPECT uniformity is helpful to determine whether there are any centre-of-rotation issues, detector-to-detector sensitivity mismatches, or failure-of-attenuation correction methods. Centre-of-rotation failures would be seen as concentric-ring artefacts on all slices, or in some slices if there is an issue with detector sag. Detector-to-detector sensitivity issues would be seen as a count gradient across the uniformity image, while attenuation correction failures could be seen as a hot or cold centre compared to the edge of the phantom. This could be because of a poor choice of linear attenuation coefficient for Chang attenuation correction, or misregistration of CT used for attenuation correction of the emission data. SPECT spatial resolution could be compromised by the same issues as SPECT uniformity.

When? Quarterly.

Acceptable values. Accurate attenuation correction should result in a centre to border ROI ratio of > 0.98 in the uniform section of the phantom. Depending on the radius of rotation, typical SPECT spatial resolution would allow between 4 and 5 balls to be seen. Around three rod segments should be visible. No concentric rings should be visible in any slice.

Figure 23.11 shows the results from a Jaszczak phantom acquisition. The uniform section shows no ring artefacts, a centre–border ratio of 0.985, integral uniformity of 7.17 per cent and differential uniformity of 6.67 per cent. Four spheres and three rod sections are visible with maximum contrast of the largest sphere being 56 per cent, and the fourth sphere having a contrast of 15 per cent.

23.3.2.3 Others

While the two tests described above could be considered an adequate test of SPECT performance, other tests could include:

SPECT uniformity. While uniformity assessed using a Jaszczak phantom is acceptable, a uniform cylindrical phantom could be used instead to assess uniformity over a wider range of the SPECT field of view. This test could be performed on a six-monthly basis.

SPECT resolution in air and with scatter. Quantitative SPECT spatial resolution is typically assessed by using a triple line source either in air or scattering material (water). The three-line sources are placed in the centre, and at the 6 o'clock and 3 o'clock positions, which allows the assessment of FWHM and FWTM central, tangential, and radial SPECT spatial resolution. Although typically performed using ^{99m}Tc, for special applications non-^{99m}Tc radionuclides could also be used. IAEA guidelines recommend an assessment of SPECT spatial resolution on a six-monthly basis.

Tomographic contrast and assess and absolute of quantification accuracy. Described in NEMA NU 1-2018, this test uses an NEMA image quality phantom to assess the recovery of different 'hot' filled spheres of various sizes; the contrast in a lung insert; and the variability in the background. For systems capable of measurements in kBq/cc, a measurement is made of the quantitative accuracy in a background region, which is unaffected by any partial volume effects. This is a test that could be performed on an annual basis.

23.3.3 Novel SPECT Systems

While many of the tests described above are achievable, some novel SPECT designs bring with them different QC challenges. In the section on planar systems, we have already described CZT systems and their QC requirements. Many CZT systems also have unique SPECT architectures. For example, the Spectrum Dynamics Veriton system uses pixelated detector arrays in a ring arrangement. While most SPECT tests would apply to this system, being a ring detector, centre-of-rotation tests are not generally applicable. Other CZT systems are specific cardiac SPECT devices with detectors arranged in an L shape. Spectrum Dynamics D-SPECT system has regional reconstruction [15] so, uniformity is a challenging metric to measure in the field of view, and spatial resolution can change significantly across the field of view. General Electric, with their 530c dedicated cardiac SPECT system, takes a different approach, using multiple pinholes to acquire projection data. Once more this brings complexity to traditional SPECT QC metrics. Many modern systems, with their complex designs, also require modelling of the system architecture to be incorporated into iterative reconstruction algorithms, which makes it challenging to compare performance with traditional gamma camera systems where characterization is according to NEMA specifications. Even using conventional SPECT systems, the use of fan beam or multi-focal collimators also requires the modification of SPECT QC protocols and different interpretation of results. Nevertheless, for most systems, some or all of the tests described in this section can be used for routine SPECT quality control. Where tests are not appropriate, the systems operator manual should provide alternative tests.

23.4 PET QUALITY ASSURANCE

With fewer moving parts and a one fundamental photon energy of 511 keV, quality assurance of PET scanners is in some ways less problematic than gamma camera and SPECT systems. Furthermore, quality control tests tend to be more heavily controlled by the system manufacturer, which creates some restraints on what can be done in an independent and user-focused manner. Nevertheless, quality assurance programmes can still exist under general themes that apply to all PET scanners. This will be the focus of this section.

23.4.1 Acceptance Testing

The acceptance tests for PET and PET/CT as defined by NEMA in NU 2-2018 [16] are given in Table 23.8.

Details on how to perform these tests are given in the NEMA document. Some sites like to repeat these tests on an annual basis as part of their routine quality control programme, but the expense of the phantoms can make this prohibitive for some. The quality control tests that can be performed are described below.

23.4.2 Quality Control Tests

For PET, quality control tests can be grouped in terms of their periodicity, which aligns with the depth of assessment being performed. As with gamma cameras and SPECT, national and international guidance is available detailing what should be done as part of a quality control programme [2], [4], [8].

23.4.2.1 Daily Tests

Daily tests typically assess the overall PET system performance and have the following components.

Detector Response

Why? It is essential to assess the PET scanner's detectors response to ensure that the scanner produces an accurate artefact-free image.

TABLE 23.8
PET and PET/CT Acceptance Tests Defined in NEMA NU 2-2018

Spatial Resolution	Image Quality, Accuracy of Corrections
Scatter Fraction, Count Losses and Randoms	Time-of-Flight Resolution
Sensitivity	PET/CT Co-registration Accuracy
Accuracy: Corrections for Count Losses and Randoms	

How? Depending on the system manufacturer, this is done in different ways, but the outcome is always the same: An assessment of whether the system at a detector level is producing an output or outputs that are within expected limits. Such tests typically provide the capacity for both qualitative and quantitative assessment. Quantitative outputs allow trend analysis to spot underlying issues, while qualitative outputs can highlight where an issue exists. This QC may require external sources or sources within the scanner, although some modern scanners can use the natural radioactivity of the scanner's scintillation crystal to assess detector performance. Examples from outputs from these daily tests are shown in Figure 23.12.

What could go wrong? Individual detector blocks could fail – typically at the level of the photomultiplier tube in analogue PET systems, or in the electronic supporting circuitry in digital Avalanche Photodiode (APD) or Silicon Photomultiplier (SiPM) systems. Energy peaking, coincidence timing, and crystal position maps could also become ineffective. On a larger scale, issues of power supply to detectors could cause problems within individual or groups of detectors and, for digital PET systems, cooling issues could also affect system performance.

Acceptable values. Outputs following data acquisition should be assessed visually for outlying patterns and quantitatively. The software run by the system will typically tell the user if the system is fit for clinical use and may also provide trends of temporal stability.

Quantitative Accuracy

Why? PET is an inherently quantitative modality with quantitative measures such as SUV being routinely used in clinical reporting. This test is used to verify the scanner's quantitative accuracy.

How? This daily test may or may not be built into the detector response test described above. An easy way to assess quantitative accuracy is to scan a cylindrical phantom filled with Fluorine-18 solution or a solid ^{68}Ge cylindrical source with known activity concentrations and compare this activity concentration with that measured by the scanner. Normally the cylinder will be scanned with a typical clinical scanning protocol over multiple axial fields of view.

What could go wrong? Detector failure could clearly affect quantitative accuracy. Assuming that this is fine, a poor or inaccurate scanner calibration could cause quantification inaccuracy (see below for details). Alternatively, a failure in correction algorithms can cause problems that may also be seen as a visual artefact in reconstructed images.

Acceptable values. Typically, activity concentrations or SUVs from the scanner should be within +/-10 per cent of the known activity concentration or SUV.

23.4.2.2 Weekly Tests

Weekly tests typically assess and perform superficial optimization of the detector. Normally, updating gains on photomultiplier tube channels and working on coincidence-timing corrections will be done on a weekly basis, generally using manufacturer specific tools. More modern PET systems using SiPMs instead of PMTs are often more stable, with these types of tests required on a less frequent basis.

23.4.2.3 Quarterly Tests

Detector optimization. On a less frequent basis, typically quarterly but sometimes monthly, more detailed detector calibration and optimization will be performed. Crystal maps, energy information, and – because of their interdependence – gains and possibly coincidence timing tests and optimization will be performed. Because of the complexity of this optimization process, these processes are often performed by manufacturer's engineers.

System calibration. Also, on a quarterly basis, calibrations of the whole system may be performed. Once such calibration is the normalization of the scanner, which accounts for differences in outputs from different line-of-response detector pairs. This test is analogous to a uniformity correction on gamma cameras. Depending on the system, well counter and sensitivity calibrations may also be updated, which relate detected events to activity concentrations and correct for the geometric effects on sensitivity. On other systems, these calibrations may be performed on a less-frequent basis. Manufacturer maintenance engineers tend to perform these tests as part of their preventative maintenance programme, although many of these tests can be performed by local staff.

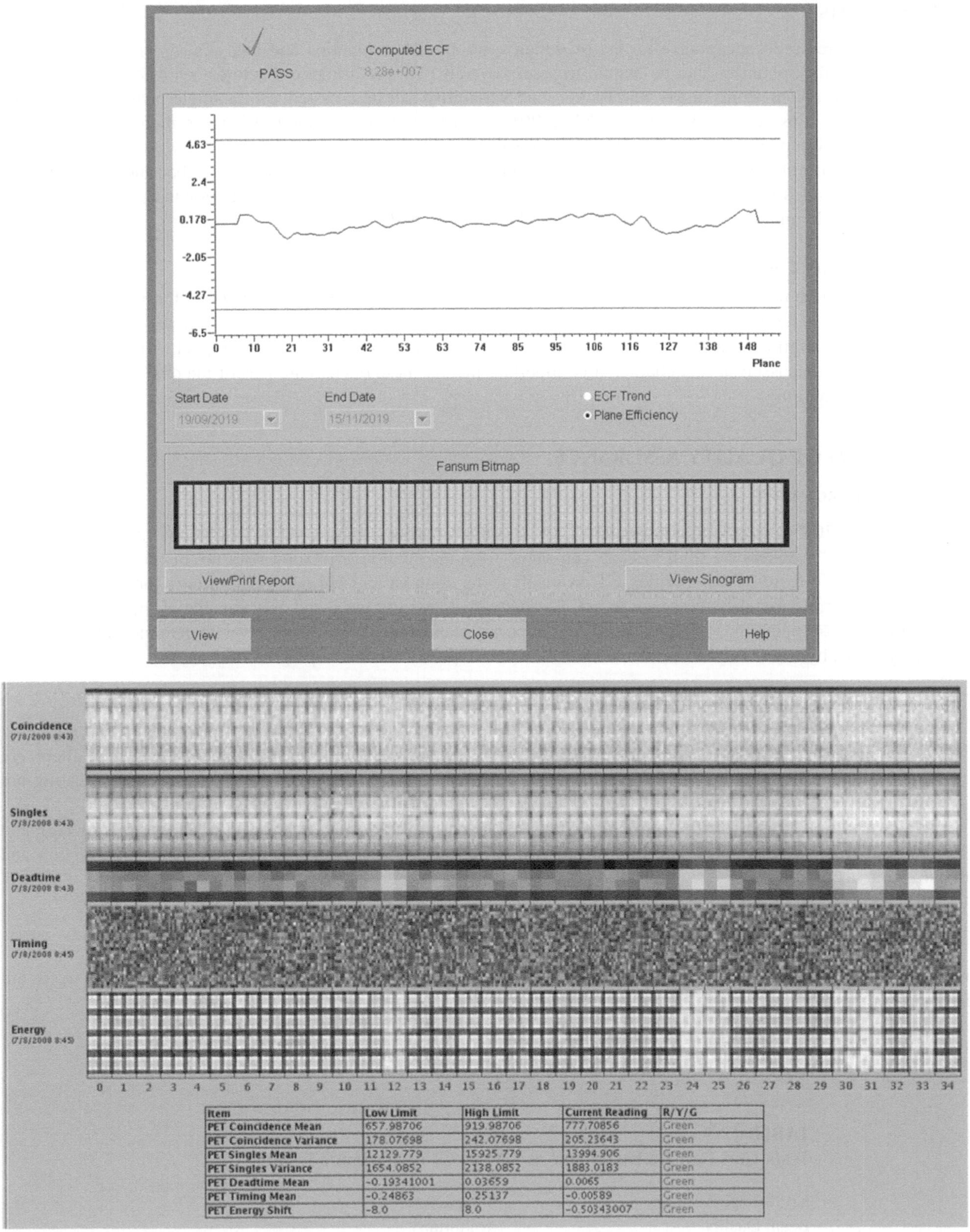

Item	Low Limit	High Limit	Current Reading	R/Y/G
PET Coincidence Mean	657.98706	919.98706	777.70856	Green
PET Coincidence Variance	178.07698	242.07698	205.23643	Green
PET Singles Mean	12129.779	15925.779	13994.906	Green
PET Singles Variance	1654.0852	2138.0852	1883.0183	Green
PET Deadtime Mean	-0.19341001	0.03659	0.0065	Green
PET Timing Mean	-0.24863	0.25137	-0.00589	Green
PET Energy Shift	-8.0	8.0	-0.50343007	Green

FIGURE 23.12 Typical outputs from daily assessments of PET detector performance.

4.2.4 Annual Tests

While some centres redo acceptance-testing procedures, an annual acquisition and analysis of a Fluorine-18 filled NEMA image quality phantom can be helpful to assess overall PET performance. Factors such as contrast recovery, scatter correction into the lung insert, and background variability can be assessed with a single phantom acquisition using the analysis steps described in NEMA NU 2-2018 document. Typically, the test is performed without the additional scatter fraction phantom that is used in NEMA acceptance testing.

Although not currently in any guidelines, more scanners are offering the ability of continuous bed-motion acquisition in addition to the standard step-and-shoot mode of PET scanning. An annual check of this motion in a manner analogous to whole-body mode imaging on a gamma camera could be considered.

23.4.2.5 Ancillary Equipment

Given the quantitative nature of PET imaging, there is a strong reliance on ancillary equipment for high-quality quantitative outputs [17]. Items such as dose calibrators, room clocks, and synchronization with scanner clocks, well or sample counters, weighing scales and height-measuring devices all contribute to PET outputs. Although only dose calibrators will be discussed in this chapter, it is important that all these devices are part of the quality assurance programme for PET imaging.

23.5 HYBRID CT QUALITY ASSURANCE

23.5.1 Background

Many modern SPECT systems and almost all PET systems now come as a hybrid system that incorporates a CT scanner. The use of the CT scanners on such systems can range from attenuation correction with no, or limited, localization of PET and SPECT features, to complex CT protocols using contrast and physiological gating. The specification of such systems typically matches their use, from CT systems that are very basic and only capable of rudimentary CT, to scanners that almost match the specification of contemporary standalone CT scanners. As with gamma camera, SPECT and PET systems, hybrid CT systems should also fall under a quality assurance programme.

23.5.2 Acceptance Testing

All CT systems, including those incorporated in hybrid systems, should be acceptance tested, and there are numerous guidance documents for this process [18–21]. Areas that could be assessed as part of this testing are given in Table 23.9.

In addition, the key test specific to hybrid systems is the co-registration accuracy of SPECT and CT, or PET and CT subsystems. Clearly, good registration of the two subsystems is very important for attenuation correction and colocalization accuracy.

23.5.3 Quality Control

Quality control of hybrid CT subsystems can be conveniently grouped based on the periodicity the tests are performed on.

TABLE 23.9
Tests That Can Be Performed as Part of CT Acceptance Testing

Kilovoltage Accuracy	Image Noise
Beam Quality	Uniformity
CT Number linearity	High Contrast Resolution
Slice thickness	Low Contrast Detectability
Accuracy of Distance Measurements	CT Dose Index (In air and Perspex)

23.5.3.1 Daily Tests

Most CT subsystems have specific daily tests that must be performed. Although the way these tests are performed are scanner specific, they typically include the following checks:

- *A tube warm up*. CT tubes need to be at an operating temperature significantly higher than room temperature to achieve optimal performance. CT tubes are warmed up to an appropriate operating temperature by performing a series of X-ray exposures over approximately one minute. A tube warm up is performed as the very first step of the daily CT quality control process.
- *Mylar window check*. Before doing any tests, the system will check for objects in the CT field of view and spillage of CT contrast onto the mylar window to the CT detectors. Any material interfering with the beam will have an impact on any following tests.
- *Detector checks*. The system will perform a series of checks to test CT detector outputs.
- *Calibration scans*. CT data is formed by taking the ratio of outputs with and without a patient within the scanner. During the standard daily QC, the scanner will acquire a number of calibration scans using various tube and collimation settings to provide the 'without patient' data that is used to form a patient CT scan.

Almost all CT scanners are provided with a CT phantom which includes various image-quality assessment components. The uniform section of this phantom should be scanned on a daily basis to assess:

- General image uniformity to see if a detector is malfunctioning.
- CT number accuracy in Hounsfield Units (HU). The phantom is typically filled with water, which should be measured as 0 HU.
- The standard deviation of voxel values in regions of interest at several positions within the phantom can be used to assess the image noise in the image. Increasing image noise could be reflective of an overall loss in system performance.

23.5.3.2 Monthly Checks

The same manufacturer-provided image phantom that is used for daily CT checks has additional components that can be used to assess parameters such as spatial resolution, image contrast, and slice thickness. Some phantoms also allow the testing of HU accuracy for different materials. These checks should be performed on a monthly basis as an addition to the daily checks of CT number accuracy, noise, and uniformity.

23.5.3.3 Six-monthly Checks

Generally, on a six-monthly basis, or at a periodicity recommended by the scanner manufacturer, co-registration of SPECT and CT, or PET and CT subsystems should be performed. It should be noted that, for mobile scanners, this test will need to be performed after each relocation and, for all scanners, it should be performed following any split of the two gantries by the system engineers. This test is typically performed with a manufacturer-specific phantom and analysis package and should be performed with the patient couch loaded in a manner that mimics a patient being on the couch.

23.5.3.4 Annual Checks

On annual basis it is sensible to repeat many of the tests performed at acceptance testing to check the overall performance of the scanner in a more detailed manner.

23.6 PET/MR QUALITY ASSURANCE

In the same way as with PET/CT, quality assurance of PET/MR scanners follows the same framework of PET and MR modality specific quality assurance and an additional co-registration check of the superimposed data. One advantage of this co-registration check is that, unlike with PET/CT where a table translation is required to move the patient between PET and CT, in PET/MR the acquisition is simultaneous with PET and MR fields of view overlapping. Consequently, there are fewer issues with co-registration checks for PET/MR compared to PET/CT, and the test can be performed less frequently.

With the exception of the co-registration test, PET quality assurance on PET/MR scanners is much more challenging than on PET/CT. This is because the PET phantoms traditionally used for PET testing are typically filled with Fluorine-18 water solution. This causes two problems.

Firstly, water-filled objects in MR scanners often produce standing wave artefacts when scanned using MR sequences [22]. This can be overcome in some instances by filtering or image processing of the resulting image data. However, the second issue of how PET/MR produces attenuation-correction information for PET data is problematic. Attenuation-correction of PET is an essential part of producing accurate and non-artefactual images. While in PET/CT or standalone PET using transmission sources, the ability to take CT or transmission source data to derive attenuation maps is relatively straightforward, in PET/MR attenuation correction is based on tissue characteristics that are not appropriate for water-based phantoms [23]. Any quality assurance assessment of PET subsystems that use water-filled phantoms therefore becomes more complex. A workaround is to use co-registered CT data of the phantom acquired on a different scanner to attenuation-correct PET data, or to use a calculated attenuation-correction approach. However, both of these methods are not ideal, and do not allow the full output of the PET/MR to be assessed. There are attempts to address the issue with PET/MR phantoms [24], but at the time of writing these are still in the development phase.

Quality assurance for MRI has other challenges. There is limited consensus for MRI quality assurance programmes – indeed quality assurance and quality control testing are not routinely performed at all sites. However, this position is changing, particularly for participants in multi-centre trials, and at centres where quantitative MR is being performed, driven by the need for consistent outputs. Given that PET/MR provides quantitative attenuation maps, quality assurance programmes of the MR component of these scanners are crucial to ensure the integrity and accuracy of PET data.

Typically, MR quality assurance starts with manufacturer engineers performing in-depth testing of an MR system following installation – mostly focusing on image quality rather than image accuracy. The same engineers will also perform fewer in-depth tests during their preventative maintenance and service programme. This can be supplemented by local sites performing more detailed acceptance testing based on factors such as signal-to-noise ratios from coil outputs, assessments of gradient uniformity, RF transmission uniformity, slice selection profiles, in plane resolution and tests for ghosting. At the same time, sites will also set baseline measurements for less in-depth parameters, which form the routine quality control programme. *IPEM Report 112* gives more details on tests that could be part of an MR Quality Assurance Programme [25].

23.7 DOSE CALIBRATOR

23.7.1 Background

A dose calibrator is the device that measures the activity of radiopharmaceuticals before, and sometimes after, injection into patients (see Chapter 5). Also known as the radionuclide calibrator, activity calibrator, or activity meter, the dose calibrator is a key part of the imaging process, given that it determines the activity and therefore the dose that is given to patients. It also helps ensure that we give an appropriate activity of radiopharmaceutical to achieve a diagnostic image. Because of its importance, the incorporation of a testing regime for dose calibrators into an equipment quality assurance program is essential.

Numerous guidelines on quality assurance programs for dose calibrators are available [26–30]. A list of recommended tests and their periodicity from these guidelines are given in Table 23.10. When assessing the performance of dose calibrators, it is normal to compare it with both the manufacturer specification and the expected level of performance given in guidelines. The following sections will describe the tests required, first for tests unique to acceptance testing, with the remaining tests explained in the quality control section.

23.7.2 Acceptance Testing

These tests should be performed before the dose calibrator is brought into clinical use. While many tests can be performed with ^{99m}Tc, which is the predominant radionuclide used in nuclear medicine, other tests should be performed using radionuclides appropriate for that site, such as ^{18}F or ^{131}I. The use of long half-life check sources such as ^{57}Co, ^{137}Cs or ^{133}Ba can also be helpful for consistency measurements, or measurements where decay correction can be an impediment. For each test, the required radionuclide will be stated.

23.7.2.1 Accuracy

Why? The dose we give to the patient is directly related to the activity we inject, with this activity being measured using a dose calibrator. It is essential, therefore, that the measurement is accurate, that is, that the instrument is recording the correct value.

TABLE 23.10
Acceptance and Periodic Quality Control Tests for Dose Calibrators

	Acceptance	Daily	Monthly	Annually
Physics Inspection	X	X	X	X
Accuracy	X			X
Reproducibility	X			X
Linearity	X			X
Subsidiary Calibrations	X			X
Background	X	X	X	X
Clock Accuracy	X	X	X	X
High Voltage	X	X	X	X
Display	X	X	X	X
Zero Adjust	X	X	X	X
Constancy	X	X	X	X

How? To know how accurate a dose-calibrator measurement is, a radioactive source of known activity is required. This known activity could come from a measurement made in a national metrology lab or from using a high-accuracy instrument with accuracy traceable to a metrology lab, for example, a secondary standard instrument. Sometimes the source will be short-lived and have an accurate measurement made on the same day as that made on the field (in the clinic) instrument, while other times a longer half-life source can be used with traceability to the metrology lab, for example, a NIST-calibrated source. More details on this process are given in Chapter 24 in this volume.

Taking the source with known activity, several measurements are made, and after background correcting, averaging and correcting for radioactive decay, the difference between the measured and known activity, that is, the accuracy can be determined. Wherever possible, accuracy should be evaluated for all radionuclides that are to be measured in the dose calibrator or, at the very least, for radionuclides that are commonly used.

What could go wrong? Calibration factors are provided by the manufacturer to convert the current collected by the instrument into activity. Although correct for the average instrument, on individual dose calibrators the calibration factor could be incorrect for that instrument and its working environment. Containers and geometry can also affect accuracy. To mitigate this, accuracy measurements should be performed using standard containers and geometries.

Acceptable values. Accuracy can depend on the radionuclide being measured, with some more challenging to measure than others because of their radioactive emissions. One would expect an accuracy within +/- 5 per cent for gamma emitters > 100 keV, and +/- 10 per cent for gamma emitters < 100 keV and beta emitters [26], [30].

23.7.2.2 Reproducibility

Why? While an average measurement may be accurate, measurements in the clinical environment are typically measured only once or twice. It is important, therefore, that measurements are consistent, that is, reproducible.

How? Using a source of clinically relevant activity and energy, 10 measurements of the source should be made going through the complete cycle of placing the source in a dipper, putting the source into the dose calibrator, and removing the source for each measurement. While a ^{99m}Tc source can be used, a long-lived source of appropriate activity and energy, such as ^{57}Co or ^{137}Cs, would avoid the need to correct for radioactive decay between measurements. The mean and the standard deviation of the mean should be calculated and recorded.

What could go wrong? A fluctuating ionization chamber power supply and/or electrometer could lead to results with low reproducibility.

Acceptable values. Within +/- 1 per cent for 1 standard deviation is an acceptable level of performance.

23.7.2.3 Linearity

Why? In addition to measurements of activity being accurate and reproducible, it is important that performance is linear – that is, accuracy is consistent across the working activity range of the dose calibrator, typically from < 10 MBq to > 20 GBq.

How? Several methods to calculate linearity can be used, although the most common is the use of the decaying source. A ^{99}Mo/^{99m}Tc generator is eluted to give a large amount of radioactivity: for example, 20 GBq. After determining the background measurement from the dose calibrator, multiple periodic readings are taken of the source until the activity of the source is less than 10 times the background. A plot is made of the background-corrected activity versus the time from the first reading. The linear slope of this plot between 100–100,000 times the background should be the negative of the decay constant of ^{99m}Tc, for example, 0.1154 hr^{-1}. Using the slope calculation, the predicted reading can be calculated for all points. The activity value where the measured value is 1 per cent and 5 per cent lower than predicted should be recorded.

This test should be performed over the range of activities used with the dose calibrator.

What could go wrong? Power-supply issues and issues surrounding any switch from lower to higher count-rate modes can show discrepancies. At very-high activities the efficiency of the counting can also degrade, causing a loss of performance.

Acceptable values. Within the used range, the one standard deviation of performance should be within +/- 5 per cent.

23.7.2.4 Subsidiary Calibrations

Why? With the tests described above, sources tend to be in standard volumes and containers. Yet it is known that for some radionuclides – for example, with low-energy gamma emissions (<100 keV) or beta emissions – because of absorption of the emissions, the measured activity will depend on the container and the source volume. Volume effects can also occur with ^{99m}Tc and ^{131}I, albeit at a much lower level.

How? Transfers by weight of activity between containing vessels can be used to assess the effects of the measuring container, while dilution with a non-radioactive solution can be used to assess volume effects.

While different calibration factors can be determined for various containers to ensure consistency in accuracy measurements across different vials and syringes, a volume correction curve can be used to correct for different activity volumes.

What could go wrong? It is important to be aware of container and volume effects, particularly with low-energy (<100 keV) gamma or beta emissions, to ensure that activity is measured accurately. Inaccurate measurements could lead to poor imaging, patient dose, or therapeutic outcomes.

23.7.3 Quality Control Tests

Some of the more-detailed and more-onerous tests have already been described in the acceptance testing described above. These should be repeated on an annual basis as part of the periodic Quality Control program. Other relatively simple tests should also be performed more regularly to ensure good consistent performance of dose calibrators (Table 23.10). These tests will now be described.

23.7.3.1 Physical Inspection

Although relatively robust instruments, dose calibrators are typically heavily used in a busy environment and have some vulnerable components. The instrument housing should be checked regularly for damage, connectors should be checked, and any linings and dippers checked for cracks, holes, and liquid contamination.

23.7.3.2 Background

Why? Non-zero values of background activity will be added to the measurement of patients' administered activity if not corrected for. The background should therefore be checked on a daily basis and dealt with if non-zero values are found.

How? A measurement of activity should be made with no known sources in the vicinity of the dose calibrator. If radioactive contamination of the device is suspected, measurements should be made with and without the dipper and well protector to isolate the source of contamination.

What could go wrong? Ideally, dose calibrators should be positioned in a stable low-background activity environment. A higher than expected measurement could be a consequence of a nearby patient or other radioactive source or, indeed, radioactive contamination of the dose calibrator dipper or well protector. A non-zero value could also be a consequence of an issue with the dose calibrator itself.

Acceptable values. The consequence of a non-zero background measurement will depend on the magnitude of the patient activity being measured. Nonetheless background activity measurements of a dose calibrator should be typically < 0.2 MBq.

23.7.3.3 Clock Accuracy

Why? For many measurements of activity, the time of measurement should also be recorded, both for the patient record, and also for measurements that may need decay correction on the scanner. Examples of the latter include PET imaging and other tests that require pre- and post-activity measurements to be made and decay corrected, such as for thyroid uptake measurements. Where the dose calibrator does not have a clock built in, a clock in the vicinity of the dose calibrator may be used. All clocks should be in agreement with the times of other clocks, instruments, and scanners within the department to ensure that decay correction is applied appropriately, and also to ensure that subsequent scans or measurements are made at the correct time.

Acceptable values. The time should be accurate to within 1 minute.

23.7.3.4 High Voltage

Why? The high voltage across the ionization chamber can have an effect on recombination effects within the chamber at high activities and, therefore, can affect the current collected and the resultant measured activity. Accordingly, it is important that this voltage is stable.

How? Many dose calibrators will include the ability to measure the high voltage – normally as part of the dose calibrators built-in daily QA routine. If such a feature is not available, then this measurement cannot be performed.

Acceptable values. This is typically given within the system's operator's manual and can be different for different systems.

23.7.3.5 Display

Before use, the display should be checked to ensure it is operating correctly. If it is not, and if it is the only dose calibrator within the department, it may not be possible to accurately determine the activity.

23.7.3.6 Zero Adjust

A drift in the zero-adjust setting may indicate that the system is in need of repair. It is typically performed as part of the system's built-in daily QA procedure and should have a value within that recommended in the system's operator's manual.

23.7.3.7 Constancy

Why? While we might know the dose calibrators accuracy on the day accuracy measurements are made, it is important to check that the performance is consistent on a day-to-day basis. This is the aim of the constancy measurement.

How? Once we are certain the background activity measurement is acceptable, a long-lived sealed check source such as ^{57}Co, ^{137}Cs or ^{133}Ba is placed in the dose calibrator and its activity measurement recorded. After accounting for radioactive decay, the measured activity of the check source should be consistent on a day-to-day basis. It should also be consistent with the expected activity calculated from the calibration certificate of the check source. In addition to making measurements with the appropriate radionuclide selected – for example, ^{57}Co with a ^{57}Co check source – it is also prudent to check the measurements for other commonly used radionuclide settings within the department, such as ^{99m}Tc and/or ^{131}I to ensure constancy across common radionuclide settings. While the measurement for one check source is adequate, given the difference in radioactive emissions and gamma photon energies in – for example, ^{57}Co (predominantly 122 keV) and ^{133}Ba (predominantly 356 keV) – it can be helpful to perform measurements with multiple check sources if they are available.

What can go wrong? Large random variations in measurements can be caused by instability issues with the electrometer, while an ongoing trend could be caused by a progressive drift in the performance of the electrometer or a leak in the pressurized ionization chamber. There is also the possibility of a corrupt calibration factor file.

Acceptable values. Measurements should be within +/- 5 per cent of the decay-corrected calibrated values of the check source.

23.8 SUMMARY

Quality assurance programmes are an essential component in the running of any high-performing medical imaging department. The processes and procedures described here represent a framework for quality assurance in nuclear medicine and hybrid SPECT/CT, PET/CT and PET/MR systems. When performing quality control, it can also be helpful to understand any underlying problems in the imaging systems. While not within the remit of this chapter, publications on gamma camera, SPECT/CT and PET/CT artefacts can be a useful aid [31–33].

REFERENCES

[1] J. R. Halama et al., *AAPM Report No. 177: Acceptance Testing and Annual Physics Survey Recommendations for Gamma Camera, SPECT, and SPECT/CT Systems*, Alexandria, Virginia, 2019.

[2] O. R. Mawlawi et al., *AAPM Report No. 126: PET/CT Acceptance Testing and Quality Assurance the Report*, Alexandria, Virginia, 2019.

[3] E. Eadie, *IPEM Report 111: Quality Control of Gamma Cameras and Nuclear Medicine Computer Systems*, York, UK, 2015.

[4] L. Pike, P. Julyan, P. Marsden, and W. Waddington, *IPEM Report 108: Quality Assurance of PET and PET/CT System*, York, UK, 2013.

[5] E. Busemann Sokole, A. Płachcínska, and A. Britten, "Acceptance testing for nuclear medicine instrumentation," *Eur. J. Nucl. Med. Mol. Imaging*, vol. 37, no. 3. pp. 672–681, Mar 2010.

[6] E. Busemann Sokole, A. Płachcínska, A. Britten, M. Lyra Georgosopoulou, W. Tindale, and R. Klett, "Routine quality control recommendations for nuclear medicine instrumentation," *Eur. J. Nucl. Med. Mol. Imaging*, vol. 37, no. 3. pp. 662–671, Mar 2010.

[7] E. Busemann Sokole, R. Z. Stodilka, A. V. Wegst, and R. E. Zimmerman, "IAEA Human Health Series No. 6: Quality Assurance for SPECT Systems," Vienna, 2009.

[8] G. El Fakhri, R. Fulton, J. E. Gray, M. Marengo, and R. E. Zimmerman, "IAEA Human Health Series No. 1: Quality Assurance for PET and PET/CT Systems," Vienna, 2009.

[9] *NEMA Standards Publication NU 1-2018 Performance Measurements of Gamma Cameras*, Rosslyn, Virginia, 2018.

[10] T. A. Hander, J. L. Lancaster, D. T. Kopp, J. C. Lasher, R. Blumhardt, and P. T. Fox, "Rapid objective measurement of gamma camera resolution using statistical moments," *Med. Phys.*, 1997.

[11] M. Dondi, S. Palm, E. Busemann Sokole, R. Z. Stodilka, a. V. Wegst, and R. E. Zimmerman, *Quality Assurance for SPECT Systems*. ISBN 978 92 0 103709 1.," *IAEA Hum. Heal. Ser. No. 6*, no. 6, pp. 1–263, 2009.

[12] K. Kacperski, K. Erlandsson, S. Ben-Haim, and B. F. Hutton, "Iterative deconvolution of simultaneous 99mTc and 201Tl projection data measured on a CdZnTe-based cardiac SPECT scanner," *Phys. Med. Biol.*, vol. 56, no. 5, pp. 1397–1414, Feb 2011.

[13] J. C. Dickson et al., "The impact of reconstruction and scanner characterisation on the diagnostic capability of a normal database for [(123)I]FP-CIT SPECT imaging.," *EJNMMI Res.*, vol. 7, no. 1, p. 10, Dec 2017.

[14] A. V. Gil, L. A. Aroche, and G. L. Poli, "IAEA NM-QC Toolkit." [Online]. Available: https://humanhealth.iaea.org/HHW/MedicalPhysics/NuclearMedicine/QualityAssurance/NMQC-Plugins/index.html.

[15] K. Erlandsson, K. Kacperski, D. van Gramberg, and B. F. Hutton, "Performance evaluation of D-SPECT: a novel SPECT system for nuclear cardiology," *Phys. Med. Biol.*, vol. 54, no. 9, pp. 2635–2649, Apr 2009.

[16] *NEMA Standards Publication NU 2-2018 Performance Measurements of Positron Emission Tomographs (PET)*, Rosslyn, Virginia, 2018.

[17] R. Boellaard, "Standards for PET image acquisition and quantitative data analysis," *J. Nuc. Med.*, vol. 50, no. SUPPL. 1. pp. 11S–20S, 1 May, 2009.

[18] P. Hiles, A. Mackenzie, A. Scally, and B. Wall, *IPEM Report 91: Recommended Standards for the Routine Performance Testing of Diagnostic X-ray Imaging Systems*, York, UK, 2005.

[19] S. Edvyean, *Type Testing of CT Scanners: Methods and Methodology for Assessing Imaging Performance and Dosimetry*, London, 1998.

[20] S. J. Shepard et al., *Quality Control in Diagnostic Radiology. Report of Task Group 12*, Published for the American Association of Physicists in Medicine by the American Institute of Physics, Madison, Wisconsin, 2002.

[21] V. D. J. Edvyean S, Gray J, Heggie J, Hiles P, Homolka P, Le Heron J, McLean D, Mutic S, Ngaile J, Strauss K, *Quality Assurance for Computed Tomography – Diagnostic and Therapy Applications*. IAEA, 2012.

[22] C. M. Collins, W. Liu, W. Schreiber, Q. X. Yang, and M. B. Smith, "Central brightening due to constructive interference with, without, and despite dielectric resonance," *J. Magn. Reson. Imaging*, vol. 21, no. 2, pp. 192–196, Feb 2005.

[23] C. N. Ladefoged et al., "A multi-centre evaluation of eleven clinically feasible brain PET/MRI attenuation correction techniques using a large cohort of patients.," *Neuroimage*, vol. 147, pp. 346–359, Feb 2017.

[24] S. Ziegler, H. Braun, P. Ritt, C. Hocke, T. Kuwert, and H. H. Quick, "Systematic evaluation of phantom fluids for simultaneous PET/MR hybrid imaging," *J. Nucl. Med.*, vol. 54, no. 8, pp. 1464–1471, Aug 2013.

[25] D. McRobbie and S. Semple, *IPEM Report 112: Quality Control and Artefacts in Magnetic Resonance Imaging*, York, UK, 2017.

[26] J. E. Carey, P. Byrne, L. DeWerd, R. Lieto, and N. Petry, *The Selection, Use, Calibration, and Quality Assurance of Radionuclide Calibrators Used in Nuclear Medicine: Report of AAPM Task Group 181*, American Association of Physicists in Medicine, One Physics Ellipse College Park, MD, 2012.

[27] E. Busemann Sokole, A. Płachcínska, and A. Britten, "Acceptance testing for nuclear medicine instrumentation," *Eur. J. Nucl. Med. Mol. Imaging*, vol. 37, no. 3. pp. 672–681, Mar 2010.

[28] E. Busemann Sokole, A. Płachcínska, A. Britten, M. Lyra Georgosopoulou, W. Tindale, and R. Klett, "Routine quality control recommendations for nuclear medicine instrumentation," *Eur. J. Nucl. Med. Mol. Imaging*, vol. 37, no. 3. pp. 662–671, Mar 2010.

[29] International Atomic Energy Agency, "IAEA TECDOC 602: Quality control of nuclear medicine instruments," Vienna, 1991.

[30] R. Gadd et al., *Report 93: Protocol for Establishing and Maintaining the Calibration of Medical Radionuclide Calibrators and their Quality Control: A National Measurement Good Practice Guide*, 2006.

[31] E. Busemann Sokole, "IAEA Quality Control Atlas for Scintillation Camera Systems, 2003," Vienna, 2003.

[32] J. C. Dickson, S. Holm, O. Mawlawi, and C. Robilotta, "IAEA Human Health Series No. 36: SPECT / CT Atlas of Quality Control and Image Artefacts," Vienna, 2019.

[33] T. Beyer, S. Holm, O. R. Mawlawi, and C. Robilotta, "IAEA Human Health Series No. 27: PET/CT Atlas on Quality Control and Image Artefacts," Vienna, 2014.

24 Calibration and Traceability

Brian E. Zimmerman

CONTENTS

24.1 INTRODUCTION

The use of radionuclides in nuclear medicine continues to expand around the world. For diagnostic nuclear medicine imaging, for example, Position Emission Tomography (PET) and Single Photon Emission Computed Tomography (SPECT), recent developments have expanded the arsenal of available agents beyond those based on just ^{99m}Tc or ^{18}F to include longer-lived radionuclides with chemistry suited to particular targeting molecules. Examples include ^{64}Cu, ^{68}Ga, ^{89}Zr, and ^{124}I. Advances in scanner technology and image-reconstruction techniques have made it possible for these modalities to be truly quantitative. In fact, PET and SPECT are increasingly used in drug development trials and in patient management to monitor patient progress [1] and inform the course of treatment [2].

In therapeutic nuclear medicine, new radionuclides are being developed for an ever-expanding array of applications. Radionuclides that emit β-particles, such as ^{67}Cu, ^{90}Y, ^{166}Ho, and ^{177}Lu, take advantage of the ability for electrons to deliver a high dose to diseased tissue while sparing nearby healthy tissue. Many of these radionuclides decay with accompanying γ-rays or emit positrons that can be used for image-based dosimetry. New radiopharmaceuticals incorporating α-emitting radionuclides take advantage of the high linear energy transfer and short range of α-particles to more precisely target diseased tissues. Some of the radionuclides being studied, such as ^{223}Ra, ^{224}Ra, ^{227}Th, and ^{212}Pb, decay in a chain that includes radioactive daughter nuclides, thereby providing an even larger dose per parent decay.

The safety and effectiveness of any nuclear medicine procedure, whether for diagnosis/dosimetry or for therapy, is highly dependent on the accuracy of the measurement of radioactivity in the injected radiopharmaceutical. The use of radioactivity standards and properly calibrated measurement equipment in the context of a well-developed quality management system (QMS) can help ensure the accuracy and dependability of those measurements. This chapter deals with the practical application of the concepts of traceability, calibration, and quality assurance.

DOI: 10.1201/9780429489556-24

24.2 TRACEABILITY AND QUALITY ASSURANCE

24.2.1 Metrology Hierarchy

Traceability is defined in the International Vocabulary of Metrology (VIM) [3] as the 'property of a measurement result whereby the result can be related to a reference through a documented unbroken chain of calibrations, each contributing to the measurement uncertainty'. Implicit in this definition is the existence of a reference or standard against which a measurement can be compared or calibrated. Developing such standards is the work of specialized laboratories called national metrology institutes (NMIs) or designated institutes (DIs) that are named by their respective governments as having the legal responsibility to develop, maintain, and disseminate standards that are traceable to the International System of Units (SI). In the case of radioactivity measurement, an NMI or DI is responsible for developing standards that are traceable to the fundamental unit of activity, known as the becquerel (Bq), which is derived from the base unit of the second.

A standard that represents a direct realization of the becquerel is termed a *primary standard*. How this realization is achieved is beyond the scope of this book, but it is sufficient to say that a primary standard represents the top metrological level of measurement in a country and usually represents the national standard for the country for that specific unit of measure. Standards that are not direct realizations of the becquerel, but which are instead calibrated against another (generally primary) standard, are termed *secondary standards*. Commercial sources purchased from source suppliers, as well as some single-dose radiopharmaceutical preparations, can be considered as secondary standards if they are calibrated in this way. While many NMIs and DIs hold both primary and secondary standards, it should be noted that some hold only secondary standards that are traceable to another country's primary standard.

Comparisons of standards amongst NMIs and DIs are carried out on a regular basis to demonstrate the degree of agreement, or 'equivalence' amongst them. Such comparisons are often organized through the Consultative Committee on Ionizing Radiation, Section II, Measurement of Activity (CCRI(II) of the International Bureau of Weights and Measures (Bureau International des Poids et Mesures, BIPM) under the guidelines established in the Mutual Recognition Arrangement of the International Committee for Weights and Measures [4], known as the 'CIPM MRA'. The equivalence established under the CIPM MRA demonstrates international comparability of standards amongst the NMIs/DIs and provides confidence to the users of standards around the world as to their validity. One of the requirements for recognition of a laboratory's performance under the CIPM MRA is that the measurements made must be covered under a QMS that undergoes periodic internal review and external audit. An important aspect of the QMS, particularly for a laboratory providing calibration services or that distributes measurement reference samples, is that the traceability chain for the measurement back to the fundamental SI unit is defined and documented.

Because the scope of the work performed by an NMI/DI is generally limited to developing and maintaining primary standards, they do not typically have the resources to interact directly with the end users, which in the case of the nuclear medicine community would be the clinics treating patients. Traceability can be extended from the NMI/DI to the clinics through *secondary standards* laboratories. For the nuclear-medicine community, an isotope producer or radiopharmacy that has established and maintains traceability to an NMI/DI for a particular radionuclide measurement can be considered secondary standards laboratories and can in turn provide products that can be considered to be traceable back to the primary standard. The ways that a laboratory can establish traceability to the SI are discussed in Section 24.22.

24.2.2 Establishing and Maintaining Traceability

According to the definition of traceability given in the VIM, it is necessary to first have a primary reference against which subsequent measurements can be made. This requires the NMI to develop such a standard and make it available. Assuming that such a standard exists, a traceable determination of the activity of a source can be made by either directly comparing the source to the standard (making any necessary corrections for geometry, differences in solution composition, etc.), or by using a series of standards to develop a calibration curve and using the calibration curve to derive the activity value. In order for the new activity measurement to be truly traceable, the process by which the measurement was made must be carefully documented, and an assessment of the uncertainty of the measurement must be made. This also includes taking into account the uncertainty of the standards that were used to derive the activity.

Maintaining traceability can generally be done through participation in a measurement assurance program or by performing and documenting regular comparisons using a standard that is traceable to an NMI/DI. Both methods fulfil the requirement of a comparison with a standard as specified in the VIM. There are no strict rules defining how long traceability is valid and how often the comparisons should be made but, in general, traceability of a result is valid as long as the calibration or comparison is valid. If a calibration does not have a definite expiration, it is incumbent on the user to ensure, through documented subsequent measurements, that both the values and uncertainties derived in the measurements have not changed over time and, if they have, to determine the maximum length of time before a new calibration should take place [5]. It is common to make this a part of a QMS and should be done for every type of source being measured.

Due to economic and geographic differences around the world, countries have adopted a number of different ways to establish and maintain traceability at the end-user level. For example, in Brazil, the Laboratório Nacional de Metrologia das Radiações Ionizantes (LNMRI)/Instituto de Radioproteção e Dosimetria (IRD), which is the DI for maintaining radioactivity standards in the country, has been operating a proficiency-testing program for activity calibrators since about 1997 [6–8]. For many years, the comparisons were run by LNMRI/IRD and served mostly clinics in the surrounding area of Rio de Janeiro. As nuclear medicine practice and the demand for traceability increased, a system of four regional laboratories (including IRD/LNMRI acting in this capacity) was set up to serve more of the country. Today, comparisons are conducted to establish and maintain traceability for ^{201}Tl, ^{131}I, ^{123}I, ^{99m}Tc, ^{67}Ga, and, more recently, ^{18}F for the almost 250 nuclear medicine centres throughout the country.

A different approach used in the United Kingdom and its NMI, the National Physical Laboratory (NPL), is to use a standard commercially available activity calibrator that that is found in clinics throughout the country. In this case, each activity calibrator is directly calibrated by (and therefore directly traceable to) NPL, which maintains a record of how each instrument compares to each other and to a master-activity calibrator maintained at NPL. In addition, traceability to the end user is encouraged through a voluntary code of practice developed jointly by NPL and the medical physicists' national professional body. This 'Measurement Good Practice Guide' [9] includes a daily stability test using a long-lived sealed source, an annual linearity test, an annual accuracy test using traceable reference materials of at least two of the radionuclides used in the clinic, and it sets forth the requirements for subsidiary calibrations for samples in non-standard containers. In this respect, it is similar to Technical Report Series 454 published by the International Atomic Energy Agency (IAEA) [10]. In both cases, guidelines are set out that define pass/fail criteria for various performance tests and distinguish between 'reference' and 'field' instruments. A 'reference' instrument has a tight specification on accuracy (2%) so it can be used to prepare reference materials to calibrate the 'field' (i.e., clinical use) instruments (to an accuracy of 5%). The approach taken by NPL also supports the clinical community through workshops to encourage adoption of best practices.

In some countries, such as the United States, the regulatory bodies overseeing the use of radiopharmaceuticals place the responsibility for ensuring accurate measurement of radioactivity content in drug preparations (or the radioactive components for drugs that will be compounded on-site) on the manufacturers of those products. In these cases, the clinical end user has the option of using the manufacturer-supplied activity value as the calibrated activity for single-use units and is merely required to apply a decay correction. For other dosages, particularly those prepared on-site, a direct measurement must be made. To help the radiopharmaceutical industry to establish the required traceability, a Measurement Assurance Program has been set up between a consortium of companies and the National Institute of Standards and Technology (NIST) that provides traceability through proficiency tests and calibrations [11].

An increasing awareness of the benefits of traceability in nuclear medicine over the past 15 years has led to more countries running proficiency tests between the NMI or DI and the clinics or an expansion of calibration services specific to the nuclear-medicine community. Recent studies involving proficiency tests at the end user level or the introduction of new services have been reported in Cuba [12], India [13], the Czech Republic [14], Indonesia [15], Switzerland [16], and Canada [16], in addition to the ongoing programs in the UK, United States, and Brazil.

24.2.3 The Role of Standards and Traceability in Quality Assurance

While in most countries traceability on its own is generally not a regulatory requirement for nuclear-medicine facilities, the majority of clinics, radionuclide production facilities, and radiopharmacies operate under some sort of a QMS or Good Laboratory Practice program. The IAEA TRS-454 [10] provides a model for how clinical facilities can implement a QMS system based on the International Standards Organization (ISO) standard 17025: General requirements for the competence of testing and calibration laboratories [17]. One of the requirements of a QMS that is compliant with the ISO 17025 standard is that 'calibration of instruments and reference sources must be traceable to a certified standards laboratory'. In addition to providing a means for improving measurement quality, the IAEA Basic Safety Standards, [18] which outlines the minimal safety requirements for the safe use of radiation, requires that 'the calibration of sources used for medical exposure [shall] be traceable to a Standards dosimetry laboratory'. While this particular guidance was obviously directed towards external beam or brachytherapy applications, it is easy to extrapolate this to the need for radioactivity measurements of injected radiopharmaceuticals to be traceable. This emphasizes the relationship between accurate and consistent clinical radioactivity measurement and patient safety.

While none of the aforementioned guidelines or standards specify how traceability is to be established and maintained, they are all in agreement that the measurements used to make the claims of traceability must be the best possible and must be meticulously documented. These documents must be kept and made available to auditors or customer requests to validate traceability claims.

24.2.4 Standards for Radioactivity

As discussed above, a primary standard generally represents the highest level of measurement quality in a measurement chain and is generally held by a national metrology institute. Because of the amount of resources necessary to prepare such a standard, they are generally not directly distributed to users. Moreover, it is common practice for most national metrology institutes to transfer a primary standard to another instrument that is less resource-intensive and that preferably enables the measurement of multiple sources at a time. One such instrument is a re-entrant ionization chamber, commercial versions of which are known as 'activity meters', 'activity calibrators', or 'dose calibrators'.

The operational principle for these devices, as well as a discussion of their use in radionuclide metrology, can be found in [19] and [20]. Briefly, these instruments generally consist of a cylindrical chamber filled with pressurized gas (usually N_2 or Ar) with a concentric well to allow for the introduction of a source, usually by means of a plastic holder. The interaction of the radiation emitted from the source with the pressurized gas results in the formation of ion pairs, the number of which is proportional to the energy absorbed by the gas. Through the application of a voltage across electrodes inside the gas-filled chamber, the ion pairs are collected and, using an electrometer, read out as a current. If the source being measured has a calibrated, traceable activity value, the measured current and the certified activity value can be used (after appropriate decay correction, if needed) to calculate a calibration coefficient in terms of pA of current per Bq of activity (pA/Bq). Most commercial activity calibrators have internal circuitry that directly converts the current to an activity value that can be displayed on a readout, thereby eliminating the need to make a direct current measurement. However, most reference-class ionization chambers and the aforementioned activity calibrators supported by NPL in the UK, utilize external electrometers for which the numerical calibration coefficient is required.

If the source used to determine the calibration coefficient is a primary standard and a proper uncertainty analysis has been carried out, then subsequent measurements made using that chamber can be considered to be secondary standards, as long as the measurement conditions are the same as those in which the chamber was calibrated, and a QMS is in place to ensure that periodic checks are carried out and properly documented.

24.3 CALIBRATION METHODS IN NUCLEAR MEDICINE

24.3.1 Activity Meters

Radionuclide calibrator settings depend strongly on the geometry of the sample to be measured, especially for low-energy photon and pure beta-emitting radionuclides. Calibrator settings provided by the instrument manufacturers are generally valid only for a given standard geometry, for which that setting was determined. For some of the major activity calibrator manufacturers, this standard geometry is a 5 mL flame-sealed thin-walled ampoule. For clinical measurements, radiopharmaceuticals are usually measured in vials or plastic syringes that could have different absorption characteristics for photon and beta radiation than an ampoule. As a result, a discrepancy of several per cent could be observed when the incorrect calibration factor is applied for a particular geometry [21], depending on the decay characteristics of the radionuclide being measured.

The response of a radionuclide calibrator for a given radionuclide is sensitive to not only the type of container used but also to the filling volume of solution in the container. For that reason, a measurement of the geometry effect should be made to ensure the quality of the measurement results for the activity of the radionuclide. This can be done easily by transferring, preferably by mass, a known amount of a calibrated solution of the radionuclide being tested into the container to be used as the new measurement geometry. With this new source in the activity calibrator, a new calibration setting can be determined by changing the 'dial setting' – the user defined calibration factor – until the correct activity is displayed. Alternatively, a correction factor can be derived by measuring the source at the original dial setting and noting the difference between the instrument reading and the expected activity (from the calibrated activity concentration and the transferred mass). Details on the application of this method and on additional ways of determining calibration settings can be found in [22].

If a system utilizes an electrometer to display units of current, the calibration coefficient can be calculated by first making a background measurement with no sources present. This background current is then subtracted from the current reading with the source in place in the activity calibrator to obtain the background-corrected current. The result is then divided by the calibrated activity of the source (corrected for radioactive decay, if necessary) to obtain the calibration coefficient in terms of current per unit activity. This process is discussed in more detail in [19].

For new radionuclides, or those for which a calibration setting is not available from the manufacturer, the setting in the measurement geometry intended to be used in the clinic should be experimentally determined. Even when a calibration setting is available in the instrument manual, it is recommended that it be verified using a traceably calibrated solution. The process for this measurement is similar to that used to investigate geometry effects and is described more completely in [22].

The use of simulated sources, that is, sources utilizing radionuclides with similar decay characteristics but longer half-lives than the nuclide of interest (for example substituting ^{57}Co for ^{99m}Tc) may be useful to estimate long-term stability and may yield useful stability information. However, using such a source for calibration cannot be justified and should not be done because of the many factors (discussed above) that that affect the activity measurement. Where necessary, calibration and calibration-setting determinations, must be done with the nuclide of interest in the measurement geometry that will be used in routine practice.

24.3.2 Gamma Well Counters

Gamma well counters consist of a solid scintillator, such as NaI doped with Tl, with a concentric cylindrical well to allow for the introduction of a sample. The geometry is similar to that of a re-entrant ionization chamber, but with several important differences. The first is that the output of a gamma well counter is typically a spectrum of the photons emitted by the source that are detected in the scintillator material, as seen in Figure 24.1. Secondly, the detection properties of most gamma well counters provide a range of measurable activities that is much lower than that of activity calibrators – lower by as much as three orders of magnitude. Lastly, most modern commercially available gamma well

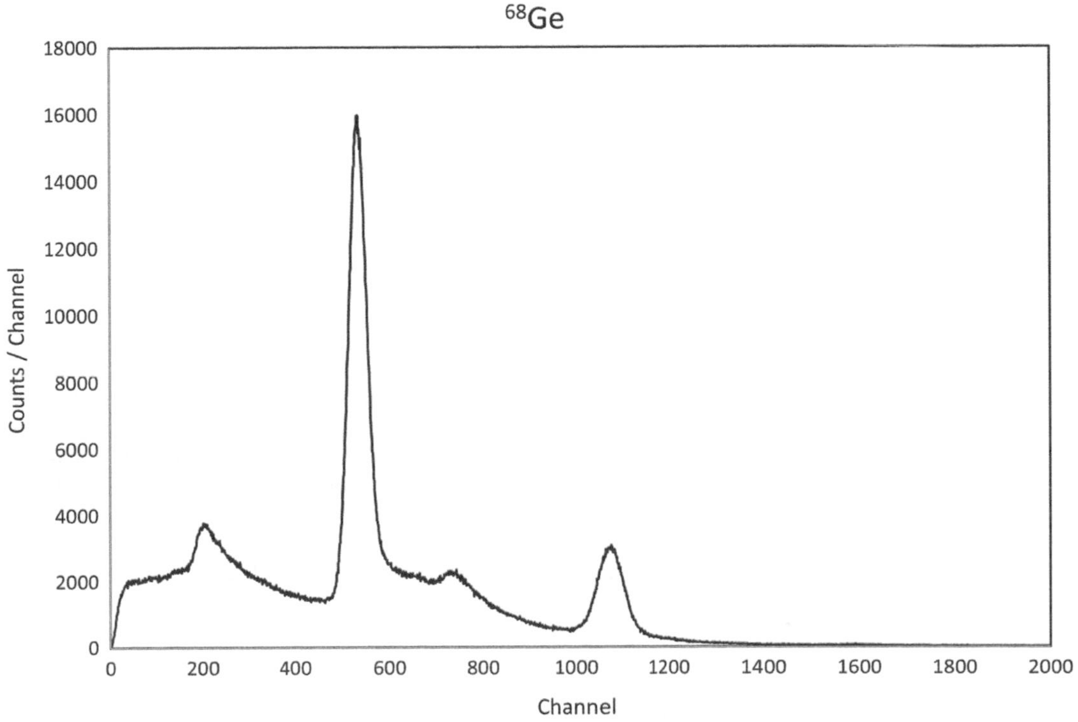

FIGURE 24.1 Spectrum of ^{68}Ge obtained from a commercial NaI(Tl) well counter.

counters incorporate sample changers with capacities of 100 samples or more, providing the opportunity to automatically measure multiple sources during a given assay.

In general, the process for calibrating a NaI well counter is similar to that for the radionuclide calibrator. The main difference is that, since the output is a spectrum instead of a single current value, a conversion factor from observed counts in the spectrum (either the total spectrum or a defined energy window) to an activity value must be experimentally derived for the specific radionuclide using a calibrated, traceable source. Consideration also needs to be given to the count rate range over which the measurements will be made. Because of the high efficiency of these types of detectors, samples are typically limited to activities on the order of a few kBq at the most, depending on the radionuclide, in order to reduce the corrections necessary to ameliorate dead-time effects. This may require that the original calibrated solution be gravimetrically diluted to a level capable of being measured in the well counter.

In general, the process of determining a calibration coefficient consists of preparing the traceably calibrated source in the correct geometry, as one would for the activity calibrator, but it is also important to prepare a blank sample (usually distilled water) in the same geometry in order to measure the background. Assuming that the spectrum energy regions have been properly set up and are applied to both samples, the total count rate in the background spectrum (total number of counts in the spectrum or energy window divided by the live time) is subtracted from the total counting rate in the spectrum (in the same region) with the radioactive sample to obtain a net counting rate (in counts per second). The result is in turn divided by the calibrated activity to obtain the calibration coefficient (in counts per second per becquerel).

REFERENCES

[1] R. Lindner Jonathan and J. Link, "Molecular imaging in drug discovery and development," *Circulation: Cardiovascular Imaging*, vol. 11, no. 2, p. e005355, 2018/02/01 2018, doi:10.1161/CIRCIMAGING.117.005355.

[2] L. D. Declercq, R. Vandenberghe, K. Van Laere, A. Verbruggen, and G. Bormans, "Drug Development in Alzheimer's Disease: The Contribution of PET and SPECT," *Frontiers in Pharmacology*, vol. 7, p. 88, 2016. [Online]. Available: www.frontiersin.org/article/10.3389/fphar.2016.00088.

[3] "2012 International Vocabulary of Metrology – Basic and general concepts and associated terms, 3rd edn., International Bureau of Weights and Measures," in Sèvres, France, 2012.

[4] "Mutual Recognition Arrangement of national measurement standards and of calibration and measurement certificates issued by national metrology institutes," International Committee for Weights and Measures (CIPM), 2019. [Online]. Available: www.bipm.org/en/cipm-mra/.

[5] "Selected Laboratory and Measurement Practices and Procedures to Support Basic Mass Calibrations" National Institute of Standards and Technology (NIST). 2019. [Online]. Available: https://nvlpubs.nist.gov/nistpubs/ir/2019/NIST.IR.6969-2019.pdf.

[6] J. A. dos Santos et al., "Implementation of a national metrology network of radionuclides used in nuclear medicine," *Applied Radiation and Isotopes*, vol. 64, no. 10, pp. 1114–1118, 2006/10/01/ 2006, doi: 10.1016/j.apradiso.2006.02.008.

[7] A. E. Oliveira et al., "Traceability from governmental producers of radiopharmaceuticals in measuring (18)F in Brazil," *Applied Radiation and Isotopes*, vol. 109, pp. 236–241, Mar 2016, doi:10.1016/j.apradiso.2015.11.051.

[8] A. E. de Oliveira et al., "Rapid and accurate assessment of the activity measurements in Brazilian hospitals and clinics," *Applied Radiation and Isotopes*, vol. 134, pp. 64–67, Apr 2018, doi:10.1016/j.apradiso.2017.07.065.

[9] R. Gadd, "Measurement Good Practice Guide 93: Protocol for establishing and maintaining the calibration of medical radionuclide calibrators and their quality control," National Physical Laboratory, Middlesex, UK, 2006.

[10] *Quality Assurance for Radioactivity Measurement in Nuclear Medicine*. Vienna: International Atomic Energy Agency, 2006.

[11] J. T. Cessna and D. B. Golas, "The NIST radioactivity measurement assurance program for the radiopharmaceutical industry," *Applied Radiation and Isotopes*, vol. 70, no. 9, pp. 2227–2231, 2012/09/01/ 2012, doi: 10.1016/j.apradiso.2012.02.110.

[12] P. Oropesa, Y. Moreno, R. A. Serra, and A. T. Hernández, "The traceability chain of 131I measurements for nuclear medicine in Cuba," *Applied Radiation and Isotopes*, vol. 70, no. 9, pp. 2251–2254, 2012/09/01/ 2012, doi: 10.1016/j.apradiso.2012.02.108.

[13] L. Joseph, R. Anuradha, R. Nathuram, V. V. Shaha, and M. C. Abani, "National intercomparisons of 131I radioactivity measurements in nuclear medicine centres in India," *Applied Radiation and Isotopes*, vol. 59, no. 5, pp. 359–362, 2003/11/01/ 2003, doi: 10.1016/S0969-8043(03)00192-1.

[14] V. Olšovcová, "Activity measurements with radionuclide calibrators in the Czech Republic," *Applied Radiation and Isotopes*, vol. 60, no. 2, pp. 535–538, 2004/02/01/ 2004, doi: 10.1016/j.apradiso.2003.11.072.

[15] H. Candra, P. Marsoem, and G. Wurdiyanto, "Performance evaluation of commercial radionuclide calibrators in Indonesian hospitals," *Applied Radiation and Isotopes*, vol. 70, no. 9, pp. 2243–2245, 2012/09/01/ 2012, doi: 10.1016/j.apradiso.2012.02.113.

[16] Y. Caffari, P. Spring, C. Bailat, Y. Nedjadi, and F. Bochud, "Activity measurements of 18F and 90Y with commercial radionuclide calibrators for nuclear medicine in Switzerland," *Applied Radiation and Isotopes*, vol. 68, no. 7, pp. 1388–1391, 2010/07/01/ 2010, doi: 10.1016/j.apradiso.2009.11.015.

[17] Standards, I. R. I. o. Malaysia, I. O. f. Standardization, and I. E. Commission, *General Requirements for the Competence of Testing and Calibration Laboratories: (ISO/IEC 17025: 2005, IDT)*. SIRIM Berhad, 2005.
[18] I. A. E. Agency et al., *Applying Radiation Safety Standards in Nuclear Medicine*. International Atomic Energy Agency, 2005.
[19] H. Schrader, "Ionization chambers," *Metrologia*, vol. 44, p. S53, 08/02 2007, doi:10.1088/0026-1394/44/4/S07.
[20] A. Pearce, C. Michotte, and Y. Hino, "Ionization chamber efficiency curves," *Metrologia,* vol. 44, p. S67, 08/02 2007, doi:10.1088/0026-1394/44/4/S08.
[21] D. E. Bergeron and J. T. Cessna, "An update on 'dose calibrator' settings for nuclides used in nuclear medicine," (in English), *Nuclear Medicine Communications,* vol. 39, no. 6, pp. 500–504, Jun 2018, doi:10.1097/mnm.0000000000000833.
[22] B. E. Zimmerman and J. T. Cessna, "Experimental determinations of commercial 'dose calibrator' settings for nuclides used in nuclear medicine," *Applied Radiation and Isotopes,* vol. 52, no. 3, pp. 615–619, 2000/03/01/ 2000, doi:10.1016/S0969-8043(99)00219-5.

25 Activity Quantification from Planar Images

Katarina Sjögreen Gleisner

CONTENTS

25.1 INTRODUCTION

Planar imaging was for a long time the most-used technique for image-based activity quantification. Owing to its simplicity and speed, the possibility to acquire dynamic image sequences, and to generate whole-body images at a reasonable acquisition time, it is still frequently used in nuclear-medicine imaging. Today, planar imaging is most commonly used for diagnostic examinations with qualitative or semi-quantitative evaluation, for instance for renal scintigraphy in static or dynamic mode using ^{99m}Tc-labelled MAG3 or DTPA, or whole-body skeletal scintigraphy using ^{99m}Tc-MDP (Chapter 15). Planar imaging is also performed for patients receiving radionuclide therapy, such as in ^{177}Lu-DOTA-TATE therapy for the treatment of neuroendocrine tumours (Figure 25.1), or ^{131}I-NaI therapy for metastatic thyroid cancer [1–3]. Activity quantification may then be applied to estimate the absorbed doses delivered to organs and tumours. It is well recognized that planar images suffer from the superposition of counts from activity located at different depths in the patient, and that quantitative SPECT is generally superior to planar-based activity quantification. However, the techniques for planar-based activity quantification preceded and, in parts, formed the basis for the subsequent development of quantification methods from tomographic SPECT images, and these techniques still carry value in terms of understanding the process with which planar image projections are formed. Notably, also when acquisition is made in SPECT mode the raw image data consist of a set of planar projections. Moreover, planar-based activity quantification is still used, for example, to estimate the total-body absorbed dose, and in combination with SPECT-based activity quantification in so-called hybrid planar/SPECT methods.

DOI: 10.1201/9780429489556-25

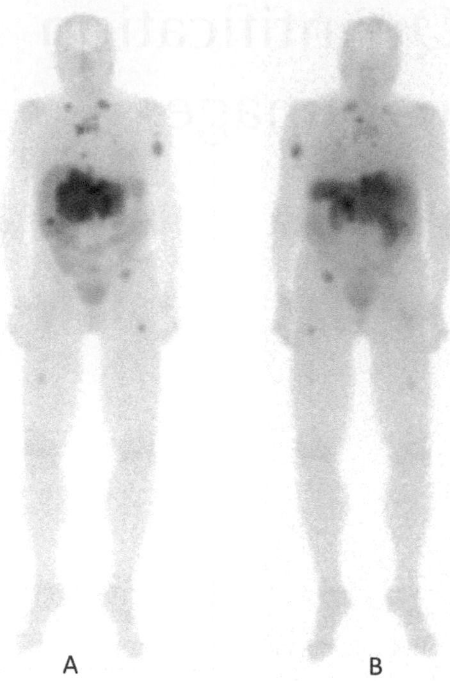

FIGURE 25.1 Examples of planar anterior (A) and posterior (B) whole-body images acquired during radionuclide therapy with ^{177}Lu-DOTA-TATE, administered for the treatment of neuroendocrine tumours [1].

25.2 IMAGE ACQUISITION AND FORMATION

Planar imaging refers to scintillation-camera imaging with the camera head positioned at a fixed angle relative to the patient, or when two opposite angles are acquired, either by sequential acquisition with a single-head camera or by simultaneous acquisition with a dual-headed camera. The camera head(s) can also scan over the patient body, which is accomplished by moving the patient bed at a constant speed relative to the camera during acquisition. In scan mode, in order to get the same effective exposure time over the entire field-of-view, the camera head stays for some time interval at the start and end positions and, gradually, electronically shuts the effective field-of-view for count detection.

The technical and physical processes that underlie the formation of a planar image are common to any gamma-camera imaging. In quantitative imaging these processes need to be considered to enable quantification of the activity in the source regions in a patient. This is generally made by application of corrections for photon attenuation and scatter, and also, depending in the radionuclide and amount of activity in the patient, corrections for septal penetration and deadtime. Usually, parallel-hole collimators are used in planar imaging, and this is the only collimator considered in this chapter. The advantage with parallel-hole collimators is that the system sensitivity (or calibration factor) is for many radionuclides independent of the source distance from the camera collimator.

25.2.1 Photon Attenuation, Scatter, and Septal Penetration

The distribution of counts in a gamma-camera image is caused by the interaction of photons in the camera detector. These photons may have travelled along different paths from the radioactive source in the patient before reaching the camera detector; they may pass directly from the site of decay into the camera, or they may undergo Compton scattering or photoelectric absorption in the tissues located along the photon trajectory. The terminology used in the field of image-based activity quantification distinguishes between primary photons, attenuated and scattered photons.

Primary photons are those that travel from the point of emission in a direction parallel to the collimator holes, without interaction in body tissues or in the collimator, and whose energy is fully absorbed in the camera-detector material (arrow 1, Figure 25.2). The positions of the image counts caused by these primary photons reflect the true position of the radioactive decay and are thus those that truly reflect the source distribution in the patient. The counts

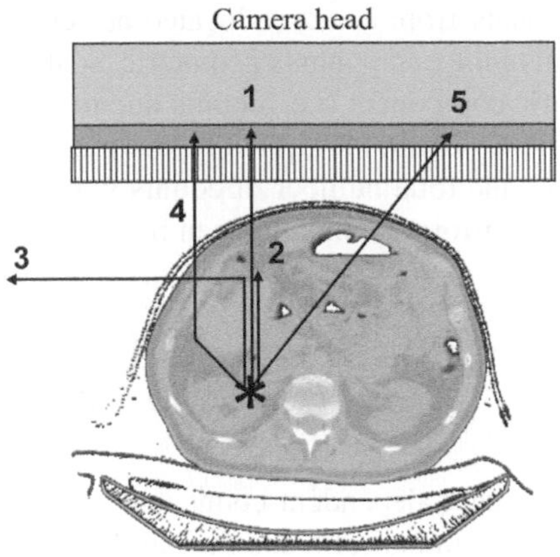

FIGURE 25.2 Schematic illustration of photon paths from the position where the radionuclide decays and emits photons in the patient (black star) to the camera detector material. Arrow 1 illustrates so called primaries, that is, photons that pass unscattered through the patient and collimator and deposit their full energy in the detector. The number of detected primaries (the geometric component) is reduced due to photo-electric absorption (arrow 2) and Compton scattering (arrows 3). Photons may scatter in an angle towards the camera and can then be detected and contribute with 'falsely' positioned counts in the image, that is, counts at positions that do not correspond to the true location of the source in the patient (arrow 4). Likewise, septal penetration (arrow 5) contributes with counts at erroneous positions.

caused by primary photons are often termed the *geometric* component since their acceptance angle is governed by the geometry of the collimator holes.

Photons that are emitted in a path parallel to the collimator holes but undergo photoelectric absorption or are scattered in an angle such that they do not reach the camera (arrows 2 and 3), are termed *attenuated* photons. Under narrow-beam conditions, the transmission fraction, that is, the number of photons $N(d)$ that transmit through an absorber of thickness d from an initial number of N_0 photons, is described according to

$$N(d)/N_0 = \exp(-\mu d) \tag{25.1}$$

where μ is the linear attenuation coefficient [4]. Photon attenuation thus decreases the count rate from primary photons.

Photons may also scatter in an angle such that they reach the camera crystal, albeit at a position away from the path of primary photons (arrow 4) thus generating counts at false positions. The resulting counts are generally termed *scatter*. The scattered photons thus add counts such that the total number of detected photons is higher than predicted from the linear attenuation of primary photons. To discriminate against scattered photons an energy window is set over the photopeak(s) at the relevant gamma-emission energy. However, due to the limited energy resolution of gamma-camera systems, these energy windows need to have a certain width, for example ±7.5 per cent or ±10 per cent of the photopeak energy. Photons scattered in small angles with low energy losses will then also be detected in the energy window.

An additional physical factor that may generate false counts in an image is septal penetration, as indicated by arrow 5. For radionuclides that include the emission of photons of sufficiently high energy to penetrate the collimator material, the count-rate contribution from septal penetration may be substantial.

The proportion of the counts that are due to primary photons, scattered photons, or septal penetration depends on a large number of factors. These include the emission spectrum of the radionuclide, the source depth, the composition and interaction properties of tissues, the collimator design, the energy-window settings, and the thickness of the scintillation crystal.

There are two similar, albeit not identical, methods to describe the proportion of the count rate that is caused by different components. The *total-to-primary ratio* describes the total number of counts with respect to the counts due to primaries (geometric component). For a medium with uniform attenuation properties, the total-to-primary ratio $\eta(\mu, d)$ may be defined as

$$\eta(\mu,d) = \frac{C_{\text{tot}}(\mu,d)}{C_{\text{g}}(\mu,d)} = \frac{C_{\text{g+s+p}}(\mu,d)}{C_{\text{g}} \cdot \exp(-\mu d)} \tag{25.2}$$

where $C_{tot}(\mu,d)$ is the totally detected counts from a source located at depth d in a medium with linear attenuation coefficient μ, which is formed as the sum of the components geometric, scatter, and septal penetration $C_{g+s+p}(\mu,d)$. The denominator $C_g(\mu,d)$ is thus the geometric component (i.e., counts due to primary photons), which is described as the distance-independent geometric count rate C_g in air, multiplied by the transmission fraction (Eq. 25.1).

The *build-up factor*, $B(\mu,d)$, describes the total number of counts with respect to the counts due to primaries and septal penetration [5] and for a medium of uniform attenuation can be described as

$$B(\mu,d)=\frac{C_{tot}(d)}{C_{g+p}(\mu,d)}=\frac{C_{g+s+p}(d)}{C_{g+p}(d)\cdot\exp(-\mu d)} \tag{25.3}$$

where the denominator is given by the distance-dependent count rate $C_{g+p}(d)$ in air, multiplied by the transmission fraction. The difference between the total-to-primary ratio and the build-up factor thus lies in the denominator. In practical applications, this difference is reflected by the way the calibration factor for activity quantification is determined and how subsequent corrections are made. For cases where septal penetration is negligible, $\eta(\mu,d)$ and $B(\mu,d)$ are equivalent. Both have values greater than or equal to unity.

A third, and more common way of describing the effects of scatter is the so-called *effective attenuation coefficient*, μ_{eff} (or broad-beam attenuation factor), in which the contribution from scatter is described as a lowered attenuation. The definition of the effective attenuation coefficient can be understood from the total-to-primary (Eq. 25.2), according to

$$C_{g+s+p}(\mu d)=C_g\cdot\exp(-\mu d)\cdot\eta(\mu,d)=C_g\cdot\exp\left[-d\left(\mu-\frac{\ln(\eta(\mu,d))}{d}\right)\right]=C_g\cdot\exp(-\mu_{eff}\,d) \tag{25.4}$$

Since $\eta(\mu,d)\geq 1$ it follows that μ_{eff} is equal to or lower than the linear attenuation coefficient. The effective attenuation coefficient thus describes the combined effects of attenuation and scatter in one value.

Figure 25.3 shows the photon transmission fraction as a function of the source depth in water for ^{177}Lu with its gamma energy of 208 keV. The transmission fraction for a narrow-beam geometry, corresponding to primary photons (geometric), has been calculated for different water depths following Eq. 25.1 [6]. Monte Carlo simulations of an elliptically shaped water phantom with a centrally located point source have been made using the software SIMIND [7]. The vertical phantom extension is varied between 6 cm and 20 cm, while the distance from the phantom surface to the

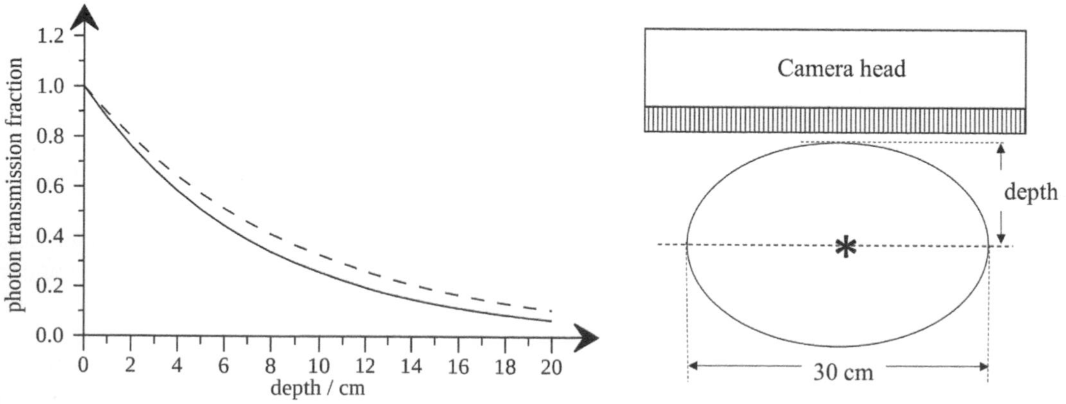

FIGURE 25.3 Right panel: Data on the total transmission fraction were obtained from Monte Carlo simulations of a point source of ^{177}Lu placed centrally in a uniform cylinder with constant lateral extension and with a varying vertical extension, as indicated 'depth'. The distance between the phantom surface and the collimator was maintained at 2 cm for all depths. Left panel: The narrow-beam transmission fraction in water calculated from the linear attenuation coefficient for 208 keV (solid line) and the total transmission fraction obtained from Monte Carlo simulation (dashed line) also including scatter.

TABLE 25.1
Examples of Values of the Effective Attenuation Coefficient, for a Selection of Radionuclides Imaged by Different Gamma Camera Systems.

Radionuclide	Considered photon energy (keV)	Linear attenuation coefficient (cm^{-1})	Effective attenuation coefficient (cm^{-1})	Camera system	Parallel-hole collimator	Energy window width (keV)	Reference
^{177}Lu	208	0.135	0.112	GE670	medium-energy	208 ± 7.5%	SIMIND [7] (calculated herein)
^{131}I	364	0.109	0.094	GE670	high-energy	364 ± 10%	SIMIND [7] (calculated herein)
^{99m}Tc	140	0.155	0.124	Picker SX-300	low-energy	140 ± 10%	[8]
^{201}Tl	68.9	0.196	0.184	Picker SX-300	low-energy	75 ± 20%	[8]
	70.8	0.194				167 ± 10%	
	80.3	0.184					
	167	0.146					
^{111}In	171	0.145	0.110	Picker SX-300,	medium-	172 ± 10%	[8, 9]
	245	0.129		GE Infinia II VC	energy	247± 10%	

Notes: Values of the linear attenuation coefficient were retrieved from [4], while photon energies were retrieved from Laboratoire National Henri Becquerel, France, www.nucleide.org/DDEP_ WG/DDEPdata.htm.

camera collimator is maintained at 2 cm. The simulated system is a Tandem Discovery 670 system (GE Healthcare System, Haifa), using a medium-energy collimator and a 15 per cent energy window centred at 208 keV.

The ratio between the total count rate and the geometric component varies with depth. For instance, in Figure 25.3 the total-to-primary ratios at depths of 6 cm, 10 cm, and 20 cm, are calculated to 1.22, 1.36, and 1.70, respectively. Thus, for a source emitting 208 keV photons located at a depth of 10 cm in water, the percentage of the total counts from scatter is approximately 26 per cent (calculated as (1.36–1)/1.36). Corresponding simulations have also been performed for ^{131}I using a high-energy collimator and an energy window of 20 per cent centred at 364 keV (data not shown). The photon transmission fraction has a similar shape as shown in Figure 25.3, although the total-to-primary ratio is higher, with values of 1.95, 2.09, and 2.47 at depths of 6 cm, 10 cm, and 20 cm. These higher values mainly reflect the larger contribution from septal penetration from the high photon energies (637 keV and 722 keV) in the emission spectrum of ^{131}I. Effective attenuation coefficients are listed in Table 25.1 for a selection of radionuclides, photon energies, and collimators.

25.3 ACTIVITY QUANTIFICATION

Internal dosimetry based on planar-image activity quantification has been pursued for a long time, and methods are summarized in Pamphlet 16 of the Medical Internal Radiation Dose (MIRD) committee [10]. The basic steps of the quantification procedure include the delineation of regions-of-interest (ROIs) in the planar image over the organs and tissues that constitute the source regions, and the application of a calibration factor and corrections for attenuation and scatter, to obtain an estimate of the activity in the source region.

25.3.1 Calibration Factor

A key parameter in image-based activity quantification is the calibration factor *CF*, a factor that is specific for the camera system, collimator, radionuclide, and energy-window settings. The calibration factor relates the measured count rate in a calibration geometry $\dot{C}_{cal}$ to a known amount of activity A_{cal}, following

$$CF = \dot{C}_{cal} / A_{cal} \tag{25.5}$$

In order to arrive at an accurate measurement of A_{cal}, traceability in activity measurements first needs to be assured by calibration of the activity meter settings towards a standard metrology laboratory. Notably, this traceability needs to be ascertained for each radionuclide used, which may be more difficult to achieve for therapeutic radionuclides such as ^{131}I, ^{177}Lu, or ^{90}Y, where standard sources are not easily available.

Usually, the calibration factor is expressed in terms of count rate per activity (in unit counts-per-second per megabecquerel). When applied to a patient image, the counts in the patient image first need to be divided by the acquisition-time interval to obtain the count rate at the time of imaging. For an appropriately uniformity-corrected camera system equipped with a parallel-hole collimator, the calibration factor is constant across the imaging field-of-view and can be applied either to the pixel values or on ROI values for the acquired image.

Despite the importance of the calibration factor, the source geometry used for its determination is not standardized. The simplest, and historically most widely used approach, is to determine the calibration factor from the first patient acquisition. A planar whole-body scan is then made early after administration, before the patient has voided, and the detected count rate over the entire image is divided by the administered activity. The resulting calibration factor is then applied for quantification of the activity in organs from images acquired at later times, when some of the activity has been excreted and the total amount in the patient is no longer known. Historically, this approach for activity quantification has been applied without other corrections, that is, by the assumption that the calibration factor takes all physical and geometric effects into account. Although attractive in its simplicity, this approach does not take into account the effects of the different tissue depths and composition that govern the probability of photon interaction. The errors in the estimated activity for different tissues such as the testes, lungs and kidneys, can thus be substantial.

When more accurate activity estimates are required and explicit corrections for attenuation and scatter are included in the quantification method, calibration can be made using a point- or disk-like source in absence of any attenuating and scattering medium (in air). This is essentially a measurement of the system sensitivity, as outlined by the National Electrical Manufacturers Association (NEMA) [11], made by use of a flat source at a source-collimator distance of 10 cm. In principle, for parallel-hole collimators the source distance from the camera collimator is not critical since the system sensitivity for many radionuclides is independent of the source-collimator distance. However, as mentioned above, some therapeutic radionuclides, such as ^{131}I, have photon emissions with a comparably high energy that produce a considerable amount of septal penetration. In contrast to the geometric sensitivity, the number of counts that are due to septal penetration is distance dependent. The effect of septal penetration can to some extent be mitigated by determining a distant-dependent calibration factor and including build-up factors in the activity determination (Eq. 25.3). However, at application to a patient image, the distance from the source to the camera is generally not known, and approximations or more elaborate methods are then required [10].

Notably, the acquisition time interval can be stated in different ways in the DICOM header, which can be of importance when determining the calibration factor for whole-body scans. For some imaging systems the acquisition time refers to the entire scan duration, whereas other systems report the exposure time per pixel, that is, the time interval that a particular position in the patient is projected on to the camera. When the entire scan time is reported, the calibration factor may need to be acquired with the same scan length, that is, by a whole-body scan (even if the calibration source is small). High count rates present particular problems, as dead time may introduce a non-linear relationship between the source activity and the observed count rate. If this is the case, then explicit dead-time corrections are required [12]. Owing to the importance of the calibration factor for accurate activity quantification, careful monitoring is required of any drifts over time, or variations across the imaging field-of-view due to insufficient non-uniformity correction [13, 14].

25.3.2 Activity Quantification from a Single Planar View

Quantification from a single planar view was the more favoured approach before two-headed scintillation cameras were available, and before the introduction of SPECT. Still today it can be a useful technique in cases where there is negligible accumulation of the radiopharmaceutical in under- or overlying tissues such that the tissue background is low.

If we consider a point source with activity A located in air, the geometric counts C_g obtained for a gamma camera equipped with a parallel-hole collimator can to a first approximation be described as

$$C_g = CF \cdot A \cdot T \tag{25.6}$$

where CF is the calibration factor determined in air, and T is the acquisition time interval. It is here assumed that the count contributions from the background and scatter are negligible. If the point source is located at a depth d in a uniform medium with linear attenuation coefficient μ, the detected number of counts can be described according to

$$C(d) = C_g \cdot \exp(-\mu d) \cdot S(\mu, d) = CF \cdot A \cdot T \cdot \exp(-\mu d) \cdot S(\mu, d) \tag{25.7}$$

Since scatter is generated when photons interact in the medium, a scatter component is included, symbolized by the function $S(\mu, d)$. For a uniform medium $S(\mu, d)$ can be obtained as either of the factors $\eta(\mu, d)$ or $B(\mu, d)$, described above. Alternatively, the effective attenuation coefficient, μ_{eff}, can be used in place of the linear attenuation coefficient (Eq. 25.4). The activity is obtained from Eq. 25.7 following

$$A = C(d) / (CF \cdot T) \cdot \exp(\mu d) / S(\mu, d) \tag{25.8}$$

It follows that the source depth needs to be estimated, which can be made by means of a CT image of the patient or a lateral gamma-camera acquisition. For some radionuclides there is also the possibility of using the double-energy peak method.

25.3.2.1 The Double-energy Peak Method for Estimating the Source Depth

Radionuclides that emit two gamma energies in the radioactive decay include, for instance, ^{111}In, ^{123}I, ^{67}Ga, and ^{177}Lu. In this case, the combination of the information of the two photopeaks can give information about the source depth. We temporarily omit the factor $S(\mu, d)$ in Eq. 25.7 and denote the geometric counts in one peak as $C_{1,g}$ for a photon energy with linear attenuation coefficient μ_1, and the counts in the second peak as $C_{2,g}$ for a photon energy with attenuation coefficient μ_2. The geometric counts are given by

$$C_{1,g}(d) = CF_1 \cdot A \cdot T \cdot \exp(-\mu_1 d) \tag{25.9}$$

$$C_{2,g}(d) = CF_2 \cdot A \cdot T \cdot \exp(-\mu_2 d)$$

The ratio between these two equations is formed and solved for d, giving

$$d = \frac{1}{(\mu_2 - \mu_1)} \ln\left(\frac{C_{1,g}(d)}{CF_1} \frac{CF_2}{C_{2,g}(d)} \right) \tag{25.10}$$

This thus gives an estimate of the source depth that can be used for subsequent attenuation correction according to Eq. 25.8. The double-peak method has been applied to quantify ^{123}I with emitted photon energies of 28 keV and 159 keV [15], ^{67}Ga with photon energies of 93 keV, 185 keV, and 300 keV, and different combinations of ^{99m}Tc, ^{67}Ga and ^{111}In (energies listed in Table 25.1) [16].

25.3.3 Activity Quantification from Conjugate Views

The conjugate-view, or geometric-mean, method was described in the late 1970s [17, 18]. The basic approach is to acquire two images with the gamma-camera heads placed at 180 degrees with respect to each other, and then take the geometric mean of the counts acquired in the two images. Figure 25.4 illustrates the imaging geometry.

The conjugate-view method relies on a distance-independent calibration factor, which implies imaging with parallel-hole collimators and radionuclides that produce a negligible count contribution from septal penetration. Moreover, the derivations that form the basis for the conjugate-view method are commonly formulated for a narrow-beam geometry. We will adhere to this tradition and omit the scatter component in the below derivations, thus giving expressions for the geometric counts affected by linear attenuation of primary photons. The scatter component is then specifically addressed in Section 25.3.3.2.

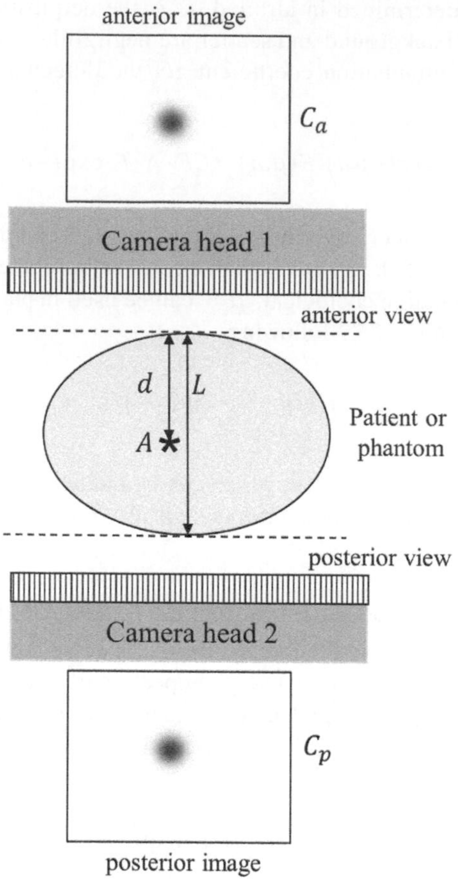

FIGURE 25.4 The conjugate-view method in which two opposing projections are acquired, for instance one anterior and one posterior view. The source depth from the patient surface in the anterior direction is indicated by *d* and the total patient thickness at the source location by *L*.

The geometric counts $C_{a,g}$ and $C_{p,g}$ obtained in the anterior and posterior views, respectively, can be described according to

$$C_{a,g}(d) = C_g \cdot \exp(-\mu d) = A \cdot CF \cdot T \cdot \exp(-\mu d) \tag{25.11}$$

$$C_{p,g}(L-d) = C_g \cdot \exp(-\mu(L-d)) = A \cdot CF \cdot T \cdot \exp(-\mu(L-d))$$

The geometric mean of the anterior and posterior counts is formed:

$$\left(C_{a,g} \cdot C_{p,g}\right)^{1/2} = A \cdot CF \cdot T \cdot \left[\exp(-\mu d) \cdot \exp(-\mu(L-d))\right]^{1/2} = A \cdot CF \cdot T \cdot \exp(-\mu L)^{1/2} \tag{25.12}$$

The application of the geometric mean thus results in an expression that is independent of the source depth, and only depends on the patient thickness L at the source location. The activity is thus obtained as

$$A = \frac{1}{CF \cdot T} \cdot \left(C_{a,g} \cdot C_{p,g} \cdot \exp(\mu L)\right)^{1/2} = \frac{1}{CF \cdot T} \cdot \left(C_{a,g} \cdot C_{p,g}\right)^{1/2} \cdot \tau(\mu L) \tag{25.13}$$

25.3.3.1 Attenuation Correction

The factor $\tau(\mu L) = \exp(\mu L)^{1/2}$ in Eq. 25.13 constitutes the attenuation correction. The value of this factor can be estimated from phantom studies performed prior to patient acquisition, as demonstrated in Figure 25.3, where the phantom is assumed to be representative of the patient geometry. A patient-specific attenuation map can be obtained by performing a transmission scan of the patient. For instance, a flat radioactive sheet such as ^{57}Co flood source can be placed on one side of the patient, and the number of counts $N_g(L)$ recorded at the location where the source is located. The measurement is then repeated without the patient in position, yielding $N_g(0)$ counts. The ratio is formed and reflects the decreased counts due to attenuation, following

$$N_g(L)/N_g(0) = \exp(-\mu_{tr}L) \tag{25.14}$$

where μ_{tr} is the linear attenuation coefficient for the photon energy of the transmission source, for example, 122 keV for ^{57}Co. The attenuation-correction factor for the radionuclide used for patient imaging can then be calculated by energy scaling, first solving for L according to

$$L = \frac{1}{\mu_{tr}} \ln\left(\frac{N_g(0)}{N_g(L)}\right)$$

$$\tau(\mu_{em} L) = \exp(\mu_{em} L)^{1/2} = \exp\left[\frac{\mu_{em}}{2\mu_{tr}} \ln\left(\frac{N_g(0)}{N_g(L)}\right)\right] \tag{25.15}$$

where μ_{em} denotes the linear attenuation coefficient for the imaging radionuclide. Often the entire patient is scanned with the transmission source, thus yielding a whole-body attenuation-correction map $\tau(\mu_{em}(x,y)L(x,y))$. As an alternative to a ^{57}Co flood source image, an X-ray scout may be used, which is otherwise acquired for positioning purposes before CT scans [19, 20]. The energy scaling (Eq. 25.15) then needs to be made with respect to the average energy of the X-ray spectrum. Figure 25.5 shows both kinds of attenuation maps. Note that the X-ray scout has the advantages of being considerably less noisy, includes less scatter, and is faster to acquire.

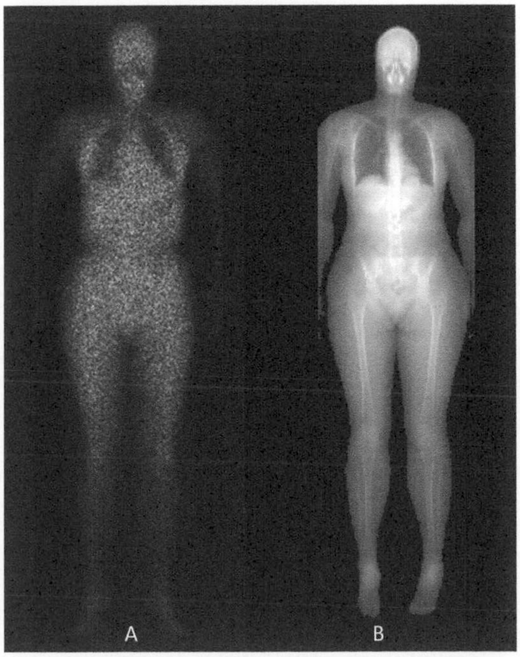

FIGURE 25.5 Whole-body attenuation correction maps obtained by use of a ^{57}Co flood source (A) and an X-ray scout image (B).

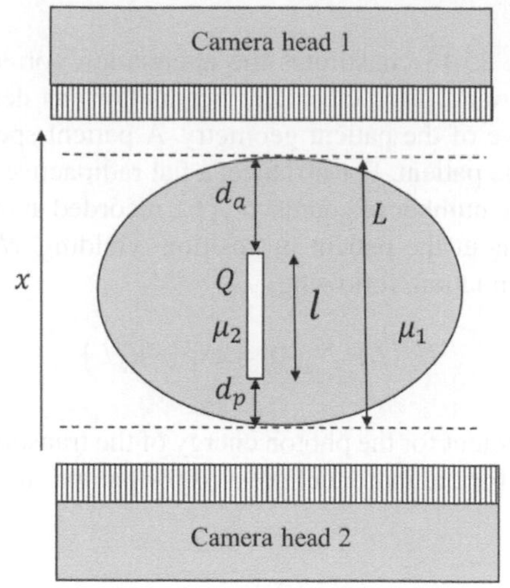

FIGURE 25.6 Geometry of an extended source (white bar) with length l and an activity per unit length of Q. The distance from the source edges to the anterior and posterior surfaces of the medium are d_a and d_p, respectively, and the attenuation coefficients of the source material and medium are μ_1 and μ_2, respectively.

When a whole-body attenuation correction map is available, the attenuation correction can be applied on a pixel-by-pixel basis. In Eq. 25.13, the geometric mean of the counts $C_{a,g}$ and $C_{p,g}$ is then applied on a pixel basis.

Instead of the point source, we can regard a source that is extended along the direction perpendicular to the camera heads. Figure 25.6 illustrates an infinitesimally thin rod with a uniform activity concentration Q in unit of activity per length along the source (such as MBq/mm, e.g.). We also let the source compartment have a different attenuation coefficient than the surrounding medium [18].

The geometric counts obtained in the anterior and posterior views can be described following

$$C_{a,g} = CF \cdot T \cdot Q \cdot \left(\int_0^l \exp(-\mu_2 x) dx \right) \exp(-\mu_1 d_a)$$

$$= CF \cdot T \cdot Q \cdot \exp(-\mu_1 d_a) \frac{1 - \exp(-\mu_2 l)}{\mu_2} \tag{25.16}$$

$$C_{p,g} = CF \cdot T \cdot Q \cdot \exp(-\mu_1 d_p) \frac{1 - \exp(-\mu_2 l)}{\mu_2}$$

As above, the geometric mean is taken

$$\left(C_{a,g} \cdot C_{p,g}\right)^{1/2} = CF \cdot T \cdot Q \cdot \exp\left(-\mu_1 \frac{d_a + d_p}{2}\right) \frac{1 - \exp(-\mu_2 l)}{\mu_2} \tag{25.17}$$

In order to correct for attenuation, an expression for the transmission fraction also needs to be derived. The transmission fraction through the patient thickness $L = d_a + d_p + l$ is given by

$$N_g(L) / N_g(0) = \exp\left(-\mu_1 (d_a + d_p)\right) \cdot \exp(-\mu_2 l) \tag{25.18}$$

The square-root of the inverse of Eq. 25.18 then gives the attenuation correction factor, $\tau(\mu_1, \mu_2, L)$. The attenuation-corrected geometric mean count is given by

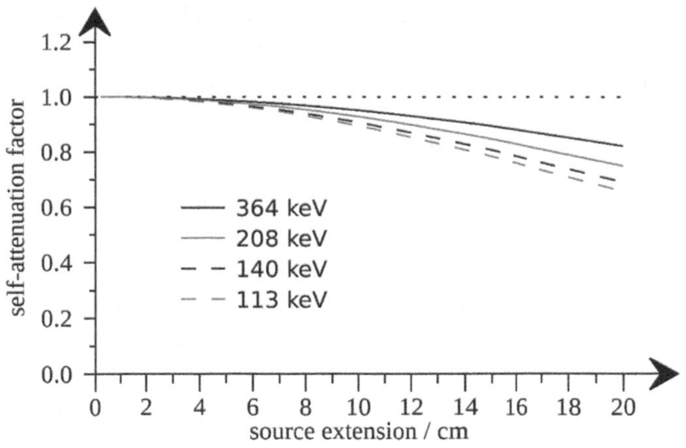

FIGURE 25.7 The self-attenuation factor as a function of the source extension for energies 364 keV (solid black), 208 keV (solid grey), 140 keV (dash black),) and 113 keV (dash grey). The dotted line indicates unity.

$$\left(C_{a,g} \cdot C_{p,g}\right)^{1/2} \tau(\mu_1,\mu_2,L) = CF \cdot T \cdot Q \cdot \exp\left(-\mu_1 \frac{(d_a + d_p)}{2}\right) \frac{1-\exp(-\mu_2 l)}{\mu_2}$$

$$\cdot \exp\left(\mu_1 \frac{(d_a + d_p)}{2}\right) \cdot \exp\left(\mu_2 \frac{l}{2}\right) =$$

$$\frac{CF \cdot T \cdot Q}{\mu_2}\left[\exp\left(\frac{\mu_2 l}{2}\right) - \exp\left(-\frac{\mu_2 l}{2}\right)\right] = CF \cdot T \cdot Q \cdot l \, \frac{\sinh(\mu_2 l/2)}{\mu_2 l/2} \tag{25.19}$$

where the hyperbolic sine function is used in the last step. The activity in the rod is given by $A = Q \cdot l$, and is thus obtained according to

$$A = \frac{1}{CF \cdot T}\left(C_{a,g} \cdot C_{p,g}\right)^{1/2} \tau(\mu_1,\mu_2,L) \frac{\mu_2 l/2}{\sinh(\mu_2 l/2)} \tag{25.20}$$

When the attenuation coefficients for the medium and the source are equal, such that $\mu_1 = \mu_2 = \mu$, we get the function $\tau(\mu, L)$. Also, the right-hand factor becomes $(\mu l/2)/\sinh(\mu l/2)$, commonly referred to as the *self-attenuation factor*. The value of this factor depends on the source extension and the energy of the emitted photons, and Figure 25.7 shows examples for ^{99m}Tc (140 keV), ^{177}Lu (113 keV and 208 keV), and ^{131}I (364 keV).

As noted, the self-attenuation factor only has importance for source regions with a considerable extension along the projection direction and is more important for low photon energies. For a source extension of 5 cm, it takes values of 0.97, 0.98, 0.98, and 0.99 for 113 keV, 140 keV, 208 keV, and 364 keV, respectively. For a source extension of 10 cm (which is uncommon in the patient body) the corresponding values are 0.89, 0.91, 0.93, and 0.95. With respect to the impact of the other corrections (or lack thereof) such as attenuation and scatter corrections, this self-attenuation factor is thus often of minor importance, although for low-photon energies it needs to be considered.

25.3.3.2 Scatter Correction

As noted, Eqns. 25.13 and 25.20 are derived from the assumption of a narrow-beam geometry and are theoretically only valid for the geometric count components in the conjugate-view images. In practice however, there are also count contributions from scatter in these images. Likewise, for the transmission fraction derived in Eq. 25.14 the scatter component may be non-negligible, especially if estimated using a ^{57}Co flood source. To estimate the geometric counts from the totally acquired counts, a similar approach as taken in Eqns. 25.7 and 25.8 can be applied. For example, in Eqns. 25.13 and 25.20 the factor $C_{a,g}$ can be obtained as $C_a / \eta(\mu,d)$ or $C_a / B(\mu,d)$, and correspondingly for $C_{p,g}$. Alternatively, the effective attenuation coefficient, μ_{eff}, can be used in place of the linear attenuation coefficient in Eqns. 25.13, 25.14 and 25.20 (see Eq. 25.4).

However, these methods for scatter correction do not account for the spatial distribution of the scatter component. Rather, they intrinsically assume that the distribution of the geometric and scatter counts is the same as for the detected counts in the photopeak energy window. For estimation of the scatter distribution there are two classes of correction methods available, energy-window based and model-based methods. Most commonly used are the energy-window based approaches in which additional scatter energy windows are set during image acquisition. The first method proposed was the dual-energy window method by Jaszczak [21], later followed by the triple-energy window (TEW) method of Ogawa [22]. The TEW method is based on setting two narrow energy windows centred just below and above the main photopeak energy window, giving counts C_{low} and C_{high}. The counts are renormalized by the width of the respective scatter energy window (ΔE_{low} and ΔE_{high}) to the width of the main energy window (ΔE_{peak}). The scatter component (C_s) and the geometric counts (C_g) are estimated according to

$$C_g = C_{\text{peak}} - C_s = C_{\text{peak}} - \left[\frac{C_{\text{high}}}{\Delta E_{\text{high}}} + \frac{C_{\text{low}}}{\Delta E_{\text{low}}}\right] \cdot \frac{\Delta E_{\text{peak}}}{2} \tag{25.21}$$

With regards to the activity quantification method outlined in Eqns. 25.13 and 25.20, the scatter is subtracted before applying the geometric mean, that is, $C_{a,g}$ and $C_{p,g}$ are obtained as C_a and C_p subtracted by their respective scatter estimate $C_{a,s}$ and $C_{p,s}$. The TEW method has been found useful but has the drawback of being sensitive to poor counting statistics in the scatter energy windows, especially in case of low activities or short scan durations.

Model-based scatter correction can be made by means of deconvolution. This approach relies on the representation of the scatter distribution as the convolution of a scatter point spread function (SPSF) and an underlying distribution of geometric counts. If the volume under the two-dimensional SPSF equates the total-to-primary ratio, deconvolution of the acquired image results in an image in which the pixel values represent the magnitude and distribution of the geometric counts. Generally, the SPSFs need to be calculated by Monte Carlo simulation of the gamma-camera system for a geometry that approximates the patient. Deconvolution has been implemented by Fourier-domain methods [23, 24] and by the Richardson-Lucy algorithm to correct for septal penetration artefacts in ^{131}I imaging [25].

When pixel-based attenuation and scatter corrections are applied to whole-body images, quantitative activity projections can be obtained. Figure 25.8 shows examples of the raw acquired counts in anterior projections, and the resulting images when attenuation correction is performed by means of an X-ray scout image, and scatter correction is made using deconvolution [23, 24]. The improved image appearance of the corrected image can be noted.

Other more elaborate methods for planar image quantification involve image reconstruction, in which the planar images are combined with the geometry of a patient obtained from a segmented CT image. Two opposite planar images are then treated as a very sparse set of projections and, by iterative methods, the mean activity in different source regions is estimated [26].

25.3.4 Corrections for Activity in Overlapping Tissues

In planar gamma-camera images, the source region for which the activity is estimated is rarely completely separated from activity uptakes in other tissues. When ROIs are drawn in the image, there will thus also be counts originating from over- or underlying tissues along the projection direction. This contribution can originate from activity circulating in plasma or blood, from activity in the extracellular fluid, and in organs and tissues situated above or below the considered source region. A number of studies have addressed background correction [10, 27–29]. The simplest method is to estimate a general background contribution by delineating an ROI at an image position in which the patient thickness is equivalent to the patient thickness over the organ, but outside the organ volume. The number of counts in this background ROI is then subtracted from the counts in the organ ROI, taking into account the possibly differently sized ROIs. Preferably, the extension of the organ should also be considered, which can be estimated from a CT image of the patient, or reference data of typical organ thicknesses [23, 28].

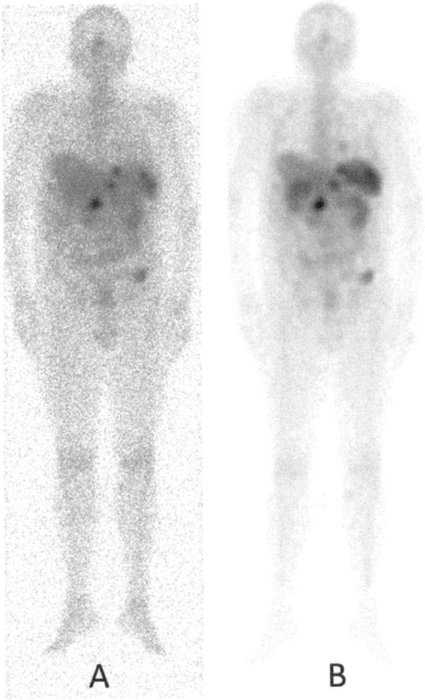

FIGURE 25.8 Whole-body images acquired seven days after infusion of ^{177}Lu-DOTA-TATE therapy for the treatment of neuroendocrine tumours. Left image is the unprocessed anterior scan, and right image shows the same image after applying deconvolution scatter correction, forming the geometric mean, and applying attenuation correction based on an X-ray scout image.

25.3.5 Hybrid Planar-SPECT-based Estimation of the Time-activity Curve

For radionuclide therapy, the aim of activity quantification is to estimate the absorbed doses delivered to patients. In this case, imaging is performed at several time points after the administration of the radiopharmaceutical, with the aim to determine the time-activity curve and the time-integrated activity [30]. When the activity distribution in the whole body also needs to be monitored, or there are other practical reasons to perform planar imaging instead of SPECT/CT at each time point, a hybrid approach can be taken [31–33]. As the accuracy of activity quantification from SPECT/CT is generally superior to planar-based quantification, the activity is quantified based on a SPECT/CT image acquired at one time point after administration. The planar images, acquired at several time points, then serve to provide relative time-activity values. Figure 25.9 illustrates a time-activity curve obtained from a planar image, and a single quantitative activity value obtained from SPECT/CT at one time point.

The magnitude of the time-activity curve is determined by scaling to the activity value from the SPECT/CT image. The scaling to the SPECT-derived activity at the j th time point is made following

$$A_j = \frac{A_{\text{SPECT},i}}{C_{\text{planar},i}} . C_{\text{planar},j} \tag{25.22}$$

where $A_{\text{SPECT},i}$ is the activity determined from SPECT/CT imaging at the i th time point, $C_{\text{planar},i}$ is the planar-derived activity value at the same time point, and $C_{\text{planar},j}$ is the planar-derived activity at time point j. Hence, by means of one SPECT/CT and sequential planar measurements, estimates of both the shape and the magnitude of the time-activity curve are obtained.

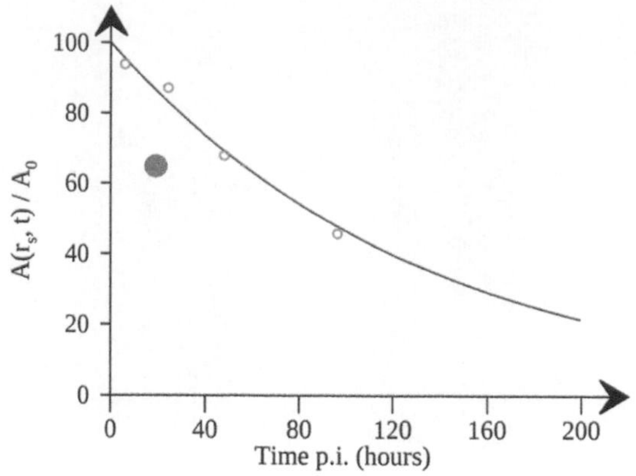

FIGURE 25.9 Open circles show the measured data of the activity $A(r_s, t)$ in a source region r_s at time t after administration relative to the extrapolated activity at time zero A_0. The line indicates the time-activity curve obtained by fitting a model to the to the data. The filled circle shows the activity quantified from SPECT/CT. In a planar-SPECT method the shape of the curve is maintained, but its amplitude renormalized such that it fits through the filled circle.

REFERENCES

[1] A. Sundlov et al., "Individualised 177Lu-DOTATATE treatment of neuroendocrine tumours based on kidney dosimetry," *Eur J Nucl Med Mol Imaging*, Mar. 22, 2017, doi:10.1007/s00259-017-3678-4.

[2] M. Lassmann et al., "EANM Dosimetry Committee series on standard operational procedures for pre-therapeutic dosimetry I: blood and bone marrow dosimetry in differentiated thyroid cancer therapy," *Eur J Nucl Med Mol Imaging*, vol. 35, no. 7, pp. 1405–1412, Jul 2008, doi:10.1007/s00259-008-0761-x.

[3] M. Luster et al., "Guidelines for radioiodine therapy of differentiated thyroid cancer," *Eur J Nucl Med Mol Imaging*, vol. 35, no. 10, pp. 1941–1959, Oct 2008, doi:10.1007/s00259-008-0883-1.

[4] M. J. H. Berger, J. H. Seltzer, S. M. Chang, J. Coursey, J. S. Sukumar, R. Zucker, D. S. Olsen, K. XCOM: Photon Cross Section Database (version 1.5).

[5] J. A. Siegel, R. K. Wu, and A. H. Maurer, "The buildup factor: Effect of scatter on absolute volume determination," *J Nucl Med*, vol. 26, no. 4, pp. 390–394, 1985. [Online]. Available: PM:0002984364.

[6] E. H. Saloman, and J. H.; Scofield, "X-ray attenuation cross sections for energies 100 eV to 100 keV and elements $Z = 1$ to $Z = 92$," *Atomic Data and Nucl. Data Tables*, vol. 38, no. 1, pp. 1–196, 1988.

[7] M. Ljungberg and S. E. Strand, "A Monte Carlo program for the simulation of scintillation camera characteristics," *Comput Methods Programs Biomed*, vol. 29, pp. 257–272, 1989.

[8] S. A. Starck and S. Carlsson, "The determination of the effective attenuation coefficient from effective organ depth and modulation transfer function in gamma camera imaging," *Phys Med Biol*, vol. 42, no. 10, pp. 1957–1964, 1997. [Online]. Available: PM:9364591.

[9] C. Chiesa et al., "Dosimetry in myeloablative (90)Y-labeled ibritumomab tiuxetan therapy: Possibility of increasing administered activity on the base of biological effective dose evaluation. Preliminary results," (in English), *Cancer Biother Radiopharm*, vol. 22, no. 1, pp. 113–120, Feb 2007, doi:10.1089/cbr.2007.302.

[10] J. A. Siegel et al., "MIRD pamphlet no. 16: Techniques for quantitative radiopharmaceutical biodistribution data acquisition and analysis for use in human radiation dose estimates," (in English), *J Nucl Med*, vol. 40, no. 2, pp. 37S–61S, Feb 1999. [Online]. Available: www.ncbi.nlm.nih.gov/entrez/query.fcgi?cmd=Retrieve&db=PubMed&dopt=Citation&list_uids=10025848.

[11] NEMA, "NU 1-2012 Performance Measurements of Gamma Cameras," in NEMA Standards Publication, National Electrical Manufacturers Association, Rosslyn, Virginia, 2012.

[12] M. Silosky, V. Johnson, C. Beasley, and S. C. Kappadath, "Characterization of the count rate performance of modern gamma cameras," *Med Phys*, vol. 40, no. 3, p. 032502, Mar 2013, doi:10.1118/1.4792297.

[13] N. Anizan, H. Wang, X. C. Zhou, R. F. Hobbs, R. L. Wahl, and E. C. Frey, "Factors affecting the stability and repeatability of gamma camera calibration for quantitative imaging applications based on a retrospective review of clinical data," *EJNMMI Research*, vol. 4, no. 1, p. 67, Dec 2014, doi:10.1186/s13550-014-0067-x.

[14] N. Anizan, H. Wang, X. C. Zhou, R. L. Wahl, and E. C. Frey, "Factors affecting the repeatability of gamma camera calibration for quantitative imaging applications using a sealed source," *Phys Med Biol*, vol. 60, no. 3, pp. 1325–1337, Feb. 7, 2015, doi:10.1088/0031-9155/60/3/1325.

[15] S. E. Strand and B. R. Persson, "The dual photopeak-area method applied to scintillation camera measurements of effective depth and activity of in vivo 123I- distributions," *Eur J Nucl Med*, vol. 2, no. 2, pp. 121–128, 1977. [Online]. Available: PM:0000891557.

[16] J. Nosil, V. Sethi, J. Bland, and R. Kloiber, "Double peak attenuation method for estimating organ location," *Phys Med Biol*, vol. 32, no. 11, pp. 1407–1416, Nov 1987. [Online]. Available: www.ncbi.nlm.nih.gov/pubmed/3423113.

[17] S. R. Thomas, H. R. Maxon, and J. G. Kereiakes, "In vivo quantitation of lesion radioactivity using external counting methods," (in English), *Med Phys*, vol. 03, no. 04, pp. 253–255, Jul–Aug 1976. [Online]. Available: www.ncbi.nlm.nih.gov/entrez/query.fcgi?cmd=Retrieve&db=PubMed&dopt=Citation&list_uids=958163.

[18] J. S. Fleming, "A technique for the absolute measurement of activity using a gamma camera and computer," (in English), *Phys Med Biol*, vol. 24, no. 1, pp. 176–80, Jan 1979. [Online]. Available: www.ncbi.nlm.nih.gov/entrez/query.fcgi?cmd=Retrieve&db=PubMed&dopt=Citation&list_uids=372956.

[19] D. Minarik, K. Sjogreen, and M. Ljungberg, "A new method to obtain transmission images for planar whole-body activity quantification," (in English), *Cancer Biother Radiopharm*, vol. 20, no. 1, pp. 72–76, Feb 2005, doi:10.1089/cbr.2005.20.72.

[20] K. Sjögreen Gleisner and M. Ljungberg, "Patient-specific whole-body attenuation correction maps from a CT system for conjugate-view-based activity quantification: method development and evaluation," (in English), *Cancer Biother Radiopharm*, vol. 27, no. 10, pp. 652–664, Dec 2012, doi:10.1089/cbr.2011.1082.

[21] R. J. Jaszczak, K. L. Greer, C. E. Floyd, C. C. Harris, and R. E. Coleman, "Improved SPECT quantification using compensation for scattered photons," *J Nucl Med*, vol. 25, pp. 893–900, 1984.

[22] K. Ogawa, Y. Harata, T. Ichihara, A. Kubo, and S. Hashimoto, "A practical method for position-dependent Compton-scatter correction in single photon emission CT," *IEEE Trans Med Imag*, vol. 10, no. 3, pp. 408–412, 1991.

[23] K. Sjogreen, M. Ljungberg, and S. E. Strand, "An activity quantification method based on registration of CT and whole-body scintillation camera images, with application to 131I," *J Nucl Med*, vol. 43, no. 7, pp. 972–982, 2002. [Online]. Available: PM:12097471.

[24] K. Sjögreen-Gleisner, "Scatter correction by deconvolution of planar whole-body scintillation-camera images using an image-based estimate of the signal-to-noise ratio," *Eur J Nucl Med*, vol. 39, no. Suppl 2, pp. S304–S353, 2012.

[25] F. Barrack, J. Scuffham, and S. McQuaid, "Septal penetration correction in I-131 imaging following thyroid cancer treatment," *Phys Med Biol*, vol. 63, no. 7, p. 075012, Mar. 27, 2018, doi:10.1088/1361-6560/aab13a.

[26] A. Liu, L. E. Williams, and A. A. Raubitschek, "A CT assisted method for absolute quantitation of internal radioactivity," *Med Phys*, vol. 23, no. 11, pp. 1919–1928, 1996. [Online]. Available: PM:0008947907.

[27] A. Kojima et al., "A preliminary phantom study on a proposed model for quantification of renal planar scintigraphy," *Med Phys*, vol. 20, no. 1, pp. 33–37, Jan-Feb 1993, doi:10.1118/1.597057.

[28] W. C. Buijs, J. A. Siegel, O. C. Boerman, and F. H. Corstens, "Absolute organ activity estimated by five different methods of background correction," *J Nucl Med*, vol. 39, no. 12, pp. 2167–2172, 1998. [Online]. Available: PM:0009867163.

[29] M. Larsson, P. Bernhardt, J. B. Svensson, B. Wangberg, H. Ahlman, and E. Forssell-Aronsson, "Estimation of absorbed dose to the kidneys in patients after treatment with 177Lu-octreotate: comparison between methods based on planar scintigraphy," *EJNMMI Research*, vol. 2, no. 1, p. 49, 2012, doi:10.1186/2191-219X-2-49.

[30] W. E. Bolch, K. F. Eckerman, G. Sgouros, and S. R. Thomas, "MIRD pamphlet No. 21: A generalized schema for radiopharmaceutical dosimetry—standardization of nomenclature," (in English), *J Nucl Med*, vol. 50, no. 3, pp. 477–484, Mar 2009, doi:10.2967/jnumed.108.056036.

[31] K. F. Koral et al., "Update on hybrid conjugate-view SPECT tumor dosimetry and response in 131I-tositumomab therapy of previously untreated lymphoma patients," *J Nucl Med*, vol. 44, no. 3, pp. 457–464, 2003. [Online]. Available: PM:12621015.

[32] M. Garkavij et al., "177Lu-[DOTA0,Tyr3] octreotate therapy in patients with disseminated neuroendocrine tumors: Analysis of dosimetry with impact on future therapeutic strategy," *Cancer*, vol. 116, no. 4 Suppl, pp. 1084–1092, Feb 15 2010, doi:10.1002/cncr.24796.

[33] A. Sundlov et al., "Feasibility of simplifying renal dosimetry in (177)Lu peptide receptor radionuclide therapy," *EJNMMI Phys*, vol. 5, no. 1, p. 12, Jul. 5, 2018, doi:10.1186/s40658-018-0210-2.

26 Quantification in Emission Tomography

Brian F. Hutton, Kjell Erlandsson, and Kris Thielemans

CONTENTS

DOI: 10.1201/9780429489556-26

26.1 INTRODUCTION

The primary objective in emission tomography is to obtain an accurate and precise one-to-one mapping of the activity distribution in the body. If this is achieved, quantification of activity concentration in tissue can be obtained directly from the reconstructed images. Quantification is an important consideration for both positron emission tomography (PET) and single photon emission computed tomography (SPECT). Although these two modalities operate on different principles with significantly different instrumentation, they have in common the ability to demonstrate the spatial activity distribution following administration of a radioactive tracer and, to some extent, its variation in time. Any image artefacts that occur will fundamentally bias regional values, and so a prerequisite is that reconstructed images should be artefact-free. This implies that the vendor corrections for non-linearity and uniformity (SPECT) and normalization of detectors (PET) result in images that accurately reflect the distribution of activity in the object or patient. Provided this is achieved, then quantification should be possible. Most of the main factors affecting quantification are common to both modalities, and the approaches to correction, often dependent on image reconstruction, are also similar for the two modalities. This chapter therefore addresses quantification for both PET and SPECT, highlighting where differences occur.

26.2 SOURCES OF ERROR IN EMISSION TOMOGRAPHY

Several factors affect quantification, and these will be outlined in this section; more details on the approaches to correction are provided in subsequent sections and other chapters.

26.2.1 Detector Calibration

Fundamental to the quantification of activity is the need for instrumentation to provide an artefact-free estimation of the activity distribution. This implies that appropriate corrections to the measured data are implemented to ensure a one-to-one relationship between the acquired data and the origin of the detected photons. The corrections that are routinely applied to acquired SPECT data include corrections for energy, linearity and uniformity to ensure spatially accurate localization and position-independent sensitivity (Chapter 13). Similarly, in PET, normalization is used to account for variations in sensitivity of individual detectors (Chapter 18). In addition, detectors are expected to have a linear response to increasing activity and any deviation due to dead-time effects needs to be corrected (Chapter 19). A primary objective of routine quality control is to ensure that performance is stable over time, flagging the need for recalibration where significant changes occur. These factors that could affect quantification are covered in other chapters and so are not further elaborated here.

26.2.2 Image Reconstruction

Image reconstruction is clearly an essential component of the processing in emission tomography and is central to many of the corrections that need to be applied to achieve quantitative measurement. There are many factors that can influence reconstructed image quality and fidelity, and the range of options in image reconstruction is a major factor contributing to cross-site variation in quantification. In theory all reconstruction algorithms aim to provide images that accurately depict the measured activity distribution but, in practice, this is not always achieved. Different algorithms perform differently and even the same algorithm can vary due to the vendor implementation. For example, in iterative reconstruction, there are many different possible algorithms where the result depends on the number of iterations used, the accuracy of the model used to define the instrument's response to emitted activity, the treatment of noise via filtering or regularization, and so forth. Any application that requires quantification needs to factor in the effects of reconstruction, usually adopting a standard procedure where the influence on quantification can be considered to be predictable and accounted for as part of the correction and calibration procedures described below. It is important to have a good understanding of image reconstruction principles and awareness of the potential effects on quantification (Chapters 16 and 20).

26.2.3 Randoms

A source of potential bias in quantification, unique to PET, is the possibility that random coincidences (randoms, also called accidental coincidences) may be acquired. If two independent annihilations occur at virtually the same time there is the possibility that one annihilation photon from each event is detected, giving rise to a false coincidence whereby the line of response is unrelated to either of the emitted positron locations. Failure to correct randoms results in a background distribution that reduces contrast. The probability of randoms occurring increases with count-rate and depends on the coincidence window width that is set in the instrument. Fortunately, random coincidences can be estimated either directly from the detector singles rates or by using a delayed timing window. In this latter case if any photons are detected in the delayed window in coincidence with photons detected in the standard window these can only be random coincidences. As the delayed counts are a noisy estimate of the randoms, ideally, they should be used after variance reduction – for example, by using a maximum-likelihood estimate based on the singles model. In both cases, the estimated randoms distribution can be subtracted directly from the acquired sinograms (appropriate for analytic reconstruction algorithms) or alternatively the random coincidences can be included in the forward projection step during iterative reconstruction (Chapter 20).

26.2.4 Attenuation and Scatter

When photons are emitted internal to the body, they will interact with surrounding tissue, mainly through the process of Compton scattering. Note that interaction in the detector is mainly via the photoelectric effect due to the high atomic number, whereas, for the emitted energies typical of SPECT and PET tracers, Compton scattering is dominant in body components. As a result of Compton scatter some photons are deflected from their initial paths with loss of energy. The probability of undergoing Compton scatter increases with distance travelled through tissue and is also dependent on the density of tissue (with angular probability of scatter defined by the Klein Nishina equation). Consequently, the detected data are subjected to two possible effects, attenuation and scatter (see Figure 26.1). Although both attenuation and scatter originate from the same interaction, corrections for these effects are usually treated independently.

Attenuation: The number of unscattered photons that reach a detector are reduced compared to the number that would be detected for the same source in air. The loss of photons is exponential given by:

$$C = C_0 e^{-\sum_i \mu_i d_i} \tag{26.1}$$

where C is the recorded counts compared to emitted counts C_0; μ_i is the linear attenuation coefficient for tissue i and d_i is the distance travelled through tissue i. The sum is calculated from the point of emission to the detector (normally patient boundary but includes any intervening structures such as the patient bed). Note that, for PET, the probability that two photons will be detected in coincidence depends on the total path through the patient, independent of the point of emission along that path. If attenuation is not corrected, the reconstructed image will appear to have reduced counts towards the centre of the subject and an apparent hot rim. This is further complicated if the attenuating medium is non-homogeneous, as this can result in misplaced counts with artificially increased counts in areas of low attenuation (Figure 26.2). Attenuation results in a significant loss of counts for both SPECT and PET. As an example, consider a 30 cm diameter water cylinder with a radioactive source placed at its centre; for ^{99m}Tc (140 keV gamma photons) the

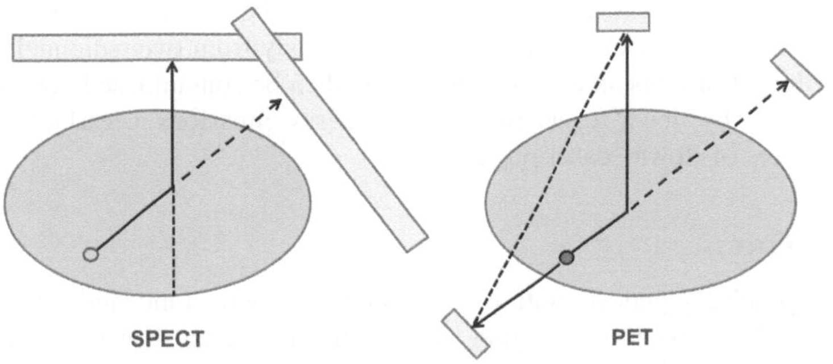

FIGURE 26.1 When an emitted photon undergoes Compton scatter it is deflected with loss of energy. The Compton interactions result in a reduction of the number of detected events (or true coincidences). This loss is considered as due to attenuation. The scattered photon may still be detected, leading to an incorrect estimate of the source location – along the trajectory of the scattered photon (SPECT) or incorrect line of response (PET).

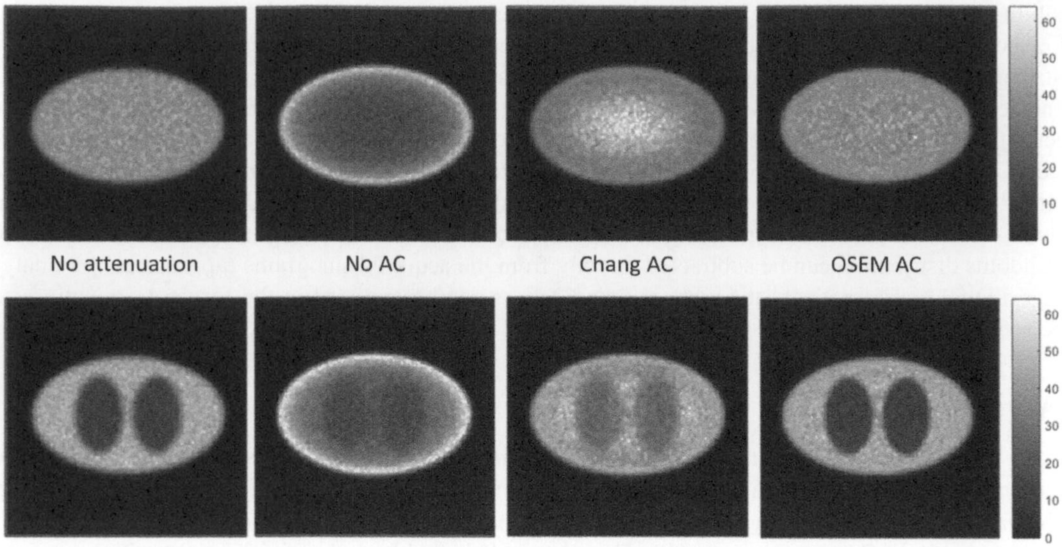

FIGURE 26.2 SPECT reconstruction for a phantom with uniform attenuation and activity (upper images) and thorax phantom with non-uniform activity and attenuation (lower images). Reduced activity concentration towards the centre without AC (No AC) compared to the perfect case, where attenuation is absent (No atten). Note apparent increased activity in the lungs relative to background. Chang multiplicative correction tends to overcorrect at the centre but does not rectify the increased activity in lungs (Chang AC). Inclusion of an attenuation map in the OS-EM reconstruction correctly accounts for the attenuation (OS-EM AC). Colour image available at www.routledge.com/9781138593268.

attenuation coefficient is 0.15 cm^{-1} resulting in an attenuation factor of 9.5 (i.e. only approximately 10% of photons reach the detector without scattering). For ^{18}F (511 keV gamma photons) in the same phantom the attenuation coefficient is lower (0.096 cm^{-1}), but the attenuation path is larger (30 cm due to the pair of annihilation photons), resulting in an attenuation factor of 17.8 (i.e. approximately 5.6% of photon pairs reach the detectors without scattering).

Scatter: A number of scattered photons will be detected in locations different from those defined by their original trajectories. An energy window is normally used to eliminate the deflected photons that have significant loss of energy, but due to the limited energy resolution of detectors (both SPECT and PET), some scattered photons will still be detected. Since the distribution of scatter extends beyond the expected source location, the effect is to reduce resolution and contrast in the reconstructed images. In the case of SPECT, scatter typically accounts for around 35 per cent of the detected photons, and the origins of scatter counts are constrained to lie within the patient boundary (including patient bed). In the case of PET, the incorrectly placed scatter coincidence (usually involving scatter of one of the annihilation photons but sometimes both) results in scattered events that can appear to originate along lines of response that lie outside the body boundary. In 3D PET the scatter fraction is typically around 40 per cent for the body (although the contrast degradation may be lower than in SPECT due to the broader distribution of scatter extending outside the body).

A further complication can occur in SPECT since some radionuclides have emissions of multiple photons with different energies (or alternatively data can be acquired simultaneously from two radionuclides of different energies). The counts recorded in a selected photopeak energy window can then be contaminated not only by the scatter from the primary energy photons but also by down-scatter from higher energy emissions. Clearly, the degree of image degradation will depend on the amount of down-scatter present.

26.2.5 Partial Volume Effects

Both SPECT and PET have relatively poor resolution compared to anatomical modalities (CT and MRI). SPECT resolution, at best, is around 7 or 8 mm for brain imaging and is typically greater than 12 mm for other applications. For PET, reconstructed resolution is typically around 6 mm in practice with better resolution demonstrated for some models, especially when designed specifically for brain imaging. In both cases the resolution can vary considerably depending on chosen reconstruction parameters. Additionally, direct modelling of the resolution during reconstruction combined with algorithms that maintain edge contrast when controlling noise can result in improvement in reconstructed resolution. The consequence of poor resolution is that the observed signal in small hot objects is reduced as a result of the blurring, with a balance between spill-out of counts to neighbouring regions and spill-in from the background regions.

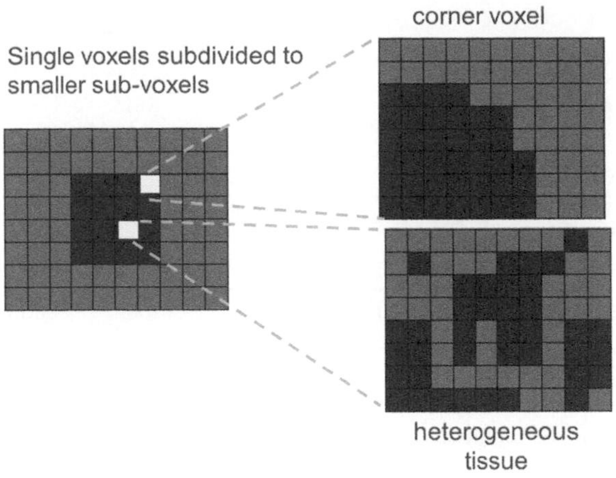

FIGURE 26.3 If a voxel volume is only partially occupied by the tissue of interest the observed activity, concentration will reflect an average rather than true concentration. This can occur on edge voxels or in heterogeneous tissue, for example, lungs contain a large volume of air.

This reduction in apparent counts due to an object partially occupying the sensitive volume of the detector is normally referred to as the partial volume effect (PVE). The effect depends on object size and, as a result, it can be difficult to interpret changes in observed uptake if volume is changing (e.g. interpretation of tumour uptake during treatment where metabolically active tumour volume may be changing). The effect is easily identified in gated cardiac perfusion studies where there is an apparent brightening of the myocardium during contraction due to the thickening of the heart wall. For large objects, the effect is much smaller since the spill-over of counts only affects the edges of the object and immediate surroundings.

There is an additional partial volume effect if the tissue of interest only partially occupies a voxel. This will occur in voxels that sample the boundary of an object but can also occur if there is heterogeneous uptake within a voxel (see Figure 26.3). This is usually referred to as a *tissue fraction effect*. An example is in the lung where most voxels have large air content, depending on the state of respiration. An observed variation in activity concentration does not necessarily reflect the true variation in tracer uptake as the air content may vary.

26.2.6 Motion

Various types of motion can occur, involuntary or voluntary. Involuntary motion refers to life-sustaining periodic motion of the heart and lungs, with associated movement of adjacent organs; voluntary movement refers to general patient movement, often prompted by discomfort. Of course, there are other causes of involuntary movement such as a cough reflex or neurological condition. In the case of a stationary detection system, motion can have an effect similar to the PVE since it introduces blurring and degradation of resolution with, consequently, reduced contrast. The non-linear nature of motion means that the blurring effect is less predictable than other PVEs and so correction is treated differently. In the case of non-stationary detectors (e.g. SPECT with a rotating gamma camera) the effect can be quite complex and in practice difficult to identify in reconstructed images, although usually quality assurance procedures are in place to check for motion occurrence. In addition to effects of motion during acquisition, there can be serious consequences of movement resulting in mis-registration between acquisitions. For example, movement during a dynamic study can result in erroneous TACs (see Figure 26.4). Also, mis-registration between sequential acquisition of transmission (CT) and emission studies can result in serious attenuation correction artefacts (see Figure 26.5).

26.2.7 Noise

Clearly noise is an inherent problem for both SPECT and PET, especially for SPECT whose sensitivity is typically a hundred times worse than conventional PET (even worse compared to PET systems with extended field-of-view). Noise imposes limits on some corrections but, in general, results in degraded precision rather than necessarily bias. Most of the work to control noise is in the domain of image reconstruction and so the reader is referred to that chapter for further discussion (Chapter 20).

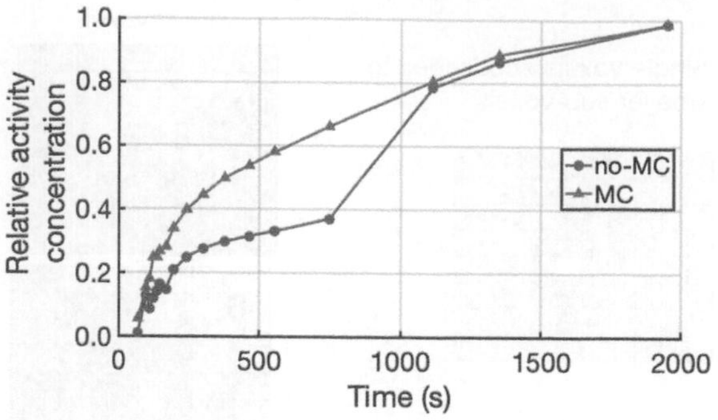

FIGURE 26.4 Time activity curves (TACs) generated for a volume of interest in the frontal lobe in a patient with epilepsy showing distortion of the TAC due to motion (no-MC) compared to the motion corrected (MC) result.

26.3 ATTENUATION CORRECTION

26.3.1 Measuring Attenuation

Accurate attenuation correction is important if the aim is to achieve quantitative reconstruction. Historically, radionuclide transmission sources were used in both PET and SPECT to directly measure attenuation [2]. However, the standard approach is now to use CT to estimate the map of attenuation coefficients for use in attenuation correction. The improved anatomical localization and complementary clinical information are compelling reasons to support the clinical use of PET/CT or SPECT/CT, with the added benefit of being able to measure attenuation efficiently. PET is no longer available as a standalone modality but usually is combined with CT [3] (or occasionally MRI). SPECT/CT is increasing in popularity, although standalone SPECT systems are still in use, with no capability to measure attenuation.

The X-rays used in CT have a range of relatively low energies, quite different to the mono-energetic photons that are emitted for radionuclides used in SPECT or PET. As a result, the reconstructed CT (usually expressed in Hounsfield units) cannot be directly used for attenuation correction. The CT data must first be transformed to produce a map of attenuation coefficients suitable for use with a specific radionuclide. In the case of PET all radionuclides decay via positron emission, resulting in pairs of 511 keV annihilation photons, so the necessary transformation is independent of radionuclide (and therefore radiotracer) used (either bi-linear or tri-linear) [4]. For SPECT an energy-specific transformation is required, usually bilinear [5]. The resulting transformed dataset is smoothed to match the poorer resolution of the PET or SPECT data.

26.3.2 Inclusion of Attenuation in the System Matrix

Assuming that an accurate map of attenuation coefficients is available, it remains to utilize this additional information to correct for the effects of attenuation. Historically, this was straightforward for PET since the integrated attenuation along a line of response could be used as a direct multiplicative correction applied prior to reconstruction. The exact correction in SPECT at that time, however, was intractable. A simple post-reconstruction approach, which is still used in standalone SPECT, is to estimate the average attenuation factor for photons originating at each specific voxel in the subject, using this as a correction factor, usually referred to as Chang attenuation correction [6]. The method is not exact but can provide an improved reconstruction in cases where attenuation can be approximated as uniform within a known subject boundary (e.g., brain). The method fails, however, if there is non-homogeneous attenuation such as in the thorax (Figure 26.2)]. The current approach that is universally applied is to use the measured attenuation map to define the probability of detection and incorporate this information in the system matrix used in iterative reconstruction (Chapter 20). This solution is applicable to both SPECT and PET reconstruction.

26.3.3 Sources of Error in Attenuation Correction

CT is subject to various artefacts as a result of motion, beam hardening or photon starvation due to presence of high-density materials or differences in the scattering probabilities for the CT low-energy photons versus SPECT/PET photons. Errors will usually propagate to the corrected emission data. Perfect attenuation correction is therefore hard

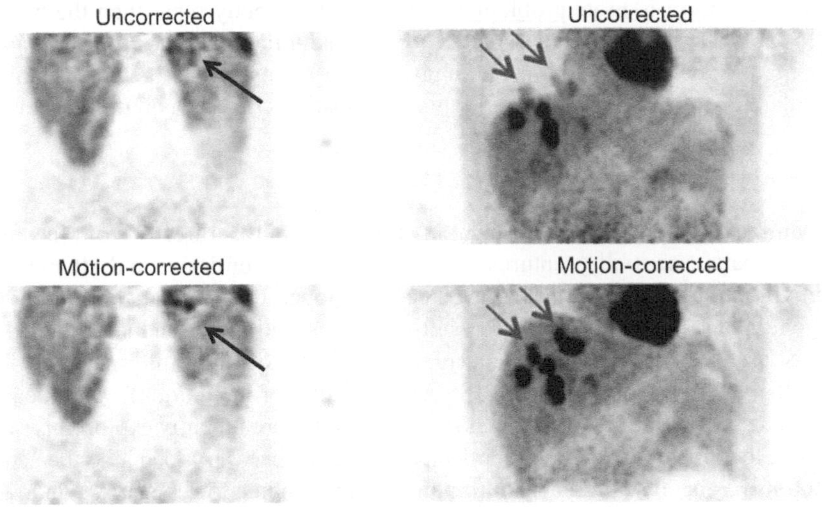

FIGURE 26.5 Respiratory motion results in blurring and loss of lesion contrast (left); respiratory mismatch of CT and PET results in apparent mis-location of the liver boundary and possible misinterpretation of lesion location (right). Correction for respiratory motion reduces both artefacts. (Adapted from [1]).

to achieve. It is important to recognize when CT artefacts occur, as failure to recognize this can lead to incorrect PET (or SPECT) interpretation [7, 8]. In general, if there are artefacts in the form of artificially increased Hounsfield units, the corresponding area in the emission study will be subject to over-correction, which can easily be misinterpreted as increased tracer concentration. It should be noted that single-energy CT does not provide sufficient information on material properties to permit exact transformation to linear attenuation coefficients for SPECT or PET. The transformation provides reasonable estimates for most tissues (including bone) but fails for metal implants and high concentrations of contrast material. This can be circumvented using dual-energy CT, although this is not commonly used.

It is assumed that the CT is perfectly registered to the emission data, but the sequential acquisition cannot guarantee this. CT involves fast acquisition where breath-holding may be feasible whereas emission tomography involves longer acquisition times with either averaging of moving structures or freezing of structures via gating procedures. A further complication arises due to the variations in attenuation during different respiratory phases; errors can therefore occur not only as a result of spatial mis-registration but also from incorrect attenuation coefficient scaling. A common artefact is visible at the boundary between lung and liver caused by acquisition of the CT data at a specific respiratory phase (e.g. with breath holding) which does not match the average position of the diaphragm during acquisition of the emission study. The so-called banana artefact is clearly visible in whole-body PET studies and leads to difficulty in locating tumours that are present near the diaphragm or, in some cases, failure to detect the lesion (see Figure 26.5). The use of time-of-flight in PET has been reported to reduce the severity of local artefacts due to respiratory mismatch [9, 10], although more recently it has been shown that this results in error propagation over a larger volume [11].

26.3.4 Attenuation Correction for PET/MRI

Attenuation correction is a problem in the case of PET/MRI where the attenuation coefficient map has to be determined from MRI rather than CT. Clearly, MRI does not measure tissue density (or electron density) and so there is not a simple conversion as in the case of CT. This has therefore been a major area for research in recent years. Much of the focus has been on correction of brain PET and a range of techniques have been developed to improve on the very basic Dixon sequence that is used to simply differentiate bone and air. A recent paper [12] provides a comprehensive overview of the possible approaches that include, for example, use of special MRI sequences (UTE or ZTE) that better distinguish tissue types (with suitable post-acquisition processing) or atlas-based methods that use a database of paired CT and MRI studies to construct a synthetic CT based on acquired MRI data. Both methods have been shown to provide acceptably low errors [13] although both are subject to limitations in their practical use in certain cases. More recently researchers have demonstrated that deep learning can be applied effectively to produce very convincing synthetic CT images [14].

The estimation of attenuation is more problematic in parts of the body other than the brain, especially the thorax. Because of the air in the chest, it is difficult for MRI to produce reliable images of the lungs (due to low signal and susceptibility artefacts) and the attenuation in the lungs can vary considerably between individuals (and even within individuals) depending on respiration and disease status. Attenuation correction in this case remains challenging [15].

26.3.5 Estimating Attenuation from Emission Data

One approach to obtaining an attenuation coefficient map is to derive this directly from the emission data; this is possible since the acquired data essentially captures information on both emission and attenuation. Algorithms for joint reconstruction of activity and attenuation maps have been developed that estimate both maps from the emission data. However, this is in general an ill-defined problem resulting in cross talk between the estimated activity and attenuation maps. This problem is largely alleviated with the availability of time-of-flight information with recent evaluation indicating promising results [16, 17] (Chapters 6 and 18). Alternatively (or in addition), benefit can be gained by including scatter information as has been demonstrated for SPECT [18] and more recently explored for PET [19, 20]. In the case of PET, modification of the normal acquisition hardware would be necessary to enable the acquisition of coincidences where one of the photons has been scattered, resulting in its detection in a lower energy window. Recently, novel reconstruction algorithms have been adapted to incorporate the additional scatter information, and these show some promise [21]; possible application to SPECT has yet to be demonstrated but the prospect of being able to correct SPECT for attenuation without the need for CT has definite appeal (recently announced by one commercial company).

26.4 SCATTER CORRECTION

For a review on scatter correction see [22].

26.4.1 Multiple Energy Windows (SPECT)

In SPECT multiple energy windows can be selected so that scatter can be estimated using narrow energy windows directly adjacent to the photopeak. Note that these can be placed below and above the photopeak to estimate any down-scatter. The measured scatter can be used to provide an estimate of the scatter that is present within the photopeak energy window. An attraction of this triple energy window (TEW) scatter correction method [23] is that it involves direct measurement of scatter and so will include scatter from activity outside the field-of-view, which cannot normally be modelled. The main limitation in using this method is that the resulting corrected images have increased noise since the scatter must be estimated from relatively narrow energy windows, with limited counts. Attempts have been made to apply similar multiple-energy-window methods in PET, but these have had limited success.

26.4.2 Scatter Models for SPECT and PET

An alternative method of scatter correction, applicable for both SPECT and PET, is to model the scatter based on an estimate of the activity distribution and a known map of attenuation coefficients. Since the activity distribution is unknown in the absence of scatter correction, an iterative process is necessary. Analytic approaches for the scatter estimation are usually restricted to single scatter events only. Monte Carlo simulation can be used to accurately model the scatter, although this ideally requires knowledge of both activity and attenuation for an extended field-of-view. Monte Carlo simulation has a reputation of being prohibitively slow, although fast computation has been shown to be possible [24]. A fast implementation for SPECT has been released by one vendor [25]. Alternatively, a simplified model can be used with some assumptions regarding the distribution of scatter; this usually assumes single scatter and ignores the possibility that photons may undergo multiple deflections [26, 27]. In PET the single scatter estimate is aided by the fact that scatter detected in lines of response that do not intersect the body can be directly measured and used to scale the scatter model [28, 29].

26.4.3 Inclusion in Reconstruction

Estimated scatter can simply be subtracted from projections (sinograms) but this does affect the noise characteristics of the raw data. It is preferable to add the estimated scatter in the forward projection step of the iterative reconstruction. This is an easily implemented modification applicable to any standard algorithm (Chapter 20).

26.4.4 Sources of Error in Scatter Correction

As mentioned above, scatter models can have limited accuracy since they usually estimate only single scatter events, although including double scatter events is becoming more practical [30] and do not account for activity out of the field-of-view, whereas methods based on use of multiple energy windows do at least partly account for these factors. Scatter models are approximate, and any errors tend to be problematic in the vicinity of hot activity (or high-contrast regions) where over-correction of scatter can lead to artificially decreased counts adjacent to the hot activity. Also, as mentioned earlier, noise can be a concern. In PET, this can affect the estimation of the scatter scale factors, with recent attempts to estimate the scale factors from all of the data as opposed to only the 'tails' [31, 32].

26.4.5 Current Research and Future Directions

There currently are no options to routinely use multiple energy windows in PET, but this may be a future possibility, which may be particularly useful if using radionuclides that have multiple emissions rather than a single energy positron (note also the potential use in estimating attenuation mentioned in section 26.3.5). This could open the possibility for alternative approaches to scatter correction but also possible use of the scatter data to improve reconstruction [33].

There are several commercially available SPECT systems that utilize solid state detectors (e.g. cadmium zinc telluride or CZT) in place of the normal NaI(Tl) detectors with photomultiplier tubes (Chapter 6). CZT operates on the basis that deposited photon energy is directly converted to an electrical signal with good intrinsic resolution (usually ~2.5 mm) and superior energy resolution (5–6% compared to 9–10% for NaI(Tl)). This results in a lower, but not negligible, scatter fraction. In the case of CZT this is somewhat complicated by the presence of non-scattered photons in a tail that extends below the photopeak due to partial charge capture within the detector elements. Distinguishing scatter and non-scatter events is therefore complicated as is dealing with down-scatter correction in dual radionuclide studies [34].

26.5 PARTIAL VOLUME CORRECTION

When analysing emission tomography images, it is important to take into account the spatial resolution of the SPECT or PET system. This factor not only affects the qualitative appearance of the images, but also the quantitative image values. The apparent reduction in signal due to resolution effects is known as the Partial Volume Effect (PVE) [35]. A number of Partial Volume Correction (PVC) methods have been developed over the years; these differ in terms of assumptions made, as well as accuracy and extent of the correction. There are several useful reviews [36–38]. A software package (PETPVC) containing a number of PVC algorithms has also been developed [39].

26.5.1 Estimating Resolution

The first step is to estimate the spatial resolution of the images. Apart from the detector geometry of the scanner, this also depends on the reconstruction algorithm used. Analytical algorithms, such as FBP, are linear and the resolution is therefore independent of the activity distribution. On the other hand, with the currently more popular iterative algorithms, such as OS-EM, the resolution will depend on both the activity distribution and the number of iterations. Other factors include the use of resolution modelling during reconstruction and any post-reconstruction filter applied to the image.

When measuring the spatial resolution with line or point sources, it is important to include a non-zero background in order to avoid artificially low values with iterative algorithms. If activity distribution specific resolution is required, this can be estimated using the perturbation method [40]. A tiny amount of data, corresponding to a point or line source, is added to the dataset in question. Images are reconstructed with and without the perturbation and then subtracted. This procedure results in an image of a point or line source representing the point-spread function (PSF) of the underlying reconstructed data.

In PET, there is a contribution to the spatial resolution from the positron range, that is, the average distance from the position of decay to the position of annihilation. This depends on the maximum positron energy, which is specific to each radionuclide. In soft tissue, the positron range is < 1 mm for ^{18}F and several mm for ^{82}Rb, while in lung tissue positron range is significantly larger.

26.5.2 Standard Assumptions

A common assumption is that the PSF is spatially invariant. This is done for reasons of computational simplicity but is not strictly necessary. In SPECT, planar resolution depends on the distance from the collimator and, like PET, the PSF becomes elongated in the radial direction with increasing distance from the centre of the scanner. In principle these effects can be taken into account, for example, by using a reconstruction-based PVC method [41].

Most PVC methods are based on a segmentation of the image into a number of regions. Usually, these regions are assumed to be uniform in terms of activity distribution, which makes the calculation of crossover between regions (spill-in and spill-out) more stable. Even with this assumption, the corrected images will not necessarily be uniform within regions. Alternatively, one could assume that the activity concentration, instead of being constant within a region, changes linearly in a certain direction [42]. This requires estimation of additional parameters.

26.5.3 Deconvolution

The PSF can be deconvolved from the reconstructed images, by either Fourier space inverse filtering or using an iterative procedure, such as the Richardson-Lucy (equation 26.2a) or the reblurred van Cittert algorithms (equation 26.2b) [43].

$$\hat{f}^{k+1}(x) = \hat{f}^{k}(x)\left\{h(x)\otimes\frac{g(x)}{\hat{f}^{k}(x)\otimes h(x)}\right\} \tag{26.2a}$$

$$\hat{f}^{k+1}(x) = \hat{f}^{k}(x)+\alpha\left\{h(x)\otimes\left(g(x)-\hat{f}^{k}(x)\otimes h(x)\right)\right\} \tag{26.2b}$$

where $\hat{f}^{k}(\cdot)$ is the estimated image after k iterations, $g(\cdot)$ is the original image, $h(\cdot)$ is the PSF, α is a scaling factor (step-length), ⊗ represents the convolution operator, and $\hat{f}^{0}(x) = g(x)$.

However, these methods cannot completely recover the true distribution, as the highest frequency components are not present. Furthermore, they will typically result in noise-amplification, and therefore need to be combined with some kind of noise-control or noise-reduction strategy. Another effect can be the introduction of ringing artefacts around sharp edges in the image, so called Gibbs artefacts.

26.5.4 Resolution Modelling

The deconvolution procedure, mentioned above, can be combined with iterative image reconstruction by incorporating the PSF into the system matrix [44]. This approach should in principle lead to a better bias versus noise trade-off, as it operates directly on the raw data. However, it suffers from some of the same downsides as the deconvolution methods, such as Gibbs artefacts [45].

Improved results can be obtained by also incorporating anatomical information into the reconstruction process. This can be done within so-called 'Bayesian' or 'penalized' reconstruction algorithms, in which prior knowledge about the image can be introduced [46] (Chapter 20).

26.5.5 Post-reconstruction Methods

PVC can be performed after image reconstruction using the PSF and a segmented, co-registered anatomical image (CT or MRI). Considering one image region at a time, it is necessary to correct for spill-out of data from within the region and spill-in to the region from the rest of the image. The correction can be performed by first subtracting the spill-in contribution and then scaling the image values to correct for spill-out (equation 26.3a) [47]. Alternatively, a single multiplicative correction can be applied (equation 26.3b) [48].

$$f_i(x) = \frac{g_i(x)-b_i(x)}{R_i(x)} \tag{26.3a}$$

$$f_i(x) = Y_i(x)\,g_i(x) \tag{26.3b}$$

where $f_i(\boldsymbol{x})$ and $g_i(\boldsymbol{x})$ are the corrected and original image values, at position $\boldsymbol{x}$ in region i, respectively, and $b_i(\boldsymbol{x})$, $R_i(\boldsymbol{x})$ and $Y_i(\boldsymbol{x})$ are the spill-in term, the spill-out factor and the correction factor, respectively. The two approaches are, in theory, mathematically equivalent, with:

$$Y_i(\boldsymbol{x}) = \frac{1 - b_i(\boldsymbol{x}) / g_i(\boldsymbol{x})}{R_i(\boldsymbol{x})} \tag{26.4}$$

The spill-in term is the sum of the contributions from all neighbouring regions:

$$b_i(\boldsymbol{x}) = \sum_{j:j \neq i} b_{ij}(\boldsymbol{x}) \tag{26.5}$$

The various proposed PVC algorithms differ in their practical implementation and in the way that the correction terms and factors are calculated [37]. PVC can be applied on a regional, or voxel-by-voxel basis. Although both regions and voxels are discrete entities, in order to emphasize the distinction between the two, here we represent regions with discrete and voxels with continuous variables.

In addition to providing sharp boundaries between regions, the anatomical image also allows for noise-reduction by means of regional averaging. Using the uniformity assumption, we can calculate the spill-in component from region j to region i at position x, as follows:

$$b_{ij}(\boldsymbol{x}) = I_i(\boldsymbol{x})\left[\left(I_j(\boldsymbol{x})\hat{f}(\boldsymbol{x})\right) \otimes h(\boldsymbol{x})\right] \approx \hat{f}_j I_i(\boldsymbol{x})\left\{I_j \otimes h\right\}(\boldsymbol{x}) \tag{26.6}$$

where $\hat{f}(\cdot)$ is an estimate of the true image, $\hat{f}_j$ is an estimate of the mean value in region j, $h(\cdot)$ is the PSF, $I_i(\boldsymbol{x})$ is the indicator function for region i, and $\otimes$ represents the convolution operator.

Again, using the uniformity assumption, the spill-out factor can be expressed simply as:

$$R_i(\boldsymbol{x}) = \left\{I_i \otimes h\right\}(\boldsymbol{x}) \tag{26.7}$$

An estimate of the true regional mean values is needed for calculation of the spill-in term, but not for the spill-out factor. To estimate the mean values, we can express the un-corrected image as follows:

$$g_i(\boldsymbol{x}) = R_i(\boldsymbol{x}) f_i(\boldsymbol{x}) + b_i(\boldsymbol{x}) \approx \sum_j \hat{f}_j I_i(\boldsymbol{x})\left\{I_j \otimes h\right\}(\boldsymbol{x}) \tag{26.8}$$

The un-corrected mean values for region i are then obtained as:

$$\bar{g}_i = \frac{\int g_i(\boldsymbol{x}) d\boldsymbol{x}}{\int I_i(\boldsymbol{x}) d\boldsymbol{x}} = \sum_j \hat{f}_j \frac{\int I_i(\boldsymbol{x})\left\{I_j \otimes h\right\}(\boldsymbol{x}) d\boldsymbol{x}}{\int I_i(\boldsymbol{x}) d\boldsymbol{x}} = \sum_j \omega_{ij} \hat{f}_j \tag{29.9}$$

with

$$\omega_{ij} = \frac{\int I_i(\boldsymbol{x})\left\{I_j \otimes h\right\}(\boldsymbol{x}) d\boldsymbol{x}}{\int I_i(\boldsymbol{x}) d\boldsymbol{x}}$$

If the union of all regions encompass the whole image, the corrected mean values can, in principle, be obtained by matrix inversion [49]. A voxel-wise correction can then be performed [50]. If the number of regions is too large, the matrix inversion operation may become unstable. A more stable alternative that avoids the need for the matrix inversion step is to use an iterative approach [51].

26.5.6 Other Factors

The concept of 'tissue fraction effect' can refer to two different effects. Firstly, because of the relatively large voxel size typically used in PET and SPECT (several mm), voxels located at the boundary between different anatomical regions may represent a mixture between the two. This problem can be ameliorated using smaller voxels, which is appropriate considering the improved resolution post-PVC.

The second tissue fraction effect occurs due to the tissue composition on a much finer scale. In dynamic imaging, the contribution resulting from blood in capillaries can be taken into account by utilizing the difference in the shape of the time-activity curves for blood and tissue [52, 53]. In lung studies, the air fraction can be corrected for using CT data [54].

26.5.7 Sources of Error

PVC can lead to large changes in the image values, so the potential for errors is also large. Several of the steps involved can be possible sources of errors, specifically: PSF estimation, co-registration, and segmentation. Co-registration errors between anatomical and functional images have a direct impact on PVC accuracy. The functional image needs to be transformed into the space of the anatomical image, or vice versa. For brain, rigid or affine transformation is adequate, while for thorax and abdomen, non-linear transformation is usually required; the latter being more susceptible to errors, as more parameters are involved. Registration errors are reduced when using combined-modality scanners, in which both functional and anatomical images are obtained within a short time-period without the need for repositioning the patient.

Segmentation can result in two kinds of errors. The first is when the boundaries of the segmented regions do not agree with those of the corresponding anatomical regions. The second kind is related to the assumption of uniformity. Some degree of bias is introduced if the segmented region is not actually uniform in terms of tracer uptake. This bias can be reduced by using smaller regions. The uniformity assumption will never be entirely correct, so the choice of segmentation represents a trade-off between bias and noise. Figure 26.6 shows an example of how the different errors affect the corrected image.

26.6 MOTION CORRECTION

Due to the duration of nuclear medicine scans, subject motion is unavoidable in the thorax (because of respiration and cardiac contraction) and often present due to voluntary motion (head or body movement), certainly in long-duration dynamic scans. As mentioned previously in this chapter, the effects of motion can be complicated: blurring because of 'displacement' of the origin of the emitted photons, errors in attenuation or even scatter correction as a result of

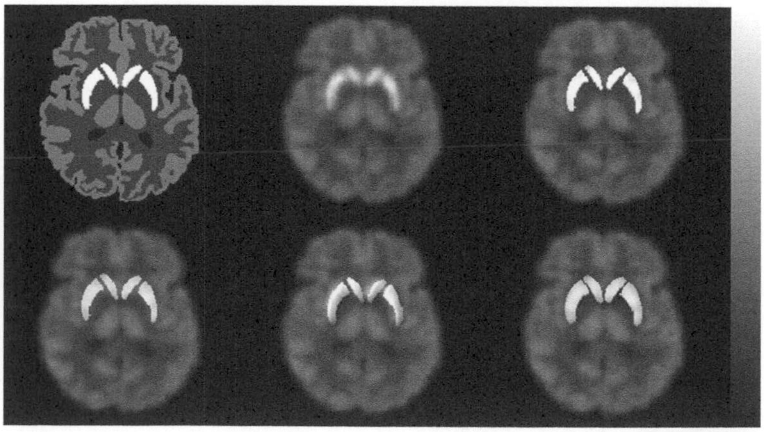

FIGURE 26.6 Examples of PVC of striatum using Single Target Correction (STC) with various errors; (a) original phantom, (b) simulated image, (c) PV-corrected, (d) reduced FWHM (80%), (e) mis-registration (2 mm in x and y), (f) mis-segmentation (1 mm dilation in x, y, and z). Colour image available at www.routledge.com/9781138593268.

mismatch between the attenuation scan and the emission data, and inconsistent data for moving detectors. Here we discuss common strategies to mitigate the effect of motion. These strategies depend on the type of movement (rigid or not) and the pattern over time (quasi-periodic or not). All strategies involve motion detection or tracking, motion estimation and motion correction. We will use the term 'motion state' when the subject's anatomy in the FOV is (nearly) in a fixed position.

Rigid movement includes translation and rotations. It is normally assumed when correcting for head movement and sometimes also for respiratory movement of the myocardium. For the latter, some authors only use translations, while others include affine transformation as it also involves scaling. An important feature of rigid and affine movement is that they preserve straight lines, and this can be exploited for motion correction, as will be clarified below. All other types of movement are called non-rigid (or non-linear).

For quasi-periodic movement, such as due to respiration or cardiac contraction, motion states get repeated over time, which is used for 'gating'. If a motion signal is available, it can be used to estimate when the subject will be in the same motion state, and data from different time points in one motion state can be added together. The simplest approach then for motion management is to select data in one single gate (often end-expiration [55, 56] and/or end-diastole) for image reconstruction. However, this method suffers from increased noise resulting from lower count statistics. Motion correction methods therefore attempt to use all (or most) of the acquired data by combining data from multiple motion states into one final image. Gating cannot be used for non-quasi-periodic movement, in which case most motion correction methods split data into different time frames (assuming little intra-frame movement).

26.6.1 Motion Detection and/or Tracking

This is currently most often done via external monitoring equipment such as cameras, belts, spirometers, and so forth [57], and ECG for cardiac motion. There is increased interest in data-driven methods that determine a motion signal from the PET or SPECT data, due to reduced cost, set-up time, and associated staff dose [58]. While initial methods needed image reconstruction of short time frame data, many methods derive a motion signal from either low spatial resolution dynamic sinograms (e.g. by using Principal Component Analysis [59, 60] or the Spectral Analysis Method [61]) or directly from list mode data (e.g. the Centre of Distribution method [62]). Commercial solutions for respiratory signal extraction using purely data-driven methods are now available for static PET, with active research on head and cardiac motion. Initial work for SPECT has shown it is possible to extend these methods for rotating scanners as well.

26.6.2 Motion Estimation Techniques

Methods to estimate the motion between the different motion states depend on the type of movement. Rigid movement has only 6 degrees of freedom (12 for affine) which can therefore be more easily estimated, for example, by tracking movement of the head via external markers attached to the head. Non-rigid movement is characterized in terms of deformation fields for every motion state, that is, we need to know where every point in the subject moved to. This is normally determined via image registration [63]. However, there can be a 'chicken-and-egg' problem as using images reconstructed for every motion state as input for the image registration suffers from two problems: noise and artefacts due to attenuation mismatch. This can be overcome using one of the following methods (sometimes used in sequence):

Registration on CT or MR images in the same motion state as the PET or SPECT data: For respiration, this has been used when 4DCT is available, although this is not common outside radiotherapy applications due to associated dose and artefacts in 4DCT images caused by irregular breathing. In contrast, there is a large body of work, including commercial implementations, for PET/MR [64].

Registration on PET/SPECT images that have been reconstructed without attenuation correction (and usually scatter correction): This is common for the head, at least with tracers with relatively wide-spread uptake. It has also been used to estimate cardiac respiratory motion, or motion of other large/high-contrast objects. For PET, the performance of this method is improved with higher time-of-flight resolution. However, for non-TOF, non-attenuation corrected images have low contrast, which makes registration of small structures in the lung challenging.

Registration on PET/SPECT images that have been reconstructed with motion-adapted attenuation: In this case, an iterative alternating process of motion estimation and reconstruction is used to gradually increase registration accuracy and image quality. Recently, it has been shown that this can be done in a joint motion-reconstruction optimization problem where the attenuation does not have to be in the same motion state as any of the emission states [65]. However, this process is computationally expensive.

26.6.3 Motion Correction Techniques

Once the motion has been determined, there are different options to obtain motion-corrected images. The methods generally fall into three categories:

Pre-reconstruction correction: As the measured data correspond in good approximation to projections along lines-of-response, if the motion is affine, the measured data can be corrected ('LOR repositioning') before reconstruction, as long as the duration that each repositioned LOR has been measured is taken into account. This strategy is very successful in brain PET imaging with continuous motion tracking [66]. It has also been used for respiratory motion of the myocardium in both PET and SPECT (for translation along the axis of the scanner). For non-rigid motion it is generally not possible to pre-correct the data, although very recently some authors have suggested that if high TOF resolution is available in PET, repositioning can give satisfactory results.

Post-reconstruction correction: Images can be reconstructed for each motion state, resampled to a reference position, and then averaged to reduce noise. This approach is often called RRA ('reconstruct, resample, average'), or RTA ('reconstruct, transform, average'). It was originally suggested for head motion as the multiple acquisition framework (MAF) [67]. It works best if each motion state is reconstructed with an appropriately positioned attenuation map. However, due to the non-linearity of iterative reconstruction (Chapter 20), the RRA image can have positive bias in low count regions [68]. Furthermore, for non-stationary SPECT, RRA is challenging due to the low quality of the gated images.

Inter-reconstruction correction: In motion-compensated image reconstruction (MCIR), the deformation is directly incorporated into the iterative reconstruction process. A separate system model that includes the warping is used for each motion state, and a single image is directly reconstructed from all the motion states. As all data are used, this approach does not suffer from the bias problem of RRA, while it has similar computational requirements. However, RRA images can have a less noisy visual appearance due to averaging effects from the interpolation used for the resampling [68, 69]. In summary, many methods have been developed for motion correction, with commercial solutions now available. It is however still an active area of research. The reader is referred to the following reviews [64, 70–72].

26.7 QUANTIFICATION IN CLINICAL PRACTICE

Quantification in nuclear medicine can refer either to the estimation of the activity concentration in an organ or tissue region or, alternatively, to the estimation of the value of some physiological or biochemical parameter, such as blood flow or metabolic rate.

Depending on the type of study performed and the outcome measure required, different calibration procedures are usually needed. Calibration can be either relative or absolute. *Relative calibration*, also known as *cross-calibration*, can be performed between a PET or SPECT scanner and a separate instrument for measuring radioactivity, such as dose-calibrator or well-counter. The objective is to ensure that the activity concentrations estimated by both instruments are directly comparable. *Absolute calibration* requires at least one of the instruments to be directly traceable to a primary or secondary standard via calibration using a known activity source from a standards laboratory (Chapter 24); in this case the objective is to measure *absolute activity* rather than just comparable estimates of activity concentration.

There are two main areas where quantification has application. The most obvious is *internal dosimetry* where one needs to know the actual distribution of activity in the body over time in order to estimate the cumulated activity and resultant radiation dose to tissue. It is clear that the absolute activity is essential for this to be feasible. With the surge of interest in radionuclide therapy and the current emphasis on patient-specific dosimetry there is an increasing demand for dose estimation in the individual, which has motivated the release of improved vendor software. Though historically mainly used in SPECT there is increasing application also in PET.

The second area where quantification is required is in *tracer kinetic analysis*, mainly in the domain of PET due to the need for rapid acquisition, though also feasible with recently developed stationary SPECT systems. By acquiring dynamic studies during the administration of a radioactive tracer the spatial distribution over time can be tracked to provide time-activity curves (TAC) for selected regions (or individual voxels). Usually a model is used (e.g. a compartmental model) where the model parameters characterize the tracer delivery, uptake and washout. However, indices of the relative tracer uptake can also be derived from either dynamic or static studies.

The simplest form of quantification in nuclear medicine does not require any form of calibration at all and can generally be referred to as *semi-quantification*. This applies to relative uptake measures for a target region in relation to a reference region, such as tissue-to-background or tissue-to-blood ratio (TBR; same acronym for both). An example of semi-quantification is the relative perfusion in myocardial studies where the regional perfusion deficit is compared with the observed peak value, which is assumed to be normal. Of course, the assumption that the reference tissue is comparable across individuals is not necessarily correct; a classic example is the incorrect interpretation of myocardial perfusion in patients with multi-vessel disease, where there may be no area of 'normal' perfusion to act as a reference tissue. Note that some more complex techniques also do not require any calibration, such as kinetic modelling of dynamic data in cases where the input function is obtained directly from the images, either from a blood pool or a reference tissue. In this case quantitative kinetic parameters are obtained without need to compare across measuring instruments or knowledge of absolute activity.

A common metric for quantification of tracer uptake is the standardized uptake value (SUV). This has been widely used in PET but is now also gaining popularity applied to SPECT. In its basic form, this metric is obtained as the maximum (or mean) activity concentration in a target region, C_T, measured a fixed time after administration, normalized to the injected activity, A_{inj}, per unit of body weight, W, of the patient:

$$SUV = \frac{C_T}{A_{inj} / W} \tag{26.10}$$

Sometimes lean body weight is used instead of body weight to account for the fact that there is not normally significant tracer uptake in fatty tissue. The result is a unit-less quantity. The activity A_{inj} is measured in a dose-calibrator, typically an ionization chamber. Therefore, a cross-calibration is needed between the scanner and the dose-calibrator. Despite this use of cross-calibration, SUV is still only a semi-quantitative parameter indicative of regional uptake, reliant on several biological assumptions, and therefore subject to several sources of error. As a result, SUV is useful to illustrate change in an individual's tracer uptake for multiple scans, but it has more limited value in comparing across individuals where the assumptions may not be valid. A further parameter commonly used in neurological analysis is the ratio of SUV values for a tissue of interest and a reference tissue. This is normally referred to as the SUV ratio (SUVR), but this in effect is simply the ratio of reconstructed tissue values and requires no cross calibration (injected dose and patient weight cancel in the ratio). It is therefore identical to the semi-quantitative indices mentioned earlier. To call it SUVR is misleading as it implies that it is somehow related to SUV, when it is not.

In some cases, blood samples are taken in conjunction with a dynamic scan for quantification purposes – for example, for estimating the arterial input function for kinetic analysis. The activity in the samples would then be measured in a well-counter, typically a scintillation detector (capable of measuring much lower levels of activity than the dose-calibrator mentioned above). In this case, cross-calibration is needed between the scanner and the well-counter, but direct traceability is not essential.

The cross-calibrations mentioned above are relative calibrations. When accurate values of activity are required, an absolute calibration is needed. As mentioned above, this is the case in dosimetry studies, performed in conjunction with molecular radiotherapy. Absolute quantification has additional merit in aiding *cross-site harmonization.* In general, achieving comparable results across sites is challenging, especially where instruments from multiple vendors are involved. This requires cross-calibration across the instruments at different sites. Alternatively, absolute quantification should, by definition, result in directly comparable values independent of system. It is clear therefore that taking the necessary steps to avoid artefacts and to perform accurate calibration to achieve absolute quantification should result in comparable results across centres.

A schematic summary of all the calibrations mentioned is shown in Figure 26.7.

26.8 VALIDATION

An important step in developing methods for quantification is validation, comparing derived quantitative values against a gold standard. Validation can involve three approaches – use of simulation, physical phantom studies, and clinical validation. Simulation can involve quite sophisticated models of the activity distribution (e.g. XCAT [73]) in combination with accurate models of the instrument. This can be achieved via Monte Carlo modelling (Chapter 29) or can involve simpler projectors. Simulation is appealing as the true values are exactly known. In addition, simulation can be used to consider separate factors that affect quantification in isolation, which is not physically possible, but which is instructive in the development of correction techniques. Of course, simulation can never fully duplicate the true clinical situation.

The next step in validation usually involves use of physical phantoms with experimental measurement, closer to the real clinical situation. The phantoms used are usually relatively simple in design, although the recent advances in 3D printing open opportunities to construct more complex phantoms, which bridge the gap between simulation and reality.

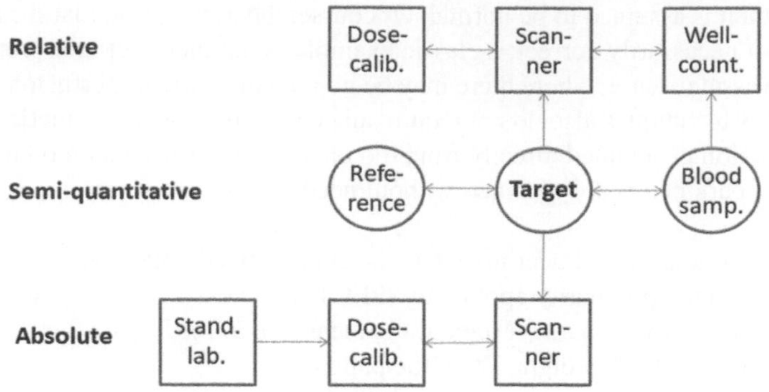

FIGURE 26.7 Diagram showing the various types of calibration required in different circumstances (see text for details). Note that semi-quantitative analysis does not require calibration.

There is wide acceptance of the quantitative capability of PET, whereas the feasibility of SPECT quantification has only recently been justified. There are several publications that demonstrate the potential accuracy of SPECT quantification in a range of phantom studies [74–76], and these illustrate that better than 5 per cent accuracy is typically achievable, though dependent on the radionuclide used.

Validation in clinical studies is much more challenging as the true values are usually unknown, with very few exceptions. Many clinical papers rely on demonstration of some secondary effect of quantification such as improvement in discrimination or improved survival. The usefulness of quantification in these studies is clear, but these usually have no means to prove that the values obtained were accurate. In theory, studies performed prior to biopsy or surgery can be used to compare in vivo measurement with tissue sample activity, but these studies are rare and are usually limited to preclinical studies where the issues facing quantification may not mirror those found in the patient population. A good example of in vivo validation is the use of microspheres in lung SPECT [74]. In this case 100 per cent of the administered dose is trapped in the lung, and so no assumption is needed regarding the in vivo distribution of activity. Other good examples involve the measurement of activity in blood or urine for direct comparison with activity concentration in heart or bladder respectively [77, 78]. The evidence is building to support the claim that SPECT can be considered a quantitative imaging modality, similar to PET [79].

26.9 APPENDIX: CROSS-CALIBRATION

Vendors of PET or SPECT systems often provide their own calibration procedures and phantoms. Below we describe an independent procedure.

In practice, a cross-calibration of a PET or SPECT scanner versus a dose-calibrator and well-counter would typically be performed using the following steps:

1. Draw up an aliquot to be used in a syringe and measure its activity (A_0) in the dose-calibrator.
2. Insert the aliquot into a cylindrical phantom of known volume (V_1), filled with room temperature water, and make sure it distributes uniformly in the phantom. Calculate the concentration, $C_1 = (A_0 - A_R)/V_1$, where A_R is the residual activity in the syringe.
3. Place the phantom in the PET or SPECT scanner and acquire data over a time t_a.
4. Reconstruct a tomographic image, including corrections for attenuation, scatter, randoms (in PET), and dead-time (if appropriate). Determine the mean image value (m_2) in a relatively large VOI in the centre of the phantom. Determine the calibration factor:

$$f_{S1} = \frac{C_1}{c_{d1} \cdot m_2 / t_a} \tag{26.11}$$

where c_{dl} is the decay correction factor.

5. Take a sample from the phantom, place it in a test tube and measure its activity (A_2) in the well-counter. The volume of the sample (V_2) can be determined either with a pipette or by weighing the test-tube before and after. Determine the calibration factor:

$$f_{S2} = \frac{C_1}{c_{d2} \cdot A_2 / V_2} \tag{26.12}$$

where c_{d2} is the decay correction factor.

The calibration factors may depend on the type of reconstruction algorithm used, (e.g. FBP, OS-EM or MAP-EM) and whether or not TOF information is utilized (PET only). On the other hand, any resolution modelling or post-smoothing applied should have no effect.

For the well-counter, the calibration factor may depend on the sample volume.

REFERENCES

[1] R. G. Manber, "PET Respiratory Motion Correction in Simultaneous PET/MR," PhD, University College London, 2016.

[2] D. L. Bailey, "Transmission scanning in emission tomography," *Eur J Nucl Med*, vol. 25, no. 7, pp. 774–787, Jul 1998, doi:10.1007/s002590050282.

[3] T. Beyer et al., "A combined PET/CT scanner for clinical oncology," *J Nucl Med*, vol. 41, no. 8, pp. 1369–1379, Aug 2000. [Online]. Available: www.ncbi.nlm.nih.gov/pubmed/10945530.

[4] S. C. Blankespoor et al., "Attenuation correction of SPECT using x-ray CT on an emission-transmission CT system: Myocardial perfusion assessment," *IEEE Trans Nucl Sci*, vol. 43, pp. 2263–2274, 1996.

[5] S. Brown, D. Bailey, K. Willowson, and C. Baldock, "Investigation of the relationship between linear attenuation coefficients and CT Hounsfield units using radionuclides for SPECT," *Appl Radiat Isot*, vol. 66, pp. 1206–1212, 2008.

[6] L. T. Chang, "A method for attenuation correction in radionuclide computed tomography," *IEEE Trans Nucl Sci*, vol. 25, pp. 638–643, 1978.

[7] A. Bockisch et al., "Positron emission tomography/computed tomography – imaging protocols, artifacts, and pitfalls," *Mol Imaging Biol*, vol. 6, pp. 188–199, 2004.

[8] T. M. Blodgett, A. S. Mehta, A. S. Mehta, C. M. Laymon, J. Carney, and D. W. Townsend, "PET/CT artifacts," (in English), *Clin Imaging*, vol. 35, no. 1, pp. 49–63, Jan–Feb 2011, doi:10.1016/j.clinimag.2010.03.001.

[9] S. Ahn, L. Cheng, and R. M. Manjeshwar, "Analysis of the effects of errors in attenuation maps on PET quantitation in TOF PET," in *2014 IEEE Nuclear Science Symposium and Medical Imaging Conference (NSS/MIC)*, pp. 1–4, 8–15 Nov. 2014, doi:10.1109/NSSMIC.2014.7430781.

[10] A. Mehranian and H. Zaidi, "Impact of time-of-flight PET on quantification errors in MR imaging-based attenuation correction," *J Nucl Med*, vol. 56, no. 4, pp. 635–641, Apr 2015, doi:10.2967/jnumed.114.148817.

[11] E. C. Emond et al., "Effect of attenuation mismatches in time of flight PET reconstruction," *Phys Med Biol*, vol. 65, no. 8, p. 085009, Apr. 20, 2020, doi:10.1088/1361-6560/ab7a6f.

[12] D. Izquierdo-Garcia and C. Catana, "MR Imaging-guided attenuation correction of PET data in PET/MR imaging," *PET Clinics*, vol. 11, no. 2, pp. 129–149, 2016, doi:10.1016/j.cpet.2015.10.002.

[13] C. N. Ladefoged et al., "A multi-centre evaluation of eleven clinically feasible brain PET/MRI attenuation correction techniques using a large cohort of patients," *Neuroimage*, vol. 147, pp. 346–359, Feb. 15, 2017, doi:10.1016/j.neuroimage.2016.12.010.

[14] T. Wanga et al., "Machine learning in quantitative PET: A review of attenuation correction and low-count image reconstruction methods," *Physica Medica*, vol. 76, pp. 294–306, 2020.

[15] J. Lillington et al., "PET/MRI attenuation estimation in the lung: A review of past, present, and potential techniques," *Med Phys* vol. 47, no. 2, pp. 790–811, 2020, doi:10.1002/mp.13943.

[16] A. Rezaei, C. M. Deroose, T. Vahle, F. Boada, and J. Nuyts, "Joint reconstruction of activity and attenuation in time-of-flight PET: A quantitative analysis," *J Nucl Med*, vol. 59, no. 10, pp. 1630–1635, Oct 2018, doi:10.2967/jnumed.117.204156.

[17] A. Rezaei, G. Schramm, S. M. A. Willekens, G. Delso, K. Van Laere, and J. Nuyts, "A quantitative evaluation of joint activity and attenuation reconstruction in TOF PET/MR brain imaging," *J Nucl Med*, vol. 60, no. 11, pp. 1649–1655, Nov 2019, doi:10.2967/jnumed.118.220871.

[18] S. C. Cade, S. Arridge, M. J. Evans, and B. F. Hutton, "Use of measured scatter data for the attenuation correction of single photon emission tomography without transmission scanning," *Med Phys*, vol. 40, no. 8, p. 082506, Aug 2013, doi:10.1118/1.4812686.

[19] L. Brusaferri et al., "Joint activity and attenuation reconstruction from multiple energy window data with Photopeak Scatter Re-Estimation in non-TOF 3D PET," *IEEE Trans Radia Plasma Med Sci*, p. 1, 2020, doi:10.1109/TRPMS.2020.2978449.

[20] Y. Berker and Y. Li, "Attenuation correction in emission tomography using the emission data – A review," *Med Phys*, vol. 43, no. 2, pp. 807–832, Feb 2016, doi:10.1118/1.4938264.

[21] L. Brusaferri, "Improving quantification in non-TOF 3D PET/MR by incorporating photon energy information," PhD, University College London, 2020.

[22] B. F. Hutton, I. Buvat, and F. J. Beekman, "Review and current status of SPECT scatter correction," *Phys Med Biol*, vol. 56, no. 14, pp. R85–112, Jul. 21, 2011, doi:10.1088/0031-9155/56/14/R01.

[23] K. Ogawa, T. Ichihara, and A. Kubo, "Accurate scatter correction in single photon emission CT," *Ann Nucl Med Sci*, vol. 7, pp. 145–150, 1994.

[24] F. J. Beekman, H. W. de Jong, and S. van Geloven, "Efficient fully 3-D iterative SPECT reconstruction with Monte Carlo-based scatter compensation," *IEEE Trans Med Imaging*, vol. 21, no. 8, pp. 867–877, Aug 2002, doi:10.1109/TMI.2002.803130.

[25] A. O. Sohlberg and M. T. Kajaste, "Fast Monte Carlo-simulator with full collimator and detector response modelling for SPECT," (in English), *Ann Nucl Med*, vol. 26, no. 1, pp. 92–98, Jan 2012, doi:10.1007/s12149-011-0550-7.

[26] Y. Du, B. M. Tsui, and E. C. Frey, "Model-based compensation for quantitative 123I brain SPECT imaging," *Phys Med Biol*, vol. 51, no. 5, pp. 1269–1282, Mar. 7, 2006, doi:10.1088/0031-9155/51/5/016.

[27] E. C. Frey and B. M. W. Tsui, "A new method for modeling the spatially-variant, object-dependent scatter response function in SPECT," in *1996 IEEE Nuclear Science Symposium. Conference Record*, vol. 2, pp. 1082–1086, Nov. 2–9, 1996, doi:10.1109/NSSMIC.1996.591559.

[28] C. C. Watson, D. Newport, and M. E. Casey, "A single scatter simulation technique for scatter correction in 3D PET," in *Three-Dimensional Image Reconstruction in Radiology and Nuclear Medicine*, P. Grangeat and J.-L. Amans (eds). Dordrecht: Springer Netherlands, 1996, pp. 255–268.

[29] J. M. Ollinger, "Model-based scatter correction for fully 3D PET," *Phys Med Biol*, vol. 41, no. 1, pp. 153–76, Jan 1996, doi:10.1088/0031-9155/41/1/012.

[30] C. C. Watson, J. Hu, and C. Zhou, "Double scatter simulation for more accurate image reconstruction in positron emission tomography," *IEEE Trans Radiat Plasma Med Sci*, p. 1, 2020, doi:10.1109/TRPMS.2020.2990335.

[31] K. Thielemans, R. M. Manjeshwar, C. Tsoumpas, and F. P. Jansen, "A new algorithm for scaling of PET scatter estimates using all coincidence events," in *2007 IEEE Nuclear Science Symposium Conference Record*, vol. 5, pp. 3586–3590, Oct. 26–Nov. 3, 2007, doi:10.1109/NSSMIC.2007.4436900.

[32] A. Rezaei et al., "Plane-dependent ML scatter scaling: 3D extension of the 2D simulated single scatter (SSS) estimate," *Phys Med Biol*, vol. 62, no. 16, pp. 6515–6531, Jul. 24, 2017, doi:10.1088/1361-6560/aa7a8c.

[33] M. Conti, I. Hong, and C. Michel, "Reconstruction of scattered and unscattered PET coincidences using TOF and energy information," *Phys Med Biol*, vol. 57, no. 15, pp. N307–17, Aug. 7, 2012, doi:10.1088/0031-9155/57/15/N307.

[34] M. Holstensson, K. Erlandsson, G. Poludniowski, S. Ben-Haim, and B. F. Hutton, "Model-based correction for scatter and tailing effects in simultaneous 99mTc and 123I imaging for a CdZnTe cardiac SPECT camera," *Phys Med Biol*, vol. 60, no. 8, pp. 3045–3063, Apr. 21, 2015, doi:10.1088/0031-9155/60/8/3045.

[35] E. J. Hoffman, S. C. Huang, and M. E. Phelps, "Quantitation in positron emission computed tomography: 1. Effect of object size," *J Comput Assist Tomogr*, vol. 3, no. 3, pp. 299–308, Jun 1979, doi:10.1097/00004728-197906000-00001.

[36] V. Bettinardi, I. Castiglioni, E. De Bernardi, and M. C. Gilardi, "PET quantification: strategies for partial volume correction," *Clin Transl Imaging*, vol. 2, no. 3, pp. 199–218, 2014/06/01 2014, doi:10.1007/s40336-014-0066-y.

[37] K. Erlandsson, I. Buvat, P. H. Pretorius, B. A. Thomas, and B. F. Hutton, "A review of partial volume correction techniques for emission tomography and their applications in neurology, cardiology and oncology," *Phys Med Biol*, vol. 57, no. 21, pp. R119–159, Nov. 7, 2012, doi:10.1088/0031-9155/57/21/R119.

[38] M. Soret, S. L. Bacharach, and I. Buvat, "Partial-volume effect in PET tumor imaging," *J Nucl Med*, vol. 48, no. 6, pp. 932–945, Jun 2007, doi:10.2967/jnumed.106.035774.

[39] B. A. Thomas et al., "PETPVC: A toolbox for performing partial volume correction techniques in positron emission tomography," *Phys Med Biol*, vol. 61, no. 22, pp. 7975–7993, Nov. 21, 2016, doi:10.1088/0031-9155/61/22/7975.

[40] J. A. Stamos, W. L. Rogers, N. H. Clinthorne, and K. F. Koral, "Object-dependent performance comparison of two iterative reconstruction algorithms," *IEEE Trans Nucl Sci*, vol. 35, no. 1, pp. 611–614, 1988.

[41] K. Erlandsson, B. Thomas, J. Dickson, and B. F. Hutton, "Evaluation of an OSEM-based PVC method for SPECT with clinical data," in *IEEE Nuclear Science Symposium & Medical Imaging Conference*, pp. 2686–2690, Oct. 30–Nov. 6, 2010, doi:10.1109/NSSMIC.2010.5874278.

[42] K. Erlandsson and B. F. Hutton, "Hyper-plane based partial volume correction for PET or SPECT," *Nucl Med Comm*, vol. 32, p. 427, 2011.

[43] J. Tohka and A. Reilhac, "Deconvolution-based partial volume correction in Raclopride-PET and Monte Carlo comparison to MR-based method," *Neuroimage*, vol. 39, no. 4, pp. 1570–1584, Feb. 15, 2008, doi:10.1016/j.neuroimage.2007.10.038.

[44] A. Rahmim, J. Qi, and V. Sossi, "Resolution modeling in PET imaging: theory, practice, benefits, and pitfalls," *Med Phys*, vol. 40, no. 6, p. 064301, Jun 2013, doi:10.1118/1.4800806.

[45] J. Nuyts, "Unconstrained image reconstruction with resolution modelling does not have a unique solution," *EJNMMI Physics*, vol. 1, no. 1, p. 98, 2014/11/30 2014, doi:10.1186/s40658-014-0098-4.

[46] B. Bai, Q. Li, and R. M. Leahy, "Magnetic resonance-guided positron emission tomography image reconstruction," *Semin Nucl Med*, vol. 43, no. 1, pp. 30–44, Jan 2013, doi:10.1053/j.semnuclmed.2012.08.006.
[47] H. W. Muller-Gartner et al., "Measurement of radiotracer concentration in brain gray matter using positron emission tomography: MRI-based correction for partial volume effects," *J Cereb Blood Flow Metab*, vol. 12, no. 4, pp. 571–583, Jul 1992, doi:10.1038/jcbfm.1992.81.
[48] J. Yang et al., "Investigation of partial volume correction methods for brain FDG PET studies," *IEEE Trans Nucl Sci*, vol. 43, no. 6, pp. 3322–3327, 1996.
[49] O. G. Rousset, Y. Ma, and A. C. Evans, "Correction for partial volume effects in PET: Principle and validation," *J Nucl Med*, vol. 39, no. 5, pp. 904–911, May 1998. [Online]. Available: www.ncbi.nlm.nih.gov/pubmed/9591599.
[50] B. A. Thomas et al., "The importance of appropriate partial volume correction for PET quantification in Alzheimer's disease," *Eur J Nucl Med Mol Imaging*, vol. 38, no. 6, pp. 1104–1119, Jun 2011, doi:10.1007/s00259-011-1745-9.
[51] P. J. Markiewicz et al., "NiftyPET: A high-throughput software platform for high quantitative accuracy and precision PET imaging and analysis," *Neuroinformatics*, vol. 16, no. 1, pp. 95–115, Jan 2018, doi:10.1007/s12021-017-9352-y.
[52] R. E. Carson, "Tracer kinetic modeling in PET," in *Positron Emission Tomography: Basic Science and Clinical Practice*, P. E. Valk, D. L. Bailey, D. W. Townsend, and M. N. Maisey (eds). London Springer-Verlag 2003, pp. 147–179.
[53] C. Tsoumpas and K. Thielemans, "Direct parametric reconstruction from dynamic projection data in emission tomography including prior estimation of the blood volume component," Nucl Med Comm, vol. 30, no. 7, 2009.
[54] T. Lambrou et al., "The importance of correction for tissue fraction effects in lung PET: Preliminary findings," *Eur J Nucl Med Mol Imaging*, vol. 38, no. 12, pp. 2238–2246, Dec 2011, doi:10.1007/s00259-011-1906-x.
[55] W. van Elmpt, J. Hamill, J. Jones, D. De Ruysscher, P. Lambin, and M. Ollers, "Optimal gating compared to 3D and 4D PET reconstruction for characterization of lung tumours," *Eur J Nucl Med Mol Imaging*, vol. 38, no. 5, pp. 843–855, May 2011, doi:10.1007/s00259-010-1716-6.
[56] C. Liu et al., "Quiescent period respiratory gating for PET/CT," *Med Phys*, vol. 37, no. 9, pp. 5037–5043, Sep 2010, doi:10.1118/1.3480508.
[57] V. Bettinardi, E. De Bernardi, L. Presotto, and M. C. Gilardi, "Motion-tracking hardware and advanced applications in PET and PET/CT," *PET Clin*, vol. 8, no. 1, pp. 11–28, Jan 2013, doi:10.1016/j.cpet.2012.09.008.
[58] A. L. Kesner, P. J. Schleyer, F. Buther, M. A. Walter, K. P. Schafers, and P. J. Koo, "On transcending the impasse of respiratory motion correction applications in routine clinical imaging – a consideration of a fully automated data driven motion control framework," *EJNMMI Phys*, vol. 1, no. 1, p. 8, Dec 2014, doi:10.1186/2197-7364-1-8.
[59] K. Thielemans, S. Rathore, F. Engbrant, and P. Razifar, "Device-less gating for PET/CT using PCA," in *2011 IEEE Nuclear Science Symposium Conference Record*, Oct. 23–29, 2011, pp. 3904–3910, doi:10.1109/NSSMIC.2011.6153742.
[60] O. Bertolli, S. Arridge, S. D. Wollenweber, C. W. Stearns, B. F. Hutton, and K. Thielemans, "Sign determination methods for the respiratory signal in data-driven PET gating," *Phys Med Biol*, vol. 62, no. 8, pp. 3204–3220, Apr. 21, 2017, doi:10.1088/1361-6560/aa6052.
[61] P. J. Schleyer, M. J. O'Doherty, S. F. Barrington, and P. K. Marsden, "Retrospective data-driven respiratory gating for PET/CT," *Phys Med Biol*, vol. 54, no. 7, pp. 1935–1950, Apr. 7, 2009, doi:10.1088/0031-9155/54/7/005.
[62] S. Ren et al., "Data-driven event-by-event respiratory motion correction using TOF PET list-mode centroid of distribution," *Phys Med Biol*, vol. 62, no. 12, pp. 4741–4755, 2017/05/18 2017, doi:10.1088/1361-6560/aa700c.
[63] F. P. Oliveira and J. M. Tavares, "Medical image registration: a review," *Comput Methods Biomech Biomed Engin*, vol. 17, no. 2, pp. 73–93, 2014, doi:10.1080/10255842.2012.670855.
[64] A. Gillman, J. Smith, P. Thomas, S. Rose, and N. Dowson, "PET motion correction in context of integrated PET/MR: Current techniques, limitations, and future projections," *Med Phys*, vol. 44, no. 12, pp. e430–e445, Dec 2017, doi:10.1002/mp.12577.
[65] A. Bousse et al., "Maximum-likelihood joint image reconstruction/motion estimation in attenuation-corrected respiratory gated PET/CT using a single attenuation map," *IEEE Trans Med Imaging*, vol. 35, no. 1, pp. 217–228, Jan 2016, doi:10.1109/TMI.2015.2464156.
[66] A. J. Montgomery, K. Thielemans, M. A. Mehta, F. Turkheimer, S. Mustafovic, and P. M. Grasby, "Correction of head movement on PET studies: comparison of methods," *J Nucl Med*, vol. 47, no. 12, pp. 1936–1944, Dec 2006. [Online]. Available: www.ncbi.nlm.nih.gov/pubmed/17138736.
[67] Y. Picard and C. J. Thompson, "Motion correction of PET images using multiple acquisition frames," *IEEE Trans Med Imaging*, vol. 16, no. 2, pp. 137–144, Apr 1997, doi:10.1109/42.563659.
[68] I. Polycarpou, C. Tsoumpas, and P. K. Marsden, "Analysis and comparison of two methods for motion correction in PET imaging," *Med Phys*, vol. 39, no. 10, pp. 6474–6483, Oct 2012, doi:10.1118/1.4754586.
[69] C. Tsoumpas et al., "The effect of regularization in motion compensated PET image reconstruction: a realistic numerical 4D simulation study," *Phys Med Biol*, vol. 58, no. 6, pp. 1759–1773, Mar. 21, 2013, doi:10.1088/0031-9155/58/6/1759.
[70] C. Munoz, C. Kolbitsch, A. J. Reader, P. Marsden, T. Schaeffter, and C. Prieto, "MR-based cardiac and respiratory motion-compensation techniques for PET-MR imaging," *PET Clin*, vol. 11, no. 2, pp. 179–191, Apr 2016, doi:10.1016/j.cpet.2015.09.004.

[71] A. Pépin, J. Daouk, P. Bailly, S. Hapdey, and M. E. Meyer, "Management of respiratory motion in PET/computed tomography: the state of the art," (in English), *Nucl Med Commun*, vol. 35, no. 2, pp. 113–122, Feb 2014, doi:10.1097/mnm.0000000000000048.
[72] S. R. Bowen et al., "Challenges and opportunities in patient-specific, motion-managed and PET/CT-guided radiation therapy of lung cancer: Review and perspective," *Clin Transl Med*, vol. 1, no. 1, p. 18, Aug. 31, 2012, doi:10.1186/2001-1326-1-18.
[73] W. P. Segars, M. Mahesh, T. J. Beck, E. C. Frey, and B. M. Tsui, "Realistic CT simulation using the 4D XCAT phantom," *Med Phys*, vol. 35, no. 8, pp. 3800–3808, Aug 2008, doi:10.1118/1.2955743.
[74] K. Willowson, D. L. Bailey, and C. Baldock, "Quantitative SPECT reconstruction using CT-derived corrections," *Phys Med Biol*, vol. 53, no. 12, pp. 3099–3112, Jun. 21, 2008, doi:10.1088/0031-9155/53/12/002.
[75] S. Shcherbinin, A. Celler, T. Belhocine, R. Vanderwerf, and A. Driedger, "Accuracy of quantitative reconstructions in SPECT/CT imaging," *Phys Med Biol*, vol. 53, no. 17, pp. 4595–4604, Sep. 7, 2008, doi:10.1088/0031-9155/53/17/009.
[76] Y. K. Dewaraja et al., "MIRD Pamphlet No. 23: Quantitative SPECT for patient-specific 3-dimensional dosimetry in internal radionuclide therapy," (in English), *J Nucl Med*, vol. 53, no. 8, pp. 1310–1325, Aug 2012, doi:10.2967/jnumed.111.100123.
[77] K. Willowson, D. L. Bailey, E. A. Bailey, C. Baldock, and P. J. Roach, "In vivo validation of quantitative SPECT in the heart," *Clin Physiol Funct Imaging*, vol. 30, no. 3, pp. 214–219, May 2010, doi:10.1111/j.1475-097X.2010.00930.x.
[78] J. Zeintl, A. H. Vija, A. Yahil, J. Hornegger, and T. Kuwert, "Quantitative accuracy of clinical 99mTc SPECT/CT using ordered-subset expectation maximization with 3-dimensional resolution recovery, attenuation, and scatter correction," *J Nucl Med*, vol. 51, no. 6, pp. 921–928, Jun 2010, doi:10.2967/jnumed.109.071571.
[79] D. L. Bailey and K. P. Willowson, "An evidence-based review of quantitative SPECT imaging and potential clinical applications," *J Nucl Med*, vol. 54, no. 1, pp. 83–89, Jan 2013, doi:10.2967/jnumed.112.111476.

27 Multicentre Studies

Hardware and Software Requirements

Terez Sera, Ronald Boellaard, Andres Kaalep, and Michael Ljungberg

CONTENTS

27.1 INTRODUCTION

The implementation of quality-assurance programs in nuclear medicine (NM) services is essential for the achievement of reliable qualitative and quantitative patient data. Quality assurance designates all those actions that are necessary to provide adequate confidence that a system will perform satisfactorily. There are recommendations in form of guidelines and standards elaborated by professional societies and different organizations (e.g.: EANM, SNM, IAEA, IEC, NEMA) based upon parameters defining quality and good clinical practice, which help NM departments to elaborate their own quality-control protocols. To ensure that the nuclear medicine instrument (Activity meter, Gamma Camera, SPECT/CT, PET/CT, dose meter, etc.) is in a proper condition for patient investigations, quality-control (QC) tests should be performed on a daily, monthly, quarterly or annual basis (see Chapter 23). The daily tests and some of the less frequent ones are usually performed by the NM staff, while the quarterly tests could be the responsibility of service which is providing the preventive maintenance. Similar to the QC programs, the patient study acquisition and processing protocols are implemented by the individual NM departments on the basis of established clinical practice in the field and taking into account the guidelines elaborated by professional organizations.

DOI: 10.1201/9780429489556-27

27.1.1 Purpose of Multicentre Comparison

On top of local quality-assurance programs, the participation in interlaboratory comparisons can add a level of reassurance of quality in addition to being a useful prerequisite for clinical multicentre trials. The aim of an interlaboratory comparison study is to evaluate the 'state of the art' in the individual laboratories, compare the performances, improve the individual practices upon what is feasible and achievable, and promote more uniform patient investigation protocols among the NM practices. The participation in the study commonly is on a voluntary basis, centres are provided with phantoms, standard operation protocols for scanning the phantoms, and processing the images, as well as standardized report forms. The test results should be sent to the core lab of the organizing body following a predefined protocol. After evaluations of the data, reports are provided to the participants, showing the individual performances of their equipment in comparison with those of the group, enabling them to identify aspects of their own practice: the fit of the technical parameters into the acceptance range defined for the specific study, the relation between the reported scan and the structure of the phantom; Respectively, the bias in the values of the quantitative parameters.

The success of an interlaboratory comparison study depends on many parameters, the most important are discussed below.

27.1.2 Organizing the Study

27.1.2.1 Financing

The cost of such a study is considerably high. The main organizers of the studies are international organizations [1, 2] or national bodies [3], often collaborating with radiopharmaceutical companies or nuclear medicine equipment-vendors.

Selection of the study: It depends on the interest of the organizers in a specific field of NM (thyroid, liver, cardiac, brain, bone, oncological, and so forth, investigations). The introduction of a new radiopharmaceutical or a new imaging technique could also attract attention in a multicentre setting.

- **Selection and procurement of the quality-control devices.** The phantoms are selected to fit the studies to be investigated. A hardware phantom for a brain protocol, for example, could be a multiple performance SPECT/CT striatal (Figure 27.1b), for the whole-body PET/CT a NEMA whole-body phantom (Figure 27.1c). Software

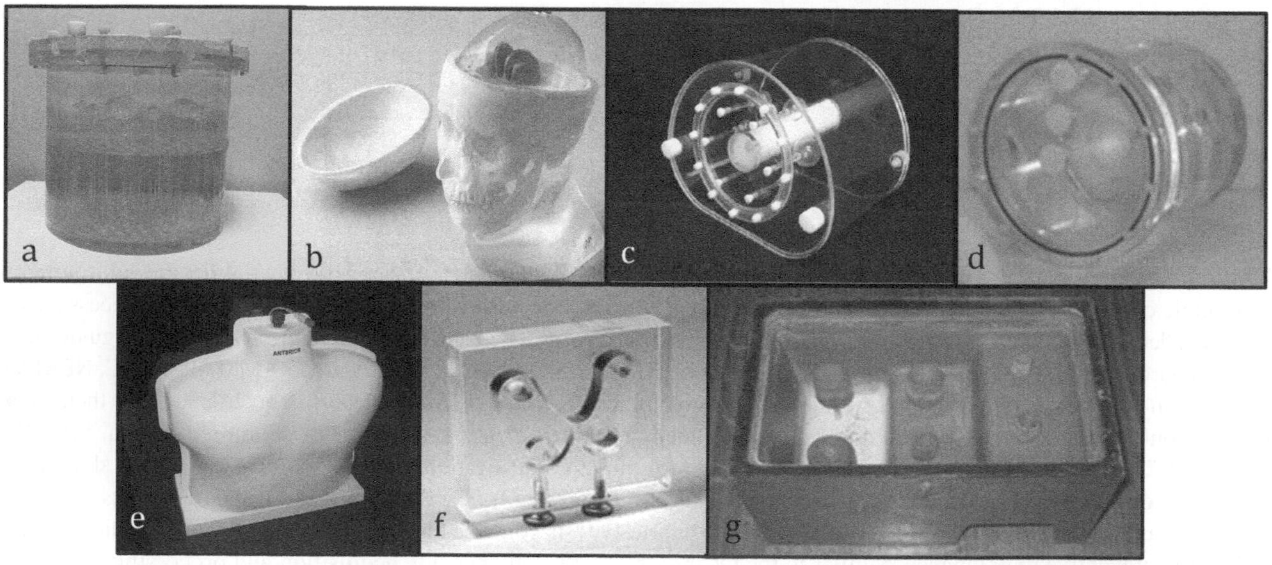

FIGURE 27.1 (a) ECT phantom, (b) striatal phantom, (c) NEMA NU-2007 phantom, (d) homogeneous phantom, (e) CTN phantom, (f) thyroid phantom, and (g) liver phantom.

phantoms are also in use. The phantoms in certain multicentre studies are procured by the core labs and provided to the centres, and in other cases participants may have/procure their own test devices.

- **Selection of participants**. The study must be advertised among the selected NM centres, whose final participation is on a voluntary basis.
- **Organizing the transport of the test device between the sites.** Each site could be responsible for shipping the phantom to the next centre; alternatively, one dedicated person could be chosen who keeps contact with the participants and organizes the mailing, or the core lab relies on a specialist who is traveling with the phantom. The advantage of the latter is that the interoperator variability is avoided.

27.1.2.2 Executing the Study

- **Elaboration of operating standard protocols for imaging/processing the data and reporting.** The protocol needs to be fitted to the parameters of the devices/software to be checked and the possibilities of the phantoms. It must be detailed and precise, defining the acquisition and processing techniques. The reporting form must be simple, and the format of the images (e.g. colour scale and its normalization) has to be standardized in advance.
- **Performing the measurements.** One option is to ship the phantom to the participating centres and ask them to execute the protocol, including the handling of the phantom, the acquisition and processing of the images, the completion of the reporting form, recording the data and sending it to the organizers. Alternatively, an appointed specialist would visit each site and help the centres perform the tests.
- **Collection and evaluation of the data.** Organizers have to create a core lab that will collect and evaluate the data coming from the participants in the study applying a common analysis for all data.
- **Informing the participants about their own performance and how it compares with the other centres**. Quick feedback of the results is vital to maintain the interest of the participants. Confidence is also an important feature. All participants receive all data, but only the personal results can be identified – the others remain anonymous. It is also important that the core lab can be contacted and provide help to identify any causes of lower performances and establish corrective actions.

The reports should be published to be accessible for the members of the NM community to propagate quality in the field.

27.1.2.3 Critical Points and Recommendations

- The study type has to be chosen to be of interest for a relatively large NM community in order to increase its value, and it is advantageous if radiopharmaceutical producing companies and/or equipment vendors are also involved.
- The phantoms must be sturdy enough to survive transportation and handling.
- The radioisotopes to be used in the tests must be ordered well in advance, and scanner time has to be secured, causing the least possible inconveniences as regards the clinical patient investigations.
- The responsible person from the participating centres has to be designated in advance in order to be involved in the local management of the study and to plan his/her time accordingly. A preferable alternative, particularly in cases of complex phantom protocols, involves the appointment of a member of the core lab to assist each site, ensuring that the study protocol is followed and data managed in a standard way.
- The scanners have to be prepared in advance. It is desirable that prior to the agreed date for the phantom studies, all relevant quality-control parameters (e.g.: intrinsic and extrinsic flood field uniformity, energy spectrum, COR, linearity, geometrical and energy resolution, sensitivity, SUV calibration) are within acceptable ranges; if they are not, then time would be needed to take remedial action and possibly re-calibration. The auxiliary equipment required for the tests should also be considered (dose calibrator, well counter, analytical scales, containers of different sizes, syringes, etc.) and -procured in time.
- The study protocols need to be clearly communicated to the participants prior the study, and all relevant documentation provided to enable them to check in advance whether the acquisition and processing parameters used on the scanners are achievable or already routine. It might happen that some parameters, such as energy window, pixel size, body contour, and so forth, required for the interlaboratory study, are different from those regularly used in clinical practice. This would then require additional setting and preparation prior to the study commencing. Taking such steps can avoid late data acquisition or failure to complete the study.

- The processing, archiving, and data transfer to the core lab is also important. The processing parameters, the file format, the archiving method, accessibility of the workstation, and so forth have to be communicated, and sites have to check them in advance.
- The core lab needs promptly to provide feedback of the individual performances: in case of low achievements the causes should be investigated and, besides the technical parameters of the scanners, attention should be paid to the quality of the interpretations/findings provided by the observers.

27.2 EXAMPLES OF MULTICENTRE STUDIES

Among the first nuclear medicine multicentre studies was a large experiment organized by IAEA involving 300 laboratories using a thyroid phantom to evaluate the quality of radioiodine uptake [4]. Only 5 per cent of the participants reported results within 5 per cent bias.

In 1971, the American College of Pathologists formed a committee with the aim to plan interlaboratory studies in the field of Nuclear Medicine [5]. Various programs were organized to assess the radioimmunoassay techniques, and check the calibration of activity meters, and other kind of studies were also initiated in which three-dimensional emission organ phantoms, dynamic organ phantoms, and transmission software phantoms were used.

From the beginning of the 1980s, the World Health Organisation (WHO) has regularly organized interlaboratory studies in Europe. In the period 1981–1983 four phantoms had been circulated, two transmission, one simulating brain and the other liver, and two emission liver phantoms, respectively [1, 6]. The transmission brain phantom was designed to comprise 13 'hot' lesions located at 8.3 cm under the surface of the device. The target background contrast ratio was constant, having a value of 1.7, the lesions diameters were in the interval of 4–20 mm. The transmission liver phantom contained 8 'cold' lesions, positioned at 3.9 cm deep. The lesions were having the same diameter of 1.6 cm, the contrast ratio lesion/background varied from 0.91 to 0.77. The two transmission liver phantoms contained two 1.0 cm, two 1.5 cm, one 2.0 cm and one 2.5 cm diameter cold lesions positioned at different depths; 12 European countries participated in the multicentre study. The size of the lesion built into the brain phantom was systematically interpreted as being larger. In the transmission liver phantom, the low contrast lesions were underestimated, the high contrast lesions overestimated. Half of the participants missed the small lesions of the liver phantom, but the larger spheres (2.0 cm, 2.5 cm diameter) were identified by most of the observers.

Between 1984 and 1986, WHO and IAEA organized multicentre studies using transmission thyroid and liver phantoms. The thyroid phantom contained 8 lesions positioned at 3.8 cm under the surface of the phantom, with the contrast ratios between 1.7 and 1.5, and diameters between 7 and 16 mm. In the liver phantom there were 10 cold lesions, with 2 cm diameter at the depths of 5 cm, with the contrast values between 0.61 and 0.94, respectively. In the multicentre study 16 European countries were involved (9). The core lab interpreting the country reports did not identify systematic errors in the case of the thyroid phantom. However, with the liver phantom, the 'Receiver Operating Characteristic' (ROC) curves [7–9] showed large differences. The areas under the curves varied between 0.77 and 0.95 (average 0.85+/ - 0.01). Better results were reported on new gamma cameras than on the old systems.

In one of the following studies, organized between 1987 and 1990, a transmission heart phantom was circulated among the participants [10]. The phantom simulated normal left ventricular wall motion and 50.7 per cent ejection fraction (EF). There were participants from 11 European countries. The core lab separately evaluated the imaging parameters and the semi-quantitative data. Regarding the acquisition parameters (type of collimator, image matrix size, number of frames per cycle), there were large variations among the sites. In some cases, errors related to the applied gating techniques could be identified, probably caused by noise of the detector-analysis system, or by erroneous interpretation of the ECG signals. Differences in the EF values were related to different ROI delineation techniques.

In 1994, the WHO, in collaboration with the IAEA, organized a multicentre study, in which a transmission bone phantom was used (Figure 27.2). The phantom simulated the thoracolumbar spine from posterior-anterior (PA) view, containing 23 'hot' lesions. 7 were positioned on the vertebra, 3–3 on the left ribs, medial and distal portions, 5–5 on the right ribs, medial and distal portions, respectively. One cold and one inter-vertebral lesion was also built in.

This study was not evaluated by a core lab; instead, results were published in country reports [11–14]. In 22 Hungarian nuclear medicine departments 31 bone phantom acquisitions were performed on gamma cameras and 7 on whole-body scanners. The participants were asked to report on the hot and cold lesions of the phantom images and to rank the evidence of the lesion on a scale of 1–4. The phantom lesion contrast values are included in Table 27.1. The observers' performance was characterized by the value of the area under the ROC curve. There were results just

TABLE 27.1
WHO/IAEA Transmission Bone Phantom Lesion Contrast Values

	Left ribs			Right ribs	
	lateral	medial	Thoracic spine	medial	lateral
5	T1	T6	1.80	T11	1.93
6	T2	1.65	T17	2.04	1.48
7	1.41	T7	P	1.32	T14
8	T3	T8	P	1.52	T15
9	T4	1.18	T18	1.25	1.76
10	2.18	T9	1.32	T12	2.3
11	1.6	1.86	T19	1.72	T16
12	T5	T10	1.7	T13	1.10
		Lumbar spine			
		1	0.58		
		2	T20		
		3	1.9		
		4	T21		
		5	T22		

Notes: T1…T22 = normal regions
P = intervertebral disc

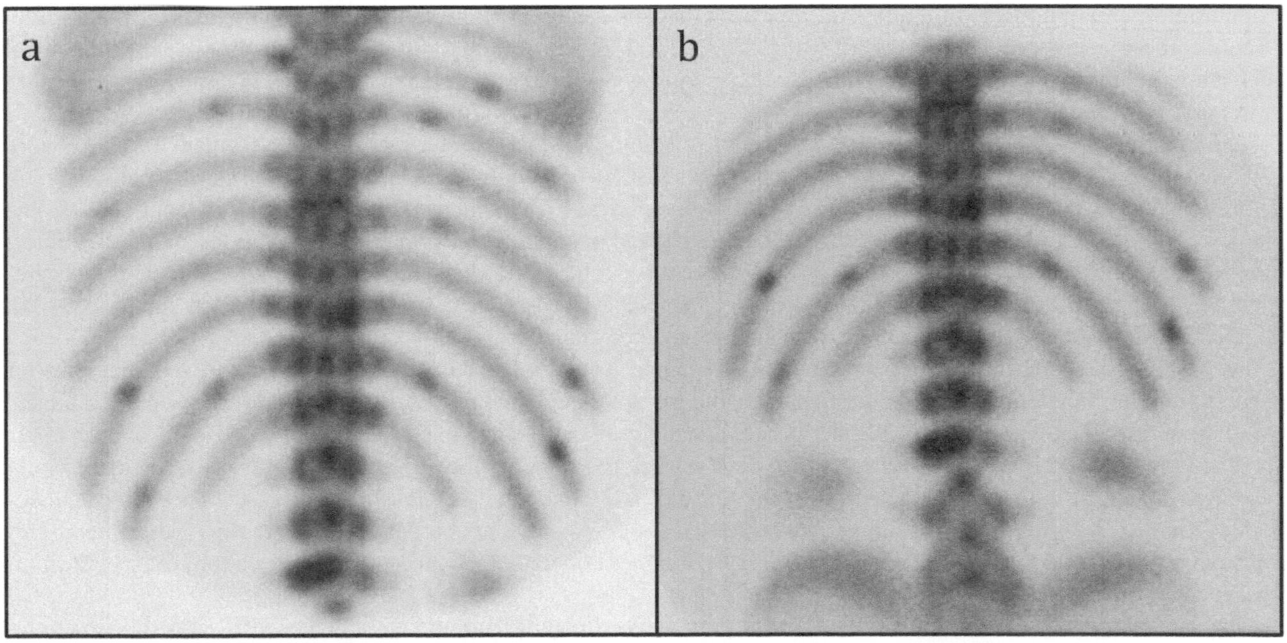

FIGURE 27.2 WHO/IAEA transmission bone phantom.

being above 0.5 (meaning decision by chance), but some very close to the maximum possible of 1. The calibration of the gamma cameras as well as the proper training of the observers had a great impact on the lesion detectability scores. By performing regular quality control and adequate preventive service maintenance of the nuclear medicine equipment, in most cases it can be avoided that the laboratory issue erroneous investigation results due to technical reasons.

27.2.1 Multicentre Studies and Standardization Programs

27.2.1.1 Multicentre Studies and Standardization Programs in ^{18}F-FDG PET/CT Investigations

18F-FDG PET/CT has been proven to be a sensitive imaging modality for detection, staging and restaging and therapy response assessment in oncology [15]. FDG PET/CT provides essential information for radiation treatment planning, helping with critical decisions when delineating tumour volumes [16, 17]. Due to the widespread use of PET/CT systems in recent years the number of multicentre trials (such as EANM/EARL, ABLE, RAPID, FALCON and many others) has increased, and quantitative performance is more frequently investigated [18–20].

PET is a quantitative imaging technique used to image biomarkers. In the modern world of evidence-based medicine a biomarker, by definition, is a characteristic that is objectively measured and evaluated as an indicator of normal biologic processes, pathogenic processes, or pharmacologic responses to a therapeutic intervention [21] where the repeatability and reproducibility of data is extremely important. The same results should be achieved not only when a patient is examined repeatedly on the same scanner, but also when the exam is performed on another scanner from a different vendor or in another institution.

Dynamic acquisition (with arterial blood sampling) is considered to be the 'gold standard' in quantitative PET imaging, when data are acquired in the area of interest during multiple time frames over the course of the accumulation of the radiopharmaceutical, providing us with information on tracer kinetics. This method however is limited due to extensive time required for the acquisition and interpretation. In clinical practice a simpler alternative is now widely used, a semi-quantitative metric, the standardized uptake value (SUV).

SUV characterizes the activity of a suspect lesion by normalizing its measured activity concentration to body habitus – typically injected activity per kg of patient weight, allowing an estimate of the tracer kinetics, without being invasive. SUV is also applied in 18F-FDG PET/CT studies and is known to be affected by multiple factors – being either technical, biological or physical [22] – which can introduce significant biases into the data.

Rapid advancement and different levels of technology in various imaging centres creates further problems to standardization as the options offered by modern technologies, such as PSF or ToF, provide increased benefit over the older technologies, creating more difficulties in comparing the data acquired on different generation systems [23].

Since there are so many factors affecting the PET/CT quantification procedures, evaluation of the individual performances in multicentre settings with the aim of finding a common feasible practice helps in improving the quality in the field.

The Society of Nuclear Medicine and Molecular Imaging (SNMMI) has been running the Clinical Trials Network (CTN) since 2008 to ensure baseline common quality-control metrics for participating PET scanners used in multicentre studies. Scanner validation requires the use of program-specific CTN phantom (Figure 27.1e) – an anthropomorphic phantom mimicking a human torso to provide qualitative and quantitative information in conditions similar to clinical imaging settings. An acceptance criterion of ±10 per cent is set for the background region SUV_{mean}, but precise sphere-specific acceptance criteria for SUV_{max} and SUV_{mean} for the various sphere sizes are currently not strictly set [24]. Scanner revalidation, including phantom scanning, is repeated annually.

From 2003, the American College of Radiology Imaging Network (ACRIN) PET core laboratory qualifies sites for participation in multicentre trials. To do this, each site is requested to submit data from one uniform cylinder scan and two patient test cases. The uniform cylinder images are analysed in the core lab by calculating the SUV_{mean} of each transverse slice while, in order to pass, the average SUV is required to be within 1.0 ± 0.1. Patient scans are qualitatively evaluated for artefacts, noise, patient positioning and PET/ CT image alignment. In addition, the core lab calculates SUVs from the ROIs comprising the livers of submitted patients scans and compares them with the reported data. This is aimed at evaluating the level of expertise of PET/CT camera operators. No acceptance criteria for size-specific contrast recovery coefficients is set [25].

The Radiological Society of North America Quantitative Imaging Biomarkers Alliance (RSNA-QIBA) provides guidelines and instructions on how to measure SUV calibration as well as uniformity between slices of a uniform cylindrical phantom. 'acceptable' and 'ideal' performance measurement methods and criteria are described for uniformity, resolution/SUV recovery and image noise. Criteria of ACRIN/EANM/SNM are frequently referred to. The results of the quality-control measurements have to be submitted annually and should be available for any site audit [25].

The National Cancer Research Institute (NCRI) PET Clinical Trials Network provides more a specific accreditation on a per trial basis, where the core lab reviews the trial protocol in the setup phase and determines specific requirements for the participating system's quality assurance. This results in a uniform distribution of scanners within a single trial, but requires multiple levels of standardization, which creates difficulties in intercomparing.

The European Association of Nuclear Medicine (EANM) published the guideline 'FDG PET and PET/CT: EANM procedure guidelines for tumour PET imaging: version 1.0' in 2010 [26], which was followed by an updated version,

2.0, in 2014 [27]. As a support for the clinical application of the European Association of Nuclear Medicine (EANM) guideline, the EANM-European Association Research Ltd. (EARL), an initiative created under the umbrella of EANM, provided an accreditation program based on regular independent review of phantom scans. The accreditation program started in 2010 and is currently still running. The basic concept was to create a network of scanners harmonized among each other to produce SUV recovery coefficients within a predefined range that accommodates most scanners in clinical use in the field of 18F-FDG PET/CT oncological studies. Participants in this network are eligible for multicentre studies, but in their local clinical work they also benefit from the independent quality-control survey. Centres are requested to send scans of a homogeneously filled cylindrical phantom (27.1d) and a NEMA NU-2007 body phantom (Figure 27.1c) to the EARL core lab, which centrally evaluates the phantom data. The uniform cylinder needs to be scanned quarterly, and data are accepted if the mean SUV is within the range of 1.0 ± 0.1. The NEMA body phantom is requested annually, and images are evaluated for contrast recovery of the six hot spheres of varying diameters (10–37 mm) built into the phantom. To achieve and then maintain the EANM/EARL accredited status, the imaging sites have to comply with harmonization (i.e. accreditation) specifications for contrast recoveries (RC) of the NEMA phantom spheres. The standard graphs of the RCs for SUV_{avg} and SUV_{max} are shown on Figure 27.3. Currently, there are 200 harmonized scanners in the EARL network [28, 29].

27.2.1.2 Multicentre Studies and Standardization Programs in DaTSCAN SPECT

Multicentre trials involving nuclear medicine imaging have been performed for many years, however, using Single Photon Emission Computed Tomography (SPECT) scanners little data have so far been reported. In 2006, a group of imaging centres from throughout Europe came together under the auspices of EARL to form a brain imaging network of excellence. The first task of this group was to produce a normal control database of ^{123}I-FP-CIT SPECT studies in a multicentre setting [30]. For this project the group used the acronym ENCDAT (European Normal Control Database of DaTSCAN). ^{123}I-FP-CIT (DaTSCAN) is a radiotracer for the dopamine transporter and has been proven to be a useful tool for neurologists, movement disorder specialists, and geriatricians in the diagnosis of Parkinson's disease, Lewy Body Dementia, and for differentiating such patients from those without loss of dopamine transporters [31–34].

The first objective when setting up this initiative was to ensure that the imaging systems produce Inter-comparable data. Criteria were set for systems before they were involved in the project, related to quality-control parameters (uniformity, centre of rotation, SPECT uniformity, SPECT spatial resolution, multiparameter phantom characteristic data) and to imaging and processing protocols [35].

Twenty-two cameras on 15 sites from 10 countries and 8 different manufacturers were evaluated for inclusion in the ENCDAT study. All scanners matched the required level of specification, so they could be connected into a network. To prepare the network for the DaTSCAN studies of the normal controls, ^{123}I SPECT images of an anthropomorphic striatal phantom (Figure 27.1b) were acquired. In order to avoid the interoperator variability the measurements were performed by one appointed physicist. The phantom was filled with nominal ratios of 10:1, 8:1, 5:1, 4:1, 2.5:1, 2:1

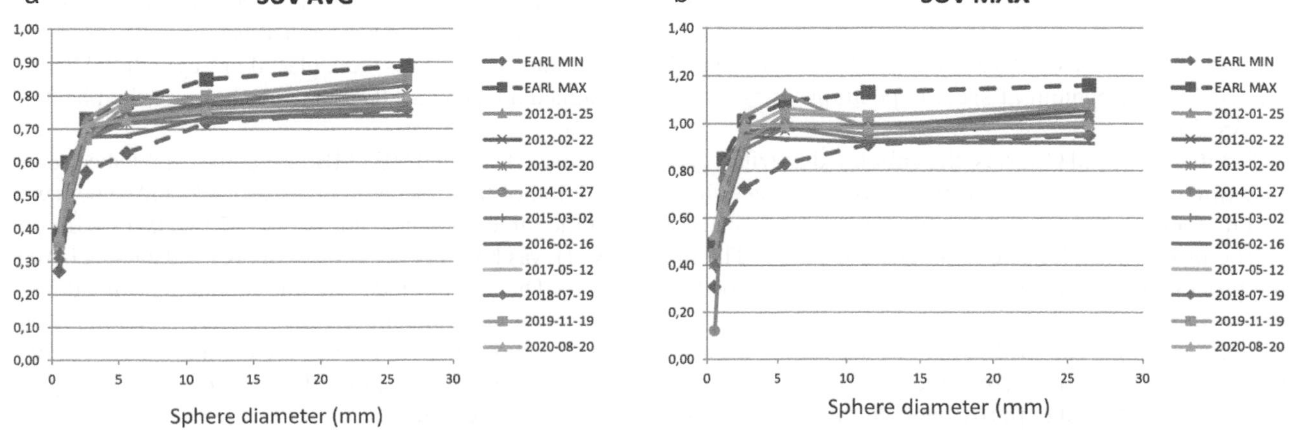

FIGURE 27.3 EARL RC standards for SUV average (a) and SUV max (b) (dotted lines). Continuous lines belong to accredited scanners.

and 1:1. The absolute ratios were measured from aliquots using well counters at the participating centres. All phantom images were acquired over 360 degrees, mostly with pixel sizes between 2–3 mm. Besides the photopeak energy window (159±10%), projections were also acquired in the lower (138±3.5%) and upper (184±3%) scatter windows to evaluate the importance of scatter and septal penetration correction (SSPC) with the triple energy window method. All images were sent to the ENCDAT study core lab for the reconstruction and analysis. OSEM (10/10 or 12/8 iterations/subsets for sets of 120 and 128 projections respectively) was performed with attenuation correction only and additional SSPC. For the quantification, the specific binding ratio (SBR) [36, 37] was measured. The aliquots provided the benchmark against which the SBR measurements were assessed. Linear regression was applied to derive calibration curves, whose slope gave the recovery coefficient (RC) for each camera.

The relationship between measured and actual ratios was linear across all SPECT systems. Variability was observed between cameras of different manufactures and, to a lesser extent, between cameras of the same type. When measured with attenuation correction only, the SBR were found to underestimate systematically the actual values for all cameras, with recovery coefficients ranging from 70 per cent to 90 per cent. The SSP correction had a significant impact on the accuracy of the measurements, bringing the RC closer to the ideal value of 100 per cent (range 92%–105%). There were generated cross-calibration factors to standardize quantitative outcome measures of DAT density obtained with different SPECT systems in a large database of healthy controls.

The findings of the study confirm that quantitative outcome measures of ^{123}I phantom images depend upon the different characteristics of the SPECT systems. The linear relationship between the actual and the measured specific binding ratio across a wide range of values would allow for standardization of different SPECT systems. The corrections for scatter and septal penetration are necessary requisite for accurate quantification and essential for the harmonization in multicentre setting.

The calibration phantom data were further analysed and works dealing with the impact of reconstruction and quantification methods on the diagnostic capability of a normal database [38–40] were published. Besides other outcomes, multicentre studies are providing lots of data for research purposes. From the ENCDAT study, an additional few papers dealing with different aspects of DaTSCAN investigation were published [41–48].

27.2.2 The International Atomic Energy Agency (IAEA) Barium Intercomparison Project

The IAEA initiated a coordinated research project titled, 'Development of Quantitative Nuclear Medicine Imaging for Patient Specific Dosimetry' to investigate the accuracy of different sites in performing quantitative imaging to determine activities in vivo. A comparison of image-quantification processes at the international level was performed for activity quantification based on a previous study that included calibrator measurements coordinated by the IAEA. The countries that participated in this study were Bangladesh, Brazil, Croatia, Cuba, Germany, South Africa, Thailand, the United States, and Uruguay. Owing to the various camera systems available at the time, the project aimed to investigate the possibility of a standardization procedure for quantitative imaging. Each site was required to have a gamma camera that can perform both planar imaging and SPECT; furthermore, they must have correction methods for attenuation and scatter (using the triple-energy-window method [49]).

The study was based on imaging ^{133}Ba (356 keV, 62%) because it mimicked ^{131}I well and afforded a long half-life of 10.5 years, thereby enabling a distribution of well-calibrated sources to the participating sites. Four cylindrical sealed ^{133}Ba sources (denoted A–D with diameters of 0.8, 1.3, 1.4, and 2.9 cm, respectively) were calibrated by the National Institute of Standards and Technology (NIST), Gaithersburg, Maryland. The sources were placed in a cylindrical homogeneous water phantom at specific positions using specially designed attachments that fitted the holes of the phantom (Figure 27.4).

For the A, B, and C series, the calibrated activity concentration was (0.199 ±0.003) MBq g^{-1} at the established reference time. For safety and transport reasons, the activity concentration of the D series sources was selected to be lower by a factor of four; hence, the calibrated activity concentration at the same reference time was (0.050 ±0.001) MBq g^{-1}.

The project was designed for separate studies. The first study (I) was based on local procedures for data acquisition and processing for the quantitative imaging developed by each site. The second study (II) was conducted using directives from the task group of the project pertaining to the acquisition and imaging of protocols, camera calibration, and the reconstruction and evaluation of the studies. Not all sites were equipped to perform SPECT/CT imaging; therefore, attenuation corrections were made via transmission estimations using other methods. In addition to local processing, all data were transferred to participants in the task group for independent processing.

This coordinated project by the IAEA indicated the necessity of staff training in addition to standardized methods and protocols to obtain accurate quantitative activities from the imaging of radionuclides.

More details and results obtained from this study (Table 27.2) have been reported by Zimmermann and colleagues [50].

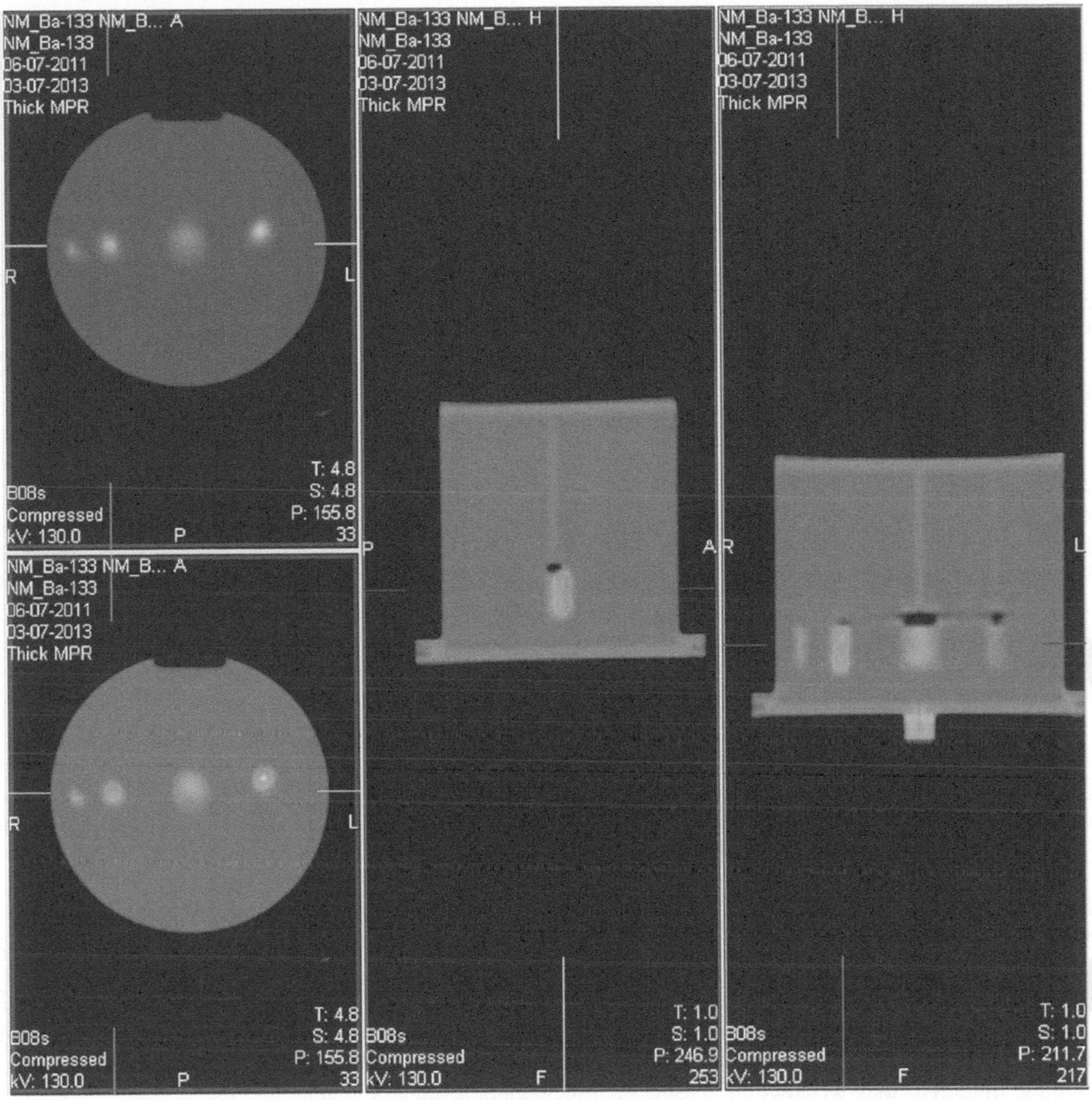

FIGURE 27.4 The phantom and the ^{133}Ba source configuration used in the IAEA intercomparison. (The image has been published from reference [50] with permission.)

TABLE 27.2
Mean and Median Values of the Ratio, *R*, of the Participants' Reported Activities to the NIST-calibrated Activity for All of the ^{133}Ba Sources as Measured by Each Technique for Each Trial.

	Study I		Study II		CDA (II)	
Method	**Mean**	**Median**	**Mean**	**Median**	**Mean**	**Median**
Planar Conjugate View	0.84(12)	0.86(7)	0.91(14)	0.88(8)	1.06(8)	1.06(6)
AC					1.09(4)	1.05(10)
SPECT Chang-AC	1.09(21)	1.10(10)	1.18(39)	0.98(10)	0.94(6)	0.94(4)
CTAC	1.08(13)	1.12(6)	1.06(8)	1.05(5)	1.00(8)	1.02(6)

Note: The Uncertainties Correspond to the Standard Deviations on the Mean Values or the Median Absolute Deviations, in the Case of the Medians.
Source: Reproduced with Permission from Reference [50].

27.2.3 Equalis National External Quality Assessment Programme in Sweden

Since 2012, a quality assessment programme has been established in Sweden and is available to Swedish hospitals. It is based on the distribution of digital images created via a Monte Carlo simulation of realistic anthropomorphic phantoms. The studies involved were coordinated by the company Equalis AB, a not-for-profit company owned by the Swedish Association of Local Authorities and Regions, the Swedish Institute of Biomedical Laboratory Science, and the Swedish Society of Medicine. The company provides support and applications for external quality for various areas, such as laboratory medicine, imaging and functional medicine, and analyses. The activities were focused on coordinating users' meetings, developing standardized terminologies, and performing activities to harmonize and improve results in the Swedish healthcare system. A number of advisory groups have been assigned to establish quality programs and arrange user meetings for discussing the associated questions and topics. Other activities aimed at harmonization and improvements may also be relevant, such as the development of recommendations. An advisory group comprising both medical physicists and nuclear medicine physicians who focused on nuclear medicine applications has been formed. Previously, this group has focused on the distribution of physical phantoms to participating clinics to audit their ability in quantifying activities in various compartments of the phantom. The possibility of using mathematically simulated studies has been posed in determining renography methods among Swedish clinical departments.

The simulations for the national study were conducted using the SIMIND Monte-Carlo program [51, 52] together with the XCAT anthropomorphic computer phantom developed by Dr. Paul Segars at Duke University in North Carolina [53]. Voxel-based images of the phantom were created from XCAT software that generated a voxel phantom with the desired properties from a file of multiple parameters that can be defined by the user. In these studies, the phantom was initially created as a code-based voxel phantom, meaning that each voxel was designed with a specific value that defined the volume-of-interest belonging to the voxel (liver, lung, spleen, etc.). Using this method, the SIMIND program can read the code-based voxel phantom images and then internally create a set of density and activity images using a table that connects the codes to the phantom organs and structures.

One of the main goals of these studies was to create images that matched the clinical evaluation procedure as closely as possible. Therefore, the participating clinical departments were initially questioned regarding their camera and acquisition-specific parameters and the administered activities such that each department can evaluate the data in a manner that closely resembled their daily protocol. The parameters included the energy resolution, energy window setting, collimator data, administered activity, and time-frame sequence.

After finalizing the simulation of the image sequences from the clinic-specific camera parameters and protocol settings, the files of the images were converted to Dicom files. To avoid potential problems with internal private Dicom tags, a template of the Dicom file that matched the camera for each vendor was used. Subsequently, the images and other related parameters in the template were exchanged to correspond to the simulated data using the program dcmodify, which is available for free from the Dicom Toolkit webpage (https://www.dcmtk.org). Subsequently, the Dicom files were exported to the local clinical workstation; this was accomplished without any major problems. Next, the local staff processed the images as they were acquired using their local gamma-camera system.

27.2.3.1 Example 1: Renal Scintigraphy

A series of national QC studies were performed to evaluate clinical renal scintigraphy studies. Because this study involved the imaging of a rapid redistribution of activity in the body, it was extremely difficult to create a physical phantom that mimicked blood flow through the kidneys into the bladder in a realistic manner. An alternative method is to create a pharmacokinetic model that mimics ^{99m}Tc-MAG3 and then simulates dynamic imaging. For this, the pharmacokinetic model developed by Brolin and colleagues [54] was referred to. Because the model maintained the total activity, the Poisson noise level added after the simulations in certain regions of the image was correct. Figure 27.5 shows simulated renography projections of six time points for a camera with perfect spatial resolution and a camera with realistic attenuation and scatter effects as well as collimator resolution with Poisson noise added.

Three cases were simulated. The first was regarded as having normal functionality, with the percentage uptake of 50/50 between two kidneys. The other two studies involved left/right relative uptakes of 33.3/66.7 and 55/45. Results for the DRF evaluation are shown in Figure 27.6.

An interesting finding in the study was that a bias occurred in the differential renal function for the left kidney; this occurred because the male who served as a template for creating the computer phantom did not have both kidneys at the same distance from the surface of the body (difference of approximately 12 mm). This resulted in a bias of

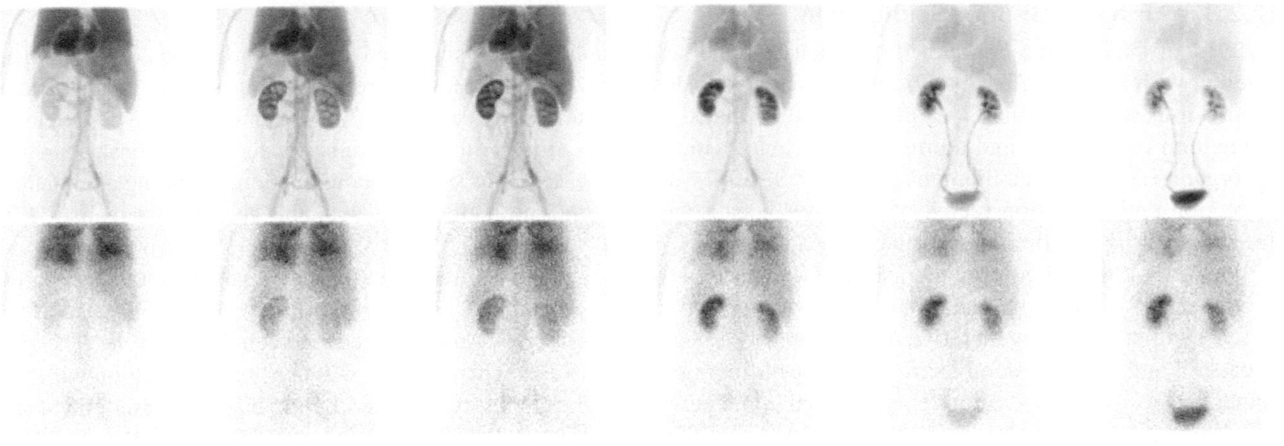

FIGURE 27.5 The top row shows images representing six time points for a scintillation camera simulated with very good spatial resolution. The lower row shows the same six time points for a camera with a realistic spatial resolution, attenuation and scatter effect, and noise level.

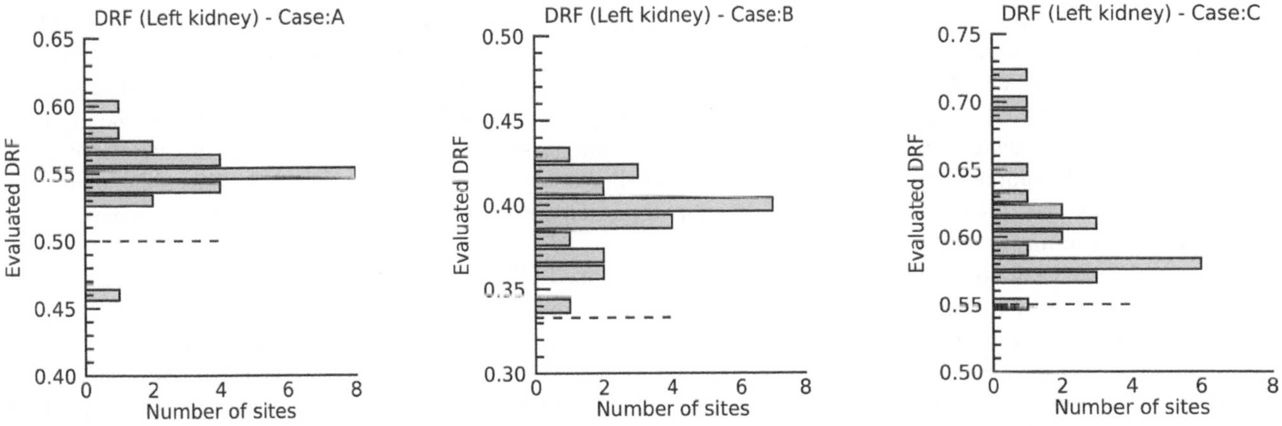

FIGURE 27.6 The graphs show the results for the three cases with the differential renal function (DRF) of the left kidney for the cases of (A) 50 per cent/50 per cent, (B) 33 per cent/67 per cent and (C) 55 per cent/45 per cent). The DRF's have been calculated by either the slope method or the integral method. The dashed line represents the true value. Data obtained from Equalis AB.

approximately 5 per cent, which can be confirmed by simulating the same geometry without photon interactions in the phantom. For this case, no bias was discovered since no attenuation was included. However, this would undoubtedly occur in real-patient studies. The study has been described in more detail by Brolin and colleagues [55].

27.2.3.2 Example 2: Myocardial Perfusion

In 2012, a similar study was conducted with the overall aim of investigating the calculated values of ejection fraction (EF), end-diastole volume (EDV), and end-systole volume (ESV) obtained from the participating departments. The simulated cases comprised images simulated with the XCAT phantom, in which the EF, EDV, and ESV were predefined. The study involved three phantoms representing patients with hearts that were small (case 1), large (case 2), and medium-sized (case 3).

In this study, however, it was difficult to tailor the simulation to the local conditions. Some departments included CT-based attenuation and different types of scatter correction methods that could not simulate each CT model and link the Dicom files of the CT to simulated SPECT images. Furthermore, some department could not participate because they used dedicated CZT SPECT systems or special types of collimators that could not be simulated. The study has been described in more detail by Trägårdh and colleagues [56].

27.2.3.3 Example 3: Bone Scintigraphy

In 2014, a survey was conducted to evaluate bone scintigraphy. Three simulated planar whole-body scintigraphs, together with two SPECT simulations, were created to detect and outline metastasis-related changes in the images. As in the previous studies, planar WB images were simulated in this study using each clinic's own camera system parameters, acquisition parameters, and administered activity. However, the SPECT images were simulated and reconstructed by the organizers based on a Siemens Symbia™ SPECT system and ML-EM/OS-EM reconstruction, including attenuation correction and resolution recovery. This was because at that time, no method was available to simulate realistic CT images and validate their appropriate usage.

For cases 1 and 3, multiple metastases were inserted (seven for case 1; three for case 3). For case 1, some metastases were clearly visible, whereas others were smaller and therefore exhibited less contrast (Figure 27.7). For case 2, an increased uptake in a degenerative manner was observed in the vertebral column located between the vertebrae. The aim of cases 4 and 5 was to demonstrate the ease of finding the metastases when SPECT was used in combination with the planar images. In case 4, an uptake occurred in the pelvis, which was partly obscured by the bladder. Case 5 involved two removals in the lumbar spine that appeared significantly better on the SPECT than on the planar images. In 2017, a new survey regarding bone scintigraphy was initiated, and 10 simulated skeletal scintigraphs were included to detect and outline increased activity uptake.

27.2.3.4 Experiences from Different Surveys

This paragraph is based on a conversation with Eva Örndahl, PhD, the scheme coordinator for the Nuclear Medicine advisory group at Equalis AB. Based on the discussion, only a few trial rounds that included measurements of physical phantoms have been performed. In 2011, three reference radiation sources were dispatched to measure and compare well counters and recently acquired images of a physical heart phantom. The dispatch of the reference sources was a major logistical challenge because the radioactive sources would first be sent for measurements to one group of participants and then transferred to a new set of participants. The production of physical phantom images required the participant's camera parameters; hence, two physicists travelled around the country to acquire images that captured all different parameter sets. Owing to time limitations, not all participants were able to use images that adapted exactly to their own camera settings.

The sample material distributed to the participants comprised clinical cases obtained from anonymized patient materials and were evaluated differently (measurements, written answers, etc.).

The digital phantom distributions required the collection of the participants' camera parameters; however, no travel was required in this case as the images can be acquired on site. As opposed to using clinical materials, access to a known value for comparison and the fact that the value is not dependent on the subjective value of any evaluator is highly advantageous. The comments that the advisory group from Equalis received from the participants regarding the clinical realism of the simulated images were predominantly positive. No participant reported that the images were unrealistic, but several participants indicated that (a) the images did not appear in the manner with which they were used in their image processing system; (b) the images were flipped or stacked in the wrong direction; or (c) they encountered problems in distinguishing between different cases (which had been loaded on top of each other as one patient). However, as indicated by Dr Örndahl, these problems can occur even when using clinical materials because of potentially different Dicom interpretations at different clinical workstations. In the trial round involving 10 skeletal cases where the task was to mark the location of the metastases, some participants were disturbed by the fact that all skeleton images were from the same phantom. Hence, in the future, the studies should be based on different phantoms such that one can learn the skeletal peculiarities. This can be achieved owing to the increased number of patient models comprising different genders, ages, sizes, and body-mass indexes developed by Dr Segars that are available to users of XCAT software [57].

A technical difficulty was encountered in the myocardial scintigraphy survey, where 7 among the 26 participants had to stand over the round because their CZT cameras could not be simulated at that time. Furthermore, the participants mentioned the lack of patient history that was required to complement their reading. In addition, they did not have any access to CT images. It is noteworthy that, as generally observed in real CT images, even if the images can theoretically be created from the XCAT phantom, the local variation in grey scales may be difficult to obtain using synthetic images compared with using XCAT images. This is because the initial density is the same for all voxels inside a specific organ.

Finally, Dr Örndahl indicated that, at the time the surveys were conducted, appropriate and thoughtful follow-ups could not be performed regarding the effect of the digital phantom on the future work of clinicians. However, based on a simple survey, most users believed that the Equalis program comprising digital phantoms contributed positively to their work quality.

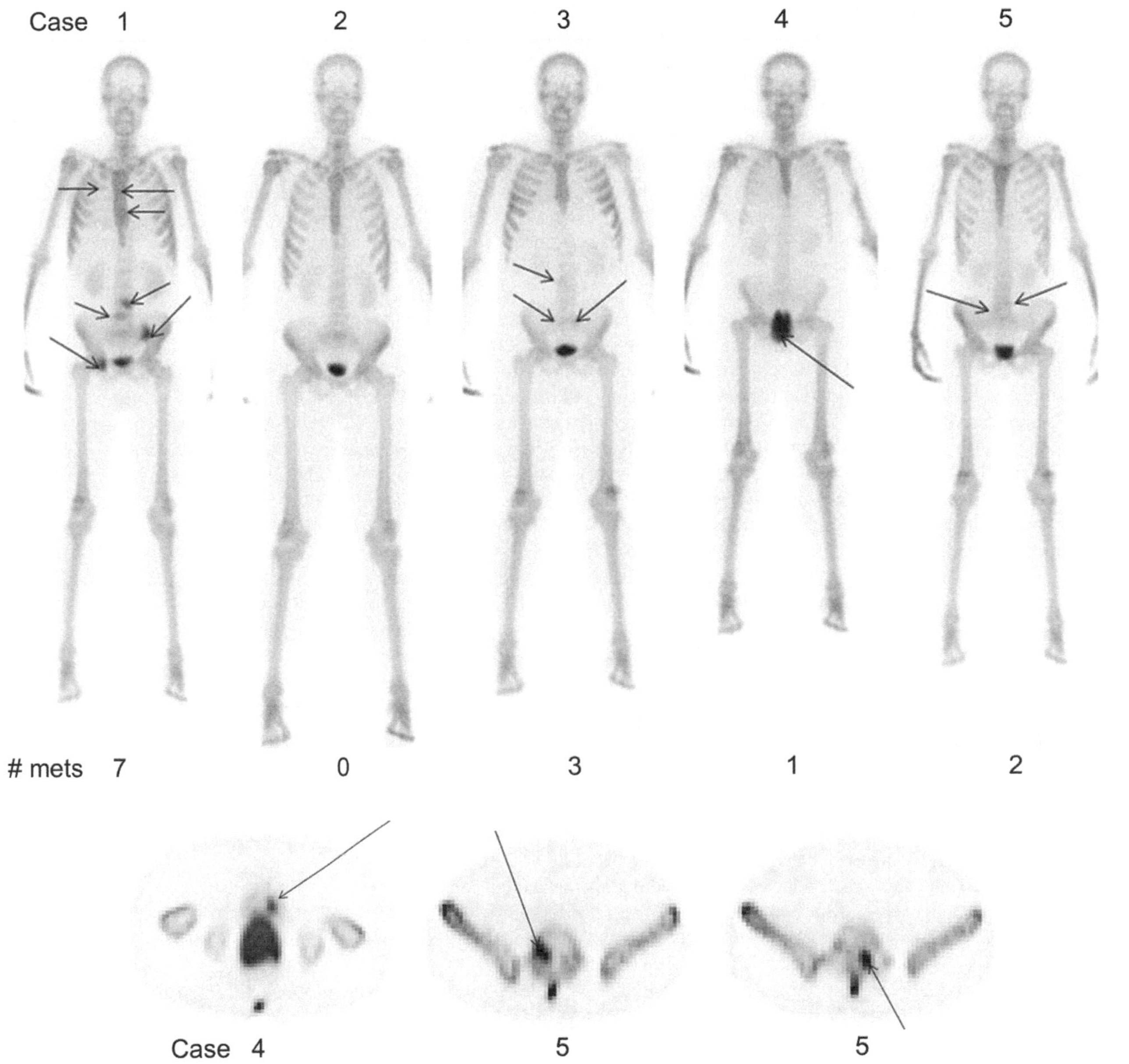

FIGURE 27.7 The image shows the five phantoms for the 2014-year bone-scan survey together with three reconstructed images. (Published with permission from Equalis AB.)

REFERENCES

[1] V. Volodin et al., "World Health Organisation inter-laboratory comparison study in 12 countries on quality performance of nuclear medicine imaging devices," (in English), *Eur J Nucl Med*, vol. 10, no. 5–6, pp. 193–197, 1985, doi:10.1007/bf00254460.

[2] G. N. Souchkevitch et al., "The World Health Organization and International Atomic Energy Agency second interlaboratory comparison study in 16 countries on quality performance of nuclear medicine imaging devices," *Eur J Nucl Med,* vol. 13, no. 10, pp. 495–501, 1988/01/01 1988, doi:10.1007/BF00256623.

[3] J. Heikkinen, J. T. Kuikka, A. Ahonen, and P. Rautio, "Quality of brain perfusion single-photon emission tomography images: multicentre evaluation using an anatomically accurate three-dimensional phantom," (in English), *Eur J Nucl Med*, vol. 25, no. 10, pp. 1415–1422, Oct 1998, doi:10.1007/s002590050317.

[4] IAEA, "Newsletter on Quality Assurance Programmes in Nuclear Medicine," IAEA, Vienna, 1987.

[5] N. Herrera. *Quality Assurance Programmes in Nuclear Medicine*. International Atomic Energy Agency. Vienna, 1987.

[6] E. Busemann Sokole, H. Bergmann, N. Herrera, R. F. Mould, and V. Volodin, "Interlaboratory comparison study of nuclear medicine imaging in 12 European countries using total performance phantoms," in *Nuclear Medicine in Research and Practice*, H. Schmidt and D. Vauramo (Eds). Stuttgart & New York: Schattauer Verlag, 1984, pp. 102–105.

[7] G. N. Souchkevitch et al., "The World Health Organization and International Atomic Energy Agency second interlaboratory comparison study in 16 countries on quality performance of nuclear medicine imaging devices," (in English), *Eur J Nucl Med*, vol. 13, no. 10, pp. 495–501, 1988, doi:10.1007/bf00256623.

[8] C. E. Metz, "Basic principles of ROC analysis," (in English), *Semin Nucl Med*, vol. 8, no. 4, pp. 283–298, Oct 1978.

[9] J. A. Hanley and B. J. McNeil, "The meaning and use of the area under a receiver operating characteristic (ROC) curve," (in English), *Radiology*, vol. 143, no. 1, pp. 29–36, Apr 1982, doi:10.1148/radiology.143.1.7063747.

[10] E. Busemann-Sokole and T. D. Cradduck, "The use of phantoms for quality control in gated cardiac studies," *J Nucl Med Technology*, vol. 13, no. 1, pp. 5–10, March 1, 1985. [Online]. Available: http://tech.snmjournals.org/content/13/1/5.abstract.

[11] G. Hart, "A UK survey of nuclear medicine imaging performance using the TransBone anthropomorphic phantom," (in English), *Nucl Med Commun,* vol. 18, no. 7, pp. 668–672, Jul 1997, doi:10.1097/00006231-199707000-00012.

[12] T. Sera, S. Medgyes, and G. Sebestyen, "Interlaboratory comparison study with WHO/IAEA transmission bone phantom in Hungary, I," *Magyar Radiologia (Journal of the Society of Hungarian Radiologists)*, vol. 68, pp. 164–167, 1994.

[13] T. Sera, S. Medgyes, and G. Sebestyen, "Interlaboratory comparison study with WHO/IAEA transmission bone phantom in Hungary, III," *Magyar Radiologia (Journal of the Society of Hungarian Radiologists)*, vol. 70, pp. 42–45 1994.

[14] T. Sera, S. Medgyes, and G. Sebestyen, "Interlaboratory comparison study with WHO/IAEA transmission bone phantom in Hungary, II," *Magyar Radiologia (Journal of the Society of Hungarian Radiologists)*, vol. 70, pp. 14–19, 1996.

[15] J. W. Fletcher et al., "Recommendations on the use of 18F-FDG PET in oncology," (in English), *J Nucl Med*, vol. 49, no. 3, pp. 480–508, Mar 2008, doi:10.2967/jnumed.107.047787.

[16] V. Gregoire and A. Chiti, "PET in radiotherapy planning: particularly exquisite test or pending and experimental tool?" (in English), *Radiother Oncol*, vol. 96, no. 3, pp. 275–276, Sep 2010, doi:10.1016/j.radonc.2010.07.015.

[17] D. Thorwarth et al., "Integration of FDG-PET/CT into external beam radiation therapy planning: Technical aspects and recommendations on methodological approaches," (in English), *Nuklearmedizin*, vol. 51, no. 4, pp. 140–153, 2012, doi:10.3413/Nukmed-0455-11-12.

[18] L. Geworski, B. O. Knoop, M. de Wit, V. Ivancevic, R. Bares, and D. L. Munz, "Multicenter comparison of calibration and cross calibration of PET scanners," (in English), *J Nucl Med*, vol. 43, no. 5, pp. 635–639, May 2002.

[19] R. K. Doot, B. F. Kurland, P. E. Kinahan, and D. A. Mankoff, "Design considerations for using PET as a response measure in single site and multicenter clinical trials," (in English), *Acad Radiol*, vol. 19, no. 2, pp. 184–190, Feb 2012, doi:10.1016/j.acra.2011.10.008.

[20] T. E. Yankeelov et al., "Quantitative imaging in cancer clinical trials," (in English), *Clin Cancer Res*, vol. 22, no. 2, pp. 284–290, Jan. 15, 2016, doi:10.1158/1078-0432.Ccr-14-3336.

[21] "Biomarkers and surrogate endpoints: Preferred definitions and conceptual framework," (in English), *Clin Pharmacol Ther*, vol. 69, no. 3, pp. 89–95, Mar 2001, doi:10.1067/mcp.2001.113989.

[22] R. Boellaard, "Standards for PET image acquisition and quantitative data analysis," (in English), *J Nucl Med*, vol. 50 Suppl 1, pp. 11s–20s, May 2009, doi:10.2967/jnumed.108.057182.

[23] A. Kaalep et al., "Feasibility of state of the art PET/CT systems performance harmonisation," (in English), *Eur J Nucl Med Mol Imaging*, vol. 45, no. 8, pp. 1344–1361, Jul 2018, doi:10.1007/s00259-018-3977-4.

[24] J. J. Sunderland and P. E. Christian, "Quantitative PET/CT scanner performance characterization based upon the society of nuclear medicine and molecular imaging clinical trials network oncology clinical simulator phantom," (in English), *J Nucl Med*, vol. 56, no. 1, pp. 145–152, Jan 2015, doi:10.2967/jnumed.114.148056.

[25] J. S. Scheuermann, J. R. Saffer, J. S. Karp, A. M. Levering, and B. A. Siegel, "Qualification of PET scanners for use in multicenter cancer clinical trials: The American College of Radiology Imaging Network experience," (in English), *J Nucl Med*, vol. 50, no. 7, pp. 1187–1193, Jul 2009, doi:10.2967/jnumed.108.057455.

[26] R. Boellaard et al., "FDG PET and PET/CT: EANM procedure guidelines for tumour PET imaging: Version 1.0," (in English), *Eur J Nucl Med Mol Imaging*, vol. 37, no. 1, pp. 181–200, Jan 2010, doi:10.1007/s00259-009-1297-4.

[27] R. Boellaard et al., "FDG PET/CT: EANM procedure guidelines for tumour imaging: Version 2.0," (in English), *Eur J Nucl Med Mol Imaging*, vol. 42, no. 2, pp. 328–354, 2015, doi:10.1007/s00259-014-2961-x.

[28] A. Kaalep et al., "EANM/EARL FDG-PET/CT accreditation – summary results from the first 200 accredited imaging systems," (in English), *Eur J Nucl Med Mol Imaging*, vol. 45, no. 3, pp. 412–422, Mar 2018, doi:10.1007/s00259-017-3853-7.

[29] N. Aide, C. Lasnon, P. Veit-Haibach, T. Sera, B. Sattler, and R. Boellaard, "EANM/EARL harmonization strategies in PET quantification: from daily practice to multicentre oncological studies," (in English), *Eur J Nucl Med Mol Imaging*, vol. 44, no. Suppl 1, pp. 17–31, Aug 2017, doi:10.1007/s00259-017-3740-2.

[30] A. Varrone et al., "European multicentre database of healthy controls for [123I]FP-CIT SPECT (ENC-DAT): Age-related effects, gender differences and evaluation of different methods of analysis," *Eur J Nucl Med Mol Imaging*, vol. 40, no. 2, pp. 213–227, January 1, 2013, doi:10.1007/s00259-012-2276-8.

[31] H. T. S. Benamer et al., "Accurate differentiation of parkinsonism and essential tremor using visual assessment of [(123) I]-FP-CIT SPECT imaging: The [(123) I]-FP-CIT study group," (in English), *Mov Disord*, vol. 15, no. 3, pp. 503–510, May 2000, doi:10.1002/1531-8257(200005)15:3<503::Aid-mds1013>3.0.Co;2-v.

[32] J. Booij et al., "[123I]FP-CIT SPECT shows a pronounced decline of striatal dopamine transporter labelling in early and advanced Parkinson's disease," (in English), *J Neurol Neurosurg Psychiatry*, vol. 62, no. 2, pp. 133–140, Feb 1997, doi:10.1136/jnnp.62.2.133.

[33] I. McKeith et al., "Sensitivity and specificity of dopamine transporter imaging with 123I-FP-CIT SPECT in dementia with Lewy bodies: A phase III, multicentre study," (in English), *Lancet Neurol*, vol. 6, no. 4, pp. 305–313, Apr 2007, doi:10.1016/s1474-4422(07)70057-1.

[34] S. A. Schneider et al., "Patients with adult-onset dystonic tremor resembling parkinsonian tremor have scans without evidence of dopaminergic deficit (SWEDDs)," (in English), *Mov Disord*, vol. 22, no. 15, pp. 2210–2215, Nov. 15, 2007, doi:10.1002/mds.21685.

[35] J. C. Dickson et al., "Proposal for the standardisation of multi-centre trials in nuclear medicine imaging: prerequisites for a European 123I-FP-CIT SPECT database," *Eur J Nucl Med Mol Imaging*, vol. 39, no. 1, pp. 188–197, January 1, 2012, doi:10.1007/s00259-011-1884-z.

[36] L. Tossici-Bolt et al., "Calibration of gamma camera systems for a multicentre European (1)(2)(3)I-FP-CIT SPECT normal database," (in English), *Eur J Nucl Med Mol Imaging*, vol. 38, no. 8, pp. 1529–1540, Aug 2011, doi:10.1007/s00259-011-1801-5.

[37] L. Tossici-Bolt, S. M. Hoffmann, P. M. Kemp, R. L. Mehta, and J. S. Fleming, "Quantification of [123I]FP-CIT SPECT brain images: an accurate technique for measurement of the specific binding ratio," (in English), *Eur J Nucl Med Mol Imaging*, vol. 33, no. 12, pp. 1491–1499, Dec 2006, doi:10.1007/s00259-006-0155-x.

[38] J. C. Dickson et al., "The impact of reconstruction and scanner characterisation on the diagnostic capability of a normal database for [123I]FP-CIT SPECT imaging," *EJNMMI Research*, vol. 7, no. 1, p. 10, 2017/01/24 2017, doi:10.1186/s13550-016-0253-0.

[39] J. C. Dickson et al., "The impact of reconstruction method on the quantification of DaTSCAN images," (in English), *Eur J Nucl Med Mol Imaging*, vol. 37, no. 1, pp. 23–35, Jan 2010, doi:10.1007/s00259-009-1212-z.

[40] L. Tossici-Bolt et al., "[123I]FP-CIT ENC-DAT normal database: the impact of the reconstruction and quantification methods," *EJNMMI Physics*, vol. 4, no. 1, p. 8, 2017/01/28 2017, doi:10.1186/s40658-017-0175-6.

[41] N. L. Albert et al., "Implementation of the European multicentre database of healthy controls for [123I]FP-CIT SPECT increases diagnostic accuracy in patients with clinically uncertain parkinsonian syndromes," *Eur J Nucl Med Mol Imaging*, vol. 43, no. 7, pp. 1315–1322, July 1, 2016, doi:10.1007/s00259-015-3304-2.

[42] S. Hesse et al., "Association of central serotonin transporter availability and body mass index in healthy Europeans," (in English), *Eur Neuropsychopharmacol*, vol. 24, no. 8, pp. 1240–1247, Aug 2014, doi:10.1016/j.euroneuro.2014.05.005.

[43] W. Koch et al., "Extrastriatal binding of [123I]FP-CIT in the thalamus and pons: Gender and age dependencies assessed in a European multicentre database of healthy controls," *Eur J Nucl Med Mol Imaging*, vol. 41, no. 10, pp. 1938–1946, October 1, 2014, doi:10.1007/s00259-014-2785-8.

[44] F. Nobili et al., "Automatic semi-quantification of [123I]FP-CIT SPECT scans in healthy volunteers using BasGan version 2: Results from the ENC-DAT database," (in English), *Eur J Nucl Med Mol Imaging*, vol. 40, no. 4, pp. 565–573, Apr 2013, doi:10.1007/s00259-012-2304-8.

[45] T. A. Soderlund et al., "Value of semiquantitative analysis for clinical reporting of 123I-2-beta-carbomethoxy-3beta-(4-iodophenyl)-N-(3-fluoropropyl)nortropane SPECT studies," *J Nucl Med*, vol. 54, no. 5, pp. 714–722, May 2013, doi:10.2967/jnumed.112.110106.

[46] G. Thomsen et al., "No difference in striatal dopamine transporter availability between active smokers, ex-smokers and non-smokers using [123I]FP-CIT (DaTSCAN) and SPECT," *EJNMMI Research*, vol. 3, no. 1, p. 39, 2013/05/20 2013, doi:10.1186/2191-219X-3-39.

[47] E. van de Giessen et al., "No association between striatal dopamine transporter binding and body mass index: A multi-center European study in healthy volunteers," (in English), *Neuroimage*, vol. 64, pp. 61–67, Jan. 1, 2013, doi:10.1016/j.neuroimage.2012.09.011.

[48] R. Buchert et al., "Reduction in camera-specific variability in [123I]FP-CIT SPECT outcome measures by image reconstruction optimized for multisite settings: Impact on age-dependence of the specific binding ratio in the ENC-DAT database of healthy controls," *Eur J Nucl Med Mol Imaging*, vol. 43, no. 7, pp. 1323–1336, July 1, 2016, doi:10.1007/s00259-016-3309-5.

[49] K. Ogawa, Y. Harata, T. Ichihara, A. Kubo, and S. Hashimoto, "A practical method for position-dependent Compton-scatter correction in single photon emission CT," (in English), *IEEE Trans Med Imag*, vol. 10, pp. 408–412, 1991, doi:10.1109/42.97591.

[50] B. E. Zimmerman et al., "Multi-centre evaluation of accuracy and reproducibility of planar and SPECT image quantification: An IAEA phantom study," *Zeitschrift für Medizinische Physik*, 2016, doi: 10.1016/j.zemedi.2016.03.008.

[51] M. Ljungberg and S. E. Strand, "A Monte Carlo program for the simulation of scintillation camera characteristics," (in English), *Comput Methods Programs Biomed*, vol. 29, no. 4, pp. 257–272, Aug 1989. [Online]. Available: www.ncbi.nlm.nih.gov/pubmed/2791527.

[52] M. Ljungberg, "The SIMIND Monte Carlo Code," in *Monte Carlo Calculation in Nuclear Medicine: Applications in Diagnostic Imaging*, 2nd edn., M. Ljungberg, S. E. Strand, and M. A. King (eds). Boca Raton: CRC Press, 2012, pp. 315–321.

[53] W. P. Segars, G. Sturgeon, S. Mendonca, J. Grimes, and B. M. W. Tsui, "4D XCAT phantom for multimodality imaging research," (in English), *Med Phys*, vol. 37, pp. 4902–4915, 2010.

[54] G. Brolin, K. Sjogreen Gleisner, and M. Ljungberg, "Dynamic 99mTc-MAG3 renography: Images for quality control obtained by combining pharmacokinetic modelling, an anthropomorphic computer phantom and Monte Carlo simulated scintillation camera imaging," *Phys Med Biol*, vol. 58, pp. 1–17, 2013.

[55] G. Brolin et al., "A national quality assurance study of 99mTc-MAG3 renography using virtual scintigraphic imaging," *Clin Physiol Func Imaging*, vol. 36, pp. 146–154, 2016.

[56] E. Tragardh et al., "Evaluation of inter-departmental variability of ejection fraction and cardiac volumes in myocardial perfusion scintigraphy using simulated data," (in English), *EJNMMI Phys*, vol. 2, no. 1, p. 2, Dec 2015, doi:10.1186/s40658-014-0105-9.

[57] W. P. Segars et al., "Population of anatomically variable 4D XCAT adult phantoms for imaging research and optimization," (in English), *Med Phys*, vol. 40, no. 4, p. 043701, Apr 2013, doi:10.1118/1.4794178.

28 Preclinical Molecular Imaging Systems

Magnus Dahlbom

CONTENTS

28.1 INTRODUCTION

PET and SPECT imaging are well-established methodologies used in a variety of preclinical research applications. Both imaging modalities have shown to be important in the development of new radiopharmaceuticals for clinical and research applications. In biological research preclinical PET and SPECT imaging can provide a better understanding of molecular pathways of disease and therapy. Using human disease models in mice and rodents, preclinical imaging can be used to facilitate the discovery and design of new diagnostic and therapeutic agents. Preclinical PET and SPECT imaging is just like human radionuclide imaging typically performed together with anatomical imaging (CT and MRI), either simultaneous or sequentially.

DOI: 10.1201/9780429489556-28

One particular strength of in vivo imaging is that it allows repeated and longitudinal studies where the same animal can be used as its own control, which reduces variability in the collected data due to inter-animal differences. This allows for studies of disease progress or regression before and after therapeutic intervention (e.g., radioligand therapy). In-vivo imaging also reduces the required number of animals to be used in a study.

In the early 1990s several research groups discussed the value of designing systems dedicated to preclinical PET and SPECT imaging [1–6]. At that time, it was clear that the spatial resolution of the existing imaging instrumentation used clinically was inadequate for small-animal imaging. The size of the detector elements used in PET systems did not provide the spatial resolution necessary to image small organs and structures in rodents. Conventional scintillation camera technology together with parallel hole collimators used in SPECT imaging had the same limitation. One of the first dedicated preclinical PET systems was designed in 1991 [7]. This system was designed using the same detector technology used in high-resolution human systems available at that time. The system had a smaller ring diameter, which reduced resolution losses due to photon non-collinearity [8, 9]. However, the system had a ring diameter that allowed imaging of larger animals such as dogs and primates. These early systems demonstrated the value and potential of preclinical imaging.

The design of the instrumentation used in preclinical imaging is dictated by the biological process to be studied and what animal model is used to answer the research question. For instance, imaging mice will in general require a system with higher spatial resolution compared to a system used to image larger animals (e.g., rats and non-human primates). Tumour xenografts may have heterogeneous uptake due to vascularity and necrosis and therefore require a system with high spatial resolution in order to detect the non-uniform uptake. Fast dynamic in vivo imaging requires high sensitivity in order to obtain enough counts in each frame to produce images with high signal-to-noise ratio.

The choice of tracer and isotope used in preclinical imaging is also dictated by the biological process to be studied. This will also determine what modality can be used for imaging. Antibodies and proteins typically have slow target uptake and plasma clearance rates. These macromolecules have to be labelled with isotopes with a relatively long half-life such as many of the isotopes used in SPECT imaging (e.g., ^{131}I, ^{123}I, ^{125}I, ^{99m} Tc, ^{111}In). The positron emitters used in PET imaging have in general a much shorter half-life (e.g., ^{18}F $T_{1/2}$ = 109.8 min, ^{11}C $T_{1/2}$ = 20 min). These isotopes are therefore better suited to study processes with fast kinetics using small molecules such as synthetic drugs. However, there are examples of long-lived positron emitters that have successfully been used for imaging of slow biological processes (^{89}Zr $T_{1/2}$ = 78.4 hrs, ^{124}I $T_{1/2}$ = 4.2 days).

There are several challenges in designing a preclinical imaging system. The spatial resolution of the detector system has to be within the same order of magnitude or higher than the dimensions of the structure to be imaged. In PET imaging, this means that the size of the detector elements has to be reduced to the same dimension as the structure to be imaged. This requires a radical change in the design of the detector technology used compared to what is used in human systems. In SPECT imaging, a conventional parallel hole collimator cannot be used due to the poor resolution properties. Instead, pinhole collimators have to be used to achieve the necessary spatial resolution [5, 6]. The magnification of the pinhole collimator reduces the limitation of the spatial resolution of the detector used (e.g., scintillation camera) and the spatial resolution is primarily limited by the size and absorption properties of the pinhole. The material and the shape of the pinhole will also affect the spatial resolution due to the penetration of the gamma rays [10, 11]. However, the efficiency of a pinhole collimator is low and, in order to improve the sensitivity, multiple pinholes have to be used [6].

Sensitivity is also a challenge in preclinical imaging. Just as in human imaging, it is necessary to collect an adequate number of counts in a study or image frame to be able to reconstruct a high-resolution image without excessive noise levels. Although animals can be imaged with higher activity levels compared to humans, this is usually not recommended. The increased activity may produce undesirable radiation-induced biological effects in the studied animals [12]. High injected activities will also increase the exposure to the personnel handling the animals, which could be significant for studies involving a large number of animals.

28.2 DETECTOR TECHNOLOGY

Preclinical imaging systems are almost exclusively based on scintillator detector technology, although there are exceptions. In these systems, the photons emitted from the animal interacts in a scintillation crystal. The scintillation light produced in the interaction process is then detected by a photodetector coupled to the scintillation crystal [8, 13]. The signal produced by the photodetector is subsequently processed by the electronics and other hardware of the system to determine the location of the photon interaction in the detector and how much energy was deposited in the interaction. The scintillation crystal material used depends on the modality (e.g., PET or SPECT) and also to a certain degree on the design of the light readout. In general, it is desirable to use a scintillation crystal with a high light yield. The high light

TABLE 28.1
Properties of Detector Materials Used in PET and SPECT Imaging

	NaI:Tl	BGO	LSO:Ce	LYSO:Ce	CsI:Tl	$LaBr_3$:Ce	$LaCl_3$:Tl	CZT
Effective Z	50	74	66	66	54	48	60	50
Density [g/cm³]	3.67	7.13	7.4	7.1	4.51	5.29	3.79	5.8
Light Yield [photons/keV]	38	8–10	35	33	54	63	46	4670*
Decay Time [ns]	250	300	42	36	680 (63%) 3340 (37%)	30	28	-
Max Emission [nm]	415	480	420	420	550	380	350	-
Hygroscopic	Yes	No	No	No	Yes	Yes	Yes	No
Application	SPECT	PET	PET	PET	SPECT	SPECT	SPECT	SPECT PET

Note: *Energy to produce one electron-hole pair.

yield results in a high energy resolution and allows for more accurate event positioning in the detector. The effective Z of the crystal material should be such that photoelectric interaction is the dominant interaction type. Photoelectric interaction of a photon will ensure that the entire photon energy is deposited in one interaction and at one location. This will minimize photon scatter due to Compton interactions in the scintillation crystal that produce spatial resolution losses or losses in detection efficiency, depending on how the events are handled by the processing electronics (e.g., rejection by energy discrimination). Minimizing detector scatter is of particular importance in preclinical imaging where it is desirable to achieve a spatial resolution around 1 mm³ or better. In SPECT imaging, where the photon energy used is, in general, relatively low (e.g., 140 keV for ^{99m}Tc), detector scatter is not a significant problem since photoelectric interaction is the dominant interaction type at atomic numbers greater than 31. For PET, where the photon energy is 511 keV, the scintillation material needs to have an effective Z greater than 79 in order for photoelectric interaction to be dominant. Even at a Z of 66 for LSO and 74 for BGO, Compton interactions in the scintillation crystal are significant. The density of the crystal material can reduce the range of these scattered photons. Table 28.1 summarizes some of the properties of detector materials used in preclinical PET and SPECT imaging.

28.3 PET SYSTEM DESIGN

In PET imaging there are three main components that limit the spatial resolution: Photon non-collinearity, positron range, detector dimensions and absorption properties of the detector material. In addition to these, effects such as detector parallax and photo-detector readout will also have a negative effect on the spatial resolution. The overall spatial resolution is the convolution of the line spread functions of the individual resolution components.

28.3.1 Photon Non-collinearity

Photon non-collinearity is a deviation in the assumed 180° emission of the two annihilation photons when the electron-positron pair annihilates. This deviation from the 180° emission happens when the momentum of the electron-positron pair is non-zero at the time of annihilation. The distribution in deviations from 180° is approximately Gaussian in shape with a FWHM of 0.5°. The effect of this is a resolution degradation that depends on the separation of the detector elements or the system diameter. For a system designed for humans where the detector separation is 80–100 cm, the resolution loss is approximately 2–3 mm FWHM. On the other hand, a preclinical system with a system diameter of about 10 cm, the resolution loss is only a fraction of a mm (~0.2 mm). Thus, photon non-collinearity has an almost negligible contribution to the overall resolution loss in preclinical PET imaging.

28.3.2 Positron Range

The emission of the two 511 keV annihilation photons that are used to localize the activity in a PET system always occur at some distance away from the actual decay location of the parent nucleus [14–16]. This distance depends on the energy that is transferred to the positron emitted in the decay. The amount of energy that is transferred is isotope dependent. For

^{18}F the average energy transferred to the positron is about 350 keV and the average range in tissue before annihilation is about 0.7 mm [14]. This translates to a resolution loss of s 0.2 mm full width half maximum (FWHM) or 1.3 mm full width tenth maximum (FWTM). For other isotopes such as ^{68}Ga, the average energy of the emitted positrons is significantly higher (836 keV and the range in tissue is about 3.5 mm. The spatial resolution is therefore substantially worse, 0.8 mm FWHM (4.7 mm FWTM) when imaging with this isotope.

28.3.3 Detector Dimensions

For an ideal PET detector, the line spread function will have a triangular shape with a FWHM equal to half the physical detector width at the midpoint between the pair of detectors [17]. At locations closer to one or the other detector, the shape of the LSF will change from triangular to trapezoidal and the resolution will degrade. This is referred as the tangential resolution loss and is primarily due to geometrical effects.

28.3.4 Detector Parallax

Resolution losses are also observed when a source is imaged at off-centre positions in the radial direction. In a PET system with a circular or polygonal detector arrangement, the detector pairs are angled relative to each other at off-centre positions. When the annihilation photon pair hit the detector elements at these positions, there is a high probability that one or both of the annihilation photons will penetrate through the detector it first enters and deposit its energy in an adjacent detector [8, 18–20]. In PET, the positioning of an annihilation event is assigned to the line connecting the front surface of the two detector elements that absorbed the two photons. If detector penetration occurs, the event will be mispositioned. This effect is referred to as detector parallax. This effect is primarily determined by the amount of radial offset and the thickness of the detectors. The thicker the detector, the greater the probability that the photon will be absorbed and detected in an adjacent detector element [8]. The effect of the detector parallax can therefore be reduced by using thinner detector elements – however, at the cost of reduced detection efficiency. Detector parallax can also be reduced by increasing the detector separation or system diameter. This will unfortunately also reduce detection efficiency. In addition, the resolution-degrading effect of photon non-collinearity will increase with an increased system diameter.

The detector parallax effect can be reduced if the actual location or detector depth of photon absorption in the detector element can be determined [21]. This effect is therefore also referred to as the depth-of-interaction effect.

28.3.5 Detector Configurations

PET detectors are usually made up of an array of scintillation-detector elements attached to an array of photodetectors. The photodetectors used to read out the scintillation light are typically an array of photomultiplier tubes (PMTs) or solid-state photodetectors. The number of photodetectors is in general much smaller than the number of scintillator elements in the array. There are several reasons for this. There are practical limitations to how small the photodetector can be made, especially when PMTs are used. Even if one-to-one coupling was possible, this would result in an extremely large number of electronic readout channels, which would be challenging both in terms of physical space requirements and power consumption. Instead, most scintillator-based high-resolution detectors use some sort of multiplexing of the light signal, which is decoded by looking at the signal amplitudes from the photodetectors and using look-up tables [22]. This detector design is usually referred to as a block detector and was first described by Casey and Nutt [23].

There are numerous variations of this design, but the basic operations are similar. The light produced in a single scintillator element following a photon absorption is guided towards the photodetectors. When the light exits the detector element, it is allowed to spread out over a number of photodetectors via a light guide. By analysing the relative signal amplitude from the photodetectors, the detector element can be identified. How accurate an event can be localized depends primarily on how much light is collected by the photodetectors and how much of this light contributes to the photodetector signal (e.g., the quantum efficiency of the photocathode in a PMT or PDE in a SiPM). This design is very cost effective since it reduces the number of required photodetectors and readout channels with little compromise in spatial resolution. Furthermore, it provides excellent light collection, which improves both energy and time resolution.

28.3.6 Segmented Detectors

Most detectors designed for PET imaging are made up of an array of narrow scintillation detector elements (Figure 28.1a). The scintillation elements are optically isolated from each other with some sort of reflective film or paint. The reflectors are needed to guide the light produced from the photon interaction toward the photodetector readout. This design is a miniatured version of the detectors used in human PET systems. There are, however, several challenges with this design as the detector elements are reduced in order to improve the spatial resolution. There are significant light losses in long and narrow scintillator elements caused by the fact that the light has to be reflected multiple times along the scintillator wall before reaching the photodetector. Each reflection on the wall means a chance of light absorption (i.e., loss of light) and results in loss in signal. This will affect both the energy and time resolution of the detector but, more important, how accurate the photodetector readout can localize the origin of the scintillation light. The light losses can be reduced by using shorter or more shallow detectors, although this will result in a loss in detection efficiency.

Manufacturing of the arrays made up of a large number of small scintillation detector elements tend to be costly due to the significant loss in scintillator material in the cutting process. Also, the placement of reflective material between the detector elements results in a significant amount of dead space (i.e., non-detector material), especially when the element size is smaller than 2x2 mm. Investigators have therefore looked into alternative manufacturing procedures using laser cutting, which results in more narrow cuts and also less material loss [24].

The use of subsurface laser engraving (SSLE) has also been applied to create an array of scintillator elements without the need of mechanical cutting. When using SSLE, the detector elements are created by laser "cutting" where the optical separation is created by cracks in the scintillation crystal [25, 26]. Due to the dispersion of the laser beam, this method is limited to crystal depths of less than 10 mm. Furthermore, the elements are not completely optically isolated from each other and cross-talk compromises detector identification, especially for elements smaller than 2 mm in width [26].

28.3.7 Monolithic Detectors

Detector modules made from monolithic scintillator crystals have also been developed [27] and have been incorporated in at least one commercial preclinical PET system [28] (See Figure 28.1b).

The event localization in these detectors is similar to the detector technology used in scintillation cameras. The spatial resolution in these detector modules is limited by the light output of the scintillator and the light collection of the photodetectors. It is therefore necessary to use a scintillator with very high light yield together with a photodetector with high quantum efficiency to maximize the signal. Using LYSO together with silicon photomultipliers (SiPM), an intrinsic spatial of around 1 mm or better has been achieved [29–31]. These modules have a two-dimensional SiPM readout, and the origin of the scintillation light is determined by the signal distribution of the photodetector array. The event positioning is not only limited in the x and y directions, but a certain level of DOI information can be extracted from the 2-D light distribution [32]. In order to achieve high spatial resolution, these detector modules have to be carefully calibrated and trained [28]. Just as in conventional scintillation cameras, positioning and spatial resolution tend to degrade near the outer edge of the crystal due to wall reflections [27].

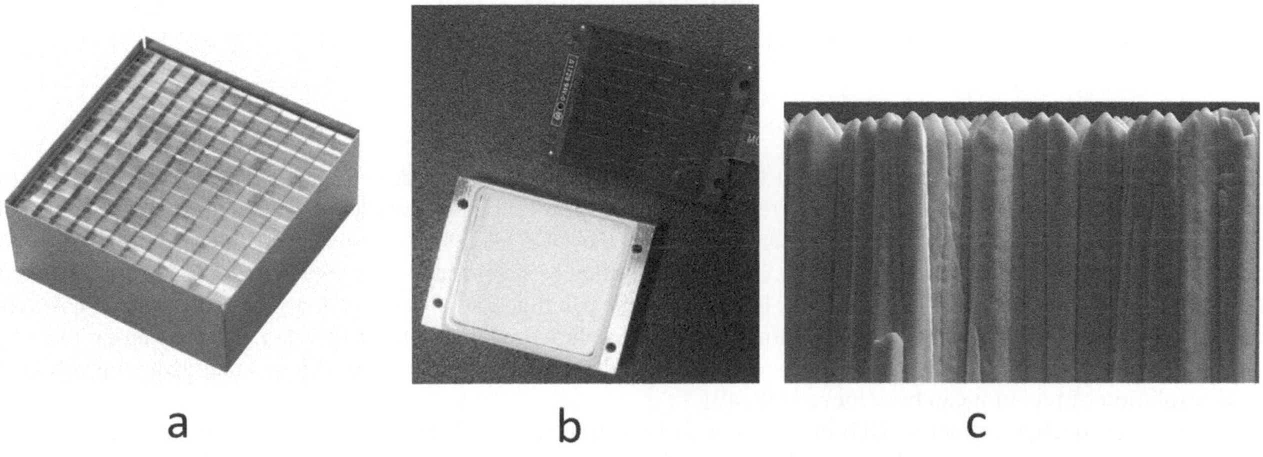

FIGURE 28.1 Scintillation detector module structures. (a) Segmented and optically isolated LYSO array. (b) Monolithic NaI (Tl) scintillator. (c) Micro-columnar CsI (Tl) scintillator. Each column is approximately 5 μm wide.

28.3.8 Scintillator Material

Lutetium-based scintillators are currently the predominant scintillator type used in preclinical PET systems. The main reason for this is the high conversion efficiency and light output, which provides accurate event localization within the detector. LSO and LYSO are also very fast scintillators, which allows for high count-rate conditions and for time-of-flight (TOF) detection [33]. TOF in preclinical PET is in most imaging situations of limited value, even at a time resolution of a few hundred ns, since the objects imaged are not large enough to take advantage of the signal-to-noise benefits TOF offers [33].

BGO has a higher density and higher effective Z compared to LSO and LYSO. These properties reduce inter-detector scatter and could potentially improve spatial resolution in high-resolution detectors [34]. The low light output of BGO may, on the other hand, compromise event localization.

28.3.9 Photodetectors

The most common type of photodetector used in nuclear imaging devices is the PMT [13]. The requirement for high spatial resolution in preclinical PET imaging does not allow for the use of an array of discrete PMTs for light detection and readout used in clinical PET systems. The reason is that the required glass envelope makes the PMTs fairly bulky and prevents close packing that is required in a high-resolution detector module. Furthermore, the performance of PMTs degrades significantly as the size is reduced. Position sensitive, multi-anode PMTs (PSPMT) have, on the other hand, been successfully used in a number of preclinical PET systems [4, 35–41].

Several types of PSPMTs have been developed, and the main difference between these designs is the electron multiplication structure and the anode readout used. Early PSPMT consisted of multiple distinct PMTs within the same vacuum enclosure, collecting photoelectrons from a common photocathode [42]. The more recent designs of PSPMTs are more compact and use metal channel–type dynodes together with a multi-anode readout. This design results in very low cross talk between channels and is less sensitive to magnetic fields compared to conventional PMTs [43]. Like conventional PMTs, the PSPMT requires a vacuum enclosure to operate, which makes them fairly large and bulky and makes it challenging to design compact high-resolution preclinical systems.

Solid-state based photodetectors are alternatives to the PMT. These include the avalanche photodiodes (APD) [44] and the SiPM [45]. Like the PMT, both the APDs and SiPM provide an internal amplification of the signal and are very compact devices. Furthermore, these devices are, in contrast to conventional PMTs, insensitive to magnetic fields. They are therefore the only photodetector that can be used in PET systems that operate within or near an MRI system.

The development of SiPMs have in recent years reached a point where these photodetectors are a viable alternative to PMTs, or even the preferred photodetector in some detector designs. In preclinical PET systems, the compactness and high performance of SiPMs has helped in improving the spatial resolution that can be achieved. The cost of SiPMs is also significantly lower compared to both APDs and PMTs. One drawback of these devices is that the gain is very sensitive to temperature and voltage fluctuation. For stable operation, the temperature has to be controlled within a fraction of a degree, which can be accomplished with Peltier cooler elements, or the bias can be adjusted to compensate for any gain change caused by temperature drifts [46].

28.3.10 Depth-of-Interaction – DOI

As described above, most detectors used in PET is made up of a matrix of detector elements with a photodetector readout attached at one end of the array. This readout does not allow for determining at what depth the photon interacted. Several approaches have been developed for determining the DOI [47]. One approach is to attach a photodetector readout at both ends of the scintillator array. The DOI can then be determined by comparing the light signal strength in the photodetectors at both ends. Since the light collection as a function of depth is non-linear, the readout has to be calibrated in order to produce the DOI information. The resolution of the DOI information will depend on the light output, light collection and quantum efficiency of the photodetector. Investigators have shown that using this design, a DOI resolution of few mm can be achieved [48, 49].

Another approach to extracting DOI information is to construct a phoswich ("phosphor sandwich") detector whereby two or more scintillation materials are sandwiched together in multiple layers with a single-end photodetector readout [21, 50]. The scintillation materials used should have significantly different properties in the decay constant of the emitted scintillation light to allow identification in which layer the photon interaction occurred. The DOI resolution of a phoswich system is limited to the thickness of the different scintillator layers. How well the scintillator signals can be separated with pulse-shape discrimination also adds to the uncertainty.

Other DOI designs include systems where multiple photodetectors are attached to the side of the photo detectors [51]. Multi-layered, stacked detector designs have also been constructed where the DOI is decoded by the pattern produced on the photodetector readout [37, 52].

28.3.11 Cadmium Zinc Telluride CZT

Solid state detectors such as Cadmium Zinc Telluride (CZT) have several properties that make them suitable as radiation detectors such as high-energy resolution and compactness. Since they are direct detection devices, they do not need a photodetector readout to extract the signal. Furthermore, CZT can operate in strong magnetic fields and could therefore be used in hybrid imaging systems (i.e., PET/MR and SPECT/MR). The main drawback is the relatively low effective Z of 50, which makes them less suitable for detection of high-energy photons and annihilation radiation. As discussed earlier, a low effective Z results in a significant amount of Compton interactions in the detector volume, which makes it difficult to differentiate primary radiation from scattered radiation. Despite this, the possibility of CZT as the detector material in preclinical PET systems has been investigated [53, 54]. In this system, several high-resolution detector modules of CZT are assembled in a box geometry (Figure 28.2f) in order to maximize the geometrical efficiency. In order to compensate for the lower stopping power of CZT, the thickness of the modules is greater than what is typically seen in scintillator-based detector systems, which may result in a significant detector parallax problem. However, CZT detectors are designed to localize events in 3-D within the module. This allows the detector module to provide DOI information for accurate positioning of the coincidence events.

Due to the low Z, there will be a significant amount of multiple Compton interactions in the detector volume, which makes accurate positioning challenging. It has been shown that by combining the multiple energy depositions in the detector volume with the Compton kinematics, the event can be positioned accurately [55].

28.4 SENSITIVITY CONCERNS

The sensitivity of a PET system depends on a number of parameters, including system diameter, FOV, and detection efficiency. Detection efficiency depends on several parameters, but one important factor is how much of the system FOV is covered by scintillation material. The detectors in both PET and SPECT systems consist of detector modules that are arranged around the object to be imaged (Figure 28.2). Since these modules typically have a cuboid shape, the arrangement of the detectors will be in the shape of a polygon (Figure 28.2b–c). The number of sides of the polygon depends primarily on the front width of the detector module and the system diameter. For a small diameter preclinical system, only a few detector modules will be required. The advantage of the modular cuboid detector design is that it allows for fairly inexpensive manufacturing and straightforward assembly. The drawback of the polygonal design is that there can be a significant amount of dead space between the detector modules that is not filled with scintillator material, especially for a small diameter system with a low number of modules. The efficiency loss due to the gaps can be significant. To reduce the number of gaps, a larger number of smaller detector modules could be used (Figure 28.2a), but this could lead to an increase in cost since more photodetectors with associated electronics would be required. An alternative approach would be to use tapered detectors [49, 56], where the back surface of the detector module is greater than the front surface (Figure 28.2d). Reducing the gaps between detector modules by increasing the volume of LSO by 38 per cent resulted in a 64 per cent increase in sensitivity for mouse-sized objects [56]. The drawback is an increase in manufacturing complexity.

Other investigators have designed systems with a box or cubic geometry with detectors placed along four sides of the FOV (Figure 28.2e–f) [54, 57]. Using this geometry, the gaps between detector modules can be reduced significantly or completely eliminated. One system that uses this geometry is the PETBox4, which uses four panel detectors, and the sensitivity was measured to be greater than 18 per cent with a wide energy window [57].

28.5 SPECT SYSTEM DESIGNS

Instrumentation for preclinical SPECT imaging has, like preclinical PET imaging, evolved from clinical instrumentation. The main challenges are to achieve a spatial resolution and sensitivity that can accomplish the imaging task. In contrast to PET, preclinical SPECT instrumentation should also be able to accommodate the measurement of a range of photon energies.

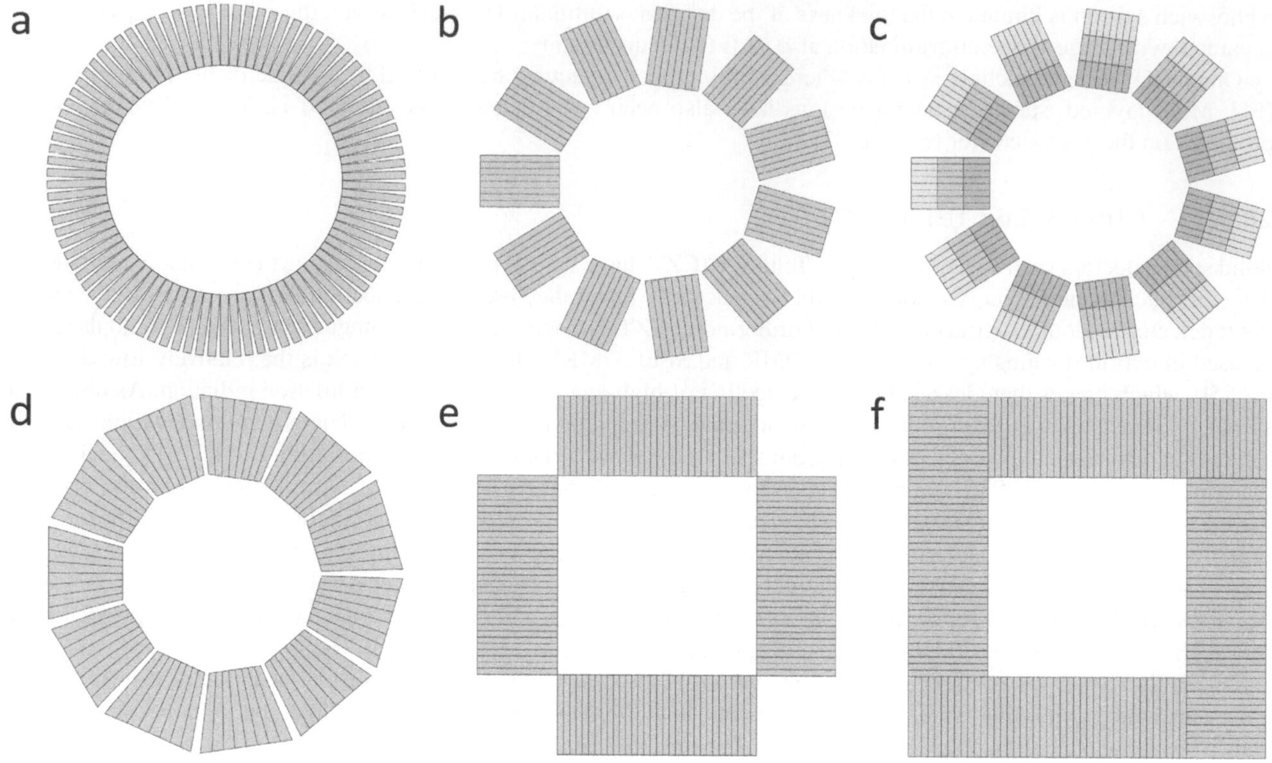

FIGURE 28.2 Different detector geometries used in preclinical imaging. (a) Circular system using detector modules with small front area. (b) Polygon system using detector modules with large front area. (c) Polygon system with multi-layered scintillators. (d) Polygon system with tapered detectors to reduce inter-module gaps. Box geometries with (e), and without detector gaps (f).

28.5.1 Detector Design and Scintillation Materials

In clinical SPECT imaging, the detector system is almost exclusively based on the original Anger design using a monolithic NaI:Tl crystal coupled to an array of PMTs. In preclinical SPECT a greater variety of scintillators and photodetector readouts are used in the instrumentation design. There are two basic scintillator designs used: monolithic and pixelated. Using a monolithic crystal, the position of the absorbed photon is localized by the light distribution projected on the photodetectors. How well the photodetector can localize the event depends on several factors, including how much light is produced, light dispersion, and spatial resolution of the photodetector. In contrast to a pixelated detector, the spatial sampling is continuous, which allows selection of arbitrary pixel dimensions.

Using a pixelated scintillator array, a large number of individual, optically isolated, scintillator elements are packed together. The purpose of using a pixelated scintillator is to minimize light spread and maximize light collection by directing as much of the emitted light towards the photodetector. The best light collection is achieved with a thin detector or a small scintillator element length-to-width ratio, which results in the best spatial and energy resolution. However, the detection efficiency degrades with decreased thickness. The detector thickness will therefore be a compromise between detection efficiency and spatial resolution. For a pixelated detector, the spatial resolution and spatial sampling are limited by the size of the scintillator elements and the resolution of the photodetector.

Since the spatial resolution is directly dependent on the light output of the scintillator, materials other than NaI:Tl with higher light output have been considered in preclinical imaging instrumentation. Both $LaBr_3$:Ce and $LaCl_3$:Ce have significant higher light output compared to NaI:Tl (see Table 28.1). Both materials are very hygroscopic and have to be hermetically sealed to prevent degradation. Manufacturing cost is also significantly higher for $LaBr_3$:Ce and $LaCl_3$:Ce compared to NaI:Tl. La-based scintillators also have a background activity originating from the decay of ^{138}La, which may interfere in SPECT imaging.

Another method to produce a scintillator with high spatial resolution is to grow the crystal material into a micro-columnar crystal array (Figure 28.1c). In the micro-columnar array, the crystals are grown together into a needle-like structure. Each crystal needle forms a light guide that channels the light towards the exit end of the scintillator. CsI:Tl has successfully been produced this way with sub-mm spatial resolution. The thickness of micro-columnar CsI:Tl crystals is limited to 2–3 mm, which limits its use to low-energy gamma imaging [58]. $LaBr_3$:Ce has, on the other hand, been successfully grown to at least a thickness of 8 mm, which allows imaging at higher photon energies [59, 60].

28.5.2 Photodetectors

The most common photodetector used in preclinical SPECT systems are PMTs and PSPMTs [61–63]. A PSPMT can accurately identify the scintillator element in a pixelated array. PMTs work especially well with NaI:Tl since the quantum efficiency of the photocathode is optimized for the emission spectrum of NaI:Tl, which has a peak emission at 415 nm. Since many scintillators have emission properties similar to NaI:Tl, conventional PMTs in general work well as the photodetectors. The emission spectrum from CsI:Tl has a peak emission at 550 nm and is poorly matched to the absorption properties of the photocathode of a conventional PMT. To take full advantage of the high light output and high-resolution properties of monolithic or microcolumnar CsI:Tl, solid state photodetectors are more suitable. These have higher QE at longer wavelengths and are therefore better matched to the emission spectrum of CsI:Tl. Several groups have reported sub-mm resolution when combining microcolumnar CsI:Tl with a Charge-coupled device (CCD) or Electron-Multiplying CCD (EMCCD) camera readout [64–66]. The absorption properties of CCD cameras are well matched to the emission spectrum of CsI: Tl, and the quantum efficiency is significantly higher compared to PMTs. Furthermore, the spatial resolution of a CCD camera is on the order of 10 μm. However, the area of CCD cameras is significantly smaller than the area of the scintillator. The light emitted from the scintillator therefore has to be projected down to the smaller CCD sensor. This can be accomplished either by lenses [60] or by tapered fibre optics [67].

SiPMs are also being used as photodetectors in preclinical SPECT systems. In one system, monolithic LYSO scintillators are coupled to an array of SiPMs [68]. LYSO is non-hygroscopic, which eliminates the need for encapsulation. Using the SiPM makes the overall detector system very compact. These detectors use the same positioning techniques as the monolithic detectors used in PET [28]. A drawback of using LYSO is that counts from the natural background activity of ^{176}Lu may interfere with the signal from the gamma photons. Although this background can be corrected for, it may add noise and interfere with event positioning in the scintillator.

28.5.3 Collimation

In order to achieve the high spatial resolution necessary to image small animals, pinhole collimation is necessary [6, 69, 70]. The system spatial resolution, R_{sys}, of a pinhole-based imaging system is given by [6]:

$$R_{sys} = \sqrt{d^2\left(1+{}^{1}\!/_{M}\right)^2 + {}^{R_{det}^2}\!/_{M^2}} \tag{28.1}$$

Where d is the effective diameter of the pinhole, R_{det} is the intrinsic resolution of the photon detector, and M is the magnification factor of the pinhole collimator for an object located in front of the collimator (Figure 28.3). M is the ratio of the pinhole-detector (f) and pinhole-object (x) distances. The magnification varies with distance from the collimator hole and is greatest for objects close to the opening of the pinhole collimator (see chapter on Collimators). M is limited by the detector FOV and the required size of the reconstructed FOV. In general, a large magnification results in a limited FOV. One important consequence of the magnification factor is that the intrinsic FWHM of the photon detector is reduced by the magnification factor. By proper collimator design and object positioning the contribution of the intrinsic resolution of the detector can be made negligible in comparison to the collimator resolution (i.e., pinhole diameter). It is therefore possible to achieve excellent spatial resolution even when using a detector with fairly nominal resolution [71].

The main drawback of the pinhole collimator is the relatively poor sensitivity which is given by [72]:

$$S_{coll} = d^2\left(\frac{sin^3\alpha}{16x^2}\right) \tag{28.2}$$

Where d is the effective diameter of the pinhole, x is the pinhole-object distance and, α is the incidence angle of the photon on the aperture plane (Figure 28.3). As can be seen from Eq. 28.2, the sensitivity drops off with the square of the distance from the pinhole opening. There are a number of approaches to compensate for the poor sensitivity. The

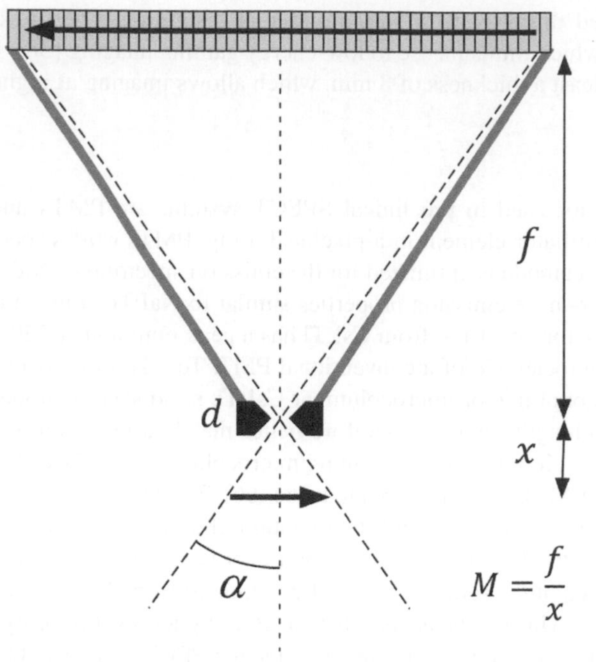

FIGURE 28.3 Schematic of the pinhole geometry indicating the key parameters affecting spatial resolution and sensitivity.

most straightforward approach is to add more detector modules around the object to be imaged [62, 73, 74]. A potential challenge for this approach is the space requirement for each detector module. Adding more detectors to the system may have the additional advantage of eliminating the need to rotate the detector assembly or object if an adequate number of projection views can be acquired simultaneously to allow artefact-free image reconstruction.

A significant improvement in sensitivity can be achieved by using multi-pinhole collimation [75–78]. Using multi-pinhole collimation, the pinholes will produce projections on different areas of the detectors (Figure 28.4). The detectors therefore have to be large enough to accommodate all the projections across its sensitive area. Multi-pinhole collimation may also allow imaging without the need for detector motion or collimator rotation. This requires an arrangement of the pinholes in such a way that enough projection views of the object are collected over 360°. Collecting a complete data set without detector motion would in turn allow dynamic imaging.

In some multi-pinhole collimation design, some overlap or multiplexing of projections is allowed. This allows for more projections to be acquired by the detector. The overlap of the projection data is modelled in the image reconstruction algorithm in order to reduce artefacts due to the multiplexing. Depending on the source distribution a certain amount of overlap can be allowed without introducing severe artefacts [76, 79]. However, in general, projection overlap should be kept at a minimum to avoid reconstruction artefacts.

There is a wide variety of designs of multi-pinholes that are designed for different detector designs, imaging situations, and photon energies [71, 76, 80–83]. For instance, an array of focused pinholes will all converge towards a common focal point. This will allow imaging of a small volume and result in a high sensitivity for that region [84]. This would be suitable for single-organ imaging, such as the brain or the heart. For whole-body imaging, a wider axial FOV is necessary, which can be accomplished by modifying the orientation of the pinholes and the opening angles of the pinholes. Depending on the design of the particular collimator, axial motion may be necessary to acquire a full body scan [84].

28.6 MULTIMODALITY IMAGING

Just like clinical radionuclide imaging, combining the functional imaging with CT or MRI adds valuable information to the study. CT adds anatomical information that can be used for localization of uptake as well as provide the necessary data for attenuation and scatter correction. MRI add high contrast anatomical information, and different MR pulse sequences can add complementary functional data to the radioisotope information [85].

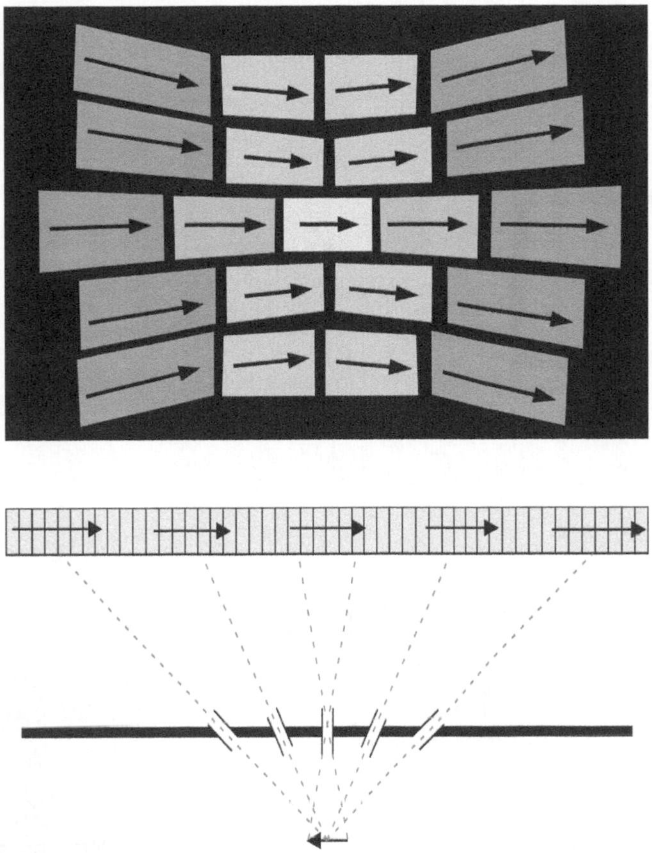

FIGURE 28.4 Schematic of the multi-pinhole geometry. Multiple pinholes are focused on the object FOV. The pinhole views of the object are projected on different areas of the detector surfaces.

There are two basic approaches to perform multi-modality imaging. The most straightforward approach is to image the animal on different instruments. The animal is first imaged on the PET or SPECT system. The animal is then moved to the second imaging modality system (i.e., CT or MRI), where the imaging continues (Figure 28.5a). This requires that the animal is immobilized on a fixture that can be accommodated by both imaging devices. A potential problem is that the animal might move in transition between imaging systems, which may result in misregistration of the images. Furthermore, images will always be acquired sequentially. An example of two mouse studies acquired on separate SPECT and CT systems is shown in Figure 28.6.

The second approach is to integrate different imaging modalities into one single instrument. Performing all imaging on one instrument reduces the amount of animal handling and reduces image registration problems. From an engineering point of view, combining X-ray CT to PET or SPECT is fairly straightforward. The design of preclinical SPECT/CT (Figure 28.5b) and PET/ CT (Figure 28.5c) systems are similar to clinical systems where a CT system is added to a PET or SPECT system and imaging is performed sequentially [86]. The CT systems in most current preclinical systems are cone-beam CTs with a low-energy microfocus X-ray source and high-resolution semiconductor panel detector, and are capable of producing images with a resolution in the range of 2–100 μm. For a CT scan for anatomical localization of activity (sub-mm resolution), the acquisition time is typically less than a minute. An example of a mouse study acquired on an integrated PET/CT system is shown in Figure 28.7.

Integrating MRI with PET or SPECT is significantly more challenging. The main problem is the magnetic field that affects the photodetectors and the electronics of the PET or SPECT system. Furthermore, the PET and SPECT detectors and associated electronics may conversely affect the performance of the MRI system. The magnetic field of the MR requires the use of solid-state photodetectors such as APDs or SiPMs [87–89]. Several MR-compatible PET systems have been produced and, unlike systems integrated with CT, these systems allow simultaneous PET and MR imaging [90].

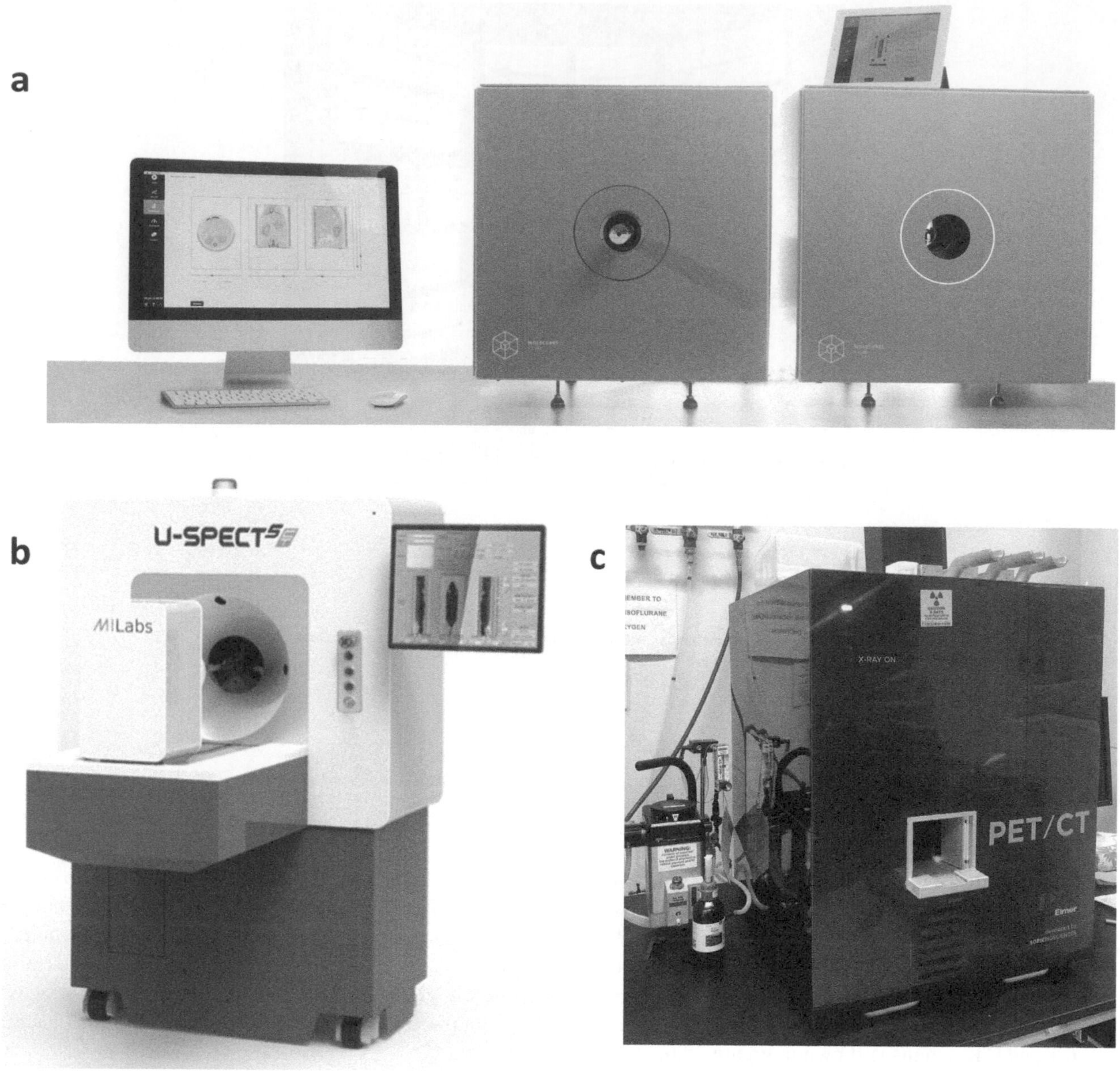

FIGURE 28.5 Example of preclinical imaging systems. (a) Separate SPECT (left) and CT (right) systems. The animal is immobilized on a fixture that is accommodated by both imaging devices (image courtesy of Dr. R. Van Holen, Molecubes). (b) Integrated SPECT/CT system (image courtesy of Dr. F. Beekman, MiLabs). (c) Integrated PET/CT system (image courtesy of C. Meyer, UCLA).

An example of an integrated SPECT/MR system is the MRC-SPECT-II [91, 92]. This system is based on compact position-sensitive CZT detectors that are insensitive to the magnetic field. One of the challenges in integrating SPECT with MRI is the need for collimation. In the MRC-SPECT system, a multi-pinhole collimator is used where an array of 8x8 pinholes projects onto the CZT detector. The complete system is made up of 8 compact detector modules arranged in an octagon around the FOV.

28.7 ANIMAL HANDLING

Proper animal preparation and handling is a very important aspect of preclinical imaging [12, 93]. Imaging animals usually means that the animal is put under general anaesthesia during the imaging session to reduce motion artefacts. Inhalant anaesthesia such as vaporized isoflurane is the most-used method in preclinical imaging, primarily due to

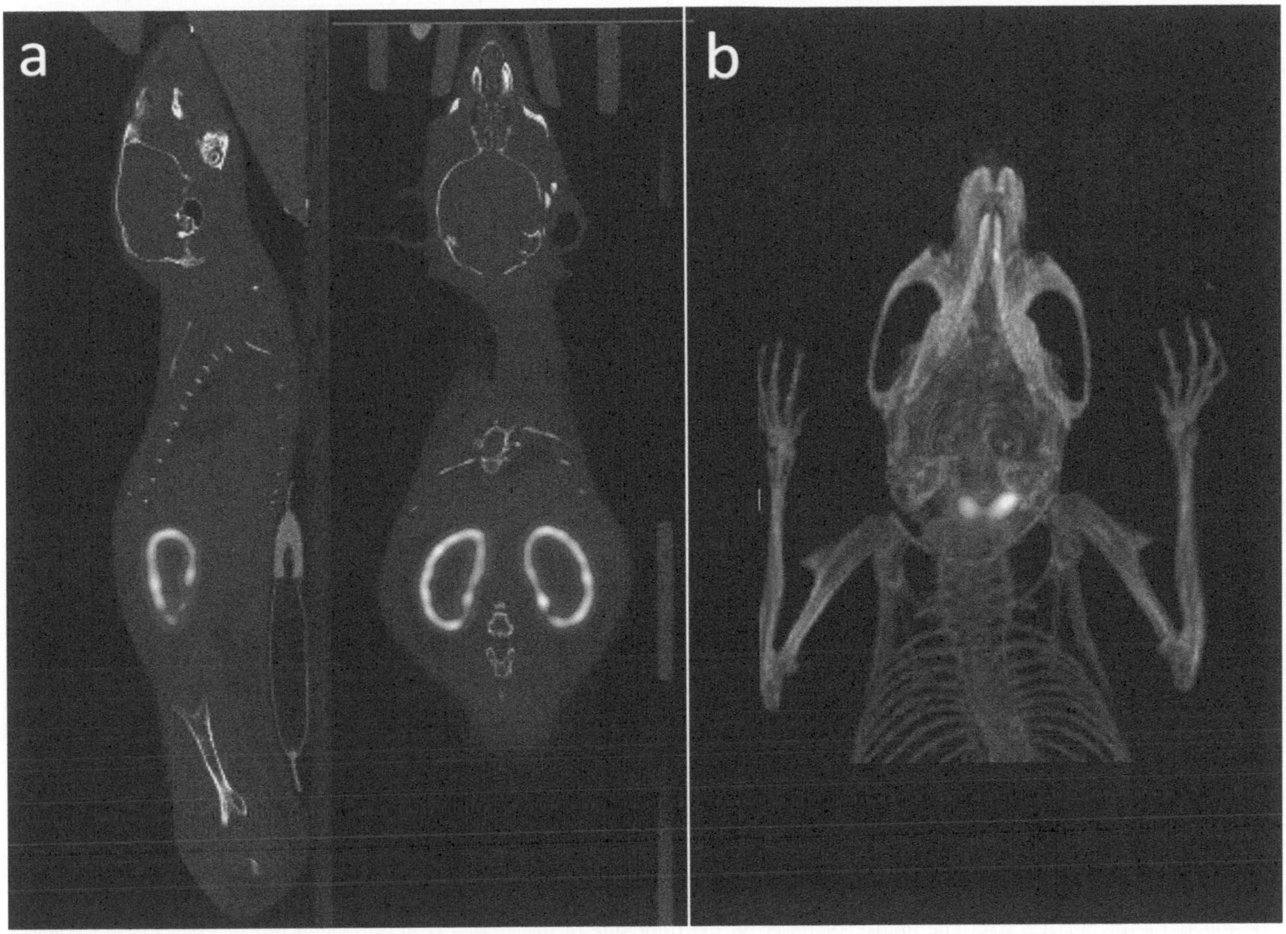

FIGURE 28.6 Example of preclinical SPECT/CT images of a mouse acquired on separate imaging system. SPECT data was acquired on a γ-cube and CT data were acquired on an X-cube (Molecubes, Ghent, Belgium). (a) Image showing a ^{99m}Tc-DMSA with uptake in the cortex of the kidneys. (b) Showing a ^{123}I-NaI study with uptake in the thyroid (images courtesy of Dr. R. Van Holen, Molecubes). Colour image available at www.routledge.com/9781138593268.

the advantages of both short induction and recovery time [93]. Many preclinical imaging procedures can be lengthy, especially multi-modality imaging procedures and dynamic imaging. For these procedures, it is necessary that the anaesthesia can be delivered continuously and at a constant level. Physiological monitoring is required to maintain safe anaesthetic levels and to ensure that good homeostasis is maintained, since this may affect tracer uptake [12]. Physiological monitoring usually includes monitoring of heart rate and respiratory rate. Since an anesthetized animal cannot maintain its core body temperature, it is necessary to provide external heating, such as the use of heating pads or integral heating elements in the imaging fixture.

28.8 PERFORMANCE EVALUATION AND QC

To evaluate the performance of preclinical PET scanners, NEMA has developed a standardized set of tests that is described in detail in NEMA NU 4-2008 [94]. This document describes in detail the various tests, specifications of required test equipment, phantoms and radioactive sources. This document also specifies how the data should be analysed and presented. The tests described include: Spatial resolution; Sensitivity; Scatter and count rate performance and randoms measurement; Image quality and accuracy of attenuation and scatter corrections.

Performing these tests allows a user to establish a system's performance under typical imaging conditions. It also allows comparison of the performance of various tomograph designs, since manufacturers of preclinical PET systems use the NEMA tests to provide performance specifications of their systems [95]. At the installation of a new system the user should repeat the NEMA tests to ensure that the system performs as specified by the manufacturer. These initial

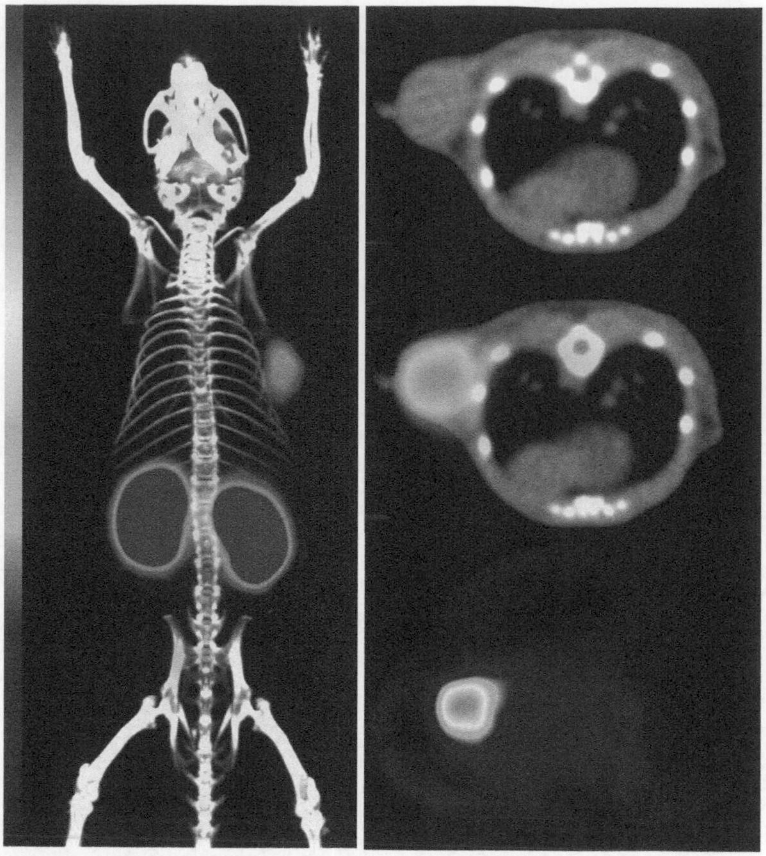

FIGURE 28.7 Example of a preclinical PET/CT images of a mouse acquired on a G8 (Sofie, Dulles, Virginia, USA). The animal was injected with ^{68}Ga-PSMA. The uptake of tracer on the right side of the animal is in implanted prostate cancer cells (images courtesy of Dr. C. Mona, UCLA). Colour image available at www.routledge.com/9781138593268.

tests also serve as the baseline to which subsequent tests are compared to as a system age. It is recommended that the NEMA tests or a sub-set of them are repeated on a regular basis (e.g., annually) to ensure consistent performance. For instrument developers, the NEMA tests are useful for comparing the performance of a new system design to existing systems.

There is unfortunately no NEMA standard defined for preclinical SPECT systems. A task group has been established to define a standard for small-animal SPECT systems. It is likely that this standard will follow the same format as the NEMA NU-4 standard with well-defined test procedures adapted to the special requirements for SPECT imaging. These are likely to include spatial resolution and image quality.

Quality control (QC) for preclinical PET and SPECT systems is similar to what is performed on clinical systems. The purpose of QC is to ensure stable and consistent performance, and the manufacturer of preclinical imaging systems will specify the necessary test for a particular system.

Phantom imaging for uniformity and quantification should be performed on a regular basis. Quantification is of particular importance to ensure that the system provides stable and quantitative values. A drift in quantification can have detrimental effects on research studies performed over longer periods. A co-registration test between other imaging systems (e.g., CT, MRI) should also be performed regularly. This is especially important if a CT system is used for attenuation correction. This test can be performed with a set of point sources that are visualized on both imaging modalities.

28.9 SUMMARY AND FUTURE DIRECTIONS

Preclinical PET and SPECT systems have undergone a remarkable improvement over the last 25 years. Much of these improvements are the results of improvement in detector technology, such as the discovery of new scintillator materials and development of new photodetector technology, and this has allowed preclinical PET systems to reach theoretical resolution limits. Innovative collimator design for preclinical SPECT systems has resulted in images with sub-mm resolution. Although not discussed in this chapter, the continuous development of image reconstruction algorithms has been a factor equally important to the improvement in image quality as is the instrument development. Preclinical PET and SPECT are now both accepted and well-established imaging technologies and integrated laboratory tools that are used to make important contributions to a wide variety of fields in medical and biological sciences.

REFERENCES

[1] J. Palmer and P. Wollmer, "Pinhole emission computed tomography: Method and experimental evaluation," *Phys Med Biol*, vol. 35, no. 3, pp. 339–350, Mar 1990, doi:10.1088/0031-9155/35/3/004.

[2] M. Ingvar, L. Eriksson, G. A. Rogers, S. Stoneelander, and L. Widen, "Rapid feasibility studies of tracers for positron emission tomography – High-resolution PET in small animals with kinetic-analysis," (in English), *J Cerebr Blood F Met*, vol. 11, no. 6, pp. 926–931, Nov 1991, doi: 10.1038/jcbfm.1991.157.

[3] S. Rajeswaran et al., "2-D and 3-D imaging of small animals and the human radial artery with a high-resolution detector for PET," (in English), *IEEE Trans Med Imag*, vol. 11, no. 3, pp. 386–391, Sep 1992, doi:10.1109/42.158943.

[4] M. Watanabe et al., "A high-resolution PET for animal studies," (in English), *IEEE Trans Med Imag*, vol. 11, no. 4, pp. 577–580, Dec 1992, doi: 10.1109/42.192694.

[5] S. E. Strand et al., "High resolution pinhole SPECT for tumor imaging," *Acta oncologica*, vol. 32, no. 7–8, pp. 861–867, 1993, doi:10.3109/02841869309096147.

[6] R. J. Jaszczak, J. Li, H. Wang, M. R. Zalutsky, and R. E. Coleman, "Pinhole collimation for ultra-high-resolution, small-field-of-view SPECT," *Phys Med Biol*, vol. 39, no. 3, pp. 425–437, Mar 1994, doi:10.1088/0031-9155/39/3/010.

[7] P. Cutler and E. Hoffman, "Use of digital front-end electronics for optimization of a modular PET detector," in *IEEE Nuclear Science Symposium*, Santa Fe, NM, 1991.

[8] S. R. Cherry, J. A. Sorenson, and M. E. Phelps, *Physics in Nuclear Medicine,* 4th edn. Philadelphia: Elsevier/Saunders, 2012, pp. xvii, 523.

[9] P. D. Cutler, S. R. Cherry, W. M. Digby, E. J. Hoffman, and M. E. Phelps, "Unique design features and performance of a new PET system for animal research," *J Nucl Med*, vol. 33, no. 4, pp. 595–604, 1992.

[10] F. van der Have and F. J. Beekman, "Photon penetration and scatter in micro-pinhole imaging: A Monte Carlo investigation," *Phys Med Biol*, vol. 49, no. 8, pp. 1369–1386, Apr. 21, 2004, doi:10.1088/0031-9155/49/8/001.

[11] F. Beekman and F. van der Have, "The pinhole: Gateway to ultra-high-resolution three-dimensional radionuclide imaging," *Eur J Nucl Med Mol Imag*, vol. 34, no. 2, pp. 151–161, Feb 2007, doi:10.1007/s00259-006-0248-6.

[12] J. L. Tremoleda, A. Kerton, and W. Gsell, "Anaesthesia and physiological monitoring during in vivo imaging of laboratory rodents: Considerations on experimental outcomes and animal welfare," *EJNMMI Res*, vol. 2, no. 1, p. 44, Aug. 9, 2012, doi:10.1186/2191-219X-2-44.

[13] G. F. Knoll, *Radiation Detection and Measurement*, 4th edn. Hoboken, NJ: John Wiley, 2010, pp. xxvi, 830.

[14] L. Jodal, C. Le Loirec, and C. Champion, "Positron range in PET imaging: An alternative approach for assessing and correcting the blurring," *Phys Med Biol*, vol. 57, no. 12, pp. 3931–3943, Jun. 21, 2012, doi:10.1088/0031-9155/57/12/3931.

[15] C. S. Levin and E. J. Hoffman, "Calculation of positron range and its effect on the fundamental limit of positron emission tomography system spatial resolution," *Phys Med Biol*, vol. 44, no. 3, pp. 781–799, Mar 1999, doi:10.1088/0031-9155/44/3/019.

[16] A. Sanchez-Crespo, P. Andreo, and S. A. Larsson, "Positron flight in human tissues and its influence on PET image spatial resolution," *Eur J Nucl Med Mol Imag*, vol. 31, no. 1, pp. 44–51, Jan 2004, doi:10.1007/s00259-003-1330-y.

[17] E. J. Hoffman, S.-C. Huang, D. Plummer, and M. E. Phelps, "Quantitation in positron emission computed tomography: VI. Effect of nonuniform resolution," *J Comput Assist Tomogr*, vol. 6, no. 5, pp. 987–999, 1982.

[18] W.-H. Wong, "Designing a stratified detection system for PET cameras," *IEEE Trans Nucl Sci*, vol. NS-33, no. 1, pp. 591–596, 1986.

[19] E. J. Hoffman, T. M. Guerrero, G. Germano, W. M. Digby, and M. Dahlbom, "PET system calibration and corrections for quantitative and spatially accurate images," *IEEE Trans Nucl Sci*, vol. NS-36, no. 1, pp. 1108–1112, 1989.

[20] S. E. Derenzo et al., "Initial characterization of a position-sensitive photodiode/BGO detector for PET," *IEEE Trans Nucl Sci*, vol. 36, no. 1, pp. 1084–1089, 1989.

[21] L. R. MacDonald and M. Dahlbom, "Depth of interaction for PET using segmented crystals," *IEEE Trans Nucl Sci*, vol. 45, no. 4 PT2, pp. 2144–2148, 1998.

[22] M. Dahlbom and E. J. Hoffman, "An evaluation of a two-dimensional array detector for high-resolution PET," *IEEE Trans Med Imag*, vol. 7, no. 4, pp. 264–272, 1988.

[23] M. E. Casey and R. Nutt, "A multicrystal two dimensional BGO detector system for positron emission tomography," *IEEE Trans Nucl Sci*, vol. NS-33, pp. 460–463, 1986.

[24] V. V. Nagarkar, S. V. Ti, K. Shah, I. Shestakova, and S. R. Cherry, "A high efficiency pixelated detector for small animal PET," *IEEE Trans Nucl Sci*, vol. 51, no. 3, pp. 801–804, 2004, doi:10.1109/tns.2004.829750.

[25] G. Konstantinou, R. Chil, M. Desco, and J. J. Vaquero, "Applications of sub-surface laser engraving on monolithic scintillator crystals: Novel pixel geometries and depth of interaction," presented at the 2016 IEEE Nuclear Science Symposium, Medical Imaging Conference and Room-Temperature Semiconductor Detector Workshop (NSS/MIC/RTSD), 2016.

[26] T. Moriya et al., "Development of PET Detectors Using Monolithic Scintillation Crystals Processed with Sub-Surface Laser Engraving Technique," *IEEE Trans Nucl Sci*, vol. 57, no. 5, pp. 2455–2459, 2010, doi:10.1109/tns.2010.2056387.

[27] D. R. Schaart et al., "A novel, SiPM-array-based, monolithic scintillator detector for PET," *Phys Med Biol*, vol. 54, no. 11, pp. 3501–3512, 2009, doi:10.1088/0031-9155/54/11/015.

[28] S. Krishnamoorthy, E. Blankemeyer, P. Mollet, S. Surti, R. Van Holen, and J. S. Karp, "Performance evaluation of the MOLECUBES β-CUBE – a high spatial resolution and high sensitivity small animal PET scanner utilizing monolithic LYSO scintillation detectors," *Phys Med Biol*, vol. 63, no. 15, 2018, doi:10.1088/1361-6560/aacec3.

[29] S. Seifert, G. van der Lei, H. T. van Dam, and D. R. Schaart, "First characterization of a digital SiPM based time-of-flight PET detector with 1 mm spatial resolution," *Phys Med Biol*, vol. 58, no. 9, pp. 3061–3074, 2013, doi:10.1088/0031-9155/58/9/3061.

[30] S. España, R. Marcinkowski, V. Keereman, S. Vandenberghe, and R. Van Holen, "DigiPET: sub-millimeter spatial resolution small-animal PET imaging using thin monolithic scintillators," *Phys Med Biol*, vol. 59, no. 13, pp. 3405–3420, 2014, doi:10.1088/0031-9155/59/13/3405.

[31] G. Borghi, B. J. Peet, V. Tabacchini, and D. R. Schaart, "A 32 mm × 32 mm × 22 mm monolithic LYSO:Ce detector with dual-sided digital photon counter readout for ultrahigh-performance TOF-PET and TOF-PET/MRI," *Phys Med Biol*, vol. 61, no. 13, pp. 4929–4949, 2016, doi:10.1088/0031-9155/61/13/4929.

[32] M. C. Maas et al., "Monolithic scintillator PET detectors with intrinsic depth-of-interaction correction," *Phys Med Biol*, vol. 54, no. 7, pp. 1893–1908, 2009, doi:10.1088/0031-9155/54/7/003.

[33] W. W. Moses, "Recent advances and future advances in time-of-flight PET," (in English), *Aip Conf Proc*, vol. 1204, pp. 119–125, 2009. [Online]. Available: <Go to ISI>://WOS:000281184400017.

[34] Z. Gu, R. Taschereau, N. T. Vu, D. L. Prout, J. Lee, and A. F. Chatziioannou, "Performance evaluation of HiPET, a high sensitivity and high resolution preclinical PET tomograph," *Phys Med Biol*, vol. 65, no. 4, 2020, doi:10.1088/1361-6560/ab6b44.

[35] S. R. Cherry et al., "MicroPET: A high-resolution PET scanner for imaging small animals," *IEEE Trans Nucl Sci*, vol. 44, no. 3, pp. 1161–1166, 1997, doi:10.1109/23.596981.

[36] S. R. Cherry, Y. Shao, R. B. Slates, E. Wilcut, A. Chatziioannou, and M. Dahlbom, "MicroPET II – Design of a 1 mm resolution PET scanner for small animal imaging," *J Nucl Med*, vol. 40, no. 5, 1999.

[37] M. Ito et al., "A four-layer DOI detector with a relative offset for use in an animal PET system," *IEEE Trans Nucl Sci*, vol. 57, no. 3, pp. 976–981, 2010, doi:10.1109/tns.2010.2044892.

[38] I. Szanda et al., "National Electrical Manufacturers Association NU-4 performance evaluation of the PET component of the NanoPET/CT preclinical PET/CT scanner," *J Nucl Med*, vol. 52, no. 11, pp. 1741–1747, Nov 2011, doi:10.2967/jnumed.111.088260.

[39] K. Sato et al., "Performance evaluation of the small-animal PET scanner ClairvivoPET using NEMA NU 4-2008 Standards," *Phys Med Biol*, vol. 61, no. 2, pp. 696–711, Jan. 21, 2016, doi:10.1088/0031-9155/61/2/696.

[40] K. Herrmann et al., "Evaluation of the Genisys4, a bench-top preclinical PET scanner," *J Nucl Med*, vol. 54, no. 7, pp. 1162–1167, Jul 2013, doi:10.2967/jnumed.112.114926.

[41] Z. Gu, R. Taschereau, N. T. Vu, D. L. Prout, J. Lee, and A. F. Chatziioannou, "Performance evaluation of HiPET, a high sensitivity and high resolution preclinical PET tomograph," *Phys Med Biol*, vol. 65, no. 4, p. 045009, Feb 12 2020, doi:10.1088/1361-6560/ab6b44.

[42] S. R. Cherry et al., "Optical fiber readout of scintillator arrays using a multi-channel PMT: A high resolution PET detector for animal imaging," (in English), *IEEE Trans Nucl Sci*, vol. 43, no. 3, pp. 1932–1937, Jun 1996, doi:Doi 10.1109/23.507249.

[43] M. Dahlbom, *Physics of PET and SPECT Imaging*, Boca Raton, FL: CRC Press, Taylor & Francis Group, 2017, p. 1 online resource. [Online]. Available: Restricted to UCLA. Limited to single concurrent user. Try again later if unavailable. http://openurl.cdlib.org?sid=UCLA:CAT&genre=book&__char_set=utf8&isbn=9781315356785 BIOMEDICALSCIENCEnetBASE. Restricted to UC campuses http://dx.doi.org/10.1201/9781315374383.

[44] R. Lecomte, D. Schmitt, A. W. Lightstone, and R. J. McIntyre, "Performance characteristics of BGO-silicon avalanche photodiode detectors for PET," *IEEE Trans Nucl Sci*, vol. NS-32, no. 1, pp. 482–486, 1985.

[45] P. Buzhan et al., "Silicon photomultiplier and its possible applications," (in English), *Nucl Instrum Meth A*, vol. 504, no. 1–3, pp. 48–52, May 21, 2003, doi:10.1016/S0168-9002(03)00749-6.

[46] P. S. Marrocchesi et al., "Active control of the gain of a 3 mm x 3 mm Silicon PhotoMultiplier," (in English), *Nucl Instrum Meth A*, vol. 602, no. 2, pp. 391–395, Apr 21 2009, doi:10.1016/j.nima.2008.12.199.

[47] M. Ito, S. J. Hong, and J. S. Lee, "Positron emission tomography (PET) detectors with depth-of- interaction (DOI) capability," *Biomed Eng Lett*, vol. 1, no. 2, pp. 70–81, 2011, doi:10.1007/s13534-011-0019-6.

[48] Y. Yang et al., "Depth of interaction calibration for PET detectors with dual-ended readout by PSAPDs," *Phys Med Biol*, vol. 54, no. 2, pp. 433–445, 2009, doi:10.1088/0031-9155/54/2/017.

[49] Y. Yang et al., "Tapered LSO arrays for small animal PET," *Phys Med Biol*, vol. 56, no. 1, pp. 139–153, Jan. 7, 2011, doi:10.1088/0031-9155/56/1/009.

[50] J. Seidel, J. J. Vaquero, S. Siegel, W. R. Gandler, and M. V. Green, "Depth identification accuracy of a three layer phoswich PET detector module," *IEEE Trans Nucl Sci*, vol. 46, no. 3, pp. 485–490, 1999, doi:10.1109/23.775567.

[51] C. S. Levin, "Design of a high-resolution and high-sensitivity scintillation crystal array for PET with nearly complete light collection," *IEEE Trans Nucl Sci*, vol. 49, no. 5, pp. 2236–2243, 2002, doi:10.1109/tns.2002.803870.

[52] T. Tsuda et al., "A four-layer depth of interaction detector block for small animal PET," *IEEE Trans Nucl Sci*, vol. 51, no. 5, pp. 2537–2542, 2004, doi:10.1109/tns.2004.835739.

[53] S. Abbaszadeh, Y. Gu, P. D. Reynolds, and C. S. Levin, "Characterization of a sub-assembly of 3D position sensitive cadmium zinc telluride detectors and electronics from a sub-millimeter resolution PET system," *Phys Med Biol*, vol. 61, no. 18, pp. 6733–6753, Sep. 21, 2016, doi:10.1088/0031-9155/61/18/6733.

[54] S. Abbaszadeh and C. S. Levin, "New-generation small animal positron emission tomography system for molecular imaging," *J Med Imaging (Bellingham)*, vol. 4, no. 1, p. 011008, Jan 2017, doi:10.1117/1.JMI.4.1.011008.

[55] S. Abbaszadeh, G. Chinn, and C. S. Levin, "Positioning true coincidences that undergo inter-and intra-crystal scatter for a sub-mm resolution cadmium zinc telluride-based PET system," *Phys Med Biol*, vol. 63, no. 2, p. 025012, Jan. 9, 2018, doi:10.1088/1361-6560/aa9a2b.

[56] S. St James, Y. Yang, S. L. Bowen, J. Qi, and S. R. Cherry, "Simulation study of spatial resolution and sensitivity for the tapered depth of interaction PET detectors for small animal imaging," *Phys Med Biol*, vol. 55, no. 2, pp. N63–N74, 2010, doi:10.1088/0031-9155/55/2/n04.

[57] Z. Gu et al., "NEMA NU-4 performance evaluation of PETbox4, a high sensitivity dedicated PET preclinical tomograph," *Phys Med Biol*, vol. 58, no. 11, pp. 3791–3814, Jun. 7, 2013, doi:10.1088/0031-9155/58/11/3791.

[58] V. V. Nagarkar, V. Gaysinskiy, I. Shestakova, and S. Taylor, "Microcolumnar CsI(Tl) films for small animal SPECT," presented at the 2004 IEEE Symposium Conference Record Nuclear Science, 2004.

[59] V. V. Nagarkar, S. Miller, R. Sia, and V. Gaysinskiy, "Microcolumnar and polycrystalline growth of LaBr3:Ce scintillator," (in English), *Nucl Instrum Meth A*, vol. 633, pp. S286–S288, May 2011, doi:10.1016/j.nima.2010.06.190.

[60] B. W. Miller, H. B. Barber, H. H. Barrett, Z. L. Liu, V. V. Nagarkar, and L. R. Furenlid, "Progress in BazookaSPECT: High-resolution, dynamic scintigraphy with large-area imagers," (in English), *Medical Applications of Radiation Detectors Ii*, vol. 8508, 2012, doi:Artn 85080f10.1117/12.966810.

[61] N. Schramm, A. Wirrwar, F. Sonnenberg, and H. Halling, "Compact high resolution detector for small animal SPECT," *IEEE Trans Nucl Sci*, vol. 47, no. 3, pp. 1163–1167, 2000, doi:10.1109/23.856564.

[62] L. R. Furenlid et al., "FastSPECT II: A second-generation high-resolution dynamic SPECT imager," *IEEE Trans Nucl Sci*, vol. 51, no. 3, pp. 631–635, 2004, doi:10.1109/tns.2004.830975.

[63] D. P. McElroy et al., "Evaluation of A-SPECT: a desktop pinhole SPECT system for small animal imaging," presented at the 2001 IEEE Nuclear Science Symposium Conference Record (Cat. No.01CH37310), 2002.

[64] M. A. Kupinski and H. H. Barrett, *Small-animal SPECT Imaging*. New York: Springer, 2005, pp. xxii, 294 p.

[65] G. A. de Vree, A. H. Westra, I. Moody, F. van der Have, K. M. Ligtvoet, and F. J. Beekman, "Photon-counting gamma camera based on an electron-multiplying CCD," *IEEE Trans Nucl Sci*, vol. 52, no. 3, pp. 580–588, 2005, doi:10.1109/tns.2005.851443.

[66] B. W. Miller, H. B. Barber, H. H. Barrett, D. W. Wilson, and L. Chen, "A Low-cost approach to high-resolution, single-photon imaging using columnar scintillators and image intensifiers," presented at the 2006 IEEE Nuclear Science Symposium Conference Record, 2006.

[67] F. J. Beekman and G. A. d. Vree, "Photon-counting versus an integrating CCD-based gamma camera: important consequences for spatial resolution," *Phys Med Biol*, vol. 50, no. 12, pp. N109–N119, 2005, doi:10.1088/0031-9155/50/12/n01.

[68] K. Deprez, R. Van Holen, and S. Vandenberghe, "A high resolution SPECT detector based on thin continuous LYSO," *Phys Med Biol*, vol. 59, no. 1, pp. 153–171, Jan. 6, 2014, doi:10.1088/0031-9155/59/1/153.

[69] B. L. Franc, P. D. Acton, C. Mari, and B. H. Hasegawa, "Small-animal SPECT and SPECT/CT: Important tools for preclinical investigation," *J Nucl Med*, vol. 49, no. 10, pp. 1651–1663, Oct 2008, doi:10.2967/jnumed.108.055442.

[70] D. A. Weber and M. Ivanovic, "Ultra-high-resolution imaging of small animals: Implications for preclinical and research studies," *J Nucl Cardiol*, vol. 6, no. 3, pp. 332–344, May–Jun 1999, doi:10.1016/s1071-3581(99)90046-6.

[71] F. J. Beekman, D. P. McElroy, F. Berger, S. S. Gambhir, E. J. Hoffman, and S. R. Cherry, "Towards in vivo nuclear microscopy: Iodine-125 imaging in mice using micro-pinholes," *Eur J Nucl Med Mol Imaging*, vol. 29, no. 7, pp. 933–938, 2002, doi:10.1007/s00259-002-0805-6.

[72] H. Jacobowitz and S. D. Metzler, "Geometric sensitivity of a pinhole collimator," *Int J Math Math Sci.*, vol. 2010, pp. 1–18, 2010, doi:10.1155/2010/915958.

[73] D. P. McElroy et al., "Performance evaluation of A-SPECT: A high resolution desktop pinhole SPECT system for imaging small animals," *IEEE Trans Nucl Sci*, vol. 49, no. 5, pp. 2139–2147, 2002, doi:10.1109/tns.2002.803801.

[74] B. W. Miller, L. R. Furenlid, S. K. Moore, H. B. Barber, V. V. Nagarkar, and H. H. Barrett, "System integration of FastSPECT III, a dedicated SPECT rodent-brain imager based on BazookaSPECT detector technology," presented at the 2009 IEEE Nuclear Science Symposium Conference Record (NSS/MIC), 2009.
[75] S. R. Meikle et al., "Performance evaluation of a multipinhole small animal SPECT system," presented at the 2003 IEEE Nuclear Science Symposium. Conference Record (IEEE Cat. No.03CH37515), 2003.
[76] K. Vunckx et al., "Optimized multipinhole design for mouse imaging," *IEEE Trans Nucl Sci*, vol. 56, no. 5, pp. 2696–2705, 2009, doi:10.1109/tns.2009.2030194.
[77] F. J. Beekman and B. Vastenhouw, "Design and simulation of a high-resolution stationary SPECT system for small animals," *Phys Med Biol*, vol. 49, no. 19, pp. 4579–4592, 2004, doi:10.1088/0031-9155/49/19/009.
[78] F. P. DiFilippo, "Design and performance of a multi-pinhole collimation device for small animal imaging with clinical SPECT and SPECT–CT scanners," *Phys Med Biol*, vol. 53, no. 15, pp. 4185–4201, 2008, doi:10.1088/0031-9155/53/15/012.
[79] G. S. Mok, B. M. Tsui, and F. J. Beekman, "The effects of object activity distribution on multiplexing multi-pinhole SPECT," *Phys Med Biol*, vol. 56, no. 8, pp. 2635–2650, Apr. 21, 2011, doi:10.1088/0031-9155/56/8/019.
[80] K. Deprez, R. Van Holen, and S. Vandenberghe, "The lofthole: A novel shaped pinhole geometry for optimal detector usage without multiplexing and without additional shielding," presented at the 2011 IEEE Nuclear Science Symposium Conference Record, 2011.
[81] K. Van Audenhaege, R. Van Holen, S. Vandenberghe, C. Vanhove, S. D. Metzler, and S. C. Moore, "Review of SPECT collimator selection, optimization, and fabrication for clinical and preclinical imaging," *Med Phys*, vol. 42, no. 8, pp. 4796–4813, 2015, doi:10.1118/1.4927061.
[82] M. C. Goorden and F. J. Beekman, "High-resolution tomography of positron emitters with clustered pinhole SPECT," *Phys Med Biol*, vol. 55, no. 5, pp. 1265–1277, Mar. 7, 2010, doi:10.1088/0031-9155/55/5/001.
[83] R. Van Holen, B. Vandeghinste, K. Deprez, and S. Vandenberghe, "Design and performance of a compact and stationary microSPECT system," *Med Phys*, vol. 40, no. 11, p. 112501, Nov 2013, doi:10.1118/1.4822621.
[84] P. E. B. Vaissier, M. C. Goorden, B. Vastenhouw, F. van der Have, R. M. Ramakers, and F. J. Beekman, "Fast spiral SPECT with stationary cameras and focusing pinholes," *J Nucl Med*, vol. 53, no. 8, pp. 1292–1299, 2012, doi:10.2967/jnumed.111.101899.
[85] S. R. Cherry, "Multimodality in vivo imaging systems: Twice the power or double the trouble?" *Annu Rev Biomed Eng*, vol. 8, no. 1, pp. 35–62, 2006, doi:10.1146/annurev.bioeng.8.061505.095728.
[86] Z. Gu et al., "Performance evaluation of G8, a high-sensitivity benchtop preclinical PET/CT tomograph," *J Nucl Med*, vol. 60, no. 1, pp. 142–149, Jan 2019, doi:10.2967/jnumed.118.208827.
[87] B. J. Pichler et al., "Performance test of an LSO-APD detector in a 7-T MRI scanner for simultaneous PET/MRI," (in English), *J Nucl Med*, vol. 47, no. 4, pp. 639–647, Apr 2006. [Online]. Available: <Go to ISI>://WOS:000236593600026.
[88] S. I. Ziegler et al., "A prototype high-resolution animal positron tomograph with avalanche photodiode arrays and LSO crystals," (in English), *European Journal of Nuclear Medicine*, vol. 28, no. 2, pp. 136–143, Feb 2001, doi:DOI 10.1007/s002590000438.
[89] M. S. Judenhofer et al., "PET/MR images acquired with a compact MR-compatible PET detector in a 7-T magnet," (in English), *Radiology*, vol. 244, no. 3, pp. 807–814, Sep 2007, doi:10.1148/radiol.2443061756.
[90] M. S. Judenhofer et al., "Simultaneous PET-MRI: A new approach for functional and morphological imaging," (in English), *Nat Med*, vol. 14, no. 4, pp. 459–465, Apr 2008, doi:10.1038/nm1700.
[91] X. Lai and L.-J. Meng, "Simulation study of the second-generation MR-compatible SPECT system based on the inverted compound-eye gamma camera design," *Phys Med Biol*, vol. 63, no. 4, 2018, doi:10.1088/1361-6560/aaa4fb.
[92] E. M. Zannoni, X. C. Lai, and L. J. Meng, "Design study for MRC-SPECT-II: The second generation MRI compatible SPECT system based on hyperspectral compound-eye gamma cameras," (in English), *J Nucl Med*, vol. 58, May 1, 2017. [Online]. Available: <Go to ISI>://WOS:000404949906112.
[93] B. A. Klaunberg and J. A. Davis, "Considerations for laboratory animal imaging center design and setup," *ILAR J*, vol. 49, no. 1, pp. 4–16, 2008, doi:10.1093/ilar.49.1.4.
[94] NEMA, "NEMA NU 4-2008 performance measurements of small animal positron emission tomographs," ed.: National Electrical Manufacturers Association, Rosslyn, VA, USA, 2008.
[95] A. L. Goertzen et al., "NEMA NU 4-2008 comparison of preclinical PET imaging systems," (in English), *J Nucl Med*, vol. 53, no. 8, pp. 1300–1309, Aug 2012, doi:10.2967/jnumed.111.099382.

29 Monte Carlo Simulation of Nuclear Medicine Imaging Systems

David Sarrut and Michael Ljungberg

CONTENTS

DOI: 10.1201/9780429489556-29

29.1 INTRODUCTION

The Monte Carlo method is a statistical method of simulating specified situations based on random numbers and probability distributions. In most Monte Carlo applications, the physical process can be simulated directly. The method requires that the system and the physical processes can be modelled from known probability density functions (pdfs). If these pdfs can be defined accurately, the simulation can be performed by random sampling from them. To obtain reasonable statistical errors, a sufficient number of histories (e.g. photon or electron tracks) must be simulated in order to get an accurate estimate of the parameters to be calculated.

Generally, simulation studies have several advantages over experimental studies. For any given model, it is quite easy to change different parameters and investigate the effect of these changes on the performance of the system under investigation. Thus, the optimization of an imaging system can be aided greatly by the use of simulations. Additionally, it is possible to study parameters and effects that cannot be measured experimentally. For example, by using a Monte Carlo technique incorporating known physics of the scattering process it is possible to simulate scattering events from an object and determine their effect on the final image. This type of study has included measurements of the scatter to primary ratios, the shape of the scatter response function, the shape of the energy spectrum, the proportion of photons undergoing various numbers of scattering events, and the effects of attenuator shape and composition in addition to camera parameters such as energy resolution and window size. However, simulations remain complementary to experimental studies and should always be contrasted, partially, to experimental measurements.

Reviews of the Monte Carlo method and its applications in different fields of radiation physics have been presented elsewhere. One of the earlier reviews was published by Raeside [1] and, for example, Andreo [2], Zaidi [3], and Buvat & Castiglioni [4] also published reviews on this topic. Ljungberg, Strand, and King edited two text books on the application of Monte Carlo calculations in diagnostic nuclear medicine imaging [5, 6], and some of the content of those books has been reproduced in this chapter with permission from the publisher. Zaidi edited a book in the same series on the therapeutic application of the Monte Carlo method in nuclear medicine [7].

29.2 PSEUDO-RANDOM NUMBER GENERATOR

A fundamental part of any Monte Carlo calculation is the pseudo-random number generator. For practical considerations, a computer algorithm is used to generate uniformly distributed (pseudo) random numbers from an initial seed number. One example of such an algorithm is the linear congruential algorithm, in which a series of random numbers I_n is calculated from an initial value I_0, according to $I_{n+1} = (aI_n + b)mod(2^k)$, where a and b are constants and k is the size of the integer dimension in the code (e.g. 32 or 64 bits). Using the same value of the initial integer value I_0 will yield exactly the same sequence of random numbers. When comparing different simulations, one must therefore change the value of the initial seed to avoid obtaining the same results in a repeated simulation. There is also a chance that the initial seed number can appear later on in the sequence and thereby create an identical replication of the following random numbers. The severity of this repetition depends on the application because, at the stage of repetition, the application must be in exactly the same state as it was when the first random number was used. If not, the random number will still be regarded as uniformly distributed with respect to the application. The period of a pseudo random number generator, that is, the expected number of generated numbers before the engine starts to repeat itself, however, is an important parameter when comparing algorithms.

29.3 SAMPLING TECHNIQUES

A priori information is needed about the physical process to be simulated and is usually expressed as probability distribution functions (pdfs). From these, different stochastic processes or parameters can be selected concerning, for example, which type of photon interaction will be simulated at a particular position or how far a particular photon will travel before its subsequent interaction. A pdf is defined over the range of $[a, b]$. Two different methods can be used to obtain a stochastic variable that follows a certain pdf.

29.3.1 Distribution Function Method

A cumulative probability distribution function *cpdf(x)* is constructed from the integral of *pdf(x)* over the interval $[a, x]$ according to

$$cpdf(x) = \int_a^x pdf(x')dx'. \tag{29.1}$$

A random sample x is then sampled by substituting the term *cpdf(x)* in Eq. 29.1 with the value of an random number, uniformly distributed in the range of [0,1], and the from this solve the equation for x. Three examples of *pdf(x)'s* and corresponding *cpdf(x)*'s are shown in Figure 29.1.

29.3.2 'Rejection' Method

If the distribution function method is not possible to use for mathematical reasons, the 'rejection method' can be used. This method works according to the steps shown in Figure 29.2a. As an example, Figure 29.2b shows a plot where the

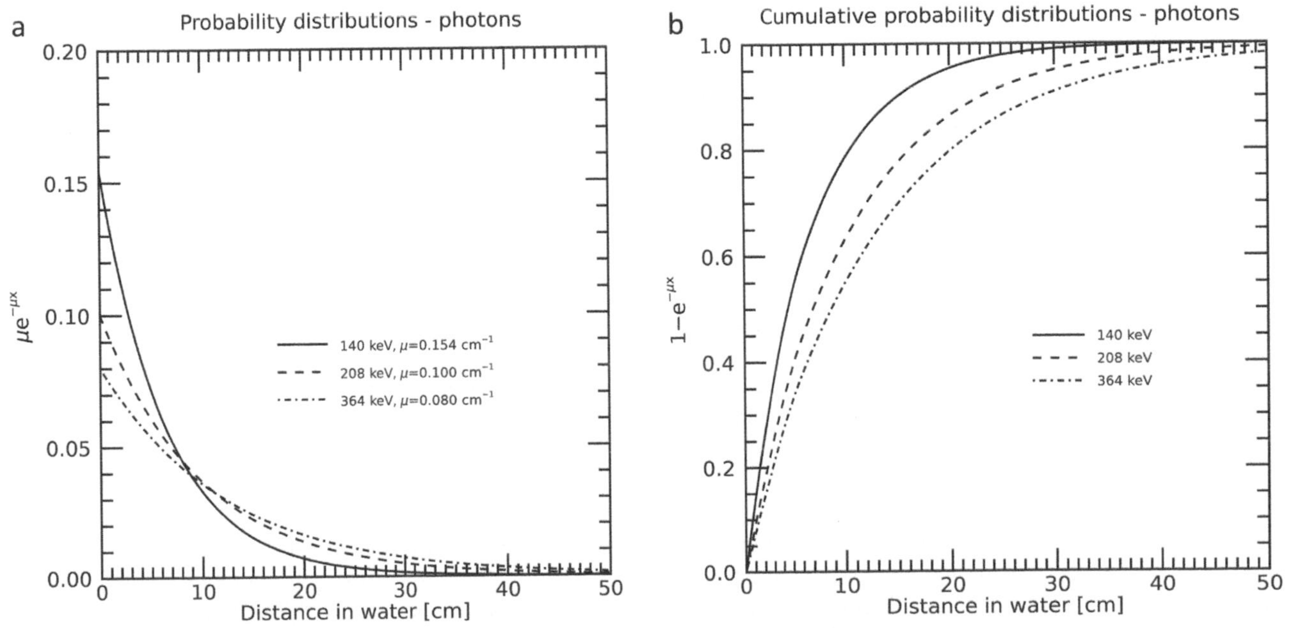

FIGURE 29.1 (a) Probability distribution functions $\mu e^{-\mu x}$ for photons of three different energies travelling in water. μ is the linear attenuation coefficient for the material and x the distance travelled by the photon (b) Cumulative probability distribution for the same three photon energies. In practical terms, these graphs show the probability that a photon travels a distance x cm or less in a volume of water.

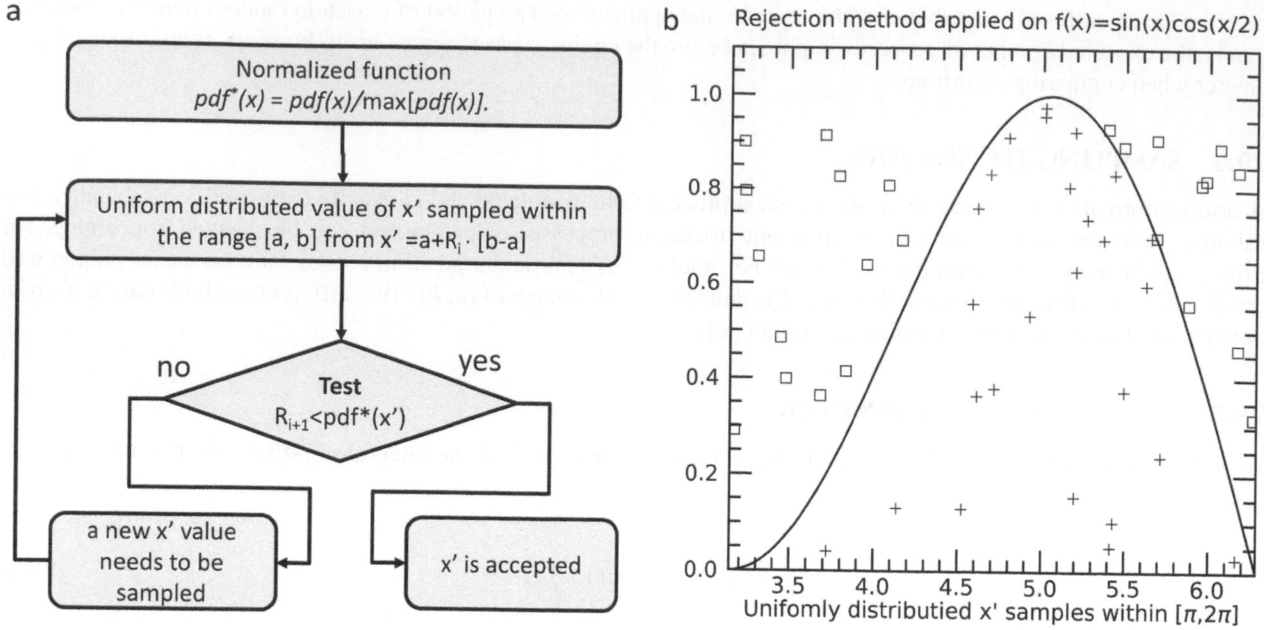

FIGURE 29.2 Flowchart (a) describing the principles of the rejection method. The graph (b) to the right shows an example of a normalized function and 100 samples. Only those samples that are below the line that describes the function are accepted. The others are 'rejected'; that is, they are not used in subsequent calculations.

purpose was to sample stochastic samples x' from a function $f(x) = \sin(x) \cdot \cos(x/2)$. One hundred samples of x' were generated randomly over the range [π, 2π]. The validity of the x' samples were tested against a normalized version $f^*(x) = f(x)/f_{\max}$. The square symbols in Figure 29.2b represent the x' values that were rejected by the test procedure, whereas the plus symbols represent those that were accepted. From this procedure, a sequence of individual x' samples can be generated that follow the distribution described by the solid line in Figure 29.2b.

29.3.3 'Mixed' Method

A combination of the two methods described above can be used to overcome potential mathematical problems that make it impossible to use either the rejection method or the distribution function method. Here, the *pdf*(*x*) may be rewritten as the product of two new probability distribution functions $pdf_A(x) \cdot pdf_B(x)$ in such a way that the two sampling methods can be used in combination. The steps of this method are illustrated in Figure 29.3.

An example of a mixed method is the sampling of the energy and scattering angle of a Compton photon from the Klein–Nishina cross-section, the procedure for which is described in detail in section 29.4.3.2.

29.4 SAMPLING OF PHOTON INTERACTIONS

As the main focus of this chapter is on Monte Carlo applications for photon transport, a description of the basic aspects of simulating photon transport in a medium is instructive.

29.4.1 Cross-section Data

Data on the scattering and absorption of photons are fundamental for all Monte Carlo calculations, because the accuracy of the simulation depends on the accuracy of the probability functions, that is the cross-section tables [8–10]. Photon cross-sections for compounds (mixtures of materials) can be obtained quite accurately (except at energies close to absorption edges) as a weighted sum of the cross-sections for the different atomic constituents. More details on cross-sections can be found in Chapter 3 of this textbook.

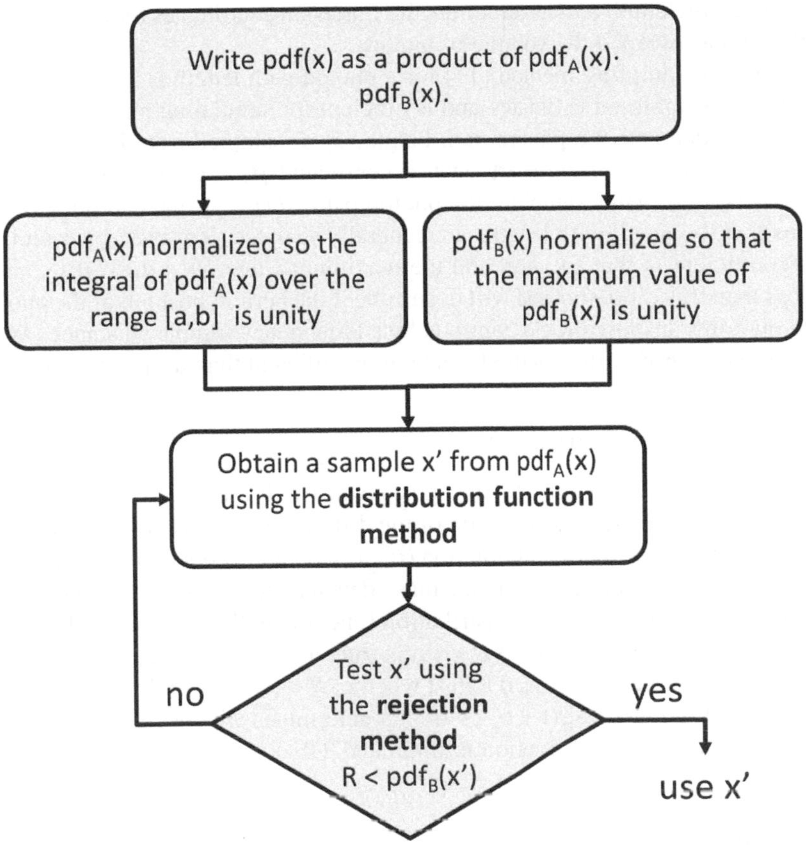

FIGURE 29.3 Flowchart describing the principles of the 'mixed' sampling method.

29.4.2 Photon Path Length

The path length of a photon must be calculated to determine whether the photon escapes the volume of interest. Generally, this distance depends upon the photon energy and the material density and composition. The distribution function method can be used to sample the distributed photon path length *x*. If the probability function is given by

$$p(x)dx = \mu e^{-\mu x}dx, \tag{29.2}$$

then the probability that a photon will travel the distance *d* or less is given by

$$cpdf(d) = \int_0^d \mu e^{-\mu x}dx = \left[-e^{-\mu x}\right]_0^d = 1 - e^{-\mu d}. \tag{29.3}$$

To sample the path length, the *cpdf(d)* is substituted by a uniform random number *R* and the equation is solved for *d*, according to:

$$R = cpdf(d) = 1 - e^{-\mu d} \tag{29.4}$$

$$d = -\frac{1}{\mu}\ln(1-R) \tag{29.5}$$

After sampling a new photon path length and its direction, the Cartesian coordinates for the end point are often calculated to check whether the photon has escaped the volume of interest.

The *Delta-scattering method* sampling method [11] for a photon path length is a statistical method involving fictitious interactions that produces unbiased estimates and is efficient for simulating photon transport in a heterogeneous volume of interest. The basic idea is that a photon path-length, d_i, is sampled from Eq. 26.5, but with the main difference that it is always the maximum value of the attenuation coefficient μ_{max}, defined in the volume where the radiation transport is simulated, that is used. After sampling the photon path, $d_i(\mu_{max})$, the coordinates (x,y,z) for the endpoint is calculated. This means that too-small path lengths are generally sampled. However, by calculating the ratio between the current attenuation coefficient at that location and the maximum value, $\mu(x,y,z)/\mu_{max}$, and checking whether a random number R_1 is less than this ratio, this test will determine if the current position of the photon should be accepted as the endpoint or if the photon transport should continue by adding a new sampled distance . When using voxel-based phantoms that consist of large matrices, this method can be more efficient than using explicit ray-tracing methods.

29.4.3 Selecting Type of Photon Interaction

The probability for a certain interaction type to occur is given by differential attenuation coefficients. These are tabulated for different energies and materials. The sum of the differential attenuation coefficients for the photoelectric effect (τ), Compton interaction (σ_{inc}), coherent interaction (σ_{coh}), and pair production (κ) is called the linear attenuation coefficient $\mu = \tau + \sigma_{inc} + \sigma_{coh} + \kappa$ or, if normalized by the mass-density, the mass-attenuation coefficient. To select a particular interaction type during the simulation, the distribution function method can be used (section 29.3.1). A uniform random number R is sampled and, if the condition $R < \tau/\mu$ is true, then a photoelectric interaction is simulated. If this condition is false, then the same value of R is used to test whether $R < (\tau + \sigma_{inc})/\mu$. If this is true, then a Compton interaction is simulated. If not, then the test $R < (\tau + \sigma_{inc} + \sigma_{coh})/\mu$ determines whether a coherent interaction is simulated. If all the conditions are false, then a pair production is simulated. Obviously, this will only occur if the photon energy is greater than 1.022 MeV.

29.4.3.1 Photo-Absorption

In this process, the photon energy is completely absorbed by an orbital electron. In the simplest terms, the photon history is terminated, and the energy and other parameters are scored. However, it is possible that secondary characteristic x-rays and Auger electrons may be emitted. The relative probability for these two emissions is given by the fluorescence yield. If a characteristic x-ray is selected, then a new photon energy and an isotropic direction are sampled. The new photon is then followed until its absorption or escape. The deposited energy is given by the incoming photon energy minus the binding energy of the rejected electron.

29.4.3.2 Incoherent Photon Scattering

Incoherent scattering, commonly denoted as Compton scattering, refers to an interaction between an incoming photon and an atomic electron, in which the photon loses energy and changes direction. The energy of the scattered photon depends upon the initial photon energy $h\upsilon$ and the scattering angle θ, with respect to the incident path. One method commonly used to sample the energy and direction of a Compton-scattered photon uses the algorithm developed by Kahn [12]. This algorithm is based on the Klein–Nishina cross-section, which is described in more detail in Chapter 3. The method is based on a mixed method sampling scheme, as shown in Figure 29.4.

The mathematical proof for this algorithm was described in an early review by Raeside [1]. The rejection method, described above, is derived assuming scattering from a free electron at rest using the Klein–Nishina cross-section. The atomic cross-section under this assumption is calculated by multiplying by the atomic number Z. For situations in which the incoming photon energy is of the same order as the binding energies of the atomic-shell electrons, the assumption of a Compton scattering against a free electron at rest becomes less valid. This is particularly applicable to the case of a Compton camera imaging system, where the so-called Doppler effect leads to reduced image resolution. A method to account for this dependence of the binding energies in a Monte Carlo simulation was described by Persliden [13].

29.4.3.3 Coherent Photon Scattering

Coherent scattering is an interaction between an incoming photon and an electron, in which the direction of the photon is changed but without energy loss. This type of interaction leads to photons that scatter mostly in the forward direction.

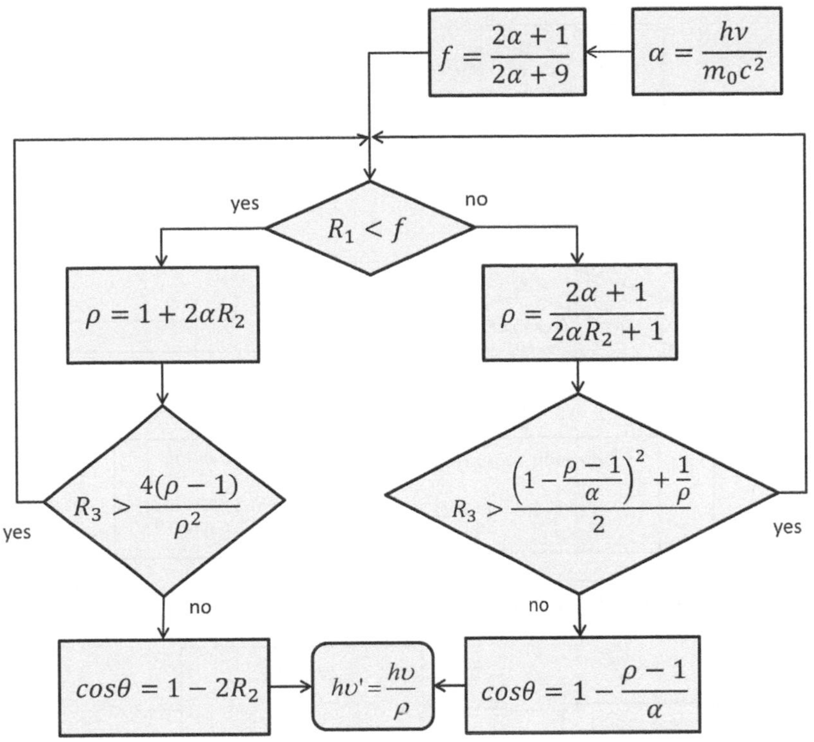

FIGURE 29.4 Flowchart describing a 'mixed method' to sample scattering angle and energy for a Compton-scattered photon. The method is based on the Klein–Nishina cross-section.

The sampling technique for coherent scattering, which was also described by Persliden [13], is based on the Thompson cross-section multiplied by the atomic form factor $F(x,Z)$ [14], and the equation for this is shown in Chapter 3 (Eq. 3.20).

29.4.3.4 Pair Production

The majority of this type of simulation concerns book-keeping parameters. In many photon-simulation programs, the range of the positron is omitted, and the point of annihilation is equal to the interaction point. An initial photon is assigned the energy of 511 keV and is emitted in an isotropically sampled direction. The location (x,y,z) and the direction vectors are stored, and the photon is within the volume until its absorption or escape. The current position is reset to the annihilation location, a second 511 keV photon is emitted in a direction opposite to the first, and then this is followed until its absorption or escape. To make the model more realistic, it may be necessary to simulate the effect of annihilation-in-flight, which results in a non-180° emission between the two photons, as well as accounting for the positron path length (range).

29.4.4 Photon Transport Calculation Scheme

Figure 29.5 shows a flowchart of a photon simulation in a volume including photo-absorption, incoherent and coherent scattering, pair production, and simulation of characteristic x-ray emission at the site of photo-absorption.

29.5 SAMPLING OF ELECTRON INTERACTIONS

In many cases in Monte Carlo simulation of nuclear medicine applications, and especially for imaging, the energy released by secondary electrons can be regarded as locally absorbed at the interaction site. However, there are some applications that this assumption does not hold. One example is absorbed dose calculations in small regions such as pre-clinical dosimetry [15]. Another example is the simulation of bremsstrahlung imaging where interacting electrons can produce photons possible to use for imaging [16–18]. In these applications, access to Monte Carlo codes that include a detailed charged-particle simulation can be necessary.

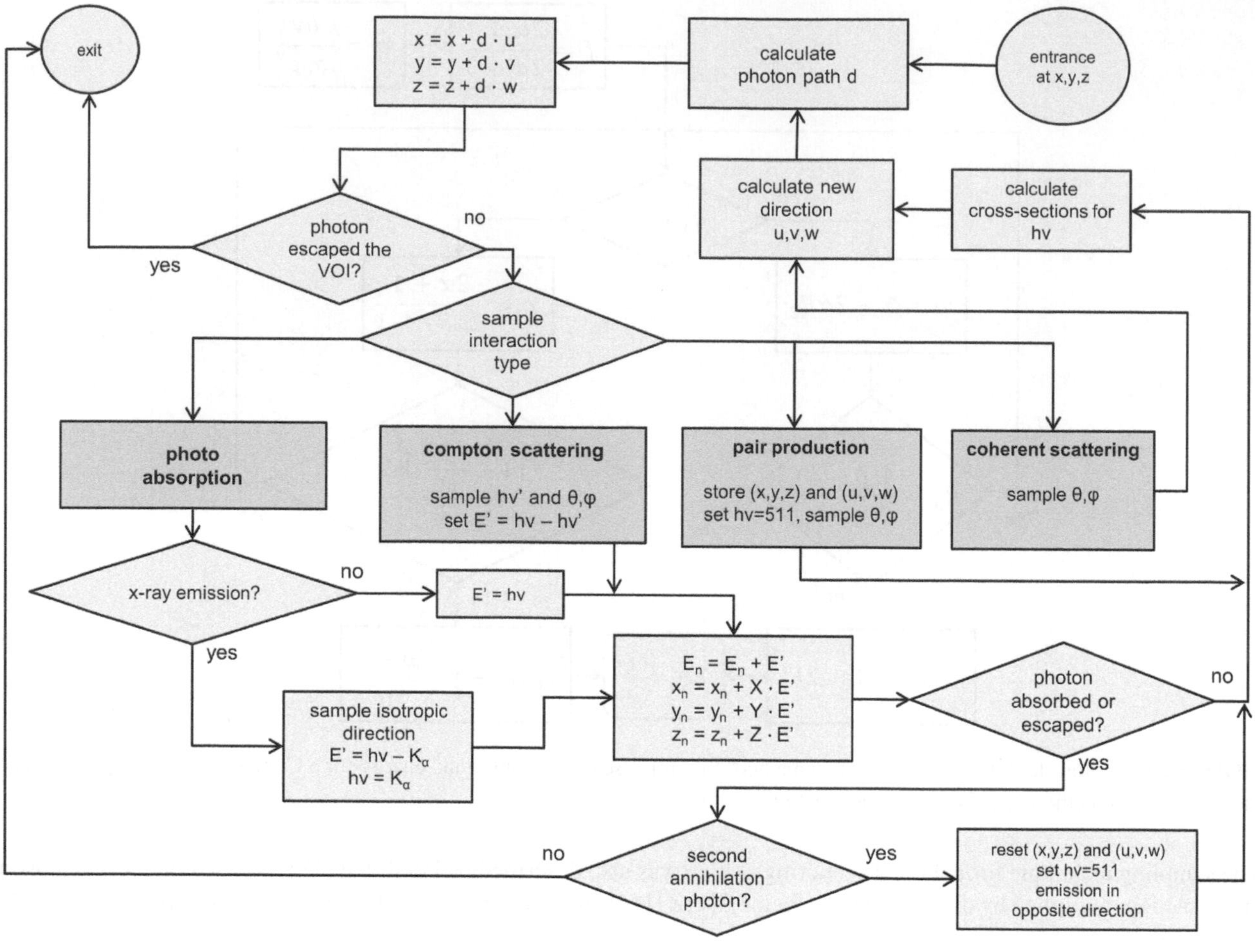

FIGURE 29.5 Flowchart describing the basic steps for simulation of photon transport in a defined volume. Four photon interaction types are simulated to calculate the deposited energy E_n and the energy-weighted coordinates (x_n, y_n, z_n). Image reprinted from [6].

A Monte Carlo simulation of charged particles, such as electrons and positrons, differ from a photon simulation in that most electrons interact by the weak Coulomb force which means that, for each electron history, interactions may typically occur in the order of millions before termination, as compared to photons that undergo relatively few interactions (order of up to 10) before absorption. Most of the electron interactions will be inelastic scatterings with atomic electrons, which results in small angular deflections at each interaction site with a very small energy loss. Only a few of the electron interactions occur by elastic scattering with the atomic nuclei, production of secondary large-energy electrons, and bremsstrahlung photon generation. Inelastic electron interactions that result in a large change in kinetic energies and directions are sometimes called 'catastrophic' events, whereas interactions resulting in only small changes in direction and energy are categorized as 'non-catastrophic' events. Since non-catastrophic events will be the vast majority, then it becomes very time-consuming to simulate the radiation transport in a detailed mode, that is, explicitly simulating each particle interaction. Therefore, in order to reduce the calculation time, it has been common to implement so-called condensed history methods whereby many electron interactions are condensed into larger steps (see Figure 29.6).

This type of transport was first suggested in 1963 by Berger [19], who proposed simulation of the diffusion of electrons by a number of 'snapshots' taken at different time or range intervals. The interactions between these snapshots were thus combined into 'large' events where changes in kinetic energy, particle direction and position are described by multiple-scattering theories. The different methods and its implementation will be discussed in detail by José M. Fernández-Varca in chapter 8, volume II of this book.

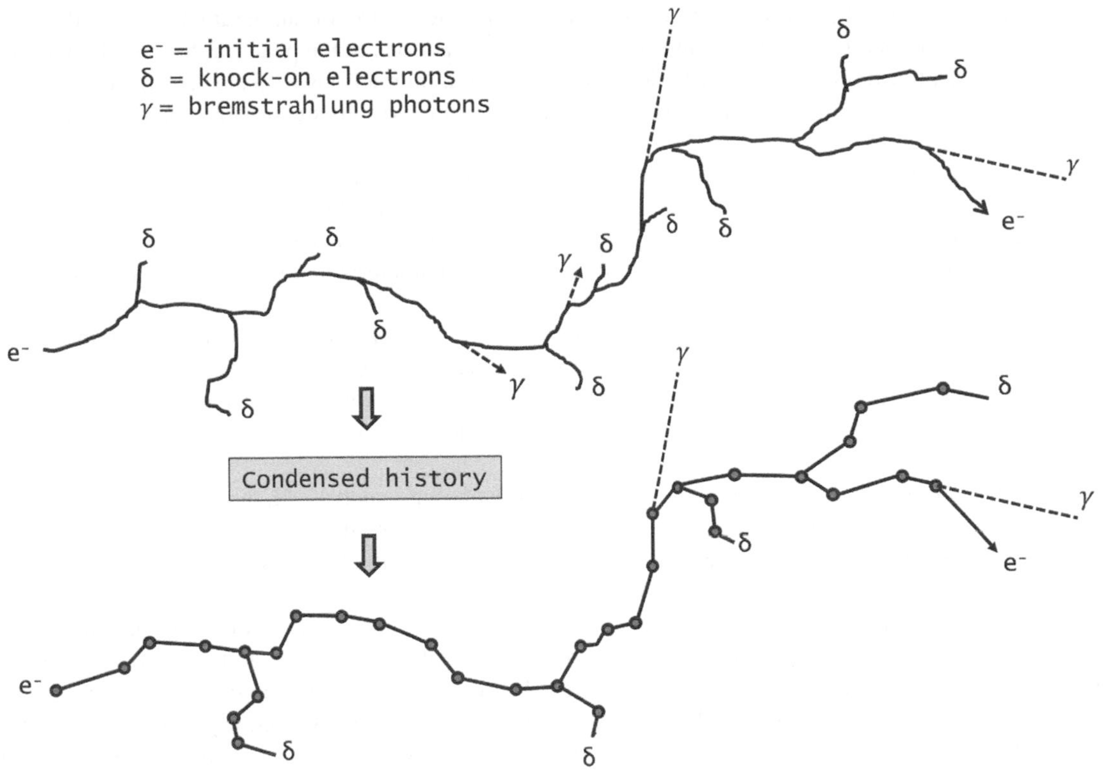

FIGURE 29.6 Schematic figure of hypothetic electron tracks and how there are discretized in small step-lengths by the application of a condensed history methods. Image reprinted from [6].

29.6 SIMIND MONTE CARLO PROGRAM

The Monte Carlo code, SIMIND [20], was designed to simulate a clinical scintillation camera and has recently been extended to solid-state CZT cameras, and could easily be modified for almost any type of calculation or measurement encountered in planar and SPECT imaging. The SIMIND system consists of two main programs, CHANGE and SIMIND. The CHANGE program provides a menu-driven method of defining the imaging system to be simulated, and the program writes these parameters to an external file. The actual Monte Carlo simulations are performed by SIMIND, which reads the input files created by CHANGE and writes the results of its calculations to the screen or to various data files. In this way, several input files can be prepared in a command file for submission to a batch queue, which is a convenient mode of working with Monte Carlo simulations.

SIMIND includes a logical flag system that provides the user a means to turn on and off different features, such as SPECT simulation, simulation of interactions in the phantom, or adding a Gaussian-based energy resolution – all without the need to redefine the input variables or the input file, as the control of the flags can be defined from the command line.

SIMIND uses the Interfile V3.3 as an intermediate file format. Although this is not a standard format, such as Dicom, it is easy to understand and to extract information. This format is defined by a list of key-value pairs located in the header of an image data file [21] or as a separate file.

The entire code was written in Fortran and includes versions that are fully operational on Linux, Windows, and MacOS operating systems. No imaging routines or reconstruction programs are provided. However, the distribution includes a conversion program, SMC2CASTOR; this program converts SIMIND projection files to be readable by the CASToR tomographic reconstruction program, which is freely available [22].

29.6.1 General Components: Geometry, Physics, Source

Briefly, the Monte Carlo program SIMIND works as follows: Photons are emitted from a simulated activity distribution (source volume) in a phantom volume, where interactions occur and are followed step by step towards the scintillation camera. Because the details of the photon history are known, important parameters that otherwise are not accessible by measurements (such as the number of scattering interactions in the phantom for a particular photon history or scatter order, scatter angles, and energy, deposited by photon interactions with electrons) can be scored during the simulation. Several variance-reduction techniques have been employed to significantly increase the efficiency of the simulation.

Differential cross-section tables for the photoelectric interactions, coherent and incoherent scattering, and pair production have been generated in steps of 0.5 keV up to 3 MeV by the XCOM program [23]. The form factors and scattering functions used for correction for bounded electrons are taken from Hubbell and colleagues [14, 24]. The SIMIND cross-section tables include energies and abundancies for characteristic x-ray simulation, which is associated with the loss of atomic-shell electrons.

Photons passing through the collimator are followed explicitly in the scintillation crystal volume until their absorption or escape. Characteristic x-ray photons, emitted from a photoelectric absorption, can be simulated for the collimator and the crystal volume. When the photon energy is above the energy threshold $2m_2c^2$, pair production is included as a possible interaction type. A layer of a different material can be defined around the scintillator volume to simulate a protecting cover.

When simulating radionuclides that emit multiple high-energy photons, such as ^{131}I, these back-scattered photons can contribute significantly to the imparted energy recorded in an energy window in the lower energy region. To include this effect in the simulation, a volume behind the crystal can be defined with a thickness estimated by the user to be equivalent to the scattering equipment behind the NaI(Tl) crystal.

A SPECT simulation can be performed for both circular and non-circular orbits. Projections are stored as floating-point matrices, because variance reduction techniques are used. Simulations of 360° and 180° rotation modes can be selected with an arbitrary start angle. Non-circular camera orbits (e.g. the radius-of-rotation (ROR) for each projection angle) can be implemented by (a) major and minor axes that define an elliptical orbit, (b) an orbit that is determined from the voxel-based phantom using a ray-tracing method, or (c) an input file that defines the ROR (extracted for example from a Dicom header). The ROR data for (a) and (b) can also be saved for further use, such as in a reconstruction program.

Statistical variations in the energy signal are simulated by convolving the imparted energy in the crystal from every photon history 'on-line' with an energy-dependent Gaussian function with a full-width at half-maximum (FWHM) that depends on $1/\sqrt{E}$. Here, the user is required to provide a reference value of the FWHM at 140 keV. An alternative is to define the energy resolution from functions fitted to measured values. The convolution procedure also affects the spatial location of each interaction site. If the simulated energy signal is within a predefined energy window, the apparent position of the event is calculated from the centroid for the imparted energy in the scintillation crystal and is stored in the projection matrix.

29.6.1.1 Source Simulations

Provided with the SIMIND code are source routines to simulate simple source shapes, such as spheres and cylinders. For the simulation of more complex activity distributions, a routine is used to generate decays according to the voxel values in a set of image matrices, where the location of an individual matrix cell is scaled to a location (x, y, z). A multiple-source routine used to set up a simulation consisting of several sources of different shapes is also included. More complex sources can be simulated by using a source map such as, for example, the digital phantoms developed by Zubal and colleagues [25] and Segars and colleagues [26].

29.6.1.2 Interactions in the Phantom

A photon history in the phantom starts by sampling a decay position within the source volume. A photon path is then calculated using the delta-scattering method, as described above. At the end of the sampled path (if inside the phantom) the scattered photon is split into two parts.

The calculation of the path length from an interaction point to the boundary is thus performed in two different ways, depending on the type of phantom being used. For simple intrinsic phantoms, the path length is analytically calculated from equations describing ellipsoids, cylinders, and rectangular phantoms. For voxel-based phantoms, a combination of the step-by-step method and the delta-scattering method is used.

Voxel-based computer phantoms are internally defined by mass-density integer matrices. The reason SIMIND works with the density distribution and not with the attenuation distribution is that the attenuation is a function of the photon energy and is calculated from the product of the density taken from the voxel maps and tabulated mass-attenuation coefficients, μ(hv,Z)/ρ, stored as differential cross-section coefficients (in steps of 0.5 keV) as a function of material and energy. The current version supports two different materials that are separated in the simulation by a density threshold.

29.6.1.3 Collimator Simulation

The CHANGE program defines the necessary parameters for simulating the properties of the collimator. The most important parameters are the hole-diameter and length and the septa thickness. A large database of parameters for most of the commercial collimators is available for both SPECT and planar imaging. These are easily defined in CHANGE or at the command-line level by a unique collimator code. Four different collimators are presently available in the program.

1. A collimator model based on an analytic formulation of the geometric response for parallel [27] and converging beam [28] collimators.
2. A collimator model, based on the delta-scattering method, which includes penetration and scattering in the collimator [29].
3. A pinhole collimator model, based on similar principles as (2), which includes wedge- or channel-shaped pinholes. Multiple pinholes are also possible to simulate.

29.6.2 User Interface

The SIMIND program can be manually run as commands in a terminal window, or as script commands. The following is an example of such a command: simind point point1/01:140/fa:12. The input file in the command line, point.smc, defines the file created by CHANGE that contains the data for the particular simulation. As SIMIND creates several result files, all files usually have the same output base name but are distinguished by different file extensions. At run time, the user can change the input parameters that were originally created by the program CHANGE, because each value is associated with an index number. By using this index, the initial value can be overridden. Below is shown an example of a script file in which point is the input file from CHANGE and point(1a,1b,...) are the names of the result files. Each row executes a simulation for a source located at three different positions relative to the origin of the coordinate system and for two radionuclides.

- Switch /01 is used to define the energy window.
- Switch /sc defines the maximum number of scatter orders in the phantom.
- Switch /53 sets collimator routine to the one that allows for scatter and penetration.
- Switch /18 define the shift of the source volume relative to the phantom.
- Switch /19 defines the acceptance angle.
- Switch /fi defines the name of the file that includes the complete photon emission from a decay.

```
C:\simind\> simind point point1a /01:140/sc:6/53:1/18:10.00/19:-5/fi:tc99m
C:\simind\> simind point point1b /01:364/sc:6/53:1/18:10.00/19:-60/fi:i131
C:\simind\> simind point point2a /01:140/sc:6/53:1/18:0.000/19:-5/fi:tc99m
C:\simind\> simind point point2b /01:364/sc:6/53:1/18:0.000/19:-60/fi:i131
C:\simind\> simind point point3a /01:140/sc:6/53:1/18:-10.0/19:-5/fi:tc99m
C:\simind\> simind point point3b /01:364/sc:6/53:1/18:-10.0/19:-60/fi:i131
```

In this example the acceptance angle is set to a small angle for ^{99m}Tc because of its low penetration effect. For ^{131}I, it is necessary to increase this acceptance angle, because the ^{131}I decay includes some high-energy photons (636 and 722 keV) that will contribute to the images acquired in the main 364 keV energy window.

29.6.3 Documentation

A manual and some tutorials can be found on the webpage www.msf.lu.se/research/simind-monte-carlo-program.

29.7 GATE MONTE CARLO PROGRAM

GATE is an open-source simulation code mainly focused on medical physics applications. It is based on the Geant4 C++ library developed by CERN for high-energy physics experiments such as the search for the Higgs boson. Geant4 includes many physics databases and physical models for both electromagnetic and nuclear interactions. By 'model', we mean that equations and data associated with a given physical interaction are translated into an algorithmic process that is used during Monte Carlo simulation to reproduce the effects of the process.

Created during the 2000s, with a first release in 2004, GATE was at first mostly dedicated to the highly realistic simulation of PET and SPECT imaging systems. Since then, several versions have been released, and GATE now contains modules for dosimetry and radiotherapy experiments, including proton therapy, carbon therapy, brachytherapy, and several imaging modalities: x-ray tomography, proton tomography, Compton camera, and so forth.

The main particularity of GATE is not the software in itself but its community-based development. Since the beginning, the OpenGate collaboration managed the code development and user community. The groups belonging to the collaboration meet twice per year and discuss the general direction and orientation of the project. Developments are performed within each group and included in the common repository provided that (1) code is considered as validated (generally associated with a publication) and (2) documentation and examples are provided. The collaboration is openly evolving, with members leaving and joining. As of 2018, more than fifty individual people had directly contributed to the code, either by correcting bugs or providing new functionalities. As a consequence of this community-driven development, the code tends to be heterogeneous and of varying quality. Some features may be nearly unused, whereas others are widely employed and tested. For example, the DoseActor is a successful module that records the absorbed dose and other associated information in matrices of voxels. This module has evolved since its creation and now has several options, such as recording dose-to-water or dose-to-media. Several variance-reduction techniques and advanced or experimental simulation methods are also available.

At the time of writing, GATE has been used for PET, SPECT, CT, proton-CT, Compton camera, spectral CT imaging systems, and absorbed dose computation in external beam radiation therapy, proton therapy, carbon therapy, brachytherapy, targeted radionuclide therapy, and radioprotection studies. A 'pubmed' search[1] revealed more than 300 papers using/contributing to GATE. Three references papers [30–32] gathered approximately 1,400, 500, and 150 citations, respectively. A new reference paper dedicated to PET and SPECT simulation has recently been published [95].

29.7.1 General Components: Geometry, Physics, Source, Actors

A GATE simulation is described by a text file containing macro commands describing four principal components. The text file, called the 'macro file', is given as an argument to the GATE engine, an executable command line that performs the simulation, prints information during the run, and stores the final results.

The four main simulation components are described as follows.

1. The *geometry* part describes all solid elements that make up the simulation: phantoms or patient image, collimator, detector, and all elements that are useful for the simulation, such as a bow-tie filter and range shifter. Various ways exist to describe objects, such as matrix of voxels, meshes, or analytical solids (e.g. box, sphere, cylinder). The material composition and density of each element are given via text files.
2. The *physics* section contains options about the physical models to be used during the simulation, such as the cross-section databases and the particle cuts. The underlying physical models and options are those of Geant4 and evolve as the Geant4 code evolves.
3. The *source* component is used to describe the initial set of particles to be created ex nihilo and tracked during the simulation. Several types of sources exist, from simple pencil beams to complex 3D distributions.
4. Finally, the *actors*' section describes the observables to be acquired during the simulation and stored at the end (dose, flux, number of counts, etc.).

29.7.2 Patient/Phantom

In PET and SPECT simulations, a patient is described by inserting a CT image in the simulation. This is a 3D matrix of voxels with HU values. The user sets the position of the image in the world coordinate system and indicates how to convert HU to material composition. Commands are provided to perform the stoichiometric calibration method of Schneider and colleagues [33]. Particle tracking in the voxelized volume is performed with a specific navigation

[1] The pubmed search was: "?term=GATE+monte-carlo"

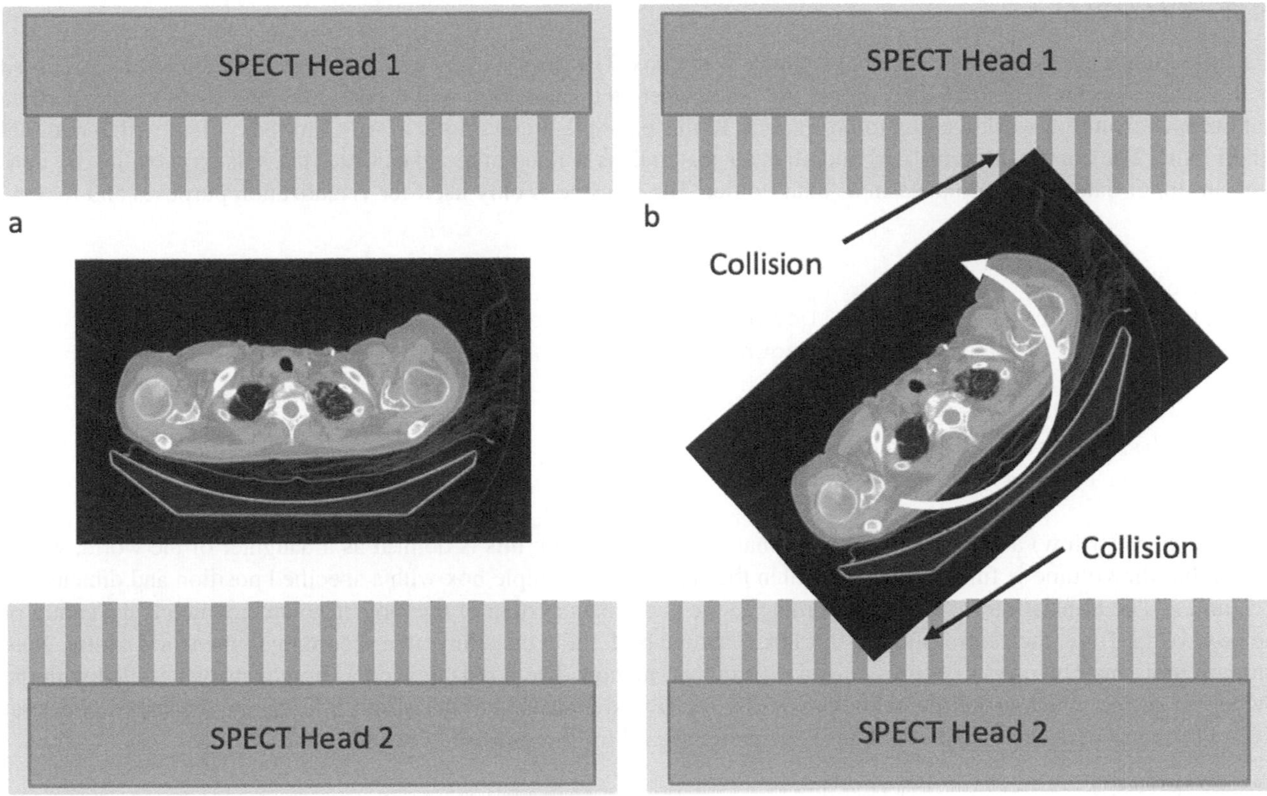

FIGURE 29.7 Illustration of potential collision between the image rectangular support and the SPECT device. Even if the collimator does not touch the patient, it overlaps certain image voxels, causing an error in the simulation.

algorithm that allows speeding up the simulation compared to standard navigation in generic volume types. Further acceleration may be achieved by an adapted tracking algorithm, called region-oriented navigation, which allows reducing the number of steps at voxel boundaries [34]. This navigator is recommended for imaging applications but may not be as efficient for dose scoring. For phantom studies, solids may be described either with a CT image or an analytical description composed of a combination of elementary geometric solids (boxes, cylinders, spheres, etc.). Usually, particle tracking along such volumes is faster than tracking in voxelized volumes, providing that the number of volumes is low compared to the number of voxels. Any number of solids may be added in the simulation.

The Geant4 mode of describing solids implies that volumes must be described hierarchically, making it explicit when volume A is fully included in volume B by declaring A as a daughter of B. The limitation is that A must be *fully* included in B, without any parts outside B: two volumes cannot have overlapping parts. This concept has implications for imaging simulations, because the detector, for example, typically the collimator for a SPECT simulation, cannot overlap the CT patient image. In practice, however, the boundaries of a CT image field of view may overlap the collimator position for certain rotation angles, which renders the direct simulation impossible (Figure 29.7). Workarounds may be performed by cropping the image to remove surrounding air pixels, or by performing simulations in two parts, using the phase-space concept. In the latter case, a first simulation without the detector is performed, and particles exiting the patient are stopped and stored in a surrounding volume (sphere or cylinder). A subsequent simulation is performed using the phase-space as a source. However, this may imply a large phase-space file (tens of gigabytes) that is not tractable. Efforts are underway to provide a more compact source model.

Several examples of patient/phantom descriptions are available among the GATE examples; usually only few command lines are required to insert them into the simulation. For advanced use, we recommend reading the Geant4 documentation to understand how geometric volumes are handled by the Monte Carlo engine. Indeed, solids are described implicitly by few functions such as 'distance to in' and 'distance to out', thereby allowing great flexibility to add new specific volumes if needed.

29.8 EXAMPLES

The following lines show how to declare a simple water box. The lines starting with a '#' are comments and are ignored by the GATE engine. The first lines define the space where the simulation will occur, called the 'world'. Any particle that escapes from the world ceases to be tracked. In this example, the world is a 3×3×3 m^3 box, filled with a material called 'Air'. The composition of 'Air' and all other materials is defined in the 'data/GateMaterials.db' text file. By convention, input data files are placed in a 'data' folder. The last line is only used for visualization purposes and may be omitted.

```
# World
/gate/geometry/setMaterialDatabase data/GateMaterials.db
/gate/world/geometry/setXLength 3 m
/gate/world/geometry/setYLength 3 m
/gate/world/geometry/setZLength 3 m
/gate/world/setMaterial Air
/gate/world/vis/setColor white
```

The next lines (below) add a volume called 'phantom' to the world; this is defined as a daughter of the world, which means that the volume is fully contained within the world. It is a simple box with a specified position and dimensions x, y, and z. The translation is defined according to the coordinate system of the mother volume, which is the centre of the box world. Here, the centre of the water box is shifted by 1 cm in the x direction according to the world centre. Note that any units may be used to express the size (mm, cm, m, etc). The box is defined as homogeneously filled with the 'Water' material, which is defined in the previously mentioned 'data/GateMaterials.db' file. Again, the last two lines set certain properties for the visualization tool but are not useful for the simulation results.

```
# Water Box
/gate/world/daughters/name                    phantom
/gate/world/daughters/insert                  box
/gate/phantom/geometry/setXLength             200 mm
/gate/phantom/geometry/setYLength             200 mm
/gate/phantom/geometry/setZLength             300 mm
/gate/phantom/placement/setTranslation        1 0 0 cm
/gate/phantom/setMaterial      Water
/gate/phantom/vis/setVisible  1
/gate/phantom/vis/setColor   blue
```

Inserting a voxelized CT image is a two-step process. First, a set of commands is provided to define how to associate materials to every HU in the image. The following lines consider two user input files (materialsTable.txt and densistiesTables.txt) that describe how to interpolate certain materials according to the HU, with a given density tolerance. The computed list of materials and HU correspondence are stored in two files (ct_HUmat.db and HU2mat.txt) that will be used in the second part.

```
# Generate materials from Hounsfield units
/gate/HounsfieldMaterialGenerator/SetMaterialTable                    data/materialsTable.txt
/gate/HounsfieldMaterialGenerator/SetDensityTable                     data/densitiesTable.txt
/gate/HounsfieldMaterialGenerator/SetDensityTolerance                 0.1 g/cm3
/gate/HounsfieldMaterialGenerator/SetOutputMaterialDatabaseFilename data/ct_mat.db
/gate/HounsfieldMaterialGenerator/SetOutputHUMaterialFilename         data/HU2mat.txt
/gate/HounsfieldMaterialGenerator/Generate
```

The second part is similar to the command used to insert a box, except that the name of the navigator used to track particles in the voxelized volume is indicated. Here, we mention 'ImageNestedParametrizedVolume', which may be considered as the default and most generic navigator. Faster navigators may be used, but they are specific to certain situations. The two previously generated files are used, and the name of the file that contains the CT image is indicated. Here, we use the so-called metaimage file format ('ct.mhd') on the conventionally used format in the medical image processing community. This is a simple format composed of two files. The first file, '.mhd', is the header that indicates the basic information about the image size, resolution, and the number of bits used to store the voxel values. The second file, '.raw', contains all voxels values in binary format. Several software packages may be used to display such file types (vv, slicer3D, ImageJ, etc.).

```
# Insert CT image
/gate/world/daughters/name                          phantom
/gate/world/daughters/insert                        ImageNestedParametrisedVolume
/gate/geometry/setMaterialDatabase                  data/ct_mat.db
/gate/phantom/geometry/setHUToMaterialFile          data/HU2mat.txt
/gate/phantom/geometry/setImage                     data/ct.mhd
/gate/phantom/geometry/TranslateTheImageAtThisIsoCenter 100 50 50 mm
```

Several volume types with different shapes (sphere, cylinder, mesh, etc.) are also available and may be inserted in the simulation.

29.8.1 Sources

In GATE, ex nihilo sources of particles may be described in several ways. The principle is for the Monte Carlo engine to sample the distributions of particle positions, direction, energy, and type. For well-known radionuclides such as ^{99m}Tc or ^{111}In, the fastest approach is to directly simulate the photo-peaks (e.g. 140.5 keV for ^{99m}Tc, 171 and 245 keV for ^{111}In) with their respective production probabilities (e.g. 89.1% for ^{99m}Tc, 90.06% and 94.12% for ^{111}In). If gammas produced by other processes need to be included in the simulation, such as photons issued from bremsstrahlung, other source particles can be included. The more generic but slower way to describe a source is to directly simulate the complete decay chain of any ion, providing that the decay model is available and validated in Geant4 (this is the case for most common radionuclides). This is important, for example, to take the positron range into account in the simulation of ^{68}Ga PET imaging. In a specific situation such as imaging ^{90}Y bremsstrahlung with SPECT, or emitted β^+ positrons from ^{90}Y with PET, ad-hoc source models may be used to accelerate the simulation [35].

The initial position distribution of source particles may be described with analytical models, from points, spheres, or any geometric volumes to be put in the simulation world. Alternatively, a voxelized source can be described as a 3D matrix of its activity distribution. Particles are sampled in all voxels according to their relative activity. Voxelized sources are usually positioned inside the patient CT image and may typically be obtained from a SPECT or PET image that have been calibrated and eventually corrected for degrading physical effects (partial volume). The directions of emitted particles are generally sampled from isotropic distributions. More complex types of sources may be described with additional ad-hoc C++ classes, such as has been done for a complex proton beam description [36].

The following box illustrates an example of a 1 MBq ^{177}Lu source that simulates the complete decay chain of this ion, hence generating electrons and gammas.

```
# add a source of Lutetium177
/gate/source/addSource                          Lutetium
/gate/source/Lutetium/gps/particle              ion
/gate/source/Lutetium/gps/ion                   71 177 0 0
/gate/source/Lutetium/gps/type                  Volume
/gate/source/Lutetium/gps/shape                 Sphere
/gate/source/Lutetium/gps/radius                3 mm
/gate/source/Lutetium/gps/angtype               iso
/gate/source/Lutetium/gps/centre                0. 0. 0. mm
/gate/source/Lutetium/gps/energy                0 MeV
/gate/source/Lutetium/setActivity               1 MBq
/gate/source/Lutetium/setForcedUnstableFlag     true
/gate/source/Lutetium/setForcedHalfLife         574300.8 s
```

GATE contains various functions to describe source of particles in a simulation. Users can select the most appropriate functions for their purpose.

29.8.2 Detector Modelling

Modelling various types of detectors in GATE is a two-step process. The user first describes the geometric elements that compose the detector, and then describes the process used to convert particle interactions in the scintillator (i.e. the crystal) into a recorded signal. This process is performed by the so-called digitizer module that models the Front End Electronics (FEE) signal-processing chain in photomultipliers (PMT), avalanche photodiode (APD), or silicon photomultiplier (SiPM), and so forth.

Geometric elements such as the head, collimators, crystal, and shielding are described as geometric volumes with shape, position, orientation, and material composition. Various examples are provided to help users with the most common geometries, such as the complex honeycomb shape of SPECT collimators. Functions are provided to repeat volumes, such as an array of crystal pixels allowing the simulation of a complete ring of complex detectors.

A digitizer module is built by the user by selecting a set of processing submodules that perform operations on the list of interaction events ('hits') occurring in the detector elements. This process chain hence allows modelling the spatial, energetic, or temporal uncertainties to calculate various energy or position thresholds, so as to finally create the output signal. The results of the digitizer module may be stored in various formats, for example, list-mode, sinograms, or projection images. Timing elements such as time-of-flight (TOF), dead-time, and coincidence windows are considered and may be user-parameterized. Under the hood, the Geant4 engine assigns timestamps to all particles and interactions, which are then used to perform the processes involving timing. In GATE, the conversion of light to electricity is thus not simulated by Monte Carlo processes, but rather by analytical models that include uncertainty.

Complex and highly realistic detector designs may be simulated with relatively few lines of code, provided that sufficient information, such as the volume size and material composition, is available from the constructor. Generally, constructors require non-disclosure agreements to share such data. The validation process that consists of comparing simulation output with experimental data remains a challenging issue, owing the complexity of the involved physical phenomena. Various publications reported successful comparisons of SPECT, PET, and radiotherapy equipment, and GATE is known to be used by various constructors to design their imaging devices.

29.8.3 Documentation

More detailed documentation and examples are available on the websites of the collaboration:

http://www.opengatecollaboration.org: main site information
https://opengate.readthedocs.io: documentation
https://github.com/OpenGATE/Gate (both text and link) code and examples

29.9 ACCELERATING METHODS

Direct (or analogue) Monte Carlo simulations follow the 'true' probability distributions according to the physics and nature of the scenario. However, in some cases, simulations may involve geometries or processes with a very low probability for detection, meaning that they require, if using an analogue simulation, many Monte Carlo histories before a sufficient number of detected (or scored) events has been reached. One example is PET imaging of ^{90}Y treatment, where only approximately 32 positrons are emitted per one million disintegrations. Another example is the low detection probability for gamma-camera imaging due to the collimator design, where only approximately 100 photons per second per unit MBq of ^{99m}Tc pass through the LEHR collimator holes and contribute to a signal or an event in the image.

In order to reduce the variance (see Volume II, Chapter 1 for more information about statistical uncertainties) in a particular analogue simulation for a given simulation time, one can improve the power of the computer by using a faster central processing unit (CPU) and introducing parallel processing using a message-passing interface (MPI) or a graphics-processing unit (GPU). However, it is also possible to reduce the variance in the result for a given computer system by introducing so-called variance-reduction methods into the code.

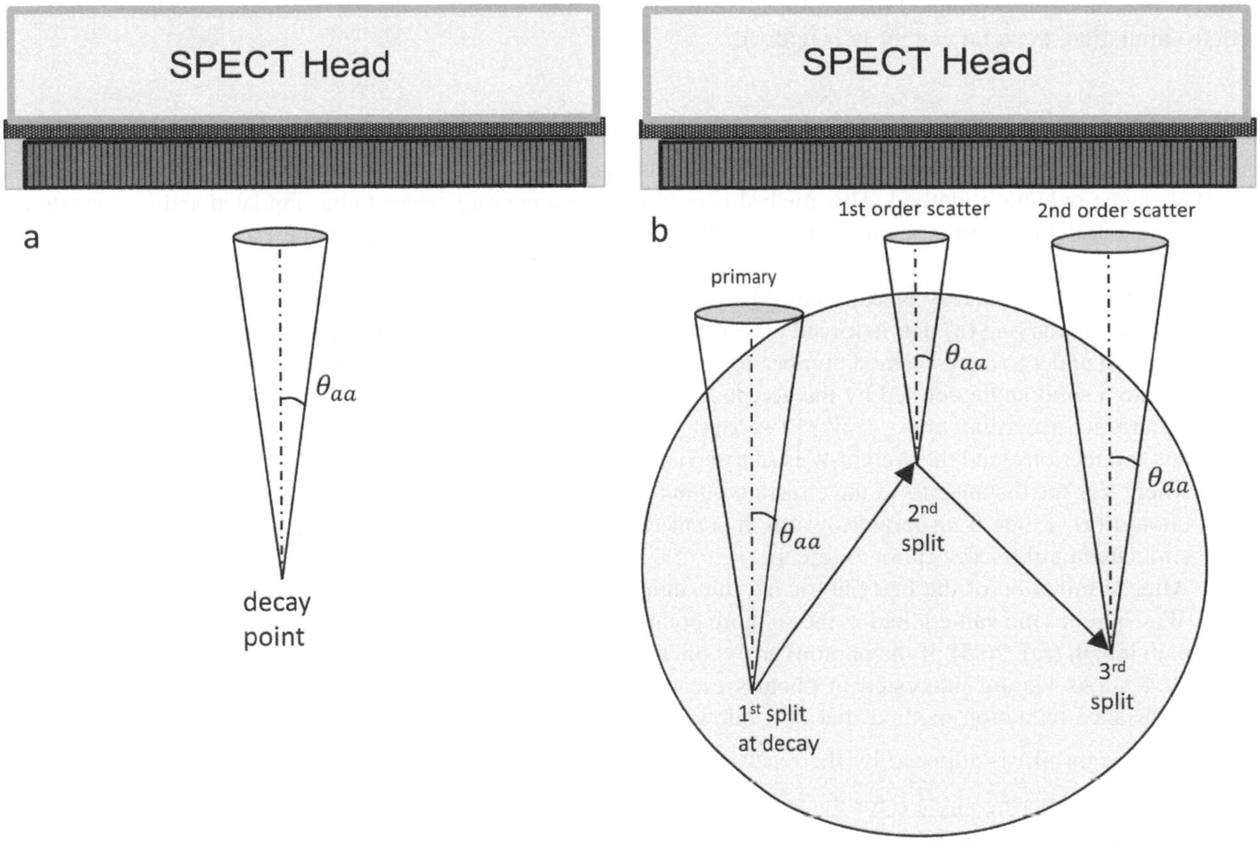

FIGURE 29.8 Schematic figures of methods for accelerating methods. Left figure show the method of forcing the directions of photons into a solid angle with a maximum acceptance angle of 0_{aa}. Right figure shows how a photon history can be split into three sub-histories to improve the speed of the calculation.

Variance-reduction methods artificially change the probabilities in order to reduce the calculation time for a given expected variance in the result or, alternatively, reduce the variance for a given CPU time. In the case of a scintillation camera simulation, one can consider the fact that, when assuming no penetration in the collimator, the probability of detection is proportional to the solid angle defined by the hole diameter and length. If we consider a point source in air, one could therefore restrict the sampling of emission directions to only those polar angles that are encompassed by this solid angle (Figure 29.8a). Then, the probability of detection would be greatly improved. However, if we want to obtain the same count level (cps/MBq) for our results as would be expected for a real measurement, we need to compensate for our 'violation' of the natural laws (the isotropic emission). This can be achieved by calculating the probability for an emission within the solid angle (as determined by an acceptance angle θ_{aa}). If we assign this value to a weight, W, we can then for a given detection combine the weight W with the energy spectrum or image matrix instead of unity. In the case described above, the weight W, assuming an initial weight $W_0 = 1$ defined at the start of the history, will be $W_1 = W_0 \cdot \left[1 - \frac{\cos(\theta_{aa})}{2}\right]$. Thus, W_n can be understood as the probability for a photon to travel along the simulated path.

In a real measurement, a detection is scored as either 1 (presence of detection) or 0 (absence of detection). For a physical detection, a value of 1 is therefore added to the multi-channel array that forms the energy pulse-height distribution as function of energy or to the image matrix as a function of the detection position. In a simulation procedure, in which variance reduction methods are used, the weight W, rather than a value of 1, is added to the energy pulse-height array and to the image array.

The same principle can be applied to force a photon to interact within a given distance by truncating the probability distribution above a certain distance d_{max}. The weight W_2 for this alteration is then equal to $\left(1 - e^{-ud_{max}}\right)$ and the weight

W for the photon being simulated is therefore $W = W_1 \cdot W_2$. As new variance reduction methods are applied during the particle simulation, the total weight W is updated.

29.9.1 Photon Splitting

Another method is to split the photon into several sub-photons. Figure 29.8b shows schematically the principle for this type of variance reduction method. This method is used to encourage rare events to be simulated and is interesting to decrease the computation time of simulations in which such rare events are the ones of interest.

i. The position of a decay is sampled and stored, and the initial weight $W_{n=0}$ is set to 1. The photon is then split into two parts: one (a) that is forced to exit the phantom towards the camera without any scattering and (b) one that will undergo a pre-defined number of scattering events. The direction of photon (a) is sampled to be emitted within a solid angle defined by the acceptance angle θ_{max}. The weight is multiplied by the probability for this occurrence according to $W_{1,a} = W_{0,a}\left[1-\cos\left(\theta_{aa}\right)\right]/2$. The photon is also forced to penetrate the phantom without any interactions, and the weight W is adjusted according to probability for this occurrence; $W_{2,a} = W_{1,a}\left(1-e^{-ud}\right)$, where d is the distance from the emission point to the surface of the phantom along the sampled direction. The photon (a) is then transported towards the camera and interactions in the collimator, crystal, and so forth, are simulated until its absorption or escape.
ii. After termination of the first photon (a), the simulation is reset back to the decay position. The photon weight $W_{0,b}$ is set to the value it had at the splitting point, and an isotropic direction is sampled together with a photon path length (Eq. 26.5). If the photons are within the phantom, the type of interaction is then determined (Section 29.4.3). As we are interested in photons escaping the phantom and contributing to the camera, we can use a variance-reduction method that only allows Compton and coherent scattering (neglecting pair production). The weight W_b is adjusted by the relative probability $\left(\left[\sigma_{coh} + \sigma_{inc}\right]/\mu\right)$ for these two types of interactions to occur. A scattering angle θ for the sampled type of scattering (Compton or coherent) is forced to be within the acceptance angle; the probability $W_{\theta,b}$ for this event to occur is then calculated and the weight W_b is adjusted again. Finally, the probability for penetration of the phantom is calculated according to the description in (i) above, but with an attenuation coefficient that is based on a possible change in the photon energy. The first-order scattered photon is then followed towards the camera.
iii. After the first order of the scattered photon history is terminated, the control is returned to the position of the first-order scattering event and the procedure is repeated until the pre-defined number of scatter orders have been simulated.

By this technique, a selection of a maximum of i scatter orders results in $i+1$ events detected by the camera for a single photon history.

29.9.2 Russian Roulette

Occasionally, the weight can be extremely small for highly unlikely paths, which thus contribute little to the results. The drawback from this is 'waisted' CPU time. One method to address this is to play 'Russian roulette'. When the current photon history weight, W, is below a certain threshold, defined by some criteria, the 'number of bullets' is decided, say for example 10. If a random number R then is less than 1/10, then the photon history weight W is increased by a factor of 10 ($W = W \times 10$) and the photon is allowed to continue. Otherwise, the history is terminated. This is essentially the opposite of the splitting method: it is used to decrease the time needed to track numerous but unwanted (low-contributing) particles.

29.9.3 ARF – Angular Response Function

A method to speed up simulations of SPECT imaging based on modelling the collimator–detector response function (CDRF) has been proposed. The CDRF combines the accumulated effects of all interactions in the imaging head and may be approximated with angular response functions (ARFs) [37–40]. The ARF method replaces the explicit photon tracking in the imaging head by a tabulated model of the CDRF. The tabulated model is derived from a simulation with a gamma source covering the energy range of the radionuclide of interest and including the complete detector head with the collimator, crystal, and digitization process. The model takes as input the direction angles and the energy of an incoming photon and determines the probability of detecting this photon in each defined energy window. This first step must be performed only once per type of SPECT head, radionuclide, and defined energy window. Once the lookup tables are computed, they can be used for every simulation with the same conditions (same collimator/detector, radionuclide energy windows), independently of the source distribution and the medium, phantom or patient. The ARF method assumes that a photon that interacts with the collimator will be detected at its geometric intersection point on the detector plane, considering the spatial uncertainty. This approximation has been shown to be sufficient. Furthermore, detector dead-time is neglected.

Recently, tabulated data have been replaced by artificial neural networks (ANNs) that are trained to learn the ARF of a collimator–detector system. The ANN is trained once from a complete simulation including the complete detector head with the collimator, crystal, and digitization process. Photons are stopped at the plane and the energy and direction are used as input to the ANN, which provides detection probabilities in each energy window. Compared to the histogram-based ARF, the ANN method is less dependent on the statistics of the training data, provides similar simulation efficiency, and requires less training data. The ARF approach has been shown to be efficient and to provide variance reduction that speeds up the simulation. The speedup compared to analogue Monte Carlo was between 10 and 3000: ARF methods are more efficient for low-count areas (speedup of 1000–3000) than for high-count areas (speedup of 20–300), and more efficient for high-energy radionuclides (such as ^{131}I) that show deep collimator penetration. This implementation of ARF with a neural network is available within the GATE platform.

29.9.4 Fixed Forced Detection

As simulations of SPECT images are slow to converge owing to the large ratio of the number of photons emitted to the number detected in the collimator, the fixed forced detection (FFD) method was proposed to accelerate photon tracking from their emission point to the detector. This idea was initially used for x-ray imaging [41–43], where the authors used forced detection onto a set of points (all detector pixels or a reduced number), and then used interpolation for the other pixels. The FFD forces photons towards each pixel of the detector for every Monte Carlo interaction, which are of type: decay, Rayleigh scattering, Compton scattering, or fluorescence. These photons are given a weight that depends on their probability to reach the SPECT head and can be separated into scattering and transmission probabilities. The scattering probability is the probability for the photon to be directed towards the pixel, according to the interaction type of the photon. The transmission probability is the probability for the photon to reach the pixel, according to the traversed medium (Beer–Lambert law). The final detector count value is the product of these probabilities for each interaction. This method was coupled with ARF detector modelling in [44]. The overall computing time gain can reach up to five orders of magnitude, but 'hot spots' can be observed because the spatial distribution of emitted primaries and a criterion to determine convergence are still lacking. This implementation of FFD is available within the GATE platform.

29.9.5 MPI – Message Passing Interface

In SIMIND, parallelization has been implemented by using the message-passing interface (MPI) standard [45] to improve the statistics of the simulated images and energy spectrum for a given calculation time. The MPI system works as follows. SIMIND is replicated to run in parallel on separate nodes automatically, where one node is selected to be a primary node that controls the execution of the others at various positions in the code. When simulating SPECT, and when the first projection has been completed, all nodes are temporarily stopped. When the master node has verified that all nodes have reached that location in the code, the primary node collects specific data from all running nodes and adds them to its own data (e.g. projection images, energy spectrum, etc.). The summed data are, if necessary, stored to a file. The primary node then releases all nodes for the simulation of the next projection.

29.9.6 GPU Coding

All the presented methods may be implemented on conventional CPU architecture or on a GPU, where the latter has large vectorization capabilities that could potentially lead to a large speedup thanks to parallelization. Through the years of increasing GPU capabilities, multiple attempts, from simple concepts to commercial solutions, have been proposed for accelerating Monte Carlo simulations in various domains (imaging, dosimetry) [39, 46–49]. One of the main difficulties is to determine the appropriate level of parallelization. For example, is it better to track one photon per GPU or to perform low-level parallel sampling computation? There are consequences in terms of memory management; it is costly to send data to and from GPU memory, which affects the overall performance. Overall, it is generally admitted that GPUs provide large computation gain for Monte Carlo simulations, as the core of the process is intrinsically parallel, with every event being independent of the others. However, there is a price to pay: Code is generally longer to write and may be more difficult to maintain and generalize. If there are few doubts that a GPU-based code dedicated to one specific task (e.g. SPECT simulation) is faster than a single CPU one, it should be compared to a multi-CPU architecture with similar costs that may have simpler code generation and maintenance, and greater generalizability to other simulations.

29.10 APPLICATIONS OF MONTE CARLO IN NUCLEAR MEDICINE IMAGING

29.10.1 Components of an Image

The Monte Carlo method is useful for understanding the measured energy pulse-height spectrum for scintillation detectors and cameras, because this spectrum provides information on the image, in both qualitative and quantitative terms. This spectrum reflects the distribution of the energy imparted in the crystal and is therefore highly dependent on the incoming photons and their energy distribution. In a measurement with a detector or a camera, one is often interested in information on the energy deposition in specific energy ranges (energy windows). However, it is possible from these events to separate photons that have been scattered in the object from unscattered photons (primaries). In a scintillation camera, the position is determined from the scintillation light as the result of an energy deposition, but their deposition could be the result of a photon scattered in the object. The quality of the image may therefore be deteriorated. In a Monte Carlo model of a detector, it is possible to 'tag' the photons as being primaries, scattered, or scattered by a certain order. The information can then be plotted as a separate curve or summed to provide a total curve that will therefore match a real measurement. A deeper analysis can then be performed to determine parameters, such as scatter-to-total ratios, for a different energy window location. This information can be useful to optimize the scatter window location for scatter correction methods commonly used in quantitative SPECT, such as the dual-energy window and triple-energy-window methods. Figures 29.9 and 29.10 show an example of such a simulation, with three radionuclides (^{99m}Tc, ^{177}Lu, and ^{131}I) displaying total, primary, and non-primary events (which can also include penetration effects and scattering in the collimator, crystal, and material behind the crystal).

29.10.2 Collimator Penetration

Some radionuclides used in nuclear medicine imaging have a complex decay scheme with multiple photon emission of extremely high photon energies. These photons are not stopped by the collimator walls and may deteriorate the images

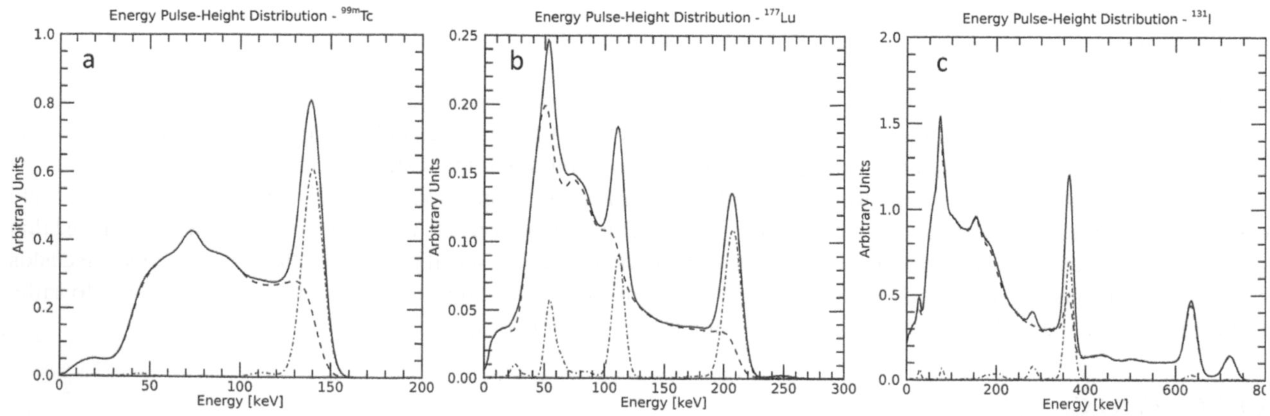

FIGURE 29.9 Simulated energy spectra for (a) ^{99m}Tc, (b) ^{177}Lu, and (c) ^{131}I from the XCAT simulation. Solid lines show the total detected spectra. Dashed lines show the component in the spectra due to photons that have been scattered in the phantom and dashed-dotted lines show the events from non-scattered photons that have passed through the phantom without interactions. The energy windows are defined as 20 per cent for all three isotopes.

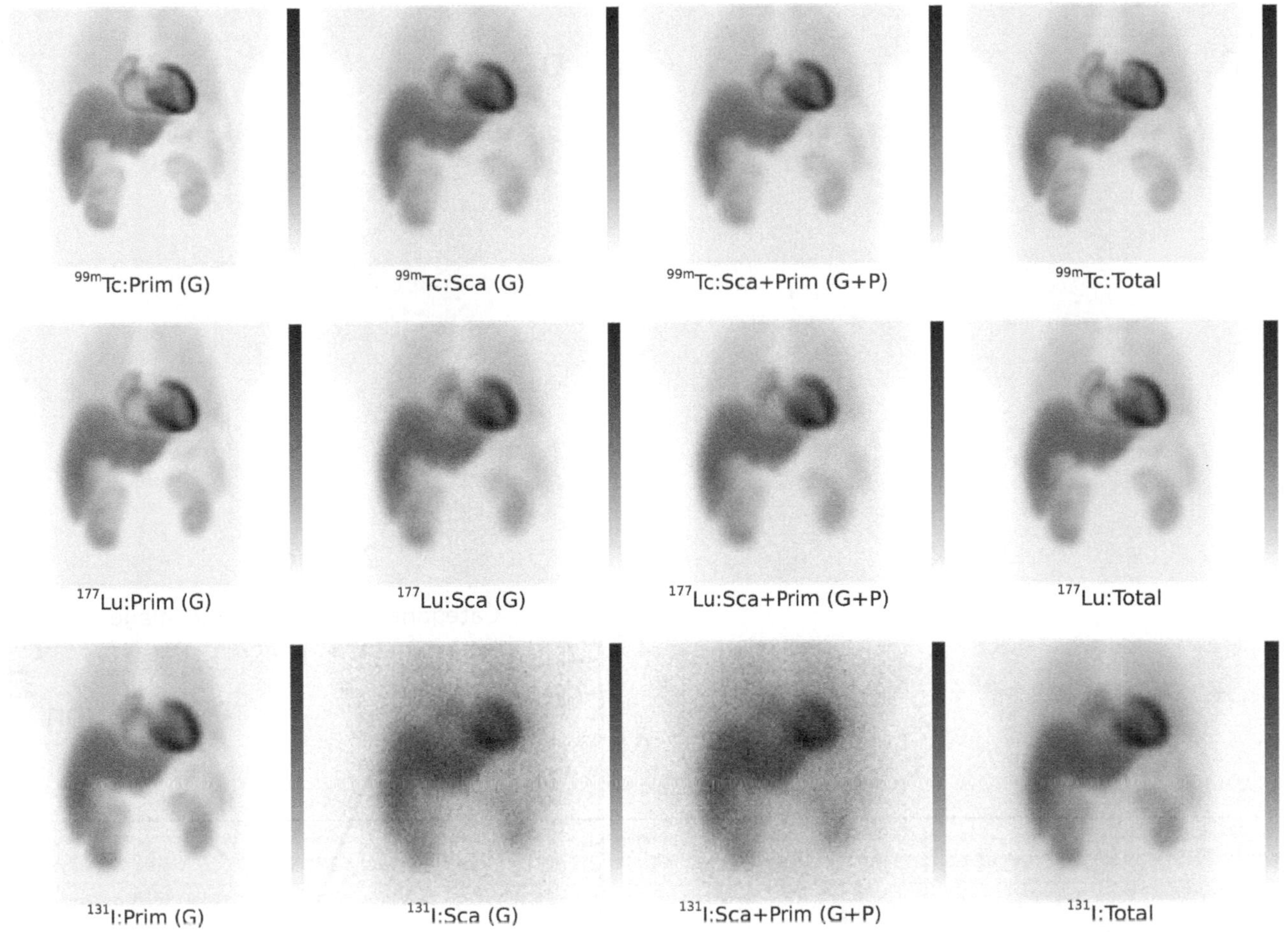

FIGURE 29.10 Simulated images for (upper row) ^{99m}Tc, (middle row) ^{177}Lu, and (lower row) ^{131}I from the XCAT simulation. Left column: The energy windows are defined as 20 per cent for all three isotopes.

from significant septum penetration. Even for HE collimators, the contribution of septal penetration from, for example, ^{131}I, can be significant because of two photons with energies 637 and 723 keV [50]. Figure 29.11 shows a simulation of a gamma camera system, equipped with a low-energy high-resolution collimator, and measuring an activity distribution emitting three different photon energies. It can here be seen that the thickness of the collimator septa is not sufficient to absorb the 200 keV and 250 keV photons and, as a consequence, the image quality is degraded (Figure 29.11a–c). Monte Carlo simulation allows also for a calculation of the relative fraction of the detected events, coming from geometrically collimated photons, penetrating photons, and photons, scattered in the collimator. These fractions are plotted in Figure 29.11d as a function of photon energy.

29.10.3 Tracking Particles for a Full SPECT Simulation: Some Numbers

To illustrate the complexity of the Monte Carlo simulation of a SPECT image acquisition, we consider a targeted radionuclide treatment with ^{177}Lu, usually for neuroendocrine tumours (NET). In this theranostic procedure, advanced midgut NETs are treated with the somatostatin agonist chelated with ^{177}Lu. The compound accumulates mostly in tumours cells, and the electrons emitted by ^{177}Lu provide, by ionization, a cytotoxic effect to lesions. At the same time, the emitted gammas could be used to acquire per-treatment images, thereby allowing the treatment to be monitored. We consider here the acquisition of this SPECT image during such a treatment.

The ^{177}Lu half-life is 6.647 days. In addition to electrons (maximum 497 keV, 78.6%), each gamma emission probability is 17.2 per cent per decay, which consists of approximately 10.38 per cent (208 keV) and 6.2 per cent (112.9 keV). The activity injected to the patient is assumed to be 7.4 GBq. We consider the SPECT image acquisition 24 h after

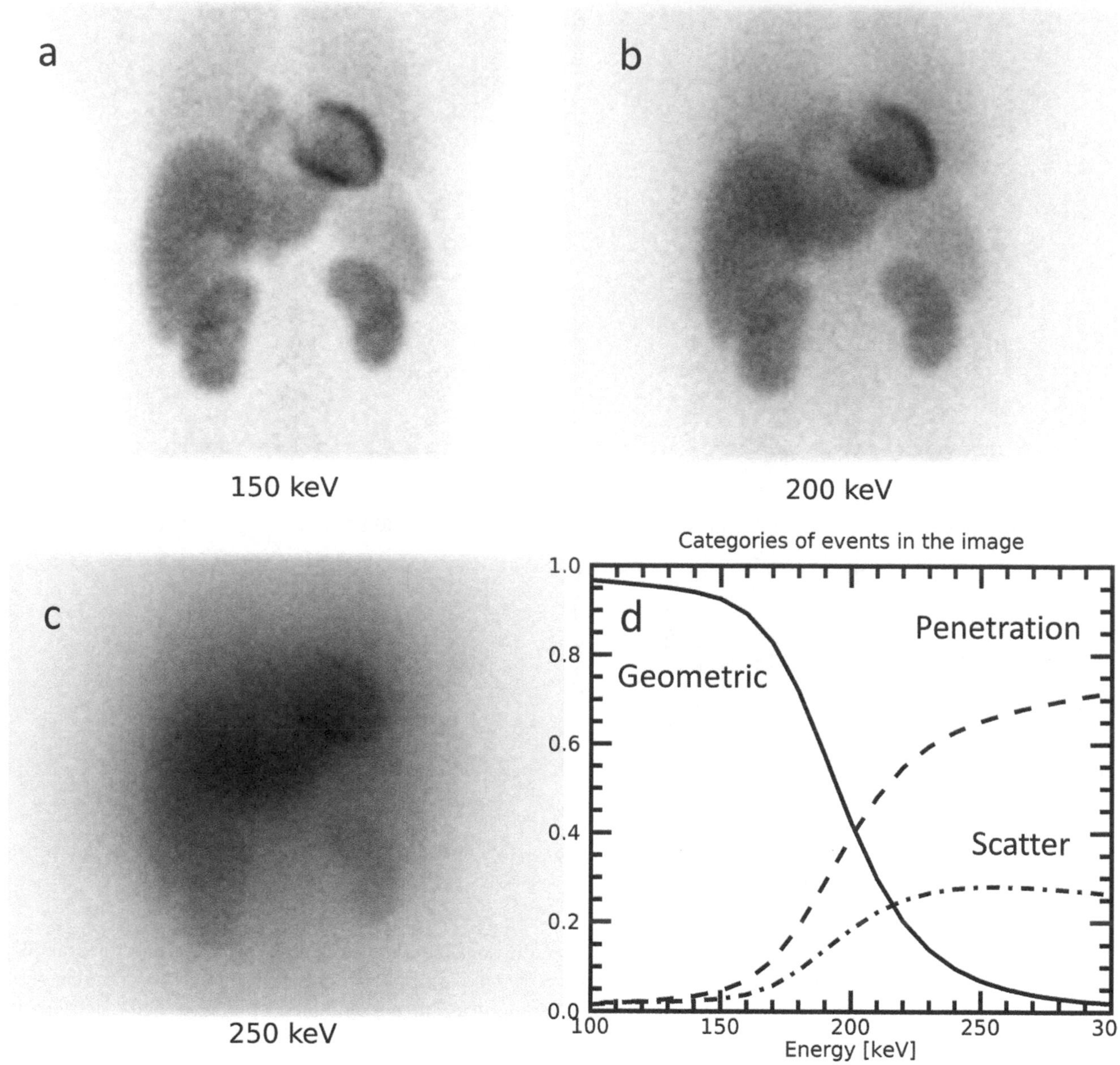

FIGURE 29.11 Images (a-c) show degradation in image quality due to collimator septal penetration and collimator scatter. Graph (d) show the relative fraction of how a photon passing though the collimator.

the injection. Considering the exponential decay, it leads to 7.07 GBq. We assume that only half of this quantity remains in the part of the patient visible from the camera head, such that 3.5 GBq (a portion clears out of the patient during miction, and a portion is in the patient but outside the field of view). One SPECT angular projection lasts 30 s. This leads to approximately 10^{11} decays or 1.8×10^{10} emitted gammas (neglecting the decay within the acquisition duration). Among those emitted gammas, we estimate that approximately 80 per cent (1.4×10^{10}) leave the patient body (see Figure 29.12). Among them, only approximately 19 per cent (2.8×10^{9}) of the gammas reach the detector head entrance plane, owing to the solid angle (or double this number if two opposite heads are used). Owing to the collimator, the crystal detection efficiency, and the choice of the energy windows, only 8.1×10^{6} and 2.1×10^{5} gammas will finally be detected in the respective energy windows of 113 and 208 keV.

Hence, the complete simulation would require the emission and the tracking of 1.8 10^{10} gammas inside the patient CT image. As an example, with Gate/Geant4, such a tracking in a voxelized geometry and in the detector may take

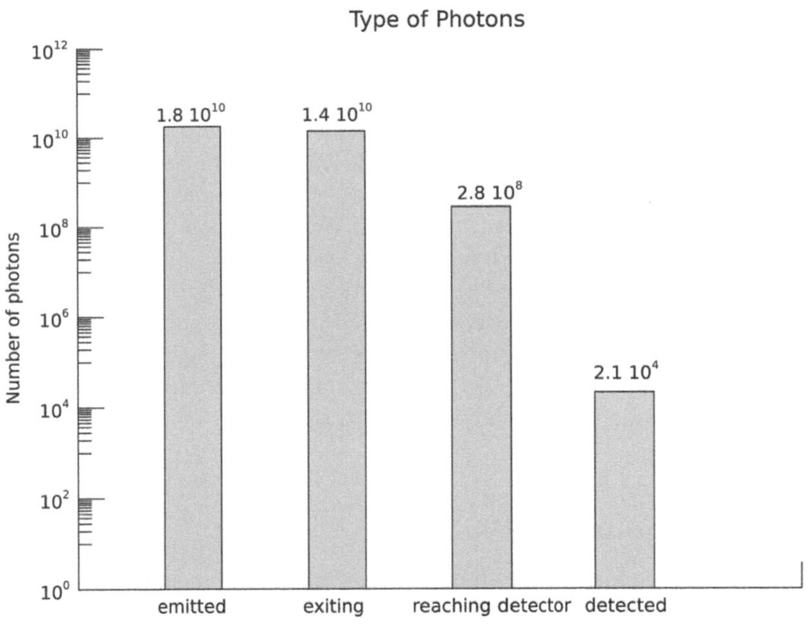

FIGURE 29.12 Number of photons from emission to detection.

approximately 5900 particles per second, leading to approximately 36 h of computation time (Intel Xeon CPU E5-2640 v4 @2.40GHz). The SPECT acquisition is performed with two heads with a 180° step-and-shoot rotation of 6° angle increment, providing 60 projection frames for a total of 15 min (900 s) acquisition time. Hence, the total computation time of the whole acquisition would require almost three years. Of course, no one would wait that long. First, parameter tuning (cuts, physics list), variance reduction techniques (e.g. restricting emission solid angle), and the use of powerful clusters significantly reduce the time to a few hours. Moreover, dedicated simulation codes, using GPUs or aggressive VRT, may reduce the time even more.

The reader should only take these numbers as an example; every simulation is different, computing time greatly varies according to the parameters, the Monte Carlo code, and the computers employed. However, these numbers give an idea of how complex a brute force simulation of a complete SPECT image acquisition could be.

29.10.4 Pinhole SPECT

Pinhole imaging uses a collimator with a single hole and acts as an ordinary camera. The pinhole consists of a small insert, of tungsten or gold, with either a knife-wedge type or a channel-type geometry (wedge with a cylindrical hole in the centre). Owing to its superior resolution when imaging smaller objects, the pinhole collimator is primarily used for small organs or imaging of small animals, such as rats or mice. A small object can be placed close to the aperture, thereby creating a large magnification. The resolution of the pinhole collimator depends on the diameter of the aperture, and the fraction of photons that penetrate the edge of the aperture. Furthermore, the number of photons that penetrate the edge of the aperture depends on the acceptance angle of the pinhole collimator, the pinhole collimator material, and the photon energy.

Smith and colleagues [51] studied the variation in sensitivity and degree of penetration by experimental measurements and a Monte Carlo program [52]. Deloar and colleagues determined, by a similar study, the relations between events coming from photons that (a) passed through the hole, (b) penetrated the collimator, and (c) were scattered in different types of knife-edge-shaped collimators for ^{201}Tl, ^{131}I, ^{123}I, and ^{99m}Tc sources and apertures. [53]. Peterson used the SIMIND code to study the Rose's metal as a potential useful material for pinhole construction and compared the sensitivity, resolution, and degree of penetration with previously published data [54]. SIMIND has also been used to study the properties of the GE Discovery Alcyone CZT SPECT Systems, which is based on 19 pinhole collimators [55].

29.10.5 Motion Artefacts

Respiratory and heart motion lead to blur and artefacts in SPECT and PET images. This uncertainty may cause incorrect lesion detection and misestimation of activity in the reconstructed image. Several methods have been proposed [56–62] to estimate and compensate for the motion during image reconstruction. The common approach consists of measuring or estimating a breathing signal synchronized to the acquisition and splitting the acquired data according to several breathing phases, thereby reconstructing the images independently for each different phase, from expiration to inspiration. This class of method is called respiration-correlated reconstruction and relies on the assumption that the respiratory motion is quasi-periodic, with all respiratory cycles being considered regular. The respiratory signal may be obtained from additional hardware on the patient, such as a belt (e.g. Siemens' ANZAI belt Respiratory Gating System) or infrared marker localization (Varian's Real-Time Position Management RPM), but the most convenient approach is to extract this signal directly from the list mode data by signal processing, such as centre of mass (COM) tracking, principal component analysis (PCA), or Laplacian Eigenmaps [63]. This type of approach is currently being investigated in clinical settings, both for PET [56, 60, 64] and for SPECT [65].

29.10.6 Reconstruction and Segmentation Algorithms

The Monte Carlo method has been useful, not only for evaluating the development of image reconstruction algorithms, but also in the implementation of the algorithm itself. A very early paper was published by Floyd and colleagues [66], which described a method in which the forward projection in an ML-EM algorithm was calculated by direct Monte Carlo methods. Although the method did not reach clinical routine application, the paper showed that it was possible to obtain good results. Monte Carlo-calculated scatter kernels in reconstruction methods were introduced, where the scatter in the photo-peak energy window were modelled and subtracted from projection data [67] or included in the forward projector and an ML-EM method [68]. A group at Utrecht University [69] developed a clinically useful Monte Carlo reconstruction code by introducing forced convolution, which significantly improved the calculation. This method is based on following the photons in the object using a Monte Carlo method to determine scattering angles and energy losses, but when an escape from the object is simulated, the photon is forced into a direction perpendicular to the detector. The impact position in the crystal is then convolved with a distance-dependent Gaussian function that simulates the collimator resolution, and the result is added to previous histories. By not simulating the collimator in an explicit manner, and thereby rejecting most of the histories, the calculation times were greatly reduced. However, by this method, additional effects, such as collimator penetration and scattering in the crystal and surrounding compartments, cannot be modelled. However, for a ^{99m}Tc application, the method worked very well. Monte Carlo-based reconstruction programs have been published by other groups for ^{90}Y bremsstrahlung imaging [70–72], ^{166}Ho imaging [73], and ^{131}I imaging [74, 75]. The group at the University of Helsinki developed a Monte Carlo-based reconstruction software package that has now been implemented in the commercial image-processing system by Hermes Medical Solutions, Sweden [76]. Rydén and colleagues [39] developed a GPU-based Monte Carlo reconstruction and generated a Monte Carlo-based SPECT/CT reconstruction of ^{177}Lu-DOTATATE treatments within a few minutes with improved resolution and contrast. Recently, work has been done to incorporate the reconstruction in the SIMIND code for parallel-hole collimation [77] and for pinhole collimation [78].

29.10.7 Simulation for Small-animal Imaging Systems

The simulation of small-animal (rodent) imaging systems with GATE is similar to the simulation of human-sized clinical systems. Indeed, the latter may be downscaled versions of human-sized systems. The differences mainly lie in the sizes and dimensions of the components and the activity quantity, but the overall principles remain identical. Recent SPECT cameras may have sub-millimetre spatial resolution with multi-pinhole detectors and reach 0.1 per cent sensitivity. PET camera resolution is approximately 1 mm with sensitivity higher than 10 per cent. Considering the animal size and the energy of the involved gamma sources, the scatter and attenuation are usually of slightly lower importance than for the human configuration. We can find several examples of GATE simulation studies for SPECT and PET systems, for example, [79, 80] among many others. More detailed information on small-animal radiation therapy and imaging platforms may be found in [81–83].

29.10.8 SiPM PET Simulation

PET technology has experienced tremendous improvements in performance over the past decades, and new trends make use of silicon photo-multiplier (SiPM) detectors, such as the Philips Vereos Digital Photon Counting (DPC) PET/CT introduced in 2013, the Siemens Biograph Vision™ PET/CT launched in 2016, and the GE Dicovery™ MI PET/CT launched in 2018. The use of TOF information that spatially constrains the location of the event on the line of response has led to improved signal-to-noise ratio in the reconstructed image. The TOF resolution improved with the use of SiPMs, which have an intrinsic timing resolution lower than conventional PMTs or APDs, with a compact electronic configuration. Detecting and processing signals using digital SiPMs bypasses the need to treat analogous signals by a direct binary count of optical photons, thereby reducing the noise in the processed output. As an example, the Philips Vereos system has been modelled with GATE [84] by comparison with experimental data obtained from the NEMA protocols NU 2-2018 [85]. Experimental and simulated data were found to be in good agreement, with differences of less than 10 per cent for activity concentrations used in most standard clinical applications.

29.10.9 Machine Learning and Monte Carlo

Deep convolutional neural networks (CNNs) have become a major and powerful methodological actors, learning statistical properties and correlation in recent years. As Monte Carlo simulation is primarily statistically driven and deals with large quantities of data, it seems quite natural that CNNs could play a role in that field. Indeed, some works are emerging. For example, [86, 87] proposed deep learning-based methods to estimate the absorbed dose distribution for internal radiation therapy treatments. The accuracy with respect to the reference Monte Carlo calculation seems quite good, and the main gain is related to the computation time, which is of the order of minutes instead of hours/days. Another work [40] proposed to use a neural network to learn the ARF of a SPECT collimator–detector system, as explained previously (see 'Accelerating Methods, ARF'). More recently, generative adversarial networks (GANs) were reported [88, 89] as deep neural network architectures that allow mimicking a distribution of multidimensional data and have gained great popularity owing to their success in realistic image synthesis. A GAN learns representations of a training dataset by implicitly modelling high-dimensional distributions of data. It has been applied in the field of Monte Carlo simulations to learn a Linac phase space [90]. At the end of the training process, the resulting network could generate particles that belong to the probability distribution of the phase space while being approximately 10 MB, compared to the initial file size of a few gigabytes.

These examples are only pioneering works in the field, and further studies are needed to better understand the advantages and drawbacks of such approaches. Even if they are promising, there are still pitfalls, limitations, and unknowns. For example, the training time is still long and requires substantial computing resources (GPU). The generalization of the learned network is still not clear. The final accuracy obtained seems to not yet reach that of the conventional Monte Carlo method, but the application in the field of Monte Carlo simulation is promising. For more information about machine-learning, see Chapter 11.

29.10.10 Compton Camera

In addition to conventional SPECT and PET, alternative nuclear imaging system principles are also being investigated by Monte Carlo simulations. Among them, Compton cameras are actively collimated gamma-ray imaging devices, which were originally designed for astronomy applications. Their application for medical imaging was first proposed by Todd and colleagues [91]. They are currently used in astronomy or for environmental measurements in homeland security and are still being studied in the medical field, in particular for proton therapy monitoring. Compton cameras potentially overcome the mechanical collimation of SPECT systems and their low sensitivity by considering several position- and energy-sensitive gamma ray detectors arranged in scatterer and absorber planes working in time coincidence. Using the kinematics of Compton scattering, the origin of the gamma-ray can be constrained to a cone surface. As a result of the intersection of different cones, information regarding the activity distribution can be obtained. To date, it has been shown that their sensitivity can be approximately twenty times better than that of a collimated SPECT system for a given resolution. However, current prototypes only reach this performance for gamma energies larger than 400 keV [92]. Moreover, they are still complex to design, and the reconstruction process is still complicated and under development [93]. Further investigations are required to achieve the clinical standards. Therefore, accurate Monte Carlo simulations are important.

Recently, an extension of GATE was proposed [94] to support the simulation of Compton cameras. This module facilitates the investigation of such systems and realistic comparisons between different configurations. As for PET systems, the main principle is to consider coincidence events, and timing is important. The interactions in the absorber and scatterer layers, named 'hits', are considered and aggregated together in 'singles' according to the detector layers response. The 'singles' are then sorted in time and associated in sequences of 'coincidences'. Numerous processing modules, also employed for SPECT and PET, are available – for example, energy resolution, time resolution, energy window, spatial resolution, and dead time. The Compton camera module has been validated against experimental data [94] and may be tailored for a wide range of detectors.

REFERENCES

[1] D. E. Raeside, "Monte Carlo principles and applications," *Phys Med Biol* vol. 21, pp. 181–197, 1976.

[2] P. Andreo, "Monte Carlo techniques in medical radiation physics," *Phys Med Biol* vol. 36, pp. 861–920, 1991.

[3] H. Zaidi, "Relevance of accurate Monte Carlo modeling in nuclear medical imaging," *Med Phys* vol. 26, no. 4, pp. 574–608, 1999.

[4] I. Buvat and I. Castiglioni, "Monte Carlo simulations in SPET and PET," *Q J Nucl Med* vol. 46, no. 1, pp. 48–61, 2002.

[5] M. Ljungberg, S. E. Strand, and M. A. King, *Monte Carlo Calculation in Nuclear Medicine: Applications in Diagnostic Imaging*. Boca Raton: Taylor & Francis, 1998.

[6] M. Ljungberg, S. E. Strand, and M. A. King, *Monte Carlo Calculation in Nuclear Medicine: Applications in Diagnostic Imaging*, 2nd edn. Boca Raton: Taylor & Francis, 2012.

[7] H. Zaidi and G. Sgouros, *Therapeutic Applications of Monte Carlo Calculations in Nuclear Medicine*. Boca Raton: Taylor & Francis, 2002.

[8] J. H. Hubbell, "Photon cross sections, attenuation coefficients and energy absorption coefficients from 10 keV to 100 GeV," *Natl Stand Ref Data Ser* vol. 29, 1969.

[9] Lawrence Livermore Lab, "Compilation of Xray Cross Sections," Livermore, CA, 1969.

[10] E. Storm and H. I. Israel, "Photon cross sections from 1 keV to 100 MeV for Elements Z=1 to Z=100," *Nucl Data Tables* vol. A7, pp. 565–681, 1970.

[11] E. Woodcock, T. Murphy, P. Hemmings, and S. Longworth, "Techniques used in the GEM code for Monte Carlo neutronics calculations in reactors and other systems of complex geometry."Conference Proceedings: Applications of Computing Methods to Reactor Problems, ANL-7050, p. 557, Argonne National Laboratory 1965.

[12] *Rand Corp.* Application of Monte Carlo, 1956.

[13] J. Persliden, "A Monte Carlo program for photon transport using analogue sampling of scattering angle in coherent and incoherent scattering processes," *Comput Meth Programs Biomed* vol. 17, pp. 115–128, 1983.

[14] J. H. Hubbell, J. W. Veigle, E. A. Briggs, R. T. Brown, D. T. Cramer, and R. J. Howerton, "Atomic form factors, incoherent scattering functions and photon scattering cross sections," *J Phys Chem Ref Data* vol. 4, pp. 471–616, 1975.

[15] E. Larsson, B. A. Jonsson, L. Jonsson, M. Ljungberg, and S. E. Strand, "Dosimetry calculations on a tissue level by using the MCNP4c2 Monte Carlo code," *Cancer Biother Radiopharm* vol. 20, no. 1, pp. 85–91, 2005.

[16] D. Minarik, K. Sjogreen Gleisner, and M. Ljungberg, "Evaluation of quantitative ^{90}Y SPECT based on experimental phantom studies," *Phys Med Biol*, Evaluation Studies Research Support, Non-U.S. Gov't vol. 53, no. 20, pp. 5689–5703, 2008.

[17] D. Minarik, M. Ljungberg, P. Segars, and K. S. Gleisner, "Evaluation of quantitative planar ^{90}Y bremsstrahlung whole-body imaging," *Phys Med Biol* Evaluation Studies Research Support, Non-U.S. Gov't vol. 54, no. 19, pp. 5873–5883, 2009.

[18] D. Minarik et al., "90Y Bremsstrahlung imaging for absorbed-dose assessment in high-dose radioimmunotherapy," *J Nucl Med* vol. 51, no. 12, pp. 1974–8, 2010.

[19] M. J. Berger, "Monte Carlo calculation of the penetration and diffusion of fast charged particles," in *Methods in Computational Physics vol. 1*, B. Alder, S. Fernbach, and M. Rotenberg (eds). New York: Academic Press, 1963, p. 135.

[20] M. Ljungberg and S. E. Strand, "A Monte Carlo program simulating scintillation camera imaging," *Comp Meth Progr Biomed* vol. 29, pp. 257–272, 1989.

[21] A. Todd-Pokropek, T. D. Cradduck, and F. Deconinck, "A file format for the exchange of nuclear medicine image data: A specification of Interfile version 3.3," *Nucl Med Comm* vol. 13, pp. 673–699, 1992.

[22] T. Merlin et al., "CASToR: A generic data organization and processing code framework for multi-modal and multi-dimensional tomographic reconstruction," *Phys Med Biol* vol. 63, no. 18, p. 185005, 2018.

[23] National Bureau of Standards, XCOM: Photon pross-sections on a personal computer. 1987.

[24] J. H. Hubbell and I. Øverbø, "Relativistic atomic form factors and photon coherent scattering cross section," *J Phys Chem Ref Data* vol. 8, pp. 69–105, 1979.

[25] I. G. Zubal and C. R. Harrell, "Computerized three-dimensional segmented human anatomy," *Med Phys* vol. 21, pp. 299–302, 1994.

[26] W. P. Segars, G. Sturgeon, S. Mendonca, J. Grimes, and B. M. W. Tsui, "4D XCAT phantom for multimodality imaging research," *Med Phys* vol. 37, no. 9, pp. 4902–4915, 2010.

[27] C. E. Metz, H. L. Atkins, and R. N. Beck, "The geometric transfer function component for scintillation camera collimators with straight parallel holes," *Phys Med Biol* vol. 25, pp. 1059–1070, 1980.

[28] B. M. W. Tsui and G. T. Gullberg, "The geometric transfer function for cone and fan beam collimators," *Phys Med Biol* vol. 35, no. 1, pp. 81–93, 1990.

[29] M. Ljungberg, A. Larsson, and L. Johansson, "A new collimator simulation in SIMIND based on the delta-scattering technique," *IEEE Trans Nucl Sci* vol. 52, no. 5, pp. 1370–1375, 2005.

[30] S. Jan et al., "GATE: A simulation toolkit for PET and SPECT," *Phys Med Biol* vol. 49, pp. 4543–4561, 2004.

[31] S. Jan et al., "GATE V6: A major enhancement of the GATE simulation platform enabling modelling of CT and radiotherapy," *Phys Med Biol* vol. 56, pp. 881–901, 2011.

[32] D. Sarrut et al., "A review of the use and potential of the GATE Monte Carlo simulation code for radiation therapy and dosimetry applications," *Med Phys* vol. 41, no. 6, p. 064301, 2014.

[33] W. Schneider, T. Bortfeld, and W. Schlegel, "Correlation between CT numbers and tissue parameters needed for Monte Carlo simulations of clinical dose distributions," *Phys Med Biol* vol. 45, no. 2, pp. 459–478, 2000.

[34] D. Sarrut and L. Guigues, "Region-oriented CT image representation for reducing computing time of Monte Carlo simulations," *Med Phys* vol. 35, no. 4, pp. 1452–1463, 2008.

[35] J. Strydhorst, T. Carlier, A. Dieudonne, M. Conti, and I. Buvat, "A gate evaluation of the sources of error in quantitative (90) Y PET" *Med Phys* vol. 43, no. 10, p. 5320, 2016.

[36] L. Grevillot, D. Bertrand, F. Dessy, N. Freud, and D. Sarrut, "A Monte Carlo pencil beam scanning model for proton treatment plan simulation using GATE/GEANT4," *Phys Med Biol* vol. 56, no. 16, pp. 5203–5219, 2011.

[37] X. Song, W. P. Segars, Y. Du, B. M. Tsui, and E. C. Frey, "Fast modelling of the collimator-detector response in Monte Carlo simulation of SPECT imaging using the angular response function," *Phys Med Biol* vol. 50, no. 8, pp. 1791–1804, 2005.

[38] P. Descourt et al., "Implementation of angular response function modeling in SPECT simulations with GATE," *Phys Med Biol* vol. 55, no. 9, pp. N253–66, 2010.

[39] T. Ryden et al., "Fast GPU-based Monte Carlo code for SPECT/CT reconstructions generates improved (177)Lu images," *EJNMMI Phys* vol. 5, no. 1, p. 1, 2018.

[40] D. Sarrut, N. Krah, J. N. Badel, and J. M. Letang, "Learning SPECT detector angular response function with neural network for accelerating Monte-Carlo simulations," *Phys Med Biol* vol. 63, no. 20, p. 205013, 2018.

[41] N. Freud, J. M. Létang, and D. Babot, "A hybrid approach to simulate multiple photon scattering in X-ray imaging," *Nucl. Instrum. Methods Phys. Res. **B*** vol. 227, no. 4, pp. 551–558, 2005.

[42] A. P. Colijn and F. J. Beekman, "Accelerated simulation of cone beam X-ray scatter projections," *IEEE Trans Med Imag* vol. 23, no. 5, pp. 584–590, 2004.

[43] G. Poludniowski, P. M. Evans, V. N. Hansen, and S. Webb, "An efficient Monte Carlo-based algorithm for scatter correction in keV cone-beam CT," *Phys Med Biol* vol. 54, no. 12, pp. 3847–3864, 2009.

[44] T. Cajgfinger, S. Rit, J. M. Letang, A. Halty, and D. Sarrut, "Fixed forced detection for fast SPECT Monte-Carlo simulation," *Phys Med Biol* vol. 63, no. 5, p. 055011, 2018.

[45] W. Gropp, E. Lusk, N. Doss, and A. Skjellum, "A high-performance, portable implementation of the MPI message passing interface standard," *Parallel Comput* vol. 22, pp. 789–828, 1996.

[46] J. Bert et al., "Geant4-based Monte Carlo simulations on GPU for medical applications," *Phys Med Biol* vol. 58, no. 16, pp. 5593–5611, 2013.

[47] M. P. Garcia, J. Bert, D. Benoit, M. Bardies, and D. Visvikis, "Accelerated GPU based SPECT Monte Carlo simulations," *Phys Med Biol* vol. 61, no. 11, pp. 4001–4018, 2016.

[48] Y. Lai et al., "gPET: A GPU-based, accurate and efficient Monte Carlo simulation tool for PET," *Phys Med Biol* vol. 64, no. 24, p. 245002, 2019.

[49] J. Lippuner and I. A. Elbakri, "A GPU implementation of EGSnrc's Monte Carlo photon transport for imaging applications," *Phys Med Biol* vol. 56, no. 22, pp. 7145–7162, 2011.

[50] Y. K. Dewaraja, M. Ljungberg, and K. F. Koral, "Characterization of scatter and penetration using Monte Carlo simulation in 131I imaging," *J Nucl Med* vol. 41, no. 1, pp. 123–130, 2000.

[51] M. F. Smith, R. J. Jaszczak, W. Huili, and L. Jianying, "Lead and tungsten pinhole inserts for I-131 SPECT tumor imaging: Experimental measurements and photon transport simulations," *IEEE Trans Nucl Sci* vol. 44, no. 1, pp. 74–82, 1997.

[52] M. F. Smith, "Modelling photon transport in non-uniform media for SPECT with a vectorized Monte Carlo code," *Phys Med Biol* vol. 38, pp. 1459–1474, 1993.

[53] H. M. Deloar, H. Watabe, T. Aoi, and H. Iida, "Evaluation of penetration and scattering components in conventional pinhole SPECT: Phantom studies using Monte Carlo simulation," *Phys Med Biol* vol. 48, pp. 995–1008, 2003.

[54] M. Peterson, S.-E. Strand, and M. Ljungberg, "Using the alloy Rose's Metal as pinhole collimator material in preclinical small animal imaging: A Monte Carlo evaluation," *Med Phys* vol. 42, pp. 1698–1709, 2015.

[55] P. H. Pretorius, C. Liu, P. Fan, M. Peterson, and M. Ljungberg, "Monte Carlo simulations of the GE Discovery Alcyone CZT SPECT Systems," *IEEE Trans Nucl Sci* vol. 62, no. 3, pp. 832–839, 2015.

[56] Y. Kitamura et al., "The efficiency of respiratory-gated (18)F-FDG PET/CT in lung adenocarcinoma: Amplitude-gating versus phase-gating methods," *Asia Ocean J Nucl Med Biol* vol. 5, no. 1, pp. 30–36, 2017.

[57] L. Le Meunier, R. Maass-Moreno, J. A. Carrasquillo, W. Dieckmann, and S. L. Bacharach, "PET/CT imaging: Effect of respiratory motion on apparent myocardial uptake," *J Nucl Cardiol* vol. 13, no. 6, pp. 821–830, 2006.

[58] C. Liu, L. A. Pierce, 2nd, A. M. Alessio, and P. E. Kinahan, "The impact of respiratory motion on tumor quantification and delineation in static PET/CT imaging," *Phys Med Biol* vol. 54, no. 24, pp. 7345–62, 2009.

[59] W. Qi, Y. Yang, C. Song, M. N. Wernick, P. H. Pretorius, and M. A. King, "4-D reconstruction with respiratory correction for gated myocardial perfusion SPECT," *IEEE Trans Med Imaging* vol. 36, no. 8, pp. 1626–1635, 2017.

[60] K. Thielemans, S. Rathore, F. Engbrant, and P. Razifar, "Device-less gating for PET/CT using PCA," in *2011 IEEE Nuclear Science Symposium Conference Record*, Oct. 23–29, 2011, pp. 3904–3910.

[61] P. P. Bruyant, M. A. King, and P. H. Pretorius, "Correction of the respiratory motion of the heart by tracking of the center of mass of thresholded projections: A simulation study using the dynamic MCAT phantom," *IEEE Trans Nucl Sci* vol. 49, no. 5, pp. 2159–2166, 2002.

[62] G. Kovalski, O. Israel, Z. Keidar, A. Frenkel, J. Sachs, and H. Azhari, "Correction of heart motion due to respiration in clinical myocardial perfusion SPECT scans using respiratory gating," *J Nucl Med* vol. 48, no. 4, pp. 630–636, 2007.

[63] J. C. Sanders, P. Ritt, T. Kuwert, A. H. Vija, and A. K. Maier, "Fully automated data-driven respiratory signal extraction from SPECT images using Laplacian Eigenmaps," *IEEE Trans Med Imaging* vol. 35, no. 11, pp. 2425–2435, 2016.

[64] L. Le Meunier, R. Maass-Moreno, J. A. Carrasquillo, W. Dieckmann, and S. L. Bacharach, "PET/CT imaging: Effect of respiratory motion on apparent myocardial uptake," *J. Nucl. Cardiol.* vol. 13, no. 6, pp. 821–830, 2006.

[65] A. Robert, S. Rit, and D. Sarrut, "Respiration-correlated 4D SPECT reconstruction " in *2019 IEEE Nuclear Science Symposium and Medical Imaging Conference (NSS/MIC)*, 26 Oct.-2 Nov. 2019 2019,

[66] C. E. Floyd, R. J. Jaszczak, K. L. Greer, and R. E. Coleman, "Inverse Monte Carlo as a unified reconstruction algorithm for ECT," *J Nucl Med* vol. 27, pp. 1577–1585, 1986.

[67] M. Ljungberg and S. E. Strand, "Scatter and attenuation correction in SPECT using density maps and Monte-Carlo simulated scatter functions," *J Nucl Med* vol. 31, no. 9, pp. 1560–1567, 1990.

[68] E. C. Frey and B. M. W. Tsui. *A New Method for Modeling the Spatially-variant, Object-dependent Scatter Response Function in SPECT*. (1996). Conference Records of the IEEE Medical Imaging Conference. Anaheim, CA, 3–9 November.

[69] H. W. A. M. d. Jong, E. T. P. Slijpen, and F. J. Beekman, "Acceleration of Monte Carlo SPECT simulation using convolution-based forced detection," *IEEE Trans Nucl Sci* vol. 48, no. 1, pp. 58–64, 2001.

[70] D. Minarik et al., "90Y Bremsstrahlung imaging for absorbed-dose assessment in high-dose radioimmunotherapy," *J Nucl Med* vol. 51, no. 12, pp. 1974–1978, 2010.

[71] M. Elschot, M. G. Lam, M. A. van den Bosch, M. A. Viergever, and H. W. de Jong, "Quantitative Monte Carlo-based 90Y SPECT reconstruction," *J Nucl Med* vol. 54, no. 9, pp. 1557–1563, 2013.

[72] Y. K. Dewaraja et al., "Improved quantitative (90) Y bremsstrahlung SPECT/CT reconstruction with Monte Carlo scatter modeling," *Med Phys* vol. 44, no. 12, pp. 6364–6376, 2017.

[73] M. Elschot et al., "Quantitative Monte Carlo-based holmium-166 SPECT reconstruction," *Med Phys* vol. 40, no. 11, p. 112502, 2013.

[74] Y. K. Dewaraja, M. Ljungberg, and J. A. Fessler, "3-D Monte Carlo-based scatter compensation in quantitative I-131 SPECT reconstruction," *IEEE Trans Nucl Sci* vol. 53, pp. 181–188, 2006.

[75] C. A. van Gils, C. Beijst, R. van Rooij, and H. W. de Jong, "Impact of reconstruction parameters on quantitative I-131 SPECT," *Phys Med Biol* vol. 61, no. 14, pp. 5166–5182, 2016.

[76] E. T. Hippelainen, M. J. Tenhunen, H. O. Maenpaa, J. J. Heikkonen, and A. O. Sohlberg, "Dosimetry software Hermes Internal Radiation Dosimetry: From quantitative image reconstruction to voxel-level absorbed dose distribution," *Nucl Med Commun* vol. 38, no. 5, pp. 357–365, 2017.

[77] J. Gustafsson, G. Brolin, and M. Ljungberg, "Monte Carlo-based SPECT reconstruction within the SIMIND framework," *Phys Med Biol* vol. 63, no. 24, p. 245012, 2018.

[78] M. Peterson, J. Gustafsson, and M. Ljungberg, "Monte Carlo-based quantitative pinhole SPECT reconstruction using a ray-tracing back-projector," *EJNMMI Phys* vol. 3, p. 32, 2017.

[79] D. Lazaro et al., "Validation of the GATE Monte Carlo simulation platform for modelling a CsI(Tl) scintillation camera dedicated to small-animal imaging," *Phys Med Biol* vol. 49, no. 2, pp. 271–285, 2004.

[80] F. D. Popota et al., "Monte Carlo simulations versus experimental measurements in a small animal PET system. A comparison in the NEMA NU 4-2008 framework," *Phys Med Biol* vol. 60, no. 1, pp. 151–162, 2015.

[81] F. Smekens et al., "Split exponential track length estimator for Monte-Carlo simulations of small-animal radiation therapy," *Phys Med Biol* vol. 59, no. 24, pp. 7703–7715, 2014.

[82] F. Verhaegen et al., "ESTRO ACROP: Technology for precision small animal radiotherapy research: Optimal use and challenges," *Radiother Oncol* vol. 126, no. 3, pp. 471–478, 2018.

[83] C. Noblet et al., "Validation of fast Monte Carlo dose calculation in small animal radiotherapy with EBT3 radiochromic films" *Phys Med Biol* vol. 61, no. 9, pp. 3521–3535, 2016.
[84] J. Salvadori et al., "Monte Carlo simulation of digital photon counting PET," *EJNMMI Phys* vol. 7, no. 1, p. 23, 2020.
[85] Performance measurements of gamma cameras. NEMA NU 1-2018.
[86] M. S. Lee, D. Hwang, J. H. Kim, and J. S. Lee, "Deep-dose: A voxel dose estimation method using deep convolutional neural network for personalized internal dosimetry," *Sci. Rep.* vol. 9, no. 1, p. 10308, 2019.
[87] T. I. Gotz, E. W. Lang, C. Schmidkonz, A. Maier, T. Kuwert, and P. Ritt, "Particle filter de-noising of voxel-specific time-activity-curves in personalized (177)Lu therapy," *Z Med Phys*, 2019.
[88] I. Gulrajani, F. Ahmed, M. Arjovsky, V. Dumoulin, and A. Courville, "Improved training of wasserstein GANs," presented at the Proceedings of the 31st International Conference on Neural Information Processing Systems, Long Beach, CA, 2017.
[89] I. Goodfellow et al., "Generative Adversarial Nets" pp. 2672–2680, 2014.
[90] D. Sarrut, N. Krah, and J. M. Létang, "Generative adversarial networks (GAN) for compact beam source modelling in Monte Carlo simulations," *Phys Med Biol* vol. 64, no. 21, p. 215004, 2019.
[91] R. W. Todd, J. M. Nightingale, and D. B. Everett, "A proposed γ camera," *Nature* vol. 251, no. 5471, pp. 132–134, 1974.
[92] M. Fontana, D. Dauvergne, J. M. Letang, J. L. Ley, and E. Testa, "Compton camera study for high efficiency SPECT and benchmark with Anger system," *Phys Med Biol* vol. 62, no. 23, pp. 8794–8812, 2017.
[93] V. Maxim, "Enhancement of Compton camera images reconstructed by inversion of a conical Radon transform," ***Inverse*** *Probl.* vol. 35, no. 1, p. 014001, 2018.
[94] E. Ane et al., "CCMod: A GATE module for Compton camera imaging simulation" *Phys Med Biol*, vol. 65, no. 5, p. 055004, 2019.
[95] D. Sarrut et al., "Advanced Monte Carlo simulations of emission tomography imaging systems with GATE" *Phys Med Biol*, vol. 66, no. 10, 2021.

30 Beta and Alpha Particle Autoradiography

Anders Örbom, Brian W. Miller and Tom Bäck

CONTENTS

30.1 INTRODUCTION

Since the discovery of radioactivity in the early twentieth century, autoradiography ("self" "radioactivity" "writing") has been an important tool in many applications. Surprisingly, even with the advancement of other imaging modalities such as SPECT and PET, its importance remains strong because it provides unmatched spatial resolutions of

DOI: 10.1201/9780429489556-30

radioactivity distributions within tissues or samples. In this chapter we discuss various technologies that are used to generate autoradiograms, ranging from early film-based methods to next-generation digital and single-particle detection methods that utilize scintillation, gaseous, or semi-conductor technologies. Our discussion focuses on systems designed for imaging beta particles and, more recently, alpha particles, with current example studies and applications.

30.2 AUTORADIOGRAPHY

Autoradiography can be said to be the first imaging technique ever used for the localization of radioactivity in a sample, and images of biological samples were produced as early as 1904 [1]. Autoradiography is actually a number of technologies and methods used to produce a two-dimensional image of the distribution of activity in, or on, a thin, solid, and immobile sample that is in close proximity to the detector or detection medium [1, 2]. The images are formed by the charged particles emitted (α or $\beta\pm$), while γ radiation often passes through the detector without interacting with it [3]. Autoradiography has been widely used in pharmaceutical development, and apart from superior sensitivity and quantification compared to non-radionuclide methods, autoradiography is vital for dosimetry at the cell and tissue levels, since it images the actual interactions delivering the absorbed dose. When the important information is the location of the biological molecule, however, it can sometimes be a problem that only the radionuclide is actually imaged, as it might be released from the radiolabelled molecule (i.e. free radionuclide) attached to a metabolite or other structure and not give the distribution of interest. This is why mass-spectrometry imaging has recently become an important tool that in some cases will be preferred over autoradiography [4]. SPECT, with long measurement times of frozen samples, is an alternative to at least macro-autoradiography when working with gamma-emitting radionuclides, where researchers have reached 120 μm spatial resolution [5].

The performance of each autoradiography instrument is only partly dependent on the instrument itself. The particular radiation characteristics of the radionuclide measured will affect the sensitivity and spatial resolution of the image with, for example, higher-energy beta particles traveling longer in a detector and reducing the precision of the position measurement. Also, the properties of the sample will matter with, for example, a thicker cryosection of a tissue allowing for more activity and better statistics but also a reduction in spatial resolution as charged particles will travel longer before reaching the detector. In the same way, sample material with high density – bone, for example – will affect the particle range and influence the resolution. Future development may counteract certain sample difficulties with reconstruction techniques that improve depth-resolution, for example, by using a form of tomosynthesis [6]. Many autoradiography systems will also have difficulty imaging low-energy beta emitters like ^{3}H since the electron does not have enough energy to traverse some protective or non-measurement layer before reaching the actual detector.

30.2.1 Applications of Autoradiography

The most common application of autoradiography in biomedical research is the imaging of the distribution of radionuclides in a tissue section from experimental animals. The sections can either be cryosections directly frozen after excision or paraffin-embedded sections. The latter often requires fixation of the tissue through incubation in fixative, which might dissolve tissue components, alter the radionuclide location, and not give a true-to-life image of the distribution. The radionuclide might have been injected into the animal at some time prior to sacrifice, either in its pure form, or often used as a label on a biological molecule or pharmaceutical of some kind. It might also have been applied post sacrifice through immersion of a tissue section in a solution containing the radionuclide-labelled tracer, so-called *in vitro* autoradiography.

Other applications of autoradiography systems are for reading out thin-layer chromatography (TLC) strips to determine the purity of the radiolabelling of a compound, although this does not require a high spatial resolution. Autoradiography can also be used in nuclear forensics, investigating labs or other areas where tracking possibly illicit nuclear materials is important. In a 2018 paper, Parsons-Davis and colleagues argued that storage phosphor imagers could be a good first-step instrument to localize tracks of nuclear material [7]. In 2020 Clarke and colleagues similarly employed a scintillator-based instrument to image environmental swipe samples and claim to be able to distinguish certain radionuclides from each other based on energy response [8]. To examine the mobility of uranium daughters after *in situ* uranium leaching, Angileri and colleagues, in a 2020 paper, used a gaseous autoradiography detector to examine stone samples [9].

30.2.2 Advanced Applications

One more advanced application of autoradiography is to employ samples containing several radionuclides and separate their distribution in different images from one or several measurement sessions. Using samples containing two radionuclides with significantly different half-lives and imaging them twice, before and after decay of the short-lived isotope, is a fairly common method of separating their contributions to an autoradiography image. In some such studies, the contribution of the long-lived radionuclide to the first image is very small, and no separation is considered necessary [10–13]. In a 2018 study, Broisat and colleagues labelled albumin with ^{125}I and red blood cells with ^{99m}Tc to investigate blood flow in rats with and without stroke [14]. Separation was done by first imaging the sample with both radionuclides on a phosphor storage imager system and then waiting for ^{99m}Tc to decay to get only the ^{125}I image, which was then used to subtract from the combined one to get one of only ^{99m}Tc. An early example of the separation of the contributions of different radionuclides according to their energy deposition was published in 1966 by Wimber and colleagues, who compared the visible tracks of ^{3}H and ^{14}C in layers of film emulsion [15]. Separation of ^{35}S and ^{32}P using storage phosphor plate screens by imaging with and without an attenuating filter between the sample and imaging plate has also been reported [16]. The separation of radionuclides by energy after a single imaging session has been demonstrated in modern digital autoradiography systems using the scintillation light from a scintillator sheet in contact with a sample, or by using a parallel-plate avalanche chamber [17]. In a 2014 study, Miller and colleagues demonstrated imaging discrimination between alpha, beta, and fission-fragment charged particles in a sample using appropriate scintillators and energy information in an autoradiography system with single-particle detection sensitivity [18]. Semiconductor detectors have a better energy resolution than scintillation detectors, and successful separation of ^{35}S and ^{32}P was shown on an early double-sided silicon strip (DSSD) system [19]. One of this chapter's authors, Örbom, has performed several multi-radionuclide studies using a more modern DSSD system, separating, for example, ^{131}I and ^{125}I labelled to different tracers targeting the same biological target in a tumour [20]. In another study using the same instrument, the same two radionuclides were used, this time labelled to two different antibodies, but also accompanied by ^{18}F-FDG in the same animal. The distributions of all three radionuclides in a mouse aorta were separated based on the difference from ^{131}I in measured energy spectra for ^{125}I, or rate of decay for ^{18}F during one single measurement session [21].

Another advanced application is to make use of the autoradiography images to calculate dosimetry. One way this can be done is by drawing regions of interest in the images and determining the activity concentrations in different organs or parts of organs and using standard MIRD dosimetry S-values [22]. In many cases, however, this can be done as efficiently using in vivo imaging modalities like PET or SPECT, which have the benefit of containing 3D information of the activity distribution. Autoradiography is more suited to investigate the small-scale distribution of absorbed dose, or often only dose-rate for the particular moment in time imaged in the autoradiography sample. For true cellular level dosimetry, micro-autoradiography will be needed to determine the distribution of radionuclides, especially for α-emitters and radionuclides with dominant Auger emission [2]. However, for the level of cell-populations, dose point kernels (DPKs) can be convolved with the activity distribution from macro autoradiography to calculate the absorbed dose-rate distributions. A DPK is either analytically calculated or simulated using Monte Carlo techniques and contains the absorbed dose at all locations within the kernel diameter around the decay of a specific radionuclide [22, 23].

30.3 FILM-BASED AUTORADIOGRAPHY

The basic mechanisms of detection of ionizing radiation by film is that the incoming radiation will interact with the film emulsion, which consist of silver bromide crystals in gelatine. When an ionization event occurs, an electron is set free and can eventually find a silver ion that will then be reduced to a silver atom. The more ionization events, the more silver atoms, and in the development step all non-reduced silver ions are washed off, leaving coloured silver grains where ionizations have occurred [24]. From this process it follows that film-based autoradiography requires an exposure step with the sample in contact with the emulsion or emulsion-covered film and then, after a specific time, the sample is removed, and the film developed [25]. Since film is sensitive to visible light, many operations of the autoradiography will have to be carried out in darkrooms. The limits of the range of linear response for film emulsion are that there is a level where the energy of the radiation is not high enough to cause an ionization and reduction of the silver ion, and once all available silver ions have been reduced you have reached over-exposure and cannot get more signal even if there are more ionizations [24]. This leads to difficulty in quantification of film autoradiography, and you are almost always required to image your sample together with a prepared calibration curve of some sort to compare "grey levels" in the final image between the standard and the sample [25]. Sensitivity and spatial resolution in a film autoradiography image will depend both on the range of the radiation emitted by the radionuclide and on the configuration of the emulsion.

Both a thicker emulsion and one with larger silver grains will increase sensitivity but decrease spatial resolution. There is also a way to amplify the signal from the ionizing radiation by using a scintillation material and converting it to visible light, but this again will reduce spatial resolution and complicate quantification [24].

30.3.1 Film Autoradiography

Films come covered in different thicknesses of layers of emulsion with various-sized grains and with a variety of cover layers which, if present, will prevent imaging of low-energy beta emitters such as ^{3}H, but specific non-covered films have been developed to perform such imaging [24, 25]. Ideal spatial resolution for ^{14}C on film would be around 2–5 µm [24]. Today, a developed film is often scanned or photographed to enable digital processing of the image.

A 2020 study making a typical use of film autoradiography is by Bogdanska and colleagues, who investigated the distribution of a ^{14}C-labelled version of the environmental pollutant Perfluorooctanoic acid (PFOA) [26]. Mice were given ^{14}C-PFOA in their diet, and biodistribution was measured using liquid scintigraphy, but a few animals were also frozen whole after sacrifice and whole-body sectioned and then exposed to autoradiographic film from Agfa while still frozen at -20° C. After exposure the film was developed and the tissue section stained for histology, as seen together with the autoradiography image in Figure 30.1. In a 2019 study from Smith and colleagues, employing both ex vivo

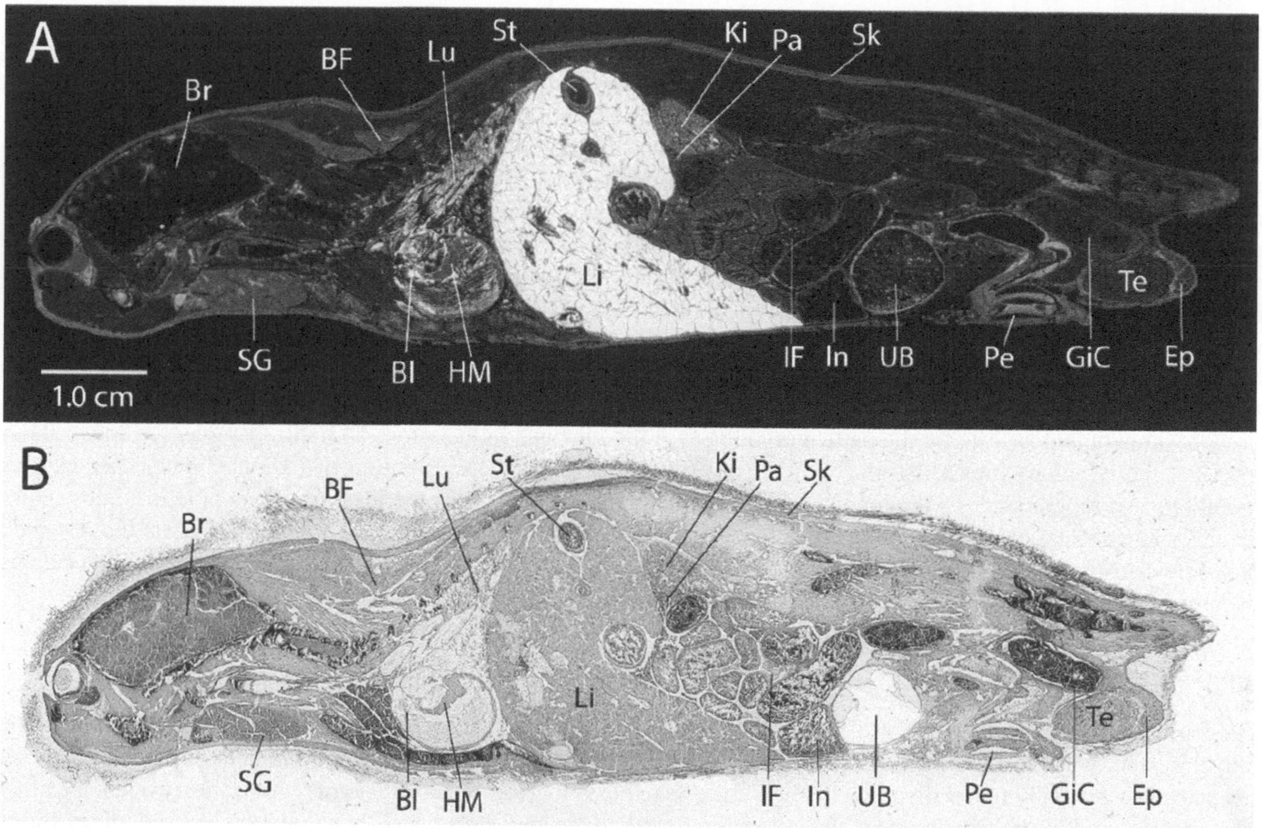

FIGURE 30.1 Whole-body autoradiography and histological staining of a mouse exposed to ^{14}C-PFOA. (A) Whole-body autoradiogram of a 40 mm-section following 5 days of dietary exposure and (B) the corresponding tape-section stained with safranin O/Fast green for cartilage and Von Kossa's staining for calcified bone. The brighter areas in (A) correspond to higher levels of radioactivity. The autoradiogram demonstrates highest level of PFOA in the liver and somewhat lower levels in blood, lungs, kidneys and skin. PFOA can also be observed in brown fat, salivary gland, and penis. The brain demonstrates low levels of PFOA. The positive autoradiogram has been inverted and in the corresponding tape-section artefacts in areas outside of the mouse have been removed. One artefact that occurred during the autoradiogram development can be observed in the brain as a bright white spot. Bl: blood, BF: brown fat, Br: brain, Ep: epididymis, GiC: gastrointestinal content, IF: inguinal fat, HM: heart muscle, Ki: kidney, Li: liver, Lu: lungs, Pa: pancreas, St: stomach, SG: salivary gland, Sk: skin, Pe: penis, Te: testis, UB: urinary bladder. (Image originally published in [26] and reprinted with permission.)

and in vitro autoradiography, the effects on neuroinflammation and microglial activation in monkeys allowed to self-administer cocaine was studied [27]. Monkeys were given ^{14}C-labelled glucose given intravenously before sacrifice and, after sacrifice, brain sections were incubated with ^{3}H-PK11195; brain sections were then exposed to either normal or ^{3}H-senistive film together with standards for quantification. Another study that used *in vitro* autoradiography was by Albores-Garcia and colleagues from 2020, which investigates the effect on opioid receptors in the rat brain by exposing the rats to lead in their diet [28]. After exposure to lead and then sacrificing the rats, brain sections were incubated with ^{3}H-D-Ala^{2}-MePhe^{4}-Gly-ol^{5} enkephalin and then exposed to ^{3}H-sensitive film for 16 weeks before development to show the receptor density. In 2017, more in vitro work was done by Hoon Yoo and colleagues, who investigated the uptake of the oxycodone metabolite oxymorphone in the brains of different mouse models [29]. Brain sections from mice with different receptors knocked out were incubated with ^{3}H-Oxymorphone and then exposed to ^{3}H-sensitive film for 7–9 weeks, along with standards before development and analysis.

30.3.2 Film Emulsion Autoradiography

When the absolute highest spatial resolution is required – for example to see whether a decay occurs inside or outside a cell – film autoradiography is not sufficient due to the physical distance between the radionuclides and the silver halide to be converted to metallic silver. Instead, the researcher can put the sample to be imaged in direct contact with the film emulsion. This is commonly done in either of two ways. One is to section a tissue onto a microscope slide and then dip that slide into emulsion to coat the tissue, store it in a light-tight environment, and then develop it to show dark grains where decays have occurred [24]. An issue with this method is that soluble compounds can be washed away during the dipping procedure and therefore an alternative method has been developed whereby the slides are pre-coated with emulsion and then dried, and the tissue sections are picked up directly on these slides [30]. While the sensitivity of the emulsion method is generally lower than when using films, the resolution when viewing the developed slides through a microscope can be very high, and with an electron microscope even as good as 0.1 μm [24].

In 2017, Asano and colleagues investigated the distribution of selective glucocorticoid receptor agonist (SEGRA) SA22465 in the eyes and eyelids of rabbits, which were sectioned and mounted on slides pre-coated with photographic emulsion [31]. Before sacrifice, the rabbits were given ^{14}C-SA22465 ophthalmic suspension into each eye. The slides containing 5 μm thick sections were exposed for 28 days and then developed and counter-stained with haematoxylin and eosin. The dark grains from autoradiography on top of the haematoxylin of different parts of the eye and eyelid can be seen in Figure 30.2. Emulsion autoradiography was also used by Aguero and colleagues on brain sections from deceased Alzheimer patients [32]. The sections were incubated with either the PET tracer ^{18}F-MK-6240 or ^{18}F-AV-1451 before the slides were covered with emulsion, one of two autoradiography methods employed, and exposed and developed.

30.4 PHOSPHOR STORAGE PLATE IMAGING

The phosphor storage plates are perhaps the most common of autoradiography systems currently found in modern radionuclide imaging labs. It has the benefits of higher dynamic range, reusability, and digital readout over film and was quickly adapted for use with autoradiography [24]. The imaging plates consist of barium fluorohalide crystals doped with europium on the surface, covered by a protective layer unless specially designed for very low (e.g., ^{3}H) energy electrons that cannot penetrate such a layer. Ionizing radiation that interacts with the crystals will further ionize the europium, taking it from Eu^{2+} to Eu^{3+}. The electrons leaving the europium will be trapped in fluorohalide vacancies [7]. The entrapment of the electrons is a metastable state with a limited half-life, which will limit the usability of phosphor storage screens for very low activity samples where new ionizations cannot generate new Eu^{3+} fast enough to compensate for the recombination of electrons and ions [24]. If the activity in the measured sample is over this limit, however, it will generate more and more trapped electrons and europium ions the more ionizing radiation interacts with the active layer. The length of this exposure step will have to be individually decided for each study depending on radionuclide and activity in the measured samples.

The read-out of phosphor storage plates is similar to the development of film in that it can be done after any chosen time of exposure, and that it will prevent any true resumption of the imaging session. The basic mechanism is that after being loaded into a light-tight instrument, individual points on the plate are exposed to red light from a laser [24]. This light gives enough energy to the trapped electrons that they can travel and recombine with an Eu^{3+} ion, reducing it to Eu^{2+}. The excess energy in this process is emitted as a photon of 390 nm wavelength, which is detected by the read-out instrument and digitized into a pixel value based on the amplitude of the light [7, 24, 33]. Before reuse of the imaging plate, all residual signal must be erased, and this is typically done by exposure to bright white light [33]. While phosphor

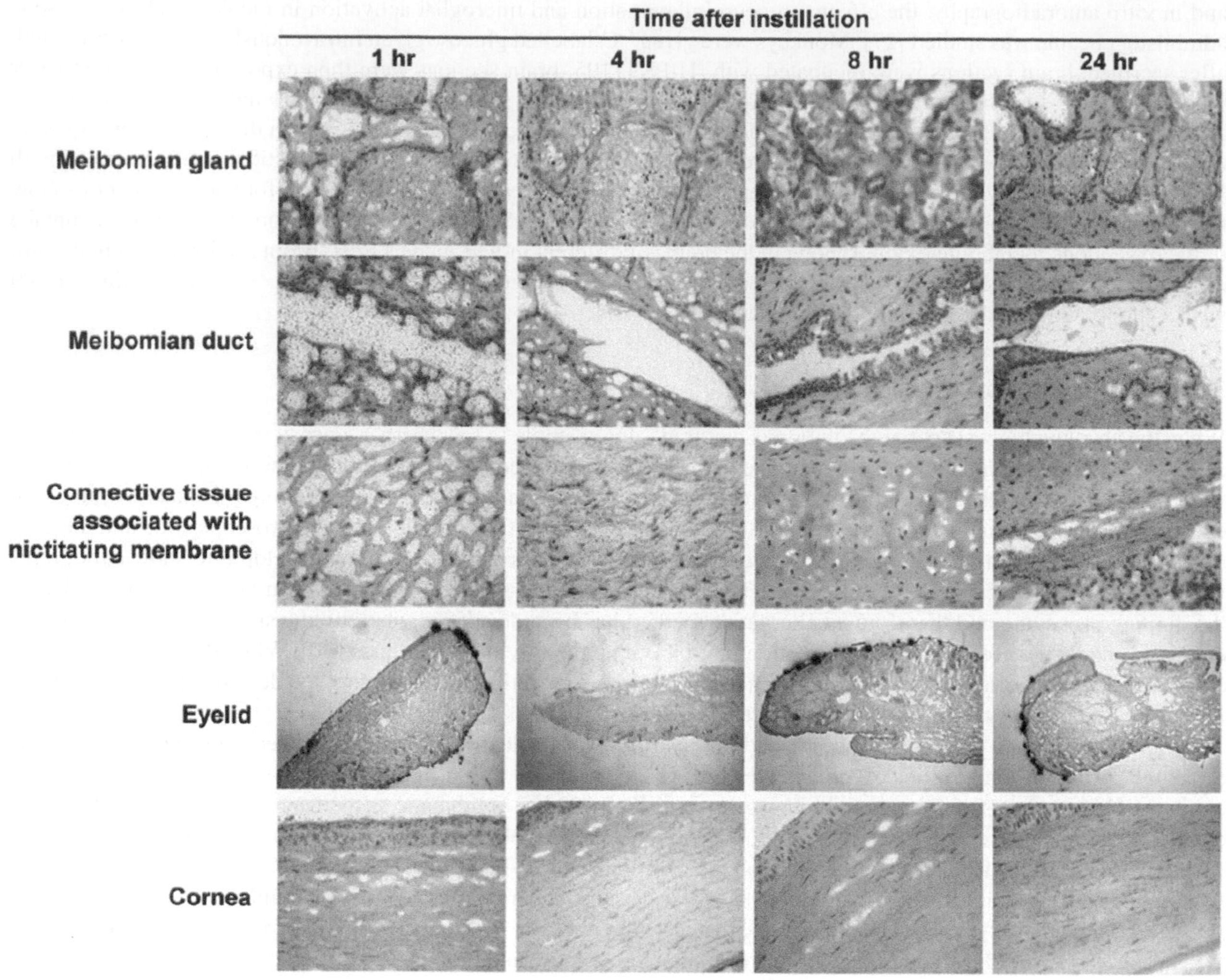

FIGURE 30.2 Time course micrographic distribution of ^{14}C-SA22465 in the meibomian gland, meibomian duct, connective tissue associated with nictitating membrane, eyelid, and cornea following a single ocular administration. Magnification: x2.5 for eyelids, x20 for other tissues. (Image originally published in [31] and reprinted with permission.)

storage plates have a greater linear range of response than film, they can be underexposed, as mentioned above, and also overexposed by high-activity samples measured for too long. This often makes pilot studies necessary to determine the exposure time for each particular study. Apart from the properties of the radiation that created the image, the spatial resolution in the final image is determined by the properties of the read-out instrument, the emulsion thickness, and grain size in the plate [7, 33]. Due to light-scattering of the read-out laser in the plate, a pixel size of much less than 50 µm is generally not seen to improve resolution even with very short-range radiation [24]. A 2015 study by Örbom and co-authors measured the FWHM in the resulting image of an 18 µm thick ^{58}Co-wire to be 344 ± 14 µm [20] in a Perkin Elmer phosphor storage plate system.

30.4.1 Applications

A typical example of a preclinical study using phosphor storage plate autoradiography is the 2020 study by Waaijer and colleagues, where they ^{89}Zr-label a bispecific antibody targeting CD3ε on T cells and glypican 3 (GPC3) on tumours. This is given to tumour-bearing mice who are PET-imaged and then sacrificed and tissues collected, embedded and paraffin-sectioned at 4 µm for autoradiography [34]. Another PET and phosphor storage plate autoradiography study was Putzu and colleagues where the cerebral distribution of ^{18}F-FDG was investigated in rats post cardiac arrest and resuscitation. Rather thick, 2 mm, sections were used for autoradiography using a 1 h exposure time [35].

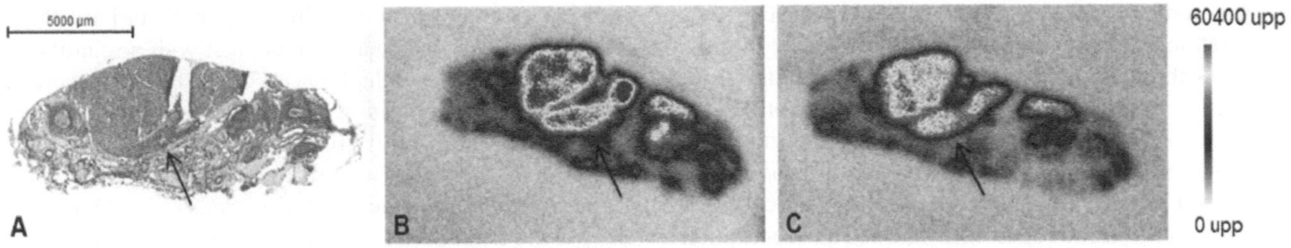

FIGURE 30.3 Salivary gland tissue cryosections (pig, 10 m). Arrows indicate glandular areas. (A) H&E staining; autoradiography after incubation for 1.5 h at ambient temperature with 80 nM [177Lu]Lu-PSMA-617 showing total binding (B) and additional incubation with 80 M 2-PMPA (highly potent PSMA-inhibitor) indicating non-specific binding (C). (Image originally published in [40] and is reprinted according to Creative Common CC BY license (creativecommons.org).) Colour image available at www.routledge.com/9781138593268.

In vitro autoradiography allows for the study of human material, for example sections of brains from deceased Alzheimer disease patients, which were incubated with ^{3}H-THK5117 and ^{3}H-deprenyl in a 2017 study by Lemoine and colleagues that, through autoradiography using ^{3}H-sensitive phosphor storage plates, showed that tau deposits and inflammatory processes are closely spatially related [36]. Brain sections from diseased Alzheimer patients were also used by Aguero and colleagues, who incubated the sections with either the PET tracer ^{18}F-MK-6240 or ^{18}F-AV-1451 before autoradiography [32]. Varnäs and colleagues used both (normal) human brain sections and non-human primate brain sections, both incubated with the PET tracer ^{11}C-UCB-J targeting Synaptic vesicle glycoprotein 2A before storage phosphor plate autoradiography [37]. Similar methods, on sections of mouse brains, were used in a study by Valuskova and colleagues [38]. Here, three ^{3}H-labelled tracers were used, ^{3}H-pirenzepine, ^{3}H-AFDX-384 and ^{3}H-QNB to show their different bindings to specific receptors in the brain using phosphor storage plate autoradiography and the MCID analysis software. Fjellaksel and colleagues in 2019 also incubated sections of mouse brain before phosphor storage plate autoradiography and analysis in ImageJ software, this time with a potential SPECT tracer, ^{125}I-[D-Trp6]-LHRH, targeting a gonadotropin-releasing hormone receptor in Alzheimers disease [39]. With cryosections of pig salivary glands, Tönnesmann and colleagues used in vitro phosphor storage plate autoradiography and analysis in the Optiquant software to investigate the distribution of the radionuclide therapy agent ^{177}Lu-PSMA-617 [40] as can be seen in Figure 30.3.

^{14}C is a common radionuclide for autoradiography, for example in this 2017 study by Asano and colleagues where both phosphor storage plates and film emulsion imaging is used to detect the distribution of ^{14}C-SA22465 in rabbit eyelids and eyes [31]. A more exotic animal model is used by Soubaneh and colleagues in their 2019 study where the distribution of ^{14}C-labelled amide multi-walled carbon nanotubes was imaged using whole body phosphor storage autoradiography of fish exposed to water containing the tracer [41].

Non-medical research applications of the phosphor storage plate imaging technique include a 2020 study by Nguyen and Thi, who expose a plant to ^{13}N gas, and the nitrogen transport in the plant is then imaged both by a real-time system and also by autoradiography [42]. Parsons-Davis and colleagues discuss using phosphor storage plate imaging for nuclear forensics of debris or fuel-cycle material in reference [7] and mention that the method could help quickly identify sources with very little preparatory work, and any discovered hotspots could later be analysed by other methods to determine the particular radionuclide and so forth.

30.5 SCINTILLATION-BASED AUTORADIOGRAPHY DETECTORS

The basic principle of scintillation is that ionizing radiation is converted to photons in the visible or at least imageable spectrum. This occurs by electrons in the scintillator material getting excited by interaction with ionizing radiation and, then, when they are de-excited, the excess energy is emitted as photons. The photons can then be imaged with, for example, a charge-coupled device (CCD) allowing for real-time registration and display of the data.

In pursuit of a tool to image tumour margins during surgery of cancer patients, Vyas and colleagues in 2018 tested a flexible scintillator sheet of 6 µm thickness and unknown composition. They found that while the sheet was less efficient than thicker firm scintillator sheets, the ability to be wrapped close to a bulky sample could provide a better signal to volume background noise ratio with a spatial resolution of 1.5 mm [43]. For a similar purpose, intraoperative imaging during radioguided surgery, Shestakova and colleagues in 2006 developed and successfully tested an imager that consisted of a micro columnar CsI(Tl) scintillator connected by fibre-optic wires to a CCD so it could more easily be moved around the patient to detect beta particle emissions [44].

Autoradiography can also be used to image live systems like a cell culture and, in their 2019 paper, Almasi and Pratx have refined this method of radioluminescence microscopy [45]. A cell culture is placed in contact with a scintillator, which is then connected by oil to a low-light microscope, which is read out by a high-speed CCD. In each recorded frame, software analysed the tracks in the scintillator lights made by interacting beta particles and determined a high-resolution original position of the decay, allowing imaging of individual cells with 10 μm spatial resolution.

Biospace Lab have introduced a scintillation-based imaging system under the name of the BetaIMAGER DFINE. It is based on the samples being put into contact with scintillating membrane containing 3 μm grains of Y_2SO_4 or ZnS which will scintillate upon ionization. The scintillation is very weak, so a light intensifier is required before the CCD that records the light of each event [24]. Biospace Lab claims 10 μm spatial resolution for ^{3}H imaging and the ability to discriminate between different radionuclides based on the shape and strength of the scintillation.

The iQID camera, recently commercialized by QScint Imaging Solutions, is based on phosphor screens, microcolumnar scintillators, or thin monolithic scintillators coupled to an image intensifier (night vision or scientific grade) [46]. iQID can be used to image alphas, betas, γ rays, fission fragments, and thermal neutrons. The charged particles interact with the scintillator, generating a flash of light that is amplified by the intensifier [46]. The output screen of the intensifier is imaged onto a fast CCD or CMOS camera, where individual events appear a cluster of signal. For imaging experiments, the iQID camera frames are processed using a graphics-processing unit (GPU) with events localized to construct a real-time counts image. List-mode data is stored for each event, including 2D positional (3D for γ rays), temporal, energy, and spatial features. Post-processing of the list-mode data is used for sub-pixel image reconstruction, charged-particle-type and isotopic discrimination, and activity estimation. Autoradiography with x-ray photons as low as 5.9 keV has been demonstrated [unpublished data] and a reported intrinsic spatial resolution up to 20 microns for alphas [47]. An important feature of the system is that it can accommodate large, irregularly shaped samples that otherwise are not compatible for imaging with other techniques. For example, betas from ^{99}Tc-emitting surfaces of cast stone pucks [48], alphas from bone surfaces [49], and electro-plated sources. To increase detector area, a fiber-optic taper is coupled to the intensifier. Large-area (~20x20 cm^2) systems have been developed by tiling fibre-optic tapers [50].

30.5.1 Application of Scintillation-based Imaging Systems

One example of an application of scintillation cameras with the iQID is evaluation of tumour uptake versus normal tissues in pre-targeted radioimmunotherapy (PRIT), a multi-step method of selectively delivering high doses of radiotherapy to tumour cells while minimizing exposure to normal tissues and organs. Frost and colleagues looked at murine biodistributions for ^{90}Y and ^{177}Lu in tumours and normal organs. The two radionuclides displayed comparable biodistributions in tumours and normal organs, but ^{90}Y delivered more than twice as high a dose. ^{90}Y -DOTA-biotin was dramatically more effective that with ^{177}LU-DOTA-biotin [51]. Results can be seen in Figure 30.4.

Hernández and colleagues in 2019 looked at different nanobodies penetration into tumour spheroids, based on whether or not they contained targeting properties. They labelled different types with either ^{111}In or ^{177}Lu and used both multi-radionuclide SPECT and then BetaIMAGER DFINE autoradiography of 10 μm cryosections where the signal from different radionuclides was separated based on half-life [52]. Results can be seen in Figure 30.5.

30.6 GASEOUS DETECTOR-BASED AUTORADIOGRAPHY DETECTORS

There are many configurations of real-time gaseous detectors. One is the parallel plate avalanche chamber, which consists of two electrodes between which there is a volume with a high electric field applied over a noble gas. Electrons ejected by ionizations by radiation from the autoradiography sample can then be accelerated to be accelerated to ionize the noble gas, creating more electrons. You get an electron avalanche that is proportional to the number of original ionization electrons and this can be read out either by electronics or, as in the BetaIMAGER tRACER by Biospace Lab, by optical means [24]. This system has a gas mixture of argon and trimethylamine, and the latter substance will scintillate in ultraviolet light when ionized and can be recorded by a high-speed CCD, marking the location, size and shape of each electron avalanche. This can be used to get real-time imaging of a sample and also to discriminate between radionuclides using the size and shape of the light emission, reaching an energy resolution of 18 percent [53]. Biospace Lab advertises that the instrument has a 200 × 250 mm maximum field of view and can detect radionuclides such as ^{3}H, ^{14}C, ^{32}P, ^{35}S, ^{125}I and more.

Recently, a new model of gaseous detector has been introduced by the French company AI4R under the name of the BeaQuant. This is a form of micropattern gas detector called a parallel ionization multiplier [24]. The instrument works

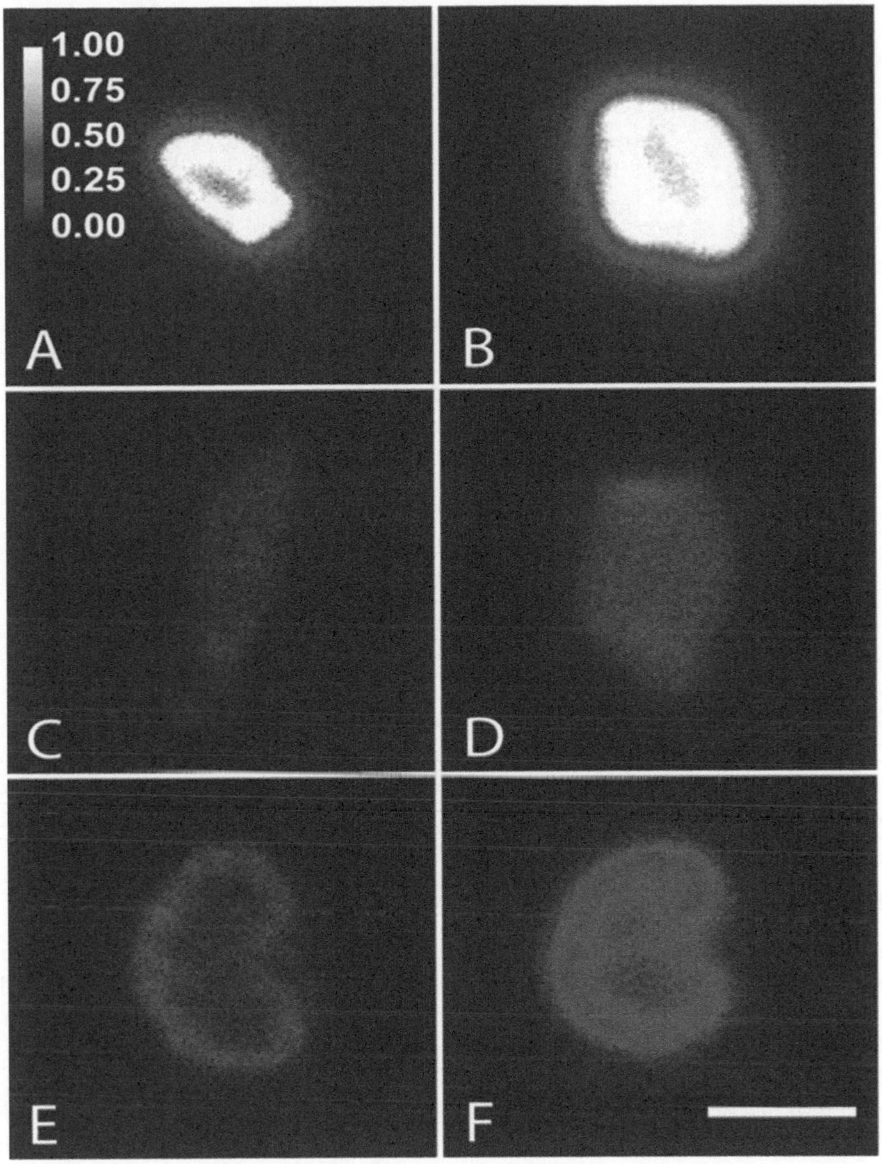

FIGURE 30.4 Autoradiographs with ^{177}Lu and ^{90}Y acquired using the iQID camera. Panels show the distribution of ^{177}Lu (A, C, and E) and ^{90}Y (B, D, and F) 24 hours after injection of labelled DOTA-biotin in 1F5-SA-pretargeted tumour (A and B), liver (C and D), and kidney (E and F) cryosections. The scales apply to all panels; the colour bar represents relative pixel intensity normalized to the mean for the tumour sections and the white scale bar indicates 5 mm. (Image originally published in [51] and reprinted according to Creative Common CC BY license (creativecommons.org).) Colour image available at www.routledge.com/9781138593268.

in the way that the microscope slides with samples are placed upon a copper base in a chamber filled with a Neon and CO_2 mixture gas. There is then an initial amplification gap of 200 μm between sample and a nickel micromesh under a high-voltage potential where radiation from the samples ionizes the gas, and electrons are accelerated in the electric field to form an electron cloud moving onwards from each point of ionization. This electron cloud then passes into a drift space of 1 cm with lower electric potential, where it is allowed to expand in size, effectively amplifying the size of the sample. This electron cloud is then passed into a final amplification gap of 50 μm between another micromesh and the pixelated anode readouts, which measure the charge and determine the coordinates of the ionization [54]. It is claimed that ionizing radiation of very different energy can be separated from each other on the strength of the detected signal, and also that different half-life radionuclides can be separated [54]. The instrument is not sensitive to gammas or x-rays [55]. Spatial resolution depends on the radionuclide, but as high as 20 μm for ^{3}H has been reported [24].

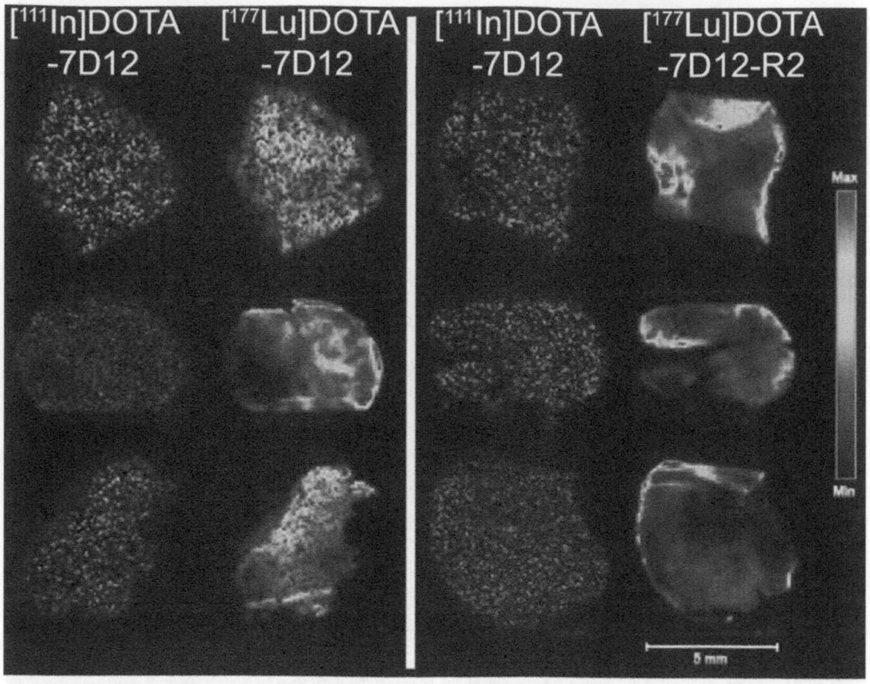

FIGURE 30.5 Ex vivo autoradiography of tumour tissue and signal quantification. A tumour-bearing mouse were co-injected with [^{111}In]DOTA-7D12/[^{177}Lu]DOTA-7D12 or [^{111}In]DOTA7D12/[^{177}Lu]DOTA-7D12-R2 and tumour sections were imaged based on decay rates. Representative tumour sections are shown. (Image originally published in [52] and is reprinted according to Creative Common CC BY license (creativecommons.org).) Colour image available at www.routledge.com/9781138593268.

30.6.1 Applications of Gaseous Detector-based Imaging Systems

One group making use of the parallel plate avalanche chamber from Biospace Lab is Alitalo and colleagues who, in 2020, injected three groups of mice with ^{3}H-2-deoxy-D-glucose, the first a control group, the second exposed to phencyclidine, and the third also exposed to the antipsychotic drug clozapine as well as phencyclidine. The mice were sacrificed 15 minutes post injection, the brains cryosectioned at 20 μm and imaged in the BetaImager Tracer for 6 hours; results are seen in Figure 30.6 [56].

Use of the micropattern gas detector from AI4R has also seen an increase. For example, the group of Lumen and colleagues in 2019 who investigated the distribution of ^{111}In-labelled dual-PEGylated thermally oxidized porous silicon nanoparticles in mice bearing breast cancer allografts [57]. They performed both ex vivo distribution studies and in vivo SPECT imaging but then used autoradiography to establish that the nanoparticle (due to its large size) did not penetrate very well into the tumours. Another study employing the AI4R BeaQuant system is from Bailly and colleagues in 2020, where authors investigated whether ^{64}Cu or ^{89}Zr-labeling was preferable for antibodies targeting CD138 in a syngeneic mouse model of multiple myeloma [58]. Both biodistribution and PET-imaging were employed and autoradiography used for studying the uptake of ^{89}Zr in bone, as can be seen in Figure 30.7.

30.7 SEMICONDUCTOR-BASED AUTORADIOGRAPHY DETECTORS

While scintillator detectors convert ionizing radiation to visible light, solid-state detectors directly register the deposited charge, allowing for real-time registration of position, energy, and time of each interaction. Semiconductor detectors also provide much better energy resolution than that obtained with scintillation-based detectors due to the large number of electron-hole pairs created when ionizing radiation interacts with the detector. These act as information carriers for the energy measurement, providing better statistics than the smaller number of photoelectrons that serve the same purpose in a scintillation detector [59].

Using CCD or CMOS detectors, not for imaging visible photons, but rather for directly registering charge left by ionizing radiation in the pixels of the detector, was employed by a University of Surrey group headed by J. Cabello and

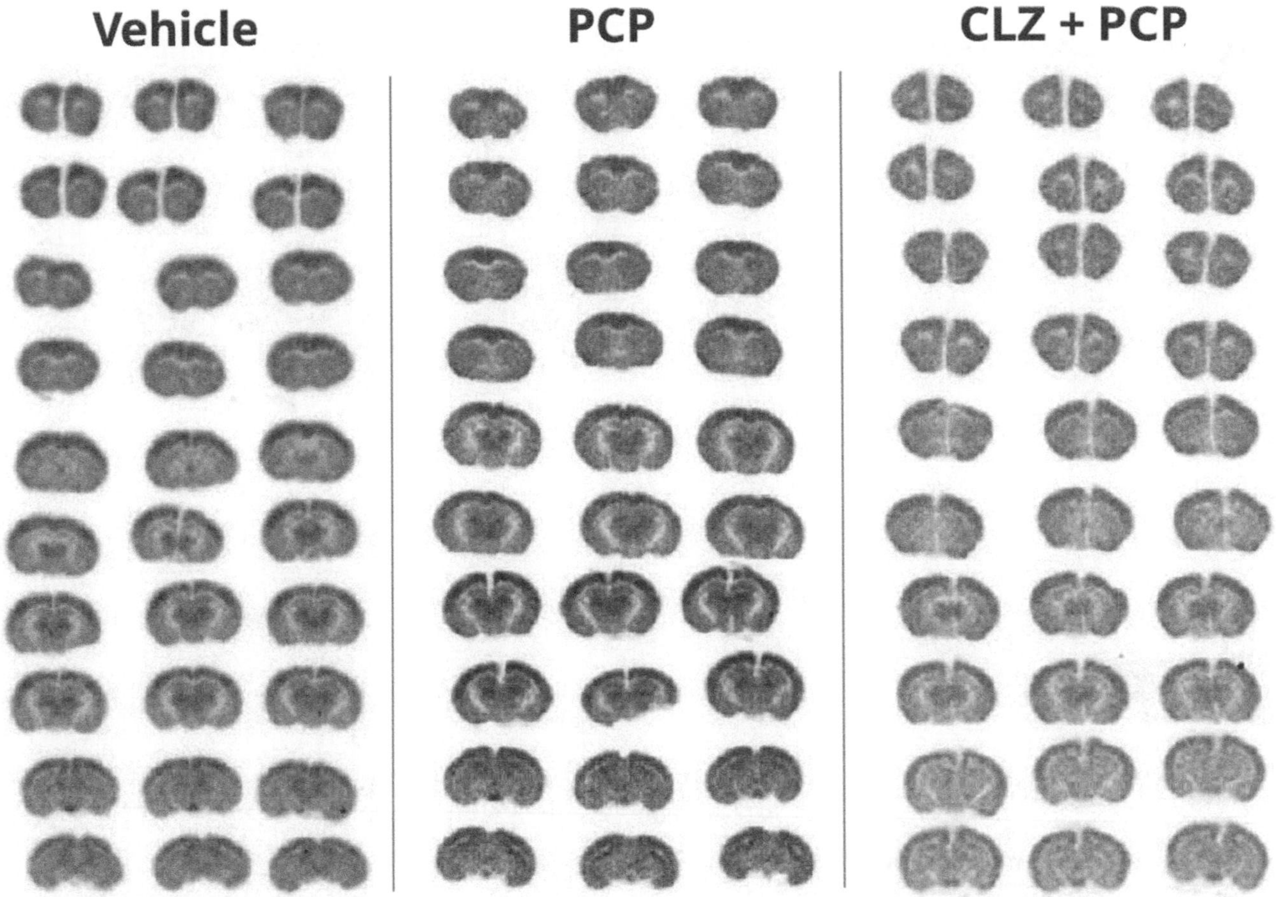

FIGURE 30.6 Representative sections from the The BetaIMAGER TRacer of most relevant regions of interest from each treatment group displaying treatment-induced effect in regional glucose utilization. As seen in the clozapine (CLZ) + phencyclidine (PCP) group, overall glucose utilization is lower due to sedative effects of CLZ. (Image originally published in [56] and reprinted according to Creative Common CC BY license (creativecommons.org).)

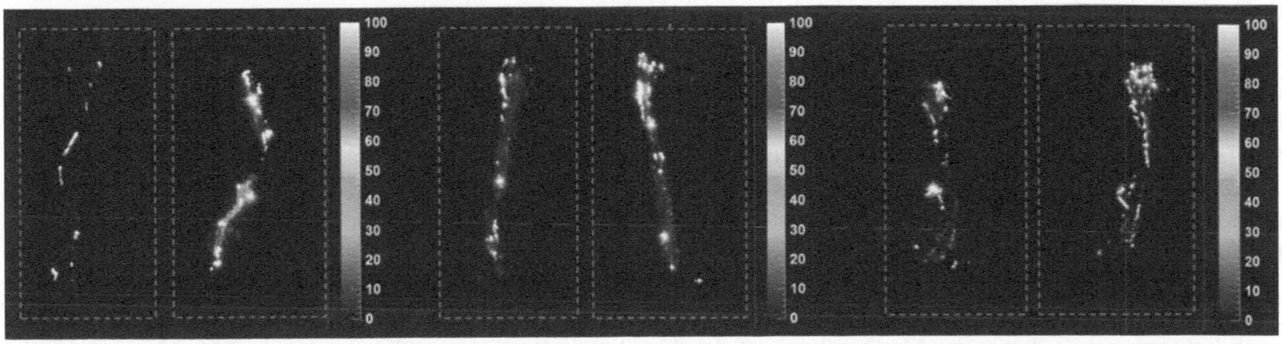

FIGURE 30.7 Digital autoradiography acquisitions performed on femurs of 3 mice. (Image originally published in [58] and reprinted according to Creative Common CC BY license (creativecommons.org).) Colour image available at www.routledge.com/9781138593268.

detailed in a 2007 paper [60]. This group used room temperature CCD and CMOS detectors to image radiation from ^{14}C and ^{35}S, both from standards and tissue samples as can be seen in Figure 30.8. The detectors have a pixel pitch of 22–25 μm and very low noise levels. The paper also details imaging of ^{3}H but, in that case, the sample was applied to the backside of a CCD that had been back-thinned using acid to reduce the inactive layer where low-energy electrons

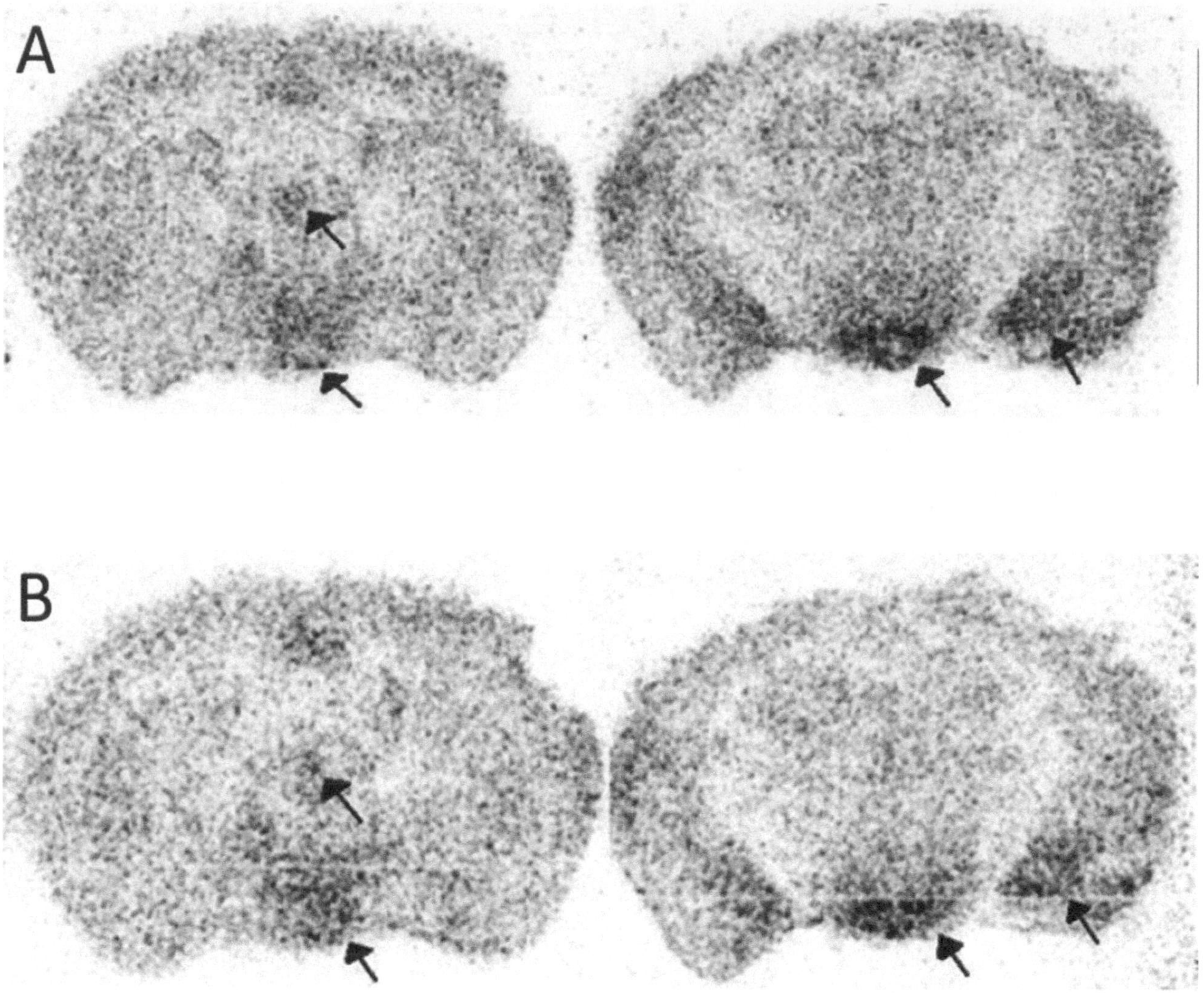

FIGURE 30.8 Coronal mouse sections, bound with ^{35}S-GTPγ S exposed to CCD sensor (A) and CMOS sensor (B) for 16 h each. (Image originally published in [60] and reprinted with permission.)

could be stopped before detection. Although energy information can be read out with these detectors, the CMOS system has such a thin detection layer that it acts mainly as a dE/dx detector, not registering the whole energy of most electrons. In a later paper, Cabello and colleagues also showed ^{3}H imaging using a similarly back-thinned CMOS detector [61].

Another more recent paper, Pham and colleagues 2020, also used a CMOS detector, this one previously employed as a detector in high-energy physics, to perform ^{18}F of autoradiography [62]. The detector had a spatial resolution of 144 μm for ^{18}F and was able to image samples with a few kBq in 4.8 hours.

A form of CMOS is the series of read-out chips developed by European collaborations, beginning with the Medipix1, developed at CERN, which could be bonded to a detection material like silicon for a 64x64 pixel readout with 170 μm pitch between pixels. Later the Medipix2 project developed a read-out chip with more pixels, 256x256, and smaller pitch, 55 μm. Mettivier and colleagues reported in 2003 of a Medipix2 bump-bonded to a 300 μm thick silicon pixel detector that could successfully image ^{14}C microscales [63]. A further development of the Medipix2 is the Timepix, which has the additional abilities to register the time of an interaction, as well as the energy deposited from an interaction with the detector. The Timepix can consist of up to eight Medipix2 style detectors [64].

One of this chapter's authors, Örbom, has done extensive work using a solid-state detector based around a double-sided silicon strip detector (DSSD), the Biomolex 700 Imager from Biomolex AS of Norway. The system uses a DSSD with 560 x 1260 parallel silicon strips, with a pitch of 50 μm, on each side of a 300 μm thick n-doped silicon wafer,

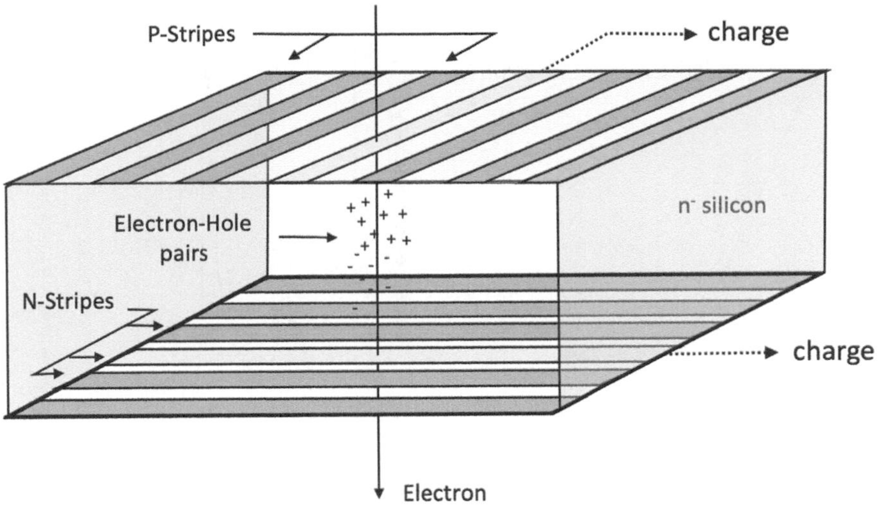

FIGURE 30.9 Schematics of the detection principle of the DSSD detector.

which acts as a depletion area. The strips on the two sides run orthogonal to each other and are p+-doped or n+-doped. They are connected to read-out electronics via aluminium contacts. When a voltage difference is applied over the detector, each strip operates as a reverse-biased diode. Electron-hole pairs are created by incoming radiation in proportion to the energy deposited, and a current can be read out from the strips closest to the point of interaction. The detector design is illustrated in Figure 30.9. The detector is cooled by a Peltier element and a fan to 20°C during operation, and the energy threshold for detection is between 15 and 30 keV, depending on the specific detector unit. The readout electronics consists of a dedicated, self-triggering CMOS chip that reads out the coordinates and the energy of a detected event in the microstrip detector. These data are sent in real time to software running on a computer connected to the system. After imaging, the data can be exported in binary format or in list-mode as a text file. A file reporting the number of registered counts each minute is also created to allow for assignment of interaction time to each event. The strip on either side of the detector reporting the highest current during the measurement period ($\geq 10\ \mu s$) determines the coordinates for each event. The preparation of samples for imaging in the autoradiography system consists of placing them on a standard 75 mm × 25 mm microscope slide. The slide is then placed in a plastic sample holder and covered with a 3 μm Mylar foil before being loaded either manually or through a 12-slot sample changer into the instrument, where it is gently held against the detector from below. Some registered energy spectra for different radionuclides on the detector can be seen in Figure 30.10; note the similarity between the spectra of pure beta emitters whereas conversion electrons give more specific peaks [65]. While the image can be observed during measurement, the final image is reconstructed using in-house software that corrects for the individual strips that might either give no signal at all or give a too-high or too-low signal.

Örbom and colleagues reported a spatial resolution of the system of 154 ± 14 μm for ^{58}Co as compared to 343 ± 15 μm for the same line-source for a storage phosphor imager system. In the same paper, the sensitivity is reported as 0.249 cps/Bq for ^{177}Lu and 0.064 cps/Bq for ^{99m}Tc [20].

30.7.1 Applications of Semiconductor-based Imaging Systems

In 2008, Russo and colleagues used a Medipix2 read-out chip bump-bonded to a silicon pixel detector to perform ^{18}F autoradiography and compare it to classic film autoradiography [66]. They used 10 μm sections from brains of rats injected with ^{18}F-FDG as can be seen in Figure 30.11. In 2011, in 2011, Esposito and colleagues employed the more advanced Timepix chip, also bonded to a silicon detector, to image ^{14}C samples, determining the spatial resolution of ^{14}C autoradiography to be 76.9 μm [67].

The Biomolex 700 Imager DSSD detector has been used in experiments in several papers, one of which is from 2013 by Örbom and colleagues, in which an antibody labelled with ^{177}Lu was injected into tumour-bearing rats and the intratumoural distribution was imaged by autoradiography at different timepoints [68]. A point-dose kernel was also used to calculate the absorbed dose rate.

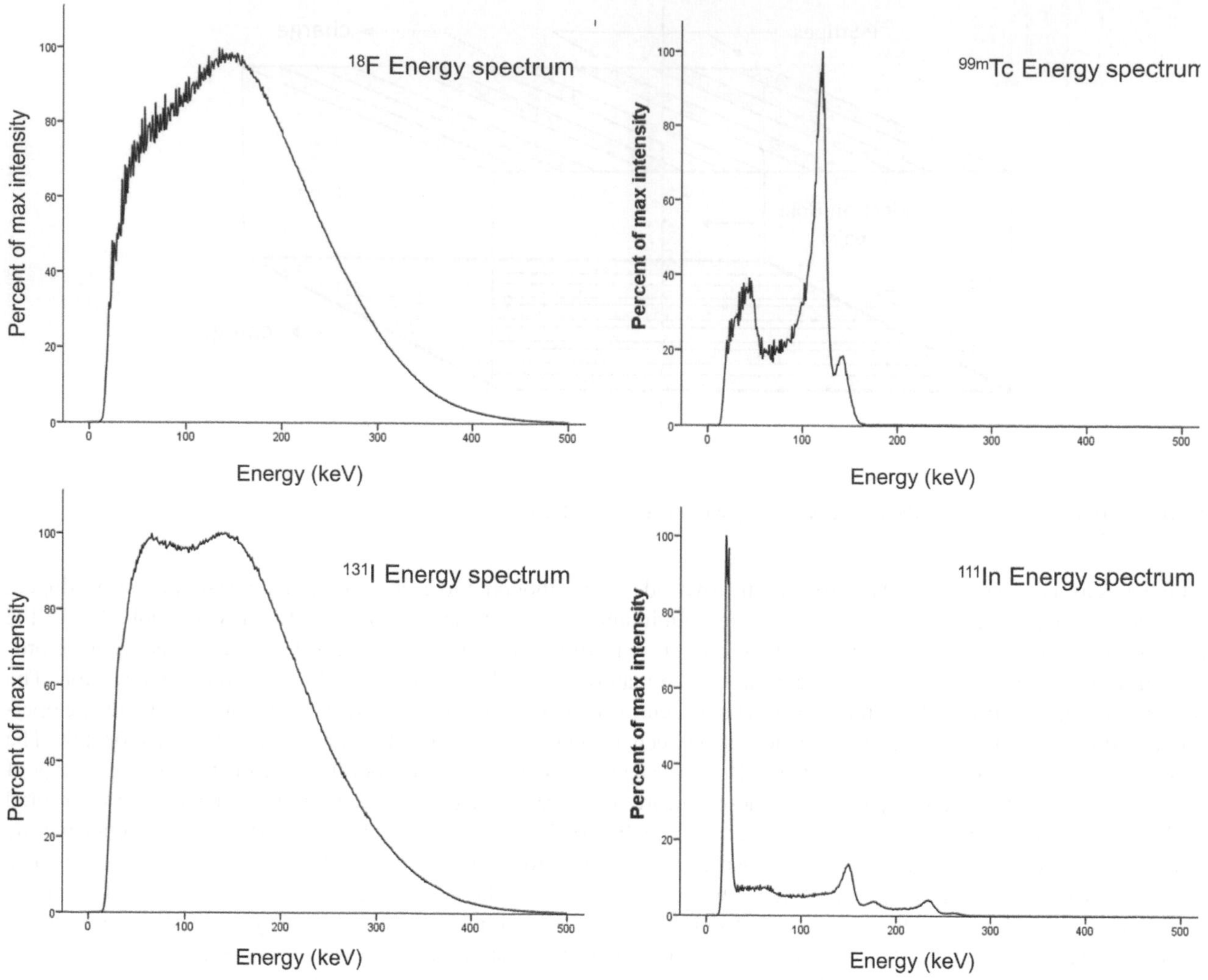

FIGURE 30.10 Energy spectra of different radionuclides as measured on the DSSD detector of the Biomolex 700 Imager [65].

30.8 ALPHA PARTICLE AUTORADIOGRAPHY

In recent years there has been growth in the need for α-particle autoradiography. This is primarily driven by the advancement of cancer therapies using alpha emitters and the need to understand and quantify the spatial distribution within tissues at high spatial resolution. This has led to new autoradiography applications and detection systems. Many of the previously discussed autoradiography systems are capable of imaging both beta and alpha particles. Here we discuss those and dedicated systems for α-particle autoradiography.

30.8.1 Dedicated Alpha Particle Autoradiography Detectors

30.8.1.1 The Alpha Camera

The first version of the Alpha Camera system [69] was presented in 2010 and employed the combination ZnS:Ag phosphor sheets and detection of scintillation photons using an optical lens connected to a high-sensitivity CCD-camera. The optical system could be altered, using different lens set-ups, so that the field of view could vary from below a millimetre up to tens of centimetres. The CCD-version is based on the principle of imaging a large number of scintillation events in the same exposure. The pixel intensity in the acquired images is linear towards activity in the imaged specimen but is dependent on the optical geometry, requiring a system calibration between pixel intensity and activity. The typical samples to be imaged are thin sections of tissue, either cryosections or paraffin-embedded. The sections are placed

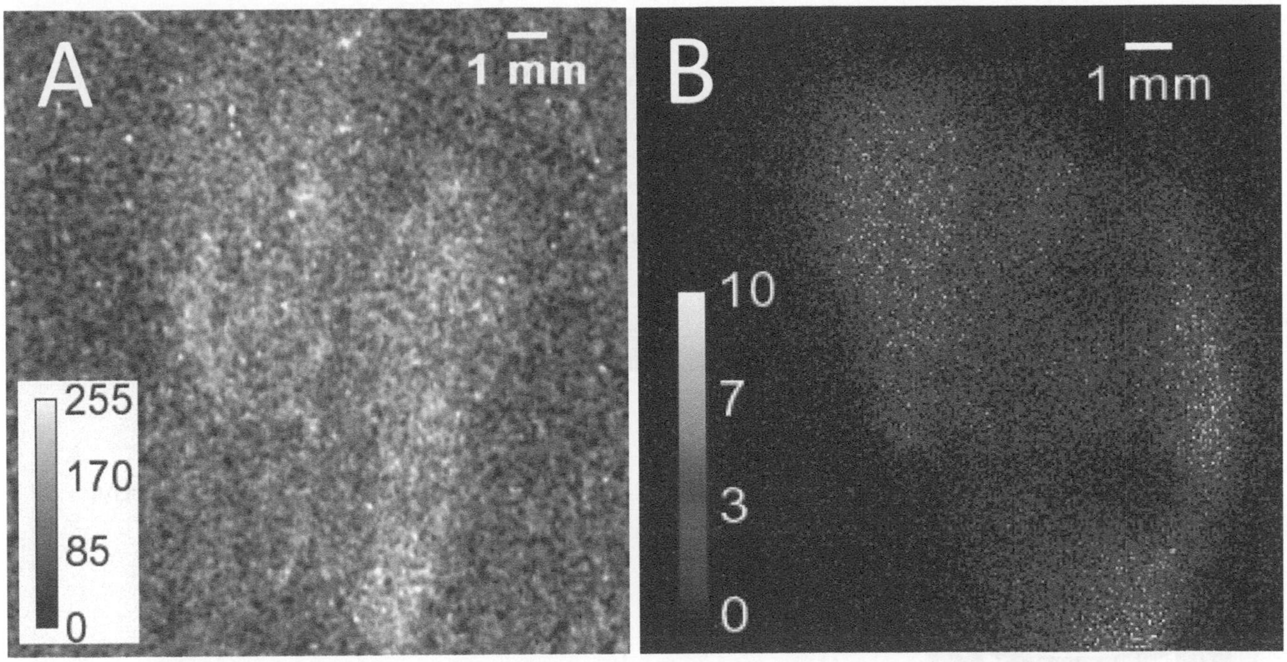

FIGURE 30.11 Autoradiography of 10 μm brain section from rat injected with ^{18}F-FDG and imaged for 20 hours with film autoradiography (A) and for 1000 seconds with the Medipix2 detector and processed by cluster analysis software to improve resolution (B). (Image originally published in [66] and reprinted with permission.) Colour image available at www.routledge.com/9781138593268.

directly on a piece of the scintillation sheet that in turn is placed under the optical lens. Depending on the field of view, one or several section samples can be imaged simultaneously. The whole camera system is set-up inside a dark box to minimize artefacts from ambient light. The principle components for the camera, together with an example of a spatial resolution measurement, are shown in Figure 30.12.

The second generation of the system utilizes an EMCCD-camera that allows for single-event imaging by which individual α-particle decays can be imaged, meaning that each image directly contains data on the activity in the imaged specimen area. Each detected α-particle decay corresponds to a single-light event in the acquired image. The spatial location of each event is recovered by digital image post-processing, and the total number of events in a series of exposures is used to form the activity image. In the initial characterization of the Alpha Camera, the FWHM was estimated to ~35 μm, but later evaluations suggest a spatial resolution in the range of 20–25 μm (unpublished data), mostly depending on features of the imaged samples.

The Alpha Camera imaging has been used for imaging of several α-emitting radionuclides, for example, ^{211}At, ^{213}Bi, ^{225}Ac, ^{227}Th and ^{223}Ra, and mostly in experimental studies of targeted alpha therapy.

30.8.1.2 Other α-particle Imaging Systems

In 2020, Yamamoto and colleagues reported the development of an alpha scintillation camera using a fibre-structured scintillator (1-μm diameter fibres) coupled to a fibre-optic taper acting as a magnifier with coupling to an EMCCD [70]. They reported the ability to distinguish tracks of particles at an imaging spatial resolution of ~11 μm. In 2020, Segawa and colleagues reported the development of an α-scintillation-camera system, similar to Bäck's Alpha Camera, using a macro lens coupled to a scientific-grade CCD [71].

30.8.2 Alpha and Beta Particle Autoradiography Detectors

The iQID system: As previously discussed, the iQID is a scintillation-based, single-particle digital autoradiography system. Figure 30.13 illustrates an experimental setup with tissue sections placed on ZnS:Ag phosphor screens. Particle discrimination is also possible with appropriate scintillators and source geometry [46]. For example, Miller recently showed, beta-gamma-blind alpha autoradiography with the iQID and ZnS:Ag phosphor screens [72].

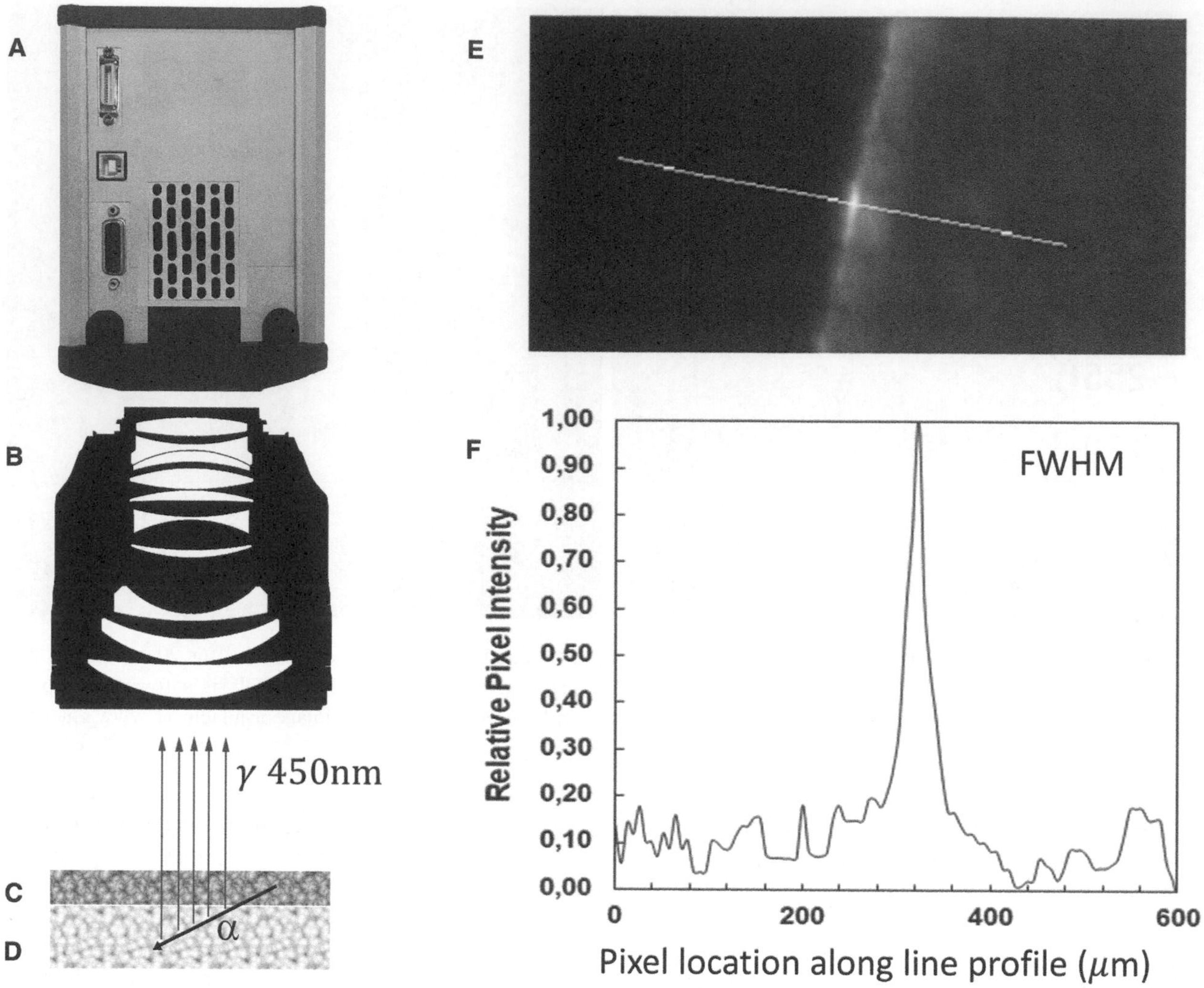

FIGURE 30.12 Schematic principle of the Alpha Camera set-up with CCD/EMCCD camera (A), optical lens system (B), thin tissue section (C) and ZnS:Ag phosphor sheet scintillator (D). Alpha particles (α) emitted from the tissue section hit the scintillator and generate light photons (γ). Radioactive line source of ^{211}At (E) and example of a line profile used to estimate the spatial resolution of the Alpha Camera system (F). The FWHM was estimated to be 20–35 μm, depending on sample geometry.

A key benefit of single-particle digital autoradiography with alphas is the ability to quantify the activity directly. Beta emitters have a continuous energy spectrum ranging from low to high energies. Accordingly, most detectors employ an energy threshold for detection above noise levels. Also, there will be an energy dependent efficiency associated with the scintillator. These factors and beta spectra differences among isotopes require the use of isotope-specific calibration standards for quantification. Alpha particles, however, have discrete energies and a short range (e.g., 20–80 μm) in the scintillator detector. This makes them easy to detect at nearly 100 percent efficiency in single-particle detectors, without the need for relative activity calibration standards. Miller showed with the iQID approximately 100 per cent detection efficiency (within the uncertainty of certified electroplated alpha sources) for alphas [47]. Imaging and activity quantification is illustrated in Figure 30.14 with ^{211}At (7.2h half-life) and estimating the activity using spatio-temporal information for each event.

The BeaQuant system: In addition to high-resolution beta-imaging, gaseous-based detectors can also be used for single-particle α-particle digital autoradiography. This is illustrated in the BeaQuant system with geological samples having α-emissions from the ^{238}U decay chain (see Figure 30.15).

The Timepix system: Besides x-rays and electrons, the semiconductor Timepix sensor can also be used in single-particle detection for a large range of charged particles with both energy and directional information [74]. It has been

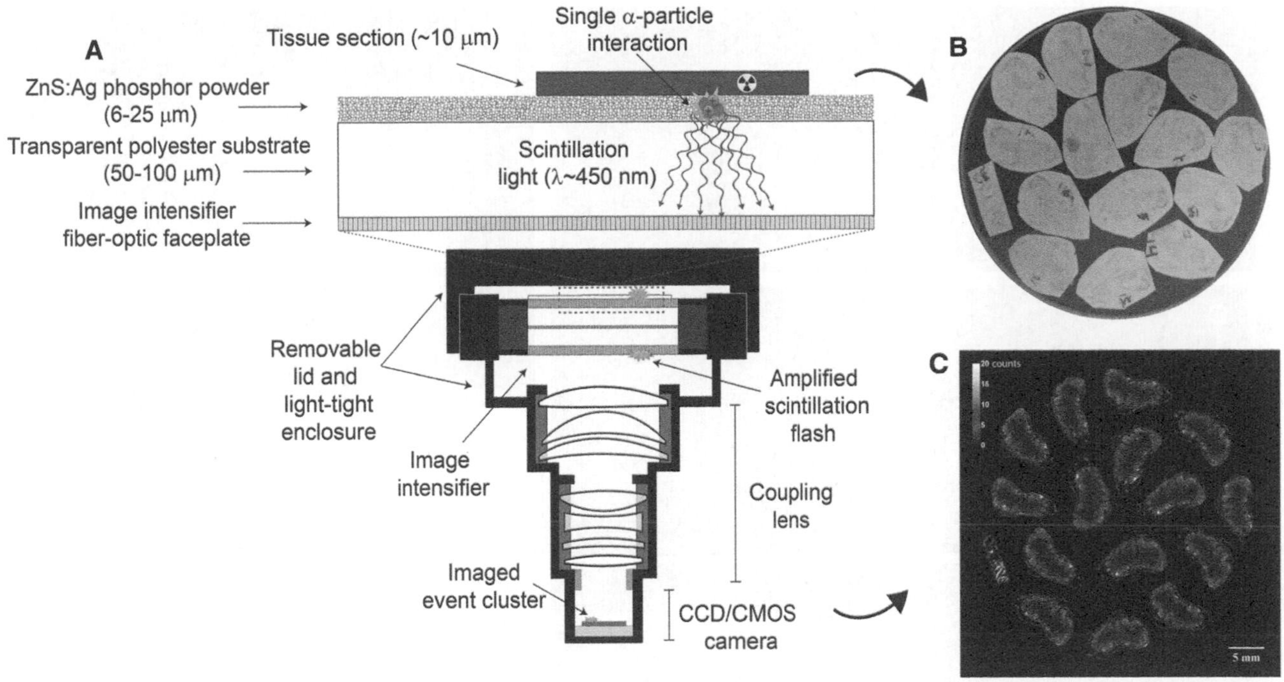

FIGURE 30.13 (A) An iQID schematic/setup for quantitative digital autoradiography with α-emitters, (B) iQID optical view of biopsy tissue sections placed on a 40 mm diameter iQID detector and (C) corresponding digital autoradiograph. (Image originally published in [73] and is reprinted according to Creative Common CC BY license (creativecommons.org).) Colour image available at www.routledge.com/9781138593268.

commercialized in single- and tiled-array configurations for charged-particle detection and imaging by Advacam. Al Granja and colleagues demonstrated α-particle digital autoradiography with the Timepix using the α-emitter ^{227}Th and for cell dosimetry studies with ^{223}Ra [75, 76]. An image of the Timepix's response to various particles [77] is shown in Figure 30.16A.

The Biomolex system: The solid-state DSSD system the Biomolex 700 imager can also be used to image α-emissions, as can be seen in Figure 30.17 of a mouse injected with a ^{211}At.

30.8.3 Applications of α-Particle Autoradiography

30.8.3.1 Targeted Alpha Therapy

Alpha-particle digital autoradiography has recently become important in the application of targeted alpha therapy (TAT). Due to their high linear energy transfer (LET) and short path length of α-particles in tissues, α-emitters are promising in cancer therapy, with recent trials showing remarkable success when other therapies have failed [78]. Alpha particles are potent at killing cancer cells, but are also toxic to normal tissues and organs, and understanding their biodistribution is key in development of new radiopharmaceuticals. Many of the α-emitters being considered for TAT are not suited for traditional imaging approaches such as scintigraphy or SPECT due to a combination of administered activities below detection limits, low photon emission efficiencies, and photon energies that are challenging or impractical to image. However, autoradiography with alphas is highly efficient with excellent spatial resolution making it an essential means for understanding biodistributions in pre-clinical TAT.

Recent example studies using alpha autoradiography in pre-clinical research include Al Darwish and colleagues demonstrating autoradiography of tumour sections using a Timepix sensor (see Figure 30.18B) in a ^{227}Th TAT study comparing mice treated with TAT and TAT combined with chemotherapy. Dekempeneer and colleagues evaluating the efficacy of single-domain antibody fragments (sdAb) labelled with the α-emitter ^{213}Bi (45-minute half-life). The iQID was used in biodistribution studies to image and quantify the radioactivity distribution within tumour sections [79].

Using alpha imaging to quantify the uptake in microtumours, the activity uptake was found to be linearly correlated to the number of cells in the microtumours (Figure 30.18).

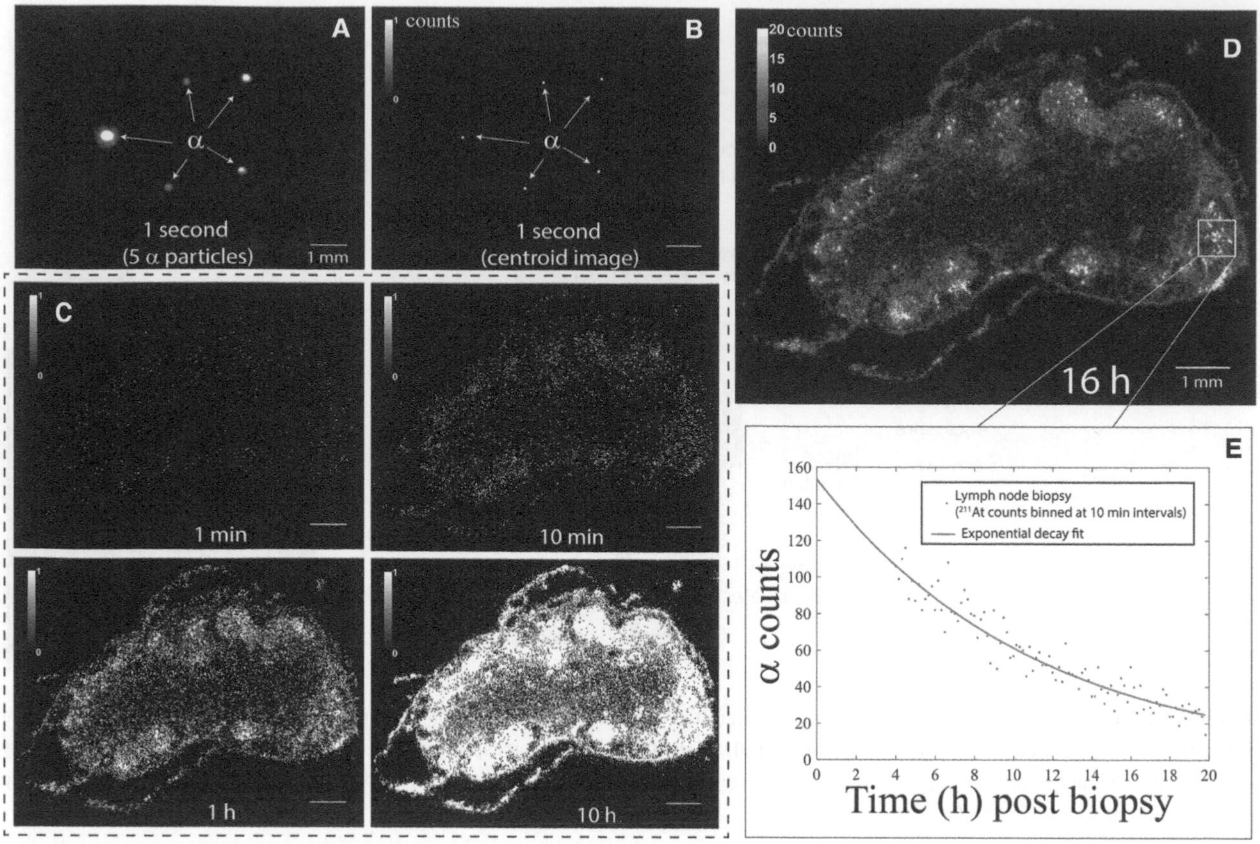

FIGURE 30.14 (A) Single-particle alpha autoradiography. One-second iQID image of ^{211}At α-particles from a biopsy tissue section and (B) corresponding centroid autoradiograph. (C) iQID biopsy autoradiographs at various time points displayed with a binary colour scale. (D) Final 16 h autoradiograph and € temporal information for a small region of interest, where the estimated total activity of the section ROI at the biopsy extraction time point is 477 mBq (approximately 4 hours prior to the autoradiography imaging experiment start). (Image originally published in [73] and is reprinted according to Creative Common CC BY license (creativecommons.org).) Colour image available at www.routledge.com/9781138593268.

For intraperitoneal treatments of ovarian cancer using ^{211}At-farletuzumab, Palm and colleagues showed a 6 to 10–fold increase (treated vs controls) in antitumoural efficacy, using the Alpha Camera to image the radionuclide spatial distribution in micro tumours [80]

The combination of the alpha autoradiograph and registration with stained microscope images makes for a powerful tool for understanding and quantifying the dosimetric effects near cellular levels for α-emitters in TAT [47]. This is illustrated in Figure 30.19 with registered iQID and H&E-stained sections from a lymph node biopsy. Additional findings related to this radioimmunotherapy study are presented in [81].

30.8.3.2 High-resolution α-Particle Scintigraphy

High-spatial resolution autoradiography requires flat contact with the sample and detector. Even a small gap between the source and detector can significantly degrade the spatial resolution. A recently developed autoradiography imaging method, similar to parallel-hole collimator imaging in γ-ray scintigraphy and SPECT, is alpha scintigraphy using microcapillary array plates. This imaging approach was recently demonstrated by Miller and colleagues using a high-activity (MBq) α-source to assess and quantify the spatial distribution of a Plutonium source [72]. Plutonium was painted on a foil for beam activation experiments. The painted surface had a texture with slight variations in height of the Pu particles, making it not possible to achieve direct contact (no gaps) between detectors. The results, Figure 30.20, show significant differences between phosphor imaging plate, iQID, and microcapillary iQID scintigraphy. The goal of the study was to estimate the uniformity of the Pu deposit where using traditional contact methods significantly underestimate the non-uniformity.

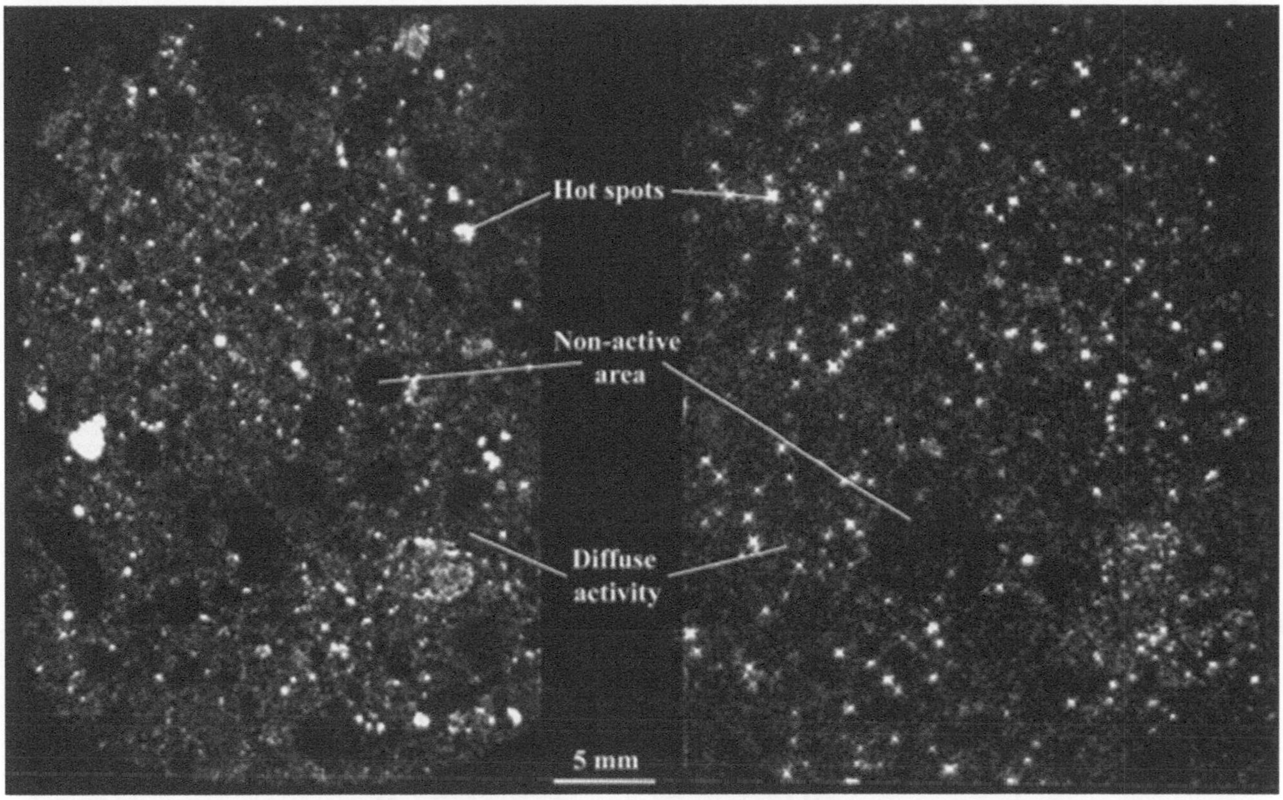

FIGURE 30.15 Alpha map of geological samples. Alpha activity from the 238U decay chain is represented in grey scale. Black areas correspond to a null alpha activity. In both samples the activity is heterogeneous in spatial repartition, with hot spots of activity and areas of diffuse and null activity. (Figure originally published in [9] and reprinted with permission.)

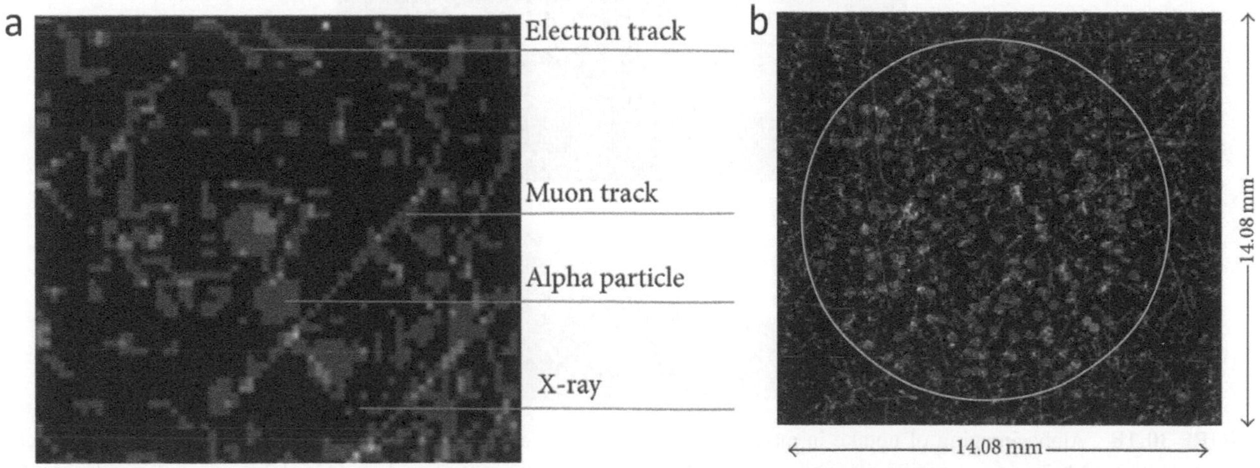

FIGURE 30.16 (Left) Timepix responses to electron Muon, x-/γ ray, and α-particles. (Right) Timepix autoradiography from a tumour section from a mouse treated with ^{227}Th-DAB4. The red circle indicates the approximate tumour section boundaries. (Images originally published in [77] and is reprinted according to Creative Common CC BY 3.0 license (creativecommons.org).)

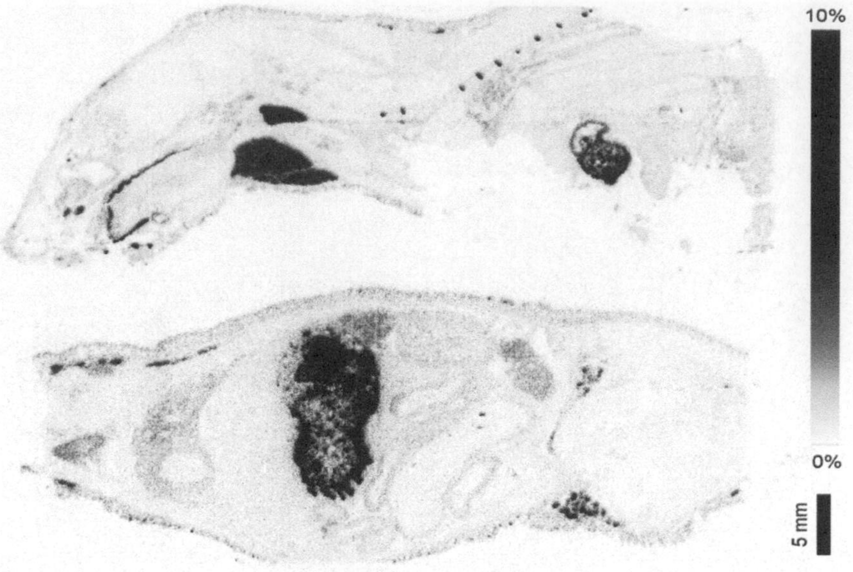

FIGURE 30.17 Example alpha autoradiography from the Biomolex 700 DSSD system of a mouse injected with ^{211}At.

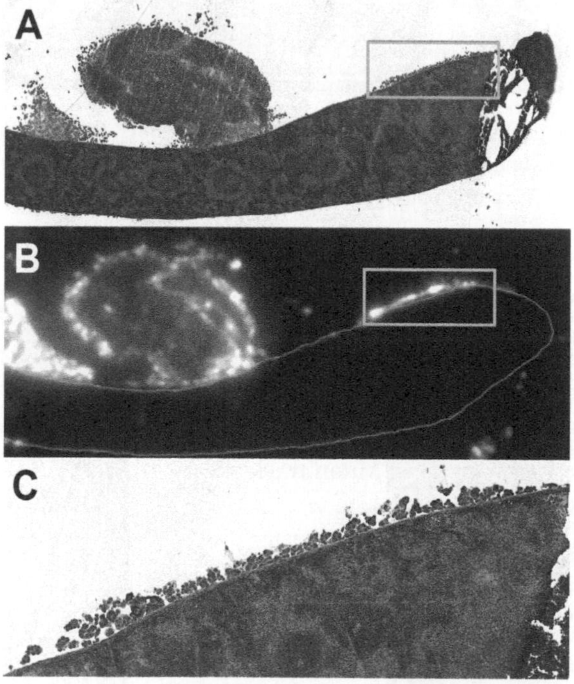

FIGURE 30.18 Alpha imaging of uptake in macro- and micro tumours on the peritoneal lining of the spleen after i.p.-injection of ^{211}At-faretuzumab in a mice model of ovarian cancer. H&E-stained section of the spleen (A), alpha image (B) with cyan ROI indicating periferum of the spleen and a zoom of an area (C, and red rectangles) with hot-spot activity in areas corresponding to small tumour cell clusters.

30.8.3.3 Radio-TLC

The α-scintillation-camera system reported by Segawa was designed for thin-layer chromatography of α-emitters. The system generates an image by integrating scintillation light using a high-sensitivity CCD camera. They reported higher sensitivity compared to phosphor imaging plates with activity estimates ranging from 56–672 Bq (uncertainty of 5%) for a 1000 s exposure of ^{211}At [71].

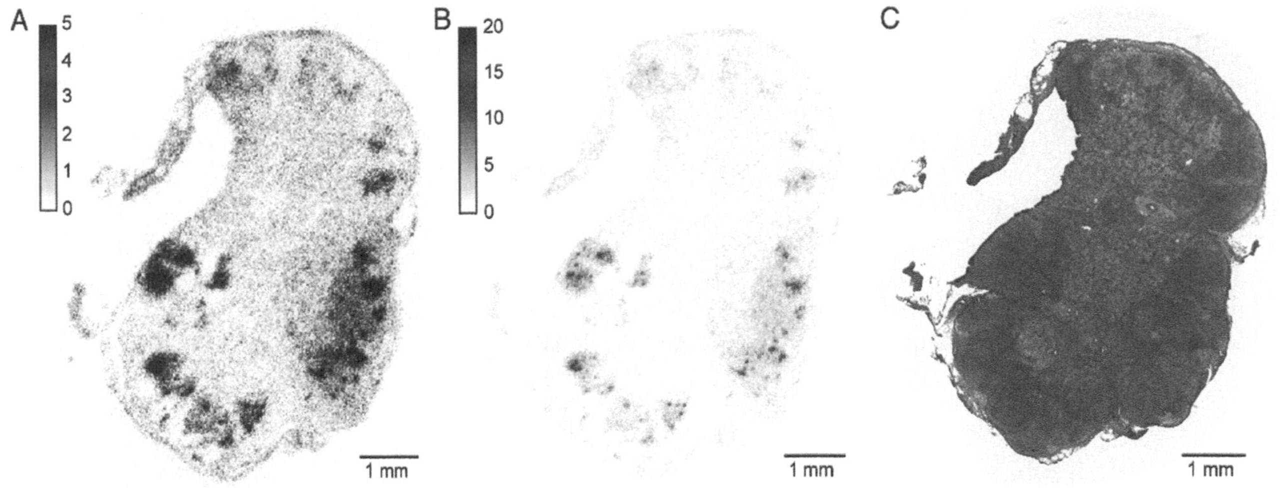

FIGURE 30.19 Images of a cryosectioned canine lymph node, biopsied 2h after ^{211}At-anti-CD45 radioimmunotherapy. [(A) and (B)] iQID α images with pixel values normalized to the mean activity concentration at the time of biopsy. The heterogeneous uptake includes activity concentrations up to 19.6× higher than the mean activity for the whole section. (C) Neighbouring H&E-stained section. Comparison of the ^{211}At distribution with the corresponding morphology demonstrated targeting of areas within the T lymphocyte-rich paracortex, with little or no accumulation in lymphoid follicles and medulla. (Image originally published in [47] and reprinted with permission.) Colour image available at www.routledge.com/9781138593268.

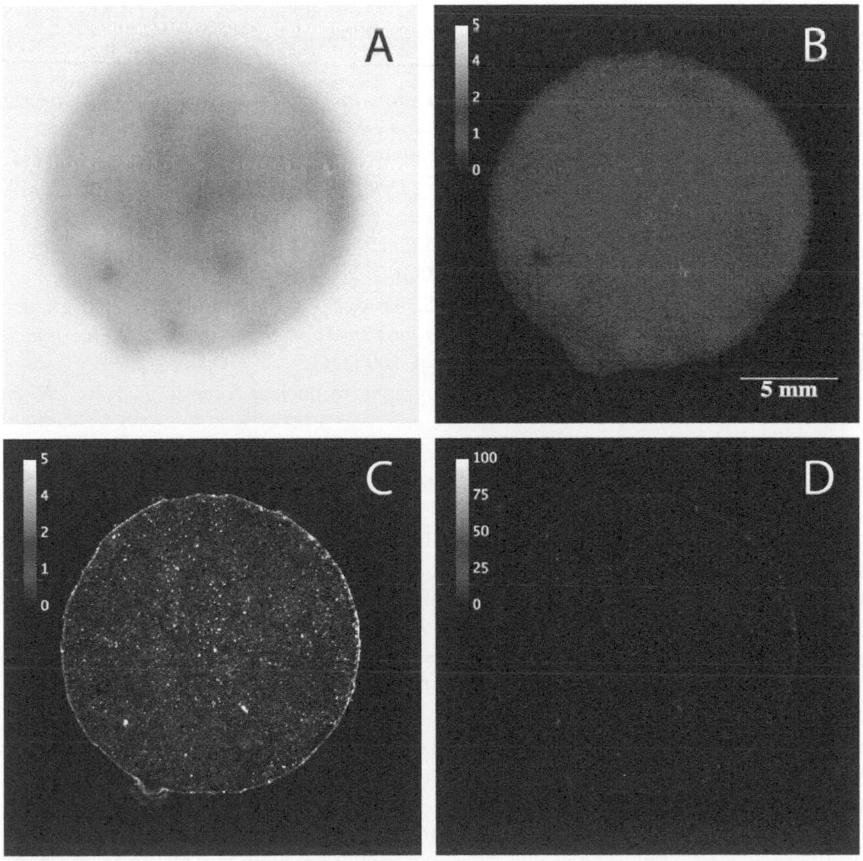

FIGURE 30.20 Autoradiographs of a 1–2 mg ^{239}Pu painted disc source using (A) phosphor imaging plate autoradiography (10 min exposure), (B) iQID contact imaging (10 min acquisition, 20 fps with 1 ms exposure per frame), and (C-D) iQID imaging using a microcapillary array collimator; 2 hr image acquisition, 20 fps with full exposure time of 50 ms per frame. Pixels are normalized to the mean non-zero activity map and displayed at different thresholds. Note the significantly higher spatial resolution provided by the collimator, which prevents obliquely emitted alpha particles from interacting with the scintillator. From Miller and colleagues 2020 [82]. Colour image available at www.routledge.com/9781138593268.

30.8.3.4 Geological Studies

As demonstrated with the BeaQuant system, alpha autoradiography is proving to be an important tool for geological studies and can be used to determine the mobility of radioactivity elements from the ^{238}U decay chain, which can otherwise be difficult to detect [9]. In 2020, Billon and colleagues used the BeaQuant to map and quantify α-emitting ^{226}Ra in uranium mine tailings [83].

30.8.3.5 Radio Toxicology

Alpha autoradiography is also a useful tool in radiation toxicology studies. Tazrart and colleagues used iQID single-particle digital autoradiography to quantify and study the absorption in skin of americium and plutonium, in various chemical forms [84]. Tabatadze, and colleagues used iQID quantitative digital autoradiography to study the spatial distribution of ^{241}Am within anatomical bone structures from individuals who received occupational exposure [49].

30.9 SUMMARY

Autoradiography is an essential tool for a variety of applications. Many detector types and configurations exist with various trade-offs in terms of activity quantification, spatial resolution, energy resolution, and detector area. Modern digital systems offer many advantages over traditional systems, but opportunities still exist for further improvement to meet the needs of growing and future applications. Especially in the field of radiopharmaceutical development, the demand for autoradiography will likely continue to grow, as there is no other technology that matches its spatial resolution and sensitivity.

REFERENCES

[1] E. G. Solon, A. Schweitzer, M. Stoeckli, and B. Prideaux, "Autoradiography, MALDI-MS, and SIMS-MS imaging in pharmaceutical discovery and development," (in English), *Aaps J*, Review vol. 12, no. 1, pp. 11–26, Mar 2010, doi:10.1208/s12248-009-9158-4.

[2] S. Adelstein et al., *ICRU Report 67: Absorbed-Dose Specification in Nuclear Medicine, J ICRU,* vol. 2, pp. 64–65, 2002.

[3] F. H. Attix, *Introduction to Radiological Physics and Radiation Dosimetry*. New York: Wiley, 1986.

[4] E. Solon, "Nonclinical applications of quantitative whole-body autoradiography, and imaging mass spectrometry in drug discovery and development," in *Visualizing and Quantifying Drug Distribution in Tissue IV*, 2020, vol. 11219: International Society for Optics and Photonics, p. 1121903.

[5] M. Nguyen, R. M. Ramakers, C. Kamphuis, S. Koustoulidou, M. C. Goorden, and F. J. Beekman, "EXIRAD-3D: Fast automated three-dimensional autoradiography," *Nucl Med Biol*, 2020.

[6] T. Mertzanidou, N. Calvert, D. Tuch, D. Stoyanov, and S. Arridge, "Tomosynthesis method for depth resolution of beta emitters," in *Medical Imaging 2019: Biomedical Applications in Molecular, Structural, and Functional Imaging*, 2019, vol. 10953: International Society for Optics and Photonics, p. 109531G.

[7] T. Parsons-Davis et al., "Application of modern autoradiography to nuclear forensic analysis," *Forensic Sci Int*, vol. 286, pp. 223–232, 2018.

[8] A. M. Clarke, B. S. McDonald, M. A. Zalavadia, D. M. Kasparek, and B. W. Miller, "Characterization of a large-area iQID imager for safeguards applications," *Nuclear Instruments and Methods in Physics Research Section A: Accelerators, Spectrometers, Detectors and Associated Equipment*, vol. 954, p. 161125, 2020.

[9] A. Angileri et al., "Mobility of daughter elements of 238U decay chain during leaching by In Situ Recovery (ISR): New insights from digital autoradiography," *J Environ Radioact*, vol. 220, p. 106274, 2020.

[10] Y. Petegnief, A. Petiet, M. C. Peker, F. Bonnin, A. Meulemans, and D. Le Guludec, "Quantitative autoradiography using a radioimager based on a multiwire proportional chamber," *Phys Med Biol*, vol. 43, no. 12, p. 3629, 1998.

[11] D. P. Holschneider et al., "Changes in regional brain perfusion during functional brain activation: Comparison of [(64) Cu]-PTSM with [(14)C]-Iodoantipyrine," (in English), *Brain Res*, vol. 1234, pp. 32–43, Oct. 9, 2008, doi:10.1016/j.brainres.2008.07.038.

[12] M. Picchio et al., "Intratumoral spatial distribution of hypoxia and angiogenesis assessed by 18F-FAZA and 125I-Gluco-RGD autoradiography," (in English), *J Nucl Med*, Research Support, Non-U.S. Gov't vol. 49, no. 4, pp. 597–605, Apr 2008, doi:10.2967/jnumed.107.046870.

[13] K. Iwanishi, H. Watabe, T. Hayashi, Y. Miyake, K. Minato, and H. Iida, "Influence of residual oxygen-15-labeled carbon monoxide radioactivity on cerebral blood flow and oxygen extraction fraction in a dual-tracer autoradiographic method," (in English), *Ann Nucl Med*, vol. 23, no. 4, pp. 363–371, Jun 2009, doi:10.1007/s12149-009-0243-7.

[14] A. Broisat et al., "Mapping of brain tissue hematocrit in glioma and acute stroke using a dual autoradiography approach," *Sci Repo*, vol. 8, no. 1, pp. 1–9, 2018.

[15] D. E. Wimber and L. Lamerton, "Cell population kinetics in the intestine of continuously irradiated mice, using double-labelling autoradiography," *Radiation Research*, vol. 28, no. 3, pp. 694–700, 1966.

[16] R. Johnston, S. Pickett, and D. Barker, "Double-label image analysis using storage phosphor technology," *Methods*, vol. 3, no. 2, pp. 128–134, 1991.

[17] N. Barthe, K. Chatti, P. Coulon, S. Maitrejean, and B. Basse-Cathalinat, "Recent technologic developments on high-resolution beta imaging systems for quantitative autoradiography and double labeling applications," (in English), *Nucl Instrum Meth A*, vol. 527, no. 1–2, pp. 41–45, Jul. 11, 2004, doi:DOI 10.1016/j.nima.2004.03.014.

[18] B. W. Miller et al., "Quantitative single-particle digital autoradiography with alpha-particle emitters for targeted radionuclide therapy using the iQID camera," *Med Phys*, vol. 42, no. 7, pp. 4094–4105, Jul 2015, doi:10.1118/1.4921997.

[19] Y. Kvinnsland and A. Skretting, "Methods for separation of contributions from two radionuclides in autoradiography with a silicon strip detector," (in English), *Phys Med Biol*, vol. 45, no. 5, pp. 1183–1193, May 2000. [Online]. Available: www.ncbi.nlm.nih.gov/pubmed/10843099.

[20] A. Örbom et al., "Characterization of a double-sided silicon strip detector autoradiography system," *Med Phys*, vol. 42, no. 2, pp. 575–584, 2015.

[21] A. Örbom et al., "Multi-radionuclide digital autoradiography of the intra-aortic atherosclerotic plaques using a monoclonal antibody targeting oxidized low-density lipoprotein," *Am J Nuc Med Mol Imaging*, vol. 4, no. 2, p. 172, 2014.

[22] Y. K. Dewaraja et al., "MIRD pamphlet No. 23: quantitative SPECT for patient-specific 3-dimensional dosimetry in internal radionuclide therapy," (in English), *J Nucl Med*, vol. 53, no. 8, pp. 1310–1325, Aug 2012, doi:10.2967/jnumed.111.100123.

[23] J. C. Roeske, B. Aydogan, M. Bardies, and J. L. Humm, "Small-scale dosimetry: challenges and future directions," (in English), *Semin Nucl Med,* Review vol. 38, no. 5, pp. 367–383, Sep 2008, doi:10.1053/j.semnuclmed.2008.05.003.

[24] N. Barthe, S. Maîtrejean, N. Carvou, and A. Cardona, "High-resolution beta imaging," in *Handbook of Radioactivity Analysis: Volume 2*: Elsevier, 2020, pp. 669–727.

[25] B. Larsson and S. Ullberg, "Whole-body autoradiography," *J Histochem Cytochem*, vol. 29, no. 1A_suppl, pp. 216–225, 1981.

[26] J. Bogdanska et al., "Tissue distribution of 14C-labelled perfluorooctanoic acid in adult mice after 1–5 days of dietary exposure to an experimental dose or a lower dose that resulted in blood levels similar to those detected in exposed humans," *Chemosphere*, vol. 239, p. 124755, 2020.

[27] H. R. Smith, T. J. Beveridge, S. H. Nader, M. A. Nader, and L. J. Porrino, "Regional elevations in microglial activation and cerebral glucose utilization in frontal white matter tracts of rhesus monkeys following prolonged cocaine self-administration," *Brain Structure and Function*, vol. 224, no. 4, pp. 1417–1428, 2019.

[28] D. Albores-Garcia, J. L. McGlothan, Z. Bursac, and T. R. Guilarte, "Chronic developmental lead exposure increases μ-opiate receptor levels in the adolescent rat brain," *bioRxiv*, 2020.

[29] J. H. Yoo et al., "Characterization of [3H] oxymorphone binding sites in mouse brain: quantitative autoradiography in opioid receptor knockout mice," *Neuroscience Letters*, vol. 643, pp. 16–21, 2017.

[30] E. Solon, "Autoradiography, imaging mass spectrometry and other preclinical imaging techniques to study tissue distribution of xenobiotics in animal models," in *Visualizing and Quantifying Drug Distribution in Tissue III*, 2019, vol. 10859: International Society for Optics and Photonics, p. 1085904.

[31] N. Asano, K. Ueda, and K. Kawazu, "Penetration route of the selective glucocorticoid receptor agonist sa22465 and betamethasone into rabbit meibomian gland based on pharmacokinetics and autoradiography," *Drug Metabolism and Disposition*, vol. 45, no. 7, pp. 826–833, 2017.

[32] C. Aguero et al., "Autoradiography validation of novel tau PET tracer [F-18]-MK-6240 on human postmortem brain tissue," *Acta Neuropathologica Communications*, vol. 7, no. 1, p. 37, 2019.

[33] A. Sorcic, "Investigation of Storage-Phosphor Autoradiography for Alpha Emitters on Different Types of Filters," Colorado State University Libraries, 2019.

[34] S. J. Waaijer et al., "Preclinical PET imaging of bispecific antibody ERY974 targeting CD3 and glypican 3 reveals that tumor uptake correlates to T cell infiltrate," *J Immunother Cancer,* vol. 8, no. 1, 2020.

[35] A. Putzu et al., "Regional differences in cerebral glucose metabolism after cardiac arrest and resuscitation in rats using [18 F] FDG positron emission tomography and autoradiography," *Neurocritical Care,* vol. 28, no. 3, pp. 370–378, 2018.

[36] L. Lemoine, L. Saint-Aubert, I. Nennesmo, P.-G. Gillberg, and A. Nordberg, "Cortical laminar tau deposits and activated astrocytes in Alzheimer's disease visualised by 3 H-THK5117 and 3 H-deprenyl autoradiography," *Sci Repo,* vol. 7, no. 1, pp. 1–11, 2017.

[37] K. Varnäs, V. Stepanov, and C. Halldin, "Autoradiographic mapping of synaptic vesicle glycoprotein 2A in non-human primate and human brain," *Synapse*, p. e22157, 2020.

[38] P. Valuskova, V. Farar, S. Forczek, I. Krizova, and J. Myslivecek, "Autoradiography of 3H-pirenzepine and 3H-AFDX-384 in Mouse Brain Regions: Possible insights into M1, M2, and M4 muscarinic receptors distribution," *Front Pharmacol*, vol. 9, p. 124, 2018.

[39] R. Fjellaksel et al., "Evaluation by metabolic profiling and in vitro autoradiography of two promising GnRH-receptor ligands for brain SPECT imaging," *J Labelled Comp Radiopharm*, vol. 63, no. 2, pp. 72–84, 2020.

[40] R. Tönnesmann, P. T. Meyer, M. Eder, and A.-C. Baranski, "[177Lu] Lu-PSMA-617 salivary gland uptake characterized by quantitative in vitro autoradiography," *Pharmaceuticals*, vol. 12, no. 1, p. 18, 2019.

[41] Y. D. Soubaneh, E. Pelletier, I. Desbiens, and C. Rouleau, "Radiolabeling of amide functionalized multi-walled carbon nanotubes for bioaccumulation study in fish bone using whole-body autoradiography," *Environ Sci Pollut Res*, vol. 27, no. 4, pp. 3756–3767, 2020.

[42] H. V. Nguyen and H. T. Thi, "Visualization of the initial fixed nitrogen transport in nodulated soybean plant using [13N] N2 tracer gas in real-time," *bioRxiv*, 2020.

[43] K. Vyas et al., "Flexible scintillator autoradiography for tumor margin inspection using 18F-FDG," in *Molecular-Guided Surgery: Molecules, Devices, and Applications IV*, 2018, vol. 10478: International Society for Optics and Photonics, p. 1047811.

[44] I. Shestakova, V. V. Nagarkar, V. Gaysinskiy, G. Entine, B. C. Stack, and B. Miller, "Feasibility studies of an emccd-based beta imaging probe for radioguided thyroid surgery," in *Hard X-Ray and Gamma-Ray Detector Physics and Penetrating Radiation Systems VIII*, 2006, vol. 6319: International Society for Optics and Photonics, p. 63191E.

[45] S. Almasi and G. Pratx, "High-resolution radioluminescence microscopy image reconstruction via ionization track analysis," *IEEE Trans Radiat Plasma Med Sci,* vol. 3, no. 6, pp. 660–667, 2019.

[46] B. W. Miller, S. J. Gregory, E. S. Fuller, H. H. Barrett, H. B. Barber, and L. R. Furenlid, "The iQID camera: An ionizing-radiation quantum imaging detector," (in English), *Nucl Instrum Meth A*, vol. 767, pp. 146–152, Dec. 11, 2014. [Online]. Available: <Go to ISI>://WOS:000344994600018.

[47] B. W. Miller et al., "Quantitative single-particle digital autoradiography with α-particle emitters for targeted radionuclide therapy using the iQID camera," *Med Phys*, vol. 42, no. 7, pp. 4094–4105, 2015.

[48] R. M. Asmussen et al., "Getters for improved technetium containment in cementitious waste forms," (in English), *J Hazard Mater*, vol. 341, pp. 238–247, Jan. 5, 2018. [Online]. Available: <Go to ISI>://WOS:000412378700026.

[49] G. Tabatadze, B. W. Miller, and S. Y. Tolmachev, "Mapping Am-241 spatial distribution within anatomical bone structures using digital autoradiography," (in English), *Health Phys*, vol. 117, no. 2, pp. 179–186, Aug 2019. [Online]. Available: <Go to ISI>://WOS:000474255200008.

[50] L. Han, B. W. Miller, and L. R. Furenlid, "LA-iQID: A novel high-resolution CCD-based gamma camera for lymphatic imaging," in *2017 IEEE Nuclear Science Symposium and Medical Imaging Conference (NSS/MIC)*, 2017: IEEE, pp. 1–3.

[51] S. H. L. Frost et al., "Comparative efficacy of Lu-177 and Y-90 for anti-CD20 pretargeted radioimmunotherapy in Murine Lymphoma Xenograft models," (in English), *Plos One*, vol. 10, no. 3, Mar. 18, 2015. [Online]. Available: <Go to ISI>://WOS:000352138500181.

[52] I. B. Hernández et al., "Imaging of tumor spheroids, dual-isotope SPECT, and autoradiographic analysis to assess the tumor uptake and distribution of different nanobodies," *Mol Imag Biol*, vol. 21, no. 6, pp. 1079–1088, 2019.

[53] V. Peskov, G. Charpak, W. Dominik, and F. Sauli, "Investigation of light emission from a parallel-plate avalanche chamber filled with noble gases and with TEA, TMAE, and H2O vapours at atmospheric pressure," *Nucl Instrum Methods Phys Res A*, vol. 277, no. 2–3, pp. 547–556, 1989.

[54] J. Donnard et al., "The PIMager: A new tool for high sensitive digital β autoradiograph," in *2009 IEEE Nuclear Science Symposium Conference Record (NSS/MIC)*: IEEE, pp. 3672–3674.

[55] S. Billon, P. Sardini, S. Leblond, and P. Fichet, "From Bq cm− 3 to Bq cm − 2 (and conversely) – Part 1: A useful conversion for autoradiography," *J Radioanal Nucl Chem*, vol. 320, no. 3, pp. 643–654, 2019.

[56] O. Alitalo, T. Rantamäki, and T. Huhtala, "Digital autoradiography for efficient functional imaging without anesthesia in experimental animals: Reversing phencyclidine-induced functional alterations using clozapine," *Prog Neuropsychopharmaco Biol Psychiatry*, vol. 100, p. 109887, 2020.

[57] D. Lumen et al., "Site-specific 111In-radiolabeling of Dual-PEGylated porous silicon nanoparticles and their in vivo evaluation in Murine 4T1 breast cancer model," *Pharmaceutics*, vol. 11, no. 12, p. 686, 2019.

[58] C. Bailly et al., "What is the best radionuclide for immuno-PET of multiple myeloma? A comparison study between 89Zr-and 64Cu-labeled anti-CD138 in a preclinical syngeneic model," *Int J Mol Sci*, vol. 20, no. 10, p. 2564, 2019.

[59] G. F. Knoll, *Radiation Detection and Measurement*, 3rd edn. New York: John Wiley, 2000, p. 802.

[60] J. Cabello et al., "Digital autoradiography using room temperature CCD and CMOS imaging technology," *Phys Med Biol*, vol. 52, no. 16, p. 4993, 2007.

[61] J. Cabello et al., "Digital autoradiography imaging using CMOS technology: First tritium autoradiography with a back-thinned CMOS detector and comparison of CMOS imaging performance with autoradiography film," in *2007 IEEE Nuclear Science Symposium Conference Record*, 2007, vol. 5: IEEE, pp. 3743–3746.

[62] T. N. Pham, P. Marchand, C. Finck, F. Boisson, D. Brasse, and P. Laquerriere, "18F autoradiography with the Mimosa-28: Characterisation and Application," *IEEE Transactions on Radiation and Plasma Medical Sciences*, 2020.

[63] G. Mettivier, M. C. Montesi, and P. Russo, "First images of a digital autoradiography system based on a Medipix2 hybrid silicon pixel detector," *Phys Med Biol*, vol. 48, no. 12, p. N173, 2003.

[64] R. Al Darwish, L. Marcu, and E. Bezak, "Overview of current applications of the Timepix detector in spectroscopy, radiation and medical physics," *Appl Spectrosc Rev*, vol. 55, no. 3, pp. 243–261, 2020.

[65] A. Örbom, "Preclinical Molecular Imaging Using Multi-Isotope Digital Autoradiography – Techniques and Applications," PhD, Dept Medical Radiation Physics, Lund University, Lund, 2013.

[66] P. Russo et al., "18F-FDG positron autoradiography with a particle counting silicon pixel detector," *Phys Med Biol*, vol. 53, no. 21, p. 6227, 2008.

[67] M. Esposito, G. Mettivier, and P. Russo, "14C autoradiography with an energy-sensitive silicon pixel detector," *Phys Med Biol*, vol. 56, no. 7, p. 1947, 2011.

[68] A. Örbom et al., "The intratumoral distribution of radiolabeled 177Lu-BR96 monoclonal antibodies changes in relation to tumor histology over time in a syngeneic rat colon carcinoma model," *J Nucl Med*, vol. 54, no. 8, pp. 1404–1410, 2013.

[69] T. Bäck and L. Jacobsson, "The Alpha-Camera: A quantitative digital autoradiography technique using a charge-coupled device for ex vivo high-resolution bioimaging of alpha-particles," (in English), *J Nucl Med*, Research Support, Non-U.S. Gov't vol. 51, no. 10, pp. 1616–1623, Oct 2010, doi:10.2967/jnumed.110.077578.

[70] S. Yamamoto, Y. Hirano, K. Kamada, and A. Yoshikawa, "Development of an ultrahigh-resolution radiation real-time imaging system to observe trajectory of alpha particles in a scintillator," (in English), *Radiat Meas*, vol. 134, Jun 2020, doi:ARTN 10636810.1016/j.radmeas.2020.106368.

[71] M. Segawa, I. Nishinaka, Y. Toh, and M. Maeda, "Analytical method for the determination of (211)At using an alpha-scintillation camera system and thin-layer chromatography," (in English), *J Radioanal Nucl Chem*, vol. 326, no. 1, pp. 773–778, Oct 2020. [Online]. Available: <Go to ISI>://WOS:000563609100002.

[72] B. W. Miller, J. M. Bowen, and E. C. Morrison, "High-resolution, single-particle digital autoradiography of actinide sources using microcapillary array collimators and the iQID camera," *Appl Radiat Isot*, vol. 166, p. 109348, 2020.

[73] W. Hofmann et al., "Internal microdosimetry of alpha-emitting radionuclides," (in English), *Radiat Environ Bioph*, vol. 59, no. 1, pp. 29–62, Mar 2020. [Online]. Available: <Go to ISI>://WOS:000503661300001.

[74] C. Granja et al., "Resolving power of pixel detector Timepix for wide-range electron, proton and ion detection," (in English), *Nucl Instrum Meth A*, vol. 908, pp. 60–71, Nov. 11, 2018, doi:10.1016/j.nima.2018.08.014.

[75] R. AL Darwish, A. H. Staudacher, E. Bezak, and M. P. Brown, "Autoradiography Imaging in Targeted Alpha Therapy with Timepix Detector," (in English), *Comput Math Method M,* 2015. [Online]. Available: <Go to ISI>://WOS:000349090800001.

[76] R. Al Darwish, A. H. Staudacher, Y. Li, M. P. Brown, and E. Bezak, "Development of a transmission alpha particle dosimetry technique using A549 cells and a Ra-223 source for targeted alpha therapy," (in English), *Med Phys*, vol. 43, no. 11, pp. 6145–6153, Nov 2016. [Online]. Available: <Go to ISI>://WOS:000387007500036.

[77] R. Al Darwish, A. H. Staudacher, E. Bezak, and M. P. Brown, "Autoradiography imaging in targeted alpha therapy with Timepix detector," *Comput Math Methods Med*, vol. 2015, p. 612580, 2015, doi:10.1155/2015/612580.

[78] C. Kratochwil et al., "225Ac-PSMA-617 for PSMA-targeted αrRadiation therapy of metastatic castration-resistant prostate cancer," *J Nucl Med*, vol. 57, no. 12, pp. 1941–1944, December 1, 2016, doi:10.2967/jnumed.116.178673.

[79] Y. Dekempeneer et al., "Therapeutic efficacy of Bi-213-labeled sdAbs in a preclinical model of ovarian cancer," (in English), *Mol Pharmaceut*, vol. 17, no. 9, pp. 3553–3566, Sep 8 2020, doi:10.1021/acs.molpharmaceut.0c00580.

[80] S. Palm et al., "Evaluation of therapeutic efficacy of 211At-labeled farletuzumab in an intraperitoneal mouse model of disseminated ovarian cancer," *Trans Onco*, vol. 14, no. 1, p. 100873, 2020.

[81] S. H. Frost et al., "alpha-imaging confirmed efficient targeting of CD45-positive cells after 211At-radioimmunotherapy for hematopoietic cell transplantation," *J Nucl Med*, vol. 56, no. 11, pp. 1766–1773, Nov 2015, doi:10.2967/jnumed.115.162388.

[82] B. W. Miller, J. M. Bowen, and E. C. Morrison, "High-resolution, single-particle digital autoradiography of actinide sources using microcapillary array collimators and the iQID camera," *Appl Radiat Isot*, vol. 166, p. 109348, Dec 2020, doi:10.1016/j.apradiso.2020.109348.

[83] S. Billon et al., "Quantitative imaging of 226Ra ultratrace distribution using digital autoradiography: Case of doped celestines," *J Environ Radioact*, vol. 217, p. 106211, 2020.

[84] A. Tazrart et al., "Skin absorption of actinides: influence of solvents or chelates on skin penetration ex vivo," *Int J Radiat Biol*, vol. 93, no. 6, pp. 607–616, 2017.

31 Principles behind Computed Tomography (CT)

Mikael Gunnarsson and Kristina Ydström

CONTENTS

31.1 INTRODUCTION

After its clinical introduction in 1971, computed tomography (CT) developed from a modality limited to axial imaging of the brain into a 3D whole-body imaging modality for a wide range of applications, including interventional radiology, oncology (dose planning) and nuclear medicine. In this chapter a wide scope of topics will be covered, from the basic principles behind CT imaging to model-based iterative reconstruction algorithms, spectral CT, and CT dosimetry.

31.2 BASIC PRINCIPLES OF X-RAY AND CT

The principle of conventional X-ray and computed tomography is to measure densities in the body. Different tissues have different densities and, therefore, they attenuate different amounts of radiation. Computed Tomography is commonly described as a process of creating cross-sectional, or tomographic, images from projections (line integrals) of an object at multiple angles and using a computer for the image reconstruction. This process involves the measurement of X-ray transmission profiles through the object for a large number of views. A profile from each view is achieved by using a detector arc consisting of multiple detector elements. The transmission profiles are collected by rotation of an X-ray tube and detector around the object (Figure 31.1) and are associated with the attenuation of the corresponding tissue, or, more specifically, to the linear attenuation coefficients μ (cm^{-1}).

DOI: 10.1201/9780429489556-31

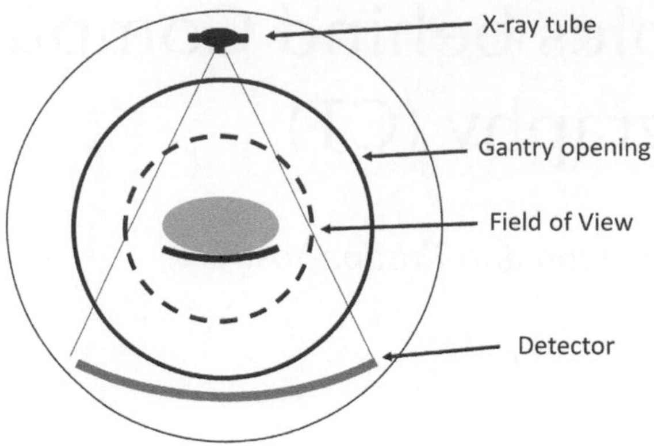

FIGURE 31.1 Geometry of a CT Scanner.

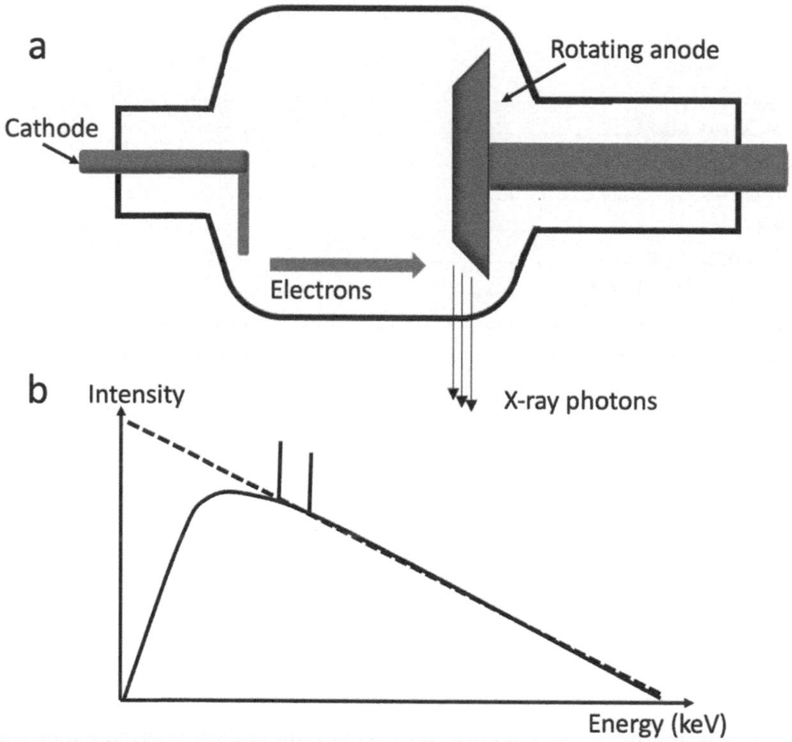

FIGURE 31.2 Schematic illustration of the X-ray tube (a) and an X-ray spectrum (b).

With conventional radiology, objects get superimposed in the image. With CT the basic principle is the same but, using a collimated beam (z-direction) and sampling round the patient, the images can be reconstructed in the cross section of the body and the problem of superimposition of objects is removed. As a result, the image contrast is also improved.

The X-ray source in a CT consists of an X-ray tube that produces X-rays by accelerating electrons from a heated filament, the cathode, to a high velocity with a high-voltage field and causing them to collide with a target, the anode. The design of a classic X-ray tube with its components is shown in Figure 31.2a. The output from the X-ray tube results in an X-ray beam with an energy spectrum that has origins in two different processes. When the electrons from the cathode are decelerated in the anode, it gives rise to rays called *bremsstrahlung* radiation. The other process is when electrons in the material of the anode get 'excited' while interacting with the incoming electrons and then 'de-excited', which results in characteristic X-rays with definite energies, due to the atomic energy levels that they transit. An X-ray spectrum is shown in Figure 31.2b.

To be able to determine the attenuation in the object and use this information to reconstruct images, some problems must be solved. Assuming a narrow-beam geometry, the attenuation of photons (decrease in the count rate due to photon absorption and photon scattering) depends on the composition of the material (Z, ρ), the photon energy $h\nu$, and the thickness of the object. In a homogeneous object one can express this as

$$\dot{N} = \dot{N}_0 \, e^{-\mu T} \tag{31.1}$$

where $\dot{N}$ is the transmitted photon rate through the object, $\dot{N}_0$ is the photon rate impinging on the object, T is the thickness of the object and μ is the linear attenuation coefficient. In a heterogeneous object, μ will vary with the composition for a fixed photon energy, so

$$\bar{\mu}T = \int_0^T \mu(s)\,ds \tag{31.2}$$

One can combine Eqs. 31.1 and 31.2 to

$$p = \bar{\mu}T = \int_0^T \mu(s)\,ds = \ln\left(\frac{\dot{N}_0}{\dot{N}}\right) \tag{31.3}$$

From this, a set of attenuation projections, $p(r,\theta)$, as function of projection angle θ and position r, can be obtained and reconstructed to transversal slices of $\bar{\mu}(x,y)$ by filtered back-projection.

31.2.1 Image Reconstruction

In order to reconstruct a CT image, numerous measurements of the transmission of X-rays through the patient are acquired; this information is the basis for reconstruction of the CT image. CT image reconstruction has traditionally relied on filtered back-projection (FBP). For further details of the FBP technique, see Chapter 16 in this book.

In a modern CT system, the image reconstruction is based on iterative reconstruction algorithms. The principle of iterative reconstruction is well known in medical imaging; it has been routinely used in nuclear medicine for decades and was also used in the pioneer work of Hounsfield in the 1970s. Iterative reconstructions are now commonly used in CT and, as the computers get more powerful, the reconstructions models that the reconstruction algorithms are based on can be more complex and larger amounts of data, that is, optics, physics, scanner, noise statistics can be handled much faster without significantly affecting the reconstruction time. In recent years, iterative reconstruction (IR) algorithms that work in both the raw data and image data domains have been developed.

In iterative reconstruction (IR), a starting image is transformed and compared to the original data. A correction is calculated. The image is corrected and then compared with the original data until a sufficiently good agreement is reached, and a final image is produced. For further details of IR technique, see Chapter 20 in this book.

When using iterative reconstruction, a degree of noise reduction can usually be selected. The example in Figure 31.3 shows how the noise in the liver is reduced with different choices of noise reduction. It is also seen that the choice affects the appearance of the image, that is, noise distribution in the image. It is, thus, a choice that can be optimized based on the image quality that is being sought.

A potential benefit of using IR in CT is the reduction of noise while the resolution is maintained. This enables the use of lower radiation dose and improves the possibilities of removing of streak artefacts (particularly when fewer projection angles are used), which means a better performance in low-dose CT acquisitions. However, iteratively reconstructed images may be affected by artefacts that are not present in filtered back-projection images, such as aliasing patterns and overshoots in the areas of sharp intensity transitions.

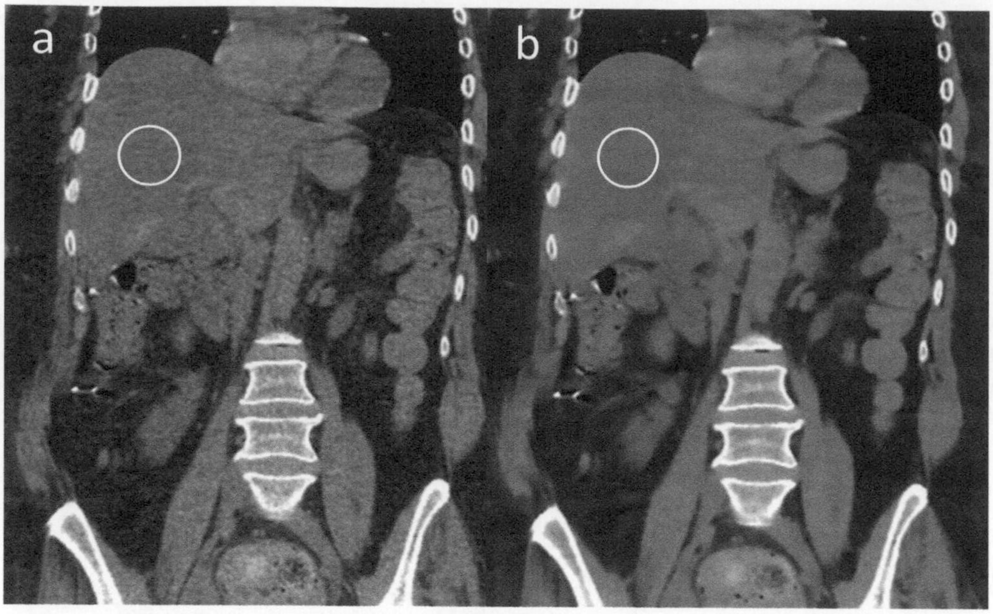

FIGURE 31.3 Iterative reconstruction. Low grade of iterative reconstruction (left) and medium grade of iterative reconstruction (right). Noise (SD 14 and 6 respectively).

31.2.2 Hounsfield Units

In the CT image, the matrix of reconstructed linear attenuation coefficients ($\mu_{material}$), is transformed into a corresponding matrix of so-called Hounsfield units ($HU_{material}$), where the HU scale is expressed relative to the linear attenuation coefficient of water at room temperature (μ_{water}):

$$HU = \frac{\mu_{material} - \mu_{water}}{\mu_{water}} * 1000 \tag{31.3}$$

A change of one Hounsfield unit (HU) corresponds to a 0.1% difference of the attenuation coefficient difference of water. Air corresponds to a HU of -1000, since $\mu_{air} = 0$ to a good approximation. The HU of water and air is independent of the energy of the X-rays and therefore constitute a fix points for the HU value scale. Calibration of a CT system includes measurement on a water phantom to verify that the values of water is correct. CT values of different tissues are shown in Figure 31.4a.

The HU for substances except water and air is however affected when the tube voltage is altered. The reason is that different substances exhibit a non-linear relationship of their linear attenuation coefficient relative to that of water as a function of photon energy. This effect is most notable for substances that have a relatively high effective atomic number, such as bone and iodine (contrast medium).

The HU is decreased for high density material (e.g. bone, iodine), zero for water and increased for low density materials (e.g. fat) and for both high and low density material the HU is approaching zero with increasing energy. (Figure 31.4b). The changes are relatively small but may limit the use of HU number for diagnostic purposes. The initial filtration of the x-ray tube changes the spectrum and due to this, the HU varies between different scanner systems. The HU value is also affected by the size of the body due to the impact of the amount of scatter in patients of different sizes.

The HU scale for medical CT scanners is typically in a range from -1024 HU to +3070 HU. Consequently, 4096 different values are available which corresponds to 12 bits per pixel. To be able to visualize bone, lung tissue or other tissues of interest to the diagnostic task one can use the concept of windowing to make a discrimination in HU values which is represented on the screen. The grayscale of the image is defined by using different settings of window 'width' and window 'level'. Typical window settings for different diagnostic tasks are shown in Figure 31.5.

The more grayscale levels an imaging system can represent, the better its contrast resolution. By changing the width of the window setting, center the grayscale levels to be presented, the contrast can be varied.

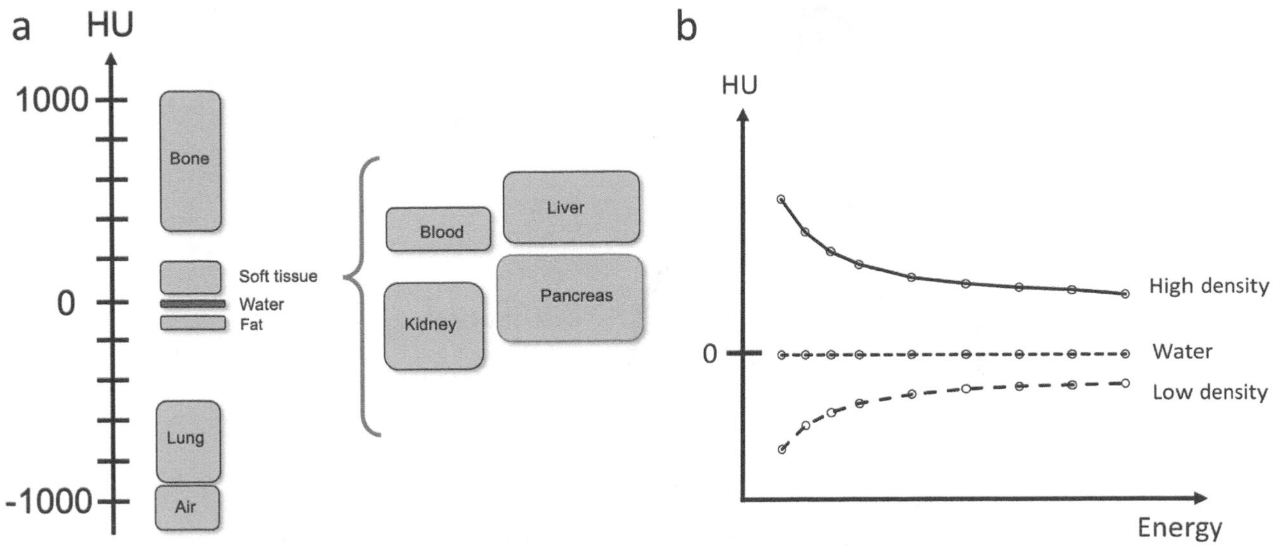

FIGURE 31.4 Panel (a) HU values for various materials/organs. Panel (b) shows how CT numbers vary with different materials and with tube voltage (kV).

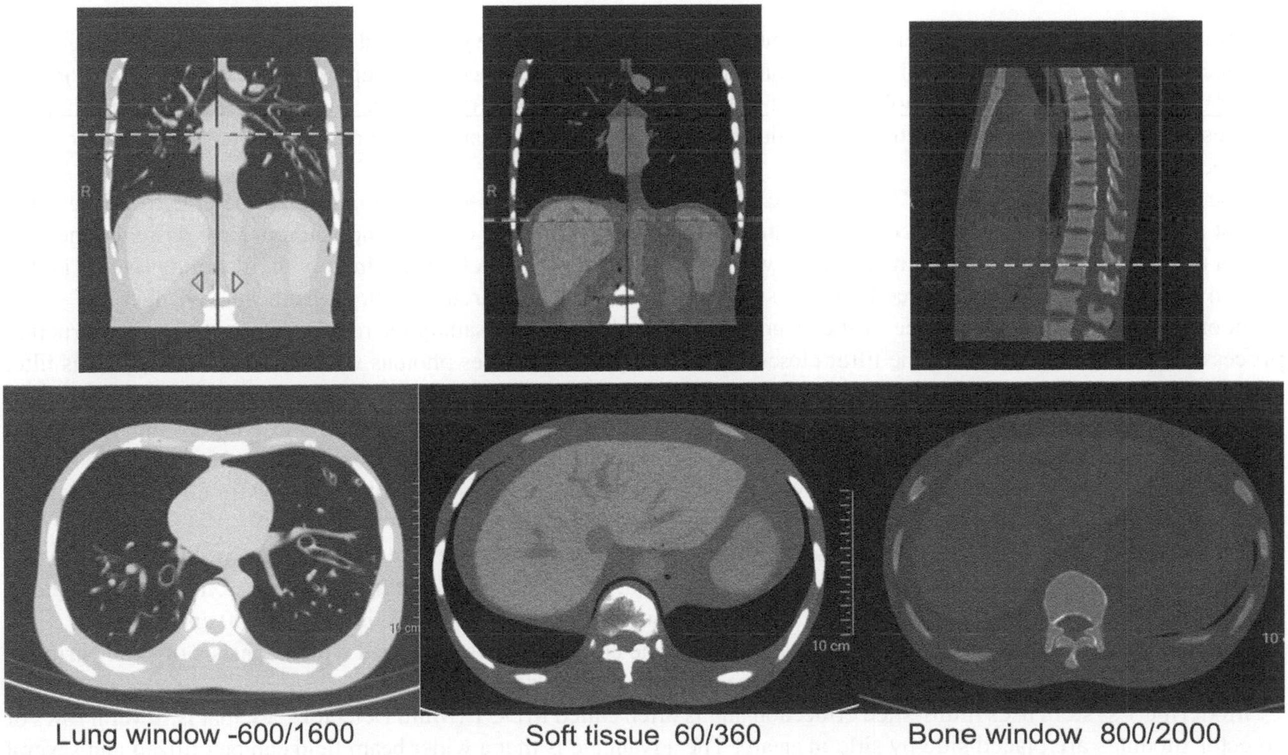

FIGURE 31.5 Window settings for different clinical tasks demonstrated in a scan of an anthropomorphic phantom.

31.3 TECHNICAL CONCEPTS

31.3.1 Scanner Configuration

In a CT, the X-ray tube with high voltage generator and tube cooling system, detector packages, collimators and beam shaping filters are mechanically interconnected in the gantry (Figure 31.6). Since transmission profiles must be recorded

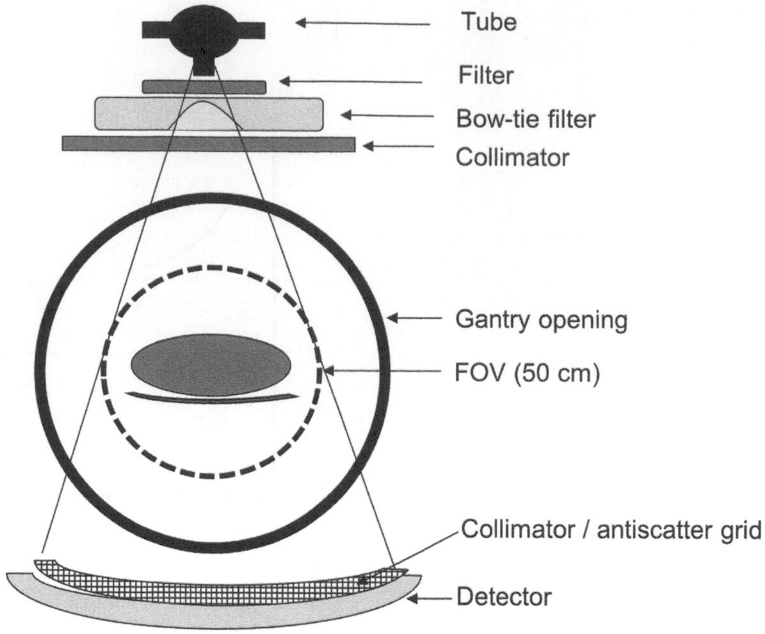

FIGURE 31.6 The principial structure of the CT scanner.

at different angles, these components are mounted on a support within the gantry that can be rotated at high speeds.

Electrical power and the recorded projection profiles data from the detectors is supplied to/from the rotating gantry through the use of slip-ring technology, which allows for continuos gantry rotation. Slip rings are electromechanical devices consisting of circular electrically conductive rings and brushes that transmit electrical energy across a rotating interface.

The X-ray tube in a modern CT must have good heat dissipation and fast cooling rates to be able to deliver high output during demanding high-speed examinations. To achieve this some of the manufacturers have designed the tube with a tungsten anode that is in direct contact with the circulating oil which results in very high cooling rates. The CT system also often consists of a forced cooling system using oil or water circulated through a heat exchanger.

In order to make the appearance of the beam more monochromatic to satisfy the requirements of the reconstruction process, a special filter is used. The filter closest to the X-ray tube removes photons with the lowest energy. This filter varies in material and thickness between manufacturers and affects the energy distribution of outgoing photons.

After the filter, the photons pass a forming filter, a so-called Bowtie filter. This results in a higher intensity in the center of the patient and lower intensity on the edge of the patient. This makes the dose more evenly distributed in the patient and thus the image quality is improved. Some manufacturers have several different bowtie filters, and they are changed depending on the type of examination and patient size chosen.

The tube also has a fixed collimator that prevents radiation outside the intended area, and a movable collimator that allows you to choose which detector width (in the z-direction, the patient's longitudinal axis) to use. The gantry opening of the computer tomograph is usually 70 cm and the maximum field of view (FOV) is normally 50 cm.

In front of the detector is usually a collimator or in some cases an anti-scatter grid which consists of small strips of highly attenuating material, which reduces scattered radiation to the detector.

Modern CT system uses multi-slice collection and is often called MDCT (Multi Detector CT), that is, several rows of detector modules are placed side by side in an arc. The advantage is that a wider beam field can be utilized and several slices are imaged in one rotation which reduce the scan time and also reduces the risk of motion artefacts.

It is important that the detectors used in a CT system have good detection efficiency and a fast response with little afterglow. Currently, solid state detectors are used, as they have a detection efficiency close to 100%. When the X-rays interact with the detector (scintillator), light is generated which is further converted to an electrical signal by photodiodes that are attached to the back of the scintillator to ensure optimal detection.

The size of the detector in the z-direction varies between different manufacturers and application. A modern CT system have detectors with a coverage of between 1 to 16 cm. The scanners with highest numbers of detector modules (320 rows) have the advantage of being able to cover organs such as the brain or the heart within one single rotation.

CT systems with a smaller number of rows must use multiple rotations to cover the same anatomical region/organ. In combined systems (SPECT-CT and PET-CT) the most common detector coverage is 1 to 4 cm and in some cases cone beam CT is used.

The components in a CT system need to be able to withstand high G forces due to the strong centrifugal force that occurs during the fast rotation, down to 0.25 s, of the gantry. The table must also be able to withstand heavy weights (>250 kg) without bending and be able to move with high accuracy during the scan.

31.3.2 Scan Modes and Scan Parameters

A CT examination is usually initiated by taking one or more overview images (AP or PA and / or lateral) of the patient. X-ray tubes and detectors stand still as the patient table moves. Based on the overview image, the examination region of the patient is selected.

It is important to position the patient correctly during a CT examination. The bow-tie filters require that the patient is centered in the gantry, that is, in the isocenter. Erroneous centering results in an inhomogeneous dose distribution in the patient, which causes areas of higher or lower radiation dose than intended. This may cause artefacts and increased noise. The overview image should be limited to the area that is intended to be investigated (Figure 31.7). However, it is important that the entire area to be investigated is included in the overview image, because the information from the overview image is used by the CT's automatic exposure control (AEC) system to adjust the tube current to the patient's size in order to achieve optimal image quality. The AEC system is described in more detail in the CT dosimetry section.

The tomographic scan is usually performed as an axial- or helical-scan. An axial CT scan involves an acquisition of transmission profiles with a rotating X-ray tube and a static table. A complete axial CT examination generally involves subsequent axial acquisitions in order to cover a clinically relevant volume. The table translation is equal to the slice thickness, so that subsequent axial acquisitions can be reconstructed as contiguous axial images.

The most commonly used scan mode is helical CT scan which involves an acquisition of transmission profiles with a rotating X-ray tube and a moving table. The advantages of helical CT scans include shorter scan time which decrease the risk of getting artefacts from patient movement, possibilities to use contrast media in a more efficient manner and results in more consistent 3-D data from the examination. The table translation is generally expressed relative to the collimated beam width and the so-called pitch factor is defined as:

$$P = \frac{\text{Table movement}}{\text{collimated beam width}} \tag{31.4}$$

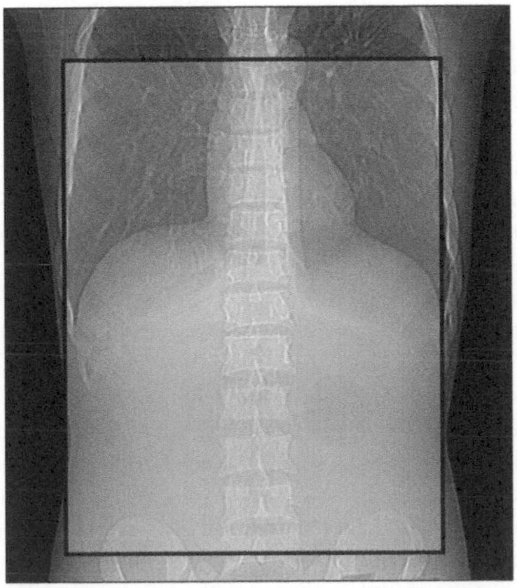

FIGURE 31.7 Overview image from a thorax/abdomen CT examination. The scan area is planned from the overview image, for example, in this case outlined by the rectangle.

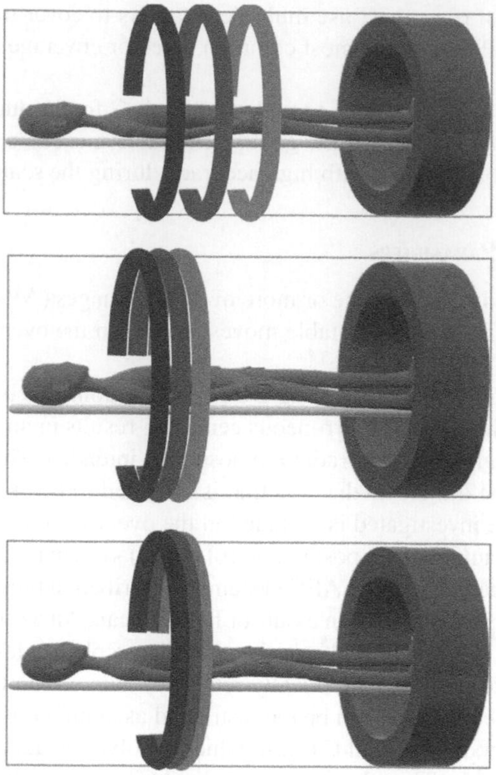

FIGURE 31.8 Examples of different pitch factors.

Three examples of different pitch factors are shown in Figure 31.8.

- Pitch > 1: High pitch, sparse collection and shorter collection time
- Pitch = 1: Scan edge-to-edge
- Pitch < 1: Low pitch, overlapping scan

31.3.3 Dual Energy

The selection of tube voltage and filtration determine the spectra and the average energy of the photons. By changing the average energy for example, by changing the tube voltage, attenuation for a particular material is affected. The principle of spectral imaging/dual energy is illustrated in Figure 31.9. The difference in attenuation for the two energies vary with the material or tissue, for example, for iodine, the difference is larger than for bone. The difference in attenuation is used to calculate different types of images such as mono energetic images, iodine images or material composition.

There are various solutions and techniques for conducting a double energy examination. The different techniques result in different dose efficiency and sensitivity to movement. To make two collections followed by one another with different tube voltages requires that you can match the images from the two collections and that no movement has taken place. Technology with alternating tube voltage basically entails simultaneous collection but may have the limitation that tube current modulation cannot be used.

Technology with two X-ray tubes with different tube voltages means that the average energy of the two spectra can be well separated by means of extra filtration. However, two X-ray tubes and two detectors result in limited field of view with dual-energy data because one of the detectors is smaller. One challenge is to deal with scattered photons from one X-ray tube that hits the other detector. Using filters to obtain a split spectrum with different mean energy is a cheaper solution than using two sets of X-ray tubes and detectors but that have limited energy separation.

Another solution is to use a detector consisting of two different scintillation materials – photons with lower energy are detected in the upper layer, and photons of higher energy in the lower layer. The energy separation is less compared to using dual X-ray tubes but has the advantage that the collection takes place in principle at the same time, and all studies carried out with 120 kV or 140 kV can be processed as double-energy studies.

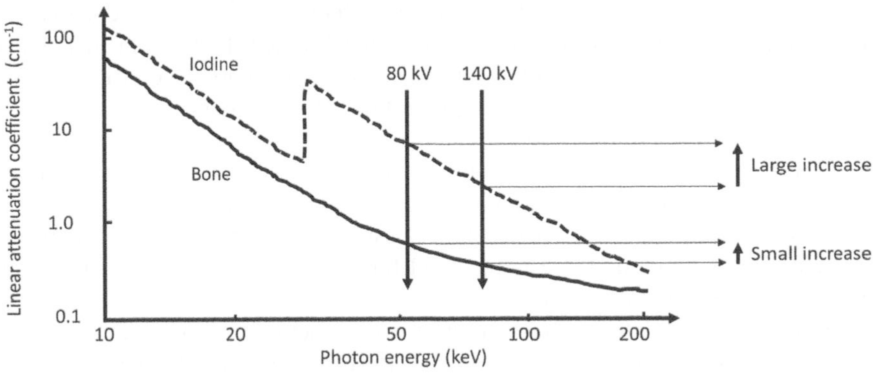

FIGURE 31.9 Linear attenuation coefficient as a function of photon energy for bone and iodine.

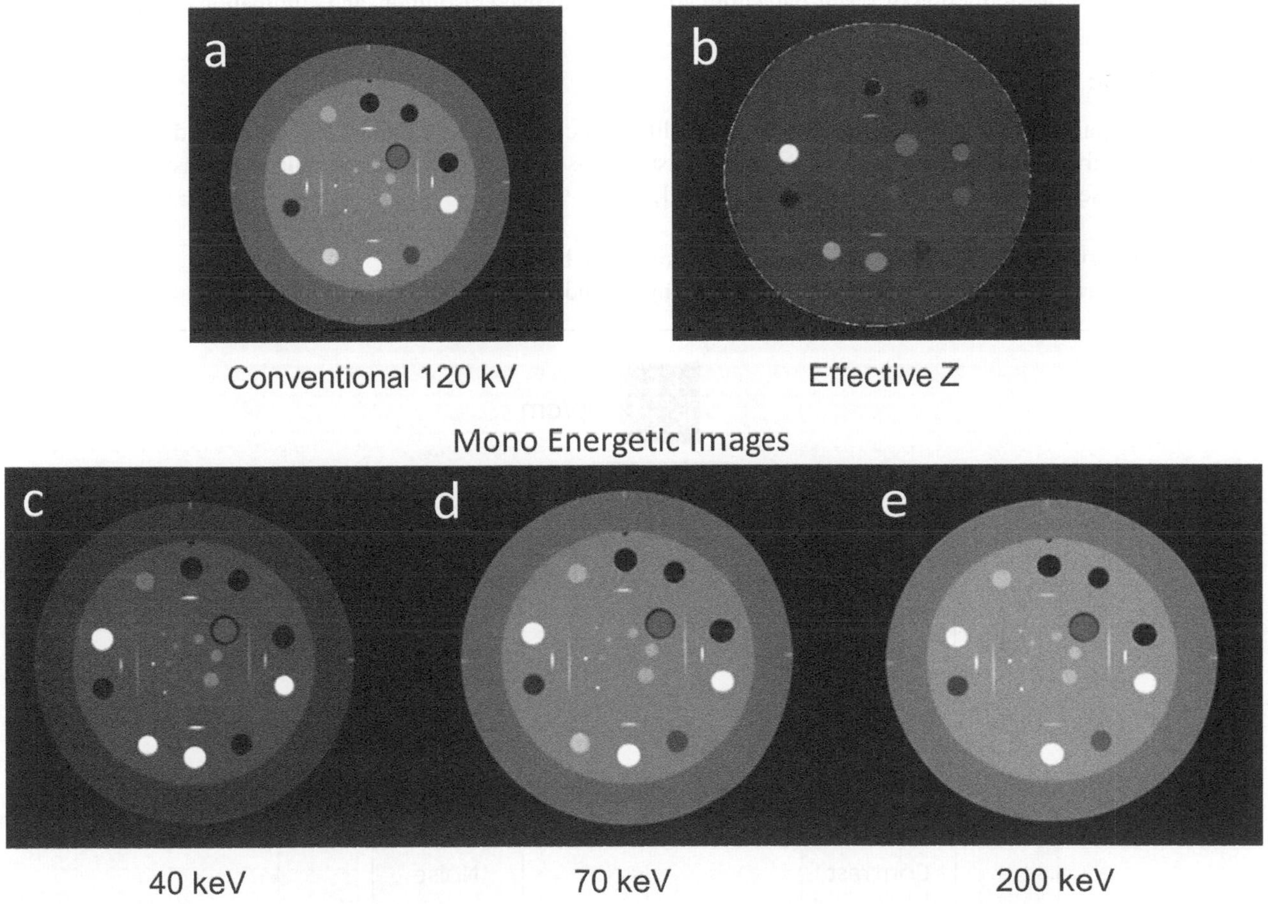

FIGURE 31.10 Illustrations of results from dual energy. Phantom scan showing conventional image, mono energetic image and effective Z.

There are several potential clinical applications for dual-energy CT. For example, in an iodine-based contrast agent study, iodine can be identified, and a virtual native series reconstructed. It is also possible to enhance the contrast charge by reconstructing a virtual monoenergetic image of lower energy.

When using a two-layer detector, spectral information is collected by the two scintillation materials. An image corresponding to single-energy 120 kV can be reconstructed. On the basis of spectral information, virtual monoenergetic images can be reconstructed. A monoenergetic image of 70 keV corresponds to approximately the average energy at 120 kV but has improved low contrast resolution due to less scattered radiation. A mono-energetic image of lower energy for example, 50 keV can be used to enhance the image contrast of iodine. By identifying iodine in the image, it is possible to produce an image without contrast. It is also possible to obtain an image of the iodine density mg / ml. In order to obtain a measure of the composition of the material, an image presenting effective atomic number (Z) can be reconstructed. Figure 31.10 illustrates some examples of dual energy results.

A common assumption is that double-energy CT results in twice the radiation dose. This is not true, but in the case of dual energy implementation, equal or reduced radiation dose should be sought compared to a conventional CT examination. However, the different technologies offer different possibilities regarding dose efficiency and the possibility of performing a dose-neutral examination. There is potential for dose savings by, for example, utilizing virtual non-contrast instead of performing a true native series.

With the exception of the two-layer detector that can separate two energies, no information is obtained on the energy of the absorbing photons with existing scintillation detectors. Research and development of semiconductor detectors with energy resolution for CT, so-called photon counters, is under development. By using different thresholds for the pulse readout, two or more energies can be determined. One of the challenges is managing and reading the high flow of photons, the so-called count rate. Potential benefits of photon counters are improved image contrast, increased contrast loading of iodine, reduction of beam-hardening artefacts, improved resolution, and elimination of electronic noise.

31.4 IMAGE QUALITY

Image-quality parameters can be divided into several different components: contrast, resolution, and noise. These are interrelated by the signal-to-noise ratio (SNR), contrast-to-noise ratio (CNR), the modulation transmission function (MTF), and noise power spectrum (NPS) (see Figure 31.11).

- SNR describes the visibility of an object in relation to the background.
- CNR describes the contrast difference between objects and background.

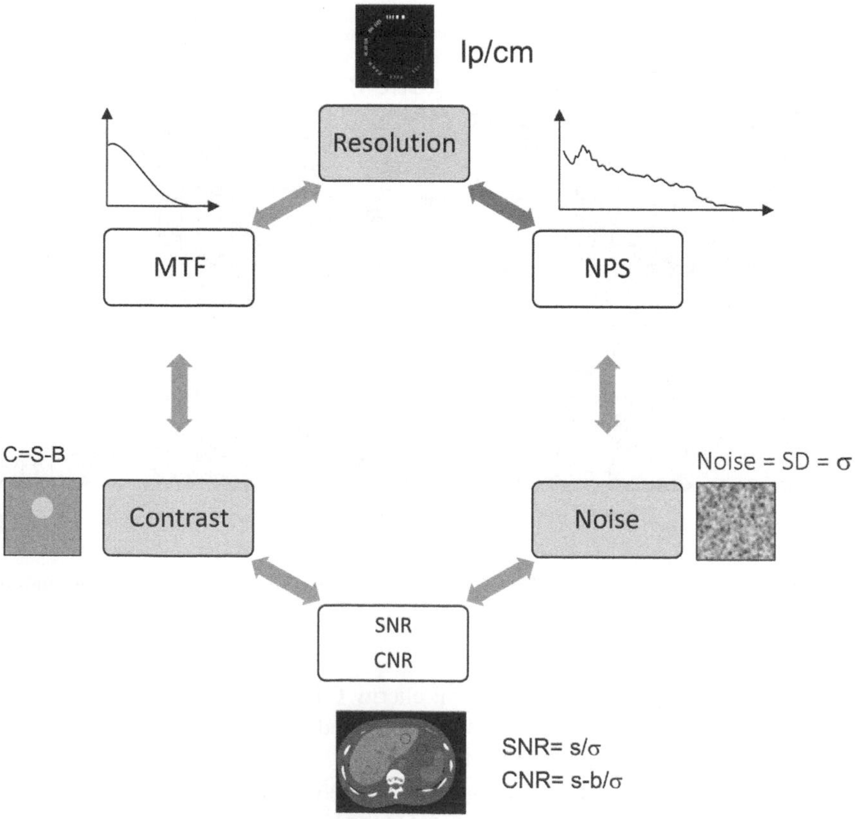

FIGURE 31.11 Image quality parameters.

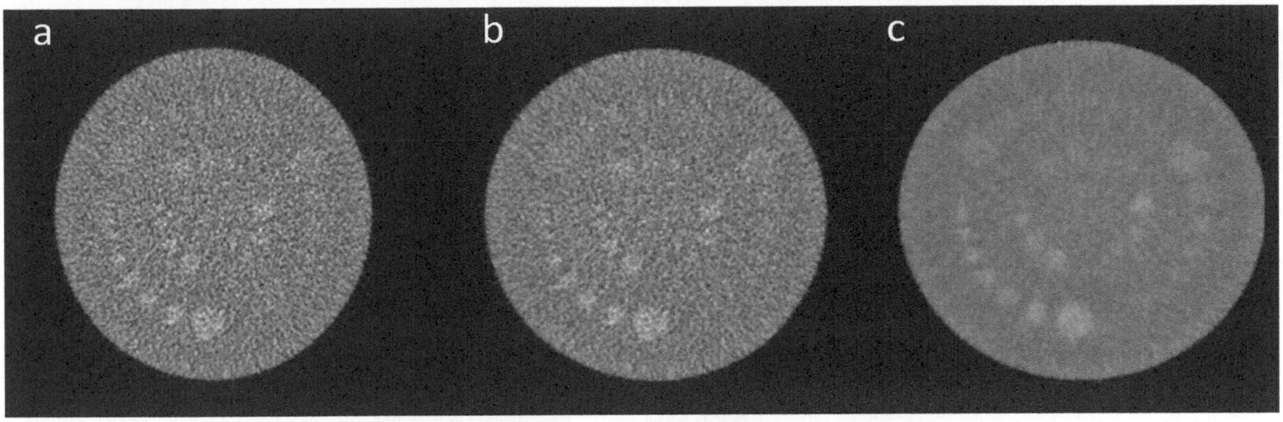

FIGURE 31.12 A comparison of image quality / noise depending on reconstructed slice thickness and reconstruction technique: (a) Filtered back-projection (FBP); (b) Iterative reconstruction; (c) Model-based iterative reconstruction.

- MTF is a measure of how well a system maintains different spatial frequencies.
- The NPS, also called the Wiener spectrum, describes the noise size (amplitude) at different spatial frequencies, that is, the 'appearance' or 'distribution' of the noise.

Noise is signal variations in the image that do not originate from the patient. The amount of noise in the image depends, among other things, on the quality of the X-ray detector, the amount of scattered radiation (size of patient, collimation), radiation dose, slice thickness and kernel / filter.

CT parameters that influence the image noise:

- Increase of tube current (mAs) results in increased dose to the patient but decreased noise. For FBP the relation is: $1/\sqrt{\text{mAs}}$. For example, to decrease the noise level by 50 per cent, the tube current must be increased by a factor of 4.
- Thinner slice collimation reduces the amount of scattered radiation but increases scan time.
- Choosing a larger reconstructed slice thickness results in more signal and thus reduced noise in the image. The disadvantage is that the resolution deteriorates.
- Choosing a softer reconstruction filter (kernel) decreases the noise, but again the resolution in the image is reduced.
- Reconstruction with an iterative process reduces the noise but can affect the noise structure.
- Reducing the irradiated volume – for example, by placing the patient's arms up and not along the body during abdominal examination – reduces the dose and results in better image quality.

Figure 31.12 illustrates how the choice of reconstruction method (FBP, IR, model-based IR) and slice thickness affects noise and, in turn, the low contrast resolution.

When reconstructing a CT image, the choice of thickness results in a balance between resolution and noise. A thicker slice consists of more information, and the reconstructed image will have less noise. The disadvantage is the deteriorated resolution and the overlaying of the tissue, which can give rise to partial so-called volume effects.

The reconstruction filter affects the noise, and a sharp filter results in a higher noise level than a soft filter. A sharp filter is a high-pass filter that amplifies high frequencies in the image and suppresses the low frequencies. The high frequencies improve the spatial resolution (details, edges) but at the same time increase the noise. It is thus a balance between resolution and noise. The image resolution is also affected by several factors such as detector characteristics, patient movements, focus size, reconstruction kernel (filter), and matrix size.

31.5 CT DOSIMETRY

31.5.1 CT Dose Index

To determine the radiation dose from a CT examination, a measurable quantity CT Dose Index (CTDI) has been introduced. CTDI is a measure of the average absorbed dose for a rotation and is measured in cylindrical plexiglass

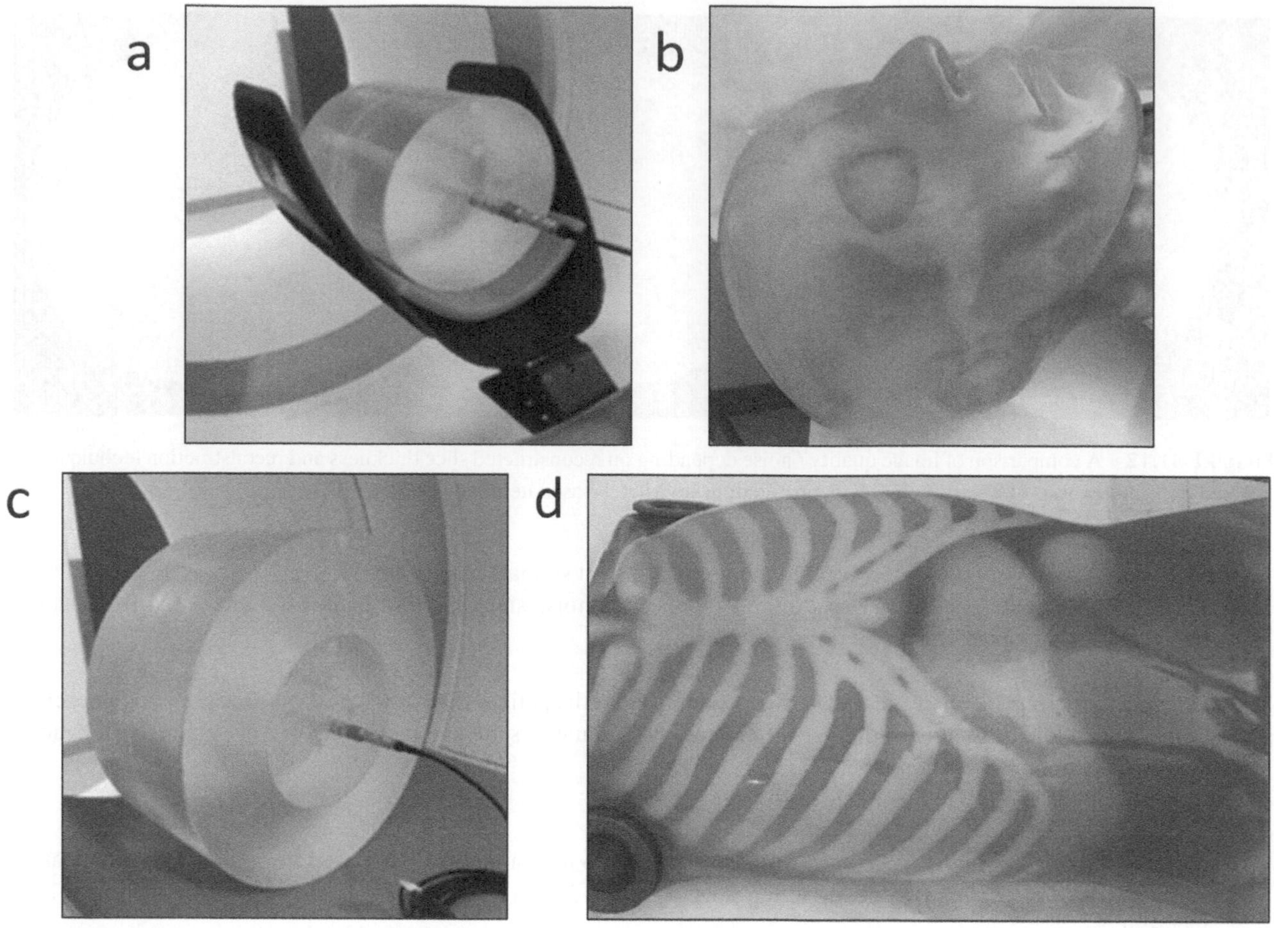

FIGURE 31.13 (a) 16 cm phantom to measure CTDI for the head, exemplified by (b) and (c) a 32 cm phantom to measure CTDI for the body (d).

phantoms with a 10 cm long ion chamber or a semiconductor detector. The unit is mGy. Depending on whether it is the head or other body part that is to be examined, two different phantoms are used. A smaller phantom with a diameter of 16 cm represents the head, and a larger phantom with a diameter of 32 cm represents the body (see Figure 31.13).

CTDI volume ($CTDI_{vol}$) is displayed on the console during a study and represents the average absorbed dose in the area under study. The unit is mGy. $CTDI_{vol}$ does not represent radiation dose to a patient but to the 16 cm or 32 cm diameter cylinder phantom. $CTDI_{vol}$ is a measure of output and is calculated based on scanner model and parameter settings. $CTDI_{vol}$ can be used to compare doses from different protocols and scanners. According to the prevailing IEC standard the stated value of the console shall correspond to the measured value with +/- 20 per cent accuracy.

31.5.2 Dose Length Product

The dose to the patient/phantom is proportional to the total scan length. The quantity dose-length product (DLP) has been introduced (see Figure 31.14). DLP is calculated as the product of $CTDI_{vol}$ and the length of the scan. The unit is mGycm.

$$DLP = CTDI_{vol} \cdot L \tag{31.5}$$

In Table 31.1 typical values for $CTDI_{vol}$ and DLP are shown. Note that $CTDI_{vol}$ is specified for the 16 cm CTDI phantom at CT brain. Other investigations refer to the 32 cm CTDI phantom.

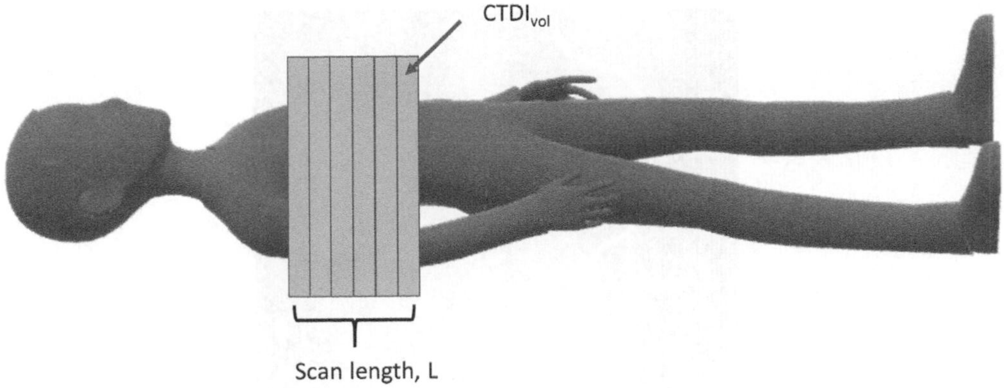

FIGURE 31.14 Illustration of the concept of dose-length product (DLP).

TABLE 31.1
Typical Values for $CTDI_{vol}$ and DLP

Examination	$CTDI_{vol}$ (mGy)	DLP (mGy cm)
CT brain	45	900
CT neck	10	300
CT thorax	5	250
CT abdomen	9	500

TABLE 31.2
Conversion Factors (Adult) for Different Anatomical Regions.

Anatomical region (adult)	CTDI-phantom (Φ cm)	k (mSv/mGy*cm)
Head	16	0.0024
Neck	32	0.0107
Thorax	32	0.0204
Abdomen	32	0.0163
Pelvis	32	0.0143

Note: The Data Have Been Taken from [1]

To make a rough estimate of the effective dose, the DLP can be multiplied by a conversion factor depending on the anatomical area examined. The factors are developed by Monte Carlo simulation, taking into account that different organs are differently radiation sensitive. Conversion factors for five different anatomical regions of the body for an adult are listed in Table 31.2.

In hybrid studies such as PET / CT and SPECT / CT, information from the CT images is used for attenuation correction for the PET and SPECT collection, respectively. For this purpose, a lower-dose CT scan can be used, a so-called attenuation correction CT scan. Corresponding examination of full diagnostic quality results in about a six-times higher radiation dose. Therefore, it is of great importance for the radiation dose to assess whether it is justified to carry out a full diagnostic quality CT examination or if a low dose of CT is sufficient.

31.5.3 Possibilities for Reducing the Dose

Automatic exposure control (AEC), or tube current modulation, is currently offered by all CT manufacturers and is used routinely. Based on the overview image, information about the patient's size is obtained. Some systems also use information from the attenuation profiles that are collected during the CT scan itself. The tube current is adjusted relative to

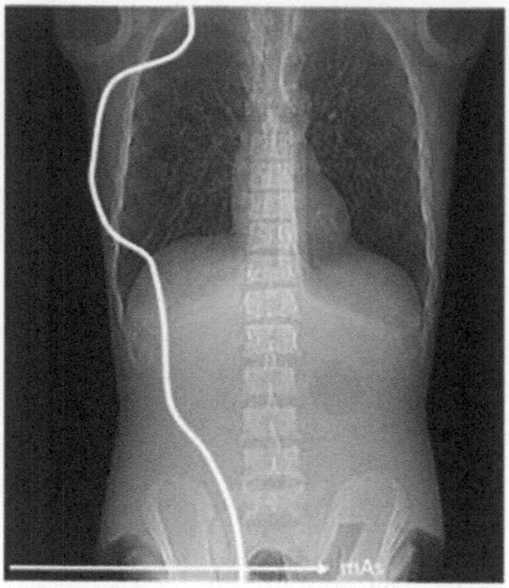

FIGURE 31.15 Illustration of how the tube current varies during a scan using AEC.

attenuation in the patient and patient size to achieve the desired image quality. The advantages are that the image quality is more consistent through the whole scan as well as between patients of different sizes. In addition, fewer artefacts are introduced. An example of how the AEC affect the tube current during a scan of an anthropomorphic phantom is shown in Figure 31.15.

Today, all CT manufacturers have AEC systems that adjust the tube current in three dimensions. They use different nomenclature, and their systems differ in how you choose to specify the desired image quality. One method is to specify a desired noise level, and the system adjusts the tube current to achieve the desired image quality. An alternative method is to specify a reference mAs for a reference patient. If the patient being examined is smaller than the reference patient, a lower tube current will be adjusted and vice versa when examining a larger patient.

Since the system uses information from the overview image to determine the size of the patient, it is important that the patient is well centered in height. If the patient is positioned close to the X-ray tube, an enlargement will be obtained. This means that the patient is assumed to be larger, and the exposure automatics will adjust a higher tube current, resulting in an increased radiation dose. If the patient is positioned further away from the detector, the patient will be assumed to be smaller, and the AEC will adjust to a lower tube current. This may result in insufficient image quality. There are systems that can correct for this effect. However, one should always be careful with patient centring to avoid artefacts and inhomogeneous dose distribution due to the bow-tie filter.

Another form of organ-based tube current modulation is offered by some manufacturers, where it is possible to adapt the image quality and dose within a certain region, for example, liver, breast, and eye.

To achieve optimum image quality at the lowest radiation dose, tube voltage should also be adjusted. A further development of exposure automatics is to also involve the choice of tube voltage. Some manufacturers today offer systems that choose tube voltage according to the patient's size and then modulate the tube current to optimize contrast to noise ratio and radiation dose. Depending on the manufacturer, issues can also be taken into account, for example, whether the study is done with or without contrast agents.

31.6 ARTEFACTS

An artefact is a structure that appears in the image but is not included in the examined object. Important causes of artefacts are patient movement, beam hardening, scattered radiation, partial volume effects, metallic implants, and the patient exceeding the limits of the field of measurement.

Ring defects can arise due to noise: for example, low dose (low mAs) in combination with larger patients. This is seen as concentric rings in the center of the image. Ring defects can also be caused by detector problems or non-calibrated detectors. To avoid ring artefacts, the detector should be calibrated and the radiation dose adjusted to patient size.

The X-rays have a spectrum of different energies. When the radiation passes through an object, low-energy photons are absorbed more than high-energy photons, that is, the average energy of the radiation increases after passage, resulting in so-called beam hardening, 'radiation hardening'. The X-ray spectra look different at the edge compared to the center and without correction; CT values will be different. Without correction it can cause dark bands / streaks in the image. If the beam passes more dense material (bone, iodine contrast), this also affects the image. Beam hardening is corrected with bowtie filters and special algorithms for beam-hardening correction.

Bowtie filters are adapted to patient size and are often selected automatically without the user being able to influence the choice. Therefore, it is important to use the correct examination protocol when the selection is controlled by certain scan parameters.

There are many causes of motion artefacts: for example, breathing, heartbeat, bowel movement. The best tools to reduce this are to fix the patient and to minimize the scan time. Partial volume effect occurs when tissues of different densities are included in the same voxel, an average value of tissue attenuation (HU value) is generated. In order to increase the possibility of distinguishing small structures and obtaining a better resolution, thinner section thickness is chosen.

Metal artefacts are due to the fact that virtually no photons pass through a compact metal object. Metal artefacts can be compensated by iterative reconstruction methods, known as MAR, metal artefact reduction. All CT manufacturers today have algorithms, and these can be added afterwards by doing an extra reconstruction.

BIBLIOGRAPHY

Comprehensive lists of suggested reading are provided in the textbooks [2–5]

[1] W. Huda, D. Magill, and W. He, "CT effective dose per dose length product using ICRP 103 weighting factors," (in English), *Med Phys,* vol. 38, no. 3, pp. 1261–1265, Mar 2011, doi:10.1118/1.3544350.

[2] International Atomic Energy Agency, *Diagnostic Radiology Physics*. Vienna: IAEA, 2014.

[3] W. Kalendar, *Computed Tomography: Fundamentals, System, Technology, Image Quality, Applications*, 2nd edn. Erlangen: Publicis Corporate, 2005.

[4] E. Seeram, *Computed Tomography: Physical Principles, Clinical Applications, and Quality Control*, 3rd edn. St. Louis: Saunders Elsevier, 2009.

[5] J. Hsieh, *Computed Tomography: Principles, Design, Artifacts, and Recent Advances*, 2nd edn. Bellingham, WA: SPIE Press, 2009.

32 Principles behind Magnetic Resonance Imaging (MRI)

Ronnie Wirestam

CONTENTS

DOI: 10.1201/9780429489556-32

32.1 INTRODUCTION AND HISTORICAL BACKGROUND

Magnetic resonance imaging (MRI) is a well-established medical imaging technique, traditionally associated with excellent soft-tissue contrast properties. Current clinical MRI systems provide not only morphological information throughout the body, but also a number of advanced techniques related to tissue and organ function, physiology, and microstructure.

MRI is fundamentally based on the phenomenon of nuclear magnetic resonance (NMR), demonstrated in nuclear beams by Rabi in 1938 and in samples of condensed matter (water and paraffin) by Bloch and Purcell in 1945–1946. The concept of NMR and the discovery of the chemical shift soon found important applications in spectroscopy in the 1950s and 1960s, primarily for determining molecular identity and structure.

The concept of using NMR for imaging, that is, spatially localized NMR signal information in a two-dimensional plane, was announced by Lauterbur in 1973, followed by important supplemental descriptions of slice selection by Mansfield in 1974. The first whole-body MRI images were presented by Damadian in 1977, and the original reconstruction of brain MRI data was reported in 1978. The first commercial whole-body MRI system was introduced in 1980. Thereafter, the number of clinical MRI installations worldwide has been steadily increased.

32.2 BASIC PHYSICS OF MAGNETIC RESONANCE IMAGING

32.2.1 Nuclear Magnetic Resonance and Signal Generation

MRI is based on registration of a radiofrequency signal originating from atomic nuclei within the object, in most cases hydrogen. In a closer look at the term NMR, the word 'nuclear' implies that relevant properties are associated with the atomic nucleus, 'magnetic' refers to the fact that nuclear magnetic fields are central, and 'resonance' tells us that the ability to affect the nucleus is restricted to certain discrete energies or electromagnetic field frequencies. Atomic nuclei with an odd number of protons and/or an odd number of neutrons exhibit a non-zero *spin*. The spin is a non-classical nuclear property (an intrinsic angular momentum) that is often compared with classical physical properties associated with a nucleus rotated around its central axis, that is, spin angular momentum ('rotating mass') and magnetic dipole moment ('rotating electrical charge'). The nuclear spin quantum number is denoted I, and the magnitude of the spin angular momentum $\overline{S}$ is thus

$$\left|\overline{S}\right| = \hbar\sqrt{I\left(I+1\right)} \tag{32.1}$$

The component of spin angular momentum measured along direction *i* is given by

$$S_i = \hbar I_i. \tag{32.2}$$

where $I_i \in [-I, -(I-1), \ldots, I-1, I]$, that is, there are $2I+1$ different angular momentum states.

The nuclear magnetic moment $\bar{\mu}$, arising from the spin, can be expressed in terms of the spin angular momentum and the so-called gyromagnetic ratio γ, that is,

$$\bar{\mu} = \gamma \bar{S}. \tag{32.3}$$

The vast majority of all MRI investigations are based on hydrogen nuclei, for which $I=1/2$ and $\gamma/(2\pi) = 42.577$ MHz/T.

In NMR, the quantum number associated with the spin component along the z-axis (in the direction of an applied external magnetic field) is often referred to as the magnetic quantum number m_I, where $m_I \in [-I, -(I-1), \ldots, I-1, I]$. Hence, $S_z = \hbar m_I$ and $\mu_z = \gamma \hbar m_I$, according to the general relationships given above (Eqs. 32.2–32.3). For ^{1}H, there are two linearly independent spin states corresponding to $m_I = \pm 1/2$, sometimes referred to as 'spin-up' and 'spin-down'. These two states are degenerate (i.e. show the same energy) in the absence of an external magnetic field, and this corresponds to an isotropic spin distribution (Figure 32.1a). However, in an external magnetic field, with magnetic flux density $\bar{B}_0$ a splitting into two (i.e., $2I+1$) different energy levels occurs, and this phenomenon is often referred to as the nuclear Zeeman effect. The potential energy E of the magnetic dipole is given by

$$E = -\bar{\mu} \cdot \bar{B}_0, \tag{32.4}$$

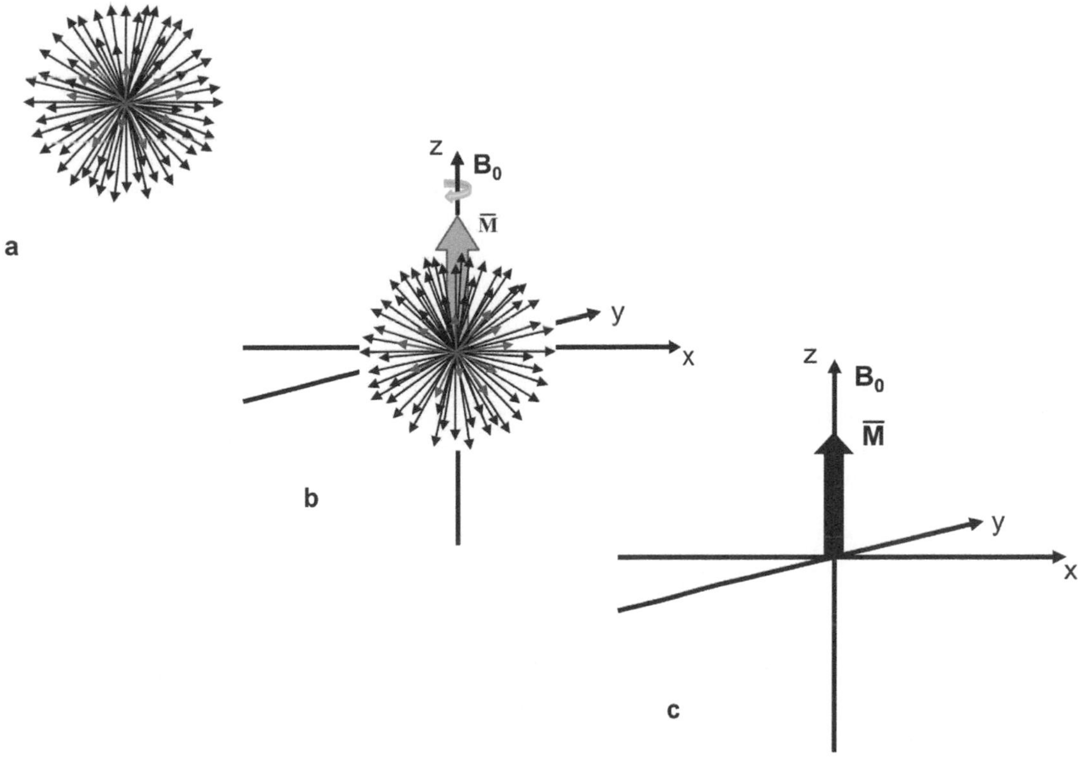

FIGURE 32.1 (a) Illustration of isotropic spin orientation in the absence of an external magnetic field. (b) The spin distribution is placed in a three-dimensional coordinate system. In the presence of an external magnetic field $\bar{B}_0$, the spin distribution is reorganized, and a net magnetization is formed. Note that precession of individual magnetic dipoles around the z-axis occurs, indicated by the curved arrow. (c) The net magnetization vector $\bar{M}$, that is, the vector sum of the magnetic dipoles of the reorganized spin distribution.

which reduces to $E = -\mu_z B_0 = -\gamma\hbar m_I B_0$, if $\bar{B}_0$ is along the z-axis. Since $m_I = \pm 1/2$, it is easily seen that the energy difference between the two states is

$$\Delta E = 2\mu_z B_0 = \gamma\hbar B_0. \tag{32.5}$$

Furthermore, an external magnetic field (denoted $\bar{B}$ for the general case) exerts a torque, which by definition equals $d\bar{S}/dt$, on the magnetic moment, that is:

$$\frac{d\bar{S}}{dt} = \bar{\mu}\times\bar{B} \Rightarrow \frac{d\bar{\mu}}{dt} = \bar{\mu}\times\gamma\bar{B} \tag{32.6}$$

The properties of the cross product imply that $d\bar{\mu} \perp \bar{\mu}$ and $d\bar{\mu} \perp \bar{B}$. This implies, in turn, that $|\bar{\mu}|$ must be constant, and that $\bar{\mu}$ precesses about $\bar{B}$. For a positively charged particle, like the hydrogen nucleus, the procession is directed clockwise. In the case when $\bar{B} = \bar{B}_0$, the precession occurs about the z-axis (Figure 32.1b), but the cross product can describe the change in magnetization also in the presence of additional external magnetic fields, as will be described in more detail later.

The angular frequency of the precession is denoted ω and is given by the so-called Larmor equation:

$$\omega = -\gamma B \tag{32.7}$$

The minus sign indicates negative (i.e. clockwise) precession direction about $\bar{B}$ (cf. Figure 32.1b). In the context of NMR and MRI, the minus sign is often omitted for simplicity, and in the presence of the main magnetic field B_0, the Larmor equation is expressed as

$$\omega_0 = \gamma B_0 \Leftrightarrow \upsilon_0 = \frac{\gamma}{2\pi}B_0, \tag{32.8}$$

where v_0 is the precession frequency (sometimes referred to as the Larmor frequency). Note that this is also the frequency of the electromagnetic field that can induce a transition between the two states with energy separation ΔE (corresponding to ΔI_i=1). When this transition is induced (and electromagnetic energy is absorbed), the electromagnetic field is said to be *in resonance* with the transition frequency. Hence, v_0 is also referred to as the resonance frequency and $\Delta E = hv_0$.

As indicated above, the two spin states of the hydrogen nucleus correspond to S_z being either parallel (m_I = +1/2) or antiparallel (m_I = -1/2) with $\bar{B}_0$, while the spin angular momentum in the plane perpendicular to the z-axis (denoted the xy-plane) is undetermined. The parallel direction corresponds to the lower energy state, and the antiparallel direction corresponds to the higher energy state. Due to the thermal energy of the system, not only the lower energy state is occupied, but the population ratio of the two spin/energy states is described by a Boltzmann distribution. Assume a population of $N = N_+ + N_-$ nuclei, where the indices '+' and '-' correspond to m_I = +1/2 and m_I = -1/2, respectively. At thermal equilibrium, the following ratio applies:

$$\frac{N_-}{N_+} = e^{-\frac{\Delta E}{kT}} = e^{-\frac{\gamma\hbar B_0}{kT}} = e^{-\frac{2\mu_p B_0}{kT}}, \tag{32.9}$$

where k is the Boltzmann constant and $\mu_p = 1.4104 \cdot 10^{-26}$ J/T, that is, μ_z for the proton. It is important to realize that a general state $|\Psi\rangle$ of an individual nucleus can be described by a so-called superposition state, that is, by a linear combination of the two spin/energy states (denoted '↑' and '↓'):

$$|\Psi\rangle = c_+|\uparrow\rangle + c_-|\downarrow\rangle, \tag{32.10}$$

where c_+ and c_- are complex numbers. Eq. 32.9 implies that there is a small excess of spins (~1–20 ppm) in the parallel direction at typical temperatures (T ~ 300 K) and magnetic flux densities (~0.1–3 T). Hence, when the external field $\bar{B}_0$ is introduced, the spin or magnetic dipole distribution reorganizes from being isotropic (Figure 32.1a) to exhibiting a net magnetization vector $\bar{M}$, parallel with $\bar{B}_0$ (Figure 32.1b–c). Note that all orientations are still possible, due to the superposition states, but the distribution is no longer isotropic (Figure 32.1b). The reorganization time, that is, the time it takes for $\bar{M}$ to establish after application of $\bar{B}_0$, is described by the time constant T1 (further discussed below). A Taylor expansion of $N_+ - N_-$ using Eq. 32.9, in combination with the fact that $N_+ \approx N_- \approx N/2$, gives an approximation of $M_0 = |\bar{M}|$:

$$M_0 \approx N \frac{\mu_p^2}{kT} B_0 \tag{32.11}$$

In this context, N is often replaced by the proton concentration (i.e. the number of protons per unit volume of tissue), and in such cases M_0 would represent the magnetic dipole density. The term proton density (PD) is closely related to the proton concentration, but PD usually refers to the concentration of MRI-visible protons in a tissue relative to that in the same volume of water at the same temperature (i.e. in percentage units).

The equilibrium magnetization M_0 is not time-varying (Figure 32.2a), and thus not detectable by Faraday induction. However, the equilibrium state can be temporarily disrupted by exposing the system to an electromagnetic field of resonance frequency, that is, a so-called radiofrequency (RF) pulse. A reorganization will gradually take place during the RF pulse, corresponding to a change in states $|\Psi\rangle$ of individual nuclei (i.e. c_+ and c_- become continuous functions of time), so that, collectively, the whole spin distribution is rotated away from the positive z-axis towards the xy-plane. Hence, the tip of $\bar{M}$ will exhibit a nutational motion pattern (precession about the z-axis in combination with rotation towards the xy-plane) (Figure 32.2b).

Effectively, it is the magnetic field of the RF pulse that causes the effect on $\bar{M}$, and this time-varying magnetic field can, equivalently, be described as the vector sum of two vectors of magnitude $|\bar{B}_1|$ rotating, with the resonance frequency, in opposite directions in the xy-plane (cf. linear polarization). These two vectors are sometimes referred to as $\bar{B}_1$ (clockwise rotation) and $\bar{B}_1'$ (counter-clockwise rotation). Often, a rotating coordinate system or rotating frame

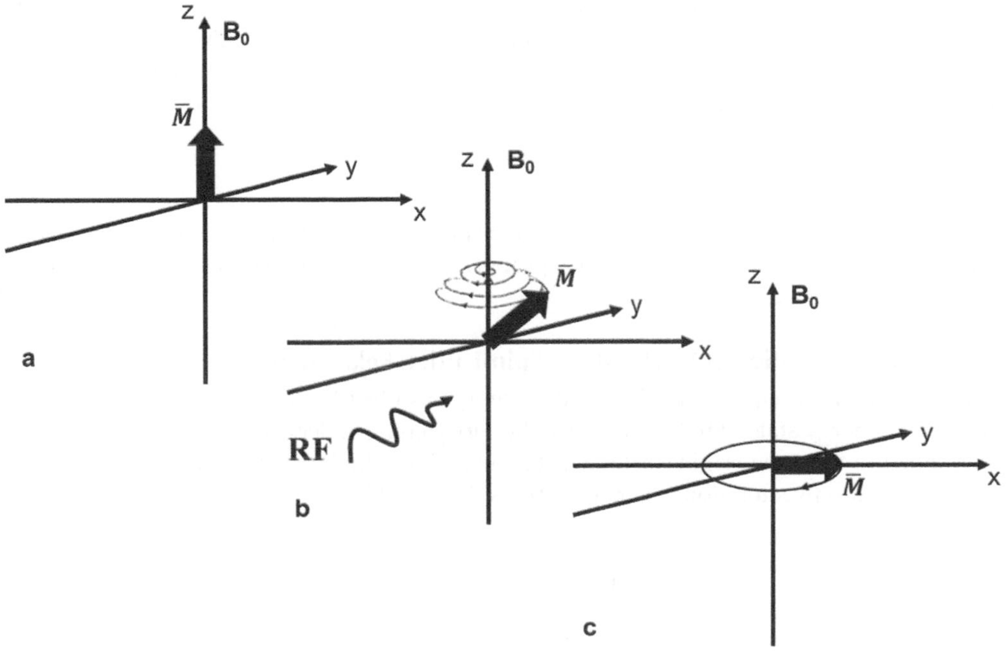

FIGURE 32.2 (a) The net magnetization vector $\bar{M}$ at thermal equilibrium. (b) Nutational motion of $\bar{M}$ caused by a radiofrequency field with resonance frequency. (c) Precession of $\bar{M}$ in the xy-plane, with resonance frequency, after a 90° RF pulse.

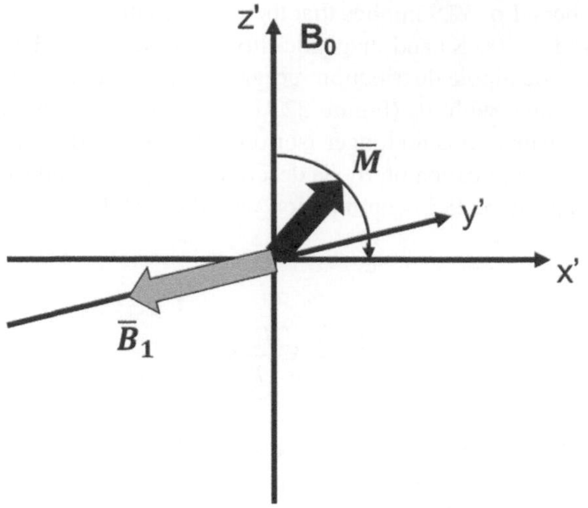

FIGURE 32.3 In the rotating frame of reference, the magnetization vector $\bar{M}$ precesses around $\bar{B}_1$ when $\bar{B}_1$ is on resonance (cf. Figure 32.2b in the laboratory frame of reference).

(with axes denoted x', y', z') is introduced, rotating with the Larmor frequency around the z-axis (i.e. the z' axis equals the z axis). In the rotating frame, $\bar{M}$ will exhibit a precession around $\bar{B}_1$ (Figure 32.3), according to Eq. 32.12:

$$\frac{d\overline{M}}{dt} = \bar{M} \times \gamma \bar{B}_1 \tag{32.12}$$

Eq. 32.12 is thus an alternative way of describing how $\bar{M}$ is rotated away from the z-axis towards the transverse plane by the RF pulse. The angle between $\bar{M}$ and the z-axis is called the flip angle, and an RF pulse that leads to a flip angle α is referred to as an α-pulse. As soon as $\bar{M}$ starts to deviate from the z-direction, a component of $\bar{M}$ occurs in the transverse plane, referred to as $\bar{M}_{xy}$, and in the laboratory frame of reference, this component can be detected by Faraday induction of an electromotive force in an RF receiver coil oriented to detect magnetization variations in the xy-plane. The generated signal, in its most basic form, is called free induction decay (FID), and it is normally observed as an alternating voltage, at the resonance frequency, with an amplitude that decreases with time over, typically, some tens of milliseconds. The maximal FID amplitude is obtained after a 90° pulse (Figure 32.4).

32.2.2 Relaxation

When the RF field at resonance frequency is terminated, the spin system gradually returns to thermal equilibrium, that is, towards the situation depicted in Figure 32.1b. This process is generally referred to as relaxation, and is divided into two main categories, as further described below.

32.2.2.1 Longitudinal Relaxation (T1 Relaxation, Spin-Lattice Relaxation)

One aspect of return to thermal equilibrium is that the energy absorbed by the system, leading to an increased spin population in the higher energy state, will be released to the surrounding molecular environment (i.e. the lattice) and the spins will be rearranged to form the original net magnetization of amplitude M_0 along the z-axis. After a 90° pulse, the regrowth of the magnetization vector along the z-axis, M_z, is described by:

$$M_z(t) = M_0\left(1 - e^{-\frac{t}{T1}}\right), \tag{32.13}$$

where T1 is the corresponding time constant, referred to as the longitudinal (spin-lattice) relaxation time (and R1=1/T1 is the longitudinal relaxation rate) (Figure 32.5a). The energy loss (and thereby the longitudinal relaxation rate) is greatly

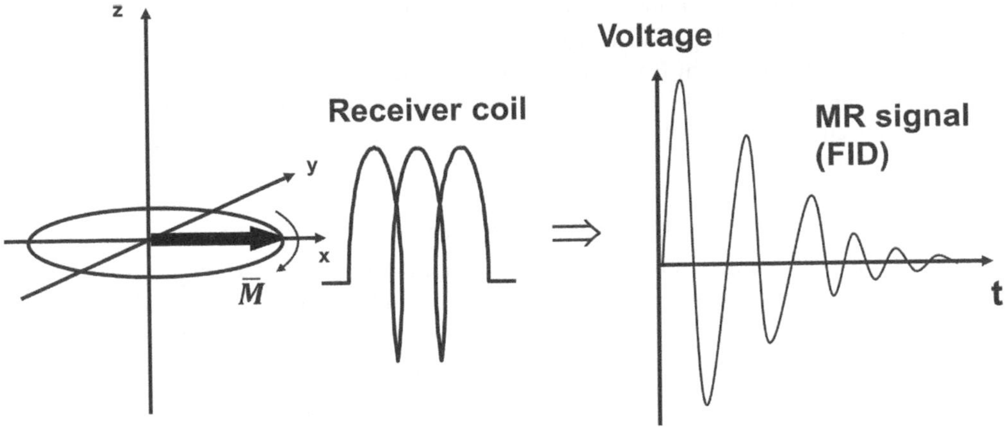

FIGURE 32.4 Simplified illustration of the signal detection in MRI, using a receiver coil in which an alternating voltage of resonance frequency is induced.

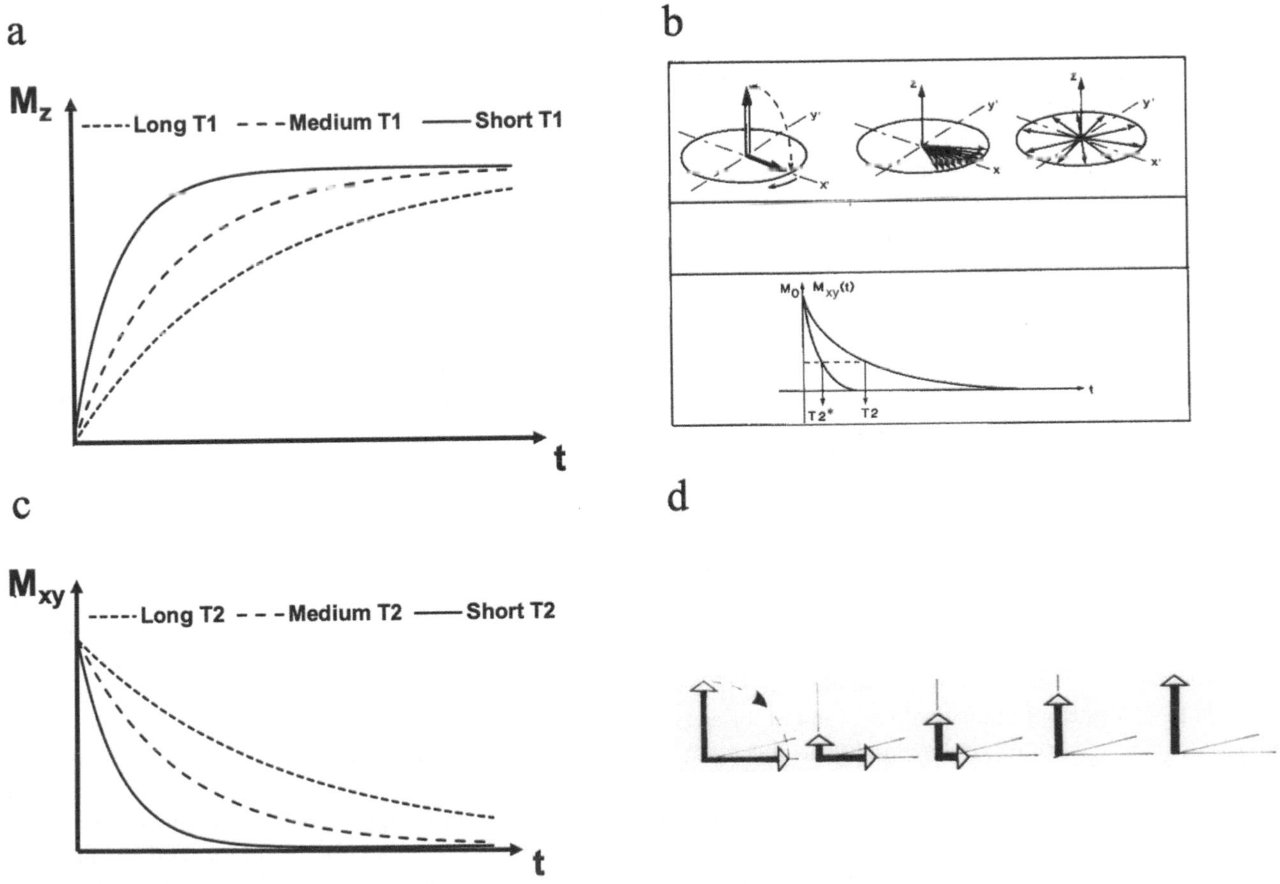

FIGURE 32.5 Illustration of relaxation processes. (a) Regrowth of M_z due to longitudinal relaxation in compartments with different T1. (b) Dephasing (phase dispersion) and corresponding decay of M_{xy} due to transverse relaxation (T2- and T2*-relaxation). (c) M_{xy} decay due to transverse relaxation in compartments with different T2. (d) Transverse and longitudinal relaxation processes occur simultaneously and are partly coupled. In biological tissue, T2 is shorter than T1.

enhanced by local magnetic field fluctuations within the surrounding environment of the spin, caused by molecular motion (rotation, vibration, translation), provided that the fluctuations occur at the resonance frequency. The local field fluctuations are more prominent at certain frequencies, and for typical tissue compartments the field fluctuations that correspond to higher frequencies tend to be less dominant. This implies that T1 is prolonged at higher B_0, because the corresponding resonance frequency (according to Eq. 32.8) will be less well-matched with the frequency spectrum of the local field fluctuations at higher field strengths.

32.2.2.2 Transverse Relaxation (T2 and T2* Relaxation, Spin-Spin Relaxation)

Return to thermal equilibrium also requires loss of transverse magnetization after the RF pulse has been turned off, that is, a so-called loss of phase coherence in the xy-plane. Short-lived spin interactions cause random local variations in the magnetic field (i.e. a spread in resonance frequencies), implying that precession frequencies vary among different spins (deviating from υ_0), and a phase dispersion or dephasing evolves over time in the x'y'-plane (Figure 32.5b). With the loss of phase coherence, the vector sum of the transverse magnetization vector will decrease, implying that (after a 90° pulse):

$$M_{xy}(t) = M_0 \, e^{-\frac{t}{T2}}, \tag{32.14}$$

where T2 is the corresponding time constant, referred to as the transverse (spin-spin) relaxation time (and R2=1/T2 is the transverse relaxation rate) (Figure 32.5c). In this context, the microscopic interactions between dipoles can be viewed upon as causing time-varying inhomogeneities in the static field, and only relatively slow field fluctuations (i.e. lower frequencies) lead to dephasing while higher-frequency fluctuations are averaged out during the measurement time (so-called motional averaging). Hence, only local field fluctuations of lower frequencies contribute to transverse relaxation, and there is consequently no dependence of T2 on the resonance frequency and, in turn, no pronounced B_0 dependence (at least not at low and medium field strengths).

An additional source of transverse phase dispersion is the time-invariant inhomogeneities of the static external B_0 field in the MRI unit, corresponding to an apparent relaxation time component sometimes denoted T2'. When this additional transverse phase dispersion is included, the transverse relaxation time is denoted T2*, that is, T2*<T2 (Figure 32.5b), and in terms of relaxation rates the relationship is R2*=R2+R2'. Note that the decay of a FID curve reflects T2* relaxation.

Transverse and longitudinal relaxation processes occur simultaneously and are partly coupled so that T2≤T1 (Figure 32.5d). Relaxation times can vary considerably among different compartments. For brain parenchyma T1 is, very roughly, of the order of a second, while T2 is approximately 0.1 seconds or less. For pure water, the T1 and T2 values are equal (around 3 seconds). Solid materials show very short T2 (of the order of milliseconds), resulting in signal void for solid compartments (e.g. bone) in conventional MR images.

32.2.3 Introduction to Conventional Pulse Sequences

In order to optimize signal and contrast in MR images, for a given diagnostic purpose, the combined effects of manipulating the magnetization by an RF pulse (Eq. 32.12) and subsequent relaxation processes (Eqs. 32.13–32.14) can be exploited. In its most basic form, a so-called MRI pulse sequence is a scheme of RF pulses and time delays, carefully designed to generate optimal signal and contrast for a given purpose. To introduce this concept, three fundamental MRI pulse sequences are briefly outlined below, at this point primarily described by the effects of RF pulses and relaxation processes.

In MRI, the execution of RF pulses and time delays are, in most sequence designs, repeated a number of times in order to obtain a two-dimensional signal data matrix for reconstruction of a two-dimensional image. This implies that a time delay between repetitions of the pulse sequence blocks, as outlined below, is also introduced, referred to as the repetition time (TR). Normally, the waiting time between repetitions, that is, between one signal registration and the subsequent excitation, is utilized for imaging of other nearby slices to maximize organ coverage, an approach referred to as multi-slice mode.

32.2.3.1 Spin Echo

The spin echo (SE) pulse sequence consists of an initial 90° RF pulse, without any signal detection immediately after it, followed by a period of delay (top row of Figure 32.6).

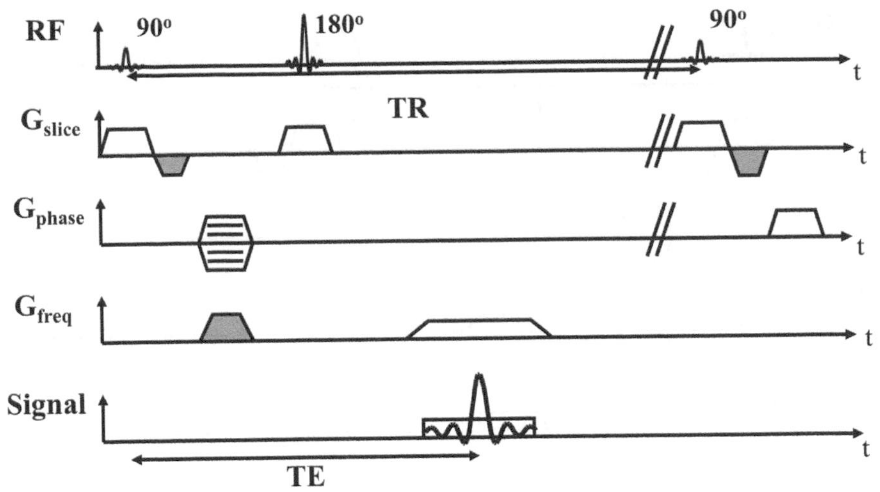

FIGURE 32.6 Complete pulse sequence scheme for a spin echo. Refocussing or balancing gradient lobes are indicated by grey filling.

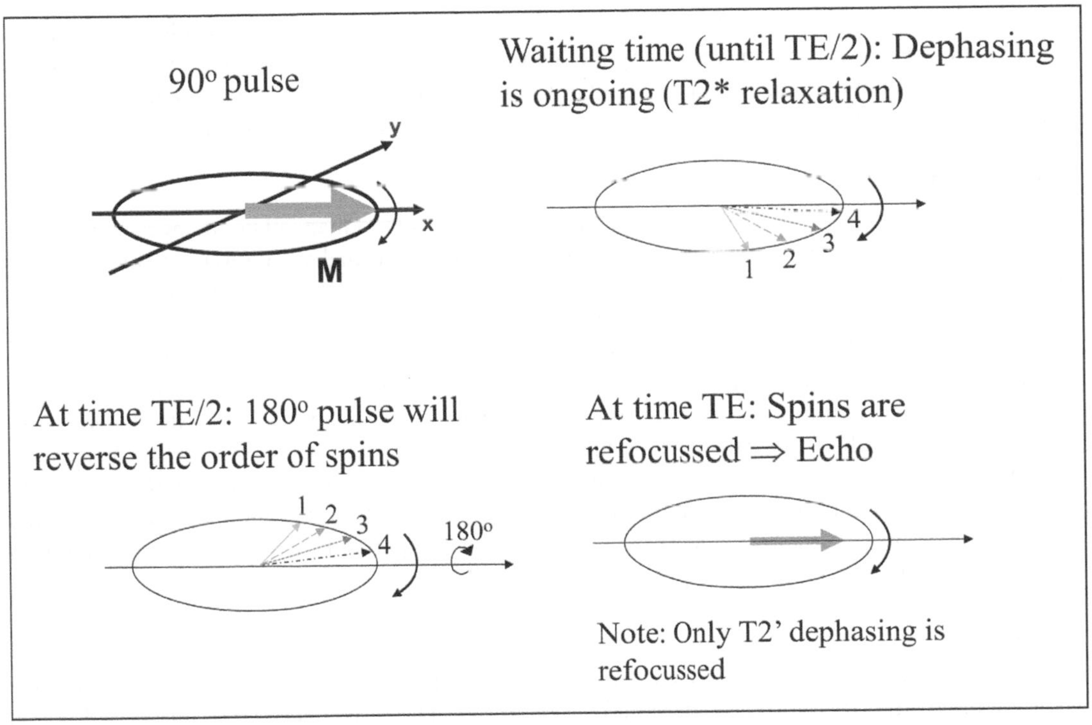

FIGURE 32.7 Principle of the spin refocussing accomplished by the 180° RF pulse in the spin echo pulse sequence.

After this first delay time, at time TE/2, a 180° RF pulse is executed. This pulse corresponds to a B_1 field applied along one axis in the xy-plane, around which the partially dephased spins (due to T2* relaxation during time zero to TE/2) will precess an angle of 180°. The spin distribution is thereby flipped from, for example, quadrant 4 to quadrant 1 in the xy-plane, leading to a reversed order of spin precession in the xy-plane (Figure 32.7).

After the 180° pulse, a partial refocusing of the spins is initiated, because the precession frequencies of the individual spin, related to the time-invariant static field inhomogeneities, are the same as before the 180° RF pulse, that is, the last spin in the line, after the 180° pulse (the component denoted #1 in Figure 32.7), still has the highest precession frequency. Hence, at time TE, called the echo time, the dephasing effects caused by time-invariant field inhomogeneities (i.e., the effects of T2' relaxation) are refocused and a detectable M_{xy} component is formed (Figure 32.7), inducing a so-called spin echo in the receiver coil (bottom row of Figure 32.6). Note that the dephasing caused by time-varying local

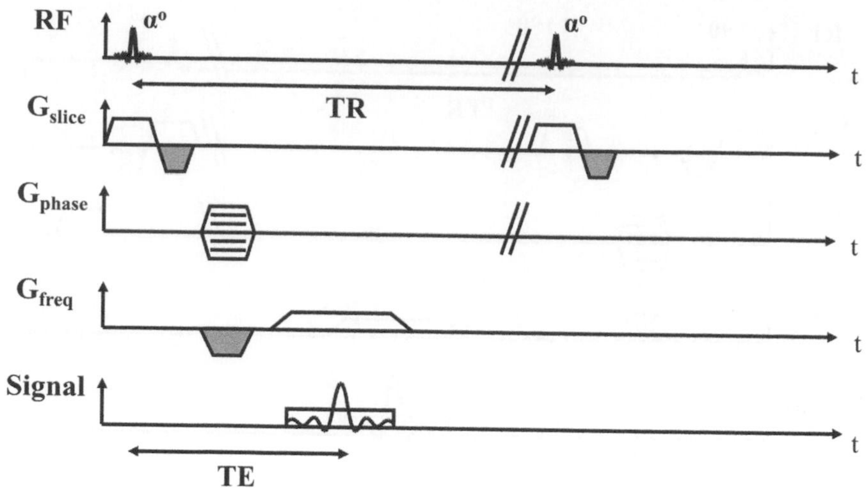

FIGURE 32.8 Complete pulse sequence scheme for a gradient echo. Refocussing or balancing gradient lobes are indicated by grey filling.

field fluctuations of lower frequencies, causing a slowly varying static field, cannot be fully refocused by the 180° pulse because the field experienced by a given spin during the interval [0, TE/2] is not exactly the same as during the interval [TE/2, TE]. As described above, the latter phase dispersion component is the one that corresponds to the true T2 relaxation. The amplitude of the spin-echo signal at time TE is thus governed by T2 relaxation (and not T2*).

32.2.3.2 Inversion Recovery

Another way of using an RF pulse in combination with relaxation effects to manipulate the measured signal is the so-called inversion recovery (IR) sequence. This type of pulse sequence is normally used to extinguish the signal from a given compartment. In this design, the pulse sequence is initiated by a 180° RF pulse, transmitted a certain period of time before the 90° pulse (which is, as in the SE, applied in order to rotate the magnetization into the transverse plane). The preceding 180° pulse leads to an inversion of the spins, that is, $\bar{M}$ becomes directed opposite to the $\bar{B}_0$ direction (i.e., along the negative z-axis). The idea is to exploit the fact that T1 relaxation occurs during the time between the 180° pulse and the 90° pulse. By adapting this waiting time (the so-called inversion time, TI) to the T1 value of the tissue whose signal is to be extinguished, one can choose to apply the 90° pulse exactly at the point in time when the regrowing magnetization M_z passes through zero (i.e. exactly when $\bar{M}_z$ changes from being opposite to $\bar{B}_0$ to, once again, being parallel with $\bar{B}_0$). At that point in time, there is no magnetization available to rotate into the xy-plane by the 90° pulse, and no signal in the MR image will be obtained from the selected compartment. From the point of the 90° pulse and onwards, the IR pulse sequence design corresponds to a common SE.

32.2.3.3 Gradient Echo

In principle, the FID, as described above, is the result of a pulse sequence consisting of a single α-pulse, without any delay between the end of the RF pulse and the signal collection. When a time delay TE is introduced in combination with pulse-sequence components related to spatial encoding (as further described below), this design is referred to as a gradient-echo (or gradient-recalled-echo, GRE) sequence (at this point, see top and bottom rows of Figure 32.8). The signal is collected at the echo time TE, but the mechanisms of echo formation differ from those of the SE, as will be further explained below. No refocusing of spin dephasing related to time-invariant magnetic field inhomogeneities occurs in GRE (no 180° pulse is present), so the amplitude of the gradient-echo signal at time TE is thus governed by T2* relaxation during the time period from zero to TE.

32.2.4 The Bloch Equation and the Concept of MR Signal Equations

A convenient way of phenomenologically describing the time variation of the net magnetization $\overline{M}$, caused by the combined effect of RF pulses (i.e., B_1 fields) and relaxation processes, is offered by the Bloch equation:

$$\frac{d\bar{M}}{dt} = \bar{M} \times \gamma \bar{B} - \frac{M_x\hat{x} + M_y\hat{y}}{T2} - \frac{(M_z - M_0)\hat{z}}{T1}, \tag{32.15}$$

where $\hat{x}$, $\hat{y}$ and $\hat{z}$ are unit vectors in the x-, y- and z-directions, respectively. Note that relaxation processes will change the magnitude of $\bar{M}$, while the precession is associated with a constant magnitude. If T2<T1, as is typically the case in tissue compartments, the transverse component can decay to zero long before the magnetization has regrown to be M_0 in the z-direction. After spin inversion, as in the IR sequence, the magnetization can even be zero at some point, prior to its regrowth along the positive z-axis. Eq. 32.15 can be reformulated as a set of differential equations (not explicitly shown here), one for each direction, and these can be solved with appropriate limiting conditions. In the simplest form, assumptions of spatially uniform magnetic field and homogeneous objects are often made.

In the SE sequence, for example, precession is first introduced by the 90° pulse. After turning off this initial B_1 field, time for relaxation is first allowed until the time point TE/2. Thereafter, another B_1 field is applied (180°) to allow the spins in the transverse plane to precess around one axis in the xy-plane, and relaxation is then accounted for until the time TE. At TE, the length of M_{xy} decides the amplitude of the echo. During the subsequent remainder of the repetition time TR, only relaxation effects are present, and the most relevant process during this (typically longer) time period is the regrowth of M_z by T1 relaxation. Mathematically, the series of events, associated with a given pulse sequence, results in a so-called signal equation, assumed to represent the amplitude of the recorded signal (i.e. the length of M_{xy} at TE). For a SE, the temporal signal development is often assumed to have reached a fairly stable level (the so-called signal steady-state situation) at the second echo. The signal S of the SE sequence, at the second echo, is given by the following equation:

$$S_{SE} \propto \left(1 - 2e^{-\frac{TR - \frac{TE}{2}}{T1}} + e^{-\frac{TR}{T1}}\right) e^{-TE/T2} \approx \left(1 - e^{-\frac{TR}{T1}}\right) e^{-TE/T2}, \tag{32.16}$$

where the approximation to the right is valid if TE<<TR. Corresponding signal equations can be formulated for other pulse sequences, for example, IR and GRE.

32.2.5 Principles of Spatial Encoding

In most medical digital imaging modalities, information originating from different internal parts (volume elements, voxels) of the object is externally detected and recorded, and the basic challenge is to assign a correct internal position to each component of externally detected information. In MRI, the detected information is the alternating voltage induced in the receiver coil by the M_{xy} vector precessing at resonance frequency. Hence, the key element of spatial encoding in MRI is to introduce linear spatial magnetic field gradients $\bar{G}$ (in mT/m) over the object, to assign different resonance frequencies to different positions within the object. The three principal spatial dimensions need to be considered, and spatial gradients are applied separately in the three directions, as will be further described below. Generally, the gradients are, in most cases, adding to B_0 over one half of the object while detracting from B_0 over the other half, that is, the field change, caused by the gradient, is zero at the centre of the object in the relevant dimension. The most common approach is to first selectively excite spins within a given slice, by so-called slice selection along one direction, and thereafter record signal components, appropriately coded in each of the two remaining directions, from the selected slice.

32.2.5.1 Slice Selection

In order to excite the spins within a given slice, a slice-selecting gradient G_{slice} is applied in the direction perpendicular to the desired imaging plane. It is important to realize that the gradient is a linear change in the magnetic flux density, along the gradient direction, and the gradient direction can be chosen in any direction, independently of the direction of the B_0 field lines. According to the Larmor equation, the presence of G_{slice} will assign each slice position a unique resonance frequency, and the idea is to adjust the frequency of the RF pulse to the resonance frequency of the slice to be selected. Hence, G_{slice} is present, in time, during the same time period as the exciting RF pulse.

Assume, for example, that G_{slice} is applied along the Z-axis in the patient's coordinate system. A slice located at position Z will show a magnetic flux density of $B_0^* = B_0 + G_{slice} \cdot Z$ and a corresponding resonance frequency of

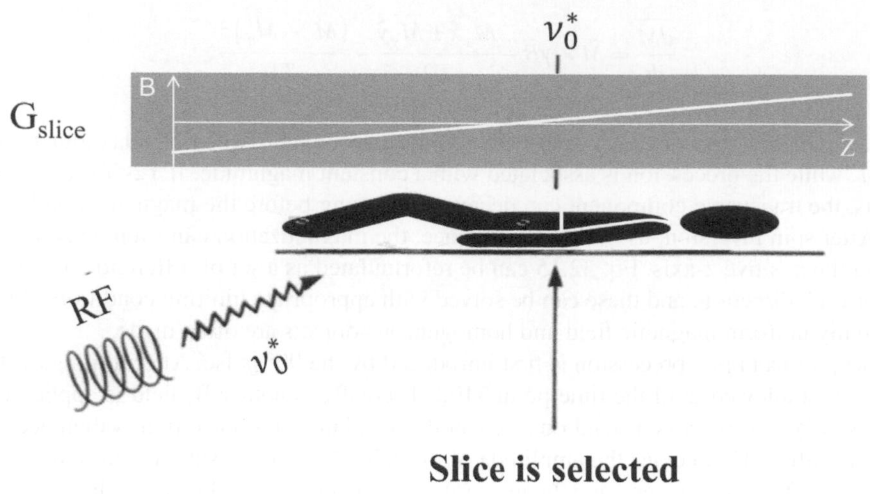

FIGURE 32.9 Slice-selection principle using a combination of a spatial magnetic field gradient and an RF pulse with a frequency adapted to the resonance frequency of the specific slice position.

$\nu_0^* = \frac{\gamma}{2\pi}\left[B_0 + G_{slice} \cdot Z\right]$ (Figure 32.9). In practice, it is unfeasible both to apply an RF pulse with a completely well-defined frequency (infinite duration) and to use an infinitely thin slice (insufficient signal). Thus, an RF pulse with a certain bandwidth, centred around ν_0^*, is applied. Additionally, at a given bandwidth, a higher gradient amplitude (steeper gradient) will result in a thinner slice, and vice versa. Not only the slice thickness ΔZ is important, but also the slice profile, and a rectangular slice excitation profile, that is, a box-car frequency profile of the RF pulse, is normally desired. In the so-called low-flip-angle approximation, the slice profile in the frequency domain and the $B_1(t)$ function in the time domain constitute a Fourier pair. This implies that the B_1 pulse, with carrier frequency ν_0^*, must be amplitude-modulated with a sinc-function in order to generate a rectangular excitation profile centred at ν_0^*. In other words, M_{xy} is supposed to show a non-zero value inside the box-car function of the slice profile, and a zero value outside it.

32.2.5.2 In-plane Spatial Encoding: The Frequency-encoding Principle

After slice selection, G_{slice} is turned off and the spins within the slice can return to a common precession rate in the transverse plane (disregarding, for the moment, differences related to the T2/T2* dephasing). The key feature of frequency encoding is to apply a magnetic field gradient G_{freq} in one of the directions perpendicular to G_{slice}, and to activate it just during the signal collection, that is, during the echo. One often talks about the 'readout' of the signal, and this gradient is sometimes referred to as the 'readout gradient'. In the presence of this gradient, the different columns (or rows, depending on the chosen gradient direction) of the selected slice will be assigned different resonance frequencies according to the Larmor equation, and column frequencies will vary continuously from one side to the other in the slice (from left to right).

Spins belonging to a given column will then generate an alternating voltage in the receiver coil, with a frequency that is specific to that particular column. As the frequency is decided by the position X, according to $\nu_0(X) = \frac{\gamma}{2\pi}\left[B_0 + G_{freq} \cdot X\right]$, the observed amplitude of the alternating voltage, with this frequency, in the receiver coil will thus reflect the amount of signal-generating spins (weighted by relaxation effects) from this particular column. Hence, column position is directly identified by signal frequency, but the problem is that the receiver coil detects signal components of all frequencies, from all columns, at the same time, and only the total sum of components of different frequencies is observed. In short, the required decoding of the recorded signal components is accomplished by a frequency-analysis technique, typically achieved mathematically by the Fourier transform. After Fourier transform, a frequency spectrum of registered signal components is obtained, and the amplitude seen at a given frequency is the amount of signal belonging to the corresponding column (Figure 32.10). This decoding procedure is, basically, what constitutes the image reconstruction method in MRI, and it will be described more comprehensively below.

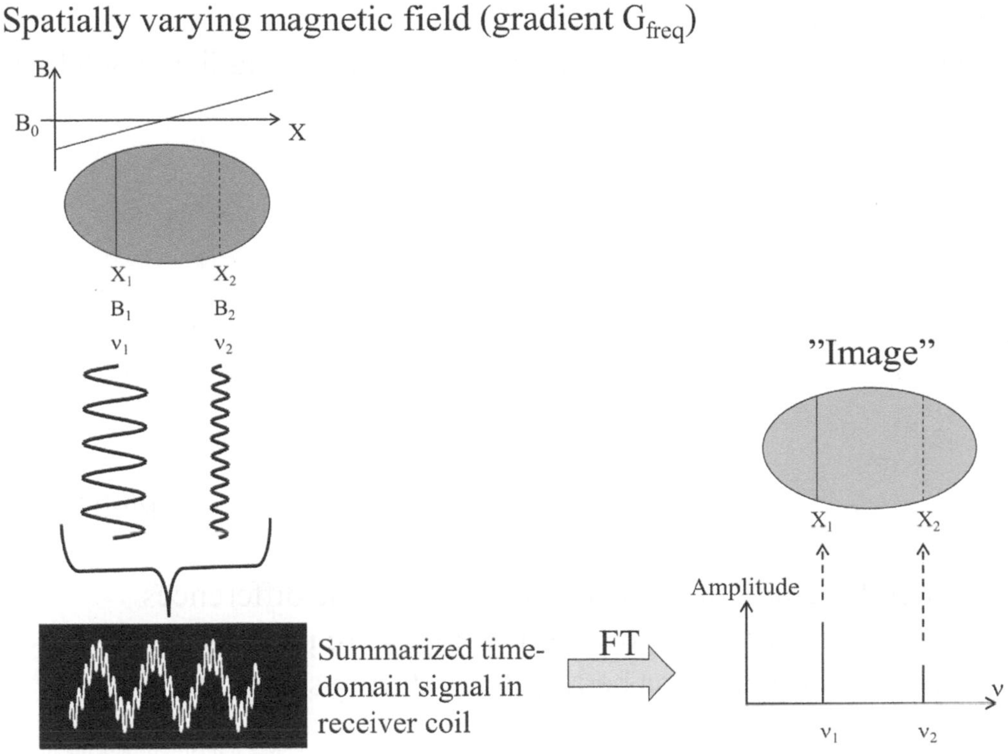

FIGURE 32.10 Frequency-encoding principle using a combination of a spatial magnetic field gradient and frequency analysis of the registered time-domain signal.

32.2.5.3 In-plane Spatial Encoding: The Phase-encoding Principle

In the phase-encoding part of spatial encoding, a magnetic field gradient G_{phase} in the direction perpendicular to both G_{slice} and G_{freq} is applied. The problem now is that decoding by frequency is already (obviously) allocated by the frequency encoding. Thus, in this dimension, the different rows (or columns, depending on the chosen direction of G_{freq}) of the selected slice will be assigned different resonance frequencies during a short time between excitation and signal readout, leading to a controlled phase shift ΔΦ in the xy-plane that will vary linearly with position along the Y-axis, according to $\Delta\Phi(Y) = \gamma\left[B_0 + G_{freq} \cdot X\right] \cdot T_{phase}$, where T_{phase} is the duration of G_{phase}. When G_{phase} is turned off, prior to signal readout, the spins will return to a common precession frequency, but the accumulated difference in phase shift between different rows will prevail during the remaining time until the echo, and can, in principle, be detected by the receiver coil (Figure 32.11).

Hence, the phase shift of each row will vary continuously from bottom to top in the selected slice. In this case, the decoding becomes more problematic, because it is less obvious how to identify different frequencies (from the frequency-encoded dimension) and different phase shifts at the same time, from the total sum of components of different frequencies and phase shifts induced in the receiver coil. In brief, the key feature of the phase decoding is to repeat the signal collection a number of times (e.g. 128, 256 or 512), referred to as phase-encoding steps, and to use different amplitudes of G_{phase} each time. If the G_{phase} amplitude increases between repetitions, in equidistant steps, the phase shift for each row will obviously increase one step for each repetition. For a given row, the difference in phase shift *between* subsequent repetitions will be constant and, most importantly, unique for this row. In the final image reconstruction of the two-dimensional dataset, it is the difference in phase shift (or, equivalently, the difference in experienced magnetic field change) *between* repetitions that identifies to which row a given signal component belongs. In fact, if one plots the signal amplitude as a function of phase-encoding step, at a given point in time of the echo, the data points will actually form a function that can be analysed by Fourier transform, independently of the frequency-encoded time curves described in the previous section. This concept is very simply illustrated in Figure 32.12, assuming that signal originates from a single point (i.e. one single frequency is seen in the time domain signal).

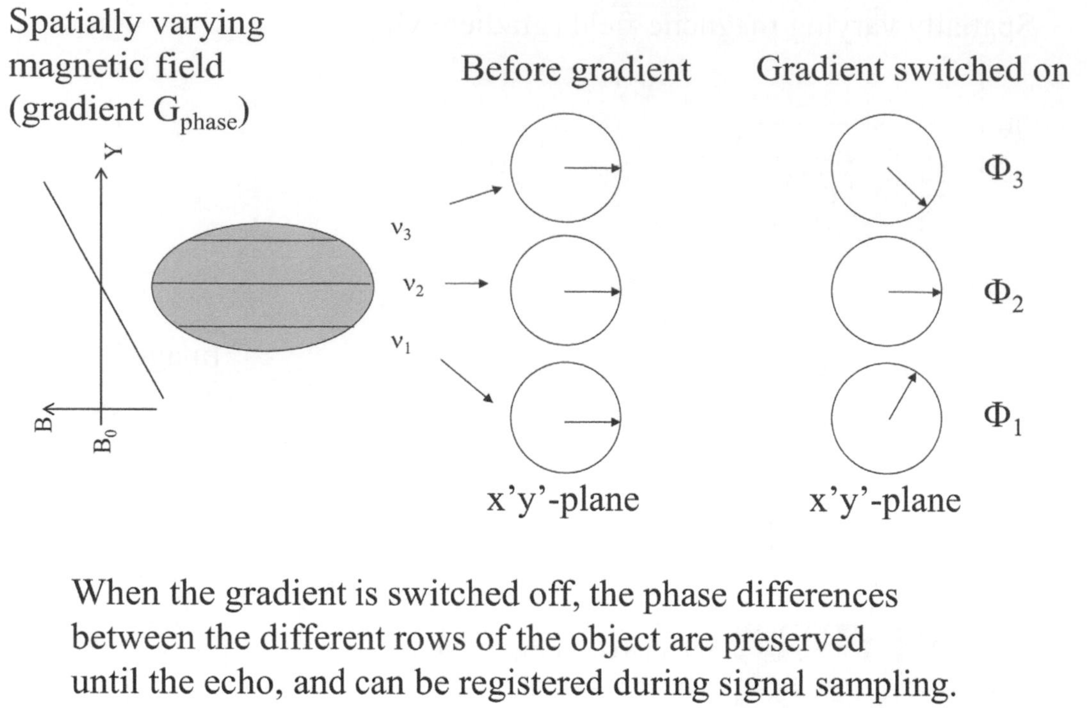

FIGURE 32.11 Phase-encoding principle using a spatial magnetic gradient that introduces a position-dependent signal phase shift prior to signal registration.

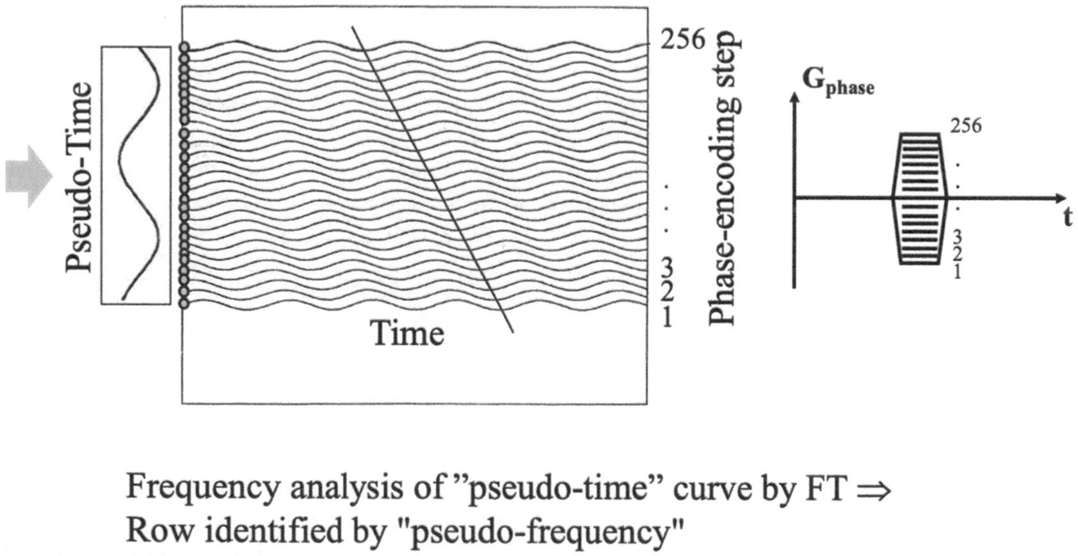

FIGURE 32.12 Simple illustration of the phase-encoding principle based on repeated measurements with different phase-encoding gradient amplitudes. Different phase-encoding steps correspond to different degrees of phase shift, for a given position (solid diagonal line indicates how curves are differently shifted at different steps). Plotting of registered data points along the phase-encoding direction reveals a 'pseudo-time curve', which can be analysed analogously to the frequency analysis of the time curve. The position of the row is identified by the so-called pseudo-frequency.

The phase-encoding steps can, effectively, be used as a pseudo-time dimension, and the resulting 'pseudo-frequency' domain, obtained by Fourier transform, can be treated in the same way as the frequency encoding, that is, a pseudo-frequency spectrum of registered signal components is obtained, and the amplitude seen at a given pseudo-frequency is the amount of signal belonging to the corresponding row. Hence, a two-dimensional Fourier transform is employed in the image reconstruction processes, that is, one Fourier transform analysing the time-domain data and one subsequent Fourier transform, applied to the result of the first Fourier transform, analysing data in the other dimension of the two-dimensional signal matrix, that is, the pseudo-time domain.

32.2.6 Conventional Pulse Sequences Revisited: The Complete Scheme

It is now clear that a pulse sequence for imaging purposes must contain gradients for spatial encoding, and these are turned on and off in a well-defined manner. Hence, it is reasonable to call them gradient pulses, in analogy with RF pulses, and complete schemes of the time courses of SE and GRE sequences are described in this section. In terms of gradients, the IR sequence does not differ from the SE, and the IR is therefore not further discussed here.

32.2.6.1 Refocusing of Spins – Balancing of the Pulse Sequence

One additional component needs to be described before the full design of the pulse sequence scheme can be understood. Gradients introduce, in practice, a static magnetic field inhomogeneity over a voxel of finite dimensions (of the order of millimetres). Although the gradient-induced field inhomogeneities are controlled, well-defined, and introduced with a purpose, they will still, in analogy with T2' relaxation (also caused by static magnetic field inhomogeneities), result in pronounced phase dispersion and associated signal loss.

Fortunately, the fact that the source of this gradient-induced field inhomogeneity is limited in time as well as spatially controlled means that the dephasing can be compensated for during another period of the pulse sequence (after the RF pulse but before the echo) by using a gradient in the opposite direction (i.e. with opposite polarity), with the same time integral as the one used for spatial encoding. In general terms, such gradients are referred to as refocusing gradients (or, less commonly, balancing gradients). The basic criteria for refocusing of spins or balancing of a sequence are:

$$\int_0^{TE} G(t)\,dt = 0, \tag{32.17}$$

or, for a SE-type sequence with a 180° refocusing pulse

$$\int_0^{TE/2} G(t)\,dt = \int_{TE/2}^{TE} G(t)\,dt \tag{32.18}$$

The latter version, for SE, is due to the fact that the reversal of the spin order in the xy-plane, introduced by the 180° pulse, is equivalent to changing the polarity of the gradient required for refocusing.

First, in the slice selection direction, the dephasing can be reversed by application of a gradient along the same dimension, but in the opposite direction (changed polarity), applied after termination of the RF pulse. However, in simplified terms, the magnetization is successively rotated towards the transverse plane, and all spin components are thus not precessing in the xy-plane directly from the initiation of the RF pulse. One can show that the desired refocusing is obtained if the refocusing gradient is assigned a time integral that is half of the time integral of the slice selection gradient. This implies that the time zero (Eqs. 32.17–32.18) is normally assigned to the time point that corresponds to the centre of the RF pulse (cf. Figure 32.6 and 32.8).

Second, for the frequency encoding direction, the situation is a bit different. Since the readout gradient is, by definition, applied during the signal collection, it would be 'too late' to apply a refocusing gradient lobe afterwards. However, the balancing criteria can instead be fulfilled by applying a dephasing gradient prior to signal readout, leading to maximal refocusing at the centre of the echo. Conventionally, the time TE is thus assigned to the time point corresponding to the centre of the echo (cf. Figure 32.6 and 32.8). Note that spins are maximally refocused only exactly at the centre of the echo, according to Eqs. 32.17–32.18, while the readout gradient is active, for a while, both before and after the time TE.

Third, the phase-encoding gradient is applied with the purpose of intentionally introducing phase differences between rows, that is, to cause a phase dispersion over the object. Hence, if this gradient were to be balanced, the phase encoding would be lost. A certain degree of intravoxel phase dispersion caused by the phase-encoding gradient is therefore a price to be paid for spatial encoding.

32.2.6.2 The Spin-echo Pulse Sequence Scheme

An overview of the time courses of the required components of a conventional spin-echo sequence (i.e. RF, G_{slice}, G_{freq}, G_{phase}) as well as the recorded signal (the echo) is given in Figure 32.6. The gradients for balancing the pulse sequence are illustrated by solid grey gradient lobes. Note that the balancing conditions, according to Eq. 32.18, can be fulfilled in several different ways, as long as the balancing gradients do not interfere with the RF pulses. For example, the balancing gradient (the defocusing gradient lobe) in the frequency-encoding direction could also, in principle, be applied immediately prior to the readout, in such a case with negative amplitude.

32.2.6.3 The Gradient-echo Pulse Sequence Scheme

The scheme of the gradient-echo sequence is displayed in Figure 32.8. In this case, it is the refocusing of the preceding gradient-induced dephasing that forms the echo (i.e., without the gradients, it would, in principle, have been an FID after the RF pulse). This is why the name gradient-induced echo is illustrative and appropriate.

32.2.7 Signal Reception and Image Reconstruction

32.2.7.1 Signal Receiver System and Mathematical Signal Representation

As pointed out above, it is convenient to view the motion of the net magnetization vector, and thus the received signal, in a frame of reference which rotates around the z-axis, with the Larmor frequency. The transition to the rotating frame of reference is, in practice, accomplished by a mathematical treatment of the signal induced in the receiver coil, often referred to as demodulation. The most common approach can, somewhat simplified, be illustrated by the following outline. In the laboratory frame of reference, a simplified signal (or signal component) is given by $A(t)\cdot\sin(\omega_1 t)$, where $A(t)$ is the amplitude and ω_1 is the angular frequency of a signal component from the selected slice. Hence, ω_1 is assumed to be close to ω_0, within the frequency interval that corresponds to the image plane, after gradient encoding, and any difference between ω_1 and ω_0 is thus caused by the applied in-plane gradient. The so-called demodulation of the signal (i.e., the transition to the rotating frame of reference) is achieved by a multiplication by $\cos(\omega_0 t)$ and using the trigonometric relationship $A(t)\cdot\sin(\omega_1 t)\cdot\cos(\omega_0 t)=\frac{A(t)}{2}\left[\sin(\omega_1-\omega_0)t+\sin(\omega_1+\omega_0)t\right]$. The first term is close to zero, while the second term is close to $2\omega_0$. The second term represents an off-resonance, high-frequency, part that can be eliminated by low-pass filtering. The first term represents the frequency interval over the image, but it has now been transferred to be located around zero (instead of around ω_0), which, by definition, is what is expected from the transition to the rotating frame of reference (cf. Figure 32.13a, upper part). In modern MRI systems, the central frequency is usually not exactly zero, but around 125 kHz. The basic principle of the approach is, however, still valid.

In the rotating frame of reference, object frequencies will be assigned both positive and negative values (with zero at the centre), corresponding to precession of M_{xy} clockwise or counter-clockwise with absolute frequency $|\delta\omega|=|\omega_1-\omega_0|$ in the rotating frame of reference. This will cause problems for the decoding, because the Fourier transform will not be able to separate $+\delta\omega$ from $-\delta\omega$. For this purpose, the so-called quadrature detection is introduced, in which the signal is split into two so-called channels, and the signal of one channel is phase shifted 90° relative the other (Figure 32.13a). It is easily realized that this is equivalent to detection of the signal along both the x' and the y' axes (as conceptually illustrated in Figure 32.13b), albeit only one single physical receiver coil is, in fact, required due to the signal split and subsequent 90° phase shift. With quadrature detection, the relative phase of the two signal components ($+\delta\upsilon$ and $-\delta\upsilon$ in Figure 32.13b) will be different in the two channels, and this allows for differentiation of the two positions relative to the centre.

It is also, in this context, convenient to view the x'y'-plane as the mathematical complex (Argand) plane (Figure 32.13b). In such a representation, the resulting M_{xy} vector of a voxel is interpreted as a complex number, with the x'-axis representing the real part and the y'-axis representing the imaginary part. M_{xy} generates the signal $S(t)$ in the time domain, so

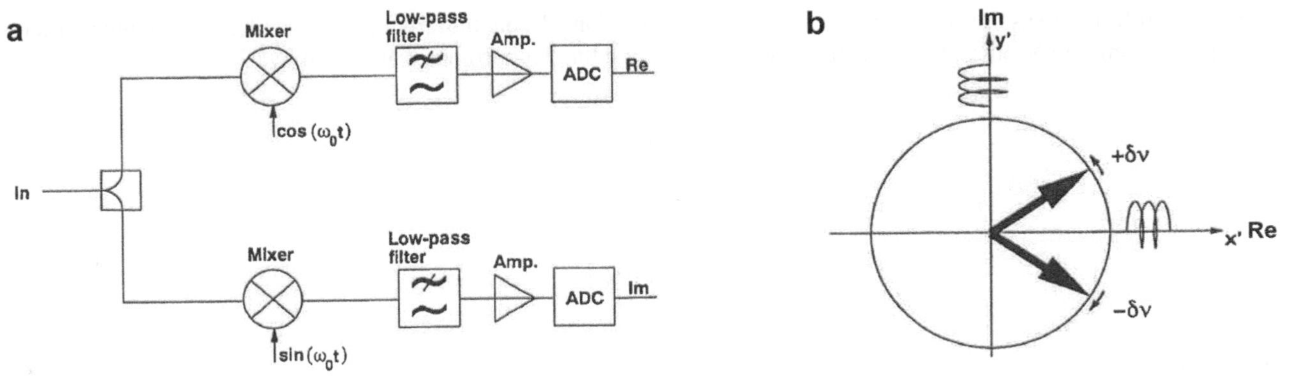

FIGURE 32.13 (a) Block diagram of quadrature demodulation, where the two channels are interpreted as the real (Re) and imaginary (Im) parts of a complex number. (b) Illustration of quadrature detection, which enables frequencies with the same absolute value, on either side of the central frequency, to be separated. (Note that two physical receiver coils are not required, due to the use of quadrature demodulation).

$$S(t) \propto M_{xy}\cos(\omega t) + iM_{xy}\sin(\omega t) = M_x + iM_y \tag{32.19}$$

Obviously, it must be considered that signal components are obtained from all positions $\overline{r}$ = (X, Y) within the selected slice, and that the total signal is the integral of all these components, that is,

$$S(t) \propto \int_{slice} (M_{xy}(\overline{r})\cos[\omega(\overline{r})t] + iM_{xy}(\overline{r})\sin[\omega(\overline{r})t])\overline{dr} \tag{32.20}$$

or, equivalently

$$S(t) \propto \int_{slice} M_{xy}(\overline{r})e^{i\omega(\overline{r})t}\,\overline{dr} \tag{32.21}$$

where $M_{xy} = \left|\overline{M_{xy}}\right|$ and $\omega t(\overline{r}) = -\gamma\int_0^t \overline{G}(\tau)\overline{r}d\tau$, according to the Larmor equation.

32.2.7.2 Image Reconstruction: Introduction of Spatial Frequency *k* and Use of the Fourier Transform Relationship

As pointed out above, the decoding of the recorded time-domain signal components is typically achieved by frequency analysis using the Fourier transform. This approach is further formalized below and displayed in mathematical form. Eq. 32.21 can be rewritten as follows:

$$S(t) \propto \int_{slice} M_{xy}(\overline{r})e^{-i\gamma\int_0^t \overline{G}(\tau)\overline{r}d\tau}\,\overline{dr} \tag{32.22}$$

At this point, a fundamentally important change of variables is introduced, stating that

$$\overline{k(t)} = \frac{\gamma}{2\pi}\int_0^t \overline{G}(\tau)d\tau \tag{32.23}$$

where $\bar{k}$ is referred to as the spatial frequency in units of m^{-1}. Considering that frequency- and phase-encoding gradients are applied along the X- and Y-axes of the slice, the corresponding k coordinates are given by:

$$k_X(t) = \frac{\gamma}{2\pi}\int_0^t G_{freq}(\tau)d\tau \quad (32.24a)$$

$$k_Y(t) = \frac{\gamma}{2\pi}\int_0^t G_{phase}(\tau)d\tau \quad (32.24b)$$

This implies that

$$S(t) \propto \iint_{XY} M_{xy}(X,Y)e^{-2\pi i[k_X(t)X + k_Y(t)Y]}dXdY \quad (32.25)$$

Comparing Eq. 32.25 with the Fourier transform of $M_{xy}(X,Y)$, it is immediately seen that $S(t)$ (or $S(k_x, k_y)$) and $M_{xy}(X,Y)$ constitute a Fourier pair. The MR image is, effectively, the pixel map of $M_{xy}(X,Y)$ (where M_{xy} is implicitly weighted by relaxation and other contrast mechanisms), and the MR image is thus reconstructed by the inverse 2D Fourier transform of the signal $S(t)$ in the time domain, detected by the receiver coil. Hence, Eq. 32.25 is the most important relationship in MR imaging.

32.2.7.3 Signal Matrix Organization and Introduction to *k*-space

In the pulse-sequence section as well as in the introduction of the Fourier transform relationship above, it was indicated that a two-dimensional matrix of signal data is built up (often referred to as the raw-data matrix) and, in conventional pulse sequences, this is accomplished by repetition of the excitation and the subsequent signal registration using different phase-encoding gradient amplitudes at each repetition (i.e. different so-called phase-encoding steps). Thus, a basic illustration of the signal matrix construction is as follows (Figure 32.14a):

1. Run the pulse sequence with a given G_{phase} amplitude and acquire echo 1. Take N_X samples from the first echo (by analogue-to-digital conversion, ADC).
2. Wait during the TR. Run the pulse sequence again, with a slightly different G_{phase} amplitude and record echo 2. Take N_X samples from the second echo.
3. Repeat the signal acquisition and sampling N_Y times, that is, in N_Y phase-encoding steps, with different values of G_{phase} each time. A $N_X \times N_Y$ data matrix is obtained (often $N_X = N_Y = N$, e.g., 128, 256, or 512).

Given the definitions of k_x and k_y (in Eq. 32.24), it is, firstly, easily realized that the different sampling points [1:N] on the echo correspond to increasing k_x values; G_{freq} is active (with positive value) during the echo and the accumulated area (the time integral) of $G_{freq}(t)$ is thus successively increasing during the sampling process (Figure 32.14c). During the first part of the echo, k_x is negative because G_{freq} is not yet balanced. At TE, k_x is zero (the sequence is balanced at that point) and after TE, when the echo decays, k_x is positive. Note that the k_x interval to be covered will be the same for each phase-encoding step. Secondly, different phase-encoding steps correspond, similarly, to different k_y values, and k_y increases from large negative to large positive values (or in the reverse order depending on how gradients are arranged) when the different pulse-sequence repetitions are executed (Figure 32.14c).

Hence, it is very convenient to view the matrix into which signal values are placed, as a mathematical space spanned by the k_x and k_y axes (Figure 32.14b–c). In general terms, the desired image properties, in terms of, for example, field of view and spatial resolution, require sufficient amounts of signal data to be collected, with appropriate sampling density and sampling window, and the signal sampling strategy is elegantly illustrated with the *k*-space formalism. The signal sampling track in *k*-space is often referred to as the trajectory. For a conventional GRE, SE or IR sequence, *k*-space is, as indicated above, covered 'line by line' over the different phase-encoding steps, and, in the standard case, a N×N signal data matrix in *k*-space (m^{-1}) results in a N×N image matrix (m) after inverse Fourier transform (Figure 32.15).

In 3D or volumetric imaging, a thick 'slab' is excited and combined with a stepwise phase encoding procedure also in the slice (i.e. in this case 'slab') direction, resulting in a 3D signal matrix reconstructed by a 3D Fourier transform.

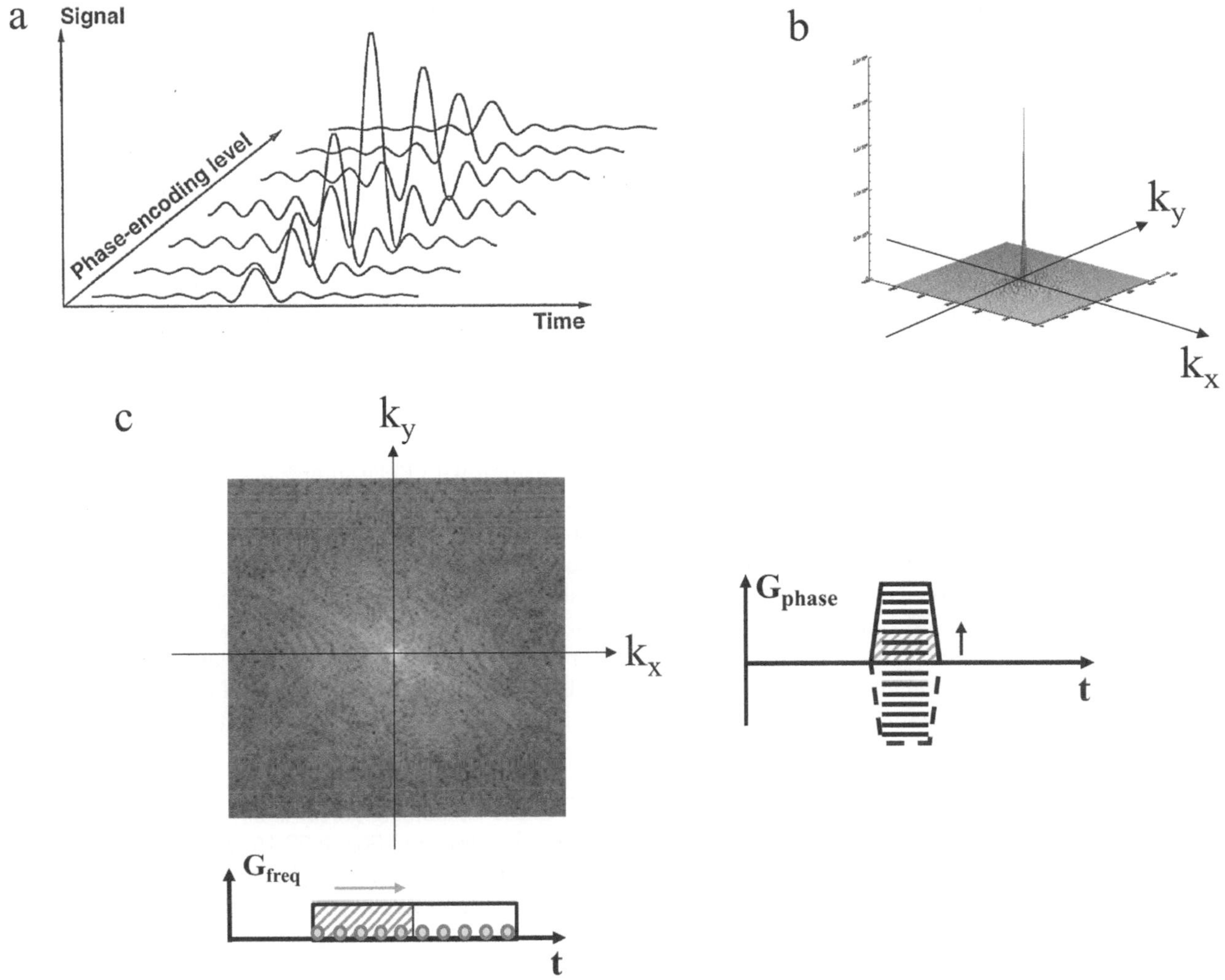

FIGURE 32.14 (a) Schematic illustration of the build-up of the signal matrix. (b) Actual MRI signal data arranged in *k*-space. (c) Actual MRI signal data, displayed as a raw-data map, arranged in *k*-space. The *k*-space coordinates k_x and k_y (reflected by accumulated gradient time integrals) are increased when signal is sampled during readout and when the phase-encoding gradient is successively stepped between repetitions, respectively. For clarity, the G_{phase} arrow and the corresponding accumulated time integral indicates the increase in positive k_y values (after completion of coverage along the negative k_y-axis).

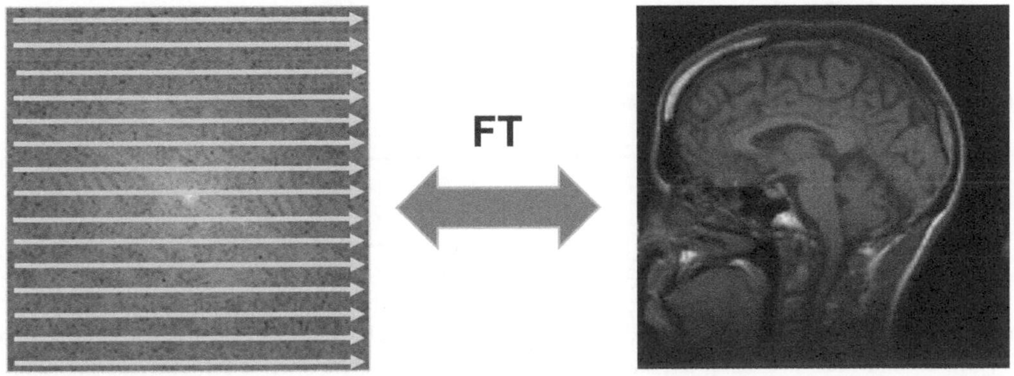

FIGURE 32.15 Fourier-transform relation between *k*-space (in m^{-1}) and image space (in m). The image is reconstructed by 2D inverse Fourier transform of the signal matrix in *k*-space. The arrows schematically illustrate that *k*-space is covered 'line by line' in conventional pulse sequences, where one line corresponds to one phase-encoding step.

Provided that the spatial resolution within the slab is more or less isotropic, arbitrary image planes can be reconstructed using the obtained 3D data set.

32.3 IMAGE ACQUISITION TECHNIQUES IN CLINICAL MRI

In clinical MRI, diagnostic accuracy as well as patient comfort and throughput are of the essence, and the complete pulse sequence signal acquisition outlined above tends to be overly time consuming, in most cases. Hence, a priority in MRI development has been to increase the imaging speed, and an alternative heading for this section might have been 'fast MRI sequences'. Contemporary MRI scanners offer a vast multitude of pulse sequences, signal readout options, and signal-sampling strategies, and it is far beyond the scope of this text to cover such a topic in detail. However, in this section, a few basic approaches to speed up the signal data collection and to shorten the image acquisition time will be briefly illustrated.

32.3.1 Fast Spin Echo

One of the major workhorses of clinical MRI is the Rapid Acquisition with Relaxation Enhancement (RARE) pulse sequence, often referred to as fast spin echo (FSE) or turbo spin echo (TSE). As the name implies, it is of the spin-echo type, but with a few modifications. First, it is a so-called multi-spin-echo, meaning that each 90° RF pulse is followed by two or more 180° pulses. Each 180° pulse will refocus the spins in the xy-plane, and several echoes can thus be obtained after each excitation. Note that the echo time, obviously, will be longer for later echoes and that T2 relaxation occurs during the time of multi-echo generation. Second, the element of time saving in FSE (known as segmentation) is that each echo in the multi-echo train is assigned an individual phase encoding, that is, an individual amplitude of G_{phase}. A third point that must be remembered is that the phase-encoding gradient needs to be balanced before acquisition of the next echo, that is, a rewinding phase-encoding gradient is played out after each echo, to restore the phase, before the next 180° pulse.

If, for example, five echoes are sampled (Figure 32.16), one excitation will generate signals at five lines in the *k*-space or raw-data matrix. In the next repetition (excitation), all phase-encoding gradient amplitudes are changed one step, and five new lines in *k*-space are covered. Hence, in the case illustrated in Figure 32.16, the total time of image acquisition is reduced by one fifth, and the so-called turbo factor or echo train length (ETL) is 5. All data corresponding to the same echo, from different excitations, constitute a partition and, as pointed out above, the different partitions will correspond to different TEs. It can be understood that the TE of the partition placed in the centre of *k*-space (often called

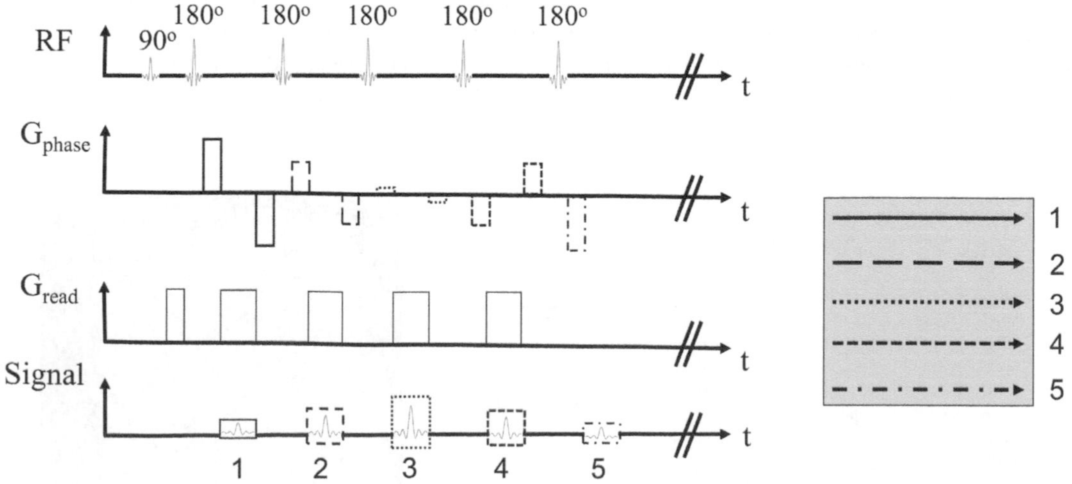

FIGURE 32.16 Left: Schematic pulse sequence diagram for a fast-spin echo with echo train length 5. Corresponding phase-encoding lobes, echoes and lines in *k*-space are indicated by the same line style. Slice selection works the same way as for a conventional spin echo and is therefore omitted. Right: Illustration of *k*-space coverage. The five differently styled arrows indicate lines in *k*-space that are acquired at one excitation. In the next excitation, all the phase-encoding lobes are changed one step, and adjacent *k*-space lines are sampled.

TE_{eff}) will, in principle, determine the image contrast. Where in *k*-space a given partition ends up is, obviously, decided by the actual k_y values (i.e. G_{phase} amplitudes) that are assigned to that particular echo. Generally, longer ETLs will result in more pronounced T2-weighting (cf. section 32.4.1.1), because the overall signal will contain more contributions from late echoes with longer TEs. The fact that T2 relaxation occurs during the process of covering *k*-space implies that different partitions will show different echo amplitudes, due to exponential T2 signal decay, and this is a complicating factor in the image reconstruction process that can cause some degree of image blurring. The extreme case of FSE is the single-shot FSE, in which only one initial excitation is executed, followed by a very long train of echoes covering all phase-encoding steps. In the so-called Half-Fourier Acquisition Single-shot Turbo Spin Echo imaging (HASTE) version of the single-shot FSE approach, a general complex conjugate symmetry of *k*-space data is used, implying that only slightly more than half of *k*-space needs to be covered by measurement while the rest can be calculated.

32.3.2 Gradient-Echo Variants

Without going into too much detail, it is safe to say that GRE sequences exist in a huge number of variants. GREs are typically very fast, with TEs down to around one millisecond, and GRE sequences are often combined with 3D imaging methods. To reduce acquisition times, TR needs to be short, and the flip angle is thereby kept fairly small in order for M_z to be reasonably large at the subsequent excitation, also after a short TR. The optimal flip angle, that is, the one resulting in the highest signal, at a given TR, for a specific T1 value of the object, is called the Ernst angle, where $\alpha_{Ernst} = arccos\left(e^{-TR/T1}\right)$. A common class of GRE sequences is said to be spoiled, that is, the remaining transverse magnetization is destroyed before the next excitation. Spoiling can be accomplished either by applying a large gradient in the slice-selection direction (sometimes also in the frequency-encoding direction) after each signal readout (referred to as gradient spoiling), or by applying the B_1 field (i.e. the RF pulse) along different angles in the x'y'-plane at each excitation (so-called RF spoiling). In ultrafast GRE imaging, TR and TE are often shortened to the extent that tissue contrast becomes very poor. Typically, an initial 180° inversion pulse is therefore applied one time, as a magnetization preparation, before the actual GRE data collection. In this way, TI can be adjusted so that one tissue of interest generates very little signal at the time of the subsequent, very rapid (~1 s), GRE data acquisition block. Note that all lines (or a substantial segment) of k-space are acquired after a single inversion pulse. The Magnetization Prepared Rapid Gradient Echo (MPRAGE) sequence is a common 3D imaging example of using magnetization preparation in fast GRE. Finally, in so-called steady-state free precession (SSFP) sequences, where short TR is used, the remaining magnetization, after signal readout, is not destroyed but instead refocused by gradients and will contribute to the signal at the next excitation step; a so-called steady-state condition of the magnetization (transverse as well as longitudinal) is achieved. Very roughly, if TR is below 100 ms, a flip angle of 45–60° would generate a reasonable steady-state level.

32.3.3 Echo-Planar Imaging

The most rapid image acquisition technique used in contemporary clinical environments is the echo-planar imaging (EPI) sequence, enabling imaging of a single slice in 50–100 ms. Obviously, this minimizes the risk of patient motion during acquisition and enables dynamic imaging to capture signal changes over time. Note that most applications require reasonable brain coverage, and the volume imaging time is, in principle, proportional to the number of included slices. It is thus realistic to expect typical temporal resolutions of the order of 1–2 seconds in dynamic EPI. The principle of single-shot EPI is to apply the frequency- and phase-encoding gradients in a temporal pattern so as to cover a sufficient region of k-space, with adequate sampling density, after one single signal preparation, that is, either a 90° RF pulse in GRE-EPI (Figure 32.17a) or a 90°/180° RF combination in SE-EPI.

During the period when MR signal is available, the readout gradient is switched with alternating polarity, and every time the spins are refocussed by the readout gradient (either when the polarity is positive or negative), an echo is formed. This switching is repeated multiple times, and a number of echoes (e.g., 64 or 128) are thus generated. During each echo, k_x is successively changed, similarly to conventional pulse sequences, but in EPI the path is alternating between 'left-to-right' and 'right-to-left' due to the alternating readout gradient polarity during sampling. After each echo, the phase-encoding gradient is active for a very short period (sometimes referred to as a 'blip') in order to increase the k_y coordinate one discrete step before the next echo. In this way the entire *k*-space is traversed in a zig-zag-like pattern after one single excitation (Figure 32.17a, right-hand panel). Very short image acquisition time inevitably leads to inferior image quality. One striking feature in EPI is related to the long time that passes between the sampling of neighbouring data points in the k_y direction, near the centre of *k*-space. A long sampling interval corresponds to low bandwidth

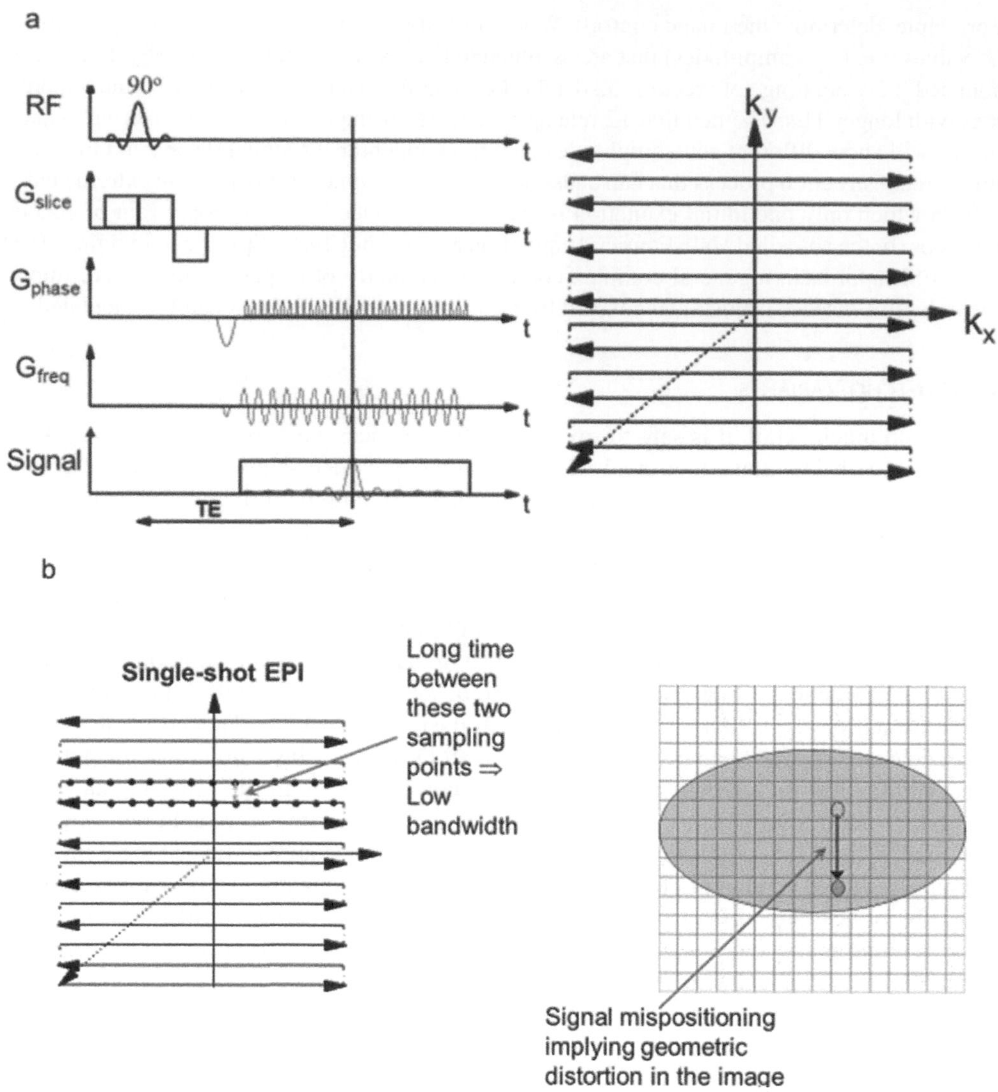

FIGURE 32.17 (a) Left: Schematic pulse sequence diagram of blipped gradient-echo echo-planar imaging (EPI). Right: The *k*-space trajectory of blipped EPI. (b) Illustration of geometric distortion in EPI caused by low bandwidth in the phase-encoding direction.

(see sections about signal-to-noise ratio and artefacts below), and EPI is thus characterized by low bandwidth in the phase-encoding direction (Figure 32.17b). This drawback can be alleviated by traversing *k*-space in a small number of excitations (referred to as multi-shot/segmented EPI) instead of only one.

32.4 IMAGE CONTRAST AND IMAGE QUALITY

32.4.1 Image Contrast

After the image reconstruction, as described above, an image with a certain tissue-specific intensity (or image signal) is obtained, depending on the length of the transverse magnetization M_{xy}. Immediately after a 90° RF pulse, M_{xy} would, in principle, reflect the proton concentration or proton density, and the image contrast would be PD-weighted. Although PD-weighting has a proven value in clinical MRI, it is fair to say that the most important contrast patterns are determined by the relaxation processes described above. For example, the echo amplitude (i.e. the image signal) will decrease with increasing TE, due to tissue-dependent T2/T2*-related dephasing, and the amount of magnetization in the z-direction that is available after regrowth due to T1 relaxation, at a given point of excitation after the waiting time between

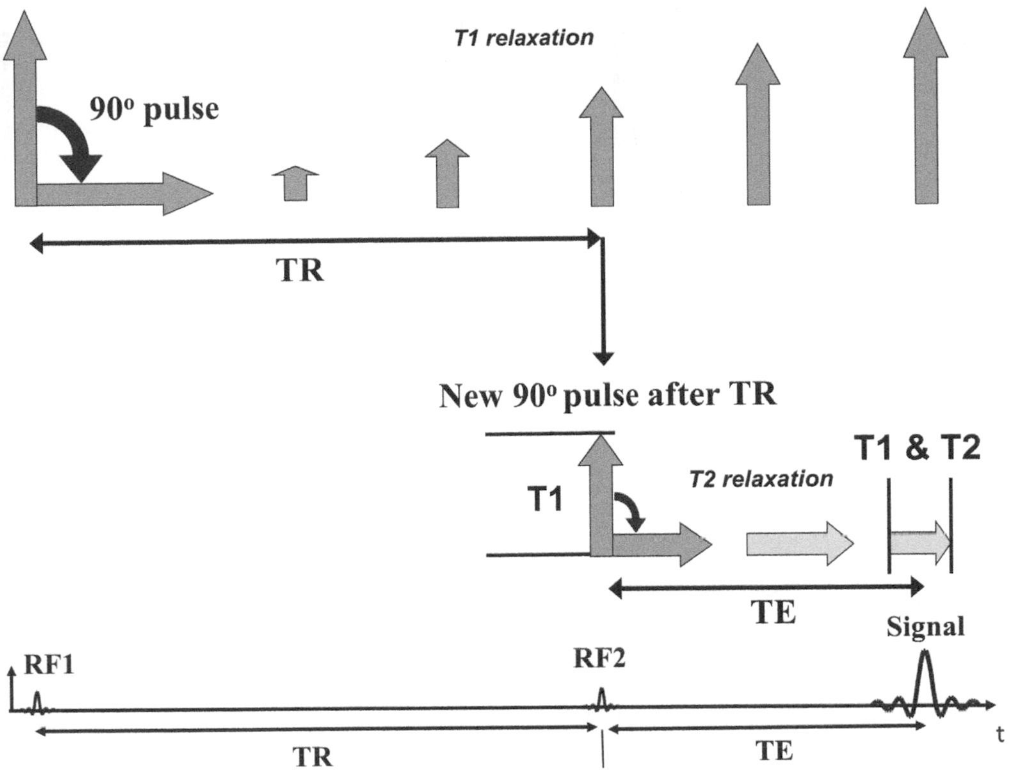

FIGURE 32.18 Illustration of basic contrast manipulation mechanisms in MRI, based on T1 relaxation (growing longitudinal magnetization) and T2 relaxation (decreasing transverse magnetization). Measured signal is proportional to the magnetization vector labelled 'T1 & T2'. It is the combination of T1, T2, TR and TE that determines the detected signal amplitude. For example, shorter TR and/or longer T1 implies that less longitudinal magnetization is available for rotation into the xy-plane at the subsequent repetition of the pulse sequence; longer TE and/or shorter T2 leads to more phase dispersion during TE and lower signal. The bottom part shows the time course of the corresponding RF excitations (RF1 and RF2) and the signal after the second excitation; the signal following RF1 has been omitted.

repetitions, will increase with increasing TR (Figure 32.18). Because T1 and T2 vary between tissue compartments, it is thus clear that signal as well as contrast will depend on the TE and TR settings in the imaging protocol of the MRI unit.

32.4.1.1 Spin-echo Contrast

Taking a SE pulse sequence as an example, it can be designed to enhance/suppress inherent differences in a given object parameter (PD, T1, T2) between tissue compartments by appropriate choice of scanner parameters (TE, TR) (Figure 32.19).

For example, at short TE, the time of dephasing is insufficient to allow differences in T2 relaxation between tissues to manifest themselves, and T2-based contrast thus becomes low. At long TR, the magnetization in all tissue compartments has regrown in the z-direction between repetitions, and T1-based contrast becomes low. On the other hand, long TE (of the order of the tissue T2 values, ~100 ms) gives sufficient time for T2-related differences in dephasing to manifest themselves, while still ensuring enough signal to be measured (i.e. if TE is too long the signal is extinguished). Moderate or 'short' TR (~500 ms) leads to pronounced differences in M_z magnetization between tissues at the time of the subsequent excitation, and the corresponding differences in M_{xy} after the 90° rotation will manifest themselves as differences in recorded signal at the time of the echo, that is, T1-based contrast is achieved. Note that if TR is too short, the signal will be deficient due to insufficient regrowth of M_z magnetization between repetitions (Figure 32.18). Finally, PD-based contrast is consequently achieved by suppressing both T1- and T2-based signal differences between tissues, that is, by using short TE and long TR. To summarize: (i) suppression of PD- and T2-based differences between tissues while enhancing T1-based differences is referred to as *T1-weighting*, and this is achieved by short TE and short TR (Figure 32.19a); (ii) suppression of PD- and T1-based differences while enhancing T2-based differences is referred to

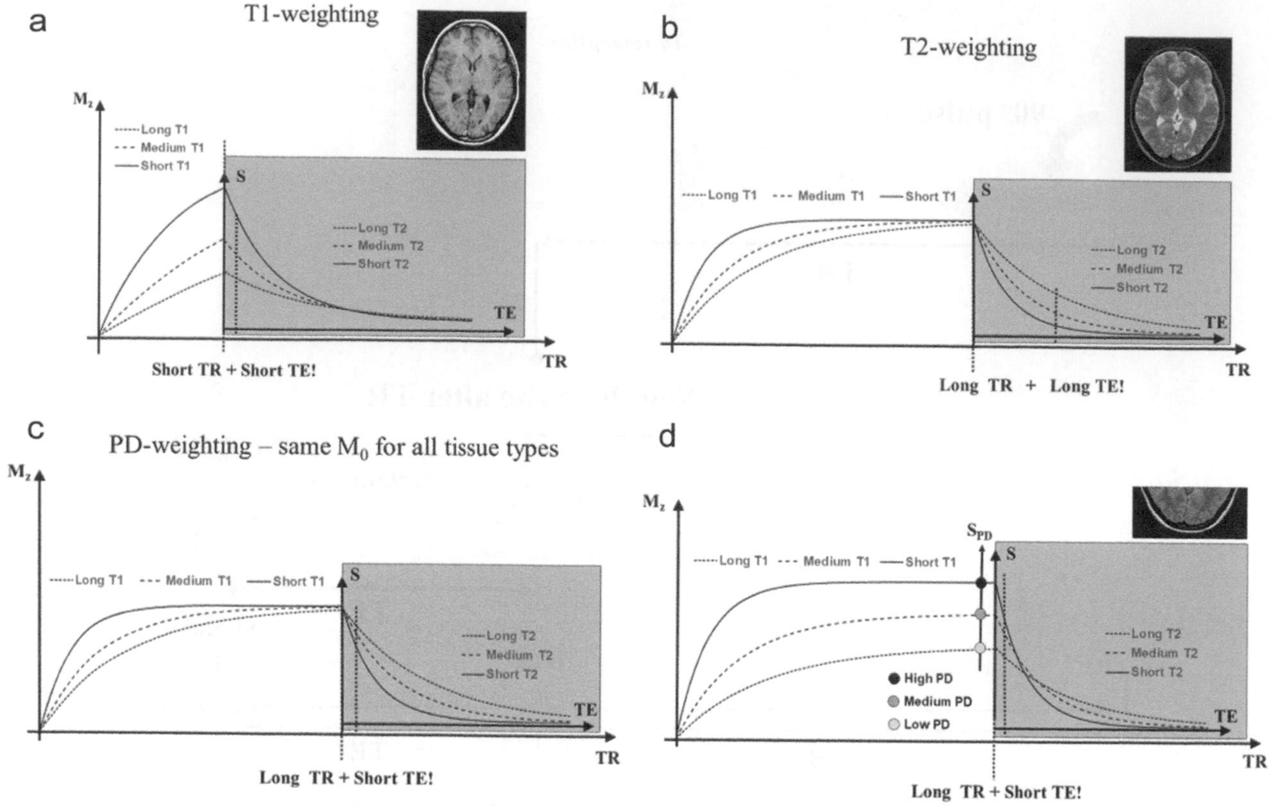

FIGURE 32.19 (a) Principle of T1-weighting, that is, how a combination of short TR and short TE leads to enhanced contrast between compartments with intrinsic T1 differences within the object. A T1-weighted image is shown for illustration (top right). (b) Principle of T2-weighting, that is, how a combination of long TR and long TE leads to enhanced contrast between compartments with intrinsic T2 differences within the object. A T2-weighted image is shown for illustration (top right). (c) Principle of PD-weighting, that is, how a combination of long TR and short TE leads to enhanced contrast between compartments with intrinsic PD differences within the object. In this simplified case, all tissue compartments show the same M_0. (d) PD-weighting in a realistic case with different compartments showing different M_0 (i.e. different PD). A PD-weighted image is shown for illustration (top right).

as *T2-weighting*, and this is achieved by long TE and long TR (Figure 32.19b); (iii) suppressing both T1- and T2-based differences between tissues, and thus obtaining a signal that reflects M_0 of the tissue, referred to as *PD-weighting*, is achieved by short TE and long TR (Figure 32.19c–d). Representative image examples of different weightings are given in Figure 32.19.

32.4.1.2 Gradient-echo Contrast

In GRE, short TR (~150 ms) is often used to realize the potential of rapid imaging in GRE and, if the flip angle is high, this leads to low amount of M_z magnetization being available at the next repetition. Hence, a flip angle well below 90° is often used in short-TR GRE, to ensure that a reasonably large M_z vector is available also after the excitation pulse to compensate for the limited regrowth between repetitions. Maximal signal at a given T1 is obtained by using the Ernst angle, but note that a given object contains tissue compartments with different T1 values, and the flip angle must, in practice, be chosen to provide optimal contrast between tissues rather than a high signal for one specific tissue type. Small flip angles will suppress T1-differences, while large flip angles will give T1-weighting. Analogous to the SE case, long TE provides T2*-weighting while short TE will suppress T2* differences. To summarize, T1-weighting is obtained by short TE (<15 ms) and large flip angle (>50°), T2*-weighting is achieved by long TE (>30 ms) and small flip angle (<40°) and PD-weighting is seen for short TE and small flip angle. TR is generally assumed to be short in all these cases.

32.4.1.3 Magnetization Preparation

Another strategy for contrast enhancement is to use some kind of magnetization preparation, and a simple example is the previously described IR pulse sequence, in which, in principle, a SE sequence module is preceded by a preparatory 180° pulse that inverts the magnetization at a time TI before the 90° pulse of the SE module. For example, the image signal of fat (showing short T1) can be extinguished by using short TI (called Short TI Inversion Recovery, STIR) and the image signal of cerebrospinal fluid (showing long T1) can be nulled by using long TI (called Fluid Attenuating Inversion Recovery, FLAIR). Magnetization preparation can also be used in numerous other applications. In ultrafast GRE acquisitions, as mentioned above, very short time intervals between the gradient echo events fail to provide sufficient time for longitudinal magnetization contrast to develop between repetitions. As also pointed out above, large flip angles are typically avoided in GRE, due to poor signal, implying that many ultrafast GRE sequences are characterized by low flip angle and short TE (i.e. PD-weighting). Improved T1-based contrast can, in such cases, be obtained by preparing the magnetization to show required differences in longitudinal magnetization only one time, just at the beginning of the GRE signal acquisition cycle.

32.4.1.4 Dixon Methods for Fat Suppression

A very common implementation in clinical MRI is so-called Dixon methods, which exploit the fact that water and fat hydrogen nuclei show somewhat different resonance frequencies. Different kinds of molecules show different electron cloud configurations, which shield the hydrogen nuclei in the respective molecules from the main magnetic field to various degrees, leading to different net magnetic fields experienced by the respective hydrogen nuclei. M_{xy} vectors corresponding to fat and water will thus show different precession rates, and different degrees of phase shift at a given point in time. Hence, over time, fat and water will periodically alternate between being in phase (parallel magnetization vectors) and out of phase (opposite magnetization vectors). In the simplest Dixon implementation, information from two images with different TE are combined; one where water and fat are in phase at TE and one where water and fat are out of phase at TE. Based on this information, maps showing either water-only signal or fat-only signal can be calculated. A variety of Dixon methods exist, and it is common to use several different echo times with varying degrees of phase angle differences between water and fat. Normally, four different images are extracted: 'in phase' image, 'out of phase' image, water-only map and fat-only map (Figure 32.20).

32.4.2 Contrast Agents

MRI contrast agents in clinical use are, in most cases, based on shortening the relaxation times, and it is generally common to exploit the shortening of the T1 relaxation time to enhance the contrast in T1-weighted images, according to the approaches outlined above. Protons inside tissues interact with contrast agent molecules available in the proximity, and thermal motion of the strongly paramagnetic metal ions of the contrast agent will generate the magnetic field fluctuations required to enhance the rate of regrowth of the longitudinal magnetization, that is, the dissipation of excess energy. A paramagnetic substance will also cause local static magnetic field inhomogeneities, and it is thus feasible also to use contrast-agent induced T2* shortening as a contrast mechanism. The most common group of MRI contrast agents is based on gadolinium (Gd), in the form of paramagnetic Gd(III) chelates. The Gd(III) contrast agent is normally administered by intravenous injection in a peripheral arm vein and applied to, for example, vessel enhancement in MR angiography or for brain tumour enhancement. Gd(III) chelates do not pass through the intact blood–brain barrier, and Gd(III) contrast agents are thus useful in enhancing tumours where the blood-brain barrier is compromised and the Gd(III) leaves the intravascular compartment. Outside the brain, a common Gd(III) contrast agent initially resides in the circulation but then distributes into the extravascular extracellular space and subsequently, for the vast majority of available contrast agent preparations, is eliminated almost exclusively by the kidneys. The normal dose (sometimes referred to as a single dose) of Gd(III) contrast agent is 0.1 mmol per kg body mass. Finally, the amount of change in relaxation rate that is induced by a given concentration of contrast agent is described by the relaxivity r_i:

$$r_i = \frac{\Delta Ri}{c}; \qquad i = 1, 2, 2^* \tag{32.26}$$

where the relaxation rate is Ri=1/Ti (in s^{-1}), ΔRi is the contrast-agent induced change in relaxation rate (in s^{-1}) and c is the concentration of contrast agent (in M).

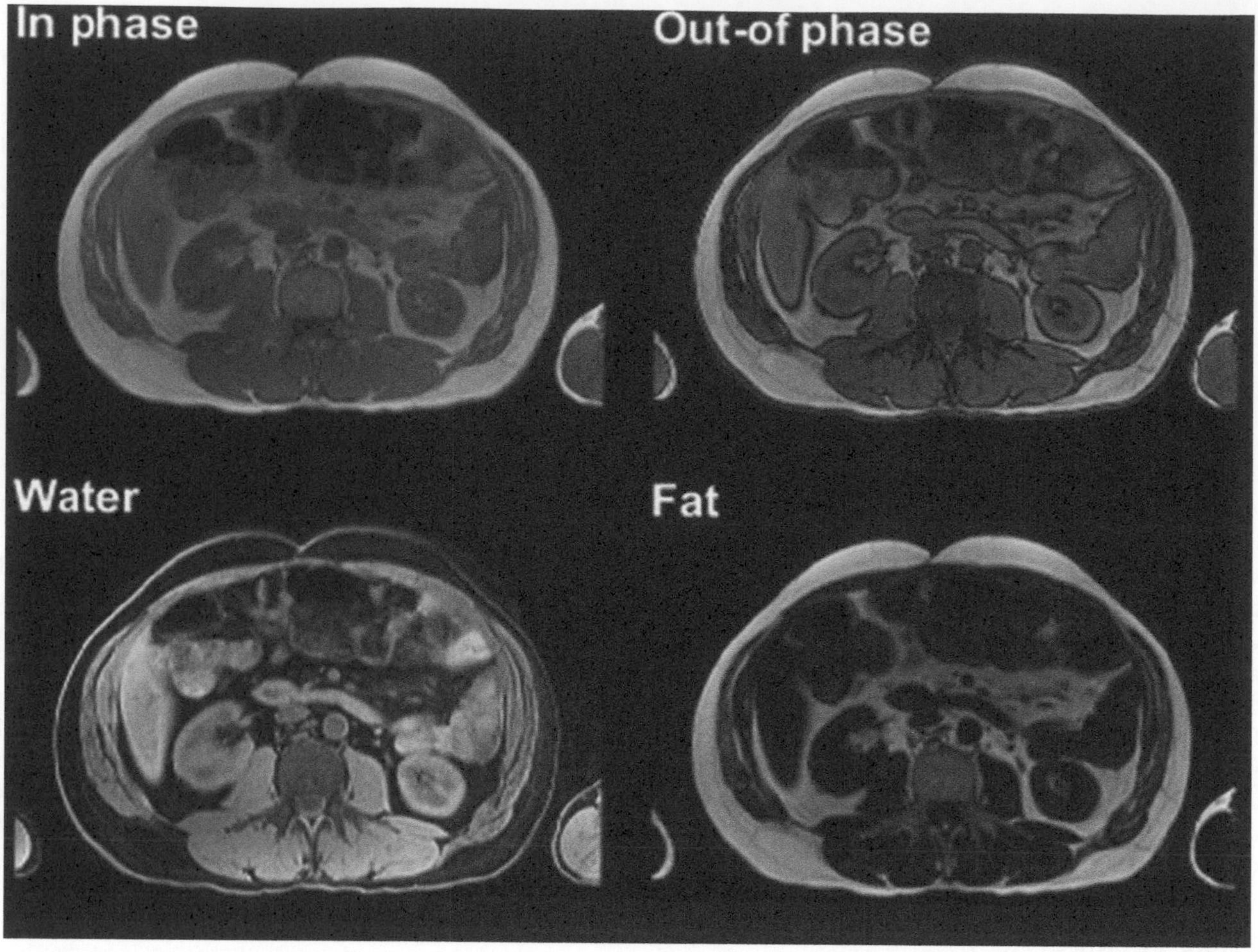

FIGURE 32.20 Axial abdominal slice illustrating the types of images typically acquired with so-called Dixon techniques. Top row: Images collected 'in phase' and 'out of phase'. Bottom row: Calculated maps of water and fat.

32.4.3 Signal-to-noise Ratio Issues

As pointed out above, there is only a small excess of spins in the parallel direction, contributing to M_0, and MRI is consequently a very noise-sensitive technique. Most of the noise that is picked up by the receiver coil is thermal white noise from the object, and this section will briefly address how the signal-to-noise ratio (SNR) is influenced by common MRI parameters.

1. *Pulse sequence type, scanner settings*: The choice of pulse sequence type (e.g. SE, GRE) and the specific settings of TE, TR and flip angle, and so forth, will influence the MRI signal, according to the respective signal equation (cf. Eq. 32.16), and thereby the SNR. For example, SE tends to give better SNR than GRE, due to the refocussing 180° pulse, short TE results in higher SNR than long TE, and so forth.
2. *Proton density:* SNR is proportional to M_0 and thus to proton density (PD).
3. *Static magnetic field strength:* SNR is, to a first approximation, proportional to B_0.
4. *Slice thickness:* SNR is proportional to ΔZ. A thicker slice means a larger number of excited spins that contribute to the signal. Noise, on the other hand, is picked up from the entire sensitive volume of the receiver coil and is not dependent on the slice thickness.
5. *Number of excitations (NEX) or number of averages n:* This refers to the possibility of repeating the signal acquisition of each echo (i.e. at each phase-encoding step). The summed signal S is then proportional to n, while the random noise N is proportional to the square root of n, that is, SNR=S/N is proportional to $\sqrt{n}$.
6. *Bandwidth (BW):* The receiver bandwidth per pixel is the signal sampling rate f during signal readout divided by the number of samples N_X. If the bandwidth is increased, the receiver coil opens up a larger frequency window,

and more noise will be included (because white noise is distributed equally over all frequencies). The power of the noise is thus proportional to the bandwidth, implying that the noise voltage N is proportional to the square root of the bandwidth, that is, $SNR = S / N \propto 1 / \sqrt{BW}$. If the time interval between sampling points is Δt, the sampling rate is $f=1/\Delta t$, and $BW = f/N_X = 1/(N_X \cdot \Delta t)$, that is, $SNR \propto \sqrt{N_X \cdot \Delta t} = \sqrt{T_{readout}}$, where $T_{readout}$ is the readout time of one echo.

7. *Field of view (FOV):* The field of view is the maximum dimension (in m) of a scanned object that can be represented in the reconstructed image, that is, one pixel represents an in-plane object area of $\Delta X \cdot \Delta Y$, where $\Delta X = FOV_X/N_X$ and $\Delta Y = FOV_Y/N_Y$. An expanded FOV with retained matrix size leads to increased voxel size and a correspondingly increased number of signal-generating spins in the voxel, that is, $SNR \propto FOV_X \cdot FOV_Y$.
8. *Number of sampling points/Matrix size:* (a) The voxel size is reduced if the matrix size is increased, with constant FOV, that is, $SNR \propto 1/(N_X \cdot N_Y)$. (b) Additionally, an increase in N_Y implies that the number of signal acquisitions is increased (more phase-encoding steps and a larger number of repetitions), which is beneficial to the SNR, that is, $SNR \pm \sqrt{N_Y}$. The combination of (a) and (b) results in $SNR \propto 1/\left(N_X \cdot \sqrt{N_Y}\right)$.

In summary, items 4–8 above can be summarized as follows:

$$SNR \propto \frac{\Delta Z \cdot FOV_X \cdot FOV_Y \cdot \sqrt{N_X \cdot \Delta t} \cdot \sqrt{n}}{N_X \cdot \sqrt{N_Y}} = \Delta Z \cdot \Delta X \cdot \Delta Y \cdot \sqrt{N_Y \cdot N_X \cdot \Delta t \cdot n} =$$

$$= V \cdot \sqrt{N_Y \cdot n \cdot T_{readout}} = V \cdot \sqrt{T_{readout,total}} \quad (32.27)$$

where V is the voxel volume and $T_{readout,\,total}$ is the total signal readout time for all echoes over all phase-encoding steps and all averages. Hence, for a given object in a given MRI unit (i.e. with a specific field strength), SNR is proportional to the voxel volume and to the square root of the total signal readout time. This is a very compact and convenient way of assessing how changes in parameter settings will influence the SNR.

32.4.4 Common Image Artefacts

MRI suffers from a number of well-known image artefacts, and the most common, or 'classical', ones are briefly explained, described and illustrated below. Note that the reader is assumed to be familiar with basic properties of the discrete Fourier transform. The list below is far from complete and should be regarded as an introduction to artefacts seen using conventional pulse sequences. Fast imaging sequences and new hardware solutions have introduced additional image quality issues, and a few of those are mentioned in other sections.

32.4.4.1 Metal Artefact

A ferromagnetic or strongly paramagnetic object will become pronouncedly magnetized by the external static magnetic field, and it will thus distort the magnetic field lines in its surroundings, implying that the magnetic flux density becomes altered in a way that is difficult to predict. For example, this implies that the magnetic flux density will not vary linearly with the X position during readout, when the frequency-encoding gradient is applied, and the signal will thus be mispositioned along the frequency-encoding direction in the image (Figure 32.21a). This occurs because the image reconstruction assumes a linearly varying magnetic field and, furthermore, the frequency encoding is, by definition, based directly on the resonance frequency at each X position and thereby on the absolute value of the magnetic flux density. In the phase-encoding direction, however, the encoding is based on the so-called pseudo-frequency, which is defined by the difference in experienced magnetic field, at a given Y position, between repetitions with different G_{phase} amplitudes. If the field distortion, caused by the metal object, is static over time, this field difference between repetitions is still the same, even if the absolute value of the magnetic flux density is altered, and no artefact (i.e. no, or very minor, signal mispositioning) is observed along the phase-encoding direction. In GRE, there is an additional effect of pronounced signal loss or signal void, because the magnetic object will also cause a magnetic field inhomogeneity in its surroundings that leads to rapid phase dispersion (cf. T2' relaxation).

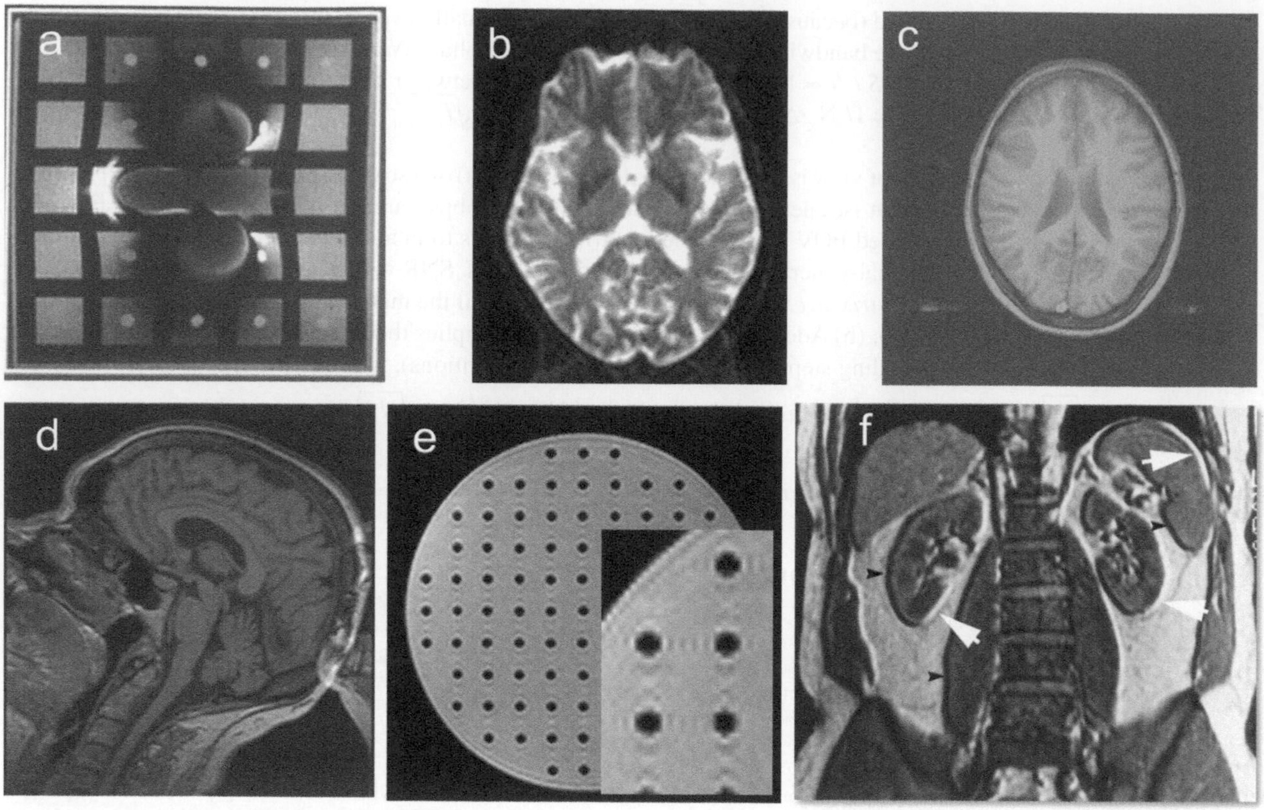

FIGURE 32.21 Examples of some common MRI artefacts. (a) Metal artefact. Note severe geometric distortion and signal displacement in the frequency-encoding direction (left-right). (b) Susceptibility artefact in EPI. Note geometric distortion in the frontal region of the brain along the phase-encoding direction. (c) Motion artefact. Ghosting and random signal distribution in the phase-encoding direction (left-right), caused by the sagittal sinus vein. (d) Aliasing (wrap-around artefact) where the nose and chin appear in the back of the head of the limited FOV. (e) Truncation artefact. Gibbs ringing (periodic signal fluctuations) close to sharp edges. (f) Chemical shift artefact. Mispositioning of fat structures (arrows) relative water in the frequency-encoding direction (left-right).

32.4.4.2 Susceptibility Artefact

This type of artefact is, in principle, the same as the metal artefact. As for the metal artefact, a difference in magnetization between materials with different magnetic susceptibility occurs (in this case, most commonly tissue and air), leading to geometric image distortions close to interfaces. If the bandwidth is low, the frequency range over one voxel is limited, and the signal of spins experiencing an erroneously altered magnetic flux density has a higher risk of falling into another pixel than anticipated. Pulse sequences with low bandwidth per pixel, such as EPI, are thus most sensitive to susceptibility artefacts of this kind (Figures 32.17b and 32.21b).

32.4.4.3 Motion Artefact

Using a somewhat simplified illustration, the Larmor equation implies, for gradient G and position r that $\omega = \gamma B = \gamma G r$. Assume that a spin, or a spin population, moves coherently a distance Δr during Δt. The phase shift of the MRI signal would then correspond to $\Delta\Phi = \omega\Delta t = \gamma G\Delta r$, that is, a coherent motion of the spin or a spin population would cause a corresponding phase shift of the MRI signal. At the same time, the phase is employed for spatial encoding, using the phase-encoding gradient. Hence, if an extra phase shift, caused by motion (e.g. bulk motion of the head or movements of internal structures), occurs at a given phase-encoding step, the phase encoding will be disturbed, and the echo will be recorded with an erroneous phase. If such additional phase shifts, caused by motion, occur randomly or periodically during the build-up of the signal matrix, artefacts will be seen in the reconstructed image. The artefact manifests itself as mispositioned signal along the phase-encoding direction. If the motion pattern is random and/or occurring during some of the phase-encoding steps, signal components are distributed more or less evenly, in a random manner, along the whole column/row (depending on the phase-encoding direction) in which the moving structure is located. If, on the other hand, the motion pattern of the moving structure is periodical (e.g. breathing or blood pulsation in vessels), one can observe blurred or smoothed replications of the moving structure in the phase-encoding direction, often referred to as ghosting (Figure 32.21c). Note that if the motion pattern, hypothetically, would generate the same extra phase shift at each phase-encoding step, there would be no artefact.

32.4.4.4 Aliasing, Wrap-around Artefact

Aliasing is a well-known phenomenon in digital signal processing. In MRI, FOV=$1/\Delta k$, and aliasing thus occurs when the signal sampling density is too low. Sampling is a multiplication of the signal function in k-space with a train of Dirac delta functions, that is, a shah function, with sampling interval Δk, which corresponds to convolution of the object function (in image space) with the Fourier transform of the shah function (which is also a shah function). In simple terms, if the shah function in *k*-space corresponds to too-sparse sampling, the shah function in the image space will be densely distributed and the repeated images resulting from the convolution will overlap. The entire object will not fit into the FOV but, on the other hand, parts of the repeated images of the convolution will show up in the FOV, appearing as if the image of the outer parts of the object has been 'wrapped around' into the limited FOV. In a sagittal slice, aliasing can, for example, be seen as 'the nose in the back of the head' and/or 'the back of the head in the nose' (Figure 32.21d). Aliasing is avoided by more dense sampling in k-space (leading to a sparser shah function in image space), and this can be accomplished, in principle, either by increasing the sampling frequency ($f=1/\Delta t$) with retained readout gradient amplitude, or by lowering the readout gradient amplitude (which leads to smaller Δk even if Δt is kept constant).

32.4.4.5 Gibbs Ringing, Truncation Artefact

Another sampling-related artefact is Gibbs ringing, and this artefact occurs when the signal is truncated in *k*-space, that is, the coverage of *k*-space is insufficient. Truncation equals multiplication of the signal function in *k*-space with a rectangular function, where insufficient coverage implies a narrow rectangle. In image space, this means that the object function is convolved with the Fourier transform of a narrow rectangle, that is, a broad sinc-function. This convolution with a sinc function manifests itself as periodic signal fluctuations in the image, typically near sharp edges or borders between different tissue types (Figure 32.21e).

32.4.4.6 Chemical Shift Artefact

As pointed out above, in connection with the Dixon methods, water and fat hydrogen nuclei show slightly different resonance frequencies, referred to as a chemical shift. In the frequency-encoding approach, signal positioning in the image is directly based on the registered frequency of the signal. Assuming that a water signal with a specific frequency ends up in the expected column of the image, fat in the same position will show another frequency, which may complicate the image reconstruction. If the bandwidth is high, both the water frequency and the fat frequency of a given position will fall into the same pixel in the image, and no artefact will occur. If the bandwidth is low, however, the frequency interval over one voxel is limited and the fat frequency will end up in another pixel than water, although the actual fat and water positions in the object, along the frequency-encoding direction, are the same. Fat will thus be mispositioned one or a few pixels along the frequency-encoding direction, relative to water, and this is referred to as a chemical shift artefact (Figure 32.21f).

32.5 ADVANCED MRI TECHNIQUES: FUNCTION, PHYSIOLOGY, AND MICROSTRUCTURE

32.5.1 Cine MRI: Imaging of Moving Structures

Rapidly moving structures, for example the heart, can be studied by dynamic MRI, often called cine MRI. The acquisition of the time series of images is synchronized, using an ECG, to take place at different points in time over the cardiac cycle, and the reconstructed images are displayed in a video sequence. The dynamic imaging is carried out in multislice mode to cover the whole heart. If rapid GRE is employed, the entire procedure can be accomplished while the patient holds his/her breath.

32.5.2 Flow and Motion, MR Angiography

The inherent motion sensitivity of the MRI signal has been known for a long time, and the disadvantages in terms of motion artefacts were described above. However, the extra phase shift caused by a displacement of the spins, in the presence of a magnetic field gradient, can also be used in a constructive way for quantitative measurements of *flow* – for example, in large blood vessels. In so-called phase maps, each pixel shows the phase angle, which is proportional to the velocity of the spins.

MR angiography (MRA) is a non-quantitative visualization of blood vessels, often utilizing 3D imaging methods (Figure 32.22a). MRA can be based on the so-called inflow phenomenon, where a signal increase occurs when spins that are previously non-excited continuously are brought into the imaging slice between pulse sequence repetitions. In phase contrast MRA, the motion-induced phase shift (mentioned above) is instead utilized for vessel visualization. In recent years, the contrast-enhanced MRA option has become more popular, particularly for regions with complex flow patterns. This method is based on signal enhancement caused by an intravenously injected Gd contrast agent. It is important that the dynamic imaging (typically 3D-GRE with short TE and TR) is temporally synchronized with the

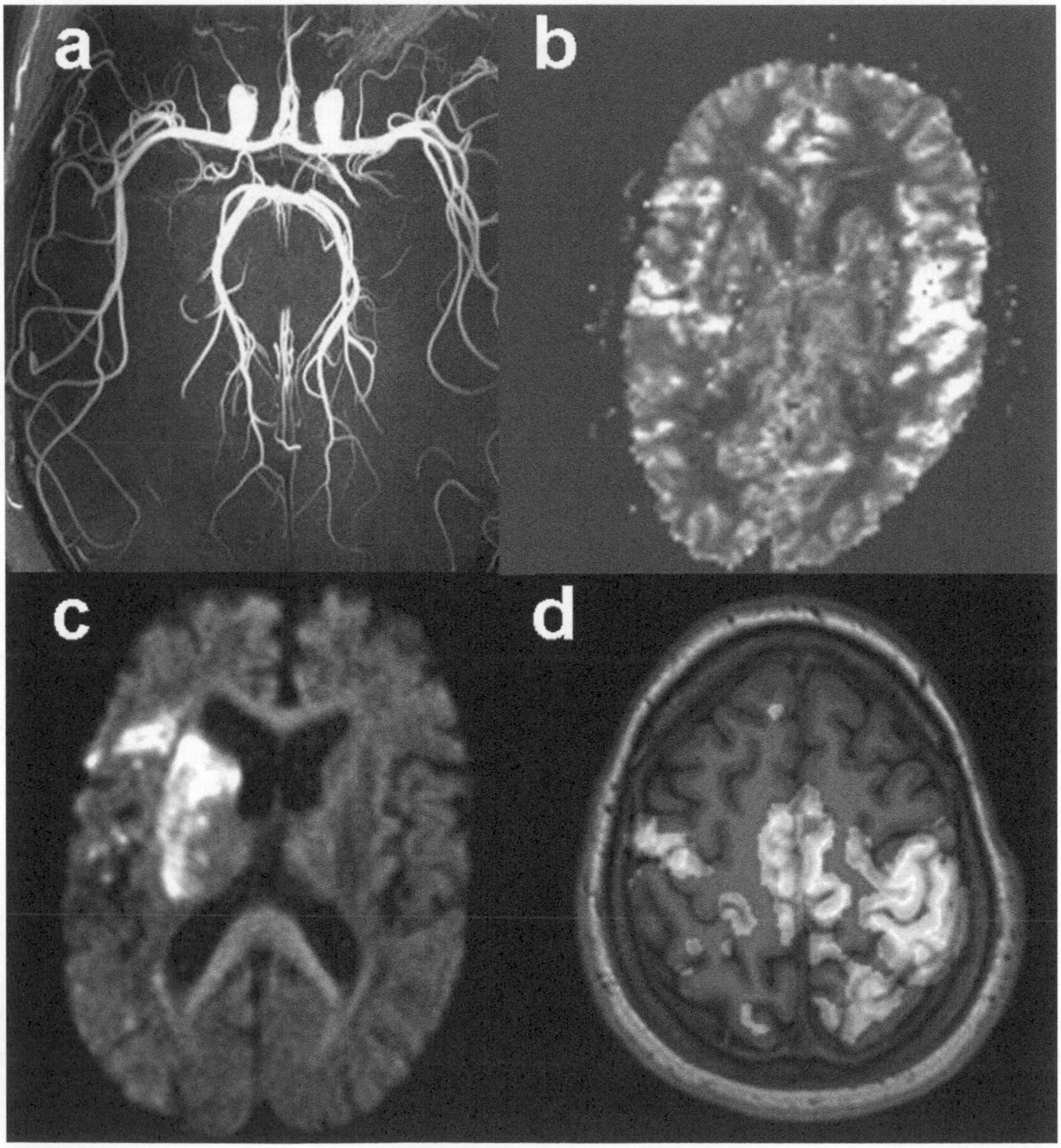

FIGURE 32.22 Examples of advanced MRI methods: (a) MR angiography; (b) Cerebral blood flow (CBF) map from T2*-weighted dynamic susceptibility contrast MRI (DSC-MRI); (c) Diffusion-weighted MRI. Note that the cerebrospinal fluid in the ventricles, with almost freely diffusing water molecules, shows pronounced hypointensity, while the ischaemic core exhibits high (i.e. more or less unaltered) signal due to low diffusion; (d) BOLD-fMRI visualizing cortical activity in an experiment where the subject has moved the right hand, arm, and elbow. Colour image available at www.routledge.com/9781138593268.

first passage of the compact volume (bolus) of contrast agent through the organ of interest; the maximal contrast agent concentration in the blood must coincide with the point in time when signal sampling in the central parts of *k*-space takes place.

32.5.3 Perfusion and Permeability

The most established MRI technique for assessment of cerebral blood flow (CBF) is *dynamic susceptibility contrast MRI (DSC-MRI)*. A Gd-based contrast agent is injected intravenously, and the signal time course is monitored by rapid T2*-weighted EPI (approximately 1 image/second for 1–2 minutes) during the first passage of the contrast agent bolus in the brain. The contrast agent concentration is quantified, as a function of time, in the tissue (pixel by pixel) as well as in a brain-feeding artery (providing the arterial input function, AIF). Perfusion-related parameters such as cerebral blood volume (CBV), CBF (Figure 32.22b), and mean transit time (MTT) can be calculated using kinetic theory for intravascular tracers, most commonly by using deconvolution of the tissue concentration time curve with the AIF to obtain CBF and MTT. The basic concept is often referred to as 'bolus tracking' (applicable also to, for example, computed tomography). Quantification in absolute terms has, so far, been problematic due to difficulties in obtaining accurate AIFs and uncertainties concerning the transverse relaxivities in tissue and blood.

Another perfusion MRI method is *arterial spin labelling (ASL)*, based on inversion (by a 180° RF pulse) of the water spin population in a brain-feeding artery upstream to the imaging slice, a procedure sometimes referred to as magnetic labelling. The labelled water molecules are transported by the arterial blood flow from the site of labelling to the blood capillary network of the tissue of interest, where they are incorporated by exchange mechanisms into the signal-generating water of the tissue. The tissue signal is then reduced in proportion to the tissue blood flow (i.e. the perfusion), because the inverted magnetization of the spins in the inflowing blood partly counteracts the original M_z magnetization of the tissue. The magnetically labelled image is finally subtracted from a control image (without labelling) and a CBF map can be calculated.

In other organs than the brain, and in conditions with an impaired blood-brain barrier (e.g. in brain tumours), T1-weighted so-called *dynamic contrast enhanced MRI (DCE-MRI)* is often applied, typically utilizing dynamic 3D-GRE imaging over several minutes after intravenous injection of a Gd contrast agent. In this scenario, the Gd chelate can diffuse from the intravascular compartment into the extravascular extracellular space (EES), to an extent that is determined by a combination of perfusion and permeability. Using pharmaco-kinetic modelling, parameters reflecting plasma volume, transcapillary permeability and/or perfusion can be extracted. K^{trans} is a commonly used composite parameter in DCE-MRI, reflecting perfusion and vascular permeability.

32.5.4 Diffusion

For visualization of water self-diffusion – that is, the random thermal motion of the water molecules in tissue – strong pulsed magnetic field gradients are employed, applied before and after the 180° pulse of a SE-EPI sequence with long TE. Each water molecule exhibits a random and unique motion pattern, causing a random phase shift of its contribution to the MRI signal of the voxel. When the summarized signal of the voxel, containing a huge number of water molecules, is registered, a phase dispersion is observed that lowers the magnitude of the signal. In a so-called diffusion-weighted image, high diffusion is characterized by reduced signal (hypointensity) due to pronounced phase dispersion, while regions with low diffusion show remaining high signal (i.e. hyperintensity relative to areas with high or normal diffusion). This phenomenon has found widespread use in acute ischaemic stroke (Figure 32.22c). Generally, diffusion MRI is often used as a biomarker of cell membrane integrity or cell density. By post-processing of image data, the diffusion coefficient or an approximation to it (often referred to as the apparent diffusion coefficient, ADC) can be calculated and mapped. More comprehensive diffusion measurements, including different degrees of diffusion sensitivity (so-called b-values) and multiple gradient directions, allow for quantification and visualization of the directional dependence of the diffusion, that is, the diffusion anisotropy, using tensor analysis. Such information can serve as the basis for neuronal fibre tracking.

32.5.5 Visualization of Cortical Activation (BOLD-fMRI)

Blood oxygenation level dependent functional MRI (BOLD-fMRI) is used for visualization of local neuronal activation, often induced by a specific task or stimulus, and the method is based on the fact that activation of a certain brain centre leads to an increased supply of oxygenated blood and, importantly, this delivery of oxygenated blood exceeds the actual metabolic needs. Hence, the relationship between deoxyhaemoglobin, which shows paramagnetic properties, and oxyhaemoglobin, which is diamagnetic, is altered in activated regions. Deoxyhaemoglobin causes a local magnetic field inhomogeneity and associated phase dispersion, and the abundant supply of diamagnetic oxyhaemoglobin means that the MRI signal increases by a few percent during activation because of reduced phase dispersion (longer T2*). Thus, an observed local signal increase in the image (often GRE-EPI) means that this area can be associated with the applied stimulus. The technique allows for studies of activated regions by statistical analysis of images acquired in a series of periods alternating between rest and stimulation/activation. Relatively straightforward stimuli as, for example, sensory/motor activation are, to some extent, used in pre-surgical planning (Figure 32.22d). Higher cognitive functions like language, memory, decision making, and so forth, are more complicated but frequently investigated by BOLD-fMRI in a neuroscientific context. In recent years, analysis of data acquired during rest only, so-called resting-state fMRI, has met considerable interest.

32.5.6 Magnetic Resonance Spectroscopy (MRS)

As indicated above, the resonance frequency of an MRI-active nuclide varies slightly, even at a given external field strength, depending on its chemical environment. Nuclei in different molecules or at separate parts of a given molecule experience different surrounding electron clouds, and varying electron configurations will shield the nuclei from the main magnetic field to various degrees in different molecular positions. If the frequency resolution is high (which requires reasonably high field strength) different types of molecules in a voxel can be identified by the obtained spectral pattern in a frequency spectrum. MRS can be accomplished with a number of nuclides of biochemical relevance, out of which ^{1}H and ^{31}P are most commonly encountered in clinically available MRS installations. Proton-MRS (^{1}H) can be used for neurological investigations and in prostate, while ^{31}P-MRS is applied to studies of energy metabolism and pH changes, for example, in muscle and liver.

32.6 MRI HARDWARE AND SAFETY

32.6.1 MRI Hardware and Technology

An MRI unit consists of a main magnet, a radiofrequency transmit-/receive system, gradient coils and a computer system for pulse sequence control and image reconstruction (Figure 32.23).

Magnets in commercial whole-body MRI units operate in the magnetic flux density range of approximately 0.1–7.0 T, and currently 1.5 T and 3 T MRI scanners are very common in clinical use. Magnets up to approximately 0.2–0.3 T can be of the permanent magnet type, where the magnetic field is induced at installation. Magnetic fields produced by running an electric current through a solenoid with an iron core are available up to approximately 0.6 T. The most common technique at higher field strengths (currently up to ~8 T in whole body units) is to use superconductive magnets, implying that the coil is cooled to approximately -269 °C (4 K) using liquid helium, whereby zero resistance is achieved, and the electric current thus requires no external voltage and generates no heat. If some part of the coil loop ceases to be superconductive, heat will be generated and spread along the wire. The electric current will subsequently drop, the magnetic field is lost, and the helium boils off rather quickly. This event is referred to as a quench, and it is typically an expensive and delicate procedure to successfully restore the magnetic field after such an incident. An important quality parameter for the B_0 field is the magnetic field homogeneity, often stated for a given volume of interest (e.g. a sphere with a given diameter). The inhomogeneity (in ppm) is the field variation over the volume of interest (in T) divided by the nominal magnetic flux density (in T), multiplied by 10^6. Measures to optimize or improve the homogeneity is referred to as shimming, and this can be achieved either passively, by adding pieces of ferromagnetic materials to improve the homogeneity, or actively, by adding compensating currents in shim coils. Shim coils can be activated for an individual examination, in order to account for inhomogeneities arising from the introduction of a specific object into the volume of an MRI unit. Finally, shielding is normally applied to reduce stray field components outside the MRI unit, either in terms of passive shielding, using steel or iron, or active shielding, achieved by introducing a reverse electric current outside the main coil to partly counteract the field components generated outside the active volume.

The RF system consists of a transmitter of RF fields, for excitation of spins, with a well-defined amplitude, phase, central frequency, and bandwidth, and a receiver in which the demodulation, described above, is accomplished. Transmit coils are designed to generate a uniform B_1 field that corresponds to the desired flip angle over the entire slice or volume to be excited. Above, the excitation was described in terms of a linearly polarized transmitter coil, but circular

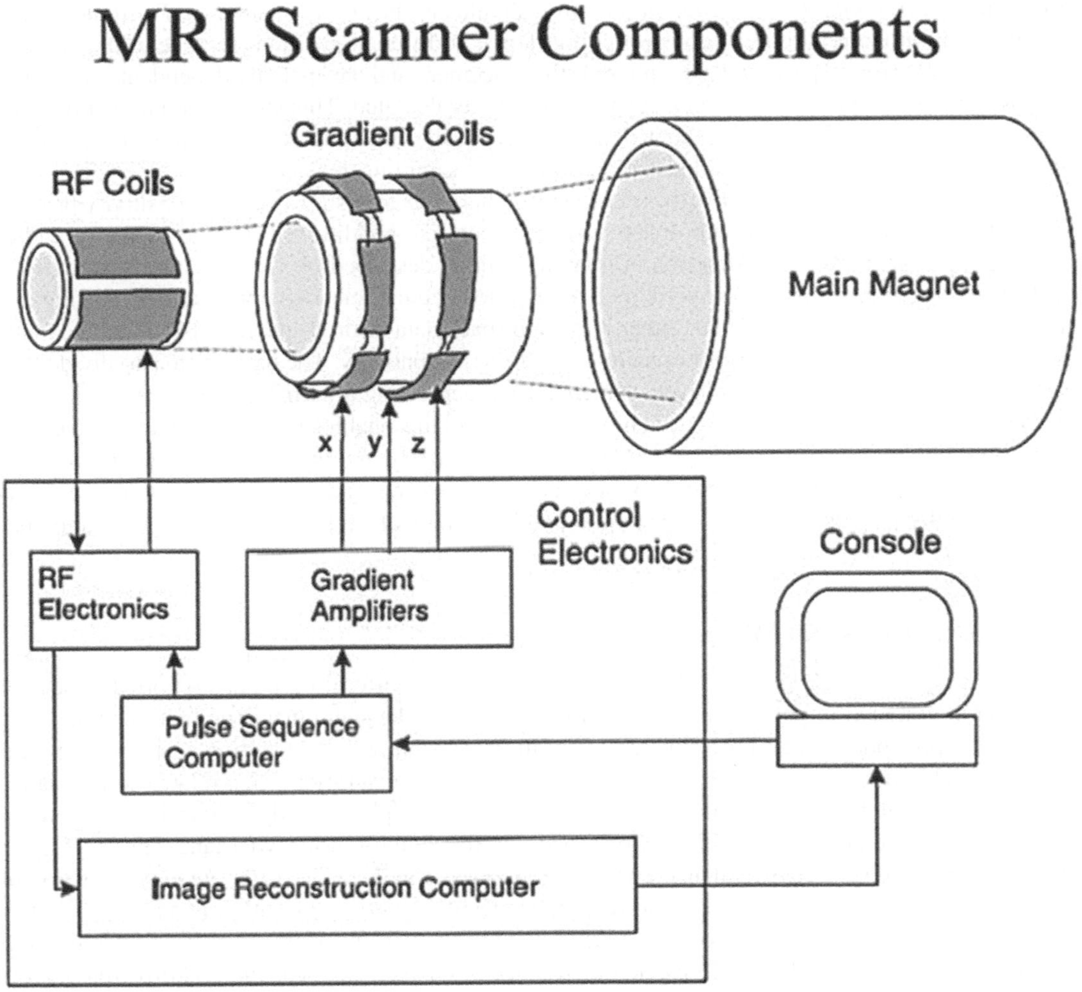

FIGURE 32.23 The basic components of an MRI system: The main magnet generates the B_0 field, and the gradient coils enable spatial encoding along three orthogonal directions. The RF coil system excites the spins and subsequently receives a Faraday induction signal from the object. All pulse sequence components can be controlled by the user via a computer system interface. (Reproduced from [1] with permission from John Wiley).

polarization is also available, based on two coils in which the supplied currents are phase shifted 90° relative to each other. With regard to receiver coils, a multitude of designs are available to optimize SNR. One important category is the volume coils, which surround the object and are often both transmitting and receiving (e.g. the head coil), and the other category is the surface coils, which tend to show a more inhomogeneous reception but show high SNR near the surface of the object. In a quadrature receiver coil system, one coil for each channel is used (cf. quadrature detection above), which improves SNR by a factor of $\sqrt{2}$. So-called phased array coils consist of a number of small single coils that are connected into one receiver coil, thereby combining the small coil's advantage of high SNR with the large sensitive volume or field of view achieved by joining a number of coils. In this context, one small coil is a coil element, one or a few coils constitute a segment, and a channel is an independent chain of electronics.

The gradient system consists of a coil system located between the main magnet coil and the RF coils, designed to create linear magnetic field variation along any arbitrary direction in space. Maxwell coils are used for gradients along the direction of the B_0 field, while either double saddle (Golay) coils or more advanced so-called fingerprint coils are used in the two principal directions perpendicular to the B_0 field lines. The maximal gradient strengths in an MRI unit are of the order of 15–80 mT/m, and the gradients can be switched on and off very rapidly. The slew rate is the maximal gradient strength (in T/m) divided by the rise time (in s), and it is typically in the range 50–200 $Tm^{-1}s^{-1}$ for contemporary MRI units. The rapid switching of gradients can induce so-called eddy currents in surrounding materials, known to cause image artefacts by affecting B_0 and introducing unwanted time-varying magnetic field gradients.

A few other important *technical and theoretical solutions* should also be briefly mentioned. *Parallel imaging* employs separation of information from individual coils (or coil groups), at different positions relative to the object within the phased-array coil to allow for *k*-space undersampling. If the placement and the spatially dependent sensitivity profile of each coil is known, the source of a given signal component can be calculated. This allows for a reduction in the number of phase-encoding steps, and thus for accelerated scan time. Parallel imaging reconstruction in the image space is performed by obtaining one aliased image (due to reduced *k*-space sampling density) for each array element and, subsequently, a full FOV image is calculated from the set of intermediate images using available sensitivity profiles and position data for each coil. The reconstruction problem tends to become ill-conditioned at some degree of undersampling, which leads to specific artefacts. The reduction in imaging time is described by the reduction factor (R), defined as the reciprocal of the fraction of *k*-space covered in an accelerated acquisition. Different parallel imaging concepts are available, and calculations can be carried out either on image space data or on *k*-space data. A natural extension of the parallel imaging concept is *parallel transmit* or *multi-transmit* RF technology. The basic idea is to divide the RF transmission between separate, independently driven elements, each generating a partial B_1 field. The sum of these partial RF fields comprises the total B_1 field experienced by the object, and this enables improved control of the RF excitation homogeneity in tissue. Another recent development is the use of *sparse sampling* and *compressed sensing* techniques to considerably reduce the amount of measured *k*-space data, using a semi-random sampling pattern, while maintaining reasonable image quality by noise-reduction techniques and by using mathematical iterative optimization processes to reconstruct the image.

32.6.2 Biological Effects and MRI Safety

With regard to the *static magnetic field*, very few significant biological effects have been reported. Documented temporary phenomena in connection with exposure to and movements in the static magnetic field include vertigo, visual sensations (magnetic phosphenes), and metallic taste sensations as well as, less commonly, stimulation of peripheral nerve cells at high field strength. It is probably fair to say that the main safety aspects related to the static magnetic field concern the effects of torque (potentially dislocating magnetic implants) and the translational force arising on magnetic objects in the spatial gradient of the magnetic field in the vicinity of the MRI unit. The projectile effect is a well-known safety issue in MRI environments. For the *RF fields*, the main issue is the Faraday induction of currents in tissue and associated heat generation by tissue resistance. The RF energy absorption is quantified in terms of the specific absorption rate (SAR), that is, absorbed energy per unit time per unit mass (in W/kg). SAR is approximately proportional to B_0^2 , but SAR distributions are generally complex, geometry-dependent, and difficult to predict. SAR-level restrictions are determined mainly based on heating effects and temperature limits. Thermal burns can occur when patient positioning creates a conductive loop pathway, for example, when hands are clasped or legs are crossed, as well as when leads and sensors are inappropriately arranged. Contact burns may occur in interaction with metallic objects acting as conductors (coils, cables, monitoring equipment, transdermal patches, metallic fibres in clothes, etc.). The exposure to *gradient fields* is relevant mainly in terms of the rapid switching (d*B*/dt), and biological effects in MRI are primarily associated with peripheral nerve stimulation; cardiac nerve stimulation is also often mentioned or discussed in this context but is only relevant at much higher stimulation threshold. A practical gradient switching issue, related to Lorenz forces in the gradient system, is the very loud acoustic noise that is produced, and patients should always wear hearing protection when undergoing MRI procedures.

32.7 CONCLUDING REMARKS

The rate of technical development in MRI is still very high, as exemplified by significant advancements in transmit-/receive system development, coil design, and *k*-space sampling strategies aiming, for example, at increased SNR, improved volume coverage, and reduced image acquisition time. Increasing magnetic field strength has been a long-term trend in MRI systems and, at this point, 7 T installations are under careful investigation with respect to clinical usefulness and diagnostic ability. A variety of advanced methods for functional/physiological and microstructural imaging is introduced and implemented to facilitate various forms of multiparametric data analysis. Interventional MRI is another topic of increasing importance, and combinations of separate medical systems, both for diagnostics and therapy, are becoming common – for example, in MR-PET and MR-Linac designs. Hyperpolarization techniques and novel contrast mechanisms (such as chemical exchange saturation transfer, CEST) are other examples of important recent developments.

BIBLIOGRAPHY

[1] B. Gruber, M. Froeling, T. Leiner, and D. W. J. Klomp, "RF coils: A practical guide for nonphysicists," (in English), *J Magn Reson Imaging*, vol. 48, no. 3, pp. 590–604, Jun. 13, 2018, doi:10.1002/jmri.26187.
[2] R. W. Brown, Y. N. Cheng, E. M. Haacke, M. R. Thompson, and R. Venkatesan, *Magnetic Resonance Imaging: Physical Principles and Sequence Design*, 2nd edn. Hoboken, NJ: John Wiley, 2014.
[3] D. McRobbie, E. Moore, M. Graves, and M. Prince, *MRI from Picture to Proton*. Cambridge: Cambridge University Press, 2017.
[4] D. G. Nishimura, *Principles of Magnetic Resonance Imaging*. Lulu: www.lulu.com, 2010.

NOTE

1 Comprehensive lists of suggested reading are provided in the textbooks [2–4].